THE CHEMISTRY AND PHYSICS OF CLAYS

THE CHEMISTRY AND PHYSICS OF CLAYS

AND ALLIED CERAMIC MATERIALS

By REX W. GRIMSHAW

B.Sc. Ph.D. F.R.I.C. F.I.Ceram. M.I.M.M.

SENIOR LECTURER · DEPT. MINING AND MINERAL SCIENCES · UNIVERSITY OF LEEDS
FORMER DIRECTOR OF RESEARCH AND DEVELOPMENT · LEEDS FIRECLAY CO LTD
FORMER LECTURER IN CERAMICS AND REFRACTORY MATERIALS · UNIVERSITY OF LEEDS AND
QUONDAM VISITING PROFESSOR OF CERAMICS · PENNSYLVANIA STATE UNIVERSITY · U S A

FOURTH EDITION

REVISED

LONDON · 1971

ERNEST BENN LIMITED

Store
5313

666.42
GRI

PUBLISHED BY ERNEST BENN LIMITED
BOUVERIE HOUSE · FLEET STREET · LONDON · EC4A 2DL

DISTRIBUTED IN CANADA BY
THE GENERAL PUBLISHING CO · LTD · TORONTO

FIRST EDITION 1923
SECOND EDITION REVISED 1933
THIRD EDITION REVISED AND ENLARGED 1959
FOURTH EDITION REVISED AND ENLARGED 1971

ISBN 0 510-47701-7

PRINTED IN GREAT BRITAIN

PREFACE TO THE FOURTH EDITION

ALTHOUGH it is only eleven years since the last printing of the third edition of *The Chemistry and Physics of Clays*, further extensive revision has proved necessary in this new publication.

The book deals with basic scientific principles pertaining to clays and allied ceramic materials and although these have not changed to any great extent in the last decade, certain new additions to knowledge warrant inclusion. Perhaps the most important single factor which has necessitated a revision has been the pronounced shift of emphasis in the clay industries and the expansion in the number of applications in which clays play a dominant role.

It was predicted in the third edition that the trends in the utilisation of clays would be towards more automatic units in which scientific control would replace the ancient arts of the craftsman.This has been realised to an extent, perhaps undreamed of twenty years ago, but the changes in industrial production methods, in brick, pottery and refractory manufacture are truly astounding and will undoubtedly continue. On the whole, such changes have been brought about only by a clearer appreciation of the underlying scientific principles which are involved. A better selection of raw materials, correct preparation of bodies from the physical, chemical and mineralogical points-of-view coupled with closer attention to the principles of forming, drying and firing have all been re-examined in the modern technologies. It is these aspects which have been more fully re-appraised in this new edition.

It is perhaps not always realised by ceramists that the vital raw material, clay, is also one of the most troublesome contaminants of so many mineral processing operations. It is an impurity which has to be removed because of its adverse influence on separation processes, but, because of its peculiar nature, this is far from easy. The coal industry of the U.K. has to remove and then safely dispose of a greater tonnage of clay waste than the total clay requirement of the ceramic industry. The new edition has been enlarged to include sections dealing in more detail with the scientific principles involved in problems of this nature. A detailed study on these lines may realise what is at present only a pipe-dream—the beneficiation and processing of waste clay into useful ceramic raw materials, either as basic clays or as aggregates.

New applications for clays are being developed. As catalysts, absorbants, binders and fillers, the useful characteristics of this most versatile material are being appreciated. The production of special grogs or aggregates, the purification, bleaching and organic modifications of clays are all acquiring new emphasis at the present time. Clays are being investigated as sources of alumina and aluminium; many of them form base materials for exciting new synthetic derivatives.

Finally, as man finds himself with greater leisure time, the art and science of clay-working is developing in home industries and hobbies. The number of enquiries which are received for more scientific information on clays and allied materials from schools, adult centres and individual operators is truly remarkable. It is an encouraging sign that whereas the larger industrial companies are being obliged to reduce their range of bodies, glazes and colours, the old arts and crafts are being rejuvenated, but now with greater scientific knowledge than ever before.

This new edition of *The Chemistry and Physics of Clays* has been designed to help all those concerned with clay problems. In keeping with previous editions it has preserved the essential characteristics of enabling the student of pure science to expand his knowledge of an exciting applied technology, whilst enabling production or managerial staff to acquire the basic scientific information to appreciate the newer skills of the industry. The information contained is sufficiently comprehensive to be of value to students reading for a ceramics or mineral processing degree in various countries throughout the world and has also been designed to help those wishing to acquire professional qualifications with various scientific bodies within the ceramic sphere.

During the course of preparation of this volume, new S.I. units were introduced. It has not been possible or even appropriate to include these in all parts, but in some of the more recent sections they have been inserted.

Readers will learn with regret of the death of Mr. Alfred Broadbent Searle on 26 November 1967, in his ninetieth year. He will be remembered by many as one of the pioneers of clay science and a great contributor to early ceramic literature. It is my sincere wish that his spirit will live on in this new edition.

It is impossible to list all those who have helped in the preparation of this volume. To the many academic and industrial associates who have provided information and guidance, I am very much indebted. I most sincerely thank the numerous post-graduate students and research assistants who have searched out new information and checked out new subject matter.

My greatest debt of gratitude is to my family, particularly my wife, for undertaking the enormous amount of copy correction, proof-reading and index compilation which a work of this size entails. I am particularly grateful to Miss Constance M. Stocks for her secretarial expertise and organisation and to my publishers and their printers for their friendly and helpful advice and assistance.

Rawdon
June 1971

REX W. GRIMSHAW

PREFACE TO THE THIRD EDITION

In the twenty-five years which have elapsed since the publication of the Second Edition of the *Chemistry and Physics of Clays*, all branches of science and technology have advanced to an almost unbelievable degree. The increased knowledge of ceramic materials has led to a new technical science, a fact recognised by many Universities and Colleges which now include 'ceramics' as part of their curriculum.

Although ceramics have been part of man's inheritance since prehistoric times, the last few decades have witnessed a remarkable advance in many aspects of their manufacture and application, due in the main to physico-chemical studies of the fundamental nature of the basic materials and their behaviour at high temperatures. The classical work of Bragg on the crystal structure of silicates, the applications of colloidal science and a more thorough understanding of the implication of phase equilibrium and reaction kinetics have all contributed to the increased knowledge of ceramic materials. The old rule-of-thumb methods of control are now being replaced by scientific processes, which although, unfortunately supplanting the craftsman, are undoubtedly opening the way to precision methods of manufacture. Despite the advances which have been made in the scientific approaches to ceramics, the older branches of the industry have, in the main, been the least to benefit. The next few years will be critical for the ceramic industry as a whole, especially that based on clays. The advent of competitive materials which are mass-produced and the changing standards of modern civilisation demand a new approach, both to the type of product and the methods of manufacture, which can be achieved only by a clearer and more detailed understanding of basic principles.

When I was approached by Mr. Searle to assist in the preparation of a new edition of the *Chemistry and Physics of Clays*, I was indeed honoured, and it has been a great personal pleasure to be so closely associated with one of the pioneers of ceramic science. Unfortunately a serious eye complaint has prevented Mr. Searle from making the contribution which he would have wished but, nevertheless, his encouragement and advice have been given unsparingly and to the full. Although the onus of compiling and the arrangement of the subject matter has had to be my entire responsibility, I sincerely hope that sufficient of the character and emphasis of the original editions have been retained.

To present an up-to-date picture of the physico-chemical aspects of clay and ceramic materials, of necessity, the subject matter must emphasise the importance of recent discoveries. Modern science, with the ramifications of crystal structure and mathematical theory is indeed a complicated subject, but, in keeping with the objects of previous publications by Mr. A. B. Searle, the present volume is designed to be of help to those with but a basic knowledge of science. At the same time, it has been compiled to cover the requirements of students of ceramics to full degree level in both Great Britain and the United States of America. The subject matter also is sufficiently comprehensive to embrace the basic science curriculum of the professional degrees of the Institute of Ceramics.

I am indebted to many present and former colleagues for their help and guidance in the preparation of this volume. To Professor A. L. Roberts of the University of Leeds and to Professors F. A. Hummel and W. A. Weyl of the Pennsylvania State University, I am particularly grateful for their stimulating discussion and advice. Much of the proof-reading and the compilation of the index has been undertaken by my wife, who, at all times, has given every help and encouragement. Amongst others who deserve particular mention are Dr. C. T. Cowan for his help with diagrams and Mr. P. Kirk who assisted with the final checking and reading of the proofs.

REX W. GRIMSHAW

Rawdon, Leeds
June 1958

A FURTHER impression having been called for the opportunity has been taken to correct a number of minor faults throughout the text.

1960 R.W.G.

PREFACE TO THE SECOND EDITION

Some years ago I endeavoured to follow Michael Faraday's method of 'selecting, co-ordinating, and publishing all the important facts that appertained to the advance of Chemistry and Physics in relation to clays and allied materials', and the first edition was one result of such effort. So rapid has been the increase in the amount of information on the Chemistry and Physics of Clays and other ceramic materials in the ten years since the first edition of this work was published that it is impossible for anyone to be engaged in active work and at the same time to read every statement on this vast subject, for the multiplicity and the kaleidoscopic diversity of chemical and physical theories constitute a phenomenon probably without parallel in recent scientific work. It has been said that the stream of serious articles and books having some bearing on the chemistry and physics of ceramic materials continues at the rate of more than five thousand a year. I need not say that I have not been able to keep abreast of this torrent; I have not even ventured to verify the estimate.

The ceramist who wishes to keep fully informed cannot simplify his task by selecting any single publication or any one theory without being liable to read too little and to miss much that would be of service. Yet, it is essential that some selection must be made in order to bring the more important facts and conclusions within the compass of a single volume. Some of the theories mentioned are of significance today, but the ceramics of the future will probably develop along lines as yet dimly foreseen. Moreover, ceramists are mainly practical men to whom theories are only of interest in so far as they are 'useful'. It seemed, therefore, both unwise and unnecessary to delay publication until there is complete agreement among leading chemists and physicists, even on such everyday subjects as the cause of plasticity, the chemical constitution of clay, and the precise nature of colouring agents in many clay products.

It is, surely, better to attempt to summarise the present state of knowledge of the subject—even if imperfectly—than to leave those in search of knowledge to flounder helplessly in an inchoate mass of articles and books, or to attempt, without any assistance, to find out laboriously for themselves what has already been found by others.

That much has been omitted is inevitable, but it is hoped that the present edition—which has been largely rewritten—will contain the greater part of the information required by those who wish to apply a knowledge of chemistry and physics to the clayworking industries, or to undertake research on these subjects. The fact that the first edition sold out rapidly encourages the author to hope that this second edition may prove to be even more useful than its predecessor.

ALFRED B. SEARLE

PREFACE TO THE FIRST EDITION

The clay-working and allied industries are exceedingly old and have reached a remarkable state of perfection in craftsmanship, aided by only very little scientific knowledge. Within the last forty years, however, and especially during the last decade, increasingly stringent demands have been made by users of electrical and chemical pottery and refractory materials, and manufacturers of paper, textiles, and other materials in which clays are used, so that inherited and artistic skill is no longer sufficient, and Science must contribute to that knowledge of causes and effects on which future developments in these industries must depend.

Only those who have devoted special attention to the subject can be aware of the importance of the applications of both physics and chemistry to the industrial uses of clays and other ceramic materials. Such applications are so extensive, that it is remarkable that no volume has previously been published in which they are dealt with in a systematic manner, for there is scarcely a branch of physics or of inorganic chemistry which is not of value when applied to the treatment of clays, allied materials, and the products made from them, and the more intense such application, the greater is, and will continue to be, the benefit to all concerned.

It may well be asked why those engaged in such extensive and important industries as those connected with the manufacture or use of ceramic materials have paid comparatively little attention to the fundamental scientific principles involved, and why so many advanced students of Science have largely neglected the study of clays and other ceramic materials? The answer to such questions is threefold: (*a*) physicists have chiefly concerned themselves with the properties of matter in mass; (*b*) students of pure chemistry, on the other hand, have chiefly dealt with the atoms and molecules of substances of a simpler character, the reactions of which can be more easily controlled, and whose properties and relationships can be studied in a more direct manner; and (*c*) most manufacturers and users have scarcely realised, as yet, the enormous importance, to them, of chemists and physicists with a highly specialised knowledge of these materials.

The difficulties experienced in obtaining perfectly pure substances, the general insolubility and apparent inertness of most ceramic materials at temperatures below a dull-red heat, and the impossibility of obtaining many of them in some readily recognisable form—such as crystals of convenient size—have hindered research, but as these difficulties are overcome, and more and more information regarding the constitution and properties of these materials and products is obtained, great technical advances will be made.

The field of research in this subject is so vast, however, that it will be many years before it is fully occupied. At present, many important fundamental principles still need to be investigated, and far too little is known of what may appear to be such simple matters as the effect of the texture of many ceramic products, the causes and control of their strength, the distribution of the water in partially dried articles, and the relief of the various strains which are produced during the drying and burning of many pieces of pottery, or of other ceramic articles used in the construction of furnaces, coke-ovens, sanitary appliances, etc. The causes and prevention of distortion afford another wide field of investigation which has, as yet, scarcely been entered, and the complexity of the problems of chemical equilibrium in relation to ceramics is such that only the simplest cases are, as yet, reasonably well understood. Far too little attention has been paid to the enormous influence of the size of the particles on the progress and nature of chemical reactions and of the physico-chemical changes which occur on heating clays and allied materials; and the study of such phenomena is, as yet, in its infancy. Colloidal phenomena also appear to offer almost endless opportunities for gaining further knowledge of the nature of clays and allied materials.

It will also be seen, in the following pages, that the chemical constitution of clays offers most fascinating problems, the complete solution of which has, so far, baffled the ablest chemists who have studied them, and a similar remark is equally applicable to increasing the plasticity of clays and to many of the other problems in physics and chemistry, with which those concerned with ceramic materials are in constant contact. Under these circumstances it will be obvious that the present volume is not intended to be an exhaustive treatment of the subject, but to provide, in a convenient form, such a description of the properties of various ceramic materials and articles, and of the application to them of the more important principles of chemistry and physics, as will be equally useful to students, manufacturers, and users. Several subjects of purely academic interest—such as the quantum theory and entropy—have been purposely omitted, as to have included them would have made the book unwieldy, without adding correspondingly to its usefulness.

Finally, the author wishes to acknowledge the zealous and skilled assistance of several members of his staff, without which this volume could not have been written. In this respect, he is specially grateful to Mr. W. L. Emmerson, who has taken a large share in collecting data, made many useful suggestions and read the proofs, and to Mr. F. Stones, who has assisted in other ways, including the compilation of the Index.

ALFRED B. SEARLE

CONTENTS

CHAPTER III

CRYSTAL STRUCTURE OF THE SILICATES

CHAPTER IV

THE CRYSTAL STRUCTURE OF NON-SILICEOUS CERAMIC MATERIALS

CHAPTER V

THE IDENTIFICATION AND ESTIMATION OF MINERALS IN CERAMIC MATERIALS

CHAPTER VI

THE CHEMICAL AND MINERALOGICAL COMPOSITION OF CERAMIC RAW MATERIALS

CHAPTER VII

PHYSICAL AND STRUCTURAL PROPERTIES OF RAW CERAMIC MATERIALS

CHAPTER VIII

THE INFLUENCE OF WATER IN CERAMIC SYSTEMS

CHAPTER IX

CHANGES IN THE PHYSICAL STATE OF CERAMIC MATERIALS BROUGHT ABOUT BY THE REMOVAL OF WATER

CHAPTER X

THE EQUILIBRIUM STATE AND PHYSICO-CHEMICAL REACTIONS

CHAPTER XI

CHEMICAL CHANGES IN CERAMIC MATERIALS

CHAPTER XII

PHYSICAL CHANGES IN CERAMIC MATERIALS

CHAPTER XIII

STRENGTH AND ALLIED PROPERTIES

CHAPTER XIV

MISCELLANEOUS PROPERTIES OF CERAMIC MATERIALS

LIST OF ILLUSTRATIONS

CHAPTER IV

CHAPTER V

CHAPTER VI

CHAPTER VII

CHAPTER VIII

CHAPTER IX

CHAPTER X

CHAPTER XIV

CHAPTER I

THE NATURE OF CLAYS AND OTHER CERAMIC MATERIALS

THE scope of materials classed as ceramics has increased widely in the last decade. Ceramic materials[1] include *clays*, which are essentially alumino-silicates with many other minor but important components, *several oxides* and other compounds valued for special purposes, *glass*, *concrete* and many *electrical* and *magnetic materials* synthesised for some useful purpose. In general they may be regarded as of a mineral nature, but the range is almost unlimited. It is no longer implicit that a ceramic material should be fired as the last stage in its manufacture, as is evidenced by concrete and refractory ramming mixes. Other ceramic materials are heated and thus shaped—as is glass, which is first liquefied and then formed to suitable shapes or products. Most materials, however, which are defined as ceramics are shaped from some mineral form and then fired to give strength and permanence.

The word *ceramics* is derived from the Greek *keramikos*, which literally translated means *earthen*. So ceramics may be defined in a general sense as 'materials and articles made from naturally occurring earths'.

Since the earliest times, man has recognised that clay can be modelled, thrown or moulded and then transformed into a hard, dense, permanent shape by the action of heat. So 'ceramics' is largely synonymous with pottery and other articles made of burned clay. In recent years, the development of a wide range of articles from natural earths and rocks[2] other than clay has extended the scope of the definition; it is still true, however, that clays are, by far, the raw material most extensively used in the manufacture of ceramic articles.

Clay cannot, at present, be defined in precise terms. The word has a popular and also a series of technical and legal meanings, which are useful but not identical; so that a farmer's or agricultural chemist's concept of clay may not be acceptable to a potter, a brick manufacturer, a judge or a ceramist.[3]

In the popular sense, clay is a finely-grained rock which becomes sticky when wet. Potters and others interested in ceramics apply the term to substances which are plastic or which can be made plastic by suitable treatment and the addition of water. Chemists and physicists define *clay* as an admixture of certain mineralogical species; such a concept is both convenient and valuable, but may be misleading in some instances.

A concise definition of clay by the American Ceramic Society is as follows: 'Clay is a fine-grained rock which, when suitably crushed and pulverised, becomes

[1] For a list of the chief raw ceramic materials and the products made from them, see p. 35.

[2] A *rock* is any mineral or aggregate of minerals which forms an essential part of the earth and shows a well-marked individuality (see p. 47).

[3] A *ceramist* is a person engaged in the use or scientific investigation of clays and other ceramic materials.

plastic when wet, leather-hard when dried and on firing is converted to a permanent rock-like mass.'

In seeking a definition of clay, it is essential to bear in mind the precise purpose for which the definition is required, the nature of the operations to be applied to the material and the use to be made of it. This may be unscientific, but any attempt at a single definition of clay which will be equally satisfactory to all who have to deal with it and other ceramic materials, may lead to confusion or to wholly erroneous conclusions.

In the present volume, the word *clay* is chiefly used as meaning the natural material either as found, or after such processes of cleaning, purification or preparation as it may have undergone. When a highly-purified clay is being considered, its essential constituent is termed *clay mineral* (or in some instances *clay substance*) according to which best suits the context.

The Historical Development of Ceramics

Some idea of the relative importance of the different types of ceramic materials may be gained from a study of their historical development.

The discovery of ceramics ranks as one of man's earliest scientific achievements.

Some enthusiastic ceramists claim that theirs was the first Science, for in Genesis there is the statement that 'God made man from clay'. Perhaps if this is not sufficient to prove that ceramics is the earliest science, it does suggest that at least it was the earliest technology.

The early history of man is traced mainly through his ceramics. All ancient settlements left traces of their existence only through indestructible material. Wood and cloth have vanished; only pottery, flint, bricks, stone products and some metal articles (chiefly gold, copper and bronze) have survived the rigours of thousands of years. From a study of these we can often visualise the modes of life of our early forebears.

The earliest traces of primeval man are to be found in the clay on which he walked. Footprints, baked hard by the sun, have remained as such up to the present day. Clay will stick to most things with which it comes into contact. An early development seems to have been the cementing of a wood and grass framework by clay, and this probably led to the manufacture of rough sun-dried bricks. These were serviceable in dry weather and could resist showers, but not prolonged heavy rain.

Then one day came the *fire*, and the clay body was converted into a hard mass, indestructible by water. Later, deliberate burning would follow and then the making of special fired-clay shapes for drinking vessels and other utensils. Ceramics had been born.

The exact date can never be known with certainty, but it is possible that simultaneous discoveries of fired clay-ware were made in several parts of the world. Crude vessels of burned clay have been unearthed in England, Belgium, Germany, North and South America, Egypt, Mesopotamia and elsewhere, dating back to about 5000 B.C.

The development from these first crude shapes to the beautiful, glazed articles of the later civilisations was slow and can be traced only with difficulty.

Egypt has produced pottery discovered on the sites of old tombs, dwellings and temples, which has proved very illuminating. Terra-cotta vases, probably used to contain provisions for the dead, can be ascribed to the Memphite period (5000–3000 B.C.). Although these are of crude clay they are of exquisite shape, and so highly polished that they appear glazed. Undoubtedly the Egyptians had learned the art of glazing by 3500 B.C., for decorative tiles have been found with a blue copper glaze covering a body of sandy clay. Some insight into the methods of these early potters has been given by the tableaux on the walls of the tombs depicting scenes from everyday life in the Theban period (3000–1700 B.C.). Later, the Egyptians also knew how to make glazed earthenware, for many beautiful decorative pieces have been found in the pyramids and temples.

In **Babylonia and Assyria** among the ruins of the cities of the ancient civilisation of Mesopotamia are to be seen many examples of the art of pottery-making, using crude clays as raw material. A feature of this period is the large single-piece ware, made with high precision. The British Museum possesses coffins made of clay which have been enamelled with different colours.

Persian pottery was a natural successor to the Assyrian. Coloured glazes were developed by adding metal oxides to an alkaline base. There are undoubted examples of ware made on an early type of potter's wheel. The early Indian races almost certainly copied Persian methods, for much of their produce is of similar type.

Early Grecian civilisation is remarkable for the contribution it has made to the Arts and Sciences; in the field of Ceramics, the Greeks departed from the conventional methods of earlier workers and specialised more on delicacy of shape than on heavy glazed ware.

They covered their ornate pieces with a thin transparent gloss, which was acidic in character. To achieve exquisite shape they must have purified the clay by washing and sieving, thus producing a fine-textured material from which the larger particles had been removed.

The **Ancient Romans** were great empire-builders and architects, and although they did produce some pottery with rich ornamentation in Grecian style, their real contribution to Ceramics is in large-scale work. They specialised in bricks and tiles, and many of their buildings (largely of brick) are preserved to this day. They made baths, aqueducts, drain-pipes and even vitrified stoneware pipes which were shaped at the ends so that they could be fitted together and then cemented into place. Wherever the Romans travelled, they took their ceramic art, and there is one remarkable fact—the ware is always similar and independent of the locality in which it is found; its colour, texture and density vary but little. Roman pottery is characterised by two different types:

(*a*) *Samian ware*, which is red;
(*b*) *Etruscan ware*, which is black and not so abundant.

On the decline of the Roman Empire, Ceramic Art in Europe also fell into a state of decay, and so remained for many centuries.

Far Eastern Ceramic Ware. The first records of bricks and tiles being made in China relate to the period of the Emperor Wu (140–88 B.C.), though legends infer a much earlier date. Historians[1] have stated that the first pottery articles produced in China were made about 2500 B.C.; these were crude, and fine Chinese pottery does not date back further than 200 B.C. Ceramic ware has been associated with the various Dynasties of Emperors:

(*a*) The *Han Dynasty* (206 B.C.–A.D. 220) is noted for its stoneware type with a brown or bluish-green glaze and for the invention of the potter's wheel in China. Attempts to make porcelain by applying imported ground glass yielded a poor product and the firing temperature was much too low.

(*b*) The *Sui Dynasty* (A.D. 581–617) witnessed the appearance of the first translucent porcelains and other ceramic wares. Opaque, glossy chinaware was made from kaolin. The early part of this Dynasty is regarded as the Great Period of production of Celadon (green) ware.

(*c*) The *Tang Dynasty* (A.D. 618–906) saw great advances in porcelain and stoneware sculptures. A much higher firing temperature was used and the quality of the porcelains produced was thereby greatly improved. About this time, pottery manufacture had a great fillip in China with the advent of ceremonial tea-drinking and the consequent necessity for many cups.

(*d*) The *Sung Dynasty* (A.D. 960–1280) produced finer bodies and a greater variety of glazes and colourings. During this Dynasty, porcelains of delicate appearance and good quality were produced and new designs in shape, colour and decoration were employed. For these reasons, this period is regarded as a golden one in the history of pottery and porcelain in China. The export of ware to Europe began and was known (from its place of origin) as *chinaware*.

(*e*) The *Ming Dynasty* (A.D. 1368–1643) represents a period in which ceramic wares attained a high degree of delicacy of product and decoration and technical perfection. Beautiful examples of overglazed porcelains made in this period are much sought after to this day.

(*f*) The *Manchu Dynasty* (A.D. 1643–1912) appears to have included two or more minor dynasties:

(1) The *K'ang-Hsi Dynasty* (A.D. 1662–1722) saw an increase in the standard; in it were developed coloured glazes of a quality which surpasses our present-day endeavours.

(2) The *Yung Cheng* and *Ch'ien Lung Dynasties* (A.D. 1723–95) are remarkable for the close similarity of manufacturing methods to those of the present day. Wood-fired kilns were used, which may account for the perfection achieved in glaze colouring.

Although the latter part of the Manchu Dynasty saw a period of marked decline, the porcelains retained their great reputation and the industry flourished into the early years of the twentieth century.

[1] Li Ch'iao-p'ing, 'The Chemical Arts of Old China', *Journal of Chemical Education* (Easton, Pa., U.S.A., 1948).

Japanese pottery was a natural development from Chinese ware. The Japanese extended the scope of 'ceramics' and became famous for 'egg-shell' and thin-sectioned translucent chinaware. They were probably the first to use lead-borax glazes.

European pottery (apart from the crudest earthenware) almost ceased to be made after the withdrawal of the Roman legions and so continued until the Arabs and Saracens brought the art to Spain. They introduced ware with a porous body of clay covered with a white, opaque glaze of high lustre. *Majolica ware* was made originally on the Isle of Majorca, but its manufacture spread throughout Europe, especially to Italy. Many Italian artists devoted themselves to Ceramics in the fifteenth century and a characteristic style emerged, known as *faience*, after the town of Faenza where the industry was centred. The French exploited this kind of ware to the full in the seventeenth and eighteenth centuries. Excellent clays for the production of porous pottery were common in many parts of France and ware with a matt glaze, containing tin, became very popular.

The development of *porcelain* in Europe about the end of the eighteenth century brought about a decline in the popularity of faience. The Dutch also mastered the art of manufacturing faience, but they, too, suffered from the competition of foreign porcelains. True porcelain was first produced in Europe by the Germans, who, after the discovery of local white kaolin, closely followed Chinese methods and established the porcelain industry at Dresden and elsewhere.

British pottery was relatively slow to develop. The 'Cistercian ware', thought to have been made by monks of the sixteenth century, has a coarsely-grained body with a heavy glaze. *Slip-ware* was manufactured in Staffordshire in the seventeenth century by the famous brothers Toft and others; although rather crude in design, this ware represents a great step forward in ceramic technique and is therefore prized by collectors. Following the introduction of the Majolica industry at Lambeth by the Dutch, Dwight produced a type of salt-glazed porcelain, which proved superior to the German counterpart. A few years later, around 1690, the Elers brothers used washed Staffordshire clays for making dense ware, but it was not until well into the eighteenth century that china clay was used to give whiter earthenware bodies. Once this step had been taken, English chinaware progressed at an amazing rate, so that at the beginning of the nineteenth century, with the introduction of bone china, it became world-renowned. In recent years, the trend has been towards more mechanisation and precision of ware than to new materials or bodies.

American pottery has developed on distinctive lines. The primitive pottery of South America does not differ materially from other primitive wares although it has distinctive styles of decoration not found elsewhere. The same may also be said of the 'Indian' earthenware, which is typified by a heavy, brightly-coloured glazing. The modern pottery industry of the United States was brought in by European emigrants, but an individual style emerged which was dictated largely by the types of clay and the needs of a rapidly-expanding populace. Ceramic ware developed along rather different lines from those in Great Britain and resembled more closely that of Central and Western Europe. In recent years, intensive research has led to a great improvement in body and glaze compositions, so that, despite the lack of suitable, easily-accessible china clay, modern American pottery is of excellent quality.

There is a very large and intensive literature on Ancient and Historical Ceramic Wares and any reader interested in this aspect of the subject should refer to it; the present volume is mainly concerned with the chemistry and physics of ceramic materials and articles. Clay, however, as a survey of the modern pottery industry shows, is still the all-important raw material.

Important Ceramic Articles and Materials

Table I.I shows the chief ceramic products and the materials from which they are made. Minerals and other substances shown in brackets are sometimes used to impart special properties or to modify undesirable ones, such as excessive shrinkage during drying or firing, low resistance to high temperatures or to thermal shock.

Structural Clay Products. Articles made chiefly of clay and used in the Building Trades and in Civil Engineering are commonly known as 'Structural Products'; they include what is generally known as The Heavy Clay Industries.

In addition to the articles mentioned in Table I.I this group includes acid-resisting bricks, tiles, tanks and pipes used in the chemical industries and to some extent in others, such as leather manufacture and in a variety of articles used in the transmission of electricity such as cable conduits and insulators.

Clay products for the retaining or exclusion of heat at low or moderate temperatures are also included, but those used at higher temperatures are included in 'Refractories'.

Refractory Materials. Firebricks and other materials, which are chiefly used because of their great durability at high temperatures, have been in use since about 2000 B.C., but those which are now known specifically as *refractory materials* (colloquially *refractories*) have only been recognised, as such, for about 300 years and some of them for less than a few decades. This is because the need for materials to resist very high temperatures arose only with the development of the metal-smelting and the glass-melting industries.

At first, stones of moderate heat resistance were utilised, then the superior quality of bricks made of certain *fireclays* was realised. About 200 years ago, *silica bricks* made of *ganister* and other silica rocks were found to be particularly suitable for furnaces used in steel-making, and still more recently basic materials, such as *magnesite*, *dolomite* and *chromite*, were introduced. With the coming of the Jet-engine and Atomic Ages, even more problems have had to be solved and the new developments include the manufacture of articles made from sintered oxides, other highly-refractory bodies and combinations of metals and ceramic materials known as *cermets*.

The introduction of these new products has, however, been on highly specialised lines, for the main bulk of ceramic products still contain clays, sand, flint, felspar and the like as their principal raw materials.

TABLE I—I. RAW MATERIALS USED IN THE PRINCIPAL CERAMIC INDUSTRIES

Industry	*Type of Product*	*Raw Materials Used*
Structural	Building bricks of all types	
	Hollow blocks; Terra-cotta	Clay (sand, chalk)
	Roofing tiles and floor tiles	
	Flower pots. Chimney pots and flues	Clay
	Unglazed (land) drain-pipes	
	Glazed pipes and accessories	Clay and glaze (often salt)
	Concrete	Limestone, clay
Refractories	Fireclay bricks and raw fireclay	Fireclay (grog or pre-calcined clay)
	Silica- and siliceous-bricks and raw stone	Quartzite rocks, ganisters, etc.
	Magnesite bricks	Magnesium-containing ores, principally carbonates, hydroxides or from sea-water
	Chrome bricks	Chrome ores
	Chrome-magnesites	Chrome- and magnesium-ores
	Dolomite bricks	Dolomite (serpentine, talc)
	Carbon bricks	Graphite (clay)
	Silicon carbide bricks and shapes	Coke, sand (salt, sawdust)
	Insulating refractories	Clay or other refractory and combustible, e.g. sawdust
		Diatomaceous earth, vermiculite, asbestos
	Sillimanite bricks	Sillimanite. Kyanite or Andalusite
	High (over 50 per cent) alumina bricks	Fireclay and Bauxite
		Bauxitic clays
	Retorts, Crucibles, Ladles, etc., for metallurgical, gas-plant, scientific and other applications	Clays, silica rock, graphite (other refractory materials)
Pottery	Glazed wall and earth tiles	Clays and Glaze, Pyrophyllite, Talc
	Domestic and Sanitary Earthenware	Clays, Felspar, Flint, Glaze, Cornish stone
	Bone china (chinaware)	Clays, Bone ash, Cornish stone
		Felspar
	Porcelain (domestic, laboratory, industrial)	Clays, Felspar, Flint
	Stoneware	Clay (fluxing agents)
	Glazes and Engobes	Clays, Flint, Felspar, Other fluxing reagents (colouring compounds)
Miscellaneous Ceramic Materials	Fused silica	High-purity quartz crystal or sand
	Special Refractories	Pure oxides, e.g. BeO, Al_2O_3, MgO, ZrO_2, ThO_2, zircon, spinels, carbides, nitrides, phosphides, etc.
	Electrical ware	Clays, Flint, Felspar
		Rutile (TiO_2)
		Steatite, Talc
		Magnesium and Barium Titanates
		Zircon
	Magnetic materials	Ferrites
	Cermets	Metals and Ceramic Oxides
	Glasses, Glazes, Enamels	Sand, Soda Ash, Limestone (felspar, sodium sulphate and nitrate, borax and boric acid, lead oxide, potassium compounds, miscellaneous colouring and opacifying agents)
	Abrasives	Silicon carbide, alumina

Pottery.[1] The term *pottery* includes many varieties of ware from the crudest vessels of prehistoric times to the most beautifully decorated porcelains, stoneware and earthenware; it also includes many articles such as large grain-jars used in ancient times for storing corn and other dry materials, wine-jars and modern sanitary ware and the large tanks for containing corrosive acids. Many kinds of earthenware, stoneware and porcelains are used for scientific and experimental purposes as well as for electrical apparatus (insulators, switch-bases, sparking plugs and as bases or frames for electrical heating appliances).

As clays are the chief constituent of all kinds of pottery, they and their products are of very great importance to many industries.

Glass. Glass is not a ceramic material in the sense that articles made of it are not produced in the same way as ceramic ones, i.e. glass articles are made from molten or semi-molten glass by moulding or blowing, and (unlike pottery and other fired ceramic bodies) heating is not the last part of the manufacturing process. Various kinds of glasses are formed in the burning of many ceramic bodies (especially in porcelain, stoneware, refractory and other clay materials, and in glazes applied to certain types of pottery, bricks, tiles, sanitary ware, etc.). Any study or treatise on ceramic materials would therefore be incomplete without due consideration being given to glass.

Glass must have had its origin shortly after the first fired ceramics were produced. The fusion of sand in the ashes of a wood fire appears to have been an early observation, but difficulties arose in the shaping of the molten mass and little is known of ancient glass. Assyria and Egypt provide some of the earliest examples, but few articles can be dated back with certainty to pre-2000 B.C. Beads made from a pottery glaze about 1600 B.C. have been found, but it was not until a much later date that the art of blowing glass was discovered.

The Romans had drinking glasses in the time of Nero (A.D. 54) and by A.D. 220 this was a flourishing industry. Window glass was in use by A.D. 422.

Venice was famed in the thirteenth century for mirrors and drinking glasses of beautiful colour with unrivalled gilding and filigree work.

The first record of glass being used in England was at Monkwearmouth, where windows were installed in the Abbey in A.D. 674. The industry really started in A.D. 1557, when the first lead-flint glass was made. By 1700, the manufacture of drinking glasses superseded in quality those of Venetian origin, which were mainly of lime glass. In 1773, a factory was established at Prescott in Lancashire and thus began what is now a great industry using local sand.

The principal raw materials used in the manufacture of glass are shown in Table I.I and are discussed in more detail in Chapter VI.

Glazes. The glazes which produce a glossy surface on pottery, stoneware and other clayware are *glasses* though they differ in composition from most commercial glasses and are used for a different purpose. (See Chapter VI.)

[1] It has been customary for many years to regard *porcelain* as distinct from *pottery* largely because of its greater beauty and the mystery which was, for a long time, associated with its production. Strictly, the term pottery includes a series of a great variety of ware with the crudest earthenware at one end, fine earthenwares and stonewares as intermediate and porcelain at the other end of the series.

Enamels may be regarded as a variety of glazes which are suitable for use on metals. The term is also applied to some opaque glazes used on pottery and porcelain for decorative purposes.

THE GEOLOGY OF CLAYS AND OTHER CERAMIC MATERIALS

NO study of clays and related minerals could be complete without an understanding of the basic geological principles underlying their formation. Ceramic materials are represented in virtually every type of rock formation from the oldest to the most recent, and in igneous and sedimentary formations of all types.

Composition of the Earth

A FULL discussion of the nature and origins of the earth is outside the scope of this volume. It may be presumed that the earth has passed through several evolutionary cycles before assuming its present form. Even to the present day, however, both the true origins and the detailed composition of the earth are purely speculative. It has been calculated that the average density of our planet is over 5·5 and that the centre must be extremely hot and probably molten. Man's knowledge of the mineralogy of his world extends barely two miles into the crust of 7,600 miles which is the total diameter. That the surface minerals are markedly different from the whole is confirmed by their average density which has been calculated at a mere 2·7. The central core must be far richer in heavy elements such as nickel, iron and other components than the surface formations.

Seismographic, gravimetric and other studies of the earth's hidden mysteries have suggested that within the relatively thin surface crust there are well-defined layers, although probably with no well-defined boundaries. It is to be expected that lighter components of the basic magma would rise to the surface, and so immediately underlying the crust is a hot, probably fluid, layer of lighter components consisting chiefly of the elements silica, alumina, alkalies, oxygen, hydrogen and sulphur. This is often referred to as the *sial* layer and is between 25 and 50 miles in thickness. From a chemical standpoint the sial is predominantly acidic in character because of its high content of silica.

Below the sial layer, another formation is recognised with a different composition, predominantly of magnesium, iron and minor amounts of heavy minerals with oxygen and some silicon. This layer, termed the *sima* layer, is basic in character and is probably hotter and more fluid than the layer above. The sima layer is also of variable thickness, but is thought to vary between 25 and 50 miles.

These two geological horizons are important in that they may be regarded as the parents of all known surface formations. Undoubtedly the first rocks to crystallise were of these layers, and, in general, all surface igneous strata have been derived from these layers at some stage in geological history.

Igneous Formations

THE earth's crust is relatively thin. Under certain circumstances such as faulting or folding, the underlying molten layers may force their way to the surface or flow into masses of solid rock strata. Subsequent cooling of such magmas results in the formation of igneous rocks which correspond roughly in composition with that predicted for either the sial or the sima layers. Because of its closer proximity to the surface, sial material more readily breaks through the solid rock crust, so that most of the present-day volcanic activity is of acid magma, characterised by violent eruptions of an explosive nature. At times of major earth upheavals, magma from the sima layer can well-up and penetrate to the surface, so causing widespread rock formations of a basic character on cooling.

Geologists can thus distinguish two major divisions of **igneous rocks** based on composition—acid materials typified by *granites*, and basic formations of which *basalts* are an example. Within the two types there are many subdivisions of mineral composition and crystallite size, a full description of which may be found in most standard books of mineralogy and structural geology. By and large, granitic and basalt rocks are of small importance to ceramists. They are hard rocks, valuable as aggregates for concretes and allied materials and some are used as components of cheap glazes.

The process of cooling of the molten magmas is usually slow and not infrequently fractional crystallisation may occur. Refractory minerals, crystallising in the earlier stages, may rise or sink in the body of the predominantly liquid magma and so become concentrated. Equally, as crystallisation becomes well advanced, the remaining molten phase becomes more concentrated in less refractory ingredients. Capillary effects may localise the final liquid which ultimately crystallises with a predominantly different composition from the magma as a whole. Such localised concentrations are termed **segregations** and these often contain minerals of great economic importance. As examples of ceramic source-materials derived from such segregations may be cited *chromites*, *magnetites*, *ilmenites* and several *felspars*, *felspathoids* and *micas*.

Deep-lying molten magmas and the hot fluid which is pushed towards the surface as sills and dykes, affect the surrounding country rock, often over extensive areas. Heat and pressure play their part in processes of compaction, sintering and partial fusing. Migrations of minerals or of chemical components of minerals are common, so that the character of the original rock may dramatically change. Such geological processes are known as **metamorphisms** and many important ceramic minerals originate in this way. *Kyanites*, *corundum*, *syenites*, *marbles*, *slates* and *quartzites* are but a few examples of products from metamorphic transformations. Not all metamorphic changes involve heat and some are apparently unrelated to igneous activity. Pressure developed either by the weight of overlying strata or by land movements such as folding or faulting can be sufficient of itself to produce marked changes in rock character, as, for example, in *slates* and *graphite*.

Igneous activity is not confined to fluid magmas, but invariably involves massive volumes of gases at high temperatures and pressures. Although water vapour, oxygen and nitrogen are undoubtedly the major components involved, other volatile components such as carbon, sulphur, phosphorus, fluorine, chlorine and boron are frequently involved, and many other elemental components may be contained either in solution or in the vapour form. Such gases readily penetrate porous strata or cracks and fissures in surface rocks and bring about new mineral entities by processes which can generally be grouped as **pneumatolytic** changes. Three main types of activity may be recognised:

(*a*) Where the pneumatolytic agencies change the chemical and physical nature of rocks or mineral components. Such changes may be widespread over a massive rock formation, an example of which is the hydrothermal transformation of the felspar in some Cornish granites to *kaolinite*. In other cases, the changes may be confined to small veins or fissures where the country rock is dissolved and then recrystallised in another form, as for example in some *asbestos* deposits derived from ultra-basic rocks.

(*b*) Where components of the gaseous phase react with the country rock to form a new mineral. This is the probable origin of *fluorspar*, calcium fluoride, derived from contact between limestone rock and fluorine-rich gases. *Tourmalines* and certain *apatites* have probably a similar origin.

(*c*) Where the gaseous phase carries components from deep-seated layers and deposits them in surface rocks either in disseminated form or as veins ranging in thickness from a fraction of an inch to several feet. This is the geological origin of many of the earth's important metal sources—*copper*, *lead*, *tin*, *zinc*, *mercury* and many other elements in either sulphide or oxide form originate in this way. *Barytes*, barium sulphate, is another example of a mineral formed in a similar fashion.

Apart from pneumatolytic activity all processes of an igneous nature lead to the formation of *primary* rocks and minerals. The first solid formations to crystallise on the earth's surface were undoubtedly of this nature and similar processes are involved in many of our present-day surface rocks. However, natural climatic agencies alter many of the igneous formations by both physical and chemical processes and give rise to entirely different types of deposits.

Sedimentary Formations

The action of ascending hot gases and solutions on surface rock formations has already been described, but surface agencies, although less severe, can also bring about dramatic changes in rocks of all types. This category of rock-forming agencies is of importance as it includes most *clays*, *quartzites* and other large tonnage ceramic materials. Such reactions which take place on the surface are called **epigenic changes** and the class comprises all changes in the composition and distribution of minerals which are brought about by the action of natural agencies such as air, water and ice, in conjunction with climatic conditions. They are also included under the general term of *weathering*. Epigenic reactions are of four types:

1. Influence of air, water vapour and other substances contained therein.
2. Action of water in the form of rain, rivers, hail, lakes or sea.
3. Action of snow and ice, as in glaciers.
4. Action of plants and animals in their growth and decay.

The action of sunlight and gases such as carbon dioxide, sulphur dioxide, etc., may also be an influencing factor in some cases.

Weathering reactions are of two types involving the mechanical and chemical breakdown of the parent rock. A common example of mechanical disintegration is frost erosion—water enters a crack or fissure in the rock and then freezes. The slight expansion resulting serves to enlarge the cavity, more water enters and the process continues until a fragment of rock breaks away. (See Chapter VII.)

Chemical attack is greatly enhanced by the increase in the surface area of the rock, resulting from this mechanical fissuring and fracturing. Water, sometimes heavily charged with substances such as carbon dioxide and humic acids, is capable of decomposing some of the primary rock-forming minerals. An example is the conversion of felspars to kaolinite or other similar minerals; micas are broken down to a hydrated variety; chlorites and many other minerals can be produced under certain conditions. Although such changes are slow they do take place even under normal climatic conditions. Changes of this type result in the formation of a wide range of *secondary* minerals. In addition, soluble salts are produced which may remain in intimate association.

Thus, under the influence of external agencies the parent rock is decomposed to form a mixture of finely-divided primary and secondary minerals which may remain *in situ* or be transported and deposited elsewhere. In the first case, the product is termed *residual*, whilst in the second case it is known as *sedimentary*. These two terms are preferable to 'primary' and 'secondary' which are often misleading.

Residual deposits are often formed under hypogenic conditions. They are frequently contaminated by large-grained primary minerals, and in many cases the alteration is present to great depths. There are reasons for believing that the Cornish china-clay deposits are over a mile in depth.

Most laterites are residual deposits but resulting from surface activity. The secondary minerals of clay, hydroxides of alumina and iron remain as replacement minerals of the original weathered rock.

Many talc deposits are also derived in this way.

Sedimentary deposits are by far the most common, and their nature is determined largely by the mode of transportation and of deposition. They can be classified under six headings:

1. *Fluviatile*—formed by sedimentation along the course of a river.
2. *Glacial*—transported by large masses of ice.
3. *Estuarine* or *deltaic*—deposited at the mouth of a river.
4. *Lacustrine*—deposited in a fresh-water lake.
5. *Marine*—deposited in a sea or ocean.
6. *Aeolian*—transported by air.

(1) **Fluviatile deposits** along the course of a river are often small in extent and of poor quality clay. They are rarely more than 20 feet thick, sometimes only a few inches. Because of this and their variable composition, such clays are of commercial value only where they have been deposited by a large river, e.g. the Nile valley and the London Basin. The formation of fluviatile clay requires a large, slow-moving, meandering river, continuously eroding its banks and bed and depositing as illustrated in Fig. I.1.

Fluviatile clays may also be formed in the flood plains of large rivers, though they are then preferably termed *alluvial clays*.

(2) **Glacial drift** is the detritus left by the large ice masses (glaciers) which

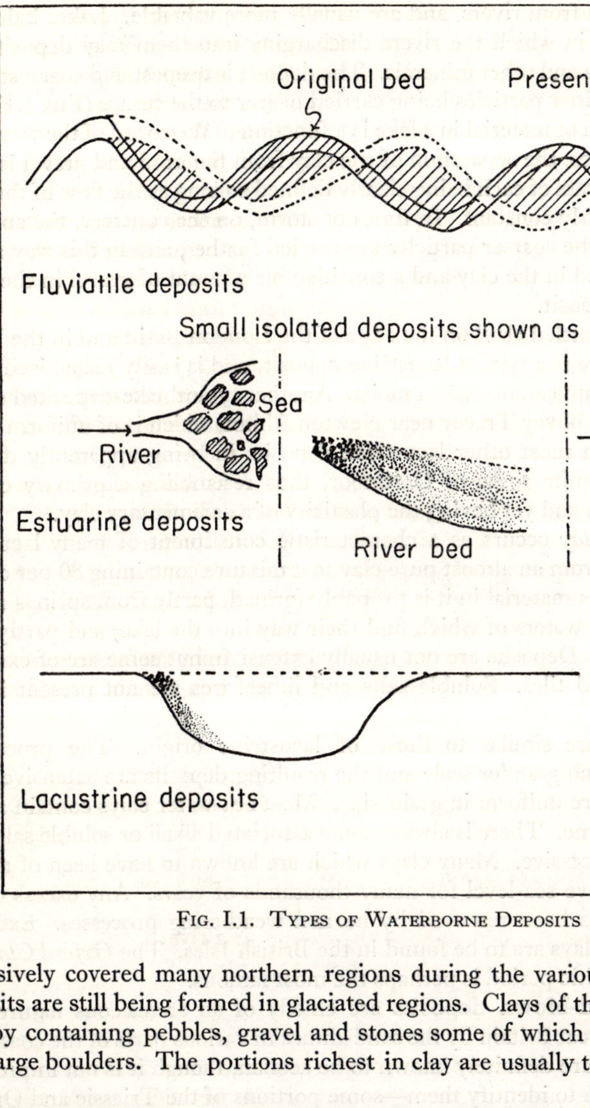

FIG. I.1. TYPES OF WATERBORNE DEPOSITS

extensively covered many northern regions during the various Ice Ages. Similar deposits are still being formed in glaciated regions. Clays of this type are characterised by containing pebbles, gravel and stones some of which may be agglomerated into large boulders. The portions richest in clay are usually termed *boulder clay* or *till*.

(3) **Deltaic formations** are usually sandy and irregular in composition. The fast-flowing river carries fragments of rocks of many sizes by virtue of the speed of flow. At the river-mouth a sudden considerable reduction in velocity takes place; hence many particles of different sizes may settle out together (Fig. I.1). The deposits show frequent cross-bedding. In Lincolnshire and Northamptonshire, estuarine clays of Jurassic formation are used for brick-making.

(4) **Lacustrine beds** are those formed by the sedimentation of matter in lakes. They are characterised by a much greater uniformity of composition than beds

formed by deposition from rivers, and are usually more valuable. Lakes behave as great 'settling tanks' in which the rivers discharging into them may deposit their burden of clays, sands and other minerals. The deposit is deepest and coarsest at the inlet of the lake, the finer particles being carried nearer to the centre (Fig. I.1).

As the deposition of material in a lake is a function of the speed of the water and the size of the particles, the separation of the clay from the sand and gravel is often very complete, the strata of each being clearly defined so long as the flow of the river feeding the lake is fairly constant. In times of storm, on the contrary, the speed of the water is greater, the coarser particles are carried further, and in this way a little sand may be deposited in the clay and a considerable amount of gravel in the sand, forming a diluvial deposit.

The 'Reading mottled clay' on the edge of the London Basin and in the Isle of Wight and Hampshire is a typical lacustrine deposit, and is easily recognised by its striking red-mottled appearance when moist. Another typical lake-deposited clay of great purity is that at Bovey Tracey near Newton Abbot, which is of uniform thickness and differs from most other lacustrine deposits in being apparently derived directly from the granite rocks of Dartmoor, thus resembling a primary clay in chemical composition and yet having the plasticity of a sedimentary clay.

Marl or *marly clay* occurs as a characteristic constituent of many lacustrine deposits, and varies from an almost pure clay to a mixture containing 80 per cent of chalk. The calcareous material in it is probably formed, partly from springs rich in lime compounds, the waters of which find their way into the lake, and partly from fresh-water shellfish. Deposits are not usually extensive, but some are of excellent quality for bricks and tiles. Soluble salts and limestones are not present in any quantity.

(5) **Marine clays** are similar to those of lacustrine origin. The process of formation is on a much grander scale and the resulting deposits are extensive, deep and are generally more uniform in grain size. Most sea-water clays contain appreciable quantities of lime. There is always some associated alkali or soluble salts, but this is not usually excessive. Many clays which are known to have been of marine origin have been above sea-level for many thousands of years. Any excess of salt originally present has been removed by normal weathering processes. Excellent examples of marine clays are to be found in the British Isles. The *Oxford Clay* laid down during the Oolitic period is perhaps the most famous.

(6) **Aeolian or wind-blown deposits** are chiefly of an arenaceous nature, and are represented in Great Britain by the sand-dunes in various parts of the coast. No clays in this country are definitely known to be aeolian, though it is not improbable that—were it possible to identify them—some portions of the Triassic and Ordovician clays are of aeolian formation. The *Loess* of some areas appears to be an air-transported silty clay, but more usually it is of subaqueous origin.

These processes have been in progress for millions of years. It is certain that in the course of geological time many minerals have been formed, transported and redeposited numerous times. Folding, faulting, metamorphism and other geological processes can fundamentally alter the nature of a particular deposit. Clays and other related deposits have been formed throughout geological time and the age of final deposition is an important feature in determining the character of the clay.

Some of the earliest deposits have been compressed by the weight of overlying strata and have been so altered by this and by earth movements that they can no longer be termed clays. Table I.II shows the geological sequence and the age of the most important deposits.

Geologically, therefore, clays are composed of various minerals of primary and secondary origin, all of which are of a comparatively 'fine' or small grain size. Some minerals are often present in grains which are larger than those of the other minerals, but, because of the nature of their formation, true clays only infrequently contain fragments visible to the unaided eye. Clays in their natural form may consist of many different minerals, not one of which predominates. It is meaningless therefore to refer to *pure clays*. There are, however, certain minerals or classes of minerals which occur in all clays and have been termed *clay minerals*. These can be separated, by virtue of their small particle size, from the coarser-grained components in a clay and in some instances can be isolated in a fairly pure state.

Clays may be found in a variety of forms. They may occur as a soft paste (mud) or as a soft solid (mudstones) or as a laminated rock (shales). Some clays are at a great depth below the surface and may be in thick beds (some Lias clay beds are more than 600 feet thick) or as narrow seams (e.g. the *fireclays* which occur below coal seams). The china-clay deposits of Cornwall are so thick that the bottom has not been reached in some places and field tests suggest that they may be over a mile in depth. Other clays lie quite close to the surface and are, in general, termed *alluvial clays*.

Deposits of clays of all kinds are described as *argillaceous* (French *argile* = clay) to distinguish them from sandy (*arenaceous*) formations; they are also classified as *clastic* to differentiate them from the igneous rocks.

Another important class of minerals directly derived by processes of erosion, transportation and deposition are **placer** deposits. In this type of formation, nature has not only segregated minerals by virtue of their particle size, but also as a function of their density. Some minerals, usually minor components of igneous rocks, are remarkably stable to weathering processes and are not chemically changed during the breakdown of the primary rock. The individual crystallite fragments are, however, released from the hard rock and transported by some agency from their original site. Many minerals which behave in this way have densities far in excess of the quartz and other secondary minerals which form the bulk of the transported material; hence they tend to become segregated and concentrated by natural processes. Numerous mineral deposits of this nature are known, and important metalliferous formations originate in this way. Perhaps the best known examples are the important *tin* deposits of Malaysia and elsewhere, but of similar nature are the beach sand deposits of *ilmenite*, *rutile*, *zircon*, *monazite*, *diamonds* and several of the precious and noble metals.

PRECIPITATION FORMATIONS

VARIOUS chemical components of primary rocks pass into a soluble form when subjected to weathering action and are carried in solution by water until they eventually enter the sea. The oceans thus become vast store-houses of mineral matter which under certain favourable conditions precipitates to form important solid mineral formations. It can be calculated[1,2] that about 5×10^{16} metric tons of mineral matter are contained in the world's oceans. At any one time enormous tonnages are being precipitated, but the amounts are being constantly replenished by the ever-eroding agencies of surface rocks and carried away by rivers.

Analyses of sea water reveal that virtually all known elements are present, but their concentrations are subject to local conditions. By far the most common elements, however, are sodium and chlorine which are present in 10,560 and 18,980 p.p.m. respectively, but other alkali elements and the alkaline earths, sulphates, carbonates, nitrates and phosphates are also present in considerable amounts.

Minerals may precipitate from sea water under a variety of conditions as follows:[3]

Evaporation, where the rate of water loss by surface evaporation exceeds the rate at which it is replaced by rainfall and rivers. Such phenomena are more likely to occur in small inland seas of which a present-day example is the Dead Sea, but in past geological eras much greater masses of water have been involved with extensive deposition. The tendency of particular minerals to precipitate from sea water depends not only on the relative concentration of the components, but also on the solubility product of respective ions in the complex. Thus calcium carbonate as *limestone* will precipitate in quantity from sea water which has been reduced in volume by about 30 per cent. Calcium sulphate, as either *gypsum* $CaSO_4\,2H_2O$ or *anhydrite* $CaSO_4$ also crystallises after moderate evaporation.

The more generally accepted *evaporites*, however, consist mainly of rock-salt or *halite*, NaCl, which requires a nine-fold concentration in sea water before it commences to crystallise. Much smaller amounts of magnesium sulphates and chloride and potassium salts, mainly *sylvite* KCl, form only after intensive evaporation involving virtually desert conditions. Other minor ingredients may concentrate under unusual natural conditions; the *borate* deposits of California and the *nitrates* of Chile are such examples.

In addition to the conventional surface evaporation phenomena involving sea water, solutions percolating through rocks often evaporate *in situ* and there is a lower level in the earth's crust below which water does not exist as a liquid phase. Any salts which such solutions contain will precipitate within the pores of the rocks through which percolation is proceeding. The 'secondary enrichment' of ores of copper and zinc is of this nature, where natural percolation dissolves minerals near to the surface and carries them downwards to precipitate and thus enrich lower levels of the deposit. The phenomenon is, however, common to many mineral species. The hard, compact and non-porous nature of many *quartzites* is due to the precipitation of a colloidal or crypto-crystalline silica phase around sand grains in a sedimentary deposit. *Opals, quartz crystals, tufa* are other examples of this nature.

Change of pH. Normally sea water is slightly acid, but local agencies can often interfere and produce changes sufficient in magnitude to cause precipitation. Decaying vegetation in mangrove swamps may induce basic characteristics resulting in the precipitation of magnesium hydroxide which later changes to the carbonate, magnesite $MgCO_3$. Decaying sea organisms and bacteria may behave similarly

[1] Dittmar, W., 'Rep. Sci. Results voyage H.M.S. Challenger 1873–6', *Phys. and Chem.* i, 1, 1884. [2] Lyman, J. and Fleming, R. H., *J. mar. Res.* 3, 134, 1940.
[3] Barnes, H., *Chemical Aspects of Oceanography*, R.I.C. Monograph No. 4, 1955.

and at the same time generate heat which is undoubtedly a significant factor in the formation of some *iron ore* and *manganese* deposits. Some bacteria are known to be instrumental in converting sulphide minerals to sulphates, or in liberating sulphuric acid from such minerals which then reacts with other components. Some *phosphate* deposits are thought to have precipitated from sea water by the change in pH conditions around coastlines.

Percolation Replacement. Closely allied with secondary precipitation are the changes which often occur in mineral formations due to chemical reaction with percolating solutions. A common mode of origin of *dolomite* is where percolating solutions, rich in magnesium, react with limestone rocks to form the double carbonate of calcium and magnesium. The latter ions are thought to replace those of calcium in the crystal lattice.

In other cases, water percolation may dissolve certain components, but leave others often in a changed form, as, for example, in bauxite.

Chemical Combinations. Reactions in nature involving the chemical union of two or more soluble phases to form an insoluble product are also known. They are most likely where two masses of water intermingle thus forming favourable precipitation conditions. In the same category might be included the mineral formation resulting from the chemical activity associated with gases dissolved in large masses of water. Carbonate rocks, for example, form only when the dissolved carbon dioxide content in the water attains a certain high level.

Some *montmorillonite* deposits are thought to be the result of chemical combination in merging water masses. Certain iron and manganese deposits have a similar origin.

Organic Derivation. Whether it be in the form of vegetable or animal matter or simple bacteria, organic material, either in a living form or as decaying matter, often has an important influence on inorganic mineral formation.

Two general types of mechanism are recognised, (*a*) fossil and (*b*) nodular formations.

(*a*) *Fossil deposits* are those derived directly from the skeletons or shells of organisms which normally live in a water environment or which by some other natural circumstance can form themselves into a mineral deposit.

Chalk is derived from the calcium carbonate fossilised remains of sea organisms which die and accumulate on the sea-bed. *Diatomaceous earths* are of similar origin but where the original organism had a skeleton of silica.

Coal, *oil* and *jet* are the result of partly decomposed vegetable matter which has decayed under anaerobic conditions. Coral reefs are also of vegetable origin where replacement reactions have also played a part.

(*b*) In many other cases, organic matter acts as a nucleation centre for the precipitation of minerals from solution. Ironstone formations are of this type, as also are *flints* and *agates*, both of which are mostly in the form of *nodules* suggesting layering formation.

The above list includes the major geological processes which are involved in mineral formation with particular reference to those of importance in ceramic studies. In many cases more than one agency may be involved in a particular formation and some minerals may be formed by more than one process.

STRUCTURAL TYPES AND STATES OF AGGREGATION

A STUDY of the nature of raw ceramic materials involves a consideration of the forms of aggregation and structure occurring both in primary rock formations and secondary deposits.

In the natural state, most ceramic materials are *solids*, but they are mostly used in the form of *muds* and *pastes* or as *suspensions* or *slurries*.

Solids

The solid ceramic materials are usually in compact masses, although several occur in a loose, powdery form (sands). There are some—particularly clays—which break down into a soft paste when mixed with a suitable quantity of water, and if a large excess of water is used, the solid particles may remain in permanent suspension and produce a viscous *slurry* or *slip*. Most clays possess some *colloidal properties* or can be partially converted into the colloidal state.

In the solid state, materials may be either crystalline or amorphous or mixtures of the two.

(1) **Crystalline.** A mass may consist wholly of crystals more or less perfectly developed and interlocking with each other to form a compact, impermeable material or a porous one; the difference lies in the shape and compactness of the crystals. Such structures are usually (*a*) the result of slowly cooling a molten material, (*b*) metamorphism, or (*c*) crystallisation from a solution; typical examples of value to ceramists are rock- and vein-quartz, quartzites, crystalline limestones, dolomite sillimanite, chromite and sometimes magnesites. Many artificially prepared ceramic materials, such as carbides and sintered alumina, are also wholly or partially crystalline.

For a substance to be crystalline it must have a characteristic form or shape. X-rays have shown that the atoms (Chapter II) of which the mineral is composed must be arranged in a symmetrical manner throughout the mass. Crystals range in size from a few Ångstrom units[1] to several inches in size, depending on the type, composition and mode of formation.

The properties of a crystalline mass are in some ways advantageous and in others detrimental; thus, when a material occurs in this form it is often fairly pure, although there is always the possibility of isomorphous substitution. Some materials have, however, the disadvantage of undergoing changes when heated, and this may cause difficulties when they are used. A typical example of this is quartz or quartzite, which is used in the manufacture of silica bricks. When these bricks are burned, the silica is changed from the form in which it normally occurs in nature to allotropic varieties of lower specific gravity, with a consequent volume expansion. This change takes place very slowly in the case of crystalline quartz, but much more rapidly when amorphous silica is heated. Thus, the cost of heat treatment is much greater for crystalline than for amorphous silica, and the conversion is not usually so complete. For this reason, an entirely crystalline structure is not so desirable in the case

[1] An Ångstrom unit (1 Å) $= 1 \times 10^{-8}$ cm.

of raw siliceous materials, but as this is the purest and most abundant form, it is the one chiefly used.

Where the crystalline form is stable at high temperatures, it has the advantage that it is less easily attacked than amorphous material, and therefore, bricks or other refractory articles with such a structure (e.g. sillimanite, mullite, tridymite, silicon carbide, corundum, periclase and spinels) are more resistant than those composed of amorphous material. When the material has an appreciable coefficient of expansion, crystals are more liable to crack and cause the disintegration of the ware than are amorphous grains. From the point of view of users of refractory materials, satisfactory structures must have low thermal coefficients of expansion.

Wholly crystalline structures are fairly common amongst some raw refractory materials, but are less frequent in the finished products.[1] In the latter, there is usually some other material between the crystals; it is generally of a glassy nature. The size of the grain enables classification into (*a*) coarse crystalline, in which the grains are readily visible to the unaided eye, and (*b*) fine crystalline, micro-crystalline or crypto-crystalline, in which the particles are much finer, the last two being so fine as to require a microscope to enable the structure to be identified. A fine crystalline texture is generally preferred both in raw and refractory fired products, as it has some of the advantages of amorphous structures, without possessing in such marked degree the disadvantages of a coarse crystalline structure already enumerated. A moderately coarse structure is, however, desirable in the case of magnesite, as on calcination the carbon dioxide escapes more conveniently than from a fine-grained or crypto-crystalline magnesite. Quartzites in which the quartz grains show clear cracks are not suitable for the manufacture of refractory bricks, as they tend to be weak along these lines. This is also true of chromites, where the texture of the spinel grains is all-important.

A single amorphous or crystalline substance is termed a *mineral*,[2] and within wide limits, minerals have a definite chemical composition and a characteristic structure. It is uncommon in nature for minerals to occur independently; they are more frequently found in association with others. Such mixtures are known as *rocks*, which may be defined as any portion of the earth which has individuality. Although a rock is often thought of as a hard, compact material this is not correct because some soft-textured materials are termed 'rocks'.

Although many rocks consist essentially of one crystalline type, there are others where several mineral species are present. An example of the latter is the acid igneous formation, granite, which consists principally of quartz, mica and felspar with many other minor ingredients. Within the various ranges of igneous rocks, numerous crystallising conditions have applied so that textures ranging from very fine to coarsely crystalline can be recognised.

[1] The present trend is to increase the proportion of crystalline matter in the more important ceramic products, especially industrial porcelains, stonewares and refractory materials.

[2] The legal interpretation of the term 'mineral' is based on the supposed origin of the word as 'something mined or only obtainable by mining'. Hence a surface clay is not legally a mineral, though if it forms part of a mass of the same material, most of which can be obtained only by mining, it is (at any rate as regards the bulk) deemed to be a mineral. The various important legal decisions on the subject have not made the matter wholly clear.

TABLE I—II. GENERAL TABLE OF THE EARTH'S STRATA AND GEOLOGICAL TIME-SCALE

(After Arthur Dendy, F.R.S., Arthur Holmes, D.Sc., F.G.S. and *British Museum Guide to Department of Geology and Palæontology*)

Era	*Epoch*	*Depth in thousands of feet*		*Age in millions of years*		*Beds, Deposits, etc.*	*Important Ceramic Deposits*
		for each epoch	*from surface*	*for each epoch*	*from surface*		
Anthropozoic	Holocene	0·25	0·25	0·02	0·02	Present Historic Neolithic	Now being formed: silt, fluviatile, alluvial clay, sand
Quaternary	Pleistocene	3·75	4	0·98	1	Palæolithic Glacial	Boulder clay
Tertiary	Pliocene	13	17	7	8	Norfolk, Norwich and Lenham beds	Brick earths, sand, pottery, and ball-clays
	Miocene	14	31	12	20	Tortonian and Helvetian Limestones	
	Oligocene	12	43	15	35	Hemstead Bembridge Osborne	
	Eocene	20	63	25	60	Bagshot London clay Lower London Tertiaries	
Mesozoic or Secondary	Cretaceous	44	107	50	110	Chalk Greensands Wealden Purbeck	Brick-clays, sand
	Jurassic	8	115	30	140	Portland clays Oolites Cheltenham Lias	Cement-clays Oxford clays

TABLE I—II. GENERAL TABLE OF THE EARTH'S STRATA AND GEOLOGICAL TIME-SCALE

(After Arthur Dendy, F.R.S., Arthur Holmes, D.Sc., F.G.S. and *British Museum Guide to Department of Geology and Palæontology*)

	Triassic	17	132	40	180	Lias, Marls (Keuper) Sandstones	Brick-clays, marl-clays, sand shales
	Permian	12	144	25	205	Limestones Red sandstone	Magnesian limestone, brick-clays
	Carboniferous	29	173	80	285	Coal, Millstone Grit Limestone	Brick-clays, fireclays ganister
Palæozoic or Primary	Devonian	22	195	45	330	Red sandstone	Brick-clays and shales
	Silurian	15	210	40	370	Ludlow Wenlock Llandovery Tarannon	Slates, grit-stones
	Ordovician	17	227	78	448	Caradocian Llandeilian Arenigian	Sands and slates
	Cambrian	26	253	77	525	Oleneidian Paradoxian Olenellian	Slates
Proterozoic	Upper Pre-Cambrian	90	343	325	850	Sudburian, etc.	Serpentines, Quartzites, Clay-slates
Archæozoic	Lower Pre-Cambrian	90	433	400	1250	Laurentian Grenville	
	Metamorphic	?	?	?	?		Mica-schist and quartzites
Azoic				1250	2500	Igneous Molten Gaseous	

(2) **Amorphous.** An amorphous solid has, by definition, neither recognisable shape nor geometrical internal structure. Many ceramic materials (especially plastic clays) appear to be amorphous even under a microscope, but the use of X-rays and an electron microscope shows that they are essentially crystalline. In some ways such materials behave as amorphous substances and in others as crystalline ores.

Modern science recognises that there is no sharp dividing line between amorphous and crystalline substances. Some silicate mixtures, when melted and rapidly cooled, solidify with a random atomic arrangement better described as amorphous or crypto-crystalline. Other materials precipitate from solutions as colloidal gels, the particles of which consist of only a few atomic groups and as such may be classified as amorphous.

Between the two extremes of crystalline and amorphous minerals there exists a wide range of materials which is referred to by a variety of names of which crypto- or pseudo-crystalline are more generally preferred. The nature of these materials can only be conjectured in many cases, but it can be inferred that they possess an ill-defined or randomly-distributed lattice. Such a condition might result on crystallisation, but it can also be induced by grinding either by natural agencies or artificial processing. In these cases, the surface layers of atoms become distorted and an apparently amorphous or non-crystalline condition results. This feature is more common in crystal lattices of open texture, such as quartz, micas, felspars, etc.

In addition to the two states (amorphous and crystalline) which relate to the fundamental nature of individual particles in a mass, many ceramic materials exist in various types of aggregation, where grains of perhaps many sizes and mineral forms are bonded together. The chief of these are:

Agglomerates in which the particles are united to form a mass with a granular structure. These are predominantly of either crystalline or amorphous matter, held together by previously applied (natural) pressure, or they are cemented together by some form of bond which may be glassy, amorphous or colloidal. In the raw materials, the principal cementing materials are clay, quartz, opal, chalcedony, iron oxides (hæmatite, limonite and magnetite), carbonates of calcium, barium, magnesium and iron (calcite, witherite, dolomite, chalybite and ankerite), sulphides of iron (chiefly pyrites), hydrated silicates (zeolites, chlorites, epidote, serpentine, talc, mica), anhydrous silicates (felspar, hornblende), sulphates of calcium and barium (gypsum, barytes), and various phosphates. Of these, the commonest cement is clay or silica, but carbonates, sulphates, silicates and iron oxides are also important.

Colloidal silica cement is usually produced by the weathering of silicate rocks; the percolating waters containing the silica in solution descend until they reach a porous rock, in which the silica, being unstable, tends to separate from the solution and is deposited as a cement between the grains of the rock through which it percolates. A siliceous cement may also be formed as a result of the silica being precipitated owing to a decrease in the temperature of the solution. Many rocks are united by a calcareous cement derived from water containing carbon dioxide which has previously percolated through limestone or chalk and has dissolved some calcium carbonate. The solution is carried through the fissures and pores of other rocks until the carbonate is deposited and forms a calcareous cement.

Ferruginous cements are common in sandstones and other rocks; they appear to have been formed by the percolation of water containing ferrous carbonate or hydroxide in solution.

Gypsum acts as cement in some shales and sands. Thus, Fontainebleau sand contains sand-calcites which consist of isolated masses composed of gypsum and sand. Some of the Northumbrian fireclays are thought to be cemented by barytes.

The precipitation of cements in rocks is usually a result of one of the following actions: (*a*) the mingling of solutions from different sources and their mutual precipitation; (*b*) the chemical action between solutions and the rocks they traverse; (*c*) a decrease in temperature causing supersaturation; and (*d*) a decrease in pressure.

A ceramic material for modern high-temperature uses in which the natural cement occurs in large proportions is generally undesirable unless it is very highly siliceous or argillaceous and is itself refractory.

In the finished goods, the cement or bond produced by the burning process is usually of a very complex nature and is generally in the form of a glass composed of various silicates, alumino-silicates, etc.; in stoneware and porcelain such a bond is essential.

The larger particles or granules in a material of granular structure may vary greatly in size from that of pebbles (as in quartzitic conglomerates which are sometimes employed for the manufacture of silica bricks) to the fine, amorphous grains which comprise the bulk of most clays, these latter being so minute as to render identification impossible, even when they are examined under a microscope. The granules may be either amorphous or crystalline or an indefinite mixture of materials in both these forms.

The principal types of granular structure are as follows:

(*a*) *Completely crystalline material.* Because the crystalline form represents the state of lowest entropy, most of nature's rocks are completely crystallised. Cooling of long duration of a molten magma, metamorphic activity and the slow evaporation of a solution, most frequently result in a rock species of total crystallinity. In many fired ceramic products, the solid forms are mainly crystalline minerals, although when silica is a component, a glassy phase often develops.

In general, the longer the processes of formation, the more crystalline is the body which forms.

(*b*) *Granular Fragments in a Glassy Matrix.* These are not very common in raw ceramic materials, though they occur frequently in igneous rocks. This structure is found in porcelain, stoneware, fired silica and in some clay bricks; these consist of crystals of mullite, quartz and other forms of silica, cemented together by a glassy mass composed of various complex silicates and alumino-silicates which sometimes include minerals such as wollastonite and anorthite. The whole mass is usually coloured by the iron compounds present. The granular matter may be either crystalline or amorphous and either coarse or fine.

(*c*) *Amorphous or Crystalline Fragments in a Non-glassy Matrix.* This structure is more frequent in the raw materials than in the finished products as, in the latter, the strength of the articles depends more often on the presence of some fused material to act as a bond. Many raw materials, however, consist of fragments which are cemented together by amorphous matter, which has been precipitated or otherwise deposited from solutions percolating through them.

The ideal structure of a quartzite for the manufacture of silica bricks is one in which the grains of quartz are very minute and are bound together with an amorphous cement. This is very important, as when a quartzite of this kind is heated, the elastic nature of the cement reduces the expansion of the silica to a minimum, whereas a quartzite which does not contain any cement has a very marked expansion and frequently cracks on account of the strains set up in the material. One of the most suitable quartzites used for making silica bricks is the amorphous or crypto-crystalline form found in flints and chalcedony. *Findlings-quartzit* consists of erratic boulders of minute grains of quartz surrounded by chalcedonic cement. Of similar composition is *silcrete*, a quartzite from South Africa, now used for the manufacture of high-grade silica bricks. In some cases, silica has been precipitated round irregular quartz grains and has apparently taken on the crystalline form of quartz. The Stiperstones quartzites occurring near Shrewsbury are a good example of silicification, the sand grains having been enlarged by secondary growth, though crystal faces have not developed. The Lickey, Bwlchgwyn and other Welsh quartzites are due to similar silicification, and will be discussed in detail in a later chapter. Many shales and slates are often silicified by the deposition of secondary silica in their fissures, and some limestones are probably cemented with a similar material.

Ganister is a fine-grained silica rock which may be sericitic in character. The term was originally applied to a portion of the Lower Coal Measure Series near Sheffield but it has been largely applied to other silica rocks, found elsewhere, which are not so fine-grained nor so angular, but can be used for some of the same purposes as 'true ganister' (the name applied solely to the South Yorkshire and Derbyshire material. The other so-called ganisters have been described by the Geological Survey as 'ganister-like sandstones' and are of a different origin.).[1]

Laterites appear to be mixtures of hydroxides, of iron, aluminium, titanium and manganese bonded together. The bond is very weak in the freshly obtained material, but, on exposure, its strength increases at the same time as the characteristic hardening of the material. Laterite may have been formed by the weathering of basalts as in the Deccan of India, or of other rocks, such as gneiss, slate, sandstone and granite, as in Africa. It has been suggested that it may have been produced by the action of bacteria which precipitated silica in the colloidal form, the silica then being removed by dilute alkaline solutions formed at the same time. Laterite may also be regarded as a terrestrial silt, indistinguishable mineralogically from the silts deposited by water. *Bauxite* is a laterite in which hydroxides of alumina predominate.

Other refractory materials, for example chromites and some magnesites, have a structure consisting of crystalline grains cemented together by a crypto-crystalline cement. Hydrous magnesium or iron silicates are of frequent occurrence.

(*d*) *Amorphous or Crystalline Grains without Cement*. Some ceramic materials

[1] 'Memoirs of the Geological Survey. Special Reports on the Mineral Resources of Great Britain.' Vol. VI. *Refractory Materials: Ganister, etc.* (London, H.M. Stationery Office, 1918).

occur in loose incoherent beds, without any cement to bind the grains together. Many clays, sands, decomposed quartzites and kieselguhr are of this type.

Most *clays* in the natural state are composed principally of finely-divided clay minerals and sand; some contain gravel and even boulders of stone embedded in the clay. Glacial drift detritus, in certain localities, is rich in limestone powder and is then not suitable for brick manufacture.

Much useful information on the composition of clays can often be obtained by mixing thoroughly a weighed quantity of the dried material with four or more times its weight of water and then passing the liquid suspension through a series of sieves of increasingly finer mesh. The residues on each sieve are washed free from finer particles, then dried and weighed to give their relative proportions. An examination of each with a hand lens or microscope will show that most of the particles of greater size than 0·005 in. consist almost entirely of sand.

Among the most important sands for refractory purposes is the *Dinas sand*—a pale-yellow material produced by the disintegration of the famous Dinas quartzite—occurring in the Vale of Neath, Glamorganshire. Several other silica sands are found in various geological formations, especially in the Lower Greensand and Estuarine beds.

Another important sand is that from Loch Aline; it is remarkably free from iron and is in great demand in the glass industry.

Dolomite sands are deposits formed by the weathering of dolomitic limestone; the calcareous material, being soluble in water containing carbon dioxide, is removed in solution, leaving the less soluble magnesium mineral behind, together with some silica sand. Such sands are seldom pure enough for use as refractory materials.

Chromite occasionally is found as a loose detrital sand, the beds of which tend to be variable.

Zirconia and *zircon* are commonly in the form of sands in river-beds and along the sea-shore. They are associated with other loose materials derived from pegmatites and syenites. *Monazite sands* are of a similar nature.

Many of the rarer refractory materials occur in individual grains disseminated through igneous rocks or in sand deposits and placers derived from the disintegration of such rocks. The principal of these are beryllium oxide, titanium oxide, zirconia, ceria, didymia, lanthana, thoria, yttria, etc.

(*e*) *Entirely Glassy Materials.* These are very uncommon in nature and can only be found in a rapidly cooled magma which is rich in silica, e.g. the mineral obsidian and basalts.

Massive Structures. Many ceramic materials occur in the form of large masses; these depend largely on the mode of formation. The principal types are as follows:

Stratified masses are those in which the materials are aggregated in the form of layers, strata or laminations. This structure is characteristic of most of the rocks laid down by the action of water and includes the majority of clays, which have been carried in suspension in rivers and have finally settled at the bottom of rivers, lakes, seas, etc. They are in beds or strata of varying thickness and often interlaid with layers of sand and other materials. For further details see *British Clays, Shales and Sands*, by A. B. Searle (Griffin).

Unstratified masses are those in which the materials show no special arrangement or lines of stratification. This type of structure is common amongst raw ceramic materials and is also characteristic of many of the finished products. In fact, in the case of the finished articles, it is the most desirable structure.

The nature of unstratified rocks comes within the province of geology and is outside the scope of the present volume. It is of interest, however, that most residual clays and some sedimentary ones are unstratified, as are also some forms of silica rock, magnesite and zirconia. Unstratified materials may be (i) homogeneous or uniform, or (ii) heterogeneous or irregular in character and in the arrangement of the particles of which they are composed.

Homogeneous structures are those which are uniform throughout the whole mass and may therefore be expected to have the same physical properties, no matter from which part a sample may be taken. They differ from stratified, laminated and other heterogeneous structures in the absence of definite planes of cleavage or other signs of being composed of 'units' of different character.

In the manufacture of articles such as pottery and brick-shapes, homogeneity is usually of great importance and any definitely laminated or segregated type of structure must be destroyed during the process of manufacture.

As homogeneity is closely allied to 'texture' further information is given in Chapter VII.

Flaky and Laminated Materials. When a rock appears to be composed of very thin *flakes*, it is said to be *laminated* and if the splitting occurs along the same plane as the bed in which it was deposited the structure is termed *shaly*. If the material cleaves in any other direction than one parallel to the bedding plane, the term *fissile* is employed to distinguish such a rock from the *shales*. In some laminated clays, each layer is silty in its lower part and carrying the finest clay at the top; these bands are known as *varves* which, in some clays, have been shown to be annual layers.

Many clays and all shales show signs of stratification; they may generally be split into thin flakes and foliations parallel to the bedding plane. When a clay has been greatly compressed in a direction at a large angle to the bedding plane and simultaneously subjected to great heat, *slate* may be produced.

Flaky clays do not readily form a homogeneous plastic mass except after an excessive amount of grinding and are very difficult to use satisfactorily. The structure of a clay is of great importance industrially; a compact, laminated type is highly objectionable for it can seldom be completely destroyed and a mass of homogeneous structure produced except at a prohibitive cost. Two clays may have the same chemical composition, but one may be useless and the other valuable simply because of the differences in their structures. The best clays have a perfectly homogeneous texture and are free from any laminated structure; but, for various reasons, laminated clays are often used and with care can be made into satisfactory articles. It is, however, essential to break them down to such an extent as to destroy all the laminations, otherwise the various laminae may not unite properly and the finished articles will be weak along the line of cleavage.

Lamination also occurs in some sandstones, ganister, bauxite, laterite, etc. Siliceous materials, other than clay, which show lamination should be avoided, as their structure is not easily destroyed in the course of making them into bricks, etc.

Sometimes a somewhat laminated structure is desirable, particularly in the case of graphite used for making crucibles and other shapes. The best graphites for this purpose consist of small flakes which adhere and form a laminated mass. Of American graphites, those from Alabama have the smallest and thinnest flakes; the flakes in Pennsylvanian graphites are slightly larger and thicker; whilst those from graphite from Canada are even larger and more irregular. The graphite from Madagascar consists of grains nearly twice as large as those in the American materials. Ceylon graphite is the most preferred; it has very thin rectangular or triangular grains, rather than true flakes; it is very satisfactory and gives most durable products, whilst articles made from a coarser and more obviously laminated structure tend to spall.

Foliated structures are produced by the recrystallisation of some of the constituents of a rock along lines parallel to the original bedding or along joints or cleavage planes, the rock being afterwards subject to lateral pressure. A foliated structure is characteristic of the schists but is not common in clays and other ceramic materials. Schistose quartz on account of its foliated nature is not satisfactory for making silica bricks, as those made from it tend to spall badly.

Cellular structures and substances are not common in clays. Of other ceramic materials the most important ones with a cellular structure are: (*a*) those such as pumice, which are too fusible to be used to any great extent as refractory materials, and (*b*) diatomaceous earths, which consist of the minute, siliceous skeletons of dead marine and freshwater plants and animals such as diatoms, radiolaria, etc. One of the most extensively used refractory materials with a cellular structure is kieselguhr. Its texture is readily seen when it is observed through a microscope, though to the naked eye it appears to be structureless. Moler consists of a similar material to kieselguhr but it is contaminated by considerable proportions of clay and often by volcanic ash. The clay acts as a bond which unites the other particles into a fairly strong mass.

A cellular structure in bricks and some other clay articles can be produced by mixing the raw material with sawdust, seeds or other organic matter. This burns away in the kiln, leaving a highly cellular and porous mass. The same effect can be achieved by introducing a gaseous component into the raw material (e.g. zinc and hydrochloric acid; calcium carbonate and acid) or by adding a foaming agent and beating air into the material.

Concretionary structures consist of a nucleus surrounded with layers of aggregations of other materials, the whole forming irregular masses and nodules. This kind of structure is often formed with nuclei of leaves, shells or the remains of plants and animals, which are deposited in water containing salts or silica in solution. Hæmatite, limonite, clay ironstone and other forms of iron oxide commonly occur in nodular masses. Ironstone pebbles frequently show traces of a fossilised nucleus when broken across. Limestone and some forms of silica are found as concretionary deposits, the former often resulting from extensive coral formations. Oolitic limestones are frequently, and dolomites more rarely, composed of concretionary masses. Flint occurs in nodular form frequently of large size associated with calcareous matter. Bauxite and laterite often occur in small round grains, varying in size from a pea to 1 in. in diameter and such nodules are sometimes separated from the matrix. Crypto-crystalline magnesite often forms nodular accretions.

Segregated structures are those in which part of the material forms veins or segregated masses in other rocks, such as vein-quartz, magnesite bands traversing masses of serpentine and hæmatite, spathic and magnetic iron ores occurring as large irregular masses in other rocks. Chromite frequently is present in massive and lenticular forms and as veins traversing serpentine and peridotite rocks and also in irregular masses in residual clays associated with serpentine.

Fibrous structures in ceramic materials are rare; the only one within the scope of this volume is asbestos. It has a fibrous structure, the crystalline grains forming long flexible fibres which interlock and produce a strong mass. Of the several varieties of asbestos, chrysotile is the most common.

Capillary structures are present in materials which behave as though they were composed of a multitude of minute tubes, each of which is of an extremely small diameter. In clays and clay products a capillary structure is somewhat analogous to cellular structure. It is best understood by observing the behaviour when the lower end of a very narrow tube is dipped into a liquid which wets it; the liquid rises in the tube to a greater height than that of the external liquid, because the surface film is in tension and the liquid in the tube gains in potential energy. The liquid rise will be greater with smaller-diameter tubes and will also depend on the tension or energy of the surface.

Porous clays and clay products contain an assemblage of small cellular units, many of which are connected through narrow necks; in this sense, they may be regarded as capillary structures.

A capillary structure has a great influence on the mechanism whereby water is taken up into a porous body, or lost from it during drying. This may be beautifully demonstrated by immersing such a body in a solution of a coloured compound allowing penetration on one face only; conversely, the migrational effect along capillaries is also brought out by drying such a body which has previously been impregnated with a coloured solution—the colour is concentrated in the outside areas.

The efflorescence which sometimes forms on bricks and other porous substances is due primarily to their capillary structure, which enables a solution of salts contained in the mass as a whole to become concentrated at the surface. When the solvent (water) evaporates, the salts are left as a 'deposit' on the surface of the article.

BIBLIOGRAPHY

(*a*) History of Ceramics

1. *Pottery and Porcelain* (3 vols.), by E. Hannover (E. Benn Ltd., London, 1925).
2. *English Pottery* (3 vols.), by J. F. Blacker (Stanley Paul & Co., 1923).
3. *Ceramic Art in Great Britain* (2 vols.), by L. Jewitt (Virtue & Co. Ltd., London, 1878).
4. *The Art of the Potter*, by W. B. Honey (Faber & Faber, London).
5. *Ceramics for the Collector*, by G. Savage (Rockliff, London, 1949).

(*b*) General Geology

1. *British Clays, Shales and Sands*, by A. B. Searle (Griffin, London, 1913).
2. *Petrology of Sedimentary Rocks*, by F. H. Hatch, R. H. Rastall and M. Black (George Allen and Unwin Ltd., London, 1938).
3. *Principles of Physical Geology*, by A. Holmes (Nelson & Sons Ltd., London, 1944).
4. *Sedimentary Rocks*, by F. J. Pettijohn (Harper & Bros., New York, 1948).

CHAPTER II

ATOMIC AND CRYSTALLINE STRUCTURE

ALL ceramic materials are composed of various units, just as a wall is built of stones or bricks. These may be regular in shape (as are bricks), or amorphous as lumps of glass appear to be. Under a microscope, these units can frequently be identified as minerals (see p. 47) with a regularity of optical property which is not visible with the unaided eye. X-rays enable other structural details and the internal arrangement of atoms within these minerals to be ascertained. Indeed, prior to the discovery of X-rays by Röntgen in 1895 and their application by the Braggs and others to the study of the structure of various minerals, the nature of many substances in the solid state was largely a matter of conjecture: since that time, great advances have been made. Much information has been available for many years on crystalline substances and various theories concerning their nature have been published, but fundamental details of atomic and ionic structure have been ascertained only recently and valuable additional knowledge is still being gained.

The importance of this knowledge of the intrinsic structure of clays and other ceramic materials and their products is steadily being recognised by manufacturers and users of these substances. Already considerable developments in ceramic industries and processes have been brought about by applying such knowledge, and the scope of the subject is almost unlimited. The basic structure of minerals is now of value in relating the properties of a ceramic material to its essential nature, whereas formerly there was only a series of apparently unrelated subjects. An understanding of the chemistry and physics of substances depends very largely on an adequate knowledge and application of their crystal structures.[1]

Some of the Ancient Greek philosophers had concluded that all material substances were composed of separate or discrete particles. This idea was further developed by Isaac Newton and John Dalton, the latter using the term *atom*[2] to represent these fundamental particles, which he imagined to be natural *units* too small to be further divided or decomposed. It was recognised that these units were not all identical, yet they represented the ultimate subdivision of matter and were the elementary units which take part in chemical reactions. They termed the complex units formed from two or more atoms *molecules* and defined a substance composed of atoms of one single type as an *element*. Their work showed that the combinations of these atoms of the elements are always in constant proportions and that the equivalent combining ratios form the basis of atomic weights.

As early as 1864, Newlands recognised that some elements are remarkably like others in some of their properties and established that a periodicity exists whereby,

[1] The fundamental conception of matter, in general, is too detailed and diverse a subject to be dealt with fully in the present volume. There are many books which treat of it to which the reader may refer if he so desires.

[2] The word *atom* means 'that which cannot be cut or divided'.

TABLE II—1A. THE PERIODIC TABLE

(incorporating the new atomic-weight values 1969)

1 H 1·008$_0$																	2 He 4·00260
3 Li 6·94$_1$	4 Be 9·01218											5 B 10·81	6 C 12·011	7 N 14·0067	8 O 15·999$_4$	9 F 18·9984	10 Ne 20·17$_9$
11 Na 22·9898	12 Mg 24·305											13 Al 26·9815	14 Si 28·08$_6$	15 P 30·9738	16 S 32·06	17 Cl 35·453	18 Ar 39·94$_8$
19 K 39·10$_2$	20 Ca 40·08	21 Sc 44·9559	22 Ti 47·9$_0$	23 V 50·941$_4$	24 Cr 51·996	25 Mn 54·9380	26 Fe 55·84$_7$	27 Co 58·9332	28 Ni 58·7$_1$	29 Cu 63·54$_6$	30 Zn 65·3$_7$	31 Ga 69·72	32 Ge 72·5$_9$	33 As 74·9216	34 Se 78·9$_6$	35 Br 79·904	36 Kr 83·80
37 Rb 85·467$_8$	38 Sr 87·62	39 Y 88·9059	40 Zr 91·22	41 Nb 92·9064	42 Mo 95·9$_4$	43 Tc 98·9062	44 Ru 101·0$_7$	45 Rh 102·9055	46 Pd 106·4	47 Ag 107·868	48 Cd 112·40	49 In 114·82	50 Sn 118·6$_9$	51 Sb 121·7$_5$	52 Te 127·6$_0$	53 I 126·9045	54 Xe 131·30
55 Cs 132·9055	56 Ba 137·3$_4$	57 La 138·905$_5$	72 Hf 178·4$_9$	73 Ta 180·947$_9$	74 W 183·8$_5$	75 Re 186·2	76 Os 190·2	77 Ir 192·2$_2$	78 Pt 195·0$_9$	79 Au 196·9665	80 Hg 200·5$_9$	81 Tl 204·3$_7$	82 Pb 207·2	83 Bi 208·9806	84 Po —	85 At —	86 Rn —
87 Fr —	88 Ra 226·0254	89 Ac —															

58 Ce 140·12	59 Pr 140·9077	60 Nd 144·2$_4$	61 Pm —	62 Sm 150·4	63 Eu 151·96	64 Gd 157·2$_5$	65 Tb 158·9254	66 Dy 162·5$_0$	67 Ho 164·9303	68 Er 167·2$_6$	69 Tm 168·9342	70 Yb 173·0$_4$	71 Lu 174·97
90 Th 232·0381	91 Pa 231·0359	92 U 238·029	93 Np 237·0482	94 Pu —	95 Am —	96 Cm —	97 Bk —	98 Cf —	99 Es —	100 Fm —	101 Md —	102 No —	103 Lr —

TABLE II—1B. REVISED TABLE OF ATOMIC WEIGHTS[1]

IUPAC Atomic Weights Commission 1969

(Based on $^{12}C=12$)

Element	*Symbol*	*Atomic No.*	*Atomic Weight*	*Element*	*Symbol*	*Atomic No.*	*Atomic Weight*
Actinium	Ac	89	(227)	Mercury	Hg	80	200·59
Aluminium	Al	13	26·9815	Molybdenum	Mo	42	95·94
Americium	Am	95	(243)	Neodymium	Nd	60	144·24
Antimony	Sb	51	121·75	Neon	Ne	10	20·179
Argon	A	18	39·948	Neptunium	Np	93	237·0482
Arsenic	As	33	74·9216	Nickel	Ni	28	58·71
Astatine	At	85	(210)	Niobium	Nb	41	92·9064
Barium	Ba	56	137·34	Nitrogen	N	7	14·0067
Berkelium	Bk	97	—	Nobelium	No	102	—
Beryllium	Be	4	9·01218	Osmium	Os	76	190·2
Bismuth	Bi	83	208·9806	Oxygen	O	8	15·999
Boron	B	5	10·81	Palladium	Pd	46	106·4
Bromine	Br	35	79·904	Phosphorus	P	15	30·9738
Cadmium	Cd	48	112·40	Platinum	Pt	78	195·09
Caesium	Cs	55	132·9055	Plutonium	Pu	94	(239)
Calcium	Ca	20	40·08	Polonium	Po	84	(210)
Californium	Cf	98	—	Potassium	K	19	39·102
Carbon	C	6	12·011	Praseodymium	Pr	59	140·9077
Cerium	Ce	58	140·12	Promethium	Pm	61	(147)
Chlorine	Cl	17	35·453	Protactinium	Pa	91	231·0359
Chromium	Cr	24	51·996	Radium	Ra	88	226·0254
Cobalt	Co	27	58·9332	Radon	Rn	86	(222)
Copper	Cu	29	63·546	Rhenium	Re	75	186·2
Curium	Cm	96	(242)	Rhodium	Rh	45	102·9055
Dysprosium	Dy	66	162·50	Rubidium	Rb	37	85·4678
Einsteinium	Es	99	—	Ruthenium	Ru	44	101·07
Erbium	Er	68	167·26	Samarium	Sm	62	150·4
Europium	Eu	63	151·96	Scandium	Sc	21	44·9559
Fermium	Fm	100	—	Selenium	Se	34	78·96
Fluorine	F	9	18·9984	Silicon	Si	14	28·086
Francium	Fr	87	(223)	Silver	Ag	47	107·868
Gadolinium	Gd	64	157·25	Sodium	Na	11	22·9898
Gallium	Ga	31	69·72	Strontium	Sr	38	87·62
Germanium	Ge	32	72·59	Sulphur	S	16	32·06
Gold	Au	79	196·9665	Tantalum	Ta	73	180·9479
Hafnium	Hf	72	178·49	Technetium	Tc	43	98·9062
Helium	He	2	4·00260	Tellurium	Te	52	127·60
Holmium	Ho	67	164·9303	Terbium	Tb	65	158·9254
Hydrogen	H	1	1·0080	Thallium	Tl	81	204·37
Indium	In	49	114·82	Thorium	Th	90	232·0381
Iodine	I	53	126·9045	Thulium	Tm	69	168·9342
Iridium	Ir	77	192·22	Tin	Sn	50	118·69
Iron	Fe	26	55·847	Titanium	Ti	22	47·90
Krypton	Kr	36	83·80	Tungsten	W	74	183·85
Lanthanum	La	57	138·9055	Uranium	U	92	238·029
Lawrencium	Lr	103	—	Vanadium	V	23	50·9414
Lead	Pb	82	207·2	Xenon	Xe	54	131·30
Lithium	Li	3	6·941	Ytterbium	Yb	70	173·04
Lutetium	Lu	71	174·97	Yttrium	Y	39	88·9059
Magnesium	Mg	12	24·305	Zinc	Zn	30	65·37
Manganese	Mn	25	54·9380	Zirconium	Zr	40	91·22
Mendelevium	Md	101	—				

[1] *Pure app. Chem.* 20 (4), 1970. 25th Compt. Rend. IUPAC Jan. 1970.

when elements are arranged in order of their atomic weight, similarity occurs at every eighth one. Mendeléeff extended the idea and in 1869 published his Periodic Table which, with but little modification, is still in general. use When this Table was first outlined, many elements had not been discovered but were predicted. The complete series, recognised at the present time, is shown in Table II.IA.[1] In this Table, the elements are arranged in horizontal series and vertical groups, in increasing order of the atomic weights. Like elements occur in the same vertical column.

In some Periodic classifications sub-groups *a* and *b* are shown together in the same vertical column. Thus copper, silver and gold would appear in the same major group as the alkali elements. In Table II.IA the more usual arrangement is shown in which the transition elements are in separate vertical columns.

The rare earth elements (Atomic Numbers 58–71) and the actinides (Atomic Numbers 90–103) are shown in separate horizontal series associated respectively with lanthanum (57) and actinium (89). The elements in these series are similar in property and do not show typical Periodic progression for reasons which will be discussed later.

At the present time elements up to atomic number 103 have been identified with reasonable certainty. Undoubtedly more will be discovered, although such elements are extremely short-lived. Seaborg[2] has predicted that elements of atomic numbers up to 155 can exist and he discusses their probable stability relationships. The atomic weight and number of each element in alphabetical order is shown in Table II.IB.

The Periodic Table enables many known physical and chemical facts to be more clearly understood. It was shown, for example, that the periodicity of valency, atomic volumes and atomic heats can all be related to the position of the elements in the Table and certain other characteristic properties can be predicted.

It is remarkable that the atomic weight of most elements is very nearly a whole number and the idea of a simple unit as the basis of all atoms was suggested by Dalton, who regarded the atom of hydrogen as the fundamental unit of atomic weight. More than a century passed before Moseley showed that the wave-length of X-rays emitted by each element when bombarded with electrons, increases in a simple ratio with the *atomic number* of the atoms as arranged in the order in the Periodic Table; this discovery led to the conclusion that atoms themselves were composed of still smaller units.

THE STRUCTURE OF ATOMS

THE present view of the structure of atoms is the result of detailed studies of electrical discharges through gases. This led to the discovery of various types of radiation, such as cathode rays, positive or α-rays and X-rays, a fuller account of which can be found in many books on Modern Physics. It was proved that the atoms of all elements consist of *electrons*, or negatively-charged particles, revolving around a core or nucleus, which has a net positive charge. The nucleus of all atoms is composed of *protons*, each of which bears a positive charge, and *neutrons*, which have no charge.

[1] Greenwood, N. N., *Chemistry in Britain*, 6 (3), 119, 1970.
[2] Seaborg, G., *Chemistry*, 43 (1), 6, 1970.

Both protons and neutrons are, by definition, of unit mass and about 1,840 times heavier than an electron. Hence the weight of an atom is largely determined by the number of particles in its nucleus.

The charge on the nucleus depends on the number of unit positive charges, or protons, which is equal to the atomic number of the element or atom. The total positive charge of the nuclear protons is exactly sufficient to neutralise the negative charges of the electrons. As each electron is of unit negative charge, it follows that there must be an equal number of electrons and protons in a particular atom.

Although the diameter of the nucleus of all atoms is very small, *circa* 10^{-12} cm., the size of the atom itself is much greater, being determined by the size of the orbits in which the electrons revolve.

The protons, neutrons and electrons of all atoms are identical in all respects and atoms differ only in the number of these ultimate particles which they contain. The simplest atom is that of hydrogen, which consists of a single proton with one electron revolving in an orbit around it. The proton is thus a hydrogen atom devoid of its electron and is thus synonymous with the free hydrogen ion or H^+. Although the electron is an even smaller particle than the proton, the hydrogen atom is more than 10,000 times larger than the ion. The atom may thus be regarded, in a simple way, as an infinitesimally-sized planetary system, with electrons in place of planets and the nucleus in the position of the centrally-located sun.

The number of atoms in the gram-atomic weight[1] of each element is always constant. This is termed the *Avogadro Number* and has the value of $6{\cdot}02 \times 10^{23}$; it is represented by the symbol N. Consequently, if the nucleus of an atom were composed only of protons, the atomic weight should, in all cases, be equivalent to the number of protons it contains, i.e. Z, or, in other words, to the atomic number, because each proton is of unit mass and almost all of the mass of an atom is centred in the nucleus. In most cases, however, the atomic weight of an element is nearer $2Z$; this is because neutrons also occur in the nucleus of atoms and these uncharged particles of the same weight as protons are present in similar numbers. However, the number of neutrons present in a particular atom is not always constant and this gives rise to the concept of *isotopes* which may be defined as atoms of the same element, having the same atomic number, but with slightly different atomic weights.

All atoms therefore consist essentially of three types of particles:

(*a*) *Electrons*, which are particles of unit negative charge revolving around a nucleus. They are of very small mass, being only $9{\cdot}1 \times 10^{-28}$ gm., or 1/1838 of the weight of a proton.

(*b*) *Protons*, which are confined to the nucleus. These are of unit positive charge, and of unit atomic weight or mass.

(*c*) *Neutrons*, which are also situated in the nucleus, but are electrically neutral; they are of similar mass to the proton.

These three kinds of particles may be regarded as the fundamental bricks in the framework of the structure of an atom.[2]

[1] The gram-atomic weight of an element is simply the atomic weight expressed in grams.

[2] Modern developments have indicated that atomic configuration is more complex, but this simple picture is sufficient for the present purpose.

Electrons are the important units which determine the nature of an atom and which take part in chemical reactions. The electrons revolve around the nucleus in definite orbits and the number of electrons in any atom is equivalent to the positive charge of the protons in the nucleus. The atomic number also represents the number of electrons present in the atom. Hydrogen, for example, with the atomic number of 1 has one orbital electron; helium with atomic number 2, has two electrons; and, so on, up to lawrencium with an atomic number of 103 and one hundred and three electrons.

A simple picture of the atom was suggested by Bohr, who visualised the electrons as separate or discrete particles moving in definite orbits and occupying certain spacial positions at any specified time.

With the development of the wave mechanical theory,[1] Bohr's suggestion has been shown to be untenable and the electron distribution is now based on a statistical wave representation. Schrodinger and others visualise the distribution around the nucleus as in the form of an electron cloud and equations have been derived to predict probable positions at a given time.

In all atoms, the electrons revolve around the nucleus in a series of specified orbits or 'shells' and the number in each separate orbit is limited. The states of the electrons in any atom are defined by four *quantum numbers* which are designated n, l, s, and j.

The quantum number 'n' refers to the number or order of the orbit in which a particular electron revolves. Wave mechanical theory predicts that electrons revolve round the nucleus in a series of orbits or shells. The number of electrons in any one orbit is limited and is given by the expression $2n^2$, where n represents the order of that shell. Thus in the first orbit there can be only two electrons; in the second orbit, eight; the third, eighteen; and so on. The first shell has its orbit nearest to the nucleus, the other shells are at ever-increasing distances from the centre. Each of these orbits is given a symbol: electrons in the innermost shell, where $n = 1$, are referred to as being in the K shell; those in the second as being in the L shell; those in the third, the M shell, etc.

The second quantum number 'l' expresses the orbital angular momentum or the 'degree of ellipticity' of the orbit. The value of l can range from 0 to $n-1$. Thus if an electron is in the M shell or where $n = 3$, then l can be 0, 1, or 2. In other words, the orbit can be spherical (when $l = 0$), elliptical if $l = 1$; and even more ellipsoidal when $l = 2$. Electrons in the K shell (when $n = 1$) can only have $l = 0$ and therefore they must all have spherical orbits. In general the ratio of the major to the minor axes of the ellipsoidal orbit increases as l increases. The electrons in any particular shell can therefore be divided into sub-shells, each with different orbits, and the number of electrons possible in each sub-group of quantum number l is given by the expression $2(2l + 1)$; thus when $l = 2$ the number of possible electrons in this orbit is 10; where $l = 1$, there are 6 electrons possible; and when $l = 0$, there cannot be more than 2 electrons. The M shell, which can contain a maximum of

[1] In 1924, de Broglie suggested that all atomic particles are associated with waves of a definite frequency. Bohr's conception of circular and elliptical orbits is now replaced by the appropriate wave-functions. The old ideas are still useful in that they provide a simple picture of the atom which can be understood by most readers.

18 electrons, will have these distributed throughout three different orbits with 2, 6, and 10 electrons in each orbit respectively; those in the first group revolve in a spherical orbit, those in the second revolve in an elliptical orbit and those in the third group revolve in an even greater ellipsoid.

The *third quantum number* 's' represents the angular momentum or spin of the electron about its own axis; s always has the value of $\frac{1}{2}$.

The fourth quantum number 'j' expresses the total angular momentum of a particular electron. In other words, it is the resultant of the orbital and spin moments, represented as vector quantities; it is equal to $l + s$.

Any further quantum numbers which need to be introduced are only extensions of those already described. Thus in a strong magnetic field an electron can develop a spin in either of two directions, the magnitudes of which can be represented by $\pm\frac{1}{2}$. This magnetic quantum number is designated m and its relation to the spin of the electron by m_s.

It is to be expected that the orbit of an electron is also subjected to magnetic influences in a strong field. The magnetic component of l is termed m_l and its value must lie between $+l$ and $-l$.

Hence, every electron in an atom, *in a strong magnetic field* can be described by *five* quantum numbers with definite values as follows:

n of value 1, 2, 3, . . . etc.
l of value 0 to $n-1$
s of value $\frac{1}{2}$
m_s of value $\pm\frac{1}{2}$
m_l of value $+l$ to $-l$

The possible number of electrons in any shell is limited by *Pauli's exclusion principle*, which states that 'no two electrons in the same atom can have an identical set of quantum numbers'. In other words, no two electrons can be in the same place at the same time and doing the same thing. Hence, for electrons in any shell a series of possible positions can be calculated. For example: where $n = 1$

	l	m_l	m_s
i.e. the K shell	0	0	$+\frac{1}{2}$
	0	0	$-\frac{1}{2}$

Only two different quantum representations are possible for the electrons in the K shell, because l and m_l have the value 0, and m_s can only be $+\frac{1}{2}$ or $-\frac{1}{2}$.

When $n = 2$ there are more possibilities

l	m_l	m_s
0	0	$+\frac{1}{2}$
0	0	$-\frac{1}{2}$
1	+1	$+\frac{1}{2}$
1	+1	$-\frac{1}{2}$
1	0	$+\frac{1}{2}$
1	0	$-\frac{1}{2}$
1	−1	$+\frac{1}{2}$
1	−1	$-\frac{1}{2}$

Thus there can be a total of eight electrons in the *L* shell, each with a different quantum representation. These two examples show why the number of electrons in each shell of quantum number n is equal to $2n^2$.

All electrons with $l = 0$ are termed *s* electrons; those with $l = 1$ are termed *p* electrons; those with $l = 2$, *d* electrons; and those with $l = 3$, *f* electrons. In any sub-shell, i.e. electrons of a certain l value, the number of electrons possible is $2(2l+1)$, because m_l can be any integer between $+l$ and $-l$, and m_s can be $+\frac{1}{2}$ or $-\frac{1}{2}$. For example, if l is 2, m_l can have the values $+2, +1, 0, -1, -2$; i.e. $2l+1$; each of these can have magnetic spins of $+\frac{1}{2}$ or $-\frac{1}{2}$; consequently, the number of possible electrons is $2(2l+1)$.

Table II.II shows the maximum possible number of electrons in each of the *K*, *L*, *M* shells, assuming that the electrons fit successively into shells, commencing with the *K* shell, then the *L* shell, etc., filling first the sub-shell *s*, then *p*, then *d* and so on.

TABLE II—II. MAXIMUM NUMBER OF ELECTRONS IN FIRST THREE ELECTRON ORBITS

Symbol	*K shell*	*L shell*		*M shell*		
n	1	2	2	3	3	3
	0	0	1	0	1	2
m_l	0	0	3	0	3	5
m_s	2	2	2, 2, 2	2	2, 2, 2, 2; 2, 2, 2, 2	
$2(2l+1)$, i.e. total electrons in sub-shell	2	2	6	2	6	10
$2n^2$, i.e. total electrons in complete shell	2	8			18	
Nomenclature*	$(1s)^2$	$(2s)^2$	$(2p)^6$	$(3s)^2$	$(3p)^6$	$(3d)^{10}$

* The number of electrons in a particular sub-shell is indicated by the number of the orbit, followed by the sub-shell number and the electrons it contains as an index figure.

The Arrangement of Electrons in Atoms

THE element with atomic number 1 is hydrogen; it has its single electron in the *K* shell, which is thus not fully filled or completed. The next element, helium (atomic number 2), has two electrons—both in the *K* shell which is now completed. As the number of orbital electrons is increased they must go into the *L* shell, filling first the 2*s* positions, then the six 2*p* positions, at which point the second shell is complete as in the inert gas, *neon*, with an atomic number of 10. Elements with higher atomic numbers must have some electrons in the *M* shell.

An electron will usually enter the sub-shell in which it will be held most firmly and with the greatest stability. The *K* and *L* shells are orbits which are relatively close to the nucleus and so the order for the electrons to enter is as expected, 1*s*, 2*s*, and 2*p*. In the *M* shell, however, the distance from the nucleus is greater, and so the attractive force on the electrons in this shell is less than in the two previous ones. The orbit of the 3*d* sub-shell is an elongated ellipsoid and this is apparently less stable than the spherical orbit of the 4*s* sub-shell of the *N* layer. Once the 3*p* sub-shell has been completed with the inert gas, *argon*, atoms of increasing atomic numbers, potassium, calcium, etc., have some electrons which enter the 4*s* sub-shell before the 3*d* orbit is filled. The irregularity is found, on an increased scale, in the *N*, *O*, *P* and *Q* shells.

Table II.III shows the electronic structure of most of the elements.[1]

Elements in which the last (outermost) shell but one is incomplete usually have characteristic properties, such as variable valency. They are called *transition elements*, and are of great value to the ceramic industry because their compounds are often coloured.

The arrangement of electrons shown in Table II.III represents those in atoms in an isolated state. In practice these are not very common; in most cases the electrons exist in combination with one another or as ions. The inert gases, helium, neon, argon, krypton, xenon and radon, are exceptions in having atoms with orbital structures capable of a stable, independent existence. A study of the electronic arrangements within these atoms has thrown considerable light on the nature of bond formation between two or more atoms.

The electrons in the various shells of the inert gases are distributed as follows:

TABLE II—IV. THE ELECTRONIC STRUCTURES OF THE INERT GASES

Element	At. No.	*K* orbit 1*s*	*L* orbit 2*s*	2*p*	*M* orbit 3*s*	3*p*	3*d*	*N* orbit 4*s*	4*p*	4*d*	4*f*	*O* orbit 5*s*	5*p*	5*d*	*P* orbit 6*s*	6*p*
Helium	2	2														
Neon	10	2	2	6												
Argon	18	2	2	6	2	6										
Krypton	36	2	2	6	2	6	10	2	6							
Xenon	54	2	2	6	2	6	10	2	6	10		2	6			
Radon	86	2	2	6	2	6	10	2	6	10	14	2	6	10	2	6

Helium and neon represent the condition where the *K* shell in the former, and the *K* and *L* shells in the latter are completely filled. In the other inert gases the electron configuration has the *s* and *p* sub-shells of the outer orbit completed with eight electrons and no others in the corresponding *d* and *f* sub-shells. It might be expected that such arrangements would be very stable, and they are, in fact, so. In all other elements one or more sub-shells of an orbit are not completely filled; there is always a deficiency or excess of electrons.

Four different types of electron structures can be recognised, each with special features:

Group 1. Atoms with outer completed shells or sub-shells. This group contains the inert gases which have the *s* and *p* sub-shells of their outer orbits complete.

[1] Elements with a higher atomic number than 92 have, in the main, only been produced in microscopically small amounts as a result of radio-active processes. There is some difficulty in determining the arrangement of the electrons in the outermost orbits, for these are weakly held and easily lost. One hundred and three elements have now been identified and occupy the positions as shown in the Periodic Table (Table II.IA).

TABLE II–III. THE ELECTRONIC STRUCTURES OF THE ELEMENTS

Symbol of Element	*At. No.*	*K orbit*	*L orbit*		*M orbit*			*N orbit*				*O orbit*			*P orbit*			*Q orbit*
		1s	*2s*	*2p*	*3s*	*3p*	*3d*	*4s*	*4p*	*4d*	*4f*	*5s*	*5p*	*5d*	*6s*	*6p*	*6d*	*7s*
H	1	1																
He	2	2																
Li	3	2	1															
Be	4	2	2															
B	5	2	2	1														
C	6	2	2	2														
N	7	2	2	3														
O	8	2	2	4														
F	9	2	2	5														
Ne	10	2	2	6														
Na	11	2	2	6	1													
Mg	12				2													
Al	13				2	1												
Si	14		Neon core of		2	2												
P	15		10 electrons		2	3												
S	16		$(1s)^2(2s)^2(2p)^6$		2	4												
Cl	17				2	5												
A	18				2	6												
K	19	2	2	6	2	6		1										
Ca	20							2										
Sc	21						1	2										
Ti	22						2	2										
V	23		Argon core of				3	2										
Cr	24		18 electrons				5	1										
Mn	25		$(1s)^2(2s)^2(2p)^6(3s)^2(3p)^6$				5	2										
Fe	26						6	2										
Co	27						7	2										
Ni	28						8	2										
Cu	29	2	2	6	2	6	10	1										
Zn	30							2										
Ga	31							2	1									
Ge	32		Copper core of					2	2									
As	33		28 electrons					2	3									
Se	34		$(1s)^2(2s)^2(2p)^6(3s)^2(3p)^6(3d)^{10}$					2	4									
Br	35							2	5									
Kr	36							2	6									
Rb	37	2	2	6	2	6	10	2	6			1						
Sr	38											2						
Y	39									1		2						
Zr	40		Krypton core of							2		2						
Nb	41		36 electrons							4		1						
Mo	42		$(1s)^2(2s)^2(2p)^6(3s)^2(3p)^6$							5		1						
Ma	43		$(3d)^{10}(4s)^2(4p)^6$							6		1						
Ru	44									7		1						
Rh	45									8		1						
Pd	46									10								

TABLE II—III. THE ELECTRONIC STRUCTURES OF THE ELEMENTS

Symbol of Element	At. No.	*K* orbit 1s	*L* orbit 2s	2p	*M* orbit 3s	3p	3d	*N* orbit 4s	4p	4d	4f	*O* orbit 5s	5p	5d	*P* orbit 6s	6p	6d	*Q* orbit 7s
Ag	47	2	2	6	2	6	10	2	6	10		1						
Cd	48											2						
In	49											2	1					
Sn	50				Silver core of							2	2					
Sb	51				46 electrons							2	3					
Te	52	$(1s)^2(2s)^2(2p)^6(3s)^2(3p)^6(3d)^{10}(4s)^2(4p)^6(4d)^{10}$										2	4					
I	53											2	5					
Xe	54											2	6					
Cs	55	2	2	6	2	6	10	2	6	10		2	6		1			
Ba	56														2			
La	57				Xenon core of 54 electrons									1	2			
					$(1s)^2 - (4d)^{10}(5s)^2(5p)^6$													
Ce	58	2	2	6	2	6	10	2	6	10	1	2	6	1	2			
Pr	59										2			1	2			
Nd	60										3			1	2			
Il	61										4			1	2			
Sm	62										5			1	2			
Eu	63										6			1	2			
Gd	64										7			1	2			
Tb	65				Xenon core as above						8			1	2			
Dy	66										9			1	2			
Ho	67										10			1	2			
Er	68										11			1	2			
Tu	69										12			1	2			
Yb	70										13			1	2			
Lu	71										14			1	2			
Hf	72	2	2	6	2	6	10	2	6	10	14	2	6	2	2			
Ta	73													3	2			
W	74													4	2			
Re	75				Hafnium core of 68 electrons									5	2			
Os	76				$(1s)^2 - (4s)^2(4p)^6(4d)^{10}(4f)^{14}(5s)^2(5p)^6$									6	2			
Ir	77													9				
Pt	78													9	1			
Au	79	2	2	6	2	6	10	2	6	10	14	2	6	10	1			
Hg	80														2			
Tl	81														2	1		
Pb	82				Gold core of										2	2		
Bi	83				78 electrons										2	3		
Po	84				$(1s)^2 - (5s)^2(5p)^6(5d)^{10}$										2	4		
At	85														2	5		
Rn	86														2	6		
Fr	87	2	2	6	2	6	10	2	6	10	14	2	6	10	2	6		1
Ra	88																	2
*Ac	89				Radon core of												(1)	(2)
*Th	90				86 electrons												(2)	(2)
*Pa	91				$(1s)^2 - (6s)^2(6p)^6$												(3)	(2)
*U	92																(4)	(2)

* The electron being added in this group of elements is thought to be filling 5*f* positions thus forming an actinide group similar to the rare earths.

Group 2. Atoms with the outermost shell only partially filled. This group includes atoms with atomic number up to 17 (with helium and neon as exceptions), the sub-group elements in the Periodic Table and also the two elements which follow each inert gas in Table II.IB, e.g. potassium and calcium following argon.

Group 3. Atoms with two outermost shells partially filled. This group includes the transition elements, the compounds of which are often coloured.

Group 4. Atoms with three outermost shells partially filled. This group includes the rare earth elements with an atomic number between 58 and 71. The *N*, *O*, and *P* shells are all incomplete. The element *lanthanum* (At. No. 57) has the electron arrangement shown in Table II.V.

TABLE II—V. THE ELECTRONIC STRUCTURE OF LANTHANUM

K orbit	*L* orbit		*M* orbit			*N* orbit				*O* orbit			*P* orbit		
1*s*	2*s*	2*p*	3*s*	3*p*	3*d*	4*s*	4*p*	4*d*	4*f*	5*s*	5*p*	5*d*	6*s*	6*p*	6*d*
2	2	6	2	6	10	2	6	10		2	6	1	2		

As the atomic number increases through cerium (At. No. 58), praseodymium (At. No. 59), etc., electrons fill the 4*f* positions, the 5*d* and 6*s* shells remaining the same until the element hafnium (At. No. 72) is reached. This group is often termed the 'Lanthanide contraction' and the properties of all elements in it are similar. There is some evidence to suggest that there may be a similar contraction beginning with actinium (At. No. 89) in which the 5*f* orbit is filled.

The Table of electron distributions can be related to the Periodic Table (p. 58). Elements which are alike in properties show remarkable similarity in the arrangement of the outermost electron orbits. The alkali elements illustrate this clearly, as shown in Table II.VI.

TABLE II—VI. THE ELECTRONIC STRUCTURES OF THE ALKALI ELEMENTS

Element	At. No.	*K* orbit	*L* orbit		*M* orbit			*N* orbit				*O* orbit				*P* orbit			*Q* orbit
		1*s*	2*s*	2*p*	3*s*	3*p*	3*d*	4*s*	4*p*	4*d*	4*f*	5*s*	5*p*	5*d*	5*f*	6*s*	6*p*	6*d*	7*s*
Lithium	3	2	1																
Sodium	11	2	2	6	1														
Potassium	19	2	2	6	2	6		1											
Rubidium	37	2	2	6	2	6	10	2	6			1							
Caesium	55	2	2	6	2	6	10	2	6	10		2	6			1			
Francium	87	2	2	6	2	6	10	2	6	10	14	2	6	10		2	6		1

In all cases the outermost shell contains a single electron in the *s* orbit.

So general is the relation between properties and the distribution of electrons in the various shells that Thomsen and Bohr were able to draw up a Periodic Table based on electronic structure in which like elements are joined by lines and the transitional elements are contained in frames (Table II.VII).

TABLE II—VII. THE PERIODIC TABLE

(Thomsen and Bohr)

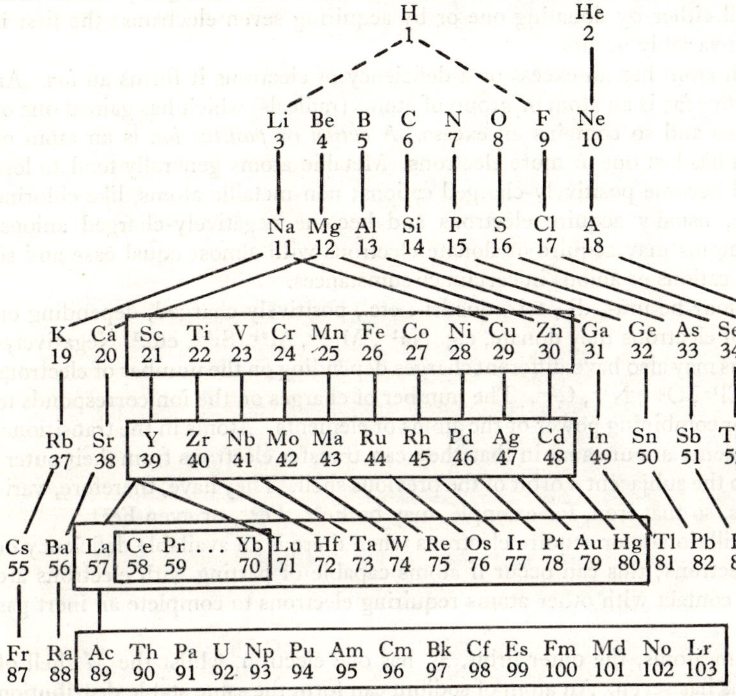

The important point established by this Table is, that if the atoms of an element have the same electron distribution in the outermost shells as the atoms of another element, then those two elements will have similar properties. The obvious implication is that many properties of an element are determined entirely by the arrangement of electrons in the outermost orbits of the atoms. The completed, internal shells are stable and, being disturbed only in exceptional circumstances, have less effect on the properties of the elements in which they are present.

ELECTRON BONDS BETWEEN ATOMS

THE manner in which electrons behave when they take part in reactions depends on the distribution of electrons in the orbits or shells. The atoms of the *inert gas* elements are very stable and unreactive, hence the electrons in these atoms must be in orbits which are complete in themselves and are not susceptible to change. Atoms of other elements appear to be constantly striving to change the distribution of electrons in their outermost shells to the arrangement found in the inert gas orbits, i.e. an external orbit $(s+p)$ of 8. This can be brought about either by 'acquiring' or 'donating' some electrons. In general, an atom will 'take the easy way out'; if fewer electrons have to be acquired than donated in order to form a distribution with completed orbits, then the electrons will be acquired and *vice versa.*

All the alkali atoms have one electron in their outermost orbit; they can form an inert gas shell either by donating one or by acquiring seven electrons; the first is 'easier' and invariably occurs.

When an atom has an excess or a deficiency of electrons it forms an *ion.* An *anion* or *negative ion* is an atom or group of atoms (radicals) which has gained one or more electrons and so contains an excess. A *cation* or *positive ion* is an atom or radical which has lost one or more electrons. Metallic atoms generally tend to lose electrons and become positively-charged cations; non-metallic atoms, like chlorine and bromine, usually acquire electrons and become negatively-charged anions. Some other atoms may acquire or donate electrons with almost equal ease and so can be either cations or anions in certain circumstances.

Cations may be uni-, di-, tri-, quadri-, etc., positively-charged, depending on the number of electrons they donate, e.g. Na^{1+}, Mg^{2+}, Al^{3+}, Si^{4+}, etc.[1] Negatively-charged anions may also have different charges depending on the number of electrons gained, e.g. Cl^{1-}, O^{2-}, N^{3-}, C^{4-}. The number of charges on the ion corresponds to the *valency*[2] or combining power of the atoms or elements. Atoms in the transitional group of elements are unusual in that they can transfer electrons from their outermost orbits to the subjacent *d* orbit of the previous shell. They have, therefore, variable valencies, so that iron, for example, may be Fe^{2+}, Fe^{3+}, or even Fe^{4+}.

Atoms will donate or acquire electrons when there is an available deficiency or surplus of electrons; this can occur if atoms capable of parting with electrons are brought into contact with other atoms requiring electrons to complete an inert gas shell.

In sodium atoms, the outer orbit, 3*s*, has one electron, whilst the *M* shell of chlorine atoms has seven. An atom of sodium can form the same stable distribution of electrons as in the inert gas, neon, by giving up one electron. An atom of chlorine can also acquire an inert gas shell (that of argon), by acquiring one electron. If a sodium atom donates its surplus electron to an atom of chlorine, two stable arrangements can result and the compound sodium chloride is formed in atomic proportions. Now the atoms are no longer electrically neutral; the sodium has lost an electron and has therefore a net positive charge whilst the chlorine atom has gained an electron and so is negatively charged. The reaction can be represented schematically by the notation suggested by Lewis:

$$\mathrm{Na}\cdot + \,.\underset{\cdot\cdot}{\overset{\cdot\cdot}{\mathrm{Cl}}}: \;\rightarrow\; \mathrm{Na}^{+}\; :\underset{\cdot\cdot}{\overset{\cdot\cdot}{\mathrm{Cl}}}:^{-}$$

This consideration may be extended to more complicated cases, e.g. the formation of calcium bromide by the union of one calcium atom with two bromine atoms. The calcium atom gives up two electrons, one to each of two bromine atoms, thus:

$$:\underset{\cdot\cdot}{\overset{\cdot\cdot}{\mathrm{Br}}}. + \cdot\mathrm{Ca}\cdot + .\underset{\cdot\cdot}{\overset{\cdot\cdot}{\mathrm{Br}}}: \;\rightarrow\; \overset{-}{:\underset{\cdot\cdot}{\overset{\cdot\cdot}{\mathrm{Br}}}:}\quad \overset{++}{\mathrm{Ca}}\quad \overset{-}{:\underset{\cdot\cdot}{\overset{\cdot\cdot}{\mathrm{Br}}}:}$$

[1] Occasionally, the + or − sign is repeated the necessary number of times, instead of preceding it by a number, e.g. Na^{+}, Mg^{++}, Al^{+++}, Si^{++++}, etc.

[2] For further information on valency see most books on Physical Chemistry.

Potassium oxide illustrates the case where two atoms each donate one electron to another atom

$$K\cdot + \cdot\underset{..}{\overset{..}{O}}\cdot + \cdot K \rightarrow \overset{+}{K}\quad \overset{--}{:\underset{..}{\overset{..}{O}}:}\quad \overset{+}{K}$$

This mechanism is termed **ionic bonding,** because the electrostatic charges or Coulomb attractive forces hold the two oppositely-charged ions together.

The product of the combination of atoms is called a *molecule*, which is the smallest quantity of a compound that can exist alone.

Another type of union of atoms which is very common, is known as the mechanism of *electron sharing* and of this the halogens provide a good example. The atoms of these elements have seven electrons in their outermost orbit. If two such atoms combine to share a pair of electrons then both of them will have effectively an inert gas configuration. Thus:

$$:\underset{..}{\overset{..}{Cl}}\cdot + \cdot\underset{..}{\overset{..}{Cl}}: \rightarrow\ :\underset{..}{\overset{..}{Cl}}:\underset{..}{\overset{..}{Cl}}:$$

This is called **covalent bonding**; by it the atoms are held together by the shared electrons which are part of the orbits of all the atoms forming the molecule. The *shared electron pair* must have opposite spins to be in accordance with the Pauli Exclusion Rule (p. 63).

More than one pair of electrons can be shared. For example, an oxygen atom must share two pairs of electrons to form an oxygen molecule.

$$:\overset{..}{O}: + :\overset{..}{O}: \rightarrow :\overset{..}{O}::\overset{..}{O}:$$

The union of atoms by either ionic or covalent bonds results in the formation of a stable molecule with the same distribution of electrons around each atom as there is in an inert gas. Molecules are capable of independent existence whereas single isolated atoms are found only in exceptional circumstances.

There are other types of bonding between atoms, two of which are described below: reference to a third and fourth (the hydrogen and hydroxyl bond) will be made later.

(i) **Van der Waal bonding.** When gases are cooled they liquefy and the atoms must therefore come together in some way. This also occurs in the inert gases, yet no electron transfer or sharing can result. The atoms exert some influence on one another but the mechanism is not fully understood. This type is termed *Van der Waal* bonding and is described more fully on p. 139.

(ii) **Metallic bonds.** There is another type of bonding between metal ions. Sodium atoms, for example, can unite with other similar atoms to form a strong bond, yet all the atoms are alike and each contains one electron in the outermost shell. It is conceived that each gives up an electron and becomes a positively-charged ion. The free electrons then wander about the structure without being attached to any particular ion.[1] As a result, metals have characteristic properties such as conductivity, malleability, etc. (see p. 74).

[1] Hume-Rothery, W., *Ann. Rep. chem. Soc.*, 46, 42, 1949.

THE BONDING OF ATOMS AND IONS IN SOLIDS

ATOMS may combine with others of similar or unlike type by one or other of the bonding mechanisms outlined in the previous section. In the simple examples which have been illustrated, the end product is a *molecule*, which is a complete and stable unit within itself. Such molecules have a definite existence in gases, liquids and in some kinds of solids and in these cases the individual particles of matter are distinct and separate and not strongly bonded to other particles in the immediate vicinity.

In crystals, however, there is a perfect continuum and X-ray analysis has shown that such substances cannot consist of discrete molecules placed either randomly or symmetrically but—except in isolated instances—of atoms, each of which is surrounded by a fixed number of other atoms, either of the same type or different. The linkages between all atoms are similar and there is no preferential bonding as would be the case if discrete molecules existed. Sodium chloride, for example, forms crystals in which each sodium atom or ion is centrally positioned with respect to six chlorine atoms, and each chlorine atom is symmetrically surrounded by six sodium atoms. The linkage between all atoms is identical throughout the crystal. Therefore, within the crystal lattice, the concept of simple electron transfer from a sodium atom to an adjacent chlorine is not tenable. The true picture is difficult to visualise unless the wave mechanical concept is introduced. The excess electron of a sodium atom must be imagined as a cloud with no discrete existence. Each chlorine atom shares a part (one-sixth) of this cloud and also acquires a further one-sixth share from each of five other sodium atoms in the immediate vicinity, so that six chlorine atoms each have a sixth share in the excess electrons of six sodium atoms. By this means, each atom acquires the same electronic structure as an inert gas and becomes an ion, the sodium ion being positively- and the chlorine ion negatively-charged. The resulting lattice is thus a three-dimensional network of positive and negative ions, and is called an *ionic crystal*, the bonds being between ions with opposite charges.

A crystal can also be formed by covalent bonds arranged in a similar fashion to ionic ones, the shared pair of electrons being distributed uniformly between all the adjacent atoms and forming *covalent crystals* with electrons, visualised only as a wave mechanical function, common to the orbits of several atoms.

Metallic crystals are simply groups of metallic ions with the excess electrons moving freely through the lattice; i.e. the extension of the metallic bond in three-dimensional space.

Crystals made up entirely of *Van der Waal linkages* are uncommon if they exist at all, but there are many examples of crystals in which they are associated with other kinds of bonds. Graphite, for example, is made up of sheets of covalently-bonded carbon atoms, the linkages between adjacent sheets being through Van der Waal linkages. In some clay minerals, the structures are held together by similar links. Bonds of this type are weak, and so both graphite and clay minerals can be sheared very easily because the sheets or plates glide readily upon each other. Van der Waal linkages appear to be due to the permanent dipoles which are set up in adjacent molecules. The attractive force is proportional to $1/r^n$ where n is usually about 6.

The four principal bonds which occur in molecules, namely ionic, covalent, Van der Waal and metallic, thus have their counterparts in crystalline structures. There are two other types of bonds, however, which occur in a limited number of lattices and involve many more atoms than there are in simple molecules. These are:

(*a*) **The Hydrogen Bond.**[1] The atom of hydrogen is unique in consisting of a single proton and one electron. When hydrogen is incorporated in a crystal lattice it cannot be regarded as a separate atom with an electron configuration. It is better to regard the proton as a dimensionless centre of force which tends to modify the electronic structure of neighbouring atoms. For example, an oxygen ion (O^{2-}) can form a hydroxyl ion (OH^-) by combining with a hydrogen atom. The resulting ion is of similar size to the oxygen, but with only half its negative charge. In some structures hydrogen can form bonds which are almost ionic in character and in these cases the proton or H^+ acts as a kind of bridge. A hydroxyl (OH) unit, for example, can link with an adjacent oxygen ion. The hydrogen is represented as continuously jumping backwards and forwards between the two oxygens, although, in fact, a 'hybrid' structure is formed.

$$-O-H^+ \text{——} O^- - \longleftrightarrow O^- \text{——} H^+O^- -$$

The resulting alternating or *resonance*[2] structure behaves as a rather weak bond, and is known as a *hydrogen bond.*

(*b*) **The Hydroxyl Bond.** When two or more hydroxyl (OH) groups occur close to each other in a crystal lattice, they may bond together by the electrostatic attraction between the negative and positive parts of the groups, thus·

$$\begin{array}{ccc} \text{—}\, O^- & \ldots\ldots\ldots & H^+ \\ | & & | \\ H^+ & \ldots\ldots\ldots & O^- \,\text{—} \end{array}$$

Water molecules may also form similar bonds with suitable hydroxyl groups in a lattice.

When bonds of the hydrogen or hydroxyl type occur they usually do so along well-defined planes in a particular crystal. They are most common in minerals which are composed of sheets or planes of atoms and such bonds provide the linkages between adjacent layers. Bernal and Megaw[3] have discussed these two types of bonds in some detail and have given several examples which will be illustrated in Chapters III and IV.

There are thus six definite types of bonds which may be present in crystals, namely, ionic, covalent, Van der Waal, metallic, the hydrogen and the hydroxyl bond. All of these can occur in crystal lattices and more than one form of bond may be present in a single crystal though ionic and covalent bonds are by far the most common.

[1] Pimentel, G. C., and McClellan, A. L., *The Hydrogen Bond* (Freeman, London, 1960).

[2] A resonance structure is one where the atoms or electrons can occupy two or more alternative positions of similar energy values. The arrangement which results does not correspond to any extreme but has a higher bond energy and the distance between the atoms or groups involved is invariably shorter than when no resonance occurs. A typical example is graphite which is described in detail in Chapter IV (p. 177).

[3] Bernal, J. D. and Megaw, H. D., *Proc. Roy. Soc.* (London), **A**151, 384, 1935.

TABLE II—VIII. PROPERTIES ASSOCIATED WITH PRINCIPAL TYPES OF ATOMIC BONDS

(Modified and enlarged after Evans*)

Property	*Ionic Bond*	*Covalent Bond*	*Metallic Bond*	*Van der Waal Bond*	*Hydrogen and Hydroxyl Bonds*
Mechanical	Strong bonding hard crystals	Strong, hard crystals	Strength of varying degrees	Weak bonding soft crystals	Weak bonding. Slightly stronger
Thermal	Fairly high melting point. Low coefficient of expansion. When melted forms ions	Variable melting point. Low coefficient of expansion. Forms molecules in melt	Variable melting point. Long liquid-forming period	Low melting point. Large coefficient of expansion	Decomposed at fairly low temperatures
Electrical	Moderate insulation. When molten conducts by ionic transportation. Occasionally soluble in liquids of high dielectric constant	Insulators in both solid and liquid form	High conductivity by electron transport	Excellent insulators	Good insulators
Optical and magnetic	Behave like individual ions in adsorption and other properties. Similar in solution	High refractive index. Adsorption different in solutions and gaseous forms	Opaque. Similar properties in liquid state	Behave as individual molecules. Similar in solution and as gas	Show cleavage and usually have a dipole
Heat transfer	Fairly high insulation. Often conductivity increases with temperature	Highly insulating at all temperatures	Conductivity usually high	Good insulation properties	Good insulators
Structural	Non-directed structures of high co-ordination	Spacially directed and numerically limited. Structures of low co-ordination and low density	Non-directed structures of very high co-ordination. High density	Formally analogous to metallic bond. Usually high spacings	Weak bonds. Lines of easy fracture

* Evans, R. C., *An Introduction to Crystal Chemistry* (Cambridge University Press, London, 1939).

The various types of bond confer certain properties on crystals which are summarised in Table II.VIII.

The actual bonding between atoms in crystal structures is not well-defined and though the principal types are ionic and covalent linkages, most bonds are non-ideal and appear to be intermediate between the two kinds. The most direct proof of the composite ionic-covalent nature of some bonds is in the dipole moment of that linkage. In a true covalent bond, each atom must be non-polar, that is, each component atom is electrically neutral. On the other hand an ionic bond is polar because each atom acquires a charge; the donating-electron atoms become positive ions, or cations, whilst those receiving electrons become negative ions, or anions. A dipole[1] is thus set up between the two ions of ionic linkages but not when the bond is covalent. Most crystals exhibit dipole moments intermediate between the two extremes of ionic and covalent bonds.

With elements lying between the alkali metals and the halogens in the Periodic Table (p. 58), ionic bonding tends to give way to covalent and most positive ions have seldom lost more than three electrons or negative ions gained more than two. Thus, although Al loses three electrons to form Al^{3+} with a symmetrical and inert-gas electron-structure, its tendency to do so is strongly opposed by the attraction of the highly-positive ion so formed for electrons in the shells of surrounding negative ions.

The transition from a covalent to an ionic bond can be regarded as a displacement of the shared electron pair or pairs towards the more electronegative atom. Alternatively, a partly covalent bond can be imagined as one in which the electrostatic fields of two oppositely-charged ions (forming an ionic linkage) have been partially neutralised by the displacement of the transferred electrons to the positive ion.

Further evidence for mixed bond formation can be gathered from X-ray studies of bond angles and length. Composite bonds of other types of linkages are also recognised.

The degree of covalency, or alternatively the ionic character of a bond, has been calculated empirically by Pauling;[2] it is related to the relative *electronegativity* of the atoms forming that bond. The electronegativity of one atom with respect to another is derived from the 'electrochemical series' or list of electrode potentials. The alkali metals are the most highly electropositive, whilst the halogens are most electronegative. When elements with the greatest possible difference in electronegativity combine together the most ionic linkages will be formed. The electronegativity can be defined as the attraction of a neutral atom in a molecule for electrons. Pauling has compiled a list of values for most elements, some of which are shown in Table II.IX.

[1] An electrostatic field is set up between two ions of opposite charge, in like manner to the magnetic field between opposite poles of a bar magnet. The strength and intensity of this field is expressed as a *dipole moment*. Use has been made of dipole moments to find the relative orientations of bonds in molecules and crystals. If the moments of the individual bonds are known, the angles between the polar bonds can be calculated.

[2] Pauling, L., *The Nature of the Chemical Bond* (Cornell Univ. Press and Oxford Univ. Press, 1960).

TABLE II—IX. ELECTRONEGATIVITY OF THE ELEMENTS

(after Pritchard and Skinner)[1]

				H			
				2·1			
Li	Be	B		C	N	O	F
1·0	1·5	2·0		2·5	3·0	3·5	3·9
Na	Mg	Al		Si	P	S	Cl
0·9	1·2	1·5		1·8	2·1	2·5	3·0
K	Ca	Sc	Ti	Ge	As	Se	Br
0·8	1·0	1·3	1·6	1·8	2·0	2·4	2·8
Rb	Sr	Y	Zr	Sn	Sb	Te	I
0·8	1·0	1·3	1·6	1·8	1·9	2·1	2·5
Cs	Ba						
0·7	0·9						

Percentage Ionic Character of Bonds

PAULING has also evaluated the approximate percentage of ionic bonding between two atoms from the difference in electronegativity between them. The values are obtained from the dipole moments and the lengths of the bonds (from the absorption spectra) and taking the value of a completely ionised compound as 100. Some of these are shown in Table II.X.

TABLE II—X. PERCENTAGES OF IONIC CHARACTER OF BONDS

	% *ionic*		% *ionic*		% *ionic*
Mg—O	71	Si—O	50	Al—O	63
Mg—F	82	Si—Cl	30	Al—F	79
Mg—Cl	59	Si—Br	20	Al—Br	35

The difference in electronegativity can be calculated from Table II.IX and then this value used in Fig. II.1 to calculate the percentage ionic character of the bonding as in Table II.X.

It is difficult to visualise these intermediate types of bonds, the composition of the linkage can be represented only as a wave mechanical function and by postulating the occurrence of resonance between the two types of bonds.

THE SIZES OF ATOMS AND IONS

THE mass of an *atom* is nearly all centred on the nucleus, but the size of the atom depends on the extent of the electron orbits. X-ray diffraction methods have enabled the size of atoms to be measured with great precision.

Although the sizes of atoms can be calculated, their shapes are difficult to assess. The outermost orbits of electrons are often elliptical (see p. 62) and so, at any one instant, an atom may have an irregular shape. The electron field is dynamic, however, and the volume that it sweeps out in space as it rotates will be approximately

[1] Pritchard, H. O. and Skinner, H. A., *Chem. Rev.*, 55, 745, 1955.

spherical, provided that the encircling, external electrical field is of uniform distribution. It is customary, although not always strictly correct, to assume that atoms (and ions) behave as spheres and their dimensions are expressed in terms of their diameters or radii. The size of an atom is defined as one-half the distance of closest approach between two adjacent atoms.

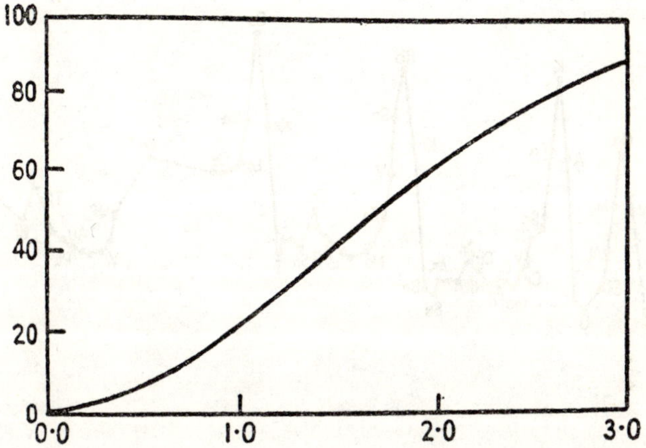

FIG. II.I. THE IONIC NATURE OF THE BOND AS A FUNCTION OF THE DIFFERENCE IN ELECTRONEGATIVITY OF THE ATOMS (after Pauling)

Fig. II.2 shows a graph of the atomic number against atomic radii. A succession of peaks occurs with the alkali elements in zenithal positions. The periodicity of these peaks is most marked and follows closely the curves of specific heat, valency, and other properties of atoms plotted against the atomic number.

The sizes of ions are of greater importance in the investigation of crystal structure because, in crystals, ions are the more common units. The measurement of the size of ions is fraught with difficulty for the ionic dimension varies with the environment. The sizes of ions have been measured by assuming that they behave as rigid spheres when in contact with one another. A series of analogous structures must be examined and the relative sizes can then be assessed.

There are great differences between the sizes of atoms and the sizes of ions of the same element. The positive ion is always smaller than the neutral atom of the same element, whilst a negative ion is larger than the corresponding neutral atom. The greater the positive charge of an ion, the smaller will be its size, all else being equal. This decrease in ionic radius with increasing nuclear charge is clearly shown by studying the series Na^{+} to S^{6+} in Table II.XI. The electronic structure is the same in all cases.

In a crystal lattice a positive ion is surrounded by a number of negative ions, the number depending on certain factors which are discussed later in this chapter. The size of the internally-contained cation varies with the number of anions surrounding it, so that great difficulty is involved in measuring the size of ions under standard conditions. This has, however, been largely overcome, and these results form the basis of Table II.XI. In all cases the values of ionic radii are based on the assumption that the ion is surrounded by six oppositely-charged ones; this may not be the stable state of some ions, but it does, at least, form a comparative basis.

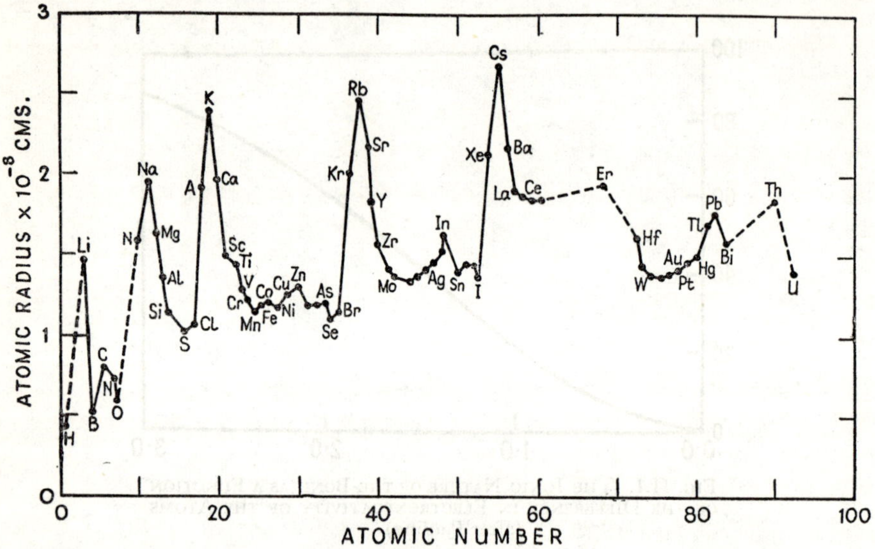

FIG. II.2. THE RELATIONSHIP BETWEEN THE ATOMIC NUMBER AND THE ATOMIC RADII OF ATOMS

THE ARRANGEMENT OF IONS IN A CRYSTAL LATTICE

A CRYSTAL lattice is composed of a large number of atoms or ions. A stable structure can be produced only if these ions are packed tightly and are retained in more or less rigid positions in space by surrounding units. Most inorganic crystals contain some proportion of ionic bonding in their atomic linkages. Therefore, the crystal lattice may be regarded as being composed of a series of positively- and negatively-charged ions. The anions are much larger than the cations (see Table II.XI) and, as would be expected, they play a larger part in determining the nature of the lattice. The cations, however, are instrumental in fixing the position which the anions can take up in space and, before the disposition of the anions can be considered, the cation-anion relationship must be established.

(1) THE PACKING OF CATIONS WITHIN ANIONS

In order to produce a stable arrangement, the ions must be packed as closely as possible. Hence, the large anions must be situated in such a way as to enclose the smaller cations tightly. The manner in which this is brought about depends almost entirely on the relative sizes of the cations and anions. If it is assumed that ions

TABLE II—XI. IONIC RADII IN ÅNGSTROM UNITS (1Å = 1 × 10⁻⁸ cm.)
(after Pauling)

Ions with Inert Gas Structures

Valency −3	−2	−1	+1	+2	+3	+4	+5	+6
		H 1·54	Li 0·60	Be 0·31	B 0·20	C 0·15	N 0·11	
N 1·86	O 1·40	F 1·36	Na 0·95	Mg 0·65	Al 0·50	Si 0·41	P 0·34	S 0·29
P 2·11	S 1·84	Cl 1·81	K 1·33	Ca 0·99	Sc 0·81	Ti 0·68	V 0·59	Cr 0·52
As 2·28	Se 1·98	Br 1·95	Rb 1·48	Sr 1·13	Y 0·93	Zr 0·80	Nb 0·70	Mo 0·62
Sb 2·66	Te 2·21	I 2·16	Cs 1·69	Ba 1·35	La 1·15	Ce 1·01	Ta 0·84	W 0·71

Ions of Sub-group Elements

Valency +1	+2	+3	+4	+5	+6
Cu 0·96	Zn 0·74	Ga 0·62	Ge 0·53	As 0·47	Se 0·42
Ag 1·26	Cd 0·97	In 0·81	Sn 0·71	Sb 0·62	Te 0·56
Au 1·37	Hg 1·10	Tl 0·95	Pb 0·84	Bi 0·74	

Ions of Transitional Group Elements

Valency	Sc	Ti	V	Cr	Mn	Fe	Co	Ni
+2					0·80	0·76	0·74	0·72
+3	0·81	0·76	0·73	0·69	0·66	0·64		
+4		0·68	0·60		0·60			
+5			0·59					
+6				0·52				

behave as solid, rigid spheres, large-sized anions can be arranged in numerous ways to enclose smaller cations in the vacant spaces between them. The size of the enclosed cation determines the situations of the anions which surround it, because these must be so positioned as to hold the cation tightly; in fact they must touch it in as many points as possible.

Case (i) When three large spheres (anions) are arranged so that they each touch the other two, their centres are at the corners of an equilateral triangle. A smaller sphere (cation) can fit into the space between the three larger spheres. (Fig. II.3.)

If R is the radius (BE) of the larger spheres, the radius r of the smaller internal sphere can be calculated. BO is equal to $R+r$ and

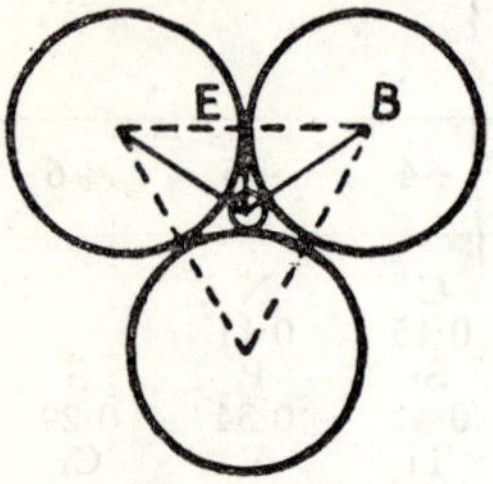

$$\cos 30° = \frac{BE}{BO} = \frac{R}{R+r} = \frac{\sqrt{3}}{2}$$

$$\text{Therefore } r = \frac{(2-\sqrt{3})R}{\sqrt{3}} = 0{\cdot}155R$$

Fig. II.3. Three-fold or Equilateral Co-ordination

Thus a sphere of 0·155R can fit between three others of radius R, when these are arranged so that their centres form an equilateral triangle.

Case (ii) For four anions to surround a central cation, the larger spheres must be arranged in the form of a regular tetrahedron.[1] Three of the spheres have their centres at the corners of the triangular base and the fourth rests at the apex, centrally above the other three, touching all of them. The small cation lies in the vacant space inside and must touch all the four anions.

The size of the central small sphere can be calculated quite simply, for the angle at the centre of a tetrahedron, i.e. $\angle$AOC is 109° 28′. (Fig. II.4.)

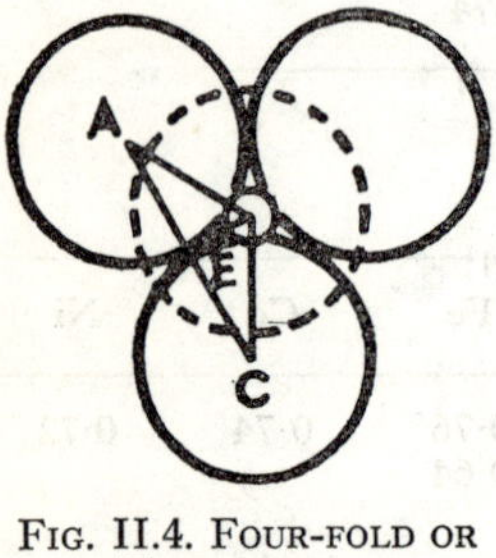

Therefore

$\angle$CAO = 35° 16′ for AOC is an isosceles triangle

$$\cos \angle CAO = \frac{AE}{AO} = \frac{R}{R+r} = 0{\cdot}8166$$

$$\therefore R - 0{\cdot}8166R = 0{\cdot}8166r$$

$$r = \frac{0{\cdot}1834}{0{\cdot}8166}R$$

$$= 0{\cdot}225R$$

Fig. II.4. Four-fold or Tetrahedral Co-ordination

Thus a sphere of 0·225R can be surrounded by four spheres of radius R, situated at the corners of a regular tetrahedron.

Case (iii) Six large spheres can surround a centrally-placed smaller sphere when each is at the corner of a regular octahedron. Four of the large spheres are on the same plane, one other lies centrally above and the other centrally below the other four as shown in Fig. II.5, in which BC = 2R and AB = $2R+2r$.

[1] A tetrahedron has three equilateral triangular sides and a base of the same shape; it is sometimes described as a three-sided or triangular pyramid.

$$\cos \angle ABC = \cos 45° = \frac{BC}{AB} = \frac{2R}{2R+2r} = \frac{1}{\sqrt{2}}$$

$$2\sqrt{2}R = 2R+2r$$

$$r = (\sqrt{2}-1)R$$

$$= 0{\cdot}414R$$

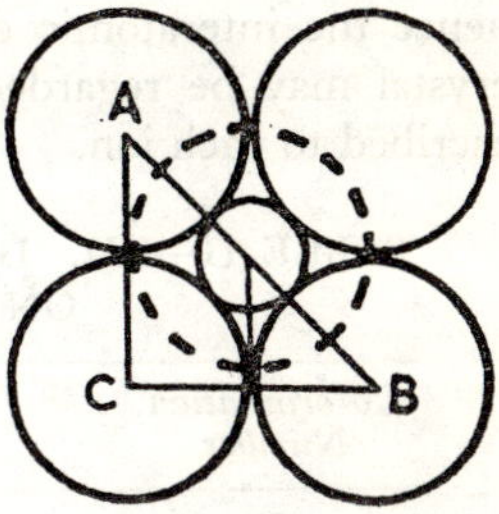

FIG. II.5. SIX-FOLD OR OCTAHEDRAL CO-ORDINATION

Therefore the largest cation which will fit in the centre of six anions arranged in the form of a regular octahedron is 0·414R where R is the radius of the anion.

Case (iv) When eight spheres surround a smaller one they occupy the corners of a cube. This is shown in Fig. II.6, where three of the front spheres A, B, C, and one back one G, are shown with the central smaller sphere.

The radius of the central sphere can be found from the two triangles ABC and ACG, in which AC = CB = 2R.

In the triangle ABC, $AB^2 = AC^2+CB^2 = 4R^2+4R^2 = 8R^2$. AB is the face-diagonal of the cube and so also is AG, which must also equal $\sqrt{8R^2}$.

In the triangle ACG, $GC^2 = AC^2+AG^2$

$$= 4R^2+8R^2 = 12R^2$$

$$\therefore GC = \sqrt{12}R$$

GC is a body-diagonal of the cube and passes from the centre of sphere C, through the diameter of the small sphere, to the centre of sphere G.

Therefore $CG = 2R+2r$ (where r is the radius of the small central sphere)

$$2r = \sqrt{12}R-2R$$

$$r = \sqrt{3}R-R$$

$$= 0{\cdot}732R$$

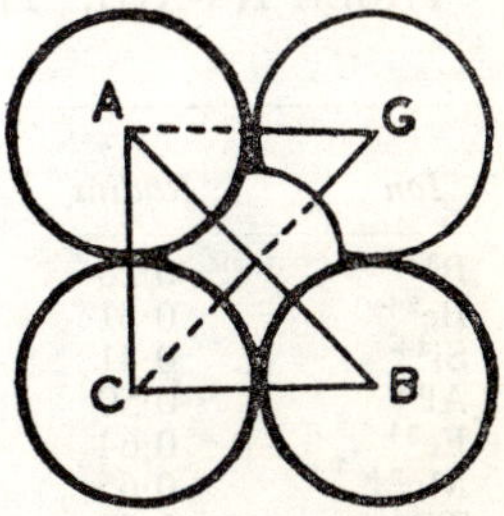

FIG. II.6. EIGHT-FOLD OR CUBIC CO-ORDINATION

Therefore, the largest size for a cation at the centre of eight anions (each of radius R), arranged at the corners of a cube, is 0·732R.

The number of oppositely-charged ions which surround a central one is called the *co-ordination number* of that ion and the value is governed entirely by the relative sizes of the ions as shown in Table II.XII.

In deducing the foregoing figures it has been necessary to assume that all ions behave as rigid, perfect spheres, and remain so even when in contact with other unlike ions. This is not entirely correct, but as a guide to a visual picture of crystal structure, it can be regarded as approximately so. In actual fact, there is no rigid boundary to an ion comprising a nucleus and orbital electrons because when two ions approach each other, a repulsive force develops with great abruptness. This arises mainly as a result of the close approach of electrons in the outermost shells, and hence the interatomic distance is limited; but, for most purposes, the ions in a crystal may be regarded as rigid spheres in contact and a characteristic radius ascribed to each ion.

TABLE II—XII. DEPENDENCE OF THE CO-ORDINATION NUMBER ON THE CATION/ANION SIZE RATIO

Co-ordination Number	*Dispositions of ions around central one*	*Critical radius ratio (cation/anion)*
8	Corners of a cube	1·000–0·732
6	Corners of a regular octahedron	0·732–0·414
4	Corners of a regular tetrahedron	0·414–0 225
3	Corners of an equilateral triangle	0·225–0·155
2	Linear	0·155–0

In the majority of structures which are of importance to ceramics, the predominating anions are those of oxygen. They surround various cations each of which is of a different size. The co-ordination number of these cations with respect to oxygen is of paramount importance in the crystal structure of all minerals.

If the radius of oxygen is taken as 1·40 Å, a series of cation/anion ratios can be calculated and hence the theoretical co-ordination of that cation with respect to oxygen. (Table II.XIII.)

TABLE II—XIII. THE CO-ORDINATION NUMBERS OF SOME CATIONS WITH OXYGEN ANIONS

Ion	*Radius*	*Cation/anion ratio*	*Predicted co-ordination*	*Observed co-ordination*
B^{3+}	0·20	0·14	3	3 (4)
Be^{2+}	0·31	0·22	4	4
Si^{4+}	0·41	0·29	4	4
Al^{3+}	0·50	0·36	4 or 6	4 (5) 6
Fe^{3+}	0·64	0·46	(4) or 6	4, 6
Mg^{2+}	0·65	0·46	4 or 6	6
Ti^{4+}	0·68	0·49	(4) or 6	4, 6
Fe^{2+}	0·76	0·54	6	6
Zr^{4+}	0·80	0·57	6 or (8)	6 (8)
Na^{+}	0·95	0·68	8	6 (8)
Ca^{2+}	0·99	0·71	8	6 (7) 8 (9)
K^{+}	1·33	0·95	(8) or 12	(6), (7), 8 (9), (10) 12

The bracketed figures are co-ordination values which are not of common occurrence. The ionic radius varies with the co-ordination number and should be corrected. According to A. F. Wells, the ionic ratios can be calculated from Born's equation which can be rearranged to read:

$$\frac{r_p}{r_m} = \left(\frac{p}{m}\right)^{\frac{1}{n-1}}$$

where r_p and r_m are the radii of cations for co-ordination numbers of p and m respectively; n is approximately equal to 9.

Thus, when the co-ordination number changes from 6 to 8 the radius is increased by 1·036 times, and reducing the co-ordination number from 6 to 4 changes the size by a factor of 0·950.

The principle stated above, that the co-ordination number is a function of the cation/anion size, forms the basis of *Pauling's first rule* of crystal structure.[1]

The *second rule of Pauling* states that the sum of positive and negative charges in a lattice must be equal if the structure is to be stable. Therefore the valency residing on a particular cation must be divided equally amongst the co-ordinating anions. An anion consequently receives from each cation with which it is in contact a contribution towards the satisfaction of its own valency.

This may be illustrated by examples:

(i) Silicon is quadrivalent, i.e. Si^{4+}; it is also in four-fold co-ordination with oxygen ions (O^{2-}). Hence, an ion of silicon contributes one valency share to each of four oxygen ions.

As each oxygen ion requires two valency units to become completely satisfied electrically, each must be in contact with two silicon ions to form a structure which is balanced electrically. This may be represented schematically as follows:

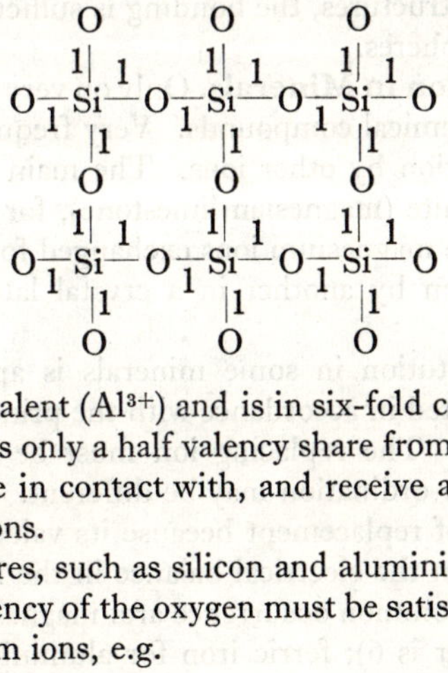

(ii) Aluminium is trivalent (Al^{3+}) and is in six-fold co-ordination with oxygen, so each oxygen ion receives only a half valency share from the aluminium ion. Each oxygen must, therefore, be in contact with, and receive a half valency share from, each of four aluminium ions.

(iii) In mixed structures, such as silicon and aluminium ions, with oxygen, the rule still holds and the valency of the oxygen must be satisfied by contributions from both silicon and aluminium ions, e.g.

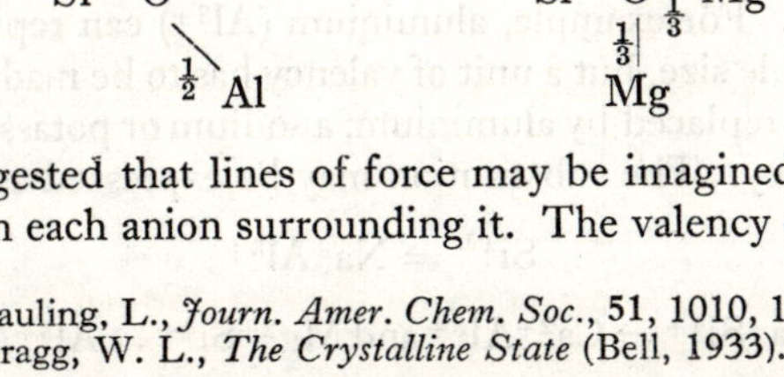

Bragg[2] has suggested that lines of force may be imagined as starting on each cation and ending on each anion surrounding it. The valency of the cation divided

[1] Pauling, L., *Journ. Amer. Chem. Soc.*, 51, 1010, 1929.
[2] Bragg, W. L., *The Crystalline State* (Bell, 1933).

by its co-ordination number provides a measure of the strength of the electrostatic bond developed between the two ions. Table II.XIV shows the values for some of the important cations.

TABLE II—XIV. STRENGTH OF BOND IN CATION-OXYGEN LINKAGES

Cation	B^{3+}	Be^{2+}	Si^{4+}	Al^{3+}	Fe^{3+}	Ti^{4+}
Co-ordination No.	3	4	4	4, 6	6	4, 6
Strength of Bond	1	$\frac{1}{2}$	1	$\frac{3}{4}, \frac{1}{2}$	$\frac{1}{2}$	$1, \frac{2}{3}$
Cation	Mg^{2+}	Zr^{4+}	Fe^{2+}	Na^{+}	Ca^{2+}	K^{+}
Co-ordination No.	6	6, 8,	6	6, 8	8	8, 12
Strength of Bond	$\frac{1}{3}$	$\frac{2}{3}, \frac{1}{2}$	$\frac{1}{3}$	$\frac{1}{6}, \frac{1}{8}$	$\frac{1}{4}$	$\frac{1}{8}, \frac{1}{12}$

The co-ordination number depends on the cation to anion size ratio only when the bonding between the atoms is essentially ionic, because only in such circumstances can the individual units be regarded as independent spheres. In groupings of atoms where the bonding is principally covalent, the co-ordination number is decided by the number of covalent bonds formed by the atom and this favours small values. In most silicate structures, the bonding is sufficiently ionic for the units to be regarded as isolated spheres.

Isomorphous Substitution in Minerals. Only on very rare occasions can minerals be regarded as simple chemical compounds. Very frequently one or more ions are replaced in some proportion by other ions. The main replacements occur in the cation component; dolomite (magnesian limestone), for example, is similar to calcite, but with some of the magnesium ions exchanged for calcium ones. This form of replacement of one ion by another in a crystal lattice is termed *isomorphous substitution.*

Although the substitution in some minerals is apparently bewildering, the amount and type are limited in accordance with the principles laid down in the two rules of Pauling (p. 83). The replacing ion must be comparable in size to the original one or else the co-ordination may be different. Even then a suitably-sized ion may not be capable of replacement because its valency contribution is not the same; if this were so, then the electrical balance in the lattice would be destroyed. Typical replacements of common occurrence are: magnesium for calcium (when the co-ordination of the latter is 6); ferric iron for aluminium; ferrous iron for magnesium.

There are other instances where an ion of lower valency can replace one of a higher valency. When this occurs, additional ions have to be introduced into vacant spaces in the lattice. For example, aluminium (Al^{3+}) can replace silicon (Si^{4+}) for they are of comparable size, but a unit of valency has to be made up. In the felspars, when a silicon ion is replaced by aluminium, a sodium or potassium ion is introduced to balance the valency. The substitution may be expressed as;

$$Si^{4+} \rightleftharpoons Na^{+}Al^{3+}$$

The substitutions $Na^{+}Si^{4+} \rightleftharpoons Ca^{2+}Al^{3+}$ and $Mg^{2+}Si^{4+} \rightleftharpoons Al^{3+}Al^{3+}$ are also common.

Complex Ions. Acid radicles, such as $CO_3{}^{2-}$ and $SO_4{}^{2-}$, are similar to co-ordinated groups in having the oxygen ions regularly arranged round the central cation. The linking, however, is mainly covalent and the atoms do not behave as rigid spheres. The outer electron orbits of the atoms merge together and as a consequence, the effective diameter is less than in electrostatic or ionic bonding. The geometrical arrangements of some of these complex units have been deduced by Bragg, and are reproduced in Fig. II.7.

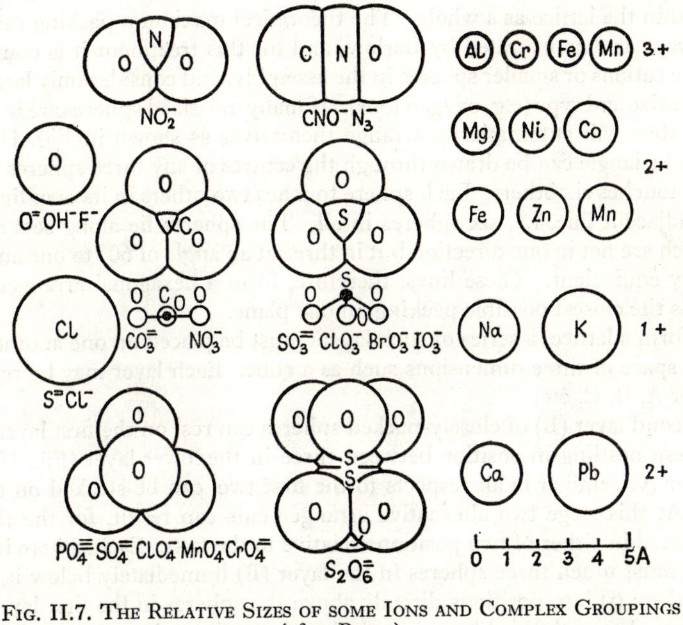

FIG. II.7. THE RELATIVE SIZES OF SOME IONS AND COMPLEX GROUPINGS (after Bragg)

(2) THE PACKING OF CO-ORDINATED UNITS IN A CRYSTAL LATTICE

All cations present in a crystal are surrounded by a certain number of anions, which are usually of larger size. From these small units of only a few ions, a crystal lattice, composed of many thousands of ions, is built up. A crystal may have cations of a single or of numerous types; so, in some cases, only one type of co-ordinated unit is present, whilst, in others, there may be several different group species. Furthermore, the linking between these units of structure must be such that the valency requirements of each ion are fully satisfied by contact with a sufficient number of ions of opposite sign.

The build-up of any crystal lattice can only be effected provided that these two principles are obeyed. There is, however, a third requirement which must be satisfied, namely, that the packing of the ions in the lattice must be tight or 'close', otherwise the crystal would not hold together.

In most silicate structures, oxygen anions predominate and are, by far, the largest ions present. These must be arranged in a crystal, so that, when they are grouped together, they must adequately and tightly contain cations; they must also be positioned so that electrical neutrality is maintained throughout the structure; and they must also fit together so that all are firmly held by other oxygen ions in the immediate neighbourhood. The tightness of packing in a lattice is thus principally determined by the arrangement of the large spheres of oxygen atoms.

The packing of ions within a crystal lattice always tends towards a maximum, because this would give a structure of the greatest stability. The larger ions which form the essential framework are so arranged that there is the minimum of void space within the lattice as a whole. The theoretical maximum packing can result in several ways as demonstrated by Barlow, and for this treatment it is convenient to ignore the cations or smaller spheres in the assembly, and consider only large spheres of a single size and type (e.g. oxygen ions). If many uni-sized spheres are in one plane or layer, they will automatically arrange themselves as shown in Fig. II.8(*a*). An equilateral triangle can be drawn through the centres of any three spheres in contact and each touches six others. Each sphere touches two others in its own line and two in each adjacent line, i.e. six spheres in all. The spheres lie along sets of parallel lines which are not in one direction but in three at an angle of 60° to one another and all exactly equivalent. These lines, therefore, form a hexagonal arrangement, and this forms the closest possible packing on one plane.

To form a lattice, a series of such layers must be placed on one another so as to occupy a space of three dimensions such as a cube. Each layer may be represented by a letter A, B, C, etc.

A second layer (B) of closely-packed spheres can rest on the first layer (A) with each sphere nestling in position between three in the lower layer (Fig. II.8(*b*)). A third layer (C), similar in all respects to the first two, can be stacked on top of the second. At this stage two alternative arrangements can result, for the third layer can be placed in either of two positions relative to the first. Each sphere in the new layer (C) must touch three spheres in the layer (B) immediately below it, but they can go either: (i) into positions directly above the spheres in the first layer, so that the structure has each third layer identical with the first. A sequence of ABABAB, etc., fully represents the structure where A is the arrangement of spheres in one layer and B that of the layer immediately above. This is known as *hexagonal close-packing*;

(ii) so that the third layer occupies entirely new positions and is unlike that of layers A and B and creates the sequence ABCABC, etc. This order would build up a pyramid and is referred to as *cubic close-packing*. Each sphere in the new layer (C) is directly above the space between three spheres in the first layer (A).

The two arrangements are in Fig. II.8 (*c* and *d*) where the arrangement of successive layers is shown.

The terms 'hexagonal' and 'cubic' close-packed refer to the shape of the resulting structures. In the former, there is a six-fold axis of rotation in a direction perpendicular to the layers so that if an imaginary axis is drawn downward through any sphere in the top layer, on rotation about this axis an identical configuration is produced six times during one complete turn of 360°. When the third layer is arranged in cubic packing there are four three-fold axes of symmetry inclined in the same way as in the body-diagonals of a cube. Rotation about an axis parallel to the

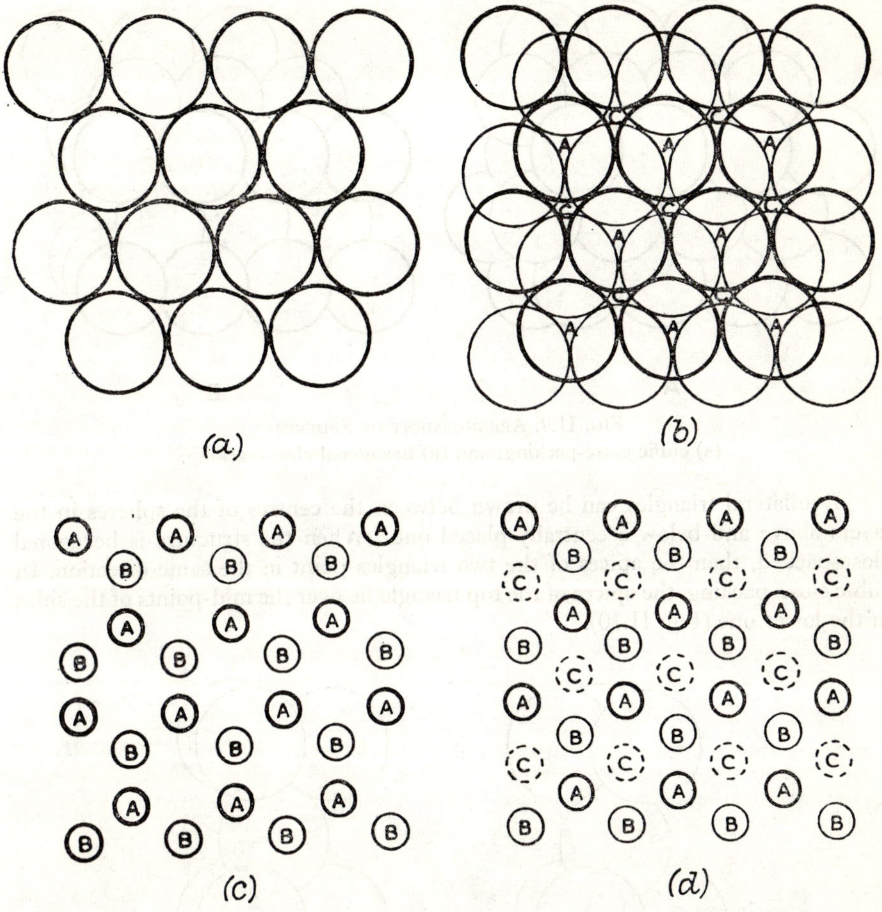

FIG. II.8. DIAGRAMMATIC REPRESENTATION OF THE CLOSE-PACKING OF SPHERES

(*a*) one layer of close-packed spheres; (*b*) two successive layers—note the available sites for the third layer marked AAA, etc., and CCC, etc.; (*c*) sequence of successive layers in hexagonal close-packing ABABAB, etc.; (*d*) successive layers in cubic close-packing ABCABC, etc.

body-diagonal brings the same structure into coincidence three times during a complete turn. When the spheres are seen in plan, the layers are perpendicular to any body-diagonal of a unit cube; they are situated at the corners and mid-points of the faces of a cube, hence the alternative term *face-centred cubic* for a cubic close-packed structure.

The diagrams in Fig. II.9 illustrate these features clearly.

In both hexagonal and cubic close-packed structures, each sphere is in contact with twelve other like spheres, six of which are in the same plane and three each in the adjacent layers above and below. They differ only in the spacial arrangements of spheres in these adjacent layers.

FIG. II.9. ARRANGEMENT OF SPHERES
(A) cubic close-packing; and (B) hexagonal close-packing.

Equilateral triangles can be drawn between the centres of the spheres in the layers above and below a centrally-placed one. When the structure is hexagonal close-packed, then the apices of the two triangles point in the same direction. In cubic close-packing, the apices of the top triangle lie over the mid-points of the sides in the lower one (Fig. II.10).

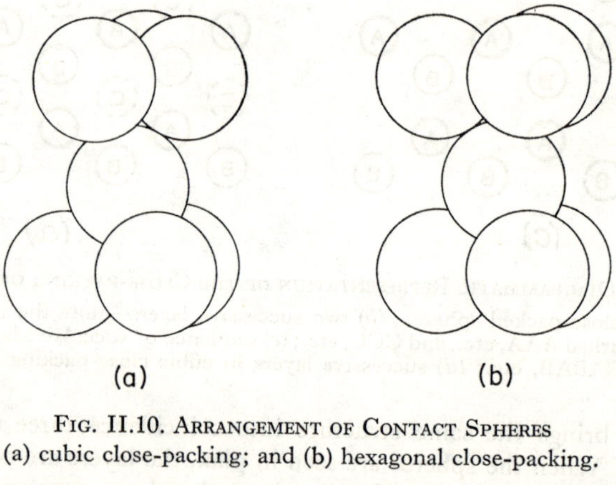

FIG. II.10. ARRANGEMENT OF CONTACT SPHERES
(a) cubic close-packing; and (b) hexagonal close-packing.

In a third type of packing, the spheres are arranged with their centres as the corners of a cube, and another one of similar size is at the centre of the cube touching all the rest. A more open structure results, for each sphere now only touches eight others. It is called *body-centred cubic packing*, and is illustrated in Fig. II.11.

Other arrangements are also possible; layers of close-packed spheres can be stacked in various ways, for example 'double hexagonal close-packing' ABACABAC, etc., but the environment of all atoms in the lattice cannot then be the same.

Some structures are not fully close-packed. The lattice may be so constructed that valency requirements could not be satisfied if there were not voids between the anions. Many silicate minerals have this structure.

Close-packing in crystal structures is an ideal rarely found in practice. The co-ordinated units of which the lattice is composed may be distorted by a too large or an unsymmetrical cation. Many crystals, however, have anions in approximately close-packed positions but more open than the ideal, whilst in others, the symmetrical distribution is considerably distorted.

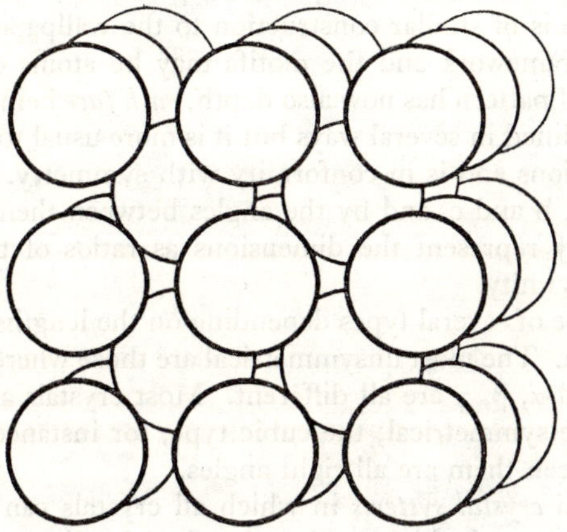

FIG. II.11. THE ARRANGEMENT OF SPHERES IN BODY-CENTRED CLOSE-PACKING

Defective Lattices. The theory of lattice structures, as described in the preceding pages, assumes that every lattice is perfect in shape and content. Many crystals show evidences of imperfect lattices and all have defects developed to a greater or less degree. In recent years, defects in the lattice have been shown to be the activation centres for solid and other reactions and will be described in detail in Chapter X.

CRYSTAL SYSTEMS

A PERFECT crystal is bounded by plane faces at constant angles to each other and the relation between such faces and angles is an important feature of the structure of all crystals. The measurement of these characteristic dimensions is one means of identifying the various compounds which occur in the crystalline form. Yet, though the angles between the faces of any crystal are constant for that substance, the shape or *habit* of those crystals may vary greatly, depending on the environmental conditions during their formation. Thus the same substance may in some occurrences be as crystals of cubic shape, but, in others, as lath- or needle-shapes and so appear to be quite different; but the various angles between the faces for all (perfect) crystals of one substance are the same irrespective of its crystalline shape.

The fundamental units of all crystals are the ions, but these are united to form larger units which give symmetry to crystals. All crystal lattices are composed of blocks of ions arranged in a specific manner. The arrangement of ions in each block is identical and any one block in the lattice has the same spacial situation and orientation as all others.

A useful analogy which has been suggested by Bragg is that of patterned wall-paper. A particular motif repeats at regular intervals in both horizontal and vertical directions. If lines are drawn joining identical motifs then parallelograms result, all of which are regular. Each parallelogram or cell might be described as the unit of pattern, for all others are exactly the same and the wallpaper has finite extension by virtue of a number of these units being joined together.

A crystal lattice is of similar construction to the wallpaper but now there is a three-dimensional framework and the motifs may be atoms or ions or groups of them; the unit cell of pattern has now also depth, *each face* being a parallelogram. A unit cell may be outlined in several ways but it is more usual to select one which has the smallest dimensions and is in conformity with symmetry. It is defined by the length of its sides a, b and c, and by the angles between them α, β, γ. Crystallographers occasionally represent the dimensions as ratios of the crystal axes; the b axis being taken as unity.

Unit cells can be of several types depending on the lengths of the sides and the angles between them. The most unsymmetrical are those where a, b, and c are all of different lengths and α, β, γ are all different. Most crystals are composed of unit cells which are more symmetrical; the cubic type, for instance, has all sides equal and the angles between them are all right angles.

There are seven *crystal systems* in which all crystals can be classified on the basis of their symmetry and which can be related to the dimensions of the unit cell. They are:

	Axes	*Angles*
1. *Cubic*—which has the highest symmetry with all three axes and angles equal	$a=b=c$	$\alpha=\beta=\gamma=90°$
2. *Tetragonal*—where one axis is of unequal length but all the angles are equal, and at right angles	$a=b\neq c$	$\alpha=\beta=\gamma=90°$
3. *Orthorhombic*—with all axes unequal but all at right angles to each other	$a\neq b\neq c$	$\alpha=\beta=\gamma=90°$
4. *Rhombohedral*—with all axes equal, but the angles between them are not right angles	$a=b=c$	$\alpha=\beta=\gamma\neq 90°$
5. *Hexagonal*—with four axes, three lateral and equal at an angle of 120° to each other, the fourth is unequal and at 90° to the plane of the lateral ones	$a_1=a_2=a_3\neq c$	$\gamma_1=\gamma_2=\gamma_3=120°$; $\alpha=90°$
6. *Monoclinic*—all axes are unequal but there are two equal angles at right angles and one other not at 90°	$a\neq b\neq c$	$\alpha=\gamma=90°\neq\beta$
7. *Triclinic*—this is the most unsymmetrical form with all axes and angles different	$a\neq b\neq c$	$\alpha\neq\beta\neq\gamma$

These systems refer to the external form of crystals and the unit cells may not be simple, i.e. with lattice-points situated only at corners. Additional lattice-points may be involved without altering the symmetry of the system. There are in all fourteen possibilities and these are termed the *Bravais space lattices* which are outlined in Fig. II.12. The cells numbered 1, 2, 4, 8, 9, 10, and 12 are simple; the others are either termed face- or body-centred cells. All crystals can be assigned to one or other of these space lattice groups.

This is not the complete story because the lattice-points may consist of several atoms grouped into a characteristic spacial configuration, and giving rise to different symmetry elements.

Symmetry Elements

A crystal which is composed of single atoms situated at the lattice-points has the greatest possible symmetry. In the simple cube (Fig. II.12), for example, the symmetry elements are as follows:

(*a*) Rotation about an axis drawn through the centre of two opposite faces brings the same configuration into coincidence on four occasions. There are three similar axes at right-angles to each other which permit this operation. Therefore, the cube has 3 four-fold axes of symmetry.

(*b*) Rotation about a body-diagonal axis results in three structure coincidences per revolution through 360°. There are four body-diagonals, hence there are 4 three-fold axes of symmetry.

(*c*) Rotation about an axis drawn through the centres of two opposite edges brings coincidence only twice in a complete revolution. There are six such axes. There are, therefore, 6 two-fold axes of symmetry.

(*d*) Planes parallel to the faces are reflection planes, for if a mirror were placed in these planes then the virtual image seen in it would be in coincidence with the actual structure behind it.

(*e*) Similar reflection planes occur parallel to the face-diagonals of the cube.

(*f*) There is a centre of symmetry or inversion because faces and edges of the cube occur in pairs on opposite sides of a central point. Any lattice-point, therefore, projected through the centre comes into coincidence with a similar point at an equal distance from the mid-point.

If the single atom or ion at the corners is replaced by a group of atoms or ions many of these symmetry elements are no longer present.

Crystals may have different groups of symmetry elements. For example, in the cubic system, spinel, rock-salt, and galena among others have all the elements outlined for the simple cube; cuprite has only a, b, and c; zinc blende and tetrahedrite have b, c, and e; pyrite has b, d, and f and in addition 3 two-fold axes; ilmenite has only b and also 3 two-fold axes, deducible from their external symmetry. Each of these groups of symmetry elements is termed a *point group* and is characteristic of a particular *crystal class*. There are five such classes possible in the cubic system and a total of thirty-two divided in the seven crystal systems.

A crystal cannot be fully defined by its point group for this refers to the symmetry of a crystal as a whole and the individual groups of atoms at a lattice-point

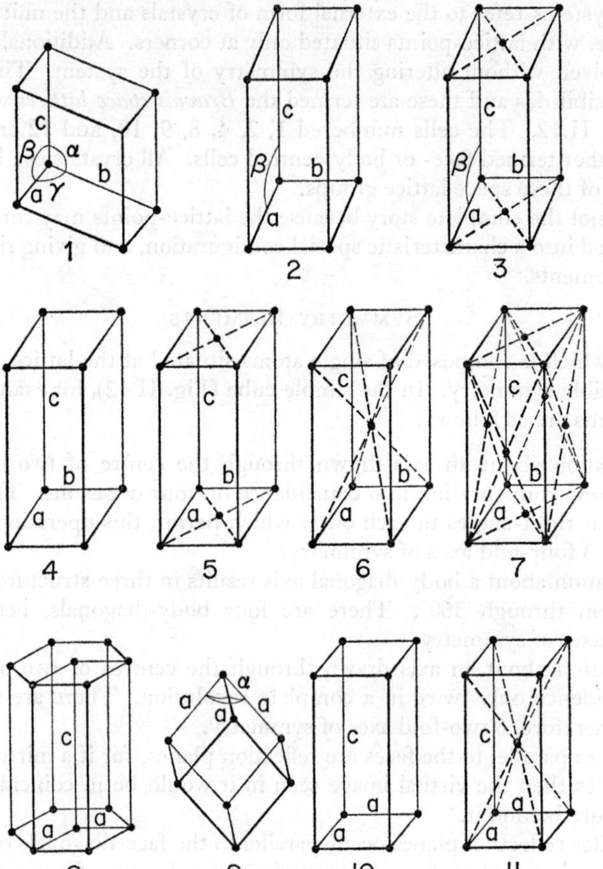

FIG. II.12. THE FOURTEEN BRAVAIS SPACE LATTICES

1. Triclinic; 2, 3. Monoclinic; 4, 5, 6, 7. Orthorhombic; 8. Hexagonal; 9. Rhombohedral; 10, 11. Tetragonal; 12, 13, 14. Cubic.

(after Bragg)

may be quite different and yet the crystal can have the same point group. This is well demonstrated by the examples above; spinel, rock-salt, and galena have all the same point group, yet they are of entirely different chemical composition, hence the atomic grouping round each lattice-point cannot possibly be the same. The symmetry of the group itself is important and essential if a full description of the crystal is desired; it is represented by the *space group*, which shows what operations are necessary to convert a unit part of a particular arrangement or motif into the complete pattern of the unit cell. These operations involve axes of rotation, reflection planes, glide planes and screw axes, and there are a possible 230. They enable a crystal to be completely described and the positions of the symmetry operations in each of them are outlined in the International Tables for the Determination of Crystal Structure.

There have been many notations for symbolising space groups, probably the most familiar being that of Schoenflies. Hermann and Maugin however have formulated an alternative method which describes the actual structure perhaps rather better. For a fuller space-group notation see *Crystal Structures of Minerals* by Bragg and Claringbull.

The Determination of Crystal Structures

A CRYSTAL lattice may be conceived as sets of atomic or ionic groups arranged with a certain symmetry so that planes of atoms occur throughout the lattice at equal spacings along the same crystallographic direction. Crystals behave as a three-dimensional grating which can diffract suitable radiation just as light is diffracted or 'scattered' by a 'grating' of closely-spaced objects or lines. Because these planes are of very small separation (a few Ångstrom Units at the most) radiation of comparable wavelength is required. Consequently, X-rays, which have a wavelength about ten thousand times as short as visible light, are used.

An X-ray beam, on passing through a set of lattice-points, will be diffracted and reinforcement will occur in accordance with W. L. Bragg's equation:

$$n\lambda = 2d \sin \theta$$

where n is a whole number, λ is the wavelength of the diffracted X-ray beam, d is the spacing of the planes and θ is the angle through which the beam is reflected. A series of reflections occurs from atoms in particular planes at different angles where $n = 1, 2, 3$, etc.

The characteristic X-ray pattern produced by any crystal thus enables the arrangement of the atoms in the crystal to be fully ascertained.[1]

The interpretation of a pattern of a complex crystal is difficult and tedious and many minerals have not yet been completely analysed in this way.

Atomic planes can be drawn in many different ways each having its own position in the lattice and with an interplanar spacing d. It is therefore important to use a recognised nomenclature in order to produce a complete description of the structure of a crystal, including the indexing of the crystal planes.

The Miller Indices. When the system of a crystal has been determined, a diagram

[1] The theory and techniques used in X-ray analysis are described fully in Chapter V.

showing the major axes can be constructed. Thus, in a simple cubic crystal (Fig. II.13) the axes are all of equal length and are at right angles, so the *a* axis can be represented by the length CG, the *b* axis by GH and the *c* axis by EG. Each of these lines makes an angle of 90° with both of the others. The face ABCD cuts the *a* axis at C, that is, one unit from the origin G, but does not cut either the *b* or *c* dimension. The intercepts on each axis are, therefore, 1, ∞, ∞. These are the parameters[1] of the face ABCD and are the lengths of the intercept of this face on the *a*, *b*, *c* axes in turn.

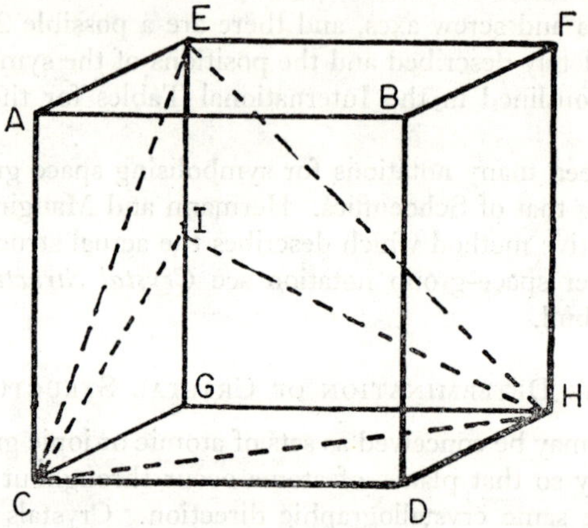

FIG. II.13. THE MILLER SYSTEM OF INDEXING CRYSTALLOGRAPHIC PLANES

In the Miller system the reciprocals of the above-mentioned parameters are used. Thus, the face ABCD has the indices 1/1, 1/∞, 1/∞, i.e. (100). A plane CEH cuts each axis at 1, therefore it has the indices (111). The plane CIH cuts the *a* and *b* axes at 1 and the *c* axis at $\frac{1}{2}$; its Miller indices are therefore (112).

As a plane face must be composed of atoms or ions, the crystal lattice must be composed of planes of this type, all of which are parallel and at a characteristic distance from each other. Many sets of such planes are to be found in any one lattice, the number and type depending on the situation of the atoms, or, in other words, on the space group. The reflections recorded on an X-ray photograph of a mineral can be related to the distribution of atoms or planes of atoms, which can be represented by their Miller indices.

Crystallographic Notation and Structural Representation. It is difficult to represent a three-dimensional crystal lattice by a two-dimensional drawing. In many cases the full space-group symmetry can be visualised only by constructing a solid model. To help those readers who wish to study crystallographic features in more detail, minerals will be classified in the text, wherever possible, according to the notation adopted by Bragg,[2] in which the space group, the crystal system and other

[1] A parameter is a figure representing a measurement or other datum, which is constant in the case being considered but may be different in other instances.

[2] Other notations are also in use, notably that of Schoenflies, but the convention used in this text is now internationally accepted as standard. Various alternative notations are listed fully in *International Tables for X-ray Crystallography*, Vol. 1952 (Kynoch Press).

pertinent features are tabulated. For instance, in a triclinic mineral such as the soda felspar, *albite*, the constants can be represented as follows:

	System: Triclinic	
Albite $NaAlSi_3O_8$	Axes	Angles
Space Group[1] $C\bar{1}$	a = 8·1Å	α = 94°
No. of molecules per unit cell[2] Z = 4	b = 12·9Å	β = 116° 30′
	c = 7·2Å	γ = 88°

The figures on the right-hand side give the unit cell dimensions and enable the crystalline form to be constructed, and those on the left determine the relative positions of the atoms in the unit cell.

When drawing and constructing atomic models it is well to bear in mind that, although the individual atoms are extremely small, they each have a finite size and are tightly packed together. There are several methods of representing the spacial arrangement. Firstly, the model can be built up of conveniently-sized spheres, the dimensions of which are in proportion to the actual size of the atoms they portray. The result is a reproduction of the actual close-packed arrangement, but the distribution of spheres and spaces within a complex structure, where small spheres are surrounded by larger ones, may be difficult to see. The alternative is to represent atoms by points, each of which is linked to others by rods of a length in proportion to the centre to centre distance of the atoms concerned; in the resulting open network the positions of atoms in the lattice may be indicated by small spheres at the junctions of the rods. These connecting links must not be confused with the valency bonds of organic molecules for they are co-ordination linkages which indicate the spacial arrangement and the structural details of the atoms in contact. The advantage of this second method is that atomic relationships in a complex structure can be clearly seen, but the idea of the close-packing within the lattice is not shown.

There is a third method of atomic model construction which is of value in certain cases, particularly in constructing solid units. The crystal lattice is built up of a number of units which contain anions surrounding smaller cations in certain co-ordinations. These units may be equilateral triangles, tetrahedra, octahedra or cubes (depending on whether the co-ordination is three-, four-, six-, or eight-fold), so arranged as to show a pattern in which these geometrical units share corners, edges and faces with each other.

These methods are used in the following chapters; where possible the most convenient and suitable representation is shown.

[1] The space group conventional representation after Hermann and Maugin.

[2] All the atoms in the chemical formula (i.e. the molecular composition) must appear in the unit cell one or more times. In albite the unit cell contains 4 atoms of sodium and aluminium, 12 silicon atoms and 32 oxygen atoms.

FURTHER BIBLIOGRAPHY

Structural Inorganic Chemistry, by A. F. Wells (Oxford Univ. Press, 1962).
Electronic Structure and Chemical Bonding, by O. K. Rice (McGraw-Hill Book Co., New York, 1940).
The Fine Structure of Matter (3 vols.), by C. H. Douglas Clark (Chapman and Hall, 1937).
Radioactivity Applied to Chemistry, by A. C. Wahl and N. A. Bonner (John Wiley & Sons Inc. New York; Chapman & Hall, London, 1951).
Chemical Crystallography, by C. W. Bunn (Oxford Univ. Press, 1961).
Crystal Structures of Minerals, by Sir L. Bragg and G. F. Claringbull (G. Bell & Sons Ltd., London, 1965).
An Introduction to Crystal Chemistry, by R. C. Evans (Cambridge U.P., 1969).

CHAPTER III

CRYSTAL STRUCTURE OF THE SILICATES

DESPITE the fact that the majority of minerals in the earth's crust are silicates, little was known about their fundamental structure and interrelationship until the present century was well advanced.

Until the work of Bragg (which was first published in the early 1930's), silicates were regarded as salts of hypothetical silicic acids, and as the present-day nomenclature still retains some of this earlier terminology, it is necessary to understand this system of classification in which silicates were arranged according to the ratio between the oxygen combined with the base and that combined in the acid radicles of the silicate as shown in Table III.I.

TABLE III—I. NOMENCLATURE OF THE SILICATES

Ratio of Acid Oxygen to Basic Oxygen	*Formula*		*Metallurgical Name*	*Mineralogical Name*
	RO *Base*	R_2O_3 *Base*		
Less than 1	$3RO.SiO_2$	$R_2O_3.SiO_2$	Sub-silicate Basic silicate	— —
1	$2RO.SiO_2$	$2R_2O_3.3SiO_2$	Singular silicate Monosilicate	Orthosilicate
1·5	$4RO.3SiO_2$	$4R_2O_3.9SiO_2$	Sesquisilicate	—
2	$RO.SiO_2$	$R_2O_3.3SiO_2$	Bisilicate	Metasilicate
3	$2RO.3SiO_2$	$2R_2O_3.9SiO_2$	Trisilicate	—
4	$RO.2SiO_2$	—	Quadrisilicate	Dimetasilicate

Orthosilicates have a ratio of acid oxygen : basic oxygen of less than 1·5–1·7. Metasilicates have a ratio greater than 1·7. In the absence of alumina the dividing line between ortho- and metasilicates is always 1·7, but where alumina is present a range of intermediate ratios occurs, due to the alumina being able to act both as an acid and as a base.

The principal theoretical silicic acids from which silicates were thought to be formed were, according to Clarke:[1]

				oxygen ratio
Dimetasilicic acid	$H_2Si_2O_5$	or	$H_2O.2SiO_2$	4:1
Trisilicic acid	$H_4Si_3O_8$	or	$2H_2O.3SiO_2$	3:1
Metasilicic acid	H_2SiO_3	or	$H_2O.SiO_2$	2:1
Diorthosilicic acid	$H_6Si_2O_7$	or	$3H_2O.2SiO_2$	1·33:1
Orthosilicic acid	H_4SiO_4	or	$2H_2O.SiO_2$	1:1

Some of these acids are unknown in nature as isolated substances but there are many compounds which appear to be salts of them.

[1] Clarke, H., *Bull. U.S. Geol. Survey*, 125, 1895.

Mellor recognised that there were numerous silicates and suggested that the corresponding acids could be arranged as in Table III.IIA.

TABLE III—IIA. NOMENCLATURE OF SILICIC ACIDS (MELLOR)

Type		1 (*mono-*)	2 (*di-*)	3 (*tri-*)	4 (*tetra-*)	5 (*penta-*)
Ortho-	$H_{2n+2}Si_nO_{3n+1}$	H_4SiO_4	$H_6Si_2O_7$	$H_8Si_3O_{10}$	$H_{10}Si_4O_{13}$	$H_{12}Si_5O_{16}$
Meta- (proto-)	$H_{2n}Si_nO_{3n}$	H_2SiO_3	$(H_4Si_2O_6)$	$(H_6Si_3O_9)$	$(H_8Si_4O_{12})$	$(H_{10}Si_5O_{15})$
Meso- (deutero-)	$H_{2n-2}Si_nO_{3n-1}$	—	$H_2Si_2O_5$	$H_4Si_3O_8$	$H_6Si_4O_{11}$	$H_8Si_5O_{14}$
Para- (tritero-)	$H_{2n-4}Si_nO_{3n-2}$	—	—	$H_2Si_3O_7$	$H_4Si_3O_{10}$	$H_6Si_5O_{13}$
Tetero-	$H_{2n-6}Si_nO_{3n-4}$	—	—	—	$H_2Si_4O_9$	$H_4Si_5O_{12}$
Pentero-	$H_{2n-8}Si_nO_{3n-4}$	—	—	—	—	$H_2Si_5O_{11}$

The following examples show the molecular composition of various natural silicates and their classification on the basis of hypothetical silicic acids:

TABLE III—IIB. MOLECULAR COMPOSITION OF SOME SILICATES

Sub-silicates (oxygen ratio, less than 1:1):			
Sillimanite	$Al_2O_3.SiO_2$		
Tricalcium silicate	$3CaO.SiO_2$		
Orthosilicates (oxygen ratio 1:1):			
Forsterite	$2MgO.SiO_2$	or	Mg_2SiO_4
Fayalite	$2FeO.SiO_2$		
Calcium orthosilicate	$2CaO.SiO_2$		
Willemite	$2ZnO.SiO_2$		
Phenacite	$2BeO.SiO_2$		
Zircon	$ZrO_2.SiO_2$		
Many other orthosilicates also occur in nature.			
Diorthosilicates (oxygen ratio 1·33:1):			
Barysilite	$3PbO.2SiO_2$	or	$Pb_3Si_2O_7$
Metasilicates (oxygen ratio 2:1):			
Sodium metasilicate	$Na_2O.SiO_2$	or	Na_2SiO_3
Calcium metasilicate (Wollastonite)	$CaO.SiO_2$	or	$CaSiO_3$
Magnesium metasilicate (Enstatite)	$MgO.SiO_2$	or	$MgSiO_3$
Trisilicates (oxygen ratio 3:1):			
Meerschaum	$2MgO.3SiO_2.2H_2O$		

This theory was useful at the time as it resolved a complicated system into some semblance of order, but there were many questions still to be answered. For example, kaolinite has the formula $Al_2O_3.2SiO_2.2H_2O$ or $Al_2Si_2O_7.2H_2O$. It would, therefore, be classified as a diorthosilicate derived from the theoretical acid $H_6Si_2O_7$, yet there is no resemblance in the properties of kaolinite and of other diorthosilicates. Furthermore in hydrated minerals like kaolinite the water molecules present are not in the same form as in most hydrates but are part of the actual crystal lattice. This was a little difficult to explain but some ingenious structures and linkages were postulated. The early workers in this field came very near the truth when they likened the crystal chemistry of silica to that of carbon, but they con- ructed their models to represent simple molecules as in organic chemistry and assumed that the crystalline silicates were composed of groups of these molecules whereas silicate minerals are obviously crystalline and the bonds between the atoms are strong and continuous throughout the lattice.

Although modern theories have refuted the idea of silicates being the salts of complex acids, the classifications of Mellor, Asch, Tschermak and others have been shown to be correct in many details and their terminology is still widely accepted.

Modern ideas of the structure of silicates are due largely to the work of W. H. and W. L. Bragg, Pauling and Taylor. They each were responsible for ascertaining the crystal structures of many silicates, but it was W. L. Bragg[1] who first presented the complete pattern which is now generally accepted.

He postulated that all silicates are built around a unit tetrahedron, composed of a silicon ion at the centre of four symmetrically placed oxygen ions. This unit is shown in Fig. III.1 and may be chemically represented as $(SiO_4)^{4-}$. The silicon ion is quadrivalent, i.e. it has four valency shares to donate to all surrounding anions. As four oxygen ions are contained in tetrahedral grouping around each silicon ion, they each receive one valency share from the silicon. Oxygen requires two valency electrons to become completely satisfied, so each has to acquire a further unit of charge from some external source.

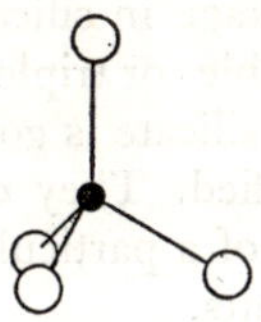

$(SiO_4)^{4-}$

FIG. III.1. THE SILICA TETRAHEDRON

The small black circle represents a silicon ion and the large circles are oxygen ions.

The completion of valency requirements of the oxygen can be secured by linking it to external cations of another type such as K^+, Ca^{2+} or Al^{3+}, or to a silicon of an adjacent identical tetrahedral group, thus (figuratively):

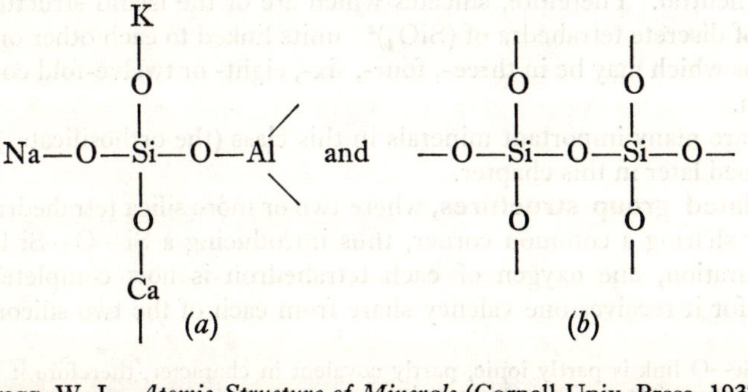

(*a*) (*b*)

[1] Bragg, W. L., *Atomic Structure of Minerals* (Cornell Univ. Press, 1937).

The structure is by no means as simple as is represented, for the external cations in (*a*) are shown as being linked to only one oxygen atom[1] of a tetrahedral silica group. In actual fact they are each co-ordinated with several oxygen anions and give a fixed valency share to each. Each oxygen ion must be linked to as many cations of one or more kinds to satisfy its divalent requirements.

When an oxygen ion is linked to two silicon ions it receives one valency share from each and is thus electrically satisfied. This

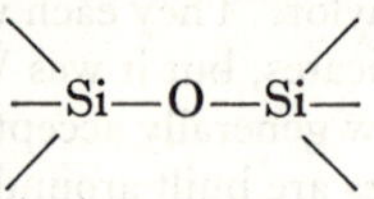

linkage permits the extension of the silicate structures throughout space and is analogous to the

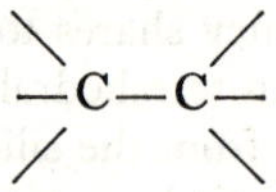

link of organic chemistry. The chemistry of the silicates can be likened in many respects to that of carbon for it is possible to have all the complex ring and group formations in both of them. The linkage in silicates however is through oxygen ions and not direct, and there is no double- or triple-bonding.

The type of structure of any one silicate is governed by the way in which the unbalanced oxygen valencies are satisfied. They can be divided into five groups depending on the number of oxygens of a particular tetrahedron which are linked directly to other neighbouring silica units.

STRUCTURAL FORMS OF SILICATES

(*a*) **Island structures** where separate $(SiO_4)^{4-}$ units exist, joined to each other only through additional cations which may be uni-, di-, tri-, or even quadrivalent. In all cases of structure formation the two rules of Pauling must apply; viz. (i) that all the cations concerned must have a certain co-ordination, governed entirely by the ratio of the size of this cation to that of the anion, and (ii) that the sum of negative charges must equal that of the positive ones; in other words, the structure must be electrically neutral. Therefore, silicates which are of the island structure type are composed of discrete tetrahedra of $(SiO_4)^{4-}$ units linked to each other only through other cations which may be in three-, four-, six-, eight- or twelve-fold co-ordination with oxygen.

There are many important minerals in this class (the orthosilicates) which are fully described later in this chapter.

(*b*) **Isolated group structures,** where two or more silica tetrahedra are linked together by sharing a common corner, thus introducing a Si—O—Si linkage. In this configuration, one oxygen of each tetrahedron is now completely satisfied electrically for it receives one valency share from each of the two silicon ions with

[1] The Si—O link is partly ionic, partly covalent in character, therefore it is correct to refer to the components as either atoms or ions.

which it is in contact. The remaining three oxygens attached to each silicon must receive their extra valency share from external cations as in island structures. Such a unit (represented in Fig. III.2A) has the chemical composition $(Si_2O_7)^{6-}$, so that in each unit of two tetrahedra there are six active oxygen atoms which can take part in complex formation with other cations.

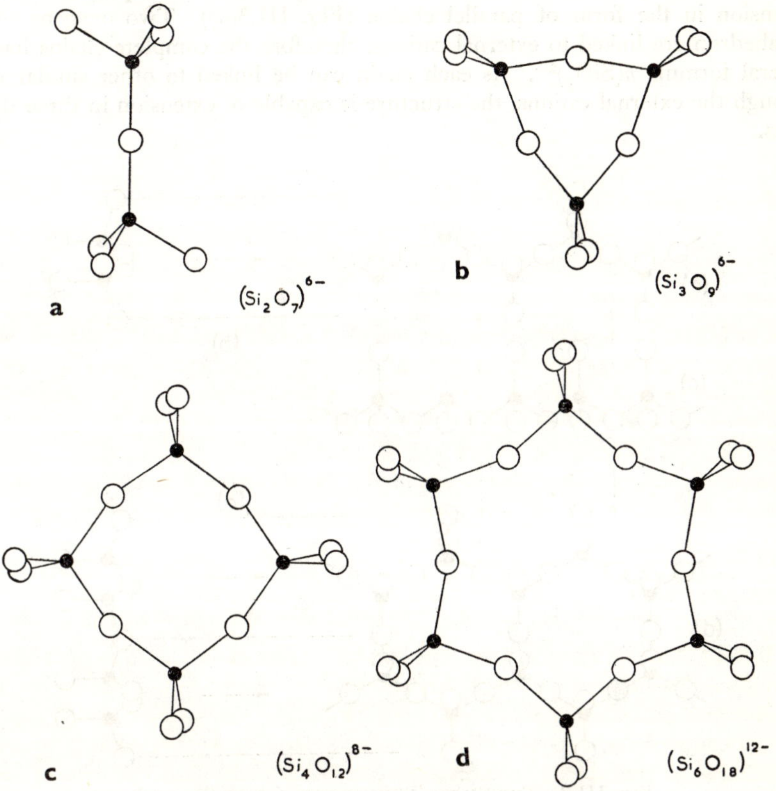

FIG. III.2. SILICA TETRAHEDRAL GROUPINGS
(*a*) two linked units; (*b*) a three-unit ring structure;
(*c*) a four-unit ring structure; and (*d*) a six-unit ring structure.

Other silicates are known which contain more complicated groupings. Thus, three tetrahedra may be linked together by sharing oxygens to form ring structures (Fig. III.2B). In this case each tetrahedron of silica $(SiO_4)^{4-}$ shares two of its oxygen atoms with similar groups, and the configuration may be expressed as $(Si_3O_9)^{6-}$, there being six oxygen ions available to combine with other cations, and the other three oxygen ions of the complex are linked to (and are electrically satisfied by) silicon atoms alone. Figs. III.2C and D show more groupings, the first being a ring of four linked tetrahedra and the second having six $(SiO_4)^{4-}$ units in the complex ring. The chemical representations are respectively $(Si_4O_{12})^{8-}$ and $(Si_6O_{18})^{12-}$

In all cases of group structure formation the individual unit (whether composed of two, three, four or six $(SiO_4)^{4-}$ linked tetrahedra) is completely isolated from others of similar construction and is joined to another only through additional cations which are not silicon ions.

(*c*) **Chain structures** are formed when two oxygen atoms of every silica tetrahedron are joined to others and the resulting structure is capable of indefinite extension in the form of parallel chains (Fig. III.3(*a*)). Two oxygens of each tetrahedron are linked to external cations, therefore the complete chains have the general formula $n(SiO_3)^{2-}$. As each chain can be linked to other similar chains through the external cations, the structure is capable of extension in three dimensions.

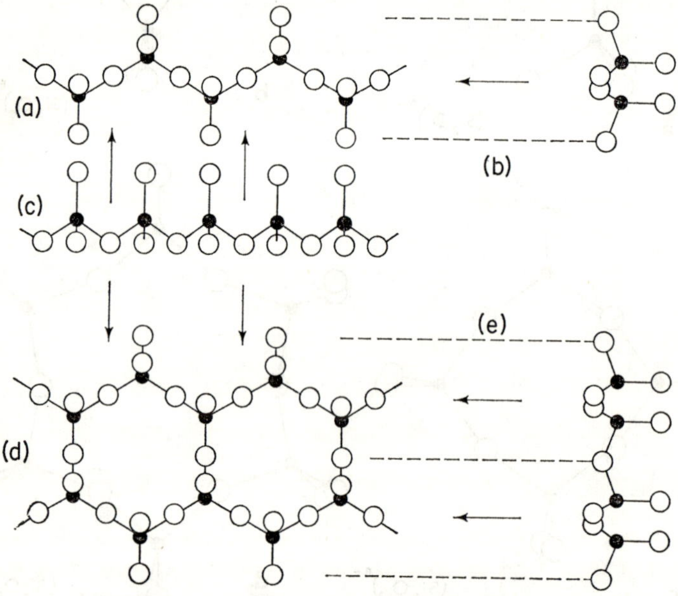

FIG. III.3. THE SILICA TETRAHEDRAL CHAIN STRUCTURES
(*a*), (*b*) and (*c*) the pyroxene single chain; and
(*d*) and (*e*) the amphibole double chain

Fig. III.3(*d*) illustrates a structure where two identical parallel chains have polymerised by sharing a further oxygen in every alternate tetrahedron. The unit is now $n(Si_4O_{11})^{6-}$ and, as shown later, it is characteristic of the *amphiboles* whereas the simple chain is exemplified by the *pyroxenes*.

(*d*) **Sheet structures** are the result of the linking of three oxygens of each tetrahedron with adjacent units. The structure may be visualised as the infinite extension of the amphibole double-chain formation described in the previous section. All oxygen ions are now electrically satisfied by silicon ions except one on each tetrahedron which is free to combine with external cations. As a consequence, the silica units are arranged in sheets composed of hexagonal rings and the structure is of indefinite extension in two planes, *a* and *b*, parallel to the plane of the paper on which the base of each tetrahedron and the electrically satisfied oxygens may be supposed to rest (Fig. III.4).

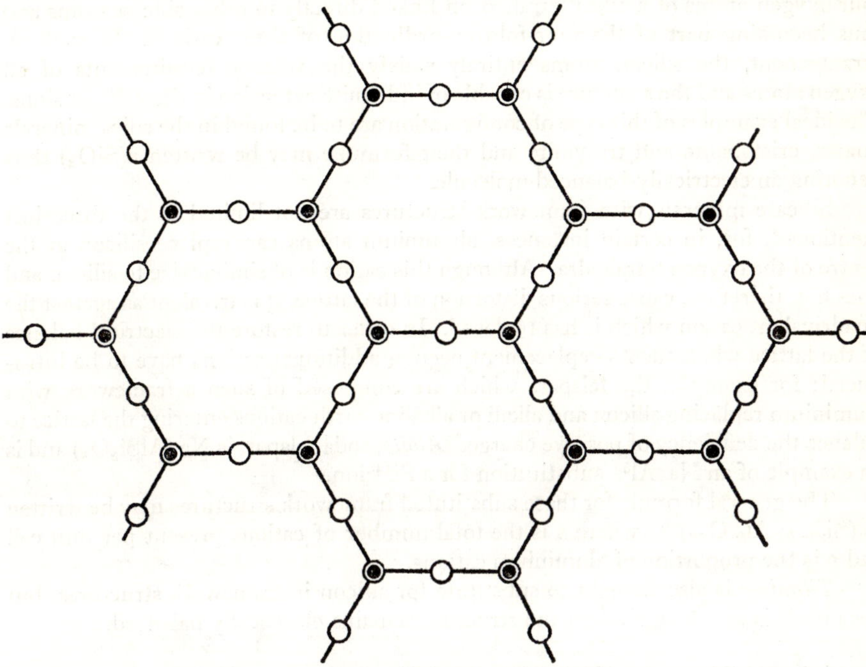

FIG. III.4. THE SILICA TETRAHEDRAL SHEET STRUCTURE

The composition of the layer can be represented by $n^2(Si_2O_5)^{2-}$; n^2 because the structure now extends in two dimensions and $(Si_2O_5)^{2-}$ is the unit expressed in its lowest terms. One hexagonal ring has six silicon atoms, although it has only a third share of each of these because they are partly contained in two other rings. Six oxygen atoms are shared between two rings, and six further oxygen atoms, which still require one valency share, are above each silicon and therefore shared between three rings. The composition of each ring is therefore

$$\frac{6Si}{3} + \frac{6O}{2} + \frac{(6O^-)}{3} \quad \text{i.e.} \ (Si_2O_5)^{2-}$$

In the ring structure of Fig. III.4 the oxygen ions which combine with other cations are shown projecting upwards in all cases. This is the assumed arrangement in most sheet-structure minerals of which kaolinite, mica and related materials are examples. The angle between —Si—O—Si— must therefore be 141° 34′. There is an alternative arrangement for some clay mineral structures which implies that every alternate unsaturated oxygen points downwards and the others upwards, thereby making the —Si—O—Si— angle 180°. This problem is discussed in more detail later in this chapter when the individual structures are considered.

Some sheet structures embody rings which are not hexagonal, e.g. *apophyllite* has a ring structure where only four tetrahedra are concerned.

(*e*) **Framework structures** represent the extreme end of the series with all four oxygen atoms of a silica tetrahedron linked directly to other silicon atoms and thus becoming part of the four-fold co-ordination of these cations. In such an arrangement, the silicon atoms entirely satisfy the valency requirements of all oxygen atoms and the structure is capable of indefinite extension in three dimensions. The ideal examples of this type of configuration are to be found in the silica minerals quartz, cristobalite and tridymite and their formulæ may be written $n(SiO_2)$ thus denoting an electrically-balanced molecule.

Silicate minerals with framework structures are not limited to the three just mentioned, for, in certain instances, aluminium atoms can replace silicon at the centre of the oxygen tetrahedra. Although this cation is of similar size to silicon and does not, therefore, cause serious distortion of the lattice, it is trivalent as against the quadrivalent cation which it has replaced. In order to restore the electrical balance of the lattice when such a replacement occurs, additional cations have to be introduced; for example, the felspars which are composed of such a framework with aluminium replacing silicon and alkali or alkaline earth cations entering the lattice to balance the deficiency of positive charge. *Albite*, soda felspar, is Na $(AlSi_3O_8)$ and is an example of an Na^+Al^{3+} substitution for a Si^{4+} ion.

The general formula for these substituted framework structures may be written as $(Si_{(n-p)}\,Al_p\,O_{2n})^{-p/2}$ where n is the total number of cations present per unit cell and p is the proportion of aluminium cations.

Titanium is also thought to substitute for silicon in framework structures, but because it is also quadrivalent the structure remains electrically balanced.

It is possible to classify all silicates within the framework of these five groups and it is interesting to compare the predicted structures with those resulting from the hypothetical silicic acid theory. *Orthosilicates* are those which belong to the 'island' or simple group structure formation; *metasilicates* are those with simple chain linkages. More complicated silicates may have aluminium ions replacing silicon within the lattice.

1. SILICATES WITH SEPARATE SiO_4 GROUPS

IN silicate structures where each unit is connected only through extra cations, the framework is largely determined by the size of these cations and their oxygen configuration.

Small cations, like beryllium, are in four-fold co-ordination with oxygen in silicates, whilst larger cations including magnesium, ferric iron and sometimes calcium are in hexagonal or six-fold co-ordination. Still larger cations are in eight-fold grouping with oxygen, for example *zircon*, $ZrSiO_4$, where the zirconium ion is surrounded by eight oxygen atoms each of which is also linked tetrahedrally to silicon. Other structures are known where the extra cations are all of one element, yet they may be co-ordinated in more than one way. Thus in the alumino-silicates of this group, *kyanite*, *andalusite* and *sillimanite*, aluminium may be present in four-, five- or six-fold co-ordination. Complex structures also exist in which more than one cation, other than silicon, is present and each of these may have different co-ordinations.

There are two further general points: (i) the anionic configuration must be such that each oxygen receives two valency shares, one of which is provided by a silicon and the other by different cations, and (ii) the oxygen atoms (which are the largest spheres in the structure) must be as closely packed as possible.

The silicates with isolated units of $(SiO_4)^{4-}$ tetrahedra may conveniently be studied in order of increasing size of cation.

The Phenacite Group

Beryllium, Be^{2+}—which is one of the smallest-sized cations with an ionic radius of 0·31 Å—forms the silicate *phenacite*, Be_2SiO_4. The beryllium ion is divalent and is situated at the centre of four oxygen atoms, so each co-ordinating anion receives a half valency-share from the beryllium. To achieve complete valency satisfaction, each oxygen must therefore be linked to two beryllium and one silicon atom. The complete crystal structure of phenacite is rather complicated, because the unit cell is very large, contains many cations and is of low symmetry.

The constants of phenacite[1] are:

Phenacite Be_2SiO_4	Crystal System:	Rhombohedral
Space Group: $R\bar{3}$	Axis:	a = 7·68 Å
Z = 6	Angle:	α = 108° 01′

The phenacite structure typifies a number of minerals, one of which is the zinc silicate *willemite*, Zn_2SiO_4. In this silicate, zinc atoms replace those of beryllium in phenacite; this is rather surprising as zinc ions are much larger and should really be in six-fold and not four-fold co-ordination with oxygen. However, zinc belongs to the sub-group of elements and the outermost electron orbits can easily be distorted, and only four oxygens are spaced around the cation. The introduction of the larger cation results in an increase in the size of the unit cell. Presumably, the oxygen atoms are not as tightly compacted and an overall increase in size results. This type of expansion is of frequent occurrence and reference will be made to numerous other examples.

The Olivine Group

When the size of the extra cations in the silicate structure is increased until the size-ratio compared to oxygen is greater than 0·414, six-fold co-ordination results. The complete structure is thus composed of silica tetrahedra linked to octahedral units. The ions of magnesium and ferrous iron are respectively 0·65 Å and 0·76 Å in size, and the ratio of these to oxygen is 0·46 and 0·54 respectively, so that they are well within the limits of six-fold co-ordination. As all oxygen ions which are co-ordinated to these cations receive only a third of a valency share, each must be linked to three

[1] Bragg, W. L. and Zachariasen, W. H., *Zeit. Krist.*, 72, 518, 1930.

divalent cations and also to a silicon atom if it is to receive two valency shares and thus be electrically neutralised. The *olivine group* of minerals has this configuration and many important ceramic minerals are of this class.

The mineral olivine itself is often referred to as a solid solution of magnesium and iron silicates, although the structure can be more clearly understood if olivine is imagined as magnesium silicate with some of the magnesium ions replaced by ferrous ions Fe^{2+}.

The structure of olivine is quite simple provided that the following points are remembered:

(1) that the oxygen atoms are hexagonally close-packed,
(2) that all silicon atoms are in tetrahedral co-ordination with oxygen,
(3) that the cations of magnesium and iron are always in six-fold co-ordination,
(4) that the valency of oxygen is satisfied by one share from a silicon and one-third share from each of three cations of magnesium or iron.

The ideal structure is illustrated in Fig. III.5. The oxygen atoms are hexagonally close-packed and the arrangement of the cations is clearly demonstrated; some oxygen ions have been omitted.

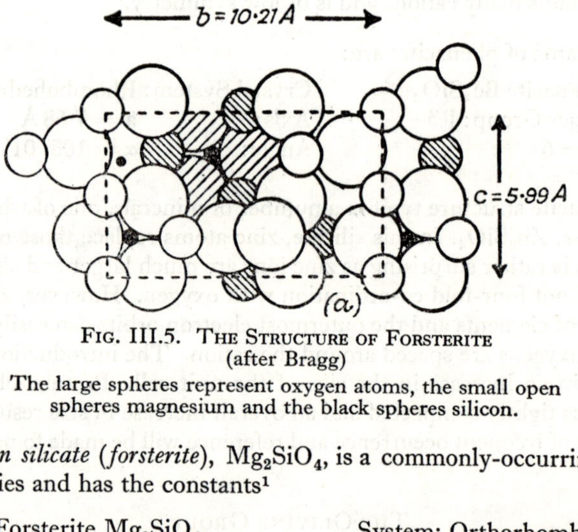

FIG. III.5. THE STRUCTURE OF FORSTERITE
(after Bragg)
The large spheres represent oxygen atoms, the small open spheres magnesium and the black spheres silicon.

Magnesium silicate (*forsterite*), Mg_2SiO_4, is a commonly-occurring mineral in basic refractories and has the constants[1]

Forsterite Mg_2SiO_4	System: Orthorhombic
Space Group: Pbnm	Axes: a = 4·77 Å
Z = 4	b = 10·26 Å
	c = 5·99 Å

Other similar minerals of the olivine group are shown in Table III.III. They all have the same spacial configuration and form orthorhombic crystals, but the dimensions of the unit cell are influenced by the size of cation taking part in the structure.

[1] Rinne, F., *Zeit. Krist.*, 59, 230, 1924.

TABLE III—III. LATTICE CONSTANTS OF THE OLIVINE GROUP

Mineral	*Formula*	*a*	*b*	*c*
Forsterite	Mg_2SiO_4	4·77Å	10·26Å	5·99Å
Fayalite	Fe_2SiO_4	4·80	10·59	6·16
Olivine*	$(Mg.Fe)_2SiO_4$	4·76	10·21	5·99
Monticellite†	$Mg.Ca.SiO_4$	4·82	11·08	6·37

* Bragg, W. L. and Brown, G. B., *Zeit. Krist.*, 63, 538, 1926.
† Brown, G. B. and West, J., *Zeit. Krist.*, 66, 154, 1927.

The mineral *olivine* is of variable composition, as iron can replace magnesium in all proportions. The lattice constants given are for a mineral with the approximate composition $(Mg_{0\cdot9}Fe_{0\cdot1})_2SiO_4$ or $9Mg_2SiO_4.Fe_2SiO_4$ expressed in more simple yet misleading terms. *Chrysolite*[1] is also an olivine; it has six times as many magnesium as iron atoms.

Monticellite, $CaMgSiO_4$, has a structure which is also based on olivine, but with calcium ions present in positions of six-fold co-ordination. Unlike the ferrous ions in olivine, calcium ions do not replace magnesium in all proportions. Monticellite has calcium ions only in certain positions in the lattice and the mineral is a definite compound with almost equal proportions of calcium and magnesium ions. Monticellites commonly have magnesium ions partially replaced by ferrous ions.

The Zircon Group

The large cation zirconium forms a silicate in which it is in eight-fold co-ordination with oxygen. Isolated $(SiO_4)^{4-}$ units are present, linked only through the zirconium ion.

The contribution of the zirconium cations (Zr^{4+}) to each surrounding anion is one-half of a valency share, so that all oxygen atoms must be in contact with one silicon and two zirconium ions in order to satisfy their valency requirements.

The structure is shown in Fig. III.6 where the unit cell is clearly tetragonal. The lattice constants[2] are

Zircon $ZrSiO_4$	System: Tetragonal
Space Group: I4/amd	Axes: a = 6·58 Å
Z = 4	c = 5·93 Å

The eight oxygen atoms surrounding each zirconium atom are not all situated at an equal distance; four are at a distance of 2·15 Å and the others at 2·29 Å.[3]

Thorite, $ThSiO_4$, has a structure similar to zircon.

[1] Whitaker, E. J. W., *Acta Cryst.*, 5, 143, 1952.
[2] Wyckoff, R. W. G. and Hendricks, S. B., *Zeit. Krist.*, 66, 73, 1927.
[3] Krstanović, I. R., *Acta Cryst.*, 11, 896, 1958.

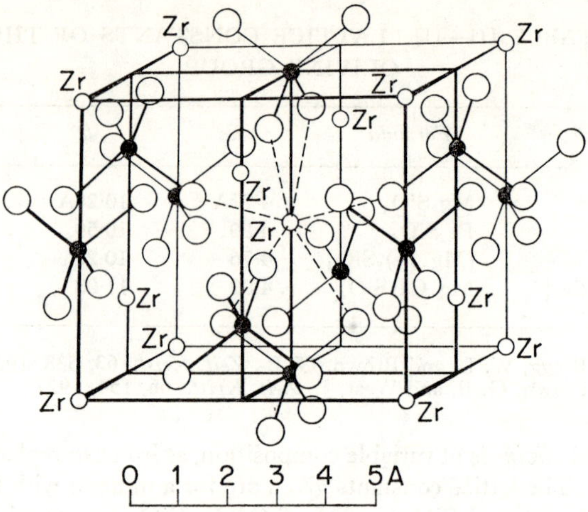

FIG. III.6. THE STRUCTURE OF ZIRCON (after Wycoff)

THE ALUMINO-SILICATE GROUP

IN the island structures previously described, the cations other than silicon have all been of one type or valency, and in any one group have always had the same co-ordination with oxygen, but when aluminium is the extra cation involved in silicate structures of this type, it may be in either six- or four-fold co-ordination, and in one case five-fold, which is most unusual.

There are three minerals which have the chemical composition $Al_2O_3.SiO_2$. These are *kyanite*, *andalusite* and *sillimanite*[1] and to these may be added *mullite*, $3Al_2O_32SiO_2$, which has a very similar structure to sillimanite, and is a decomposition product of many alumino-silicates, and commonly occurs in fired alumino-silicate bodies.

The structure of these minerals consists of isolated units of $(SiO_4)^{4-}$ linked together by aluminium cations which, in kyanite, are all in six-fold co-ordination; in andalusite they are in both six- and five-fold and in sillimanite they are in six- and four-fold linkages.

The *kyanite* structure has only two sets of co-ordinated groupings, tetrahedrally-linked oxygen atoms round each silicon and hexagonally-positioned oxygen atoms about the aluminium ions. Furthermore, the oxygen atoms are (ideally) in a close-packed arrangement as illustrated in Fig. III.7.

A feature of the structure is the parallel chains of aluminium-oxygen linkages joined by silica units. The valency satisfaction of oxygen atoms is not simple on the basis of the proposed structure by Naray-Szabo, Taylor and Jackson[2]; although some are linked to aluminium and silicon atoms, others are satisfied by aluminium ions

[1] Mullite so closely resembles sillimanite in properties that it was, at one time, mistaken for it and many references to 'sillimanite' in firebricks and other forms of heated clay really relate to mullite and so do many references to 'artificial sillimanite'.

[2] Naray-Szabo. St., Taylor, W. H. and Jackson, W. W., *Zeit. Krist.*, 71, 117, 1929.

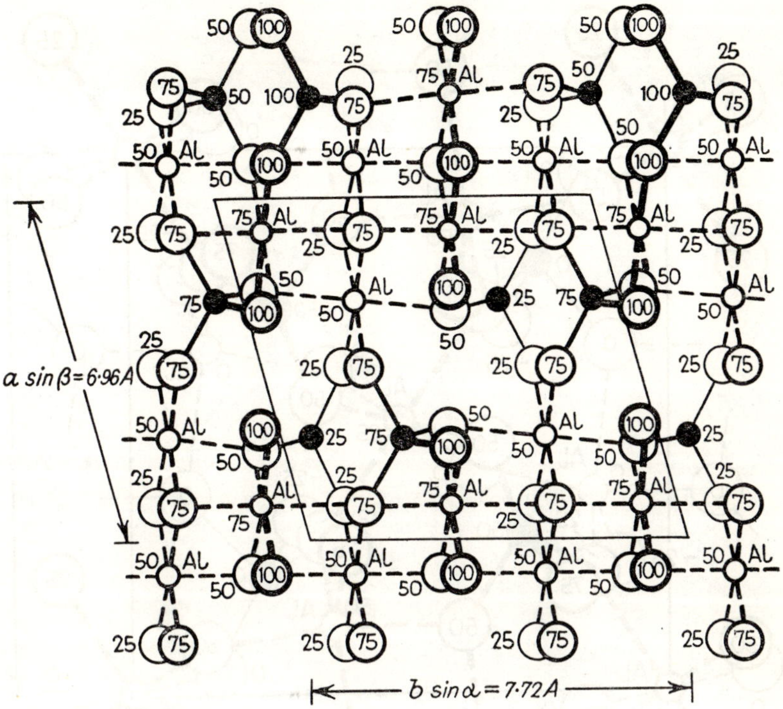

FIG. III.7. THE STRUCTURE OF KYANITE
(after Bragg)
The numbers in the circles refer to the heights of the atoms above a zero plane.

alone. Some oxygen atoms, therefore, do not belong to $(SiO_4)^{4-}$ units and the composition is best represented by the formula $Al_2O.SiO_4$.

The structural constants[1] are

	System: Triclinic	
Kyanite $Al_2O.SiO_4$		
Space Group: $P\bar{1}$	Axes:	Angles:
Z = 4	a = 7·12 Å	α = 89° 59′
	b = 7·85 Å	β = 101° 07′
	c = 5·57 Å	γ = 106° 00′

The structures of *andalusite* and *sillimanite* are somewhat similar to each other and are shown in Figs. III.8 and III.9; the unit cell, in both cases, is orthorhombic. Aluminium atoms are at the corners and centre of the unit cell in the projection along the *c*-axis. These aluminium atoms are linked to other similar atoms immediately below and above them and so there is a series of chains of aluminium-oxygen linkages running parallel to the *c*-axis and linked laterally by silica tetrahedra and, in andalusite, by five-fold co-ordinated aluminium and, in sillimanite, by four-fold or tetrahedrally co-ordinated aluminium.

[1] Burnham, C. W., *Zeit. Krist.*, 118, 337, 1963.

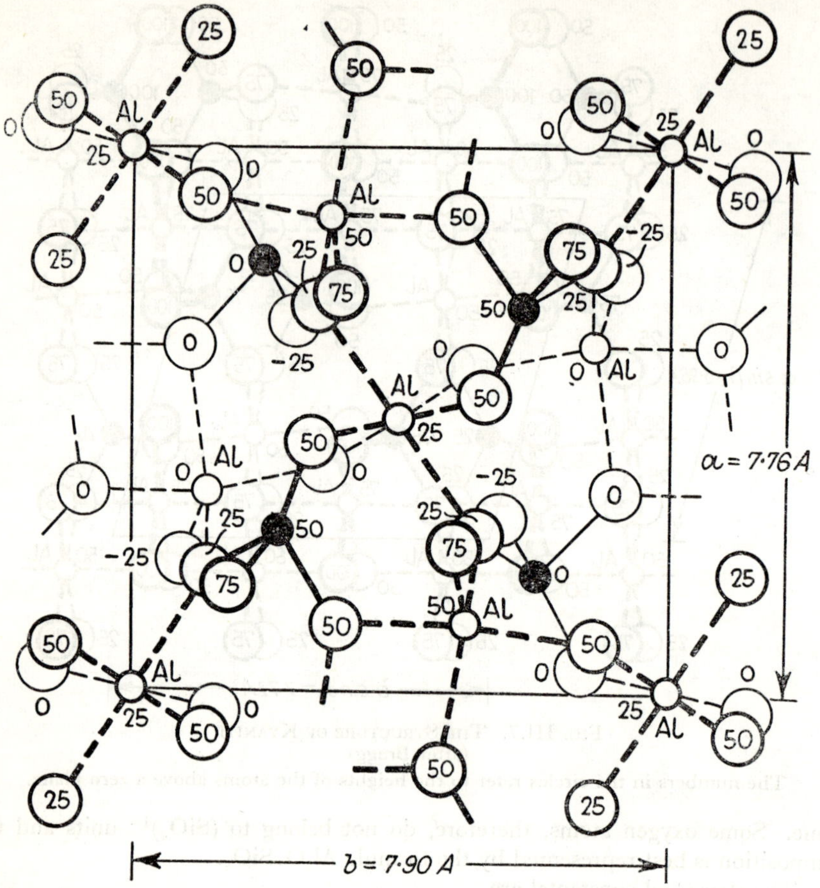

FIG. III.8. THE STRUCTURE OF ANDALUSITE
(after Hey and Taylor)

The packing of the oxygens in andalusite and sillimanite is less close than that in kyanite. The structure of sillimanite is, in fact, slightly more open than that of andalusite.

The structural constants of andalusite[1, 2] are:

Andalusite $Al_2O.SiO_4$	System: Orthorhombic
Space Group: Pnnm	Axes:
Z = 4	a = 7·79 Å
	b = 7·90 Å
	c = 5·56 Å

[1] Hey, J. S. and Taylor, W. H., *Zeit. Krist.*, 80, 428, 1931.
[2] Burnham, C. W. and Buerger, M. J., *Zeit. Krist.*, 115, 269, 1961.

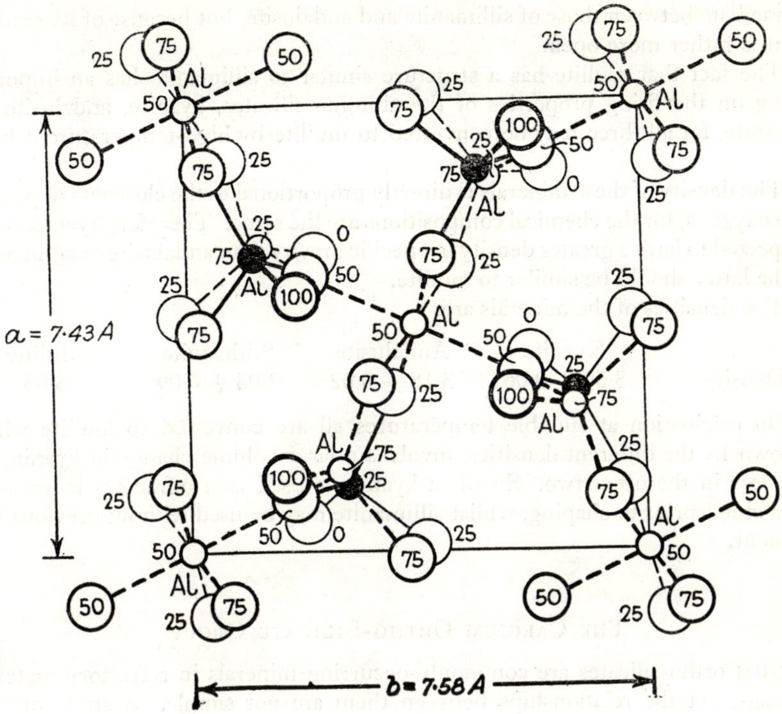

FIG. III.9. THE STRUCTURE OF SILLIMANITE
(after Taylor[1])

and those of sillimanite[2] are:

Sillimanite $Al_2O.SiO_4$	System: Orthorhombic
Space Group: Pbnm	Axes:
Z = 4	a = 7·49 Å
	b = 7·67 Å
	c = 5·77 Å

The structure of *mullite* has presented considerable difficulty because true single crystals which give a satisfactory X-ray photograph rarely occur. The problem has been further complicated because the mineral appears to have variable composition within the range $2Al_2O_3.SiO_2$ and $3Al_2O_3.2SiO_2$. Sadanaga, Tokonami and Takéuchi[3] have evolved a structure based on a synthetic crystal of composition $2Al_2O_3.SiO_2$ and showed that the structure is closely related to that of sillimanite. However the silicon and aluminium atoms are more randomly positioned and one-fifth of the oxygens in sillimanite are removed, and more aluminium replaces silicon to give a change in valency. In some respects the structure of mullite is

[1] Taylor, W. H., *Zeit. Krist.*, 68, 503, 1928, and ibid., 71, 205, 1929.
[2] Burnham, C. W., *Zeit. Krist.*, 118, 127, 1963.
[3] Sadanaga, R., Tokonami, N. and Takéuchi, Y., *Acta Cryst.*, 15, 65, 1962.

intermediate between those of sillimanite and andalusite, but because of its randomness it is rather more open.

The fact that mullite has a structure similar to sillimanite has an important bearing on the firing properties of the alumino-silicates, kyanite, andalusite and sillimanite, for all three may be converted to mullite by high-temperature calcination.

The density of these minerals is directly proportional to the closeness of packing of the oxygens, for the chemical compositions are the same. Therefore kyanite might be expected to have a greater density or specific gravity than andalusite or sillimanite, and the latter should be similar to mullite.

The densities of the minerals are:

	Kyanite	Andalusite	Sillimanite	Mullite
Density	$3{\cdot}59 \pm 0{\cdot}06$	$3{\cdot}18 \pm 0{\cdot}02$	$3{\cdot}08 \pm 0{\cdot}09$	3·03

On calcination at suitable temperatures all are converted to mullite which, as shown by the different densities, involves a large volume change in kyanite but much less in the other two. So when kyanite is used as a refractory it has to be pre-calcined prior to shaping, whilst sillimanite may be used without previous heat treatment.

The Calcium Ortho-Silicate Group

Calcium ortho-silicates are commonly-occurring minerals in refractory materials and slags, yet the relationships between them are not simple. A study of their crystal structures has explained many of the perplexing changes which take place when they are heated.

There are two chemical varieties of orthosilicate of calcium, tricalcium silicate, $3CaO\,.\,SiO_2$, and dicalcium silicate, $2CaO\,.\,SiO_2$. The latter is capable of existing in at least three modifications named α, β and γ varieties.

If lime and silica are mixed in the molecular proportions of 2:1 and heated to a sufficiently high temperature both α- and β-*dicalcium silicate*, which have similar properties, are formed. On cooling, however, these forms convert slowly to the γ-variety with a large increase in volume. The temperature stability ranges of the different forms are given as[1]

$$\alpha \overset{1420^\circ\,C}{\rightleftharpoons} \beta \overset{675^\circ\,C}{\rightleftharpoons} \gamma$$

Tricalcium silicate which is formed by heating lime and silica in the correct, molecular proportions is stable under normal conditions, but can be converted to β-dicalcium silicate and lime by slowly cooling between 1250° C. and 600° C. which at lower temperatures, in turn, changes into γ-dicalcium silicate.

The relationship between γ- and β-dicalcium silicate depends upon the co-ordination of calcium with oxygen at low and high temperatures. Under normal conditions, calcium forms a silicate which has a similar structure to olivine, so cal-

[1] Day, A. L. and Shepherd, E. S., *J. Amer. Chem. Soc.*, 28, 1104, 1906.

cium is in octahedral linkage with oxygen and the formula may be written as Ca_2SiO_4; its lattice constants are[1]

γ-Dicalcium Silicate Ca_2SiO_4	System: Orthorhombic
Space Group: Pbnm	Axes:
Z = 4	a = 5·06 Å
	b = 11·28 Å
	c = 6·78 Å

At high temperatures, however, calcium has a different co-ordination; O'Daniel and Tscheischwili[2] consider that it is surrounded by four oxygens but Bredig[3] has postulated that 9 or 10 anions are associated with a calcium ion, and that the structure of high-temperature dicalcium silicate is rather similar to K_2SO_4.

The mineral *merwinite*, $Ca_3Mg(SiO_4)_2$, is considered by Bredig to have a similar structure to β-dicalcium silicate and may not be a definite compound but 'a high-temperature isomorphous mixture of forsterite in the high-temperature form of Ca_2SiO_4'; the forsterite component may act as a stabiliser. Goldschmidt and Rait[4] have suggested that merwinite is based on the perovskite structure (p. 194) where each calcium is surrounded by twelve oxygen atoms.

Tricalcium silicate also exists in the α, β and γ modifications but all three are similar in property and structure. The crystal structures of this mineral have not been established with certainty. O'Daniel and Hellner[5] have suggested that the structure is based on rings of $(Si_3O_9)^{6-}$ units, but Jeffery[6] indicates that the mineral is similar to *alite*, $54CaO.16SiO_2.Al_2O_3.MgO$, which is a common constituent of cement clinker. Isolated tetrahedra of silica exist in the structure and explain the easy conversion to β-dicalcium silicate.

Jeffery has given the lattice constants of tricalcium silicate as

Tricalcium Silicate $3CaO.SiO_2$	System: Pseudo-rhombohedral (hexagonal)
Space Group: R3m	Axes:
Z = 9	a = 7·0 Å } for hexagonal cell
	c = 25·0 Å } for hexagonal cell

The Garnet Group

SILICATE minerals are known where isolated $(SiO_4)^{4-}$ units are linked to more than one type of cation. The best example is the mineral group, *garnet*, which has the general formula $R_3^{2+}R_2^{3+}(SiO_4)_3$ where R^{2+} is a divalent cation such as Ca, Mg, Fe^{2+} or Mn, and ^{3+}R is a trivalent cation like Al, Cr, Fe^{3+}.

The structure is rather complex as there are tetrahedrally co-ordinated groups of oxygen atoms around each silicon, octahedral units around trivalent cations and

[1] O'Daniel, H. and Tscheischwili, L., *Zeit. Krist.*, 103, 178, 1941.
[2] O'Daniel, H. and Tscheischwili, L., *Zeit. Krist.*, 104, 124, 1942.
[3] Bredig, M. A., *Amer. Miner.*, 28, 594, 1943.
[4] Goldschmidt, H. J. and Rait, J. R., *Nature*, 152, 356, 1943.
[5] O'Daniel, H. and Hellner, E., *Neues. J. Min. Geol. Paläent.* 5, 108, 1950.
[6] Jeffery, J. W., *Acta Cryst.*, 5, 26, 1952.

oxygen atoms in eight-fold co-ordination round divalent ones. Every oxygen atom is linked to one silicon atom, one trivalent ion, and to two divalent cations. The valency contribution from each is thus 1, $\frac{1}{2}$ and two $\frac{1}{4}$'s respectively, making a total of 2. The unit cell is cubic and varies in size with the cations present.

The structural constants of the group are[1]

Garnet $R_3^{2+}R_2^{3+}(SiO_4)_3$	System: Cubic
Space Group: Ia3d	Axis: a—variable but about 12 Å
Z = 8	

An unusual feature of the garnet group is that hydroxyl units can apparently replace $(SiO_4)^{4-}$ tetrahedra. Artificial hydrogarnets[2] have been prepared, one of which has the composition $Ca_3Al_2(OH)_{12}$, but is structurally similar to *grossularite*, $Ca_3Al_2(SiO_4)_3$.

Titanite, $CaTiO.SiO_4$, also contains isolated $(SiO_4)^{4-}$ tetrahedra which are linked to titanium ions in octahedral co-ordination with oxygen and to calcium ions which are within groups of seven. Titanite is a commonly occurring mineral, although it is sparsely distributed. Titanite (or *sphene*) is often the source of titanium in volcanic and metamorphosed rocks. The crystals are monoclinic with the structural constants[3]

Titanite $CaTiO.SiO_4$	System: Monoclinic	
Space Group: C2/c	Axes:	Angle:
Z = 4	a = 6·55 Å	
	b = 8·70 Å	β = 119° 43′
	c = 7·43 Å	

The mineral *topaz* is an alumino-silicate containing fluorine and is sometimes used as a refractory material after the fluorine has been driven off by heat and mullite formed.

In topaz, isolated units of silica tetrahedra are present linked to aluminium ions which are in six-fold co-ordination, but the surrounding anions are not all the same. Four are oxygen atoms and the remaining two are fluorine atoms which are shared between two adjacent octahedra. The oxygen atoms are close-packed, but not in a simple form.

The structural constants[4/5] are

Topaz $(AlF)_2SiO_4$	System: Orthorhombic
Space Group: Pbnm	Axes:
Z = 4	a = 4·64 Å
	b = 8·78 Å
	c = 8·37 Å

Other minerals exist which have separate $(SiO_4)^{4-}$ tetrahedra but they are not of importance as ceramic materials.

[1] Menzer, G., *Zeit Krist.*, 63, 157, 1926, and 69, 300, 1928.
[2] Flint, E. P., McMurdie, H. F. and Wells, L. S., *J. Res. Nat. Bur. Stand.*, 26, 13, 1941.
[3] Zachariasen, W., *Zeit. Krist.*, 73, 7, 1930.
[4] Pauling, L., *Proc. Nat. Acad. Sci.*, 14, 603, 1928.
[5] Alston, N. A. and West, J., *Zeit Krist.*, 69, 149, 1928.

2. SILICATES WITH COMPLEX SILICON-OXYGEN GROUPS

COMPLEX silicon-oxygen groups may be of many types, so the structures which are included in this group may differ widely. Although there is no important family of minerals with structures of this type, some compounds frequently associated with slags and high-fired ceramic products undoubtedly contain isolated complex groupings.

As explained in the earlier part of the chapter, the complex groups most commonly occurring are (i) those with two linked tetrahedra $(Si_2O_7)^{6-}$, (ii) those with three tetrahedra arranged in the form of a ring giving the group $(Si_3O_9)^{6-}$, (iii) those with four tetrahedra in a ring $(Si_4O_{12})^{8-}$, and (iv) those with hexagonal rings or groups of $(Si_6O_{18})^{12-}$.

Structures with $(Si_2O_7)^{6-}$ Groups

SIMPLE structures are known where one type of cation satisfies the valency requirements of the groups and provides a linking ion for a three-dimensional network. Such an example is the rare mineral *thortveitite*, or scandium silicate, $Sc_2Si_2O_7$, but in general, greater complexity in structure is more usual.

The Melilite Group includes minerals which are the result of slag attack at a high temperature on silicate refractories. A range of solid solutions exists with *gehlenite*, $2CaO.Al_2O_3.SiO_2$, as one end-member and *akermanite*, $2CaO.MgO.2SiO_2$, as the other. Alkali ions and ferrous iron ions may also replace other cations and enter the complex.

The structure is based on groups of two silica tetrahedra sharing a common corner, but as aluminium ions can replace silicon ions at the centre of four oxygen atoms, the groups have the general formula $(Si,Al)_2O_7$. Magnesium ions are also within a tetrahedral group of oxygen atoms and calcium is in eight-fold co-ordination. The full structural formula may be represented as $(Ca,Na)_2(Mg,Al)_1.(Si,Al)_2O_7$, so on this basis gehlenite is $Ca_2Al(Si_1Al_1)O_7$ and akermanite is $Ca_2.Mg.Si_2O_7$. The replacement may be imagined as Al + Al for Mg + Si. This is a common form of substitution in other silicates. The crystal form of the melilites[1] is tetragonal with a space group of $P\bar{4}2_1m$.

Structures with Isolated $(Si_3O_9)^{6-}$ Rings

THE well-known example of structures with isolated rings of $(Si_3O_9)^{6-}$ units is the mineral *benitoite*, $BaTiSi_3O_9$. Titanium and barium atoms lie within groups of six oxygen atoms.

Structures with $(Si_4O_{12})^{8-}$ Rings

THE best known examples of minerals with four silica tetrahedra arranged in the form of a ring are certain *zeolites*. Aluminium ions are present also in tetrahedral

[1] Warren, B. E., *Zeit. Krist.*, 74, 131, 1930.

units, therefore it is better to regard zeolites as framework silica structures with aluminium ions replacing some silicon and to include them in the appropriate section of this chapter.

STRUCTURES WITH $(Si_6O_{18})^{12-}$ RINGS

THE important source of beryllia for refractory purposes is the mineral *beryl*, $3BeO.Al_2O_3.6SiO_2$, which is composed of symmetrically-placed rings of six silica tetrahedra linked to other similar groupings through beryllium and aluminium ions. The structure is illustrated in Fig. III.10; beryllium atoms are in tetrahedral co-ordination whilst aluminium atoms are at the centre of six oxygen atoms. Each 'valency' oxygen is linked to one silicon, one aluminium and one beryllium atom.

The lattice constants[1] are

Beryl $Be_3Al_2Si_6O_{18}$	System: Hexagonal
Space Group: P6/mcc	Axes:
$Z = 2$	$a = 9{\cdot}19$ Å
	$c = 9{\cdot}17$ Å

The mineral *cordierite*, $2MgO.2Al_2O_3.5SiO_2$, which is of frequent occurrence in steatite-clay bodies, has been shown by Bystom[2] to be based on the beryl structure. One silicon atom in every ring of six is replaced by an aluminium atom and the chemical representation is

Cordierite	$Al_3Mg_2Si_5Al\ O_{18}$
cf. Beryl	$Be_3\ Al_2\ \ Si_6\ \ O_{18}$

There are three forms α, β and μ of cordierite, all of which form fibrous crystals. The β- and μ-forms revert to the α-variety when heated to 830–1050° C. The α-form (which is the commonest) is pseudo-hexagonal. Ferrous iron can replace magnesium in the structure to form iron cordierite, which occasionally occurs in imperfectly-fired fireclay materials and in blast-furnace hearth bricks.[3]

The structure of the mineral *tourmaline* is also based on isolated units of hexagonal rings of six silica tetrahedra. Tourmaline is a silicate of boron and aluminium, although replacements by other cations are frequent. The mineral is commonly associated with china clays as it is a primary mineral of granites and pegmatites.

3. SILICATES WITH CHAINS OF SILICON-OXYGEN LINKAGES

MINERALS with chain structures are limited entirely to the pyroxenes and amphiboles, with the exception of a few examples which are intermediate between true amphiboles and layer structure minerals.

Within the pyroxene and amphibole group many minerals are classified which vary widely in composition. Their relationships can be established by their optical properties and the type of X-ray pattern.

The *pyroxene group* of minerals may be represented by the general formula

[1] Bragg, W. L. and West, J., *Proc. Roy. Soc.*, **A**111, 59, 1926.
[2] Bystom, A., *Arkiv. Kemi. Min. Geol. Stockholm*, 15**B**, 12, 1942.
[3] Rigby, G. R., *The Thin-section Mineralogy of Ceramic Materials* (B.C.R.A., 1953).

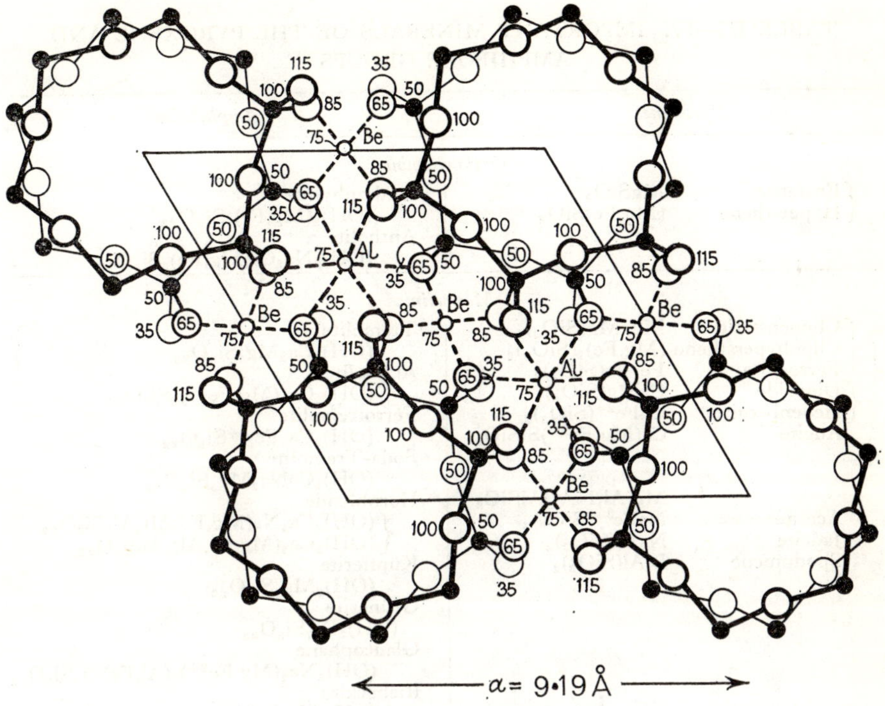

FIG. III.10. THE STRUCTURE OF BERYL
(after Bragg and West)

$R^{2+}SiO_3$,[1] but the degree and type of substitution is complex. The R^{2+} cation, for example, can be replaced by Al^{3+} and Fe^{3+} and aluminium ions can take the place of silicon in any ratio up to about 1 in 3. Pyroxenes crystallise in several classes, some of which have not been fully analysed. Pyroxenes are present in slags and refractory materials, and they can invariably be identified in magnesites or chrome-magnesites which have been exposed to a siliceous iron-bearing slag at a high temperature.

Amphiboles are similar in composition and it is impossible to differentiate them from pyroxenes by chemical analysis alone. This is to be expected as amphiboles are based on a unit of $(Si_4O_{11})^{6-}$ and pyroxenes on a unit of $(SiO_3)^{2-}$; aluminium may replace silicon, thus adding further confusion. Some guide to the type of mineral is given by the loss of water on calcination, because hydroxyl groups are essential constituents of the structure of amphiboles, although fluorine atoms sometimes replace them. The main difference between the groups is the angle between the cleavage planes of the crystals which in pyroxenes is 93° and in amphiboles, 56°. This point is dealt with in more detail in Chapter V.

Crystallographically, pyroxenes are made up of single chains of linked silica tetrahedra. Each of the units in the chain has two oxygens out of four shared with

[1] The group is known as the metasilicates, a term which dates back to the days when silicates were classified as salts of hypothetical silicic acids (p. 97).

TABLE III—IV. IMPORTANT MINERALS OF THE PYROXENE AND AMPHIBOLE GROUPS

Pyroxenes		*Amphiboles*	
		Orthorhombic	
Enstatite	$MgSiO_3$	Anthophyllite	$(OH)_2(Mg,Fe)_7Si_8O_{22}$
Hypersthene	$(Mg,Fe)SiO_3$	Antholite	$(OH)_2Na_2Ca(Mg,Fe)_5Si_8O_{22}$
		Monoclinic	
Clinoenstatite	$Mg.Mg(SiO_3)_2$	Tremolite	$(OH)_2Ca_2Mg_5Si_8O_{22}$
Clinohypersthene	$(Mg,Fe)_2(SiO_3)_2$	Actinolite	$(OH)_2Ca_2(Mg,Fe^{2+})_5Si_8O_{22}$
Ferrosilite	$Fe.Fe(SiO_3)_2$	Ferrotremolite	$(OH)_2Ca_2Fe_5^{2+}Si_8O_{22}$
Diopside	$CaMg(SiO_3)_2$	Soda-Tremolite	$(OH)_2CaNa_2Mg_5Si_8O_{22}$
Hedenbergite	$CaFe^{2+}(SiO_3)_2$	Hornblende	$(OH)_2Ca_2Na(Mg,Fe,Al)_5Al_2Si_6O_{22}$
Augite	$Ca(Mg,Fe^{2+})SiSiO_6$ ↕ $Mg(Al,Fe^{3+})AlSiO_6$		$(OH)_2Ca_2(Mg,Fe,Al)_5AlSi_7O_{22}$
Acmite	$NaFe^{3+}(SiO_3)_2$	Küpfferite	$(OH)_2Mg_7Si_8O_{22}$
Jadeite	$NaAl(SiO_3)_2$	Grünerite	$(OH)_2Fe_7Si_8O_{22}$
Spodumene	$LiAl(SiO_3)_2$	Glaucophane	$(OH)_2Na_2(Mg,Fe^{2+})_3(Al,Fe^{3+})_2Si_8O_{22}$
		Riebeckite	$(OH)_2Na_2Fe_3^{2+}Fe_2^{3+}Si_8O_{22}$

other similar tetrahedra; the other two oxygen atoms are connected to other cations and receive a full valency share to be electrically neutral. The chains may be of indefinite length and are linked to others through the extra cations. The bond angle between the Si—O—Si linkage is about 141°, hence the 'unsatisfied' oxygen atoms of each tetrahedron lie in the same planes.

In the amphibole structure two parallel pyroxene chains are directly joined through an oxygen of every alternate tetrahedron. Once again the double chain is continuous from left to right, but is limited in the directions at right-angles where linkage must be through other cations.

A comparison of the two units is given in Fig. III.3 (*a*) and (*b*). The upper diagram is the pyroxene chain viewed in plan, in elevation and end on; the lower shows the amphibole unit from the same aspects.

The crystals of both pyroxenes and amphiboles are lath- or needle-shaped, and in some cases fibrous. This is a direct result of the preferential development along the chain length.

The two groups can be conveniently divided into sub-sections on the basis of their crystal form. Both pyroxenes and amphiboles crystallise mainly in either the orthorhombic or monoclinic classes, and Table III. IV summarises some of the main types.

The table includes only a few known pyroxenes and amphiboles, but it does indicate their comparative complexity of composition. Most of the end members form solid solutions with others; the main series are indicated by brackets.

THE PYROXENES

THE solid solution series enstatite-hypersthene is probably the most common group of naturally occurring pyroxenes. The minerals are orthorhombic at normal temperatures, but at high temperatures transform to the monoclinic clino-varieties. The structure of orthorhombic *enstatite* is shown in Fig. III.11, *hypersthene* and *ferrosilite* are similar but with some iron replacing magnesium ions. The three-dimensional nature of the crystal is illustrated, because the magnesium ions, which are in six-fold co-ordination with oxygen, link four chains, the top two of which have silicon atoms at positions 65 and 85 above a zero plane. The lower two chains have silicon atoms at 15 and 35. The uppermost chain has its vertices lying in opposite directions to those in the lower one.

The structural constants of a *hypersthene* which contains 30 per cent $FeSiO_3$ and 70 per cent $MgSiO_3$ are, according to Warren and Modell:[1]

Hypersthene $(Mg,Fe)SiO_3$	System: Orthorhombic
Space Group: Pbca	Axes:
Z = 16	a = 18·20 Å
	b = 8·86 Å
	c = 5·20 Å

Diopside in solid solution with clinoenstatite is of common occurrence in dolomites, particularly when they have been in contact with silica slags. Ferrous iron may also be present in solid solution.

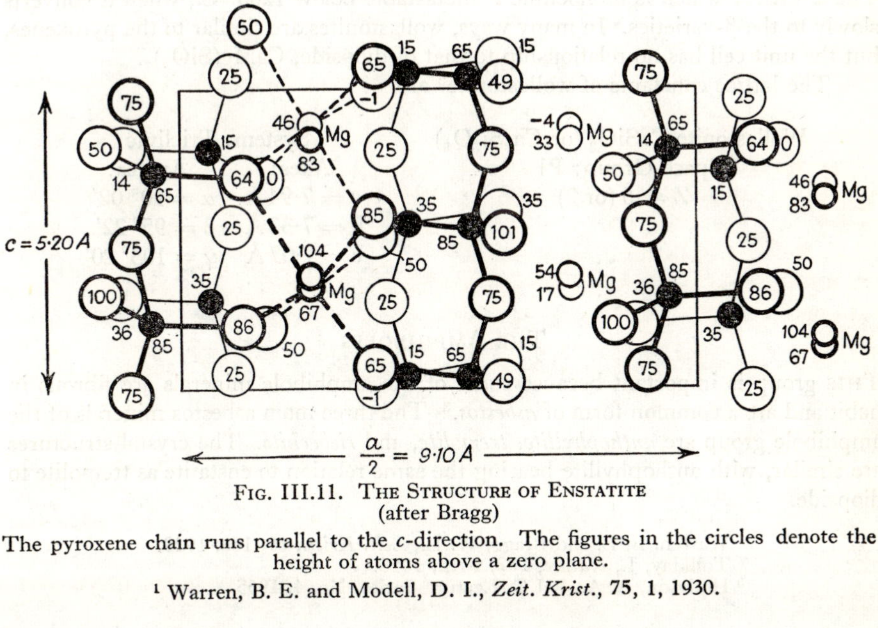

FIG. III.11. THE STRUCTURE OF ENSTATITE
(after Bragg)

The pyroxene chain runs parallel to the *c*-direction. The figures in the circles denote the height of atoms above a zero plane.

[1] Warren, B. E. and Modell, D. I., *Zeit. Krist.*, 75, 1, 1930.

The crystal structure of diopside is similar to enstatite, except that the chains are not so symmetrically positioned, and the structure becomes monoclinic. Calcium and magnesium ions link adjacent chains, but the former is within a group of eight oxygens as against six around magnesium. Calcium is in contact with six 'active' oxygens and two 'inactive' anions which are links in silicon-oxygen chains.

The structural constants have been given by Bragg and Warren[1] as

Diopside $CaMg(SiO_3)_2$	System: Monoclinic	
Space Group: C2/c	Axes:	Angle
Z = 4	a = 9·71 Å	
	b = 8·89 Å	$\beta = 105° 30'$
	c = 5·24 Å	

The other pyroxenes *augite*, *acmite* and *jadeite* are of similar structure to diopside, but the lithium pyroxene *spodumene* has apparently a slight distortion due to the small size of the alkali ion.

In pyroxenes, typical isomorphous replacements are Fe^{2+} for Mg^{2+}, Al^{3+} (oct.) for Mg^{2+} and Al^{3+} (tetr.) for Si^{4+}, with some Fe^{3+} replacing Al^{3+} (oct.).

Calcium metasilicate or *wollastonite*, $CaSiO_3$, is classified as a silicate with a chain formula. Three modifications of this mineral have been recognised, *pseudo-wollastonite* or α-CaO . SiO_2 and two forms of β-CaO . SiO_2, *wollastonite* and the less common *para-wollastonite*. The last variety has not been studied in detail but as the normal beta-variety closely resembles it in its properties, it is probable that the lattice configurations are similar.

Wollastonites are common in lime-slags and often crystallise from lime glasses. The α-variety which is monoclinic is metastable below 1200° C., when it converts slowly to the β-varieties. In many ways, wollastonites are similar to the pyroxenes, but the unit cell has no relationship to that of diopside, $CaMg(SiO_3)_2$.

The lattice constants of wollastonite[2] are

Wollastonite $CaSiO_3$ (or $Ca_3Si_3O_9$)	System: Triclinic	
Space Group: $P\bar{1}$	Axes:	Angles:
Z = 6 (or 2)	a = 7·94 Å	$\alpha = 90° 02'$
	b = 7·32 Å	$\beta = 95° 22'$
	c = 7·07 Å	$\gamma = 103° 20'$

The Amphiboles

This group is important because some of the amphibole minerals are fibrous in habit and are a common form of *asbestos*.[3] The three main asbestos minerals of the amphibole group are *anthophyllite*, *tremolite*, and *riebeckite*. The crystal structures are similar, with anthophyllite bearing the same relation to enstatite as tremolite to diopside.

[1] Warren, B. E. and Bragg, W. L., *Zeit. Krist.*, 69, 168, 1928.
[2] Tolliday, J., *Nature*, 182, 1012, 1958.
[3] Hodgson, A. A., R.I.C. Lecture Series, No. 4, 1965.

The structural constants of tremolite[1] are

Tremolite $(OH)_2Ca_2Mg_5Si_8O_{22}$	System: Monoclinic
Space Group: C2/m	Axes: Angle:
Z = 2	a = 9·84 Å
	b = 18·05 Å $\beta = 104° 42'$
	c = 5·28 Å

If the unit cell dimensions are compared with those of diopside it is obvious that the only material difference is a doubling along the *b*-crystallographic axis.

In the tremolite structure[2] Fig. III.12 there are large voids unoccupied by any cations; this enables one of the two Ca^{2+} ions to be replaced by two Na^+ ions, one-half of the Na^+ replacing the Ca^{2+} and the other half of the Na^+ ions entering the voids.

This is an ideal picture of the structure of an amphibole; the position of the cations and hydroxyl units with respect to the silica chain is also of great importance. In order that the cations may have the correct co-ordination and that the structure shall be electrically balanced, hydroxyl (OH) groups must be linked to the magnesium ions and receive a third of a valency share from three of them.

Isomorphous replacement occurs as in pyroxenes,[3] so that in a typical amphibole, such as hornblende, there are the pyroxene replacements and also $Na^+ + Al^{3+}$ (oct.) for $Ca^{2+} + Mg^{2+}$, $Ca^{2+} + Al^{3+}$ (tetr.) for $Na^+ + Si^{4+}$ and F^- for $(OH)^-$.

In the amphibole hornblende, aluminium ions replace silicon in the chains, thereby creating a valency deficiency. Although this is counterbalanced to some extent by further aluminium substituting for magnesium ions, balancing ions, such as Na and K, enter the lattice and are situated at the points marked AA′.

Chrysotile, the serpentine mineral, was thought until quite recently to be an amphibole, but it is now known to have a layer-silicate lattice (see p. 137).

There is one group of clay minerals which has a similar structure to the amphiboles. This is the mineral group which was originally termed *palygorskite* by Russian workers. More recently, minerals of this same group, of American origin, have been named *attapulgites*.

Palygorskites, which are usually fibrous in texture, are complex hydrated magnesium silicates often with considerable amounts of aluminium, calcium and iron. Bradley[4] has described the structure as amphibole-like which is more feasible than the mica-like structure originally suggested.

Fig. III.13 is a diagrammatic representation of the structure of palygorskite viewed along the *c*-axis. The amphibole double chains run along the *c*-direction, that is, downwards through the plane of the book, therefore the chains are shown in the end-on position. Palygorskite differs from a true amphibole in that each double chain is linked directly to similar ones in the lateral direction, but each alternate chain has its vertices (valency electrons) pointing in opposite directions. Thus, there

[1] Zussman, J., *Acta Cryst.*, 12, 309, 1959.
[2] Warren, B. E., *Zeit. Krist.*, 72, 12, 1929.
[3] Whittaker, E. J. W., *Acta Cryst.*, 13, 291, 1960.
[4] Bradley, W. F., *Amer. Min.*, 25, 405, 1940.

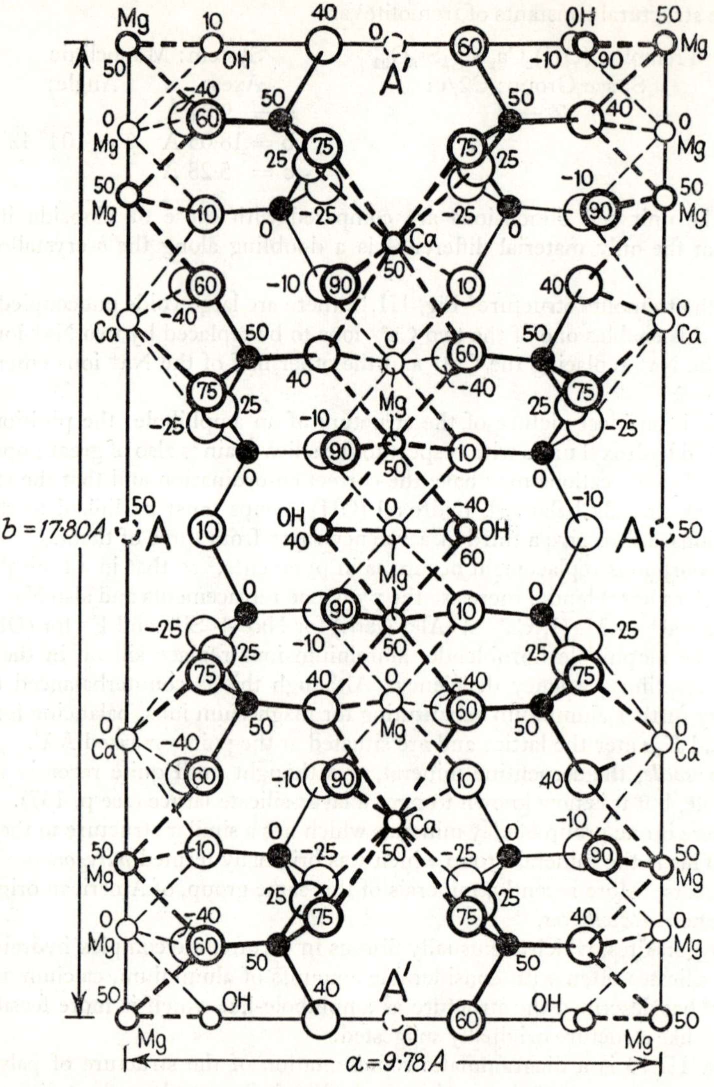

FIG. III.12. THE STRUCTURE OF TREMOLITE

The amphibole double chains run parallel to the *b*-dimension. The small numbers refer to the height of atoms above a zero plane.

are continuous sheets of silicon-oxygen linkages, but they may be regarded as linked amphibole chains rather than the symmetrical sheet structure of the next section.

The 'valency' oxygens are limited to one in each tetrahedron so the structure is based on $(Si_2O_5)^{2-}$ (see sheet structures, p. 124), rather than on the amphibole unit

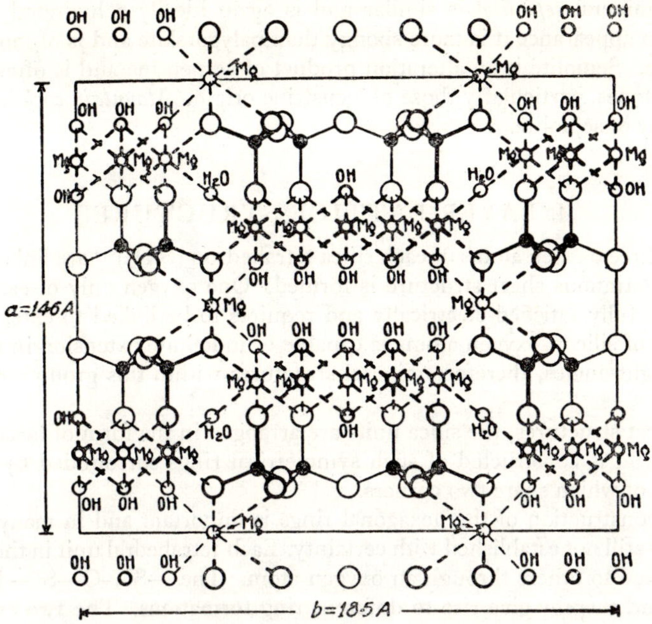

FIG. III.13. THE HYPOTHETICAL STRUCTURE OF PALYGORSKITE

$(Si_4O_{11})^{6-}$. The linkages between chains in the vertical direction of the diagram are through extra cations of magnesium or aluminium which are in six-fold co-ordination with oxygen and hydroxyl units. A feature of the palygorskite structure is that series of holes are present which arise as a result of the alternating chain elements of the structure. Water molecules and, possibly, hydrated cations are contained in these channels, which resemble those of zeolite structures. Aluminium atoms replace silicon in tetrahedral co-ordination within the chains giving a unit composition which is complex. Bradley has suggested that the ideal chemical formula would be $Si_8O_{20}Mg_5(OH)_2(OH_2)_4 . 4H_2O$; which represented on a balanced layer concept would be $(Si_2O_5)_4{}^{8-}(Mg_5(OH)_2)^{8+}(OH_2)_4 . 4H_2O$, i.e. units of $(Mg_5(OH)_2)^{8+}$ linking $n(Si_2O_5)^{2-}$ sheets.

The concept of the two sets of water molecules arises as a result of dehydration studies, which indicate that two molecules per unit are held more firmly than the others. It is suggested that these are directly linked to the cations which connect silica sheets as indicated in the diagram.

The structural constants of palygorskite are

Palygorskite (Attapulgite)
Space Group: C2/m
Z = 2

System: Pseudo-orthorhombic
Axes:
a sin α = 12·9 Å
b = 18·0 Å
c = 5·3 Å

The mineral *sepiolite*[1] is similar and is again ideally a hydrated magnesium silicate. In appearance it is more spongy than palygorskite and is of more common occurrence. Sepiolite is an alteration product of serpentine and is often associated with limestones, particularly those of lacustrine origin. *Mountain cork* is thought to be a variety of sepiolite.

4. LAYER LATTICE STRUCTURES

WHEN three oxygen atoms of each silica tetrahedron $(SiO_4)^{4-}$ are linked to similar units, a continuous sheet structure is formed. One oxygen only of each group of four is not fully satisfied electrically and requires to be linked to external cations. The sheet of silicon-oxygen atoms is capable of indefinite extension in two dimensions at right angles, therefore mineral structures within this group are invariably platey.

In most structures, the silica units are arranged in the form of hexagonal rings and the sheet is constructed of such symmetrical rings surrounded by six similar ones, each of which share two corners.

The construction of the hexagonal rings is important and in many structures the form is still not established with certainty. Each tetrahedral unit in the hexagonal ring is linked to others through an oxygen atom. The —Si—O—Si— bond angle can vary and thereby give rise to different ring formations. The two extremes are when the angle is 141° 34′ and when it is 180°. In the first case, a ring can be constructed where each silicon atom is at the same level above a zero plane and the valency oxygens all point vertically upwards. In the second case there are two possibilities: (*a*) every alternate valency oxygen points in opposite directions and the silicon atoms are on two levels; this is termed the 'rocking chair' or trans-form; and (*b*) the valency oxygens point in many directions and the silicon atoms are on three levels giving the so-called 'boat' or cis-form.

The majority of sheet structures are based on the concept that the Si—O—Si bond angle is about 141° 34′ but alternative structures are suggested, based on an angle of 180°, for montmorillonites. Ring structures of the trans-form occur in framework silicates but the cis-form of ring is not known, although it is common in organic chemistry.

Unlike all silicates which have so far been considered, sheet structures do not build up an extensive three-dimensional network by the simple link-up of silicon-oxygen units through additional cations. Definite layers are formed which are stacked one above another, and the bonds between each layer are relatively weak. Such minerals can, therefore, be sheared with comparative ease, and the crystal form is invariably plate-like.

Each mineral layer or plate is an independent unit, so Pauling's rules must be applied within each sheet. The layers must each be electrically balanced and each cation co-ordinated with anions in a way dependent on their size ratio. In some cases, the layers are not electrically balanced within themselves, so additional ions have to be absorbed on the surface of the layers.

[1] Brauner, K. and Preisinger, A., Tschermak's *Min. Pet. Mitt.*, 6, 120, 1958.

Sheet structures can be simple where layers of tetrahedral silica units are linked to cations of one type with a specific co-ordination, or they may be more complex with substitutions of many types.

The principles governing the structure of layer silicates were established by Pauling[1] following a detailed study of large mica flakes suitable for single crystal X-ray investigation. Similar lattice configurations have now been ascribed to most clay minerals and well-defined groups have been established.

Naturally-occurring clays are usually composed of an admixture of many minerals. Quartz and other crystal forms are present, but, in addition, there is always a hydrated silicate of aluminium, magnesium or iron sometimes of small grain size. Such minerals are termed *clay minerals*, but the group also includes *micas*, *chlorites*, *serpentines* and other minerals, for these are similar in structure.

The simplest structure is that of the kaolin-type layer where the valency oxygen of a silica sheet is satisfied by one type of cation in octahedral co-ordination with oxygen or hydroxyl units. This is known as a single-layer structure because only one sheet of silicon-oxygen linkages is involved. Many minerals are examples of this class and this group has probably been studied in more detail than any other silicate type.

Single-layer Group

In general terms, the structure of this group is a series of layers, each with a sheet of linked silicon-oxygen tetrahedra, the excess valencies of which are satisfied by cations in octahedral co-ordination. The cations which can be contained in a group of six symmetrically-placed oxygen atoms are limited in numbers; there are in fact only four common ones, Al^{3+}, Fe^{3+}, Mg^{2+} and Fe^{2+}.

The valency contribution of a trivalent cation to each surrounding anion in six-fold co-ordination is one-half, whilst that of a divalent cation in a similar environment is only one-third. The lattice configuration of minerals which have trivalent cations must obviously be different from those which contain cations of lower charges. The single-layer minerals may, therefore, be further subdivided into structures with differently charged cations.

1. **Trivalent cations.** The important *kaolin*[2] group of minerals is based on layers of a single sheet of silicon-oxygen linkages bound through common anions to aluminium atoms arranged as in a gibbsite sheet and surrounded by six anions of oxygen and hydroxyl units.

Kaolinite is the name given to a specific mineral which is found in certain localities and has certain properties which enable it to be distinguished from others. *Cornish china clay* is composed largely of kaolinite which has been freed from impurities of micaceous, felspathic and quartzitic types. Several other minerals have been isolated which are identical in chemical composition, yet are different from kaolinite in properties and structure. These minerals include *nacrite*, *dickite*, *livesite*, *halloysite*, *anauxite* and possibly *allophane*.

[1] Pauling, L., *Proc. Nat. Acad. Sci.*, 16, 123, 1930.

[2] The Clay Minerals Group of the Mineralogical Society has proposed the term *kandites* as a general name for kaolin group minerals.

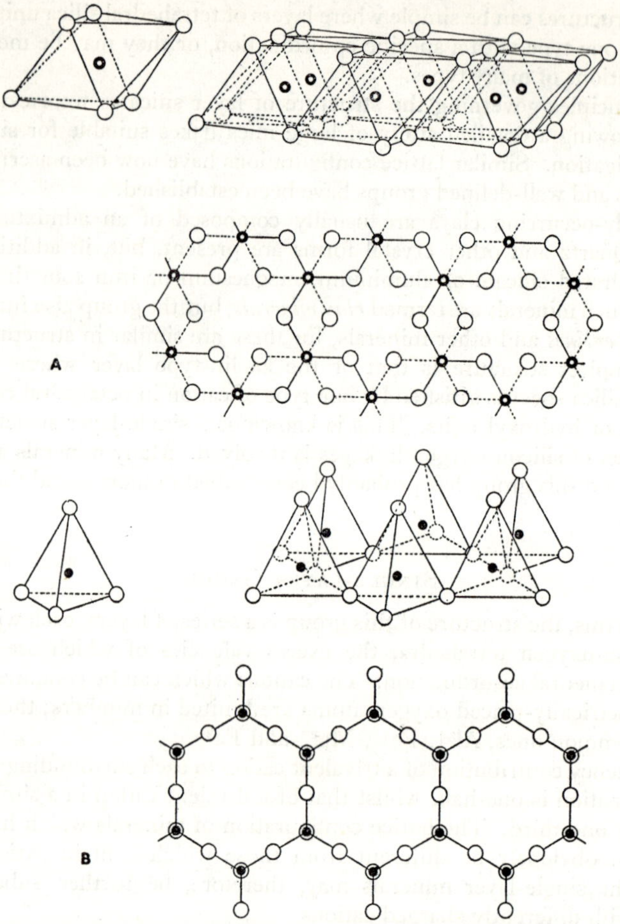

FIG. III.14. THE STRUCTURE OF A KAOLIN LAYER
(A) The gibbsite unit showing an octahedral $Al(OH)_6$ unit and the planar construction from side and top elevations; (B) the silica plane showing a tetrahedral unit and the planar construction from side and top projections

Nacrite and *dickite* are not common, although the latter is occasionally found in ironstone beds as a white powder in cavities and at the centre of nodules. Both nacrite and dickite are more coarsely crystalline than other kaolin minerals.

Livesite is probably the commonest kaolin mineral as it is associated with many sedimentary clays. *Fireclays*, certain *bond-clays*, many *brick-earths* and some *ball-clays* are rich in livesite which is present as a finely-grained component, thereby endowing plasticity or stickiness to the clay.

Halloysite is a commonly-occurring kaolin mineral but is rare in the British Isles. Like kaolinite itself, halloysite is thought to be of hydrothermal origin.

Anauxite contains less aluminium than kaolinite yet has all the other characteristics of a kaolin group member—it is probably not common.

Allophane also differs from kaolinite in composition. It is a finely-grained mineral rather like an amorphous mixture of colloidal silica and alumina and may be a constituent of some sedimentary clays.

All kaolin minerals are composed of identical layers of composition $Al_2O_3.2SiO_2.2H_2O$, and differ only in the way in which these layers are arranged relative to each other. The fundamental unit consists of a kaolin layer, based on a hexagonal *silica network* (Fig. III.14B) which is linked to a layer of aluminium atoms through common oxygen linkages.

Aluminium hydroxide or *gibbsite*, $Al(OH)_3$, has also a layer structure (Fig. III.14A) and the kaolin layer is essentially such a gibbsite layer condensed on the hexagonal silica-sheet. As the valency contribution of each aluminium ion to each co-ordinating anion is one-half, the 'valency' oxygens of the silica sheet, which all point vertically upwards, have to be linked to two aluminium atoms. Each oxygen atom thus receives one valency-share from a silicon and half from each of two aluminium ions and is electrically satisfied. Each aluminium ion is co-ordinated with three oxygen atoms linked to silicon atoms and to three hydroxyl units, one of which is on the same plane as the 'valency' oxygens and the other two are at a higher level.

The structure of kaolinite viewed from the side is shown in Fig. III.15A. The lowest plane in the layer is of oxygens linking adjacent silicons in the sheet above; then comes a layer of close-packed anions with a hydroxyl unit $(OH)^-$ at the centre of six oxygen atoms. The next layer is of aluminium atoms, above which is a sheet of close-packed hydroxyl ions. As hydroxyl units are similar in size to oxygen atoms and all anions are relatively much larger than the cations, the layer structure can be imagined ideally as a close-packed system of spheres.

The thickness of the layer is 4·37 Å and the height of each plane is given in the diagram. The aluminium atoms are not situated directly above each 'valency' oxygen of the silica sheet, otherwise the structure would not be electrically balanced. The positions of aluminium atoms can be more clearly seen in Fig. III.15B, which is a diagrammatic representation looking downwards on to a layer.

Two aluminium atoms lie immediately above each hexagonal ring of silicon atoms. There are, however, three possible positions over each ring where octahedrally co-ordinated cations could fit, but if each were filled with trivalent cations the structure could not be electrically neutral. Only two-thirds of the available cationic positions are filled in the kaolin layer so the structure is termed *dioctahedral.*

The chemical representation of a kaolin layer is $(Si_2O_5)Al_2(OH)_4$ implying sheets of $n(Si_2O_5)^{2-}$ linked to $(Al_2(OH)_4)^{2+}$ units. This unit is of indefinite length in two directions along the *a* and *b* crystallographic axes.

This simple representation of a kaolin layer is not an absolutely accurate interpretation of present-day thought. The gibbsite layer does not fit perfectly over the silica sheet, so that distortions must occur. Both the octahedral and tetrahedral units are considered[1,2,3] to rotate and become slightly distorted to accommodate the

[1] Drits, V. A. and Kashaev, A. L., *Dokl. Sobr. Mez. S.S.S.R. Akad. Nauk.*, 15, 1960.
[2] Newnham, R. E., *Min. Mag.*, 32, 683, 1961.
[3] Radoslovich, E. W., *Amer. Min.*, 48, 76, 1963.

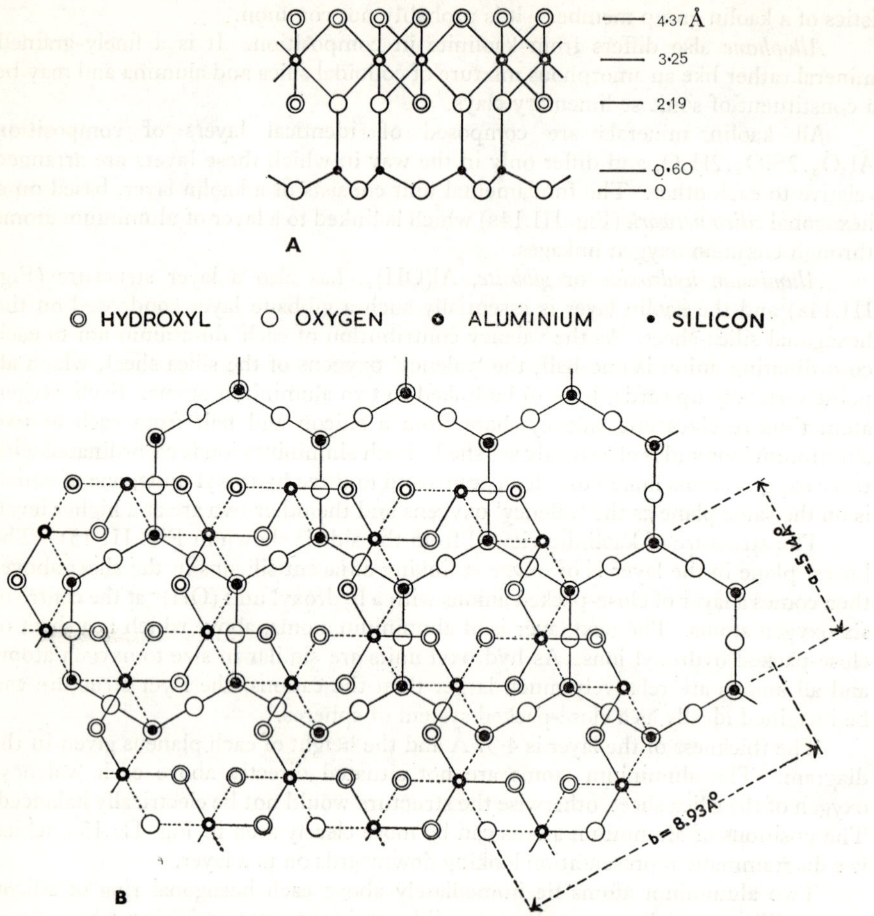

FIG. III.15 THE DIAGRAMMATICAL STRUCTURE OF A KAOLIN LAYER
(A) The kaolin layer (side elevation); (B) the kaolin layer (top elevation)
(after Brindley[1] and Newnham[2]).

bond positions, thus resulting in a puckering in both the basal oxygen plane and in the upper hydroxyl layer.

Polymorphism in kaolin minerals occurs because the individual layers can be stacked upon each other in different ways. Although within one layer the electrical valency forces are balanced, weak bonds develop between the basal oxygen atoms of one plane and the upper hydroxyl groups of the next. A hydrogen or O-OH bond (p. 73) forms, which, although inherently weak, is sufficiently strong to lead to crystal build-up in the *c*-dimension at right angles to the basal section. There is

[1] Newnham, R. E. and Brindley, G. W., *Acta Cryst.*, 9, 759, 1956.
[2] Newnham, R. E., *Min. Mag.*, 32, 683, 1961.

close contact between atoms in adjoining layers with a close-packed system throughout and a 7·2Å separation distance between similar levels in adjacent sheets.

It is generally accepted that the stacking of kaolin layers to form a specific mineral requires the closest approach of adjacent O and OH groups in successive layers (Fig. III.16), and furthermore each O-OH linkage must be equalised. Newnham is of the opinion that elevated hydroxyl and oxygen units in successive layers are involved in sharing the common hydrogen.

A stable configuration is formed when the OH-groups on the top of one kaolin layer are 'paired' with oxygens in the lowest sheet of the unit immediately above as shown in Fig. III.16. This symmetrical arrangement can result only if one layer is placed upon another in a limited number of ways. For example, if one unit is arranged above another so that equivalent atoms lie immediately above each other, then the relationship of OH to O groups in successive layers will be as illustrated. The *c*-direction in the resulting structure will then be at right angles to the basal section and the crystallographic angle $\alpha = 90°$.

However in the idealised layer structure (Fig. III.15), each layer would have to be shifted one-third of a unit cell in the *a*-dimension to conform to the O-OH bonding requirement, although the distortion of the lattice and the rotation of the octahedral and tetrahedral groups usually imply a slightly higher displacement. This results in an ideal regular shift of $\pm a/3$ from one layer to another implying a crystallographic angle $\beta = 103° 30'$. There is not a regular progression in all kaolin minerals and some show a sequential right and left shift alternation.

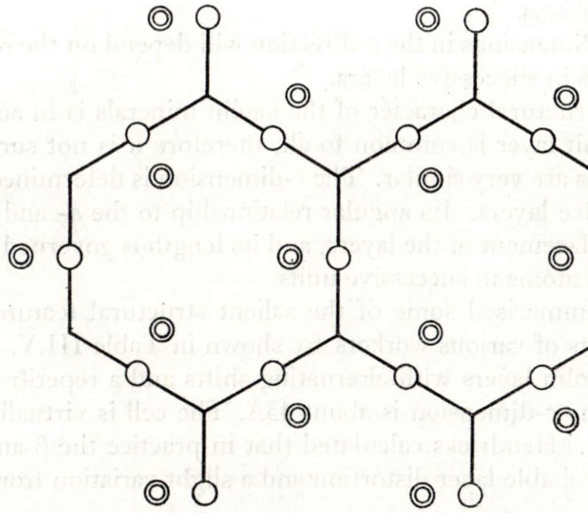

FIG. III.16. THE STABLE ARRANGEMENT OF OH GROUPS AND O ATOMS IN SUCCESSIVE LAYERS OF KAOLIN MINERALS

There is another complication which arises when kaolin layers are stacked above each other. There are only two aluminium atoms associated with each hexagonal silica ring, yet there are three possible positions into which they may be placed.

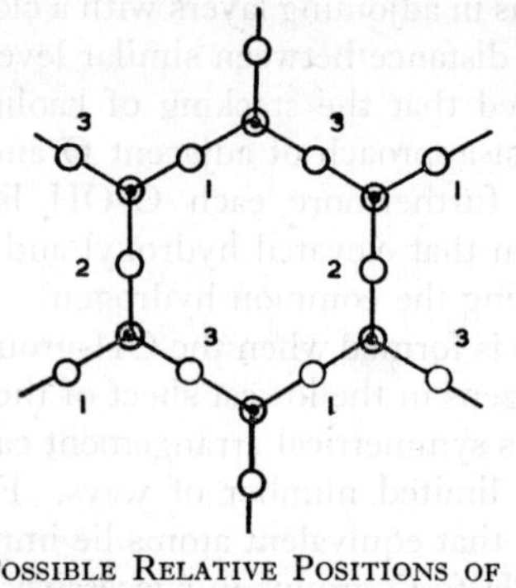

FIG. III.17. THE POSSIBLE RELATIVE POSITIONS OF ALUMINIUM ATOMS TO A SILICA HEXAGONAL RING IN THE KAOLIN LAYER

Fig. III.17 shows the hexagonal ring with the positions for two aluminium atoms. They may be at the points 1 and 2, or 1 and 3, or 2 and 3. In a single layer, all these combinations are identical, but when two successive layers are considered, the relative positions of the aluminium atoms influence the dimensions of the unit-cell. If, for example, one kaolin layer is placed immediately above another, the oxygen atoms in the lowest sheet of one layer are directly below similar ones in the top layer; silicon atoms in adjacent layers must also be along vertical lines for they are symmetrically arranged. Aluminium atoms need not be above one another, for in the lower layer they may be in the positions 1 and 2, but in the upper unit in any of the three alternatives 1–2, 1–3 or 2–3. Three or more layers provide additional complications.

The unit-cell dimensions in the *c*-direction will depend on the relative positions of aluminium atoms in successive layers.

The general structural character of the kaolin minerals is in accord with these principles. The unit layer is common to all, therefore it is not surprising that the *a*- and *b*-dimensions are very similar. The *c*-dimension is determined by the type of stacking of successive layers. Its angular relationship to the *a*- and *b*-axes depends on the relative displacement of the layers, and its length is governed by the arrangement of aluminium atoms in successive units.

Bailey[1] has summarised some of the salient structural features of the kaolin minerals, and results of various workers are shown in Table III.V.

Nacrite has kaolin layers with alternating shifts and a repetitive unit cell with six layers so that the *c*-dimension is about 43Å. The cell is virtually orthorhombic with $\alpha=\beta=\gamma=90°$. Hendricks calculated that in practice the β angle was $90°\ 20' \pm 10'$, indicating probable layer distortion and a slight variation from orthorhombic symmetry.

Dickite is perhaps the only kaolin mineral of which perfect crystals can be isolated. Newnham and Brindley have carried out a detailed structural analysis which has contributed greatly to the knowledge of this group of minerals. Their careful X-ray compilations have proved the interatomic associations and indicated the type and extent of distortions and rotations.

[1] Bailey, S. W., *Amer. Miner.*, 48, 1196, 1963.

The stacking of individual layers to build up the unit cell probably involves regular alternation of right- and left-handed shifts, but to accommodate the O-OH linkages the β angle is 96° 50′.

TABLE III—V. STRUCTURAL SCHEMES OF SOME KAOLIN MINERALS

Mineral	*Crystal form*	*Crystallographic dimensions*					
		axes Å			*angles*		
		a	*b*	*c*	α	β	γ
Nacrite[1]	Orthorhombic	5·15	8·96	43	90°	90° 20′	90°
Dickite[2]	Monoclinic	5·15	8·96	14·45	90°	96° 50′	90°
Kaolinite[3]	Triclinic	5·14	8·93	7·37	91° 36′	104° 48′	90°
Livesite[4] (fireclay)	Pseudo-monoclinic	5·1	8·9	7·4	(90°)	104° 30′	90°
Halloysite[5]	Pseudo-hexagonal	5·1	8·9	7·2	(90°)	(90°)	90°

Kaolinite. Gruner[6] analysed kaolinite in 1932 and deduced a monoclinic cell. The *c*-axis was inclined only in one direction to give a β-angle of about 104° 30′ corresponding to a displacement of layers in the *a*-direction of $-\frac{1}{3}$. More recent work by Brindley and Robinson[3] has indicated that the unit-cell is really triclinic with $\alpha = 91° 48'$. The *c*-dimension of 7·37 Å indicates that there is only one layer per unit-cell.

The structure is illustrated in Fig. III.18.

Livesite. That a previously unrecognised kaolin mineral was present in fireclays and in some other sedimentary clays was first established by Grimshaw, Heaton and Roberts[7] using mainly thermal methods. The structure of this mineral has been deduced by Brindley and Robinson from X-ray measurements and they have suggested that the chief clay mineral in fireclays is intermediate between kaolinite and halloysite. The β-angle is 105°, which is near to that of kaolinite, but the α-angle is 90°, therefore the unit-cell is monoclinic. The displacements of layers along the *a*-direction are regular and equal to $-\frac{1}{3}$. The X-ray pattern indicates a high degree of disorder along the *b*-axis where layers are apparently randomly displaced by multiples of $\pm\frac{1}{3}$. The net result of considering several layers stacked upon each other is to give $\alpha = 90°$.

Some of the lines in the X-ray diagram of the clay mineral from fireclays are banded, as in halloysite, and it has been suggested by Brindley and Robinson that a series of minerals exists with kaolinite as one extreme and meta-halloysite the other. but the ultimate crystallites are so minute that pure livesite gives a diffuse pattern thereby making structure analysis most difficult.

1 Hendricks, S. B., *Zeit. Krist.*, 100, 509, 1938.
2 Newnham, R. E. and Brindley, G. W., *Acta Cryst.*, 10, 88, 1957.
3 Brindley, G. W., *The X-ray Identification and Crystal Structures of Clay Minerals*, p. 51, Mineral Soc. G.B. Monograph, 1961.
4 Brindley, G. W. and Robinson, K., *Trans. Brit. Ceram. Soc.*, 46, 49, 1947.
5 Bates, T. F., Hildebrand, F. A. and Swineford, A., *Amer. Miner.*, 35, 463, 1950.
6 Gruner, J. W., *Zeit. Krist.*, 83, 75 and 394, 1932.
7 Grimshaw, R. W., Heaton, E. and Roberts, A. L., *Trans. British Ceramic Society*, 44, 20, 1945.

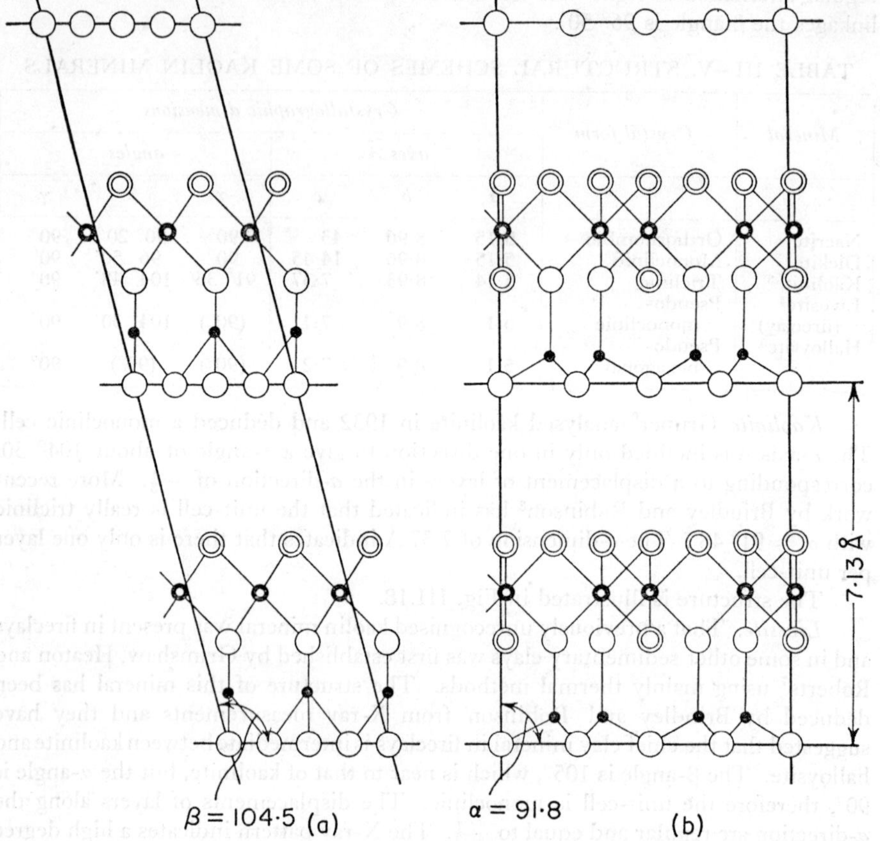

FIG. III.18. THE STRUCTURE OF KAOLINITE
(*a*) Looking down the *a*-direction; and (*b*) looking down the *b*-direction.

A feature of all samples of purified livesite from many localities has been the presence of a small amount of iron (1–2 per cent) which cannot be removed without disruption of the clay mineral. Chemical analysis suggests that it is in the ferrous state; it is not attributable to any known hydrate or oxide of iron. The clay mineral has a characteristic, pale yellow colour which turns faintly pink at about 450°C. If ferrous or ferric ions were contained in the crystal lattice of livesite, strains would be set up and the crystal size would be expected to be small.

Halloysite. Unlike the kaolin minerals previously described, halloysite exists in more than one form. The terminology applied to these varieties is not yet standardised. McEwan has proposed a nomenclature in which halloysite is retained as a general term and the particular form referred to as hydrated or dehydrated halloysite. The latter is commonly termed *metahalloysite*, although this may lead to confusion particularly among ceramists, for meta-kaolin has been defined as china clay which has lost its water of crystallisation. There seems to be no valid reason why the two forms should not simply be called (*a*) *hydrated halloysite* when the mineral with the high water content is implied, and (*b*) *normal halloysite* when referring to the type which resembles other members of the kaolin group in chemical composition.

Hydrated halloysite has the approximate composition $Al_2O_3.2SiO_2.4H_2O$, i.e. two more molecules of water per unit of composition, whilst normal halloysite has the usual kaolinite formula. The hydrated variety has a basal spacing or distance between successive layers of 10·1 Å which falls to 7·2 Å on dehydration. Hendricks[1] has shown that the decrease of 2·9 Å corresponds to a molecular sheet of water molecules and on this basis suggests that in hydrated halloysite an oriented layer of water is present between layers of kaolin.

Some natural deposits of halloysite are composed chiefly of the hydrated variety (Angleur halloysite is an example), other samples contain little, if any, and may have been dehydrated by natural processes. Grimshaw recently examined an American halloysite where two forms could be clearly recognised. A soft, easily water-dispersed portion gave a strong 10·1 Å X-ray line, whereas hard nodular pebbles were mainly the normal variety contaminated with trihydrate of alumina, gibbsite.

Hydrated halloysite changes to the normal form at about 50° C., but the basal spacing does not fall to 7·2 Å until the mineral is heated to about 400° C.

There has been more scientific discussion on the structure of halloysite than on all other kaolin minerals but there is as yet no general agreement on the finer details. For reasons which are so far unexplained the kaolin layers in halloysite are curved—in fact, they are probably rolled up within themselves to produce distinct tubular crystals. Bates[2] considers that halloysite is similar to kaolinite but in the hydrated form water molecules are interposed between adjacent layers, thus raising the basal spacing to about 10·1 Å. The inclusion of the water band prevents the normal *c*-axis bonding of O-OH units so that the layer distortion is manifested as an actual curvature.

The resulting structure has random layer shifts in both the *a* and *b* crystallographic directions so that the X-ray pattern is banded and diffuse, and extremely difficult to interpret.

There might be a complete series of minerals between kaolinite and halloysite differing in their degree of lattice distortion and in the amount of interstratified water.[3]

Anauxite has a similar X-ray pattern to those of other kaolin minerals, yet its molecular ratio $SiO_2:Al_2O_3$ is 3:1 as against a normal 2:1. Hendricks[4] has suggested that anauxite may have a structure consisting of alternate layers of kaolin and neutral silica layers of composition $4SiO_2.H_2O$. If so, this silica unit would be of

1 Hendricks, S. B. and Jefferson, M. E., *Amer. Min.*, 23, 863, 1938.
2 Bates, T. F., *Amer. Miner.*, 44, 78, 1959.
3 Brindley, G. W., de Sousa Santos, P. and de Sousa Santos, H., *Amer. Miner.*, 48, 897, 1963.
4 Hendricks, S. B., *Jour. Geol.*, 50, 276, 1942.

similar size to a kaolin layer, and the X-ray pattern would resemble that of a true kaolin mineral.

Allophane is classed as a kaolin mineral on rather inconclusive evidence. The molecular ratio of alumina to silica varies between 1:1 and 1:2, although it is usually about 1:1·5. The mineral appears amorphous and it gives no X-ray pattern. Grimshaw has analysed a sample from the south of England which had a molecular ratio of alumina to silica of 1:1·56; the differential thermal curve was quite characteristic and unlike a mixture of hydrated silica and alumina, which some authorities suggest allophane may be.

One feature of the kaolin minerals is the differences in their crystal sizes. Nacrite and dickite occur in comparatively large fragments, easily visible with an ordinary microscope; kaolinite crystals are much smaller and, for the most part, are only a few microns (1 micron $= 10^{-4}$ cm.) in size; halloysite tubes are also small, very rarely larger than 1 micron; livesite is invariably in the form of smaller crystallites which are not perfectly resolved under magnifications of 35,000 times in the electron microscope. These differences may be related to crystal structure or more particularly to the configuration along the *c*-dimension.

The stacking of kaolin layers may be compared to that in a pack of playing cards. Nacrite represents the state when each card lies directly above lower ones and a large structure can build up. Dickite is the condition where a progressively slight shift (6–7°) is produced in one direction. Kaolinite is similar, but with a larger displacement (14–15°) from the vertical in one direction. Livesite has displacements in two directions, randomly in one, regularly in the other; a pack of cards stacked in such fashion would be liable to collapse before many units could be added.

The thickness of the mineral plates undoubtedly influences the cross-sectional area of plate-like crystals. Unfortunately, detailed information of the thickness : size ratio for all kaolin minerals is not available but R. W. Grimshaw and T. E. Currie have examined dickite crystals under the microscope, and kaolinite and livesite by electron microscopy. In each, the ratio of the thickness of the crystal to the average length of its sides was closely similar and varied from 1:12 to 1:20.

Possibly all crystals which are composed of kaolin layers have maximum stability at a certain thickness: size ratio and each mineral can build up to a certain size, governed largely by the stability of stacking of the successive layers.

Most kaolin minerals are of extremely small particle size and are difficult to separate from associated mineral impurities. Consequently reliable X-ray diffraction patterns are not easily produced or interpreted. Some of the proposed structural configurations of these minerals are, at the best, inspired guesswork—perhaps stretching the observed results beyond sensible limits. In some of the interpretations, physical concepts have been allowed to override significant chemical evidence. Infra-red studies of these minerals suggest that some aspects of their structure require modification and it may well be that with the development of this and other techniques, the true structural relationships within the kaolin minerals will ultimately be established.

Kaolin minerals must be one of the few silicate structures in which isomorphous substitution of cations is not extensive. This is all the more remarkable, because ferric iron can freely replace aluminium atoms in most structures (see pp. 113, 120, 145 etc.). It may be that a gibbsite layer just fits on to a hexagonal network of silica units and the introduction of ions of different sizes would be sufficient to distort the whole layer catastrophically.

At the present time there is no positive evidence of substitution in kaolin minerals, but there is no obvious reason why a small amount should not occur.[1]

2. **Divalent cations.** Although the replacement of aluminium atoms in kaolinite with other trivalent cations is uncommon (even if it takes place at all), very similar structures with divalent ions are of frequent occurrence. Two important types are recognised, (*a*) those with magnesium and (*b*) those with ferrous iron as the additional cation. Both structures are similar, for they are composed of layers stacked above one another in the *c*-dimension and held together by weak bonds.

The composition of a single layer is not unlike that of kaolin, because a silica hexagonal sheet unit is linked to cations in octahedral or six-fold co-ordination with oxygen and hydroxyl groups. Yet, as the cations are divalent, they donate to each surrounding anion only one-third of a valency-share. Therefore the valency-oxygens of each silica tetrahedron, which once again are all directed upwards, require to be linked to three additional cations for valency satisfaction. Each hydroxyl group, of which there is one between each ring of six 'valency oxygens', and also a close-packed assembly forming the uppermost sheet of the layer structure, are linked to three divalent cations. To satisfy these requirements *three* cations must be placed symmetrically over each hexagonal ring of silicon atoms. By this means, all available octahedral positions are filled, whereas only two in every three are occupied by aluminium atoms in the kaolin structure. The minerals may be termed *trioctahedral*, to distinguish them from those in the dioctahedral kaolin series.

The composition of the unit layer is shown in Fig. III.19 A, B, and the diagram may be compared with that of kaolin in Fig. III.15.

Whereas the layer in kaolinite may be imagined as a gibbsite ($Al(OH)_3$) layer condensed on a sheet of hexagonally arranged silica units, so may the trioctahedral minerals be visualised as a condensation of the corresponding hydroxide with a similar silica sheet. This is shown by comparing the kaolin unit with that of the magnesium- and iron-bearing analogue on the basis of their lattice dimensions.

Figs. III.15 and 19 show the diagrammatical representation of dioctahedral and trioctahedral layers and the hexagonal ring structure of the silica sheet beneath; arbitrary *a*- and *b*-directions are marked.

The ring structure of silica is common to all minerals and the dimensions of this unit can be calculated quite simply by assuming that the Si—O distance is 1·54 Å and that the Si—O—Si angle is 141° 34′. The length along *a* is 5·031 Å and along *b* is 8·715 Å. On to this unit a hydroxide layer has to be fitted, the *a*- and *b*-dimensions of which have been calculated for the three main types and are as follows:

Gibbsite $Al(OH)_3$		Brucite $Mg(OH)_2$		Ferrous hydroxide $Fe(OH)_2$	
a	*b*	*a*	*b*	*a*	*b*
5·060 Å	8·624 Å	5·38 Å	9·33 Å	5·50 Å	9·72 Å

[1] Youell, R. F., *Nature*, 181, 557, 1958.

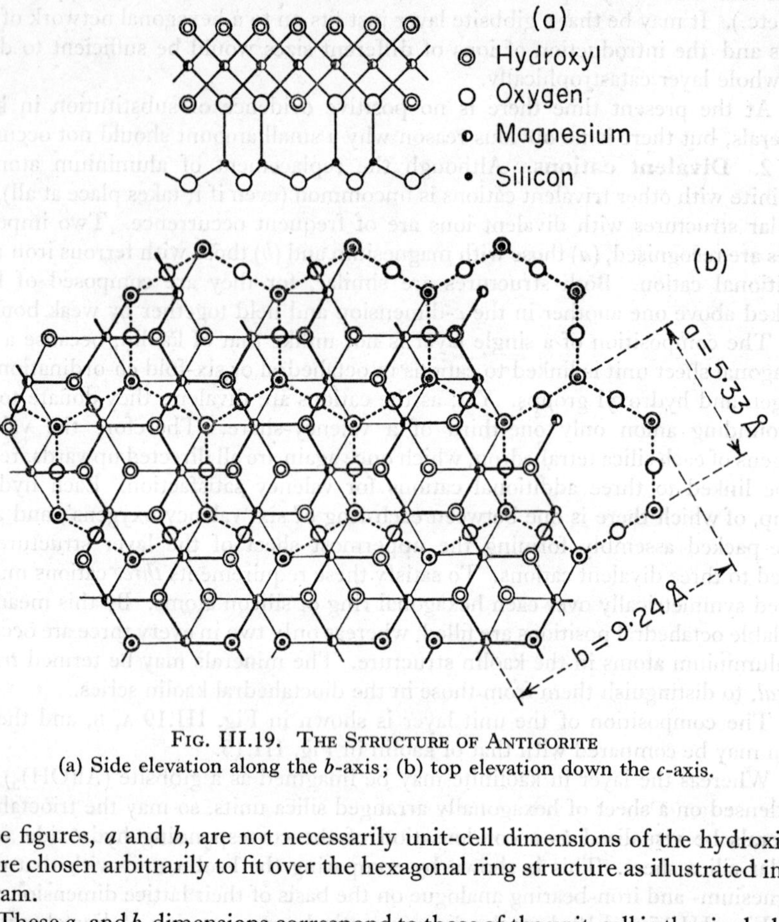

FIG. III.19. THE STRUCTURE OF ANTIGORITE
(a) Side elevation along the *b*-axis; (b) top elevation down the *c*-axis.

These figures, *a* and *b*, are not necessarily unit-cell dimensions of the hydroxides, but are chosen arbitrarily to fit over the hexagonal ring structure as illustrated in the diagram.

These *a*- and *b*-dimensions correspond to those of the unit-cell in the single-layer silicate minerals, and the effect of condensing different hydroxide structures with a silica sheet may be compared directly as in Table III.VI.

TABLE III—VI. COMPARISON OF THE DIMENSIONS OF THE HYDROXIDE LAYER AND CORRESPONDING LAYER SILICATE

	Al mineral		*Mg mineral*		*Fe^{2+} mineral*	
	a	*b*	*a*	*b*	*a*	*b*
Hydroxide	5·06Å	8·62Å	5·38Å	9·33Å	5·50Å	9·72Å
Layer silicate	5·14Å	8·93Å	5·33Å	9·24Å	5·39Å	9·31Å

The ideal dimensions of the silica hexagonal network are $a = 5{\cdot}031$ Å and $b = 8{\cdot}715$ Å so that considerable distortion must take place if this sheet is to accommodate the magnesium and ferrous hydroxide layers.

(*a*) *Magnesium-bearing Minerals*

Minerals which contain magnesium ions in trioctahedral single-layer silicate structures are important in ceramic studies for they are present in serpentines and the more common forms of asbestos.

Two distinct mineral entities have been identified, *chrysotile* and *antigorite*, both of which are similar, being composed of unit layers of $Mg_3Si_2O_5(OH)_4$.

Chrysotile is a common constituent of asbestos minerals, and because of its fibrous nature it was thought to be an amphibole (see p. 121). However Warren and Hering[1] have established that the structure is really a layer-lattice very similar to that in the kaolin minerals.

The lattice distortion in the *b*-direction of the individual layers must be very great if a brucite layer is to fit over a silica network, and so the supposition that a considerable curvature occurs along this dimension is reasonable.[2] The fibrous structure of chrysotile may be compared to the tubular form of halloysite where lattice distortion also results in curvature. Chrysotile crystals are capable of extension in the *a*-direction and long fibres result. Layers are stacked above one another, but in random fashion in one direction as in livesite. The structural constants of the common clino-form are:

Chrysotile $Mg_3Si_2O_5(OH)_4$	System: Monoclinic	
Space Group: Cm	Axes:	Angle:
Z = 2	a = 5·34 Å	
	b = 9·25	β = 93° 16′
	c = 14·65	

Chrysotile is the most common form of asbestos; other types include tremolite, anthophyllite, etc., which are true amphiboles and have been mentioned in the appropriate section. Hodgson[3] has presented a comprehensive study of the asbestos minerals.

Antigorite has virtually the same unit layer configuration as chrysotile, but it does not possess complete curvature of the sheets, possibly because the orientation or polarity of the layers reverses at regular intervals.[4] The layers are thought to be corrugated in some way. Cell dimensions are similar to chrysotile but the unit length of the *c*-axis can be up to 45 Å (6 units).

(*b*) *Iron-bearing Minerals*

Hydrated ferrous silicates are usually classified as *chlorites*, which are double-layer minerals to be described in the next section. Brindley[5] has, however, shown

[1] Warren, B. E. and Hering, K. W., *Phys. Rev.*, 59, 925, 1941.
[2] Whittaker, E. J. W., *Acta Cryst.*, 8, 571, 1955.
[3] Hodgson, A. A., 'Fibrous Silicates', R.I.C. Lecture Series No. 4, 1965.
[4] Zussman, J., *Min. Mag.*, 30, 498, 1954.
[5] Brindley, G. W., *Min. Mag.*, 29, 502, 1951.

that some chlorites, particularly those of the variety known as chamosite, have a single-layer construction rather like that of antigorite or chrysotile.

Chamosites are minerals which are common in the ironstone deposits associated with many of the Coal Measures and in some Irish (Antrim) bauxites. Such minerals have been found in associated fireclays, and some chlorites give X-ray powder diagrams strongly resembling that of livesite, which suggests that the stacking of layers is somewhat similar.

Antigorite-like chamosites appear to be of two forms, one of which has a monoclinic unit-cell (cf. livesite, p. 131) and the other with an orthorhombic cell suggesting a high degree of symmetry in the stacking of successive layers.

Apparently magnesium atoms can replace ferrous iron in the chamosite lattice as some samples contain appreciable proportions.

Another iron-bearing mineral based on a single-layer lattice is *cronstedtite* which ideally is $2FeO.Fe_2O_3.SiO_2.2H_2O$. Hendricks[1] suggests that this mineral has ferric iron replacing silicon in the hexagonal network and that the trioctahedral positions are filled with Fe^{2+}, Fe^{3+} and Al atoms. The full structural representation is, therefore, $(Fe^{3+}, Fe^{2+}, Al, Ca)(Fe^{3+}SiO_4)(OH)_5$. The replacement of a silicon by a ferric ion must result in the substitution of a hydroxyl unit for an oxygen atom to balance valencies.

Amosite is another mineral based on antigorite, but with aluminium ions replacing silicon. Magnesium and ferrous ions fill the octahedral positions and some aluminium atoms are also present.

DOUBLE-LAYER FORMATION

WHEN a hydroxide layer such as *gibbsite* is condensed on a hexagonal silica network, a single-layer structure is formed. Gibbsite (also brucite and ferrous-hydroxide) layers consist of two sheets of hydroxyl units with linking cations; the top sheet is similar to the bottom one and there can be condensation of a further silica network, thereby completing a sandwich structure of two hexagonal silica units with a gibbsite layer between. The structure is capable of indefinite extension in two dimensions by covalent/ionic linkages.

This type of unit is illustrated in Fig. III.20. The uppermost silica network is displaced relative to the lower and the unit layer is bounded by two sheets of close-packed oxygen atoms.

When aluminium ions are present, only two out of three of the octahedral lattice sites are filled. The result is a dioctahedral double-layer structure and this fundamental unit is called the *pyrophyllite layer*. Expressed chemically the ideal formula of this unit is $Al_2O_3.4SiO_2.H_2O$ or structurally $Al_2(Si_2O_5)_2(OH)_2$.

There is also a trioctahedral series when magnesium, iron or other divalent cations replace aluminium and the structure is equivalent to a further silica hexagonal network condensed on the single-layer structure of antigorite. This unit is termed a *talc layer* and has the chemical formula $3MgO.4SiO_2.H_2O$ or, structurally, $Mg_3(Si_2O_5)_2(OH)_2$; it is shown in Fig. III.20.

[1] Hendricks, S. B., *Amer. Min.*, 24, 529, 1939.

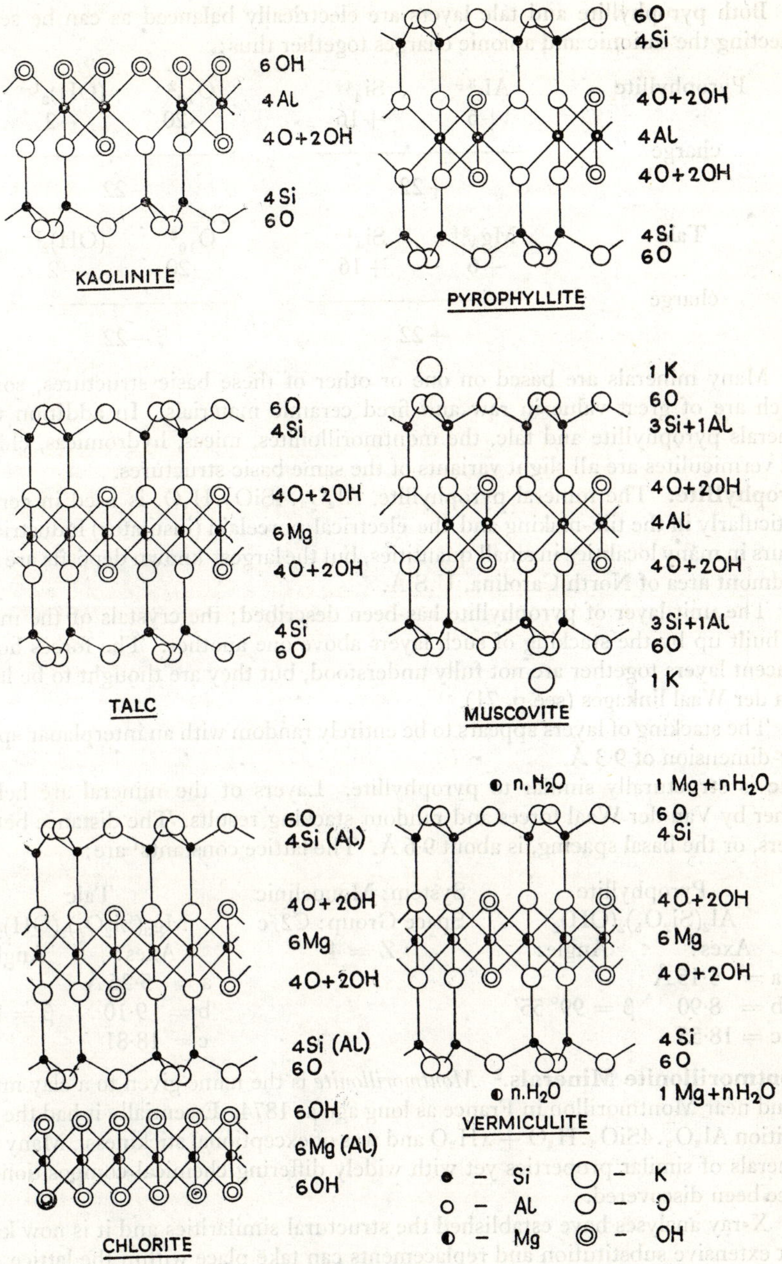

FIG. III.20. THE STRUCTURES OF SOME LAYER LATTICE MINERALS

Both pyrophyllite and talc layers are electrically balanced as can be seen by collecting the cationic and anionic charges together thus:

Pyrophyllite	Al_2^{3+}	Si_4^{4+}	O_{10}^{2-}	$(OH)_2^{1-}$
	$+6$	$+16$	-20	-2
charge	$+22$		-22	
Talc	Mg_3^{2+}	Si_4^{4+}	O_{10}^{2-}	$(OH)_2^{1-}$
	$+6$	$+16$	-20	-2
charge	$+22$		-22	

Many minerals are based on one or other of these basic structures, some of which are of great value in raw and fired ceramic materials. In addition to the minerals pyrophyllite and talc, the montmorillonites, micas, hydromicas, chlorites and vermiculites are all slight variants of the same basic structures.

Pyrophyllite. The mineral pyrophyllite, $Al_2O_3.4SiO_2.H_2O$, is used in ceramics particularly in the tile-making and the electrical porcelain (insulator) industries. It occurs in many localities in small quantities, but the largest-known deposits are in the Piedmont area of North Carolina, U.S.A.

The unit layer of pyrophyllite has been described; the crystals of the mineral are built up by the stacking of such layers above one another. The forces holding adjacent layers together are not fully understood, but they are thought to be largely Van der Waal linkages (see p. 71).

The stacking of layers appears to be entirely random with an interplanar spacing or *c*-dimension of 9·3 Å.

Talc is structurally similar to pyrophyllite. Layers of the mineral are held together by Van der Waal forces and random stacking results. The distance between layers, or the basal spacing, is about 9·6 Å. The lattice constants[1] are:

Pyrophyllite		System: Monoclinic	Talc	
$Al_2(Si_2O_5)_2(OH)_2$		Space Group: C2/c	$Mg_3(Si_2O_5)_2(OH)_2$	
Axes:	Angle:	Z = 4	Axes:	Angle:
a = 5·15 Å			a = 5·26 Å	
b = 8·90	$\beta = 99° 55'$		b = 9·10	$\beta = 100°$
c = 18·55			c = 18·81	

Montmorillonite Minerals.[2] *Montmorillonite* is the name given to a clay mineral found near Montmorillon in France as long ago as 1874. Essentially it had the composition $Al_2O_3.4SiO_2.H_2O + xH_2O$ and was of exceptional stickiness. Many other minerals of similar properties yet with widely differing chemical composition have since been discovered.

X-ray analyses have established the structural similarities and it is now known that extensive substitution and replacements can take place within the lattice.

[1] Gruner, J. W., *Zeit. Krist.*, 88, 412, 1934.

[2] The Clay Minerals Group of the Mineralogical Society has proposed the terms *smectites* as a general name for montmorillonite group members.

A wide range of minerals exists which might be included in the 'montmorillonite group', but the term *montmorillonite* now has a more restricted meaning, and implies a compound which is essentially a hydrated alumino-silicate with only little substitution.

Many clays which contain montmorillonite type minerals have been termed *bentonite*. This name has become common usage for commercial grades of montmorillonites.

The proposed structure of all minerals of the montmorillonite group is based on that of pyrophyllite or talc, but with the essential difference that the unit layers are not electrically neutral.

Montmorillonites have the property of absorbing cations and holding them strongly. As cations are positively charged it is assumed that the layers have a net negative charge, the result of ionic replacements within the lattice. Thus, the replacement of an Al^{3+} by a Mg^{2+} cation in octahedral co-ordination creates a unit deficiency of charge. In the mineral montmorillonite itself magnesium appears to replace aluminium in the approximate ratio of 1 to 6 and in other minerals of the montmorillonite group ferrous ions can replace, and aluminium (Al^{3+}) can take the place of silicon (Si^{4+}) in the hexagonal network. The atoms in successive sheets of a layer are arranged as follows. Pyrophyllite is included for comparison.

TABLE III—VII. THE ATOMIC ARRANGEMENT IN SUCCESSIVE SHEETS OF MONTMORILLONITE AND PYROPHYLLITE LAYERS

Montmorillonite	*Net charge*	*Pyrophyllite*	*Net charge*
3 O^{2-}	−6	3 O^{2-}	−6
2 Si^{4+}	+8	2 Si^{4+}	+8
2 O^{2-} + $(OH)^{1-}$	−5	2 O^{2-} + $(OH)^{1-}$	−5
1·67 Al^{3+} + 0·33 Mg^{2+}	+5·67	2 Al^{3+}	+6
2 O^{2-} + $(OH)^{1-}$	−5	2 O^{2-} + $(OH)^{1-}$	−5
2 Si^{4+}	+8	2 Si^{4+}	+8
3 O^{2-}	−6	3 O^{2-}	−6
Sum of positive charges = 21·67		Sum of positive charges = 22	
Sum of negative charges = 22		Sum of negative charges = 22	

In all other respects, the structure of pyrophyllite shown in Fig. III.20 is indicative of montmorillonite. (Fig. III.21*a*.)

These are the views of such eminent authorities as Hofmann, Endell, Wilm, Maegdefrau, Marshall and Hendricks[1] who have all made important contributions to the structural relationships of montmorillonite minerals. There is also an alternative structure, proposed by Edelman and Favejee[2] which is claimed to account for certain properties of montmorillonites unexplained by the pyrophyllite analogy, and a similar structure is suggested for halloysite. They conceive that the silica hexagonal network is based on one of the alternative arrangements (p. 124) where the

[1] (*a*) Hofmann, U., Endell, K. and Wilm, K., *Zeit. Krist.*, 86, 340, 1933; (*b*) Marshall, C. E., *Zeit. Krist.*, 91, 433, 1935; (*c*) Maegdefrau, E. and Hofmann, U., *Zeit. Krist.*, 98, 299, 1937; (*d*) Hendricks, S. B., *Jour. Geol.*, 50, 276, 1942.

[2] Edelman, C. H. and Favejee, J. C. L., *Zeit. Krist.*, 102, 417, 1940.

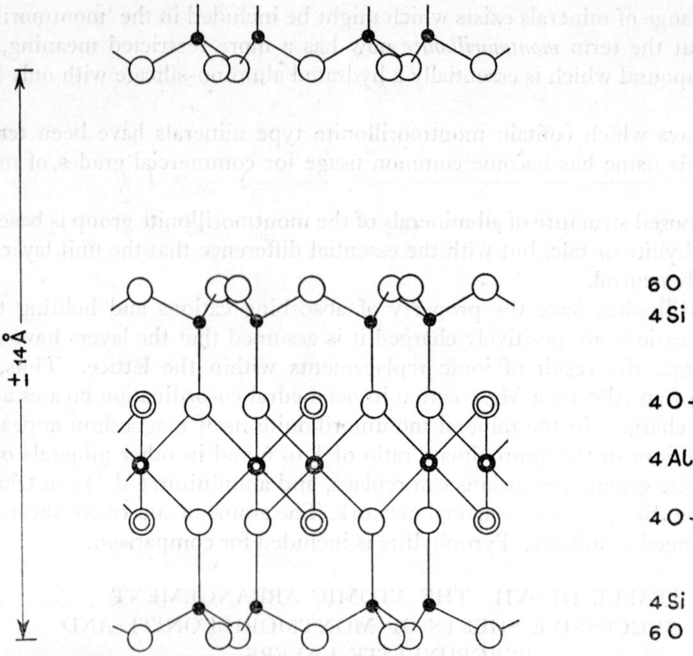

FIG. III.21(*a*). THE PROPOSED STRUCTURES FOR MONTMORILLONITE (IDEALISED) (Hofmann, *et al.*)

Si—O—Si bond angle is 180° and the rings are of the 'rocking chair' or *trans*-form, so that three alternate valency-oxygens point upwards and the other three downwards. Two of these sheets are linked to a gibbsite unit as illustrated in Fig. III.21*b*. Hydroxyl units project outwards from each layer and it is alleged that these behave as true hydroxide units and can combine with cations (M) to form O—M groupings. Other adsorption properties can be explained in a similar manner. The interesting feature of the Edelman and Favejee concept of montmorillonite is that the layers are electrically neutral and isomorphous replacements resulting in a charge-deficiency are not needed to account for the cationic adsorption. The layer may be represented as sheets composed as follows:

1 $(O—H)^{1-}$	net charge	−1	Silica hexagonal ring unit (with −11 and +8 charges)
1 Si^{4+}	,, ,,	+4	
3 O^{2-}	,, ,,	−6	
1 Si^{4+}	,, ,,	+4	
1 O^{2-} + 2$(OH)^{1-}$	,, ,,	−4	
2 Al^{3+}	,, ,,	+6	
etc., in reverse order			Silica hexagonal ring unit

Sum of negative charges = 22
Sum of positive charges = 22

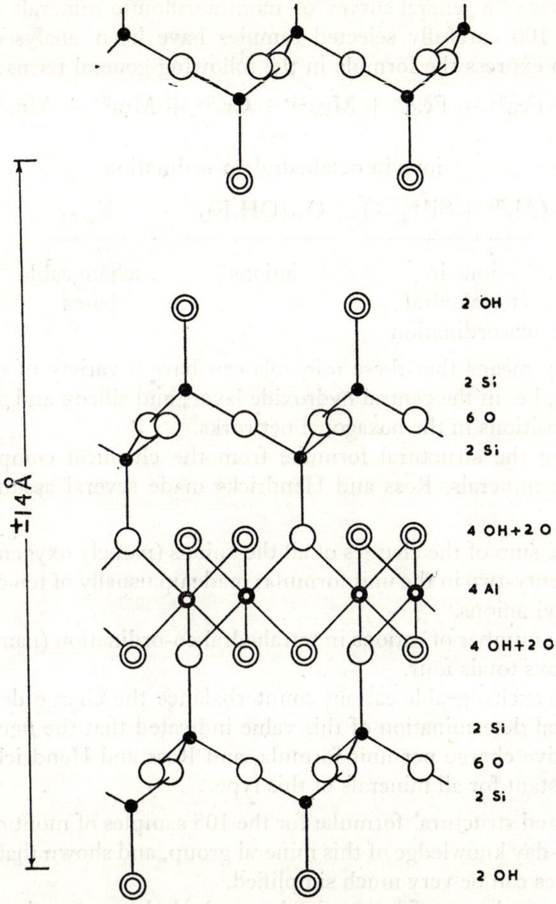

FIG. III.21(*b*). THE PROPOSED STRUCTURES FOR MONTMORILLONITE (IDEALISED)
(Edelman and Favejee)

The resulting total formula is $Si_4Al_4O_8(OH)_8(O—H)_4$. This structure would give rise to a much higher cation-adsorption than would actually take place if all (O—H) groups were ionisable. For the formula to be correct it must be assumed that all the (O—H) ions are not dissociated or that the silica units in the hexagonal networks are not inverted regularly.

Ross and Hendricks,[1] however, have accepted the structure based on pyrophyllite and have classified all the montmorillonite group members on this basis. Their ideas have met with fairly general approval, but the Edelman-Favejee concept has not been fully disproved.

[1] Ross, C. S. and Hendricks, S. B., *U.S. Geol. Survey Prof. Paper*, No. 205–B, 1945.

In the course of a general survey of montmorillonite minerals from numerous localities, over 100 carefully selected samples have been analysed by Ross and Hendricks, who express the formula in the following general terms:

$$\underbrace{(Al^{3+}_{a-y} + Fe_b^{3+} + Fe_c^{2+} + Mg_d^{2+} + Cr_e^{3+} + Mn_f^{3+} + Mn_g^{2+} + Li_h^{1+})}_{\text{ions in octahedral co-ordination}}$$

$$\underbrace{(Al_y^{3+} + Si^{4+}_{4-y})}_{\text{ions in tetrahedral co-ordination}} \quad \underbrace{O_{10}(OH,F)_2}_{\text{anions}} \quad \underbrace{X_{0\cdot 33}}_{\text{exchangeable bases}}$$

This simply means that these minerals can have a variety of cations in octahedral positions, i.e. in the central hydroxide layer; and silicon and aluminium ions in tetrahedral positions in the hexagonal networks.

In deducing the structural formulæ from the chemical composition of the montmorillonite minerals, Ross and Hendricks made several assumptions. These include:

(*a*) That the sum of the charges of all the anions (namely oxygen and hydroxyl) always totals twenty-two in the unit formula; made up usually of ten divalent oxygen and two hydroxyl anions.

(*b*) That the number of cations in tetrahedral co-ordination (namely silicon and aluminium) always totals four.

(*c*) That the exchangeable cations counterbalance the charge deficiency on the layers. The actual determination of this value indicated that the figure was usually about 0·33 positive charge per unit formula, and Ross and Hendricks suggest that this value is constant for all minerals of this type.

The calculated structural formulæ for the 105 samples of montmorillonites has clarified present-day knowledge of this mineral group, and shown that an apparently bewildering series can be very much simplified.

The number and type of cations in the octahedral layer can be calculated and on this basis alone there is a major distinction between the minerals of the montmorillonite group, because the number of cations in the central sheet was either between 2·00 and 2·22 or between 2·88 and 3·00. Both a *dioctahedral* and a *trioctahedral* series are thus clearly established.

The amount of substitution in the silica (tetrahedral) layer and in the central (octahedral) layer of cations can be calculated and as a result it was established that montmorillonites form continuous series of minerals with end members which could be predicted with some certainty. Although many cations are present in the structure of montmorillonite minerals, those which occur in appreciable amounts are limited to silicon (Si^{4+}), aluminium (Al^{3+}), magnesium (Mg^{2+}), iron in the ferric Fe^{3+}) and ferrous (Fe^{2+}) forms and lithium (Li^{1+}).

Many of the earlier analyses have been revised by Ross[1] and the original classification modified slightly.

[1] Ross, C. S., *7th Nat. Conf. Clays and Clay Minerals* (Pergamon Press, London, 1960).

Most of the samples which were analysed fell within a narrow range of composition which can be represented by the structural formula[1]

$$\begin{array}{c} Na_{0\cdot 33} \\ \uparrow \\ (Al_{1\cdot 67-2\cdot 00}Mg_{0-0\cdot 66})\,(Al_{0-0\cdot 33}Si_{3\cdot 67-4\cdot 00})O_{10}(OH)_2 \end{array}$$

Ferric ion may replace aluminium in the octahedral layer.

The name given to this group is *montmorillonite.* The number of cations in octahedral or six-fold co-ordination in this class varies between 2·00 and 2·22 so they are of the dioctahedral series.

Montmorillonites of higher alumina content are thought to be impure and the name *beidellite* is not so generally accepted. Synthetic high alumina derivatives have been termed *aluminian montmorillonites.*

Wyoming bentonite contains a mineral truly of the montmorillonite type with the structural formula

$$\begin{array}{c} Na_{0\cdot 33} \\ \uparrow \\ (Al_{1\cdot 53}Fe^{3+}{}_{0\cdot 18}Mg_{0\cdot 33})\,(Al_{0\cdot 13}Si_{3\cdot 87})O_{10}(OH)_2 \end{array}$$

The mineral from Montmorillon, France, is

$$\begin{array}{c} Na_{0\cdot 33} \\ \uparrow \\ (Al_{1\cdot 51}Fe^{3+}{}_{0\cdot 07}Mg_{0\cdot 60})\,(Al_{0\cdot 28}Si_{3\cdot 72})O_{10}(OH)_2 \end{array}$$

The so-called *Beidellite* from Beidell, Colorado, is

$$\begin{array}{c} Na_{0\cdot 40} \\ \uparrow \\ (Al_{1\cdot 46}\,Fe^{3+}{}_{0\cdot 50}\,Mg_{0\cdot 08})\,(Al_{0\cdot 36}\,Si_{3\cdot 64})\,O_{10}\,(OH)_2 \end{array}$$

When a montmorillonite mineral is of the above type, i.e. when the sum of octahedral cations is between 2·00 and 2·22, and yet a considerable amount of iron is present, the mineral is termed *nontronite.* Ferric cations can replace aluminium in all proportions in the octahedral layer, so there is really a complete range of solid solutions between the end members montmorillonite, beidellite and nontronite. Nontronite has to be represented by two extreme type formulæ:

$$\begin{array}{c} Na_{0\cdot 33} \\ \uparrow \\ (Fe^{3+}{}_{1\cdot 67}\,Mg^{2+}{}_{0\cdot 33})\,Si_4\,O_{10}\,(OH)_2 \end{array}$$

and

$$\begin{array}{c} Na_{0\cdot 33} \\ \uparrow \\ (Fe^{3+}{}_{2\cdot 22})\,(Al_1\,Si_3)\,O_{10}\,(OH)_2 \end{array}$$

[1] The exchangeable cation is usually represented in the formula, linked to the cationic group which is deficient in charge.

The former is not entirely suitable as no nontronites have been reported which contain so much magnesium, and the charge-deficiency is invariably centred mainly in the tetrahedral layer.

Of the remaining montmorillonite type minerals analysed by Ross and Hendricks a small number were entirely unrelated to the dioctahedral series. These minerals have structural formulæ where the sum of cations in octahedral positions is very nearly 3. Therefore the structure must be based on that of trioctahedral talc rather than on pyrophyllite which is the basis of the minerals of the montmorillonite-beidellite-nontronite group.

There is a series of minerals of the trioctahedral series with two end members represented by the formulæ

$$\overset{Na_{0\cdot 33}}{\uparrow}$$
$$(Mg^{2+}_{2\cdot 67}\ Li^{1+}_{0\cdot 33})\ Si_4\ O_{10}\ (F,OH)_2 \qquad \text{called } \textit{hectorite}$$

and

$$\overset{Na_{0\cdot 33}}{\uparrow}$$
$$Mg_3^{2+}\ (Al_{0\cdot 33}\ Si_{3\cdot 67})\ O_{10}\ (OH)_2 \qquad \text{called } \textit{saponite.}$$

Apparently there is a complete range of solid solutions between these two extremes, but it is unlikely that dioctahedral and trioctahedral montmorillonite types form intermediate members.

In all the theoretical structures which have been represented, the exchangeable base has been given as sodium. This is not necessarily the natural form of the mineral, for other cations such as Ca^{2+} and K^+ may be absorbed. The structural formula is not influenced in any way by the inclusion of other cations, but the properties of the mineral may be considerably affected.

Montmorillonite minerals are often expressed chemically as having the formula $Al_2O_3.4SiO_2.H_2O$. This can be misleading as the aluminium content may be very small as in nontronite or hectorite. It is not even accurate to express the formula as

$$\left.\begin{matrix} R_2^{3+}O_3 \\ 3R^{2+}O \end{matrix}\right\}.4SiO_2.H_2O$$

because the molecular ratio of metallic oxide to silica may approach 1:2, as in beidellite, nontronite and saponite. Only in ideal montmorillonite is the oxide ratio about 1:4.

In the foregoing pages only the composition of the layers which make up the montmorillonites has been described; there remains the problem of extension in the *c*-dimension, i.e. the stacking of successive layers.

Alkali or alkaline earth cations which are absorbed by the layers in montmorillonites must play a large share in the linking of adjacent units. Such ions penetrate between the layers and, although not held firmly in any way, they probably form a bridge between two negatively-charged units.

Apparently, in montmorillonites, stacking of successive layers is completely random; some minerals are fibrous in habit, e.g. hectorite, suggesting that the layers may be distorted as in halloysite or chrysotile.

When montmorillonites are completely dried at temperatures of about 150° C., the basal spacing, or distance between the lowest sheet of one layer and the equivalent sheet in the layer above, is about 10 Å and varies with the size of adsorbed cation. When water is added, this spacing increases to about 14·0 Å indicating that water has penetrated between the layers. As more water is introduced this spacing increases to 20 Å or more and in a dilute montmorillonite suspension the individual plates are probably completely isolated. This theory has been substantiated by adsorption measurements[1] by which the surface area can be determined.

The mechanism of water adsorption by montmorillonites is still a matter for conjecture. There is conflicting evidence as to whether the basal spacing of montmorillonite increases uniformly with water adsorption, or whether the increase is step-wise, thus indicating the formation of hydrate layers round the cations.[2] Experiments with minerals in which the adsorbed cations were Ca^{2+} or H^+ showed, undoubtedly, a non-uniform increase whilst Na^+-montmorillonites have a smooth expansion with increasing water content. This increase in the size of montmorillonite crystals on the addition of water is termed 'swelling' and has many important practical applications.

Montmorillonites are known to exhibit great plasticity, stickiness and strength in the wet and dry states. This is mainly due to the extremely large specific surface area which is exposed on dispersion in water.

In addition to their ability to absorb inorganic cations and water, montmorillonites are capable of taking up organic molecules in interlamellar positions. Large organic cations such as amines and proteins can replace the exchangeable balancing cations. Mono- and, sometimes, di-molecular layers are absorbed in this way depending on the size of cation and charge-deficiency in the layers. Organic liquids or solids may also be adsorbed where the mechanism is not attributable to base-exchange phenomena. Glycerols, resorcinol, catechol and pyrogallol are only a few examples and it was partly to explain the mechanism by which such compounds are held between montmorillonite plates that Edelmann and Favejee postulated their structural representation. Absorptions of this type are diagnostic in that montmorillonites are the only minerals which give a normal 14 Å X-ray spacing increased to about 18 Å by glycerol and some similar compounds.

Dyestuffs are also adsorbed by montmorillonites and, in some cases, this is of value in the microscopic examination of these fine-particled minerals. This topic will be discussed more fully in Chapter V. Fullers' earths, which are usually montmorillonites of one type (commonly *saponite*), are of particular importance in the paint industry, as a direct result of their ability to absorb colouring agents.

The unusual capacity of montmorillonites for absorbing various components from aqueous media has many interesting aspects in commercial developments. There is none more so than the ability of certain montmorillonites to absorb organic amines and similar derivatives and, in so doing, to change their entire physico-chemical nature.

[1] Schofield, R. K., *Nature*, 160, 408, 1947.
[2] Norrish, K., *Trans. Farad Soc.*, 18, 120, 1954.

Bentones is the commercial name given to this class of treated montmorillonite-type minerals. Wyoming bentonite was the prototype source, hence the etymological derivation of the name. In these materials, the clay minerals have absorbed amines from solution to such an extent that the product has acquired a hydrophobic character, i.e. it is water repellent. As such, bentones have an affinity for organic non-polar species such as oils and greases which they absorb freely and often preferentially.

The whole field of organic derivatives of montmorillonites offers exciting possibilities. By their means, the extraction of essential oils from natural products is easier, more efficient and cheaper than distillation procedures which were hitherto necessary. Bentones are finding increasing applications in a wide range of industries ranging from industrial effluent disposal, oil refining, preparation of foodstuffs, to perfume extraction.

Mica Group. *Micas* may be regarded as naturally following the montmorillonites as their structure (Fig. III.20) is also based on those of pyrophyllite or talc. The layers are not electrically balanced and require the presence of extra cations in interlamellar positions.

There is one important difference, however, which results in a fundamental change in the physical character. The layers in mica have a greater deficiency of charge and the balancing cations are held more strongly; in fact, there is ionic bonding, linking successive sheets through these additional cations. Unlike montmorillonite type minerals the layers of micas cannot be expanded by water, nor can the cations between them be exchanged or removed in any way other than by decomposition of the mineral.

Micas are primary igneous minerals; they crystallise in large platelike masses which are easily cleaved along their basal section. They were the first layer-lattice silicates to be examined by X-rays and their structure, as defined by Pauling,[1] has formed the basis for other minerals of similar layer construction.

The essential difference between montmorillonite and the mica minerals is that whereas the former has its charge-deficiency in the central octahedral layer, the deficiency in micas is entirely in the outer tetrahedral networks. From this it is to be expected that the cations in micas would be held more firmly than those in the montmorillonite minerals. The average net residual negative charge on a unit of montmorillonite is 0·33 whilst in mica it is about 1·00.

The structural formula of Muscovite mica can be expressed as

$$\begin{array}{c} K_{1\cdot 00} \\ \uparrow \\ Al_2\,(Al_1\,Si_3)\,O_{10}\,(OH)_2 \end{array}$$

Potassium cations in muscovite are symmetrically placed over each hexagonal ring in the external sheets of the layers. In this way they are linked to six oxygen atoms of one sheet and six in the adjacent layer, making twelve in all. The interlayer distance is about 9·8 Å.

Micas are also of complex chemical composition being mainly hydrated

[1] Loc. cit., p. 125.

alumino-silicates but always containing notable proportions of alkalies, alkaline earths, iron and, sometimes, fluorine.

The structural formula of a particular mica may be calculated from the chemical analysis as for montmorillonites. Within the mica group there is a further subdivision into a dioctahedral and trioctahedral series based on pyrophyllite and talc respectively.

Muscovite mica is a true dioctahedral mica and has the chemical composition $K_2O.3Al_2O_3.6SiO_2.2H_2O$.

One silicon atom in every four in the outermost tetrahedral positions of each layer is replaced by an aluminium atom, thereby causing a unit deficiency of charge. This is not a general rule, however, for isomorphous replacements in the octahedral layer may also cause at least part of the net negative charge.

There are three important groups of micas:

(*a*) A dioctahedral series when trivalent cations occupy the octahedral positions and the structure is based on pyrophyllite;

(*b*) A trioctahedral series, with trivalent and monovalent cations (Lithium Li^+) in octahedral positions;

(*c*) A normal trioctahedral series with divalent cations arranged as in talc.

Examples of each series are summarised in Table III.VIII.[1]

TABLE III—VIII. THE MICA SERIES

Series	*Mineral*	*Composition**
Dioctahedral		
Muscovite system	Muscovite	$K_2(Al_4)(Al_2Si_6)O_{20}(OH)_4$
	Phengite	$K_2[(Mg^{2+}, Fe^{2+})Al_3](Al_1Si_7)O_{20}(OH)_4$
	Paragonite	$Na_2(Al_4)(Al_2Si_6)O_{20}(OH)_4$
Trioctahedral		
Lepidolite system	Lepidolite	$K_2(Li_3{}^+Al_3{}^{3+})(Al_2Si_6)O_{20}(OH, F)_4$
	Polylithionite	$K_2(Li_3{}^+Al_3{}^{3+})Si_8O_{20}(OH, F)_4$
	Protolithionite	$K_2(Li^+, Fe_4{}^{2+}Al^{3+})(Al_2Si_6)O_{20}(OH,F)_4$
	Zinnwaldite	$K_2(Li_2{}^+Fe_2{}^{2+}Al_2{}^{3+})(Al_2Si_6)O_{20}(OH,F)_4$
	Phlogopite	$K_2(Mg_6)(Al_2Si_6)O_{20}(OH)_4$
Biotite system	Biotite	$K_2[(Mg^{2+},Fe^{2+})_6](Al_2Si_6)O_{20}(OH)_4$
	Annite	$K_2(Fe^{2+})_6(Al_2Si_6)O_{20}(OH)_4$

* These compositions are ideal, as natural micas are complex in composition. Extensive solid solution occurs, but there is a sharp contrast between the dioctahedral and trioctahedral series.

Winchell[2] has termed the two series *heptaphyllite* and *octaphyllite* on the basis of the number of cations per structural unit; thus *muscovite*, $K\,Al_2(Al.Si_3)\,O_{10}\,(OH)_2$, has seven cations, whilst *phlogopite*, $K\,Mg_3(Al.Si_3)O_{10}(OH)_2$, has eight. This

[1] In this table the structural formulæ have been doubled to avoid fractional representation.

[2] Winchell, A. N., *Amer. Min.*, 13, 267, 1927.

terminology is common in mineralogical text-books but is not of such general application in other mineral groups.

The minerals in Table III.VIII are called the *elastic* or true micas, but there is another type known as the *brittle* micas which are of similar composition. The distinguishing feature of the brittle type is that calcium or magnesium ions are always present as 'linking' or 'bridging' ions. The charge-deficiency lies mainly in the tetrahedral layer but it is usually greater than in true mica.

Three minerals *margarite*, *chloritoid* and *clintonite* are thought to belong to this group, but their structural relationships are still in doubt. *Margarite* is considered to be $CaAl_2(Al_2Si_2)O_{10}(OH)_2$; with every alternate silicon in the tetrahedral layer replaced by aluminium. There is extensive isomorphous replacement and solid solution.

The stacking of the successive layers in micas is of great complexity. Each cation linking successive sheets is in twelve-fold co-ordination with oxygens, six of which are part of the lower sheet of one layer and the other six are part of the upper sheet of the layer-unit below. There is a multiplicity of ways in which this arrangement can come about and, consequently, the polymorphism in micas is highly complex. This subject has been carefully analysed by Smith and Yoder[1] who have shown that the unit-cell dimensions of many micas are exceptionally large along the *c*-dimension.

Most of the mica minerals are monoclinic, as a result of the non-symmetrical arrangement of the two silica sheets in each layer, in order to accommodate potassium ions in twelve-fold co-ordination. The structural constants of muscovite are[2]

Muscovite $KAl_2(Al.Si_3)O_{10}(OH)_2$ System: Monoclinic

Space Group: C2/c	Axes:	Angle:
Z = 4 (normal)	a = 5·19 Å	
	b = 9·00 Å	β = 95° 11′
	c = 20·10 Å (normal equivalent to two sheets)	

The *elastic* or *true micas* have predominantly potassium cations linking successive sheets. Although the sodium mica, *paragonite*, has been described, Winchel considers that there is no conclusive evidence that such a mineral exists. Undoubtedly, micas which contain a high proportion of sodium are rare and this may be related to the spacial configuration within this mineral group. As sodium ions are smaller than potassium ions, the former may not be held quite as tightly between layers of mica units. If so, sodium micas would presumably not form so readily and would be more easily decomposed.

Hydrous Micas and Illites. In addition to the true micas of primary formations there is another type which is of secondary origin. Many micas are known in sedimentary clays which do not correspond in composition to the true micas. These secondary minerals consist of grains of variable size ranging from readily visible crystallites to particles of the same dimensions as the clay minerals—a range of 4 mm. or more to less than 0·005 mm.

[1] Smith, J. V. and Yoder, H. S., *Min. Mag.*, 31, 209, 1956.
[2] Radoslovich, E. W., *Acta Cryst.*, 13, 919, 1960.

These secondary micaceous minerals are of common occurrence in fireclays and some other clays and are the principal alkali-bearing constituents in them. It seems certain that many of these micas are secondary products and not formed by crystallisation from magmas. The larger-grained samples are most likely of *autigenic* origin, i.e. they are the result of a breakdown of true primary mica by weathering or chemical agencies. On the other hand, some of the finely-grained minerals may be *digenic*, i.e. synthesised from colloidal constituents by natural processes.

These secondary micas, though of general occurrence, are known by a variety of names and no standard terminology has yet been accepted.

The first specific name proposed for a mineral of the micaceous type, having only half the normal content of potassium, was *bravaisite*. This mineral has, however, been proved to be a mixture of montmorillonite and a micaceous mineral. The name bravaisite is, however, still retained by some authorities.

In 1937, Grim[1] proposed the general name *illite* for a group of mica-like minerals with less alkali ions than muscovite, of small particle size, having a 10 Å spacing and showing no expanding lattice characteristics. Maegdefrau and Hofmann[2] have termed similar minerals *glimmerartigen tonmineralen* or more simply *glimmerton*. Other mineralogists have adopted the name *hydrous mica* for larger-sized fragments which are common in sedimentary deposits and this is probably the best and most descriptive term for the whole group. These micaceous minerals are characterised by having a larger water-content together with a smaller amount of alkali than true micas. Another name which is in general use for this class of minerals is *sericite*.

In addition to this group-terminology, specific minerals have been described and named, e.g. *brammalite* which is a sodium-bearing hydrous mica.

Although the mineral group of hydrous micas is widely recognised, the structural relationships have not been fully elucidated. Finely-grained samples are invariably contaminated with impurities which are difficult to remove, and this makes the true composition of the mineral uncertain. Larger-sized flakes have been isolated[3] in what appears to be a fairly pure state, but cracks and fissures may contain impurities which cannot be removed.

Typical chemical analyses of hydrous micas and illites are given in Table III.IX: an analysis of true muscovite mica (No. 7) is included for comparison.

In general, the secondary micas contain less alkali and more water than true muscovite. All the minerals in the table are of the dioctahedral series, but it is probable that a trioctahedral series exists, based on biotite.

The mineral *glauconite*, which is a common constituent of marine clays, may also be classified as a hydrous mica or illite, with more iron than is usual.

There are large variations in the compositions of minerals in the hydrous mica or illite group. A wide range of minerals is probable with gradually decreasing amounts of alkali ions and a corresponding increase in water-content. In addition, as the alkali content is lowered, there is a tendency for the silica: alumina ratio to increase beyond 2:1, which corresponds to ideal muscovite. These facts have led

[1] Grim, R. E., Bray, R. M. and Bradley, W. F., *Amer. Min.*, 22, 813, 1937.
[2] Maegdefrau, E. and Hofmann, U., *Sprechsaal*, 74, 369, 1941.
[3] Carr, K., Grimshaw, R. W. and Roberts, A. L., *Min. Mag.*, 30, 139, 1953.

TABLE III—IX. CHEMICAL COMPOSITIONS OF HYDROUS MICAS

	(1)	(2)	(3)	(4)	(5)	(6)	(7)
SiO_2	46·75	50·30	47·55	50·10	46·54	45·65	44·85
TiO_2	1·15		0·64	0·50	0·17	1·95	—
Al_2O_3	32·43	32·80	32·45	25·12	36·37	25·05	36·95
Fe_2O_3	2·98	0·00	0·76	5·12	0·72	5·41	1·54
FeO	n.d.	0·00	1·85	1·52	0·36	n.d.	1·36
MgO	1·00	1·95	1·70	3·93	0·50	3·40	0·00
CaO	1·04	0·55	0·06	0·35	0·22	1·59	0·00
Na_2O	0·94	0·52	1·05	0·05	0·46	0·51	1·78
K_2O	5·72	6·72	6·22	6·93	8·06	6·50	9·13
H_2O^-	0·04	—	—	1·90	0·52	0·60	0·00
H_2O^+	8·01	6·98	7·73	7·18	6·31	8·53	4·08
Total	100·06	99·82	100·01	102·70	100·23	99·19	99·69
Molecular ratio of SiO_2 to R_2O_3	2·32	2·66	2·45	3·00	2·14	2·72	1·91
Molecular proportion of alkalies	0·76	0·79	0·81	0·74	0·93	0·77	1·25

(1) Purified hydrous mica from a Yorkshire fireclay.[1]
(2) Hydrous mica from Sarospatak (Maegdefrau-Hofmann).[2]
(3) Illite from South Wales shales (Nagelschmidt).[3]
(4) Illite from Fithian Shale, Illinois (Grim).[4]
(5) Hydromuscovite (Brammall, Leech and Bannister).[5]
(6) Illite (Bates).[6]
(7) Muscovite (Jakob).[7]

Grim[8] to suggest that a continuous series of minerals exists with muscovite as one end-member and montmorillonite as the other. The change might be visualised by comparing the two end-members:

$$\underset{\text{muscovite}}{K\,Al_2(Al_1Si_3)O_{10}(OH)_2} \rightarrow \underset{\text{pyrophyllite or ideal montmorillonite}}{Al_2(Si_4)O_{10}(OH)_2x.H_2O}$$

This suggestion is supported to some extent by the fact that many secondary micas occur in association with montmorillonites, but it does not account for all the facts; e.g. the increase in the water-content of these micaceous minerals is not in keeping with a gradual sequence from muscovite to montmorillonite, which both have about the same amount of combined water.

[1] Carr, K., Grimshaw, R. W. and Roberts, A. L., *Min. Mag.*, 30, 139, 1953.
[2] Maegdefrau, E. and Hofmann, U., *Sprechsaal*, 74, 369, 1941.
[3] Nagelschmidt, G. and Hicks, D., *Min. Mag.*, 26, 297, 1943.
[4] Grim, R. E., Bray, R. M. and Bradley, W. F., *Amer. Min.*, 22, 813, 1937.
[5] Brammall, A., Leech, J. G. and Bannister, F. A., *Min. Mag.*, 24, 507, 1937.
[6] Bates, T. F., *Amer. Min.*, 32, 625, 1947.
[7] Jakob, J., *Zeit. Krist.*, 62, 443, 1925.
[8] Grim, R. E., *Clay Mineralogy* (McGraw-Hill, N.Y., 1968).

Norrish and Brown[1] have suggested that hydroxonium ions (H_3O^+) replace potassium ions between the layers of muscovite and they have deduced structural formulæ on this basis. This is a not unreasonable supposition, for these two cations are of comparable sizes and such substitution is known in other mineralogical systems.

Grimshaw, Sadler and Roberts[2] have examined a wide range of so-called hydrous micas from British fireclays. They showed that these represented stages in the alteration of true muscovite mica to livesite. This alteration commenced on the outside of crystallites and progressed more rapidly between the layers. In most single crystals this sequential breakdown could be detected and followed by X-ray diffraction patterns. Differential thermal analysis tests indicated the presence of an intermediate, transitory form of mineral but its nature could not be established. They suggest that this type of hydrous mica is not a true identity but an intimate association of mica and livesite.

Chlorites. *Chlorites* form a class of platey minerals, occurring in many sedimentary deposits, which are similar to mica in appearance but much softer and of quite different compositions. Essentially they are hydrated silicates of magnesium and aluminium with a proportion of iron in either the ferrous or ferric form. The crystals, which are usually of small grain size, are greenish in colour whence the name *chlorites*. Some mineralogists term them '*flexible*' *micas*; they are extremely variable in composition.

Winchell considers that chlorites may be expressed as solid solutions of four end members which are:

$$\begin{array}{ccc} \text{Ferroantigorite } (H_4Fe_3Si_2O_9) & \longleftrightarrow & \text{Daphnite } (H_4Fe_2Al_2SiO_9) \\ \updownarrow & & \updownarrow \\ \text{Antigorite } (H_4Mg_3Si_2O_9) & \longleftrightarrow & \text{Amesite } (H_4Mg_2Al_2SiO_9) \end{array}$$

Mauguin,[3] McMurchy[4] and Brindley *et al.*[5] have shown that some chloritic minerals have a characteristic structure which is not shared by antigorite and amesite (i.e. single-layer silicates). Their structure is based on that of *talc* but with a layer of *brucite*, $Mg(OH)_2$, interposed between each talc layer. Isomorphous replacements are present in both layers; e.g. aluminium replaces silicon in the tetrahedral layers and aluminium may also be substituted for magnesium or ferrous ions in the octahedral positions. Ferric iron (Fe^{3+}) and chromium (Cr^{3+}) may also be present. As a result of such replacements the talc layers acquire a net *negative* charge (because of the substitution of Al^{3+} for Si^{4+}), which is balanced by a residual *positive* charge on the brucite layers (where Al^{3+} replaces Mg^{2+} ions). It is therefore probable that the bonding between all the layers is partly ionic in character.

The *stacking of layers* in the *c*-dimension is complex and many units may be

[1] Norrish, K. and Brown, G., *Min. Mag.*, 29, 929, 1952.
[2] Grimshaw, R. W., Sadler, A. G. and Roberts, A. L., *Clay Min. Bull.*, 5, 110, 1962.
[3] Mauguin, C. H., *C.R. Acad. Sci.*, 186, 1852, 1928.
[4] McMurchy, R. C., *Zeit. Krist.*, 88, 420, 1934.
[5] Brindley, G. W., Oughton, B. M. and Robinson, K., *Acta Cryst.*, 3, 408, 1950.

involved.[1] The separation between equivalent sheets of atoms in successive planes in this direction is 14·2 Å. The structural constants of a typical chlorite, *penninite*, are

Penninite—approx. $Al_2Mg_5Si_3O_{10}(OH)_8$	System: Monoclinic
Space Group: C2/m	Axes / Angle:
$Z = n \times 2$	a = 5·32 Å
	b = 9·21 Å $\beta = 97°$
	$c = n \times 14{\cdot}2$ Å

This is one of the simplest forms; other minerals show lower symmetry. Penninite is the most abundant true chlorite and the arrangement of atoms in the various layers is shown in Fig. III.20. If the chemical formula is represented as follows the structure may be more clearly appreciated:

$$\underbrace{Mg_2(Al,Fe^{3+})(OH)_6}_{\text{brucite layer}} \cdot \underbrace{Mg_3(Al_1Si_3)O_{10}(OH)_2}_{\text{talc layer}}$$

Other chlorites with a similar structure and differing only in the cations present include *clinochlore*, *corundophilite* and *prochlorite*, as well as some forms of *chamosite*.

Most chlorites are of the trioctahedral series and are based on talc; there are chlorites with a structure related to pyrophyllite.[2]

Mineralogists recognise a series of minerals from *antigorite* (see p. 140) with a single-layer trioctahedral kaolin-like structure, to *penninite*, the true chlorite. Antigorite has virtually no replacements of aluminium for silicon in tetrahedral co-ordination or for magnesium in octahedral positions. It cannot form a chlorite type structure because the hypothetical layers of talc and brucite would be electrically balanced and no ionic linking could result. Thermodynamically, the single-layer structure is probably more stable and therefore antigorite has this configuration.

Vermiculites. Another distinct group of minerals which form large mica-like flakes are the *vermiculites*. When heated, these minerals *exfoliate* in an amazing fashion with an increase in volume up to 100-fold. As a result they are valuable in ceramics for thermal insulation. These minerals derive their name from their peculiar property after the Latin *vermiculare*—to breed worms.

Vermiculites are mainly hydrated magnesium silicates although considerable amounts of aluminium and iron may be present. Gruner[3] derived the average molecular composition of vermiculite as

$$22MgO.5Al_2O_3.Fe_2O_3.22SiO_2.4OH_2O$$

although undoubtedly some of the magnesium is present in the form of an exchangeable base.

The structure may be compared with that of chlorite (Fig. III.20); the resemblance is shown in the structural formulæ:

Chlorite	Mg_3	(Al,Si_3)	$O_{10}(OH)_2$.	$[Mg(Al,Fe^{3+})]_2$	$(OH)_6$
Vermiculite	$(Mg,Fe^{2+})_3$	$(Al,Si)_4$	$O_{10}(OH)_2$.	$x[Mg + H_2O]$	$x(OH)$

[1] Brindley, G. W., *X-ray Identification and Crystal Structure of Clay Minerals*, Min. Soc. G.B., 1961.
[2] Sudo, T., Takahashi, H. and Matsui, H., *Zeit. Krist.*, 109, 379, 1957.
[3] Gruner, J. W., *Amer. Min.*, 24, 428, 1939.

In vermiculites, hydroxyl units or water molecules take the place of the brucite layers of chlorites and there is extensive hydration. The analogy between the two structures is complete when the exchangeable magnesium cations, which are fully hydrated,[1] are included between the electrically-unbalanced talc-like layers.

Normally, the basal spacing of vermiculite is 14·2 Å, but on heating to moderately high temperatures the lattice collapses to 9·26 Å. This takes place in two stages; an intermediate spacing of 11·8 Å results after dehydration at 110° C.

Mathieson and Walker[2] have suggested that those magnesium ions which are loosely held between the talc layers are hydrated normally in octahedral co-ordination; they lose part of their water at 110° C. and then finally collapse after the removal of all the water. They give the structural constants as:

$$\underset{\uparrow}{\overset{Mg_{0\cdot 33}}{}}$$

Vermiculite $(Mg,Fe)_3(AlSi)_4O_9(OH)_3 3{\cdot}5H_2O$ System: Monoclinic

Space Group: Cc

Z = 4

Axes:	Angle:
a = 5·33 Å	
b = 9·18 Å	β = 97°
c = 28·85 Å	

The *exfoliation* in vermiculite is apparently of two types, because it can be produced by two entirely different agencies.

(i) Rapid heating is one cause and presumably the explosive release of water molecules serves to push the layers of mineral apart. Slow calcination with gradual increase in temperature does not cause exfoliation and there is little expansion when the vermiculite particles are of small size.

(ii) Some vermiculites exfoliate on treatment with oxidising agents, such as hydrogen peroxide. Drosdoff and Miles[3] suggest that extraneous material may be occluded between the plates which may decompose the peroxide and liberate oxygen. Such compounds include manganese dioxide, carbonaceous matter and various oxides.

Vermiculites are often confused with *hydrobiotites* which are really hydrous micas. Hydrobiotites always have associated alkalies and have a lower basal spacing.

Mixed-layer Structures

Many layer-silicate minerals which occur in nature are composed of mixed layers of more than one type.[4] Some are known which consist of units of mica, chlorite and vermiculite; in others, hydrous mica and montmorillonite layers may be included.[5] When mixed-layer structures occur, the interpretation of X-ray diffraction data is difficult, especially when the arrangement of layers is complex.

[1] Walker, G. F. and Milne, A. A., *Clay. Min. Bull.*, 1, 171, 1951.
[2] Mathieson, A. McL. and Walker, G. F., *Amer. Min.*, 39, 231, 1954.
[3] Drosdoff, M. and Miles, E. F., *Soil Sci.*, 46, 391, 1938.
[4] Weaver, C. E., *Amer. Min.*, 41, 202, 1956.
[5] Brindley, G. W. and Gillery, F. H., *Nat. Acad. Sci. Pub.* 327, 349, 1954.

The Synthesis of Layer Lattice Mineral Groups

With the present-day refinements and modern developments in high pressure/temperature techniques, there has been intensive research into the synthesis of and the transformations between various clay mineral types. Such work has added substantially to the knowledge of the origins and structures of this class of minerals although there is much conflicting evidence.

One of the main problems is that the reactions which occur are very much dependent on experimental conditions which are difficult to control, such as the presence of trace impurities, physical state of the reactants, maturing time. Furthermore, the products are invariably fine-grained and hence difficult to analyse with precision, and more than one mineral type may form.

Classical work by Noll,[1] Roy and Osborn,[2] and Brindley and De Kimpe[3] has shown that there is a consistent pattern in the Al_2O_3—SiO_2—H_2O system. Working mainly with mixed gels, these authors proved that at high temperatures and pressures, in excess of 400° C. and 300 atmospheres pressure, pyrophyllite was the most likely phase to develop, sometimes in association with forms of alumina or aluminium hydroxides. This conclusion was valid over a wide range of Al_2O_3:SiO_2 ratios. Below the temperatures and pressures quoted above, a kaolin mineral, probably halloysite,[2] formed, although at a slower rate.

The introduction of alkali or alkaline earth components into the reaction system produces bewildering results in fine detail, although a regular pattern does seem to be emerging. Montmorillonite minerals usually form if the reaction conditions are maintained alkaline and if high pressures and temperatures are involved,[4] whilst acid conditions seem to favour kaolin minerals or pyrophyllite.[1]

However under specific conditions of cation additions, changes in the alumina/silica ratio and variations in reacting temperature and pressure, felspars, talc,[5] micas,[6] chlorites[7] and vermiculites[8] can all be synthesised.

In addition to experiments at high temperatures and pressures, syntheses under more normal conditions have been reported, although higher reactant concentrations are required and the establishment of equilibrium is slow. The products resulting from such syntheses are complex, but in general follow the pattern of high temperature work; acid or neutral conditions result in kaolins or pyrophyllite;[9] alkaline conditions favour montmorillonites,[5] although zeolites can also develop.[10]

It is to be expected that much research has been undertaken to effect laboratory breakdown of primary minerals which are thought to be the parent materials of clay minerals. Felspars have been extensively investigated and their reactions at high temperatures and pressures and with different added chemical components were

[1] Noll, W., *Min. Pet. Mitt.*, 48, 210, 1936.
[2] Roy, R. and Osborn, E. F., *Amer. Min.*, 39, 853, 1954.
[3] Brindley, G. W. and De Kimpe, C., *Nature*, 190, 254, 1961.
[4] Ames, L. L. and Sand, L. B., *Amer. Min.*, 43, 641, 1958.
[5] Stresse, H. and Hofmann, U., *Z. anorg. Allg. Chem.*, 247, 65, 1941.
[6] Yoder, H. S., *Proc. 6th Nat. Clay. Conf.* (Pergamon, N.Y., 1959), p. 42.
[7] Nelson, B. W. and Roy, R., *Amer. Min.*, 43, 707, 1958.
[8] Roy, R. and Romo, L. A., *J. Geol.*, 65, 603, 1957.
[9] Hénin, S., *Nat. Acad. Sci. Pub.* 456, 54, 1956.
[10] Barrer, R. M., *Chemistry in Britain*, 2, 380, 1966.

reported by Gruner[1] in 1944. Kaolin and pyrophyllite were the main phases to form and their respective appearances follow generally the pressure/temperature stabilities of synthesis from gels. Muscovite mica also formed under certain conditions. More recent work has confirmed Gruner's findings but it would appear that several intermediate phases, at present not established, may be important to the course of the reactions.

Transformations between the Layer Lattice Mineral Groups

Transformations within clay mineral groups are known to occur, but whether these are direct changes or whether they occur through a solution or gel phase is a matter for speculation. Caillère, Hénin and Meriaux[2] and Aleshin[3] have shown independently that certain montmorillonites when treated with potassium-containing solutions at 110° C. became 'fixed', developing a micaceous form. Treatment of the same mineral with magnesium or aluminium ions gave a chlorite.[4,5]

When acid conditions are employed, it may be concluded tentatively that most clay minerals would alter to a kaolin form, provided reaction conditions were sufficiently prolonged.

The conditions for the formation of the main clay mineral types are known with reasonable certainty and inferences can be made as to the changes which take place under varied conditions. As an example, the degradation of muscovite is thought to proceed through a hydrous mica or illite to a kaolin type of mineral, possibly always livesite, although the mechanism of the final stage in this transformation is not fully understood. It is easy to visualise the gradual decrease in alkali ion, but the process causing the conversion of a double-layer silicate to a single-layer one, and at the same time bringing about the elimination of charge deficiencies in the external silica network, must be more complex. Yet the weathering of mica appears to proceed so smoothly and easily in nature that one is tempted to speculate whether the mineral configurations are so rigid and inviolable as structural physicists believe.

Within the kaolin group, the only change which has been confirmed is that between kaolinite and halloysite. In the region of mineral springs, rich in sulphate or chloride ions, kaolinite is known to form halloysite, but whether this is a true transformation or whether the clay mineral is dissolved and then reprecipitated has not been established.

Detailed studies of the transformations between clay minerals have been presented by Correns.[6]

The Identification of Clay Minerals

Although there are many layer silicates, the various groups are well defined and can be recognised from various diagnostic data.

[1] Gruner, J. W., *Econ. Geol.*, 29, 578, 1944.
[2] Caillère, S., Hénin, S. and Meriaux, S., *Compt. Rend.*, 226, 680, 1948.
[3] Aleshin, S. N., *Dokl. Akad. Nauk. S.S.S.R.*, 61, 693, 1948.
[4] Sawhney, B. L., *Nature*, 187, 261, 1960.
[5] Barnhisel, R. I. and Rich, C. I., *Soil Sci. Am. Proc.*, 27, 632, 1963.
[6] Correns, C. W., *Proc. 10th Nat. Clay Conf.* (Pergamon, N.Y., 1963), p. 443.

All the single-layer minerals have a basal spacing of about 7 Å whether they are of the dioctahedral series (like kaolinite) or the trioctahedral series (like antigorite or amesite).

Double-layer minerals may also be distinguished from each other by chemical and physical tests and the main differences may be summarised in Table III—X. Warshaw and Roy[1] have published more comprehensive diagnostic data.

All layer silicates in this section are based on the hexagonal ring structure of silica. This is by far the most common configuration, but minerals are known where the silica layer is of rings with less than six units. *Apophyllite* is an instance where four silica groups are linked into a ring formation extending in two dimensions.

5. FRAMEWORK STRUCTURES

TO complete the possible silicate structures, those must be included where all four of the oxygens tetrahedrally linked to a particular silicon are also shared with other silicon atoms. In these, every oxygen is linked to two silicon atoms and receives one valency share from each. By this means, electrical balance is maintained throughout the resulting three-dimensional framework.

Theoretically, all structures of this type must have the chemical composition SiO_2, but substitution can occur and many important minerals are of this group.

The Silica Minerals

EVEN among the minerals which are composed entirely of silica there is complexity of structure. Silica occurs in many forms in nature; three distinct *crystalline* varieties are recognised, *quartz*, *tridymite* and *cristobalite*. All the three polymorphic forms are capable of existing in more than one modification, with different crystal structures.

The interrelationship between these minerals is of great importance in the manufacture and use of silica and siliceous refractory materials.

The temperature stability ranges of the various modifications of silica may be summarised as:

β-Quartz	870° C. ⇌	$\beta_1 + \beta_2$-tridymite	1470° C. ⇌	β-cristobalite
573° C. ⇅		117° C. ⇅ ⇅ 163° C.		220° C. ⇅ −280° C.
α-quartz		$\alpha_1 + \alpha_2$-tridymite		α-cristobalite

Each form has a limited temperature stability range.

There are two distinct types of transformations in silica minerals:

(*a*) The *conversions*, where quartz changes to tridymite and then to cristobalite. These changes involve a fundamental structural rearrangement; β-quartz is hexagonal whilst β-tridymite is trigonal and β-cristobalite is cubic. Although these

[1] Warshaw, C. M. and Roy, R., *Bull. Geol. Soc.* (*Amer.*), 72, 1455, 1961.

TABLE III—X. DISTINGUISHING FEATURES OF DOUBLE-LAYER SILICATES

	Main Minerals	*Composition*	*Basal Spacing*	*Other Distinguishing Features*
Electrically Balanced Layers	Pyrophyllite Talc	$Al_2Si_4O_{10}(OH)_2$ $Mg_3Si_4O_{10}(OH)_2$	9·25 Å 9·40 Å	
Electrically Unbalanced Layers				
1. Montmorillonite-type Minerals	Montmorillonite	$Na_{0\cdot33}$ ↑ $(Al_{1\cdot67}Mg_{0\cdot33})Si_4O_{10}(OH)_2$	circa 14 Å	The 14 Å spacing may be expanded by treatment with water or glycerol
	Beidellite Nontronite	$Na_{0\cdot33}$ ↑ $(Al_{2\cdot22}(Al_1Si_3)O_{10}(OH)_2$ As above but with Fe^{3+} ions		
	Hectorite	$Na_{0\cdot33}$ ↑ $(Mg_{2\cdot67}Li_{0\cdot33})Si_4O_{10}(OH)_2$		
	Saponite	$Na_{0\cdot33}$ ↑ $Mg_3(Al_{0\cdot33}Si_{3\cdot67})O_{10}(OH)_2$		
2. Micas	Muscovite Lepidolite Biotite	$K\ Al_2(Al_1Si_3)O_{10}(OH)_2$ $K(Al, Li)_3(Al_1Si_3)O_{10}(OH)_2$ $K(Mg, Fe)_3(Al_1Si_3)O_{10}(OH)_2$	9·9–10·1 Å	High alkali content
3. Hydrous Micas		Variable	9·9–10·1 Å	Lower alkali, higher water content than true mica
4. Chlorites	Penninite	$Mg_5(Al, Fe^{3+})(Al, Si)_4O_{10}(OH)_8$	14 Å	No change with water or heat treatment[1]
5. Vermiculite		$(Mg, Fe^{3+})_3(Al, Si)_4O_9(OH)_3 3\cdot5H_2O$	14 Å	Heat treatment reduces basal spacing to 11·8 Å

[1] Some chlorites are thought to have an expanding lattice when treated with glycol.

transformations are reversible, the rate of change is normally exceedingly slow. Pure quartz crystal can be raised to the melting point at 1680° C. and eventually melted completely without any appreciable conversion into cristobalite or tridymite taking place. Similarly, cristobalite, which is theoretically metastable at normal temperatures, would revert to quartz only in a length of time equivalent to a geological age. Some of the changes may, however, be accelerated by catalytic agents.

(*b*) The *inversions*, which are the transitions between the α and β varieties of the minerals, involve only slight structural rearrangements and the changes are instantaneously reversible.

Both conversions and inversions are accompanied by a change in the density of the silica.

The β-modifications of all these minerals are stable at the higher temperatures and these have the most symmetrical structures. As the structure of all the silica minerals is based on linked silica tetrahedra $(SiO_4)^{4-}$, any variations in structure must be the result of the way in which these tetrahedra are combined. If two tetrahedra are linked together through one common oxygen there are three possibilities as shown in Fig. III.22. In the first two diagrams (A) and (B) the Si—O—Si angle is 180° in both cases, but in (A) the top three oxygen atoms are the mirror image of the lower ones, whilst in (B) the upper layer has been rotated through 180° with respect to the lower one. The third alternative (C) is when the units are not arranged linearly and the top unit is inclined at an angle to the lower one.

These three arrangements illustrate the essential differences between the β-modifications of tridymite, cristobalite and quartz. The (A) type of linking, or the *cis*-form as it is called, is the structural scheme of tridymite; the *trans*-form (B) is that of cristobalite; whilst quartz is unit (C) extended in space.

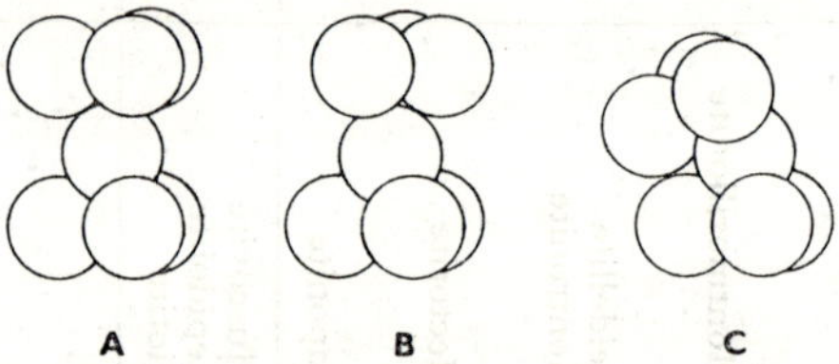

FIG. III.22. THE IDEAL DISPOSITIONS OF ADJACENT SILICA TETRAHEDRA IN THE SILICA MINERALS
(A) Tridymite; (B) cristobalite; and (C) quartz.

The complete structures are shown in Fig. III.23 A, B and C.

It is difficult to illustrate the structure of quartz in a way which brings out all the details. The relative positions of adjacent tetrahedral units produce a spiral formation throughout the crystal. The direction of twist of the spirals determines whether the crystal is left- or right-handed.

The crystal structures of both tridymite and cristobalite are more open than that of quartz; this accounts for the much lower density of these forms of silica.

The conversions of the silica minerals involve a complex rearrangement within the crystal. Bond angles have to change and considerable rotational and transla-

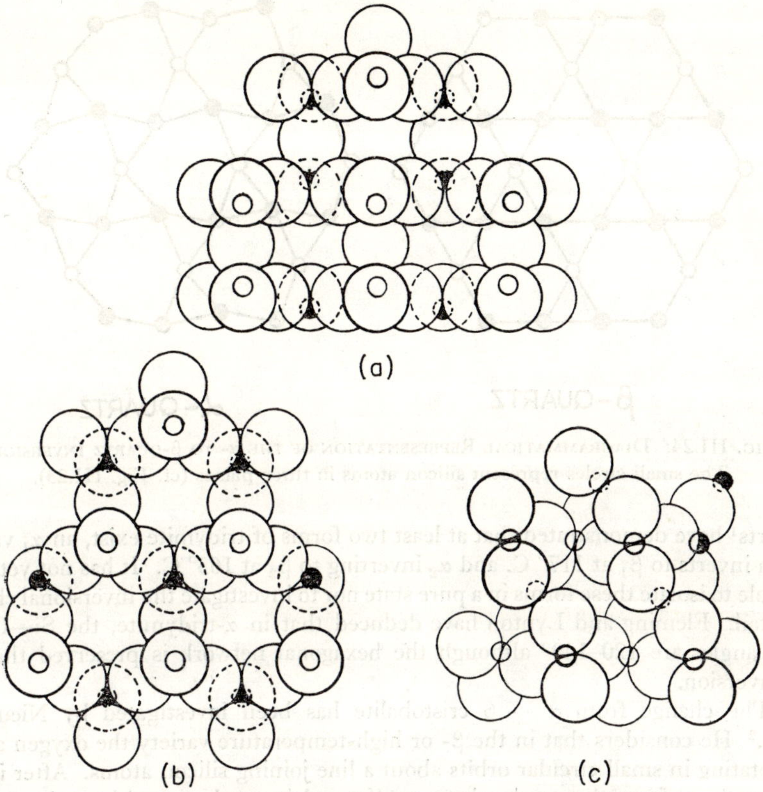

FIG. III.23. THE IDEALISED STRUCTURES OF THE SILICA MINERALS
(a) Tridymite; (b) cristobalite; and (c) quartz.

tional displacements are necessary for the change from one form to another. It is not surprising, therefore, that such changes are sluggish and take place only with difficulty.

The inversional change is much simpler. The β-form of the minerals, which is stable at high temperatures, is symmetrical. When the inversion occurs the crystal is deformed and assumes a lower symmetry. Cristobalite, tridymite and sometimes quartz are formed at high temperatures, therefore the β-modification is produced, but as the crystals cool, contraction occurs and eventually distortion results. Bond angles are altered and the structure becomes more dense. The structural change from α- to β-quartz has been deduced by R. E. Gibbs[1] and the atomic configuration of the two forms is shown in Fig. III.24. It will be appreciated that although the α-modification is a distorted form of β-quartz, the same general atomic arrangement is maintained and there is no bond breaking and no ionic migration.

The inversion of tridymite has not been deduced with certainty and is complicated because it appears to take place in two stages. Grimshaw, Westerman and

[1] Gibbs, R. E., *Proc. Roy. Soc.*, A110, 443, 1926.

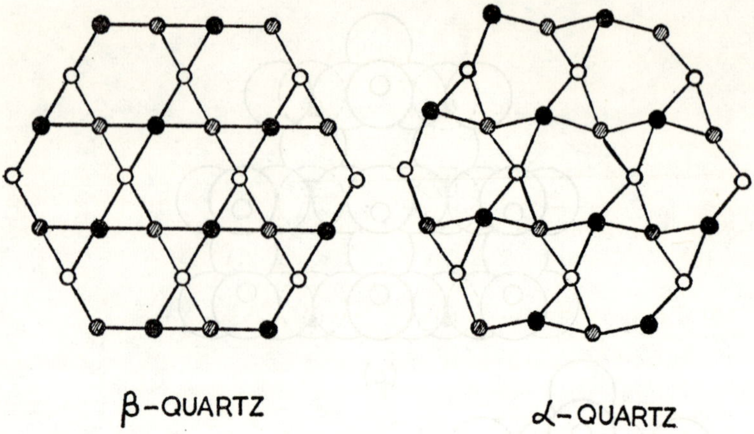

FIG. III.24. DIAGRAMMATICAL REPRESENTATION OF THE α- TO β-QUARTZ INVERSION
The small circles represent silicon atoms in three planes (cf. Fig. III.23).

Roberts[1] have demonstrated that at least two forms of tridymite exist, an α_1 variety which inverts to β_1 at 117° C. and α_2 inverting to β_2 at 163° C. It has not yet been possible to isolate these forms in a pure state nor to investigate the inversional change in detail. Fleming and Lynton have deduced that in α-tridymite, the Si—O—Si bond angles are 120–140° although the hexagonal network is preserved through the inversion.

The change from $\alpha \rightarrow \beta$ cristobalite has been investigated by Nieuwenkamp.[2] He considers that in the β- or high-temperature variety the oxygen atoms are rotating in small circular orbits about a line joining silicon atoms. After inversion to the α-form the oxygen ions are 'frozen' in random positions about these circular orbits, but always maintaining a Si—O—Si bond angle of about 150°. In the high-temperature form of cristobalite, the distance from the centre of one silicon atom, through an oxygen to the centre of the next silicon atom is 3·08 Å, but in the α-form, the separation distance is 3·067 Å.

The structural constants of the silica minerals are shown in Table III.XI.

The Mechanism of the Transformations of Silica Minerals. As explained on p. 170, when silicates are heated, the tetrahedral units of silica $(SiO_4)^{4-}$ appear to dissociate and to be capable of migration. On cooling, they reform around other ions or groups in such a manner as to form the most stable configuration.

When quartz is heated, it is converted slowly into tridymite or cristobalite but the change can be accelerated by the addition of other cations, some of which catalyse the formation of cristobalite (examples are Ca^{2+}, Mg^{2+}, Fe^{2+} and Ba^{2+} ions), whilst others (mainly alkali ions) promote the formation of tridymite. Some other cations (principally Al^{3+} and Ti^{4+}) do not appear to assist the conversion of quartz, indeed, alumina tends to retard it.

[1] Grimshaw, R. W., Westerman, A. and Roberts, A. L., *Trans. Brit. Ceram. Soc.*, 47, 269, 1948.

[2] Nieuwenkamp, W., *Zeit. Krist.*, 96, 454, 1937.

It is probable that when such cations are present, the silica tetrahedra which are dissociated at high temperatures, nucleate around them. The larger the sphere of influence of the cation, the more open will be the silica configuration around it and hence the less dense will be the resulting silica form. Tridymite has the most open lattice of the silica minerals with spaces as large as 2·5 Å, cristobalite has also an open structure but with voids of smaller size, and quartz is much more dense. Cations with the largest range of influence may nucleate tridymite; but as the size is reduced the formation of cristobalite would be favoured and, finally, very small spheres will produce the closest packing of the quartz lattice.

TABLE III—XI. STRUCTURAL CONSTANTS OF SILICA MINERALS

Mineral	*Crystal system*	*Space group*	*Cell dimensions Å* *a*	*c*	*Density*
α-quartz[1]	Trigonal	$P3_221$	4·90	5·39	2·65
β-quartz[2]	Hexagonal	$P6_422$	4·99	5·45	
α-tridymite[3]	Hexagonal	–	5·01	8·18	2·28
β-tridymite[4]	Hexagonal	P62c	5·03	8·22	
α-cristobalite[5]	Tetragonal	$P4_12_12$	4·97	6·92	2·32
β-cristobalite[6]	Cubic	Fd3m?	—	7·14	

The different effects of various cations does not depend entirely on their ionic radii, for although a large ion, like potassium (K^+) with a radius of 1·33 Å, facilitates the formation of tridymite, lithium (Li^+) with a radius of 0·78 Å behaves similarly, but barium (Ba^{2+}) (radius 1·43 Å) catalyses the formation of cristobalite. Dietzel[7] considers that the activity of cations in glasses is related to their field-strength at the separation distance of surrounding anions, i.e. the charge per unit surface area of a sphere equivalent in size to the radius of cation and anion. If cations are regarded as spheres, the field-strength will be proportional to Z/a^2 where Z is the charge on the cation and a is the radius of cation (r_c) plus that of the anion (r_a).

Table III.XII shows that the value of Z/a^2 is related to the ability of the various cations to catalyse the formation of tridymite and cristobalite from quartz. Aluminium and similar cations which do not accelerate the rate of conversion to any great extent may be classified as *quartz-formers*.

Another theory advanced to explain the catalytic conversion of silica is that the oxides form silicate compounds which 'seed' the formation of one of the various forms. This is really only another way of expressing the field-strength, for the cations present govern the configuration of the surrounding silica tetrahedra.

Considerable support for these theories is found in a study of compounds which are isostructural with the silica minerals.

[1] Gibbs, R. E. *Proc. Roy. Soc.*, **A110**, 443, 1926.
[2] Wyckoff, R. W. G., *Amer. J. Sci.*, **11**, 101, 1926.
[3] Gibbs, R. E., *Proc. Roy. Soc.*, **A113**, 351, 1927.
[4] Fleming, J. E. and Lynton, H., *Phys. Chem. Glasses*, **1**, 148, 1960.
[5] Nieuwenkamp, W., *Zeit. Krist.*, 92, 82, 1935.
[6] Smith, J. V., *Carnegie Inst. Wash. Year Book*, **52**, 61, 1951.
[7] Dietzel, A., *Zeit. Elektrochem.*, **48**, 1, 9, 1942.

TABLE III—XII. EFFECTS OF VARIOUS CATIONS IN CATALYSING SILICA

Cation	*Ionic radius*	$a = (rc + ra)$	$\frac{Z}{a^2}$	*Effects*
K^+	1·33 Å	2·73 Å	0·13	Tridymite-forming catalysts
Na^+	0·95	2·35	0·18	
Li^+	0·60	2·00	0·25	Cristobalite-forming catalysts
Ba^{2+}	1·35	2·75	0·25	
Ca^{2+}	0·99	2·39	0·35	
Fe^{2+}	0·76	2·16	0·43	
Mg^{2+}	0·65	2·05	0·47	
Be^{2+}	0·31	1·71	0·68	Quartz-forming catalysts
Al^{3+}	0·50	1·90	0·83	
Ti^{4+}	0·68	2·08	0·92	

Minerals Isostructural with Silica Minerals

Hummel,[1] in 1949, described the structural relationships of *aluminium phosphate*, $AlPO_4$, which exists in three forms resembling quartz, cristobalite and tridymite.

Of more importance to ceramists are the minerals *nephelite*, $KNa_3Al_4Si_4O_{16}$, and *disodium calcium silicate*, Na_2CaSiO_4, which have been described by Buerger and his colleagues.[2]

Nephelite is an important mineral in glaze manufacture and is used extensively in the ceramic industry as a substitute for felspar. It has a crystal form which is similar to β-tridymite, with the difference that every alternate silicon atom (Si^{4+}) is replaced by aluminium (Al^{3+}), thereby causing a unit-charge deficiency which is balanced by the introduction of sodium (Na^+) and potassium (K^+) ions in the ratio of 3 to 1.

The replacement is $Na^+ Al^{3+} \rightleftharpoons Si^{4+}$.

Sodium ions are contained in the vacant sites within the lattice and are held firmly by ionic bonding. The structure does not undergo any change resembling inversion but, at 1248° C., nephelite changes to carnegieite which converts back very slowly to the original form on cooling. *Carnegieite* is cubic at high temperatures and is identical with β-cristobalite except for the isomorphous substitutions.[3] Furthermore, on cooling, its high-temperature form inverts to a distorted form at 690° C. Sodium ions occupy only part of the vacant sites so inversion can take place, through a change in bond angles.

Disodium calcium silicate, Na_2CaSiO_4, is cubic at all temperatures and, like carnegieite, has the same structure as cristobalite. Calcium ions (Ca^{2+}) replace silicon, so there are two unit-charge deficiencies for each replacement; these are balanced by two sodium ions. Sodium ions fill all vacant lattice-sites and so the structure is close-packed and no inversion can take place. Further studies of these and similar isostructural minerals may throw more light on the conversions of silica.

[1] Hummel, F. A., *J. Amer. Ceram. Soc.*, 32, 320, 1949.
[2] Hahn T. and Buerger, M. J., *Zeit. Krist.*, 106, 308, 1955.
[3] Barth, T. F. W. and Posniak, E. *Zeit. Krist.*, 81, 376, 1932.

They do establish that alkali ions can enter the lattice of these minerals and influence the crystal form.

According to Buerger[1] (1954) there are many minerals which have the same structure as the silica minerals, but with isomorphous substitutions. The replacement of a silicon ion by aluminium or a similar-sized cation involves no spacial disarrangement, but the resulting positive charge deficiency has to be made up by the inclusion of other cations in the voids in the lattice. The size of the balancing cation influences the structure of the so-called 'stuffed' derivatives.

If the structure is that of quartz, then the included cations must be small, e.g. Li, Be, or B, to fit the voids; hence, *eucryptite*, $LiAlSiO_4$ and $LiAlSi_2O_6$ are of this type.

When the voids in the structure are larger, as in the tridymite or cristobalite lattice, then larger cations can be incorporated. As would be expected, very large cations can fit only into the tridymite structure and so most potassium derivatives are of this form, e.g. $KAlSiO_4$, $K_2MgSi_3O_8$ and $K_2FeSi_3O_8$. Glasser and Glasser[2] have examined several synthetic minerals containing large alkaline earth cations, such as $BaAl_2O_4$, $SrAl_2O_4$ and $SrGa_2O_4$. These are structurally related to β-tridymite, where two trivalent ions have replaced two quadrivalent silicon ions per unit of structure and the resulting charge deficiency is balanced by the introduction of a divalent cation. They term the structures 'stuffed derivatives of tridymite'. The inclusion of sodium cations is most usual in cristobalite-type structures, some examples of which have already been quoted.

Isomorphous replacement in silica minerals is probably the rule rather than the exception. Keith and Tuttle[3] have reported that the cause of anomalous inversion temperatures in quartz, when prepared under different conditions, might be due to solid solution or replacements. Quartz containing incorporated lithium and aluminium gave a low inversional change, but if germanium (Ge^{4+}) ions were incorporated, the temperature of the change was much increased (see also p. 756).

Other Forms of Silica

In addition to the different crystalline modifications already mentioned, silica exists in nature in other forms; these include *flint*, *chert*, *chalcedony*, *agate*, *opal*, *geyserite* and *silica gel*. Two other crystalline forms, namely coesite and keatite have been synthesised.

Many of these varieties are amorphous or crypto-crystalline in character, but some have a pseudo- or distorted-type of quartz lattice.

Recent work on the grinding of quartz has shown that a distorted layer is formed on the surface and the crystalline form of quartz is destroyed.[4] Most samples of natural silica are sedimentary in origin and, in many instances, mechanical erosion during transportation has distorted the minute crystallites.

[1] Buerger, M. J., *Amer. Min.*, 39, 600, 1954.
[2] Glasser, F. P. and Glasser, L. S. D., *J. Amer. Ceram. Soc.*, 46, 377, 1963.
[3] Keith, M. L. and Tuttle, O. F., *Amer. J. Sci.*, 203, 1952.
[4] Dempster, P. B. and Ritchie, P. D., *Nature* (London), 169, 539, 1952.
Nagelschmidt, G., Gordon, R. L. and Griffin, A. G., *Nature* (London), 169, 539, 1952.

TABLE III—XIII. THE FELSPAR GROUP
(After Taylor)

Crystal habit	*Mineral*	*Composition*	*Space group*	*Crystallographic dimensions*					
				a	*b*	*c*	α	β	γ
Monoclinic	Orthoclase	$KAlSi_3O_8$	C2/m	8·56 Å	12·99 Å	7·19 Å	90°	116°	90°
	Hyalophane	{ $KAlSi_3O_8$ / $BaAl_2Si_2O_8$	,,						
	Celsian	$BaAl_2Si_2O_8$	,,	8·63	13·10	14·41	90°	116°	90°
	Microcline	$KAlSi_3O_8$	,,	8·44	13·00	7·21	90° 65′	115° 50′	87° 70′
Triclinic	Albite (Ab)	$NaAlSi_3O_8$	Cl	8·14	12·86	7·17	94°	116° 30′	89°
	Oligoclase	Ab_6An to Ab_3An	,,						
	Andesine	Ab_3An to Ab An	,,	8·14	12·86	7·17	93° 23′	116° 28′	89° 59′
	Labradorite	AbAn to $AbAn_3$	,,						
	Bytownite	$AbAn_3$ to $AbAn_6$	,,						
	Anorthite (An)	$CaAl_2Si_2O_8$	,,	8·21	12·95	14·16	93° 13′	115° 56′	91° 12′

Opal is a variety of hydrated silica with a composition $SiO_2.nH_2O$. It is the main constituent of kieselguhr or diatomaceous earth.

Coesite, a dense crystalline form of silica, was synthesised by Coes[1] by high-pressure/high-temperature treatment of silica gels.

Keatite is of similar origin but not as dense as coesite; it was synthesised and described by Keat.[2]

Both the above forms represent silica minerals in which the hexagonal ring configuration of the normal forms has been broken and compressed to a more close-packed form by pressure. In coesite, the silica tetrahedra are arranged in four-membered rings[3] whilst in keatite it is suggested that four- or five-membered loops characterise the structure.[4]

Stishovite is another crystalline form of silica, which can be formed only at very high pressures and temperatures. It has been described by Ostrovsky[5] and confirmed by several other workers, but its crystal lattice has not been established.

FELSPARS

THE most common constituents of igneous rock formations are the *felspars*. These important minerals are used extensively in the manufacture of porcelains, stoneware, earthenware and many kinds of engobes and glazes and are common constituents of alumino-silicate slags.

There are two clearly-defined groups: (i) based on *orthoclase*, $KAlSi_3O_8$, (ii) termed the *plagioclase* group, which are solid solutions of *albite*, $NaAlSi_3O_8$, and *anorthite*, $CaAl_2Si_2O_8$; *celsian*, $BaAl_2Si_2O_8$, which is not of common occurrence is similar to orthoclase. There are also the *felspathoids*, such as *leucite*, $KAlSi_2O_6$, and *nephelite*, $NaAlSiO_4$.

Orthoclase and celsian are monoclinic crystals which are soluble in each other in virtually all proportions, yet they have a limited solubility in triclinic plagioclase. The important felspars are classified in Table III.XIII which also gives their crystallographic details.

Orthoclase and albite undergo inversion at high temperature, changing to *sanidine* and *barbierite* respectively.

The structure of all felspars is based on a framework of SiO_2, but one which is more open than that of the silica minerals. There is replacement of silicon by aluminium ions however, which results in charge-deficiencies, balanced by cations of K^+, Na^+, Ba^{2+} or Ca^{2+} entering the lattice thus:

$$4SiO_2 = Si_4O_8 = \underset{\displaystyle K^+Al^{3+}.Si_3O_8}{\overset{\displaystyle Si^{4+}.Si_3O_8}{\updownarrow}} = \underset{\displaystyle Ca^{2+}Al_2^{3+}Si_2O_8}{\overset{\displaystyle Si_2^{4+}.Si_2O_8}{\updownarrow}}$$

[1] Coes, L., *Science* (*N.Y.*), 118, 131, 1953.
[2] Keat, P. P., *Science*, 120, 328, 1954.
[3] Zoltai, T. and Buerger, M. J., *Zeit. Krist.*, 111, 129, 1959.
[4] Zoltai, T. and Buerger, M. J., *Zeit. Krist.*, 114, 1, 1960.
[5] Ostrovsky, I. A., *J. Geol.*, 5, 325, 1967.

Albite, $NaAlSi_3O_8$, is isomorphous with *anorthite*, $CaAl_2Si_2O_8$, therefore Na^+ Si^{4+} must be equivalent to and capable of being substituted by Ca^{2+} Al^{3+}, and so albite and anorthite may be regarded as the end members of a complete series of solid solutions which are termed the plagioclase felspars.

The framework structure of the felspars is based on condensed rings of four tetrahedra as shown in Fig. III.25. A type of chain structure results which is linked to similar formations on all sides, thus forming a three-dimensional framework. The resulting structure contains voids large enough to hold extra cations which electrically balance the lattice.

In orthoclase, one silicon atom in every four is replaced by aluminium, but it is not possible to show by X-ray studies whether such substitution is regular or random.

The structural relationships were established in 1934 by Taylor and his co-workers,[1] and later workers have confirmed their findings. The alkali or alkaline earth ions dictate the type of structure in that they are the units around which the Al/Si framework is constructed. Potassium and barium cations, being of large size, confer a different structural arrangement from the smaller cations of sodium and calcium. So in orthoclase-type felspars the ring structure is larger, with a consequent increase in unit cell dimension and a lower density than in the plagioclase forms.

Claringbull and Bragg have an excellent discourse on the types and structures of felspars in their *Crystal Structures of Minerals* (G. Bell & Sons Ltd., London, 1965).

Felspars change to kaolinite under suitable weathering conditions yet the structural mechanism remains a complete mystery. The two configurations are very different, and if present concepts are correct, then extensive rearrangement would be necessary.

Zeolites

The zeolites are also based on framework structures of $(Si,Al)O_2$ units, but the lattice is more open than that of the felspars, and cavities and wide channels are present.

The framework has a net charge-deficiency which is electrically balanced by large-sized cations contained in the voids in the lattice. Unlike those in the felspar structure, these cations are not held firmly, but can move freely within the more open cavities and channels. Consequently they can be readily exchanged and, in addition, large quantities of water may be contained which can be driven off by heat without destroying the structure. In addition to absorptions of this type, other substances such as ammonia, carbon dioxide, alcohol, iodine and mercury may penetrate along the channels. Zeolites are of three types with cubic, fibrous or lamellar form.

Analcite, $NaAlSi_3O_8.H_2O$, is an example of a cubic zeolite; *natrolite*, $Na_2(Al_2Si_3O_{10})2H_2O$, and *edingtonite*, $Ba(Al_2Si_3O_{10})4H_2O$, are fibrous and *heulandite*, $Ca(Al_2Si_7O_{18})6H_2O$, has a sheet-structure.

The zeolites are commonly employed as water softeners as a result of their

[1] Taylor, W. H., Darbyshire, J. A. and Strunz, H., *Zeit. Krist.*, 87, 464, 1934.

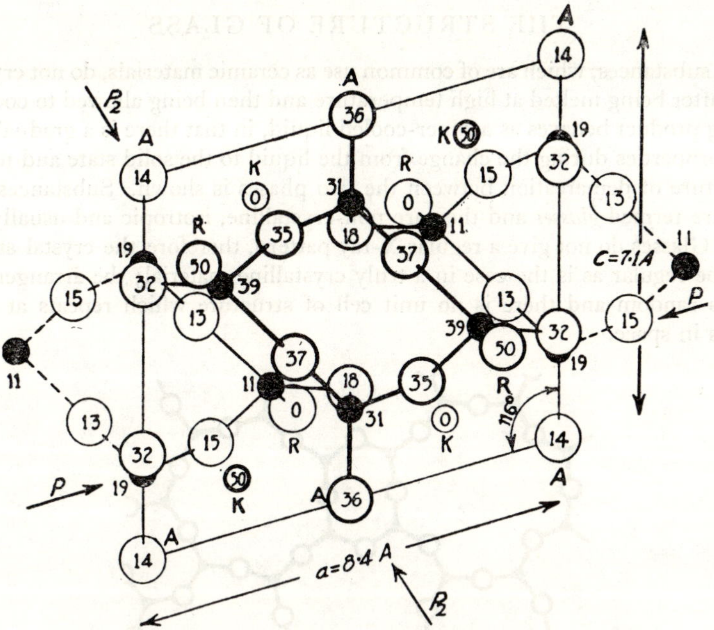

FIG. III.25. THE IDEALISED STRUCTURE OF ORTHOCLASE FELSPAR (after Taylor)

The small numbers refer to the heights of atoms above a zero plane. The chain formation along the *a*-direction is clearly shown.

capacity for cationic exchange. A sodium zeolite will readily adsorb calcium ions from solution thus:

$$2NaAlSi_2O_6 . H_2O + Ca^{2+} \rightleftharpoons Ca^{2+}(AlSi_2O_6)_2 2H_2O + 2Na^+$$

The original zeolite can be regenerated by treatment with a solution of sodium chloride.

In recent years, great interest has been shown in zeolites both from an academic and an industrial standpoint. The interesting feature of these minerals is that the frameworks can be large and the 'gate-ways' or points of entry to the ring configuration can be controlled in size. Zeolites thus have a large internal surface area for absorption, but at the same time the 'gates' permit certain sizes of molecular units to pass through them whilst excluding larger ones. The materials behave as molecular sieves and have important uses in filtration and separation processes.

Zeolites with desirable properties can be synthesised more or less at choice by caustic alkali treatment of silica and alumina gels, or of kaolinite. Some of the important synthetic forms are *faujasite*, *phillipsite*, *Linde Sieve A-type*, each with a characteristic structure and properties.

R. M. Barrer and his co-workers have played a prominent role in developments of this nature.[1]

[1] Barrer, R. M., *Chemistry in Britain*, 2, 380, 1966.

THE STRUCTURE OF GLASS

SOME substances, which are of common use as ceramic materials, do not crystallise readily after being melted at high temperature and then being allowed to cool. The resulting product behaves as a super-cooled liquid, in that there is a gradual alteration in properties during the change from the liquid to the solid state and no sharp temperature of demarcation between the two phases is shown. Substances of this nature are termed *glasses* and they are non-crystalline, isotropic and usually transparent. Glasses do not give a regular X-ray pattern, therefore the crystal structure cannot be regular as is the case in a truly crystalline material; the arrangement of atoms is random and there is no unit cell of structure which repeats at regular intervals in space.

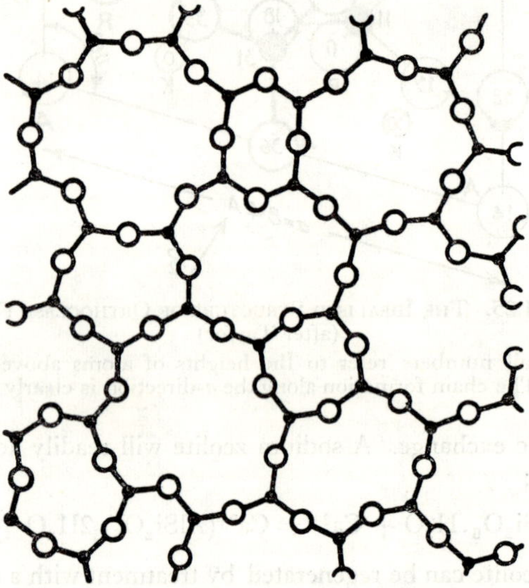

FIG. III.26. THE STRUCTURE OF SILICA GLASS
(after Zachariasen)

In all the crystal structures described in the preceding pages, certain laws must be obeyed in the building-up of units, and such units must 'repeat' at frequent intervals throughout a three-dimensional space. In accordance with modern theories of the structure of glass, the same laws apply to the unit construction, but there is no regular order or arrangement in the building-up of such units into the glass structure.

This may be more clearly appreciated by considering *silica*, which is the most commonly-employed glass-former.

It is well known that, at high temperatures, ordered silicate lattices dissociate, mainly because the individual silica tetrahedra $(SiO_4)^{4-}$ become detached and are free to migrate. On cooling, the units come together, bonds are reformed and usually a symmetrical arrangement is produced before the material becomes completely

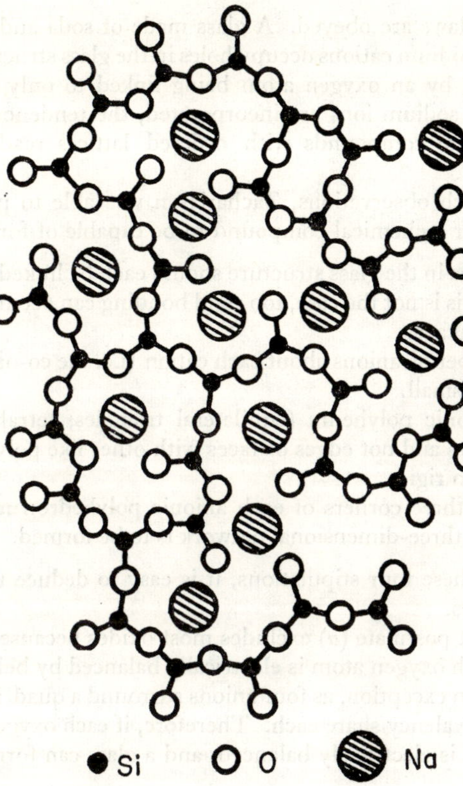

FIG. III.27. THE STRUCTURE OF ALKALI-SILICA GLASS

solid. If the temperature is lowered rapidly, there is a tendency for the liquid to become too viscous before the units can form into an ordered lattice. The resulting solid behaves as a super-cooled liquid and is termed *glass*.

The difference between a symmetrical crystal formation and a randomly arranged lattice of glass is represented diagrammatically in Fig. III.26. The true picture is, of course, an extension of such structures in three dimensions. An important point is that in silica glass, discrete $(SiO_4)^{4-}$ units are present with tetrahedrally co-ordinated groups. These are linked to other similar units through common oxygen atoms so that the structure is electrically balanced. The Si—O distance remains the same as in true crystalline structures, but the Si—O—Si angle is variable and capable of distortion through wide angles.

Zachariasen[1] first suggested this structure for glass in 1932 and his ideas have been subsequently confirmed by Warren[2] and co-workers following careful analytical X-ray studies.

When other cations are added to silica, the glass structure is more complex,

[1] Zachariasen, W., *J. Amer. Chem. Soc.*, 54, 3941, 1932.
[2] Warren, B. E., *J. Applied Physics*, 13, 602, 1942.

but the same general laws are obeyed. A glass made of soda and silica serves as a useful example. The sodium cations occupy holes in the glass structure and electrical balance is maintained by an oxygen atom being linked to only one silicon as in Fig. III.27. As more sodium ions are incorporated, the tendency to form glass is diminished and definite compounds with ordered lattices readily crystallise on cooling.

As a result of such observations, Zachariasen was able to postulate that the conditions required for a chemical compound to be capable of forming a glass are:

(*a*) That the anions in the glass structure should each be linked to not more than two cations. Where this is not the case, too rigid bonding can occur and there can be no distortion of the linkages.

(*b*) That the number of anions about each cation (i.e. the co-ordination number of the cation) must be small.

(*c*) That the anionic polyhedra (equilateral triangles, tetrahedra, octahedra, etc.) must share corners and not edges or faces with other like polyhedra, otherwise the bonding will be too rigid.

(*d*) That at least three corners of each anionic polyhedron must be linked to other similar units if a three-dimensional network is to be formed.

On the basis of these four stipulations, it is easy to deduce those compounds which will form glasses.

Zachariasen's first postulate (*a*) excludes most oxides because very few cations form oxides where each oxygen atom is electrically balanced by being linked to only two cations. Silica is an exception, as four anions surround a quadrivalent silicon ion (Si^{4+}) and receive one valency share each. Therefore, if each oxygen (O^{2-}) is linked to two silicon atoms it is electrically balanced, and a glass can form.

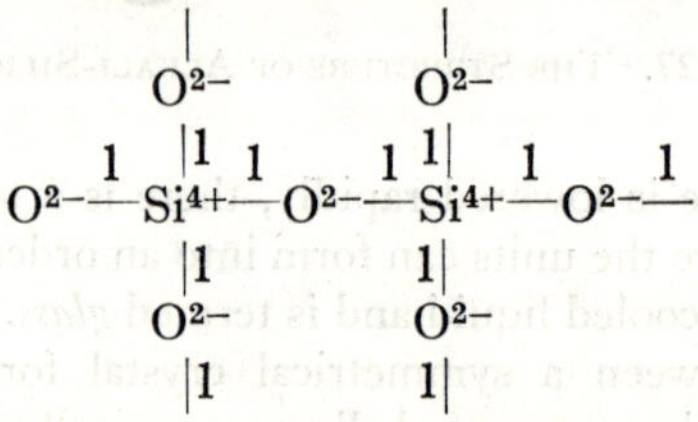

Aluminium (Al^{3+}) may form six-fold (octahedral) or four-fold (tetrahedral) co-ordinated groups with oxygen. In the former case, each oxygen receives only a half-valency share from an aluminium cation and would require to be linked to three others and could not form glass. In the second case (tetrahedral grouping), each oxygen around an aluminium ion receives three-quarters of a valency share. To achieve electrical balance, each oxygen must be linked to an average of 2·66 aluminium ions and under these conditions glass can again, theoretically, not be formed.

The glass-forming capacity of an oxide is thus dependent on the size and charge of the cation. The former determines the co-ordination number and the latter the valency-contribution to each oxygen atom. It may be concluded that:

A *monovalent oxide* (M_2O) cannot form a glass.

A *divalent oxide* (MO) cannot form a glass because the cation would have to be linked to only two oxygens, and a three-dimensional network could not be formed.

A *trivalent oxide* (M_2O_3) can form a glass if the cation is in threefold co-ordination. By reference to Table II.XIII (p. 82) it will be seen that B_2O_3 is the only trivalent oxide which can be a glass-former, although As_2O_3 and Sb_2O_3 may be so in certain circumstances.

A *quadrivalent oxide* (MO_2) is a glass-former if the cation is in three- or four-fold co-ordination, or has a radius less than 0·5 Å. Silica (SiO_2) and germania (GeO_2) fulfil these requirements.

A *pentavalent oxide* (M_2O_5) again cannot have more than tetrahedral co-ordination if it is to be a glass-former, and phosphorus pentoxide (P_2O_5) and arsenic oxide (As_2O_5) are the only possibilities, although Sb_2O_5, Cb_2O_5 and Ta_2O_5 are border-line cases.

Oxides of higher valency cations do not form glasses.

In general the only oxides which can form glasses are those with a co-ordination of three or four and with one of the following formulæ: M_2O_3, MO_2 or M_2O_5. There are five such oxides, *boron oxide* (B_2O_3), *silica* (SiO_2), *germania* (GeO_2), *phosphorus pentoxide* (P_2O_5), and *arsenic oxide* (As_2O_5). *Arsenious oxide* (As_2O_3) and *antimony oxide* (Sb_2O_3) will form glass, but with more difficulty than the others, and P_2O_3, V_2O_5, Sb_2O_5, Cb_2O_5 and Ta_2O_5 are claimed to be glass-formers in exceptional circumstances. Arsenic and antimony are interesting because they can form glasses even when in six-fold co-ordination with oxygen. Apparently three of their oxygens are nearer the cation than the others and the unit behaves almost as though there were three-fold co-ordination.

Some other compounds may form glass, e.g. *beryllium fluoride*. Fluorine atoms are similar in size to those of oxygen, but are monovalent. Although in beryllium fluoride, the beryllium ions (Be^{2+}) are surrounded by four fluorine atoms, they give a half-valency share to each, and therefore each fluorine atom need be linked to only two cations and the Zachariasen laws are obeyed.

Some other cations can take part in the network structure of the true glass-formers, e.g. aluminium can replace silicon ions over a wide range of composition and form alumino-silicate glasses. Titanium, zirconium and tantalum can behave in similar fashion.

Other ions which play a large part in the structure of glass, yet can take no part in the network system, include the alkali metal cations Li^+, Na^+, K^+, Rb^+, and Cs^+ and the alkaline earth cations Ca^{2+}, Sr^{2+}, Ba^{2+}. Beryllium (Ba^{2+}) and magnesium (Mg^{2+}) are of this group, though they may also, to some extent, enter the network system, as does aluminium. These large cations occupy holes in the network and serve to balance valency deficiencies which may arise through isomorphous substitutions in the network, or from unsatisfied oxygen atoms. These cations are important, for they act as nuclei around which the network forms and so they determine the size of the holes in the lattice.

One anomalous feature of glass structures is that lead (Pb) can take part in the network structure of silica or boron oxide. Stanworth[1] has suggested that the reason is attributable to the electron configuration in the outermost electron orbits

[1] Stanworth, J. E., *J. Soc. Glass Tech.*, 32, 154, 1948.

of lead atoms. Table II.III (p. 67) shows that there are (18 + 2) electrons in the penultimate orbit of the valency electrons, and such a unit is capable of easy deformation and forms covalent bonding. Thallium and bismuth behave similarly and impart refractivity to the resulting glass. Zinc and cadmium have an outer completed orbital of 18 and can also form part of the glass network.

Stanworth has also pointed out that the linkages in all glass structures are mainly covalent and he suggests that this covalency is essential before a lattice can be formed.

The cations which take part in oxide glass structures can be divided into three groups:

(i) Those which form the network structures;

(ii) Those which cannot form networks by themselves, but may replace other true glass-forming cations in the lattice;

(iii) Those which occupy vacant holes in the network.

These are summarised in Table III.XIV.

TABLE III—XIV. ROLE OF CATIONS IN GLASS STRUCTURES

Glass lattice-forming cations		*Cations capable of replacing lattice-forming cations in network*		*Other cations capable of entering network*		*Cations which occupy holes in lattice*	
Cation	*Valency*	*Cation*	*Valency*	*Cation*	*Valency*	*Cation*	*Valency*
		Ta	5	Bi	3	Ca	2
P	5	Ti	4	Pb	2	Sr	2
As	5	Zr	4	Tl	1	Ba	2
Si	4	Al	3			Li	1
Ge	4			Cd	2	Na	1
B	3	(Be)	2	Zn	2	K	1
(As)	3	(Mg)	2			Rb	1
(Sb)	3					Cs	1
V	5						
Sb	5					(Be)	2
Cb	5 } ?					(Mg)	2
Ta	5						
P	3						

The concept of glass as presented in the previous pages is over-simplified and scientists consider that the idea of a three-dimensional random distribution of ions and atoms is not completely valid.

The ease of preferential leaching of groupings or elements from certain glasses suggests that localised molecular complexes, probably intergrowing with other species, is a more likely explanation. The interlayering of several glass forms has been suggested by Kumar and Maitra;[1] Sawai[2] has discussed some anomalous leaching characteristics of glasses. The reader is also referred to *Properties of Glass* by G. W. Morey (Reinhold, 1954) and *Modern Aspects of the Vitreous State* by J. D. Mackenzie (Butterworth, 1968).

[1] Kumar, S. and Maitra, A. K., *Trans. Indian Ceram. Soc.*, 27, 69, 1968.
[2] Sawai, I., *Glass Ind.*, 38, 197, 1957.

That the types of glasses which exist in fired alumino-silicates have an imposed sub-structure has been demonstrated by Babcock.[1] It is well known that several properties, such as refractive index, density, thermal expansion and fluidity, can be represented by a mathematical expression of the form

$$P = A\,SiO_2 + B\,Na_2O + C\,CaO + D\,Al_2O_3 + \ldots$$

where P is the property and the various letters A, B, C, D . . . are numerical values to be multiplied by the respective molar concentrations of the oxide components.

It is not easy to apply such calculations to systems of more than two components, but Babcock has developed computer techniques to solve this problem with considerable success.

He has shown that the values of A, B, C . . . of a glass, change significantly when calculated for composition ranges which cross a theoretical solid-state boundary. It may be inferred that the glass has a sub-structure resembling the equilibrium solid-state phases of that composition (see Chapter X).

As examples, Babcock quotes the numerical values derived for oxides in terms of refractive indices and fluidities of glasses of silica with alkali and alkaline earths as in Table III.XV.

TABLE III—XV. OXIDE CONSTANTS IN SILICATE GLASSES

A. Refractive Index Measurements (Na_D line)	*Constants*				
Glass type	*SiO_2*	*Na_2O*	*Al_2O_3*	*CaO*	*Primary solid field*
SiO_2—Na_2O	1·458	1·619	—	—	quartz/tridymite/ cristobalite
	1·477	1·563	—	—	$Na_2Si_2O_5$
	1·483	1·553	—	—	Na_2SiO_3
SiO_2—Na_2O—CaO	1·456	1·634	—	1·782	tridymite/ cristobalite
	1·474	1·566	—	1·782	$Na_2Si_2O_5$
	1·481	1·548	—	1·756	$2Na_2O.CaO.3SiO_2$
	1·472	1·574	—	1·764	$3Na_2O.CaO.6SiO_2$
	1·468	1·562	—	1·781	α $CaSiO_3$
	1·468	1·604	—	1·752	β $CaSiO_3$
SiO_2—Na_2O—Al_2O_3	1·464	1·598	1·563	—	quartz/tridymite
	1·476	1·564	1·544	—	$Na_2Si_2O_5$
	1·483	1·551	1·541	—	Na_2SiO_3
B. Fluidity $\left(\frac{1}{\text{viscosity}}\right)$ Measurements at 1500° C.					
SiO_2—CaO—Al_2O_3	1·196	—	1·261	5·358	$2CaO.Al_2O_3.SiO_2$
	1·089	—	0·247	5·748	α $CaSiO_3$
	−0·088	—	0·265	7·297	$CaO.Al_2O_3.2SiO_2$
	−1·058	—	0·726	9·616	tridymite

Glazes, Vitreous Enamels, Frits and some *Fluxes* resemble glasses in their general structure and properties. They are described in Chapter VI.

[1] Babcock, C. L., *J. Amer. Ceram. Soc.*, 51, 163, 1968.

CHAPTER IV

THE CRYSTAL STRUCTURE OF NON-SILICEOUS CERAMIC MATERIALS

FOR many years, the term *ceramic* was used almost exclusively for clays and clay products, but as the demands increased for furnaces and crucibles which could withstand higher temperatures and more prolonged heating than those made of *fireclay*, more attention was paid to other materials—particularly those of a high refractoriness which could be shaped and fired like clays.

Silica was the first of these non-argillaceous substances to be used, then *plumbago* or *graphite* was mixed with fireclay to increase the heat-resisting property. Various oxides of a highly refractory nature, such as *magnesia*, *lime*, *zirconia*, *titania*, *beryllia*, etc., were introduced, as also were *carbides* of such elements as silicon and boron. In recent years, *nitrides*, *borides* and other unusual compounds of rare elements have been manufactured for highly specialised application where extremely high temperatures are encountered, so that today large quantities of non-clayey materials are used industrially and must be included under the term *ceramic materials*.

The chief property of most of these non-siliceous materials is their great resistance to heat (refractoriness) which enables them to be used at very high temperatures. Other factors may also be involved, such as resistance to molten metals and slags or to their capability of withstanding high thermal or mechanical stresses. New ceramic bodies have been developed where other properties are the main consideration; these include materials of low thermal conductivity or resistivity, those with special electrical properties and those with unusually low expansion characteristics. In many cases, the valuable properties of the materials can be explained from their crystal structures and such studies have often led to improvements in performance.

Various non-siliceous materials are of great importance as opacifying and colouring agents and for the decoration of porcelain, chinaware and some domestic earthenware. Others are used as fluxes in bodies, engobes and glazes.

There are other non-siliceous minerals which, although of little value as ceramic materials, are present as impurities in some otherwise useful materials. These are among the most serious causes of defects in bricks, tiles, pottery, refractory materials and other ceramic articles, and as they are difficult if not impossible to eliminate, only a thorough knowledge of their nature and structure can provide means for reducing their deleterious properties to a minimum.

It is obviously impossible and undesirable to consider in detail all the minerals which are present in unfired and fired ceramic material, as it is a safe presumption that these would include most of the minerals so far discovered. Only those which occur most commonly in important ceramic materials will be considered.

ELEMENTS

THE number of *elements* which are to be found as such in ceramic materials is limited. The only one of importance is carbon, although gold, silver, platinum and some others are used to a limited extent for decoration and as colourisers for glazes.

CARBON

CARBON occurs in the free state in two varieties.

(*a*) **Amorphous.** As a soft, black mass or powder in *lampblack*, *soot*, *retort-scurf*, etc. Very little is known of its fundamental structure.
(*b*) **Crystalline** as **diamond** and **graphite.**

The *chief forms* of carbon as a ceramic material are:

(i) As a *refractory material*, in a graphitic form and used under reducing conditions.

(ii) As an *abrasive*, as diamond which is unequalled in hardness by any other substance.

(iii) As a constituent of the *carbonaceous matter* of many clays and sedimentary deposits.

Diamonds have long been known as crystals of exceptional beauty and hardness and were one of the first minerals to be analysed structurally by X-rays.[1] The tetrahedral arrangement of the bonds linking one carbon atom to its near neighbours was clearly demonstrated as early as 1913. There is a remarkable similarity between the structure of diamond and that of high or β-cristobalite as postulated by Wyckoff (cf. Fig. III.23B). In diamond, carbon atoms occupy the positions held by silicon in cristobalite, and there are, of course, no oxygen atoms. Each carbon atom is linked to four other similar atoms arranged at the corners of a regular tetrahedron. The bonding is wholly covalent as shown by the structure, hardness, brittleness, appearance and electrical resistance. Four of the eight corners of a cubelet in the cubic unit cell are occupied and by placing atoms at the centre of such cubelets every carbon atom can form four tetrahedral covalent bonds (see Fig. IV.1(*a*)).

The structure is based on a face-centred cube, but there are extra carbon atoms conferring cubic holosymmetry. The constants are given by W. H. and W. L. Bragg as

Diamond C	System: Cubic
Space Group (Mauguin): Fd 3m	Axis:
Number of atoms per unit cell = 8	a = 3·56 Å

The carbon to carbon distance in diamond is 1·54 Å.

Graphite is a rather soft greyish material which occurs in nature as *graphite* or *plumbago* and artificially as coke and in some carbon electrodes. The contrast between the softness of graphite—which is used as a lubricant and for *black-leading*—and the hardness of diamond is very great.

[1] Bragg, W. H. and W. L., *Proc. Roy. Soc.*, **A**89, 277, 1913.

The carbon atoms in *graphite* are arranged in sheets and the atoms in each layer are nearly co-planar. The linking in graphite layers may be likened to that in the benzene ring C_6H_6 where the bonds between the carbon atoms are pictured as alternating double and single bonds with six atoms uniting to form a complete hexagonal ring. The modern view is that there is resonance and each link is about 50 per cent double-bond and consequently, in benzene the carbon to carbon distance is between that of C—C (1·54 Å) and that of C = C (1·34 Å), being 1·39 Å. The remaining valency of each carbon atom in benzene is satisfied by a hydrogen.

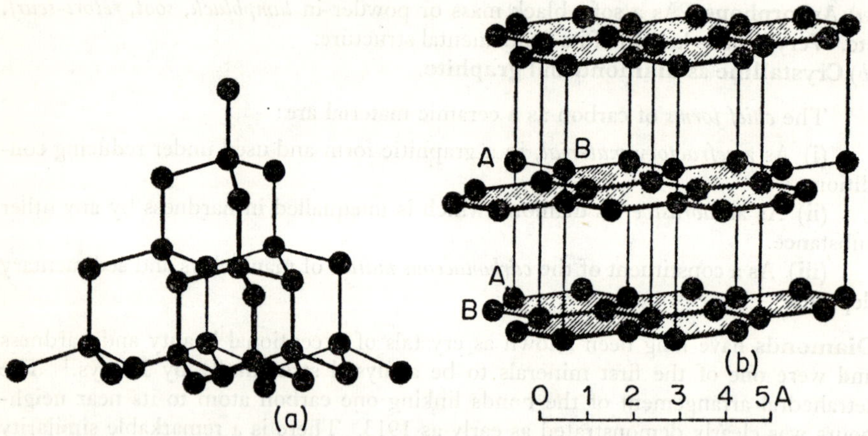

FIG. IV.1. THE CRYSTAL STRUCTURES OF (*a*) DIAMOND, AND (*b*) GRAPHITE

Graphite, on the other hand, consists of hexagonal carbon-linked rings extending indefinitely in two dimensions; so that any particular carbon atom is bound to two other similar atoms in the same ring and to another atom in an adjacent ring. Each atom resonates between three positions, thus:

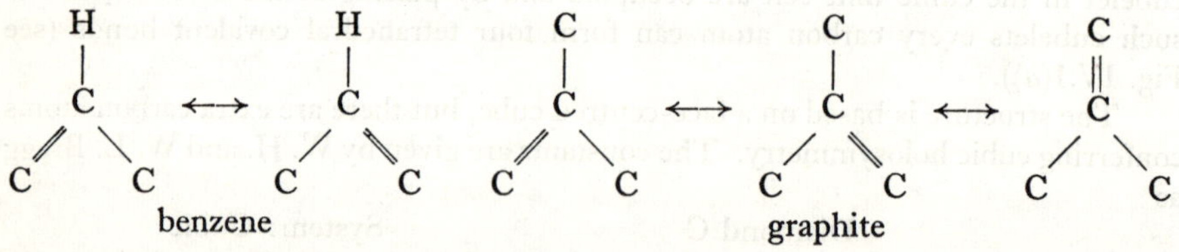

It may be concluded that each bond in graphite has a third of a double-bond character and this is borne out by the increase in length of the carbon to carbon linkage to 1·42 Å. The resonance energy strengthens the bonds and reduces the separation of the atoms to below that theoretically expected.

The bonding between the sheets is rather similar to that between the plates in montmorillonites. Van der Waal linkages have been suggested and the distance between each layer of carbon atoms is 3·40 Å. The complete structure is given in Fig. IV.1(*b*).

Graphite has the following constants:[1]

Graphite C	System: Hexagonal
Space Group: C6/mmc	Axes:
Z = 4	a = 2·46 Å
	c = 6·80 Å (i.e. two layers to one unit)

The similarity of graphite to montmorillonite is also seen on comparing their swelling properties. If graphite is treated with strongly oxidising agents such as potassium chlorate and nitric acid, the interlayer distance increases to between 6 and 11 Å. An oxide layer is most probably formed, as the expansion of the lattice may be related to the increase of combined oxygen. Molecules of water are also absorbed between the layers. Sulphuric acid and fluorine have somewhat similar effects.

The linkages between the atoms in diamond have the characteristics of aliphatic compounds where carbon is in tetrahedral co-ordination; on the other hand, the bonding of the carbon atoms in graphite closely resembles that in aromatic compounds.

The black forms of carbon (e.g. **soot** and **lampblack**) including those produced by charring vegetable and other carbonaceous matter in clays are very different in appearance from commercial graphites, but they are composed of minute crystals of it and have a similar atomic structure. The chief structural difference is that in black carbon the distances between the layers are relatively larger than in coarser graphites and there is sometimes less regularity in the distribution of the atoms.

In many carbon blacks the carbon atoms are arranged regularly in hexagonal rings and the various planes are parallel in the *a*-direction but wholly random in the *c*-direction. Only after heating to 2800° C. are crystals of true graphite produced.

Some commercial graphites have crystals extending 50–100 Å in the *a*-direction and 20–50 Å in the *c*-dimension.

Carbon in the combined state occurs as:

(i) The *gases* carbon monoxide, CO, carbon dioxide, CO_2, acetylene, C_2H_2, methane, CH_4, etc.

(ii) Various *solid* and *liquid components*; those of chief interest in ceramics being (*a*) the *carbonaceous matter* (chiefly of vegetable origin) which occurs in clays, etc., and (*b*) various fuels such as coal, anthracite, coke and various oils.

The nature of the carbon in carbonaceous material associated with sedimentary clays is still largely conjectural. Much valuable and interesting work on the minerals in coals and shales has been carried out by Riley[2] who has suggested that small crystallites of graphitic-like minerals exist, which appear to be composed of sheets of co-planar carbon atoms alternating with undulating layers of complex groupings. There is little orientation in the stacking of layers and the interplanar distance is not constant. *Fusain* is a coal mineral with a structure resembling graphite and it is readily converted to true graphite on carbonisation at a moderately high temperature

[1] Lipson, H. and Stokes, A. R., *Nature*, 149, 328, 1942.
[2] Riley, H. L., *Quart. Reviews*, 1, 59, 1942.

under reducing conditions. *Clarain*, *vitrain* and *durain* are other minerals which contain a higher proportion of complex groupings.

Coke prepared at relatively low temperatures appears to be a mixture of amorphous carbon and graphite. Coke prepared at temperatures above 700° C. consists mainly of crystallites of graphite with a considerable proportion of border carbon atoms. On heating to about 1400° C. recrystallisation of the crystallites occurs; this is the basis of the Acheson method of making graphite from coke or anthracite. The chemical reactivity of coke is at a minimum when it is recrystallised into graphite.

Some clays contain a proportion of carbonaceous material in the form of algæ or leaf remains in which the carbon molecules appear to have discrete existence.

OXIDES

MANY oxides are used as ceramic materials, as refractories, fluxes, opacifiers and colouring agents. The crystal structures of most of these have been accurately worked out and Wells[1] has given an excellent summary and classification of the main types.

TABLE IV—I. THE CRYSTAL STRUCTURE OF SOME OXIDES*
(after Wells)

Co-ordination numbers of M and O and formulæ	*Name of structure*	*Examples*
4:8 M_2O	Anti-fluorite	Li_2O, Na_2O, K_2O
4:4 } MO	Wurtzite	BeO, ZnO
6:6 } MO	Rock-salt	MgO, CaO, BaO, FeO, etc.
6:4 } M_2O_3	'A' rare-earth sesquioxides	La_2O_3, Ce_2O_3, etc.
} M_2O_3	'C' rare-earth sesquioxides	Mn_2O_3, and other rare earths
6:4 } M_2O_3	Corundum	α-Al_2O_3, α-Fe_2O_3, Cr_2O_3, V_2O_3
4:2 } MO_2	Silica	SiO_2
6:3 } MO_2	Rutile	TiO_2, VO_2, MnO_2, MoO_2, WO_2, etc.
8:4 } MO_2	Fluorite	ThO_2, CeO_2, PrO_2, UO_2, (ZrO_2)
6:2 MO_3		ReO_3
Complex oxides		
ABO_3	Perovskite	$CaTiO_3$, etc.
ABO_3	Ilmenite	$FeTiO_3$, etc.
AB_2O_4	Spinel AB_2O_4	$FeAl_2O_4$, $ZnAl_2O_4$, etc.
	$B(AB)O_4$	$FeMgFeO_4$, $MgTiMgO_4$

* All these structures are infinite three-dimensional complexes.

Although this table is an excellent guide to the relationships between the oxides, it must be remembered that the structures are idealised and represent the oxygen atoms in some form of close-packing, which they are certainly not in many instances. Furthermore, some oxides are capable of existing in more than one modification, each of different spacial arrangement. Certain oxides undergo transformations at high temperatures, thereby forming new modifications, which are

[1] Wells, A. F., *Quart. Reviews.* 2, 185, 1948.

often instantly reversible on cooling to normal temperature, yet in ceramic reactions it is sometimes the high-temperature form which is important, especially in the study of firing reactions.

The oxides, however, are conveniently arranged in order of type.

The M_2O group is unusual and for reasons stated later they are best considered along with MO_2 forms.

OXIDES MO

METALLIC oxides with the formula MO form two groups, depending on the different co-ordinations of the cations and oxygen ions.

(*a*) **With 4:4 co-ordination.** Oxides of composition MO with this low co-ordination must be divalent and with cations of very small ionic radii; *beryllia*, BeO, is the only oxide of importance which conforms to this configuration, because beryllium has an ionic size of 0·31 Å. Each beryllium ion is surrounded by four oxygen atoms, which each require to be in contact with four metal ions to satisfy their valency requirements. The structure resembles that of *wurtzite*, ZnS, but the bonds are probably more covalent than in the alkaline earth oxides.

The constants of beryllia[1] are:

Beryllia BeO	System: Hexagonal
Space Group: C6mc	Axes:
Z = 2	a = 2·67 Å
	c = 4·39 Å

and the structure of wurtzite is shown in Fig. IV.2; in beryllia, beryllium ions replace the zinc, and oxygen the sulphur in the wurtzite structure.

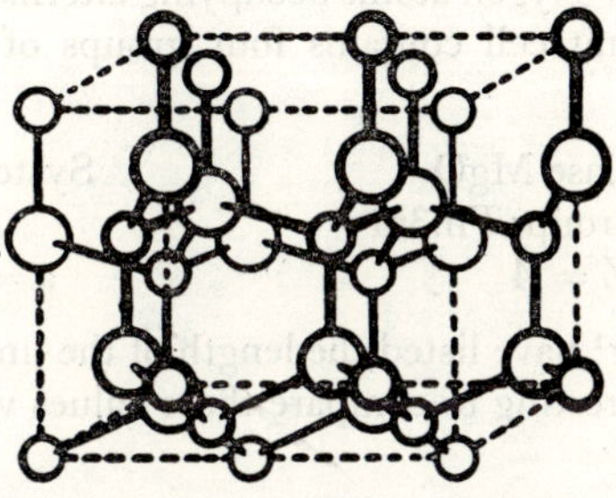

FIG. IV.2. THE STRUCTURE OF BERYLLIA (BASED ON WURTZITE)
(Small circles represent the cations, larger circles the anions)

In oxides of the beryllia type, the oxygen atoms are ideally hexagonally close-packed. The metallic cations occupy all vacant lattice sites and the sequence of layers is ABABAB, etc. If in Fig. IV.3 the centre unit (b) is moved to the left above unit (a) until each oxygen is directly over a cation, then the structure of beryllia results. The third layer of oxygen atoms must then be arranged to give four-fold co-ordination in the beryllia structure.

Zinc oxide (ZnO) has a similar structure to beryllia although the tetrahedral bonding is probably more covalent.

[1] Nitka, H., *Nature*, 29, 336, 1941.

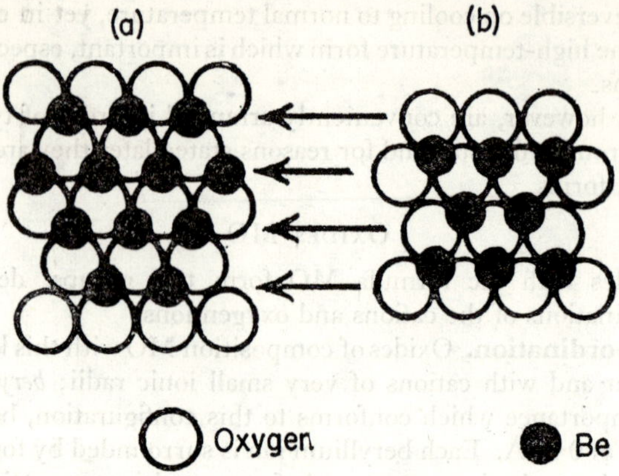

FIG. IV.3. THE LAYER ARRANGEMENT OF BERYLLIA
(b) superimposed on (a) gives beryllia.

(*b*) **With 6:6 co-ordination.** This group includes the alkaline earth oxides MgO, CaO, SrO, BaO, etc., along with the transitional metal oxides CdO, VO, MnO, FeO, CoO and NiO.

The structure of all oxides in this group is based on that of rock-salt, NaCl. Each cation is in six-fold co-ordination with oxygen atoms, each of which receives one-third of a valency share. As a consequence, all oxygen atoms must, in turn, be in contact with six cations. The structure of all oxides in this group is *cubic* with the corresponding cation and oxygen atoms occupying alternate corners of a set of cubes (see Fig. IV.4). The unit cell contains four groups of MO. The constants for periclase, MgO, are:

Periclase MgO	System: Cubic
Space Group: Fm3m	Axis:
Z = 4	a = 4·20 Å

Huber and Wagener[1] have listed the length of the unit cell of each oxide of the rock-salt type. It is interesting to compare these values with the size of the corresponding cation.

TABLE IV—II. LATTICE CONSTANTS OF ALKALINE EARTH AND RELATED OXIDES

Oxide	*Axis*	*Radius of cation*	*Oxide*	*Axis*	*Radius of cation*
MgO	4·20 Å	0·65Å	CdO	4·70 Å	0·97 Å
CaO	4·80	0·99	MnO	4·42	0·80
SrO	5·15	1·13	FeO	4·28	0·76
BaO	5·53	1·35	CoO	4·25	0·74
			NiO	4·17	0·72

[1] Huber, H. and Wagener, S., *Z. tech. Physik.*, 23, 1, 1942.

The length of the unit cell in each instance is almost in direct ratio to the ionic size of the respective cation.

An unusual feature of the oxides of the *alkaline earth metals* is that only magnesia can form a stable crystal which is not liable to hydration when exposed to normal atmospheric conditions. No matter how lime, strontia or baria are heat-treated, it is impossible to convert them into a form with the permanent character of periclase. The answer lies partly, if not entirely, in the crystal configuration of these oxides. If reference is made to Table II.XIII (p. 82), the stable co-ordinations (calculated as the ratio of the cation/anion radii) of each alkaline earth metal with respect to oxygen may be compared. Magnesium is well within the limits of six-fold co-ordination, but calcium and, to an even greater extent, strontium and barium are too large. The resulting octahedral configuration of oxygens around each cation must, therefore, be highly distorted except in the case of magnesium. The close-packed idealised structure must be replaced by a more open structure, permitting the free access of water molecules and the consequent ease with which all alkaline earth oxides, except periclase, can be hydrated.

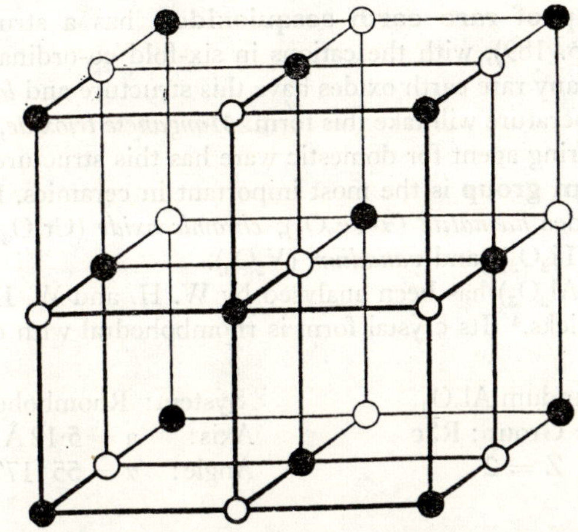

FIG. IV.4. THE PERICLASE STRUCTURE, BASED ON ROCK SALT
(Black dots represent cation, and circles, anion locations)

The *transition group oxides* have ions of suitable sizes to conform to the requirements of six-fold co-ordination with oxygen. Cadmium is a little large and is the least resistant to hydration. The mineral FeO (*ferrous oxide* or *wustite*) is not very common in nature, but is considered to be the normal form of iron oxide at temperatures above 1200° C. *Wustite* very readily oxidises to the ferric form under normal conditions.

(*c*) **Other forms of MO minerals.** Compounds are known where the composition corresponds with the general formula MO, but where simple co-ordinations are not present.

Lead oxide (PbO) is thought to exist in three forms, (*a*) tetragonal, (*b*) orthorhombic, and (*c*) monoclinic, but only the structure of the first has been established with certainty. Each lead atom is linked to four oxygen atoms, but these all lie to one side of the cation and are each linked to two lead atoms and receive a valency share from each. Layers of such linkages are built up and are linked to other similar layers by bonds, the nature of which has not been established.[1]

Cupric oxide (CuO) has a structure in which each copper atom is linked to four co-planar oxygen atoms.[2]

Oxides M_2O_3

This group includes those compounds which have long been known as the sesquioxides. Wells has suggested that there are three separate sections with distinct structures.

(i) **The 'A' group of rare earth sesquioxides** is of importance in ceramics, and includes *lanthana* and *ceria*. These are unusual in that the metal cation is co-ordinated with seven oxygen ions.

(ii) **The 'C' group of rare earth sesquioxides** has a structure resembling *fluorite*, CaF_2 (see p. 189), with the cations in six-fold co-ordination but not in a simple manner. Many rare earth oxides have this structure and *lanthana* if crystallised at a high temperature will take this form. *Manganese trioxide*, Mn_2O_3, which is an important colouring agent for domestic ware has this structure.

(iii) **The corundum group** is the most important in ceramics, for in addition to *α-alumina*, it includes *hæmatite* (α-Fe_2O_3), *chromic oxide* (Cr_2O_3), and the *sesquioxides of titanium* (Ti_2O_3), and *vanadium* (V_2O_3).

Corundum (α-Al_2O_3) has been analysed by W. H. and W. L. Bragg,[3] and by Pauling and Hendricks.[4] Its crystal form is rhombohedral with constants of:

Corundum Al_2O_3	System: Rhombohedral
Space Group: $R\bar{3}c$	Axis: $a = 5{\cdot}12$ Å
$Z = 2$	Angle: $\alpha = 55° 17'$

The structure consists of layers of oxygen atoms arranged in hexagonal close-packing. Each aluminium atom is surrounded by six oxygens. The valency contribution to each co-ordinating anion is one-half, so that each oxygen must be linked to four cations. If all possible positions were filled by aluminium atoms there would be as many cations as anions, as in periclase, and each oxygen would be receiving three valency shares—a half from each of six cations. Only two-thirds of the available octahedral lattice sites are in fact filled by aluminium and the structure is electrically balanced. The idealised form of packing of two oxygen layers is shown in Fig. IV.5; the third layer is immediately above the first.

[1] W. J. Moore and L. Pauling, *J.A.C.S.*, 63, 1392, 1941.
[2] G. Tunell, E. Posnjak and C. J. Ksanda, *Zeit Krist.*, 90, 120, 1935.
[3] Bragg, W. H. and W. L., *X-rays and Crystal Structure* (Ball and Sons, 1916).
[4] Pauling, L. and Hendricks, S. B., *J. Amer. Chem. Soc.*, 47, 781, 1925.

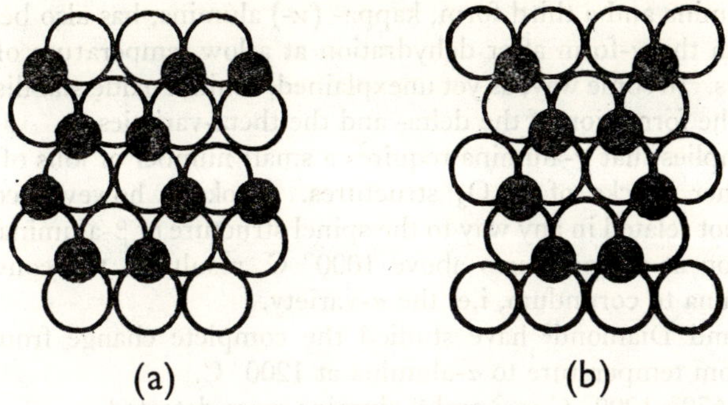

FIG. IV.5. THE STRUCTURE OF CORUNDUM
Unit (b) is superimposed on (a) and the progression of layers is ababab, etc.

The dimensions of other minerals which have a similar configuration are tabulated below:

TABLE IV—III. LATTICE CONSTANTS OF THE CORUNDUM GROUP

		Space group	Axis a	Angle α	Radius of cation
Corundum*	Al_2O_3	R3c	5·118 Å	55° 17′	0·50 Å
Chromic oxide*	Cr_2O_3	—	5·351	55° 06′	0·69
Hæmatite	Fe_2O_3	—	5·420	55° 17′	0·64
Ilmenite†	(Fe,Mg)TiO_3	$R\bar{3}(C_{3i})$	5·520	54° 16′	—

* Jay, A. H. and Wild, W. T., *Min. Mag.*, 27, 56, 1944.
† Ilmenite is structurally similar, but the introduction of two different cations destroys the symmetry of the space group and results in one of lower order.

Aluminium oxide is known to exist in other crystal modifications, viz. γ- and β-alumina: γ-*alumina* may be prepared by heating any hydrated aluminium oxide to 200–500° C. at which temperature it loses water of crystallisation and produces small crystallites of γ-alumina. This mineral is also a decomposition product of many clays and similar materials. The crystal structure is cubic with a lattice parameter of a=7·88 Å. Aluminium ions are in both four- and six-fold co-ordination in what is apparently a defect spinel lattice (see p. 193) with $21\frac{1}{3}$ metal ions in a unit cell as against a normal 24.

The effect of heat on this form is most interesting; Rooksby[1] has identified at least two other modifications by maintaining the crystal between 600 and 950° C. for long periods of time. The crystallite size increases under these conditions and the structure assumes a lower symmetry. X-ray photographs suggest that there is a change from cubic to tetragonal symmetry, but the full structure has not yet been ascertained. Rooksby has named these two new modifications as delta- (δ-) and

[1] See *X-ray Identification and Crystal Structures of Clay Minerals* (Mineralogical Society, London, 1951), p. 250.

theta- (θ-) alumina and a third form, kappa- ($\varkappa$-) alumina, has also been identified, associated with the γ-form after dehydration at a low temperature of certain synthetic gibbsites. In some way, as yet unexplained, lithium oxide stabilises γ-alumina and restricts the formation of the delta- and the theta-varieties.

Wells[1] implies that γ-alumina requires a small number of ions of another type to link together blocks of Al_2O_3 structures. Rooksby however considers that γ-alumina is not related in any way to the spinel structure of β-alumina.

Calcination at temperatures above 1000° C. results in the conversion of all forms of alumina to corundum, i.e. the α-variety.

Dragoo and Diamond[2] have studied the complete change from amorphous alumina at room temperature to α-alumina at 1200° C.

Between 670°–1200° C γ, δ and θ alumina were detected.

Beta- (β-) *alumina* is another modification which is of frequent occurrence when a hydrate containing an alkali impurity is heated to a moderately high temperature. It has unquestionably a spinel structure but the alkali ions are part of the lattice. The mineral is discussed in more detail in the appropriate section of this chapter (p. 193).

Iron oxide may also exist in many forms, two of which are frequently associated with clays. These are *hæmatite* or α-ferric oxide, Fe_2O_3, and *magnetite*, Fe_3O_4.

The crystal structure of the former is similar to that of α-alumina or corundum. The dimensions have been given on page 185.

Ilmenite, $FeTiO_3$, is also similar but every alternate iron is replaced by titanium.

Magnetite, Fe_3O_4, has a cubic structure and has been classified as a *spinel* (see p. 193). The iron atoms in the magnetite lattice are present in both ferrous[3] and ferric forms. The formula is best expressed as $Fe^{2+}Fe_2{}^{3+}O_4$. Under conditions of careful oxidation, magnetite can be oxidised to *maghemite*, γ-Fe_2O_3. The structure is a type of 'defect' spinel structure and may represent the condition where each iron atom has been oxidised, without any rearrangement within the crystal having taken place. γ-ferric oxide is readily converted into the stable α-form by heat treatment at 400° C. Wells has reported that this change is accompanied by a pronounced colour change from light brown to Turkey red.

Vanadium oxide, V_2O_3, crystallises with the corundum structure so that the vanadium ion is octahedrally co-ordinated.

Lead trioxide, Pb_2O_3, is a black compound which in the presence of a little alkali is analogous to β-alumina. The lead atoms (Pb^{4+}) are surrounded by four oxygen atoms at the corners of a deformed square and the Pb^{2+} atoms by two oxygen atoms at the apices of a distorted octahedron.[4]

Bismuth oxide, Bi_2O_3, is used in lustre-wares. When fused with an MO oxide the metal content in a unit cell is constant but the oxygen content varies.[5] Its structure is not known with certainty.

Boric oxide, B_2O_3, has a structure of which little is known. It appears to consist of two BO_3 groups sharing all their oxygen atoms between them. The crystalline

[1] Loc. cit., p. 180.

[2] Dragoo, A. L. and Diamond, J. J., *J. Amer. Ceram. Soc.*, **50**, 568, 1967.

[3] *Wustite*, or ferrous oxide, is described on p. 183, in relationship to the similar mineral, periclase, MgO.

[4] A. Bystrom, *Arkiv. Kemi. Min. Gerl.*, 1944, 18A, No. 23.

[5] B. Aurivillius, *Arkiv. Kemi. Min. Gerl.*, 1943, 16A, No. 17.

oxide is difficult to obtain as it fuses so readily and its structure is still rather doubtful. There is a similar lack of knowledge about the structure of *borax*, $Na_2B_4O_2 + 10H_2O$ and other *borates*, though it seems clear that each B—O link is a full valence and that each B^{3+} ion is at the centre of a triangle of oxygen ions and that these BO_3 triangles can link through their vertices in the same way as SiO_4 tetrahedra but with a greater tendency to a random network. As each boron ion has only three linkages, boric oxide has a much lower softening point than silica.

Borates can be classed in three groups:

(i) Isolated triangles, as boric acid $B(OH)_3$.
(ii) Rings composed of three triangles as $K_3B_3O_6$.
(iii) Chains of triangles—$n.(BO_2)^{2-}$ as calcium borate CaB_2O_4.

More complex borate ions are also known[1] (see p. 198).

Antimony trioxide, Sb_2O_3, and *arsenic trioxide*, As_2O_3, each exist in two polymorphic forms. After heating to high temperatures and cooling, both these oxides appear to have a chain structure.

Oxides MO_2 and M_2O

In this section, oxides with an MO_2 formula are grouped with those of M_2O formula because, crystallographically, they are similar. Oxides of the type M_2O such as Li_2O, Na_2O and K_2O have the same lattice configuration as fluorite, CaF_2, except that the positions of the cations and anions are reversed. These groups may be distinguished:

(*a*) **MO_2 with 6:3 co-ordination.** *Rutile*, TiO_2, is the typical mineral of this group, which includes *germania*, GeO_2, *cassiterite*, SnO_2, *plattnerite*, PbO_2, *pyrolusite*, MnO_2, and the *oxides of molybdenum*, *tungsten*, *ruthenium*, *osmium* and *iridium*.

Some of these oxides are capable of existing in more than one form and this is well-illustrated by titania, TiO_2, which occurs in nature as rutile, anatase and brookite. The titanium atoms in all three minerals are in six-fold co-ordination with respect to oxygen, but the linking of the octahedral groups in each is different.

The valency contribution of a titanium atom to each oxygen atom surrounding it is two-thirds; consequently, every oxygen must be in contact with three titanium atoms to fulfil its valency requirements.

Rutile and *anatase* were first analysed structurally by Vegard[2] in 1916. Their constants are as follows:

Rutile TiO_2	System: Tetragonal
Space Group: P4/mnm	Axes:
Z = 2	a = 4·58 Å
	c = 2·95 Å

Anatase TiO_2	System: Tetragonal
Space Group: I4/amd	Axes:
Z = 4	a = 3·73 Å
	c = 9·37 Å

[1] P. C. Carman, *Constitution and Properties of Eng. Materials* (Edward & Co., 1949).
[2] Vegard, L., *Phil. Mag.*, 32, 65, 1916.

The main difference in the structures lies in the orientation of the three titanium atoms surrounding each oxygen. In rutile, they are situated approximately at the corners of an equilateral triangle, whilst in anatase the angle between two of the titanium atoms is nearer 180° than 120°.

Brookite is more complex and Pauling and Sturdivant[1] have suggested that its constants are:

Brookite TiO_2	System: Orthorhombic
Space Group: Pbca	Axes:
$Z = 8$	$a = 9{\cdot}14$ Å
	$b = 5{\cdot}44$ Å
	$c = 5{\cdot}15$ Å

There is no evidence that the other oxides of the rutile group have any modification of the anatase and brookite type. The lattice dimensions are based on the rutile cell, and are as follows:

TABLE IV—IV. LATTICE CONSTANTS OF RUTILE-LIKE OXIDES

	Radius of cation	*Axis a*	*Axis c*		*Radius of cation*	*Axis a*	*Axis c*
GeO_2	0·53 Å	4·39 Å	2·86 Å	OsO_2	0·67 Å	4·51 Å	3·19 Å
MnO_2	0·60	4·44	2·89	MoO_2	0·62	4·86	2·79
TiO_2	0·68	4·58	2·95	WO_2	0·71	4·86	2·77
RuO_2	0·65	4·51	3·11	SnO_2	0·71	4·72	3·16
				PbO_2	0·84	4·93	3·37

This table shows that as the size of the cation increases, so in general do the lattice dimensions.

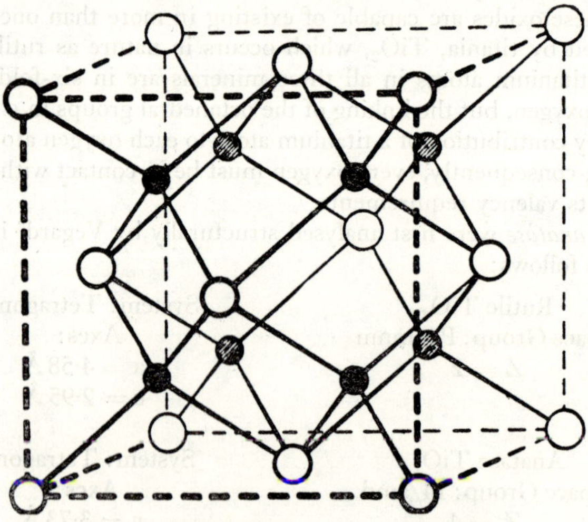

FIG. IV.6. THE STRUCTURE OF FLUORITE
(after Bragg)
Shaded circles are anions, open circles are cations.

[1] Pauling, L. and Sturdivant, J. H., *Zeit. Krist.*, 68, 239, 1928.

(*b*) **MO_2 with 8:4 co-ordination.** When quadrivalent cations are too large for six-fold co-ordination, the surrounding oxygens are arranged in the form of a cube; thus the cation is in eight-fold co-ordination. Each oxygen receives a half valency share from the cations with which it is in contact and consequently requires to touch four.

The type-structure of this group of oxides is the mineral *fluorite*, CaF_2, which is based on a face-centred cube, and shown in Fig. IV.6. The oxides of thorium, cerium, uranium and zirconium are of similar structure with oxygen replacing fluorine, and the metallic ion replacing calcium.

Zirconia (*baddeleyite*) occupies an anomalous position, for it has a monoclinic structure formerly thought to be a distorted form of the fluorite structure. McCullogh and Trueblood[1] have concluded that the structure of zirconia is anomalous because each zirconium atom in it is co-ordinated with seven oxygens, which are not all at the same distance. The structure possesses some characteristic features of the cubic fluorite and the tetragonal rutile.

When the cationic radius and the cation/oxygen size ratio of some of these quadrivalent elements are compared, very interesting facts emerge (see Table IV.V).

TABLE IV—V. DEPENDENCE OF STRUCTURE ON CATIONIC SIZE OF THE OXIDES RO_2

Cation	*Radius*	*Cation/oxygen radius ratio*	*Co-ordination No.*	*Type of structure of oxide*	
Titanium	0·68Å	0·485	6	Rutile	Tetragonal
Tungsten	0·71	0·507	6	,,	,,
Tin	0·71	0·507	6	,,	,,
Zirconium	0·80	0·572	6 (8)	Distorted	Monoclinic
Cerium	1·01	0·722	8	Fluorite	Cubic
Uranium	1·05	0·751	8	,,	,,
Thorium	1·11	0·794	8	,,	,,

The *titanium*, *tin* and *tungsten* cations fall well within the size range for six-fold co-ordination with oxygen, whilst the *cerium*, *thorium* and *uranium* cations are within the eight-fold range. The *zirconium* cation falls near to the borderline of the two groups. It might, therefore, be expected to fall in either class, or to form a distorted grouping of one of them, which, in fact, it does. At high temperatures pure zirconia is transformed into a tetragonal structure; the zirconium ion presumably increases in size less than the oxygen anion, and the cation/anion ratio falls into the stable six-fold co-ordination range of rutile. On cooling, the mineral reverts to the monoclinic form, but if it is fired to high temperatures with other oxides, it becomes stabilised in a cubic modification resembling fluorite.

Ruff and Ebert[2] originally reported that cerium oxide was the most effective stabilising oxide. Crystallographically, this is to be expected, for the fluorite structure of this oxide would 'seed' the crystallisation of a cubic zirconia. Geller and Favorsky,[3] however, have investigated the stabilising effect of other oxides and report that the alkaline earth oxides are superior even to ceria. Solid solution may occur

[1] McCullogh, J. D. and Trueblood, K. N., *Acta Cryst.*, 12, 507, 1959.
[2] Ruff, O. and Ebert, F., *Z. anorg. Chem.*, 180, 19, 1929.
[3] Geller, R. F. and Favorsky, P., *J. Res. Nat. Bur. Standards*, 35, 87, 1945.

and the combined effect is to produce a stable cubic form. The packing in the cubic type of zirconia will be very tight and would account for the high coefficient of thermal expansion observed in the stabilised form.

(*c*) **M_2O with 4:8 co-ordination.** Minerals with this configuration include the alkali oxides Li_2O, Na_2O and K_2O. The structure is similar to that of fluorite, but with the cation and anion positions reversed, the alkali metals occupying the F positions and the oxygen atoms the Ca positions. The structure-type is therefore called *anti-fluorite*.

(*d*) **The cuprite M_2O Group with 2:4 co-ordination** contains *cuprous oxide*, Cu_2O, and *argentum oxide*, Ag_2O. Both these oxides have a structure quite different from that of the anti-fluorite one (*c*), in consisting of two completely penetrating and identical frameworks which are not crossed by any primary M—O bonds.[1]

OXIDES MO_3

THE simple cubic ReO_3 structure is related to that of perovskite (p. 194) with large (A) ions and O^{2-} ions forming a close-packed lattice with the smaller (B) ions in the interstices.

According to Wells,[2] the structures of CrO_3 and WO_3 are only slightly distorted versions of the ReO_3 structure.

OXIDES M_3O_4

IN this small group four oxides are of ceramic interest—*red lead*, Pb_3O_4, *magnetite*, Fe_3O_4, *cobaltic tetroxide*, Co_3O_4, and *manganese tetroxide*, Mn_3O_4.

Red lead is usually a mixture of several lead oxides with an average composition corresponding to Pb_3O_4 so that no single crystalline structure can be assigned to commercial red lead. The pure tetroxide consists of PbO_6 octahedra sharing opposite edges and linked by Pb atoms each with a pyramidal arrangement of three oxygen atoms. According to Wells,[1] the constants are:

System:	Bond lengths:		Angles:
Octahedral	$Pb^{4+}O$	6 at 2·15 Å	2 of 95·3°
	$Pb^{2+}O$	2 at 2·23 Å	1 of 91°
		1 at 2·15 Å	

Several mixed oxides have the same structure.

Magnetite, Fe_3O_4, is included among the Complex Oxides (p. 193).

Cobalt tetroxide is obtained by calcining cobaltous oxide and is the essential constituent of *prepared cobalt*—one of the most extensively-used potters' colours.

Manganese tetroxide, Mn_3O_4, is the *calcined manganese* used as a colouring agent by potters. Like other M_3O_4 oxides it contains the metal in two different states of valency and corresponds to the AB_2O_4 formula for complex oxides except that in the case of the four above-mentioned oxides the A and B elements are the same and could be written AA_2O_4, though this would not so clearly show that they are of different valencies.

[1] A. F. Wells, *Structural Inorganic Chemistry*, 2nd ed. (Clarendon Press, Oxford, 1950).
[2] Ibid.

Complex Oxide Structures

Wells states that complex oxide structures of the general formula $A_xB_yO_n$ may be of three types:

(*a*) where one ion (B) simply replaces another (A) without affecting the spacial configuration;

(*b*) where blocks of oxides are joined by other cations; and

(*c*) where complete new units are formed unlike any of the component oxides in the structure.

Type (a) requires little description, and is best represented by the mineral *ilmenite*, $(FeMg)TiO_3$. The basic structure is *hæmatite* (α-Fe_2O_3), which is structurally similar to *corundum*. Titanium ions (Ti^{3+}) replace ferric ions in the lattice, for they are of comparable size and valency. There may be further replacement of iron by magnesium, for many ilmenites contain this element. $MgTiO_3$, $MnTiO_3$, $CoTiO_3$ and $NiTiO_3$ are all isomorphous with hæmatite.

Type (b) includes many compounds composed of blocks of a simple structure linked by ions of other types. In this group, of importance in ceramic studies, is β-alumina and possibly γ-alumina, and γ-Fe_2O_3. In the former, the structure is composed of blocks of a spinel type, linked together through alkali ions. They can be studied as an adjunct to the last group.

Type (c), where new structure types are formed, includes the important group of minerals—the *spinels*. The well-known form of spinel—the gem-stone $MgO.Al_2O_3$—was first analysed by W. H. Bragg[1] and by Nishikawa,[2] and shown to have a cubic structure. Other workers have identified minerals of variable composition in this group, and it is now customary to write the spinel form in general terms $R^{2+}O.R_2{}^{3+}O_3$ where R^{2+} is a divalent cation, such as Mn, Mg, Fe^{2+}, etc., and R^{3+} is a trivalent cation such as Al, Fe^{3+}, Cr, etc. The chrome ores, for example, contain a high proportion of spinel where the trivalent cation is chromium; these ores are used extensively in the manufacture of basic refractories (see Chapter VI).

The structure of all spinels is based on a unit cell of 56 atoms including 32 oxygen atoms arranged in cubic close-packing. In the interstices between the oxygen atoms are two types of cations, 8 of which are in four-fold co-ordination and 16 in six-fold co-ordination (see Fig. IV.7). There are essentially two different cation/oxygen layers, the first (*a*) with a close-packed layer of oxygen atoms and cations of four- and six-fold co-ordination. The second layer (*b*) is also a close-packed layer, but with six-fold cations only; if (*b*) is moved directly to the right until the points XX and YY coincide, the arrangement of atoms in the first two layers is produced. The third layer (*c*) is like the first, but to satisfy the co-ordination requirements of cations B it has to be stacked in such a way that cubic close-packing results. This may be shown by moving unit (*c*) directly over the superposed (*a*) and (*b*) units so that XXX and YYY now lie immediately above each other.

The formula of a spinel may be represented by the general formula AB_2O_4, i.e. it contains twice as many B atoms as A atoms. Therefore it is logical to place the

[1] Bragg, W. H., *Phil. Mag.*, 30, 305, 1915.
[2] Nishikawa, S., *Proc. Tokyo Math. Phys. Soc.*, 8, 199, 1915.

A atoms in the eight tetrahedral or four-fold co-ordinated groups, and the B atoms in the sixteen octahedral or six-fold co-ordinated groups. Barth and Posnjak[1] showed, by careful measurement of the X-ray line-intensities of certain spinels, that there could be an alternative arrangement whereby eight B-type atoms occupy the tetrahedral positions and the octahedral positions are filled by eight A type and eight further atoms of the B type, randomly arranged. This leads to two general formulæ: Group 1 or *normal* spinels represented by AB_2O_4, and Group 2 or *inverse* spinels represented by $B(AB)O_4$.

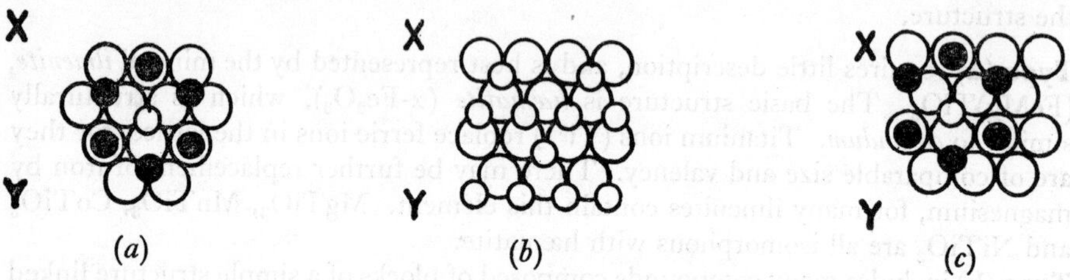

FIG. IV.7. THE STRUCTURE OF SPINELS

Small open circles represent cations in octahedral co-ordination. Small black circles represent cations in tetrahedral co-ordination, large circles represent oxygen ions.

The structure is obtained by moving (*b*) on to (*a*) and then (*c*) on to (*b*) in such a way that the points X and Y are superimposed.

Dunitz and Orgel[2] consider that the important governing factor influencing the type of spinel structure which forms is the relative inclination of the various cations present to enter the octahedral groupings. On the basis of the theory of crystal field they have adduced the excess octahedral stabilisation energy of various cations which can take part in spinel lattices. The order of preference (diminishing to the right) of some common cations is as follows:

Cr^{3+}, Mn^{3+}, Ni^{2+}, Cu^{2+}, Al^{3+}, Co^{2+}, Mg^{2+}, Ti^{3+}, Fe^{2+}, Fe^{3+}, Mn^{2+}, Zn^{2+}.

On this basis, all chromium-based spinels are of the *normal* type and those based on Fe^{3+} are of *inverse* form (except $Mn^{2+} Fe_2^{3+} O_4$). This is true in practice. Most of the aluminium-containing spinels are normal except those of nickel and copper.

The structure constants for the spinel gemstone $MgAl_2O_4$ is:

Spinel $MgAl_2O_4$	System: Cubic
Space Group: Fd3m	Axis:
Z = 8	a = 8·10 Å

The types of spinel associated with chrome ore will be considered in more detail in Chapter VI, but the wide range in chemical composition of this group may be judged from the Table IV. 6. These minerals represent end-members of the spinel group, for extensive solid solution may occur between all of them.

[1] Barth, F. W. and Posnjak, E., *Zeit. Krist.*, 82, 325, 1932.
[2] Dunitz, J. D. and Orgel, L. E., *J. Phy. Chem. Solids*, 3, 318, 1957.

It is probable that only the last two, *magnesioferrite* and *magnetite*, belong to Group 2 and have the inverse structure, although some doubt is attached to spinel $MgAl_2O_4$ itself. The unit cell parameter figures shown are the mean of values obtained by several workers.

More detailed consideration must be given to the spinel structures of gamma (γ)-alumina Al_2O_3, gamma (γ)-Fe_2O_3 and magnetite Fe_3O_4.

TABLE IV—VI. SOME MINERALS WITH SPINEL STRUCTURE

Name	*Chemical Composition*	*Length of Unit Cell*
Spinel	$MgAl_2O_4$	8·10 Å
Gahnite	$ZnAl_2O_4$	8·075
Hercynite	$FeAl_2O_4$	8·115
Chromite	$FeCr_2O_4$	8·346
Picrochromite	$MgCr_2O_4$	8·308
Magnesioferrite	$MgFe_2O_4$	8·368
Magnetite	$FeFe_2O_4$	8·38

Magnetite, Fe_3O_4, is a true spinel, for in a unit cell there are 32 cubic close-packed oxygen atoms and 24 cations, 8 of which are in the ferrous form and 16 ferric ions. The structure may then be written as $Fe^{2+}Fe_2^{3+}O_4$ or better still, for it is an inverted form, as $Fe^{3+}Fe^{2+}Fe^{3+}O_4$. Gamma (γ)-Fe_2O_3 has only $21\frac{1}{3}$ cations to a unit cell and all of these are in the ferric form distributed in a random fashion throughout the 24 available lattice sites. Wells[1] has made a careful study of the relationships between FeO, Fe_3O_4 and γ-Fe_2O_3. The former mineral also is built up of oxygens in cubic close-packing configuration, but there are as many ferrous ions as oxygens; it may be considered as the extreme end of the series with 32 oxygen and 32 iron ions all of which are in the ferrous form. The oxidation of ferrous oxide, FeO, to magnetite, Fe_3O_4, may be visualised as the addition of extra layers of oxygens and the migration of some cations to new lattice sites making the ratio of O^{2-} to Fe ions 32:24, and two-thirds of the iron is as Fe^{3+}. Further oxidation results in a continuation of this process until the oxygen/iron ratio is $32{:}21\frac{1}{3}$ and all the iron is as Fe^{3+}.

In the spinel structure of Fe_3O_4, Fe^{2+} and Fe^{3+} ions are distributed over a few octahedral and tetrahedral voids whereas in γ-Fe_2O_3 the Fe^{3+} ions are distributed at random over eight-ninths of the spinel positions.

The large series of solid solutions between Fe_3O_4 and γ-Fe_2O_3 can be regarded either as Fe_3O_4 with an oxygen excess or as γ-Fe_2O_3 with an excess of metal.

γ-alumina has a similar structure to γ-Fe_2O_3, but with aluminium cations in place of ferric iron.

Gamma iron oxide and gamma alumina, are not very stable and are readily converted into haematite or corundum respectively; both can be stabilised, however, by the addition of ions such as lithium, and it may be that in these forms the alkali ions act as a bridge between structural blocks. This is the case in *beta* (β)-*alumina*,

[1] Ibid., p. 192.

which has the chemical composition $Na_2O.11Al_2O_3$. Beevers and Ross, showed that it is composed of spinel units, linked together through sodium ions with aluminium in both four- and six-fold co-ordination. The lattice constants are:

β-alumina $Na_2O.11Al_2O_3$	System: Hexagonal
Space Group: C6/mmc	Axes:
Z = 2	a = 5·584 Å
	c = 22·45 Å

A structural representation is given in Fig. IV.8. The mineral may also be prepared with potassium replacing sodium as the 'bridging' ions for the spinel units.

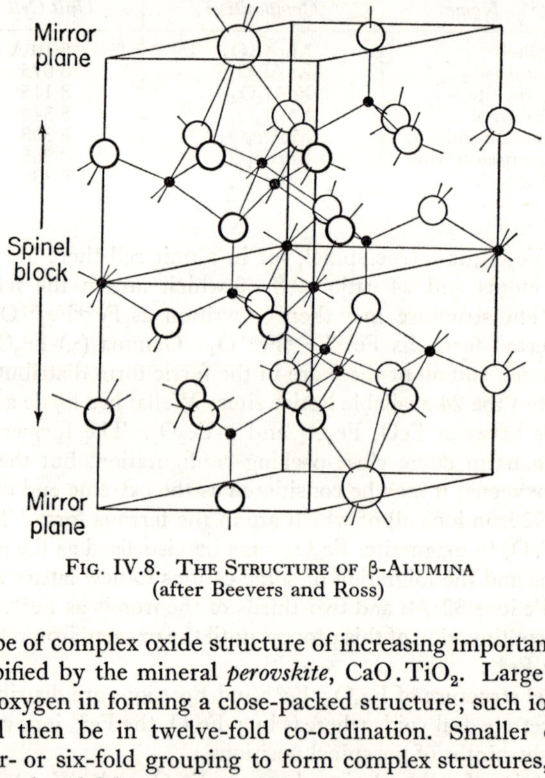

FIG. IV.8. THE STRUCTURE OF β-ALUMINA
(after Beevers and Ross)

Another type of complex oxide structure of increasing importance in ceramics[1] is the group typified by the mineral *perovskite*, $CaO.TiO_2$. Large cations can be associated with oxygen in forming a close-packed structure; such ions, e.g. Sr, Ba, K and Na, will then be in twelve-fold co-ordination. Smaller cations occupy positions of four- or six-fold grouping to form complex structures, which include some *tungstates* and *molybdates*, e.g. $NaWO_3$. *Barium titanate*, which has valuable electrical properties, is of this structure and Goldschmidt and Rait[2] suggest that *merwinite*, $3CaO.MgO.2SiO_2$, may be of this type.

Where cations of variable valency are present in a compound, there is the possibility of complex structures; in some instances a simple chemical formula disguises a complicated configuration.

Chrysoberyl, $BeAl_2O_4$, is of similar composition to the spinels, but is structurally

[1] Duncan, J. H., and Creyke, W. E. C., *Trans. Brit. Ceram. Soc.*, 68, 137, 1969.
[2] Goldschmidt, H. J. and Rait, J. R., *Nature*, 152, 356, 1943.

different. The oxygens in this case are hexagonally close-packed and the structural constants are:

Chrysoberyl $BeAl_2O_4$	System: Orthorhombic
Space Group: Pbmn	Axes:
Z = 4	a = 4·24 Å
	b = 9·39 Å
	c = 5·47 Å

Cobalt oxide, Co_3O_4, is also a spinel, but is very often impure, like chromite, as the result of replacements.

HYDROXIDES

IN order to understand the structure of the important hydroxides it is necessary to have a clear idea of the unusual features of the linkage in hydroxyl (OH) groups. This is all the more necessary because the simple hydroxides range from strongly basic and alkaline compounds such as caustic soda, NaOH, through the amphoteric hydroxides, such as $Al(OH)_3$, to the hydroxy-acids of boron, phosphorus, silicon, etc.

The oxygen atom has six electrons in the outermost orbit (p. 66). If a hydroxyl unit is formed, one electron is added to the oxygen orbit and the eighth can be acquired either by forming a covalent linkage, (—O—H), or by acquiring an electron from a metal element and forming an ion which will be the hydroxyl or $(OH)^-$ anion.

The hydrogen atom is very small compared with oxygen and consequently the hydroxyl ion can be imagined as a sphere of similar size to an oxygen ion, but of only unit negative charge.

The greater the difference in electronegativity between the atoms forming a linkage, the greater will be the degree of ionic bonding (see p. 76). Consequently alkali atoms such as sodium, potassium, etc., will be linked largely in ionic fashion to hydroxyl units and molecules of this type may be represented by $M^+(OH)^-$.

As the difference in electronegativity becomes less, the ionic nature of the linkage decreases through the amphoteric hydroxides like Al and Fe to the hydroxy-acids of B, P, and Si. In these latter elements the covalent union of oxygen is so strong that the hydrogen can be *ionised* (i.e. split up into separate ions) and an acid complex formed, thus:

$$R\text{—}(OH)_n \rightleftharpoons RO_n^{n-} + nH^+$$

Boric, phosphoric and silicic acids are of this type.

When these principles are applied to the arrangement within a crystal lattice, there can be two types of bonds: (i) the hydroxyl-$(OH)^-$ ion can behave as a rigid sphere and form close-packed ionic-linked structures rather similar to the oxides, or (ii) covalently bonded atoms may have the distorted configuration of complex ions (see p. 85), with the hydroxyl unit behaving as a dipole (p. 75). Other negatively-charged oxygen ions in the vicinity can influence the dipole and attract the hydrogen ion, thereby setting up the resonance structure (see p. 73):

$$\text{—}O^-\text{—}H^+ \cdots\cdots O^-\text{—} \rightleftharpoons O^- \cdots\cdots H^+\text{—}O^-\text{—}$$

which represents a 'hybrid' of two states of equal energy. Most bonds are intermediate between the extremes of ionic and covalent types, consequently intermediate types of hydroxyl linkages may be expected.

The alkaline earth hydroxides are examples of hydroxyl bonding which is virtually ionic in character:

Magnesium hydroxide (brucite, $Mg(OH)_2$) has a layer-lattice structure with hydroxyl units in positions of hexagonal close-packing with the Mg ions in octahedral co-ordination and each OH ion linked to three Mg cations. Each layer consists of two sheets of OH groups with a layer of magnesium atoms between, arranged so that each cation is in contact with six larger ions. Another exactly similar layer lies above the first situated in such a way that all OH groups in it touch three identical groups in the layer beneath. Because each layer is electrically balanced there can only be weak secondary forces between opposed OH layers and these hold the structure together. Aminoff[1] was the first to determine the atomic configuration of brucite and the constants are:

Brucite $Mg(OH)_2$	System: Trigonal
Space Group: $P\bar{3}ml$	Axes:
Z = 1	a = 3·12 Å
	c = 4·73 Å

The structural features can be seen in Fig. IV.9; (*a*) shows the configuration of $(OH)^-$ and Mg^+ ions in one layer and (*b*) the packing of hydroxyl units between adjacent layers.

Aluminium hydroxide, $Al(OH)_3$, is known as *gibbsite* or *hydrargillite* and unlike

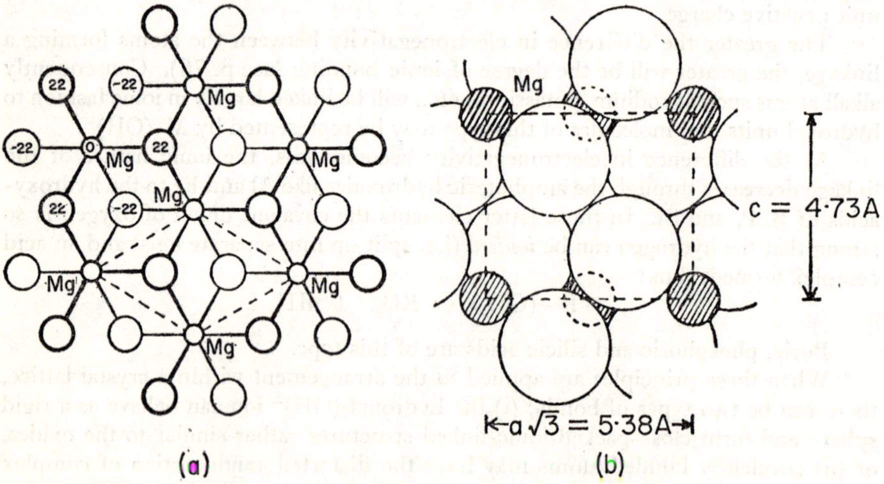

FIG. IV.9. THE STRUCTURE OF BRUCITE, $Mg(OH)_2$
(after Megaw)
(*a*) Schematic; (*b*) the relative position of atoms in adjacent layers.

[1] Aminoff, G., *Geol. Fören. Föhr. Stock.*, 41, 407, 1919.

brucite is not basic, but *amphoteric*, i.e. it may be weakly electronegative (acid) or electropositive (basic) under certain conditions.

The hydroxyl groups are not entirely ionic and non-directed but are slightly polar as a result of the dipole caused by the greater degree of covalency in the bonding. This structure is made up of layers composed of two close-packed hydroxyl sheets with aluminium atoms lying in six-fold co-ordination between them. Aluminium is trivalent, therefore only two-thirds of the possible vacant lattice sites are occupied. The essential difference between gibbsite and brucite lies in the relative stacking of adjacent layers. The hydroxyl groups in the gibbsite structure are sufficiently polar to exert an attractive force on others in the immediate vicinity and as a result each OH group lies directly over another in the adjacent layer. It is postulated[1] that hydroxyl bonds link up the individual OH groups which are in contact. The structural constants are:

Hydrargillite or gibbsite $Al(OH)_3$	System: Monoclinic	
Space Group: $P2_1/n$	Axes:	Angle:
Z = 8	a = 8·624 Å	
	b = 5·060 Å	β = 85° 26′
	c = 9·699 Å	

and the idealised atomic arrangements are illustrated in Fig. IV.10 and may be compared with brucite.

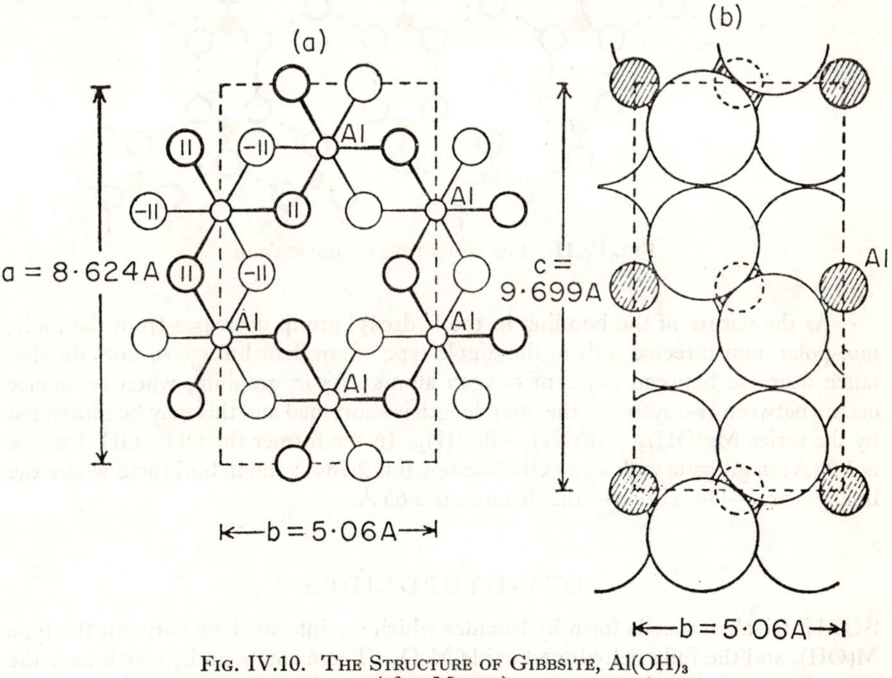

FIG. IV.10. THE STRUCTURE OF GIBBSITE, $Al(OH)_3$ (after Megaw)

(*a*) Schematic; (*b*) the relative positions of atoms in adjacent layers.

[1] Megaw, H. D., *Zeit. Krist.*, 87, 185, 1934.

Gibbsite is a naturally-occurring mineral, but when the tri-hydrate of alumina is prepared artificially by precipitation from a salt, the resulting form, *bayerite*, has a rather different atomic configuration. The structure has been described by Montoro[1] as having hexagonal symmetry.

The hydroxide of *beryllium* is similar to gibbsite but all available octahedral positions in the lattice have to be filled, because the cation is divalent.

The extreme (acid) end of the hydroxyl series is typified by *boric acid*, H_3BO_3. The structure appears to consist of discrete molecules of $B(OH)_3$ linked to each other through hydroxyl bonds (—O····H····O). A figurative representation is given in Fig. IV.11. *Silicic acid* may have a similar configuration.

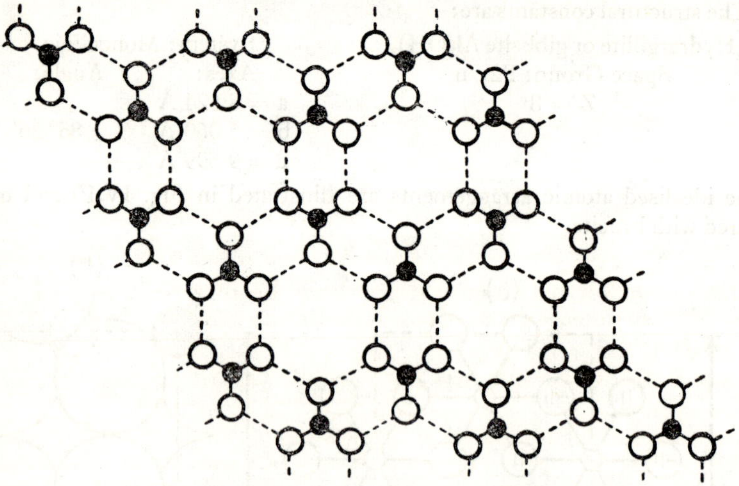

Fig. IV.11. The Structure of Boric Acid

As the nature of the bonding in the hydroxyl group decreases from the ionic, non-polar, non-directed unit to the dipole type of covalent linkage so does the distance decrease between adjacent oxygen atoms. As in graphite, when resonance occurs between two systems, the bond length is shortened and this may be illustrated by the series $Mg(OH)_2$—$Al(OH)_3$—$B(OH)_3$. In the former the OH—OH distance is 3·22 Å; in gibbsite with some OH bonding it is 2·78 Å; and in boric acid where the linkages are —O····H····O— the distance is 2·65 Å.

OXY-HYDROXIDES

SOME trivalent metals form hydroxides which are intermediate between the type $M(OH)_3$ and the fully dehydrated oxide M_2O_3. These compounds, which have the general formula $MO(OH)_n$, are termed *oxy-hydroxides*. The oxy-hydroxides of

[1] Montoro, V., *Ricerca Sci.*, 13, 565, 1942.

aluminium and iron are isodimorphous, i.e. they are capable of existing in two forms as follows:

Diaspore	α-AlO.OH	Goethite	α-FeO.OH
Boehmite	γ-AlO.OH	Lepidocrocite	γ-FeO.OH

The structure of *diaspore* has been deduced by Takané[1] and Deflandre;[2] it is based on hexagonal close-packing of oxygens, with aluminium atoms lying between six oxygen atoms and every oxygen touching three cations. The contribution which each anion receives from the aluminium atoms in contact with it, is therefore three half valencies. The structure can be electrically balanced only if the hydrogen atom indicated by an asterisk behaves as a small cation with planar configuration between two oxygens, thus donating a half-valency share to each. The structure is illustrated in Fig. IV.12, and the constants are:

Diaspore $HAlO_2$	System: Orthorhombic
Space Group: Pbnm	Axes:
Z = 4	a = 4·43 Å
	b = 9·36 Å
	c = 2·80 Å

By using neutron diffraction techniques, Busing and Levy[3] have confirmed the hydrogen bonding and located the position of the proton precisely.

Goethite is of similar structure but with dimensions of a = 4·54 Å; b = 10·0 Å; and c = 3·03 Å.

Boehmite and *lepidocrocite* have a different configuration from the α-modifications but are similar to each other. The structure is shown in Fig. IV.13. The configuration is a little complicated at first sight as the lattice is composed of layers each of which has three planes of oxygen atoms. Cations, either of aluminium or iron, are in distorted six-fold co-ordination within these layers. Half the oxygens of a layer touch four cations and thus receive their full valency share; the other half are in contact with only two cations but are linked to oxygens in other layers through two hydrogen atoms, thereby completing their valency requirements. Therefore it is better to represent the formulæ of boehmite and lepidocrocite as RO(OH) whereas diaspore and goethite are more logically described as HRO_2.

The constants of lepidocrocite are:

Lepidocrocite FeO(OH)	System: Orthorhombic
Space Group: Amam	Axes:
Z = 4	a = 3·87 Å
	b = 12·51 Å
	c = 3·06 Å

Boehmite has the same structure but with a = 3·69 Å, b = 12·2 Å and c = 2·86 Å.

[1] Takané, K., *Proc. Imp. Acad. Tokyo*, 9, 113, 1933.
[2] Deflandre, M., *Bull. Soc. Franc. Min.*, 55, 140, 1932.
[3] Busing, W. R. and Levy, H. A., *Acta Cryst.*, 11, 798, 1958.

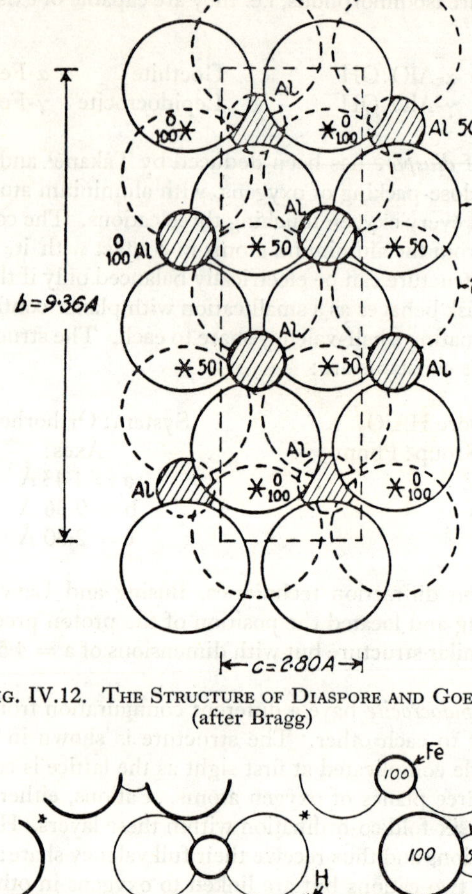

FIG. IV.12. THE STRUCTURE OF DIASPORE AND GOETHITE
(after Bragg)

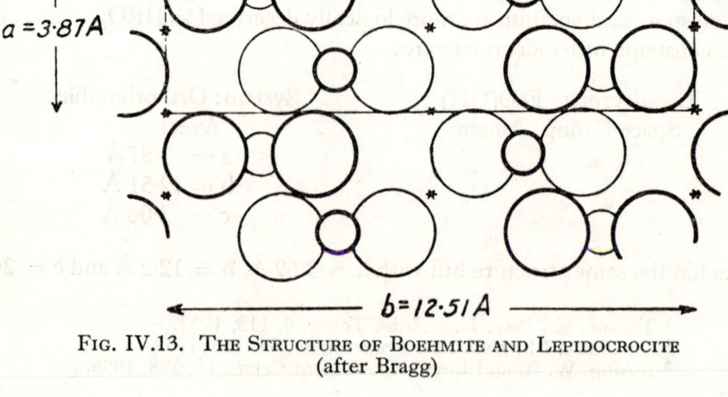

FIG. IV.13. THE STRUCTURE OF BOEHMITE AND LEPIDOCROCITE
(after Bragg)

HYDRATES

MANY compounds containing the elements of water are not definite chemical entities, but have the molecules of water attached very loosely to the structure. Water is lost at comparatively low temperatures without complete disruption of the compound and involves little rearrangement within the crystal itself. The term 'hydrates' is often used to distinguish compounds of this type from the *hydroxides* in which the water is held firmly as an integral part of the structure.

Some hydrate structures will be considered in the following pages.

CARBONATES

THE carbonates of the alkaline earth elements form a very important series of possible basic refractories. The mineral magnesite and the composite calcium-magnesium carbonate, dolomite, are used as such but the others fail to form stable oxides and cannot, *as yet*, be used on an extensive scale.

The atomic structure of these carbonates is quite simple despite the fact that the complex unit or ion CO_3^{2-} is present. Bragg[1] has shown that the structure of *calcite* (calcium carbonate, $CaCO_3$) closely resembles that of *rock-salt*, NaCl, in which calcium ions replace sodium and the carbonate radicle takes the place of chlorine. But, because the carbonate radicle is non-symmetrical, it lowers the symmetry and destroys the cubic arrangement of the unit cell. Calcite is rhombohedral with constants of:

Calcite $CaCO_3$	System: Rhombohedral
Space Group: $R\bar{3}c$	Axis: $a = 6{\cdot}361$ Å
$Z = 2$	Angle: $\alpha = 46° 07'$

An interesting feature of the alkali carbonates is the change-over of structural type with an increasing ionic size of cation. Elements which have smaller ions than calcium form crystals similar to calcite itself, but those of larger sizes have a different configuration. Calcium carbonate has the distinction of being able to form both modifications, the first of which is rhombohedral *calcite*, which has already been mentioned, and the other is *aragonite* of orthorhombic symmetry.

The two series of minerals are compared in Table IV.VII, which also gives the ionic sizes of the cations and the lattice parameters.

The essential difference between the calcite and aragonite structures lies in the relative positions of the carbonate ion and the calcium ion. In both cases, the calcium ions are in hexagonal close-packing, but in the calcite group, each oxygen of the carboxyl group touches two cations, whereas in the aragonite group they are in contact with three cations (see Fig. IV.14). The $(CO_3)^{2-}$ ion must therefore be placed asymmetrically in aragonite with respect to two layers of calcium ions.

[1] Bragg, W. L., *Proc. Roy. Soc.*, A89, 468, 1914.

TABLE IV—VII. LATTICE CONSTANTS OF THE ALKALINE EARTH CARBONATES

	Mineral	*Composition*	*Size of ion*	*Space group*	*Parameters* *Axis*		*Angle*
Calcite Group (rhombohedral)	Magnesite	$MgCO_3$	0·65 Å	R3̄c	a = 5·66 Å		α = 48° 10′
	Siderite	$FeCO_3$	0·76	,,	5·84		47° 45′
	Smithsonite	$ZnCO_3$	0·74	,,	5·87		48° 20′
	Rhodochrosite	$MnCO_3$	0·80	,,	5·84		47° 20′
	Calcite	$CaCO_3$	0·99	,,	6·75		47° 07′
	Dolomite	$(CaMg)CO_3$	—	R3̄	6·00		47° 30′
	Ankerite	$Ca(Mg,Fe)(CO_3)_2$	—	—	—		—
					Axes *a*	*b*	*c*
Aragonite Group (orthorhombic)	Aragonite	$CaCO_3$	0·99	Pcmn	4·94 Å	7·94 Å	5·72 Å
	Strontianite	$SrCO_3$	1·13	,,	5·12	8·40	6·08
	Cerussite	$PbCO_3$	1·32	,,	5·14	8·45	6·10
	Witherite	$BaCO_3$	1·35	,,	5·31	8·83	6·54
	Bromlite	$(Ca,Ba)CO_3$	—	—	—	—	—

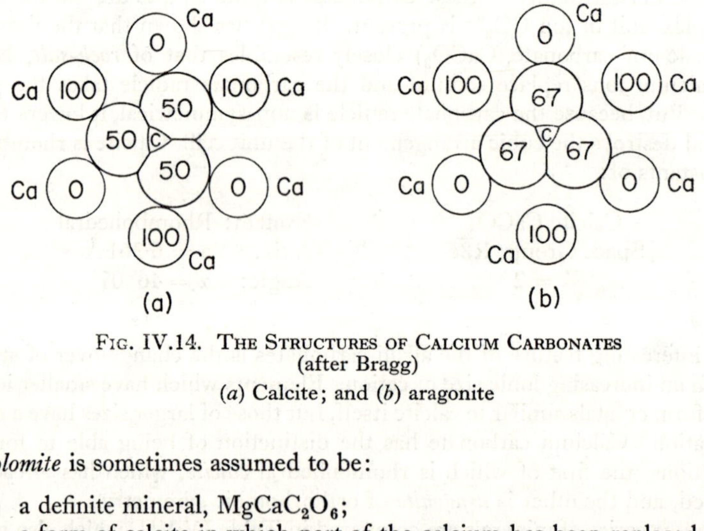

FIG. IV.14. THE STRUCTURES OF CALCIUM CARBONATES (after Bragg)
(*a*) Calcite; and (*b*) aragonite

Dolomite is sometimes assumed to be:

(*a*) a definite mineral, $MgCaC_2O_6$;

(*b*) a form of calcite in which part of the calcium has been replaced by magnesium;

(*c*) a mixture of calcium carbonate and magnesium carbonate.

X-ray examinations show that it is a single substance,[1] but of variable composition.

Some deposits known as *magnesian limestone* may be mixtures of calcium carbonate and magnesium carbonate and should not be described as 'dolomite'.

The *simple alkali carbonates* (soda, potash and lithium carbonate) are, ceramically, of interest mainly because of their occurrence as impurities in clays, etc., and

[1] Bradley, A. L. and Jay, A. H., *Iron and Steel Inst. Sp. Rep.*, No. 33, 48, 1946.

their use as constituents of glaze-frits. Their atomic and crystal structure is of small importance in ceramics, and for that reason is not described in this volume.

SULPHATES

SULPHATES are also examples of complex ion configurations, and once again two types of structure are associated with the sulphates of the important group of divalent elements.

The elements with large cations form structures of the *barytes* type; those with small ions have the *anhydrite* structure; calcium sulphate can only have the latter structure.

Both barytes and anhydrite crystallise with orthorhombic symmetry, but their space groups are different. In *barytes*, $BaSO_4$, a barium ion lies between twelve oxygens of sulphate radicles; in *anhydrite*, $CaSO_4$, a calcium ion is linked to eight oxygens, not all of which are at the same distance from it.

The constants of minerals in the sulphate group are given in Table IV.VIII.

TABLE IV—VIII. LATTICE CONSTANTS OF THE ALKALINE EARTH SULPHATES

	Name	*Chemical composition*	*Size of ion*	*Space group*	*Parameters*		
					a	*b*	*c*
Anhydrite Group (orthorhombic)	Anhydrite	$CaSO_4$	0·99 Å	Bbmm	6·22 Å	6·96 Å	6·97 Å
Barytes Group (orthorhombic)	Celestite	$SrSO_4$	1·13	Pnma	8·36	5·36	6·84
	Anglesite	$PbSO_4$	1·32	,,	8·45	5·38	6·93
	Barytes	$BaSO_4$	1·35	,,	8·85	5·44	7·13

Gypsum, $CaSO_4.2H_2O$, is monoclinic in structure and like some hydroxides has a layer-lattice. Each layer contains sulphate radicles $(SO_4)^{2-}$ linked to calcium ions as shown in Fig. IV.15. A calcium ion is surrounded by eight units, six of these are oxygen ions, of which there are two from each of three sulphate groups, and the other two co-ordinating groups are water molecules. The linking between layers is through these water molecules which have weak bonds and give rise to cleavage along these planes.

Plaster of Paris, $2CaSO_4H_2O$, has a structure akin to zeolites in that water may be removed without destroying the crystal lattice. Thus on heating to about 300° C. it forms *soluble anhydrite*, $CaSO_4$, which has the same structure and reacts in the same way with water. On heating to a higher temperature a further change in structure occurs and *normal anhydrite* is formed.

Alkali sulphates occur chiefly as adventitious impurities in clays and other ceramic materials. Their structure is of small importance to ceramists and is therefore omitted from this volume.

Alunite, $KAl_3(OH)_6(SO_4)_2$. The mineral *alunite* is sometimes associated with clays. Its structure is rhombohedral with each aluminium atom surrounded by four OH groups and two oxygen atoms from sulphate ions. Potassium ions are in twelve-fold co-ordination with six oxygen atoms and six OH groups.

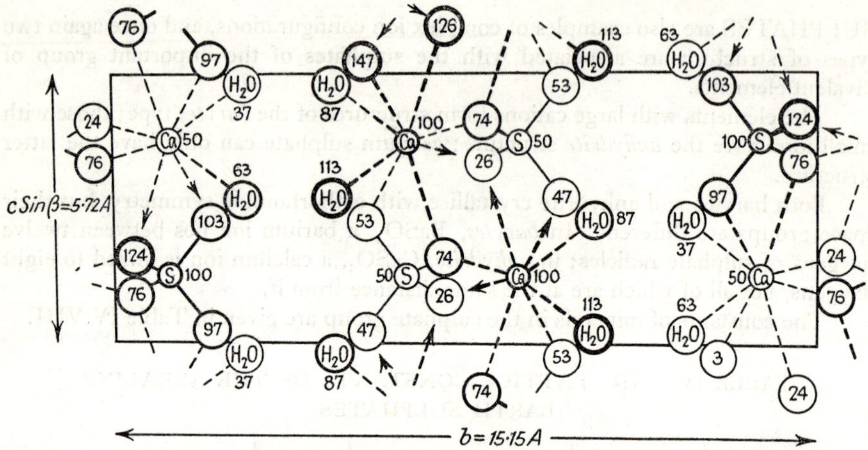

FIG. IV.15. THE STRUCTURE OF GYPSUM (after Bragg)

There are many minerals in the alunite group but only three need be mentioned; these are *alunite* itself, the iron isomorph, *jarosite*, $KFe_3(OH)_6(SO_4)_2$, and the sodium analogue, *natro-alunite*, $NaAl_3(OH)_6(SO_4)_2$.

The structural constants of alunite[1] are:

Alunite $KAl_3(OH)_6(SO_4)_2$	System: Rhombohedral
Space Group: R3m	Axis: a = 7·05 Å
Z = 1	Angle: α = 59° 02′

The oxides and complex salts of the alkaline earth elements provide an interesting study of the value of crystal structure interpretation in explaining the properties of the minerals.

TABLE IV—IX. CATION/ANION RATIOS AND STABILITIES OF ALKALINE EARTH SALTS

	Cation radius Å	*Oxide* (*anion radius 1·40Å*)	*Carbonate* (*anion radius 2·40Å*)	*Sulphate* (*anion radius 3·30Å*)
Magnesium	0·65	0·465	0·271	0·197
Calcium	0·99	0·705	0·414	0·301
Strontium	1·13	0·808	0·472	0·342
Barium	1·35	0·964	0·563	0·427

[1] Gossner, B., *Zeit. Krist.*, 96*A*, 488, 1937.

All the structures may be regarded as co-ordination groupings of the cations with anionic forms of differing sizes. In every case the co-ordination number is six, so that the most stable form results when the size ratio of cation : anion is close to the ideal for octahedral grouping, namely 0·414. Table IV.IX. shows these values and the compounds of greatest stability are underlined.

OTHER OXYGEN SALTS

Phosphate minerals, particularly *apatite*, occur in some clays and so do *coprolites* and *phosphorite* produced from organic remains, but these are only 'impurities' and their structure is of little importance. The most important phosphate used in chinaware is *bone ash*, made by calcining bones. Little is known of its structure but it appears to resemble that of silicates, the PO_4-ions in bone ash occupying the same positions as the SiO_4-ions in calcium silicates.

In a PO_4 tetrahedron, three oxygen atoms form normal covalencies and each possesses a free valency whilst the fourth is inactive. The P—O—P bond angle is fixed at about 126° but sufficient ionic character is retained for this angle to be variable and to enable glasses to be formed. As the PO_4 tetrahedron has only three linkages a phosphate glass has a low softening point.

Other groups which occur in minerals include phosphates, tungstates, borates, etc. Many of these have not been investigated fully and others are beyond the scope of this volume. An interesting feature of some borates is that they have a framework structure based on boron and are comparable with the silicates (see also p. 187).

SULPHIDES

THE sulphides are a highly complex group of minerals because the anion of sulphur is capable of existing in structures in two forms. In one series it behaves as a single ion S^{2-}, larger than oxygen, but in certain circumstances capable of being polarised and distorted from a spherical shape. In the other series, two sulphur atoms link together by a single covalent bond to form a double unit S_2^{2-}; this has a dumb-bell shape and in crystal structures behaves rather like a complex grouping (p. 85).

The two series are known as the mono- and di-sulphides respectively.

MONOSULPHIDES

MINERALOGICALLY the monosulphides form an important group although few are of interest in ceramic studies. Many of the important metallic sulphides are of this form and include galena PbS, zinc blende ZnS, pyrrhotite FeS, chalcopyrite $CuFeS_2$.

Chalcopyrite, $CuFeS_2$, is the well-known copper pyrites. Its structure is based on that of *zinc-blende*, ZnS, which crystallises in the tetragonal system. Each sulphur atom lies between two copper and two iron atoms which are situated at the corners of a tetrahedron. The structure is shown in Fig. IV.16, and the constants are:

Chalcopyrite $CuFeS_2$	System: Tetragonal
Space Group: I42d	Axes:
Z = 4	a = 5·24 Å
	c = 10·30 Å

The unit cell of chalcopyrite is twice as large as that of zinc-blende and corresponds to two cubes one above another.

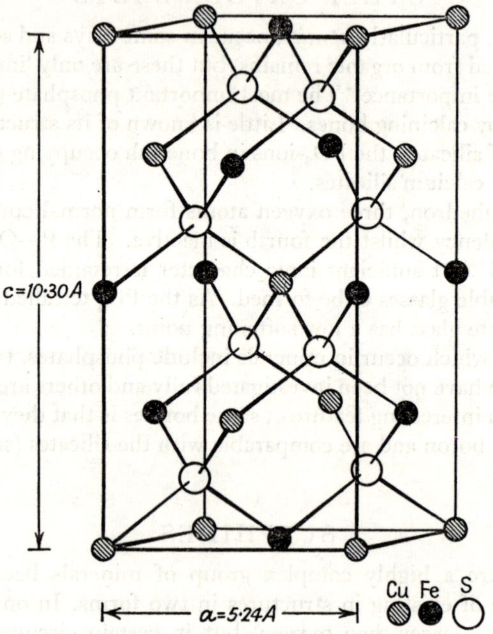

FIG. IV.16. THE STRUCTURE OF CHALCOPYRITE
(after Pauling and Brockway)
Dark-shaded circles, iron atoms; light-shaded circles, copper atoms; open circles, sulphur atoms

Embescite, $CuFeS_2$, is a variety of chalcopyrite found in some clays.

Galena, PbS, is chiefly of interest as one of the main sources of lead and the lead oxides. It has a cubic structure like rock-salt but with six equidistant sulphur atoms at 2·97 Å.

DISULPHIDES

THE more typical iron sulphides, particularly those found in clays, are of this group.

Pyrite,[1] FeS_2, is the most commonly occurring form of iron sulphide. Its crystal structure may be compared to that of rock-salt, NaCl, in which the sodium

[1] The term *pyrites* is a general one for all iron sulphides, and its use for *pyrite*, FeS_2, is incorrect, but frequent. The letter *s* should be omitted when pure, crystalline pyrite is intended.

ion is replaced by iron and the chlorine ion by a dumb-bell group of two sulphur atoms. The structure is illustrated in Fig. IV.17, and the constants are:

Pyrite FeS_2	System: Cubic
Space Group: Pa3	Axis: a = 5·40 Å
Z = 4	

The sulphur to sulphur distance is only about 2·10 Å, thereby suggesting that a considerable amount of covalent linking is present.

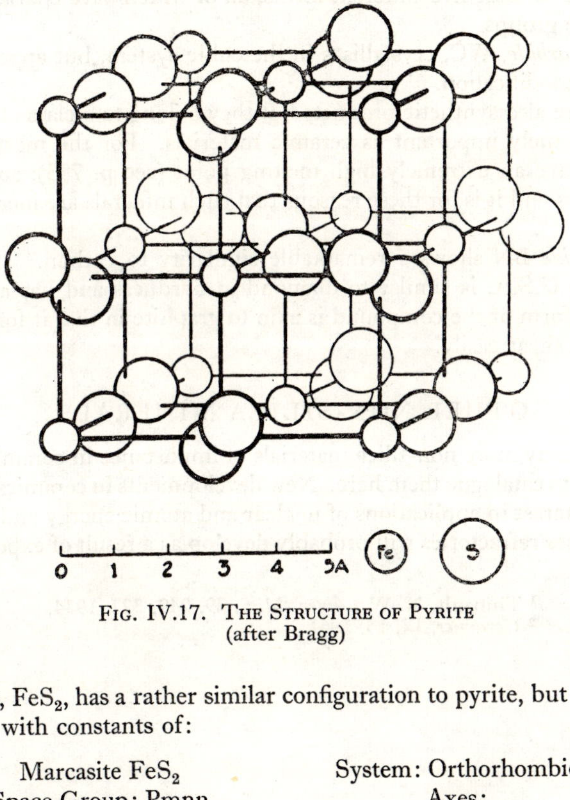

FIG. IV.17. THE STRUCTURE OF PYRITE
(after Bragg)

Marcasite, FeS_2, has a rather similar configuration to pyrite, but the unit cell is orthorhombic with constants of:

Marcasite FeS_2	System: Orthorhombic
Space Group: Pmnn	Axes:
Z = 2	a = 3·37 Å
	b = 4·44 Å
	c = 5·39 Å

CARBIDES AND NITRIDES

CARBIDES do not occur in nature in commercially useful quantities. Those used industrially are all manufactured and are usually known by trade names.[1]

[1] *Carborundum* is the registered trade mark of carbides manufactured by the Carborundum Co. and should not be applied to carbides made by other firms though this is often done.

Those of ceramic interest are *silicon carbide* and *tungsten carbide*. These commercial products are not usually very pure crystals; some of them vary considerably from the 'ideal' formula, M_xC_y, and probably consist of mixtures of carbides, carboxides and unaltered raw materials.

Silicon carbide, SiC. The crystal structure of silicon carbide was first determined in 1944 by Thibault,[1] who found that among a profusion of types, there are two main modifications—an α-form with a hexagonal structure and a β- or cubic variety. The former, which predominates in commercial abrasive powders, may be subdivided into at least five different forms, all of which have characteristic structures and space groups.

Tungsten carbide, WC, crystallises in the cubic system, but appears to exist as more than one modification.

Nitrides are also synthetic products but they belong to a class of new minerals which are extremely important as ceramic materials. For the most part nitrides and carbides have an extremely high melting point (see p. 783); some behave as semi-conductors and it is for these reasons that such minerals are increasing rapidly in importance.

Boron nitride BN shows a remarkable similarity to carbon.[2] 'Borazon' synthesised in the U.S.A. is similar to diamond in hardness and has a cubic lattice, whilst another form of the compound is akin to graphite in that it forms hexagonal platelets which shear easily.

OTHER NON-SILICA MINERALS

THERE are many more non-silica materials of importance in ceramic studies, but it is impossible to catalogue them here. New developments in ceramics include those of particular interest in applications of nuclear and atomic energy and a new field in high-temperature refractories will probably develop as a result of experiments along these lines.

[1] Thibault, N. W., *Amer. Min.*, 29, 249, 327, 1944.
[2] *Ceramics*, 12, 15, 1961.

CHAPTER V

THE IDENTIFICATION AND ESTIMATION OF MINERALS IN CERAMIC MATERIALS

MANY of the properties of clays and other ceramic materials are dependent on the nature and amounts of the various minerals in them, so that their identification, determination and estimation[1] are of very great importance. The problem is made more difficult because ceramic materials are rarely composed of one pure crystal type. Usually, several minerals are present in appreciable quantity, and many others as minor constituents; whilst the identification of a particular pure mineral may be comparatively easy, there is usually much more difficulty when others are present especially if they are of similar type.

In some cases, one mineral cannot be differentiated from others with which it is associated unless there is some prior purification. This is particularly true of many clays, and details of a useful method for separating the minerals from such materials are given later in this chapter. Clays are very diverse in composition and some contain many mineral types. In addition to the clay minerals (see Chapter III), quartz, micas and other components may be present in considerable amounts.

Furthermore, many ceramic materials are of small grain size. This is again exemplified by clays where particles as small as 100 Å (i.e. 10^{-6} cm.) are of frequent occurrence. The identification of such minerals is extremely difficult for they are too small to be observed except with an electron microscope.

The mineralogist is also faced with the difficult problem of isomorphous substitution, for many minerals vary within wide limits of chemical composition. In many instances different grains of a particular mineral may show pronounced differences in property and composition; this is particularly true of micaceous flakes in sedimentary deposits.

Therefore, in general, methods of analysis must be capable of detecting and estimating minerals which may not be of constant composition, which are usually admixed with other types and sometimes of very small grain size.

The identification of a mineral must depend on some fundamental characteristic which is always the same, independent of the mode of occurrence or the nature of the locality or environment. Methods of mineral identification have been devised which make use of a particular property or class of properties. They can be summarised as:

[1] A *determination* is the result of definitely ascertaining the position, nature, amount, etc., of anything by measurement, weighing or counting.

An *estimation* is the result of approximately ascertaining numbers, quantities, magnitudes, etc., without actual enumeration, weighing or measurement or by using such limited means as are insufficient or unsuitable for producing an accurate and reliable result. For many purposes an estimation is sufficiently accurate but from its nature it cannot (except by coincidence) be as positive as a determination. This distinction is important but is not always realised.

1. Those dependent on the chemical nature of the mineral.

2. Those dependent on the external crystallographic features of the mineral, i.e. the microscopic properties.

3. Those dependent on the internal atomic or ionic arrangements within the crystal lattice; this group includes X-ray, infra-red and electron diffraction analysis.

4. Those dependent on some physical or chemical change within a mineral which can be measured under controlled conditions, such as differential thermal analysis (i.e. the measurement of heat changes which occur during reactions).

Other methods are known and used to identify and even roughly estimate minerals present in ceramic materials. Most of these use an inferred property of a mineral, and provided that only one mineral in a mixture exhibits this property the method can be of value. The magnetic properties of a substance may sometimes be indicative of the presence of iron-bearing minerals. The cation exchange capacity (see p. 264) may differentiate between montmorillonite and kaolin-containing clays. Properties such as plasticity, thixotropy, etc., also are of value in some cases in suggesting the presence or otherwise of certain minerals.

Sampling. The identification of minerals in a ceramic material does not usually require great care in the selection of samples. However, when it is necessary to ascertain the proportion of any mineral present, it is essential to procure one or more samples of ample size which are truly representative of the whole. If the total quantity of the material is large, it is highly important that any sample shall be typical of the whole and not of one (possibly abnormal) portion. Unless very great care is taken in obtaining a 'fair sample' the results of testing may be seriously in error.

A single piece, selected haphazardly from the mass, will be unlikely to be fairly representative, and even a large number of such casually selected samples may not be satisfactory. Very frequently large variations occur, so that it is most important that proper care should be taken to obtain a truly representative sample. When a large, natural deposit of clay or other material is to be analysed, the samples should be taken from many different parts of the bed, both horizontally and vertically. It is often desirable to have several distinct selections from different parts of the deposit, so as to determine whether the material is constant in composition throughout the whole bed. Where the material appears to be fairly uniform, about 1 per cent of the whole of the material may suffice for the preliminary sample from which the final is to be taken, but with a very heterogeneous material, 2 per cent or even 5 per cent may be necessary to obtain a reasonably typical sample. The preliminary sample should be composed of numerous smaller ones selected systematically from every part of the deposit and not merely from those which are the most readily accessible.

If the material to be examined is contained in trucks or in a heap, it is equally necessary to take an ample quantity in the form of small samples from many parts. Samples from large quantities of material contained in trucks or shiploads are often taken by means of mechanical appliances.

The preliminary sample is next treated as an original material from which a representative sample is to be obtained by the process known as 'quartering', which is usually effected as follows:

The material should be thoroughly mixed and placed in a symmetrical and rather flat pile and divided into four equal parts. Two opposite parts are removed, the material in them is again mixed and then quartered, this being continued until a sample weighing about 28 lb. is obtained. This is coarsely crushed and again quartered, the final half (about 7 lb.) being sent to the analyst, who should then obtain a representative sample by still further crushing and quartering until a suitable quantity is obtained. When very small samples are required for analysis, fine grinding and thorough mixing are very essential.

The samples sent to the analyst should not be ground too finely before transit, otherwise some compounds may be oxidised and so render the results inaccurate; thus, ferrous oxide may be oxidised into ferric oxide, and pyrite (FeS_2) to iron sulphate or oxide. Grinding is also undesirable because some minerals, especially those of platey character, may be destroyed. Phosphor-bronze instruments should be used as far as possible instead of iron or steel ones in sampling, grinding and sifting, so as not to include any adventitious iron in the sample. Other useful information relating particularly to the sampling of clays is to be found in a publication of the British Standards Institution No. 1902 (1967).

In analysing clays and other ceramic materials, techniques to determine the overall composition on an elemental or oxide basis may be employed, otherwise methods to identify and estimate individual mineral components may be utilised. Both approaches are useful but they require very different experimentation and interpretation of results.

ELEMENTAL ANALYSIS METHODS

THE determination of the total types and amounts of elements present in a sample is probably the most favoured control tool of ceramists. Often, however, taken by itself, such an analysis has little meaning. Whilst chemical entities frequently determine the firing characteristics and reactivity in chemical changes, they are of little significance in explaining physical features.

Nevertheless, elemental analysis can be regarded as an exact science, provided that due care is taken, whilst many mineral analytical techniques are fraught with errors and uncertainties. In certain circumstances an accurate knowledge of the total amount of each element present can be used to compile an exact mineral balance sheet in which the mineralogical tests are used mainly for identification and semi-quantitative measure.

The role of trace amounts of minor elements in clays, glasses and glazes is just beginning to be appreciated. The analysis of these components is being developed.

Elemental analysis involves many techniques mostly designed to obtain quick results. Modern laboratories tend to use several methods to combine accuracy with rapidity of analysis.

CHEMICAL METHODS

THE chief purpose of a full chemical analysis is to reveal all the elements present in a sample and the proportion of each. In some instances it may reveal the presence of an unsuspected mineral which might, otherwise, be overlooked.

The results of a chemical analysis of any substance are usually expressed in terms of the various oxides present, such as silica, alumina, titanium oxide, ferric and ferrous oxides, lime, magnesia, soda, potash, carbon dioxide, sulphur dioxide, water, etc. Such an analysis does not indicate the forms in which the various elements are combined and, when sulphides are present, the reporting of the results in terms of oxides introduces errors (usually small, but sometimes important) for which allowance must be made.

Prior to commencing a chemical analysis, the sample should be dried and ground as finely as possible—usually in an agate mortar, which does not add any appreciable amount of silica to the sample. Insufficient grinding may cause serious errors, especially when the substance is slow to react.

Many silicates and other naturally-occurring minerals are not readily attacked or broken down by common laboratory agents. The first step, therefore, in chemical analysis is to convert the mineral entities into a form in which the individual elements or oxides can be easily separated from each other. The following methods are in fairly common use:

(*a*) The silica may be evaporated by heating with hydrofluoric and sulphuric acids in a platinum dish; the residue of sulphates or fluorides of the other elements, which are mostly readily soluble, may then be analysed. For many purposes, this removal of the chief constituent (silica) is particularly convenient. Some minerals are slow to react under this treatment, particularly those which have been produced at high temperatures (e.g. mullite). For this reason it is advisable to repeat the evaporation until a constant weight of residue is produced.

(*b*) The sample may be fused with molten sodium and potassium carbonates, sodium hydroxide or, for some purposes, with a mixture of calcium carbonate and ammonium chloride. Virtually all minerals are decomposed by this treatment and dissolve in the molten mass. Soluble carbonate salts of the non-acidic oxide components are dissolved in dilute hydrochloric acid, to convert them to chlorides, and silica is precipitated from sodium silicate.

Whichever method is used to decompose the original sample, the common procedure of analysis is to precipitate, gravimetrically, compounds of the elements present. Use is made of the different solubilities of salts under certain conditions; thus silica is insoluble in hydrochloric acid whereas chlorides of aluminium, iron, calcium, etc., are completely soluble; hydroxides of aluminium, iron, chromium and titanium may be precipitated in a weak alkaline solution provided that the chloride ion concentration is high, yet under these same conditions, calcium, magnesium and the alkalies remain in solution. These latter elements may be separated by utilising the different solubilities of their oxalates and phosphates.

The alkali elements, sodium and potassium, form very few insoluble salts and consequently are difficult to estimate gravimetrically. Probably the best method is to use a flame spectrometer (see p. 216), although reliable chemical techniques have been developed. It is important to estimate each alkali element separately rather than to assess one by difference.

The original techniques for silicate analysis as described by A. W. Groves,[1]

[1] Groves, A. W., *Silicate Analysis* (Allen & Unwin, 2nd edition 1951).

which have become known as the classical method, were extremely tedious in the filtration, washing and ignition of precipitates.

Bennett and Reed[1] have evolved a comprehensive form of analysis which, whilst retaining all the accuracy of former practice, enables a full analysis to be completed in a few hours. Instead of systematically plodding through a complete analysis on one sample, they recommend that several aliquots be treated for the determination of individual or small groups of elements. Various solution techniques are advised for these individual operations and the authors make full use of modern instrumental analytical techniques, such as absorption spectrophotometry (p. 217), flame photometry and EDTA titrations.

These methods have become standard for chemical analysis throughout the ceramic industry and form the basis of a British Standard.[2]

Chemical analysis as a means of mineral identification and estimation will obviously be of the greatest value when one particular mineral predominates in the sample, for if that mineral has a simple formula, it can be accurately assessed and the nature and amounts of the minor, accessory minerals may also be indicated. However, if a mixture of minerals is analysed, the result will give the total of each element or oxide, but no information about the nature and type of the actual minerals containing those elements. So, in a typical clay containing quartz, clay minerals, mica, etc., the total silica content on analysis will come in part from each silicate mineral present and the actual contribution of each cannot be determined. Taken by itself, therefore, a chemical analysis of a composite rock, mineral deposit or fired product is of little value in the estimation and detection of the minerals present, but when used in conjunction with other methods of analysis, it can be of very great importance.

To a ceramist who is not specifically concerned with the derivation of mineralogical composition, a chemical analysis is often very useful. For example, the alumina/silica ratio of a clay may indicate the approximate value of that material as a refractory. Similarly, the total content of alkalies in a clay will show whether the material is likely to have a high melting point, but it would not be safe to predict the vitrification temperature on this result alone. Soda and potash, when in the form of felspar, are more detrimental to the refractoriness of a clay than when they are in the form of mica, because the former melts more readily. Sodium silicate is still more easily fusible.

A chemical analysis is also valuable in showing the presence of some deleterious ingredients, such as (i) lime, which may destroy bricks and tiles made of clay containing it, or (ii) ferric oxide, which may either discolour or enhance the beauty of the product, depending on whether a white or red material is desired, or (iii) alumina, which in siliceous refractories greatly reduces the load-bearing capacities at high temperatures.

A complete analysis, including those elements which are present in very small proportions (less than 0·1 per cent), is very rarely required but the possible need for it

[1] (*a*) Bennett, H. and Reed, R. A., *Analyst*, 92, 466, 1967; (*b*) Bennett, H. and Reed, R. A., B.C.R.A. report, Special Publication No. 61, 1969.

[2] (*a*) British Standards Institution, B.S. 1902, Part 2A, 1964; (*b*) British Standards Institution, B.S. 1902, Part 2B, 1967.

should never be ignored, as the omission of a small percentage of some constituents may make adequate interpretation of the results of the analysis very difficult. For this reason, no constituents should be estimated 'by difference' in order to make the analysis complete, though this is frequently done. It is especially important that the alkalies should not be 'estimated' in this way, as a small error in the proportion of them may have a serious effect in the interpretation of some of the properties of the material. Duplicate determinations should always be made and, if careful and accurate procedures are adopted, they should not differ by more than 0·1 per cent.

Selective Chemical Analysis. The separation of one or more minerals from others in a mixture by a chemical agent is, where practical, of great importance. Such separation may be achieved in either of two ways: (i) where a single mineral entity is attacked by a specific reagent and converted into a form in which it can be extracted from the other components, and (ii) where all the ingredients of a composite material are destroyed except one, which can then be estimated.

Probably the best example of the first method is the estimation of carbonates in rocks by measuring the carbon dioxide evolved on treatment with dilute hydrochloric acid. The type of carbonate may then be determined by analysing the cations passing into the acid solution. Also of value are the methods of estimating free oxides of iron and sometimes aluminium in clays. These minerals are present in many cases, not as discrete particles, but as a colloidal film surrounding grains of other minerals and may be troublesome if a complete mineral balance is required. The method of Dion[1] has been used with considerable success, where, under carefully controlled conditions, goethite can be quantitatively removed from clays. Iron oxide dissolves in a weak alkaline solution in the presence of nascent hydrogen, conditions which are secured by treating the clay with a boiling solution of ammonium tartrate in an aluminium or porcelain vessel. The rate of reaction is accelerated by the addition of spirals of aluminium which should be freshly cut from sheet metal at the start of the experiment.

Hallimond[2] has outlined a method for the preferential solution of chlorites in dilute acid solutions, but the raw material must be free from other acid soluble components.

A chemical method[3] has been described which is claimed to be capable of distinguishing between four-fold and six-fold co-ordinated aluminium in layer structures.

The chemical estimations of quartz or free silica involve destroying all other minerals by violent acidic reagents, which have little or no effect on the silica component. Trostel and Wynne[4] favour the use of fused potassium bisulphate, whereas Shaw and Skinner[5] prefer to use a mixture of sulphuric and hydrochloric acids. In both methods, any silicates that may be present are decomposed, thereby leaving some colloidal silica in suspension, which can be removed by careful treatment with sodium hydroxide.

These methods attempt a compromise because too violent a treatment unquestionably results in the solution of some free silica; on the other hand, some

[1] Dion, H. G., *Soil Sci.*, 58, 41, 1944.
[2] Hallimond, A. F., *Spec. Rep. Min. Res. Mem. Geol. Survey Gt. Britain*, vol. 29.
[3] Brindley, G. W. and Youell, F., *Acta Cryst.*, 4, 495, 1951.
[4] Trostel, L. J. and Wynne, D. J., *J. Amer. Ceram. Soc.*, 23, 18, 1940.
[5] See Nagelschmidt, G., *The Analyst*, 81, 210, 1956 (Appendix).

silicates—e.g. felspar and mullite—require prolonged treatment before they are completely decomposed. Correction factors have recently been proposed to allow for the errors that are introduced.

An excellent chemical technique for determining the sulphide content of clays has been evolved by Neglia.[1]

The total colloidal matter of clays can also be determined chemically.[2]

Alexiades and Jackson[3] have used selective dissolution analysis (S.D.A.) for quantitative mineralogical determinations by chemical means.

SPECTROGRAPHIC ANALYSIS

If atoms are excited by a sufficiently high external energy source a condition will be reached where electrons in normal orbit will be raised to one of higher energy. As the electrons fall back to their normal position, the energy is released as radiation of a characteristic wavelength most usually in the range 2000–8000 Å. The atoms of each element have different radiation spectra and if the emission from a sample is suitably recorded, the spectral lines will indicate all the elements present. Some elements excite readily but others need to be raised to extremely high temperatures before the electron orbitals can be disturbed. It is customary, therefore, in techniques designed to identify every element present in a sample, to use an electric arc or a series of sparks from an induction coil or a bank of condensers and to select spectra throughout the whole wavelength range.

Thus, if a sample of the powder to be analysed is placed between carbon rods and a high-frequency current is applied, the radiation emitted by the electron excitation within the atoms can be projected through a prism on to a photographic plate. As the radiation from each element is of a characteristic wavelength, a series of lines is produced on the plate, each of which is a result of an electron change in the particular element. These lines may be calibrated and the elemental composition of any particular powder determined. The amount of a particular element may often be estimated, as it is proportional to the intensity of its characteristic line or lines in the spectrographic pattern, provided that the test has been carried out under standard conditions. The method is rapid and is of especial value in the identification and estimation of elements present in small amounts. The line-pattern produced by a single element may contain a profusion of lines; hence, if many elements are present, the resulting pattern is complicated and difficult to analyse. Increasing use is being made of spectrographic methods, and their value as a means of estimating alumina associated with silica in raw quartzites and in fired bricks has been demonstrated by Mason and Speight.[4] A comprehensive survey of spectrochemical methods and their application in the analysis of silicates has been published by Ahrens.[5]

[1] Neglia, S., *Clay Min. Bull.*, 243, 1961.
[2] (*a*) Mitchell, B. D. and Mackenzie, R. C., *Soil Sci.*, 77, 179, 1954; (*b*) Wada, K. and Greenland, D. T., *Clay Minerals*, 8, 241, 1970.
[3] Alexiades, C. A. and Jackson, M. L., 'Clays and Clay Minerals', *Proc. 14th Nat. Confer.* (Pergamon, 1966).
[4] Mason, G. L. and Speight, G. E., *Trans. Brit. Ceram. Soc.*, 50, 35, 1951.
[5] Ahrens, L. H., *Quantitative Spectrochemical Analysis of Silicates* (Pergamon, 1955).

Flame Photometry

Most students of chemistry are aware that sodium compounds when introduced into a flame, colour it yellow, and the violet and brick-red colorations of potassium and calcium respectively are equally well known. These characteristic emissions are again the result of disturbances in the electron orbits of the elements, but where the wavelength of the emission is in the visible region of the spectrum.

Within the last few years, the development of commercial flame photometers has greatly relieved the tedium of many of the processes in silicate analysis and, at the same time, increased the accuracy with which many elements can be estimated. The essential features of the instruments are a suitable flame and a means of ejecting a solution of the sample into it at a predetermined rate. A filter, designed to transmit only the characteristic emission of a particular element, is placed between the source and a photocell so that the intensity of radiation may be measured and compared with standard samples tested under the same conditions.

The content of calcium, potassium and sodium in a material can be rapidly and accurately assessed in most spectrometers and others can be adapted to estimate lithium, barium and other elements.

Atomic Absorption

If a compound can be dissociated into its elemental forms, the electron orbits will absorb radiation normally emitted by that element when in an excited state. When this effect is utilised, accurate determinations of the nature and amounts of elements can be deduced.

Unfortunately the elements to be analysed have firstly to be prepared quantitatively in solution form. The aqueous phase must then be injected into a flame sufficiently high in temperature to dissociate all compounds of the element being analysed and to volatilise them without exciting the atomic form. Emission of a characteristic wavelength of the element from a standard lamp is passed through the flame so that a proportion of it, corresponding to the amount of element present, is absorbed by the element. In practice only one element is analysed at a time because it is preferable to use a single-element lamp. Multi-element lamps are available but they have certain limitations.

Different flame temperatures have to be selected for elements depending on their dissociation and volatilisation characteristics. Some elements, such as arsenic, selenium, tin, volatilise readily in a gas/air flame, but others, including those of greatest interest in ceramic practice, require to be heated in a nitrous oxide/acetylene flame. Table V.I, due to T. S. West,[1] lists the flames required for a range of common elements.

The method has potential for the analysis of clays and other ceramic materials, but only the most sophisticated units capable of utilising high-temperature flames are suitable for the wide range of elements.

[1] West, T. S., *Chem. and Ind.*, 12, 387, 1970.

TABLE V—I. ELEMENTS CAPABLE OF BEING DETERMINED BY ATOMIC ABSORPTION

Flame	*Elements detectable*
Air/H_2	As, Se, Sn.
Air/propane	Ag, Cs, K, Li, Na, Rb.
Air/acetylene	Ag, Au, Ba, Bi, Ca, Cd, Co, Cr, Cs, Cu, Fe, Ga, Hg, In, Ir, K, Li, Mg, Mn, Na, Ni, Pb, Pd, Pt, Rb, Ru, Rh, Sb, Sr, Tl, Zn.
Nitrous oxide/acetylene	Al, B, Be.

Absorption Spectrophotometry

A similar principle to that of atomic absorption is employed in colorimetric absorption spectrophotometry but the radiation is absorbed by a solution of the element which has been complexed to give a coloured solution, usually in the visible region, but a wider range into the infra-red and ultra-violet can be utilised. Most commercial instruments cover a wavelength range from 800 to 8000 Å.

Provided that an element can be dissolved from the mineral form in which it is contained, and provided also that it will complex with some agent to give a stable characteristic lake, it can be determined accurately by absorption spectrophotometry.

The method is particularly applicable to some of the components commonly found in clays, such as silica, alumina, iron oxide, titania, and the method is a useful accessory technique to supplement chemical methods.

X-ray Fluorescence

In this technique, a mineral is bombarded by a wide wavelength range of X-rays (p. 232). If the intensity is sufficiently large, the inner electron orbitals of the elements present become excited and emit their own characteristic X-ray wavelength —usually the K orbits give the most distinctive lines. In theory, X-ray fluorescence can be applied to a powdered sample directly, but errors can arise when more than one mineral type is present, each of which may have different absorption characteristics to X-rays. Clays and similar materials are best analysed as a glass produced by fusing them with a suitable flux. This technique is most useful for analysing elements with atomic numbers in the middle of the Periodic Table.

Electron Excitation Analysis

A similar type of effect to X-ray fluorescence can be produced by using an electron beam as a source of bombardment. The inner electron orbits of elements emit characteristic X-ray wavelength radiation under these conditions but there is less penetration into the sample. From many points of view this is an advantage and the technique is particularly valuable in determining elements of lower atomic numbers, as are most of those in clays and other alumino-silicates. In the author's experience, much more reproducible results and greater accuracy can be achieved by using a sample made up as a uniform glass. In electron excitation techniques the sample has to be intimately admixed with carbon to provide conductivity for the electron beam.

Both X-ray fluorescence and electron excitation methods are rapid and a simultaneous analysis of a wide range of elements is feasible on the same sample. Where multiple analyses are required in a very short time, the methods are to be recommended but high capital cost is involved, particularly in X-ray fluorescence methods, and skilled and dedicated operators are essential if accurate and meaningful results are to be obtained.

Electron Probe Analysis

The tendency for elements to absorb an electron beam and then emit radiation of characteristic wavelength is utilised in electron probe methods which are extremely valuable not only in elemental analysis but in locating the positions of various elements in crystals, glasses, across reaction boundaries and in other applications.

A powerful beam of electrons from a hot wire source is allowed to impinge on the surface of a section of the material under investigation. A particular wavelength of a suspected element is selected and transmitted to a receptive measuring unit. The complete section, often greatly magnified, can be scanned in this way and the distribution of the element in it thus ascertained. The procedure can be repeated for other elements.

The technique is especially valuable for studies of solid/solid reactions or of slag penetration mechanisms. Along with I. A. White, the author has studied the weathering breakdown of mica flakes and the consequent removal of potassium ions. Fig. V.I shows the potassium scan of such a flake; the dark areas are where the alkali ion has been more severely leached and is present in lower concentration.

Polarography

The establishment of the principle and techniques of polarography is due to J. Heyrovsky.[1] Basically a solution of a particular element or series of elements, in an appropriate base electrolyte, is subjected to a continuously increasing electrical potential. If conditions are correctly chosen, an element will accept an electron at a characteristic potential and undergo a reduction in its valency state, which is detected and its amount measured accurately by a conductivity change. The method is capable of wide application but it is more generally used for estimating elements of the transition group which change valency readily and tend to form complex ions in solution. It is also a useful accessory technique to be used in conjunction with conventional chemical analysis.

A useful summary has been published by Davis.[2]

Chromatographic Analysis

This method of analysis involves the use of a comparatively new technique, but which is likely in the coming years to supersede many of the conventional analytical

[1] Heyrovsky, J., *Polarographischen Praktikum* (Lange and Springer, Berlin, 1960).
[2] Davis, J., R.I.C. Publ., 1964.

procedures.[1] Chromatographic analysis depends on the fact that when different substances are absorbed, suspended or dissolved in a medium, they travel through that medium at different rates, so that a separation of them is effected. The method has been used to a considerable extent in gas analysis, but it is equally effective in the estimation of mixed solutions. A common technique is to allow the solution to run slowly down an inclined paper impregnated with a suitable resin; after a certain period of time, the components of the solution will have reached different levels on the paper, where they can be estimated, usually by colorimetric techniques.

Most analytical techniques of the instrumental variety give results which require to be compared with calibrated standards before accurate determinations can be obtained. This can present serious difficulty because the nature of the standard and the 'environment' of the element in it can produce anomalous results. In solution analysis, calibrated amounts of the dissolved element can be tested under similar conditions to the unknown provided that impurity elements do not interfere.

The problem of standards for solid sample analysis has been mitigated by the Bureau of Analysed Samples Ltd., Middlesbrough, which supplies several materials containing a wide range of elements, all of which have been accurately calibrated using various analytical techniques.

Pecsok and Shields[2] have produced an excellent summary of the modern analytical techniques and for further information the reader is referred to this and other similar works of reference.[3]

Recalculated Analysis. Although the methods of analysis so far described indicate only the type and amounts of elements present in a sample, it is sometimes convenient to recalculate the results to give an indication of the mineralogical composition. At the best, only an approximate mineral balance will be given, and where the material is complex, containing several components, so many assumptions have to be made that the recalculation may be of little value. In samples where the major mineral entities are known or can be identified by other means, recalculation of the chemical analysis can be most informative. Fireclays, for example, contain predominantly three mineral entities, quartz, mica and the kaolin clay mineral, livesite; the total alkali content will be present as mica (apart from a small proportion of exchangeable cations associated with the clay mineral) so that the proportion of this mineral can be assessed; the alumina content comes from both mica and livesite, but knowing the content of the former, the latter can be calculated; the remaining silica, after subtracting that present in the mica and livesite, will be as quartz or some other form of free silica. Recalculated analysis used in this way can give a mineralogical balance which is in close agreement with that obtained when each mineral is assessed separately by other methods. It can be extremely misleading, however, if a major component is overlooked.

[1] Williams, T. I., *The Elements of Chromatography* (Blackwell Ltd., Oxford, 1954).
[2] Pecsok, R. L. and Shields, L. D., *Modern Methods of Chemical Analysis* (John Wiley & Sons, London, 1968).
[3] Ewing, G., *Instrumental Methods of Analysis* (McGraw-Hill, 1969).

MINERALOGICAL ANALYSIS METHODS

Microscopic Examination and Analysis[1]

THE observation of the external form of minerals either by eye or under magnification has been one of the principal methods of identification and analysis from the earliest days of mineralogy. It was recognised that particular mineral species always formed crystals of exactly the same type and shape, and with definite characteristics whereby they could be differentiated from others. The use of the *polarising* microscope establishes not only the external crystal form but also many other important optical properties of minerals.

A microscope is also invaluable for observing many other facts about ceramic materials (both raw and manufactured) in either their natural, polished or powdered state. It may be used to study defects in ceramic bodies and to investigate the texture of both raw materials and manufactured products.

For many purposes, a hand-magnifier is sufficient, for others a low-power microscope magnifying 20–50 diameters, but for an exhaustive examination a good, modern, petrological microscope with a magnification of up to one thousand times is needed. Useful accessories to such an instrument include a universal stage, wedges and plates of different optical properties, sub-stage and analyser diaphragms, a Bertrand lens for observing interference figures, micrometer scales and camera equipment for obtaining permanent records.

In clay mineralogy, great difficulty is encountered because particles may be present which are too small to be seen even under the highest magnification of an optical microscope. Under certain conditions, however, many clay minerals can be detected by measuring the optical properties of an oriented aggregate and where large fragments of other minerals are present in a clay, the microscope is an invaluable aid to their identification and often to their estimation.

Ceramic materials and the minerals in them can be examined under a microscope in either transmitted or reflected light. In the former, a beam of light is projected upwards through the specimen but in the latter case the light is shone downwards on to the specimen and only the surface contours are viewed through the microscope. Although the former method is more generally applicable and reveals most information about the optical properties of the minerals present, reflected light microscopy is becoming increasingly important, especially when the material examined is heterogeneous or not entirely transparent, and where its surface is flat or has been polished.

Preparation of the Specimen or Sample. The specimen may be prepared in several ways, but if it is to be viewed in transmitted light it must always be of thin section, i.e. less than two-thousandths of an inch in thickness, or in the form of a powder. If the material is sufficiently hard, the customary way of preparing sections is to saw a small piece of about a half of an inch square section and a tenth of an inch

[1] The microscopical examination of minerals has been described in great detail by numerous authors. The reader is recommended to consult (i) Winchell, A. N., *Elements of Optical Mineralogy* (Chapman & Hall, London, 1947); (ii) Rigby, G. R., *The Thin-section Mineralogy of Ceramic Materials* (British Ceramic Research Association Pub., Stoke, 1953). (iii) Hartshorne, N. H., and Stuart, A., *Crystals and the Polarising Microscope* (Arnold, London, 1970).

thick. One face must then be ground perfectly flat by means of carborundum or other suitable abrasive powder. After drying, this section may be mounted on a glass slide with Canada balsam which has been previously heated to ensure that it will harden on cooling. The ground face of the specimen must be in contact with the glass at all points, and there must be no entrapped air bubbles. The rough upper surface of the mounted specimen may then be ground, again using carborundum powder. The final stages of the grinding have to be carried out carefully, using the finest grade of abrasive powder. Whenever practicable, the thickness of the rock slice should be one and a half thousandths of an inch, so that comparative results may be obtained from each section to be examined.

The final stage is to fix a cover glass over the section, to prevent the scattering of light when using high magnification. A small drop of balsam can be melted on to the rock slice and the cover glass may be pressed down on to it using a small cork. Care must be taken to ensure that all air bubbles are removed before the slide is set in a warm place to hasten the hardening of the balsam.

When making sections of ceramic articles or of porous or fragile rocks, it is advisable to impregnate the specimen prior to grinding. A simple way is to add a small amount of warm bakelite to a fragment of the specimen contained in a small cardboard pill-box. Heating on a hot-plate expels the air from the voids in the specimen and the bakelite can penetrate on cooling. Further heat treatment hardens the bakelite and the fragment can then be ground in the usual way.

An alternative method is to add the bakelite as a concentrated solution in acetone. The air may then be removed by evacuation in an enclosed vessel and, at the same time, the acetone will evaporate. The bakelite impregnates the specimen and can be hardened by baking in an oven.

Powder specimens may also be prepared for examination under the microscope by heating a little Canada balsam on a microscope slide until it would harden if allowed to cool: whilst it is still soft a small sample of the powder may be stirred into the warm balsam with a glass rod. When completely cold the section can be ground as before, and a warm cover glass applied on completion. Another method which is of great value if many powders have to be examined rapidly, is to add the sample (powder) to a drop of water or other liquid on a slide and then apply a cover glass. It must be examined before the liquid evaporates, but the method is useful because no grinding is required and platey minerals tend to be oriented along their basal section. The powder must naturally be composed of small-sized particles.

For the examination of specimens in reflected light only the top face need be ground flat, and in many instances better results are obtained by etching this surface with some chemical reagent prior to examination.

Examination of Sections in Transmitted Light

The *Polarising* or *Petrological* microscope enables the thin sections of minerals or rocks to be examined under a variety of conditions, and the salient features of such an instrument are shown in Fig. V.2.

It may be used as a means of high magnification, the magnitude of which is governed by the selected combination of *objective* (O) and *eyepiece* (E) lenses. The *mirror* (M), situated below the *stage* (S) of the microscope, is usually flat on one side and concave on the other, so that a beam of either parallel or convergent light may be projected upwards through the section and instrument; in examinations in ordinary light the flat side is invariably used so that a parallel beam of light is employed.

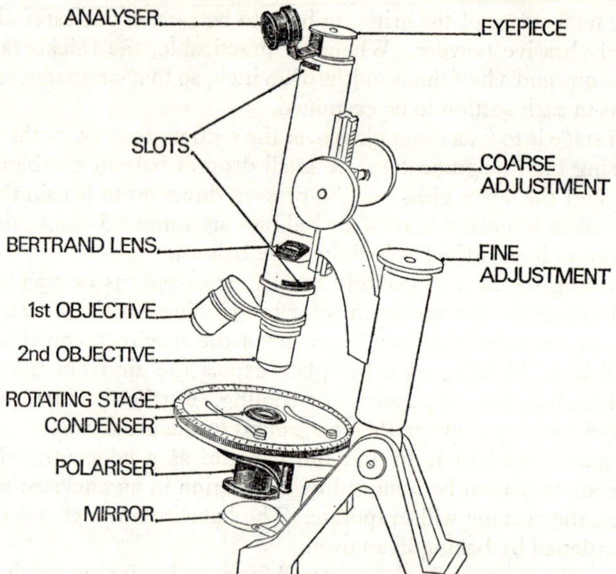

FIG. V.2. FEATURES OF THE PETROLOGICAL MICROSCOPE
The location of the analyser and the Bertrand lens may be different in some types of microscopes, but their function is the same.

Two nicol prisms or other polarising media are provided on all petrological microscopes—one below the stage, which is called the *polariser* (P), and another above the objective, which is termed the *analyser* (A). If the lower one only is inserted in the beam of light then the emerging rays vibrate in only one direction or plane, and the section can thus be examined in *plane polarised light*. If, however, both the polariser and the analyser are placed in position, the light is completely *polarised* provided that the nicol prisms are in the 'crossed' positions, where the vibration direction of one is at right angles to that of the other.

Alternatively, the section can be examined in convergent light in conjunction with crossed nicols when *interference figures* are produced. These can be seen quite clearly either by using the subsidiary *Bertrand lens* (B), or by removing the eyepiece (E). Useful information may be derived from an examination under all these four conditions.

(*a*) **Under Ordinary Light.** The first examination is best made using a low-power magnification. The composition of the section as a whole will then be revealed; many different mineral species may be observed, some of which may be transparent, others opaque and occasionally some may be coloured. The particle size of each mineral may be compared, and if the eyepiece is fitted with a calibrated scale then the actual size of each crystal can be measured.

The texture and general appearance of the section may indicate whether the material is of igneous or sedimentary character.

After the preliminary investigation has been completed, high-power magnification ($\frac{1}{4}$ in., $\frac{1}{8}$ in. and $\frac{1}{12}$ in. objectives in turn) may be used to study each different mineralogical variety in more detail. Sometimes the *crystal habit* or form of a mineral may be revealed in a well-developed fragment; in other cases, the angles between adjacent sides or faces can be measured using the rotating, calibrated stage of the microscope to bring first one side and then the other parallel to the cross-wires in the eyepiece. Care must be taken in such measurements, as the angle between faces is dependent on the inclination of the crystal with respect to the flat face of the section.

In most cases, the minerals present in a rock slice will be imperfect and the external characteristics will offer little indication of the crystal form or habit, but some minerals show pronounced *cleavage*, or a tendency to fracture along definite planes. Mica is an example of a mineral with perfect cleavage parallel to the basal section, for, as shown in Chapter III, it is composed of layers or plates, each held together by weak bonds, and it is along such planes that cleavage will develop. Other minerals may show more than one set of cleavage lines. The angle between these planes may be diagnostic of that particular mineral.

One of the most useful determinations which is carried out in ordinary light is the measurement of the *Refractive Index*[1] of the mineral. Briefly, the refractive index of a material is defined as the ratio between the velocity of light in that material and its velocity in air.

The refractive index is characteristic for a particular mineral although its value may vary when measured along different optical directions within the crystal.

The measurement of refractive index is by far the simplest and most direct method of distinguishing between some transparent substances—particularly when they have the same composition but different physical properties, such as quartz, cristobalite, and tridymite, calcite and aragonite, lightly-calcined magnesia and periclase. It is an important means of ascertaining the proportion of unaltered quartz in silica bricks.

The *Becke method* is the most usual way of determining the refractive index of a mineral. A sample of the crushed mineral or rock is immersed in a liquid of known refractive index on a microscope slide. The barrel of the microscope is adjusted until the fragments are in focus and it is then raised slowly. A halo of light either surrounds or is enclosed by the particles when they are slightly out of focus, and this either expands or contracts as the barrel is raised further. The general rule is that when the microscope barrel is raised, the light halo moves from the medium of higher refractive index to the medium of lower index. Thus, if the immersion liquid is of lower refractive index than the mineral, the light halo will move outwards

[1] For the general theory of the refractive index of minerals the reader is referred to the standard textbooks on mineralogy, petrology and optics.

Care must be taken to avoid confusing the term refractive (which relates to the behaviour of light) and *refractory* (which is the capacity of a substance to withstand heat).

from the fragments, and the opposite will occur when liquids of greater refractive index are used.

When the index of the immersion liquid is similar to that of the mineral under examination the haloes are replaced by coloured bands, and if the fragments are transparent and colourless, they will be almost indistinguishable against the background. Provided that a range of liquids with slight variations in refractive index is available, accurate determinations can be made. An alternative method is to mix a liquid with a higher refractive index than the specimen with a liquid of lower index until the correct value is reached. The refractive index of the mixed liquids, and hence that of the specimen, can then be measured by means of a refractometer.

If a specimen, consisting of two minerals of different refractive index, is immersed in a liquid with the same index as one of them, that mineral will be almost or quite invisible, but the other will be clearly observed under the microscope. This method devised by A. B. Dick is a very delicate one and has been used for many years to distinguish tridymite from cristobalite in silica bricks. The solution used is mercuric-potassium iodide which has a refractive index of 1·477, the same as that of tridymite, whilst that of cristobalite 1·484.

A useful set of liquids of value in the determination of refractive indices is given in Table V.II.

TABLE V—II. REFRACTIVE INDICES OF SOME IMMERSION LIQUIDS

Material	*Refractive Index*	*Material*	*Refractive Index*
Methyl alcohol	1·330	Ethyl salicylate	1·523
Ethyl alcohol	1·362	Ethylene dibromide	1·538
Ethyl acetate	1·373	Nitro-benzene	1·552
Amyl acetate	1·396	Benzyl benzoate	1·569
Ethyl bromide	1·424	Bromoform	1·598
Ethylene dichloride	1·445	Iodo-benzene	1·619
Liquid paraffin	1·470	Quinoline	1·625
Xylene	1·490	α-Bromo-naphthalene	1·658
Benzene	1·500	Methylene iodide	1·727

Care should be taken, when using mixtures of liquids, that the components are miscible and have a similar vapour pressure. Suitable combinations which cover the range of refractive indices normally required in mineral studies are as follows:

1·330–1·362 Methyl alcohol and ethyl alcohol
1·362–1·445 Ethyl alcohol and ethylene dichloride
1·445–1·538 Ethylene dichloride and ethylene dibromide
1·538–1·598 Ethylene dibromide and bromoform
1·598–1·658 Bromoform and α-bromo-naphthalene
1·658–1·727 α-bromo-naphthalene and methylene iodide
1·727–1·84 Methylene iodide and various sulphides and sulphur

Once proficiency has been acquired in measuring refractive indices, values can be determined rapidly and with great accuracy. In addition, a composite mixture

Fig. V.1 (*a*)

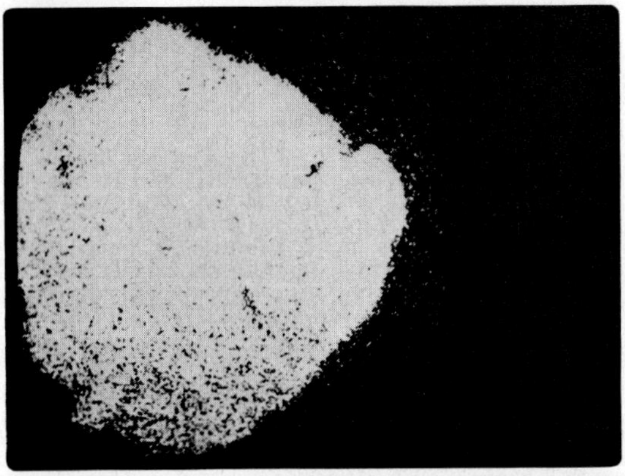

Fig. V.1 (*b*)

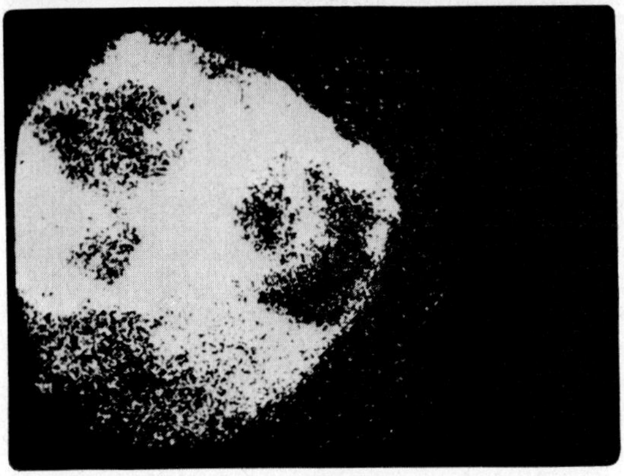

Fig. V.1 (*c*)

Fig. V.1. Photograph of Hydrous Mica by Electron Microprobe
Field = 300 μ

(*a*) Silicon X-ray Image (*b*) Aluminium X-ray Image (*c*) Potassium X-ray Image

[N.B. *The clay content of this paper is approx.* 20 *per cent* *See p.* 923.]

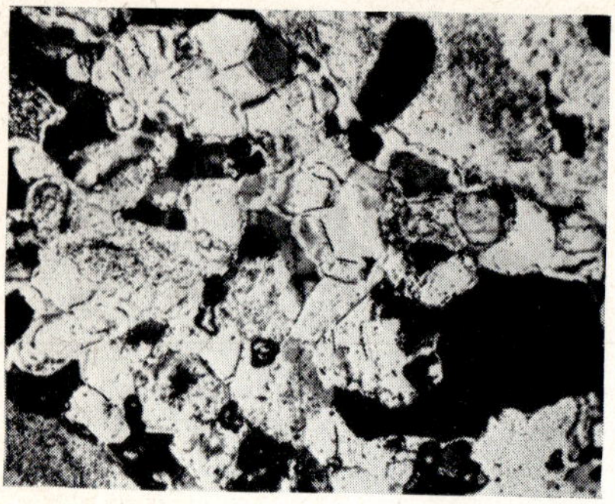

FIG. V.3 (*a*)

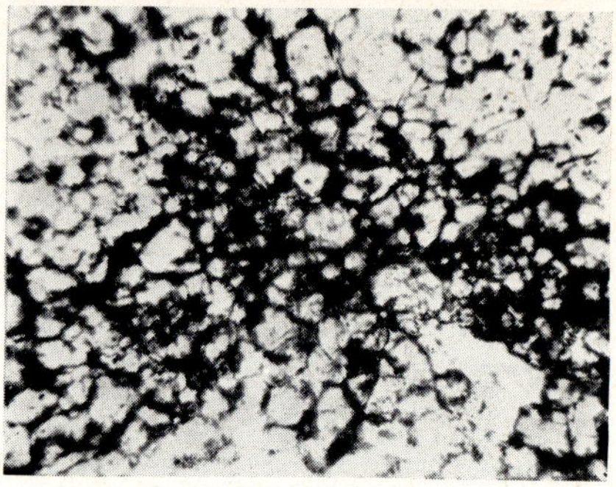

FIG. V.3 (*b*)

FIG. V.3 (*c*)

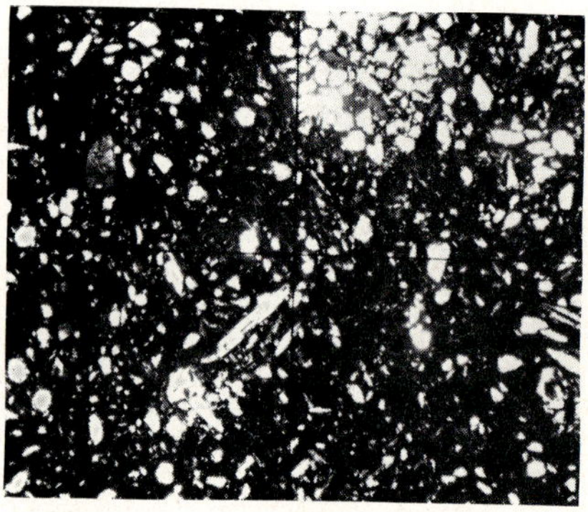

FIG. V.3 (*d*)

FIG. V.3. PHOTOMICROGRAPHS OF THIN SECTIONS OF SOME CERAMIC MATERIALS UNDER POLARISED LIGHT

(*a*) a typical Welsh quartzite. (× 150 magnifications)
(*b*) a silica brick showing cristobalite. (× 800 high contrast technique)
(*c*) tridymite arrow-head twinned crystals in a silica brick. (× 800)
(*d*) a siliceous firebrick. (× 150)

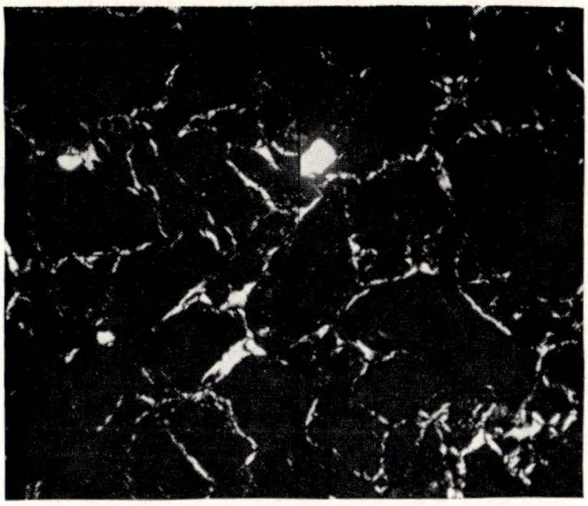

Fig. V.3 (*e*)

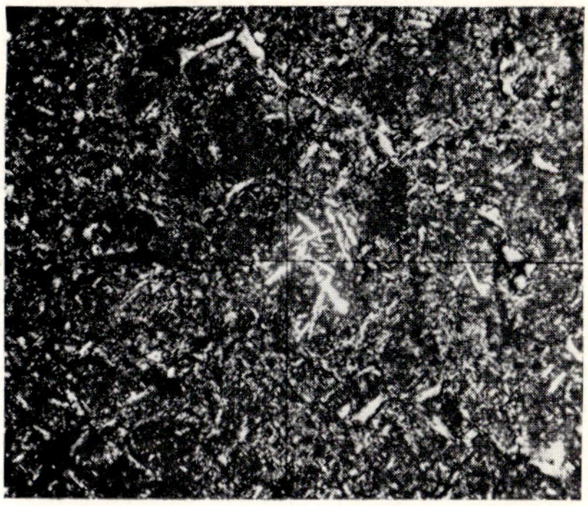

Fig. V.3 (*f*)

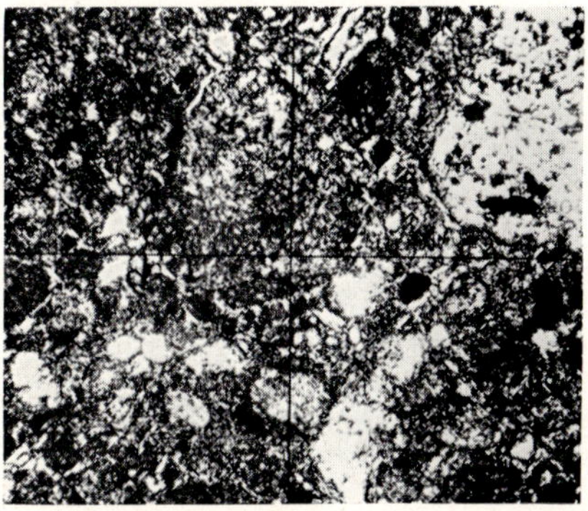

Fig. V.3 (*g*)

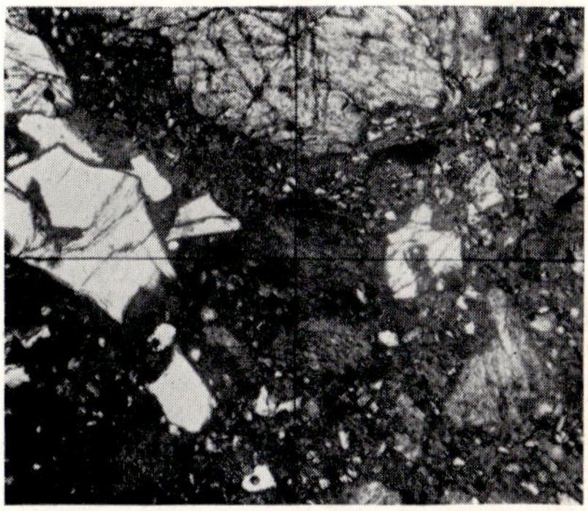

Fig. V.3 (*h*)

Fig. V.3. Photomicrographs of Thin Sections of some Ceramic Materials under Polarised Light

(*e*) a high grade firebrick. (×150)
(*f*) an aluminous firebrick. (×150)
(*g*) a sillimanite refractory. (×150)
(*h*) an alumina refractory. (×150)

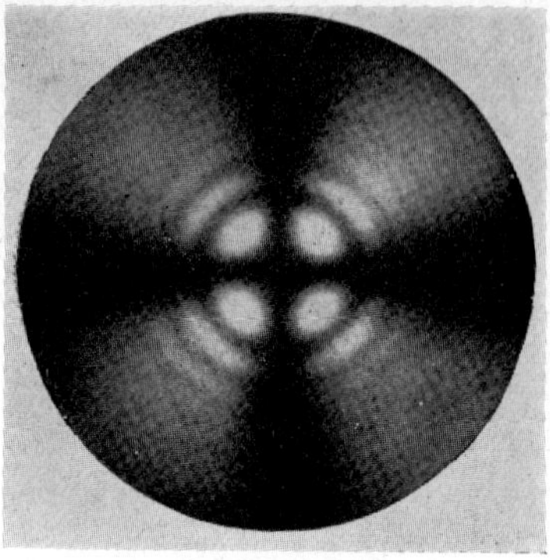

Fig. V.4 (*a*)

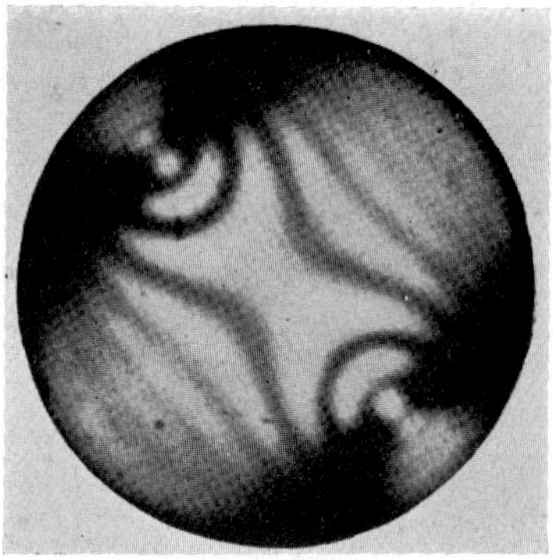

Fig. V.4 (*b*)

Fig. V.4. Ideal Interference Figures of Minerals
(*a*) Uniaxial and (*b*) Biaxial

Fig. V.5 [*opposite*]. Photomicrographs of some Kaolin Minerals
(*a*) nacrite (*b*) dickite and (*c*) kaolinite. (× 500)

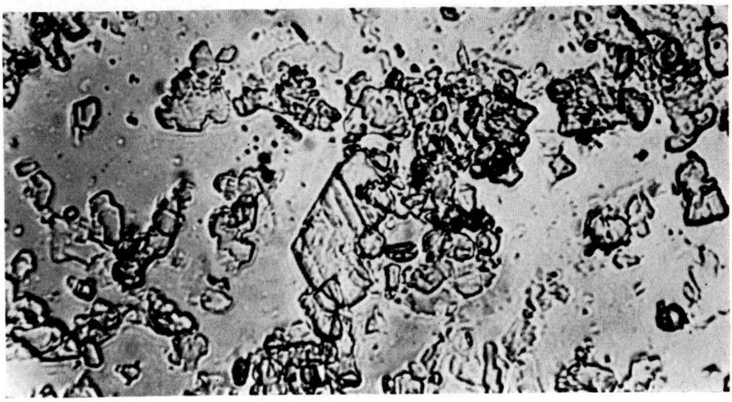

Fig. V.5 (*a*)

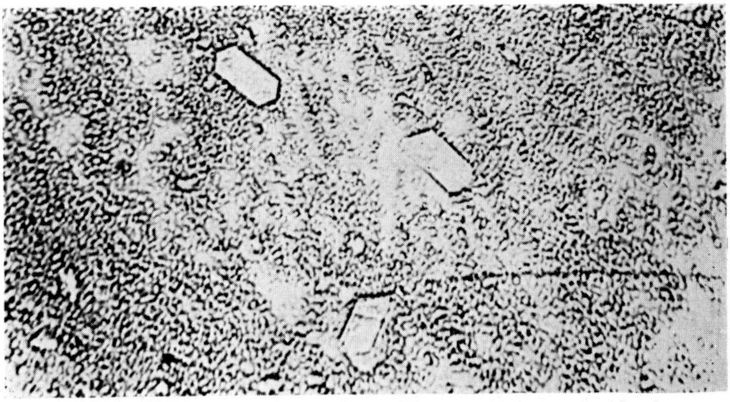

Fig. V.5 (*b*)

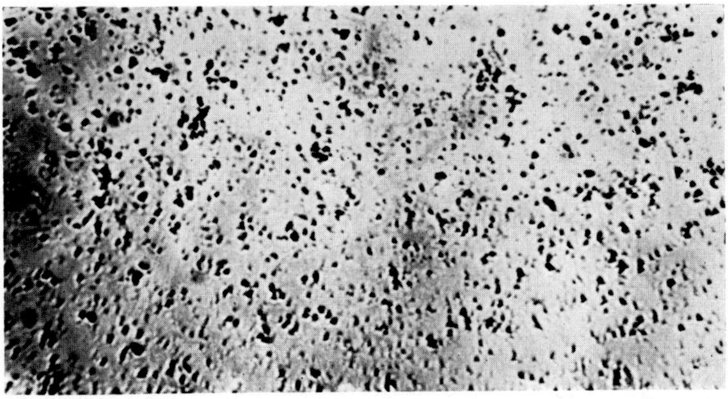

Fig. V.5 (*c*)

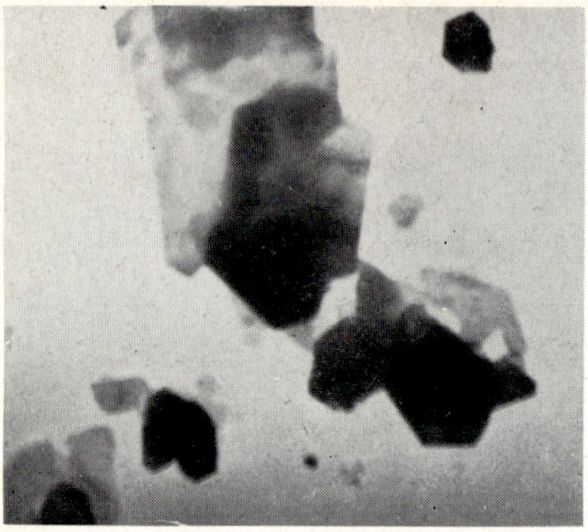

Fig. V.6 (*a*) (*i*)

Fig. V.6 (*b*) (*i*)

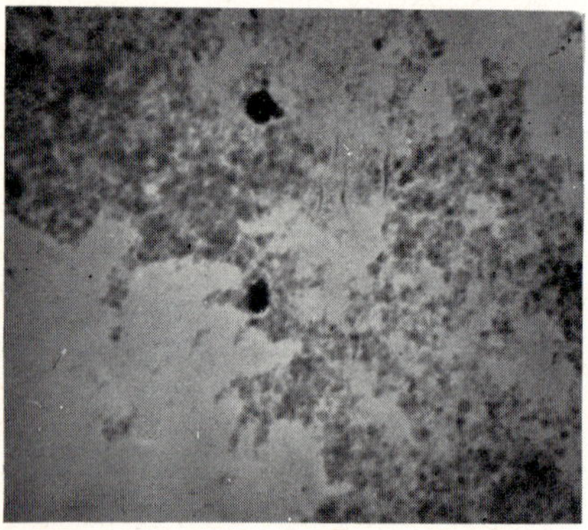

Fig. V.6 (*a*) (*ii*)

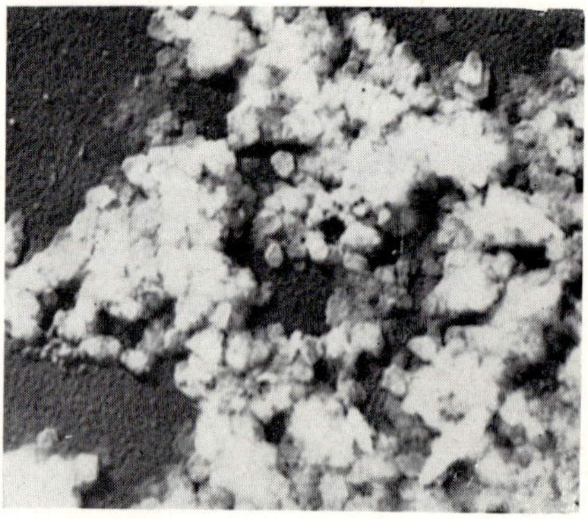

Fig. V.6 (*b*) (*ii*)

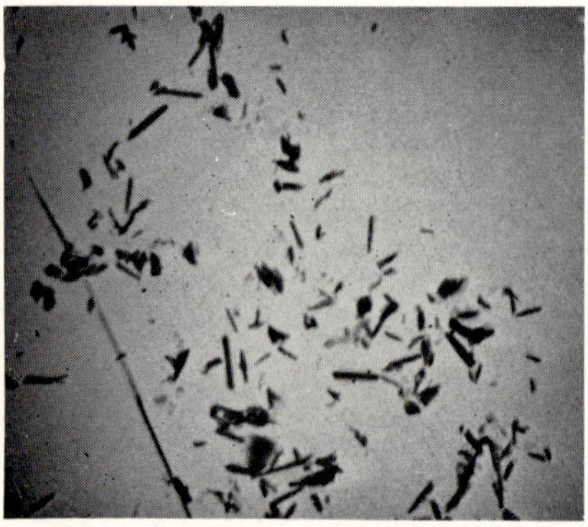

Fig. V.6 (*a*) (*iii*)

Fig. V.6 (*b*) (*iii*)

FIG. V.9

FIG. V.9. X-RAY POWDER PHOTOGRAPHS OF THE SILICA MINERALS
(*a*) quartz (*b*) cristobalite and (*c*) tridymite

FIG. V.6. [*opposite*]. ELECTRONMICROGRAPHS OF SOME KAOLIN MINERALS
(*i*) kaolinite (*ii*) livesite and (*iii*) halloysite. (× 20,000)
(in every case *a* indicates normal, and *b* gold shadow treatment)

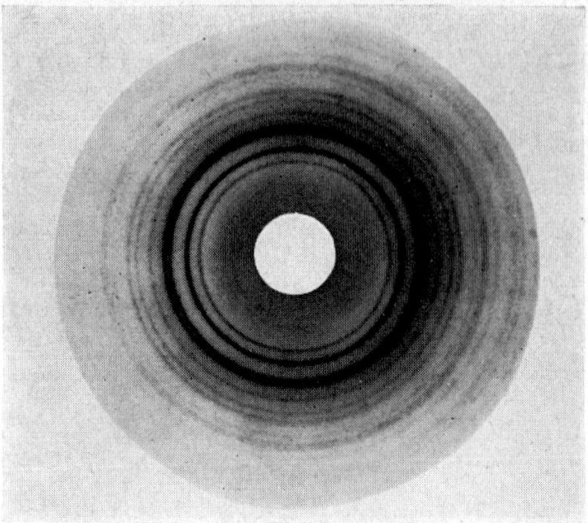

Fig. V.11 (*a*)

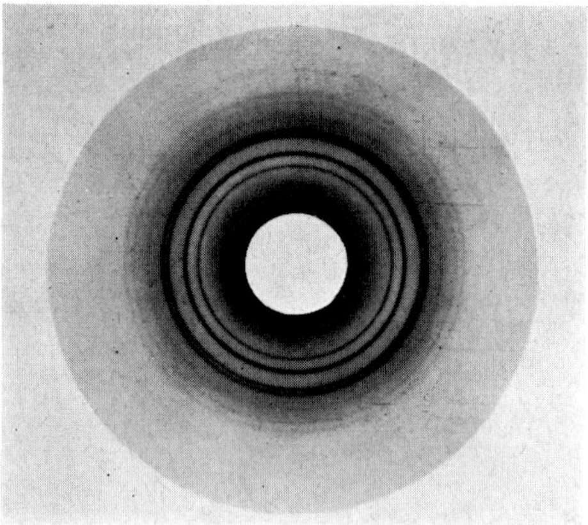

Fig. V.11 (*b*)

Fig. V.11. Back Reflection X-ray Photographs of Compacted Alumina Powders

(*a*) particles of 5 microns (*b*) particles of 0.5 microns

may be analysed without prior separation of the components, and the refractive indices parallel to different crystallographic directions in a mineral can be assessed.

Among other useful information which may be revealed by examination in ordinary light is the *degree of alteration* and the *inclusions* within a particular mineral. Felspars are frequently changed by weathering action to clay minerals, and traces of the alteration product may be visible under the microscope. Quartz crystals often contain occluded fragments of other minerals and, more particularly, liquid or gas bubbles.

(*b*) In **plane-polarised light** some materials show a colour change as they are rotated upon the stage of the microscope. This phenomenon is termed *pleochroism* and is well illustrated by the micaceous mineral, biotite, which exhibits a colour change from a pale to a dark brown.

Some other minerals, which are colourless, exhibit twinkling or rapid changes of relief as the section is rotated on the stage.

(*c*) **Between crossed nicols** or **under polarised light,** minerals fall into one of two classes, (i) *isotropic*, or (ii) *anisotropic*. If a mineral fragment is examined under these conditions and it remains dark during a complete rotation of the section, it is said to be *isotropic*. If, however, the mineral is alternately dark and illuminated as the stage rotates, it is *anisotropic*. Cubic crystals and certain sections of other crystals are isotropic, all others are anisotropic. Minerals of the latter group show darkness (extinction) and illumination four times in succession during a complete rotation of 360° of the stage. If the angle at which the extinction occurs can be referred to a principal crystallographic direction in the mineral, then the *extinction angle* can be measured.

At maximum illumination, a fragment may be white or coloured, depending on the optical orientation of the mineral, its birefringence (see below) and the thickness of the section. The plane-polarised light passing through the sub-stage polariser is resolved into two vibration directions at right angles to each other within the fragments of minerals in the section. In anisotropic crystals, light travels at different velocities along the main crystallographic directions, and therefore, the refractive index is different along each of these axes. If light entering the section is resolved into two components vibrating along two directions at right angles, it follows that one of these beams will be retarded relative to the other. If a certain mineral is oriented along one of its crystallographic axes, the amount of retardation will be directly proportional to the difference in the refractive index along this direction and that at right angles to it. The retardation is characterised by the development, at maximum illumination, of a specific spectrum colour of one of seven orders, and the Newton Colour Scale has been drawn up enabling the retardation to be calculated directly. The difference in the refractive indices in two major crystallographic directions is termed the *birefringence* and is often used in the identification of minerals. It may be calculated from the retardation and the thickness of the section by graphical means. (For further details the reader should refer to the books recommended on p. 220.)

The examination of sections under crossed nicols is of great value, for anisotropic minerals are thrown into relief, especially if they are contained in an isotropic, glassy matrix. In addition, if more than one mineral is present in a section, they may often be distinguished by the differences in birefringence. This is strikingly illustrated by the silica minerals quartz, cristobalite and tridymite. The first is often present as small mineral fragments agglomerated together, as in quartzites, where its comparatively high birefringence bestows a characteristic coloured mosaic when a section is examined under polarised light. Cristobalite and tridymite (which are present in silica bricks) are clearly distinguished between crossed nicols, the former because of the weak anisotropic nature of its small, globular crystals, and the latter by its characteristic twinned arrow-heads. Photomicrographs of these three minerals under polarised light are shown in Fig. V.3*a*, *b*, and *c*, facing p. 224.

Other alumino-silicate refractories show characteristic features in thin sections which can be distinguished under the microscope. Siliceous fireclays usually contain a high proportion of unchanged quartz, which is readily recognised in polarised light (Fig. V.3*d*). Less siliceous fireclays appear amorphous or glassy, although occasionally finely-divided mullite can be observed (Fig. V.3*e*). As shown in Fig. V.3*f*, mullite is the predominant mineral in aluminous firebricks (circa 72% Al_2O_3); this mineral has an unmistakable needle-like structure clearly revealed in polarised light. Sillimanite refractories, manufactured from calcined kyanite, are also rich in mullite and, as Fig. V.3*g* shows, the original grains of kyanite can be observed, criss-crossed with mullite. Corundum, α-Al_2O_3, is the predominant mineral in alumina refractories; it is present as angular fragments of high relief (Fig. V.3*h*).

Many minerals, some of which are commonly associated with ceramic materials, exhibit *twinning*, i.e. crystals exist in which one part is aligned at a different crystallographic direction from another. When such a crystal is observed under the microscope, two adjacent parts have different positions of extinction. Tridymite may be cited as an example, because the 'arrow-head'-like crystals are twins, and the microsection in Fig. V.3*c* shows that when one half of the crystal is in the extinction position, the other half is illuminated.

(*d*) **When convergent light is used in conjunction with crossed nicols,** optical *interference figures* are often shown by mineral sections. These figures can be related to the difference in refractive index along the main crystallographic axes. Uniaxial crystals are those where the refractive index along only one axis is different; in biaxial crystals it is different along all three axes. The interference figures of these two groups are different, and they may be examined under high-power magnification either by inserting the Bertrand lens into the microscope or by removing the eyepiece. The ideal figures are shown in Fig. V.4. The uniaxial figure is composed of two isogyres, or black bands, intersecting centrally to form a cross, and radiating from the centre are a series of circles which are coloured unless monochromatic light is used. The biaxial figure again contains two isogyres, but these are now in the form of hyperbolæ situated in opposite quadrants of the figure. Once again circles are formed but now there are two sets, one at the centre point of each isogyre. Rings with large radii tend to fuse together to form lemniscates and finally ellipses. As the section is rotated on the stage, the biaxial interference figure changes and the isogyres form hyperbolæ of varying angles.

The interference figures of the uniaxial and biaxial crystals may be further differentiated into a negative and positive type, and they can be distinguished by observing the colour change produced in the interference figures when a mica or gypsum plate is inserted in the slot in the barrel of the microscope.

In biaxial crystals, the *axial angle* can be determined from the shape of the hyperbolæ made by the isogyres when they are at a maximum separation. When the isogyres remain almost completely together in the form of a cross during a rotation of the stage, the axial or 2V angle is near to 0°. With a little practice the axial angle can be estimated with a fair degree of accuracy.

The production of interference figures is not always easy, and satisfactory measurements can be made only when the mineral to be examined is present in large fragments which are not too thin. The ideal figures shown in Fig. V.4 are produced only when the crystal is inclined along a main crystallographic direction. If this is not the case, the figure may be difficult to identify, because the centre of the figure will be displaced away from the mid-point of the image. A microscope fitted with a universal stage is of great value, for by its use the crystal can be inclined until the figure is centralised.

Phase Contrast Microscopy is a technique which is proving of the greatest value in the examination of the thin sections of minerals. Briefly, a thin section of the sample is viewed under a single or crossed nicols, a specially designed plate of mica or similar material being inserted to alter the phase of the light. The internal features of some minerals are more clearly revealed and in many cases the crystal form of fragments is brought out. The value of the method in mineralogy has been demonstrated by Zernike[1] and Smithson[2] and undoubtedly the technique will have great application in the field of ceramic materials. Schoblik[3] has applied phase contrast microscopy in studies of glaze discoloration.

A general review of new techniques in microscopy has been published by Vickers.[4]

Examination under Reflected Light

The examination of opaque minerals is not possible by transmitted light, but the use of reflected light applied to the microscopic examination of materials has yielded information of very great value. The method is of particular importance in ceramic studies and has been the subject of a paper by C. Schouten.[5]

The section to be examined can be of any convenient thickness and only the top surface need be polished. Excellent results are sometimes obtained by cementing a thin sheet of clear glass to the section with Canada balsam so as to secure complete optical contact and a smoother surface than that produced by polishing, especially if the specimen is friable. Many modern microscopes are fitted with an attachment which projects a beam of light vertically downward on to the section; this light is then reflected upwards through the objective and measurements may be made in both ordinary and polarised light. The reflectivity of minerals is in direct proportion to their refractive index and this often provides a valuable means of identifica-

[1] Zernike, F., *Physica, Eindhoven*, 9, 686 and 974, 1942.
[2] Smithson, F., *Min. Mag.*, 28, 384, 1948.
[3] Schoblik, A., *Sprechsaal*, 85, 95, 1952.
[4] Vickers, A. E. J., *Research, Lond.*, 8, 17, 1955.
[5] Schouten, C., *Trans. Brit. Ceram. Soc.*, 50, 185, 1951.

tion. Reflected light examination offers additional advantages, but is complementary to, and not intended to replace the use of, transmitted light. More surface detail is revealed by its use, but the standard techniques of transmitted light microscopy are probably of more value in mineral identification.

In reflected light, some minerals in polished sections of ceramic materials can be readily recognised. Corundum (Al_2O_3), for example, always exhibits clear relief and the rounded grains of periclase (MgO) in basic materials are equally clearly shown. Spinel minerals, by contrast, show little relief and many silicates show none. Mullite crystals may often be recognised by their peculiar cross-sections. Defects in glazes and other ceramic bodies are often more conveniently studied in reflected light.

An excellent account of the examination and identification in reflected light of minerals in refractory materials has been presented by Rutter.[1]

More detail may be revealed in a section to be examined in reflected light by etching the surface with a suitable chemical agent. Hydrofluoric acid, for example, attacks some silicates more rapidly than others and helps to differentiate various minerals.

The measurements of the optical properties of minerals, outlined in the previous pages, often enable those species to be identified. Tables have been compiled by various authors which contain the crystallographic data of all minerals so far identified. The appendix to this volume contains a list of the optical properties of some minerals occurring in ceramic bodies, which aid in their identification.

The Estimation of Minerals under the Microscope

In the examination of ceramic materials by means of a microscope, a little of the powder or a thin section of the specimen is used; it normally consists of fragments of several minerals. The area of the thin section which is occupied by the particles of a particular mineral is in direct relation to the volume of that mineral in the original substance. Hence, the estimation of minerals under the microscope depends on the accurate measurement of the relative area of the fragments of the mineral. A rough estimation can be secured by using a squared micrometer plate in the eyepiece. The area occupied by the fragments of the mineral to be estimated can then be measured relative to the area of the total field under the microscope.

More accurate measurements are possible with the use of an integrating stage. The thin section is held in a carrier fitted with screws capable of moving the section in two directions at right angles. The screws are geared to a vernier graduated scale, so that the distance travelled by the stage can be accurately recorded. One edge of the section is set up along one of the cross-wires of the microscope, and the section is moved across the field of view by operating one of the screws. Most integrating scales are designed so that the graduated scale may easily be slipped in or out of gear. Consequently, when the edge of a fragment of the mineral which is being estimated coincides with the junction of the cross-wires, the scale can be geared to the screw thread. The scale records the distance traversed until such time as the mineral passes beyond the junction of the cross-wires, when it may be degeared. The same

[1] Rutter, F., *J. Brit. Ceram. Soc.*, 7, 85, 1970.

process is repeated when all other fragments are encountered whilst the section is passed across the field of view. The fraction of the total traverse which is occupied by the mineral gives the amount of that mineral by volume in the section across the field. A series of determinations at several positions along the length and breadth of the section must be made, and the mean result will be an accurate measure of the volume content of the mineral in the sample. Where the material being examined is of irregular composition throughout, several sections which are representative of the whole must be examined.

The estimation of minerals by microscopic means can be carried out with accuracy, provided that the grain size is large enough to be capable of reliable measurement and that the grain boundaries are clearly distinguishable. Needle-shaped minerals, such as mullite, are difficult to estimate, particularly if the particle size is small. In some cases, crystals are oriented along one particular direction in a sample, e.g. mica in slates and in some clays; reliable microscopic estimations can then only be made by taking samples along three directions in the material at right angles to each other and making determinations on thin-sections of these.

The identification and estimation of minerals by microscopic means depends largely on the size of the ultimate particles. The degree of magnification in an optical system is limited by the resolution of the lenses, and in practice, amplification factors of over 1,500 times are not easily attainable without resolution being lost. Consequently the optical microscope is not capable of identifying, let alone estimating, many of the minerals in clays, when particle sizes of less than 10μ are numerous.

Examination of Clays. When clay minerals can be obtained in a sufficiently pure state, some of their optical properties can be determined by preparing an oriented specimen. The individual crystallites of clay and allied minerals are platey, consequently if they are allowed to fall freely in water, they tend to align parallel to their largest face. If a microscope slide is inserted into a suspension of clay, the particles settle on to it with the basal sections of all fragments along the plane of the slide. Hence, a section of appreciable thickness and size can be built up, which behaves as though it were a single crystal. The mineral can be examined in ordinary and polarised light, and birefringence effects produced in the usual way. The results obtained are not so accurate as those using true single crystals, for the section is composed of a mosaic agglomeration rather than a continuous medium.

The optical properties of the kaolin minerals determined on oriented aggregates are given in Table V.III.

TABLE V—III. SOME OPTICAL PROPERTIES OF KAOLIN MINERALS

Mineral	*Optical Character*	*Mean Refractive Index*	*Birefringence*
Allophane	Isotropic	1·494	—
Halloysite	Biaxial negative	1·522	Small
Kaolinite	Biaxial negative	1·558	Small
Livesite	Biaxial negative	1·540	Greater than kaolinite
Nacrite	Possibly biaxial negative	1·560	Very small
Dickite	Possibly biaxial negative	1·562	Very small

The last two minerals, nacrite and dickite, form crystals which are large enough to be identified by high-power magnification. Microphotographs of these two are shown in Fig. V.5 along with kaolinite for comparison under the same conditions, facing p. 224.

Staining. Another useful technique for identifying small-sized particles of clay under the microscope involves the use of dyes or staining fluids.

Certain silicate minerals become *pleochroic* (see p. 225) when treated with dyestuffs. The adsorptive properties of clay minerals are well known, and a special method has been evolved for differentiating between the main types.[1]

The dyestuffs (of which crystal violet, methylene blue, malachite green, safranine 'O' are the most useful) are dissolved in nitrobenzene and added to the powdered clay. Pre-treatment of the clay is essential to remove carbonates and iron oxides which might influence the results; a satisfactory method is to add two or three drops of dilute hydrochloric acid to a few milligrams of the sample on a microscope slide and, after a short interval, the liquid may be evaporated by gentle heat. When cool the dye solution may be added and a cover glass applied. Kaolin-type minerals adsorb all the dyes very readily and the pleochroic effect produced may be observed under high-power magnification.

Montmorillonite-type minerals not only adsorb the dyestuffs, but in certain cases change the colour of the dye. Malachite green, for example, is similar to litmus, and under acidic conditions undergoes a colour change to yellow. It must be emphasised that natural montmorillonites do not always exhibit this colour change, and it is important to treat the sample first with acid as described previously. If standard conditions are observed, it is easy to distinguish between kaolinite and montmorillonite clays by stain tests. The method is also claimed to be of value in differentiating between some other clay minerals.[2]

Although the optical microscope can be used with advantage in clay minerals studies, the extremely small particle sizes introduce severe difficulties. To observe some of the smallest of these particles a degree of magnification is required which is unattainable with optical systems.

Electron Microscopy

The Electron Microscope provides a means of high magnification without using an optical lens system. Furthermore, the source of radiation is not light but an electron beam of short wavelength; much better definition is obtained.

In this form of microscope a beam of electrons, emitted by a heated filament, is accelerated by an electric field and impinges upon a specimen. If the specimen consists of a composite powder or small particles of minerals which vary in thickness, the electron beam is attenuated differently over the whole field. Where there are no crystals, the beam is transmitted unaltered, but on passing through mineral fragments, the electrons will be adsorbed in varying degrees, and carry an image of the specimen which can be greatly magnified by passing the beam through a magnetic or electrical field, in which magnetic poles or electrically-charged plates replace the

[1] Faust, G. T., *Report Invest. U.S. Bur. Mines*, RX, 3522, 1940.

[2] Mielenz, R. C., King, M. E. and Schieltz, M. C., Report No. 7, Amer. Petrol. Inst. Project 49 (Columbia Univ., N.Y., 1950).

glass lenses in the optical microscope. The degree of magnification can be varied by increasing the charge on the plates, and magnifications of up to 60,000 times are possible.

The magnified image may be viewed directly by projecting the electron beam on to a fluorescent screen. Alternatively, a permanent record can be produced on a photographic plate or film.

The specimen has to be inserted in the barrel of the instrument, and is therefore at reduced pressure throughout; furthermore, the electron beam is intense, especially when high magnifications are being used, so great care must be taken to avoid disrupting or melting the specimen.

An electron microscope cannot be used to examine all the minerals in a raw clay, for these include particles of a wide grain-size range. The larger particles have to be removed by elutriation, and only those of less than 2μ examined. Clay mineral sections are normally prepared by dispersing a little powder in water, allowing the large grains to settle and applying a drop of the suspension to a collodion sheet. The water may then be evaporated and the sample is ready for analysis.

The electron microscope has revealed the fine structure of the smallest-particled clay minerals and is an excellent method for the identification of the principal types.[1]

Clay minerals exist as plate-like or fibrous crystals and the method of specimen preparation automatically orients the particles along their basal section, so the thickness cannot be judged until they have been shadowed with a film of fine gold particles 'spluttered' from a filament under the influence of a strong electrical field. The particles moving at high velocities are allowed to impinge on to the surface of the section at a very small angle, so that a layer is built up against the side of the crystal, directly in the path of the beam, thus forming a kind of shadow. If the angular projection of the particles is known, the thickness of the crystal can be calculated from the magnification factor of the microscope and the extent of the gold shadow. This 'gold-shadow' technique represents an important step forward in electron microscopy, as it enables sections to be examined with almost three-dimensional clarity (Fig. V.6*b* facing p. 224).

Photographs of kaolinite, halloysite and livesite are shown in Fig. V.6; those on the right were the result of the shadow technique, whilst those on the left were taken without any such treatment of the specimen.

Kaolinite is typified by hexagonal plates with a thickness to size ratio of about 1:12; halloysite is apparently lath-like in character; livesite is in the form of small plate-like particles with no well-defined crystal habit, and montmorillonite has a characteristic fibrous habit.

Recent investigations[2] into the crystal habit of halloysite have revealed that the lath-like crystals shown in Fig. V.6*b* (iii) are in reality rolled-up plates. Electron micrographs have been taken in the end-on position, and the hollow cylinder nature of the crystal established. Other samples show the tubular form of the plates along only a part of their lengths.

Montmorillonites exist as particles which are virtually single-layer; as more magnification power becomes available, the details of these individual units will be

[1] Kay D., *Techniques for Electron Microscopy* (Blackwell, Oxford, 1967)
[2] Bates, T. F., Hildebrand, F. A. and Swineford, A., *Amer. Miner.*, 35, 463, 1950.

established, and, perhaps, the perplexing problem of the structure of these minerals may be finally solved.

T. F. Bates[1] has presented excellent electron micrographs of a wide range of clays and associated minerals.

X-RAY ANALYSIS

THE development and use of X-ray analysis has brought order into all branches of structural chemistry. Few have gained so much as the field of silicate minerals, which, prior to 1930, was a difficult and complex branch of science.

X-rays consist of electromagnetic waves, similar to light, but of much *shorter* wavelength (visible rays have a wavelength of $4–7 \times 10^{-5}$ cm. from violet to red, X-rays are between 10^{-7} and 10^{-9} cm. long). They are produced when a fast-moving stream of electrons impinges on a metal surface or target—termed an anti-cathode. They result from the energy liberated when the electrons of the metal atoms are 'excited' out of their normal orbits by the bombarding particles. The wavelength of the X-ray radiation so produced depends on the material of which the target is composed. Most metals emit a wide band of X-rays but with intense peaks at two characteristic wavelengths, termed K_α and K_β radiation, the value of which is related to the atomic number of the element (see p. 60).

Monochromatic X-rays can be produced if a material (termed a filter) is selected which transmits the K_α rays and absorbs the K_β. Copper K_α radiation, for example, can be produced by introducing a sheet of nickel foil into the beam, which effectively blocks the K_β rays. Table V—IV indicates the filters which are used to give monochromatic radiation from other metal targets. When it is impracticable to use the filtering element in the form of foil, a sheet or film of a suitable salt is often satisfactory.

TABLE V—IV. CHARACTERISTIC WAVELENGTHS OF X-RAYS EMITTED BY SOME METALS ($\times 10^{-11}$ cms.)

Metal	*Aluminium*	*Chromium*	*Iron*	*Cobalt*	*Nickel*	*Copper*	*Molybdenum*	*Tungsten*
K_α	8320	2289	1936	1789	1658	1541	713	213
K_β	7965	2081	1753	1617	1497	1389	631	184
Filter	—	Vanadium	Manganese	Iron	Cobalt	Nickel	—	—

When a parallel beam of either ordinary light or X-rays is passed through a very fine grating or a series of lines placed very close together, each line becomes the centre of an emerging spherical wavefront. As these cross similar wavefronts radiating from other lines, regions of darkness and brightness are produced depending on the phase relationships in the coinciding beams. This is known as a *diffraction pattern* and the spectral colours sometimes seen through a fine grating or a dusty window are given by white light with a wide range of wavelength radiation. When monochromatic radiation (i.e. of one wavelength) is used the colours are replaced by alternate regions of darkness and brightness. These patterns are produced only if the spacing of the grating is of a similar size to the wavelength of the radiation.

[1] Bates, T. F., Circ. No. 51, *Min. Ind. Exp. St.*, Penn. State Univ., 1960.

The atoms or ions in a crystal are arranged symmetrically (see Chapter II) and form planes or groups of planes within which there is a constant separation distance. These constitute a type of grating which will diffract radiation of a wavelength comparable to the distance between the planes. Most crystals have spacings between planes of their atoms corresponding to the wavelength of X-rays; consequently, characteristic patterns are produced and if these are suitably measured, the dimensions of the atomic planes and hence the structure of the crystal can be predicted.

In Fig. V.7 atoms, represented by dots, are arranged along planes *p* separated by a distance *d*. A monochromatic beam of X-rays, XY, impinges on these planes and makes an angle of θ with them.

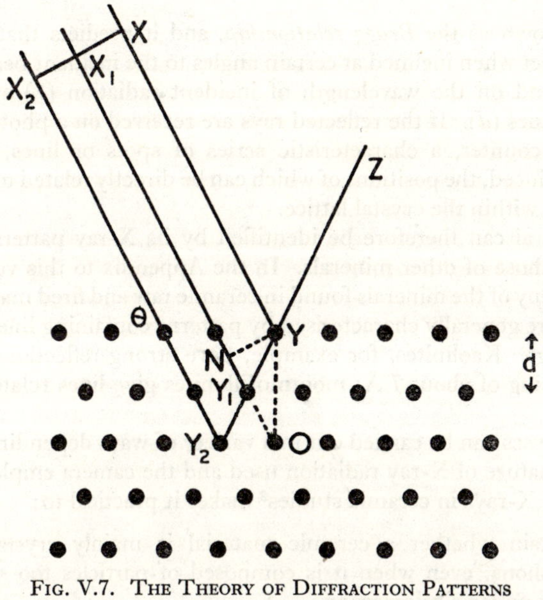

FIG. V.7. THE THEORY OF DIFFRACTION PATTERNS

The beam is made up of a series of waves which can be represented by X, X_1, X_2, X_3, . . ., etc. Each of these is reflected at different lattice points, and join together to move along YZ. The path lengths of each ray are different, and can be represented by XYZ, X_1Y_1Z, X_2Y_2Z, X_3Y_3Z, etc. If YN is drawn to cut X_1Y_1 perpendicularly at N, and X_1Y_1 is produced to O, where Y_1O is equal to Y_1Y, then $OYN = \theta$.

The difference in path length between rays XYZ and X_1Y_1Z is clearly NO, because $XY = X_1N$ and Y_1Y equals Y_1O.

$$\sin \angle OYN = \sin \theta = \frac{NO}{YO} = \frac{NO}{2d}$$

$$\text{therefore} \qquad NO = 2d \sin \theta$$

By similar calculations, the path difference between X_2Y_2Z and X_1Y_1Z is also equal to $2d \sin \theta$ and so on.

If the distance NO is exactly equal to the wavelength (or to a whole number multiple of the wavelength) of the incident radiation, all reflected waves along YZ will be of the same phase and the amplitude will be the sum of all the individual rays added together. Should NO not be exactly equal to the wavelength of the X-ray beam, each beam will be of different phase, and the net resultant amplitude from many reflections will be zero.

Only when θ has certain values will each ray be in phase and therefore, as the crystal is rotated, reflections will occur only when the relationship

$$n\lambda = 2d \sin \theta$$

is fulfilled, where $n = 1, 2, 3$, etc.

This is known as the *Bragg relationship*, and it predicts that all planes in a crystal will reflect when inclined at certain angles to the incident beam, and that the angles (θ) depend on the wavelength of incident radiation (λ) and the distance between the planes (d). If the reflected rays are received on a photographic film or Geiger-Müller counter, a characteristic series of spots or lines, or electrostatic impulses is produced, the positions of which can be directly related to the positions of planes of atoms within the crystal lattice.

Each mineral can therefore be identified by its X-ray pattern which will be different from those of other minerals. In the Appendix to this volume, the principal lines of many of the minerals found in ceramic raw and fired materials are given. Clay minerals are generally characterised by patterns containing lines corresponding to large spacings. Kaolinites, for example, have strong reflections resulting from their basal spacing of about 7 Å; montmorillonites give lines related to their 14 Å spacing.

X-ray analyses can be carried out in a variety of ways depending on the type of specimen, the nature of X-ray radiation used and the camera employed.[1]

The use of X-rays in ceramic studies[2] makes it practical to:

(i) Ascertain whether a ceramic material is mainly crystalline or largely amorphous, even when it is composed of particles too small to be seen clearly under a microscope.
(ii) Determine the structure of the lattice in any crystalline substance.
(iii) Identify most crystalline minerals in a ceramic body.
(iv) Estimate the proportions of one or more minerals in a substance under certain favourable conditions.
(v) Measure (indirectly) the grain size of crystalline particles in a sample.

Preparation of the Specimen or Sample. When a sample to be analysed is in the form of a single crystal uncontaminated by impurities, much more information about the nature of its atomic configuration and the composition of the mineral can

[1] Klug, H. P. and Alexander, L. E., *X-ray Diffraction Procedures* (John Wiley, London, and N.Y. 1967).

[2] Peiser, H. S., Rooksby, H. P. and Wilson, A. J. C., *X-ray Diffraction by Polycrystalline Materials* (Blackwell Ltd., Oxford, 1954).

be deduced. The crystal may be aligned in the X-ray beam in such a way that reflections parallel to certain crystallographic axes can be recorded. The number of lines is then reduced, for only a limited group of planes within the crystal is able to satisfy the Bragg relationship. For example, a flake of mica oriented along its basal section will give only those reflections which correspond to the distance between the mineral sheets. The number of reflections may be increased by oscillating the crystal through a small angle. Eventually, the complete atomic configuration can be deduced by calculation from the planes of atoms which reflect at different incident angles of the X-ray beam.

The positions of lines or spots which appear in an X-ray pattern are determined by the arrangement of planes of atoms in the lattice, but the intensities of the reflections are dependent largely on the scattering power of the atoms which make up these planes. Therefore, the location of different atoms in the lattice can be established with great precision.

In most ceramic materials, the individual crystals are not large enough for single-crystal examination. This is particularly so with clay minerals, and most fired ceramic materials, because single crystals which can be isolated for analysis are not common. As in microscopy, oriented specimens can be prepared which may be analysed as a single crystal, but this method is restricted to clay minerals which are in a pure state.

The most common method of specimen preparation when the particle size is small, is to compact or mould the powdered material into a long fibre or rod. Satisfactory results are obtained only when the specimen is less than 0·5 mm. in diameter.

The powder may be packed into a thin-walled capillary tube of glass, but care has to be taken that the X-ray beam is not absorbed or scattered by the container. Silica glass tubes are the most satisfactory, but even this material is not entirely suitable when low-angle lines are to be measured. One advantage of using capillary tubing is that the sample can be kept perfectly dry throughout; the tube may be sealed after filling, and the specimen thereby kept free from rehydration. This is of value when studying the decomposition products of certain minerals which lose water or carbon dioxide on heating, and readily reabsorb it on exposure to the atmosphere.

When such precautions are unnecessary, the specimen may be prepared by adding a little gum tragacanth to the powder, slightly wetting with water and then rolling it into a cylindrical shape between two glass plates. An alternative method is to attach the powder to a fine hair or fibre with a suitable adhesive. Successful specimens can often be made by extruding the material mixed with gum from a hypodermic syringe fitted with a needle of large bore.

A powder specimen is an aggregation of fine crystallites, usually with random orientation. An X-ray beam striking such a specimen will be reflected by all planes of atoms which are present in the minerals in the powder. To ensure this, the specimen should be rotated throughout the exposure, especially if the individual particles are relatively of large size. Except where only simple crystals are present, powder analysis photographs are difficult to interpret. It is impossible to restrict reflections to a small group of planes inclined at a certain angle to the beam. The lines which are produced by a mineral are, however, characteristic, and an X-ray photograph of a powder can establish the nature of the minerals present.

The X-ray pattern of a powder consists of rings and not spots, as is that produced by a single crystal. The reason is that whereas a certain set of planes in one crystal is at the same orientation to the incident beam throughout, the identical planes in crystals in the powder are inclined at many angles to the beam. Thus, for a given wavelength of radiation, a set of planes in a single crystal will produce one spot on a photographic plate (or set of spots corresponding to different angles of reflection for various orders), whilst those in a powder produce a ring formed from a merging series of spots, each with the same inclination to the direct beam, but radiating in the form of a cone from the specimen. The various orders of reflection give rise to other rings with different inclinations to the direct beam.

The X-ray Radiation. The wavelength of the radiation emitted from a target is determined by the nature of that target and the electrical potential applied between the filament and target. If that potential is low, only X-rays of a continuous range of wavelength are emitted. At a high potential, the inner electron orbits of the atoms of the target are disturbed and superposed on the continuous spectrum are peaks of high intensity at characteristic wavelengths.

The continuous wavelength form of X-rays is known as *white radiation*, and was used by the earlier workers in the field of X-ray crystallography. The peaks of maximum intensity are termed the *characteristic wavelength* of the target, and a list of the principal values has been given in Table V.IV (p. 232). The use of suitable filters or a monochromator reduces or eliminates the background or continuous spectrum and all but one characteristic wavelength, thereby giving *monochromatic radiation*.

The analysis of single crystals using 'white radiation' is useful, for if measurements are made along one crystallographic axis, reflections will occur whenever $n\lambda = 2d \sin \theta$ (p. 234). A family of spots or lines will therefore result at characteristic positions on a film placed behind the specimen; the resulting pattern is called a *Laue photograph*.

It is more usual, however, to use monochromatic radiation of a fixed wavelength. Copper K_α ($\lambda = 1541 \times 10^{-11}$ cm.) is commonly used, because a high intensity beam can then be used without serious deterioration of the target. If the specimen is rich in iron, considerable scattering takes place, and it is better to use a cobalt target in such cases. Molybdenum, iron and other targets are used according to the wavelength which is most suitable for the type of specimen being examined.

X-ray Cameras, Structure Determination and the Detection of Minerals. Many kinds of camera are designed specifically for structure analysis. Large crystals may be examined by means of a single crystal or Weissenberg goniometer. Attachments to these instruments enable a mineral to be set up along any crystallographic axis and the X-ray reflections are received on a photographic plate which is usually bent to form an arc of a circle with the specimen at the centre. Many modern cameras can be fitted with a Geiger-Müller counter which replaces the photographic plate. The accurate positions of spots can thus be located and the relative intensities determined directly.

The type of photograph obtained with a single crystal contains many spots and sometimes rings. It often shows a symmetrical pattern which is characteristic of the crystal form of the mineral. Powder samples may also be used in cameras of this type; the resulting photograph is a series of rings or ellipses radiating from the centre of the film.

There are special types of cameras designed specifically for powder specimens, which are also satisfactory for mineral identification purposes. Such cameras are circular in shape, the specimen being located at the centre and rotated by a small electric motor built into the instrument. The X-ray beam passes through a collimator, impinges on the specimen, and the resulting reflections are recorded on a strip of film inserted round the perimeter of the camera. The diameter of the cameras may be of any size, although the two most generally used are 5·73 cm. and 11·46 cm. The smaller the camera the less will be the exposure time required, but the separation of lines along the length of the film will not be as great as with those of larger diameter. The film is in the form of strips, thereby saving photographic paper and enabling the camera to be reduced considerably in bulk. The photographic pattern produced is a series of lines which are, in reality, segments of rings. This line pattern is related directly to the spacings between the planes and the *d* value of a particular line can be calculated quite simply.

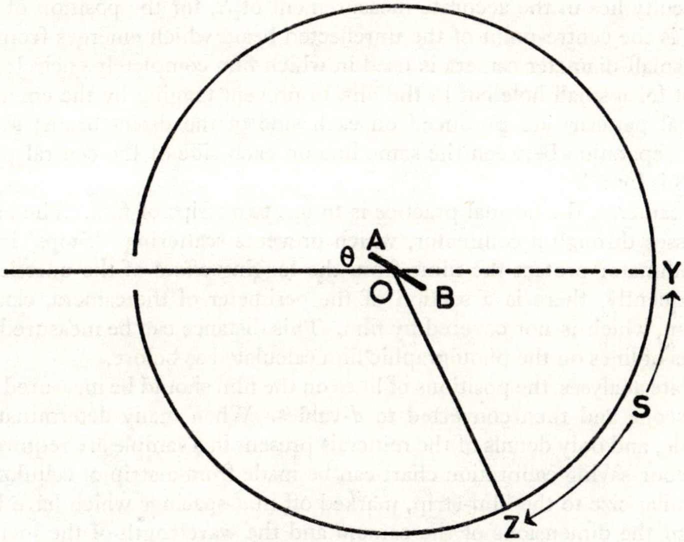

FIG. V.8. METHOD OF CALCULATING THE SPACING OF AN X-RAY LINE

In Fig. V.8, AB represents the inclination of a set of planes in a crystal to the incident beam of X-rays. If the angle of incidence is θ and the reflected ray is along OZ, the angle ZOB must equal θ because the angle of reflection must equal the angle of incidence. Furthermore, the reflected ray OZ must satisfy the Bragg relationship $n\lambda = 2d \sin \theta$ where λ is the wavelength of the incident X-ray beam, and *d* is the spacing between planes parallel to AB.

By geometry the angle YOB must also equal θ, therefore the angle YOZ equals 2θ.

The distance YZ is S or the separation of the line at Z from the direct beam, and OY is the radius of the camera.

So $$\frac{YZ}{OY} = \frac{S}{r} = 2\theta \text{ (in radians)}$$

$$\therefore \theta = \frac{S}{2r} \text{ or } \frac{S}{D} \quad \text{where } D \text{ is the diameter of the camera.}$$

Now $$n\lambda = 2d \sin\theta$$

$$\therefore d = \frac{n\lambda}{2 \sin\theta} = \frac{n\lambda}{2 \sin \frac{S}{D}}$$

All values on the right-hand side of the equation are known, or can be measured, therefore d (the spacing of the planes giving rise to the line at Z) can be calculated. The only difficulty lies in the accurate measurement of S, for the position of Y is uncertain. (Y is the centre-point of the unreflected beam which emerges from the camera.) If a small-diameter camera is used in which film completely encircles the camera (except for a small hole cut in the film to prevent fogging by the emergent beam), identical patterns are produced on each side of the direct beam; so, by measuring the separation between the same line on each side of the central point, the value of $2S$ is found.

In larger cameras, the normal practice is to use two strips of film. The emergent beam passes through a collimator, which prevents scattering. 'Stops' in the ends of the collimator protect the films from the fogging effect of the unreflected beam. Consequently, there is a section of the perimeter of the camera, close to the direct beam, which is not covered by film. This distance can be measured and hence positions of lines on the photographic film calculated as before.

For accurate analyses, the positions of lines on the film should be measured by a vernier microscope and then converted to d-values. When many determinations have to be made, and only details of the minerals present in a sample are required, a very useful labour-saving calibration chart can be made from a strip of celluloid or perspex, of similar size to the film-strip, marked off in d-spacings which have been calculated from the dimensions of the camera and the wavelength of the incident radiation. The lines on a photograph of an unknown mineral may then be estimated directly in d-values by comparison with the calibrated strip.

The identification of minerals by X-ray analysis is comparatively easy and rapid, and the method is of value in many ceramic investigations. Fig. V.9 shows the characteristic X-ray patterns of the three silica minerals, quartz, cristobalite and tridymite. Although some lines are common to all, certain reflections are unique to each of the crystallographic types (see facing page 224).

One feature of most X-ray powder cameras is that the collimator at the emergence side is so designed that lines corresponding to a d-spacing of greater than 7 Å cannot be recorded. X-rays are scattered to some extent by air particles, therefore the unreflected rays may be deflected slightly from their normal parallel path; the emergent beam would be spread over a comparatively large area, and reflections at low angles would be masked or obliterated by fogging if a collimator were not used. This is usually of little consequence as most minerals do not have spacings greater than 7 Å, but it is a serious drawback in clay mineral analysis where planes with separations up to 25 Å may be present.

Cameras for studying the X-ray patterns of clay minerals must be capable of revealing reflections which are at small angles to the direct beam. The removal of the exit collimator, or the design of a camera without one, is a simple matter, but the problem of eliminating low angle scattering still remains. A commercial 11·46 cm. camera from which the exit collimator has been removed is suitable if the scattering caused by air molecules is reduced by completely evacuating the camera. Spacings of 35 Å can be detected without any noticeable contamination by random radiation, provided that the pressure in the camera is lower than 1 mm. Cameras of 5·73 cm. and 11·46 cm. diameter have been modified in this way, but the larger size is preferred because this gives a greater separation of lines.

Low angle scattering may also be reduced by replacing the air in the camera with hydrogen.

Because the arrangement of the atoms in all layer lattice silicates is similar, most of the lines in the patterns of these minerals are common to all. Ionic replacements may cause a slight shift in the line from a particular *d*-spacing, but the same characteristics can usually be observed in all minerals. The basal spacings of layer lattice minerals, i.e. the distance between identical planes of atoms in two successive layers, are different, and reflections from such planes are characteristic of the mineral type. The principal line corresponding to the *d*-spacing of the basal section for the layer lattice minerals is as follows:

7 Å (invariant)	Minerals of the kaolin group
10 Å (invariant)	Micas and hydrous micas
14 Å (invariant)	Chlorites
14 Å (reduced to 11·8 Å by heat treatment at 110° C.)	Vermiculites
14 Å (increased to 17 Å or above by treatment with glycerol)	Montmorillonites

The values given are only approximate and differ slightly for particular minerals in any one group.

The best means of studying minerals with these long spacings is with a camera of the Brindley focusing or semi-focusing type. Details of the technique are given in *X-ray Diffraction by Polycrystalline Materials* by Peiser *et al.* (see p. 234).

From the examples shown, it is obvious that a complex line pattern is produced by any one mineral. Each line is characteristic of a set of planes in the crystal, and there may be as many as fifty of these recorded on one photograph. The identification of any mineral does not present much difficulty, provided that it is in a fairly pure state. Ceramic materials, however, may contain many minerals, each of which has its own characteristic line pattern. The X-ray pattern of a composite mixture of minerals may therefore be highly complex and difficult to interpret. Some minerals are of variable composition, and isomorphous substitution and solid solution are common phenomena in minerals associated with ceramic mixtures. The X-ray pattern of such minerals may vary considerably between different samples, and although such variation may be indicative of the degree of substitution when the pure mineral is examined, in a composite mixture the change in pattern may be confusing.

Despite these difficulties, X-ray analysis is the quickest and most positive method available for the detection of many minerals. A complex ceramic material would usually be analysed using the powder technique described in the preceding pages, so that a line pattern is obtained containing the characteristic reflections from each mineral present. A comprehensive index of the *d*-spacing for all known minerals has been drawn up by the American Society for Testing Materials, Philadelphia, U.S.A., and it is by reference to this, or to a similar Table, that the minerals present in an unknown material may be identified. The *d*-spacings corresponding to the three strongest lines of many minerals in ceramic bodies are given in the Appendix.

If one mineral predominates in the sample, the strongest lines in the pattern will probably be characteristic of this mineral, hence its identification is a comparatively easy matter. The remaining lines can then be catalogued and other minerals established by referring to the Index.

When no one mineral predominates, the indexing of the lines in a pattern is complicated and is largely a matter of trial and error. The task can often be simplified if the approximate composition or the previous history of the sample is known. The characteristic line patterns of only those minerals which can be present, need be examined and compared with the lines from the unknown sample.

If many mineral analyses by X-ray methods are contemplated, it is advisable at the outset to prepare a set of line patterns of the important minerals occurring in ceramic bodies and comparing these, in turn, with the pattern of the substance being examined. Much tedious measurement can be avoided by this procedure.

An important development in the use of X-rays in ceramic studies has been the introduction of *high temperature techniques*. The progress of reactions as they take place can be observed and the phases present in a fired mixture can be identified without prior cooling and the consequent chance of further reaction.

High temperature cameras are usually of conventional type, designed to take powder specimens and fitted with a heating element. Johnson[1] has described a simple tantalum resistance heater capable of maintaining a specimen at 2200° C. A further modification is a detachable film holder which enables a series of photographs to be taken of a specimen whilst it is maintained at a high temperature.

Jay[2] was one of the early pioneers in the field of high temperature X-ray techniques applied to ceramic materials. He studied the changes in the lattice constants of quartz from room temperatures to above the inversion point. Johnson and Andrews,[3] using the same method, have studied the inversion and thermal expansion of cristobalite.

[1] Johnson, J. R., *J. Amer. Ceram. Soc.*, 37, 360, 1954.
[2] Jay, A. H., *Proc. Roy. Soc.*, **A**.142, 237, 1933.
[3] Johnson, W. and Andrews, K. W., *Trans. Brit. Ceram. Soc.*, 55, 227, 1956.

High temperature X-ray techniques are now firmly established and several commercial cameras are capable of being operated at temperatures up to 1500° C., but the interpretation of the results which they give is often not an easy matter. One great difficulty is the correct assessment of the temperature of the specimen being analysed and no entirely satisfactory method, of general application, has yet been found. A critical survey of high temperature X-ray methods has been published by Brand and Goldschmidt.[1]

The Estimation of Minerals by X-ray Methods

The identification of minerals by X-ray analysis is based upon the spacings of the characteristic line pattern. The intensity of the reflections depends upon the arrangement of atoms in the crystal, the number of reflecting planes and the intensity of the X-ray beam. If the radiation can be controlled, the intensity of lines, for a certain crystal, will be in proportion to the number of reflecting planes, or, in other words, to the amount of that mineral present in the sample.

The intensity of radiation is difficult to regulate, for although it is largely dependent on the electron stream impinging on the target and the potential across the X-ray tube, other factors (such as the condition of the surface of the target) have an influencing effect. Furthermore, minerals vary considerably in their relative scattering or absorptive power to X-rays, so the nature of the contaminant may alter the intensity of the line-pattern of a particular mineral.

The *estimation* of minerals is not possible, therefore, by comparing intensities produced under normal exposure conditions.

There are two methods, however, each widely used, which permit the quantitative analysis of certain minerals by X-ray means. If quantitative methods of X-ray analysis are to be successful and accurate, the first essential is to measure the total intensity during an exposure; this may be achieved either (*a*) directly, or (*b*) relatively, and both methods are used extensively.

(*a*) The Geiger-Müller counter is a type of electrostatic voltmeter, which can be used for measuring the intensity of X-ray radiations. Many cameras are designed so that the counter can be mounted on a moveable arm, and used in place of a photographic film. A motorised drive moves the counter through an arc of a circle centred on the specimen. As each reflected ray is encountered, the intensity of the beam is recorded directly on a microvoltmeter and the positions of the reflections and their absolute intensity are determined in one operation.

The intensity measured by the Geiger-Müller counter is, in the simplest terms, the X-ray flux falling on the instrument per unit time. A particular reflection which is characteristic of the mineral to be estimated, is allowed to fall on the counter, and the resulting deflection on the voltmeter is measured. The counter is then adjusted to measure the intensity of the direct beam, or a small fraction of it. Duplicate measurements may be carried out and the relative intensity of the reflection to that of the direct beam estimated.

If a series of standards is compiled containing different known percentages of

[1] Brand, J. A. and Goldschmidt, H. J., *J. Sci. Instrum.*, 33, 41, 1956.

minerals to be estimated, the amount of a particular mineral in an unknown sample can sometimes be determined.

(*b*) When a Geiger-Müller counter is not available, quantitative measurements can be made by measuring the intensity or density of the corresponding lines on a photographic film by means of a microphotometer, which is usually of the photoelectric cell type. A beam of light is projected through the line on the film and the density determined by measuring the microvoltage produced in a photocell. Unfortunately the density of a line on a photographic film, as measured by a microphotometer, is not directly related to the intensity of the reflected ray which produced it. This is due to a variety of reasons but it depends mainly on the nature of the film and the development conditions. The difficulties can be largely overcome by the use of a calibration strip which compares the intensity of radiation with the density of the film.[1] This strip is usually printed on the margin of each film exposed, so that any variations in the developing process may be automatically corrected. These standard exposure strips are produced by rotating a specially-designed sector wheel in the X-ray beam, and allowing the transmitted beam to fall only along the edge of the film.

Once the calibration strip has been made, the density of a line in the X-ray pattern can be related directly to the intensity of radiation which produced it.

Small variations of X-ray intensity cannot be avoided during long exposures because of fluctuations in operating conditions and irregularities in target emissions, so that it is not possible to compare the intensity of radiation of a characteristic line with a series of standards, and so deduce the amount of mineral present. This difficulty is not serious when a Geiger-Müller counter is employed, as the intensity of a reflected ray can be assessed in a few seconds and compared immediately with the direct beam, but with a photographic film, comparatively long exposures are required to produce a line pattern of sufficient intensity. For example, a crystalline material such as quartz, requires an exposure of about 45 minutes using a 9 cm. camera and copper K_α radiation at 50 kV and 20 milliamperes. Clay minerals require 4 hours in a 19 cm. camera under the same conditions to produce a suitable photograph.

A neat method of compensating for these variations is to use an *internal standard*. The sample of powder to be analysed is mixed with a prearranged amount of some known mineral. The resulting X-ray pattern will contain the reflections of the mineral superimposed on that of the sample. The relative intensities of characteristic lines in the mineral to be analysed may then be compared with the intensity of a line or lines in the added mineral. As an example, in the estimation of quartz using fluorspar as the internal standard described by H. M. Richardson,[2] standards are first prepared by mixing quartz with fluorspar in different proportions. Quartz has a strong characteristic X-ray reflection at 3·35 Å and fluorspar one at 3·15 Å. The relative densities of these two lines in a series of photographs of quartz-fluorspar mixtures of known proportions can be assessed and plotted graphically and a calibration graph prepared.

An unknown sample of quartz may be analysed by adding a known amount of

[1] Bunn, C. W., *Chemical Crystallography* (Oxford U.P., 1961).
[2] Richardson, H. M. and Wild, F. G., *Trans. Brit. Ceram. Soc.*, 51, 398, 1952.

fluorspar and comparing the relative densities of the characteristic lines. The amount of quartz can be estimated directly from the calibration graph.

There are several factors which may either reduce the accuracy or even prevent a quantitative determination being made:

(i) For instance, no other mineral must be present in the sample which gives a reflection near to the characteristic line of the mineral being estimated and that of the standard mineral. Mica, cristobalite and tridymite, when associated with quartz, restrict the use of this method when fluorspar is used as the internal standard.

(ii) Another source of error is that the line-pattern of a mineral is influenced by the size of the crystallites. As the particle size decreases, the X-ray reflections broaden and the photometer curve of a line-pattern altered in this way shows an increase in the base of the 'peak', and a reduction in magnitude. For this reason, it is better to measure the relative areas of the peaks in the intensity curve rather than the heights.

This method proved useful for estimating the mullite content of mixtures of amorphous silica and alumina fired to different temperatures. Quartz was used as the internal standard, and the 4·21 Å line of this mineral compared with the mullite line of 5·4 Å. Neither of the components of the mixtures was contaminated with quartz and no other minerals were produced which interfered with the lines to be measured.

Fig. V.10 shows the type of photometric curve produced by measurement of the X-ray pattern in the region of the two lines. Superimposed on the mullite line is the trace of a 'peak' which appeared after light firing when presumably the mullite was of small grain size. The broad base line and the shallowness may be compared with that of the full curve, which represents a more crystalline mullite.

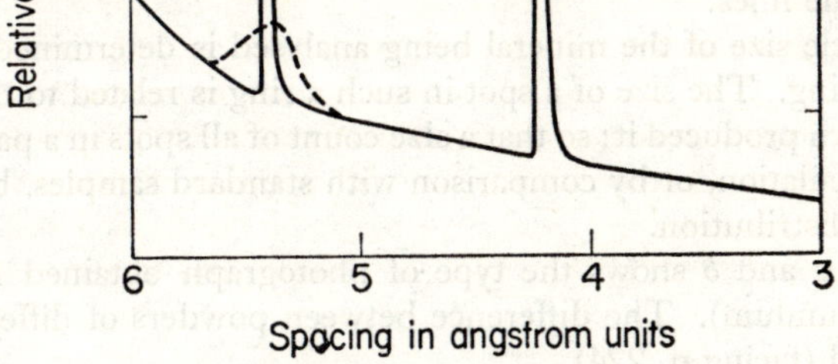

FIG. V.10. A PHOTOMETRIC CURVE OF THE X-RAY LINE PATTERN OF A MIXTURE OF MULLITE AND QUARTZ

The quantitative estimation of minerals by X-rays is accurate to about $\pm$ 5 per cent when the conditions of testing are ideal; i.e. when the mineral to be estimated and the internal standard used have characteristic reflections far removed from those of contaminants.

A further requirement is that the mineral being estimated should be available in a pure state so that a calibration graph can be constructed. Only reflections with a strong intensity should be selected for quantitative measurement. Weak lines of a pure mineral will be fainter still when diluted with impurities, and are then of little value.

The Determination of Crystallite Size by X-rays

If the X-ray powder photographs of crushed specimens of the same mineral but of different crystallite size are compared, a change is observed as the size decreases, from a set of spots, through lines, partially continuous and partially spotty, to complete lines of uniform intensity. The degree of 'spottiness' of the pattern is an index of the crystallite size and if uni-sized particles are present, a fair measure of their dimensions may be calculated. With small particles, the sharp rings characteristic of particular reflections of the mineral become banded or diffuse (see Fig. V.11) and the spread of the lines can also be related to the crystallite size. Exceptionally fine particles give rise to diffuse bands or haloes at low angle, which may be observed surrounding the direct beam and at a small separation from it.

The limits of crystallite size which produce spots, rings, diffuse bands and low-angle haloes, vary with the type of mineral and the degree of crystallinity. Wherever possible, standard photographs should be compiled from materials of known size.

Particle size determinations by X-ray methods are more conveniently made by using a *back-reflection technique*. The specimen to be tested can be in almost any form, but the best results are obtained on a large fragment or on a compressed powder with one face freed from irregularities. A narrow monochromatic beam of X-rays is projected on to this face, and is reflected *back* at characteristic angles on to a photographic plate or film. By this means, a pattern is produced by reflections from the planes of atoms in the minerals of the specimen, and the *d*-spacings can be calculated. The film, which is set parallel to the reflecting face of the specimen, is usually flat and the incident X-rays pass through a small hole in it. The distance between the specimen and the film can be adjusted to give the most convenient separation of the lines.

The particle size of the mineral being analysed is determined by examining a characteristic ring. The size of a spot in such a ring is related to the dimensions of the crystal which produced it; so that a size count of all spots in a particular ring can, by suitable calculation, or by comparison with standard samples, be converted into a particle size distribution.

Fig. V.11 *a* and *b* shows the type of photograph obtained from samples of α-alumina (corundum). The difference between powders of different particle size is most marked (facing p. 224).

Back reflection photography may also be used for small particles because the line broadening is well-defined, as also is the low-angle scattering. This is illustrated by Fig. V.12 A, B and C, where samples of silica of different crystalline development are compared.

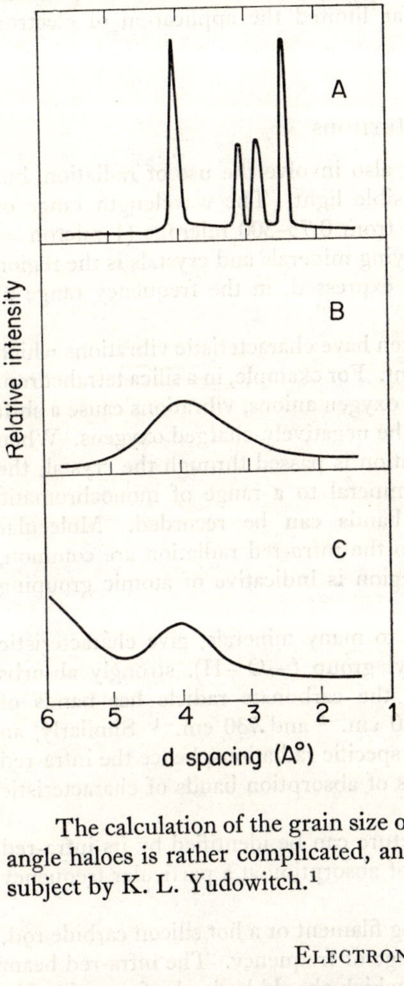

FIG. V.12. PHOTOMETRIC CURVES OF THE X-RAY LINE PATTERNS OF (A) CRISTOBALITE; (B) FUSED SILICA; (C) SILICA GEL (showing the line broadening and the development of low-angle haloes as the crystallite size decreases)

The calculation of the grain size of a powder from the line broadening and low-angle haloes is rather complicated, and the reader should refer to an article on the subject by K. L. Yudowitch.[1]

ELECTRON DIFFRACTION

THE technique of electron diffraction is similar to that of X-ray diffraction, but a beam of electrons is substituted as a source of radiation. Most modern electron microscopes are readily adapted for this purpose.

The modern concept of the electron is expressed as a wave mechanical function. Instead of a rigid particle, a wave motion is postulated but the wavelength is very short. The atoms in a crystal diffract such radiation, and the Bragg relationship $n\lambda = 2d \sin\theta$ applies. As a consequence, a certain *d*-spacing will give rise to a reflected ray at a much smaller angle than when X-rays are used.

[1] Yudowitch, K. L., *J. Applied Physics*, 20, 174, 1949.

Electron diffraction is of great value in the structure analysis of complex crystals, because reflections corresponding to smaller planar spacings can be detected. More precise positioning of atoms in a complex lattice is therefore possible, although other technical factors have so far limited the application of electron diffraction.

Infra-red Methods

INFRA-RED spectroscopy and spectrometry also involve the use of radiation, but of much *longer* wavelength than that of visible light. The wavelength range of the spectrum classified as infra-red, ranges from 0·75–300 microns (1 micron = 10^{-4} cm.), but the important region for studying minerals and crystals is the region 2–25 microns, or as it is more commonly expressed, in the frequency range of 4000–400 cm.$^{-1}$

Molecular groupings within a crystal often have characteristic vibrations which are accompanied by a change in dipole moment. For example, in a silica tetrahedron, where a silicon cation is at the centre of four oxygen anions, vibrations cause a shift of the positively-charged silicon, relative to the negatively-charged oxygens. When radiation of the same frequency as the vibration is passed through the crystal, the radiation is absorbed, so by subjecting a mineral to a range of monochromatic radiation frequencies, several absorption bands can be recorded. Molecular vibrations with frequencies corresponding to the infra-red radiation are common, and hence the absorption spectra in this region is indicative of atomic grouping within a crystal.

Certain groupings, which are common to many minerals, give characteristic absorption bands. For example, a hydroxyl group (—O—H), strongly absorbs radiation of frequency about 3500 cm.$^{-1}$; the carbonate radicle has bands of absorption at frequencies of 1450 cm.$^{-1}$, 880 cm.$^{-1}$ and 730 cm.$^{-1}$ Similarly, an Al—O and a Si—O group will also absorb at specific radiations, hence the infra-red spectrum of a mineral is made up of a series of absorption bands of characteristic frequencies.

Quite often a mineral in a complex mixture can be identified by its infra-red absorption and, in certain cases, the degree of absorption at a particular frequency may be indicative of the amount present.

The source of infra-red may be a glowing filament or a hot silicon carbide rod, both of which emit radiation over a wide range of frequency. The infra-red beam passes through a collimator to the specimen which should be in the form of a thin film. The transmitted rays then pass through a system of mirrors on to a rock salt prism which can be rotated, so as to vary the frequency of radiation received on the detector. The magnitude of the spectra over the wavelength range is measured on a thermopile or bolometer, and is recorded continuously on a potentiometer.

Most infra-red spectrometers work on the split-beam principle, for the intensity of incident radiation is apt to vary over the wavelength range. One half of the beam passes directly through the sample, whilst the remainder is transmitted directly on to the mirror system. The two beams are recorded simultaneously and the difference in intensity at any point represents absorption by the specimen.

Rock salt is commonly used for specimen mounts and the reflecting prisms because it is one of the few materials which do not absorb infra-red radiation over most of the wavelength band.

Clays have proved most difficult materials from which to obtain satisfactory infra-red absorption pattern. Because the crystallite size of such materials is small, specimens for analysis have to be composed of agglomerates, with the result that considerable internal reflections occur and a consequent substantial reduction in the intensity of the transmitted beam. Satisfactory transmissions are possible if the clay samples are mounted in a cold-pressed potassium bromide pellet. The absorption region in the hydroxyl range, i.e. 2·7–3·2 microns, is best studied by using either a lithium fluoride prism or a diffraction grating in place of a prism.

Infra-red spectra are difficult to interpret and can be further complicated by slight variations in sample preparation. Nevertheless, the technique has great potential in inorganic structural analysis and many workers are operating in this field.

The two important spectral regions for clay and associated minerals are:

(*a*) Frequency range 3500–3750 $cm.^{-1}$ (wavelengths 4–2 microns) where the main absorptions due to hydroxyls and water molecules occur.

(*b*) Frequency range 1150–400 $cm.^{-1}$ (wavelengths 10–25 microns) where lattice vibrations between main cations occur.

Fig. V.13 due to Stubičan and Roy[1] shows the important regions of infra-red absorptions of many hydrated silicate minerals.

Thermal Analysis Methods

Thermal analysis methods are based on the observation or measurement of the heat changes which occur when a material undergoes physical or chemical changes on being heated or cooled. All reactions involve an energy change which is usually manifested as a thermal effect, and are termed *exothermic* when heat is evolved, and *endothermic* when heat is absorbed. This principle has been known for many years, and is a recognised means of studying the structural changes which take place in metal systems at different temperatures. Early workers in the field of thermal analysis closely copied metallurgical methods and observed the inflections in the temperature rise when a material was heated under uniform conditions.

The technique was first applied to clays in 1887, when H. Le Chatelier[2] studied the thermal characteristics of alumino-silicates and attempted to classify them by this means.

J. W. Cobb and H. S. Houldsworth[3] were amongst the first to apply the method of *differential* thermal analysis, in which the temperature of the test material is measured relative to that of an adjacent inert material, instead of to that of the furnace. A thermocouple embedded in the test-piece and another in the inert material are connected in opposition, so that any appreciable E.M.F. set up during the heating results from the evolution or absorption of heat in the test sample. This technique forms the basis of all modern methods of thermal analysis.

[1] Stubičan, V. and Roy, R., *J. Amer. Ceram. Soc.*, 44, 625, 1961.
[2] Le Chatelier, H., *Bull. Soc. Min.*, 10, 204, 1887.
[3] Houldsworth, H. S. and Cobb, J. W., *Trans. Eng. Ceram. Soc.*, 22, 111, 1923.

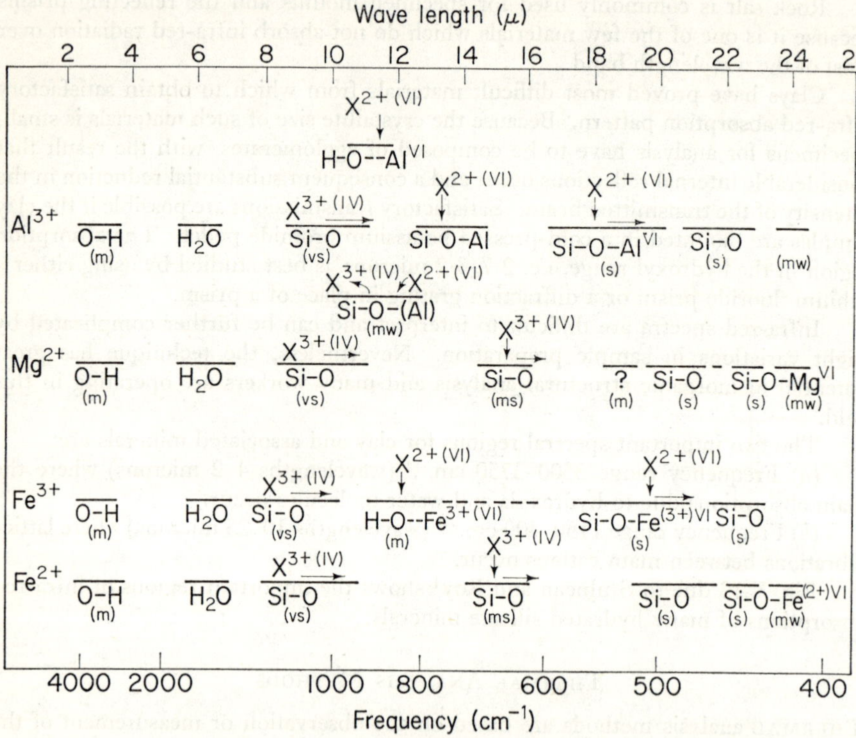

FIG. V.13. INFRA-RED ABSORPTION BANDS IN LAYER-LATTICE SILICATES

$X^{3+(IV)}$ refers to a trivalent cation (3+) in tetrahedral co-ordination (IV)
$X^{3+(VI)}$ refers to a divalent cation (2+) in octahedral co-ordination (VI)

The intensities within a band-spread are indicated by vs = very strong, s = strong, m = medium and w = weak.

The group giving rise to absorption is shown by chemical symbols, e.g. Si—O; Si—O—Al etc. Where isomorphous substitution is likely in these groups, the position of the corresponding peaks or the shift which occurs in the band is shown by arrows.

During the last two decades, D.T.A. has become an important analytical method ranging from a routine control of raw materials to a precise scientific measure of such properties as specific heat, heats of reaction, latent heat of phase changes. It is a technique which can be readily moulded to suit the requirements of a particular research exercise and it is small wonder that apparatus and methods of operating are numerous. To attempt to list all the techniques and applications of thermal analysis would be impossible in the present context, but the reader is referred to an excellent publication under the editorship of R. C. Mackenzie entitled *Differential Thermal Analysis* (Academic Press, London, 1970).

Apparatus for D.T.A. ranges from the simple home-made unit, capable of being constructed and producing valuable results for a few tens of pounds, to the sophisticated custom-designed models, embodying all the latest control and recording accessories, but costing several thousand pounds. Each has its merits and demerits, but for ordinary routine identification and estimation of typical ceramic minerals and for undergraduate teaching and demonstration, the author finds the design of Grimshaw and Roberts[1] quite adequate. The essential features are described below.

Apparatus. A fixed weight of the sample (about 0·8) is packed into one cubical compartment of a thin-walled refractory container (Fig. V.14), the other compartment being filled with an inert material which has no thermal reaction over the range of temperature of the test. This specimen holder, which is made of sintered alumina, fits into the lower half of a refractory block, the lid of which completes the cylindrical shape, thus enabling it to be used in a tubular wire-wound furnace. The square end-section facilitates the handling of the block and provides a means for its location in the furnace.

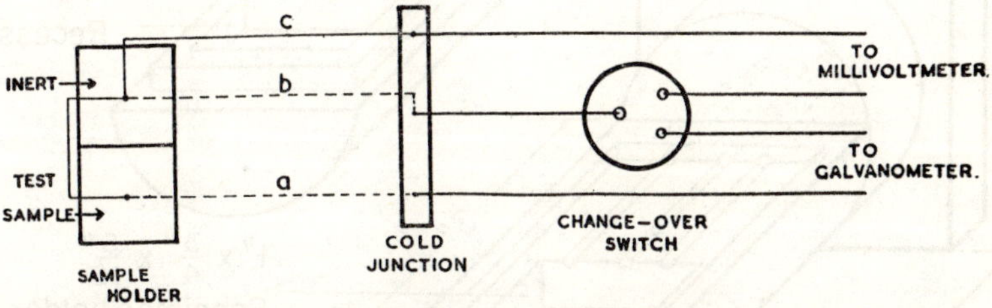

FIG. V.14. DIFFERENTIAL THERMAL ANALYSIS APPARATUS. THE SPECIMEN HOLDER AND THERMOCOUPLE ARRANGEMENT

Thermocouples, which are usually of fine-gauge (28 s.w.g.) chromel-alumel wire,[2] are arranged in the specimen holder as shown in Fig. V.14. A chromel-to-alumel junction is located in the centre of both the test and the inert sample compartments. If these are regarded as separate thermocouples, the principle of differential thermal analysis can be clearly appreciated. The junctions in the two compartments are identical in all respects, therefore if they are both at the same temperature an E.M.F. will be generated in both circuits, the voltages of which will be exactly the same and current will flow in the same direction, i.e. chromel to alumel, in each. These conditions apply independently of the actual temperature of the junctions. When the junctions are connected as shown in the diagram with the alumel wire common to both, the potential difference generated at one junction will be equal to but in opposition to that at the other and the net reading on a galvanometer connected across the circuit-leads *a* and *b* will be zero at all temperatures of the specimen holder. If, however, there is a difference in temperature between the sample compartment and that of the inert, the E.M.F. in the two thermocouples will differ and a net potential across the leads *a* and *b* will result, which will be a measure of *the difference in temperature* between the two junctions. The actual temperature at any time is given by connecting the leads *b* and *c*, which are respectively chromel and alumel, to a millivoltmeter through a change-over switch.

[1] Grimshaw, R. W. and Roberts, A. L., *Trans. Brit. Ceram. Soc.*, 44, 69, 1945.

[2] Chromel-alumel thermocouples are preferred because of the large thermo-E.M.F. of this metal combination. Platinum-platinum/rhodium couples are sometimes employed because they can be used to higher temperatures (1250–1300° C.) with greater stability although their sensitivity is five or six times less.

The refractory block is of alumina or mullite fired to a high temperature to give strength to the moulded shape. The specimen holder rests in a depression cut near the end of the lower half, and three holes are drilled in the body of the block through which pass the thermocouple leads. The edges of the lower half are raised slightly and a complementary taper is removed from the lid in order to give a better fit and to ensure that the specimen holder is totally enclosed (see Fig. V.15).

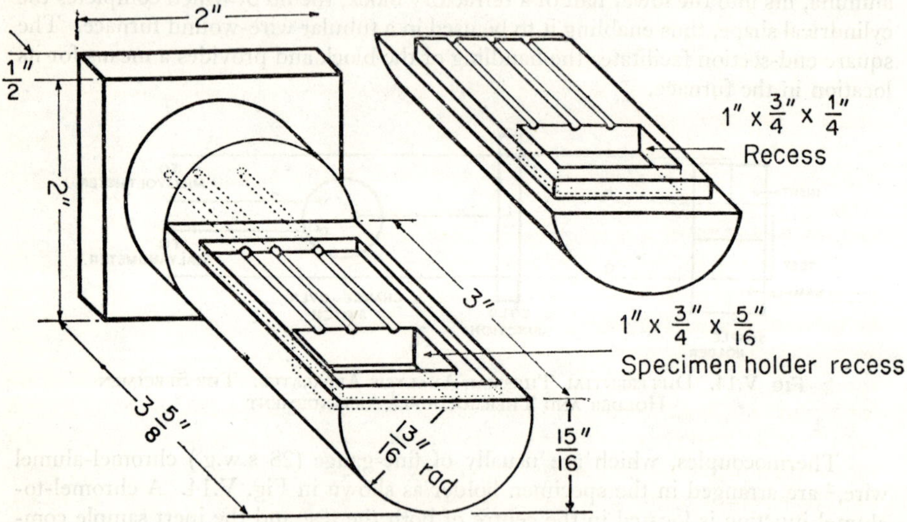

FIG. V.15. DIFFERENTIAL THERMAL ANALYSIS APPARATUS. REFRACTORY BLOCK ASSEMBLY (after Grimshaw and Roberts)

Both the block and specimen holder are now made commercially, and consequently identical sets can be used by different laboratories and results may be directly compared.

The use of blocks and specimen holders made of refractory material confers many advantages over the metal type used by some workers. The temperature differences when exo- or endo-thermic changes take place in the test sample are greater and are defined more sharply, because the heat transfer between the test sample and the surrounding material is slower as a result of the lower thermal conductivity of the refractory block. Furthermore, a block and specimen holder made from alumina or mullite is free from corrosion or oxidation and permits tests to be carried out at temperatures well above 1000° C., which is difficult or impossible with metal blocks. On the other hand, the rate of rise in temperature is usually more uniform throughout when metal blocks are used, especially at low temperatures. When thermal effects are to be measured which occur below 200° C., a nickel steel block is preferred.

The furnace design most suitable for the above apparatus consists of a mullite or similar refractory tube (1·8 in. internal diameter, 0·1 in. wall thickness, and 7 in.

long), the central 5·5 in. of which is wound with nichrome wire[1] 0·05 in. in diameter, in such a way that the winding has a total resistance of 11 ohms. The wire is anchored at both ends, coated with a thick slip of alumina or other suitable refractory, dried, and placed in a furnace case 9 in. × 6 in. × 9 in., made of sheet metal. The space between the tube and the case may be packed with diatomaceous earth or calcined china clay, but immediately around the tube it is better to have a layer of alumina or sillimanite powder to withstand the high temperatures.

Furnaces constructed in this way are excellent in service and are very durable when used up to 1000° C. There would probably be no disadvantage in using a wire-wound furnace with a higher resistance, although if it is greater than 40 ohms it would not be suitable for tests requiring a controlled rate of temperature rise unless a step-up transformer is also used.

The standard thermal analysis block fits rather loosely into a tube of 1·8 in. diameter, but this is an advantage because it is better to support the block in a refractory ring and thereby prevent direct contact with the furnace tube. Heating is entirely by radiation and troublesome conduction effects are avoided.

The **electric heating current** to the furnace is supplied from a variable transformer. If greater simplicity is desired, a fixed 3:1 step-down transformer may be used, with a heavy-gauged rheostat of about 12 ohms in the primary circuit for control purposes. In the choice of heating rate there is a great divergence of view in the design of thermal analysis equipment. Some investigators prefer to control the input to the furnace to give a heating rate which falls steadily as the temperature is raised; other workers favour a linear rate of rise, which is secured by continuously increasing the electrical input to the furnace. There is little to choose between the two methods, provided that a reproducible rate of rise of temperature can be ensured.

When a controlled electrical input is used no elaborate equipment is necessary, small fluctuations in mains-voltage and change in resistance of the furnace wire can be compensated by manual adjustment of the variable transformer or rheostat. When a furnace of the type described is used, an input of 275–300 watts raises the temperature from 0–1000° C. in 100 minutes. If constant conditions are maintained the heating period required is reproducible to ± 20 seconds.

One disadvantage of the constant input method is that a time against temperature graph has to be plotted, and peaks in the thermal curve related to temperature by reference to this graph. A further disadvantage is that reactions taking place at different temperatures cannot be compared directly because the heating rate is not constant. Such difficulties are overcome, to some extent, by adopting a linear rate of temperature rise. The most usual method is to record the temperature on a potentiometric controller fitted with a cam wheel which is contoured to give the rate of rise required. The electrical input is in two parts; (i) the primary source supplying the furnace directly, and (ii) the secondary current regulated by the controller, so that it boosts the primary current when the rate of heating is too slow, or switches off when the heating-rate is too rapid. In practice, the primary source has to be

[1] When thermal tests at temperatures higher than 1000° C. are required, a nichrome tape winding will give long life up to about 1200° C., but for temperatures up to 1300° C. a platinum-wound furnace is essential.

increased as the temperature is raised, therefore it is necessary to use a variable transformer, with a motorised drive. A heating rate produced in this way is sensibly linear over the complete range of temperature of test, but it is difficult to avoid a step-wise increase which arises as a result of the constantly changing electrical supply. Under these conditions it is almost impossible to reproduce thermal curves quantitatively.

The use of proportional controllers and modern high speed recorders has largely solved the problems, and such instruments are now frequently employed; unfortunately, the initial cost of installation of such elaborate equipment is very high.

Another point to be borne in mind is that a linear rate of rise is secured in the inert reference sample only. When endothermic or exothermic reactions take place in the specimen holder, the heating rate cannot be linear, so there is some doubt whether a constant rate of temperature rise is a necessity in the qualitative and quantitative determination of minerals.

Whether the method of constant input or the linear rate of rise method is chosen, a suitable heating rate must be selected to bring out the reactions in the sample to the best advantage. For general work on the minerals in clays, a temperature rise of about 10° C. per minute is most useful, but, it is often advisable to reduce this rate, especially when low temperature effects are of importance or if two minerals are present with closely separated reaction temperatures. The lower the rate of rise of temperature, the less pronounced will be thermal effects from endothermic reactions, but if the rate is too rapid several reactions may be superimposed.

The **methods of recording thermal effects** depend largely on the reactions which occur in the sample. In thermal analysis tests the difference in temperature between the sample and the inert reference material is measured continuously as the temperature is raised. If no reaction is taking place, there will be zero current across the differential thermocouple and no deflection on the galvanometer to which the thermocouple is connected. When a reaction takes place, a difference in temperature between the test sample and the inert material results which creates a net E.M.F. in the thermocouple system, and causes the galvanometer to be deflected. Two opposing effects occur, (i) the heat of reaction tending to increase the temperature difference, and (ii) the restoring influence of the furnace environments. As long as the first effect predominates, the galvanometer will be further deflected from the zero position, but when the reaction rate declines, the sample is restored to the temperature of the inert and the galvanometer returns to zero.

The thermal curve therefore consists of a series of peaks superimposed on a base line. The peaks may be either above or below the base-line, depending on whether the reaction is endo- or exo-thermic, and will have magnitudes in proportion to the heat of reaction.

The recording galvanometer used for measuring the differential temperature must be sensitive enough to record the effects produced, and yet must have sufficient stability and rapidity of action. When natural clays are being examined, temperature differences of up to 50° C. are developed using 0·8 g. of material. On the other hand, the minerals in siliceous samples have endothermic reactions which produce temperature differences of only about 2·5° C.

When the thermal effects are very small, a more sensitive system of recording is necessary. One method is to use a D.C. amplifier which magnifies the differential voltage sufficiently to allow it to be recorded on a suitable ammeter or galvanometer. Alternatively, a mirror galvanometer may be used for measuring the differential temperatures.

The *advantages of differential thermal analysis* lie in its relative simplicity and the rapidity with which results may be obtained. Furthermore, certain minerals can be identified (and often the amount of each may be estimated) even when they are present in complex mixtures. The method is of particular value in the analysis of clays, for the principal impurity, quartz, has only a small endothermic effect which does not interfere with the peak of the clay minerals. Moreover, the endothermic change in quartz is reversible with temperature, whilst that in clay minerals is not. Thermal tests on calcined clay using a sensitive method of recording will indicate the presence of quartz.

One difficulty which is encountered in thermal analysis is the variability of results obtained from one apparatus to another. Other factors which need strict attention if good reproducibility of results is to be attained are the position of the thermocouple junctions and the packing density of both the test-sample and the reference material. If these requirements are observed, the reproducibility of results is generally good.

The principal difficulty encountered in differential thermal analysis is that appreciable temperature differences may occur during the test which are not due to the evolution or absorption of heat in the test-sample. As a result, the zero or base-line deviates from the horizontal. This 'drift' can be caused by:

(i) Unsymmetrical location of the sample-container in the block, or of the block itself in the furnace. These are constructional faults and easily corrected.

(ii) Poor fit between the lid and lower half of the block causing localised over-heating. This may be overcome by fitting the lid with a tongue which engages in a groove in the lower half of the block (cf. Fig. V.15).

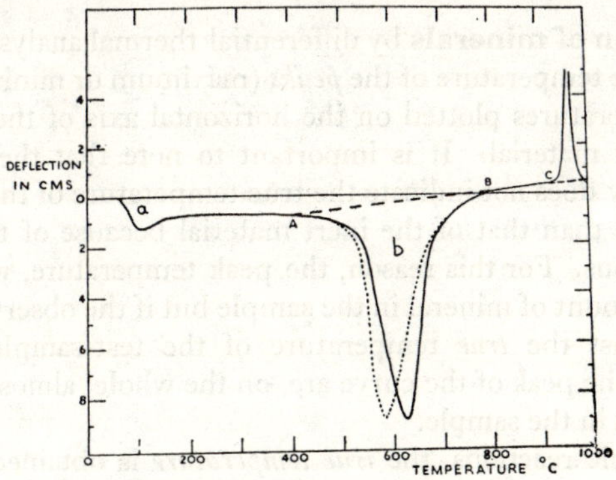

FIG. V.16. DIFFERENTIAL THERMAL CURVE SHOWING BASE LINE CORRECTION

There is a third cause of drift which is more difficult to eliminate. Since thermal analyses are conducted on a rising-temperature basis, the rate of temperature rise in both the test and reference materials depends on their respective coefficients of thermal diffusivity. Not only may these be different at the start, but both will vary as the temperature is raised and not necessarily at the same rate. In addition, the thermal diffusivity of the test material may alter considerably when the mineral decomposes.

The drift of base-line from this cause may be greatly reduced by choosing a reference material with thermal characteristics approaching those of the test-sample as closely as possible, and it is usually advisable to use precalcined test-material for this purpose, and the curves finally corrected as shown in Fig. V.16.

To attempt to catalogue all the refinements and modifications to the apparatus used in differential thermal analysis by the many workers in the field would be impossible.[1] Units range from the simplest, costing a few pounds and occupying little space, to the most complex, comprising the latest electronic devices and requiring most of the space in a large room for housing. It is debatable whether the latter has the advantage that the extra cost would suggest. Of prime importance is that the operator should appreciate the limitation of the method and his own apparatus in particular and design his experiments accordingly.

The apparatus described in the previous pages represents a basic type of comparatively low cost, but nevertheless one which has proved eminently satisfactory for most applications of differential thermal analysis, with, at the most, minor refinements.

Applications of Differential Thermal Analysis

This method of analysis has several important applications of which the chief are:

(i) The identification of minerals.
(ii) The estimation of the content of some minerals in composite mixtures.
(iii) Assessing the principal reactions and the temperatures at which they occur in ceramic bodies on firing.

The identification of minerals by differential thermal analysis is based mainly on observations of the temperature of the *peaks* (maximum or minimum) of the thermal curve. The temperatures plotted on the horizontal axis of the observed curve are those of the inert material. It is important to note that the 'peak' temperature derived in this way does not indicate the true temperature of the test-sample, which is lower or higher than that of the inert material because of the endothermic and exothermic reactions. For this reason, the peak temperature, when so measured, is affected by the amount of mineral in the sample but if the observed curve is redrawn and plotted against the *true* temperature of the test-sample, the temperatures corresponding to the peak of the curve are, on the whole, almost independent of the amount of mineral in the sample.

In endothermic reactions, the *true temperature* is obtained by subtracting the

[1] For fuller details see *Differential Thermal Analysis*, edited by R. C. Mackenzie (Academic Press, London and New York, 1970).

observed differential temperature from the corresponding temperature of the inert material, as shown by the dotted curve in Fig. V.16.

The peaks in D.T.A. curves represent heat changes associated with specific reactions. In clays and allied materials the following major effects can give characteristic and diagnostic results.

(1) Loss of free, adsorbed or absorbed water. Endothermic peaks in the temperature region 100–150° C. are usually due to these causes. Broken bond water may not be lost until the temperature reaches 200–300° C.

(2) Breakdown of hydrates, hydroxides and other minerals which contain lattice water is an endothermic effect which often characterises specific minerals. The changes may take place between 200 and 800° C. depending on the type of mineral.

(3) Breakdown of other oxy-salts, such as carbonates, sulphates. These are also endothermic changes and may occur between 500 and 1200° C.

(4) Oxidation reactions involving exothermic changes and including the conversion of sulphides to sulphates or oxides, the burning of carbonaceous material.

(5) Recrystallisation reactions from amorphous or disordered phases often result in an overall exothermic change of extreme rapidity. These are discussed more fully in a later chapter, but many hydrated silicate minerals show characteristic peaks due to this cause.

(6) Changes of state either from solid to liquid or liquid to gas show endothermic effects often at diagnostic temperatures.

(7) Phase changes, such as those in the silica minerals, may be either endothermic or exothermic changes, but are frequently small in magnitude. They are often reversible and can be measured also on the cooling cycle.

Many of the characteristic peaks of minerals commonly found in ceramic materials are shown in the Appendix to this volume. Fig. V.17 contains reproductions typical of the thermal curves of the clay minerals kaolinite, illite and montmorillonite. Both the maximum temperature and the shape of peaks can be diagnostic.

The characteristic peaks of many of the minerals identified by differential thermal analysis which have been compiled by R. W. Grimshaw and his colleagues, using the apparatus described in the previous pages, are given in the Appendix.

From the description given above, it will be noted that the method is applicable to the principal clay minerals and to most of the accessory components, with the notable exception of alkali-bearing constituents such as *micas* (other than the hydrous micas or illites) and felspars, which appear to be thermally inert.[1]

The *silica minerals* are also capable of identification by thermal means,[2] but because the thermal changes which take place are the result of the reversible inversions with a small heat change (see p. 662), a sensitive method of recording is essential.

The Quantitative Determination of Minerals by Differential Thermal Analysis. The peaks in differential thermal analysis are the direct result of the heat

[1] Köhler, A. and Weiden, P. (*Neues. Jb. Miner. Mh.* 249, 1954) have suggested that some felspars invert to high temperature varieties in the temperature range 700–900° C., and that such changes are accompanied by a thermal effect of small magnitude.

[2] Grimshaw, R. W., Westerman, A. and Roberts, A. L., ibid., 48, 269, 1948.

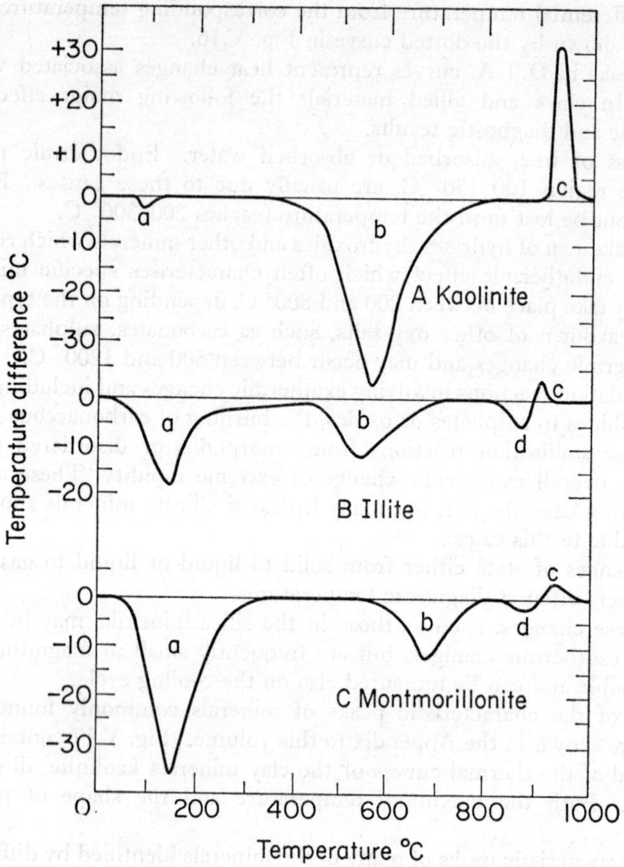

FIG. V.17. DIFFERENTIAL THERMAL CURVES OF (A) KAOLINITE; (B) ILLITE; (C) MONTMORILLONITE

change during a reaction. Because this heat change corresponds to the energy of reaction, it can be predicted theoretically that the area under a characteristic peak of a mineral is related to the energy change during the reaction; hence if that reaction is specific in both magnitude and temperature to all samples of the mineral, the amount of the mineral present in a composite sample will be related to the peak area.

The first and prime essential for the quantitative estimation of minerals by differential thermal analysis is that the technique for carrying out the tests should be rigorously standardised; only under these conditions are accurate measurements possible.

Thermal analysis measurements are relative in that conclusions can be drawn only by comparing the curves of an unknown sample with those of known materials. So, with quantitative measurements, the thermal curves of different proportions of pure minerals are required for an analysis. To be capable of estimation, minerals must give well-defined thermal curves with at least one peak sufficiently distinctive and capable of accurate measurement, e.g. the kaolinite decomposition peak at 590° C. The peak corresponding to the loss of hygroscopic water would not be suitable for quantitative work, nor would a peak which is liable to interference by commonly associated impurities, e.g. the exothermic peak in kaolin minerals at 980° C., which is modified or suppressed by certain soluble salts particularly sodium sulphate.

Grimshaw and Roberts[1] have shown the value of the method for the quantitative determination of some ceramic minerals and have demonstrated the limits of accuracy. They point out that the main source of error in such measurements is due to differences in thermal characteristics (specific heat, thermal diffusivity, etc.) between the test-sample and inert reference material. It can be eliminated or largely reduced by diluting the test-sample to 25 per cent with the inert material and increasing the sensitivity of the recording equipment.

Under these conditions, kaolinite, halloysite, montmorillonite, calcium carbonate and many other minerals can be measured under favourable conditions with an accuracy of $\pm$ 1 per cent.

One difficulty encountered in the estimation of minerals, not only by thermal analysis, but by all other methods, is the isolation of a pure mineral sample which is representative of that mineral wherever it is found. Some minerals present no difficulty; calcite, for example, occurs in a pure state in nature and apparently has a similar thermal effect in all cases. Kaolinite also is well-defined and can be purified from crude china clay by centrifugal separation. Other kaolin minerals may be intermixed and difficult to isolate; however the characteristic endothermic peaks at about 550–600° C. of kaolinite, livesite and halloysite are all of similar magnitude so that the total content of kaolin minerals in a mixture can be estimated, even though the individual proportions cannot be ascertained. Other minerals, such as the montmorillonites are variable in composition (see p. 145), and their estimation by thermal means should always be undertaken with caution. In any one particular deposit an 'average mineral' may be obtained by suitable methods of purification, and differential thermal analysis may then be used to estimate the content of that mineral in other samples from the same deposit.

The *silica* minerals present many anomalies in their detection and estimation by differential thermal analysis.

Quartz may be identified by the heat change of its inversion at 573° C. Usually this change is rapid and reversible and for a variety of reasons it is better to measure the change on the cooling cycle,[2] recording the thermal effects by a sensitive mirror galvanometer. For most samples of quartz the maximum of the differential peak occurs at the same temperature and the form of the curve is similar. Keith and Tuttle[3] have demonstrated that in exceptional cases, quartz can show an anomalous inversion and in many instances, a particular sample may invert over a range of temperature (amounting to 2–3° C.) rather than at one definite point. However, the analysis of many quartz samples by normal differential thermal methods has shown that the peak area is virtually constant in amount and independent of the sample.

[1] Grimshaw, R. W. and Roberts, A. L., *Trans. Brit. Ceram. Soc.*, 52, 50, 1953.
[2] Grimshaw, R. W., Westerman, A. and Roberts, A. L., loc. cit., 1948.
[3] Keith, M. L. and Tuttle, O., *Amer. J. Sci.*, 1, 203, 1952.

The method of using peak height as a basis of mineral content, whilst satisfactory for well-crystalline quartzites, is less accurate than area measurements when the quartz is in the form of fine crystallites.

Both *cristobalite* and *tridymite* invert at a low temperature, hence the method of cooling as in quartz cannot be used without considerable modification to the apparatus. The changes are therefore best measured on the heating cycle, but with a reduced rate of rise (6–8° C. per min.) than that normally employed. Less drift is encountered if a metal block is used instead of the usual refractory one, although the thermal peaks are reduced as a consequence.

The inversion temperature of *cristobalite* varies between 224–269° C.; it appears to depend on the temperature at which the cristobalite is prepared and on the type of catalytic agent. Differential thermal curves of cristobalite samples from different sources show this variability of inversional temperature, and further indicate that the shape of the peak is not identical in all samples.

Despite these differences, the peak area of pure cristobalite is reasonably constant and is independent of the maximum temperature and shape.

Tridymite also provides a complication. Although every sample of this mineral so far tested gives thermal peaks at 117° C. and 163° C., the ratios of the heights of these peaks show considerable variation. The time and temperature of firing and the nature of the fluxing oxide both influence the form of curve. Two forms of tridymite probably exist, one with an inversion at 117° C., and the other at 163° C., and not one form with two inversions.

It has been shown by Gaskell and Grimshaw that the combined areas of the two inversional peaks of the thermal curve of tridymite give a good measure of the content of that mineral in a sample.

Carbonaceous Matter is a common constituent of clays and shales and under oxidising conditions it burns off between 250–600° C. giving a pronounced exothermic peak in a differential thermal analysis test. As such, it is the most serious deterrent to mineralogical identification by this means.

The best solution to this problem is to carry out differential thermal analyses of carbonaceous clays in an atmosphere of nitrogen. The oxidation of the organic matter is thus suppressed and the thermal effects recorded are those of the inorganic minerals. This technique is applicable to estimating some of the inorganic constituents of coals, but if the ash content is less than about 10 per cent, the peaks of distillation and pyrolysis reactions of the coal minerals themselves tend to predominate.

Differential thermal analysis can also be applied to assessing the nature and amounts of carbonaceous material in clays and also to determining the rank and approximate calorific value of fuels.

Influence of Grain Size on Thermal Analysis. Provided that the mineral is not crystallographically distorted or altered as the grain size is reduced, the differential thermal curve does not appear to be seriously influenced by the sizes of the particles in the sample. The thermal curves of many minerals are however, seriously affected by grinding. Many alumino-silicates conform to a general pattern and give results which are remarkably similar. Short-term grinding invariably gives rise to a product with an endothermic peak at about 550° C. when tested by thermal means. In addition, the peak resulting from the loss of hygroscopic water is prolonged and accentuated and may persist up to 300° C. As the grinding is continued, the endothermic effect at 550° C. increases to a maximum and then declines, whilst the lower temperature peaks increase in size continuously. The probable explanation is that mechanical breakdown occurs by the fracture of corners of crystals, thereby exposing free valencies. Water molecules are absorbed which are held more firmly than hygroscopic water. It is of interest to record that more alumina and silica may be extracted from a ground sample by moderate leaching with dilute acid or alkali than from the original material (see p. 708).

When preparing a mineral sample for thermal analysis, it is therefore advisable to minimise grinding whenever there is a possibility of its causing any breakdown of the mineral.

Double Differential Technique

Natural raw materials occasionally give thermal curves containing several peaks, some of which may mutually interfere. When a composite thermal curve containing two or more clay mineral peaks is obtained by the usual method of testing, it is often possible to decide upon the identity of at least one of the minerals, and to estimate its amount approximately. The test may be repeated, after adding to the inert material the appropriate amount of the mineral already identified in the test-sample. Thermal effects due to this particular constituent thus occur in both the test-sample and in the reference material, and so almost or completely cancel each other out if the correct amount has been added to the latter. By successive approximations it is possible to obtain a thermal curve showing only the thermal effect due to the second mineral. This method is illustrated by Fig. V.18, in which curve A relates to a laboratory-prepared mixture of 40 per cent of kaolinite and 25 per cent of montmorillonite when tested in the usual way against an inert reference material. In this particular instance there is no difficulty in recognising the presence of both kaolinite and montmorillonite, but mutual interference between the two peaks is sufficient to make quantitative interpretation uncertain. Curve B shows the results obtained by repeating the test against a reference material containing 40 per cent of raw kaolinite the rest being pre-calcined material; the principal peak due to kaolinite is almost completely suppressed, and only the montmorillonite peak remains as a major effect. In this case, the sensitivity of the equipment was increased in order to magnify the montmorillonite peak area. The residual kaolinite effect corresponds to an apparent excess of less than 1 per cent of this mineral.

This technique has been termed *double-differential thermal analysis*, and is of great practical value. One particular application which deserves special mention is the determination of small amounts of kaolin clay minerals in ganister and silica rocks. Usually less than 3 per cent of clay is present in such minerals, and consequently the endothermic $\alpha \rightleftharpoons \beta$ inversion of quartz, which occurs at a similar temperature to the breakdown of the clay, is superimposed and prevents accurate measurement. The percentage of quartz may be found by testing a sample of pre-calcined material; this amount of pure quartz may then be included in the standard reference material. The resulting thermal curve of the ganister tested against the new reference is that of the clay mineral alone.

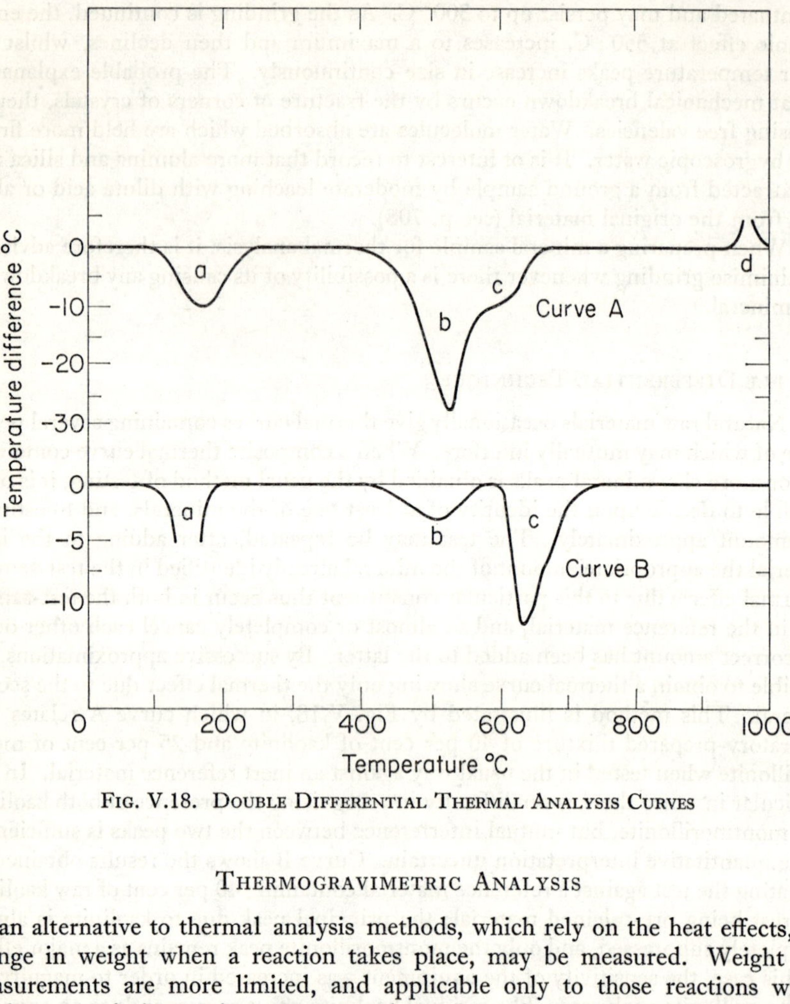

FIG. V.18. DOUBLE DIFFERENTIAL THERMAL ANALYSIS CURVES

THERMOGRAVIMETRIC ANALYSIS

As an alternative to thermal analysis methods, which rely on the heat effects, the change in weight when a reaction takes place, may be measured. Weight loss measurements are more limited, and applicable only to those reactions which involve an evolution, or in some cases, absorption, of a gaseous component. They are valuable in clay studies, for many of the principal minerals decompose on heating with the evolution of water or carbon dioxide. Some examples are:

			Loss of weight
$Al_2O_3.2SiO_2.2H_2O$ (kaolin)	→	$Al_2O_3 + 2SiO_2 + 2H_2O$	13·9 per cent
$Al_2O_3.4SiO_2.H_2O$ (ideal montmorillonite)	→	$Al_2O_3 + 4SiO_2 + H_2O$	5·06 per cent
$CaCO_3$ (calcium carbonate)	→	$CaO + CO_2$	44·0 per cent

$2FeS_2 + 4O_2$ (pyrites)	$\rightarrow$	$Fe_2O_3 + 4SO_2$	33·0 per cent
$4FeCO_3 + O_2$ (siderite)	$\rightarrow$	$2Fe_2O_3 + 4CO_2$	31·0 per cent

The reactions take place at well-defined temperatures and, as with thermal tests, a series of standard curves may be drawn up which help in the identification and estimation of many minerals.

The **loss on ignition** at a temperature of about 900° C. is regarded as an important characteristic of many clays. In addition to being a vital determination in the routine chemical analysis of ceramic materials, the total weight loss on calcining is also used as a means of identifying some minerals and as a rough means of estimating the percentage of some of them in various samples. Thus in a sandy clay, in which the clay mineral is almost entirely kaolinitic, the loss on ignition gives a good measure of the clay content. (The weight loss of the pure clay mineral is 13·9 per cent, so if 6·95 per cent loss were measured, the sample would contain half its weight of clay). The type of clay mineral in an unknown sample can often be conjectured if the loss on ignition is determined. It is important to determine the loss on ignition on a sample which has previously been oven-dried, because, otherwise, hygroscopic or absorbed water would contribute to the total.

The loss of moisture, i.e. the loss in weight after drying at 120° C., can itself be an important and an informative determination. Some minerals take up large quantities of moisture on being exposed to the atmosphere; montmorillonites, for example, absorb up to 20 per cent, whilst kaolinite rarely holds more than 2 per cent. Livesite, the kaolin mineral in many sedimentary clays loses about 8 per cent at 120° C. provided that it has been allowed to hydrate fully before testing.

The total weight loss of a sample includes:

(i) loss of hygroscopic and adsorbed water;
(ii) loss of colloidal water;
(iii) loss of water of hydration and absorbed water.

These three are, for the most part, removed by heating to 120° C. for long periods and are usually classified as the loss of moisture.

(iv) loss of combined water, which is part of the structure of minerals;

(v) loss of carbon dioxide or monoxide from the burning of carbonaceous material or the decomposition of carbonates;

(vi) loss of sulphur dioxide and trioxide from sulphur compounds, such as pyrites, sulphates, etc. In exceptional circumstances, hydrogen sulphide, H_2S, may sometimes be evolved;

(vii) loss of various other gases and volatile substances present in the material.

For an accurate and complete analysis, the individual proportions of water, carbon dioxide and sulphur dioxide liberated should be estimated separately.

Although it is usual to determine the total loss on ignition by heating at a high temperature (above that at which all the minerals will decompose) much information can be obtained from tests in which the sample is raised in temperature at an extremely slow rate and weighed at several temperatures. Such tests can be extremely tedious, especially if the sample is allowed to reach equilibrium at each temperature; for example, a kaolinite sample reached a sensibly constant weight only after two months at 410° C., and even then there was a small yet appreciable weight loss at slightly higher temperatures!

Excellent differentiation between different minerals may be secured by such careful measurements. Livesite, for example, loses almost all its combined water at 405° C., kaolinite reacts at 410° and halloysite at 400° C. Montmorillonites, micas and other minerals can be identified in a similar way.

A neat method of recording the *rate of weight loss* in a sample has been described.[1] Reactions involving an evolution of a gas give peaks with a maximum which corresponds to the temperature at which the reaction is proceeding at its fastest rate. The form of typical curves for clays obtained from this apparatus are not unlike those from differential thermal analysis, except that there is no differentiation between endo- and exothermic effects.

In recent years apparatus has been developed to carry out simultaneous thermogravimetric (TG)/differential thermal analysis (D.T.A.) tests. The results have been described by Wendlandt,[2] but as yet the potential of the dual technique has not been fully realised. Nevertheless, this type of complementary experiment has often led to the nature of changes being established. When gas evolution is involved, the nature of the volatile component can often be established by chromatographic methods (p. 218).

Rehydration Methods

The property of *ad*sorbing and *ab*sorbing water from the atmosphere is possessed by all clay minerals to a greater or less extent, the amount being determined by (*a*) the surface area of the material, and (*b*) its chemical and physical nature. For example, montmorillonite, which has a much smaller grain size, and hence a larger specific surface area than kaolinite, adsorbs much more water; but when saturated with various bases it will absorb different amounts of water depending upon the ease of hydration of the cations.

In the kaolinite group, where little ionic effect can be expected, water will be adsorbed only and the amount of adsorption will depend on the relative surface area of each mineral.

A simple rehydrator can be constructed from a desiccator with water substituted for the drying agent. Small petri dishes containing the sample of clay dried at 250° C. are placed in the rehydrator and weighed from time to time. The test is best carried out at constant temperature in order to standardise humidity conditions.

Berkelhamer[3] has stated that it is necessary to dry the samples at temperatures of about 250° C. to eliminate all absorbed water which in the case of montmorillonite may not be released much below 200° C.

Results obtained on samples of kaolinite, halloysite, montmorillonite (Wyoming) and a foundry bond clay rich in livesite are shown in Fig. V.19. Although the test is not sufficiently critical to be diagnostic of a particular mineral, it does indicate the

[1] Bird, R. T. H., *Claycraft*, 30, 340, 1957.
[2] Wendlandt, W. W., *Chem. Inst. Canada*, 101, 1965.
[3] Berkelhamer, L. H., *J. Amer. Ceram. Soc.*, 26, 130, 1943.

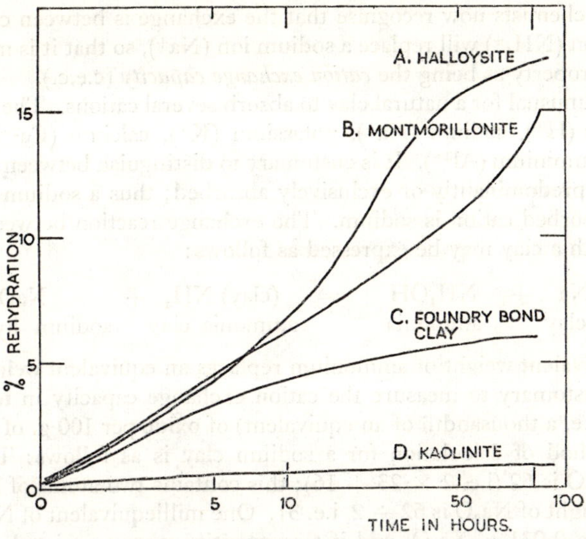

FIG. V.19. THE RATE OF REHYDRATION OF SOME CLAYS

mineral type, and is also of value in assessing the behaviour of a clay under atmospheric conditions.

IL/MA METHOD

KEELING[1] has shown that the ratio of the ignition loss and moisture adsorption of clays is a valuable parameter for predicting the nature and proportion of clay minerals which are present. Care has to be taken to determine the values under controlled conditions which take account of impurities such as hydrates, carbonates and carbonaceous material.

BASE EXCHANGE CAPACITY

ALL clay minerals adsorb positive cations by virtue of the unsaturated negative ions in the surface layers. In addition, montmorillonites, and to a lesser extent some finely grained illites absorb cations because they have a charge deficiency within the lattice. These cations are held by only feeble electrical forces and are easily replaceable by others.[2]

That soils were able to remove ammonia from aqueous solution was first recognised by J. T. Way, who reported his finding to the Royal Agricultural Society in 1850 and 1852. The absorption of bases in this way does not involve any structural change in the minerals, and as one base is taken up from solution, another is generally released in its place. Hence this property of soils and clays became known as the *base exchange capacity* (b.e.c.).

[1] Keeling, P. S., *Trans. Brit. Ceram. Soc.*, 60, 217, 1961.
[2] Carrol, D., *Bull. Geol. Soc. America*, 70, 749, 1959.

Physical chemists now recognise that the exchange is between cations, e.g. an ammonium ion (NH_4^+) will replace a sodium ion (Na^+), so that it is more correct to refer to the property as being the *cation exchange capacity* (c.e.c.).

It is not unusual for a natural clay to absorb several cations. The common ones are hydrogen (H^+), sodium (Na^+), potassium (K^+), calcium (Ca^{2+}), magnesium (Mg^{2+}) and aluminium (Al^{3+}). It is customary to distinguish between clays in which one cation is predominantly or exclusively absorbed; thus a sodium clay is one in which the absorbed cation is sodium. The exchange reaction between two cations associated with a clay may be expressed as follows:

$$\underset{\text{sodium clay}}{(\text{clay})\,Na} + \underset{\text{ammonia}}{NH_4OH} \rightleftharpoons \underset{\text{ammonia clay}}{(\text{clay})\,NH_4} + \underset{\text{sodium hydroxide}}{NaOH}$$

One equivalent weight of ammonium replaces an equivalent weight of sodium, hence it is customary to measure the cation exchange capacity in terms of milli-equivalents (i.e. a thousandth of an equivalent) of oxide per 100 g. of clay.

The method of calculation for a sodium clay is as follows: The molecular weight of Na_2O is 62 (i.e. $2 \times 23 + 16$); this contains two atoms of Na, hence the equivalent weight of Na_2O is $62 \div 2$, i.e. 31. One milliequivalent of Na_2O therefore corresponds to 0·031 g. Na_2O, and if this quantity were associated with 100 g. of clay, then the *cation exchange capacity* would be 1 milliequivalent per 100 g.

The cation exchange capacity of clays is known to have a pronounced influence on their properties (see Chapter VIII). The fertility of soils is partially dependent on the inorganic salts held in this way. In addition, a measurement of the cation exchange capacity may provide useful information to the ceramist as to the nature of the minerals in a clay. It may be determined by (i) electrodialysis, or (ii) by a chemical method.

In electrodialysis a cell is used which consists of three compartments, the two end ones being separated from the centre by a semi-permeable membrane or a suitable porous plate. The end compartments are filled with distilled water and in the centre one is placed a known weight of the sample, suspended in distilled water. A direct electrical potential (100 volts) is applied between platinum electrodes in the two end cells and this causes the cations absorbed on the clay to migrate through the membrane to the negative electrode. After a period of time, when the test is judged to be completed, the alkalinity of the solution in the cell containing the negative electrode may be measured by titration with a standard solution of acid. From this, the total cation exchange capacity (other than hydrogen ions) can be calculated. If it is required to know the types of cations associated with a clay, a quantitative analysis must be carried out on the solution.

The chief effect in electrodialysis is to replace the exchangeable cations by hydrogen cations, but it is alleged that some clay minerals, montmorillonites in particular, tend to be broken down and release cations which are normally part of their structure. Some iron compounds may also be removed by this treatment.

In the chemical methods, the sample is treated with a cation which readily replaces all others and at the same time is capable of easy estimation.[1] Ammonium (NH_4^+) is probably the most convenient ion to estimate, but it is not absorbed by

[1] Worrall, W. E., *Trans. Brit. Ceram. Soc.*, 62, 447, 1963.

clays in preference to such ions as calcium, magnesium or hydrogen. However, if a salt of ammonia is used, the anion of which forms an insoluble compound with most other ions, quantitative replacement can be effected. Ammonium acetate is most commonly employed, because not only are the acetates of calcium, magnesium, etc., insoluble, but any hydrogen ion is removed as undissociated acetic acid.

The usual procedure is to treat a known weight of the sample with a normal (1.ON) solution of ammonium acetate. Five or six such treatments are normally required to ensure complete replacement and between each, the samples should be allowed to sediment and the clear liquid decanted. If a centrifuge is available, this stage of the determination can be completed rapidly. Excess ammonium acetate is removed following the final treatment by washing with alcohol.

The sample is then transferred to a long-necked (Kjeldahl) flask, an excess of sodium hydroxide added and the ammonia distilled off—preferably on a steam bath—and collected in a known volume of standard (N/10) sulphuric acid. When the distillation is complete, the ammonia which has been liberated is estimated by back-titrating the remaining acid with a standard solution of alkali. Further details of the method have been published by Truog[1] and a useful micro-method has been described by Mackenzie.[2]

In many cases, the type and amount of the cations associated with a particular clay in its natural state are equal to or greater in importance than the total exchange capacity. For the most part, these can be identified and estimated by carrying out a standard analysis on the decanted liquids which are collected after each washing with ammonium acetate. Exchangeable hydrogen is best estimated on a separate sample, using barium acetate[3] which replaces hydrogen more rapidly and completely than does ammonium acetate. The free acetic acid which forms can be titrated against a standard solution of alkali (NaOH), with phenol-phthalein as an indicator.

Other methods of determining the cation exchange capacity of clays have been described, but none is of such general application as those outlined.

In a rapid method of determining the cation exchange capacity devised by Robertson and Ward,[4] a known volume of a standard solution of methylene blue is shaken up with a weighed quantity of the sample which is suspended in water. After several hours, an aliquot part of the clear liquid is withdrawn and the colour compared with other samples of the dye solution diluted with known quantities of water. From this comparison, the amount of colour adsorbed by the sample can be calculated. This method is particularly useful for industrial tests, but it is not as accurate as the ammonium acetate technique. The type of ion associated with the clay may affect the colour of the methylene blue and give erroneous results, but the error is not serious unless precise measurements are required.

The cation exchange capacities of some clay minerals and natural clays are shown in Table V.V. The actual cations present are shown in some cases, but it must be remembered that these apply only to the samples analysed, and may differ considerably in other specimens from the same locality. This is particularly true

[1] Truog, E. and Jackson, H. L., *Proc. Soil Sci. Amer.*, 4, 136, 1939.
[2] Mackenzie, R. C., *J. Colloid Sci.*, 6, 219, 1951, and *Clay Min. Bull.*, 1, 203, 1952.
[3] Parker, M., *Soil Sci.*, 30, 391, 1930.
[4] Robertson, R. H. S. and Ward, R. M., *Pharmacol.*, 3, 27, 1951.

TABLE V—V. THE CATION EXCHANGE CAPACITIES OF CLAYS

Clay mineral	*Total cation exchange capacity (m.e./100g.)*	*Cations originally present*				
		Ca	*Mg*	*Na*	*K*	*H*
Kaolinite (china clay)	5·74	3·12	0·61	0·41	0·40	1·20
Halloysite (Eureka)	12·76	0·50	1·25	1·74	1·56	7·71
Livesite (from bond clay)	39·3	—	—	—	—	—
Montmorillonite (Wyoming)	76·46	17·72	14·78	38·75	3·31	1·90
Hydrous Mica (from fireclay)	4·50 (variable)	0·00	0·00	0·62	3·43	0·45
Natural Clays						
Ball Clay (Dorset)	14·45	4·62	0·32	5·23	0·31	4·00
Bond Clay	21·5	—	—	—	—	—
Fireclay (weathered)	10·65	1·75	3·70	0·68	0·84	3·68
Fireclay (unweathered)	11·32	7·45	1·72	0·81	0·54	0·80
Brick Clay (Jurassic bed)	7·32	3·62	2·26	0·05	0·02	1·37
Bentonite	56·80	18·86	2·55	28·04	3·89	3·46

when comparing weathered and unweathered clays; the former are always richer in hydrogen ions although the total capacity is not usually altered substantially.

OTHER METHODS OF ANALYSIS

MANY other methods of analysis of the minerals in ceramic materials may be used, but they are not of such general application as those outlined. However, in certain specific instances, one of them may be employed with perhaps greater success than obtained by standard methods.

SPECIFIC GRAVITY OR TRUE DENSITY

THE true specific gravity of a substance is the relation between its weight and that of an equal volume of water. Its value is determined by the type of atoms or ions and the way in which they are stacked in the crystal lattice, hence the specific gravity of a pure mineral is constant for all samples and is a positive means of identification.

The specific gravity can be measured rapidly and accurately by one of the methods described in Chapter VII (p. 428) and if the material is composed of a single mineral, then it can often be readily identified. Many ceramic minerals are very similar in density, so that it is usually advisable when examining an unknown material to use specific gravity measurements in conjunction with other tests.

Different minerals of the same chemical composition can often be differentiated by measuring their specific gravity; such examples are (i) the silica minerals, quartz, cristobalite and tridymite which have values of 2·65, 2·32 and 2·28 respectively, and (ii) the alumino-silicates, kyanite, andalusite and sillimanite with the chemical composition $Al_2O_3 . SiO_2$ and specific gravities of 3·59, 3·18 and 3·08. The pertinent figures for many of the minerals found in ceramic bodies are given in the Appendix.

Most ceramic materials are composed of many minerals and components, so that the specific gravity is the resultant of the various ingredients. In these cases, the method is not of great value, unless the ingredients can be separated and determined singly.

Hardness

The hardness of a mineral can be defined as its resistance to scratching or abrasion. It is a property which is related to the crystal structure, so, in a sense, the hardness of a mineral can be used as a means of its identification.

The hardness of a substance is a purely relative term, and it can be assessed only by comparing its abrasion resistance with that of standard minerals. Diamond, the hardest substance known, has the index 10 and others are graded down the scale to the softest, talc, with the index 1. The simplest measure of the hardness of a substance is its capacity to scratch less hard materials and, in turn, to be scratched by others less soft than itself. Other, more precise, methods are described in detail in Chapter XIV (p. 865).

In ceramic bodies composed largely or entirely of a single mineral type, the measurement of the hardness may be a valuable guide to the composition. However, in composite bodies, as a means of mineral identification, the determination is usually of little value.

Colour

Many ceramic materials are white—some like china clay, magnesia, lime, alumina, zirconia and titania are impressively so when pure. With experience it is often possible to predict the impurity content of such materials by their appearance and deviation from the pure white colour. The compounds of some elements have a characteristic colour, whereby they can be readily recognised. Iron compounds, for example, tend to impart a yellow, buff, red, brown or black coloration to bodies which contain them; cobalt and copper compounds in ceramic bodies usually give a blue and green coloration respectively. A stain or discoloration on a body may be indicative of the presence of certain undesirable minerals and should never be disregarded.

There are other properties also which can be a useful guide to the minerals in a material. These include the *odour*, the *feel* and even the *taste* of a substance which to the expert may be all-important. Properties such as the *refractoriness*, the *plasticity*, the *firing* and *drying shrinkages* may also be indicative of the content and amount of particular mineral types, but they are not usually employed for this purpose.

THE CALCULATION OF THE MINERALOGICAL ANALYSIS OF A CLAY

THERE are many methods which are of value in the detection and estimation of minerals in ceramic materials, all of which are useful in specific instances. No one method can, by itself, give a complete analysis of the minerals in any mixture, and it is only by considering the results from several that a true picture can be secured.

One difficulty which is not easy to overcome in mineralogical analyses is that naturally-occurring minerals are prone to vary in composition and properties especially if they are from different localities, and as all the methods of mineral analysis outlined are comparative, it is essential to secure a standard pure sample of each mineral to be analysed from the source being investigated. Even then, it is impossible to guarantee that each fragment of mineral will be identical, but some idea of the extent of the variation can often be obtained.

The only method which is likely to be successful in the estimation of minerals in a complex sample is to separate the individual components, to select a suitable technique from the ones outlined, which is accurate for the quantitative determination of this component and then to apply the method to the raw material.

Purely chemical methods of separation are of limited value, and satisfactory isolation of pure minerals is, in general, only achieved by physical or mechanical means, and even then treatment must not be so severe as to cause breakdown of any component.

Grimshaw and Roberts[1] have outlined a method by which the composite minerals of a natural clay may often be isolated or their nature inferred.

The clay deposit should always be examined *in situ* if possible, for the type of clay, whether it be residual or sedimentary, may indicate the minerals likely to be present. The geological history and environments may be informative, for it is often possible to tell whether a clay deposit is of marine, lacustrine, fluviatile or other origin. In addition, a study of the surrounding strata may indicate the nature of impurities in the clay; e.g. in sandstone areas, the clays will usually have a high free silica content, whereas clays from a predominantly limestone area will be rich in calcium carbonate.

The examination of the clay in mass sometimes reveals stratification and micaceous flakes are often observed along the bedding planes. Ironstone nodules are sometimes revealed, and occasionally large fragments of other minerals. The presence of carbonaceous material can be inferred from the colour, the remains of roots and plants, and sometimes particles of shale or coal. A hand-lens is useful in such examinations.

The next stage is to break down the clay into individual crystallites without disrupting or decomposing any mineral particles, and then to try to separate them in some way. The mildest agency is water, in which many clays break down with gentle stirring, although others may need more vigorous treatment. The addition of a dispersing agent, such as sodium carbonate or sodium hexametaphosphate, helps in the division and, occasionally, in particularly tenacious clays, the process is quickened by heating the water. Stirring must be continued until no undispersed lumps remain on decantation. At least twenty litres of water should be used for every kilogram of clay.

The resulting suspension is poured through a nest of sieves, the largest one being 60 B.S.S., the smallest 240 B.S.S. The residue on the sieves will be appreciable in many cases. The 60-mesh sieve retains a high proportion of ironstone nodules and occasionally mica flakes. The latter, however, usually predominate at lower sieve sizes along with much of the quartz and most of the heavy minerals.

The suspension which passes the sieves should be allowed to settle for 24 hours, and after decantation almost the whole of the remaining quartz and heavy minerals

[1] Grimshaw, R. W. and Roberts, A. L., *Trans. Brit. Ceram. Soc.*, 51, 327, 1952.

remain in the sediment. Washing this residue removes any retained clay or fine-particled mineral, which can be returned to the suspension.

The decanted liquid is then separated into fraction ranges by centrifugal means. Two types of centrifuge have been used. The first is a model capable of sedimenting large volumes of suspension batchwise in a large bowl. The advantage of this apparatus is that considerable quantities of liquid can be dealt with in a single operation but its range of separation is restricted as the maximum g is 490 (i.e. 490 × the force of gravity).

Later stages of the fractionation are best carried out in a super-centrifuge operating under continuous feed conditions capable of producing a g of 50,000. Particles down to 0·05μ can thus be separated from suspension.

In practice it is better to split the sediment into eight or nine fraction sizes, each consisting of particles with an ever-decreasing maximum grain size. Various minerals tend to predominate in narrow particle size ranges, often with Gaussian distribution, and under controlled conditions fairly pure samples have been obtained. By re-centrifuging, the particles can be separated into closely monodispersed fractions along the wall of the centrifuge rotor. Introducing a liner into the rotor enables the sample to be withdrawn and the grain size at any particular point calculated from nomograms as outlined by Saunders.[1]

In this way the various minerals in clays can be separated into grain size zones, which by suitable treatment can be further purified.

Although sedimentation and centrifuging methods have gone a long way towards completely separating the minerals in a clay, it is frequently impossible to isolate pure specimens by these means alone. This is particularly the case in the fractions of larger grain-size where three or four minerals sometimes predominate over a specific particle size range. Other separation methods must therefore be employed.

(i) **Selective Chemical Action.** This can be used where one component of a mixture of minerals is unattacked by a certain chemical reagent which effectively removes the rest. The chemical method for estimating quartz is such an example.

Other methods employ the reverse principle of dissolving one component preferentially to others, as, for example, dilute acid solution of carbonates, and the Hallimond method of estimating chlorites.[2]

(ii) **Heavy Liquid Separation.** The use of organic liquids to separate minerals of different densities is well known. In general only comparatively large-sized particles are efficiently isolated by density difference methods. Smaller sizes tend to flocculate in organic liquids and so restrict free sedimentation. Bush *et al.*[3] claim that some swelling clay minerals can often be separated centrifugally because of their different hydrated densities.

(iii) **Froth Flotation.** By the use of a suitable frothing agent some separation of minerals can be obtained. The most suitable are amino-acids of high molecular weight, and with these mica and quartz can be separated. A full account of froth-flotation methods is given by Perry.[4]

[1] Saunders, E., *Ind. Eng. Chem.* (*Anal.*), 20, 379, 1948.

[2] Loc. cit., p. 215.

[3] Bush, D. C., Jenkins, R. E. and McCalb, S. B., *Clays and Clay Minerals. 14th Nat. Conf.* (Pergamon Press, 1966).

[4] Perry, J. H., *Chemical Engineer's Handbook* (McGraw-Hill, N.Y., 1952).

(iv) **Electrostatic Separation.** The principles of electrostatic separation have been illustrated by A. L. Johnson.[1]

Briefly, many minerals if allowed to fall freely between two oppositely charged plates, will move towards one or the other. Muscovite mica, for example, can be deflected appreciably by comparatively low applied voltages (about 3,000 V.) and the electrostatic separation of felspar from quartz is carried out commercially in the United States.

Unfortunately many of the minerals commonly associated with clays behave similarly in an electric field and a very high voltage is required to achieve satisfactory separation.

(v) **Magnetic Separation.** This is useful in a limited field. Should there be any large-sized particles of magnetite or similar magnetic mineral, easy separation is secured.

(vi) **Elutriation.** Elutriation has proved effective for separating fine-grained fractions from the centrifuge into very closely monodispersed ranges, with subsequent purification of mineral species.

A stream of water or air is passed upwards through a bed of dispersed particles and, if the flow is carefully regulated, particles within a narrow size range can be extracted. The separation of large particles is not very effective because the liquid flow rate has to be high; consequently turbulence influences the separation and the volume of liquid is excessive. This method will be described in more detail in a later chapter.

Although a sample of the pure minerals can be obtained with the methods outlined, it is usually impossible to effect quantitative separation of all fragments of that mineral. Therefore a test must be selected which is suitable for the estimation of each mineral present in the presence of the impurities. The isolated pure sample is used for standardisation purposes.

In many cases, all the principal and some of the accessory minerals in a clay can be estimated in this way. Another important conclusion which can be derived is the grain-size range of any mineral in the raw material.

A mineral balance sheet should be drawn up to compare with the full chemical analysis which is an essential. If the two do not agree then some important mineral constituent has not been detected or has been assessed incorrectly. Frequently some mineral defies estimation, in which case it can be established only by difference.

The above basis for the separation of components in a clay has proved excellent in assessing the mineralogical composition and the grain-size distribution of different species. Certain minor modifications have been necessary in some cases, but the same general methods have been of value for clays of many types.

[1] Johnson, A. L., *Amer. Inst. Min. Met. Tech. Paper*, 677, 1938.

CHAPTER VI

THE CHEMICAL AND MINERALOGICAL COMPOSITION OF CERAMIC RAW MATERIALS[1]

CERAMIC materials can seldom be termed pure. Although, in some, one mineral type may predominate, the 'impurities' present may be all-important in determining its properties. Clays, in particular, contain many mineral types and are extremely variable in composition. So complex is their study and so great are the difficulties involved, that, although clays are the most extensively used of all ceramic materials, very few have been scientifically examined and the complete details of their mineralogical and chemical composition derived.

For many centuries it has been possible to manufacture beautiful pottery and many useful articles such as bricks, tiles, crucibles, stoneware vessels and furnace linings without it being necessary to have any formal knowledge of their constitution. This is still true of many clays used for making common bricks, tiles, terra-cotta, coarse earthenware and some of the simpler forms of pottery (e.g. peasant ware).

Present day manufacture, with its precision methods and bulk production techniques, necessitates a complete understanding of all the factors involved in the processing of ceramic ware of most types. Not least of these requirements is a knowledge of the raw materials themselves, the mineral types that are present and how their physical state and chemical nature can influence the properties during manufacture and in use.

In the refractories industry, increasing demands are being made by the user in the metallurgical, carbonising and other industries for materials capable of withstanding more severe conditions. To meet this need, great attention must be paid to the raw material and rigid control is essential. Tremendous improvements in the quality of bricks have been made over the last few years, simply by appreciating the role of the various components present in the raw material on the properties of the fired product. The elimination or reduction in amount of deleterious impurities has done much to improve the quality.

The scientific approach to clay problems has not, as yet, been quite as fruitful as it has in other ceramic fields, principally because of the greater complexity encountered. Because clays consist essentially of many different mineral types, most of which are in the form of minute particles, their examination is difficult even when using modern experimental methods. A further cause of difficulty is the influence of the particle size distribution of the various minerals on the properties of the clay. This is often as important as the type and amount of the minerals themselves.

[1] For convenience some materials which are not strictly 'raw' (e.g. bodies, engobes glazes) but are compounded from several ingredients, are included in this chapter.

THE COMPOSITION OF CLAYS

IN their natural state, clays consist of:

(i) Minerals of primary origin, i.e. those present in igneous rocks and which have undergone no significant alteration in composition. These may be described as *primary minerals* (see p. 39).

(ii) Minerals of secondary origin produced by the action of chemical and physical agents on primary minerals resulting in their breakdown and conversion to *secondary minerals*.

Both primary and secondary minerals occur in clay deposits in many different proportions. The *residual clays*, which have undergone little or no transportation, usually contain a much larger proportion of primary minerals, whilst the *sedimentary clays*, which have been transported and deposited perhaps many times, contain few, if any (except quartz), of the minerals associated with igneous rocks.

In composition, clays consist of *clay minerals* (described in Chapter III), together with varying proportions of other substances commonly known as *impurities* —a term which is not always appropriate because the presence of some so-called impurities is of great technical importance in the use of certain clays. For instance, the pleasing red colour of some *terra-cotta*, the imperviousness of *stoneware* and its resistance to corrosion, the delicate translucency of *chinaware* and *porcelain*, the enormous resistance to crushing of some *engineering bricks*, and the extraordinary resistance of *electric insulators* are all due to the presence of a suitable proportion of certain 'impurities' which impart to these materials their valuable characteristics.

A pure clay mineral may be highly refractory, but it may be very weak when burned, owing to the absence of suitable 'impurities' which would, if present, form a vitrifiable bond to unite the clay particles into a mass of great strength.

The proportion of impurities allowable in a clay depends on the purpose for which it is to be used. Thus for the manufacture of *whiteware*, the clay must not contain a harmful proportion of colouring impurities, such as iron oxides; similarly, where a clay is to be used as a refractory material it must be as free as possible from fluxes (p. 275), as these would reduce its resistance to heat.

As a chemical analysis does not reveal the state in which the various substances are combined, it is very important when considering the results of analyses of clays, silica rocks and other materials containing alumino-silicates, to remember that the proportion of fluxes shown by such analyses does not represent the total amount of impurity present; the silica and alumina with which the fluxes are combined (or with which they will combine when the material is heated) will be included in the total amounts of silica and alumina, so that the presence of 1·6 per cent of potash may represent 10·0 per cent of impurity in the form of felspar, the 1·8 per cent of alumina and the 6·6 per cent of silica being included in the total silica and alumina present. Hence, a clay which contains a very small percentage of lime, magnesia, potash, and soda may actually contain a large proportion of minerals other than clay. For this reason, it is often more important to know the mineralogical composition of a clay or other ceramic material than its composition as shown by chemical analysis.

The effects of impurities in a clay depend upon:

(*a*) Their nature and behaviour when the clay is used.
(*b*) The proportions in which they occur.
(*c*) The size and shape of the grains of clay and of the impurities.
(*d*) The conditions under which interaction takes place, including (i) the temperature reached, (ii) the duration of the heating, (iii) the atmosphere of the furnace or kiln, and (iv) the effect of any other substances which may be present.

The principal chemical components in clays may be classed as follows: (*a*) silica; (*b*) alumina; (*c*) alkali-bearing minerals; (*d*) iron compounds; (*e*) calcium compounds; (*f*) barium compounds; (*g*) magnesium compounds; (*h*) titanium compounds; (*i*) manganese and other compounds which occur in small proportions in some clays; (*j*) complex alumino-silicates often containing other elements; (*k*) carbonaceous matter; (*l*) moisture and colloidal water (i.e. water of constitution and of crystallisation); (*m*) exchangeable bases.

Various other impurities which occur in some clays are usually of minor importance.

Silica occurs in clays and allied minerals (*a*) in the *free* state, as quartz or other form of crystalline silica and as amorphous, hydrated, or colloidal silica, and (*b*) *in combination* (i) with alumina in the form of clay and other clay minerals, (ii) with fluxes and alumina in the form of felspar, mica, or other alumino-silicates, or (iii) with various bases forming simple silicates, such as wollastonite ($CaOSiO_2$), etc.

The effects of free silica in clay are as follows:

1. It reduces the plasticity.
2. It lessens the shrinkage on drying and firing.
3. It reduces the tensile and crushing strengths, unless it is of small particle size.
4. It reduces the refractoriness in many cases, though not in all.

The size of the grains of silica is also important, as very small particles will often react and act as a flux under conditions where larger particles of silica increase the refractoriness of the mass.

Free silica occurs in clays of most types. In residual clays, it is present mainly as crystalline *quartz*, for this is a common constituent of igneous rocks and is unaltered by normal weathering processes.

Sedimentary clays also often contain appreciable amounts of quartz which has been transported and deposited along with the clay minerals. In them, the content of free silica is very rarely entirely quartz, because during transportation the original crystalline material may have been subjected to mechanical breakdown which not only reduced the grains in size, but may have converted the quartz into distorted, crypto-crystalline or amorphous varieties. Furthermore, some finely-divided silica may have been dissolved and later precipitated again under more favourable conditions. Such processes are common, for *flint*, *agate*, '*drusy*' *quartz* and pockets of well-developed quartz crystals are formed in this way. When reprecipitation is rapid, the silica formed will be amorphous or crypto-crystalline, and although it may be compacted and metamorphosed at a later stage in its geological history, it may never develop the crystallinity or lattice configuration of quartz. Material of this type is of common occurrence, *chalcedony*, *flint*, *opaline silica* are but a few examples, of what may be present as 'free silica' in clays. These minerals do not give the same X-ray or thermal pattern as quartz crystal, and it is always unwise to accept the quartz content, determined by these methods, as the total amount of free silica.

Silica which is not in the form of quartz is much more reactive than is crystalline quartz; it also converts more readily to cristobalite and tridymite, and will combine with fluxing agents at appreciably lower temperatures.

Alumina. Alumina occurs in clays not only in the form of clay minerals, but also in felspars, mica, hornblende, tourmaline, and other similar alumino-silicates, all of which are moderately fusible; free alumina is also found in some clays, derived from bauxites and laterites. Halloysite clays frequently contain associated free alumina in the form of *gibbsite* ($Al(OH)_3$). Such clays are highly valued for refractory purposes, especially if the total alumina content is high.

Other clays may contain free alumina either as *gibbsite*, *diaspore* (AlO. OH) or as a *colloidal* variety. Some fireclays, particularly of Scottish origin, often have a higher alumina content than would be indicated from the rational analysis.

Aluminous compounds, other than clay minerals, have the following **effects on clays** in which they occur:

1. They reduce the plasticity of the clay, as they are non-plastic.
2. They increase the refractoriness of a clay, provided that the total proportion of alumina is greater than 5 per cent. The refractoriness of a silica brick is drastically reduced by alumina. For each 0·1 per cent of alumina which is present in silica, the maximum operating temperature of the refractory is reduced by about 10° C.

Alkalies and other fluxes are usually associated with alumina compounds and these usually have a serious influence on the refractory properties.

Alkali Compounds. The chief alkalies in clays occur as:

1. Silicates or alumino-silicates of which *felspars*, *micas* or *hydrous micas* are the chief examples. The last-mentioned minerals which are the principal alkali-bearing constituents in many fireclays are classified as a clay mineral, but, unlike those of the kaolin and montmorillonite group they contain alkali as an integral part of their structure. The influence of such minerals on the properties of clays is discussed in Chapter VIII, but the presence of an alkali-bearing mineral always reduces the vitrification temperature and the refractoriness.
2. Adsorbed cations on the surface of clay mineral crystals. The degree of adsorption depends on the available surface area and on the nature of the clay minerals. Montmorillonites may have up to 3 per cent of alkalies from this source, but kaolin clays rarely have more than 0·3 per cent (see also p. 289).
3. Soluble salts, such as *potassium sulphate*, *sodium sulphate* and *sodium chloride*.

In stoneware, porcelain, and other vitrified ware, the higher the molecular proportion of potash or soda, the more fusible is the mass. In incompletely sintered ware, such as faience and earthenware, the larger the proportion of potash or soda the more solid will be the body after firing at a certain (finishing) temperature. **The chief effect of alkalies** in clays is to reduce the refractoriness or vitrification temperature. They combine with alumina and silica to form liquids at low temperatures (*c.* 700° C.) and consequently are termed *fluxes*. The liquid which forms on firing in clays containing alkalies does not crystallise readily on cooling but forms a glass, which behaves as though it were a super-cooled liquid.

Because of the readiness of liquid formation, alkalies impart imperviousness and strength to fired clay bodies and are a major constituent of earthenware, stoneware and porcelains.

Soluble alkali salts may have an important influence on the colloidal and plastic properties of clays (see Chapter VIII).

Most alkali-bearing minerals are non-plastic and, for this reason, they reduce the green and dry strengths of bodies which contain them. They reduce the drying shrinkage of clays and, in general, facilitate the drying process by preserving an open texture.

Alkali minerals in clays can usually be identified by either microscopic or X-ray methods.

Iron Compounds. The various iron compounds which may occur in clays may be classed as (*a*) ferric oxide (Fe_2O_3); (*b*) ferrous oxide (FeO); (*c*) magnetic iron oxide (Fe_3O_4); (*d*) iron sulphides (FeS and FeS_2); (*e*) iron carbonates ($FeCO_3$); (*f*) ferrous and ferric hydroxides; (*g*) ferro-silicates and ferro-alumino-silicates; (*h*) ferrous aluminate (*hercynite*); (*i*) soluble iron salts (chiefly ferrous sulphate); and (*j*) chlorites (complex silicates, see p. 153).

The principal iron-bearing minerals in clays are:

Magnetite, Fe_3O_4, which occurs as black cubic crystals, with a hardness of 5·5–6·5 and a specific gravity of 4·9–5·2. Magnetite is seldom present in raw clays, but is produced by the partial reduction of ferric oxide during the burning of some clay wares. It is common in some bricks and is readily detected in the ground material by its magnetic properties.

Haematite, Fe_2O_3, occurs as steel-grey, iron-black or reddish hexagonal crystals, or as red amorphous particles, with a hardness of 5·5–6·5 and a specific gravity of 4·5–5·3. It is not often present in clays which have been much exposed to the weather, as it is readily converted into the hydrated form—*limonite*. In burned clays it is one of the most important colouring agents, especially in red bricks, terra-cotta, etc. Haematite is a constituent of some clays, however, and colours them characteristically red or salmon-pink.

Limonite, $Fe_2O_3xH_2O$, is a hydrous iron oxide occurring as yellow or brownish amorphous particles, with a hardness of 5–5·5 and a specific gravity of 3·6–4·0. The term limonite is not confined to a single, definite, chemical compound, but includes an apparently large number of compounds of variable composition, which are probably colloidal in nature and contain varying proportions of water. The formula by which limonite is usually expressed is $2Fe_2O_33H_2O$, but the term is applied to almost any hydrous iron oxide. It is very common in clays and forms one of the principal yellow or buff colouring agents in the raw material. Limonite occurs frequently in clays which have been weathered, and many surface clays and weathered outcrops owe their colour to it. It may be present as (*a*) a film over the grains of clay or sand, or (*b*) as irregular nodules scattered through the mass.

On heating, limonite evolves water and forms red ferric oxide. Hence, like the latter, it is a very important source of the red colour to which many clay products owe their characteristic appearance.

Goethite, FeO(OH) or $Fe_2O_3 . H_2O$, is a hydrous iron oxide which forms brownish-black, yellowish or reddish orthorhombic crystals, or occurs in the massive or fibrous state with a hardness of 5–5·5 and a specific gravity of 4–4·4. It is of common occurrence in many clays, where it is often distributed in a colloidal form round particles of other minerals. Much of the so-called limonite in clays is finely-divided goethite.

Lepidocrocite, or γ-FeO.OH, is not thought to be of common occurrence in clays, but it may be a minor constituent in some fireclays.

Turgite, $2Fe_2O_3 . H_2O$, is a similar hydrous iron oxide, which produces a red streak as distinct from the brownish or yellowish streak of goethite. It also contains less water.

Ferrous oxide, FeO, does not occur in raw clays, but it may be produced during the burning of the clay in a reducing atmosphere.

Pyrites—a general term for all iron sulphides—includes three important minerals, *pyrite*, *marcasite* and *pyrrhotite.*

Pyrite, FeS_2, is a sulphide of iron occurring as yellowish cubic crystals, having a hardness of 6–6·5 and a specific gravity of 4·8–5·1. It is sometimes disseminated in minute grains through the raw clay, or it may form large nodular or fibrous masses, or it may occur in masses resembling small petrified roots; the last-named are sometimes known as *race.*

Marcasite, FeS_2, is similar in composition to pyrite, but forms radiating needles of orthorhombic crystals. It occurs chiefly in the form of fibrous masses or nodules in the same manner as pyrite. Marcasite is readily oxidised to ferrous sulphate on weathering, and later forms a whitish scum on the dried clay. Pyrite, on the contrary, is not so easily weathered and oxidised.

Pyrrhotite, Fe_nS_{n-1}, is a sulphide of iron forming red, brown, bronze or copper-coloured hexagonal crystals or occurring in the massive state, with a hardness of 3·4–4·5 and a specific gravity of 4·4–4·65. It is not of frequent occurrence in clays, except in some residual forms.

Iron sulphides are most common in the Coal Measure fireclays and in ball clays, though they also occur to a variable extent in other clays. Unlike iron oxides, carbonates, and sulphates, they do not form a pleasant red colour on the ware in the kilns, but produce small black spots if the grinding has been sufficiently fine; otherwise, they form relatively large masses of fused slag.

Copper-iron sulphides, chiefly in the form of *chalcopyrite*, $CuFeS_2$, *bornite* and *erubescite*, occur in some fireclays, the principal occurrences in this country being in Northumberland, Durham, South Scotland, North Staffordshire and Shropshire. These copper compounds act in a similar manner to iron pyrite, except that they produce greenish-black slag patches instead of black ones.

Ferrous sulphate, $FeSO_4 . 7H_2O$ (*copperas*), occurs in some clays as a soluble salt. It is often produced by the weathering of sulphides of iron and it has the same general properties as ferrous carbonate but is more soluble and can cause *scum* or *efflorescence* (see p. 925).

Chalybite or *siderite*, $FeCO_3$, is an iron carbonate occurring as yellow, brown, or red hexagonal crystals, with a hardness of 3·5–4·5 and a specific gravity of 3·7–3·9. It is present in large proportions in the Clay Ironstone of British coalfields, in the Staffordshire marls and in the Coal Measure fireclays. Chalybite is found in three forms in clays: (*a*) as a film coating the other mineral grains; (*b*) as minute crystals; and (*c*) as concretionary masses, sometimes of large dimensions consisting of chalybite or clay iron-stone. Nodular fragments of siderite and of pyrites are the principal source of 'iron-spotting' in fireclay bricks and in other ceramic ware made from these types of clay.

Vivianite, $Fe_3P_2O_88H_2O$, is a hydrous phosphate of iron in the form of blue, white or green monoclinic crystals, with a hardness of 1·5–2 and a specific gravity of 2·66. Vivianite occurs to a very small extent in boulder clays as minute blue crystals, but the proportion is usually negligible.

Many other minerals, including silicates and alumino-silicates, such as hornblende, hypersthene, augite, olivine, fayalite, glaucophane, glauconite, etc., which occur in clays also contain a variable proportion of iron compounds.

The ferruginous minerals in clays occur as (*a*) large masses scattered irregularly through the mass; (*b*) minute grains uniformly distributed; and (*c*) stains on the surface of the grains. Those in group (*a*) may often be removed fairly readily by hand-picking, but the other two forms are almost impossible to remove.

In thin sections of articles made of burned clay, the decomposition products of limonite, haematite, and many other iron compounds may be recognised under a microscope (the section being illuminated by reflected light) as brown or reddish-brown specks where the material has been fired in an oxidising atmosphere, and as bluish-black films where a reducing action has occurred. The other minerals present are frequently enveloped by a film of iron oxide.

Pyrite is never seen in microsections of properly burned clayware, as it is decomposed at a comparatively low temperature, forming the vitrified, slag-like material (fayalite).

Chlorites, which are complex iron silicates containing magnesium and aluminium, as well as iron, as cations in a layer-lattice silicate, are present in many brick shales. Chamosites are regarded by some scientists as being the iron analogue of kaolinite and they would therefore have the formula $Fe_2Si_2O_5(OH)_4$. These minerals are of common occurrence in ironstone deposits associated with Coal Measures and Jurassic formations, so it is not unlikely that these varieties of chlorite will be present in adjacent clay strata.

The identification of iron compounds in clays is not always easy, especially in those cases where they are present as a colloidal film. Hydrated varieties give a characteristic peak in differential thermal analysis and goethite is readily identified and estimated by this means. Large particles of iron-bearing minerals can often be identified under the microscope. The total amount of iron in a clay is readily determined by chemical methods; occasionally, iron can be quantitatively extracted by treatment with dilute hydrochloric acid. Chemical analysis does not reveal the identity of the iron-bearing compounds nor differentiate between colloidal and large-particled material.

The **principal effects of iron compounds** in clays are:

(i) They effect an alteration in the colour.

(ii) They may reduce the refractoriness of the clay.
(iii) Soluble iron compounds may form a scum on the ware.
(iv) They may form iron spots which are clearly visible in the burned clay.

Localised concentrations of iron are usually the result of nodular fragments of ironstone. They seriously influence the behaviour of the ware in an atmosphere of carbon monoxide at certain temperatures, for they catalyse the precipitation of carbon which may disrupt the brick. This is thought to be the cause of *scurfing* in alumino-silicate refractories used in gas retorts; it is only noticeable when the iron is present in localised centres of high concentration.

So serious is the contamination of iron nodules in clays that numerous attempts[1] have been made to remove them by some means. Goodson[2] has suggested a physical process in which an initial separation is achieved by differential comminution in a crushing unit, followed by the separation of associated clay on a fluidised deck.

Another suggested process is to heat the entire clay to 400–450° C. which converts the siderite of the ironstone nodules to magnetite, Fe_3O_4. This oxide form is magnetic so that it can be removed by passing the clay under a suitable electromagnet.

Wet methods for removing ironstone are the most effective, although they are costly. A compromise is sometimes possible if the clay is allowed to weather until it is soft and well-broken down. If it is then lightly crushed and passed over a coarse screen ($\frac{1}{2}$ in. size) the clay passes through but the weather-resistant ironstone along with coal and hard stone are retained on the deck and can be rejected.

Even very small quantities of iron oxide are undesirable where white ware is required, unless there is a comparatively large proportion of calcium carbonate in the clay. The colour normally produced by the iron compound will then be neutralised on heating and a cream product formed instead of the usual red colour due to ferric oxide.

Cobalt oxide is frequently added in small amounts to china bodies so that the slight blue coloration which it imparts will 'kill' the cream or buff-staining associated with small amounts of iron.

If iron is present as surface-staining in clays, it can often be substantially reduced by chemical treatment. Oxidising and bleaching agents can be effective, but the most successful reagent is zinc dithionite which is used successfully in the china clay industries (see also p. 920).

Ferric oxide does not greatly reduce the refractoriness of a burned clay, provided it is always maintained in an oxidising atmosphere, for it is highly refractory. In a reducing atmosphere, on the contrary, it acts as a powerful flux.

Magnetic and ferrous oxides are very undesirable, as they are powerful fluxes, and combine with clay to form viscous, fusible silicates, ferro-silicates, and alumino-silicates, the most fusible silicate being *fayalite* (melting-point, 1150–1210° C.) and the most fusible alumino-silicate, corresponding to the formula $5FeOAl_2O_35SiO_2$, which melts at 1073° C.

[1] Nicholls, G. D., *Amer. Min.*, 47, 34, 1962.
[2] Goodson, F. J., B.C.R.A. Research Paper No. 520, 1962.

Ferrous carbonate, when present in a clay, may either be reduced in the burning process to ferrous oxide with the evolution of carbon dioxide (the oxide then acts as a flux, combines with the silica in the clay and produces black, slaggy masses of fusible silicates which are very undesirable), or it may be oxidised to ferric oxide, and be comparatively harmless unless white ware is required.

Ferric sulphide (pyrite), when heated, loses half its sulphur at 400–600° C. and the rest at a higher temperature. In a reducing atmosphere it produces ferrous oxide which acts as a powerful flux, but if the burning is carried out entirely in an oxidising atmosphere, the iron may be completely oxidised to form ferric oxide, which does not greatly affect the refractoriness of the clay.

Ferro-silicates and ferro-alumino-silicates behave in a manner similar to felspar, i.e. they are moderately fusible and increase the amount of vitrified matter or 'bond' in the fired ware and so slightly increase the strength of the ware. According to their colour they may improve or spoil the appearance of the ware.

Soluble iron compounds often produce a light-coloured *scum* on the surface of the dried ware. On burning, this is usually converted into unpleasant brown or black patches.

If small amounts of iron compounds are uniformly distributed throughout a clay they have little influence on the refractoriness and are deleterious only in so far as they affect the colour. They react with the clay to form silicates or alumino-silicates on prolonged firing.

Calcium Minerals. The *chief* calcium compounds in clays are (*a*) *calcite* ($CaCO_3$), which occurs as white or grey hexagonal crystals, with a hardness of 3 and a specific gravity of 2·71, or as amorphous grains; (*b*) *aragonite* ($CaCO_3$), which occurs as white, grey or greenish orthorhombic crystals, with a hardness of 3·5–4 and a specific gravity of 2·94, or as amorphous grains; calcium carbonate may be present in the form of chalk or other form of limestone and occasionally fragments of shells may be observed, and (*c*) various *calcium silicates and alumino-silicates* which include some felspars, micas and other primary minerals or their alteration products. Residual clays derived from an igneous rock are more likely to contain such mineral impurities, for most of them are readily decomposed by weathering and water transportation.

Calcium compounds also occur in much smaller proportions in clays in other forms, including:

Gypsum (*calcium sulphate*, $CaSO_4 2H_2O$) occurs in some clays and shales in the form of colourless, white, grey, yellow or reddish monoclinic crystals, with a hardness of 1·5–2 and a specific gravity of 2·3. *Selenite*—another variety of gypsum—is more common in clays, especially in the London and Oxford clays, and in some shales.

Anhydrite ($CaSO_4$) is often regarded as a variety of gypsum. It rarely occurs in clays, but is sometimes found in those of the Lias formation, usually at considerable depths.

Apatite, a crystalline *calcium phosphate*, occasionally occurs in clay, as also may coprolites and other calcium phosphates of organic origin. In burned clays various other complex calcium compounds may occur, such as calcium ortho- and meta-ferrates.

The **chief effects of calcium compounds** in clays are:

(i) They act as fluxes, i.e. they combine with alumina and silica to form low melting-point liquids, and so they reduce the vitrification temperature and refractoriness of the clay.

(ii) They produce a mobile, fluid liquid of great corrosive power. When cooled, the liquid readily forms a glass which acts as a strong bond and also produces impermeability and resistance to acids and other chemicals. The glass softens at a comparatively low temperature and may cause serious distortion of the body.

(iii) At temperatures below that at which they react, calcium compounds may reduce the shrinkage of a clay and facilitate its drying.

(iv) At a dull red heat, calcium compounds may combine with iron minerals and 'bleach' the red colour which they normally produce on firing to a white cream, or buff. To be most effective, the calcium minerals must be finely divided and disseminated throughout the clay.

(v) Calcium carbonate, when heated to about 900° C. is converted into lime, which, if it remains uncombined on cooling, may absorb moisture from the air and swell so much as to crack or burst the body. This is a serious defect in some clays used for brick-making where it is termed 'lime-blowing'.

(vi) Calcium sulphate is stable on firing and may remain as such in low-fired bricks. It is one of the most frequent causes of scum and efflorescence in common and facing bricks. Although not of common occurrence in natural clays, calcium sulphate is often formed on firing in clays containing calcium carbonate and pyrites.

(vii) In some brick-making clays, calcium compounds, in small amounts, increase the vitrification range.

(viii) As a fluxing agent in bodies and glazes, calcium compounds reduce the expansion but raise the maturing temperature, when they replace alkalies.

(ix) Calcium silicates and some other stable calcium compounds melt at a comparatively low temperature, and then act as fluxes in a manner similar to lime, but much more slowly. The calcium alumino-silicates produce a tougher and more viscous bond than the simple silicates, and are, therefore, preferable in some types of bonding.

Acid soluble calcium salts, e.g. carbonates are readily detected and sometimes estimated by selective chemical analysis. Carbonates and sulphates of calcium may be identified by differential thermal analysis, but complex alumino-silicates, e.g. felspars, micas, can be detected only by microscopic or X-ray techniques. The total calcium content of a material can be rapidly assessed by spectrographic analysis.

Barium Compounds are not of frequent occurrence in clays, the only one of any importance being *barytes* (*barium sulphate*, $BaSO_4$), which may be present as colourless, white, yellow, red, brown or blue orthorhombic crystals with a hardness of 2·5–3·5 and a specific gravity of 4·5. Barytes sometimes occur as the cementing medium in clays and shales, as at Seaton and other districts in Northumberland.

Barium compounds in clay behave similarly to those of calcium on firing.

Strontium minerals seldom occur in appreciable quantities in clay, the only one likely to be recognised being *celestine* (strontium sulphate, $SrSO_4$), which exists as white orthorhombic crystals, with a hardness of 3–3·5 and a specific gravity of 3·96.

Magnesium minerals occur to a small extent in most clays, chiefly as *magnesite*, $MgCO_3$, *dolomite*, $MgCa(CO_3)_2$ *chlorites*, *spinel*, *cordierite* and as various magnesium silicates and alumino-silicates. *Magnesite* is sometimes found in clays as fibrous crystals, but generally the grains of magnesium compounds are too small to be recognisable as they are difficult to separate mechanically and the proportion is almost invariably less than corresponds to 3 per cent of magnesia.

Magnesium sulphate, $MgSO_4.7H_2O$ (Epsom salts), is present in small amounts in some clays for brickmaking, but even 0·05 per cent can spoil ware, because it produces a white efflorescence, which in time disrupts the body. It can be rendered harmless by adding finely-divided precipitated barium carbonate to the clay before moulding. Magnesium sulphate is sometimes formed during firing by the combination of other magnesium salts with sulphurous fumes. In this case, barium compounds are ineffective in eliminating efflorescence.

Because they are usually of fine particle size and present in small amounts, magnesium compounds in clays are difficult to identify. The carbonates and sulphates can be detected by selective chemical methods or by differential thermal analysis; other forms can occasionally be identified by optical or microscopic methods.

Magnesium compounds act as fluxes and reduce the refractoriness of clays in which they occur, though they are less powerful than the corresponding lime compounds and act more slowly. As a consequence, they are usually less harmful and less likely to cause loss of shape when heated. The slag produced by the action of magnesium compounds on clay is also more viscous than that produced by calcium compounds and this further restricts the occurrence of a rapid loss of shape.

Zinc compounds do not occur in significant proportions in clays. Their action on clays (e.g. in the distillation of zinc in fireclay retorts) can be serious (see Chapter XIV).

Titanium minerals being very resistant to the action of the weather, are fairly common in most clays, the chief ones being *rutile*, *anatase*, and *brookite*, all of which have the formula TiO_2. Occasionally *sphene* (*calcium titanite*) crystals, $CaTiSiO_5$, and sometimes *ilmenite*, $FeTiO_3$, are present. The total content seldom exceeds the equivalent of 3 per cent of titanium oxide, and in many highly plastic clays less than 1 per cent of titanium oxide is present. A few fireclays contain as much as 10 per cent of titanium oxide, and in the U.K. these are mostly found in Scotland (Ayrshire).

The oxides of titanium are the usual form of mineral. Brindley[1] has reported that *anatase* is the common form in some sedimentary clays, but *rutile* predominates in many residual clay samples.

Titanium compounds are very persistent and often remain quite unaltered in clays and other materials which have been fired at 1300° C.; they act as feeble fluxes, although they are not active below 1500°. In glasses and glazes they are powerful opacifiers.

Chromium minerals sometimes occur to a very small extent in clay, usually in the form of *chromite*.

[1] Brindley, G. W. and Robinson, K., *Min. Mag.*, 28, 244, 1947.

Tin-bearing minerals are seldom present in clays, though occasionally *cassiterite* (SnO_2) is found as black or brown tetragonal crystals, with a hardness of 6–7 and a specific gravity of 6·4–7·1.

Manganese minerals occasionally occur in clays, chiefly as a thin film of oxide on the other mineral grains.

They act as fluxes and in many respects resemble ferrous compounds but their amount is usually too small to be serious. The chief effect of manganese is in staining articles made from clays which contain them, the colours of the stains ranging from a medium brown to dark brown or black. Sometimes manganese dioxide is added to clays used for making facing or blue engineering bricks to produce dark colours, low porosity and an increase in crushing strength. They are often added to sand for facings and to veneer compositions.

Vanadium compounds in clays are rare, but, as they sometimes cause a green discoloration in buff-burning clays and pipe clays and occasionally give rise to a green efflorescence in bricks, their presence must not be overlooked.

The minerals of titanium, chromium, tin, manganese and vanadium are usually of high density and they can often be isolated by heavy liquid separation and then identified.

Sulphur compounds in clays are chiefly *pyrites*, and various sulphates, particularly of calcium, magnesium and sodium. In many instances, the precise form of the sulphur compounds in a clay may be of little importance but the total percentage (found by chemical analysis) may be very useful in indicating the source of discoloration, scum or efflorescence.

Small amounts of sulphur compounds may be a major factor in the bloating of low-grade fireclays and other clays at temperatures about 1150° C. In clays rich in iron, sulphur compounds are often responsible for producing 'black-hearts' by maintaining the iron in a reduced state. This is often confused with the residue from undecomposed carbonaceous matter (see p. 289), but it can be distinguished by its steely appearance and the sharp line of demarcation between the black zone and the surrounding red colour of the fired clay.

Phosphate minerals are found in some clays, the most important is *apatite*, a complex calcium phosphate containing proportions of chlorine and fluorine. The formula is usually expressed as $Ca_5(F,Cl)P_3O_{12}$ and it occurs as blue, grey, red, brown, green or yellowish hexagonal crystals, with a hardness of 5 and a specific gravity of 3·17–3·23.

Coprolites and *phosphorite* are natural phosphates produced by the accumulation of organic remains or fossilised animal excreta. Phosphatic *nodules* of this nature are common in the Greensand beds, and true coprolites are found in the Oxford clay and Gault formations.

Fluorine minerals are of frequent occurrence in clays derived directly from primary rocks, such as granites. The main mineral form in which fluorine may be present in clays is as *fluorspar*, calcium fluoride CaF_2, but *topaz*, $(AlF)_2SiO_4$, is also known in some china clays.

Fluorides, if they are present in appreciable amounts, are a serious contaminant in clays. At high temperatures they behave as powerful fluxes and by combining with the clay body they readily volatilise as silicon tetrafluoride, SiF_4. This product is highly reactive; it will reduce the gloss on glazed ware and attack the refractory linings of kilns.

Fluoride compounds can also be present in felspars and Cornish Stones. For most purposes, they have to be removed by froth flotation before such minerals can be used in ceramic practice.

A total fluorine content exceeding 0·2 per cent cannot be accepted in ceramic ware without serious problems arising either to the articles themselves or to the kiln structure.

Complex Alumino-silicates. The principal impurities in clays are complex silicates or alumino-silicates usually containing other elements as principal ingredients. Clays are secondary products which have resulted from the breakdown of igneous rocks. Many primary minerals in these rocks are relatively easily attacked by epi- or hypo-genic agents and converted into some other form. Clays will, therefore, contain minerals of many types, some of which will be unaltered *primary* forms, whilst others will be the *secondary* breakdown products of such minerals. The so-called *residual* clays are more liable to be contaminated with true primary minerals than those of *sedimentary* character where successive transportations and depositions have increased the possibilities of breakdown.

Primary Minerals in Clays

The **three most important primary minerals** in clays derived from igneous rocks are quartz, felspars and micas, but olivines, pyroxenes and other complex minerals may also be present.

Quartz is a common mineral in clays and for a full description see p. 316.

Felspar is the name given to a group of alumino-silicates containing one or more of the following metals as bases: potassium, sodium, calcium and barium. The felspars (p. 167) form isomorphous mixtures and, consequently, may vary greatly in composition, especially as regards the metals present. They are common in residual but seldom in secondary clays for they break down very easily under normal weathering conditions. The principal types have been outlined in Chapter III (Table, III–XIII).

When pure, the felspars are colourless or white, but most specimens are tinged reddish-grey, green or blue because of occluded or dissolved impurities. They have a hardness of about 6 and a specific gravity of 2·5–3, and may usually be recognised by their characteristic cleavage or twinning.

Felspars are frequently found in residual kaolins and, as china stone (p. 323) is fairly rich in unchanged mineral they are present in small amounts in crude English china clays. It is most unusual to find them in sedimentary clays for they are readily changed to other minerals as described on p. 290.

Felspars contain large proportions of alkali or alkaline earth elements; consequently they melt at comparatively low temperatures (between 1200° and 1250° C.) and act as powerful fluxes in ceramic bodies. It is not uncommon to find well-developed felspar crystals in fired ceramic bodies which are rich in calcium or alkalies, for they are a common product of the crystallisation of molten alumino-silicates.

Mica is a term used to designate a group of minerals crystallising in the mono-clinic system in characteristic plate-like masses or particles, and consisting of alumino-silicates of potassium, sodium, lithium, iron and magnesium.

The principal micas have been outlined on p. 149, they are:

Muscovite micas, including *muscovite*, $K_2O.3Al_2O_3.6SiO_2.2H_2O$, or potassium-mica, *paragonite* or sodium-mica, and *lepidolite* or lithium-potassium-mica. *Muscovite* consists of white, black, brown, yellow or green crystals, with a hardness of 2–2·5 and a specific gravity of about 2·85. It usually occurs in thin flat plates or scales. *Paragonite* is similar, but has a hardness of 2·5–3 and a specific gravity of 2·9. *Lepidolite* is of a white, grey, lilac or rose-red colour, with a hardness of 2·5–4 and a specific gravity of 2·8–2·9. Muscovite micas are fairly stable and, consequently, are often found in clays.

Biotite micas, including *biotite*, $K(Mg,Fe)_3AlSi_3O_{10}(OH)_2$, or iron-magnesium mica; *zinnwaldite*, or lithium biotite, and *phlogopite*, $AlMg_3KH_2Si_3O_{12}$, or magnesium mica. *Biotite* occurs either as scales or in large massive crystals of a black or dark green colour. In transmitted light the flakes appear brown, green or blood-red. It has a hardness of 2·5–3 and a specific gravity of 2·7–3·1. On weathering, it alters to *chlorite* and so is not often found in clays. *Zinnwaldite* is pale yellow or brown in colour (sp. gr. 2·9–3·1). *Phlogopite* occurs as white, colourless, brown or copper-red crystals or scales, with a hardness of 2·5–3 and a specific gravity of 2·75.

Hydrous micas are an alteration product of true micas from which some of the alkali ions have been leached away and more water has entered the lattice (see p. 152). They are the predominant alkali-bearing constituents of many sedimentary clays and are probably the parent of the kaolin-type clay mineral *livesite*.

Hydrous micas are often known as *illites*, and *sericites* are in this class.

Glauconite is another variety of hydrous mica, and is a hydrated silicate or alumino-silicate of magnesium and calcium with a varying proportion of iron. It has a yellowish to blackish-green or grey-brown colour, a hardness of about 2, and a specific gravity of 2·2–2·4. It occurs chiefly in marine deposits and is a characteristic constituent of the Greensand clays and sands, to which it sometimes imparts a greenish or brown colour; on burning the colour of the glauconite is changed to that of red ferric oxide, which is characteristic of clays burned in an oxidising atmosphere.

Any kind of mica, but especially *biotite*, is objectionable in clays which are required to burn to a good white, as the iron in the mica is liable to produce a buff or brown colour. Mica frequently occurs in unaltered scales and flakes in lightly burned goods, these scales being readily recognised by their bright glistening appearance when examined by reflected light. In transmitted light, dark micas are generally citron-coloured or light-brown, whilst muscovite mica is often quite white and transparent.

Olivines are primary minerals but alter so readily that they are rarely found in natural clays. They have the composition $(MgFeCa)_2SiO_4$, and are of common occurrence in fired ceramic ware. The usual alteration product is *serpentine* in the form of *antigorite* or *chrysotile*, $3MgO.2SiO_2.2H_2O$, which is found in some clays as predominantly green, and occasionally yellow, red, brown or almost black needle-shaped crystals with a hardness of about 3·5 and a specific gravity between 2·5 and 2·6.

The Pyroxenes are principally silicates of magnesium, iron and calcium with the general formula $(MgFeCa)SiO_3$. They are closely related to the *amphiboles* (see p. 120), which are also constituents of igneous rocks. Both mineral species alter readily on weathering and yield a variety of products, chiefly *serpentines* and *chlorites*. Pyroxenes and amphiboles are found to a limited extent in residual clays and the principal types are:

Enstatite, $MgSiO_3$, which occurs as colourless, grey, green or brown orthorhombic crystals, with a hardness of 5·5 and a specific gravity of 3·1–3·3. It is so readily decomposed that it is found only to a limited extent in clays.

Hypersthene, $(FeMg)SiO_3$, forms brownish-green, green, brown or black orthorhombic crystals, with a hardness of 5–6 and a specific gravity of 3·5. It is not as readily decomposed and, consequently, is more common in some clays derived from basic rocks.

Augite, $Ca(MgFe)(SiO_3)_2$ with $(MgFe)(AlFe)_2SiO_6$, is a silicate or alumino-silicate of calcium, magnesium and iron, which occurs in greenish-black or black monoclinic crystals, having a hardness of 5–6 and a specific gravity of 3·2–3·5. It is not easily decomposed and, consequently, is often recognised in clays.

The following pyroxenes also occur in some clays: *Bronzite*, $(MgFe)SiO_3$, a silicate of magnesium and iron, *Diopside*, $CaMg(SiO_3)_2$ or $Ca(MgFe)(SiO_3)_2$, a silicate of magnesium with some iron; *Diallage*, a mineral similar to diopside and augite; *Wollastonite*, $CaSiO_3$, a silicate of calcium.

The *amphiboles* include *hornblende*, a variable silicate or alumino-silicate of calcium, magnesium and iron, which sometimes contains silicates and alumino-silicates of sodium and potassium; it occurs as green or black monoclinic crystals, having a hardness of 5–6 and a specific gravity of 3·0–3·47.

Glaucophane is a silicate or alumino-silicate of sodium, iron and magnesium, which occurs as bluish particles, with a hardness of 6–6·5 and a specific gravity of 3·0–3·1.

Among the less common amphiboles are *anthophyllite*, $(OH)_2(MgFe)_7Si_8O_{22}$, a silicate of magnesium and iron, *tremolite*, $(OH)_2Ca_2(MgFe)_5Si_8O_{22}$, a silicate of calcium and magnesium.

All amphiboles are readily decomposed and converted into secondary products and occur only in small proportions in clays.

Pyroxenes and amphiboles are not usually present in clays in quantities sufficient to influence the properties. They would tend to discolour fired ware, act as slow fluxes, and reduce the refractoriness.

Tourmaline is an accessory mineral in many granites and as it is very stable, it is a common constituent of many crude china clays where it may be recognised in the form of blue needles. It is an aluminium borosilicate or a boroalumino-silicate which occurs as black or blue-black hexagonal crystals, with a hardness of 7·0–7·5 and a specific gravity of 2·98–3·2.

Secondary Minerals in Clays

Secondary minerals are of much more common occurrence in clays and they include the clay minerals, of the kaolin and montmorillonite group, chlorites, vermiculites and hydrous micas.

The principal minerals of granitic igneous rocks, which break down under the action of natural agencies are the felspars and micas. In the presence of weak acids (e.g. carbonic, boric), felspar changes to kaolin-type minerals. The micas give a variety of products, depending upon their composition; muscovite mica yields hydrous mica and, finally, the kaolin mineral, livesite; biotite micas change to chlorites or vermiculites and it is speculated that kaolin minerals may be the end product.

The general alteration reactions under alkaline conditions and the products from basic igneous rocks, e.g. basalts, give different products from those formed through the medium of acids. Rocks rich in ferro-magnesian minerals are the principal sources of the montmorillonite minerals; the olivines, pyroxenes and amphiboles transform through a variety of intermediate forms and the final product depends to a great extent on the composition of the original minerals. Montmorillonite minerals may also be formed by crystallisation from solutions containing aluminium, silicon, magnesium and iron, and also by the co-precipitation of gels of the hydroxides of those elements.

Kaolin minerals of composition $Al_2O_3.2SiO_2.2H_2O$ are the most prevalent mineral group found in natural clays, although hydrous mica or illite is also common. The mineral, *kaolinite*, is not of frequent occurrence and may be restricted to those clays which have been formed under hypogenic action, i.e. sub-surface reaction at high temperatures and pressures (e.g. china clay) and to the direct sediments of those clays (e.g. ball clays).

Halloysite forms under conditions of strong acid attack (e.g. sulphuric and sulphates) in the region of mineral springs or intense volcanic activity. *Nacrite* and *dickite*, the most well-formed of all kaolin minerals, have probably crystallised from solution under specific acid conditions. The latter mineral is occasionally associated with nodular ironstone in fireclay seams.

The most common clay mineral in sedimentary clays is *livesite*, a disordered kaolin form which has a fine grain size. It may be derived from a micaceous component (see Chapter III) and have a small proportion of cations, other than aluminium or silicon, as part of its lattice.

Allophane is another kaolin type mineral but of variable composition. It is not of common occurrence.

Montmorillonite minerals are less common than kaolins, but numerous examples are known. They are of variable chemical composition containing aluminium, silicon, magnesium, iron and other elements. *Montmorillonite*, *beidellite* and *nontronite* form one series of minerals where silica, alumina, magnesia, iron oxide and water are the principal components. *Saponite* and *hectorite* are other types of montmorillonite containing chiefly magnesia, lithia, silica and water. The general name given to clays which contain appreciable quantities of montmorillonite minerals is *bentonite*.

Attapulgite or **palygorskite** is another secondary mineral present in some clays. It is not of common occurrence, but, in a pure state, it is particularly valuable for drilling fluids and as a filler. It is essentially a hydrated magnesium silicate, but other elements may be present.

Chlorites are a group of hydrous alumino-silicates with iron and magnesium as essential ingredients; they occur as greenish flakes resembling mica. They have a hardness of 1·5 and a specific gravity of 2·5–2·8. Chlorites are the most usual alteration products in recent clays derived from minerals rich in magnesium and iron and they are common constituents in many sedimentary clays.[1]

Vermiculites are not uncommon in clays of recent origin. They are hydrated magnesium silicates but other elements, chiefly iron, are frequently associated. They owe their name to their property of exfoliating on heating. They occur as plate-like flakes, are softer than mica and are usually red, brown or black.

Clay minerals usually impart plasticity to clays; montmorillonite is extremely plastic and sticky, kaolin types less so, and chlorites and vermiculites not at all. Kaolin minerals are the most refractory because they are essentially alumino-silicates and contain far less fluxing elements than the other types. Kaolin minerals are white burning (except livesite) and are common constituents of whiteware bodies. The principal properties of clay minerals are summarised in Table VI.I.

TABLE VI—I. SOME PROPERTIES OF CLAY MINERALS

	Kaolin	*Montmorillonite*	*Chlorites, Vermiculites, Hydrous micas*
Particle size	Small	Extremely small	Variable, but comparatively large
Plasticity	Good	Very good	Poor
Fired colour	White (mostly)	Cream, buff or red	Brown, red or black
Refractoriness	1710° C.	1350–1450° C.	1250–1500° C.

Clay minerals are difficult to detect because they are usually of small particle size. Chlorites, vermiculites and hydrous micas are frequently of a sufficiently large size to be identified by the optical microscope; kaolins are readily observed under the electron microscope, but montmorillonites are not easily discerned. Differential thermal analysis affords a simple and effective method of detecting and estimating many of the clay minerals, from the nature and temperature of the characteristic peak or peaks corresponding to the loss of structural water. The main clay mineral groups can be differentiated by X-ray analysis and most individual types are recognised.

Another class of minerals which occurs in many clays and has an important bearing on some of their properties is referred to as **organic or carbonaceous matter**; it is present in variable proportions according to the mode of formation and the conditions to which the clays have been exposed. It is derived chiefly from (*a*) the percolation of water containing humic acids and matter in solution or suspension, the organic matter being absorbed by the pores of the clay and so retained; (*b*) the deposition of clay upon organic matter or the deposition of organic matter on clay beds, the carbonaceous material being gradually incorporated into the clay; or (*c*) the admixture of bituminous or carbonaceous rocks with the clay during the deposition of the latter.

The carbonaceous matter may be disseminated either in large masses, minute

[1] Brough, J. and Robertson, R. H. S., *Clay Min. Bull.*, 3, 221, 1958.

grains, or as films over the surfaces of the other grains. Ball clays usually contain 3–4 per cent, and sometimes as much as 10 per cent of carbon in the form of an organic jelly derived from peat, lignite, coal or other organic matter—usually of vegetable origin—or in the form of oleaginous matter derived from fossilised fishes and creatures having shells, which, on heating, produce 'shale oil' and give them a dark colour. Sometimes the plants from which the organic matter has been derived may be identified but the greater part of it has usually decomposed too much to be definitely recognisable. All the organic carbon present in a clay may be burnt off when the clay is heated to redness in a current of air, but the process requires care, or the mass may be bloated and superficially fused in such a manner as to prevent the oxidisation of all the carbon.

In primary clays, such as *china clays* and *kaolins*, organic matter is practically absent; it is most abundant in plastic clays, such as those used for the manufacture of bricks, tiles, terra-cotta, etc.

In fireclays and shales associated with Coal Measures, the carbonaceous material closely resembles the coal seams and ranges from lignitic, through bituminous to anthracitic in character.

The **chief effects of carbonaceous matter** are (i) it imparts a dark colour to the raw material (p. 921), (ii) it may produce reducing conditions during kiln firing and so affect the colour and the vitrification behaviour, (iii) it may reduce the amount of fuel required for burning the clay, (iv) it may burn too rapidly in the kiln and cause over-vitrification and loss of shape, (v) it may give rise to a 'black-heart', (vi) its burn-out may increase the porosity.

Carbonaceous matter is sometimes deliberately added to a clay to produce desirable effects: (*a*) large particles of coal (up to pea-size) provide localised reducing conditions and, in some clays, give a pleasing appearance to the outside of facing bricks made from them—about 5 per cent of coal by weight is normally added; and (*b*) finely-divided breeze incorporated into a clay can provide sufficient heat on controlled burning to burn bricks completely without the need for additional fuel. This is the usual procedure in *clamp-firing*, where about 15 per cent of combustible is used—bricks of excellent colours are thereby produced if the clays have a sufficient proportion of iron compounds.

The presence of a proportion of low-rank carbonaceous matter, e.g. algae or oily constituents, is an advantage in some clays, because if the firing is carefully controlled, the organic components decompose steadily over a wide temperature range, thereby reducing the total fuel required and giving a more uniformly-fired brick.

High-rank carbonaceous matter, e.g. bituminous and anthracitic coals, is not so desirable in clays, because it commences to burn only at high temperatures and then very rapidly, so that ware containing it tends to become over-vitrified.

If the carbonaceous matter is in the form of fine grains uniformly disseminated through the clay, it is not usually harmful, and is readily burnt out by slowly heating the clay in an oxidising atmosphere, care being taken to avoid surface fusion which seals the pores and prevents all the organic matter from being fully oxidised. Where larger masses of organic matter occur, however, it may cause trouble because of the amount of heat and gas evolved, causing superficial fusion, and because of the large cavities which are left when the particles of organic matter have been burned away. In all cases where a clay contains sufficient organic matter especially of a coaly or graphitic nature and sufficient fluxing materials, such as lime or soda compounds, the firing process is difficult to control and often results in the production of 'black hearts', and sometimes of a highly bloated mass. These objectionable changes are due to the fact that the very close texture of some clays prevents the ready access of air to the interior of the articles. If much carbonaceous matter is present, before it can be oxidised and removed, the exterior pores of the article become closed with fused material and so effectually prevent the removal of the carbonaceous matter.

Water may also be conveniently regarded as a mineral and it is present in clays in a variety of forms. These are:

(*a*) *Hygroscopic water* or *moisture*, the amount of which depends on the surface area of the clay.

(*b*) *Absorbed water*, which is associated with the exchangeable cations in certain clay minerals.

(*c*) *Broken-bond water*, which is linked to the unsatisfied valencies occurring at the edges of crystals, and

(*d*) *Combined water* or *water of constitution*, an essential part of the structure of some minerals.

The various forms of water are described in Chapter VIII.

Exchangeable Cations. In addition to the elements which form part of their crystal structure, some minerals can absorb cations (or bases) from solutions. Clay minerals have this property to a marked degree; montmorillonite, for example, may have up to 3 per cent by weight of sodium or calcium oxide retained in this way. Kaolinite does not usually absorb more than 0·3 per cent, but livesite may contain up to 1·0 per cent.

The cations absorbed are many, but hydrogen, sodium, potassium, calcium and magnesium predominate, although aluminium and iron are common in some clays. Only small-sized minerals have the capacity for cation absorption, and it is probably due to a charge deficiency on the particles. Although the cations are closely associated with the mineral crystallites, they may be readily exchanged by others under suitable conditions—hence the property is known as the *cation-exchange capacity*. It is considered in more detail in Chapter V (p. 263), where methods for its measurement are outlined.

The type of cations which are present in natural clays depends principally on the mode of formation and the environment. Clays associated with lime strata, for example, will contain principally calcium ions; those derived from rocks containing alkali have predominantly sodium or potassium ions; in clays which have been extensively leached or weathered, the exchangeable ions are chiefly hydrogen.

The properties of the clay may be greatly influenced by the type of associated cation. The behaviour of clay in water is most markedly changed when cations of different types are present (see Chapter VIII). The exchangeable cations may influence the plasticity, drying shrinkage and firing properties of clays if they are present in appreciable quantities.

THE VARIETIES OF CLAYS AND THEIR COMPOSITION

THE materials known as *clays*[1] differ so greatly in composition and properties that it is impracticable to class them accurately and in great detail. The following are some of the best-known and widely-used clays; other clays of geological or mineralogical interest but of no commercial importance are not included, though they may be used locally in small quantities for making bricks, roofing tiles and crude earthenware vessels.

All clays are composed of (*a*) one or more clay minerals, and (*b*) other minerals, the proportions of each being varied according to the origin and subsequent (natural) treatment of the clay.

In most materials described as clays, the *clay minerals* are the predominant and essential constituent, and this should always be borne in mind though in some commercially-useful brick- and tile-clays the total clay minerals may be only 30–35 per cent of the whole material, and in some workable clay-pastes it is even less. For more than a century, the existence of some essential *clay-substance* has been assumed, though the recognition of definite 'clay minerals' with a distinct individual crystalline structure has been established only within the last 30–40 years. The lattice structures of the clay minerals are described in Chapter III and the other minerals are described in Chapter IV.

China Clays. *China clays* and *kaolins* are mainly residual clays, formed *in situ*, by the breakdown of primary minerals. The principal deposits of this type of clay are the result of hydrothermal reactions in large granitic masses, and the resulting *kaolinisation* is found to great depths in some igneous rocks.

The principal mineral which undergoes alteration is the felspar component which loses its alkali and some of its silica and gains water of constitution. The reaction may be represented simply by:

$$\underset{\text{orthoclase felspar}}{K_2O.Al_2O_3.6SiO_2} \xrightarrow{H_2O} \underset{\text{kaolinite}}{Al_2O_3.2SiO_2.2H_2O} + \underset{\text{potash}}{K_2O} + \underset{\text{silica}}{4SiO_2}$$

The alkali is usually removed as a soluble salt.

The true nature of the reaction is not fully understood; several intermediate stages may form before the final product is reached.

In the Cornish granites the reaction has probably proceeded through the agency of water containing free acids (such as hydrofluoric, hydro-boric and carbonic) at high temperatures and under great pressures, the solutions having been forced up from below and penetrating the granite through fissures and jointings. As the felspar breaks down the granite mass becomes porous and more powdery and yields

[1] Many clays have names which are likely to cause confusion—particularly in the United States and Germany. The fact that a clay bears an English place-name does not necessarily mean that it is found in England. Several such clays occur only in the United States and the place-name is often a trade-name. Many clays are named after a mineral species which they contain; others, after the type or character of the clay, e.g. flint clays; some are called by a name intended to indicate the use of the clay, e.g. pot-clays.

Much of the confusion which exists in clay terminology has been outlined by Robertson and his survey of existing trade-names greatly clarifies the position. (Robertson, R. H. S., *A Glossary of Trade Names* (Mineralogical Society, London, 1954)).

what is termed *china-clay rock* or *carclazite*. This material varies greatly and contains only about 20 per cent of kaolinite on the average, the remainder consists of unchanged quartz, mica and other primary minerals, including some unaltered felspar. It is a curious fact that, broadly speaking, the greater the depth at which the altered rock occurs, the richer is the material in kaolinite.

The complexity of the decomposition and other changes is great, because in addition to (i) *kaolinisation*, two other alteration processes are recognisable in the Cornish granite; (ii) *tourmalinisation* which occurs in regions where the ascending solutions, rich in boron, have converted the micaceous and felspathic constituents into tourmaline (p. 285), and (iii) *greisenisation* or a process of alteration mainly involving solutions containing fluorine; the felspars break down forming quartz and hydrous mica; topaz $(AlF)_2SiO_4$ and fluorite (CaF_2) may also be present. The three processes are often superimposed so that the resulting china-clay rock may have a complex composition.

Fortunately, the kaolinite produced consists essentially of much smaller particles than most of the associated minerals and, as a result, it can be concentrated by a process in which the clay mineral is extracted with jets of water or steam; the slurry so formed flows to the foot of the quarry-face and settles in large tanks. Most of the coarsely-grained quartz, unchanged felspar, mica and other components are thus removed and the purified kaolinite extracted by evaporation or by filter-pressing.

The theory that kaolinisation is the result of the action of ascending hot acidic solutions is widely accepted to explain the origin of the Cornish china clay and similar deposits in France and Germany. However, kaolinite may also be produced from felspar under normal atmospheric weathering conditions and some American, Asiatic and possibly some European kaolins may have been formed in this way. Such kaolin deposits differ in thickness and texture from those of hypogenic origin.

True kaolinite deposits are not as widely spread as was formerly thought. The only ones of any economic importance are those where a large mass of igneous rock has been altered by hydrothermal action, as in Cornwall and Devon, the Massif Central in France, Saxony and Bavaria in Germany, Czechoslovakia, China, Australia and in a few localities in the Americas particularly North Carolina.

The best kaolins for refractory and ceramic use are those purified from residual deposits. Sedimentary or re-deposited kaolins are found in the United States and elsewhere and are of economic value, but they are not as pure as kaolin prepared from china-clay rock.

Kaolinite, $Al_2O_3.2SiO_2.2H_2O$, when pure, has the composition (by weight) alumina 39·56 per cent, silica 46·54 per cent, water 13·90 per cent. Commercial kaolins contain between 80–90 per cent of the clay mineral, contaminated mainly with quartz and fine-grained mica or hydrous micas but seldom with appreciable proportions of felspar. As the particle-size distribution of all the minerals present is similar, it is difficult to deduce a mineralogical analysis with much degree of accuracy. An analysis of a purified sample of Cornwall china-clay is shown in Table VI.II. From this, if an average value for a mica component is accepted, a mineral balance sheet can be computed. An accurate estimate of the kaolinite content can also be secured by thermal analysis. In this sample, the free silica content was exceptionally low and only a trace was revealed by X-ray methods. The amounts of other minerals detected by microscopic and X-ray methods have to be inferred from the chemical analysis.

TABLE VI—II. MINERALOGICAL BALANCE OF A CORNISH CHINA CLAY

Mineral	*Per cent*	SiO_2	Al_2O_3	TiO_2	Fe_2O_3	*CaO*	*MgO*	K_2O	Na_2O	*Loss on ignition*
Kaolinite	89·00	41·44	35·13							12·43
Mica	7·10	3·24	2·75					0·58	0·18	0·35
Quartz	2·04	2·04								
Goethite (FeO.OH)	0·91				0·69					0·22
Sphene	0·09			0·09						
Fluorite (CaF_2)	0·55					0·28				0·27
Tourmaline	Trace									
Mg mineral	Trace									
Total	99·69	46·72	37·88	0·09	0·69	0·28	—	0·58	0·18	13·27
Actual chemical analysis	100·04	46·72	37·88	0·09	0·69	0·28	0·24	0·58	0·18	13·38

Other kaolinites show similar results on analysis although they usually contain a greater proportion of impurities. One important point emerges from the mineralogical analysis which cannot be stressed too strongly. It is that the total amount of impurity (i.e. oxides other than alumina and silica) in the kaolinite total only 2·06 per cent, but when this is calculated on a mineralogical basis the non-kaolinitic material is as high as 11·0 per cent principally due to the proportion of silica and alumina in mica.

The high purity Cornish china clays and some others from various world sources are prepared by careful size separation using hydrocyclones or centrifuges. An upper cut-point of 5 microns at the most is normally accepted to eliminate coarse-grained impurities. However, the finest fractions—the material below 1 micron—frequently contain non-kaolinitic material also. Montmorillonites are suspected to occur in Cornish materials, but there is also associated colloidal matter, perhaps resembling allophane. The removal of this fine fraction is of advantage in the commercial product but it involves a difficult and expensive process. It is important that fine-grained impurities are kept to a minimum in china clays used in paper-making which is one of the most important applications of the mineral.

Pure kaolin or china clay is white-burning and has a refractoriness of about 1785° C., although if much impurity is present the softening temperature will be reduced. Kaolins which have been formed by surface weathering agents invariably contain a higher proportion of iron-bearing minerals than those formed by hydrothermal actions. Such clays tend to burn brown, red or pink on firing and are not as valuable for some ceramic purposes.

Halloysite-bearing Clays. Clays which contain the clay mineral *halloysite* resemble the china clays in appearance and in some properties, but are more plastic and produce a denser fired body. No halloysite-bearing clays are definitely known in the British Isles although some authorities claim that this mineral is present in fireclay in the Glenboig district in Scotland.

Halloysite probably has a similar origin to kaolinite but the hydrothermal agents were more acidic and may have contained sulphate or chloride radicles. The fact that *alunite* ($K_2O.3Al_2O_3.4SO_3.6H_2O$) is commonly associated with halloysite clays strongly supports this theory. Callaghan[1] considers that some halloysites may be crystallisation products from solutions or derived from allophane or diaspore, for those minerals are also frequently present. There is geological evidence which suggests that, under favourable conditions, halloysite may recrystallise and form kaolinite and an example of the reverse change is to be found in the halloysites at Cadouin in France, where kaolinite is altered to halloysite in the region of mineral springs, rich in sulphate.

Halloysite is a common constituent of some American flint clays, particularly in Utah and Indiana where it is found in both the normal and hydrated form. An exceptionally pure sample was worked at Angleur, near Liège in Belgium, but this supply has apparently now been exhausted.

A potentially good source of halloysite has been found in New Zealand and it is remarkable for the fact that the principal impurity is cristobalite.

Steinmark or *Lithomarge* are local names given to a variety of china clay.

Myelite is regarded as a white, conchoidal variety of lithomarge, and *carnalite* as a pink and more compact form. *Tuesite* is a bluish variety found at the mouth of the Tweed.

Other minerals similar to clay, but much contaminated by other substances, also occur. Thus *bole*, *umber*, and *sienna* consist of ferruginous clayey material of a brownish-yellow or red colour and of very variable composition. When these various minerals are heated they behave in a manner similar to clay, though the other minerals usually present as impurities enter into combination with them and their decomposition products, and so produce so complex a mixture that it is almost impossible to ascertain its mineralogical composition.

Ball Clays. Clays which are plastic or 'sticky' yet burn white, are termed *ball clays*. They are of value because they can be added to true china clays to give strength to the moulded shape and yet maintain the white colour after firing. One essential feature of ball clays is their fine grain size (much less than that of china clay) which is the main reason for the high plasticity.

Large deposits of ball clays occur in the south-western counties of England, and they are widely used in the British Isles particularly in the white-ware industry. The three important producing areas are in Dorset, North Devon and South Devon.

Unlike the residual true kaolins, ball clays are sedimentary in character, i.e. they have been transported by rivers and deposited at some place removed from the site of formation. The English ball clays were most likely formed in the Cornwall granite mass, and afterwards removed by rain, carried by rivers and probably deposited in fresh water lakes at a time when the coast-line was very different from that of the present day. Scott[2] has suggested that the Dorset deposits were laid down

[1] Callaghan, E., *Indiana Dept. Conserv. Geol. Surv. Bull.*, 1, 1948.

[2] Scott, A., *Memoirs of the Geological Survey*, vol. 31 ('Ball Clays', H.M. Stationery Office, 1929).

in the Miocene period of the Tertiary era of geological formation (see p. 48) and those of Devon at a later period. It is probably more correct to ascribe the Dorset beds to an earlier epoch, as the fossil remains resemble those in the Brown Coal deposits of Germany which belong to the Oligocene period.

Although ball clays are undoubtedly fresh-water deposits, the diverse nature of some of them suggests that they are, in part at least, of fluviatile origin. High-quality clays are usually found in small pockets associated with lignite and coarse-particled gravels.

The South Devon deposits, which occur in the vicinity of Bovey Tracey and Newton Abbot, are truly lacustrine. These clays are fairly well bedded and although there is much associated lignite and gravel, they can be worked either by open pit or underground mining methods. The beds are characteristically lenticular and there is a complicated sequence of clay, lignite and gravel which in some places is known to extend to depths of 1000 feet.

There is considerable support for the theory that the ball clays are derived from the Cornwall granite, for in the South Devon deposits there is a tendency for the clays to become finer towards the more eastern edge of the area. This would represent the position of maximum transportation if the source were centred in Cornwall.

Ball clays are invariably dark in colour, due to associated lignite, but in those of the best quality this readily burns away and produces a white body; others are suitable for coloured ware, stoneware or other purposes. Some gravel and sand occur in the clays, but can usually be removed by blunging with water and sieving or allowing the coarser particles to settle and pumping off the 'fines' in suspension.

The lignite and gravel contamination is a result of the frequently changing conditions whilst the clays were being transported and deposited. Some other minerals present in the primary rock have been transported along with the clay itself and are of comparable grain size; they form a part of the impurities in the ball clays. *Hydrous mica* or *sericite* is usually the commonest accessory mineral and as it is of similar particle size to that of the kaolin mineral itself, it is difficult to remove. Ball clays invariably contain a larger percentage of impurities than purified china-clays, and so have a lower refractoriness. They tend to shrink appreciably on firing and produce a denser body.

The ball clays from South Devon localities are composed chiefly of the clay mineral *kaolinite*, but the particle size of this mineral is much less than in china-clay. Some of the Dorset clays appear to contain a kaolin mineral more closely related to *livesite*. A sample from the vicinity of Torrington in North Devon contained a clay mineral of the livesite type and was rather richer in iron than those from South Devon. It may have resulted from surface weathering and not be of true hydrothermal origin.

The difference in clay mineral type does not seriously influence the properties of the ball clays. Both kaolinite and livesite have the same chemical composition although the latter has inherently a finer grain size, which for the same content of clay mineral should make it more plastic.

In the United States excellent ball clays occur in Kentucky and Tennessee where they are extensively worked. The Westerwald clays of the Central Rhineland of Germany are similar to ball clays and fulfil the same purposes although they have a different geological origin and generally are not so pure white-burning as the British varieties.

Ball clays are found in other countries but they are not of common occurrence and few are of such good quality.

Chemical analyses of ball clays show that they are richer in silica than china-clays and there is a correspondingly smaller alumina figure. They usually contain more alkalies, iron oxide and carbonaceous material. The percentage composition figures vary considerably, but usually fall within the following limits:

Silica	45–60 per cent
Alumina	25–35 per cent
Loss on ignition	7–15 per cent
Iron oxide	less than 2 per cent
Lime and Magnesia	less than 1 per cent
Potash and Soda	less than 3 per cent.

The mineralogical composition is also very variable but the three main components are always the clay mineral (either kaolinite or livesite), hydrous mica and quartz with some primary mica and titania, but no felspar. The carbonaceous material in ball clays is present in a finely-divided state (probably in colloidal form) surrounding grains of other minerals; it is possible that the plasticity of the clay is thereby increased.

The compositions of many ball clays from Devon and Dorset have been investigated by Holdridge.[1] He found that the essential mineral components were kaolinite (in Dorset samples it was a 'disordered' form), quartz, mica, organic matter with TiO_2 present as anatase. He calculated the mineralogical composition by means of 'rational analysis' (see p. 219) and by assuming that the principal alkali-bearing mineral was mica, he concluded that the range of mineral contents in the important British ball clay areas was as given in Table VI.III.

TABLE VI—III. MINERALOGICAL COMPOSITION OF BALL CLAYS
(after Holdridge)

	North Devon	*South Devon*	*Dorset*
Kaolin-type	33–68 per cent	20–90 per cent	20–83 per cent
Quartz	15–48 per cent	0–60 per cent	5–60 per cent
Mica	0–22 per cent	0–40 per cent	0–30 per cent
Organic Matter	0–3 per cent	0–16 per cent	0–8 per cent

This table includes the siliceous ball clays which have a total silica content of 60–80 per cent, due to a higher content of quartz than is usual.

It is evident from the manner of their formation and subsequent transportations and depositions, that the composition of ball clays must depend largely on that of the original altered rock and almost equally on the accidental impurities which have been mixed with them. In all ball clay pits, there is a great variety of clays, differing considerably from each other, so that it is necessary to exercise skill and care in excavating and in keeping the various grades separate. Serious losses in spoiled

[1] Holdridge, D. A., *Trans. Brit. Ceram. Soc.*, 55, 369, 1956.

ware may occur if ball clay from another source or, sometimes from another seam in the same area, is substituted without the possible effects of the change being realised. Changes in the physical properties of a ball clay are often quite as serious in their effects as changes in the chemical composition.

Much importance is attached to the smallness of the grain-size and the best clays are expected to contain 80 per cent of particles less than 0·001 mm., with 50 per cent between 0·0002 mm. and 0·0006 mm.

The chief uses of ball clays are to increase the plasticity of other white-burning clays, flint, etc., in the manufacture of whiteware bodies. They are important ingredients of stoneware, earthenware, engobes, dips and spray-slips and are used in the manufacture of abrasive wheels, insulators and moulding sands. Some of the cheapest ball clays are used as *fillers* in the rubber, paint and plastics industries.

Refractory Clays. Refractory clays are those with a high resistance to heat without loss of shape. This group includes many clays, but the most important are the fireclays.

Fireclays are clays which have a high refractoriness and although there is no definite stipulated lower temperature limit, a softening point of 1500° C. is usually regarded as a minimum and one greater than 1600° C. is generally required. Fireclays are widely distributed and are of many types, the principal of which are (*a*) *flint clays*, where the clay material has been compacted by pressure of overlying strata into a dense, hard, rock-like mass; (*b*) *plastic clays* which are much softer and are readily broken down by water into a mouldable mass of high plasticity, and (*c*) *shales* where a refractory clay has again been compacted by pressure, in such a manner that the mass readily cleaves along well-defined planes.

The important fireclay deposits[1] in the British Isles are associated with the Coal Measures which were laid down in the Carboniferous era of geological deposition. They are the oldest clay deposits which are worked on an extensive scale in this country. In the course of geological time, further deposition has taken place and the weight of this overlying strata has compressed the Coal Measure clays into narrow bands or beds of extremely hard material.

The true geological origins of fireclays and the influence of the overlying coals on their properties have been matters of conjecture for some time. In a recent thesis, I. A. White[2] has studied several British samples in great detail, analysing chemically and mineralogically at small intervals throughout the strata section.

He has proved conclusively that fireclays were not the seat-earths of growing vegetation, nor has there been any marked leaching of the clay strata from above at any time in its geological history. The fireclays were laid down in comparatively deep fresh water and the overlying coal seam may itself also have had a sedimentary origin.

The presence of chlorite minerals in some fireclays and more particularly in overlying shale beds as reported by Robertson,[3] proves that there has been little or no leaching nor major mineralogical changes since the deposits were laid down.

The conditions under which the Coal Measure fireclays were formed is not

[1] Some refractory clays at Ewell, Surrey, have long been used to make fireclay bricks, though they are not obtained from the Coal Measures.

[2] White, I. A., Ph.D. thesis, Leeds University, 1970.

[3] Brough, J. and Robertson, R. H. S., *Clay. Min. Bull.*, 3, 221, 1958.

known with certainty, but it is likely that the sediments were carried by large rivers from a vast land mass which extended from Greenland to Northern Scandinavia. Deposition took place over a large area, in conditions which resemble fresh water rather than marine formation.[1]

The formation of the Coal Measures extended over approximately 80 million years (see table, p. 48) and, during the period, frequent land movements took place which changed the level of the bed and altered the conditions of deposition. Swampy, deltaic, lacustrine and sometimes marine conditions followed, sometimes in regular sequence, giving rise to coal seams, sandstones and gritstones, fireclays and shell bands.

Coal Measure fireclays occur in many localities in the British Isles, the United States, Germany and elsewhere. They are worked wherever valuable seams outcrop, and important producing areas in Great Britain are the Central Lowlands of Scotland, Northumberland and Durham, West and South Yorkshire, Cumberland, parts of Lancashire, Staffordshire, Shropshire, North and South Wales and parts of South-west England. The outcropping seams in each locality are usually different, and a productive seam in one area may be of no value in another. The measures may vary in thickness from a few inches to several feet and only a limited number can be worked profitably. The usual methods of winning are by drift mining, by outcrop quarrying and by deep mine workings.

The clay seams vary considerably in composition and properties and there is no progressive relationship to geological age. In addition, clays extracted from one particular seam, but at different points along it, may show considerable variations.

Fireclays are the most extensively used refractory material and the nature of the minerals present in them and their influence on the property of the fired bricks have been the subject of much research for many years.

The behaviour of a clay at high temperatures is entirely dependent on the minerals which are present and on their grain-size distribution. This is well-illustrated by the fireclays, so that a chemical analysis can, at the most, give only the first indication of the value of a clay as a refractory material. Considerable variation in chemical composition is found in fireclays from different sources, the range of composition being:

Silica	40–80 per cent
Alumina	10–40 per cent
Loss on ignition	5–14 per cent
Iron oxide	1–5 per cent
Lime, Magnesia	Less than 5 per cent
Alkalies	Less than 3 per cent (low-grade samples may have up to 4 per cent)

In general, the higher the alumina content of a fireclay, the greater is the refractoriness. Chemical analyses of fireclays from the principal British localities are

[1] Similar conditions but on a smaller scale are to be found at the present day at the mouth of the Amazon where the water is comparatively fresh to a distance of over 300 miles into the Atlantic.

TABLE VI—IV. CHEMICAL ANALYSES OF FIRECLAYS FROM THE MAIN PRODUCING AREAS

	Cumberland		*Derbyshire and Warwickshire*		*Durham and Northumberland*		*Lancashire and Cheshire*		*North Wales*	
	Range	*Average*	*Range*	*Average*	*Range*	*Average*	*Range*	*Average*	*Range*	*Average*
SiO_2	45–69	57·5	46–70	56·0	48–65	57·0	45–70	57·5	43–78	63·0
Al_2O_3	22–36	26·5	18–41	27·5	22–37	27·0	18–36	25·0	15–37	22·5
Fe_2O_3	1·2–2·9	2·0	0·1–2·8	1·7	1·1–2·1	1·6	1·1–7·0	3·6	1·0–8·0	3·3
TiO_2	0·8–2·2	1·2	0·9–2·7	1·3	1·0–1·3	1·1	0·8–1·5	1·1	1·1–1·5	1·3
CaO	0·2–0·4	0·3	0·2–0·7	0·4	0·1–0·3	0·2	0·1–0·8	0·4	0·2–1·4	0·5
MgO	0·4–1·0	0·6	0·0–0·8	0·6	0·5–1·1	0·7	0·5–1·8	1·1	0·1–1·4	0·7
Na_2O	0·2–0·4	0·3	0·2–0·6	0·4	0·0–0·3	0·1	0·0–0·4	0·2	0·1–0·3	0·2
K_2O	1·0–4·3	2·4	0·8–3·3	2·4	1·3–4·5	2·2	0·4–3·9	2·2	0·8–3·4	1·8
Loss on ignition	7–15	9·0	7–14	9·5	9–12	10·4	7–14	9·6	5–12	7·5

TABLE VI—IV. CHEMICAL ANALYSES OF FIRECLAYS FROM THE MAIN PRODUCING AREAS

	South Wales		*Scotland*		*Staffordshire, Worcestershire, Shropshire*		*Yorkshire*	
	Range	*Average*	*Range*	*Average*	*Range*	*Average*	*Range*	*Average*
SiO_2	53–76	64·0	31–64	50·5	43–74	59·0	44–72	58·0
Al_2O_3	12–28	21·5	25–52	34·0	13–36	25·0	19–43	25·5
Fe_2O_3	0·6–6·3	2·6	0·5–6·4	2·3	1·0–3·3	2·1	1·0–4·0	2·2
TiO_2	0·9–1·4	1·1	1·0–4·3	1·7	0·5–1·7	1·1	1·0–2·3	1·5
CaO	0·1–1·5	0·5	0·1–1·0	0·4	0·0–0·6	0·3	0·1–0·5	0·3
MgO	0·2–1·2	0·7	0·1–1·2	0·5	0·1–0·9	0·4	0·0–1·5	0·7
Na_2O	0·0–0·5	0·2	0·0–0·8	0·2	0·0–0·7	0·2	0·0–0·8	0·3
K_2O	1·7–3·4	2·1	0·2–1·9	0·7	0·1–2·2	1·1	0·4–1·5	1·8
Loss on ignition	5–11	7·2	7–20	11·2	6–19	9·9	7–16	10·3

shown in Table VI.IV, but these may not be typical of the locality, particularly where more than one seam is worked.

The results given in Table VI.IV are the analyses of the most productive seams in each locality, but do not include minor workings, where clays with analyses outside the stated limits may be produced.

The wide range of composition in clays from each area is to be expected because numerous seams are mined which have very different characteristics. In the Yorkshire area, for example, clays range from those of the Lower Coal Measures to Middle and even Upper Coal Measure strata.

In general, however, the fireclays with the highest alumina contents are found in Scotland and Shropshire; North and South Wales deposits are less refractory, on the average, than those from the other producing areas.

The most refractory fireclays of a particular region are more likely to be found in the lowest geological levels. Many of the high alumina Scottish clays occur in the upper levels of the Gritstone series which underlie the Lower Coal Measures; Yorkshire fireclays of high refractoriness are invariably found only in the Lower Coal Measures.

The figures for the average analyses of fireclays from each area are the statistical mean of many samples. Of especial interest are:

(*a*) the small but fairly constant proportion of titanium dioxide;

(*b*) the ratio of magnesia to lime which is always about 2:1, but the total amount of alkaline earth oxides rarely exceeds 1·5 per cent of the whole;

(*c*) the content of alkalies (i.e. sodium and potassium oxides) usually lies between 1·3 and 3·0 per cent in the refractory clays, but in all cases the proportion of potassium to sodium oxide is between 3 and 10 to 1;

(*d*) the loss on ignition figures include carbonaceous material in addition to water of constitution. In some samples, the organic matter amounted to 10 per cent by weight.

Grimshaw, Carr and Roberts[1] have deduced the mineralogical composition of several fireclays in the Yorkshire area. This work has been extended and samples from many British localities have been analysed by standard methods. Although there is considerable diversity in chemical composition among the clays examined, the minerals present in all the samples are similar; it may be concluded that all the fireclays differ only in the relative amounts and grain size of three main constituents.

Some clays contained particles which were of coarse grain; this was particularly so in the case of the micaceous component and sand where in some fireclays, large flakes were retained in the sieves.

Of all clays examined three minerals were common to all, and together these amount to more than 90 per cent of the total clay. These minerals are:

(*a*) *Free silica* or *sandstone*, which, in most clays, is comparatively coarse-grained (upwards of 10μ). It has a high content of total silica which, in many samples,

[1] Grimshaw, R. W., Carr, K. and Roberts, A. L., *Trans. Brit. Ceram. Soc.*, 51, 334, 1952.

is largely crystalline quartz, although some flint or chalcedonic particles have been identified under the microscope. The content of this sandy component is variable, ranging from 0 to as much as 50 per cent in some samples. All fireclays which have a high total silica content on analysis are rich in free silica.

(*b*) A *clay mineral* of the kaolin group which, in all samples so far examined, has been *livesite* (see p. 131). This mineral has the composition $Al_2O_3 . 2SiO_2 . 2H_2O$, and is inherently so finely grained that it imparts plasticity to the clay. The content of livesite in fireclays is again variable, but is rarely less than 15 per cent.

(*c*) A *micaceous mineral*, which is classed as a hydrous mica because it contains less potash and more water of constitution than true muscovite (see p. 150). This mineral is of variable composition and is, therefore, difficult to estimate in clays, especially when it is present in fine grain-sizes. An average molecular composition of the micaceous component is

$$2{\cdot}1(K,Na)_2O . 1{\cdot}1(Mg,Ca)O . 8{\cdot}7(Al,Fe)_2O_3 . 24SiO_2 . 11{\cdot}2H_2O$$

(cf. muscovite mica $4K_2O . 12Al_2O_3 . 24SiO_2 . 8H_2O$), but in deriving a mineralogical balance of a particular clay it is advisable to isolate a pure sample for analysis. Hydrous mica is the principal alkali-bearing mineral in fireclays, and in many other sedimentary clays.

Some idea of the total amount of hydrous mica in a clay is shown by the total alkali figure, determined by chemical means, after first deducting the exchangeable bases in the clay. In the average hydrous mica, the alkali content is about 6·7 per cent, therefore a clay with 2 per cent total alkali may contain as much as

$$\frac{2 \times 100}{6{\cdot}7} = 30 \text{ per cent}$$

of hydrous mica. If the proportion of alkali in the hydrous mica is lower, the relative amount of this mineral in a clay will be higher still.

It is usually permissible to ascribe the *total* alkalies to the hydrous mica as the amount of exchangeable bases, soluble salts and other alkali-bearing minerals is generally insignificantly small.

Other minerals which may be present in fireclays include *ironstone*, *chlorites*, *anatase* (TiO_2), *carbonaceous material* and *amorphous material*, consisting of silica and alumina which appear to be present as a cementing medium in the clay. Traces of the kaolin mineral *dickite* have also been found in some clays, possibly derived from the ironstone nodules which frequently contain the mineral encrusted around the central nucleus. Traces of many other minerals may be present and occasionally some fireclays contain a considerable proportion of an unusual mineral: these are exceptional, however, and the vast majority of fireclays conform to the general pattern.

As livesite and hydrous mica in fireclays have average alumina contents of 39 per cent and 31 per cent (average) respectively prior to ignition, it is reasonable to assume that fireclays rich in silica contain appreciable quantities of free silica, whilst a high alumina content indicates a predominance of livesite or hydrous mica. Some differentiation between these two may be found from the alkali figure which will be comparatively large (above 2 per cent) if there is excess of hydrous mica.

A typical mineral balance sheet is shown in Table VI.V, for a fireclay from the Yorkshire area.

The constitution of the colloidal amorphous material has been deduced by difference from the chemical balance, although its amount is in accordance with that given by thermal analysis.

The mineral compositions of typical fireclays from some other localities are shown in Table VI.VI.

TABLE VI—V. COMPARISON OF CHEMICAL COMPOSITION AND MINERALOGICAL ANALYSIS OF A FIRECLAY

Mineral	*Per cent*	SiO_2	Al_2O_3	TiO_2	Fe_2O_3	*MgO*	*CaO*	K_2O	Na_2O	*Wt.* +120°	*Loss* −120°
Livesite	43·0	19·9	17·1							6·0	
Free silica	24·5	24·5									
Hydrous mica	22·0	10·5	7·0	0·2	0·7	0·1	0·1	1·3	0·3	1·8	
Ironstone	1·0				1·0						
Anatase	1·1			1·1							
Amorphous material	5·0	1·7	2·8							0·5	
Organic matter	1·2									1·2	
Adsorbed water	1·3										1·3
Total	99·1	56·6	26·9	1·3	1·7	0·1	0·1	1·3	0·3	9·5	1·3
By chemical analysis	99·3	56·6	26·9	1·3	1·7	0·3	0·2	1·26	0·22	9·5	1·3

TABLE VI—VI. MINERALOGICAL COMPOSITION OF SOME FIRECLAYS

	Scottish		*Yorkshire*		*Stourbridge*
	1	2	1	2	
Livesite	68·0	62·3	43·5	39·1	42·4
Free silica	1·2	0·3	24·8	12·3	7·1
Hydrous mica	23·4	16·6	22·6	43·0	46·3
Ironstone	2·2	1·6	1·0	0·8	1·0
Anatase	1·6	1·5	1·1	1·6	1·0
Carbonaceous material	1·0	12·3	1·2	0·3	0·2
Colloidal matter	2·2	4·9	4·9	1·6	1·1
Total	99·6	99·5	99·1	98·7	99·1

The *size distribution* of minerals in fireclays can be deduced by analysing the mineralogical composition of each sedimented fraction. The grain size range of the principal minerals in a clay is at least as important as the total mineralogical constitution in influencing the properties of that clay. This is demonstrated in other sections where actual grain size distributions are quoted. The hydrous mica may be present either as comparatively coarse grains (above 30μ) or it may be in the form of particles as small as the kaolin clay mineral itself. At high temperatures the small-sized micaceous flakes react more rapidly than large ones and therefore the behaviour of the clay can be predicted only from a knowledge of the amount *and* grain size distribution of each mineral.

High quality fireclay bricks are manufactured from clays with a low free silica content and the chief components are essentially a kaolin mineral and hydrous mica. Some Scottish clays and some from Northumberland are of suitable composition, except for their iron content which is usually too high. This would not be too serious a limitation if the iron were disseminated uniformly throughout the clay, but in many cases it is present as nodular ironstone which gives rise to *black* or *brown* spots and localised *bloating* at high temperatures.

Excellent high alumina bricks are made from calcined china clay, commercially termed *molochite*, although they are rather expensive. Other high alumina bricks are manufactured from raw materials which are hydroxides of alumina mixed with a little clay.

For further information on refractory clays see A. B. Searle's *Refractory Materials: their Manufacture and Uses* (London, C. Griffin and Co. Ltd.).

Flint Clays. Peculiar pockets of clay are known to occur in certain localities. These are of hard, virtually non-plastic material, but high in alumina. They are composed essentially of a kaolin mineral—probably kaolinite—and diaspore, the monohydrate of alumina. Although the titanium content is frequently high, the iron, alkalies and other impurities are usually low. The consequence is that the clays are extremely dense and hard and of high refractoriness. They are most commonly used as a calcined grog in bricks, castable and ramming compositions where greater volume and heat stability coupled with high resistance to abrasion is essential.

Clays of this character are not common but commercial supplies are available in Missouri, U.S.A., South Africa and Israel. Typical analyses are shown in Table VI.VII.

TABLE VI—VII. COMPOSITION OF SOME FLINT CLAYS

	Missouri, U.S.A.	*Negev, Israel*	*South Africa*
SiO_2	43·0	36·0	45·0
Al_2O_3	40·7	47·0	37·0
TiO_2	1·5	3·0	3·0
Fe_2O_3	0·3	1·2	0·5
CaO/MgO	0·1	0·2	0·1
K_2O/Na_2O	0·2	0·3	0·3
Loss on ignition	14·2	13·9	14·0

The geological origin of these clays is not fully understood but Keller[1] considers that they are the products of direct crystallisation from colloidal solutions and that the small kaolinite crystallites so formed intergrow to a mass of great hardness.

[1] Keller, W. D., *Clays Clay Mineral.*, 16, 113, 1968.

Not all clays of this type are pocket formations and some Carboniferous era seams have similar qualities. These are highly consolidated, kaolinitic clays often with diaspore or gibbsite to yield a clay of high alumina content. Some Ayrshire bauxites and clays from the Pennsylvania, Ohio and Illinois lower level coal strata are of this type. In some of the later formations, lenses of diaspore/gibbsite result in clays analysing to over 70 per cent Al_2O_3 on a calcined basis.

Flint clays are characterised by being virtually non-plastic unless they are ground very finely and allowed to weather for long periods of time. Their main use therefore is as grog or pre-calcined materials, and they are constituents of *super-duty* or *high-duty* fireclay articles.

High Alumina Clays. The presence of free alumina or more usually hydroxide of alumina in clays may result in an alumina content in excess of that of pure kaolinite. Such clays may be termed siliceous or impure bauxites and they are often the result of laterisation or the sedimentary products of laterites. The weathering of many rocks under wet tropical conditions usually leads to laterisation so that high alumina clays are common in such regions. Many of these formations are rich in iron oxides or hydroxides so they are not particularly valuable raw materials, but some are reasonably free of iron and are useful in refractory practice.

Bonding Clays. Some clays are of value as bonding agents. They are used for increasing the strength of moulding sands employed in the metal-casting industries. In order not to reduce the desirable properties of the sand, only small quantities of clay may be added, therefore, it must be of high binding power. Montmorillonite-type clays are frequently employed, but plastic kaolin clays are sometimes preferred. This latter clay is not of common occurrence, but a few have been developed commercially in the British Isles.

Such bonding clays are soft-textured and have been formed under geological conditions where the separation of most of the undesirable minerals has been achieved and only fine-grained components are present. The two principal minerals are *livesite* and *hydrous micas*, but the grain size of these is very much finer than in fireclays.

Bonding clays of high plasticity are sometimes added to non-plastic material in the manufacture of high refractory bricks, e.g. aluminous bricks.

Building Brick Clays[1] depend for their value on their physical properties rather than their chemical composition and, consequently, the permissible range of composition of a brick clay is very wide. The principal properties required in a clay for brickmaking purposes are that (*a*) it shall be present as a consistent deposit sufficiently extensive and accessible to warrant large-scale development,[2] (*b*) it shall be plastic enough to mould easily into shape and retain that shape in both wet and dry states, and (*c*) it shall vitrify sufficiently at a temperature of 950–1100° C. to form hard bricks without excessive shrinkage or deformation.

[1] Many brick clays are also suitable for making roofing tiles, terra-cotta and coarse earthenware ('brownware').

[2] For an output of ten million bricks per annum for twenty years there must be at least 800,000 cu. yds. of suitable *excavatable* material; this, at an average thickness of 3 yds., would occupy 55 acres of land (i.e. 266,200 sq. yds.), apart from that needed for margins and for the works.

A clay seam 3 feet in thickness will yield about 5,000 tons of clay per acre.

The clays which satisfy these requirements *perfectly* are limited, and although there are many which can be used quite satisfactorily, it is wrong to assume that any deposit of clay can be used for making building bricks.

Most quarries in which brick clays are mined are extensive and deep. It is only rarely economical to work a face which is less than 10 feet thick and most are in the region of 30–50 feet. Over vertical heights of this magnitude, clay strata usually vary between quite wide limits of composition. Layers of stone frequently alternate with shales and plastic seams. Nodules of iron, shell bands, small coal seams may be included in the working face. In the methods of working usually adopted in the brick industry, the feed material to the plant can and does vary widely and quality control often rests with the driver of the mechanical shovel.

The position can be further complicated by the geological dip of the strata, so that different beds or levels may be exposed and worked throughout the lateral extent of a quarry. It is only in recent years, and then only in a few larger or more progressive works, that the problems of clay variation have been realised and steps taken to ensure a more consistent regular quality.

Chemical analyses of brick clays have been described as worthless, because of the variations which can occur in any one deposit and also because there is such a wide range of compositions in use in brick-making practice. This is not an entirely valid criticism because chemical analyses can give much information on the nature of the minerals which are present, and a series of analyses on separated size fractions of a clay can often be used to predict making, drying and firing characteristics.

Table VI.VIII shows chemical analyses which are typical of particular quarries in specific brick-making areas of the United Kingdom. The samples were selected carefully to be representative of a working face. In the author's experience there is a greater consistency in the raw materials used in particular areas than is generally believed, but this observation should be limited only to the nature of minerals present throughout an area and not to the amounts of those minerals.

Holdridge[1] has analysed in detail the various strata levels in a large vertical section of the Etruria Marls. His results illustrate the type of variation to be expected if careful selection and blending are not carried out.

Although the range of composition is wide in terms of chemical ingredients, there are certain basic requirements which emerge from Table VI.VIII.

1. There must always be a proportion of clay minerals in brick clays to confer all the essentials of plasticity (see Chapter VIII).

2. A good proportion of non-plastic, non-fluxing material is also essential, so that the shapes do not shrink excessively on drying and firing. Quartz is usually present in considerable quantity, but felspar, calcite, particles of granite can fill a similar role, if they are present as large fragments.

3. Fluxing agents, commonly the alkali elements, sodium and potassium must be present but magnesium, calcium and ferrous iron can behave as fluxes at higher temperatures.

These elements are combined in mineral form as micas or felspars, but to be effective the particle size of such minerals must be small.

Other minerals include those containing iron—these are chiefly oxides or

[1] Holdridge, D. A., *Trans. Brit. Ceram. Soc.*, 58, 301, 1959.

TABLE VI—VIII. CHEMICAL ANALYSES OF SOME BRICK-MAKING CLAYS

	Coal Measure Shales			*Etruria Marl*		*Keuper Marl*			*Jurassic*	*Gault*	*London Clay*	*Glacial*
	York-shire	*Lanca-shire*	*North-umberland*	*Staffordshire a*	*b*	*Lincoln-shire*	*Leicester-shire*	*Devon*	*Oxford-shire*			*York-shire*
SiO_2	55·0	63·2	62·2	57·2	52·1	44·2	56·1	65·2	50·2	42·2	61·2	60·1
TiO_2	1·0	0·7	0·6	1·1	1·2	0·3	0·8	0·5	0·4	0·7	0·8	1·2
Al_2O_3	20·2	18·2	19·7	23·1	24·0	15·8	10·0	10·3	17·4	15·7	12·2	17·1
Fe_2O_3	7·1	4·1	3·5	8·4	9·2	6·7	4·1	5·8	3·0	5·7	8·1	5·4
CaO	0·1	0·7	0·4	0·6	1·8	9·2	5·3	3·2	8·9	14·1	2·2	3·7
MgO	1·4	0·8	1·2	1·1	1·0	6·4	12·1	4·3	1·0	1·3	2·2	2·8
K_2O	2·9	3·3	3·1	1·7	1·9	3·7	2·9	3·0	3·1	2·9	3·0	2·7
Na_2O	0·7	1·1	0·7	0·3	0·2	1·2	1·0	1·1	0·4	0·4	0·7	0·7
Loss on ignition	10·9	6·4	8·2	6·2	8·2	12·8	7·2	6·1	15·7	17·2	8·2	6·4

hydroxides; calcium carbonate as limestone or chalk; and minor impurities such as carbonaceous matter, gypsum, etc.

The analyses quoted are all British clays, but U.K. brick manufacturers use a wider range of clays, from a geological point of view, than those of most other countries. Chemical and mineralogical compositions of brick-making clays from the U.S.A., from the Continent of Europe and from Australia which have been analysed by the author show very similar characteristics to the British varieties.

The compounds most likely to cause trouble in the manufacture of building bricks are (*a*) limestones, and (*b*) soluble salts. The limestones produce lime which slakes in moist air and may crack or 'blow' the bricks. This may be largely avoided by grinding the limestones to a fine powder, as particles less than 0·02 in. (500 microns) in diameter seldom crack bricks containing them, even when present in amounts up to 15 per cent (see also p. 745).

Calcium compounds confer certain advantages on a clay for building brick purposes provided that they are finely divided. They are of value in some clays rich in iron, for they produce a pleasing buff or fawn colour on firing, and prevent the formation of a 'red brick'. Calcium compounds are also of value as a binder or glass-producer on firing, but in excess they tend to cause premature fusion and distortion of the bricks.

Soluble salts are chiefly objectionable because they form a white efflorescence or 'scum'; consequently, brick clays containing an appreciable proportion of soluble salts can be used only where the appearance of the fired bricks is of no importance.

Building-brick clays are of many types and the principal ones,[1] classified on the basis of their geological age, are:

1. *Carboniferous* or *Coal Measure* clays, which are of poorer quality than would be used for refractories. Bricks from such clays are manufactured in many parts of the country. The Etruria marl clays are of this classification. In 1968, about 33 per cent of all building bricks produced in the U.K. were from Carboniferous clays.

2. *Triassic* clays outcrop extensively along a band between 3–20 miles wide from the mouth of the Tees through the Midlands, Gloucestershire, Somerset and Devon. Another outcrop can be traced from the coast of Lancashire to its mergence with the former in Staffordshire and in this region the deposit is extensive. The most valuable clay is the Keuper Marl.

3. *Jurassic* clays—chiefly of marine origin—which include the well-known Oxford and Kimeridge clays, so highly prized by brickmakers. These clays outcrop along the length of the British Isles from Yorkshire to Dorset, but it is in the Peterborough and Bedford areas where the deposit is most extensively worked. Bricks made from Jurassic clays make up more than one-third of the total production in this country.

4. *Cretaceous* clays are worked in the South-east of England, the principal being the Hastings, Weald and Gault beds. Although the total production of bricks from clays of these beds is comparatively small, the quality and colour of the ware is excellent.

[1] Bonnell, D. G. R. and Butterworth, B., *Clay Building Bricks of the United Kingdom* (H.M.S.O., London, 1950).

5. *Tertiary* clays include those of the Reading beds and the London clay.

6. *Recent* clays are found in the valleys and flood-plains of large rivers and are termed alluvium or brickearths. The important ones for brickmaking are argillaceous soils and although such deposits are not extensive, they are comparatively easy and cheap to work.

7. *Glacial* clays are the detritus left by the receding ice-pack following the Ice Ages. They may be in the form of moraines, but much more commonly as Boulder Clay, so called because of the large rock fragments which are always present. Many deposits are of good commercial value, but the clay is extremely variable in composition and properties.

The geological formation of British building-brick clays has been described by Keeling.[1]

There is little published data on the constitution of building brick clays, but it is certain that they are generally more complex than the clays previously considered.

It may be inferred that the *Coal Measure* clays will be similar to fireclays, but with a higher proportion of silica and alkali impurity. Many surface outcrops are worked and it is customary to extract the softer clays by quarrying. Consequently, more than one clay seam is often excavated along with intervening strata of grits, shales and sometimes coal. Chlorites are common constituents of these deposits.

The geology of the *Jurassic* clays has been worked out in detail, but the complete mineralogical analyses are not known with certainty. The clays are of marine origin, so that shells and other fossils are of common occurrence. Carbonaceous material is also present, sometimes in considerable quantity, and is of value in the burning of the clay. Unlike Coal Measure clays which frequently contain coal, the *Oxford clay* is rich in algae or oil shale which burns readily at comparatively low temperatures. It is likely that the principal mineral in the clay is of the kaolin-type, but there is an associated alkali-bearing mineral which may be glauconite. Quantities of sand are also present. Oxford bed clay is of value because the associated carbonaceous material reduces the fuel required in the firing, and the high lime content, derived from the marine fossils, produces a pleasing colour, although it tends to shorten the vitrification range of the clay.

Clay of the later geological periods has an extremely complex mineralogical composition. Honeyborne[2] has reported that clays from the *Hastings bed*, which outcrops in Kent and Sussex, contain kaolinite which is not well-crystalline (livesite?). Hydrous mica (illite) and a chloritic mineral along with iron hydroxide as lepidocrocite, hæmatite and considerable free silica sand are also present. Some samples also contain calcite and small proportions of gypsum. The same author[3] has also described clays from the *Keuper Marl* which is another important and extensive building brick clay deposit. The principal mineral component is of the hydrous mica or illitic type and, in addition, a complex mixed-layer mineral of chloritic and montmorillonite-type is present. Additional minerals include calcium carbonate (in the form of fossil remains), hydrated iron oxide, free silica sand, magnesium carbonate and occasionally sepiolite ($2MgO.3SiO_2.H_2O$ approx.).

[1] Keeling, P. S., *Trans. Brit. Ceram. Soc.*, 52, 465, 1953.
[2] Butterworth, B. and Honeyborne, D. B., *Trans. Brit. Ceram. Soc.*, 51, 211, 1952.
[3] Honeyborne, D. B., *Clay Min. Bull.*, 1, 150, 1951.

Alluvium or *recent* clays have not been examined in detail largely because they occur as isolated pockets which are unrelated to other parts of the same area. Samples from small workings in the flood plain of the Humber contain hydrous mica, sand and hydrated iron oxide (goethite), along with a kaolin-type mineral.

Glacial drift is also used for making building bricks and few analyses of this material have been made. The minerals present in them are derived from the areas which were covered by the ice sheet. The impurities or non-clay minerals are usually greater in amount and of larger grain size than those in a water-sedimented clay.

Engineering brick clays are of similar type to those used in making building bricks. A larger proportion of fluxing agents is desirable so as to produce a stronger, harder, more vitrified and impervious body than that of other bricks, although these properties can often be achieved by harder firing.

Clays for Ceramic Ware.[1] The clays used in the manufacture of different wares are selected chiefly on the basis of physical properties, though the importance of chemical composition must not be ignored. *White ware* of all types—including porcelain, bone china, fine earthenware and vitreous table ware is made from white-burning clays, principally kaolinite, with additions of ball clay to promote strength in the moulded shapes. *Permeable hollow-ware*, which includes heavy clays, refractories, terra-cotta, coarse earthenware, faience, majolica and many others, is made from clay which does not vitrify excessively at the temperature of firing. Colour is usually of secondary consideration in these bodies, so that many types of clay can be employed. *Impermeable ware*, includes vitreous china, porcelain and stoneware which have dense non-porous bodies; the two former are white or light cream coloured, but stoneware varies from almost white to buff or dark brown. Clays with a high proportion of fluxing elements are required for the manufacture of stoneware, and it is common practice to add extra ones during manufacture. The clay should be fine-grained and free from large particles of sand or mica. The composition of ceramic bodies is discussed more fully on pp. 333 ff.

Montmorillonite-type Clays. Clays consisting largely of montmorillonite minerals are not common in the British Isles, but are of frequent occurrence in many other parts of the world.

The earliest reported description of a montmorillonite-type clay was by W. C. Knight[2] in 1896 from Fort Benton, Wyoming, U.S.A.; not unnaturally he termed the material *bentonite* and this name persists to the present day and is often used to describe any impure commercial montmorillonite clay.

All the details of the geological formation of montmorillonite clays have not been established with certainty but many authorities consider that they are the breakdown products of alkaline volcanic ash often in a marine environment. The parent ash must be rich in magnesium or ferrous iron. Clays which contain montmorillonite minerals are of comparatively recent geological origin and there are very few reports of these minerals being present in deposits older than the Cretaceous Period.

[1] The term *pot clay* or *pottery clay* has so wide an application as to be almost useless. It is applied to the refractory clays used for crucibles employed for melting steel and other alloys and the 'covered pots' used in glass manufacture and to the clays used for coarse and fine earthenware and for the various kinds of porcelain.

[2] Knight, W. C., *Eng. Min. Journ.*, 66, 491, 1896.

Montmorillonite clays have been discovered in many countries; they are often a feature of soil profiles or sedimentary deposits in tropical regions with a comparatively dry climate.[1] The clay mineral has formed directly from volcanic ash or alkaline lava. The most important commercial source is still the bentonite area of Wyoming where beds or seams of a few inches up to a thickness of about 30 ft. occur.

Clays containing montmorillonites are valued for their binding power and the strength which they impart to other materials. They also have a remarkable capacity for swelling when water is added to them and they readily form suspensions which exhibit unusual changes in viscosity with different cationic additions. Suspensions are prone to thixotropic characteristics (p. 473) which are frequently utilised in drilling fluids. Some samples have a high absorption capacity and are useful in purifying liquids, bleaching oils and in the paint industry where they are known as *fullers earths.*

Fullers earth is mined in two localities in England, although similar clays occur in other places. The principal deposits are near Nutfield in Surrey and Bath in Somerset. The origin of the British fuller's earths is still unknown; they present the anomaly of an apparently altered volcanic ash occurring in the middle of arenaceous beds of sedimentary formation. It has been suggested by Newton[2] that the clay has its origin in chemical precipitation in marine waters where normal sedimentation was almost non-existent. Apparently at some stage in the formation of the Lower Greensand deposits, land movements occurred which produced sheltered marine conditions for a comparatively short duration. Newton's theory is likely, for no volcanic material was associated with the geological periods in question.

Chemical analyses of samples of Wyoming bentonite and Nutfield fullers' earth are given in Table VI.IX.

The mineralogical composition has not been deduced in detail. The main mineral in both samples is montmorillonite (see p. 145). Kerr, Main and Hamilton[3]

TABLE VI—IX. CHEMICAL ANALYSIS OF MONTMORILLONITE CLAYS

	Wyoming bentonite	*Nutfield fullers earth*
	per cent	per cent
SiO_2	49·40	54·36
Al_2O_3	20·40	12·22
TiO_2	0·40	0·72
Fe_2O_3	3·70	2·91
FeO	n.d.	3·02
MgO	2·55	2·32
CaO	1·80	2·48
Na_2O	0·43	0·21
K_2O	1·12	0·61
H_2O (moisture)	5·00	5·60
H_2O (lost at 900° C.)	15·15	15·91
Total	99·95	100·36

[1] Barshad, I., *Proc. Int. Clay Conf.*, 1966.
[2] Newton, E. F., *Geol. Assoc. Proc.*, 48, 175, 1937.
[3] Kerr, P. F., Main, M. S. and Hamilton, P. K., *Amer. Petroleum Inst. Res. Proj.* **49**, Part 5 (Columbia Univ., N.Y., 1951).

state that the impurity content in the Wyoming sample is about 10 per cent, principally quartz and felspar with some calcite, limonite and glauconite. The Nutfield sample is similar but also contains traces of pyrites and sphene. Cristobalite is a common impurity in some montmorillonite clays.

The nature of the principal minerals in a clay deposit can, to some extent, be predicted from the geological details of the deposit. Hydrothermal alteration of an igneous rock will usually produce a kaolin mineral and hydrous micas if acid conditions are prevalent, but montmorillonites if the attack is under alkaline conditions. Surface weathering also favours the formation of kaolin minerals unless excess magnesium or iron in an alkaline medium is readily available. Hydrous micas or illites are of common occurrence particularly in marine-deposited shales, and it is a remarkable fact that these minerals are frequently associated with the kaolin-mineral livesite in a way which suggests inter-convertibility between the two forms.

Many deposits of early geological periods contain only hydrous mica and a kaolin mineral, along with free silica and minor constituents. More recent formations may contain more complex forms such as *chlorites* and, in some cases, *vermiculites*. These latter minerals are known to be less stable than other clay minerals and they may, in the course of geological time, revert to the hydrous mica or kaolin types.

It is important to realise that when a bed of clay extends over a large area, it may vary considerably in properties in different localities. Thus, the Middle Lias clay of Oxfordshire is very different from clay from the same seam which outcrops in Dorset; similarly, a valuable bed of fireclay in one area may be worthless when it is found in other regions.

Even when taken from two closely-separated areas, a particular clay may show considerable variation. The difference between a deep-mined clay and the same seam under shallow overburden is so well marked that it is often impossible to use the two materials for the same purpose. A striking change in the plastic behaviour of a clay often occurs when it is taken from land on which there has been intensive cultivation. The fertilisers used in the soil are sufficient to alter the exchangeable cations even in clays at a considerable depth and so change their character.

SILICA AND SILICEOUS MATERIALS

FREE silica is second only to clay as an important ceramic material. It is used on a large scale: (i) in the glass industry; (ii) for making bricks, roofing tiles, refractories and ceramic wares; and (iii) in the form of sand or aggregate for a variety of industrial processes. As with clay, the forms in which free silica is present together with the nature of the impurities have a profound influence on the properties of the material both in its raw and fired states.

The principal siliceous materials (other than clays) which occur in nature and are used in the ceramic industry include (*a*) crystalline quartz, (*b*) silica rocks, including quartzites, sandstones, ganisters, silcrete, etc., (*c*) silica sands, and (*d*) amorphous or 'organic' silica including flint, diatomaceous earth, etc.

Free silica occurs in a variety of forms and makes up a large portion of the earth's crust, something like 60 per cent of which consists of this mineral, either in bulk or disseminated through other rocks in particles of varying sizes. Free silica is present in both crystalline and amorphous conditions, the principal forms being: **Amorphous silica** occurs as (*a*) precipitated silica formed by the deposition of silica from solution; (*b*) siliceous skeletons of diatoms, etc., in such forms as kieselguhr, tripoli, etc.; and (*c*) silica glass, which is of infrequent occurrence in nature, but common in manufactured clay products and formed by rapid cooling of molten silica.

Precipitated silica occurs in various forms in nature, the commonest being flint, chert, chalcedony, opal and geyserite. *Flint* consists of silica which has been precipitated around a nucleus of either organic or inorganic matter, and consequently it often contains a variable proportion (usually about 5 per cent) of impurities, chiefly of carbonaceous and calcareous matter. It occurs as grey or black nodules of varying size, of great hardness, which break with a conchoidal fracture. Sosman[1] considers that it consists of an intimate mixture of amorphous and crystalline silica. Flint is found usually in the form of pebbles, especially on the sea-shore, where they often extend over large areas. They also occur in beds of clay or in chalk or limestone deposits—especially those of Cretaceous age—one bed being known as *clay-with-flints*. Small grains of amorphous silica occur to a varying extent in clays, sands, and similar detrital deposits.

Chert or *hornstone* is similar to flint, being formed by the deposition of sponge spicules producing greyish, brown or black grains of almost crystalline appearance, though in reality it is truly amorphous. The grains occur in beds of limestone, clay, sand, and sand-rocks, especially those of the Carboniferous Age, in which they frequently form large masses of stone.

Chalcedony is a somewhat fibrous variety of precipitated silica formed in the same manner as flint and chert. Some specimens may have a deformed quartz network structure with considerable micropores running through them. Chalcedony varies in colour from white to blue or brown and has a hardness of 7 and a specific gravity of 2·55–2·58, a refractive index of about 1·54 (rather less than that of quartz), and a slightly lower birefringence than quartz. *Onyx*, *agate*, and various other minerals resemble chalcedony so closely that for the purpose of the present volume they need not be distinguished.

Opal is a hydrated form of silica with the composition $SiO_2\ nH_2O$ where the percentage of water is usually less than 10. It is found in rocks of many kinds, but is of secondary origin behaving as a dried-up gel filling cracks and fissures and cementing grains of other minerals. It commonly occurs in sandstones and shales, and is readily converted into chalcedony on heating. In some massive forms, opal is highly prized as a gem-stone, but it is difficult to identify when disseminated throughout a rock mass. The specific gravity of opal is 2·2. It is isotropic between crossed nicols.

Diatomaceous earth or *kieselguhr* and *radiolarian earths* are mainly opaline silica and are of value as thermal insulators. Diatoms and radiolaria are minute organisms which have siliceous skeletons; they thrive in situations where there is abundant available silica in solution, as in the Miocene beds of California where volcanic emanations of a siliceous character have promoted growth. The remains of such

[1] Sosman, R. B., *The Phases of Silica* (Rutgers Univ. Press, N.J., 1965).

organisms have accumulated as large deposits on sea- and lake-beds in some places. The intrinsic nature of the skeletons of silica results in the formation of a mass containing appreciable volumes of entrapped air, so that these earths form almost ideal thermal insulators, and in addition, are of value as an absorbent or filtering medium and as surface catalysers. The chief deposits are in the United States, Germany, Denmark and North Africa, but workable seams are found in the Isle of Skye, Cumberland and Northern Ireland.

The *impurities* in diatomaceous earths and similar materials are chiefly silt, clay, sand, volcanic ash and decayed organic matter. Many samples are red in colour indicating a high proportion of iron oxide, although some American deposits are almost white. Unfortunately, the refractoriness of diatomaceous earths is low and it is inadvisable to expose them to a temperature of greater than 900° C.

Geyserite is another form of opal and occurs as a loose, porous form of silica precipitated from solution around hot springs, as in Yellowstone Park, U.S.A., in Iceland and elsewhere.

Le Chatelierite is the only form of silica glass found in nature and is of very rare occurrence.

Colloidal silica may occur either (*a*) in bulk as a horny mass in forms such as those mentioned above; (*b*) as a cementing material in other rocks, such as quartzites, sandstones, etc.; or (*c*) as an active colloidal substance possessing highly absorbent properties.

Crystalline silica occurs in nature in three forms, the commonest being quartz, whilst tridymite and cristobalite are relatively rare. The two latter forms are produced by heating silica to high temperatures, as in the manufacture of silica bricks and other siliceous articles.

Quartz occurs as hexagonal or rhombohedral crystals, which are colourless when pure, but are often stained a variety of colours on account of the impurities present.

The purest qualities are highly valued industrially either to manufacture various electronic devices based on their piezo-electric character, or in the manufacture of fused silica ware. Some varieties are coloured and highly prized as semi-precious gemstones; amethyst, citrine, rose quartz are examples.

Quartz has a conchoidal fracture, a hardness of 7 and a specific gravity of 2·65 Its refractive index is 1·55 and its birefringence 0·009. It is often readily recognised under the microscope by having numerous inclusions which cause it to be cloudy or opaque.

Quartz occurs also (i) in massive form, (ii) as small grains disseminated through other rocks, and (iii) as sand. The most useful forms of massive silica are (*a*) *quartzites*, which consist of an irregular mosaic of grains of quartz cemented by a siliceous cement, (*b*) *silica rocks*, which are grains of quartz united by a cement, preferably of a siliceous nature, but generally containing more impurity than quartzites, and (*c*) *ganisters*, which consist of small, highly angular grains of quartz.

Tridymite is not of common occurrence in nature, but is has been reported in rocks of recent volcanic formations, and it is a constituent of some rhyolites. It forms pseudo-hexagonal crystals with characteristic arrowhead twinning, with a specific gravity of 2·28 and a refractive index of 1·477. Tridymite is the stable silica modification between 870° C. and 1470° C.; it is therefore a common constituent of fired siliceous materials.

Cristobalite is another crystalline variety of silica which again is rarely found in nature. Occasionally cristobalite is present in volcanic rocks and is named after the locality San Cristobal, Mexico, where it is a constituent of andesite. It forms pseudo-cubic crystals with a specific gravity of 2·32 and a refractive index of 1·484. It is the crystalline form of silica which is stable at temperatures above 1470° C.

Silica Rock Formations

The various silica rock formations which are of value in ceramics are composed of crystalline or amorphous varieties, or usually admixtures of the two types. They vary considerably in character and property.

Crystalline quartz is the essential raw material for the manufacture of silica glass, where the impurities must be at a minimum. The purest form of silica in nature is rock crystal or crystal quartz, which occurs in veins or geodes penetrating other rock formations. The crystals, which are often perfect in form and range in size from a few millimetres to several centimetres, are the result of the slow evaporation of solutions rich in silica, or of the slow cooling of siliceous magmas. In only a few localities do such crystals occur in quantity, the main sources of supply for British industries being Madagascar and Brazil. Crystal quartz which can be washed free from any adherent associated impurity provides a material which is at least 99·98 per cent silica on analysis. Occasionally liquid or gaseous inclusions within the crystals are present, and also other minerals such as rutile (TiO_2), tourmaline, tremolite or chlorite, which cannot be removed without destroying the crystal.

Pure quartz melts at 1710° C. to form a highly viscous transparent liquid which can be supercooled to a glass without forming any crystals.

Silica sands are used in the manufacture of glasses of all descriptions; a material of high purity is required. Some sands, of which that from Loch Aline is an example, are pure enough to be used in the manufacture of silica glass; others can be treated by physical or chemical means to produce the desired qualities.

The role of silica in glasses is that of an acid medium to combine with the alkaline constituents, and it is important to employ sands of great purity. The particles of sand must be small so as to ensure rapidity of reaction. Best quality glass sands contain more than 99·5 per cent of silica, but commoner ware, such as some bottle-glass can be manufactured from material with as little as 65 per cent of total silica.

The principal *impurities* which occur in glass sands are:

(*a*) Alumina, in the form of felspar, muscovite or clay; the two latter minerals can be removed fairly easily by washing. In most glasses a small proportion of alumina is not serious and for some purposes alumina is added to the batch to impart tenacity to the glass ware.

(*b*) Iron compounds, which may be present as a film coating quartz particles, or as discrete particles of haematite, magnetite or ilmenite. Iron is most undesirable as an impurity, as it has a profound influence on the colour of the glass as shown in Table VI.X.

TABLE VI—X. EFFECT OF IRON OXIDE ON THE COLOUR OF GLASS

Colour of glass	*Ferric oxide added*
Colourless	Less than 0·02 per cent
Nearly colourless	0·02–0·05 ,, ,,
Faintly green	0·05–0·10 ,, ,,
Yellowish green	0·1–0·3 ,, ,,
Green	0·3–1·0 ,, ,,
Dark green	1·0–2·0 ,, ,,

These figures are only approximate indications of the relation of colour to iron content. The effect can be mitigated, to some extent, by the addition of a decolor-iser such as manganese dioxide, cobalt oxide or selenium.

(*c*) Lime and alkalies, mainly in the form of felspar or micas, are not serious as impurities, as large amounts of these constituents are purposely added in the production of commercial glasses.

Many sands of good quality are found in the British Isles although most have to be pre-treated in some way before they can be used for making glass.

Raw Materials for Silica Refractories. Silica bricks[1] are usually made from rock containing at least 97 per cent of silica, but the following range of compositions is usually permissible:

Silica	95–99 per cent
Alumina	0·1–2·8 per cent
Iron oxide	0·3–1·3 per cent
Lime	0·2–2·4 per cent
Soda and Potash	0·2–1·5 per cent

The raw materials used are mainly sedimentary sand deposits which have been cemented with precipitated silica and have formed a hard, dense rock mass.

The conversion of the siliceous raw material to cristobalite or tridymite must be completed in a reasonable time at temperatures of about 1450° C. For this reason, pure quartz crystal would not be suitable because its reaction is extremely slow even in the presence of catalysts. Amorphous silica, on the other hand, converts too rapidly, sometimes with disastrous effects on the brick shape. The ideal raw material is one composed of small grains of quartz which contains a proportion of a non-crystalline, flint or chalcedonic type of silica, which will be readily converted and 'seed' the slower reacting crystalline material.

Lower grade silica bricks are made from sandstones and ganisters with a total silica content between 87–96 per cent. The impurities are mainly mica or hydrous mica, limonite and occasionally a kaolin-type clay mineral. The first silica bricks were made in the early 1820's from Dinas rock found in South Wales, and although this material is no longer used for high quality bricks, similar material is in frequent use on the Continent at the present time for what are described as *Dinas bricks*.

[1] Sand-lime bricks are made chiefly of silica (sand) but the term *silica bricks* is commercially restricted to highly refractory bricks made from selected silica rocks. Sand-lime bricks are not refractory.

For many purposes, various silica rocks, which form part of the Millstone Grits, have been used for many years and are still in use although on a much smaller scale.[1] With the ever-increasing demands of metallurgists for furnace linings which will withstand greater temperatures, some of these silica rocks are unsatisfactory because they contain small amounts of impurity, particularly alumina, which reduces very seriously the refractory properties of the bricks. It is also advantageous to use material with a low porosity and one in which the individual crystallites of quartz are as small as possible.

The principal *sources of high-grade silica* have been summarised by Davies[2] as follows:

1. *Coarse-grained quartzites* including the Welsh quartzite mined in the region of Bwlchgwyn, and incoherent pebble deposits, as near Aberdeen, both of which are mainly composed of comparatively large fragments of quartz grains which have been cemented by secondary precipitated silica. The Welsh rock is contaminated with hydrous mica (*sericite*), which is the main source of alumina. Davies has stated that the cementing silica crystallises as quartz which develops on the surface of the original particles, thereby producing a mass of interlocked crystals. Grimshaw, Westerman and Roberts,[3] however, have pointed out that Welsh quartzites rarely contain more than 80 per cent α-quartz as determined by thermal analysis. They suggest that free silica in the form of crypto-crystalline or dense chalcedonic varieties may be present; this would explain the comparatively rapid rate of conversion. Flint is undoubtedly present in many pebble deposits.

2. *Chalcedonic Silica*, which is found as *chert* or *flint*, has normally an impurity content which is too high for silica brick manufacture.

3. *Fine-grained Crystals in a Chalcedonic Matrix*. One of the best known and highly prized of European silica rocks is the *Findlings Quartzite* which has been used in Germany for many years. Similar material is found in Italy, but the most extensive deposits occur in South Africa. The formation of this fine-grained aggregation or *silcrete* is the result of chemical alteration of siliceous sub-soils or sandstones. Non-acidic components in the original rock have been dissolved and leached away and the resulting rock-form consists of fine-grained quartz particles cemented with chalcedonic silica. A feature of silcretes is a high content of titania (TiO_2) principally rutile, which is concentrated along with the silica. The natural rock has a much lower porosity than other quartzitic raw materials.

Table VI.XI shows the chemical analyses and other properties of some siliceous materials.

Semi-silica Material. Refractory bricks and other shapes are usually manufactured from sand/clay mixtures with a total silica content of 78–92 per cent. Many low-grade quartzites are of this composition and contain more impurities than those used in the manufacture of silica bricks of the highest grades. The chief source of contamination is hydrous mica, which introduces appreciable quantities

[1] For a description of the historical development of silica bricks see Lynam, T. R., *Ceramics, a Symposium* (The British Ceramic Society, Stoke-on-Trent, 1953), pp. 448–73.
[2] Davies, W., *Trans. Brit. Ceram. Soc.*, 51, 95, 1952.
[3] Grimshaw, R. W., Westerman, A. and Roberts, A. L., *Trans. Brit. Ceram. Soc.*, 47, 269, 1948.

of alkali as well as alumina. Artificial mixtures of a good grade of quartzite with a fireclay low in alkali are more commonly employed.

TABLE VI—XI. CHEMICAL ANALYSES OF SILICA ROCKS

	Bwlchgwyn Rock	*Silica Pebbles Aberdeen*	*Chalcedonic Silica, Spain*	*Findlings Quartzite*	*Silcrete S. Africa*
	per cent	per cent	per cent	per cent	per cent
Silica	97·39	99·86	98·7	97·8	96–98
Alumina	0·73	0·06	0·12	0·4	0·2–0·7
Ferric oxide	0·78	0·02	0·71	0·3	0·3–1·0
Titania	0·11	0·03	Trace	0·8	1·2–2·0
Specific gravity	2·66	2·65	2·65	2·64	2·65
Porosity (per cent)	1·6	4·4	1·0	0·3	very low
Average size of quartz grains (microns)	150	1000	unresolvable	40	15

The value of volume stable semi-silica refractory has been widely recognised in recent years. Some American manufacturers have been successful with a product synthesised from a pure quartz and a high-quality kaolinite. A pyrophyllite-based composition is extensively used in the Japanese steel industry in such applications as soaking pits.

Refractory Silica Sands. Silica sands are used for many purposes in the ceramic industry, (i) in the manufacture of ware, (ii) as a setting or lining material, (iii) in the manufacture of refractory cements, and (iv) as a ramming composition in steel ladle refractories. Very pure sands are not usually essential but the percentages of lime, iron oxide, magnesia and alkali should be as low as practicable.

Moulding Sands. The casting of metal shapes is invariably carried out in moulds made from sand which is strengthened with a little clay or organic binder. Sand is the most serviceable material for this purpose as it: (*a*) is economical in use; (*b*) has a high refractoriness; (*c*) can be formed into shapes, which not only prevent the leakage of metal, but also allow generated gases to escape; and (*d*) is indestructible and can be used repeatedly.

To be of value in moulds, the sand must be free from impurities which would reduce the refractoriness, but the shape and grain size distribution of the individual grains are of greater importance. The choice of sand must be a compromise between angular particles which develop maximum strength with the binder, and rounded particles which are easily and more uniformly moulded and produce moulds with a lower porosity.

The particle size distribution of moulding sands has been carefully investigated and standard testing methods have been developed. A sieve analysis is carried out on the dried sand and a cumulative curve of the particle size distribution plotted (Fig. VI.I). Two important values are derived from this curve:

A. The *centre grain size* (*a*) which is the value in microns (10^{-4} cm.) at 50 per cent retention on the sieves.

B. The *distribution-coefficient* which is the ratio between the particle size at 75 per cent and that at 25 per cent retention. This is the value b/c as shown in Fig. VI.1.

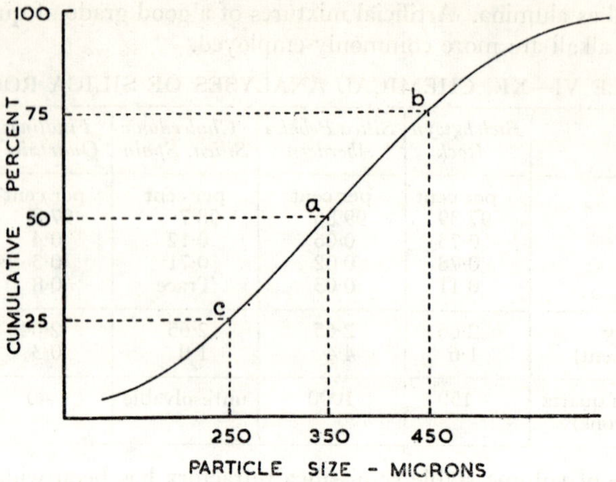

Fig. VI.1. Graph of the Grain Size Distribution of a Sand

Showing the calculation of (A) the centre grain size (*a*); and (B) the distribution coefficient (*b*/*c*)

The former value determines the size of castings which can be made successfully from the sand; a coefficient of about 400 (microns) would be suitable for large castings and, for small ones, a value about 200 is better. The ratio of b/c must be between 2–8 otherwise the permeability of the mould to gases will be unsuitable.

Ramming sands are increasing in favour as lining materials for ladles in steel practice. The best quality consists of a pure silica sand with an intimate coating of highly plastic clay. Sands with a natural bond of clay are generally preferred, although synthetic mixtures are in use, but the clay content should be as low as possible.

Similar physical properties to those of moulding sands are required, but for large ladles the quartile ratio should be in the 10–15 range.

Other Sands. Many other sands are used in ceramic practice, and their value in a particular application depends on their physical and chemical properties. Sands for brick-making purposes must be relatively cheap and if they are required in large quantities they must occur near the site on which the bricks are manufactured. Some clays from the London Basin shrink excessively on drying and firing, but, because they are highly plastic, a proportion of sand from nearby deposits can be incorporated to improve their quality.

A special quality of sand is required in the manufacture of *sand-lime* bricks. Five per cent of lime is added to a sand which contains a proportion of fine particles. The moulded shapes are treated under steam pressure in an autoclave, so that reaction occurs between the lime and silica forming cementing substances. About 12–15 per cent of the sand should pass a 100-mesh sieve, and a third of this quantity should pass a 200-mesh sieve. The rest of the sand should be coarse with a centre grain size of about 30 B.S.S. Quite often two sands are blended to give the required particle size distribution. Flint or chalcedonic silica are particularly valuable.

ALUMINA

ALUMINA occurs in a variety of forms. It is a common constituent of many rocks and minerals, but it is also found in the free state as the oxide or hydroxide.

Alumina is present in nature in massive form and as grains disseminated through other materials principally as *corundum* (α-Al_2O_3), which occurs as grey, green, reddish or yellow hexagonal crystals, having a hardness of 9 and a specific gravity of 3·9–4·1. Corundum is mined only on a small scale because all the known deposits are limited in extent. The principal sources are Asia Minor, South Africa, Greece, Canada and the United States. *Emery* is an impure form.

Artificial corundum is prepared by fusing bauxite, and is being increasingly used as a refractory material on account of its high softening point, inertness to corrosion and constancy in volume when heated.

Alumina is more prevalent in the hydrous form as *bauxites* and *laterites*, which are apparently amorphous, granular materials of a white or brownish colour according to the amount of iron oxide they contain; the main minerals in these types of deposits are: (*a*) *diaspore*, $AlO.OH$ or $Al_2O_3.H_2O$, which consists of white orthorhombic crystals, with a hardness of about 7 and a specific gravity of 3·5; and (*b*) *gibbsite*, $Al(OH)_3$ or $Al_2O_3.3H_2O$, which usually occurs in concretionary masses and rarely as monoclinic crystals. It has a hardness of 2·5–3·5 and a specific gravity of 2·3–2·4.

Bauxite is not a simple mineral, but a term used to describe a massive formation which is rich in alumina. Most bauxites are mixtures of the mono- and tri-hydrates of alumina with some clay, but some samples which are loosely termed bauxite are almost entirely kaolinite. Bauxitic deposits are fairly widespread; some have been derived from basaltic rocks and others from rhyolites. Extensive valuable areas occur in the West Indies, Guyana and Surinam, Australia, Yugoslavia, France, Hungary, the United States, Russia and elsewhere, but the most important world sources are of fairly recent tropical formations.

Smaller (commercial) quantities occur in Co. Antrim, Northern Ireland and the high-alumina fireclays of Ayrshire appear to be of similar origin.

Laterites are usually formed by the chemical alteration of certain igneous rocks, such as basalt lavas and rhyolites. They sometimes occur as residual deposits but much more commonly as a secondary formation. They consist chiefly of *gibbsite*, with some *boehmite*, *kaolinite* and occasionally some *chamosite*. The composition of laterites is very varied; they show every gradation from a highly ferruginous hydroxide with little alumina to almost pure aluminium hydroxide.

Lenses of lateritic material occur in some American fireclays and thus give a clay with a high alumina content, although of somewhat variable composition.

High alumina bricks are manufactured from natural bauxites which contain between 60 and 90 per cent alumina on a calcined basis. The main impurity is clay, usually of a kaolin type, which serves as a binder. Variable amounts of iron (0·5–12 per cent) and free silica (3–25 per cent) may also be present. Materials which are rich in diaspore are generally preferred to those containing gibbsite, because the lower content of water of constitution in the former results in considerably less shrinkage on firing. This can be mitigated by pre-calcining the gibbsite.

Aluminous refractories cover bricks ranging from about 44–99 per cent Al_2O_3. Because natural aluminous clays are uncommon in this country, the range of refractories of this type which are produced is limited. Synthetic mixtures of impure bauxites and clays are being used on an ever-increasing scale to produce good quality bricks. Commercial refractories corresponding to the ideal composition of *mullite* (i.e. *circa* 72 per cent) and others with about 90 per cent alumina are manufactured on a large scale. Aluminous clays are quite common in the United States and a wide range of refractories is made from them.

Alumina materials of great purity have been developed for many purposes. The raw material is bauxite and the alumina is extracted in acid solution in an autoclave, then reprecipitated and the final traces of impurity are removed by leaching. The product, which contains upwards of 99·5 per cent alumina is shaped and fired at 1750–1800° C. It is termed *sintered* or *fused* alumina and it has virtually zero porosity. The high cost of processing restricts the use of such material to small scale specialised articles, although a refractory material containing 96 per cent of alumina is being used commercially for making standard-sized bricks.

High-grade fired alumina is also used in abrasive wheels, laboratory ware, cutting tools, turbine blades, sparking plugs and for making guides which form the threads of synthetic fibres. Alumina powder has largely replaced silica sand as a setting material for china biscuit.

ALUMINO-SILICATES OTHER THAN CLAYS

AN important group of high alumina refractories is made from the three alumino-silicate minerals, *kyanite*, *andalusite* and *sillimanite*, all of which have the composition $Al_2O_3.SiO_2$, and on calcination at about 1650° C. are converted to *mullite*, $3Al_2O_3.2SiO_2$.[1]

Kyanite occurs the most frequently; large deposits, usually in massive rock form, metamorphosed by heat and pressure, are found in Kenya, South Africa, Australia, Spain, India and the United States. Pure crystals of up to 3 in. in length may be present in an interlocking aggregate contaminated with small amounts of pyrophyllite and quartz. Kyanite is also present in schistose materials, where kaolinite has been hydrothermally altered. On calcination to produce a stabilised form, kyanite undergoes a volume expansion of about 20 per cent.

Andalusite is less common, but deposits are worked for refractory purposes in Spain, South Africa and California. The principal form is a massive, metamorphosed rock, but fairly pure samples are occasionally found which have been weathered from such rocks and deposited elsewhere as silt or sand. The volume expansion on calcination is only about 4 per cent.

Sillimanite is also uncommon but large deposits are being developed in Assam, India. Unlike kyanite, sillimanite undergoes scarcely any change in volume (see p. 112), and the raw material can be used directly in furnace settings without any pre-firing. Sillimanite results from the metamorphism of pure clays; the Assam deposits are particularly pure, containing only minor amounts of quartz, biotite and cordierite, although there may be small amounts of excess alumina as

[1] Varley, E. R., *Sillimanite* (Chemical Publ. Co. Inc., N.Y., 1968).

corundum (α-Al_2O_3). The chemical composition of typical samples of the three minerals is shown in Table VI.XII, along with some characteristic properties.

Topaz, $Al_2SiO_4(F,OH)_2$, is converted into mullite when heated to 1300° C. and so can be used in the manufacture of high-alumina refractories. The chief sources are Southern Rhodesia, Brazil and South Carolina, U.S.A.

Pyrophyllite, $Al_2O_3.4SiO_2.H_2O$, is an important raw material used for refractories, electro-ceramic bodies and in the manufacture of tiles. It is not of frequent occurrence in Britain, but workable deposits have been found in the Transvaal, Japan and Australia to supplement those of Carolina, U.S.A. The mineral is thought to have been formed by the hydrothermal replacement of acidic volcanic glasses. It occurs in massive flake-like form, not unlike mica, but softer and more brittle; although in colour it is silver-grey, it is often tinged with green as a result of chlorite inclusions. Other common impurities include quartz and hydrous mica, but many more are present as minor constituents. Pyrophyllite is valued for its low shrinkage on firing—it loses its water of constitution at about 600° C. and forms mullite and cristobalite above 1100° C. An analysis of a typical sample from North Carolina shows SiO_2 77·4 per cent; Al_2O_3 16·5 per cent; Fe_2O_3 0·5 per cent; MgO 0·05 per cent; CaO 0·41 per cent; Na_2O 0·22 per cent; K_2O 1·26 per cent; and loss on ignition 2·87 per cent.

TABLE VI—XII. COMPOSITION AND PROPERTIES OF SOME ALUMINO-SILICATES

	Kyanite (*S. Africa*)	*Andalusite* (*California*)	*Sillimanite* (*Assam*)
	per cent	per cent	per cent
SiO_2	35·5	33·9	35·7
Al_2O_3	61·2	57·6	62·3
Fe_2O_3	0·9	1·6	1·7
TiO_2	1·1	1·5	0·2
CaO	0·1	1·3	Trace
MgO	0·2	0·8	0·1
Na_2O	0·1	{ —	{ Trace
K_2O	0·1	{ 0·6	{
Loss on ignition	1·2	3·5	0·7
Refractive Indices	1·713–1·728	1·623–1·645	1·657–1·684
Specific Gravity	3·6	3·1–3·2	3·08
Hardness	7·5	7·5	7·5

Alunite, $K_2O.3Al_2O_3.4SO_3.6H_2O$, is used to a small extent for the manufacture of tiles and other ceramic materials. It decomposes at a moderately high temperature to give a product rich in alumina (85 per cent) but with the remainder being virtually all alkali oxide—the shrinkage on firing is very large. It is formed as an alteration product of igneous rocks which are rich in alkalies and it occurs as seams or pockets. The principal deposits are in the Rocky Mountains of the U.S.A., Egypt and Australia.

FELSPARS, MICAS, PEGMATITES AND CORNISH STONE

FELSPARS and micas are minerals which are commonly used in the manufacture of glazes and in the preparation of vitrified bodies. Pegmatites, Aplites and Cornish Stone are natural rock formations which are rich in felspars and micas and are frequently used in place of the pure minerals.

Felspar is a constituent of pegmatites, such as granite and other igneous rocks which are predominantly alkali-bearing; it may be present in amounts up to 60 per cent. Felspar is also found in a reasonably pure state in Norway, Sweden, Russia and the United States, and it is readily purified from many rocks which contain it.

Felspars are the most valuable fluxes used in the ceramic industries, both in 'bodies' and 'glazes', particularly in the manufacture of fine earthenware, china ware, porcelain and stoneware. Orthoclase and albite are preferred, but anorthite is sometimes used; for many purposes, felspars can be replaced by the cheaper china stone.

None of the commercial felspars is quite pure and their selection requires great care. One essential is that the felspar should melt at a temperature not exceeding 1260° C. to form a smooth, glossy, vitreous mass, free from any discolorations, such as iron.

Mica is also present in igneous rocks and occurs in massive form in some localities. Muscovite mica is a valuable by-product in the production of china clay, because it is largely unaltered by the hydrothermal processes which formed kaolinite; the comparatively large plate-like grains of mica along with quartz particles are easily sedimented from the finely-divided clay mineral. The separation of the quartz is effected by sieving and screening. Mica is a valuable raw material for the electrical industry, but is rarely used as a constituent of glazes.

Both felspar and mica are rich in alkalies, as shown by the theoretical analyses for the potassium-bearing minerals (Table VI.XIII).

TABLE VI—XIII. COMPOSITION OF FELSPAR AND MICA

	Orthoclase felspar ($K_2O.Al_2O_3.6SiO_2$)	*Muscovite mica* ($K_2O.3Al_2O_3.6SiO_2.2H_2O$)
	per cent	per cent
Silica	64·8	45·3
Alumina	18·3	38·4
Potash	16·9	11·8
Ignition loss	0·0	4·5

Nepheline or **nepheline syenite** is a rock which contains mainly nephelite $(Na,K)_2O.Al_2O_3.2SiO_2$ and is used in place of felspar in pottery bodies and as a fluxing mineral in glasses. There are no workable deposits in this country, but supplies are imported from Canada and the United States where it is extensively employed.

An extensive occurrence is being exploited in North Cape, Norway; the material is of excellent quality and has become the most important source of alkali-bearing mineral for British manufacturers.

Pegmatites are coarsely-crystalline rocks containing principally felspar, mica and quartz. The common form is granite, but the proportions of the minerals present in pegmatites are very variable.

They are usually very hard rocks, having crystallised from a molten magma. They are mined in some localities, crushed and used as a foundation in making roads and for other purposes where a hard, durable material is required.

China stone may be described as a partially decomposed granite which, had the reaction proceeded further, would have produced a mixture of china clay, mica and quartz, with a little felspar. The best known samples in England are found in Cornwall and bear the names of *China stone*, *Cornish stone*, *moor stone* and *growan*. It contains china clay, felspar, quartz and unaltered granite in variable proportions, together with accessory minerals such as tourmaline, mica, topaz, beryl, zircon and fluorspar; the last-named is the source of colour of *purple stone*. It is not a definite mineral, but a heterogeneous mixture of the minerals just mentioned. The best qualities consist of felspar crystals in a quartz matrix, roughly in equal proportions, with a total alkali-content of about 7 per cent though the proportions of the various minerals vary greatly in the different grades.

Cornish stone in Cornwall is quarried, broken into lumps of convenient sizes and is then sorted into four principal grades:

(i) *Hard purple* which is the most fusible, contains the most felspar and is regarded as the most valuable grade.

(ii) *Mild purple* which contains less felspar and more china clay, and has a higher melting point.

(iii) *Dry white* contains still less felspar and more china clay and has the highest melting point. It is much softer than (i) and (ii).

(iv) *Buff stone* is slightly yellowish (due to iron compounds) and is unsuitable for white wares but satisfactory for some coloured ones.

The principal difference between the first three grades is in the degree of alteration. The kaolinite content is increased at the expense of the felspar, and there is a progressive increase in the content of secondary micas.

Typical analyses of the four grades as supplied by the manufacturers are in Table VI.XIV.

TABLE VI—XIV. CHEMICAL ANALYSES OF CORNISH STONES

	SiO_2	Al_2O_3	Fe_2O_3	TiO_2	CaO	MgO	Na_2O	K_2O	*Ignition Loss*
Hard purple	71·7	15·6	0·2	0·1	2·0	0·1	4·0	4·7	1·4
Mild purple	72·4	15·8	0·2	0·1	1·8	0·2	3·5	4·0	1·9
Dry white	73·2	16·9	0·1	0·1	1·1	0·1	2·0	3·8	2·7
Buff stone	71·9	16·5	0·5	0·2	0·9	0·2	2·2	3·8	3·8
Defluorinated	80·6	11·1	0·1	0·0	0·1	0·1	3·8	3·9	0·3

The changes in mineralogical composition in the samples can be followed by thermal analysis. The kaolinite and hydrous micas give characteristic thermal peaks. The approximate 'rational' mineralogical analysis of the above samples is shown in Table VI. XV. A more accurate method of measuring the mineral composition of Cornish stone has been described by Keeling[1] using staining and point-counter techniques of microscopy.

It is customary in best quality Cornish stone to remove most of the fluorspar by froth flotation, thus reducing the fluorine content from 1·0 per cent to less than 0·1 per cent.

TABLE VI—XV. APPROXIMATE MINERALOGICAL COMPOSITIONS OF CHINA STONES

	Quartz	*Hydrous mica*	*Kaolinite*	*Unchanged micas and felspar*	*Other minerals**
Hard purple	30·1	21·2	0·3	45	*circa* 4·0
Mild purple	28·2	23·5	2·7	41	,, 5·0
Dry white	36·4	26·0	9·6	25	,, 3·0
Buff stone	26·4	32·7	5·8	30	,, 5·0

* Other minerals include fluorspar, topaz, apatite, tourmaline, rutile, all of which are present in small amounts in most samples.

Cornish stone is an important substitute for part of the felspar in the manufacture of porcelain, chinaware, fine earthenware and stoneware. Being less fusible than the best felspars, it is less liable to cause distortion in the kiln and both Cornish stone and felspar are often used simultaneously, particularly in glazes. It is a convenient source of silica, alumina and potash in suitable proportions to ensure a good glass-former which will, in the kiln, slowly dissolve other parts of the body and greatly increase its strength and density when cold.

The *pe-tun-tse* rock used by the Ancient Chinese porcelain-makers appears to have been a kind of china stone, resembling that found in Cornwall, but prepared for use by careful grinding and meticulous sedimentation in order to produce the most uniform material obtainable. The term *pe-tun-tse* (little brick) should not be applied to the crude rock.

MAGNESIUM-BEARING MATERIALS

DEVELOPMENTS in the iron and steel industry have been largely limited by the types of refractories which have to be used to line furnaces and other containers. The necessity of producing ferrous alloys free from acid contaminants led to the introduction of non-silicic refractories which were at the same time resistant to molten basic slag. The choice of a truly basic material which has a high refractoriness and is not unduly expensive is limited to compounds of magnesium and lime.

The principal *magnesium minerals* which occur in nature are:

Magnesium carbonate or *magnesite* ($MgCO_3$), forms white, grey, yellow or brown amorphous grains or hexagonal crystals having a hardness of 3·5–4·5 and a specific gravity of 2·8–3. Two varieties occur; a spathic or coarsely crystalline form and a dense crypto-crystalline variety.

The spathic type of magnesite is found in Austria, Czechoslovakia, Russia, Washington, U.S.A. and Brazil. Compact magnesites occur mainly in Greece,

[1] Keeling, P. S., *Trans. Brit. Ceram. Soc.*, 53, 67, 1954.

Yugoslavia, Russia, Canada, the United States of America, and Manchuria, and are usually purer than the spathic type.

The principal impurity is free silica in the form of quartz, but there is always some associated lime and iron oxide, as a result of the ease with which the carbonates of magnesium, calcium and iron form solid solutions.

When the iron content is high the mineral is termed *breunnerite*, which implies a solid solution of magnesite ($MgCO_3$) with siderite ($FeCO_3$).

Magnesium hydroxide or *brucite*, $Mg(OH)_2$, deposits occur in Canada and Nevada, but require purification before they can be used for the production of refractory materials.

Sea water magnesia. An ever-increasing amount of high-purity magnesia is being manufactured from sea water. The world's oceans contain an average of 0·7 per cent magnesium in solution which can be precipitated quantitatively as the insoluble hydroxide if alkaline conditions are produced.

This is most readily achieved by treating sea water with a slurry of calcined limestone or calcinated dolomite to produce the following reaction:

(*a*) $$MgO + CaO + 2H_2O \rightarrow Mg(OH)_2 + Ca(OH)_2$$

(*b*) $$\begin{matrix} Ca(OH)_2 \\ Mg(OH)_2 \\ \text{(slaked, calcined dolomite)} \end{matrix} + \begin{matrix} MgCl_2 \\ \text{or} \\ MgSO_4 \\ \text{(in sea water)} \end{matrix} \rightarrow \begin{matrix} Mg(OH)_2 \\ Mg(OH)_2 \\ \text{(insoluble)} \end{matrix} + \begin{matrix} CaCl_2 \\ \text{or} \\ CaSO_4 \\ \text{(soluble)} \end{matrix}$$

The hydroxide of magnesium is then calcined at about 1500° C. to produce the stable oxide *periclase.*

Sea water magnesia is produced in California, U.S.A., Hartlepool, U.K., Sardinia, Italy, Japan and Eire.

The analyses of some raw materials used in the manufacture of magnesia refractories are given in Table VI.XVI.

TABLE VI—XVI. ANALYSES OF CALCINED MAGNESITES

	Austrian	*Grecian*	*Rhodesian*	*Indian*	*Russian*	*Sea water magnesites*
	per cent	per cent	per cent	per cent	per cent	per cent
Magnesia	80–90	90–95	85–95	85–95	85–95	80–90
Alumina	0–4	0–2	0–1	0–1	1–4	1–6
Ferric oxide	4–9	0–1	0–1	0–1	2–5	2–6
Lime	1–6	1–7	0–1	0–2	2–4	2–5
Silica	1–10	0–4	5–10	0–5	0–3	2–4

Dolomite or *Magnesian Limestone*, as the British deposits should be more correctly termed, is a mixed magnesio-calcium carbonate $MgCa(CO_3)_2$, which forms extensive rock formations. In its geological formation, two modes of origin are generally recognised. The primary form is the result of precipitation from solutions rich in both calcium and magnesium carbonates; such rock formations are uniform in composition over large areas and are *true dolomites.* A much more common mechanism is where solutions rich in magnesium salts have filtered through a limestone ($CaCO_3$) rock, and caused extensive alteration in it, because magnesium ions are able to replace calcium in the lattice in considerable proportions, thereby producing true crystals which may be of variable composition, but are not a simple admixture of two carbonates. Secondary magnesian limestone may vary considerably in chemical composition, for the degree of alteration depends on the availability of magnesium solutions and the ease of penetration. Ferrous iron can also enter the lattice and replace calcium and magnesium ions, although if the percentage is not too high, the value of the rock as a refractory material is not unduly influenced.

A belt of magnesian limestone outcrops in England and runs parallel to the Pennine Chain from Durham to Nottinghamshire. Similar samples occur in North and South Wales. The rocks are composed of interlocking crystals which are variable in size from a crypto-crystalline variety with particles of 0·01 mm. or less, to rocks with grains visible to the naked eye. Those portions of this rock which are rich in magnesium and low in impurities are valued for refractory purposes; less pure rocks are lightly calcined and used as lime for building purposes. So great and localised is the variation in chemical composition that refractory and inferior quality magnesian limestone are often extracted in the same quarry. Chemical analyses of some selected samples used for refractory and building purposes are shown in Table VI.XVII.

TABLE VI—XVII. ANALYSES OF MAGNESIAN LIMESTONES

	Durham	*Yorkshire*	*Derbyshire*	*North Wales*	*South Wales*
	per cent	per cent	per cent	per cent	per cent
CaO	32·0	29·5	30·3	31·7	30·2
MgO	19·7	21·0	19·8	19·0	19·5
FeO	—	0·4	—	0·1	—
Fe_2O_3	0·04	0·4	0·6	0·2	1·0
SiO_2	0·30	0·2	0·3	1·2	1·1
Al_2O_3	0·05	1·4	1·0	0·8	1·2
Loss on ignition (mainly CO_2)	46·8	46·1	47·1	46·5	46·8

Other magnesium-bearing materials include the important groups of magnesium silicates; serpentines, talc, forsterite, spinel, cordierite and vermiculite.

Serpentine is essentially a mixture of fibrous needles of *chrysotile* and *antigorite* which have the formula $3MgO.2SiO_2.2H_2O$. It is of frequent occurrence, as it is a common alteration product of rocks rich in magnesium and iron. In only a few localities, however, does it form extensive deposits free from excessive quantities of other minerals. A particularly valuable source is in the Shetland Isle of Unst, which has been called the 'green' isle because of the colour of the outcropping serpentine.

The mineral is of importance as a stabiliser of dolomite for refractory purposes and as a raw material in the manufacture of forsterite refractories.

Talc or *steatite* ($3MgO.4SiO_2.H_2O$), is of value in the manufacture of ceramic insulators of the electrical low-loss type, also of forsterite refractories and in the tile industry. It occurs in many localities although 65 per cent of the world's supply comes from Canada and the United States. Two forms are recognised, (*a*) soapstone, steatite or crypto-crystalline talc, and (*b*) true talc which is composed of much larger crystallites.[1]

Talc deposits, mined on a commercial scale, are extensive and may be up to 800 ft. thick. The *principal impurities* are carbonate rocks (magnesite, breunnerite), chlorites and quartz. Egyptian steatite is used in considerable quantities in this country and Gad[2] gives the following analysis for a talc-carbonate rock from this locality:

	per cent		per cent
SiO_2	31·04	TiO_2	0·08
MgO	35·15	CaO	0·23
Al_2O_3	0·54	K_2O	0·01
Fe_2O_3	3·02	Na_2O	0·01
FeO	4·32	Cr_2O_3	0·33

Loss on ignition 25·18 per cent

The relatively high figure for magnesia and the low content of alkalies suggest that this sample would be of value as a refractory, but in general, talc for the electrical and tile industry should have less iron and a much lower loss on ignition.

Asbestos is a form of magnesium silicate which occurs in a fibrous form. The principal types of minerals in asbestos have been detailed by Hodgson[3] who has also outlined the properties of each type. The most common form of asbestos contains fibres of *chrysotile*, $3MgO.2SiO_2.2H_2O$, but other deposits contain a true *amphibole*. These latter may be of three types: (*a*) the *tremolite-actinolite* series $(OH)_2Ca_2(Mg,Fe)_5Si_8O_{22}$; (*b*) fibrous *anthophyllite* $(OH)_2(Mg,Fe)_7Si_8O_{22}$; and (*c*) *riebeckite* or *crocidolite* $(OH)_2Na_2Fe_3{}^{2+}Fe_2{}^{3+}Si_8O_{22}$. Minerals of this last group form longer fibres than the mineral chrysotile, but, on the whole, they tend to be less refractory because of adsorbed alkali ions. The chief producing countries of asbestos are Canada, Southern Rhodesia, South Africa and Russia.

Vermiculites are essentially hydrated magnesium silicates although other cations are usually present in the mineral lattice (see p. 154). This group of minerals resemble soft mica in appearance, but exfoliate when heated. The expanded mass is of value as a thermal insulator and is becoming important as a light-weight filler. Vermiculite is frequently present in small amounts in soil clays, but extensive workable deposits are limited. The principal source is in Montana, U.S.A., but potentially good formations have been reported as existing in Tanganyika, Africa, Australia and India.

Olivine is a naturally occurring magnesium-iron orthosilicate $(Mg,Fe)_2SiO_4$ containing varying amounts of magnesium and iron. It is a constituent of some igneous rocks, e.g. peridotite and dunite, which is almost a pure olivine; it is also a major component in some basalts and dolerites.

[1] Steatite is dehydrated at a lower temperature than talc; it also forms clino-enstatite (p. 763) more rapidly.
[2] Gad, G. M., *Trans. Brit. Ceram. Soc.*, 51, 429, 1952.
[3] Hodgson, A. A., R.I.C. *Lecture Series*, No. 4. 1965.

Olivine rocks are found in Cornwall, the Shetland Isles, Northern Ireland and many localities abroad, but it is usually extensively altered to serpentine.

Those minerals which contain less than 5 per cent of ferric oxide are used in the preparation of *forsterite* refractories. Forsterite is the magnesium end-member of the olivine series but it rarely occurs in the pure state in nature. Serpentine and talc are also used in the manufacture of refractories, but, unlike olivine, they shrink considerably on firing, even after being pre-calcined. They are, however, usually much purer than natural olivines especially with regard to their iron content.

Cordierite, $2MgO.2Al_2O_3.5SiO_2$, is rare as a natural mineral, but artificial products of the same composition are of great commercial interest because their coefficient of expansion is less than that of vitreous silica and they possess great thermal endurance. Artificial cordierite is made by heating a mixture of china clay or fireclay with magnesite, steatite or olivine. The iron content should be kept as low as possible.

Spinels form a group of mixed oxides of which the mineral spinel $MgO.Al_2O_3$ is typical. They are described in detail in the section on chrome-ores and their structure has been considered in Chapter IV.

Most magnesium compounds decompose on heating and either combine with other substances of an acidic character (e.g. $2MgO + SiO_2 = Mg_2SiO_4$) or crystallise as *periclase MgO*.

CALCIUM-BEARING MINERALS

CALCIUM carbonate, $CaCO_3$, is a widely distributed rock formation for it is the essential component of chalk, limestone and marble. In addition, calcium carbonate occurs as Iceland Spar, in crystals of calcite and aragonite and as shells and allied organic structures.

Pure calcareous limestone is used to a very small extent as a refractory material —because of the instability of the calcined product (lime)—although a method of stabilising it has been described by Williams.[1] Lime is, however, an important base in the manufacture of glass, sand-lime bricks and mortar; chalk is used in the manufacture of some kinds of bricks and pottery, and whiting (finely-ground calcium carbonate) is an important ingredient in some glazes.

Calcium sulphate (*gypsum*, *anhydrite* and *selenite*) occurs in massive form near Newark. Gypsum is chiefly of importance as the source of plaster of Paris, from which are made the *plaster moulds* so extensively used in pottery manufacture. The gypsum ($CaSO_42H_2O$) is heated to a temperature of about 107° C. with constant stirring until it 'boils' and evolves some of the combined water present, forming plaster ($2CaSO_4H_2O$ or $CaSO_4\frac{1}{2}H_2O$). This material is ground to a 'superfine' powder and is then ready for use.

CHROME ORES

CHROME ores have been used in the manufacture of neutral fire-bricks to separate courses of basic and acidic refractories in metal-melting furnaces. The development and increasing use of chrome-magnesite bricks has drawn attention to

[1] Williams, A. E., *Trans. Brit. Ceram. Soc.*, **50**, 215, 1951.

the composition of these rocks and they are now important refractory materials. The mineral chromite, $FeCr_2O_4$, is the only significant source which forms lenses or bands which are either solidified intrusions or a highly crystallised, compacted sediment. The principal impurities are iron and titanium minerals, but the chromite grains are often of very variable composition due to solid solution and may contain appreciable amounts of aluminium and magnesium. Chrome ores occur in massive form in many localities; and the main chemical ingredients of suitable rocks are:

Chromic oxide	Cr_2O_3	between	33–45 per cent
Aluminium oxide	Al_2O_3	,,	14–27 per cent
Magnesium oxide	MgO	,,	10–20 per cent
Iron oxide	FeO and Fe_2O_3	,,	10–15 per cent
Silica	SiO_2	less than	7 per cent
Calcium oxide	CaO	,,	1·5 per cent

Despite the apparent complexity of their composition, chrome ores contain one principal mineral type, *spinel* and only small amounts of impurities.

The compositions and types of various spinels have been described in Chapter IV (p. 192), where it was shown that a complex system of solid solutions exists with the general formula $R^{2+}O.R_2{}^{3+}O_3$ where R^{2+} is any divalent cation, such as Mg or Fe^{2+} but not Ca, and R^{3+} is a trivalent cation such as Al, Cr, or Fe^{3+}.

The spinel grains which occur sometimes as octahedral crystals, but more often in massive form with a granular and compact structure, are usually brownish in colour due to the presence of iron. They are brittle with a hardness of 5·5, a specific gravity of 4·3–4·5 and a high melting point.

The *impurities* are present in the form of gangue which holds the spinel grains together. These impurities include *quartz*, *pyroxenes* and *felspars*, but are chiefly *serpentine*, $3MgO.2SiO_2.2H_2O$.

The principal sources of chrome ore are Rhodesia, Finland, India, Albania, Turkey, Greece, Yugoslavia, Russia and the Philippine Islands.

A material with a low porosity and large unfissured grains of spinel is preferred for refractory purposes.

TITANIUM-BEARING MINERALS

ALTHOUGH a small percentage of titanium compounds is invariably present in most clays and sedimentary rocks, there are only a few known sources of rich titanium-bearing rocks.

The chief titanium minerals are as follows:

Rutile, TiO_2, is a titanium oxide occurring as reddish-brown, yellowish or black tetragonal crystals, with a hardness of 6–6·5 and a specific gravity of 4·2. As it is practically indestructible, it is one of the commonest heavy detrital minerals in clays and is of very wide occurrence. It also occurs in massive form. It is the stablest form of titania and is readily formed from other oxides on heating.

Brookite is a titanium oxide of more limited occurrence; it is found as brown, red, or blackish orthorhombic crystals, with a hardness of 5·5–6 and a specific gravity of 4. It is almost as durable as rutile and has been isolated on a commercial scale from Arkansas red clays.

Anatase is a titanium oxide occurring as slender brown, blue or black tetragonal pyramids, with a hardness of 5·5–6 and a specific gravity of 3·83–3·95. It is of somewhat limited occurrence, although Brindley[1] has reported that it is the principal variety of titania in British fireclays.

Ilmenite, $FeO.TiO_2$, is a very common detrital mineral, consisting of a mixed oxide of iron and titanium; it occurs as black tetragonal crystals, with a hardness of 5–6 and a specific gravity of 4·5–5·0. In the massive form it is one of the principal sources of titanium oxide for use in refractory materials.

Sphene, $CaTiOSiO_4$, is a calcium titanite and silicate which occurs as brown, green, yellow, or black monoclinic crystals, having a hardness of 5–5·5 and a specific gravity of 3·54. It also occurs in massive form.

Leucoxene is a variety of sphene produced by the alteration of ilmenite and other titanium-bearing minerals.

Titania is the principal ingredient of bodies when a low dielectric loss is required, as in radio work; it is also used to some extent as a refractory, as a high temperature lute or cement and as an opacifier in glasses and glazes.

Titania is extracted mostly from naturally-occurring sands, found principally in India and Australia, where it occurs as ilmenite, or rutile. The minerals may be purified by heavy medium separation followed by electrostatic and electromagnetic methods. Glaze quality is chemically prepared.

ZIRCONIUM-BEARING MINERALS

ZIRCONIUM occurs in two forms in nature, as:

(*a*) the oxide ZrO_2 (*baddeleyite*) which is found principally in Brazil in the massive form or as concretionary pebbles or nodules. It forms monoclinic crystals, unstable at high temperature, with a melting point of about 2700° C., a high refractive index (*circa* 2·15), and a specific gravity of 5·7–6·0.

(*b*) The silicate, *zircon* ($ZrSiO_4$), which occurs as grey, yellow, green or reddish-brown tetragonal crystals with a hardness of 7·5, a specific gravity of 4·7, and a melting point of about 2550° C. in detrital deposits, principally monazite sands.

The entire British supplies of zircon come from the monazite beach sands in Travancore, India or Australia. The sands contain only small amounts of zircon, but when purified yield a fine-grained sample up to 99·5 per cent pure. The chief impurities are titanium compounds, quartz and felspar.

Zircon is used as a refractory material; it is very resistant to the attack of coal ash and molten metals at high temperatures. Although it begins to dissociate at about 1600° C. into zirconia and free silica, it can be used at considerably higher temperatures without serious deterioration.

Monazite sands are also important sources of other ceramic oxides; monazite itself is a phosphate of cerium, lanthanum, ytterbium with thoria, ThO_2. The various components can be purified by heavy medium separation, and other techniques of a chemical nature.

[1] Brindley, G. W., loc. cit., p. 281.

BERYLLIUM-BEARING MINERALS

Beryllia, BeO, has many advantages as a refractory material and is being used on an increasing scale in the United States. Largely because it is considered a health hazard, it has had only limited application in the U.K.

The principal source is beryl ($3BeO.Al_2O_3.6SiO_2$), which is associated with the massive felspars in Brazil, India and the United States. The natural mineral contains 10–12 per cent of BeO, so that purification is necessary; the mineral is first concentrated by froth-flotation methods and after the ore has been partially decomposed by heating it above 1000° C. with lime, the beryllia is extracted by hot sulphuric acid.

Beryllia is highly refractory (M.P. 2570° C.), yet it readily sinters at about 1300° C. to form a mass of low porosity with excellent resistance to thermal shock and with high electrical resistance at elevated temperatures. It is chiefly used for laboratory crucibles and gas-turbine blades, but it has great potential value as a super-refractory and it also confers desirable properties when it is a component of glazes and porcelains.

CARBON

ALTHOUGH most of the carbon used in the ceramic industries is now manufactured, naturally-occurring *graphites* are still highly valued for special applications.

Graphite probably originates through the metamorphism of carbonaceous material of sedimentary origin and is frequently found in coal-bearing regions where there has been volcanic intrusion. There is a small amount of graphite mined in Cumberland, but the principal sources are in Ceylon, Norway, Korea, Austria, Mexico and the United States. The quality of the graphite is of major importance and is governed principally by the size of the flakes. Ceylon graphite with large flakes is eminently satisfactory for the manufacture of crucibles and some other refractory shapes.

For most purposes, however, a synthetic form of graphite is nowadays preferred, and this is prepared by heating coke of low ash-content mixed with tar or pitch, in the absence of oxygen. The temperature of firing determines the texture of the product and the size of the crystallites. Graphite for carbon refractories is treated at about 1000° C., but super-quality for the manufacture of electrodes is prepared in an electric arc.

Diamond, in addition to being a valuable gem-stone, is commonly employed as an abrasive. It is found in ultra-basic rocks and it has been suggested that the molten magma acted on enclosed fragments of carbonaceous materials. Gem diamonds are found principally in South and Central Africa, but the most important source of abrasive material is *bort*, which is a compacted variety of diamond occurring in Brazil, South Africa and the Congo.

Silicon Carbide, SiC, is also an extremely hard material, but it does not occur naturally. It is made by fusing a mixture of sand and coke in an electric arc furnace and selecting the crystalline product from the mass. Silicon carbide is also a valuable refractory material and it has a high thermal conductivity.

BORON COMPOUNDS

THE chief boron compounds used in the ceramic industries are boric acid, borax and calcium borate.

Boric acid, H_3BO_3, is used only when it is not desirable to introduce alkalies into a body, in which boron is a necessary constituent.

Borax, $Na_2B_4O_7.10H_2O$—hydrated borax. On heating this to about 62° C. it loses half its water and forms $Na_2B_4O_7.5H_2O$ which, on further heating to 130° C. gives $Na_2B_4O_73H_2O$; at 180° C. $Na_2B_4O_7H_2O$ is produced, which changes to anhydrous borax $Na_2B_4O_7$ at 318° C. Borax occurs chiefly as the mineral *rasorite*, $Na_2O.2B_2O_3.4H_2O$, but it can also be obtained by mixing *colemanite*, $2CaO.3B_2O_35H_2O$, with sodium carbonate, boiling the solution and, after filtration, allowing the borax to crystallise.

Borax Consolidated Ltd. (Vitreous Enamels) have published the following typical analyses:

TABLE VI—XVIII. CHEMICAL COMPOSITION OF SOME BORATES

	Rasorite Concentrates	*Anhydrous Rasorite*	*Anhydrous Borax*	*Hydrated Borax* ($10H_2O$)
B_2O_3	44·5	62·2	68·5	36·5
Na_2O	20·1	28·3	30·4	16·2
H_2O	25·9	1·0	1·0	47·2
SiO_2	5·1	4·4	0·04	0·02
Al_2O_3	1·2	1·3	0·03	—
Fe_2O_3	0·3	0·4	0·01	—
SO_3	0·1	0·3	0·02	0·06
CaO	1·4	0·7	—	—
MgO	0·8	1·4	—	—
CO_2	0·6	—	—	—

Borates are formed in nature mainly as (i) deposits from waters in the neighbourhood of mineral springs or other volcanic emanation—*sassoline* (boric acid) may form under these conditions, or (ii) residues from the evaporation of lake waters rich in borates.

The principal sources of borates are in California, U.S.A., Turkey, South America and Asia Minor. Sassoline is found in small amounts in the volcanic regions of Northern Italy.

Boron compounds are used for a great variety of purposes, but their chief value as a ceramic material is as a constituent of glazes.

Boron Nitride can be synthesised into forms which resemble diamond or graphite. It is a valuable mineral in that it has a very high melting point and is amongst the best low-loss dielectric ceramic materials.

RAW MATERIALS USED IN THE MANUFACTURE OF CERAMIC BODIES

THE substance of which bricks, tiles, pottery, porcelain and other ceramic articles are composed is known as the *body* and its composition and treatment require careful consideration. Different kinds of articles require to be made from different raw materials or to be treated in a different manner. The more important bodies are as follows:

EARTHENWARE

Earthenware, as its name implies is made of earth, and of the various materials available, *clays* are the most abundant and, for many purposes, the most suitable. The term earthenware is, therefore, applied to any articles made chiefly of clay—either alone or after being mixed with some other materials. Unlike china-ware, true porcelain, stoneware and other dense wares, earthenwares have a porous body, which is usually covered with a transparent or translucent glaze.

Coarse earthenware or *Brown-ware* (including *Terra-cotta*) is made of a crude clay which may contain 50 per cent or more of sand or chalk and up to about 10 per cent of iron compounds together with small amounts of accessory minerals. Some clays can be used without any preliminary purification but others need to be washed and any excess of sand or gravel removed. For common bricks and roofing tiles, little blending is required with some clays but others must have their objectionable properties modified by the addition of sand, chalk or other clays. For hollow-ware (bread-pans, wash-basins, flower-pots, etc.), a carefully selected clay or blend of clays is used, including one or more which have previously been washed or finely ground. The crude clays used for the manufacture of coarse earthenware generally contain sufficient fluxing agents (alkali and similar impurities) to produce fired ware which will ring well when struck and will be durable under ordinary conditions of use.

Many articles of coarse earthenware are *unglazed*, but for certain purposes, a coating is applied either to the exterior or the interior of the article (or to both) as may be required.

Faience is intermediate between coarse and fine earthenware and is often made by applying a coating of white- or cream-coloured *engobe* (see p. 341) or *slip* to coarse earthenware. A special variety, known as *majolica ware* has a dense opaque coating of glaze, which completely hides the coloured ware to which it is applied, though sometimes portions of the latter are exposed as a means of decoration.

Fine earthenware, when fired, has a white or almost white body and is made of clays which are white when burned. The clays for the best, fine, white earthenware are usually *china clay* and *ball clay*, the former for the texture and translucency it imparts and the latter to give strength to the unfired shapes. *Flint* is added to reduce the shrinkage and to increase the porosity and the silica content. A *flux*, such as *china stone* or *felspar* is also introduced, which forms a glass in the fired body and therefore imparts strength, density and the power to ring when struck. In recent years *talc*, *pyrophyllite* and various other silicates have been included in earthenware bodies used for glazed wall- and floor-tiles.

For a slightly inferior fine earthenware, a mixture of *red-burning clay* and *chalk* is sometimes used. The chalk combines with the iron compounds in the clay and, if skilfully proportioned and fired, it produces a white body. This kind of hollow ware is seldom made in England, but the process is used in the manufacture of some white bricks.

Various blue colours may also be used to destroy the yellow colour of some burned clay and in this way some otherwise unsuitable clays may be utilised for fine earthenware manufacture.

The average composition of the raw materials in British fine earthenware bodies to be fired between 1100° C. and 1160° C. is:

	Range	*Average content*
Ball clay	20–45 per cent	25 per cent
China clay	25–40 per cent	25 per cent
Flint	20–35 per cent	35 per cent
China stone or felspar	7–15 per cent	15 per cent

Sanitary earthenware is of two kinds:

(i) That with a porous body, sometimes covered with a white or coloured engobe and a glaze, or with a salt glaze. This is a true earthenware.

(ii) That with a vitreous body like stoneware but not so dense. It is usually made of *china clay*, *ball clay* and *flint* with about 15 per cent of *Cornish stone* and 10–15 per cent of *felspar* (see *Stoneware*, p. 340).

Chinaware and Porcelains

Chinaware and Porcelains are made of mixtures of various white-burning clays and fluxes of such a nature and in such proportions as will produce a dense, vitreous and translucent body. The temperature of firing of these bodies is usually much higher than for earthenware, being in the range 1260–1400° C. The term *porcelain* includes a great variety of wares all of which are translucent and are, in this way, distinguished from stoneware. What is usually termed *true* or *hard porcelain* contains a large proportion of clay and forms one end of a series, the other end being *soft porcelain*, which consists chiefly of an opaque glass and may contain very little clay. *Bone china* differs from other porcelains in containing bone ash, and is almost exclusively made in England. It is true porcelain though not usually regarded as such. The paste used for bone china bodies is difficult to shape because of its low plasticity and great care is required.

Chinaware is made of *china clay*, *ball clay*, *flint*, *Cornish stone*, *bone ash*, *felspar* and other fluxes. It owes its translucency largely to the formation of a glassy binding material formed by the combination of the fluxes and the silica. Such ware usually varies in composition between the following limits:

$$1{\cdot}15 \text{ to } 8{\cdot}33\ (R_2O + RO + Al_2O_3);\ 1{\cdot}97 \text{ to } 9{\cdot}08\ SiO_2;\ 0{\cdot}35 \text{ to } 2{\cdot}67\ P_2O_5$$

True (or *Ideal*) *porcelain* should consist of a mixture corresponding to a core or skeleton of matted crystals of *mullite* ($3Al_2O_3 . 2SiO_2$), enclosed in a fusible glass rich in silica, which binds the other particles together.

There are at least six distinct classifications of porcelain with many intermediates and the clays used for one type are not necessarily suitable for another.
(1) **Hard-paste porcelain,** including *porcelaine dure*, *Halle porcelain*, and *Berlin porcelain*,[1] corresponds approximately to the composition

$$0{\cdot}2 \text{ to } 0{\cdot}3\ (R_2O+RO+R_2O_3);\ 4{\cdot}2 \text{ to } 4{\cdot}8\ SiO_2$$

It is made of *kaolin* (china clay), *orthoclase felspar* and *flint* or *quartz*. Hard porcelains are resistant to sudden changes in temperature and to most acids.
(2) **Soft porcelain** or *porcelaine tendre*, corresponding to the formula

$$0{\cdot}4 \text{ to } 0{\cdot}45\ (R_2O+RO+R_2O_3);\ 4{\cdot}8 \text{ to } 5{\cdot}3\ SiO_2$$

This material is really a complex glass and not a true porcelain. The soft porcelains are so rich in alkalies that a considerable portion of the raw material is *fritted*, i.e. pre-melted and then crushed, before being mixed with the rest. Because of the excessive amount of liquid formation at the temperature of firing, serious distortion of the ware may occur unless close control is maintained. The manufacture of soft porcelain in several well-known factories was never permanently profitable; the 'loss on firing' was so great. This product was developed in France where it became known as *porcelaine de Reaumur* and is still manufactured in the Limoges area.
(3) **Sèvres porcelain** is made essentially of a mixture of clays and fluxes which are intended to form a special kind of glass, the chief materials being *white-burning kaolins*, *felspar* and *quartz*. Sèvres porcelain corresponds approximately to the formula

$$0{\cdot}30 \text{ to } 0{\cdot}35\ (R_2O+RO+R_2O_3);\ 2{\cdot}8 \text{ to } 3{\cdot}5\ SiO_2$$

The *Meissen* and *Viennese* porcelains are similar in composition to that of Sèvres.
(4) **Chinese** and **Japanese porcelains** vary so greatly in composition and are so complex in character owing to the very impure materials used, that they can be scarcely said to correspond to any formula, though they are chiefly within the following limits:

Japanese	$0{\cdot}3$ to $0{\cdot}4\ (R_2O+RO+Al_2O_3);\ 6{\cdot}2$ to $7{\cdot}4\ SiO_2$
Chinese	$0{\cdot}4$ to $0{\cdot}45\ (R_2O+RO+Al_2O_3);\ 5{\cdot}5$ to $6\ SiO_2$
Sèvres imitation of Oriental porcelain	$0{\cdot}37\ (R_2O+RO+Al_2O_3);\ 5{\cdot}15\ SiO_2$
Seger's imitation of Oriental porcelain	$0{\cdot}36\ (R_2O+RO+Al_2O_3);\ 8{\cdot}55\ SiO_2$

They are made of *kaolins*, *pe-tun-tse* (felspar) and crushed *silica rock* (quartz), but these were, originally, so very impure that porcelains of varying properties were produced—some having so slight a translucency as to be more correctly described as stoneware.

One of the remarkable features of the older Chinese porcelains is the manner in which very crude and impure materials have been used to produce a ware of extraordinary technical and artistic quality. The very serious complexities thus introduced add greatly to the aesthetic value of such ware.

[1] Most chemical and laboratory porcelain resembles Berlin porcelain and is based on the same formula with slight modifications due to the composition of the materials.

(5) **Dental porcelains** vary greatly in composition, some being highly refractory and others very fusible. The following typical formulae are due to A. S. Watts:[1]

$$(a)\quad \left\{\begin{matrix} 0{\cdot}74\ K_2O \\ 0{\cdot}26\ Na_2O \end{matrix}\right\} \ 1{\cdot}14\ Al_2O_3 \ \left\{\begin{matrix} 7{\cdot}830\ SiO_2 \\ 0{\cdot}002\ TiO_2 \end{matrix}\right.$$

$$(b)\quad \left\{\begin{matrix} 0{\cdot}638\ K_2O \\ 0{\cdot}141\ Na_2O \\ 0{\cdot}221\ CaO \end{matrix}\right\} \ \begin{matrix} 0{\cdot}565\ Al_2O_3 \\ \\ 0{\cdot}090\ Fe_2O_3 \end{matrix} \ \left\{\begin{matrix} 5{\cdot}90\ SiO_2 \\ 0{\cdot}023\ B_2O_3 \\ 0{\cdot}011\ TiO_2 \end{matrix}\right.$$

$$(c)\quad \left\{\begin{matrix} 0{\cdot}295\ K_2O \\ 0{\cdot}650\ Na_2O \\ 0{\cdot}055\ CaO \end{matrix}\right\} \ 0{\cdot}086\ Al_2O_3 \ \left\{\begin{matrix} 4{\cdot}544\ SiO_2 \\ 0{\cdot}706\ B_2O_3 \end{matrix}\right.$$

The materials used for dental porcelains are *china clay* or *kaolin*, *felspar* (or some equivalent mineral such as china stone or pegmatite), and *free silica* in the form of crushed quartz or calcined flint.

(6) **Electrical porcelains** are chiefly used as insulators and other parts of electrical appliances. Up to a short time ago, these were made exclusively as are true porcelains, but in recent years, other materials have been introduced which have enormously improved the electrical properties. Electrical porcelains include:

(i) *True* or *hard porcelains* (p. 335) made from *kaolins*, *flint* and *fluxes*.

In present-day practice, it is usual for porcelains to be manufactured from fine quality raw materials and to contain a high proportion of alumina. Cristobalite porcelains for electrical purposes have also been developed. The essential requirement of modern electrical porcelain for insulators, spark-plugs, etc., is high tensile strength (see also p. 964).

(ii) *Cordierite bodies* which are essentially fired mixtures of *clay* (china clay, ball clay or fireclay) and *talc*, which is a hydrated magnesium silicate; *alumina* may also be added to ensure that the chief fired ingredient will be *cordierite*, $2MgO.2Al_2O_3.5SiO_2$. Cordierite bodies have an extremely low coefficient of thermal expansion. The ideal composition is 70 per cent talc with 30 per cent of china clay, although in practice it is usual to work with a slight excess of one of these or to add a little *felspar* in order to increase the vitrification range. *Zircon*, $ZrSiO_4$, is sometimes added to improve the qualities of cordierite bodies.

(iii) *Steatite bodies* are made of *talc* or *steatite* to which a small amount of *clay* has been added as a bond. Amounts of *felspar* and *alkaline earth carbonates* are introduced, principally to produce a vitrified body at the firing temperature of about 1400° C. The principal mineral after firing is *clino-enstatite*, $MgO.SiO_2$, and such bodies are excellent electrical insulators and of value as low loss ceramics. Their range of composition is:

Talc	60–90 per cent
Felspar	5–8 per cent
Clay	5–20 per cent
Magnesium barium or calcium carbonate	7–20 per cent

[1] Watts, A. S., *Trans. Amer. Ceram. Soc.*, 17, 1915.

(iv) *Zircon porcelains* are used as high frequency electrical insulators and as they easily produce a dense, non-porous body, they have, to some extent, superseded cordierite bodies. *Zirconium silicate* is the principal ingredient, with *ball clay* being added as a bond. A proportion of *lime* is also incorporated. A typical composition is:

Zircon $ZrSiO_4$ 75 per cent; Ball clay 13 per cent; Lime 12 per cent

(v) *Rutile bodies* based on titanium oxide, are also used for low-loss electrical insulators. Some *zirconia* is usually added to such bodies to improve their properties. Titania and titanate bodies have largely replaced mica in condensers. They vary greatly in composition, but are chiefly of two kinds: (*a*) those composed mainly of *titania*, TiO_2, with 8–10 per cent of *bentonite* or *ball clay* added as a bond; and (*b*) those composed of *magnesium*, *calcium* or *barium titanate* either with or without a *clay* bond. This group of bodies has very high dielectric constants and low power factors (see p. 963).

For all kinds of porcelain, the materials must be very finely divided and extremely well mixed so as to ensure the steady and regular progress of the chemical reactions between the constituents and the uniform progress of the physical changes which take place when they are heated; otherwise, distortion of the ware will occur. The materials must be sufficiently uniform in composition to enable successive batches to produce ware having the same characteristics, though some adjustment is possible by varying the proportions of each material in the mixture. The materials must be sufficiently free from iron and other discolouring impurities to enable white ware to be produced. Slight differences in the materials from various sources, as well as slight variations in the proportions in which they are used, largely account for the differences between specimens of porcelain and china ware from different factories.

The chief purpose of the clay and some of the free silica in all porcelains is to produce the 'felted' mass of *mullite* crystals which form the 'skeleton' of the finished ware and to provide a 'reinforcement' which will prevent undue distortion of the ware. The other ingredients have two purposes: (*a*) part of them commence to fuse at a comparatively early stage in the firing and so produce a solvent for the remainder; and (*b*) these ingredients, together with the clay and free silica, combine to form a glassy fluid—which must be produced slowly and progressively to avoid loss of shape in the ware—from which the *mullite* may crystallise, the remaining material forming the vitrified magma to which—with the *mullite*—porcelain owes its peculiar characteristics.

The nature of the fluxes in porcelains, etc., is very important, as upon them the nature of the finished products largely depends. It is almost impossible to produce good porcelain ware from a mixture of the various oxides—silica, alumina, potash and soda—as the reactions between them take place too rapidly and serious distortion occurs. If these oxides are used in various states of combination as clay, felspar, etc., and great attention paid to the grain size of the components, the reactions proceed at suitable rates and enable satisfactory results to be obtained repeatedly with hard porcelain, but with much greater risk of loss when soft porcelain wares are made.

Soda and *potash*—preferably in the form of felspar or Cornish stone—are the best fluxes for porcelain.

Table VI.XIX shows the chemical composition, and Table VI.XX the mixtures used for various porcelains.

The *Bone ash* used in the manufacture of bone china is prepared by calcining the bones of oxen, and consists essentially of tricalcium phosphate. The addition of bone ash to clays imparts translucency, even in porous bodies which have not been fired to the vitrification point.

The composition of porcelain depends largely on the temperature to which it is fired. For example, 28 per cent of felspar is sufficient if the body has to be fired at 1400° C., but this has to be increased to 35 per cent at the expense of other ingredients if the firing temperature is reduced to 1300° C.

TABLE VI—XIX. CHEMICAL COMPOSITION OF PORCELAINS

Porcelain	*Silica*	*Alumina*	*Ferric oxide*	*Potash*	*Soda*	*Lime*	*Magnesia*	*Titanium oxide*	*Borax*	*Loss on ignition*
Sèvres (1848)	59·2	35·2	—	3·2	—	3·2	—	—	—	—
Dental A	68·2	16·7	trace	10·1	2·3	—	—	0·23	—	2·5
Dental B	68·1	2·2	—	6·9	10·1	0·8	—	trace	10·6	1·2
Dental C	69·6	11·3	0·3	11·8	1·7	2·4	—	0·2	0·3	—
Meissen	58·5	35·1	0·8	5·0	—	0·3	0·6	—	—	—
Vienna	59·6	34·2	0·8	2·0	—	1·7	1·4	—	—	—
Berlin	64·3	29·0	0·6	3·6	2·7	0·3	0·5	—	—	—
Chinese	70·0	22·2	1·3	3·6	2·7	0·8	—	—	—	—
Marquardt's	61·6	30·0	1·56	3·26	—	3·56	—	—	—	—

Parian ware, though usually included as a porcelain is, strictly, a very fine stoneware as the articles made of it are vitreous, but not translucent. The Parian ware originated by Copeland and Garret of Stoke-on-Trent about 1845 was composed of one part of *china clay* and two parts of *felspar*. The French Parian ware is largely composed of *frit*, *kaolin* and *felspar* and sometimes contains *flint glass*.

Stoneware

Stoneware resembles porcelain in texture, being vitreous and glassy, but *plastic clays* are used in its manufacture instead of kaolins. Consequently, stoneware bodies are much easier to shape by moulding than are porcelain ones; they are also less fragile and less brittle. The chief difference between stoneware and porcelain is the translucency of the latter, although porcelains vary in themselves in their degree of translucency. Many articles described as fine stoneware are really non-translucent, under-fired porcelains, but are not regarded as such because of their colour and general appearance.

Chemical stoneware has a particularly high resistance to many chemicals and is of great industrial and technical importance.

Sometimes the term stoneware is, wrongly, applied to *glazed earthenware* used for sanitary purposes such as drain-pipes, gullies and traps. This differs from stone-

TABLE VI—XX. MIXTURES USED FOR VARIOUS PORCELAINS

Porcelain	*Sèvres*	*Berlin*	*Chinese*	*Seger*	*Bone China*			*Dental*			*American China**		*Chemical†*	*Parian Ware*		
					A	*B*	*C*	*A*	*B*	*C*	*Hotel*	*House-hold*		*A*	*B*	*C*
Kaolin	38	77	47	13·0	33–35	26	30	4	—	—	35	24	} 80	33	16	40
Ball clay	—	—	—	15·5	—	—	—	—	—	—	7	28		—	—	—
Felspar	38	23	15	30·0	15–19	—	—	81	61	12	22	13	10	67	33	58
Cornish stone	—	—	—	—	—	30	35	—	—	—	—	—	—	—	—	—
Flint	—	—	—	—	10–14	—	—	15	29	60	35	35	—	—	—	—
Quartz	24	—	38	41·5	—	—	—	—	—	—	—	—	10	—	51	2
Bone ash	—	—	—	—	32–42	44	35	—	—	—	—	—	—	—	—	—
Sodium carbonate	—	—	—	—	—	—	—	—	2	8	—	—	—	—	—	—
Borax	—	—	—	—	—	—	—	—	1	11	—	—	—	—	—	—
Calcium carbonate	—	—	—	—	—	—	—	—	5	1	1	—	—	—	—	—
Potassium carbonate	—	—	—	—	—	—	—	—	2	8	—	—	—	—	—	—

* American china ware is of slightly different composition from that of British ware. More alkali in the form of felspar or nepheline syenite is used along with a little lime or magnesia.

† Most chemical and laboratory porcelain resembles Berlin porcelain and is based on the same formula with slight modifications due to the composition of the materials.

ware in having a porous body and relies for its impermeability solely on the glaze, whereas true stoneware has an impermeable body with or without a glaze.

One of the most interesting kinds of stoneware is Josiah Wedgwood's *Jasper ware* in which barium sulphate (barytes) is the chief flux: the colour is due to chromium oxide (green ware) or to cobalt oxide (blue jasper). This ware is not glazed but has a slightly glossy surface.

The chief requirement of stoneware in general use is its impermeability to liquids. Its colour is usually of minor importance and this enables cheaper and less pure clays to be used in its production; but, for fine stoneware, white-burning clays, such as *plastic kaolins* and *siliceous ball clays* must be used, though occasionally a mixture of fireclay and china clay is quite satisfactory, particularly for some vitreous sanitary ware.

Stoneware may, therefore, be regarded as similar to earthenware which has more natural fluxes or to which additional fluxes have been added, to produce a dense, vitrified body when fired at a suitable temperature.

Coarse stoneware is chiefly made of bastard (siliceous) fireclays which contain sufficient fluxes to need no further addition; in some cases, a small proportion of a suitable flux (usually chalk but sometimes magnesia or a cheap felspar) is added.

Fine stoneware and **chemical stoneware** is made of kaolin, ball clay or one or other of these with a coarser refractory clay (e.g. fireclay), sufficient china stone or felspar and flint or fine sand. These are added in proportions sufficient to produce the kind of body required.

When a stoneware is required to have a high resistance to sudden changes in temperature (e.g. some chemical ware) it is usual to include a moderate proportion of *talc*, *mica* or *magnesia* to act as a flux, reduce thermal expansion and to impart thermal stability.

The outstanding advantages of stoneware over porcelains are: (i) ease in moulding; (ii) toughness of the fired ware; (iii) less loss (rejects) in manufacture; (iv) cheaper raw materials, especially clays; and (v) smaller total cost. These advantages also apply to many technical porcelains which, as previously stated (p. 335), are better described as stoneware than as porcelain.

Coloured bodies may be produced:

(i) By using clays which, when suitably fired, have the desired colours, e.g. red-burning clays for terra-cotta; these usually owe their colour to the presence of iron-bearing minerals such as haematite.

(ii) By adding a suitable colouring agent (or stain), such as those used for engobes and glazes (p. 352).

(iii) By firing under reducing conditions which reduce ferric compounds to the ferrous state and produce dark blue, purple or very dark brown coloured substances.

The addition of a colouring agent to the whole of the body is usually too expensive and, where practicable, it is cheaper and, in many cases, equally satisfactory to apply a coloured engobe, glaze or veneer to those parts of the article which are visible when it is in use. Such examples are sand-faced or veneered building bricks where the surfaces to be exposed are covered with a coloured sand or clay; coloured, glazed sanitary ware and glazed fireclay drain-pipes are also of this type. The colouring agents are chiefly cheap oxides which, in the kiln atmosphere, produce a pleasing appearance on the fired body.

Engobes

Engobes are intermediate in composition between clay bodies and glazes. They are vitreous but less glassy than glazes and are used principally:

(*a*) To hide the colour of the body and to prevent the top glaze, which is much more fusible, dissolving some of the coloured body and thus becoming coloured itself.

(*b*) As a material with a thermal expansion intermediate between that of the body and the glaze. A highly desirable glaze which would not otherwise adhere to the body may be used if one or more layers of engobe are fired on to the body previously.

Engobes vary considerably in composition but usually consist of *china clay*, *ball clay*, *flint*, *Cornish stone* or *felspar* and *whiting* (crushed *calcium carbonate*). The thermal expansion can be adjusted between wide limits by varying the proportions of these constituents. Coloured engobes are produced by adding suitable oxides to the mixture (see p. 352).

Glazes

Glazes are applied to many kinds of clay ware to give a glossy finish which seals the body and at the same time provides a smooth and matt or a brilliant surface. They resemble glasses in structure and texture, but their composition is adjusted to ensure a suitable but not complete amount of vitrification and a more viscous fluid. Glazes must also adhere closely, strongly and uniformly to the ware or engobe.

Glazes are made for use at all temperatures from about 600° C. to 1500° C. according to the articles to which they are applied. They are also made to have glossy, matt, dull or other surfaces, according to requirements, and in some cases to be highly resistant to acids and other liquids and corrosive gases and in others to be semi-conducting.

Broadly, glazes are composed of groups of materials:

(i) Bases corresponding to alkali and alkaline earth oxides.
(ii) Alumina.
(iii) Silica and other 'acid' radicles, such as boron oxide.
(iv) Other oxides added to produce certain effects, such as opacifiers, colorisers, etc.

It is customary when representing the formula of the glaze to do so on the basis of the molecular proportions of the various elemental oxide components. Convention usually decrees that the formula is expressed with the molecular content of fluxing oxides (including PbO and ZnO) adding up to unity and the alumina and silica (+ boric oxide) listed as separate items, thus:

$$1{\cdot}0\,(R_2O + RO).xAl_2O_3.y\,(SiO_2 + B_2O_3).$$

Instead of using the simple oxides, it is more satisfactory to use compounds or frits containing the components (e.g. felspar and china stone) which melt readily and act as solvents for the other ingredients. In most glazes, the larger the proportion of alkalies and other bases, the more fusible will be the glaze and the brighter the colours produced by pigments, although this does not apply to all glazes.

It is important in glazed ware to ensure that the coefficient of thermal expansion of the body should be similar to that of the glaze, otherwise *peeling* (when the body has a greater expansion than the glaze) or *crazing* (when the glaze has the greater expansion) will occur. A high content of alkali, particularly sodium, usually produces a glaze of large thermal expansion but replacement by calcium or magnesium reduces this tendency. The ratio of $RO:R_2O$ has a particularly great influence on the nature and many properties of glazes. Variations in it cause great differences in the fusibility, transparency, opalescence as well as the liability to craze. In this respect, the $RO:R_2O$ ratio often has a much greater effect than a similar change in the proportions of alumina and silica.

Glazes may be divided—so far as their chemical composition is concerned—into:

(*a*) **Alkaline glazes,** consisting chiefly of silicates of potash and soda; the most important of these are (i) salt glaze for stoneware; and (ii) some of the unstable glazes occasionally used for ornamental ware and enamels. These, although easy to apply, are usually slightly soluble and not so resistant to wear.

(*b*) **Felspathic glazes,** consisting of silica and alumina, with alkaline or alkaline earth bases. The main component in this type of glaze is felspar, but flint, whiting and sometimes a little clay is also added; in this way the thermal expansion can be controlled. Felspathic glazes are relatively cheap, but require extreme care in use. One of the best-known examples is *Bristol glaze* which is used on cheap earthenware and on some white sanitary ware. Its composition corresponds to:

$$\left.\begin{array}{l}0{\cdot}25\ (K,Na)_2O\\ 0{\cdot}55\ CaO\\ 0{\cdot}20\ ZnO\end{array}\right\}\ 0{\cdot}6\ Al_2O_3;\ 3{\cdot}0\ SiO_2$$

The zinc oxide is added as an opacifier.

(*c*) **Lead glazes** consist of a glaze similar to felspathic glazes, but having the melting point reduced and the appearance improved by the addition of lead compounds.

Lead glazes may be made from red lead and flint, but are often improved by the addition of a little clay, whiting or other flux.

Because of the highly poisonous nature of soluble lead compounds, it is now a legal requirement that the element can be incorporated into a glaze only if it is in an insoluble form. This is most usually achieved by 'fritting' the lead in oxide form with silica to compound a molecular composition of $PbO.2SiO_2$. It is in this form that lead is normally incorporated into glazes; even safer and better formulations include small additions of alumina or titania. Because lead has such a high atomic weight normal frits contain 65–70 per cent PbO on a weight analysis.

In lead glazes containing soda, potash and lime (or any of these) the maximum amount of alumina which can be accepted depends on the $RO:R_2O$ ratio, but with the alumina equivalent between 0·01 and 0·05 good clear glazes can usually be obtained. With a larger amount of alumina, semi-glazes and matt glazes are produced.

Like other glazes, those containing lead must be adjusted to suit the ware to which they are applied. For a reason not fully understood, however, lead glazes do not require such rigid control in use as do other types.

A commonly-used lead glaze compounded from a lead frit, felspar, whiting and a borax frit is:

$$\left.\begin{array}{l} 0{\cdot}13\ PbO \\ 0{\cdot}40\ Na_2O \\ 0{\cdot}20\ K_2O \\ 0{\cdot}27\ CaO \end{array}\right\} 0{\cdot}01\ Al_2O_3 \left\{\begin{array}{l} 0{\cdot}93\ SiO_2 \\ 0{\cdot}78\ B_2O_3 \end{array}\right.$$

Lead compounds impart high lustre to glazes in which they are present. This is because lead atoms can take part in a network structure of silica and in so doing become greatly distorted or polarised; in this form they increase the refractive power of the glaze enormously. Because lead can and does take part in network structures it is not strictly correct, when calculating glaze formulæ, to regard it as a RO type oxide.

(*d*) **Leadless boric acid glazes,** were developed during the Second World War, when lead supplies were in short supply. Borax, sodium borate, and boric acid are valuable constituents of glazes, because they have low melting points, readily form a glassy network and boron oxide can act as either an alkaline or as a basic component. Unlike silica, the greater the proportion of boric acid in a glaze, the lower will be its melting point and as this component also reduces the thermal expansion, it is a valuable additive.

Borates are usually highly soluble and for this reason they cannot be used, as such, in glazes which are applied in a watery medium. The soluble components of a glaze are *fritted* prior to use to convert them into an insoluble form. The *frit* is prepared by fusing together the soluble components and other low melting-point compounds; these include felspar, whiting, Cornish stone and flint. After being quenched, the frit is crushed and finely-ground before being added to the other components which make up the glaze composition.

A typical composition of a borax frit is:

$$\left.\begin{array}{l} 0{\cdot}11\ K_2O \\ 0{\cdot}44\ Na_2O \\ 0{\cdot}45\ CaO \end{array}\right\} 0{\cdot}30\ Al_2O_3 \left\{\begin{array}{l} 3{\cdot}43\ SiO_2 \\ 0{\cdot}88\ B_2O_3 \end{array}\right.$$

Borax glazes are often used as a replacement, in total or in part, for lead glazes. Although they are not so brilliant in appearance, they are accommodating in 'glaze-fit', permit a wide maturing range and are considerably cheaper.

Borax is sometimes used alone as a glaze in a similar manner to salt (see p. 348). Some cheap, glazed clay-ware is prepared by throwing borax into the kiln during the final stages of the firing process. The borax combines with the clay body to form a complex sodium boro-silicate, which has the advantage of being much more insoluble than the compounds produced during salt-glazing.

(*e*) **Loam glazes** are composed of a fusible clay to which felspar, whiting, flint or similar fluxes are added. They are chiefly used for slip-ware, art-ware and 'peasant' ware and are of common use in some countries.

(*f*) **Enamels,** which are ordinary lead or felspathic glazes rendered opaque by the addition of tin oxide or other opacifying agent. (Unfortunately, the term *enamel* is also used for transparent glazes applied to metals, but in the pottery industry its use is largely confined to opaque glazes.)

In order that it may be satisfactory, a glaze must possess three characteristics: (i) it must have the desired transparency or opacity and colour; (ii) it must be perfectly adapted to the body or ware to which it is applied so as to avoid crazing or peeling; (iii) it must possess a suitable fusibility in order that it may mature at a convenient temperature which is also suitable to the ware to which the glaze is applied.

These characteristics are secured by suitably modifying the chemical composition of the glaze mixture. All these characteristics are equally important and yet, in some cases, they are incompatible, so that various compromises must be effected. An increase in transparency is usually obtained by making the glaze more fusible or by reducing the proportion of non-fusible matter in it. The adhesion of a glaze to the ware may sometimes be increased by adding clay (though this may make it less transparent), by making it more fusible (though this may cause 'crazing'), or by varying the proportions of the various minerals in such ways as previous experience may suggest.

The different constituents which are used in glazes and engobes play a large part in determining the nature of the fired mixture. They are as follows:

Silica is used in glazes to supply the chief acid radical and also to regulate the temperature at which a glaze will mature. An excess of silica which remains unaltered in the glaze, decreases its transparency and prevents complete fusion except at a higher temperature. On the other hand, a deficiency of silica reduces the ductility of a glaze and imparts a tendency for it to boil and blister.

Silica reduces the expansion of a glaze, but at the same time increases the vitrification temperature. It is added in the form of (*a*) free silica, usually in the form of crushed flint, or (*b*) a silicate combined with alumina in the form of kaolinite $Al_2O_3 . 2SiO_2 . 2H_2O$ or as felspar $K_2O . Al_2O_3 . 6SiO_2$.

Alumina sometimes acts as a base in glazes, but it does not influence the fusing-point to any great extent. Its chief value is to permit considerable variations in the composition and firing of the glaze without greatly altering the physical properties; it is particular useful in preventing crystallisation. It also acts as a clarifier and is especially necessary in lime glazes, where it is required in considerable proportion to prevent turbidity.

In alkaline glazes, including some lime glazes, one effect of alumina is to make the soluble silicates less soluble. In lead glazes, alumina increases the solubility and enables them to contain more silica, but an excess of it should be avoided as it causes turbidity.

When the ratio $RO:R_2O$ is not less than 1:1, alkaline and lime glazes can take up 0·5–0·6 molecular equivalents of alumina but no more. The $RO:R_2O$ ratio is very important when alumina is present.

In earthenware glazes, the molecular ratio of alumina to silica is best kept between 1:6 and 1:10, whereas in crystalline and matt glazes the molecular proportion of alumina should be much lower, e.g. 1:5.

In opaque glazes, alumina tends to reduce the amount of opacifier (tin or titanium oxide) required and to increase the temperature range.

Alumina tends to reduce crazing in fritted glazes and acts similarly to silica and boric acid together, though more powerfully than either of these two when used alone.

Keeler has observed the following effects of varying the ratio of alumina:silica in terra-cotta glazes:

Alumina: Silica Ratio	Product
High alumina, low silica	Immature glazes in most cases. If they do fuse, they tend to flaw and craze
Low alumina, low silica	Crazing, pinholes, immaturity
High alumina, high silica	Beading, immaturity, waviness, but no crazing
Low alumina, high silica	Fair bright glazes, with tendency to waviness

Alumina is used in the form of (i) china clay or ball clay, (ii) felspar or Cornish stone. The two latter materials being fluxes, usually predominate, the clay being added, when required, to give the requisite adhesion to the ware. On account of its alumina content, clay is often effective in preventing crazing. Its chief value is that, being composed of very small particles, it combines more rapidly with some of the fluxes than the coarser forms of silicates. An excess must be avoided, as it makes glazes opaque, and insufficiently fusible.

Fluxes are employed in glazes to produce a material having the required fusing-point and transparency. The choice of the fluxes is very important.

It is desirable to have at least two bases in a glaze, and preferably three. The fact that several fluxes produce a more fusible mixture than an equivalent proportion of one flux is due to the well-known law that one gram-molecule of a substance dissolved in any solvent causes a constant depression of the fusion-point, so that if several bases are present, each acts independently and the fusion-point is depressed to a correspondingly greater extent than that caused by an equivalent proportion of only one flux. In addition, a combination of fluxes increases the vitrification range and liquid formation progresses more smoothly.

In most leadless glazes, the fluxes are lime, soda and potash, with the addition, in some cases, of magnesia, baryta or zinc oxide.

Seger gives the following as the order of strength of various fluxes in glazes:

TABLE VI—XXI. FLUXING POWER OF GLAZE COMPONENTS

Fluxes in Glazes (*Order of Activity*)	*Fluxing Power of Colouring Oxides* (*Order of Activity*)
Lead oxide	Manganese oxide
Baryta	Cobalt oxide
Soda	Iron oxide
Potash	Uranium oxide
Zinc oxide	Chromium oxide
Lime	Nickel oxide
Magnesia	
Alumina	

Soluble bases should, as far as possible, be avoided, as they necessitate the fusion of the base with silica, known as *fritting*, which is both troublesome and costly, in order to produce an insoluble material. Soluble salts do not remain distributed uniformly through a glaze or engobe, but rise to the surface by capillary attraction and so are unable to react properly with the ingredients below the surface.

Potash is usually added in the form of felspar or Cornish stone, but potassium nitrate (nitre) and carbonate are sometimes used. Felspar adds the desired amount of flux in a more concentrated form than does Cornish stone, but if in excess it tends to cause peeling and may also cause crazing if the other constituents are not in the right proportions.

Both felspar and Cornish stone also add alumina and silica to a glaze, and this must, if necessary, be corrected by reducing the proportion of clay or flint used. Care should also be taken to ensure that the iron content is small, otherwise the glaze will be stained.

Soda is added in the form of a soda felspar or as a soluble salt, such as sodium carbonate, in which case *fritting* (p. 343) with some of the other constituents of the glaze is necessary. The action of soda is almost identical with potash and the two are often interchangeable. Sometimes soda produces a slightly more 'fusible' glaze and potash a rather more glossy one.

Glazes high in alkalies are generally suitable for bodies rich in quartz, whilst glazes low in alkalies are more suitable for bodies rich in alumina.

Lime is usually added as whiting or other form of a calcium carbonate, though calcium sulphate (plaster of Paris) is sometimes used. The addition of any lime compound increases the glossiness and fusibility of fine earthenware and stoneware glazes and slightly reduces the tendency to crazing. Lime glazes with no other base usually require 0·5 equivalents of boric oxide to give the required fusibility, but more than one equivalent tends to cause turbidity.

For the reasons stated on p. 345 it is better to use a mixture of soda (or potash) and lime in glazes than either of these substances separately. The usual permissible range of composition of these two substances (expressed as oxides) is

$$\left.\begin{array}{l}0{\cdot}6\ K_2O\\0{\cdot}4\ CaO\end{array}\right\}\ \text{to}\ \left\{\begin{array}{l}0{\cdot}2\ K_2O\\0{\cdot}8\ CaO\end{array}\right.$$

Soda may be partially or completely substituted for potash in the above molecular proportions. Lime glazes usually require at least four equivalents of silica and at least 0·5 equivalents of alumina to give a satisfactory gloss.

Magnesia can usually replace an equivalent of lime in a glaze. The chief effect is to make the molten glaze less viscous and so reduce the tendency to excessive 'running'. It sometimes produces a clearer glaze with brighter colours, but neither magnesia nor lime should be the sole base in a glaze.

Baryta (BaO), behaves similarly to lime in a glaze, but as barium has a much larger atomic weight than calcium, it tends to produce a more brilliant lustre.

Zinc oxide acts as a base and increases the fusibility of glazes. It also stabilises some colours and reduces the tendency to 'craze'.

Lead compounds or frits give great mobility and a high refractivity to glazes, and consequently impart a great lustre and brilliance unobtainable by any other element, although barium compounds sometimes give similar results in leadless glazes containing soda and boric acid. Small variations in the composition of leadless glazes are important, and more likely to cause trouble than when lead is present.

Boric acid and **boric oxide** are used in some glazes in place of part of the silica to lower the fusing-point without materially altering the constitution of the glaze.

The dual behaviour of boric oxide—sometimes as an acid and sometimes as a base—appears to be closely related to the general composition of the glaze. Thus, Stull and Radcliffe found that when the base oxygen: acid oxygen ratio is less than 1:2, boric oxide behaves as an acid, whilst if the ratio is greater the boric oxide behaves as a base, but if the ratio is exactly 1:2, the boric oxide behaves as a free acid. When boric oxide behaves as an acid, it increases the tendency to crazing and when it behaves as a base it decreases the crazing.

Typical Glazes

THE following are typical formulae for glazes suitable for various purposes; they are by no means exhaustive, as the number of different compositions which can be used for each purpose is enormous:

Salt glaze is produced by throwing damp common salt into the kiln when its contents are at a sufficiently high temperature. The salt is decomposed by the heat and moisture, forming soda (Na_2O) and hydrochloric acid (HCl). The former combines with the silica and alumina, of which the articles are made, forming complex silicates, the composition of which appears to vary considerably. According to Barringer, the limits of the alumina:silica ratio in the ware within which it is commercially possible to produce a good salt glaze are 1:4·6 to 1:12·5, and Mackler—who analysed pieces of glaze carefully chipped from the ware—found that salt glazes vary in composition from

$$1\text{–}2Na_2O . Al_2O_3 . 3{\cdot}3\text{–}7{\cdot}5SiO_2,$$

the average being

$$1{\cdot}5Na_2O . Al_2O_2 . 5SiO_2.$$

If an article is made of a highly aluminous clay deficient in free silica, it is usually very difficult to produce a good salt glaze on it, but the assumption that salt glazes are silicates and not alumino-silicates cannot be correct, as bricks made of pure silica do not form a good salt glaze when treated in the ordinary manner.

Terra-cotta glazes differ widely in composition, as the clays used in the manufacture of terra-cotta vary so greatly. A glaze which has long been used with success corresponds to the formula:

$$\left.\begin{array}{l} 0{\cdot}1\text{–}0{\cdot}7\ CaO \\ 0{\cdot}5\text{–}0{\cdot}2\ K_2O \\ 0{\cdot}4\text{–}0{\cdot}1\ ZnO \end{array}\right\} 0{\cdot}55\ Al_2O_3 . 3{\cdot}3\ SiO_2$$

Stoneware consists of a vitrifiable body and can, therefore, be glazed at a higher temperature than most terra-cotta and faience ware. The composition of the glazes used, naturally varies with the clays of which the articles are made.

A wide range of compositions is quoted which fall within the formulæ

$$\left.\begin{matrix} 0{\cdot}15\text{–}0{\cdot}45\ K_2O \\ 0{\cdot}15\text{–}0{\cdot}60\ CaO \\ 0{\cdot}20\text{–}0{\cdot}60\ ZnO \end{matrix}\right\} 0{\cdot}50\text{–}0{\cdot}55\ Al_2O_3.2.5\text{–}4{\cdot}0\ SiO_2$$

Those which mature at the lowest temperature contain most fluxs and have the highest silica values.

Sanitary ware is largely made of fireclay covered with a felspathic engobe and a leadless glaze, which are fired at a temperature corresponding to Seger Cones 8 or 9.

An engobe and a glaze which may be regarded as the bases of most of those used in this country for sanitary ware—the variations being largely due to local circumstances—correspond to the formulae:

Engobe, $1{\cdot}0\ K_2O.22Al_2O_3.65SiO_2$

$$\text{Glaze} \left\{\begin{matrix} 0{\cdot}16\ K_2O \\ 0{\cdot}64\ CaO \\ 0{\cdot}20\ ZnO \end{matrix}\right\} 0{\cdot}22\ Al_2O_3.2{\cdot}5\ SiO_2$$

Glazes containing boric acid are seldom applied to sanitary ware.

Earthenware and Faience. Earthenware glazes free from lead and boric acid for temperatures between Cones 1 and 6 should have compositions within the following limits:

$$\left.\begin{matrix} 0{\cdot}3\left\{\begin{matrix} K_2O \\ Na_2O \end{matrix}\right. \\ 0\text{–}0{\cdot}15\ ZnO \\ 0{\cdot}1\text{–}0{\cdot}2\ MgO \\ 0{\cdot}5\ BaO \end{matrix}\right\} 0{\cdot}14\text{–}0{\cdot}18\ Al_2O_3.1{\cdot}4\text{–}1{\cdot}8\ SiO_2$$

A typical glaze for English earthenware (tableware) corresponds to:

$$\left.\begin{matrix} 0{\cdot}29\ PbO \\ 0{\cdot}39\ CaO \\ 0{\cdot}10\ K_2O \\ 0{\cdot}21\ Na_2O \end{matrix}\right\} 0{\cdot}37\ Al_2O_3 \left\{\begin{matrix} 2{\cdot}62\ SiO_2 \\ 0{\cdot}58\ B_2O_3 \end{matrix}\right.$$

As the terms 'earthenware' and 'faience' include many kinds of ware, from those made of crude red-burning clay covered with a roughly-made glaze composed of galena and crushed quartz, flint, or fine sand, to some of the finest examples of the potter's art, no single formula can possibly prescribe the limits of composition of glazes and engobes for such a variety of wares.

Porcelains, like earthenware, vary greatly in composition and the glazes vary correspondingly. Seger has stated that the following are the maximum and minimum compositions for hard porcelain glazes:

Maximum: $RO.1{\cdot}25\ Al_2O_3.12SiO_2$
Minimum: $RO.0{\cdot}5\ Al_2O_3.5SiO_2$

but he recommended that such glazes should be based on the composition of some Seger cones, and Mellor[1] has suggested that the compositions shown below (which approximate to Cones 4–11) may be used as glazes for felspathic porcelain.

Cones Nos. 4–6	$RO.0{\cdot}4\ Al_2O_3$.	$3{\cdot}5\ SiO_2$	Used as a glaze at Cone	7
	$RO.0{\cdot}5$ „	$4{\cdot}5$ „		9
	$RO.0{\cdot}6$ „	$5{\cdot}5$ „		10
Cones Nos. 7–8	$RO.0{\cdot}7$ „	$6{\cdot}5$ „		11
	$RO.0{\cdot}8$ „	$7{\cdot}5$ „		12
Cones Nos. 9–11	$RO.0{\cdot}9$ „	$8{\cdot}5$ „		13
	$RO.1{\cdot}2$ „	$11{\cdot}0$ „		16

Opaque[2] and opalescent glazes are made of the same materials as clear glazes with the addition of one or more opacifying agents. The opacity may be produced:

(*a*) By adding a substance, such as tin oxide, which remains suspended in the glaze and renders it opaque and white.

(*b*) By adding a substance, such as alumina, which can dissolve in the molten glaze but crystallise out on cooling and so produce opacity. Titanium oxide, zirconia and alumina are of this kind.[3]

(*c*) By so composing the glaze that two immiscible products are formed in the kiln, one being uniformly distributed through the other, so that on cooling an opaque glaze is formed. Calcium borate, calcium phosphate and calcium fluoride are used in this manner.

The maximum opacity is obtained with particles of about $0{\cdot}25 \times 10^{-3}$ mm. Such particles are much too small to be produced by grinding but are readily obtained when sufficient of the opacifier is dissolved in a molten glaze and the latter is then allowed to cool slowly.

Glazes rich in alkali, boric acid or lead tend to dissolve an excess of the opacifier and produce irregular results. Fluorides, on the contrary, increase the fluidity of the molten glaze and also its opacity.

Opacity depends upon the crystallised phase having a refractive index very different from that of the main clear glassy phase. The latter has an index in the range 1·5–1·6 so that tin oxide 2·04, zirconia 2·40, antimony oxide 2·60, titania 2·50 are all effective opacifiers.

Matt glazes are chiefly made from transparent glaze mixtures by:

(i) Reducing the proportions of base or flux.

(ii) Adding zinc oxide, calcium carbonate, magnesia or alumina (or china clay).

(iii) Reducing the finishing temperature or the duration of the heating at or near that temperature, and so underfiring.

The change which occurs is usually due to the saturation of a clear glaze with a substance, usually a crystalline silicate, which fuses at the oven temperature but is precipitated or crystallises when the glaze is cooled slowly.

[1] Mellor, J. W., *Trans. Eng. Ceram. Soc.*, 14, 176, 1914.

[2] For further information on opacity as a property of some ceramic materials, see Chapter XIV.

[3] Styhr, K. H. and Beals, M. D., *Bull Amer. Ceram. Soc.*, 37, 480, 1958.

Great skill is needed in order to obtain a pleasing effect as small errors in the composition of the glaze or the use of a substance in too coarse a powder may have a disastrous effect.

For many purposes, a semi-matt surface is nowadays preferred, with a beautiful satin-vellum finish. This does not collect dirt like the true matt glaze and can be more readily cleaned. It is produced by adding about 4 per cent each of zinc oxide, titania and tin oxide and firing to a fairly low temperature (*circa* 1050° C).

Matt glazes are difficult to fit to a body without faults occurring; consequently a high lead glaze with some boric acid replacing silica is generally preferred.

A typical composition is

$$\left.\begin{matrix} 0{\cdot}09\ K_2O \\ 0{\cdot}20\ CaO \\ 0{\cdot}58\ PbO \\ 0{\cdot}13\ ZnO \end{matrix}\right\} 0{\cdot}35\ Al_2O_3 \left\{\begin{matrix} 1{\cdot}40\ SiO_2 \\ 0{\cdot}20\ B_2O_3 \end{matrix}\right.$$

Crystalline glazes produce beautiful and fascinating effects but they are most difficult to control and to reproduce. As the name implies, these glazes contain a high percentage of crystalline phase which develops on cooling so they can be regarded as a type of matt glaze.

True crystalline glazes are those in which zinc silicate is induced to form on cooling. This is achieved by selecting a glaze composition in which the alumina is either low in amount or entirely absent. The content of zinc must be high but silica should not predominate and may be replaced by boric acid; titania in small amounts is added as a seed. A well-known composition has the formula:

$$\left.\begin{matrix} 0{\cdot}10\ K_2O \\ 0{\cdot}06\ Na_2O \\ 0{\cdot}18\ CaO \\ 0{\cdot}66\ ZnO \end{matrix}\right\} 0{\cdot}041\ Al_2O_3 \left\{\begin{matrix} 0{\cdot}90\ SiO_2 \\ 0{\cdot}04\ TiO_2 \\ 0{\cdot}03\ B_2O_3 \end{matrix}\right.$$

A second group of crystalline glazes are those in which various oxides or compounds are added which crystallise on cooling.

The chief oxides which give good crystals and may be regarded as crystallising agents are: zinc oxide, titanium oxide, tungstic oxide, molybdic oxide, vanadic oxide, and bismuth oxide. Oxides of manganese, uranium, cobalt, iron, copper, and nickel may also be used for coloured glazes. Bismuth oxide and uranium oxide produce crystalline glazes at as low a temperature as Cone 010–09, and tungstic oxide produces beautiful fernlike and star-shaped formations, showing partly iridescent reflexes quite unlike the effects obtained with zinc and titanium oxides.

Purdy and Krehbiel[1] found that manganese dioxide had the greatest tendency to cause crystallisation; zinc oxide is next, but with it the crystals tend to segregate in local areas, leaving others devoid of crystals. If a fusible silicate (glaze) is coloured (as by adding cobalt) and is then saturated with zinc silicate whilst in the molten state, the zinc silicate will crystallise out as willemite and will be stained blue, the background remaining colourless if there is no excess of colouring agent.

[1] Purdy and Krehbiel, *Trans. Amer. Ceram. Soc.*, 9, 319, 1907.

The composition of the 'foundation glaze' from which crystalline glazes may be prepared by suitable additions is very important; the following are typical formulae for this kind of glaze:

To mature at Cone 09–2, $\left.\begin{matrix} 0{\cdot}1\ Na_2O \\ 0{\cdot}2\ CaO \\ 0{\cdot}7\ PbO \end{matrix}\right\} 0{\cdot}05\ Al_2O_3 \left\{\begin{matrix} 2{\cdot}1\ SiO_2 \\ \\ 0{\cdot}2\ B_2O_3 \end{matrix}\right.$

To this is added 18–28 per cent of a crystallising agent.

To mature at Cone 08–4, $\left.\begin{matrix} 0{\cdot}25\ Na_2O \\ 0{\cdot}35\ K_2O \\ 0{\cdot}40\ CaO \end{matrix}\right\} 0{\cdot}05\ Al_2O_3 \left\{\begin{matrix} 2{\cdot}0\ SiO_2 \\ \\ 0{\cdot}5\ B_2O_3 \end{matrix}\right.$

To this is added 2–25 per cent of rutile or 0·25–2 per cent of oxides of chromium, cobalt, or copper, or 5–10 per cent of the oxides of iron, manganese, uranium or nickel.

Aventurine glazes are a special form of crystalline glaze of great beauty in which iron oxide crystallises in a radiating form as magnificent golden spangles in a red or brown matrix. Its alternative names are sunspot and tiger's eye, both very descriptive. About 20 per cent of metallic iron or finely-divided iron oxide is added to a typical foundation glaze for the best aventurine but careful control of particle size, kiln atmosphere and the rate of cooling is of vital importance for the best effects.

The Colouring of Glazes

The colouring of glazes and glazed bodies is a science in itself and has occupied the attention of ceramists from the earliest days of glazed-ware manufacture.

Glaze ware is usually coloured in one of three ways:

(*a*) *Under-glaze decoration*—where the colours are applied to the ware after it has been 'bisque' fired. The colours are usually applied in the form of a spray or as a paste and are finely-ground mixtures of metallic compounds with china clay and felspar. Some underglaze colours which have an oil base have to be hardened at about 600° C.; the ware is then covered with a transparent glaze.

Underglaze decoration has the advantage of being completely protected from abrasion and so it is permanent. It is, however, flat and dull in appearance and the range of colour and design is limited because of the risk of attack by the glaze during its maturing.

(*b*) *On-glaze decoration*—where the colours are applied on top of the glaze. In this form of decoration the ware is refired to harden the colours at a temperature which is insufficient to soften the glaze. Consequently, on-glaze colours are compounded from low-melting components—usually frits—with a coloured metallic compound.

Because on-glaze decoration is a surface application, it is not as resistant to chemicals and abrasion as is under-glazed ware, but the colours are more brilliant and lustrous and because the temperature at which the on-glaze is fired is low (750–850° C.) a greater variety of colours is available.

On-glaze decoration is usually applied to ware which has been covered with an opaque glaze or to a white body with a transparent glaze.

(*c*) *Coloured or majolica glazes.*—Metallic compounds are frequently added to glazes which are then applied to ware to give an overall coloured covering.

Such glazes are similar in composition to normal transparent glazes—apart from the addition to produce the colour—and a complete range with different softening temperatures is available.

Colouring Compounds. The ingredients which are used to produce a particular colour for glazed ware are mostly closely-guarded secrets of the manufacturer. Although certain metallic compounds are known to produce a colour, the shade or tint can often be influenced by additives or by the type of flux which is used.

Most of the compounds which are used for glaze coloration are salts of metals of the transition groups or those with relatively unstable outermost orbits of electrons (see Chapter II). By a mechanism of electron transfer within the orbits of their atoms, such elements preferentially absorb light of certain wavelengths, thereby appearing coloured.

The elements used to produce colours in ceramic glazes are many, but the most common have been summarised by Henry[1] and are as follows:

Yellow and Orange

1. A mixture of *lead oxide* and *antimony oxide* with small amounts of zinc oxide, tin oxide, calcium carbonate, aluminium oxide and iron oxide produces varying shades of yellow from lemon to orange depending on the proportion of the ingredients which are used. This is an excellent colour combination provided the temperature of firing is not too high. It is called *Naples yellow.*

2. *Cadmium sulphide* also produces a bright yellow and is suitable for overglaze decoration but will not withstand a high temperature.

3. Mixed oxides or other salts of *titanium*, *antimony* and *chromium.* This gives a stable colour at most temperatures and under most glaze conditions.

4. A mixture of *tin oxide* and *vanadium pentoxide* with small amounts of alumina and titania produces a very stable yellow, suitable for majolica and terra-cotta glazes. It is extensively used to produce yellow sanitary ware or tiles.

5. A combination of *zirconia* and *vanadium pentoxide* gives a similar colour to the above. Although it is a cheaper product it is not so stable.

6. *Uranium salts* give yellow colorations, but are not widely used.

Red

Red coloration in glazes is produced almost exclusively from a combination of *cadmium sulphide* and *selenium.* The higher the proportion of the latter ingredient, the deeper is the shade.

Pink

Gold produces the most beautiful shades of pink, especially when used with titanium frits. *Chromium salts* combined with *tin oxide* or *alumina* will also produce pinks but these mixtures must be chosen carefully as they are not stable in all glaze compositions.

[1] Henry, P., *Bull. Amer. Ceram. Soc.*, 36, 431, 1957.

Blue

Cobalt oxide is used almost exclusively as a source of blue coloration. The shade may be modified by different additions; with silica a purplish blue is given, whilst with a combination of zinc oxide and alumina a greenish blue is developed.

Green

1. *Copper salts* form beautiful green or bluey-green colorations which unfortunately are not very stable and are particularly prone to vary with different kiln atmospheres. In an alkaline glaze, copper gives a transparent turquoise coloration, but under acidic conditions it is green.

2. *Chromium oxide* is the principal agent for producing green colorations. It is very stable under most conditions.

Black

1. A mixture of the *oxides* or *salts* of *cobalt*, *chromium*, *iron* and *manganese* gives a stable neutral black and is of common use in glaze ware.

2. The above components, without chromium salts, give a characteristic metallic bluish black.

3. *Chromium oxide* with *copper oxide* gives a black colour which is stable only at low temperatures.

4. The *oxides* of *chromium*, *iron* and *manganese* produce a brown-black which is stable to high temperatures.

Brown

Iron oxide, *chromium oxide* and *zinc oxide* form the basis of most brown colours for glazes. Manganese dioxide when added to the mixture gives a dark brown whilst alumina produces a lighter, yellow brown.

White

Tin oxide is the component most commonly used to give a white glaze. To produce the purest shades all the ingredients of a glaze must be chosen carefully because a minute proportion of iron oxide gives a yellow tint. A small amount of cobalt oxide is occasionally added to mask the yellowing effect of iron oxide.

Titanium dioxide, in the form of rutile or anatase, is finding increasing use as a white pigment of greater permanence than tin oxide in various glaze compositions.

The colour of a body or glaze is dependent on factors other than the type of colouring oxide which is added. These include:

(*a*) The amount of colouring agent added.

(*b*) Its reaction with the other components of the body or glaze. Iron compounds, for example, normally burn red, but if appreciable lime is present, the colour of the fired body is white, cream, buff or brown, depending on the relative proportions of the iron and lime.

(*c*) The temperature and duration of the firing.

(*d*) The atmosphere in the kiln during burning. A reducing atmosphere gives rise to brown, grey and blue colorations in ware containing iron which otherwise would burn red.

(*e*) The particle size and state of combination of the colouring agent. Both these factors influence the rate of reaction with the body.

Several manufacturers specialise in the production of colours for most types of ware and they offer a wide range to the intending user.

Compounding Ceramic Glazes

Although complex mixtures are often used for ceramic bodies and glazes their effective composition can be greatly simplified by regarding them (especially after they have been fired) as mixtures of simple oxides, corresponding to a general formula

$$1{\cdot}0(RO+R_2O).xR_2O_3.yRO_2$$

It is better to divide the bases into two groups and to write the formula as

$$aR_2O.bRO.xR_2O_3.yRO_2$$

(The total of $a+b$ equals 1·0 as before)

The division of the bases enables the ratio of the R_2O and RO groups to be clearly shown, and sometimes this ratio is very important (p. 346).

The chief oxides may then be arranged as follows:

R_2O	RO	R_2O_3	RO_2
Na_2O	CaO	Al_2O_3	SiO_2
K_2O	MgO	B_2O_3	TiO_2
Li_2O	ZnO		ZrO_2
	PbO		

Ceramic materials cannot have strict chemical compositions which correspond to chemical formulae as they are subject to too many influences on firing. It is, however, very convenient to make use of chemical formulae as guides even though they do not wholly represent the facts.

It is customary to ignore elements present in very small proportions, but care must be taken, as they sometimes have important effects. Another important precaution is to use accurate chemical analyses of the materials actually used and not 'general figures', or serious errors may occur.

As explained on p. 343 boric acid can have a dual role in ceramic glazes. It is more usual for it to behave as an acidic component and to fulfil a similar function to that of silica. In such cases, it is more correct to include boric oxide in the total of RO_2 components.

Calculation of Molecular Formulae. A chemical analysis of a ceramic body indicates the composition in terms of oxides calculated on a *weight percentage* basis. To convert this to a *molecular* basis, the values have to be divided by the molecular weight of the corresponding oxide, which is found by adding together the atomic weights (see Chapter II, Table Ib) of the components; it is sufficiently accurate to calculate to the first place of decimals.

The molecular weight of alumina, Al_2O_3 (atomic weights—Al = 27·0; O = 16·0) is

$$2 \times 27{\cdot}0 + 3 \times 16{\cdot}0 = 102{\cdot}0$$

Similarly silica (SiO_2), is

$$1 \times 28{\cdot}1 + 2 \times 16{\cdot}0 = 60{\cdot}1$$

and water (H_2O), is

$$2 \times 1{\cdot}0 + 1 \times 16{\cdot}0 = 18{\cdot}0$$

Example: The ideal chemical composition of kaolinite is Al_2O_3 = 39·50 per cent; SiO_2 = 46·56 per cent; H_2O = 13·94 per cent. The molecular composition is

$$\begin{aligned} Al_2O_3 &= 39{\cdot}50 \div 102{\cdot}0 = 0{\cdot}387; \\ SiO_2 &= 46{\cdot}56 \div 60{\cdot}1 \;\;= 0{\cdot}774; \\ H_2O &= 13{\cdot}94 \div 18{\cdot}0 \;\;= 0{\cdot}774; \end{aligned}$$

i.e. 0·387 Al_2O_3; 0·774 SiO_2; 0·774 H_2O.

For convenience the fractions are eliminated by dividing throughout by 0·387, giving:

$$Al_2O_3 \quad 2SiO_2 \quad 2H_2O$$

The total molecular weight of kaolinite is

$$102{\cdot}0 + 2 \times 60{\cdot}1 + 2 \times 18{\cdot}0 = 258{\cdot}2$$

The molecular composition of orthoclase felspar and muscovite mica can be calculated similarly from the data in Table VI.XIII (p. 322).

Some useful molecular weights are given in Table VI.XXII.

TABLE VI—XXII. MOLECULAR WEIGHT OF IMPORTANT COMPONENTS OF CERAMIC BODIES

Oxide	*Molecular Weight*	*Compound*	*Molecular Weight*
SiO_2	60·1	Flint, quartz SiO_2	60·1
Al_2O_3	102·0	Kaolinite $Al_2O_3.2SiO_2.2H_2O$	258·2
TiO_2	79·9	Orthoclase felspar $K_2O.Al_2O_3.6SiO_2$	556·8
Fe_2O_3	159·8	Albite felspar $Na_2O.Al_2O_3.6SiO_2$	524·6
FeO	71·9	Anorthite felspar $CaO.Al_2O_3.2SiO_2$	278·3]
CaO	56·1	Muscovite mica $K_2O.3Al_2O_3.6SiO_2.2H_2O$	796·8
MgO	40·3	Mullite $3Al_2O_3.2SiO_2$	426·2
K_2O	94·2	Whiting $CaCO_3$	100·1
Na_2O	62·0	Cornish stone (average)	
B_2O_3	69·6	$\left.\begin{array}{l}0{\cdot}4\,Na_2O\\0{\cdot}5\,K_2O\\0{\cdot}1\,CaO\end{array}\right\}1{\cdot}25\,Al_2O_3.9{\cdot}25\,SiO_2.0{\cdot}8\,H_2O$	770·1
PbO	223·2		
ZnO	81·4		
H_2O	18·0		
CO_2	44·0		

When calculating the composition of a body or when compounding a glaze, the molecular composition of the actual ingredients used must be determined, paying attention to minor components as well as those of major importance, and taking into account any constituents which may be volatile on heating.

The body or glaze to be compounded must also be calculated in terms of the molecular proportions of its oxides, and its components attributed to those in the ingredients to be used. This is best demonstrated by considering an example.

Example: to compound a Bristol glaze of composition

$$(0{\cdot}35\ K_2O\,.\,0{\cdot}35\ CaO\,.\,0{\cdot}3\ ZnO)\,.\,0{\cdot}6\ Al_2O_3\,.\,3{\cdot}5\ SiO_2$$

from felspar ($K_2O.Al_2O_3.6SiO_2$), china clay ($Al_2O_3.2SiO_2.2H_2O$), flint (SiO_2), whiting ($CaCO_3$) and zinc oxide (ZnO). (For convenience, the ideal compositions of the components have been taken, but in practice they must be determined.)

The CaO and ZnO in the glaze come from the whiting and zinc oxide respectively. The K_2O content comes from the felspar but this ingredient also contributes some alumina and silica. The remainder of the alumina comes from the clay and any excess silica required will be provided by the flint.

The proportions of each ingredient may then be calculated as in Table VI.XXIII.

GLASSES

TO some extent, a glass may be likened to an overvitrified glaze, and indeed many of the general rules applicable to glazes, hold also for glasses.

A glass is a true super-cooled liquid and although it appears hard and solid at room temperature, it has all the properties of a highly viscous liquid; it will flow if a load is applied to it and, on heating, it becomes progressively more fluid, finally becoming free-flowing.

Glasses differ from glazes in having a higher proportion of flux, i.e. a greater amount of alkali and alkaline earth elements. This is mainly so that melting will take place at a lower temperature—in fact, the initial softening point of many glasses is about 600° C.

The principal components of glasses are (1) soda, potash, lime, magnesia and baria as bases, of which soda and lime are the most common; (2) silica and boric oxide, which may be considered as acid components, although the latter can have dual behaviour; (3) alumina, which helps to prevent crystallisation on cooling and has other desirable effects; (4) other ingredients added to impart colour, appearance, resistance to chemical attack, low expansion characteristics, high melting point, hardness, etc. These include iron oxide, manganese dioxide, lead oxide and zinc oxide.

The main ingredients used in the manufacture of glasses are similar to those used in compounding ceramic bodies and glazes. Silica sand is the principal source of silica, felspar of potassium and alumina, limestone of calcium oxide, borax of boric oxide. Quantities of soda ash (crude sodium carbonate), sodium sulphate and sodium nitrate are also incorporated to provide a high content of alkali. Of major importance in glass-making, is that the ingredients should be free from iron contamination, because even small quantities of this impurity cause serious discoloration of the product (see p. 315).

TABLE VI—XXIII. THE COMPOSITION OF A GLAZE

Components	Molecular weight	Molecular Prop.	CaO 0·35	ZnO 0·30	K_2O 0·35	Al_2O_3 0·60	SiO_2 3·50	H_2O —	CO_2 —	Weight required Prop. × molecular weight	Weight Proportion per cent
Whiting ($CaCO_3$)	100·1	0·35	0·35	—	—	—	—	—	0·35	0·35 × 100·1 = 35·035	10·2
Zinc oxide (ZnO)	81·4	0·30	—	0·30	—	—	—	—	—	0·30 × 84·1 = 24·230	7·0
Felspar ($K_2O.Al_2O_3.6SiO_2$)	556·8	0·35	—	—	0·35	0·35	2·10	—	—	0·35 × 556·8 = 167·040	48·4
Kaolinite ($Al_2O_3.2SiO_2.2H_2O$)	258·2	0·25	—	—	—	0·25	0·50	0·50	—	0·25 × 258·2 = 64·550	18·7
Flint (SiO_2)	60·1	0·90	—	—	—	—	0·90	—	—	0·90 × 60·1 = 54·090	15·7
		Total	0·35	0·30	0·35	0·60	3·50	0·50 (volatile)	0·35 (volatile)	344·945	100·0

The proportions required to compound the glaze of formula 0·35 CaO, 0·30 ZnO, 0·35 K_2O, 0·60 Al_2O_3, 3·50 SiO_2 are 10·2 parts of whiting, 7·0 parts of zinc oxide, 48·4 parts of felspar, 18·7 parts of kaolinite and 15·7 parts of flint.

The chief effects of the principal oxide components in glasses are as follows:

1. *Silica* is the major acidic oxide and combines with the basic components. Large proportions of silica increase the melting and softening point of glasses and also lower the thermal expansion.

2. *Boric oxide* can replace silica to some extent in glasses and has the advantage of lowering the melting point, so that by a suitable combination of silica and boric oxide, a glass of low expansion, but with a reasonable working temperature can be manufactured.

3. *Alumina* reduces the melting point, prevents crystallisation when present in certain proportions, and reduces the solubility of many glasses in water and other chemicals.

4. *Sodium and potassium oxides* are the principal fluxes in glasses, and with silica form a low melting-point liquid. Such glasses have a high thermal expansion.

5. *Lime* is another major source of flux, but it does not form a liquid phase with silica as readily as does sodium. However, when it replaces sodium, the glass has an appreciably lower thermal expansion.

6. *Magnesia* is used as a replacement for lime but the liquids which it produces are more viscous, although they crystallise less readily.

7. *Baria* behaves rather like magnesia, but also imparts a high lustre to the glass. *Lead oxide* behaves similarly.

As in glazes, it is advisable to have at least two fluxing oxides in glasses, to produce a more satisfactory melt with a longer working range.

The compositions of several types of glasses are given in Table VI.XXIV.

TABLE VI—XXIV. THE COMPOSITIONS OF SOME GLASSES

Constituent	*Window*	*Flint*	*Crown*	*Bottle*	*Pyrex*	*Electric lamps and valves*
SiO_2	72·0–72·5	45·0–70·0	58·0–74·0	72·0–74·0	80·4	48·8
Al_2O_3	0·5–1·7	0·0–2·0	—	1·0–3·0	2·2	17·7
Fe_2O_3	—	—	—	0·0–2·0	—	—
CaO	10·0–8·0	0·1–0·4	0·0–2·5	3·0–5·0	0·4	12·4
MgO	4·0–3·2	—	—	0·0–4·0	0·1	—
Na_2O	13·5–16·0	1·0–16·0	3·0–5·0	12·5–17·0	4·0	—
K_2O	—	7·0–0·0	9·0–15·0	—	0·6	—
Li_2O	—	—	—	—	—	6·7
B_2O_3	—	—	3·0–7·0	—	12·3	—
PbO	—	4·8–10·0	—	—	—	—
BaO	—	—	20·0–0·0	—	—	—
ZnO	—	0·0–4·0	5·0–0·0	—	—	—

CEMENTS, MORTARS AND OTHER BONDING AGENTS

A CEMENT or bonding agent is used to impart strength to a material which otherwise would be too weak mechanically for the purpose for which it is intended. Many substances can be considered as cements and they can be divided into two general classes:

1. Amorphous materials which form a continuous film around the non-cementing material and so hold it together, and

2. Crystalline cements, which form a felted mass of interlocking and intergrowing crystals.

Amorphous Cements

Amorphous cements include gums, resins and glasses. The mechanism by which such materials act as adhesives is not known with certainty. It is believed by some authorities that they form a continuous phase of high viscosity throughout the mass and completely enclose the other materials which may be present. Water can behave similarly, because some non-plastic bodies, e.g. sand, form a stronger mass when green (i.e. wet) than when dry. Surface tension effects may play a predominant part in their bonding action. Many examples of amorphous cements are known in ceramic practice.

(*a*) **Organic Binders.** Various gums, of which *cellulose sulphite lye* is the most common, are frequently added to non-plastic materials so that they can be moulded into shapes which have a reasonable strength on being dried. Other materials used include linseed oil, fat, gelatin, dextrin, various gums, rubber latex, resins, starch, tar and pitch.

All these substances decompose at a comparatively low temperature, so that their cementing action is destroyed. For many purposes, this is not a serious disadvantage, because the strength which they confer is required only to allow the ware to be handled and used at ordinary temperatures. The loss of strength on low temperature firing in bricks of non-plastic material is disadvantageous, however, in that it limits the numbers which can be stacked together in the kiln or furnace.

(*b*) **Inorganic binders**—some substances form gels which behave as gums or resins in forming a continuous film in a compacted body, but which are not destroyed by heat (see p. 51).

Crystalline Cements

Of greater effectiveness and permanence, are the materials which react (usually with water) to form a felted mass of needle-shaped crystals which has great tenacity. The setting properties of *calcined lime* were well-known to the Romans and this material is still in common use. Calcined limestone—heated at about 900° C. to form the oxide —is mixed with sand in the proportions of about 1:3. In the presence of water, hydrated calcium silicates are formed and also some calcium hydroxide and water. *Lime mortars*, as they are termed, set slowly and are not very strong, but they are cheap to manufacture and easy to use, so they were commonly employed until recent years as cementing materials for brick structures. Excess lime was invariably present and undoubtedly many severe cases of efflorescence could be attributed to this source.

Many calcium compounds have remarkable cementing properties.

Plaster of Paris, $2CaSO_4.H_2O$, is an ideal material for making ceramic moulds. It is made into a slip with water and poured into a suitable former and on standing crystallises as gypsum, $CaSO_4.2H_2O$, which forms a strong mass of interlocking needles. Such bodies are porous and they 'filter' clay and other slips which are poured into them. The manufacture of Plaster of Paris has been described on p. 328.

The most widely-employed cements, however, are *calcium alumino-silicates* which set to a hard, strong mass with water. These include Portland and high-alumina cements; the former are the active ingredients in concrete and other structural materials and the latter are commonly used in refractory cements and *castable* ceramic bodies.

Portland cement is usually manufactured by heating limestone and clay to about 1600° C. in a rotary kiln. The cycle of reactions is highly complex and the nature of the final product is not fully understood, but it contains lime, alumina, silica, magnesia and iron oxides as components. The predominant compound is tri-calcium silicate ($3CaO.SiO_2$), but dicalcium silicate ($2CaO.SiO_2$), tetra-calcium aluminium ferrite ($4CaO.Al_2O_3.Fe_2O_3$), tricalcium aluminate ($3CaO.Al_2O_3$) and alite (see Chapter III, p. 113) are also present. Small amounts of impurity may have an important influence on the setting properties. Portland cement is the cementing agent in *concretes* where it is mixed with an aggregate of stone or pebbles in the ratio of about 1:6. Water is essential to the setting process, which involves the hydration of the calcium compounds. Portland cement is not suitable for use above about 250° C. and is not very resistant to solutions containing acids. Its composition is shown in Table VI.XXV.

Aluminous cements contain calcium aluminates as their active components which hydrate when mixed with water. The manufacturing methods of these types of cement are still largely trade secrets but the principal ingredients are limestone and alumina with other minor additives to promote the most suitable product. The temperature of calcination in the rotary kiln is usually much higher than that used in the manufacture of Portland cement; in fact, the materials are heated almost to their fusion point.

The range and scope of aluminous cements are very great. They are of advantage because they set rapidly, are resistant to sea water and withstand greater temperatures than Portland cement.

The aluminous cements can be divided into two general classes depending on the amount of alumina which they contain.

(1) **38-40 per cent** alumina by weight comprises the range of cements known as *ciment fondu*. The active compound in such cements is mono-calcium aluminate ($CaO.Al_2O_3$), but other ingredients, particularly iron oxide, are present in large amounts and are essential to the economical preparation of the cement. The melting point of pure monocalcium aluminate is about 1600° C., but, in the commercial materials, an upper limit of 1300° C. should never be exceeded. Ciment fondu is widely used as the active constituent in refractory concretes. It is incorporated with a suitable 'grog', mixed with water, shaped and allowed to set, which it does in a matter of a few hours. One of its chief uses is in the manufacture of fireplaces and their surrounds, when it is mixed with sand or fireclay grog. Robson[1] has described the Continental practice of making brick kilns of ciment fondu and a crushed refractory aggregate usually of well-burned chamotte. Such material will withstand temperatures of up to 1350° C., provided that the refractory aggregate contains over 40 per cent alumina. If the refractoriness of the aggregate is increased by using crushed sillimanite or corundum, the maximum temperature at which the concrete

[1] Robson, T. D., *The British Clayworker*, 62, 302, 1954.

can be used is correspondingly increased. Ciment fondu is also valuable in the manufacture of light-weight refractory concretes, using crushed insulating bricks, vermiculite, diatomite, or bloated clay as aggregate.

Vast monolithic units of concrete can be cast and these have obvious advantages over the conventional brick-work structures.

Robson[1] has also indicated the value of ciment fondu in the ceramic and brick industries and in the manufacture of refractory concrete for gas-works and coke-oven plant. He recommends that a fireclay grog be used as aggregate and suggests the following mixtures to meet all purposes.

	Mix A	Mix B	Mix C
Coarse aggregate ($\frac{3}{4}$–$\frac{1}{8}$ in.)	3 cu. ft.	2 cu. ft.	—
Fine aggregate ($\frac{1}{8}$ in. to dust)	2 cu. ft.	2 cu. ft.	$2\frac{1}{2}$ cu. ft.
Alumina cement	1 cu. ft.	1 cu. ft.	1 cu. ft.

(2) **60-75 per cent** alumina cements, although they are of high cost, are of great potential value in refractory concretes and the so-called *refractory castables.* The active ingredient of such cements is tricalcium penta-aluminate ($3CaO.5Al_2O_3$)[2] which has a melting point in excess of 1700° C., and, because the impurity content can be kept low, commercial cements of this type, when admixed with a suitable grog, can withstand temperatures up to 1700° C. Many alumina cements (with up to 75 per cent alumina) have been manufactured and 'Secar 250' is a common type used in Europe and a similar type is of wide application in North America. According to Robson,[3] it has a setting time of 3–4 hours, and so rapidly does it develop strength that castable mixes made from it can be used within 24 hours. The strength of mixtures is between 2,500–3,500 lb. per sq. in. depending on the proportion of cement and this strength is maintained to high temperatures even under heavy loads.

The rapid setting characteristics of high alumina cements can give rise to inherent problems. The cements tend to be hygroscopic, hence they must be stored under dry conditions; furthermore, the mixing water must be thoroughly incorporated as rapidly as possible otherwise the maximum strength is not developed. The addition of a small percentage of calcium oxalate, or similar salt, is beneficial. Maintaining the cast article in an atmosphere of saturated steam for a few hours also increases the strength to a remarkable degree.

One of the great advantages of high alumina cements is that they maintain a considerable proportion of their strength when heated to so-called intermediate temperatures (400–1000° C.). Other inorganic cements usually lose most of their binding power when heated to these temperatures.

The chemical compositions of some calcium alumino-silicate cements are shown in Table VI.XXV.

Sorel cements are basically magnesium oxychloride and are made by mixing caustic magnesium hydroxide with a solution of magnesium chloride. The production of magnesium oxychloride is normally slow, but it can be accelerated by high

[1] Robson, T. D., *Claycraft*, 28, 244, 1955; *Coke and Gas*, 1952, p. 241.

[2] Dr. E. Hammond, in a private communication, has indicated that the principal mineral in Secar 250 is calcium di-aluminate ($CaO.2Al_2O_3$) and that the mineral $3CaO.5Al_2O_3$ does not exist.

[3] Robson, T. D., *Refract. J.*, 30, 203, 1954.

temperatures or by suitable additives. Although Sorel cements are most commonly employed in the manufacture of interior floors they have been used as the cementing agent in chemically-bonded basic refractory bricks, with calcined magnesia or dolomite as aggregate.

TABLE VI—XXV. CHEMICAL COMPOSITION OF SOME CEMENTS

	Portland	*Ciment Fondu*	*High-alumina*
	per cent	per cent	per cent
CaO	60–66	35–38	26–29
Al_2O_3	5–9	38–40	70–72
SiO_2	19–25	4–6	less than 1
Fe_2O_3	2–4	14–18	less than 1

Chemical Bonds

As a means of cementing to form a strong ceramic body without firing, chemical bonding has been developed to a remarkable degree. Whilst a saving on firing cost is obviously of importance, chemical bonding confers additional advantages of considerable technical merit. Bodies with a high content of non-plastic or 'grog' components can be prepared to close limits of particle-size control. They can thus be made dense or open as desired but remarkably resistant to many of the faults of conventional fired shapes. Refractoriness-under-load, resistance to slag attack and to spalling can often be improved, but perhaps most of all, chemically-bonded ceramic ware can be manufactured to a precision of size and shape unrealised in bodies which have to be fired to develop strength.

Numerous examples of chemical bonding are known and new types are constantly being developed. They may conveniently be sub-divided into (*a*) gel structures and (*b*) crystalline development.

(*a*) *Gel structures.* The precipitation of gels within the pores of a particulate compact is a widely used technique to develop strength. Nature has many examples of this type of structure, perhaps the best-known being quartzites where sand or quartz grains have been cemented by precipitating silica. Colloidal gels of aluminium and iron hydroxides are also known to behave in a similar fashion.

This same process can be used in ceramic practice; in fact, as early as 1912, a process for bonding magnesia bricks with colloidal magnesium hydroxide was patented. By far the commonest gel bond is colloidal silica and this can be achieved simply by hydrolysing with acid a solution of sodium silicate which has been intimately mixed with the aggregate.

The principal difficulty with this type of bond is to obtain the correct incorporation of the non-plastic aggregate with the cementing agent, and the most suitable method is to precipitate the gel in contact with the particles of the body. A solution of sodium silicate forms silica gel when acid is added to it or when it is treated with carbon dioxide and this is a common form of cement used in ceramic practice. Sodium silicate itself forms very viscous solutions, is a very effective bond and is the basis of most acid-resisting cements.

The drawback of cementing material containing alkalies is that they reduce the refractoriness of bodies which contain them. The use of an *organic silicate* compound has largely obviated this difficulty and commercially-manufactured *ethyl silicate* or one of the proprietary *silicones* are being increasingly applied. These materials are supplied as liquids which decompose upon hydrolysis or at a comparatively low temperature to give amorphous silica of great purity. Not only do such materials impart strength to a compacted powder but their reactions at high temperature with other ingredients can be put to advantage. Williams[1] has described how a high-grade mullite refractory may be synthesised by bonding alumina with ethyl silicate and firing to a temperature at which reaction occurs. Similarly zircon (zirconium silicate) can be synthesised from zirconia and the organic silicate.

(*b*) *Crystalline development*. In some ceramic compositions, compounds can be induced to form and grow by chemical reaction in the solution stage. If such crystalline growth can be of an interlocking or combination character the resulting mass can be extremely hard.

The commonest example of this type of chemical bonding is in the formation of sand/lime bricks. In this process, lightly calcined limestone or chalk is thoroughly admixed with crushed silica, either as flint or quartz. The wet mixture is pressed to desired shapes and then autoclaved. During this process the highly active lime reacts with the silica, to develop a felted mass of hydrated calcium silicate crystals of great strength. This bond is permanent and weather-proof and sand/lime bricks are highly valued as building products.

In basic compositions used for refractory purposes chemical bonding is of great importance. Many acids will combine with alkaline magnesia to form crystal phases which interlock and intergrow to develop strength. Such crystals originate on the particles of aggregate and often join them to others to give a continuous chemical bond throughout a compacted mass.

The acid bonds most commonly used with basic refractory compositions are phosphoric, sulphuric, hydrochloric, chromic; in each case a salt of magnesium is formed.

Phosphate bonds are also powerful chemical reagents in bonding high alumina compositions, and the refractoriness of such mixes is not impaired. Phosphate cements are also used in ceramic dentistry for fillings and restorations. Phosphoric acid is mixed with an oxide, usually zinc, and the meta-phosphate, which forms slowly, sets to a hard mass which is very resistant to mouth fluids. Euginates behave similarly and are used for the same purposes. Although they do not set as hard as phosphate cements, euginates are strong antiseptics and are used where the risk of mouth infections is serious.

The important advantage of chemical bonding is that the body can be prepared in much the same way as a concrete and cast into any required shape, complex or otherwise. The bond develops slowly and its rate can be controlled by adjusting the particle-size distribution. The product is a monolith of perfect shape and fit without the disadvantages of similar material bonded with cements. Chemical bonds do not usually break down on heating nor do they reduce the refractoriness of mixtures as do some cements.

[1] Williams, A. E., *Refract. Journ.*, 28, 453, 1952.

Lutes

A LUTE is a material which acts as a bond but which does not set to a rigid body. They are used where some plasticity is required in the body to permit movement without cracking or deformation.

Lutes are used to seal structures and to prevent the escape of gas or liquors. Because of their non-rigid condition, they can be used as a seal between two parts of a structure which have different thermal expansion coefficients.

In general, luting materials do not undergo chemical change, so that any strength which they develop must be purely physical in effect. The most common example known in ceramic practice is *clay* which is mouldable when wet but hardens on drying. Most finely-divided powders can be used as lutes and they should be selected according to the purpose for which they are being used. Finely-powdered *zircon* is a valuable lute for high temperature use; *asbestos* is also commonly employed. Acid-resisting lutes are usually based on *bitumen* and such materials are used to form a cushion between ducts and pipes carrying corrosive liquids and the surrounding foundation material.

Bonds and Cements for Brickwork. Although it is often not realised, the strength of a brickwork structure is largely dependent on the cementing material which binds the bricks together. It is also true that many furnaces designed to withstand high temperatures with perhaps a high degree of slag attack, fail because an unsuitable bond is employed. It is equally as important to select a serviceable bond as it is to choose a suitable brick and to bear in mind the conditions during use.

For *building bricks and for other architectural purposes*, the prime, but by no means the only requirement in the cement or mortar, is strength. This can usually be achieved by using a suitable mixture of sand (composed of angular grains) and either lime or Portland cement. A more pleasing effect is often produced by adding a colouring agent to the mortar so that it blends (or contrasts) with the bricks with which it is used.

The presence of free lime in a mortar can have serious and unpleasing consequences on bricks. Such material is a frequent source of efflorescence (p. 926) and in a sulphurous atmosphere, small amounts of free lime in the mortar may lead to bursting or powdering in the bricks. Mortars composed of sand and lime or cement are usually porous and not very serviceable when exposed to severe conditions of weathering. They can be improved considerably by incorporating with them small amounts of sodium silicate, tar or bitumen which are water-resistant and provide an effective weatherproofing.

Ordinary mortar should never be used to bond the brickwork in kilns or muffles where the temperature may exceed 200° C. or where there is any possibility of contact with sulphurous gases or with liquids of an acidic nature.

Bonds for use at moderately high temperatures (up to 1000° C.) must contain a cement which is serviceable up to those temperatures, and the one most commonly employed is *ciment fondu*, which is mixed with sand or other cheap non-plastic material. Mixtures of sand or fireclay grog, plastic clay and sodium silicate are also suitable under these conditions and if the grog and clay are carefully chosen, such cements can be used up to 1400° C.

Cements containing ciment fondu are resistant to acids in moderate concentrations and they are not unduly attacked by sea water or by acid soils; they can be further improved by using a small amount of bitumen with them.

For bonding brickwork which has to withstand *high temperatures*, the non-plastic component must be refractory and the cement must not break down under service conditions. Many proprietary cements are available, some of which use an aluminous cement, others a plastic clay and/or sodium silicate or other bond and others employ a low refractory clay which forms a glass at the temperatures of operation. The main disadvantage of most types is that they shrink excessively at high temperature and are readily eroded by slags or high velocity gases, with the result that the brickwork is weakened and corners and edges exposed to mechanical and chemical erosion. It is thus important to restrict the amount of cement between the bricks to a minimum by using shapes which are accurately sized. *Zircon cements* are particularly valuable, because they are highly resistant to most slags at all temperatures; cements containing a proportion of *kyanite* have also proved effective at high temperatures. The advantage of the latter type of material is that it expands on heating and hence prevents the bond from shrinking away from the bricks.

CHAPTER VII

PHYSICAL AND STRUCTURAL PROPERTIES OF RAW CERAMIC MATERIALS

MANY properties of ceramic materials, both in the raw and fired states, are related to their composition, but of almost equal importance are the particle size and the shape of the minerals which are present, and, also, the way in which they are agglomerated or situated with respect to each other so as to form the *structure* or *fabric* of the material. In many ceramic bodies, the treatment and processing during manufacture produces many changes in the appearance, colour, hardness and other physical properties, but, in most cases, the original arrangement of the grains persists and can be traced in the final product. The textural properties of the final article are, thus, often largely dependent on the macroscopic or bulk-properties of the raw materials.

Most ceramic materials of natural occurrence require to be reduced in size after they have been extracted from the seam, vein, or rock form. Processes of size reduction are important in the handling and transportation of the material and in its fabrication into shapes of all types by various techniques. The general term *comminution* is given to processes of this nature.

COMMINUTION PROCESSES

THE preparation of a mass of particles to a suitable size distribution from original mine material is one of the most important considerations in ceramic practice. Far too little attention is paid to the processes of size reduction and to the products from such processes in relation to the packing density and other textural properties.

PRINCIPLES OF SIZE REDUCTION

COMMINUTION processes involve applying sufficient energy in the form of pressure to a particle to cause cracks to develop, thereby resulting in sub-division of the particle. The theory of crack formation and propagation is highly complex and outside the scope of this volume. Much of the original theory of this subject was developed by Griffith[1] and subsequently enlarged and improved by McClintock and Walsh[2] and latterly by Bieniawski[3] amongst many others.

Basically the ease or otherwise of comminution is related to certain physical characteristics of the rocks or minerals of which crushing strength and hardness are important. The most significant property is, however, the Young's Modulus which

[1] Griffith, A. A., *Phil. Trans.*, A221, 163, 1921.
[2] McClintock, F. A. and Walsh, J. B., *Proc. 4th U.S. Congress App. Mech.*, 1015, 1963.
[3] Bieniawski, Z. T., *Int. J. Rock Mech. Min. Sci.*, 4, 395, 1967.

is related both to ultimate strength and degree of elasticity of the rock. The significance of these factors is discussed more fully in Chapter XIII.

Size reduction of particles can be achieved in a variety of ways, but in ceramic practice three important principles are involved.

(*a*) *Crushing or compression* of the particles in which the grains of material are squeezed between plates until fracture occurs. Theoretically, such comminution processes should be related to the compressive strength of the rock or mineral, but in practice the grains are usually irregular in shape, so that tensile and shear forces are invariably involved. Nevertheless, the characteristic breakdown associated with crushing is a disruption of large particles into two, four or more grains, each of which is of similar size. The product is thus of a narrow size range with only a small quantity of fines. This is a feature of such units as jaw and cone crushers and the mode of fracture is as illustrated in Fig. VII.1.

(*b*) *Grinding* of particles involves different principles, in that size reduction proceeds by scouring, paring or chipping the surface of large grains, thus slowly reducing their size but at the same time producing an amount of relatively fine particles. The size distribution of a ground product is therefore much wider than that resulting from crushing. In general, grinding processes are more wasteful in energy, but if clay is involved the body so produced is vastly superior in most characteristics.

Various roller pans, either with perforated base-plates or with rim discharge, and various types of rolls crushers, are the grinding units most frequently employed in ceramic practice and the principle and construction are illustrated in Fig. VII.2.

(*c*) *Impact fracture* involves a third set of principles where fracture planes develop within a grain as a result of shear forces on an edge or corner. Breakage usually takes place along a natural plane of weakness or along a contact plane between two different mineral species. In fired bodies or hard rocks, particularly those of a glassy nature, impact crushing often causes characteristic conchoidal fracture patterns.

There is a variety of impact crushers in commercial production in which a combination of fixed and swinging hammers is used to impel particles against a series of impactor plates. They produce an intermediate type of particle-size curve, but they have considerable potential in the clay industries because of their valuable propensities for differential grinding. Certain designs, when crushing composite materials, will crush selectively soft minerals, e.g. clays, to fine size, whilst leaving harder ones, such as quartz, unchanged. This property can be used to separate undesirable impurities or to prepare a well-balanced body. Some models of impact crushers can incorporate a dryer unit, so that wet clays can readily be processed. The principle and a typical design are shown in Fig. VII.3.

Typical differences in the particle size distribution of products from the different methods of comminution are shown in Fig. VII.4.

The three principles outlined above are ideal and most comminution units involve part crushing, part grinding and part impact although usually one aspect predominates in a particular unit.

The propensities for size reduction of a comminution unit are limited. As small particles build up, the efficiency reduces dramatically so that it is usual to

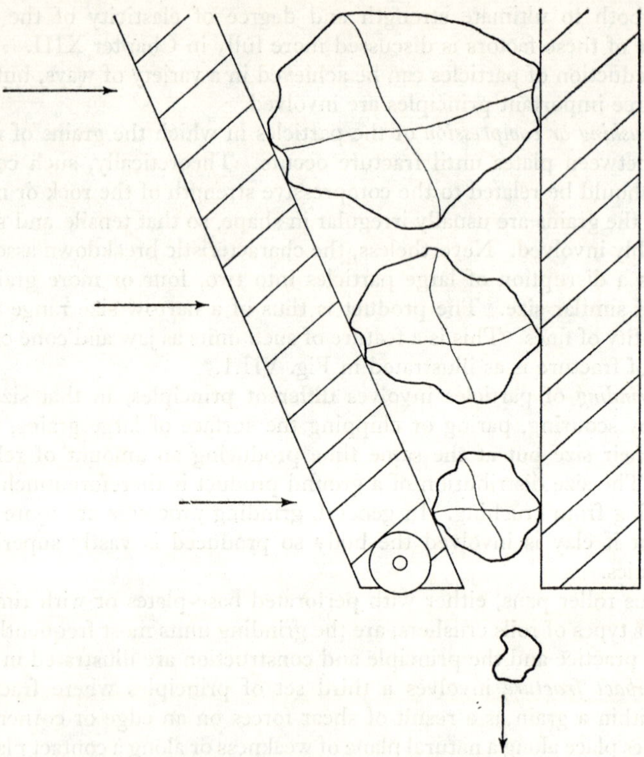

FIG. VII.1. PARTICLE SIZE REDUCTION BY CRUSHING
The lower diagram illustrates a Broadbent Jaw Crusher reproduced by permission of W. C. Holmes & Co. Ltd., Huddersfield, England.

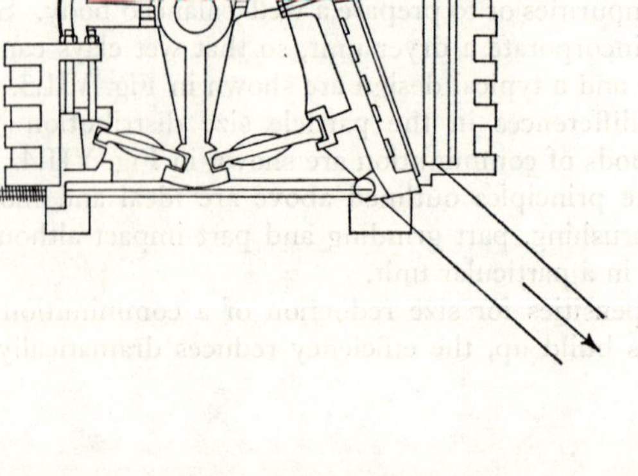

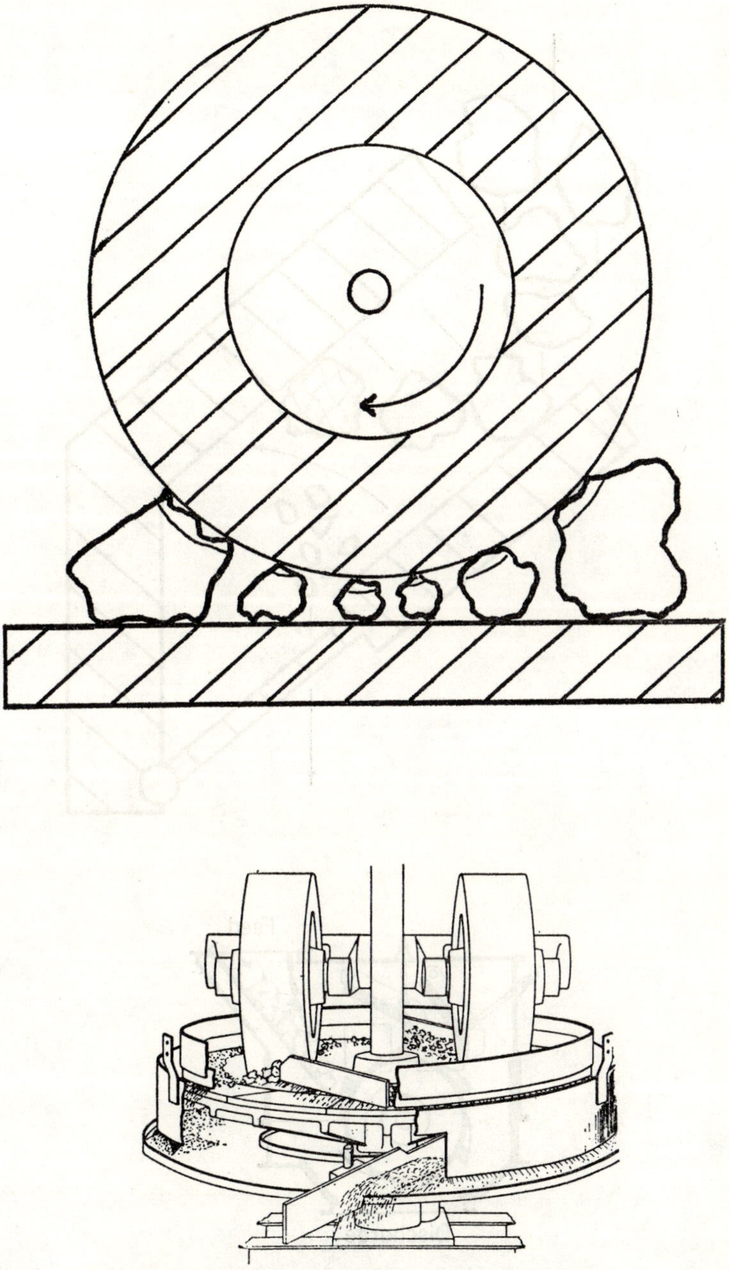

FIG. VII.2. PARTICLE SIZE REDUCTION BY GRINDING

The lower diagram shows a rim discharge pan and the diagram is reproduced by permission of Bradley & Craven Ltd., Wakefield, England.

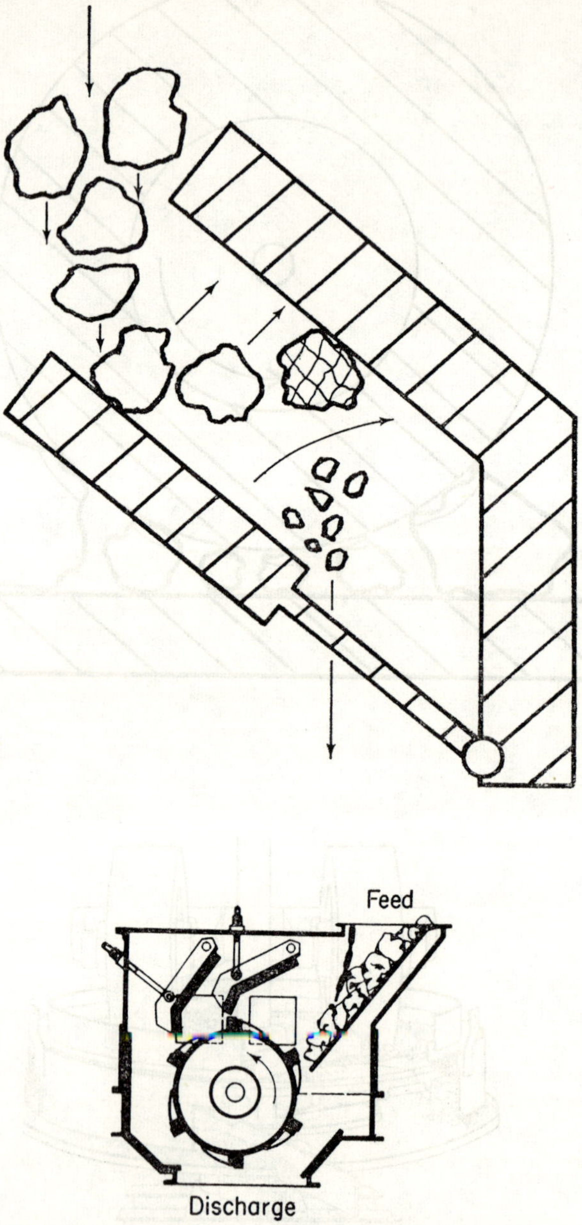

Fig. VII.3. Particle Size Reduction by Impaction

A Hazemag fixed hammer unit is shown in the lower diagram which was kindly supplied by W. C. Holmes & Co. Ltd., Huddersfield, England.

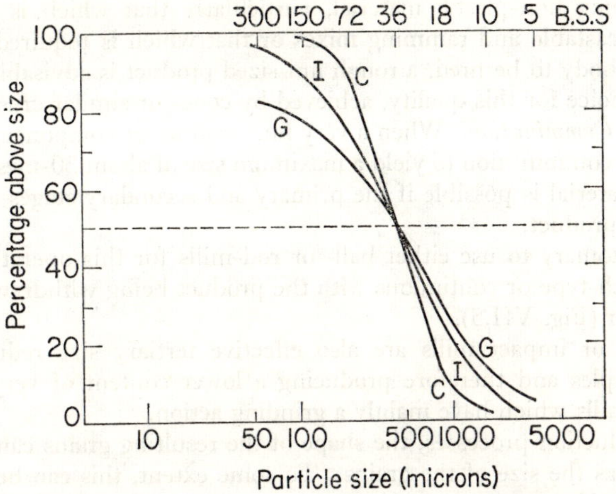

FIG. VII.4. TYPICAL PARTICLE SIZE DISTRIBUTIONS PRODUCED BY DIFFERENT METHODS OF COMMINUTION

In each case a hard rock specimen was used (Welsh Quartzite) and conditions were so adjusted as to give the same centre grain size on 36 mesh. The feed material was −2 in. lumps. Crushing was carried out in a laboratory-type cone crusher. Grinding was effected in an edge-runner pan with solid bottom. Impaction was achieved in a laboratory-size Holmes Hazemag unit

grind in closed circuit with a screen, thus progressively removing fine material from the system and returning coarser, uncrushed grains to the unit. Even under the best conditions it is inadvisable to expect a centre grain-size reduction ratio of greater than 8 : 1 with a single unit.

Because of this fact, the reduction of run-of-the-mine material to a size suitable for manufacture into ceramic shapes has to be carried out in more than one stage. Most brick plants involve two stages; the material for precision shapes or those requiring fine particles normally require three comminution units.

Primary Comminution. The initial comminution process to which mined material is subjected is to reduce it to fragments with a maximum between walnut and fist size (about 1–3 in. diameter). Harder materials are commonly treated in jaw crushers, but for softer clays, simple breakers, chains or cutters will usually suffice.

Secondary Comminution. For many ceramic materials, this is the important size reduction stage which yields a product suitable for manufacture into shapes. The 1–3 in. maximum-sized material is reduced to a maximum of about $\frac{1}{4}$ in. In the case of clays, pans with heavy rollers provide an efficient grinding operation with a well-distributed particle size leading to a dense-packed body. Pan-grinding usually assists in the development of plasticity in a clay when it is subsequently mixed with water. Conventional grinding operations normally produce rounded particles which are not good for bonding; impact crushers are better in this respect and are often preferred when the clay has an inherently high plasticity.

In crushing non-plastic material, particularly that which is to be used as aggregate in castable and ramming mixes or that which is required as a 'grog' or skeleton in a body to be fired, a rough uni-sized product is advisable. Crushing is an obvious choice for this quality, achieved by cone- or similar crushers.

Tertiary Comminution. When a very fine product or component is required, a third stage of comminution to yield a maximum size of about 30-mesh is employed. Even finer material is possible if the primary and secondary stages have produced a better feed product.

It is customary to use either ball- or rod-mills for this operation[1] which can be either batch-type or continuous with the product being withdrawn in a current of air or water (Fig. VII.5).

Attritors or impact mills are also effective tertiary size reducers, utilising impact principles and therefore producing a lower content of very fine material than do the mills which have mainly a grinding action.

In all reduction processes, the shape of the resulting grains can frequently be as important as the size of the grains. To some extent, this can be controlled by the process itself.

Crushing normally produces rough angular grains ideal for concretes and as 'grog'.

Grinding is characterised by smooth, rounded fragments—in fact, if the material being processed is a little wet, the fine grains often agglomerate to form almost spherical pellets.

Impact crushing is more inclined to produce angular particles although smoother in surface appearance than those from true crushing. Impact crushing may develop characteristic shapes when particular minerals are processed. Fractures along lines of weakness predominate and a shaley clay or slate gives rise to flaky fragments after impact crushing.

The product of size reduction is a mass of particles of various shapes and sizes which together make up the body of the material to be processed into ceramic articles. The nature and interrelationship between the various grains determine many of the physical characteristics of the body, such as texture, porosity and bulk density.

PARTICLE-SIZE DISTRIBUTION

A RAW material composed of grains of various shapes and sizes can be termed a particulate system. The size-distribution of the grains in such a system is of great importance, but it is a difficult property to measure because it involves so many factors.

In a set of perfectly spherical ball-bearings it would be feasible to separate them into various sizes or weights and hence represent a distribution on either a numerical or weight basis. In clays and other ceramic materials the position is far more complicated—in fact it is not easy to define exactly what is meant by a particle.

As shown in earlier chapters the ultimate mineral crystallites in clays are

[1] For a detailed description of fine grinding processes see North, R., *Trans. Inst. Quarrying*, October 1969.

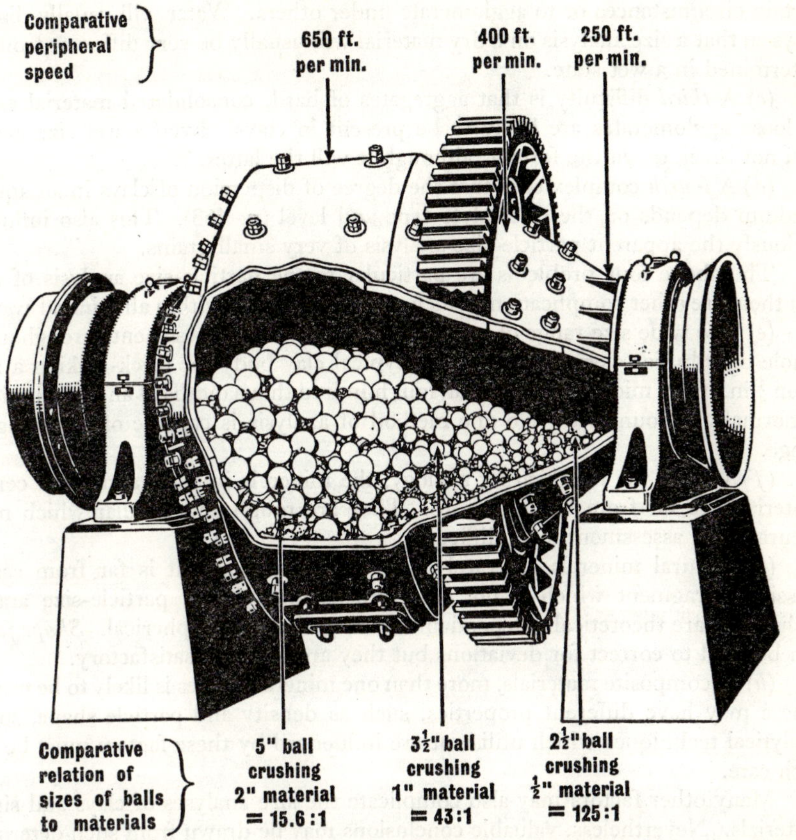

FIG. VII.5. A SCHEMATIC REPRESENTATION OF A BALL MILL

The model shown is a Hardinge Mill manufactured by International Combustion Products Ltd., Derby, England. The unit is so designed that the balls or pebbles are segregated by size as the mill rotates. Consequently the mill can be used for progressive size reduction as the charge travels from the entrance to the exit ends of the conical mill. The crushing ratios in various parts of the ball mill are shown in the diagram which also shows the estimated peripheral speeds in various sections.

extremely small. In a clay body these particles are aggregated and agglomerated so that it is often extremely difficult to isolate them in a free form. The grains produced by grinding of clays are mostly composed of many very small units which are the true crystallites of the system. In such properties as the texture, bulk density and porosities of bodies, the grains or particles following comminution are the main influencing factors but in colloidal behaviour and in firing characteristics, the crystallite size has greater significance.

(*a*) The *first* problem in the measurement of particle size of clays is to decide whether it is a *grain size* or a *crystallite size* which is in question.

(*b*) A *second* aspect is that the grains of clay are prone to breakdown under certain circumstances or to agglomerate under others. Water will usually disperse clays so that a size analysis on a dry material will usually be very different from that determined in a wet state.

(*c*) A *third* difficulty is that aggregates of hard, consolidated material as well as loose agglomerates are likely to be present in clays. Even a wet size analysis will not break down the former, although it will the latter.

(*d*) A *fourth* complexity is that the degree of dispersion of clays in an aqueous medium depends on the electrolyte type and level (p. 468). This also influences seriously the apparent particle-size analysis of very small grains.

The above four problems are particular to the particle-size analysis of clays, but there are other complications which apply to analyses within all mineral systems.

(*e*) The wide size range of particles makes accurate assessment throughout the whole distribution very difficult. In a typical clay body for brick-making a range from $\frac{1}{4}$ in. (6000 microns) to less than four hundred thousandths of an inch (1 micron) is normally encountered. No one method of analysis is capable of analysing this range.

(*f*) Unlike the ball-bearing analogy, the minerals in clays and other ceramic materials are far from being smooth. They are rough and angular which makes accurate size assessment most difficult.

(*g*) Natural minerals are rarely if ever spherical and it is far from easy to classify a fragment which is not equi-dimensional. Some particle-size analysis techniques are theoretically dependent on the grains being spherical. *Shape factors* can be used to correct for deviations but they are not very satisfactory.

(*h*) In composite materials, more than one mineral species is likely to be present. These may have different properties, such as density and particle shape, so that analytical techniques which utilise or are influenced by these factors must be used with care.

Many other factors may also complicate the size analyses of clays and similar materials. Nevertheless, valuable conclusions may be drawn from such determinations. It must always be remembered, however, that particle-size measurements can be absolute only under ideal conditions.

The Measurement of Particle Size

The measurement of the size of a single grain of irregular shape is a difficult problem of itself. The linear dimensions vary from point to point, so that a total volume or weight is the only acceptable descriptive parameter. From this an *equivalent spherical diameter* can be adduced relating the average cross-sectional length or the total volume or weight to that of an equivalent sphere and hence calculating a diameter.

When the problem involves very many irregular particles covering a wide size range, most of which cannot be inspected except under a microscope, the situation cannot be resolved except by assumption and compromise.

The only general satisfactory method of treating such a system is to relate the size of each particle to an equivalent sphere which under the conditions of experiment would behave in precisely the same manner. The concept of an *equivalent spherical diameter* (e.s.d.) to define the critical dimension of any particle is generally accepted and is a reasonable compromise for most systems.

Where particles are highly irregular in shape, such as those of a fibrous or platey nature, the concept of e.s.d. is less tenable although even in these materials, *shape factors* may be evolved to give a more meaningful picture.

In the measurement of particle-size distributions, therefore, some technique of differentiation into sizes has to be adopted. The sizes selected are defined in terms of perfect spheres and some calibration may be necessary before useful results can be obtained.

There are three general methods of assessing particle-size characteristics of particulate systems. Unless precautions are taken to break down the materials to be tested into ultimate crystallites, all methods will measure the grain size, and aggregates and agglomerates will be represented as single units.

A. Methods based on geometrical symmetry.
B. Methods based on hydrodynamic symmetry.
C. Methods based on surface area measurement.

In choosing the most suitable technique for the particle-size analysis, account must be taken of the potential significance of the results. The limitation of each method must also be appreciated particularly in the interpretation of the results. It must also be realised that all methods have a certain range of size within which they are suitable and outside of which they can be valueless or give misleading results. Many methods can be used over a wide size range but in any one analysis some of them are suitable for analysing only a narrow distribution.

The limits of particle-size measurement generally accepted for the principal methods of analysis are as shown in Table VII.1.

TABLE VII—I. THE LIMITS OF PARTICLE-SIZE ANALYSIS FOR VARIOUS METHODS OF MEASUREMENT

Method	*Approx. range of application (microns)*
Sieving	+50
Micropore sieving	0·001–50
Optical Microscope	0·25–250
Electron Microscope	0·0005–1
Sedimentation	1–50
Elutriation	2–50
Centrifugal sedimentation	0·05–5
X-ray scattering	0·005–0·1
Gas permeability (air)	2–200
Gas permeability (low pressure)	0·1–10
Gas adsorption	– 20
Organic complex adsorption	– 20

A. METHODS BASED ON GEOMETRICAL SYMMETRY

(1) **Microscopic Methods.** To examine a sample under the microscope is the most direct method of particle-size analysis.

It is the only satisfactory technique for judging the crystallite size of minerals in a hard rock sample. A thin section of such a sample will reveal the minerals present and give an indication of their size. Care in interpretation is necessary in that a section of rock prepared for either transmitted or reflected light observation will cut only one plane of each mineral fragment which is intersected.

A powder sample can also be analysed under the microscope and each particle observed as a projected image of its maximum cross-sectional area. Each particle must be observed and measured separately although much of the tedium can be alleviated by using modern automatic counting devices. There are various ways of selecting parameters to define each particle. The four most commonly used are:

(*a*) Martin's statistical diameters in which the intercept of each particle, which divides it into two equal halves in a particular plane of the microscope, is recorded as the section is traversed.

(*b*) Heywood's method of measuring two dimensions of every particle in two arbitrary directions at right angles to each other.

(*c*) Measuring the area of each particle on a calibrated projection screen.

(*d*) Using a comparison graticule of the type described by Fairs.[1]

Microscope methods have a limited application in particle-size analysis. Only a small sample can be examined so that multiple determinations are necessary if reliable results are to be obtained; the analysis is tedious no matter what technique is used; the ratio of sizes which can be analysed at any one time is small because of the magnification factor and the limited depth of focus. Phase contrast microscopy is claimed to be an excellent technique for particle size analysis of small grains.[2]

The electron microscope can be used to measure the size of very small particles down to 0·001 micron.[3]

(2) **Sieving, sifting or screening** is the most widely used technique of size analysis for materials down to about 50 microns. It involves passing the material—either in its dry state or in suspension in water—over a series of meshes each having apertures of known dimensions. The particles larger than these apertures remain on the sieves, whilst the smaller particles pass forward to sieves with successively smaller apertures until the last one is reached, through which the smallest grains eventually pass. This method is very convenient for separating all particles larger than 0·002 in. (53 microns) diameter (300-mesh).

In recent years, the development of certain plastic formulations has enabled sieves to be manufactured capable of separating extremely small particles. Commercial filters or sieves of this nature are manufactured to separate particles throughout the range from normal sieve sizes (53 microns) down to a few millimicrons. These are made with great precision and an extraordinarily high effective screen area can be secured. Naturally such units are delicate and incapable of handling large quantities of material, but they are invaluable additions to the screening range.

[1] Fairs, G. L., *Chem. & Ind.*, 62, 375, 1943.
[2] Stein, F., *Powder Tech.*, 2, 377, 1969.
[3] Hall, C. E., *Introduction to Electron Microscopy* (McGraw-Hill, N.Y., 1953).

Sieves, riddles, and screens are made with apertures of various shapes some of which are circular, others square, and still others are rectangular. The last-named are useless for investigating the texture or fineness of a material but are useful for industrial screening.

Sieves with circular apertures are simply perforated sheets of metal or other suitable material and are not as reliable for testing angular materials as those with square apertures made from wire-gauze.

Gauze-sieves are made of various metals, including iron, steel, brass, bronze and occasionally copper. In some cases, gauzes made of silk (*lawns*) are employed for the finest materials, but they are very delicate and not to be recommended for bulk separation. Those made of phosphor-bronze are usually the most satisfactory. All sieves used for testing should be examined frequently to ensure that the apertures remain of the desired sizes, as otherwise the results obtained may be erroneous. This applies particularly to sieves with the finest apertures which tend to stretch and sag with frequent use.

Standard sieves are in use in both this country and abroad; they are classified by numbers which designate the number of apertures per linear inch, although the actual sizes of holes can be derived only provided that the diameter of the wire used is known. The six types commonly employed are: (*a*) British Standard Test Sieves (B.S.S.) conforming to the British Standard Institute Specification (No. 410, 1931); (*b*) Institution of Mining and Metallurgy (I.M.M.) standard; (*c*) the German DIN standard; (*d*) the French AFNOR standard; (*e*) Tyler Standard Screens; and (*f*) the United States Standard Sieve Series. The last two are those commonly used in the United States and in all the ratio of the aperture of any particular screen to that of the succeeding one is equal to $\sqrt{2}$ or 1·414.

The aperture-size (or length of square side) of the sieves of some of the series is shown in Table VII.II.

The grain-size distribution of a sample may be estimated by sieving either in the dry state or after it has been mixed with a large amount of water to form a slip or slurry. Whilst the results obtained may be quite different, the method of using the sieves is similar in each case.[1]

A weighed quantity of the sample is placed in the largest-aperture sieve and stirred or shaken until as much of it as will pass through that sieve has done so. The material which has passed through the first sieve is similarly treated on the next finer sieve in the series and this process is continued either until all the powder is removed or no smaller sieves are available. The particle-size distribution may then be calculated by plotting the cumulative curve of the percentage of sample retained by each sieve against the size of aperture.

The dry method of sieving will indicate the fineness of division after crushing or grinding, although some particles retained on the sieves will be aggregates of much finer crystallites. When the size-distribution is being measured to assess the textural properties of the moulded powder it is usually advisable to regard such agglomerated particles as single grains. If, however, the particle size of the crystal or mineral fragments is required then such composite grains must be broken down and for this reason the 'wet' method of sieving is preferable when an ultimate size

[1] British Standard Specification No. 1902 1A (Addendum 1), 1967.

TABLE VII—II. SIZE OF APERTURES IN STANDARD SIEVES SERIES

Number of Series	*B.S.S.*	*I.M.M.*	*Tyler Series*	*U.S. Standard*
		in.	in.	in.
	From 2 in. to ¼ in. in ⅛ in. steps then $\frac{3}{16}$ in., ⅛ in., $\frac{1}{16}$ in. and $\frac{1}{32}$ in.		1·050	
			0·883	
			0·742	
			0·624	
			0·525	
			0·441	
			0·371	
2½	—	—	0·312	0·315
3	—	—	0·263	0·265
3½	—	—	0·221	0·223
4	—	—	0·185	0·187
5	0·1320	0·1000	0·156	0·157
6	0·1107	—	0·131	0·132
7	0·0949	—	0·110	0·111
8	0·0810	0·0630	0·093	0·0937
9	—	—	0·078	—
10	0·0660	0·0500	0·065	0·0787
12	0·0553	0·0417	0·055	0·0661
14	0·0474	—	0·046	0·0555
16	0·0395	0·0313	0·0390	0·0469
18	0·0336	—	—	0·0394
20	—	0·0250	0·0328	0·0331
22	0·0275	—	—	—
24	—	—	0·0276	—
25	0·0236	—	—	0·0280
28	—	—	0·0232	—
30	0·0197	0·0167	—	0·0232
32	—	—	0·0195	—
35	—	—	0·0164	0·0197
36	0·0166	—	—	—
40	—	0·0125	—	0·0165
42	—	—	0·0138	—
44	0·0139	—	—	—
45	—	—	—	0·0138
48	—	—	0·0116	—
50	—	0·0100	—	0·0117
52	0·0116	—	—	—
60	0·0099	0·0083	0·0097	0·0098
65	—	—	0·0082	—
70	—	0·0071	—	0·0083
72	0·0083	—	—	—
80	—	0·0063	0·0069	0·0070
85	0·0070	—	—	—
90	—	0·0055	—	—
100	0·0060	0·0050	0·0058	0·0059
115	—	—	0·0049	—
120	0·0049	0·0041	—	0·0049
140	—	—	—	0·0041
150	0·0041	0·0033	0·0041	—
170	0·0035	—	0·0035	0·0035
200	0·0030	0·0025	0·0029	0·0029
230	—	—	—	0·0024
240	0·0025	—	—	—
250	—	—	0·0024	—
270	—	—	0·0021	0·0021
300	0·0021	—	—	—
325	—	—	0·0017	0·0017
350	0·0017	—	—	—

analysis is required. The residue retained on each sieve can be readily freed from adhering finer particles by washing with a jet of water.

Even with prolonged treatment, some large particles of particularly hard clays remain as agglomerates; these are extremely difficult to break up and as they would certainly behave as single particles when being moulded into shape they are usually recorded as such in the particle-size analysis. As a result, fireclays, clay-shales and similar materials appear to contain a larger proportion of coarse particles than would be expected. When the ultimate crystallite sizes are required, all agglomerates must be broken down but without disrupting any individual crystal.

Many raw materials, particularly clays, contain particles which pass through the smallest-aperture sieve. The estimation of the grain-size distribution of such materials is often assessed by water sedimentation or elutriation methods.

(3) **Particle Size Counters** of which the Coulter Counter[1] is the best known. In this method a suspension of the sample in an electrolyte is drawn through a small aperture between charged plates. When particles pass through, they cause an interruption to the flow of electrical current of a magnitude proportional to the cross-sectional area of the particle. The effects are amplified and fed directly to an electronic counter. It is usual in determining a size distribution to count the number of particles at different threshold values.

In practice only a limited range of sizes can be determined with one aperture but the unit is of value in assessing material smaller than sieve sizes. It is often used in conjunction with a series of millipore screens.

B. METHODS BASED ON HYDRODYNAMIC SYMMETRY.

In the analysis of particles below sieve size (*circa* 50 microns) it is usual to employ a technique which depends upon sedimentation laws. When a particle is allowed to fall freely in a medium, it will accelerate under gravity until it attains a specific velocity. It will then continue to fall with the same terminal velocity unless there is a change in sedimentation conditions.

With large-sized particles the terminal velocity is not reached for a considerable time and the flow characteristics are complicated. Particles of 50 microns or less, however, attain their terminal velocity in less than one second and for all practical purposes the whole of their sedimentation cycle can be regarded as being at constant velocity.

The rate of fall of particles in a liquid medium was investigated by Stokes in 1845.[2] He was able to derive an equation which related the terminal velocity of fall to the square of the diameter of perfect spheres when subjected only to the influence of gravity and free from interference by neighbouring particles.

This full expression which is called Stokes' law is

$$V = \frac{2}{9} \frac{(\rho_1 - \rho_2)gr^2}{\eta}$$

[1] The Coulter Counter is an apparatus manufactured and supplied by Messrs. Coulter Electronics Ltd., St. Albans, Herts., U.K.

[2] Stokes, *Camb. Phil. Trans.*, 8, 287, 1845 and 9, 8, 1851.

where V is the velocity of fall in cm. per sec.,

η is the viscosity of the liquid $\left(\begin{array}{llll} & 10^\circ\ C. & 20^\circ\ C. & 30^\circ\ C. \\ \text{water} = & 0{\cdot}013, & 0{\cdot}010, & 0{\cdot}008\ \text{poises}\end{array}\right)$

ρ_1 and ρ_2 are respectively the density of the particles and the liquid medium,

g is the gravity constant (981 cm./sec.^{2}),

and r is the radius of particles in cm.

Stokes' equation is applicable only when it is assumed that:

(*a*) The particles of solid matter are much larger than the particles of liquid so that the effects of Brownian movement (see p. 456) are not serious.

(*b*) The liquid is homogeneous and of infinite extent in comparison with the particles.

(*c*) The particles are smooth and rigid.

(*d*) No slipping occurs between the particles and the liquid.

(*e*) The velocity is not too great or else free settling may be restricted.

(*f*) The particles are small, but not excessively so.

(*g*) The particles are free to fall. If the settling vessel is narrow and the suspension is concentrated, the walls and the presence of neighbouring particles will affect the rate of falling. The influence of other particles is most serious, because, when suspended in a water medium, each particle becomes charged and complex repulsive forces may prevent free fall (see Chapter VIII). Oden[1] has shown that particles fall freely only if the concentration does not exceed 1 per cent, but serious error does not arise provided that less than 5 per cent by weight is present.

(*h*) The particles are true spheres. This is rarely the case and the shape factor introduces serious error. Some attempt at a compromise has been introduced by the term *equivalent spherical diameter* (e.s.d.). This refers the dimension of a particle to the radius of a sphere which would settle with the same velocity.

(*i*) The temperature of the medium is constant, as the viscosity diminishes with an increase in temperature. The velocity of fall is 30 per cent greater at 20° C. than at 10° C.

(*j*) There are no currents in the medium.

(*k*) There is no overlapping in sizes due to excessive concentration which might result in finer particles being carried down underneath larger ones. This may be largely avoided by repeated sedimentations of each fraction.

Other factors which also influence the correct determination of the particle-size distribution by sedimentation methods, are the presence of particles of different densities (e.g. sand, chalk or clay) and associated soluble salts, organic and inorganic colloids.

From the foregoing it will be realised that the use of Stokes' Law may give results which have little relation to the absolute particle-size distribution. However for comparative determinations, methods based on the standard equation are invaluable and are used extensively.

Sedimentation tests may be carried out in several ways but they usually involve allowing a mixture of particles to settle for certain periods of time and then measuring

[1] Oden, S., *Int. Mitt. Bodenk.*, 5, 257, 1915.

either the amount removed from or remaining in suspension. Stokes' equation can be modified by assuming that the velocity of fall (V) is equal to the height (s) through which the particles (of radius r) fall in unit time.

i.e. $$V = \frac{s}{t}$$

The equation then becomes $$\frac{s}{t} = \frac{2}{9} \frac{(\rho_1 - \rho_2)gr^2}{\eta}$$

For a particular powder sedimenting at a constant temperature ρ_1, ρ_2, η and g are constants. Therefore the equation may be written in terms of the diameter of particles as

$$\frac{s}{t} = Cd^2$$

where d is the diameter and C is a constant, or

$$d = k\sqrt{\frac{s}{t}}$$

where k is a constant.

For material with a density of 2·6 settling in water at 25° C. (when $\eta = 0{\cdot}01$), k has the value of $13{\cdot}84 \times 10^{-4}$ (where d is in cm. and t in minutes). Table VII.III shows the time taken to fall a fixed distance and also the distance fallen in one hour for differently-sized particles.

TABLE VII—III. SEDIMENTATION DATA FOR SPHERICAL PARTICLES OF DENSITY 2·6 IN WATER AT 25° C.

Size of Particles		*Time to fall 20 cm.*	*Distance fallen in 1 hr.*	
mm.	in.			
0·1	0·004	15 secs.	3139	cm.
0·05	0·002	1 min. 31 secs.	784	cm.
0·01	0·0004	38 min. 15 secs.	31·39	cm.
0·005	0·0002	2 hr. 33 min.	7·84	cm.
0·001 (1 micron)	0·00004	63 hr. 45 min.	0·314	cm.

Sedimentation Methods Particle-size distributions may be measured by either of two *sedimentation* methods.

(*a*) Involves making up a slip of known concentration, carefully pouring a small volume of it on to the top of a column of water and then allowing the particles to settle for pre-arranged periods of time. The amount settling or that remaining in suspension can be estimated and related to the particle size by using Stokes' equation.

(*b*) A more commonly-used procedure is to make up a suspension of the sample in a cylindrical vessel and then allow it to settle. Samples of the liquid at a fixed distance from the surface are withdrawn by a pipette at suitable intervals. An apparatus designed by Andreasen and Lundberg[1] to simplify this procedure is shown in Fig. VII.6.

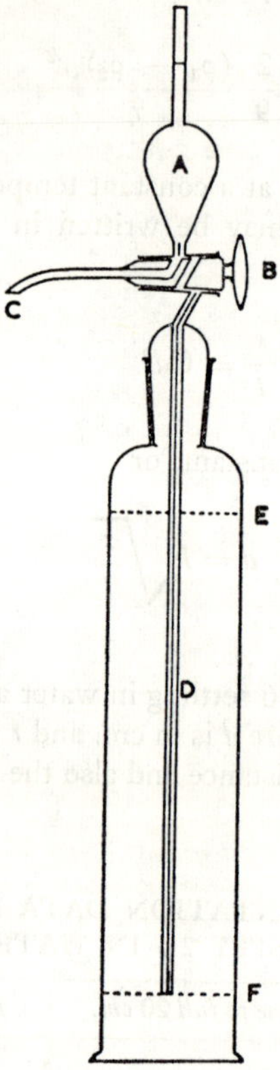

FIG. VII.6. THE ANDREASEN PIPETTE

A dilute suspension which should be entirely dispersed is contained in the graduated cylinder and filled up to the mark E. A 10 ml. pipette A is connected through a two-way tap B to a stem D by which liquid can be drawn from the level F which is exactly 20 cm. below E. The alternative outlet C to the tap allows a sample which has been collected in the pipette to be transferred to a weighed dish.

Samples should be withdrawn from time to time and the concentration at the point F determined by weighing the dried residue in the collecting dishes. The maximum grain-size which remains in suspension can be calculated by applying Stokes' Law, and a graph of the percentage of original sample remaining in suspension below each maximum grain-size may be drawn. More information may be

[1] Andreasen, A. M. H., *Ing. Skrifter*, No. 3, 1939. Andreasen, A. M. H. and Lundberg, J. J. V., *Ber. deut. keram. Ges.*, 11, 249, 1930.

derived by constructing a histogram from the curve, thereby indicating the concentration of particles in several size ranges.

Each sample collected contains particles of all sizes, from the largest remaining above the level F, to the smallest which were present originally. Particles of one particular size (d_1) will, at the start of the experiment, be distributed uniformly through the volume of the suspension. They each must fall with the same velocity, hence their concentration round the level F will be constant until such time as those which were at level E have fallen through 20 cm. No more particles of this size can then be drawn into the pipette on sampling, hence the overall concentration will be lowered, for only particles less than d_1 will remain in suspension.

The results are best represented on a graph plotting the concentration of the withdrawn samples relative to the original concentration, against the maximum size of particles which can still be in suspension at the time of sampling. For accurate results, a correction must be applied to allow for the reduction in the length of the settling column when a sample is removed. A graph of the percentage weight of the sample finer than a particular size (plotted on a cumulative basis) against the particle diameter is shown in Fig. VII.7, but a clearer picture of the size-distribution of particles is given by constructing a histogram. This shows the percentage by weight of particles in a certain size-range and is a measure of the slope of the line. It is readily constructed by determining the percentage by weight of particles over small increments of size. Thus, in the cumulative curve in Fig. VII.7, the percentage of the particles between 0–2 microns is 7 per cent, between 2–4 microns it is 29 per cent and so on, and from these values the block diagram can be constructed.

Curves of this type show that whilst clays contain particles of a wide size range, the major portion is concentrated between narrow limits.

The principle of sedimentation is used in a simple apparatus for determining

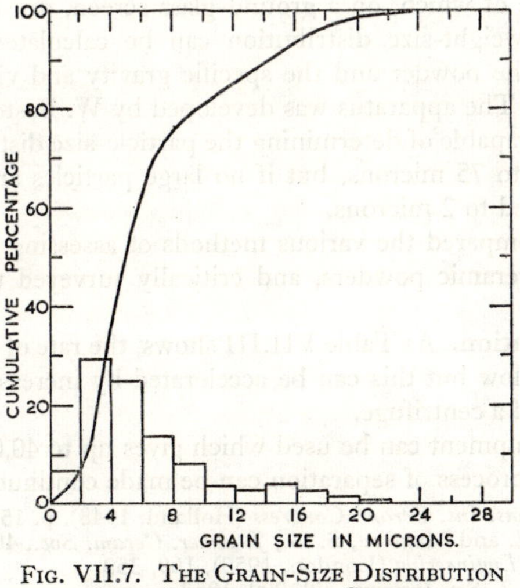

FIG. VII.7. THE GRAIN-SIZE DISTRIBUTION OF A CLAY, SHOWING THE CONSTRUCTION OF A DISTRIBUTION HISTOGRAM

the clay content of sands. The weighed sample of sand is agitated in a specified volume of water in a beaker-like container. When the stirrer is removed the particles fall freely and are allowed to do so for a period of time sufficient to settle all grains above 0·010 mm. The suspension is then carefully siphoned off, more water added and the process repeated a sufficient number of times to give a clear liquid after the specified settling period. The sand which remains as a sediment is thoroughly dried and reweighed; the loss in weight is a measure of the clay content. Apparatus of this type is manufactured commercially and the makers usually specify the most suitable conditions for the test.

Interesting and valuable refinements to the pipette method of sedimentation analysis are:

(i) Instead of withdrawing samples of the suspension at a fixed depth, a hydrometer or small 'diver' of suitable density is inserted into the liquid, where at some point it reaches equilibrium and then proceeds to fall steadily at the rate of sedimentation of the largest particles at that level. The use of small divers, made of hollow glass spheres weighted with metal shot to give the required density, was developed by Berg[1] and appears to be of more value than other methods using large hydrometers, although the latter are of frequent use for control in the clay industries.[2]

(ii) By measuring the optical density of the suspension at certain points. A narrow beam of light is passed through the liquid and the intensity of transmitted light measured by a photo-electric cell. As large particles settle out from the section being examined the optical density decreases and a graph may be plotted of size-distribution once the apparatus has been suitably calibrated.[3]

The *sedimentation balance* has been devised to measure the particle-size distribution in the sub-sieve range. The powder in suspension is allowed to fall through a column of water and settle on the pan of a torsion balance where it is continuously and accurately measured, because the torsion wire carries a pointer fitted with a graticule, the image of which, on a ground-glass screen, can be observed through a telescope. The weight-size distribution can be calculated provided that the specific gravity of the powder and the specific gravity and viscosity of the liquid (water) are known. The apparatus was developed by W. Bostock[4] and is produced commercially; it is capable of determining the particle-size distributions of powders within the range 5 to 75 microns, but if no large particles are present, the lower limit can be extended to 2 microns.

Johnson has compared the various methods of assessing the particle size distribution of some ceramic powders, and critically surveyed their limitations and advantages.[5]

Centrifugal Separation. As Table VII.III shows, the rate of settling of small particles is extremely slow but this can be accelerated by increasing the gravity constant (g) by means of a centrifuge.

Small-scale equipment can be used which gives up to 40,000 g and by using a controlled feed the process of separation can be made continuous. The calculation

[1] Berg, S., *Trans. Int. Ceram. Congress* (Holland, 1948), p. 152.
[2] Phelps, G. W. and Maguire, S. G., *J. Amer. Ceram. Soc.*, **40**, 399, 1957.
[3] Rose, H. E., *Engineering* (London, 1950), **169**, 350.
[4] Bostock, W., *J. Sci. Instrum.*, **29**, 209, 1952.
[5] Johnson, R., *Trans. Brit. Ceram. Soc.*, **55**, 237, 1956.

of the grain-size distribution of finely-divided clays is made possible by using a super-centrifuge, the details of which are shown in Fig. VII.8. The suspension enters through a nozzle at the lower end of the rotor at a controlled rate. A steam-driven turbine rotates the bowl at speeds up to 40,000 revs. per min., resulting in the separation of particles along the walls of the cylindrical rotor. The clarified liquid passes out through holes in the top of the bowl into the extracting funnels and then into a suitable receptacle. Under carefully-controlled conditions, the material in suspension can be separated into layers of gradually-decreasing particle size along the length of the rotor, so if a removable liner is inserted beforehand the grain-size distribution of the sample under test can be more readily assessed by cutting the liner into strips of calculated thicknesses and weighing the amount on them.

The calculation of the size of particles separating out in a centrifuge is complicated. The increase in g is the result of the angular velocity of rotation ($\omega^2 r$) which in turn depends on the distance between the particle itself and the axis of rotation. If the column of liquid through which the particle settles is not very small with respect to the diameter of the bowl or rotor, then r will be constantly changing and an ever-increasing gravitation force will be acting on each individual particle.

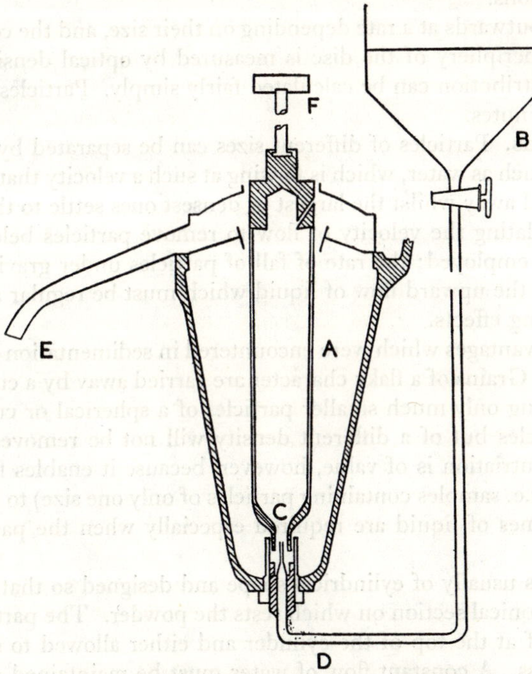

FIG. VII.8. THE CONTINUOUS SUPER CENTRIFUGE

(A) The outside protective case
(B) The feed reservoir
(C) The rotor
(D) The lead in for the suspension
(E) The outlet for the liquid
(F) The driving-head for the rotor

Even greater complexity is introduced into the calculation of the settling position of particles sedimenting under conditions of continuous flow in a supercentrifuge. In addition to being under the influence of a constantly varying *g*, each particle is moving upwards with a velocity governed by the rate of flow. Saunders[1] has prepared nomograms whereby the particle-size distribution of finely-divided clays can be determined and he has shown that fairly mono-dispersed size fractions can be separated.

A quick method of determining the particle-size analysis by centrifugal techniques is known as the Packing Volume Technique. A sediment is allowed to settle in a constricted section of a centrifuge tube. The packing volume can be assessed quickly, so that the technique affords a rapid method of a full size analysis. Conley has described the apparatus and methods of calculation.[2]

The *disc centrifuge* designed by R. H. Kaye[3] is of value when the sizes of particles of typical clays are to be measured rapidly. The principle involves sedimenting a dilute suspension of clay in a transparent hollow disc which can be rotated at a speed to give a predetermined centrifugal force. The unit is filled with water initially and the suspension inserted coaxially when the apparatus is operating under steady conditions.

Particles settle outwards at a rate depending on their size, and the concentration at any time on the periphery of the disc is measured by optical density methods. The particle-size distribution can be calculated fairly simply. Particles of 1 micron settle in about 40 minutes.

Elutriation Methods. Particles of different sizes can be separated by suspending them in a medium, such as water, which is flowing at such a velocity that the smallest or lightest are carried away whilst the largest or densest ones settle to the bottom of the vessel. In calculating the velocity of flow to remove particles below a certain size, Stokes' Law is employed; the rate of fall of particles under gravity has to be slightly exceeded by the upward flow of liquid which must be regular and uniform and free from eddying effects.

The same disadvantages which were encountered in sedimentation analysis also apply in elutriation. Grains of a flaky character are carried away by a current which is capable of removing only much smaller particles of a spherical or cubical form; similarly-sized particles but of a different density will not be removed under the same conditions. Elutriation is of value, however, because it enables fairly mono-dispersed fractions (i.e. samples containing particles of only one size) to be obtained, although large volumes of liquid are required especially when the particle size is comparatively large.

The elutriator is usually of cylindrical shape and designed so that water flows upwards through a conical section on which rests the powder. The particles in suspension are taken off at the top of the cylinder and either allowed to settle or are evaporated to dryness. A constant flow of water must be maintained until a clear liquid is being extracted at the exit end of the elutriator; the rate of flow may then be increased and a larger-sized fraction collected.

[1] Saunders, E., *Ind. Eng. Chem. (An.)*, 20, 380, 1948.
[2] Conley, R. F., *J. Amer. Ceram. Soc.*, 46, 1, 1963.
[3] Kaye, R. H., *Analyst*, 1963.

The particle size of material being removed at a particular rate of flow can be calculated provided that the bore of the elutriator is known and also the density of the sample.

The velocity of flow of water (V) = rate of flow (i.e. volume (v) per unit time (t) $\times$ area of cross-section (A)

$$\text{i.e.}\quad V = \frac{dv}{dt}.A$$

The velocity at which particles fall when suspended in a medium is predicted by Stokes' equation (p. 381), and for a particular size of particle to be extracted in the elutriator, its rate of fall must be slightly less than the velocity of flow of the liquid. Hence:

$$\frac{dv}{dt}.A = \frac{2.(\rho_1 - \rho_2)gr^2}{9\eta}$$

Therefore, as all other factors are known or can be calculated, the grain size distribution of a powder can be derived. The rate of flow of water is easily determined by measuring the quantity which is collected at the take-off of the elutriator over a certain period of time. It should be expressed in millilitres per second.

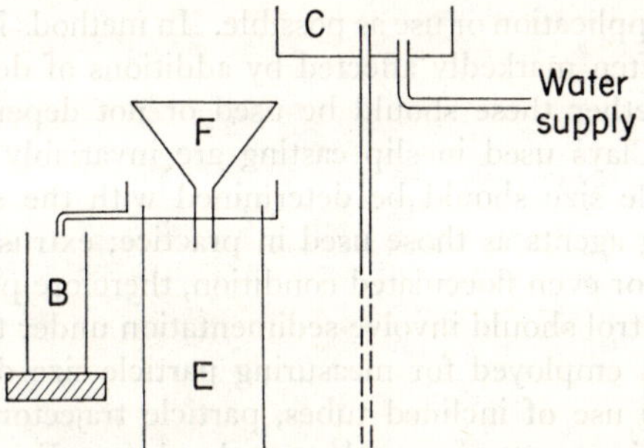

FIG. VII.9. DIAGRAM OF A SIMPLE ELUTRIATOR

A basic type of elutriator is shown in Fig. VII.9. Water from a constant head supply (C) passes into the conical flask (D) and upwards through the elutriator (E) which is so shaped at its lower end (T) to provide teeter column conditions and this keeps all particles in the system in a constant state of agitation.

The constant head level is adjusted initially to give a low rate of flow up the column, so that only the finest particles will be removed. The rate of flow (V) through the elutriator column in cm./sec. can be accurately measured by weighing the overflow into beaker (B) over a known time interval (*a* g./sec.).

Then $V = \frac{a}{\pi r^2}$ where *r* is the radius of the elutriator column.

Once steady flow conditions have been established the powder is introduced slowly through funnel (F). Coarse particles will settle whilst those which would fall with a smaller Stokes velocity than that of the upward-rising water current will be carried over into the collecting beaker. Once the liquid being collected becomes clear, the rate of water flow through the elutriator can be increased and a larger size collected.

The products in the beaker can be filtered or flocculated to remove excess of water but the larger-sized material (+5 microns) settles so quickly that decantation is usually sufficient.

There are many other techniques for measuring the particle-size distribution of fine powder, some of which are of particular value in the study of certain systems. In measurements of this nature it is important to appreciate that the results obtained are rarely absolute values and that it is best to select a comparison method of testing as closely akin to the application or use as possible. In methods involving sedimentation the results are often markedly affected by additions of deflocculating or flocculating agents. Whether these should be used or not depends on the ultimate industrial practice. Clays used in slip casting are invariably fully deflocculated, therefore their particle size should be determined with the same type and proportions of dispersing agents as those used in practice; extrusion of clays is most usual in their natural or even flocculated condition, therefore particle-size distribution as a means of control should involve sedimentation under the same conditions.

Other techniques employed for measuring particle-size distributions include hindered settling, the use of inclined tubes, particle trajectory analysis involving inertial effects, radiation scattering and several others. For further details the reader is referred to *Fine Particle Measurement* by C. Orr and J. M. Dallavalle (Macmillan Co., N.Y., 1960).

Of all the methods described, only sieving and elutriation techniques afford a completely satisfactory means of separating particles into their specific grain-size ranges. It is often important to effect such a separation particularly in scientific studies of the influence of different particle fractions on a specific property.

If large quantities of specific size ranges below the smallest sieve size are required, sequential sedimentation and decantation is reasonably satisfactory. The sample is made into a suspension and allowed to settle in a large cylindrical vessel. From a knowledge of the density of the material, the time can be calculated at which particles above a certain size will have settled below a specific point in the suspension. After this prearranged time interval the material still in suspension is siphoned into another container.

The sedimented residue is restirred with more water and again allowed to settle for the same time interval as before. Once again the fines are decanted or siphoned and added to the first extract. Three operations of this nature are sufficient to reduce the undersize in a fraction to less than 5 per cent.

The procedure can be repeated at different size intervals in the same manner. Excessive volume concentrations can be avoided by flocculating the suspension at intermediate stages, decanting the clear liquid after settling and then redispersing. Useful size ranges can be chosen on a logarithmic basis; the author prefers the following sequence which links with a 300-mesh British Standard screen (53 microns).

53–27 microns; 27–13·5 microns; 13·5–6·8 microns; 6·8–3·4 microns; 3·4–1·7 microns; less than 1·7 microns.

Material below 1·7 microns can be separated by centrifugal means.

Air-separation or **air-elutriation** methods, in which air is used instead of water, are based on the same general principles, but accurate results are difficult to obtain, because convection and eddy effects are not easy to eliminate, and these destroy the steady flow conditions so necessary for quantitative determinations. Apart from this, however, air has the advantages of separating much finer particles and delivering them in a much more convenient condition. It can be used only for dry powders.

Methods of determining the grain-size distribution of a powder which are based on the settling of particles in a liquid are subject to errors which arise mainly as a result of the shape factor and the mutual interaction between liquid molecules and particles of small size.

The influence of the shape of particles on sedimentation behaviour has been studied by several workers in recent years. Davies[1] has shown that under certain conditions, factors can be applied to correct the standard equation, although if more than one type or shape of particle is present in a heterogeneous mixture a complete solution is impossible.

C. METHODS BASED ON SURFACE AREA MEASUREMENT. Although not strictly a technique for measuring particle-size distribution, the total surface area of a particulate system may reveal many of the important characteristics associated with size. Thus the intensity of chemical reactivity, the degree of absorption, and some plasticity characteristics are related to surface area so that its measurement provides an important parameter.

Some methods of determining surface area are rapid, whilst those for particle-size distributions are invariably slow and by their very nature take several hours. Consequently the measurement of surface area can be a most useful control tool and by its use a mean particle size can be calculated.

As in all sizing techniques it is necessary to assume that the particles can be related to equivalent spheres of the same surface area.

It is usual to measure the *specific surface area*, S, which is defined as the total surface area per gram.

[1] Davies, C. N., *Symposium on Particle Size Analysis* (Inst. of Chem. Ind. and Soc. Chem. Ind., 1947), p. 25.

This can be represented mathematically as

$$S = \frac{4\pi r_m^2}{\frac{4}{3}\pi r_m^3 \rho}$$

where r_m is the mean equivalent radius of particles and ρ is their density.

$$= \frac{3\pi}{r_m \rho} \text{ or } \frac{6\pi}{d_m \rho}$$

where d_m is the mean e.s.d.

$$\therefore d_m = \frac{6\pi}{S\rho}$$

The determination of surface area of a mass of particles presents great complexity in interpretation. Surface area involves (*a*) the external surfaces of particles as determined by their geometrical shape, (*b*) the internal surfaces of cracks and pores within the grains and (*c*) internal surfaces within the lattice of the crystals themselves due to a departure from a truly closely-packed arrangement. This latter point is often ignored but in some layer lattice and framework silicates, e.g. zeolites, it can be a major contributory factor to high surface area values.

Most methods for measuring surface areas of powders depend upon the principle of surface ad- or ab-sorption. Unsatisfied electrical charges in the region of surface layers of atoms attract atomic and molecular groupings and retain them on the surface. Both physical and chemical bonding are involved and it is important although difficult in surface area measurements to differentiate between them. At very low temperatures and pressures, more simple physical adsorption is the predominant mechanism and consequently gas adsorption methods are usually carried out under these conditions.

One other problem is that surfaces may adsorb multi-molecular layers, the number of which can vary with the pressure of gas and with surface features. Micro-cracks or channels or voids in the lattice may adsorb different levels of gases from those of the outer surfaces. In some methods, large fibrous molecules are used for adsorption where the chains stick out from the surface and these cannot penetrate all channels as readily as small gaseous molecules.

Perhaps the greatest problems in adsorption techniques are in assessing what features of surface area contribute to the value measured and how results from different adsorbants should be correlated.

It is not surprising therefore that different methods of measuring surface area give widely varying results, and caution is required in interpreting values.

The surface area is usually expressed as the *specific surface area* which is defined as the total surface area in square metres per gram, and it is usual to indicate the method used in its determination.

There are three major techniques—with several variants—used in the measurement of surface area:

(1) gas adsorption
(2) adsorption of a compound or dye from solution
(3) permeability methods.

Gas adsorption Methods. The principle of these methods is that molecules can be adsorbed on to surface layers by virtue of unsatisfied valencies and the amount adsorbed depends on the molecules available (or their partial pressure) and the contact surface area. Langmuir developed a mathematical treatment of such a condition and his equation—the *Langmuir adsorption isotherm*—is the basis for calculation (see also p. 680 for more details).

This may be written in the form:

$$\frac{P}{V} = \frac{1}{V_m b} + \frac{P}{V_m}$$

where V is the volume of gas adsorbed at pressure P, V_m is the volume adsorbed if the whole surface were covered with a mono-molecular layer and b is a constant.

So by plotting P/V against P, it should be possible to calculate V_m.

This equation is valid provided that adsorption is of a mono-molecular nature which is not always the case. Brunauer, Emmett and Teller[1] have developed a generalised treatment which can be applied for all cases but only at very low temperatures and over a small range of gas pressures. This may be written as:

$$\frac{P}{V(P_s - P)} = \frac{1}{V_m C} + \left(\frac{C-1}{V_m C}\right)\frac{P}{P_s}$$

where the new terms are P_s the saturation pressure of the gas and C a constant.

The important factor established by this equation is that a graph relating the volume of gas adsorbed to the relative partial pressure of that gas will show a sensible linear relationship over part of the curve usually from values 0·05–0·35 (see Fig. VII.10). The volume adsorbed (V_m) if the whole surface were covered with a mono-molecular layer of the gas can then be calculated from the expression:

$$V_m = \frac{1}{\text{slope}}$$

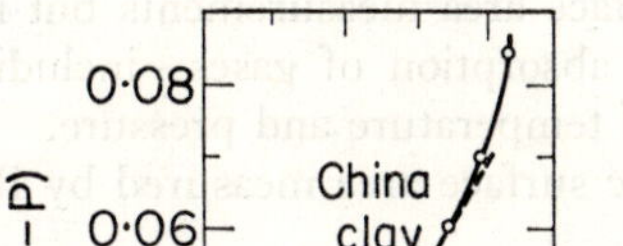

FIG. VII.10. TYPICAL BET CURVES FOR THE ABSORPTION OF NITROGEN ON A CORNISH CHINA CLAY AND A DORSET BALL CLAY

[1] Brunauer, S., Emmett, P. H. and Teller, E., *J. Amer. Chem. Soc.*, 60, 309, 1938.

This treatment is known as the BET method, so called from the first letter of the surnames of the authors who developed it.

In calculating the surface area from the value V_m, the area occupied by one adsorbed molecule must also be known. Livingston[1] has collected published data of many gases—for nitrogen, the important gas used in surface area measurement, the value generally accepted was derived by Joy[2] as 15·8Å^2.

The apparatus used in the BET method is complex in that gas volumes and pressures have to be established with great precision. In principle, however, it is quite simple. The sample must firstly be degassed and freed from other adsorbed phases by being heated in a tube at moderately high temperatures under very low pressures. It is then immersed in liquid nitrogen or oxygen and a known volume of nitrogen at a known pressure is admitted to the system. Adsorption occurs which causes a reduction in pressure. The determination is repeated at several pressures of gas and an isotherm constructed from which the surface area can be calculated.

Precision equipment for BET determinations is elaborate but simplified units have been evolved which although not as accurate, are excellent for obtaining relative results.[3,4] Commercially constructed equipment is also available.

Nelsen and Eggertsen[5] have developed an extremely elegant unit for measuring surface area by gas adsorption. They pass a gas stream of mixed nitrogen and helium over a sample at liquid nitrogen temperatures. Nitrogen is adsorbed so that the composition of emerging gases changes; this can be measured by thermal conductivity methods. The quantity of nitrogen can be calculated on both the adsorption and desorption cycle.

Another interesting and useful technique involving gas adsorption is the McBain–Bakr balance in which the adsorbed gas is measured gravimetrically.[6] The sample to be measured is suspended from a sensitive spring balance usually manufactured from fused silica. The adsorbed gas can then be weighed by observing the depression of the pan in which the sample is contained. Such a technique is useful for surface area measurements but it is more widely used for measuring adsorption and absorption of gases—including water vapour—over a wide range of conditions of temperature and pressure.

Some values of specific surface area measured by BET techniques are shown in Table VII.IV.

Values for montmorillonites are usually lower than expected probably because nitrogen does not penetrate readily between platelets.

Adsorption from a liquid phase. In a liquid medium, such as water or benzene, mineral surfaces adsorb or absorb molecular species contained in solution in a characteristic manner. The absolute mechanism is not always easy to establish but undoubtedly both physical and chemical sorption is involved. Fatty acids are absorbed by means of their carboxyl group whilst the hydrocarbon chain projects outwards from the surface. It can be calculated that each molecule thus occupies

[1] Livingston, H. K., *J. Colloid. Sci.*, 4, 450, 1949.
[2] Joy, A. S., *Vacuum*, 3, 254, 1953.
[3] Bugge, P. E. and Kerlogue, R. H., *J. Soc. Chem. Ind.*, 66, 377, 1947.
[4] Starkweather, F. M. and Palumbo, D. T., *J. Electrochem. Soc.*, 287, 1957.
[5] Nelsen, F. M. and Eggertsen, F. T., *Analytical Chem.*, 30, 1387, 1958.
[6] McBain, J. W. and Bakr, A. M., *J. Amer. Chem. Soc.*, 48, 690, 1926.

TABLE VII—IV. SURFACE AREA VALUES BY GAS ADSORPTION

Material	*Surface area sq.mt/g.*
Kaolinite (Cornwall)	13·2
Montmorillonite (Wyoming)	85·0
Fullers earth	310·0
Halloysite	14·2
Ball clay (Dorset)	19·3
Better Bed fireclay	10·1
Silica gel	520·0
Diatomaceous earth	54·0
Micronised alumina	230·0
Crushed firebrick (−30 mesh)	4·1

20·5 $Å^2$ of surface, hence if the total absorption can be measured in molecular terms, the surface area can be deduced.

Liquid absorption as a means of surface area measurement can be carried out with many complexes. Fatty acids are the best known, but radio-active isotopes and certain dyes can also be used.

Liquid phase absorption does not require the same complication of equipment as that used in gas adsorption methods, but care must be taken in the purity of the absorption complexes. By and large, good agreements in surface area values are obtained between liquid and gas absorption techniques except with materials possessing fine cracks or a microporous structure. It is found that with diatomaceous earths, zeolites, certain glasses and other similar materials, surface area values reduced between 20–50 per cent are obtained with stearic acid absorption in comparison with nitrogen methods.

Surface area by permeability. If a stream of gas or liquid flows through a compacted mass of particles, a resistance to flow is encountered resulting from friction of the molecules of the medium with the surface layers. Thus the technique can be used to measure the external surface area of a powder or the surfaces around which the medium can be induced to flow.

The mathematical expression relating the flow characteristics to the packing and surface area of the powder was derived by Kozeny[1] in 1927 and is the basis of calculation. Poiseuille has modified the equation to be of general application.

In principle, the method relates the surface area to the rate of flow of air or other gas through a compacted powder or porous body. The powder is lightly pressed on to a filter paper or perforated disc in a 1 in. diameter steel cylinder. A fixed volume of air is passed through the specimen and the time of flow or the pressure differential is measured. Provided that conditions are standardised, the specific surface area (S) of the powder is related to the rate of flow by the expression:

$$S^2 = 5{\cdot}65 \times 10^6 \left(\frac{\varepsilon^3 T}{(1-\varepsilon)^2 \rho^2 d} \right)$$

where T is the time of flow in secs. of a fixed volume of air or can be related to pressure differential;

[1] Kozeny, J., *Sitzber. Acad. Wiss. Vien.*, 136, 271, 1927.

ρ is the specific gravity of the powder;

d is the depth of powder in cm.;

and ε is the fractional volume of voids in the compacted powder and is calculated from the equation:

$$\varepsilon = 1 - \frac{W}{A\rho d}$$

where W is the weight of the powder (gm.) and A is the cross-sectional area of the cylindrical test sample.

This equation applies only if viscous flow conditions exist, which they will, provided that the pore size of the compact is appreciably greater than the molecules of gas. As such, the method is accurate for measuring particles down to about 5 microns in size and reasonably good results are obtained down to an average micron size of 1, provided that the size distribution is not large.

At smaller sizes, molecular streaming becomes an important factor and Knudsen[1] has deduced flow characteristics under these conditions. Although the calculation is complex, the surface area of particles down to 0·1 micron can be measured accurately but it is necessary to work under low pressure conditions.

A method for measuring surface area by this technique was evolved by Lea and Nurse[2] and subsequently modified by Rigden[3] to be capable of general application to ceramic bodies. A commercial instrument called the Fisher sub-sieve sizer[4] enables results to be obtained and calculated with great rapidity—in fact, the tedious part of the experiment is the prior determination of the true density of the powder being examined. The principle of the method is illustrated in Fig. VII.11.

As a means of control of particle size, the air permeability method is excellent and although it will not indicate a size distribution it gives results of centre grain-size values in close accord with those derived from tedious calculation from a sedimentation curve.

Air permeability methods indicate only the external surface area of particles. Whilst results are generally much less than those from gas adsorption methods they are more in line with those from sedimentation methods.

There are many other methods suggested for measuring the specific surface area. Two which have been applied to clays on a limited scale are the heat of absorption on wetting[5] and the measurement of negative absorption.[6]

An excellent survey of various methods of measuring the surface area of ceramic materials has been carried out by Haynes.[7]

[1] Knudsen—see Loeb, L. B., *Kinetic Theory of Gases* (McGraw Hill, N.Y., 1934).
[2] Lea, F. M. and Nurse, R. W., *J. Soc. Chem. Ind.*, 58, 277, 1939.
[3] Rigden, P. J., *J. Soc. Chem. Ind.*, 62, 1, 1943.
[4] The instrument is marketed in the U.K. by Kek Ltd., Manchester, and is the design of Fisher Scientific Inc., U.S.A.
[5] Calvet, E., *J. chim. phys.*, 48, 579, 1953.
[6] Schofield, R. K., *Trans. Brit. Ceram. Soc.*, 48, 201, 1949.
[7] Haynes, J. M., *Trans. Brit. Ceram. Soc.*, 60, 691, 1961.

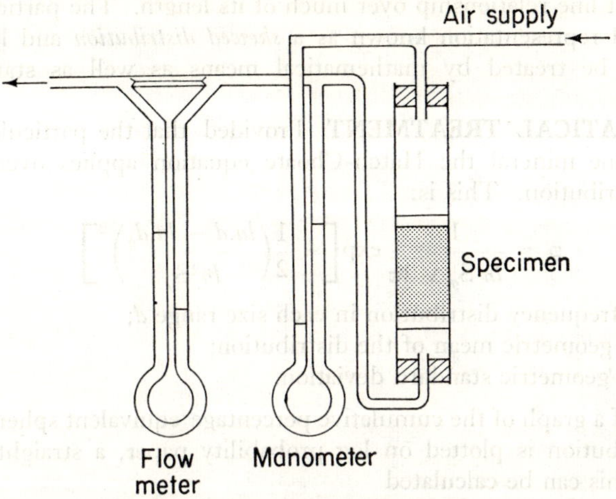

FIG. VII.11. SURFACE AREA MEASUREMENT BY GAS PERMEABILITY

The pressure drop across the specimen is measured on a manometer and the rate of flow of air indicated on the meter gauge. The height, cross-sectional area and packing porosity of the specimen need to be determined in order to calculate the surface area.

INTERPRETATION OF PARTICLE-SIZE ANALYSES

MANY attempts have been made to find a single figure or a simple diagram to represent different size distributions and so form a means of comparison. No single *fineness figure* can be wholly accurate and no diagram can yield all the desired information 'at a single glance' but both are of great importance when clays and other materials containing particles of many different sizes are to be compared. These are commonly understood as showing the 'distribution' of the different sizes and, though this term is not wholly suitable, it is widely used.

The proportions of the various particles of different sizes in a clay or other ceramic material are known collectively as the *particle-size distribution.* As such materials usually contain particles of many different sizes the size-distribution can be expressed only approximately but its importance is so great that even an arbitrary figure is of value.

The particle-size distribution may be represented in several ways of which the chief are:

(i) A *graph* or *curve* in which the distribution is represented by a single line.
(ii) A *surface factor* in which the distribution is represented by a single figure.

Typical particle-size distributions from various methods of comminution are shown in Fig. VII.4.

In the representation of the size-distribution of single mineral species which have been subjected to certain comminution processes, a graph of the cumulative percentage passing a certain size plotted against the logarithm of particle size often shows a straight line relationship over much of its length. The particle-size graph has a statistical representation known as a *skewed distribution* and hence the information can be treated by mathematical means as well as statistically and graphically.

A. MATHEMATICAL TREATMENT. Provided that the particulate system is composed of one mineral the Hatch-Choate equation applies over the greater part of the distribution. This is:

$$\varphi = \frac{1}{ln\, S_g \sqrt{2\pi}} \exp\left[-\frac{1}{2}\left(\frac{ln.d - ln.d_g}{ln^2 S_g}\right)^2\right]$$

where φ is the frequency distribution in each size range d;

d_g is the geometric mean of the distribution;

S_g is the geometric standard deviation.

Consequently if a graph of the cumulative percentage/equivalent spherical diameter of a size-distribution is plotted on log probability paper, a straight line should result. From this can be calculated

d_g the geometric mean size at 50% retention and

S_g the standard deviation which is defined as $\frac{d_1}{d_g}$

where d_1 is the size at 84·13 per cent retention.

These two factors adequately define a size-distribution and give useful parameters for control.

Pearson derived an empirical equation to fit particle-size distributions and his observations have subsequently been modified by various authorities to represent special cases.

Martin, Heywood and Gaudin have evolved laws or equations which are applied, but perhaps the best known is the Rosin-Rammler expression which is of particular value in the analysis of coal and other brittle minerals.

The expression simplifies to

$$\log \log \varphi = \log b + \log \log e + n \log d$$

where b and n are critical constants determined by plotting frequency distribution against size on special graph paper.

B. STATISTICAL TREATMENT. Rather than applying a complicated mathematical formula to the results, the important parameters of mean size and standard deviation can be calculated by statistical means. Table VII.V shows a size analysis of a crushed firebrick grog obtained by sieving followed by an Andreasen pipette determination on the −300 mesh size. The sieves and sub-sieve size fractions have been so chosen that there is the same logarithmic difference between their sizes. This is important because the interval between fractions must be the same for the analysis to apply.

In this particular example a logarithmic base has been chosen because this material has a skewed distribution as have most ceramic bodies. It would be equally correct to calculate the results on an arithmetical basis but less convenient in this case.

TABLE VII—V. STATISTICAL TREATMENT OF A PARTICLE-SIZE DISTRIBUTION (assume log d_0 = 1·880)

B.S. mesh	*Aperture range (microns)*	*log. aperture range*	*log. interval*	*% fraction in range*	*Aver. log. size fraction*	*$d-d_0$*			
				f	*d*	*u*	*uf*	*u^2f*	*u^2f^2*
5–10	3400–1696	3·53–3·23	0·30	1	3·38	+1·50	+1·50	2·25	2·25
10–18	1696–848	3·23–2·93	0·30	3	3·08	+1·20	+3·60	4·31	12·96
18–36	848–424	2·93–2·63	0·30	5	2·78	+0·90	+4·50	4·05	20·25
36–72	424–212	2·63–2·33	0·30	12	2·48	+0·60	+7·20	4·33	51·84
72–150	212–106	2·33–2·03	0·30	19	2·18	+0·30	+5·70	1·71	32·49
150–300	106–53	2·03–1·73	0·30	21	1·88	0·00	0·00	0·00	0·00
	53–27	1·73–1·43	0·30	17	1·58	−0·30	−5·10	1·53	26·01
	27–13·5	1·43–1·13	0·30	10	1·28	−0·60	−6·00	3·60	36·00
	13·5–6·8	1·13–0·83	0·30	6	0·98	−0·90	−5·40	4·86	29·16
	6·8–3·4	0·83–0·53	0·30	3	0·68	−1·20	−3·60	4·32	12·96
	3·4–1·7	0·53–0·23	0·30	2	0·38	−1·50	−3·00	4·50	9·00
	−1·7	−0·23		1	0·08	−1·80	−1·80	3·24	3·24
						Totals	−2·40	38·70	236·16

From this type of table the mean size $\bar{d}$ can be calculated from the formula

$$\bar{d} = \frac{1}{n}\Sigma df$$

where n is the number of samples—in this case 100;

d is each particular size and f the frequency of occurrence of each particular size.

Similarly the standard deviation—or square root of the variance—is given by

$$S = \sqrt{\frac{1}{n}\Sigma(d-\bar{d})^2 f}$$

These two important values can be calculated, although much of the tedium can be eased by introducing a further number d_0 which could be defined as a 'guessed mean size'. In the table this value could be 1·88 (a log. number). It can be introduced in the mean size equation to give

$$\bar{d} = \frac{1}{n}\Sigma(d-d_0)f + d_0$$

and the size values to be multiplied by the frequency are now smaller numbers.

If $d - d_0 = u$

$$\bar{d} = \frac{1}{n}\Sigma uf + d_0$$

Now $S = \sqrt{\frac{1}{n}\Sigma(d-\bar{d})^2 f}$

Substituting for $\bar{d}$

$$S = \sqrt{\frac{1}{n}\Sigma\left\{d - \left(\frac{uf}{n} + d_0\right)\right\}^2 f} \text{ but } d - d_0 = u$$

$$\therefore S = \sqrt{\frac{1}{n}\Sigma\left(u - \frac{uf}{n}\right)^2 f}$$

$$= \sqrt{\frac{1}{n}\Sigma\left(u^2 f - \frac{2u^2 f^2}{n} + \frac{u^2 f^3}{n^2}\right)}$$

The third term is insignificantly small and for most purposes can be ignored so

$$S = \sqrt{\frac{1}{n}\Sigma u^2 f - \frac{2}{n^2}\Sigma u^2 f^2}$$

The values of uf, u^2f and u^2f^2 are easily derived and substituted in the equations. The calculation can be followed from Table VII.V.

In the example under consideration

$$\bar{d} = \frac{1}{n}\Sigma uf + d_0$$

$$= \frac{1}{100} \cdot -2{\cdot}40 + 1{\cdot}880$$

$$= -0{\cdot}024 + 1{\cdot}880 = 1{\cdot}856$$

$$S = \sqrt{\frac{1}{n}\Sigma u^2 f - \frac{2}{n^2}\Sigma u^2 f^2}$$

$$= \sqrt{\frac{1}{100} \cdot 38{\cdot}70 - \frac{2}{10{,}000} 236{\cdot}16}$$

$$= \sqrt{0{\cdot}387 - 0{\cdot}047} = 0{\cdot}588$$

The critical values, therefore, in terms of real numbers are

$\bar{d}$ (mean size) = 71·8 microns

and S (the standard deviation) = ±3·87

These two values adequately describe a size analysis.

C. GRAPHICAL TREATMENTS. The simplest and most common method of expressing a particle-size distribution is as a graph of cumulative weight retained on or passing through a particle screen or size against the diameter corresponding to that size. Fig. VII.12 shows this type of information from the results shown in Table VII.V. The vertical axis represents the percentage *greater* than a particular equivalent size.

Such a curve is valuable but not easy to compare with others if they are to be used for quality control.

These curves can be broken down to two significant parameters which are:

(*a*) the *centre grain size* (c.g.s.) which is the particle size in microns at 50 per cent retention—d_m.

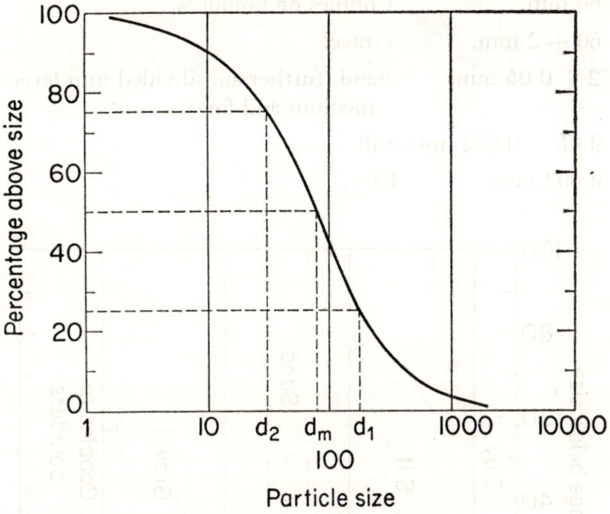

FIG. VII.12. GRAPHICAL REPRESENTATION AND INTERPRETATION OF A PARTICLE SIZE CURVE

(*b*) the *quartile ratio* which is the ratio between the particle size at 25 per cent retention and that at 75 per cent retention—d_1/d_2.

From the information given in the graph the calculated values are

$$d_m = 73 \text{ microns}$$
$$d_1/d_2 = 176/30 = 5{\cdot}83$$

These values describe a size analysis curve over the central 50 per cent; as such they are useful although they do not take account of anomalies in the extremely coarse or fine regions.

The statistical calculation in the previous section covers the range from 15·3 to 84·7 per cent of fraction values.

Despite their limitations, the determination of mean size and the standard deviation or quartile ratio are excellent control parameters for size-distribution curves. They provide a continuous record for quality control and they have a bearing on many textural properties of particulate systems.

There are other graphical representations of size-distributions which are valuable. Boswell's classification is still widely used by geologists, soil chemists and civil engineers to describe soils and sub-strata.

In cumulative percentage distribution diagrams, the horizontal scale is the equivalent diameter usually plotted on a logarithmic basis. The vertical axis is the percentage *smaller* or *larger* than a particular size. The diagram is shown in Fig. VII.13 with the horizontal ordinate divided into equal logarithmic intervals which define the type of material which makes up that fraction.

Thus + 60 mm.	Cobbles or Boulders
− 60 + 2 mm.	Gravel
− 2 + 0·06 mm.	Sand (further sub-divided into large or coarse, medium and fine)
− 0·06 + 0·002 mm.	Silt
− 0·002 mm.	Clay.

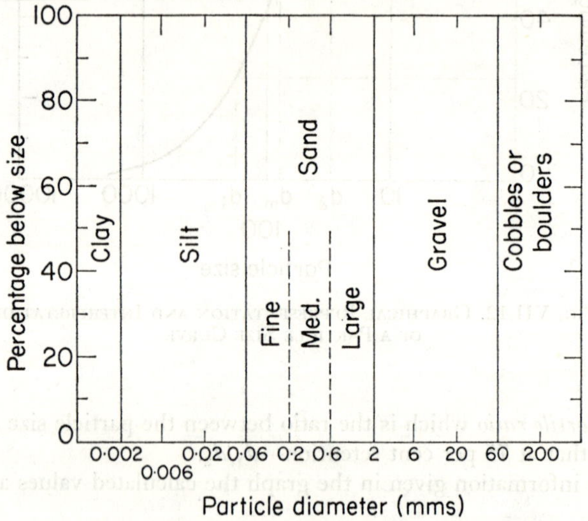

FIG. VII.13. BOSWELL'S CHART METHOD OF SIZE CLASSIFICATION

When a size-distribution graph is plotted in the usual way the amounts of material falling into each of the above five categories can be easily calculated. The various values and ratios are important in defining the behaviour of soils and sands.

A material with its centre grain size in the range 0·06–2·0 mm. is sandy and termed *arenaceous*; if the c.g.s. is below 0·002 mm. it is a clay or *argillaceous deposit.*

A sand of good drainage characteristics must have the silt + clay content less than 20 per cent.

A sand of good packing characteristics should have a sand content 60–80 per cent and the ratio silt + clay : gravel : : 15–30 : 1.

A c.g.s. of less than 0·02 mm. and at least 10 per cent clay fraction will indicate a material of good water-retention characteristics.

The principle of sub-dividing size-distribution graphs into several components has been developed in ceramic practice where a three-component basis is preferred. Thus comparisons may be made on a triangular diagram—often called a Feret diagram (Fig. VII.14).

Three size ranges are chosen, designated large (G), medium (M) and small (F); they can be of any convenient size ranges.

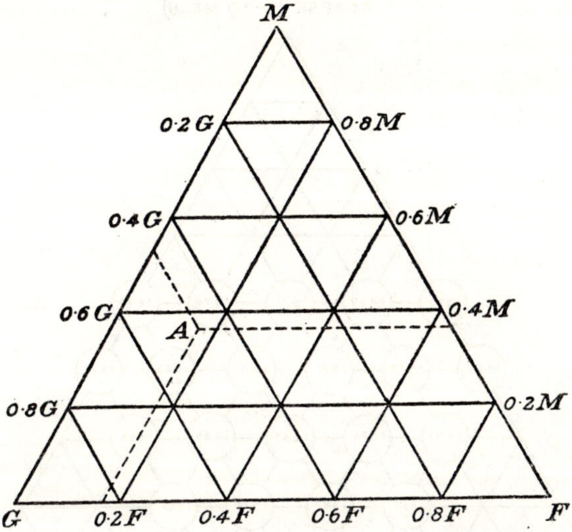

FIG. VII.14. FERET'S TRIANGULAR DIAGRAM

The proportions of each grade in a mixture may be indicated if each side of the triangle is sub-divided and three sets of parallel lines constructed by joining equivalent points. For example, the proportion of M at the apex is 1·0 but at points along the line GF it is zero. At any point on the line 0·6G – 0·4M, the proportion of M is 0·4.

The point A corresponds to a material containing 48 per cent of size G, 16 per cent, of size F and 36 per cent of size M. A large number of samples may be plotted on such a diagram and the particle-size distribution of all of them can be clearly shown. Any three desired ranges of particle-sizes may be used and if a material contains a very small percentage of each of several sizes this may either be omitted or included with another size.

It was by means of such a diagram that Hugill and Rees[1] were able to forecast the most suitable blends of graded quartzite to give an optimum bulk density in the fired product. Their triangular diagram is shown in Fig. VII.15, and the figures within the circles indicate the bulk density of the fired material of that particular blend. The best ratio of coarse : medium : fine particles was 45 : 10 : 45 which gave a maximum bulk density of 1·70.

Surface factors are based on the total surface area of all the particles in a given weight of material. So far as the ceramic industries are concerned the term was devised by W. Jackson to express the size distribution of particles by means of a single figure. He calculated the arithmetic average of each of the size-groups suggested by Boswell (p. 400), assumed all the particles to be true spheres and deduced the average surface area of equal weights of grains of various sizes. From these figures, the *surface factor* was formed by multiplying the percentage weight of each

[1] Hugill, W. and Rees, W. J., *Trans. Brit. Ceram. Soc.*, 30, 337, 1930.

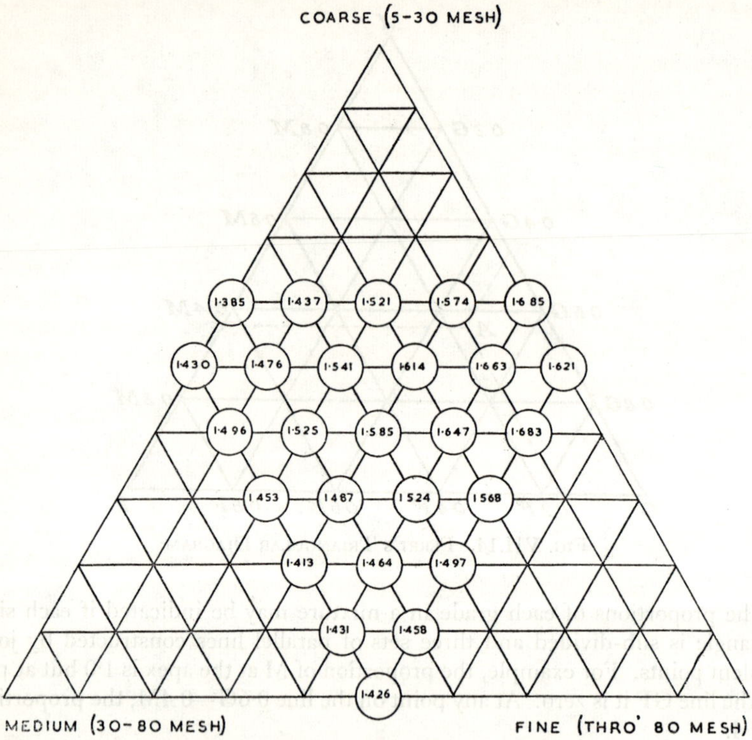

FIG. VII.15. THE RELATIONSHIP BETWEEN THE APPARENT DENSITY AND THE GRADING OF FIRED MIXTURES OF SILICA ROCK (after Hugill and Rees)

grade by the average surface area of the grains included in that grade, adding the products so obtained and dividing the result by 100.

J. W. Mellor[1] showed that the true average diameter of a series of particles of different sizes is not the arithmetic mean but corresponds to

$$\sqrt[3]{\frac{D^4 - d^4}{4(D - d)}}$$

where D is the diameter of the largest and *d* that of the smallest particles in the group. Mellor assumed three groups of sizes:[2]

(*a*) Particles below 0·010 mm.
(*b*) Particles between 0·01 and 0·063 mm.
(*c*) Particles between 0·063 and 0·107 mm.

[1] Mellor, J. W., *Trans. Eng. Ceram. Soc.*, 9, 94, 1910 and 29, 251, 1930.
[2] The extension to particles of smaller diameter than the above (200-mesh) greatly increases the error in the surface factor and is inadvisable though sometimes used.

and obtained three factors: (*a*) 359; (*b*) 53·9; and (*c*) 26·0 which, when multiplied by the respective weights of the different grades and the products added together, produces the surface factor.

No such surface factor can be fully accurate because it assumes that all the particles are truly spherical and are of the same weight—which is incorrect—and also because the conception of an 'average diameter' is inadequate. In spite of these disadvantages, such a surface factor is often convenient and is very useful in the works control of clays and other materials used in the manufacture of earthenware and porcelain.

The surface factor can also be calculated from a cumulative grain-size curve as illustrated in Fig. VII.12. Whittaker[1] has shown that if the shape of particles is constant and is known—and hence the surface area of an individual grain of a certain size—the total surface area of a powder can be calculated by summing increments of weights corresponding to a small size-range on the cumulative curve.

Many other methods used to determine the particle size of powder can be applied to give a measure of the surface factor. Webb and Ratcliffe[2] use a hydrometer technique which is of wide application in the routine control of slips and slurries. Moore[3] has shown that the light absorption method (p. 384) gives satisfactory results for the surface factors of crushed flints. Leech, Ratcliffe and German[4] compared a light absorption method using a Spekker photometer, with the Rigden specific-surface apparatus (see p. 394) for determining surface factors of flint, colours and glazes, and showed that the results from both methods were closely related.

As the surface factor is directly related to the sizes of the various particles it affords a means of representing the size-distribution of these particles, but the result may be seriously affected by the fact that several materials each with widely differing proportions of particles of different sizes and densities (and therefore very different in their behaviour in a kiln) may all have the same surface factor. For this reason, this factor is chiefly of value in certain works where the kinds of materials used are few and where their composition and physical properties are reasonably constant. Thus, in a factory manufacturing fine earthenware, any change in the surface factor indicates a change in one or more of the ingredients used and so may act as a warning.

Industrial Applications of Particle Sizing

ALTHOUGH it is outside the scope of this book to discuss the techniques of industrial sizing, it is opportune to classify the various methods on the scientific premises outlined in the previous sections.

In industrial ceramic practice, size classification and separation are essential for four prime reasons: (*a*) to remove undesirable oversized material from a comminution process; (*b*) to remove undesirable undersize, as, for example, clay from a quartzite or a sand; (*c*) to remove a valuable coarse or fine ingredient from an

[1] Whittaker, H., *J. Amer. Ceram. Soc.*, 22, 16, 1939.
[2] Webb, H. W. and Ratcliffe, S. W., *Trans. Brit. Ceram. Soc.*, 41, 51, 1942.
[3] Moore, A. E., *Pottery Gazette*, 75, 849, 1950.
[4] Leech, L. G., Ratcliffe, S. W. and German, W. L., *Trans. Brit. Ceram. Soc.*, 52, 145, 1953.

undesirable matrix; and (*d*) to produce a material having a satisfactory texture for the purposes for which it is intended.

Because of the high cost of subsequent drying, particle-size control of most low-priced clay components is carried out in the dry state with sieving for coarser sizes and air elutriation or sedimentation for smaller material.

As explained in an earlier section, clays break down into their ultimate crystallites only when they are suspended in water, hence when clay beneficiation is involved the techniques of water separation have to be employed.

Sieving is an important stage in all clay and ceramic processing. The principles have been described on p. 376 but in industrial practice, screening has to be a continuous process. It is usual to allow material to flow over a screen deck inclined at an angle, so that the particles progress under gravity. From several points of view this is an unsatisfactory technique.

(*a*) Particles tend to slide over the deck and do not make sufficient contact points with the screen grating.

(*b*) If the material flow is large, small particles tend to remain on top of the moving mass and never make contact with the screen.

(*c*) An inclined angle of screen lowers the size of particle which will pass through. The effective screening diameter (d_1) is reduced in relation to the nominal diameter of the screen aperture (d_2) by the relation

$$d_1 = d_2 \cos \varphi$$

when φ is the angle of inclination from the horizontal.

(*d*) Particles tend to stick or 'peg' in the aperture thus reducing efficiency.

These problems can be minimised by using vibrating screens which agitate the particles and require a much smaller angle of inclination.

Wet clay is troublesome material to screen industrially as it tends to stick in the apertures and so causes 'blinding'. This effect can be reduced by electrically heating the sieve deck, although this is possible only with metal screens.

In practice, clay with up to 8 per cent water can be sieved through a 6-mesh screen quite satisfactorily; it will pass a 30-mesh screen only if the water content is less than 4 per cent; 80-mesh screens are about the lowest limit employed industrially for bulk processing, but screens down to 200 mesh are in use where the oversized material is both small in amount and in size.

When a graded aggregate is to be prepared, multi-deck screens are employed with coarser units overlying those of smaller size. Products that are retained on each screen are withdrawn separately.

In the processing of large quantities of material which is to be smaller than 80 mesh, *air elutriation* within the comminution unit is the usual practice. Whilst the raw material is being reduced in size an upward circulating current of air carries away fine particles which can be collected elsewhere. The rate of flow of air can be adjusted to remove any predetermined size range. This type of unit often carries warm dry air into the comminution unit to act as a dryer as well as a classifier.

Such a system can operate only if it is completely enclosed, so that it is restricted to impact mills, attritors and ball mills.

Wet size analysis is effected by firstly dispersing the raw material in water in a blunger. The resulting slurry can then be screened through sieves or lawns to remove oversized particles. Wet sieving is highly effective and the oversize slides from the screen if it is inclined at a small angle.

Sedimentation methods are used in the separation of small particles from a slurry. It was customary to allow the suspension to flow slowly through a series of sedimentation tanks arranged in cascade fashion. Each tank was made larger than the one preceding it so that the residence time increased progressively thereby separating finer particles at each stage of the operation. The finer material from each tank remained in suspension and passed over a weir to the next separation stage or to rejection (see also p. 536).

Sedimentation tanks are still widely used but many have been replaced by units in which high gravity factors can be employed.

Hydrocyclones in which the suspension is separated by whirling it tangentially in a small diameter tube are most effective and economical to operate. They will provide a 'cut-point' down to about 5 microns and they can be used for dewatering slips (p. 538).

For smaller sizes *centrifuges* are preferred which are capable of operating at higher gravity factors. Particles of less than 1 micron can be separated by these means.

PROPERTIES DEPENDENT ON PARTICLE SIZE

THE physical aspects of the fabric of a material are best considered from the point of view of the properties which are a direct result. The principal of these are: (*a*) texture; (*b*) homogeneity; (*c*) porosity; (*d*) specific gravity; (*e*) bulk density; (*f*) permeability; and (*g*) penetrability.

Many of these properties are of equal or greater importance in the fired products and in this connection will be considered in Chapter XII.

Texture

THE term 'texture' is difficult to define for it has a wide variety of meanings, and is subject to many variables, which together impart a characteristic structure to the material. These include: (*a*) the shapes of individual grains; (*b*) their sizes; (*c*) the grading or packing of such grains; and (*d*) the density or closeness of packing of the mass.

The texture is said to be *coarse* when the particles are large or loosely-spaced and *fine* when the particles are small. *Open-textured* materials are those where the individual particles are not closely packed and many large voids are present; conversely, *close-texture* implies very small voids or, in some cases, a minimum of vacant spaces.

The texture[1] of a material has a very important influence on many of its

[1] The texture of ceramic bodies, both raw and fired, can be examined by impregnating the surface with a slip of plaster of Paris and allowing it to set. If a thin slice is cut from the body and the face polished, the particle shape and the size and shape of void space (filled by plaster) can be clearly seen. A low-power microscope is an advantage when fine-textured materials are to be examined.

When the body is light-coloured, better results are given by adding a colouring agent (e.g. iron oxide) to the plaster of Paris prior to impregnation.

properties, including shrinkage, porosity, fusibility, and, in the case of clays, plasticity. Notwithstanding its importance, the texture is not considered or controlled to anything like the extent it should be, and many failures in the production and use of articles from clay and other ceramic materials are due to the lack of proper attention to this property.

The **shapes of the grains** is a prime factor in determining the texture of a material. They may be classed as: (*a*) flaky or platelike; (*b*) needle-shaped; (*c*) angular; or (*d*) rounded.

(*a*) **Flaky particles.** The micas and clay minerals consist of layers, which result in crystals of these minerals being typically flaky: i.e. the area of cross-section is considerably greater than the thickness. A sample of clay, especially if it has been purified from associated quartz grains, is composed of an aggregated mass of plate-shaped crystallites. When suspended in a perfectly still liquid, particles will settle in such a way as to present the greatest possible surface area to the direction of fall. Therefore clays, micaceous and similarly shaped grains will, under conditions of free sedimentation, form a mass in which all flakes are oriented in the same way, with their flat faces at right angles to the direction of fall. All sedimented clay deposits have particles arranged in this way, although the presence of quartz or other angular-grained minerals in the clay may locally disturb the ordered arrangement.

During forming processes, the flaky particles in clays can become orientated and produce characteristic effects in bodies manufactured from them. When subjected to initial flow under pressure, particles tend to orient themselves to present the greatest resistance to flow. That is to say, grains will be turned so that their greatest cross-sectional area will be at right angles to the direction of applied pressure. Large flakes of mica which are often present in clay bodies can frequently be observed in pressed bodies to be lying with their major face at right angles to the pressing direction.

When conditions for true streamlined flow are reached in ceramic bodies, as, for example, in extrusion, flaky particles turn through right angles as they slip and glide along planes of failure. This preferred orientation may persist throughout the manufacture of clay bodies and contribute to characteristic lamination faults.

Frequently, clay deposits exhibit characteristic *bedding* or *lamination* which is often horizontal, but if faulting or folding has taken place, the beds may be inclined at an angle. Such clays easily cleave along these directions and close inspection will usually reveal the presence of a layer of micaceous flakes arranged parallel to the line of fracture.

Because the particle size is invariably small, clays are fine-textured. However, many micropores are present which are the result of the evaporation of water from between the deposited particles. Consequently, most clays were of open texture at some stage in their geological history, but were subjected to compression, metamorphism or secondary precipitation which increased their density.

Laminated masses are troublesome when being manufactured into ceramic shapes, for there is a tendency for such material to split along the natural lines of weakness, during firing. It is, therefore, essential to grind clays sufficiently before use so as to reduce this undesirable feature of flaky materials. Excessive grinding must be avoided as it will cause other defects in some articles.

The cleavage, on firing, of laminated clays is serious only when the flake-size is large or when the degree of compactness is great. Slip-cast clayware is composed of oriented mineral plates but flaking is uncommon except when the size of the cast body is large.

When particles of the same grain size are compared, flaky or platelike particles develop greater strength in the mass than do rounded or equi-proportioned grains. This is because the specific surface area (or surface area per unit weight) is greater and in an oriented aggregate where plates are virtually in contact along their basal section, the binding force is relatively greater than with rounded grains.

(*b*) **Needle or lath-shaped crystals** are common in some naturally-occurring ceramic raw materials such as halloysite clays and dolomite rocks, but the best-known example is *mullite* which is one of the main products when alumino-silicate mixtures are fired. The texture of bodies composed of such particles depends on the orientation of individual grains, but most fibrous or needle-shaped crystal masses are of open texture. *Asbestos* which consists of regularly-arranged needles of chrysotile or amphiboles is of value as a thermal insulator largely as a result of the micropores which separate each fibre.

When lath-shaped crystals are arranged in more random fashion the resulting structure is usually strong. The crystals of mullite in fired clays are in the form of a network or felted mass of needles which interlink and intergrow throughout the body.

(*c*) **Angular grains** are the common form of particles in ceramic raw materials (other than clays) for they are the automatic result of crushing and grinding processes. The texture of such an agglomeration is difficult to predict, for it depends on the size distribution and the methods of packing. Where the particles approximate to cubes, the way in which they may be packed together has much in common with the arrangement of rounded or spherical particles. However, as a result of the angularity, they do not pack as readily because the 'ease of flow' is considerably less.

(*d*) **Rounded grains** occur in many sedimentary deposits as a result of mechanical erosion during transportation. They are a special feature of sands and gravels of marine origin where virtually all the particles are roughly spherical.

PACKING

THE packing[1] and texture of masses composed of rounded particles have been

[1] The ratio of the weight of a powder to the minimum volume which it can occupy is known as the *packing density*. Well-graded materials with a minimum of voids, occupy a relatively small volume, so that their packing density is not very different from their true specific gravity (see p. 427).

The packing density can be measured by placing a weighed sample of dry powder into a measuring cylinder and shaking it until a constant volume is reached. The packing density is the weight divided by this volume. A less accurate figure, although one which is of more value when the powdered material is to be moulded under pressure, is given by ramming a weighed sample in a cylinder with a plunger of definite weight (between 150 and 300 gm.) which is allowed to fall on the specimen a certain number of times (usually 50) from a fixed height in rapid succession. The volume of the shape is then measured and the packing density derived. This value is sometimes known as the *rammed density* and is important in the manufacture of ceramic materials from a semi-dry or almost dry powder. This density is expressed in gm. per cc. (or in lb. per cu. ft.).

studied in detail and the results obtained have provided a valuable basis for improving the texture and quality of ceramic wares.

Where spheres of equal size are packed together, it is impossible to produce a mass with no voids. A mass with maximum porosity or open spaces will result if the spheres are arranged so that each sphere is in contact with six others (Fig. VII.16a). It will be reduced if one particle in each group of four rests equally on three others in the form of a tetrahedron (Fig. VII.16b). The extension of this unit will result in each sphere touching twelve similar ones and give rise to one of the close-packed arrangements mentioned and illustrated in Chapter II (p. 85).

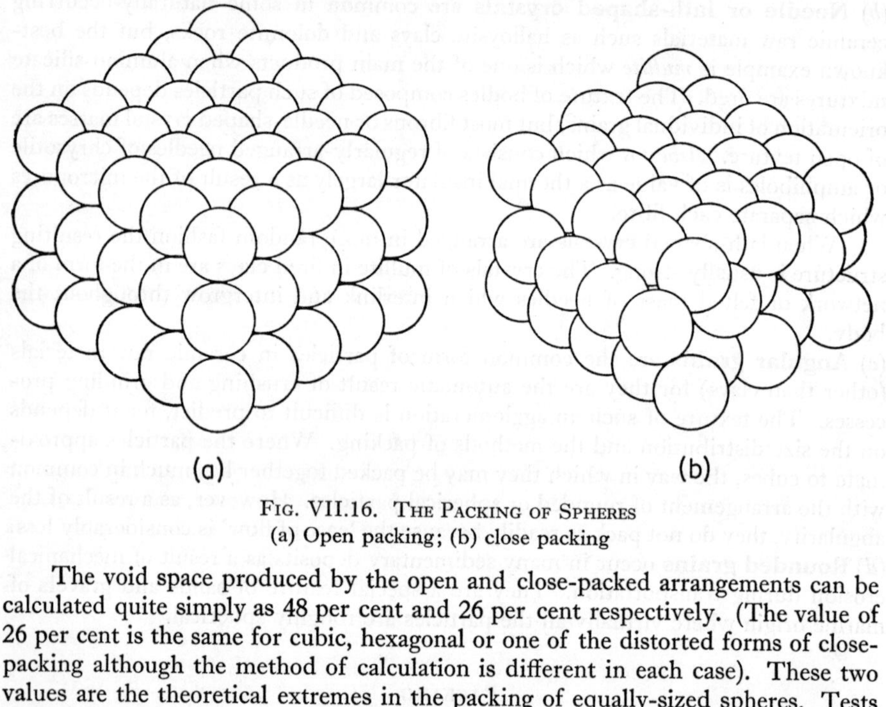

FIG. VII.16. THE PACKING OF SPHERES
(a) Open packing; (b) close packing

The void space produced by the open and close-packed arrangements can be calculated quite simply as 48 per cent and 26 per cent respectively. (The value of 26 per cent is the same for cubic, hexagonal or one of the distorted forms of close-packing although the method of calculation is different in each case). These two values are the theoretical extremes in the packing of equally-sized spheres. Tests carried out by Westman and Hugill,[1] using mono-sized spheres of different dimensions and materials have shown that the packing is virtually the same in all cases and that the system has a void space of 38 ± 1 per cent. This is intermediate in value between that of the open and close-packed arrangements and it is suggested that the particles have an open structure in the vertical direction, but horizontally they are almost ideally close-packed.

The value of 38 per cent void space is now known to depend on other factors. Sempere[2] has shown that a value of 35 per cent can be achieved by tamping and by vibrational compaction. The same author and also Koerner[3] and Gray[4] have

[1] Westman, A. E. R. and Hugill, H. R., *J. Amer. Ceram. Soc.*, 13, 767, 1930.
[2] Sempere, R., *Powder Tech.*, 2, 1, 1969.
[3] Koerner, R. M., *Powder Tech.*, 3, 210, 1970.
[4] Gray, W. A., *The Packing of Solid Particles* (Chapman and Hall, London, 1968).

related the minimum void space in systems of uni-sized grains to other factors such as the degree of sphericity and to the ultimate size. Spheres of about 10 micron size are reported to give a maximum packing density.

The void space can be reduced by adding to the original batch of large spheres a number of smaller ones of a size which can fill the vacant interstitial positions between the large grains.

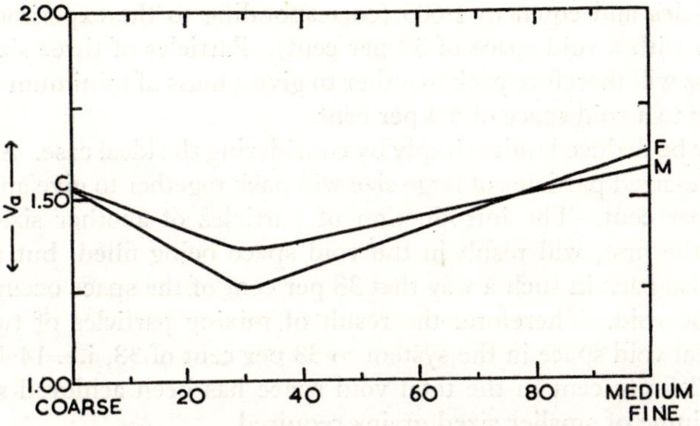

Fig. VII.17. Graph Showing the Packing Volume of Mixtures of Spheres of Two Different Diameters (after Westman and Hugill)

Westman and Hugill also proved that, in a system composed of spheres of two different diameters, maximum packing is secured when the larger particles are infinitely large compared with the smaller ones. This is shown graphically in Fig. VII.17, where the composition (in terms of the true volume of each size of sphere added), is related to the ratio of the actual volume of the mixture (V_a) to the true volume, which for convenience is regarded as unity. The apparent volume is related to the percentage of void space by the equation

$$V_a = \frac{100}{100 - \text{void space}}$$

Two experimentally determined graphs are shown representing the bulk volume of mixtures of differently-sized particles. As the size ratio is increased from 8:1 to 50·5:1 the minimum point in the curve is reduced, thus illustrating that, in a two-component system, maximum packing or minimum void space is reached when the size ratio is large. The greater is the difference in the size ratio of the spheres, the smaller is the value of the bulk volume attained at maximum packing. It is possible to calculate, theoretically, the variation in bulk volume on mixing particles of two sizes when one is infinitesimally smaller than the other. The minimum percentage of voids which could result from such a mixture is 14·4, corresponding to a bulk volume of 1·19 referred to a true volume of 1.

The effect of introducing particles of other sizes can be predicted by applying similar theoretical reasoning and Westman and Hugill have shown that maximum packing is produced by selecting spheres with very large diameter ratios. They have deduced that the minimum apparent volume (V_a) of a mixture of particles of three different sizes may be calculated from the equation

$$V_a = \frac{a^3}{(1-a)^3 + a^3}$$

where a is the bulk or apparent volume of a system containing spheres of only one size. The assumption is made that the value of a is the same for the coarse, medium and fine particles and equal to 1·605 (corresponding to the experimentally-determined system with a void space of 38 per cent). Particles of three sizes with large diameter ratios will therefore pack together to give a mass of minimum volume 1·057 corresponding to a void space of 5·4 per cent.

This may be deduced quite simply by considering the ideal case. A system composed of single-sized particles of large size will pack together to give an average void space of 38 per cent. The introduction of particles of another size, very much smaller than the first, will result in the void space being filled, but those smaller spheres will also pack in such a way that 38 per cent of the space occupied by them will in turn be void. Therefore, the result of mixing particles of two sizes is to reduce the total void space in the system to 38 per cent of 38, i.e. 14·4 per cent. A decrease of 23·6 per cent in the total void space has been achieved so this is the theoretical volume of smaller sized grains required.

Even smaller spheres can be added to the mixture of two grades of spheres and these will occupy the remaining void space, and closest packing will result if the third size of spheres is infinitesimally smaller than the second grade, which for convenience will be called the medium fraction. The overall packing can be calculated because the smallest-sized particles occupy the 14·4 per cent of void space in the mixture of coarse and medium particles, but because they are of one size, 38 per cent of the total space which they occupy will be void. Therefore, in a system containing spheres of three sizes, the total void space will be 38 per cent of 14·4, i.e. 5·4 per cent. A further reduction in the voids of 9·0 per cent has taken place, hence this must be the volume occupied by the finest particles.

The theoretically predicted ideal packing of a mixture of spheres of three different sizes where the size ratio is very large, is where the mixture contains 62·0 parts of coarse, 23·6 of medium and 9·0 of fine spheres (by volume). The percentage void space in such a mixture would be 5·4 per cent.

The nearest experimental approach to this value was obtained with a mixture containing 70 per cent coarse, 20 per cent medium and 10 per cent fine spheres by volume, where the diameter ratios of the spheres were 50·5 : 8 : 1. The bulk-volume which resulted was 1·202 giving a calculated void volume of 16·8 per cent.

The classical representation of good packing using mixtures of spheres of different diameters is different. The ideal system is presumed to consist of a framework of large-sized spheres with interstices filled by slightly smaller ones which are of such a size that they touch each of the surrounding large-sized balls. The picture is completed by imagining that smaller spheres occupy the remaining void spaces which remain as shown in Fig. VII.18a.

Such a system would ideally have a very small void space, but it would be impossible to attain in practice, because the smaller-sized grains are rigidly con-

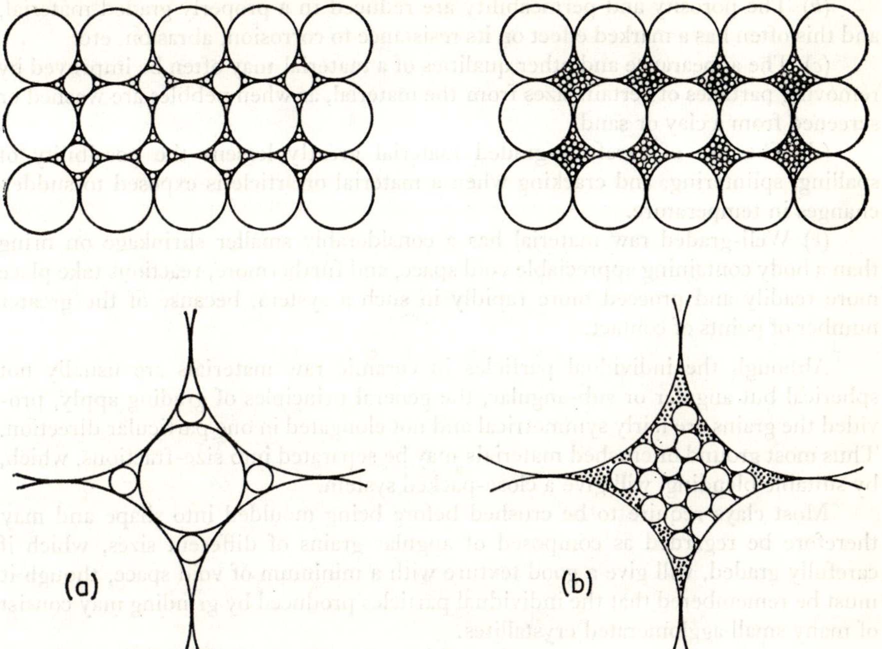

FIG. VII.18. THE PACKING OF SPHERES OF DIFFERENT SIZES
(a) Theoretical representation; (b) practical representation

tained in the space between large ones and are not free to move. Consequently no amount of tapping or pressing could produce the ideal arrangement from an assorted mass of randomly-arranged spheres. The work of Westman and Hugill has established that, only when the small-sized spheres are of a diameter which enables them to move freely through a network of larger grains, will there be any approach to close packing on a practical scale. The true pictorial representation of a close-packed system of spheres of two different diameters is shown in Fig. VII.18b. When even smaller-sized grains are added they occupy a similar position relative to those of medium size.

GRADING

THE mixing of spheres or particles of different diameters in various proportions so as to improve the closeness of texture and reduce the void-space is known as *grading* and has an important application in the manufacture of most ceramic bodies.

The use of graded materials is desirable for several reasons:

(*a*) If only coarse material were employed, it would require a very large proportion of bond to fill the voids and cement the grains together, and the resultant mass could not be so strong as a better-graded material, on account of the small number of points of contact between the 'cement' and the larger particles.

(*b*) The porosity and permeability are reduced in a properly graded material, and this often has a marked effect on its resistance to corrosion, abrasion, etc.

(*c*) The appearance and other qualities of a material may often be improved by removing particles of certain sizes from the material, as when pebbles are washed or screened from a clay or sand.

(*d*) The use of carefully-graded material greatly lessens the possibility of spalling, splintering, and cracking when a material or article is exposed to sudden changes in temperature.

(*e*) Well-graded raw material has a considerably smaller shrinkage on firing than a body containing appreciable void space, and furthermore, reactions take place more readily and proceed more rapidly in such a system, because of the greater number of points of contact.

Although the individual particles in ceramic raw materials are usually not spherical but angular or sub-angular, the general principles of grading apply, provided the grains are fairly symmetrical and not elongated in one particular direction. Thus most ground or crushed materials may be separated into size-fractions, which, by suitable blending, will give a close-packed system.

Most clays require to be crushed before being moulded into shape and may therefore be regarded as composed of angular grains of different sizes, which if carefully graded, will give a good texture with a minimum of void space, though it must be remembered that the individual particles produced by grinding may consist of many small agglomerated crystallites.

Although angular grains may be graded by mixing particles of different sizes in suitable proportions, they do not pack as readily as rounded grains. Some degree of interlocking occurs, however, which results in an overall lower porosity.

The application of the principles of grading on an industrial scale has necessarily to be a compromise. The ideal system would be one where several grades of particles are selected with large size-ratios and the proportions of each suitably adjusted, but such a mixture could not be used in practice because of the wastage in time and material.

Nevertheless a suitable compromise can often be achieved by utilising the principles deduced by Westman and Hugill. They predict that maximum packing would result if the composite material were to consist of

65·6 per cent of large material
24·9 per cent of medium grains
9·5 per cent of fine sizes

It is implied that there are the greatest possible size ratios between the large and medium and the medium and small grains.

Supposing size ratios of 100 : 1 were assumed to be sufficiently large for good packing; a cumulative histogram would result as shown in Fig. VII.19. The lines AB, CD, EF represent the values of total percentage material passing the particular size fractions.

To select such a system of single-sized particles would be impractical, but a compromise solution is where each uni-sized fraction is replaced by a range extending up to the next size with uniform distribution. The histogram becomes complete

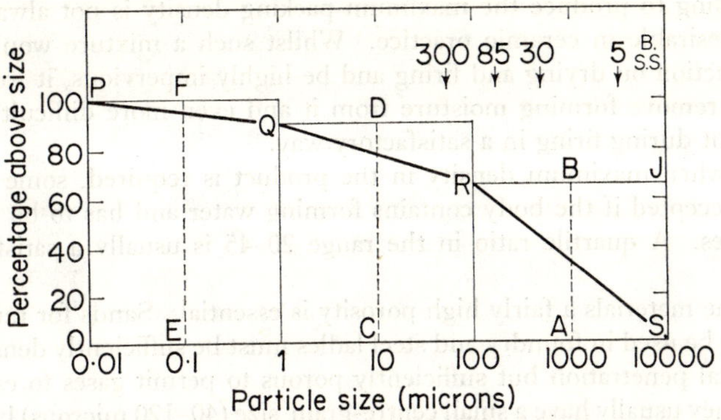

Fig. VII.19. Illustration of Good Packing Distribution under Practical Conditions of Sizing

but drawn out as shown in the diagram. A curve PQRS could be drawn representing the cumulative distribution of particles throughout the whole range.

If the assumption is correct that a spread of a size range is an adequate substitute for a specific size, then the curve PQRS should closely approximate to the best packing density given by a smooth size distribution of the type attainable in practice. This can be proved to be correct.

Although this type of grading cannot be produced by a single comminution process, a combination of crushing and grinding followed by reblending can yield a product with sizing characteristics similar to the ideal curve. Such mixtures give extremely dense products on compaction.

It is interesting that the empirically-derived best-packing of Hugill and Rees (see p. 401) conforms to the general pattern if actual values are given to the horizontal axis and chosen to fit their values.

Sieve size	+30	30−85	−85
Hugill/Rees weight per cent	45	10	45
Theoretical values weight per cent	42	14	44

The important difference is that the theoretical curve indicates the importance of the size distribution in the fine fraction. 30 per cent of material should be less than 300 B.S. mesh and 10 per cent should be less than 1 micron in size in order to ensure the best packing density.

Such a high percentage is impossible to achieve by fine grinding alone, but if clay is a component of the system and water is to be used in forming, a high percentage of very fine material results. Many brick clays have excellent packing densities on extrusion or pressing.

Unfortunately most ceramic mixtures contain too high a percentage of 'middles' —the inevitable result of most grinding processes. Good grading can be achieved only by screening out undesirable particle sizes although this can be wasteful.

The ideal grading graph can be selected with any maximum size limit but it should run parallel to PQRS. The quartile ratio of this composition is about 150 and for the best packing mixtures this ratio must be as high as possible.

A grading to produce the maximum packing density is not always necessary nor even desirable in ceramic practice. Whilst such a mixture would have very little contraction on drying and firing and be highly impervious, it would be most difficult to remove forming moisture from it and even more difficult to complete the burn-out during firing in a satisfactory way.

Even when maximum density in the product is required, some compromise has to be accepted if the body contains forming water and has to be fired to high temperatures. A quartile ratio in the range 20–45 is usually a satisfactory compromise.

In some materials a fairly high porosity is essential. Sands for ramming compositions to be used in foundry and steel ladles must be sufficiently dense to prevent molten metal penetration but sufficiently porous to permit gases to escape during casting. They usually have a small centre-grain size (40–120 microns) but a medium quartile ratio (8–15).

Particle Size of Ceramic Materials

As defined at the beginning of the chapter, particle-size distribution can imply either the inherent crystallite size of the minerals present or the grain distribution produced by a comminution process.

Natural clays are best defined in terms of their crystallite composition.

Kaolins and china clays, after separation from the parent altered granite rock and associated impurities, such as quartz and mica, usually consist of particles less than 0·01 mm., but some coarser varieties are known with only 20–25 per cent of particles less than 0·01 mm. Cornish and Devon china clays contain only a small percentage of particles less than 0·0003 mm., i.e. 0·3 microns. Purified china clay has a very limited grain size range and about 95 per cent of the particles are between 0·002 mm. (2 microns) and 0·0003 mm. (0·3 microns) in size.

Sedimentary Kaolins, which have been removed from their original site and re-deposited consist of smaller particles and are much more plastic. Norton[1] has reported that Florida plastic kaolin is over 30 per cent finer than 0·0005 mm. (0·5 microns) and that particles as small as 0·05 microns may be present.

Ball clays of the best qualities contain 60 per cent or more of particles less than 0·0005 mm. and up to 90 per cent less than 0·001 mm., but many are much coarser. The larger particles in most ball clays are usually quartz, mica or some other impurity present in small amounts. The variable and often large proportion of organic matter, which is the cause of the dark colour of the raw clay, is mostly present as a film surrounding the clay particles.

The particle size of ball clays is notoriously variable and strict control over the raw material has to be exercised, because slight changes can alter its properties considerably. For this reason it is inadvisable to use a different type of ball clay in the manufacture of a ceramic body, without first assessing its behaviour.

Bentonites and montmorillonite clays are, for the most part, of extremely small size. The larger proportion of clay particles are less than 0·05 microns and those that

[1] Norton, F. H., *Elements of Ceramics* (Addison-Wesley Press Inc., Cambridge, Mass., 1952).

are above this size are probably agglomerates. In a watery suspension, the ultimate particles of montmorillonite clays are probably only a few Ångstrom units in size (1 Å unit = 1×10^{-8} cm.). Commercial bentonites often contain a proportion of non-plastic material which is of a comparatively large particle size.

The particle size of clay minerals is best measured by means of the electron microscope.

Rock clays, shales, fireclays and some other clays would have an extremely fine texture if the individual particles could be separated, but unless special methods are adopted, each 'particle' of clay separated by the action of water will be found, on examination, to consist of an aggregation of smaller grains. Articles made of such clays have a coarser and more open texture than those made from clays in which the individual grains are readily separated from each other.

Freshly-dug clays may be classified according to their texture as: (*a*) shales; (*b*) marls; (*c*) loams; (*d*) stony clays; and (*e*) plastic clays.

Shales have a laminated or stratified texture; they may be either fine or coarse and either rich or poor in clay as the term 'shale' has no connection with the size of the grains nor with the composition of the material.

Marls or **Malms** are natural mixtures of clay and chalk and may usually be recognised by their friable nature, their texture being quite different from that of other clays. The term *marl* is often used for materials having a similar texture, but which contain very little, if any, chalk, e.g. Staffordshire marls. The greater part of most true marls can usually be washed through a 100-mesh sieve, though coarser marls are also found.

Loams are mixtures of sand and clay; they are largely used for the manufacture of bricks, roofing tiles, floor tiles, terra cotta, agricultural pipes, and coarse (red) pottery. Their texture is similar to that of marls. Loams and marls are often confused, but they may be readily discriminated by adding a little hydrochloric acid, when a marl will effervesce, whilst a loam will not do so. The texture of many loams is such that they are excellent for the manufacture of the articles just mentioned. The sand in them prevents loams from shrinking excessively and enables articles made from them to be dried and burned without undue risk of cracking or warping. Some loams contain so much sand that they cannot be used unless they are mixed with plastic clay to produce a stronger mass. Any very coarse, non-plastic material such as pebbles and gravel in a loam must either be removed or crushed so finely that they will not do any harm when the loam is made into articles.

Stony clays usually consist of fine clay or loam mixed with stones, pebbles, or similar non-plastic material of any size from $\frac{1}{8}$ in. in diameter to large boulders. Much of the *boulder clay* which forms part of the Glacial Drift is of this nature. The coarse material must be separated from the clay or loam before the latter can be used, even for brickmaking.

Plastic clays, when suitably moist, possess a smooth oleaginous texture which is often so fine that no definite 'structure' can be seen with the naked eye. If the material is rather drier it can be seen to consist of adherent granules consisting of aggregations of the much smaller particles of clay. Most plastic clays also contain a considerable proportion of sand and some (such as boulder clay) contain gravel and large stones.

Although natural clays are composed of inherently fine-grained crystallites, the grains produced by grinding are usually aggregates or agglomerates of such crystallites. The texture of the manufactured article is usually more dependent on the grain-size distribution than on crystallite size.

An excessively fine texture is objectionable, because it is usually associated with a tendency to twist or warp when the articles are being dried or burned and with an undue sensitiveness to sudden changes in temperature. In order to eliminate the less desirable features of fine-textured clays amounts of coarse non-plastic material are often added. This may be sand or alumina but more usually it is previously-calcined raw material, called 'grog', which forms a rigid skeleton within the body and thus prevents undue shrinkage or warping during drying and firing.

For *brick manufacture*, the ground clay should be free from particles larger than No. 6-mesh.

If sand is to be added to the clay it should usually consist of grains between 30- and 120-mesh. Much finer sand is seldom satisfactory though a small proportion of it is not likely to be detrimental.

For *roofing tiles* and *drain pipes* a similar clay can be used but it is much more finely ground. The material will usually all pass a 22-mesh screen.

For *refractory materials*, clays which will not pass through a No. 10-mesh sieve are usually too coarse, and a large proportion should pass through a No. 52-mesh.

If *grog* is to be added it should be of three grades: (*a*) *coarse grog* consists of particles between 0·125 in. and 0·375 in.; (*b*) *medium grog* consists of particles between 0·06 in. and 0·125 in.; and (*c*) *fine grog* consisting of particles between 0·03 in. and 0·06 in. in diameter.

Some manufacturers use a small proportion of much finer grog, but with some clays this creates difficulties by rendering the goods more sensitive to sudden changes in temperature.

For *fine earthenware*, *stoneware* and *china ware* the whole of the materials should pass completely through a No. 120-mesh sieve and a large proportion should pass through a No. 200-mesh sieve. For some of the finest porcelains the whole of the materials will pass readily through a No. 200-mesh sieve.

Glazed sanitary ware may be made of clays containing a moderate proportion of particles 0·02 in. in diameter, provided that there is at least 50 per cent fine enough to pass through a No. 72-mesh sieve.

All these suggested sizes are necessarily approximate and the grades of the various materials must always be selected with regard to their type and quality and to the requirements of the individual user.

Ceramic raw materials other than clays usually have a medium or coarse texture, unless they are ground excessively finely. Such mixtures are non-plastic and careful grading is essential if the body is to be strong on moulding and firing. Quite often a finely-grained plastic material, such as clay, or an organic substance, e.g. sulphite lye, is added as a bond. Thus, *silica bricks* are chiefly made of a mixture of three grades of silica: (*a*) particles between 0·125 in. and 0·185 in.; (*b*) particles between 0·04 in. and 0·125 in.; and (*c*) particles less than 0·04 in. in diameter but free from those less than 0·003 in. Some firms prefer to use no particles larger than 0·06 in., and include up to 30 per cent of particles less than 0·006 in.

For silica bricks accurate grading is very important because only $1\frac{1}{2}$–2 per cent of lime is used as the binding agent.

For *magnesite* and *dolomite* bricks, about 75 per cent of the particles may be less than 0·125 in. diameter and 25 per cent less than 0·006 in.

Glass-melting Tank Blocks require a coarse texture of low penetrability such as is produced by a mixture of

(*a*) Particles 0·125 in.–0·25 in.
(*b*) Particles 0·062 in.–0·125 in.
(*c*) Particles 0·031 in.–0·062 in.
} 1 part by volume

for the bottom blocks, whilst for replacement flux-line blocks the proportions may be 2:2:1.

Blocks for *muffles* and *retorts* should be made of the coarsest practicable material so graded as to be impermeable to gases. For gas-retorts a coarse grog and moderately fine fireclay is best.

In castable and mouldable compositions, the grain size of the components must be chosen with great care, usually to achieve the best possible packing density. Some of the commercial mixtures approximate closely to the ideal packing distribution (Fig. VII.19).

The texture of such mixtures is determined by the maximum size of the grog used. In compositions likely to encounter severe thermal shock conditions the grog fraction is prepared through a 3-mesh screen. Where slag or metal penetration is likely a 6-mesh product or finer is desirable. When low gas permeability is essential the maximum size of grog can be as low as 30 mesh.

Grog is usually prepared by crushing and its distribution curve represented on a logarithmic basis is virtually linear with a quartile ratio of 6–10. Some examples are shown in Fig. VII.20. This form of curve is preferred because fine clay or cements which are inherently fine-grained are normally added to such mixtures and this usually gives an overall good packing factor.

Porosity

No matter how carefully the various size grades of a raw material are mixed, it is virtually impossible to produce a body free from void space. All ceramic raw materials and products with the exception of glasses and similar bodies, are to some extent porous, therefore some fraction of the total volume is composed of air in the form of pores or voids.

The porosity of a material is defined as the total proportion of the air space contained between the solid particles of which the body is composed. Porosity should not be confused with permeability (p. 435) for the latter implies interconnected voids whilst the former embraces void space of all kinds.

According to E. W. Washburn[1] there are six types of pores (Fig. VII.21): (*a*) closed or 'sealed' pores; (*b*) channel pores connecting separate pores; (*c*) blind-alley pores; (*d*) loop pores; (*e*) pocket pores—some with narrow necks; and (*f*) micropores, which are so small as not to be filled with water or other liquid in

[1] Washburn, E. W., *J. Amer. Ceram. Soc.*, 4, 918, 1921.

Fig. VII.20. Typical Gradings of Grogs used in Castable and Mouldable Compositions

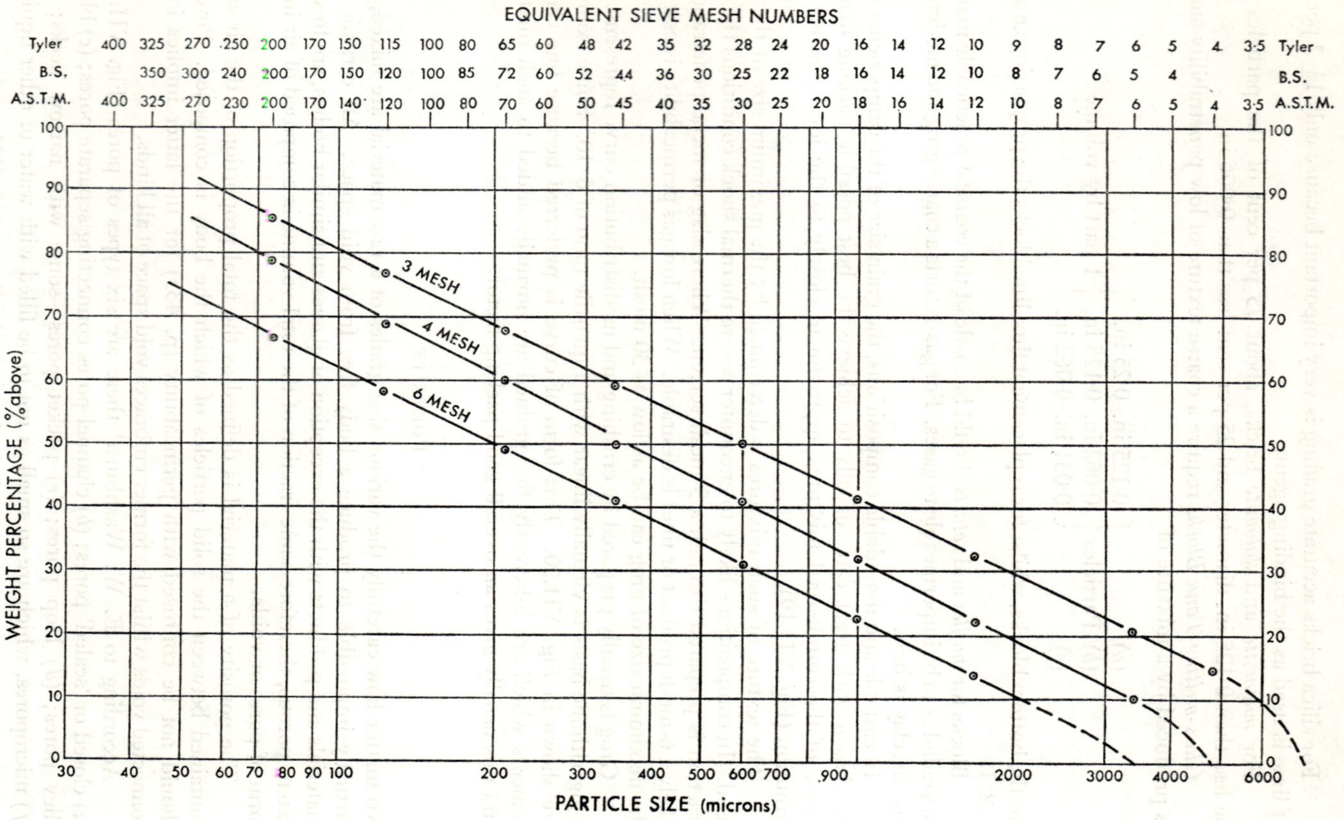

any ordinary period of soaking. To these must be added the continuous pore which runs, perhaps with a complex path, from one external face of the body to another.

These different kinds of pores produce two sets of results when the porosity of a material is determined, viz. (*a*) the true porosity, and (*b*) the apparent porosity.

The *sealed* or *closed pores* are of two kinds: (i) those which are naturally closed; and (ii) those which were originally open, but have been closed by molten matter formed when the articles were heated. Those of the first class may contain air, water, sulphur dioxide or trioxide, carbon monoxide or dioxide, oxygen or hydrocarbons commonly termed gaseous inclusions.

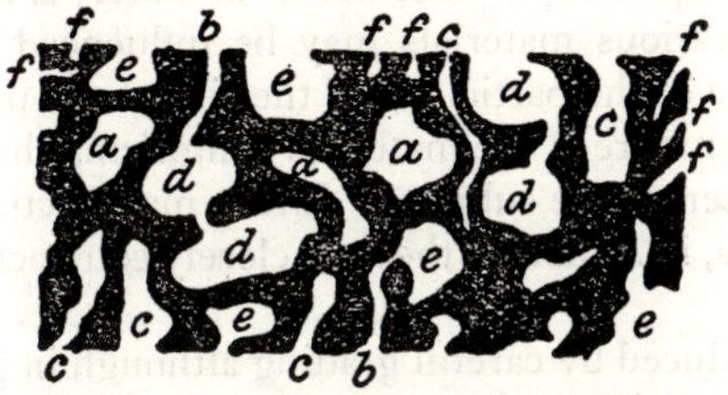

FIG. VII.21. VARIOUS KINDS OF PORES

When some clays are overheated the formation of bubbles of gas in them is greatly increased and the material swells. The bubbles each form a closed pore, and the whole mass, eventually has a vesicular structure and is frequently bloated and misshapen.

Sealed pores (with or without bloating) may also be formed by improper control of the kiln at a temperature of 700–950° C. They may be prevented by heating the articles more slowly and with more air at this range of temperature.

Pinholes and *blisters* may be regarded as forms of sealed pores. The first is usually due to the decomposition of calcium sulphate or other salt in the clay. Blisters may be due to a sudden overheating, or to mechanical defects in the shaping of the articles.

The *open pores* are those with an extremity on at least one surface and are the result of (*a*) the original packing of individual grains, (*b*) the escape of water during drying or of gases during firing, and (*c*) cracks and fissures which develop during drying and firing as a result of shrinkage (or expansion) of the material. They vary in size from that of a fine capillary to 1 mm. or more; the size depending largely on the coarseness and openness of the texture.

The **true porosity** is the ratio between the volume of all voids, both open and closed pores, and the total volume of the article.

The **apparent porosity** is defined as the ratio between the volume of water or liquid capable of being absorbed into it and the total volume of the article. Any closed pores, or for that matter, any open ones which are so fine that liquid cannot penetrate into them, will not be included as void space in determining the apparent porosity.

Both the true and the apparent porosity may be expressed as (*a*) percentage by weight, and (*b*) percentage by volume. The percentage of porosity by weight shows the weight of water or other liquid absorbed by 100 units of weight, whilst the percentage of porosity by volume is the volume of water absorbed in 100 volumes of the material or article. It is customary to use the latter as this indicates the volume of the pores. The volume of articles of simple shape may be calculated from direct measurement of the dimensions, but where the shape is complex the volume must be determined by finding the difference between the weight when immersed in water and the weight in air. The expression 'porosity by weight' is unsatisfactory because all materials have not the same density (e.g. clay particles may have a density of 1·4 to 2·6), so that the 'porosity' of different samples cannot be compared accurately. The expression of porosity wholly in terms of volume is, therefore, the most satisfactory as well as being the most consistent. Nevertheless in the brick industry the water absorption (per cent *weight* of water) is invariably quoted.

The porosity of various materials may be influenced by some or all of the following: (*a*) the shape of the particles; (*b*) the size of the particles; (*c*) the grading of the particles; (*d*) the nature of the materials comprising the mixture; (*e*) the treatment to which the materials are subjected during manufacture; and (*f*) the relative position of the particles, i.e. whether they are closely compacted or lie loosely on one another.

Porosity can be reduced by careful grading although in general, the coarser the maximum size of the particles the less will be the porosity; in other words an ideally well-graded mixture of fine particles does not approach the theoretically-predicted minimum porosity figure as closely as a coarser mixture.

A mixture of angular grains readily compacts under pressure to give a low porosity as shown in Table VII.VI.

TABLE VII—VI. EFFECT OF PRESSURE ON THE POROSITY OF CLAY

Pressure lbf. in.^{2}	*Porosity (per cent)*	*Pressure lbf. in.*^{2}	*Porosity (per cent)*
A 500	38·71	I 2500	26·87
B 750	34·73	J 2750	26·31
C 1000	32·59	K 3000	25·58
D 1250	30·83	L 3250	25·70
E 1500	30·65	M 3500	24·94
F 1750	29·83	N 3750	24·90
G 2000	28·83	O 4000	24·13
H 2250	27·47		

A fact which is often overlooked is that if the individual grains in a body are themselves porous, the body itself will have a porosity higher than that of a similarly graded mixture of non-porous particles. Thus, if the raw material is an aggregated clay, the porosity of the body will be related to the porosity of the grains. For this reason, a clay which contains a proportion of large-sized sand particles will have, all other things being equal, a lower porosity than a body composed of a purer clay.

The porosity is also greatly influenced by the methods used for moulding and the reactions which occur on firing. These will be discussed in more detail in the appropriate section (Chapter XII).

In some cases, a high porosity is desirable, particularly where a high thermal insulation is required. A large amount of air-space entrapped within the particles of the body presents the maximum resistance to the passage of heat particularly if the voids are discontinuous. Materials may be added to increase the porosity and the most usual example is carbonaceous material. When the mass is fired, the carbonaceous matter burns away, leaving corresponding pore-spaces, so that the porosity of the fired mass is roughly proportional to the volume of carbonaceous matter added and is, in some cases, very high. The chief combustible materials added to clays, etc., to increase their porosity are hardwood sawdust, cork, seeds, naphthalene, and, occasionally, finely-ground coke. The maximum proportion of carbonaceous matter which can be added depends on the bonding power of the other ingredients. The porosity of a mass may also be increased by the addition of a porous material, such as grog, kieselguhr or vermiculite.

Gas bubbles may also be artificially introduced by adding finely-divided zinc powder to the raw material and then mixing dilute hydrochloric acid with the water added during moulding. Hydrogen gas is liberated which under certain conditions can be retained as gas bubbles. A froth-stabilising agent may also be added to a clay and bubbles introduced by agitating the plastic body.

One of the best-known ceramic thermal insulating materials is diatomaceous earth where a high porosity body is produced by a natural process. The origin and mode of formation of diatomaceous earth has been described in Chapter VI (p. 312) and as would be expected the skeleton structure is capable of entrapping large volumes of air. The porosity of such bodies is very high and the void space is mainly in the form of micro- or pocket pores. The result is that the heat transfer is low and at the same time there is not an excessive amount of gas permeability.

For most purposes, the size of pores should be less than 0·04 in. and in 'thin ware' it should not exceed 0·02 in.; otherwise, the strength of the mass is reduced and the heat insulation capacity is lowered largely as a result of the greater permeability.

The porosity of a raw material or ceramic body has a great influence on many properties. These include the property of absorption, spalling resistance, conductivity, mechanical and chemical erosion, refractoriness, strength and many others, all of which will be considered in other chapters.

The *porosity of raw clays* varies greatly according to their nature. Some plastic clays appear to be devoid of porosity, some fireclays have only about 3 per cent porosity, whilst some sandy loams have a porosity of 25 per cent or more.

The Measurement of Porosity

As explained previously there is a distinction between the true and apparent porosity, but in most cases the term 'porosity' is applied to the apparent porosity as estimated from the amount of water or other suitable fluid absorbed by a given weight or volume of the sample. Hence the *absorption* or *apparent porosity* is a measure of the unsealed pores.

The usual way of carrying out the determination[1] is to take a conveniently-sized test piece, if possible a 2 in. cube or a whole brick, which is weighed (W_1)

[1] See British Standard B.S.S. No. 3921, 1965.

after being thoroughly dried. To ensure that liquid completely fills all unsealed pores the sample should be placed in a vacuum vessel and evacuated. Without releasing the pressure, the immersion liquid must then be introduced to such a level that the piece is entirely covered. The saturated sample may then be weighed in air (W_2) by suspending it from a thread attached to the arm of the balance; and finally the weight of the sample when immersed in water (W_3) has to be measured.

The weight of liquid absorbed is given by $W_2 - W_1$ g. and provided that water (density = 1) is used for immersion this figure is also the volume of the open pore spaces in cubic centimetres. The total volume of the test piece is $W_2 - W_3$, i.e. the weight 'lost' on immersion in water, so the percentage apparent porosity by volume (P) is given by the equation

$$P = \frac{W_2 - W_1}{(W_2 - W_3)\rho}$$

(where ρ is the density of the liquid which will be 1 if water is used).

If a vacuum equipment is not available an alternative technique capable of giving results comparable in accuracy has been described by Cross and Young.[1] The method is to immerse the specimen heated to just above boiling point in boiling liquid. As cooling proceeds the air space is completely filled and the procedure is then as above.

An accurate method which is suitable for small test specimens and those which break up on immersion in liquid is shown in Fig. VII.22.

The specimen holder (H) is usually cylindrical, although any size may be selected depending on the shape and dimensions of the samples being tested. A glass plate closes the lower end of the holder, the seal being made by vaseline or vacuum grease.

To calibrate the apparatus before use the glass plate is securely fixed to the specimen holder and tap T_1 closed to isolate the round-bottom flask (F) which should be of comparable volume to the specimen holder. The assembly is evacuated by turning tap T_2 to the appropriate position—the pressure can be observed on the manometer (M). When evacuation is complete, tap T_2 is closed, tap T_1 opened and the pressure (P_1) calculated from the difference in levels in the two manometer arms. The reading of the scale on the specimen-holder side of the manometer should be carefully recorded (S). The assembly may then be restored to atmospheric pressure by adjustment of tap T_2.

A bar of metal of known volume is then placed in the specimen holder and the test repeated, but after tap T_1 has been opened the level of mercury in the left-hand manometer arm must be adjusted to S by running in mercury from the reservoir (R)—this is to ensure that the volume in the assembly is maintained constant. The pressure is again measured (P_2).

If V_1 is the volume of the assembly comprising the specimen holder (H), round-bottomed flask (F) and connecting tubing to S; and V_2 is the volume of the metal bar, from Boyle's law:

$$P_1V_1 = P_2(V_1 - V_2)$$

hence V_1 can be calculated.

[1] Cross, A. H. B. and Young, P. F., *Trans. Brit. Ceram. Soc.*, 47, 121, 1948.

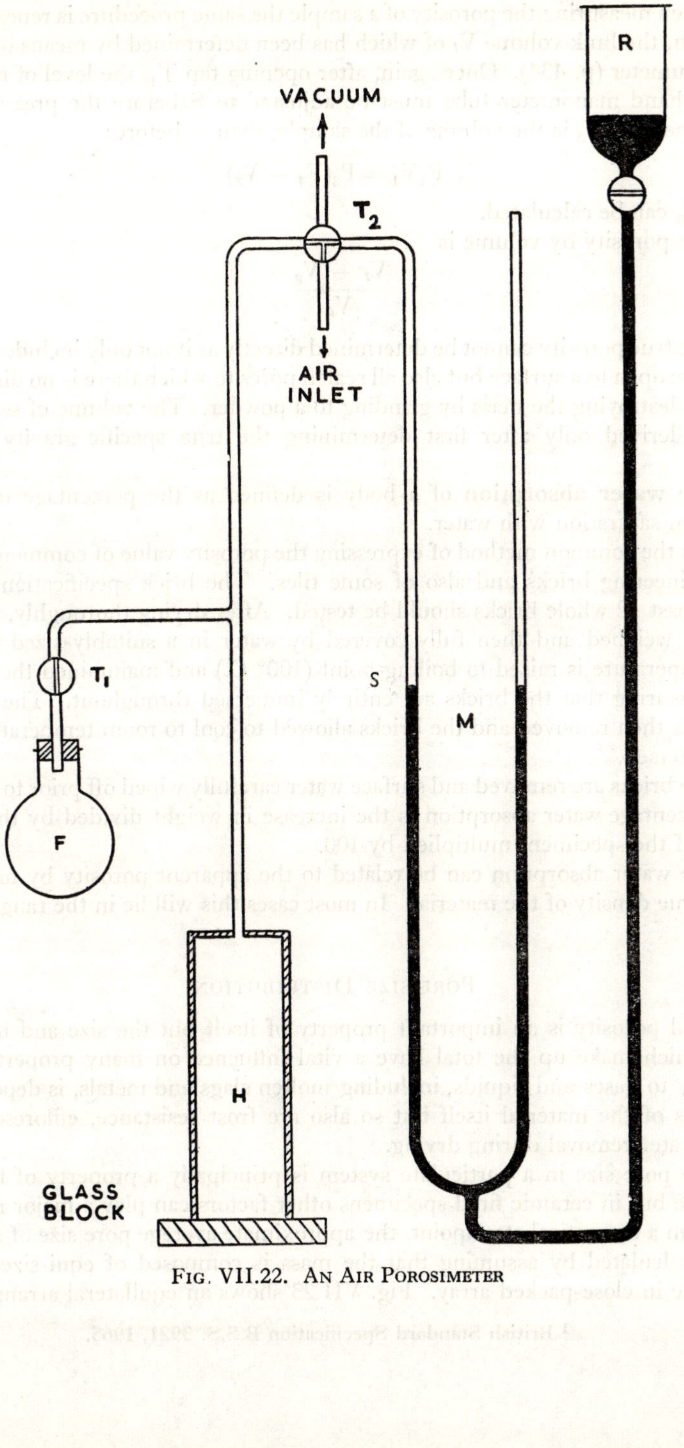

FIG. VII.22. AN AIR POROSIMETER

When measuring the porosity of a sample the same procedure is repeated with a specimen, the bulk volume V_t of which has been determined by means of the mercury volumeter (p. 434). Once again, after opening tap T_1, the level of mercury in the left-hand manometer tube must be adjusted to S before the pressure (P_3) is determined. If V_3 is the volume of the sample, then as before:

$$P_1V_1 = P_3(V_1 - V_3)$$

hence V_3 can be calculated.

The porosity by volume is

$$\frac{V_t - V_3}{V_t}$$

The true porosity cannot be determined directly as it not only includes the pores which are open to a surface but also all sealed pores to which there is no direct access without destroying the mass by grinding to a powder. The volume of sealed pores can be derived only after first determining the true specific gravity and bulk density.

The **water absorption** of a body is defined as the percentage increase in weight on saturation with water.

It is the common method of expressing the porosity value of common, building and engineering bricks and also of some tiles. The brick specification[1] requires that at least 12 whole bricks should be tested. After drying thoroughly, the bricks must be weighed and then fully covered by water in a suitably-sized container. The temperature is raised to boiling-point (100° C.) and maintained there for five hours ensuring that the bricks are entirely immersed throughout. The source of heating is then removed and the bricks allowed to cool to room temperature whilst still immersed.

The bricks are removed and surface water carefully wiped off prior to weighing. The percentage water absorption is the increase in weight divided by the original weight of the specimen, multiplied by 100.

The water absorption can be related to the apparent porosity by multiplying by the true density of the material. In most cases this will lie in the range 2·6–2·7.

Pore-size Distribution

The total porosity is an important property of itself but the size and number of pores which make up the total have a vital influence on many properties. Permeability to gases and liquids, including molten slags and metals, is dependent on the pores of the material itself but so also are frost resistance, efflorescence and rate of water removal during drying.

The pore size in a particulate system is principally a property of the centre grain size but in ceramic fired specimens other factors can play a major role.

From a theoretical standpoint, the approximate average pore size of a compact can be calculated by assuming that the mass is composed of equi-sized spheres which are in close-packed array. Fig. VII.23 shows an equilateral arrangement of

[1] British Standard Specification B.S.S. 3921, 1965.

spheres of radius R which would have a pore size of a sphere, r, sufficiently large to be contained in the space between them.

The size of the aperture can be calculated by constructing an equilateral triangle from the centre of each large sphere and then joining the apices to the centre of the small spherical aperture.

Then
$$\cos 30^\circ = \frac{R}{r+R}$$

from which
$$r = 0{\cdot}154\ R$$

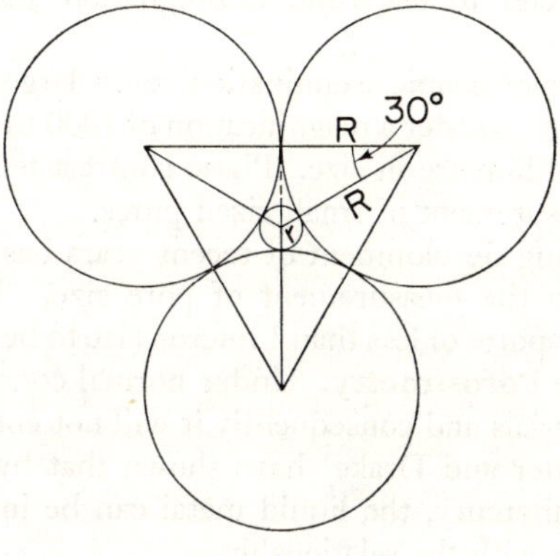

FIG. VII.23. METHOD OF CALCULATING THE APPROXIMATE PORE SIZE UNDER IDEAL CONDITIONS OF MONO-SIZED SPHERICAL PARTICLES

This is a useful guide to the average pore radius of a compacted mass and it can be applied to give an approximate value when particles are non-spherical and when a size distribution is involved. In this case the centre grain size (c.g.s.) replaces the value of R (note that radius is involved and not diameter).

Thus a mass of particles with a c.g.s. of 100 microns equivalent spherical diameter will have an average pore size roughly equivalent to $r = 0{\cdot}154 \times 50 = 7{\cdot}70\ \mu$.

The measurement and subsequent assessment of pores in raw and fired ceramic materials have received considerable attention and there are various techniques in use.

1. **Direct Measurement.** If the pores are large they can be examined and measured under a microscope at a suitable magnification. Mostly, however, the pores in ceramic bodies are too small to be observed directly except by special techniques.

Sample preparation is of the greatest importance and a specimen prepared for reflected light viewing has many advantages. In order to prevent surface disintegration during polishing and to highlight the pores, impregnated specimens are most frequently employed.

When pores of several microns in size are present, plaster of Paris impregnation is successful and for better relief the plaster may be coloured. The specimen is roughly polished and placed in a small cardboard container which is filled with a fluid plaster suspension. The whole is placed in a suitable receptacle and evacuated to a very low pressure. When air bubbles have ceased to be emitted, the vacuum is slowly released to permit the plaster to enter the pores of the body and there to set.

Polishing of the face of the section may then be completed prior to examination in reflected light. The plaster of Paris assumes the replica of the pore system and the sizes and distribution can be assessed quite accurately.

Some workers prefer to use resin impregnation and others some form of staining compound.

Under ordinary microscopic examination, only large pores can be observed and measured accurately. Under a magnification of 1000 times, a pore of 10 microns will appear as only 1 millimetre in size. Phase contrast techniques are helpful but do not permit the measurement of small-sized pores.

The most promising development of recent years has been the application of electron microscopy to the measurement of pore size. The extra magnification which it affords enables pores of less than 1 micron size to be observed and measured.

2. **Mercury Pressure Porosimetry.** Under normal conditions mercury does not wet most ceramic materials and consequently it will not enter pores unless they are of very large size. Ritter and Drake[1] have shown that by applying pressure to a material immersed in mercury, the liquid metal can be induced to enter pores of the body in accordance with the relationship

$$Pr = -2\sigma \cos \theta$$

where r is the radius of the pore penetrated at pressure P;

σ is the surface tension of mercury which can be taken as 480 dynes/cm.; and

θ is the contact angle between mercury and ceramic which for most purposes is 140°.

This equation shows that whilst mercury will not penetrate at zero pressure, at 21·5 lbf/in.^{2} pores of radius 10·0 μ will be entered, 215 lbf/in.^{2} is needed to fill pores of 1 μ, whilst if a pressure of 2150 lbf/in.^{2} can be applied, pores of 0·1 μ are penetrated.

To operate effectively, a bomb-type unit is necessary capable of withstanding the high pressures involved. The usual technique is to measure the reduction in level of mercury in a bulb containing the sample as pressure is progressively applied.

Watson, May and Butterworth[2] have shown the value of this method in application to building bricks and related materials. Some of their results are summarised in Table VII.VII.

3. **Gas Adsorption Methods.** The average pore size of a material can be measured by using the principle of gas adsorption as described on p. 391. For this determination, it is necessary to calculate the volume of gas adsorbed at total saturation pressure.

The difference between this value and the volume V_m adsorbed as a monolayer on the surface can be assumed to be the total pore volume (V_p).

[1] Ritter, H. L. and Drake, L. C., *Ind. Eng. Chem.* (*Anal. Ed.*), 17, 782, 1945.
[2] Watson, A., May, J. O. and Butterworth, B., *Trans. Brit. Ceram. Soc.*, 56, 37, 1957.

TABLE VII—VII. PORE-SIZE DISTRIBUTION IN SOME BUILDING MATERIALS

	Total porosity % *vol.*	*Porosity* (% *vol.*) *in the size range* (*e.s.d. in microns*)				
		>10·7	10·7–1·8	1·8–0·5	0·5–0·12	<0·12
Building Bricks						
London Brick Earth	42	10	24	7	1	0
Glacial Clay	37	2	1	24	11	0
Oxford Clay	37	7	4	14	11	1
Coal Measures a	38	9	6	4	15	3
b	21	8	1	3	5	3
Wall Tile	31	0	0	0	16	15
Portland Stone	15	2	0	3	8	1
Bath Stone	25	4	11	6	4	0

The average pore radius is then given by the equation

$$r = \frac{2V_p}{S_w}$$

where S_w is the surface area per g. of the solid phase.

A pore-size distribution can be calculated by desorption methods in which the volume of gas removed from a fully-saturated specimen is measured as a function of pressure. The smaller the cracks or pores in the specimen the more difficult it is to remove the adsorbed gas phase from them. The relationship is not unlike that given for the mercury pressure porosimeter.

Nitrogen adsorption and desorption methods are claimed[1] to be satisfactory for measuring pore radii down to 20 Å.

True Specific Gravity

The definition of the true specific gravity of a substance is the ratio between its weight and that of an equal volume of water. The term *density* is now considered to be synonymous with specific gravity although its use may lead to confusion, particularly when applied to ceramic bodies which are often porous. The true specific gravity of a *crystal* may be expressed in terms of the dimensions of the atoms or ions which make up the lattice. The weight of the atoms in a crystal and the way in which they are arranged determine the specific gravity and the introduction or removal of atoms or a rearrangement within the lattice are the only agencies which can alter its value. For example, rock-salt (NaCl) crystallises in the cubic system where the unit-cell dimension is 5·628 Å and there are four molecules of NaCl in each unit cell. The four molecules of NaCl imply four atoms each of sodium and chlorine, the atomic weights of which are, respectively, 22·997 and 35·457. The weight of each atom is the atomic weight divided by Avogadro's Number ($6{\cdot}02 \times 10^{23}$) (see p. 61).

[1] Joyner, L. G., Barrett, E. P. and Skold, R., *J. Amer. Chem. Soc.*, 73, 3158, 1951.

Therefore, the weight of the atoms in a unit cell is

$$\frac{4 \times 22{\cdot}997}{6{\cdot}02 \times 10^{23}} + \frac{4 \times 35{\cdot}457}{6{\cdot}02 \times 10^{23}} = \frac{233{\cdot}816}{6{\cdot}02 \times 10^{23}} \text{ gms.}$$

This weight is contained within the boundaries of the cubic unit cell of side 5·628 Å (where 1 Å ≡ 10^{-8} cm.). The total volume occupied by the atoms is therefore $(5{\cdot}628 \times 10^{-8})^3$ cubic centimetres.

The specific gravity of rock-salt is, therefore,

$$\frac{\text{weight}}{\text{volume}} = \frac{233{\cdot}816 \times 10^{24}}{6{\cdot}02 \times (5{\cdot}628)^3 \times 10^{23}} = 2{\cdot}179$$

Replacement of sodium ions by potassium gives the mineral sylvine, KCl, which is also cubic with a unit cell length of 6·277 Å. The atomic weight of potassium is 39·096 and calculating as before

$$\text{Specific Gravity} = \frac{\text{weight}}{\text{volume}} = \frac{4.(39{\cdot}096 + 35{\cdot}457)10^{24}}{6{\cdot}02.(6{\cdot}277)^3.10^{23}} = 2{\cdot}002$$

The determination of the true specific gravity of all perfect crystals can be calculated in this way provided the dimensions of the unit cell and the number of molecules in it are known. These data for most ceramic minerals are given in Chapters III and IV and in the Appendix.

The simple examples considered above indicate how both the absolute weight and the size of atoms contribute to the specific gravity of a compound. All other things being equal it is obvious that atoms where the ratio of atomic weight to atomic size is large will form compounds of high specific gravity.

Another factor which influences the true specific gravity of crystals is the way in which the individual atoms are arranged. Thus quartz, cristobalite and tridymite are all pure SiO_2 yet the specific gravities are respectively 2·65, 2·32 and 2·28. The individual atoms in the crystal lattices of each mineral are arranged differently so that the packing in quartz is closer than that in cristobalite which in turn is not as open as that of tridymite. For more details see Chapter III (p. 158). The true specific gravity is difficult to determine accurately for crystals are rarely, if ever, perfect and are liable to contain micro-pores, inclusion and other imperfections. In most cases, the errors which arise as a result of crystal flaws are not serious except when values correct to several places of decimals are required.

A ceramic body or raw material may be made up of many crystals and perhaps colloidal or glassy material; the specific gravity is the resultant of the various components. It may be determined by several methods depending on whether the material is a solid mass, a powder or a liquid.

The **true specific gravity of a non-porous solid** may be determined by:

(*a*) Weighing a piece of the solid and then suspending it in water and reweighing it. The specific gravity is found by dividing the weight of the material by its loss of weight in water, i.e. specific gravity $= W_s \div (W_s - W_e)$ where the weight of the dry sample $= W_s$ and the weight of the sample immersed in water $= W_e$.

(*b*) Weighing a piece of the material, measuring its volume in a volumeter and dividing the weight (in grams) by the volume (in c.c.).

The **specific gravity of a porous solid** may be determined by:

(i) By method (*a*) above, or

(ii) By method (*b*) above, but in both cases using the volume of the sample, after deducting the volume of the pores which can be filled by saturating the sample with water as described on p. 421. Neither of these methods will be accurate if the material contains any fine or sealed pores.

(iii) Grinding the material to a fine powder, the individual grains of which are too small to contain any pores and determining the specific gravity in the manner described below.

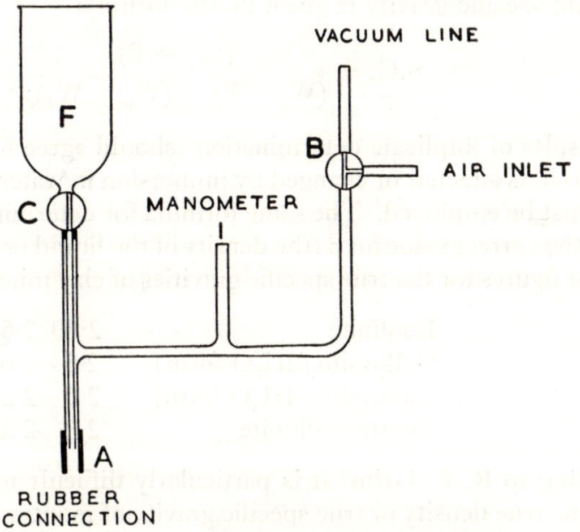

FIG. VII.24. AN APPARATUS FOR FILLING SPECIFIC GRAVITY BOTTLES WITH DISTILLED WATER UNDER VACUUM

Vitrified materials are anomalous, because they may appear to be non-porous and yet may contain many sealed pores. Hence, they should be ground to a very fine powder before their true specific gravity is determined, as, unless the sealed pores are broken open, low results will be obtained.

The **true specific gravity of a powder** may be determined by means of a specific gravity bottle, pycnometer or a Rees-Hugill flask.[1] The bottle and stopper are cleaned, dried and weighed (= P); a sufficient quantity of the dry powder (crushed to pass a 120 B.S.I. sieve) to half fill the bottle is placed in it and the bottle is again weighed (= W_1); the weight of powder is thus given by ($W_1 - P$). The bottle must now be filled with water and freed from all air. Satisfactory results can be ensured only by evacuating the air and then filling under vacuum with distilled water, which has been boiled to eliminate dissolved air bubbles and then cooled in the absence of air. A suitable apparatus is shown in Fig. VII.24. The specific

[1] See British Standard B.S.S. No. 1902, Part 1A, 1966.

gravity bottle without stopper is attached by a small piece of rubber tubing to A, and is evacuated by opening tap B which leads to a vacuum pump. When the pressure indicated by the manometer has become constant, distilled water is added from the funnel (F) by opening tap C. The system may be returned to normal pressure by opening tap B to air when the filling has been completed.

As the specific gravity of water varies with temperature, the filled bottle which must now be stoppered should be placed in a thermostat until its temperature is constant. Excess water overflowing through the stopper can be removed with blotting paper and the bottle and contents weighed (W_2). The bottle is then emptied, cleaned, refilled with distilled water and again brought to constant temperature prior to weighing again (W_3). The volume of the bottle is $(W_3 - P)/\rho$ where ρ is the density of water or other fluid used at the temperature of the test.

The true specific gravity is given by the formula

$$\text{S.G.} = \frac{(W_1 - P)\rho}{(W_3 - P) - (W_2 - W_1)\rho}$$

The results of duplicate determinations should agree to 0·005.

If the solid is affected or changed by immersion in water, some inert liquid such as xylene must be employed. The same formula for determining the specific gravity is used but the correct value for ρ (the density of the liquid used) must be substituted.

Typical figures for the true specific gravities of clay minerals are:

Kaolinite	2·60–2·68
Halloysite ($2H_2O$ form)	2·5 –2·6
Halloysite ($4H_2O$ form)	2·0 –2·2
Montmorillonite	2·3 –2·8[1]

According to R. E. Grim[2] it is particularly difficult to obtain really accurate figures for the true density or true specific gravity of montmorillonite, halloysite and illite clays because they contain a variable amount of water as an integral part of their structure. Moreover, as isomorphous substitution is common, the density of different samples varies greatly, even in the same locality, and no single figure is truly typical. Samples with less than 1 per cent of moisture range in density from 2·0–2·8—depending partly on the iron content in samples with a density greater than 2·5.

The specific gravities of other minerals in ceramic materials are characteristic and their determination may be informative. The raw materials used in the preparation of *basic* refractory materials have high values and can thereby be distinguished. The specific gravities of many of the minerals are to be found in the appendix.

The **specific gravity of slips or suspensions** offers an important means of control in the obtaining of uniform results in slip-casting, engobing or slip-coating and glazing. A glazing-slip of too low a specific gravity will form too thin a coating on the ware, whilst one which is too high will cause defects by forming too thick a coating.

[1] According to water—and iron—content.

[2] Grim, R. E., *Clay Mineralogy* (McGraw-Hill Publishing Co. Ltd., N.Y., 1968).

When the quality of a suspension is being controlled by specific gravity methods, care must be taken to ensure that the components are not changed in amount or obtained from a different source. Thus variations in the specific gravity of a flint for use in fine earthenware or china ware may cause the weight of flint in a slip to change by as much as 13 per cent and a potter who is unaware of this may unwittingly produce badly crazed ware.

The **true specific gravity of a liquid** may be determined as follows:

(*a*) By immersing a suitable hydrometer in the liquid and noting the figure on the stem which coincides with the surface of the liquid; such an instrument is frequently used in the control of body-slips and glazes and to determine the weight of dry material in them; or

(*b*) By means of a Westphal balance in which a glass plummet is immersed in the liquid and weighed in that position. Special weights are used which enable the specific gravity to be read off directly without calculation; or

(*c*) By using a specific gravity bottle or pycnometer. The weight of liquid to fill the bottle is compared with the weight of an equivalent volume of water. The specific gravity of the liquid is given by the expression:

$$\text{S.G.} = \frac{W_l - P}{(W_w - P)\rho}$$

where P is the weight of the bottle;
W_l is the weight of bottle and liquid;
W_w is the weight of bottle and water;
and ρ is the specific gravity of water.

Normally the specific gravity of water (ρ) may be taken as 1 but for accurate work a correction to allow for the temperature of test must be used.

	10° C.	20° C.	30° C.
Specific gravity of water	0·9997	0·9982	0·9956

A rapid method of assessing the **weight of solid matter in a slip** is to determine the specific gravity in ounces per pint and then apply the formula

$$\text{Weight of solid matter in a pint of slip} = (D - 20)\frac{G}{G - 1}$$

where G is the true specific gravity of the solid matter and D is the specific gravity of slip in ounces per pint.

If G is 2·6, which is the approximate value for many ceramic materials the formula above reduces to

$$\text{Concentration of slip (ounces per pint)} = (D - 20) \times 1{\cdot}625$$

A rough means of measuring the *specific gravity* (D) *of a slip* is to fill a weighed or counterpoised cylindrical vessel with the slip and then with water. The weight of slip (in ounces) divided by the weight of water (in ounces) gives the specific gravity.

Apparent Specific Gravity

The term **apparent specific gravity** or **bulk specific gravity** is used to indicate the relation between the weight of a mass of material as a whole and that of a volume of water equal to the total volume of the mass less the volume which can be filled with water (i.e. the volume of the solid material *plus* the sealed pores).

Many ceramic materials—especially when fired—contain 'sealed pores' into which water cannot penetrate. These reduce the apparent specific gravity of the material, so that the true specific gravity cannot always be found by subtracting the volume of the pores (as ascertained from the volume of water absorbed by the material) from the total volume of the mass; the result so obtained is not necessarily the volume of the solid matter, but the volume of the solid matter plus that of the sealed pores. When a material is ground to a fine powder, however, the sealed pores are destroyed and the sum of the volume of the individual grains forming the powder is the true volume of the solid matter, and from this the *true specific gravity* is obtained. The difference between the true and apparent specific gravities depends on the nature of the material examined.

Apparent Density or Bulk Density

With the exception of air-free glasses or single crystals, ceramic bodies are invariably porous to some extent. Air or gases are entrapped between and within the grains of the body or they may fill pores within the structure with the result that the external volume of the body is much greater than the actual volume of the material of which it is composed. The body has thus an *apparent density* which is simply the ratio of the total weight to the external volume.[1] The smaller the space into which the solid matter can be packed the greater will be its apparent density, and the closer the approach to the true specific gravity, so that the shape and size of the grains and the grading of the mixture greatly influence this figure. If a material has a porous texture it will necessarily have a low apparent density, because a considerable portion of the volume of the mass will be occupied by the light-weighing air in the pores. On the other hand, a material of a glassy or vitrified nature will have a similar apparent density and true specific gravity, because there are few pores in it.

The true specific gravity of the solid portion of the material will affect the apparent density, because the higher the true specific gravity, the greater (other things being equal) will be the apparent density. Thus, a piece of silicon carbide having the same proportionate volume of pores as a piece of fireclay brick will have a greater density because, whilst both pieces have the same total volume and the same volume of solid grains, the silicon carbide has a true specific gravity of 3·17–3·21, whilst the fireclay brick has a true specific gravity of only 2·6.

Apparent density is thus a function of the true porosity and true specific gravity, and factors which influence those properties will also affect the apparent density. In

[1] If the bulk density is expressed in grams per c.c. the bulk specific gravity will be identical, as water has a specific gravity of 1·0, 1 c.c. of water weighing 1 gm. If the bulk density is expressed in 'lb. per cu. ft.' the corresponding bulk specific gravity can be found by dividing by 62·3 which is the weight (in lbs.) of 1 cu. ft. of water.

calculating stress factors in brickwork construction, the apparent density is an important figure.

It may be measured quite simply in several ways:

(*a*) A first indication of the value can be obtained by calculating the volume of a symmetrical shape from the external dimensions and dividing into the total weight.

(*b*) A more accurate method is to use a mercury volumeter which is illustrated in Fig. VII.25. The sample container (A) is usually of cylindrical shape and is capable of being split to insert the specimen. The external volume is given by the difference in mercury levels in the burette (B) on filling to a mark (M) firstly with and secondly without the specimen in place. If the sample is exceptionally porous, mercury may enter some of the pores; in such cases, it is advisable to coat the specimen with a thin film of paraffin wax before testing.

(*c*) The apparent density can be determined along with the apparent porosity by the liquid impregnation method described on p. 421, and by its use the apparent specific gravity, true porosity and percentage of sealed pores may also be calculated.

The specimen is weighed dry (W_1), impregnated with water or other liquid, weighed again in air (W_2) and finally weighed suspended in water (W_3).

The properties to be measured are related to three important volumes which may be expressed as:

Vm = volume of solid material
Vo = volume of open pores
Vc = volume of closed pores

From the weights above

(*a*) The total external or bulk volume $Vt = Vm + Vo + Vc = W_2 - W_3$.

(*b*) The volume of open pores $Vo = W_2 - W_1$ (provided water is the immersion liquid used).

The *percentage apparent porosity* (P_A)

$$= \frac{Vo}{Vt} = \frac{W_2 - W_1}{W_2 - W_3} \times 100 \text{ per cent} \qquad (1)$$

The *apparent density* or *bulk density* (D_A)

$$= \frac{W_1}{Vt} = \frac{W_1}{W_2 - W_3} \qquad (2)$$

The *apparent specific gravity* (S_A)

$$= \frac{W_1}{Vm + Vc} = \frac{W_1}{Vt - Vo} = \frac{W_1}{(W_2 - W_3) - (W_2 - W_1)}$$

$$= \frac{W_1}{W_1 - W_3} \qquad (3)$$

The *true specific gravity* (S_T) $= \dfrac{W_1}{Vm}$ (4)

(cannot be determined except by the powder method p. 429).

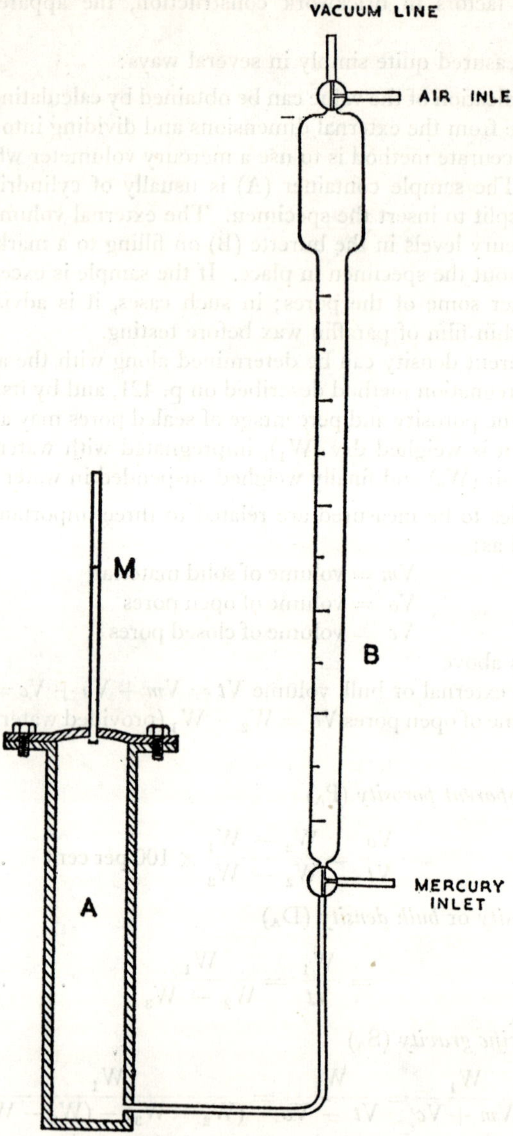

FIG. VII.25. A MERCURY VOLUMETER

The *true porosity* (P_T) $= \dfrac{Vo + Vc}{Vt} = \dfrac{Vt - Vm}{Vt} = 1 - \dfrac{Vm}{Vt}$ . . (5)

From equations 2 and 4:

$$\frac{\text{apparent density } (D_A)}{\text{true specific gravity } (S_T)} = \frac{W_1/Vt}{W_1/Vm} = \frac{Vm}{Vt} \quad . \quad (6)$$

Substituting this value in equation 5:

$$P_T = 1 - \frac{D_A}{S_T} \quad . \quad . \quad . \quad . \quad (7)$$

The percentage true porosity is given by multiplying by 100. Therefore the true porosity can be derived if the values of the apparent density and true specific gravity are known.

The *percentage of closed pores* (Pc) is

$$Pc = \frac{Vc}{Vt} = \frac{Vt - (Vm + Vo)}{Vt} = 1 - \frac{Vm}{Vt} - \frac{Vo}{Vt}$$

substituting values from equations 1 and 6

$$Pc = 1 - \frac{D_A}{S_T} - P_A \quad \left(\text{but } 1 - \frac{D_A}{S_T} = P_T\right) \quad \text{(equation 7)}$$

Therefore $$Pc = P_T - P_A \quad . \quad . \quad . \quad . \quad (8)$$

i.e. the percentage of closed pores is equal to the difference between the percentage true and apparent porosities.

The number of closed pores in a ceramic body depends on many factors, but is mainly the result of vitrification having taken place and the liquid produced effectively seals off air and gas bubbles.

To eliminate much of the tedium involved in calculating the various values when many samples have to be analysed, Lakin[1] and Cross and Young[2] have devised rapid calculator charts and nomograms which enable the results to be read off directly once the weighings have been completed.

The values of the true specific gravity, apparent density, and apparent porosity for some raw materials are shown in Table VII.VIII. The figures are only approximate and some samples give very different values when they are collected from other localities. The values for fired products are given in Chapter XII (p. 820).

Permeability

Permeability may be defined as the readiness with which a substance permits a fluid to flow through it and is measured by the rate at which water, air, or other gas flows through a mass of unit area and unit thickness at unit pressure. The permeability is due to the pores, voids, cracks or fissures which either run completely through the material or are connected with others to form a series which

[1] Lakin, J. R., *Trans. Brit. Ceram. Soc.*, 50, 208, 1951.
[2] Cross, A. H. B. and Young, P. F., *Trans. Brit. Ceram. Soc.*, 49, 191, 1950.

TABLE VII—VIII. PHYSICAL CHARACTERISTICS OF CERAMIC MATERIALS

Type	*True Specific Gravity*	*Apparent Density*	*Apparent Porosity (per cent)*
Raw materials			
Raw china clay	2·60–2·68	1·10–1·25	50–55
Plastic ball clay	2·40–2·80	1·40–1·60	35–45
Fireclay	2·60–2·70	1·70–2·10	30–35
Brick clay	2·30–2·80	1·60–2·10	20–40
Quartzite	2·65	2·60–2·65	1–2
Flint	2·61–2·63	2·52–2·6	—
Raw kyanite	3·5–3·7	3·5–3·7	1
Dolomite	2·85–2·90	2·75	—
Chrome ore	3·8–4·6	4·30 (av.)	—
Magnesite ($MgCO_3$)	2·9–3·1	2·90	—

extends through it. The fluid which fills sealed or closed pores without passing through them is not included in measurements of permeability.

The permeability (P) is most conveniently expressed by the use of an equation in which

$$P = \frac{\text{volume of air or water passed (ml.)} \times \text{thickness of specimen (cm.)}}{\text{area of cross-section of specimen (cm.}^2) \times \text{pressure (cm. of water)} \times \text{time of flow (sec.)}}$$

As pressure is necessary to cause the flow, it is customary to specify the pressure or head of the fluid. Water is occasionally employed, but air or other gas, being much more mobile, gives more accurate results.

The permeability of a ceramic article made by mechanical pressure is usually greatest in a direction at right angles to that at which the pressure was applied. This is not so noticeable in articles made by hand. In those made by extrusion (e.g. wire-cut bricks) the permeability is usually greatest in the direction in which the pressure was applied during manufacture.

To be wholly impermeable, a material must be very compact and close in texture, at any rate on its exposed surface, and for this reason such surfaces are often covered with a glaze.

In paving bricks and tiles, porcelain and some forms of stoneware, the requisite impermeability is produced by the interaction of fluxes and the material of which the goods are made, with the result that the greater part of the pores present in the early stages of firing are later filled with a glassy mass of fused silicates which renders the article impermeable. The production of impermeable articles is costly and in many cases both unnecessary and undesirable, but in others it is essential.

The permeability of an article depends on (*a*) its thickness; (*b*) the kind, number, distribution and sizes of the pores; (*c*) the presence or absence of cracks, fissures and holes; (*d*) the presence or absence of a superficial glaze or slag; (*e*) the difference in pressure on opposite sides; (*f*) the nature of the penetrating fluid; (*g*) the difference in the temperature of opposite sides of the article; and (*h*) the duration of the test.

In several of these respects permeability differs greatly from porosity, for whereas the latter is a measure of the total volume of the pores, the permeability depends on the extent to which they penetrate through the mass. A glazed brick may have a high porosity, but it should, when in use, be impermeable so long as the glaze is undamaged. In other words, permeability depends chiefly upon the number of channels or connected pores which enable the permeating fluid to pass through from one side of the article or test-piece to the other. Pores which are closed at one end, or do not penetrate far into the mass, take no part in producing a permeable mass; a single small crack penetrating right through the mass may increase the permeability more than will a thousand short or disconnected pores. In short, whilst highly porous articles are often highly permeable, there is no definite relation between the porosity and permeability of a material.

The Permeability of ceramic materials is usually determined on a cylindrical sample about 3 in. long and 2 in. diameter (Fig. VII.26).[1]

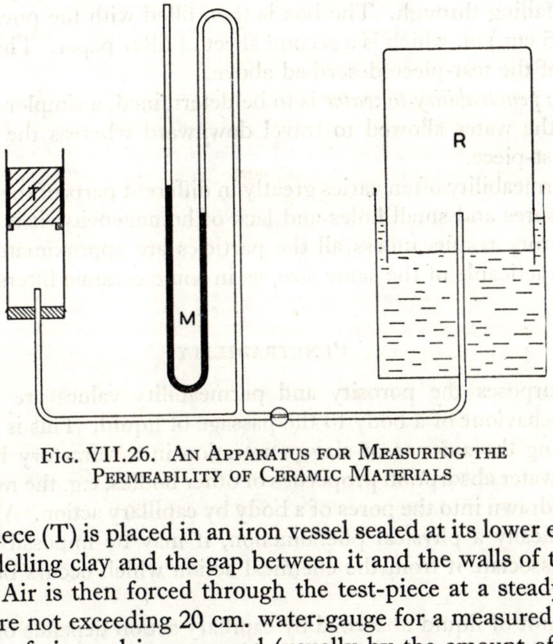

FIG. VII.26. AN APPARATUS FOR MEASURING THE PERMEABILITY OF CERAMIC MATERIALS

The test-piece (T) is placed in an iron vessel sealed at its lower end with a rubber ring or modelling clay and the gap between it and the walls of the vessel filled with mercury. Air is then forced through the test-piece at a steady rate and at a constant pressure not exceeding 20 cm. water-gauge for a measured period of time and the volume of air used is measured (usually by the amount of water used to displace it from an air-holder (R) which forms part of the apparatus). Frequent readings of the manometer (M) indicating the water-gauge pressure of the air should be noted and the average used.

The permeability (P) is measured in terms of the volume (ml.) of air[2] passed

[1] British Standard No. 1902, 1A, 1966 contains a detailed description of the British Standard method: it permits either a square or cylindrical testpiece and specifies the size as 'at least 7 cm. high and at least 25 sq. cm. cross-section'.

[2] Helium is also used particularly for glasses and porcelains (see F. J. Norton, *J. Amer. Cer. Soc.*, 30 (3), 90, 1953).

vertically through the test-piece in unit time and reduced to an imaginary test-piece of unit cross-section and stated height.

Thus, if P = permeability, V = volume of air (ml.) passed through the test-piece in t seconds under a pressure of p cm. of water, A is the cross-sectional area of the test-piece in sq. cm. and H is the height of the test-piece in cm.

$$P = \frac{V \times H}{t \times A \times p}$$

A slightly different procedure is described in British Standard 1069 (1953) which includes the determination of the maximum pore-diameter, the uniformity of pore-diameter and the relationship between pore-diameter and permeability.

The permeability of loose powders is determined in a similar manner using a circular metal box with an internal diameter not less than 5 cm. in which the bottom is formed of wire gauze. On this is laid a sheet of thin filter paper to prevent the powder from falling through. The box is then filled with the powder to a suitable height (e.g. 2·5 cm.) on which is a second sheet of filter paper. This arrangement is used instead of the test-piece described above.

When the *penetrability to water* is to be determined, a simpler arrangement can be used and the water allowed to travel downward whereas the air usually rises through the test-piece.

As the permeability often varies greatly in different parts of the same sample, on account of fissures and small holes and lack of homogeneity, it is often difficult to obtain satisfactory results unless all the particles are approximately spherical and as nearly as practicable of the same size, as in some ceramic filters.

Penetrability

For many purposes the porosity and permeability values are not sufficient to describe the behaviour of a body to the passage of liquid. This is particularly true when estimating the amount of slag penetration in a refractory brick, but it also applies to the water absorption properties of other bodies, e.g. the manner in which a liquid may be drawn into the pores of a body by capillary action. Although penetrability is specifically a physical phenomenon, it may be impossible and even undesirable to dissociate it from the chemical action which occurs between the body and liquid.

The amount of liquid absorbed by capillary action depends on the nature and viscosity of the liquid, its wetting action on the solid, the surface tension, the temperature, the time of contact and the position of the surface of the articles with respect to that of the liquid level.

All else being equal, the velocity at which a liquid penetrates a porous mass is approximately proportional to the cube of the radii of the pores and inversely related to the height to which it ascends. Hence, a coarse-grained brick is more penetrable to water than a fine-grained one.

OTHER PROPERTIES ASSOCIATED WITH TEXTURE AND PARTICLE-SIZE DISTRIBUTION

1. **Behaviour of Particles under Pressure.** The field of study which embraces the various aspects of the flow of dry particles under pressure is highly complex and forms an important subject of rock mechanics. Three different aspects are involved, each of which has an important bearing in ceramic studies:

(*a*) pressure due to the inherent weight of the particulate material itself which thus affects gravity flow, flow from bunkers, angle of repose of stock-piles,

(*b*) point-load applied pressure causing shear fracture in compacted masses. This aspect is involved in mining practice, in quarrying and in load-bearing properties of pressed bodies,

(*c*) overall load pressure affecting powder compaction, strata consolidation, strength of bodies.

The basic premise which governs all conditions of flow of powders under pressure is the well-known Reciprocal Theorem. This specifies that in a continuous mass of particles, if pressure is applied at any point it will be transmitted in diminishing amounts to all particles immediately in contact with that point. The pressure will thus be dissipated through the mass and there will be an equivalent resulting displacement, provided that there is no slipping between particles or permanent deformation or fracture.

Angle of repose. In a pile of dry particles, the stability depends on the mutual friction between grains. When the weight of overlying particles exceeds the coefficient of friction the mass will slide.

In an open stock-pile the sides of the particulate mass will attain the *angle of repose* where the coefficient of friction is equal to the tangent of that angle.

The angle of repose is related both to particle-size distribution and to ultimate size. In a mono-sized system of particles above about 120 B.S. mesh size, the angle of repose is constant at about 34°.

Smaller sizes of particles develop inter-molecular forces which increase the angle of repose considerably, thus:

Particle size (mean) (microns)	Angle of repose (φ)
1000	34°
100	37°
10	44°
1	58°

A spread of particle sizes in a mass can cause an increase in angle of repose over and above that expected of the contribution from the smaller particles. This is probably because there is a greater number of contact points in a graded mass thus increasing the coefficient of friction.

A ground mass of clay particles of a dry c.g.s. 500 μ with a quartile ratio of approximately 20 can have an angle of repose as high as 48°.

The presence of water, even in small amounts, can markedly affect the angle of repose, as plant operators who have struggled with the problem of 'hanging-up' in hoppers and bunkers know to their cost.

Other aspects of the behaviour of particles under pressure are considered in later chapters.

2. **Electrical Properties of Powders.** A moving cloud of dust particles will pick up a charge from contact with surrounding surfaces but more particularly from internal friction within the system itself.

This charge is associated with the atomic nature of the material of which the particles are composed, because metallic and some basic dusts develop positive charges whilst non-metals and acid dusts usually become negatively charged. The sign of the charge can be changed and if particles are passed between plates, one charged, the other earthed, they will pick up the charge of the plate of high potential.

The total charge picked up by a dust cloud is a function of particle-size distribution as well as of the material itself. A system of mono-sized particles will not develop a high charge density no matter what the ultimate size may be, although larger particles appear to acquire a slightly higher value.

When both large- and small-sized particles are present, very high charge densities often develop, where the larger particles acquire the greatest share. It is theorised that the smaller particles collide with larger ones whilst in motion, pick up a charge by friction contact but immediately lose it to the large particles which have a greater capacitance.

This is an important principle in pneumatic methods of clay conveying because the presence of charged particles presents problems of electrical discharge and of precipitation. Fortunately, although clays and associated minerals will acquire a charge, they lose it readily and are not as troublesome as some other minerals.

The principle of electrostatic charge pick-up is of important industrial application in some ceramic systems. Certain minerals can be separated from others by virtue of the charge they acquire—rutile from beach sands is an example; electrostatic precipitation of dusts, by attracting the charged particles to a surface of opposite charge, is a well-known separation and air-cleaning process.

3. **Optical Properties** of small particles are manifested in ceramic systems in opacifiers for glazes and glasses.

Particles suspended in a transparent medium will both reflect and diffract light. As a beam of light is passed through a dust cloud its intensity (I) is weakened relative to its original amount (I_0) and the relationship can be expressed as

$$\mathrm{I} = \mathrm{I}_0 \exp. (-\mathrm{KN}ld)$$

where K is a constant incorporating the wavelength of light;

N is the number of particles per unit volume;

l is the path length and d the mean diameter of particles.

With monochromatic light (radiation of single wavelength) the transmission decreases with particle size but with white light there is wide variation and complexity of characteristic.

Most materials have a maximum extinction coefficient to white light at about 0·5 micron size; conversely at this size the particles have the greatest reflectivity or *opacity*.

These principles are important in the selection of materials for white fillers for glazes and paints and for light reflection and are discussed more fully in Chapter XIV.

4. **Sonic Properties.** The properties of particles in relation to vibrations in the audible and ultra-sonic range may be considered from two points of view:

(*a*) in terms of coagulation and dispersion,

(*b*) as sound insulation.

(*a*) When particles are subjected to high-intensity vibrations in the frequencies of sonic and ultra-sonic regions, interesting phenomena occur.

At high frequencies (*circa* 500 cycles/sec.), the ultra-sonic vibration is sufficient to keep particles in violent agitation. In fact they readily break up and disperse, especially when they are suspended in water. The optimum frequency is quite critical and the principle is employed on a commercial scale.

At lower frequencies (*circa* 120 cycles/sec.) particles tend to congregate at nodal points and can be coagulated and agglomerated.

(*b*) Acoustical insulation depends very much on particle size and on the nature of the solid insulant itself. Sound is transmitted in the form of waves which are very long in relation to other types of radiation. Lower basal notes are 30 feet in length whilst high notes are only a few inches—ultra-sonic waves are even shorter.

Such long waves are easily reflected by solid surfaces and are not readily absorbed; in other words there is a low radiation transfer coefficient between sound waves carried in air to an interposed solid object.

For sound insulation, therefore, a solid object free from pores is best but other factors are involved. A completely smooth solid object would reflect sound waves in much the same way that a mirror reflects light so that it would give rise to echoes. A wall of perfect sound insulation would cause unpleasant reverberations within.

To overcome this problem a rougher, porous wall would be preferred; the rougher the better because only a slight departure from a smooth surface will still reflect low notes although transmitting or dissipating the ones of shortest wavelength.

For the best acoustics, therefore, large-sized particles, preferably porous within themselves, and a high void space, give the most successful results. Walls need to be thick to give good insulation and if the material composing the wall or screen is elastic, the insulation is improved. Large-sized expanded vermiculite, coke breeze, bloated shales, perlite and asbestos compositions are used but if there is no attendant fire risk, expanded plastic compositions are preferred.

5. **Electrical and Thermal Conductivity.** All forms of radiation conduction of short wavelength are governed by the same principles; the different wavelengths involved lead to minor orders of variation.

Basically the conductivity of most frequencies of radiation is considerably lower in air than in virtually all solids so that conductivity of all types is intimately associated with the total porosity and the pore size of the solid mass.

Conductivity across a porous body takes place by five mechanisms all of which may contribute:

(*a*) Conductivity through the solid phase, which is of high value, but in a particulate system it will be limited by the points and surface area of contacts.

(*b*) The conductivity in the air space, which is low and for most purposes can be ignored.

(*c*) Radiation transmission across a void space which becomes significant only when there is a high temperature difference between two surfaces.

(*d*) Convection currents in the gaseous medium which depend on movement which can occur. This is greatest in masses with large open pores and again is largely associated with hot bodies.

(*e*) Radiation transfer coefficient which determines whether radiation will be reflected from the surface or transmitted into the body itself.

The principles of electrical insulation differ from thermal insulation in that ceramic and other non-metallic materials are relatively poor conductors of electricity so there is not the same requirement of high porosity. However, when there is a possibility of heat being generated within or around the electrical insulation, porosity is of advantage, but in order to preserve compactness the pores are usually small. Ideal electrical insulators are micas and graphite, across their basal section. As explained in Chapter IV, these materials have a platey crystal habit with non-continuity of electrical bonding in their *c*-dimension.

Thermal insulation presents additional problems because at high temperatures radiation transmission increases, but also convection effects are greater in magnitude. These factors are discussed more fully in Chapter XIV, but in the present context the textural requirements for good thermal insulation are high porosity with the maximum number of sealed non-connected pores. Small pore size is therefore implied, a condition favoured by small single-sized particles. Diatomite, with its characteristic micro-cell structure, is a classical example of a good thermal insulator.

6. **Chemical Properties.** Chemical activity in gaseous, liquid or solid reactions depends principally upon the surface area exposed to attack. The solubility of a material in a liquid is simply related

$$ln\mathrm{S} = k\,\frac{\text{surface area}}{\text{mass}} \;\alpha\; k\,\frac{d^2}{d^3}$$

where S is the solubility in g./lt; d is the mean particle diameter and k is a constant.

Solubility is therefore in inverse ratio to particle size or to the centre-grain-size of a ground mass.

There is one complicating factor which must not be overlooked. In reactions involving a gas or liquid with a particulate mass, the rate of penetration can be significant in determining the rate of chemical activity. A well-graded mass has a low filtration rate and hence should be avoided if rapid reaction is required; in a ramming sand for ladle practice where reactions with slags and metals are to be avoided, a good grading is desirable.

The texture, particle-size distributions and shape of the grains have pronounced effects on many other properties of ceramic bodies in both the unfired and fired conditions. Some of these are dealt with in later chapters but by studying the basic principles involved, their probable effect can be predicted but not always their magnitude.

CHAPTER VIII

THE INFLUENCE OF WATER IN CERAMIC SYSTEMS

THE role of water in ceramics is of the greatest importance. Most raw materials are sedimentary in character and have therefore been laid down in a watery medium; furthermore, many of the principal mineral types are secondary products, resulting from the breakdown of primary rocks through the agency of water and, in addition, water is added to most ceramic bodies to facilitate shaping during manufacture. Hence the physical processes which are involved in the conversion of a rock-like body to a slip or plastic paste must be carefully studied.

Although the properties of all ceramic materials are affected to some extent by the addition of water, clays are by far the most susceptible and for this reason, the present Chapter deals principally with clay/water relationships.

If a piece of clay is placed in a large volume of water, the latter will gradually permeate the former, loosening and separating the particles and causing a *swelling* of the mass, until it *slakes*, or falls to pieces. The disintegration may be hastened by grinding or stirring the clay and water and, if they are sufficiently well mixed together, a *slip* or *slurry* is obtained. If the amount of water used is only about one-fifth of the weight of the clay and the mixing is sufficiently thorough, a plastic paste is produced. This paste can be hardened by drying, or it can be converted into a slip by mixing it with a further quantity of water.

When clays are mixed with water, the properties of the resulting suspension are those of a true colloid and the study of clays from the standpoint of colloid chemistry has led to a more complete knowledge of their behaviour.

PHYSICO-CHEMICAL CONCEPTS OF CLAY/WATER RELATIONSHIPS

PERHAPS the least understood of ceramic problems and that posing the most formidable barrier to absolute technical control in manufacture, is the true inter-relationship between clay particles and water molecules.

It is not difficult to appreciate why this should be. Water, although represented by a simple chemical formula H_2O, is in reality a highly complex entity; clays and associated minerals also are complicated in structure and possess many unusual groupings and configurations which affect associated water molecules.

In considering the influence of water on clay bodies, so many factors are involved and each must be given due regard for the particular conditions which apply.

Nature of Water. Although water in all its forms is composed simply of atoms of hydrogen and oxygen, it is known to behave in numerous different and bewildering ways. At normal temperatures it is a liquid, but only just, because at 0° C.

it changes state to become solid ice, and at 100° C. it undergoes an equally major change to become a gas, namely steam. Over a relatively short temperature interval, therefore, water changes from a well-defined, three-dimensional crystal lattice to a completely dissociated molecular form. In its liquid state, water has a transitional nature varying from a pseudo-ice lattice to individual molecules and its characteristic can change with temperature, pressure and with other external conditions.

When in molecular form, the arrangement of oxygen and hydrogen atoms is far from simple and the resultant charge distribution gives the unit an electrical unbalance or dipole moment. In other words, there are centres of positive and negative charge occurring at discrete points in the structural unit.

Although the forces associated with dipole moments are not strong they are sufficient to cause inter-molecular attraction and structural build-up. Particularly at temperatures near to its freezing point, water has a definite, although loosely-arranged structure, caused by hydrogen bonding between the protons and the electrostatic attraction of the dipoles. Not only can such a structure build up within water itself, but these same principles probably apply when water molecules become associated with a mineral surface. Unsatisfied bonds and electric charges on the outer layers of the crystal lattice induce a structural configuration within the dipoles of the water molecules in immediate contact with the surface.[1,2]

The presence of ions in solution in water plays an important role in aligning the dipoles of the HOH molecule. Cations of positive charge attract the negative centres of the water molecule and hold them with sufficient bonding power to develop a hydrosphere.

Water possesses another important property; it readily ionises so that in some circumstances it behaves as a grouping dissociated into hydrogen ions (H^+) and hydroxyl ions (OH^-). This behaviour is responsible for acidic and basic properties of solutions and also, in part at least, for the colloidal characteristics of suspensions.

Nature of Mineral Surfaces. Because of the complex character of water many aspects of the structure of minerals can be of influence, but those associated with the surface, both externally and internally, are the most important.

Basically, three features of crystal surfaces may influence water molecules which come into contact with them:

(*a*) Distorted ionic groupings in surface layers.
(*b*) Broken bonds at fracture or cleavage planes.
(*c*) Unscreened ions (usually cations) associated with the lattice.

Further complications can occur when the minerals possess lattice holes or channels large enough to permit water units to enter and to be retained. This is the case in zeolites and some clay minerals.

The change in electrical balance and equilibrium in lattices at surfaces or along broken edges is illustrated in Fig. VIII.1. If the crystal lattice is figuratively represented as a system of alternating positive and negative charges in three dimensions, as shown in Fig. VIII.1, an ion well-removed from the surface will be symmetrically

[1] Terzaghi, K., *M.I.T. Tech. Eng. News*, 9, 36, 1928.
[2] Low, P. F., *Advan. Agron.*, 13, 269, 1961.

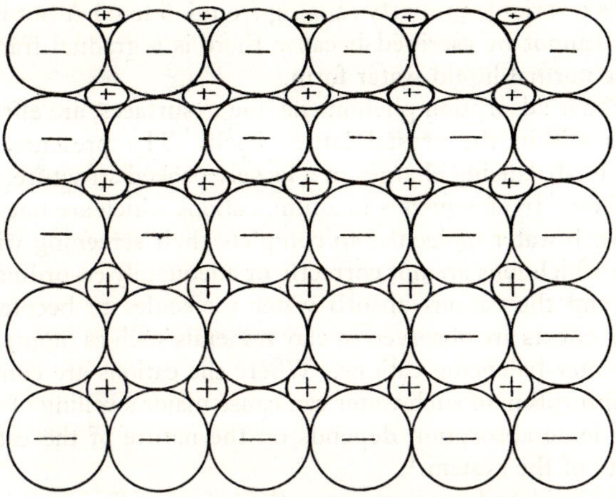

FIG. VIII.1. FIGURATIVE REPRESENTATION OF THE ATOMIC OR IONIC DISTRIBUTION IN THE SURFACE REGIONS OF CRYSTALS

surrounded in three dimensions by six ions of opposite charges, and thus will be in equilibrium. At the crystal edge, however, or along the line of a fracture the exposed ions are not surrounded on all sides, and the electrical balance is destroyed. It is on such centres, which, in most ceramic crystals, will be mainly the larger negative ions of oxygen and hydroxyl units, that broken-bond water is held.

Such water is known to be in a different physical state from that of liquid water, but opinions differ as to its true nature. It is not improbable that more than one mechanism might apply particularly when the mineral surfaces involved are clay minerals. A basal plane of a layer lattice mineral is theoretically in electrical balance, but the surface ions must be polarised because they are not fully screened. Fairly long-range but uniform attractions will therefore influence water molecules. When a fractured surface is involved, valency bonds are most likely to be unsatisfied, thus producing localised areas of high residual charge density. The attraction for water dipoles could thus be of a different type and order.

Some surfaces, particularly those of a planar nature, with oxygen or hydroxyl groupings in characteristic, co-ordinated array are thought by Hendricks and Jefferson[1] to link up geometrically with the water units to form a bonded layer capable of extension to several molecular thicknesses.

Whatever the basic underlying mechanism, it is generally accepted that surfaces can attract water molecules to themselves to form layers where a type of structure exists very different from the condition in liquid water. The density of such water layers is higher than in normal water and according to Low and Anderson[2] approaches that of ice. The viscosity of the layer is much higher than that of normal water, possibly because of the increase in surface tension.[3] The thickness of the

[1] Hendricks, S. B. and Jefferson, M. E., *Amer. Mineral.*, 23, 863, 1938.
[2] Low, P. F. and Anderson, D. M., *Soil. Sci. Soc. Amer. Proc.*, 22, 22, 1958.
[3] Leonard, R. A. and Low, P. F., *Proc. 12th Nat. Clay Conf.* (Pergamon Press, N.Y., 1964), p. 311.

adsorbed layer is a variable property ranging from 7·5 to 60 Å but in all probability a definite limit cannot be ascribed because there is a gradual transition from the structural to the normal liquid water form.

Over and above adsorption phenomena due to surfaces, are effects attributed to ionic inclusions within the crystal lattice itself. The presence of weakly-held cations, as counter-balancing charges in a loosely packed structure, invariably leads to hydration effects. In an aqueous medium, cations which are not fully electrically bonded will absorb water molecules to complete their screening within the lattice. Some lattices in which ions are not correctly or adequately co-ordinated possess this same property and the cations absorb water molecules to become hydrates (see p. 201). Similar effects are observed in clay minerals such as montmorillonites with loosely-held counter-balancing cations. Where the cations are contained in a specific layer, the absorption of such water can cause major swelling of the lattice itself. The amount of ionic absorption depends on the nature of the cation and on the relative humidity of the system.[1]

It is not easy to identify or to measure the nature of the water layers in association with mineral structures, surfaces and ions. Infra-red methods (p. 246) reveal many details in the appropriate part of the spectra and suggest that great complexity exists in the nature of water molecules. As yet, the results cannot be fully interpreted, but undoubtedly many forms and modifications can exist.

Forms of Water in Clays, Etc.

WATER is known to exist in many forms in clays and similar materials, but the boundaries of each form are not clearly defined and there is considerable overlapping. The principal types are as follows:

1. **Combined water** is present in clays and many secondary alteration products. The usual form is as hydroxyl groupings (cf. the structure of clays, Chapter III), but isolated water molecules are occasionally present. Combined water is often termed *water of constitution*, and sometimes *water of crystallisation*, although the latter is incorrect in some instances. The combined water in clays and related minerals is a definite part of the crystal lattice and is only liberated at a comparatively high temperature when the mineral decomposes. If hydroxyl units are lost, the reaction may be represented as:

$$—OH^{1-} + —OH^{1-} \rightarrow H_2O + —O^{2-}$$

Hydrogen ions may combine with oxygen or hydroxyl units to generate water and this may be the reaction involved in the decomposition of goethite:

$$\underset{\text{(goethite)}}{2HFeO_2} \rightarrow \underset{\text{(ferric oxide)}}{Fe_2O_3} + H_2O$$

Once water of constitution has been removed, the mineral from which it was released is virtually destroyed and no longer possesses the characteristics of the original form. Thus, kaolinite, $Al_2(Si_2O_5)(OH)_4$ decomposes under conditions of slow rate of temperature rise at about 410° C. The resulting product is an intimate mixture of free silica and alumina, and although partial rehydration can be effected

[1] Méring, J., *Trans. Farad. Soc.*, **42B**, 205, 1946.

in an autoclave, the original form of the kaolinite has been destroyed. The reaction is as follows:

$$\underset{\text{(kaolinite)}}{Al_2(Si_2O_5)(OH)_4} \rightarrow \underset{\text{alumina}}{Al_2O_3} + \underset{\text{silica}}{2SiO_2} + \underset{\text{water}}{2H_2O}$$

2. **Water of crystallisation** or **hydrate water** is that which is involved when substances of a crystalline nature are heated, usually at a comparatively low temperature. It differs from the combined water or water of constitution inasmuch as the original substance can be readily re-formed by treating the dehydrated material with water and then allowing it to recrystallise. The precise mechanism of the bonding of the hydrate water within the crystal is not known with certainty in all cases, but in many hydrates with ionic linkages water molecules are associated directly with the cation. The amount of water of hydration depends on the co-ordination around the cation, although the presence of anions introduces additional complication; a certain salt may form several types of crystals with different amounts of water of crystallisation, depending on the conditions of formation. Thus, hydrated sodium silicate occurs in three crystalline forms:

Formula	*Crystalline Form*
$Na_2SiO_3.9H_2O$	rhombic
$Na_2SiO_3.6H_2O$	monoclinic
$Na_2SiO_3.4H_2O$	hexagonal

When a hydrate loses its water of crystallisation the atomic configuration is not entirely destroyed for it simply reverts to a non-hydrous form.

3. **Broken-bond water** is absorbed by crystals where there is a charge deficiency at a fractured edge. Kelley, Jenny and Brown[1] in 1936 were the first to introduce the term broken-bond water, and showed by dehydration experiments that it could be distinguished from true planar or adsorbed water.

That water is absorbed by fractured surfaces can be readily illustrated by grinding many alumino-silicates. The product invariably gives a characteristic thermal pattern, which becomes more pronounced as the reduction in size is maintained. Apparently in the case of micas, felspars and quartz, fracture takes place across the crystal, thereby exposing cations and anions which will not be electrically balanced. Hydroxyl or water molecules are absorbed from the surrounding atmosphere and strongly held. So firmly is the bonding between water of this type and the crystal edges, that it is only eliminated at high temperatures, and a type of compound formation must develop. Alumino-silicates which have been subjected to ball mill grinding show two thermal peaks at 550° C. and 225° C. The first is similar to that produced by the hydrates of alumina and the latter is associated with colloidally precipitated silica. On prolonged grinding, the thermal peaks become more diffuse and water liberation takes place over a range of temperature.

4. **Colloidal water** may be regarded as the extreme case of broken-bond water in which the particle size of the solid to which water is linked is extremely small. The unsaturated surface charges on the crystallites are numerous and able to influence not only those water molecules which are in contact, but also some removed

[1] Kelley, W. P., Jenny, H. and Brown, S. M., *Soil Sci.*, 41, 272, 1936.

a small distance from the surface. As the distance from the surface is increased, the binding forces decrease rapidly and consequently water molecules may be lost over a range of temperature. This is a characteristic of many colloidal gels, including silica and alumina, and, in addition, is a feature of some plastic clay pastes.

5. **Absorbed water** is water which is taken up within the crystal lattice on exposure to humid air. The most common example of a mineral with this capacity is montmorillonite, which has the property of absorbing moisture between the layers which make up the crystal. As a result, the lattice expands by virtue of the forcing apart by the water molecules.

The corresponding absorption of water in halloysite and possibly hydrogen montmorillonite, is not likely to be due to the same mechanism, for there are no exchangeable bases in the former mineral, and the hydrogen ion is too small to form co-ordinated groupings. It may well be that absorbed water does take on a specific structure which is determined largely by the 'fit' on to adjacent layers of the absorbing mineral, but where cations are present, they must exert some influence on the surrounding water molecules.

True *ab*sorption within the lattice is possible only in minerals which contain lattice holes of sufficient size to permit water molecules to enter (e.g. zeolites), or those which are capable of expanding to accommodate the molecule (e.g. montmorillonites). Such water is held more firmly than that resulting from simple surface *ad*sorption and requires a higher temperature before it is driven off; it is rapidly re-absorbed on exposure of the mineral to a humid atmosphere, provided that the mineral has not been decomposed by over-heating.

6. **Adsorbed water** is related only to the exposed surface-area of the mineral, and may thus be differentiated from absorbed water, which enters the crystal lattice. Because adsorption is restricted to the surface of minerals, it follows that its amount will be directly proportional to the surface-area of a material, and will be independent of its nature. This is not strictly so, although, for any particular mineral, the amount of water adsorbed is in direct relation to the degree of fineness of division, i.e. to the specific surface area. The amount of adsorbed water also depends on the humidity of the atmosphere to which the material is exposed and a change in humidity conditions is sufficient to alter its value.

7. **Free water** is that which is not combined in any way with the particles of the solid, and includes all excess water which may be present naturally or added during processing. It is frequently present as so-called *pore-water* which has been drawn into the mass of the material by capillary action. The maximum amount of pore-water is sometimes termed the *imbibitional moisture capacity*. When the raw material has been exposed to rain or excess water in another form, an even larger amount of water may be present, and in the case of clay, 'stickiness' develops. The physical concept of this state is that sufficient water is present in the interstices between the individual grains to cause partial separation. Deformation of the mass is, therefore, easy, as the water acts as a free-flowing cushion.

As more free water is introduced, the clay mass becomes plastic and will flow under its own weight. Finally, the clay/water mixture assumes all the characteristics of a true liquid as the water content is further increased, until a slip or colloidal suspension is formed.

Free water in all its forms is readily lost on exposure to normal atmospheric conditions, for the partial pressure of the water present is invariably greater than the vapour pressure of water in the surrounding air.

The presence of free water, however, may change the complete character of the clay, and it is one of the principal processes involved in the 'weathering' of ceramic raw materials.

Weathering Processes

The changes in the physical state of clays and other ceramic materials, due to the application or removal of water in the ordinary course of nature, are commonly regarded as the result of 'weathering'. They include the action of rain, snow and frost in winter, and of sunshine and rain in summer—the former is sometimes known as 'wintering' and the latter as 'summering' or 'sunning', but it is convenient to regard all under the term 'weathering'.

It is sometimes desirable to increase the action which would take place during weathering and especially to increase its velocity. Such treatment may be regarded as *artificial weathering*. This includes such simple processes as 'watering' the clay with a hose-pipe, or turning the material over to expose fresh surfaces. When the action of cold water is not sufficiently intense, hot water or steam may be substituted.

Changes effected by Weathering. When clays and other raw materials used in the ceramic and allied industries are exposed for a sufficient time to the action of the weather, they may undergo one or more of the following changes:

(i) Mechanical disintegration due to the freezing of water in the material.

(ii) Physical changes which are dependent on the behaviour of water absorbed by the material.

(iii) Chemical changes.

Most fired ceramic materials are highly resistant to the action of the weather and for that reason many of them are of great value for building and constructional work.

(i) *Mechanical Disintegration.* Dry clays and some other ceramic materials have the power of absorbing moisture from the atmosphere into their pores. If the weather becomes sufficiently cold the water is frozen, and as ice has a greater volume than the water from which it is produced, it exerts an enormous pressure on the ceramic material and may cause it to burst or disintegrate. Such an action is beneficial in the case of clays and some other materials, because it not only reduces the cost of crushing and grinding the materials, but it has the advantage over disintegrating machines, such as stone-breakers, crushing rollers, and grinding mills, in that the frost works from the inside outwards, instead of externally from the surface of the material towards the centre, and, consequently, produces effects which cannot be obtained by such machines. If sufficient time can be allowed most materials used in the ceramic industries may be disintegrated by frosty weather, but the porous substances are most readily and rapidly disrupted, as water can more easily penetrate into the pores of such masses.

(ii) *Physical Changes on Weathering.* The principal *physical changes* wrought in clay by exposure to the action of the weather are:

(*a*) Moisture is absorbed up to the limit of the hygroscopicity and sometimes the imbibitional capacity of the clay.

(*b*) The moisture is more uniformly distributed.

(*c*) The cohesion of hard masses is decreased by the formation of water films which surround the individual particles and the consistency of the clay is changed.

(*d*) The plasticity is increased.

(*e*) Oxidation of some of the mineral and other constituents of the clay is a chemical change, but when it occurs the physical character is altered, e.g. a very sticky clay may become easier to work. As some iron minerals change their colour on exposure they may be more easily recognised and picked out of the weathered clay. Pyrites, on the contrary, may break down and be more difficult to separate.

Chemical action may change the type of clay or alter the amount of soluble salts and absorbed cations with a consequent change in the physical nature of the clay (see *Base Exchange*, p. 263).

Steaming has a similar effect on clay to exposing it to the action of weather. Its chief actions are to:

(*a*) Supply the water required by the colloids present in a form in which it can act more intensely than is possible with cold water.

(*b*) Secure the uniform distribution of moisture throughout the mass.

(*c*) Increase the temperature of the clay and so ensure a greater amount of fermentation of the organic matter present.

Souring is a form of artificial weathering in that a clay body is allowed to mature in the presence of added water for a period of time before being processed. In the time interval, the important change taking place is that of a more uniform distribution of the moisture so vital for satisfactory processing and subsequent drying. Some clays imbibe water readily and need little or no souring whilst others profit by treatments of 48 hours or more (see also p. 505).

By these means, the plasticity and binding power of the materials are increased and the clay so treated is more easily shaped. When only a small amount of moisture is to be added to a clay it can be distributed more uniformly if applied in the form of steam. Some fireclays are much more readily disintegrated by steam or hot water than by cold water.

The time taken to weather different materials properly varies greatly according to their nature. Thus, a comparatively soft clay may merely require to remain exposed to the weather for a few days or weeks. Other materials, such as hard shales, require two years or even more.

Some clays are very peculiarly affected by weathering; some even become less plastic the longer they are exposed; others are adversely affected by weathering in wet weather. The best conditions are obtained when a clay is immersed in water and thereby thoroughly saturated. Such a procedure is seldom practicable, so that the method generally available is to spread it out in comparatively thin layers and expose it directly to the action of rain, frost and snow and sun.

(iii) *Chemical Changes on Weathering*. The chief *chemical effects* which occur on weathering are caused principally by the agencies of air and water, although some reactions are catalysed by sunlight. These are as follows:

(*a*) The solution and replacement of some substances after oxidation and hydrolysis.

(*b*) The oxidation of carbonaceous matter, sulphides and some ferrous compounds.

(*c*) The hydration or hydrolysis of substances which are ordinarily insoluble.

Solution and replacement by water has an important effect in changing the chemical nature of rocks through which it percolates, especially when it contains dissolved substances in solution. Thus, the percolation of water containing carbon dioxide through clay beds may cause the solution and removal of any soluble substances present. Calcium carbonate may be removed in this way from a highly calcareous clay or clayey limestone, and it is generally thought that the pocket clays of Derbyshire are composed of the insoluble residuum from a clayey limestone, the greater part of which has been removed by the solution of the calcium carbonate. Some clay beds containing calcium carbonate are richer in this impurity near the bottom of the bed than at the top, this latter portion having been partially purified by carbonated water descending from the surface and dissolving the calcium carbonate in its passage.

Even silica—though normally regarded as insoluble in water—is not wholly so.

The disintegration of rocks and clay beds by weathering is largely, though not entirely, due to the removal, in solution, of soluble substances originally present as bonding material.

Conversely, water containing silica in solution, when in contact with siliceous rocks, tends to precipitate silica in the pores of the rock, with the result that the particles of aggregate are still more firmly united with a siliceous cement. As water can contain both silica and calcium bicarbonate simultaneously in solution, a limestone may have part of its calcium carbonate removed in solution by percolating waters and replaced by silica—chiefly in the form of chert. Thus, near Carlow, in Ireland, a bed of chert 30 to 40 feet thick has replaced the original carboniferous limestone. The flints and chert in some English chalk and limestone beds are due to metasomatic replacement of this character. Sometimes quartz is replaced by calcite brought in solution by percolating waters, but this type of replacement is not common.

Another typical rock formed by partial replacement is dolomite, which has been formed by the substitution of some of the calcium carbonate in limestones by magnesium carbonate. Some dolomites, however, appear to have been produced by simultaneous crystallisation of both magnesium and calcium carbonates (see p. 325).

Weather has a bleaching action on clays and some other minerals containing suitable carbonaceous matter, because the latter, when decomposing as a result of exposure to air and water, forms humic and other acids which dissolve some of the iron and other compounds and remove them in solution.

Prolonged exposure to the weather also results in the solution and removal of selenite and gypsum. Any other soluble salts in a clay or other ceramic material may also be dissolved and carried away by water percolating through the mass.

Cementation, or the binding of the particles of aggregate into a strong mass of rock by means of a binding agent or 'cement', is caused, by (*a*) the mingling of solutions from different sources, as when a solution containing oxygen comes into contact with one containing iron compounds in solution, the result being the precipitation of iron oxide and the cementation of the rock in which the precipitation occurs; (*b*) the chemical reactions which occur between rocks and solutions percolating through them. Thus, the action of water upon anhydrous rocks causes the formation of hydrates or hydroxides, which fill up the pores and cement the rock particles. For example, felspar may be replaced by zeolites or haematite by limonite. Temperature and pressure play an important part in modifying the effect of chemical reactions in cementation.

Oxidation. Exposure to weather (i.e. to water and air) converts ferrous compounds into the ferric state, limonite being the chief product. Iron sulphides are, to some extent, converted into sulphates, marcasite being more readily oxidised than pyrite. Copper sulphides are oxidised in a similar manner, erubescite being readily broken down and chalcopyrite with more difficulty.

Sunlight appears to favour oxidation and has therefore an important influence on the weathering of clays. Remarkable differences in the character of some clays occur even after a few hours' exposure to sunlight and air, whilst other clays are scarcely affected. The difference is apparently due to some oxidation processes, possibly associated with colloidal changes accompanying the hydration of the material, but the precise nature of these changes can only be surmised, as they have not been fully investigated.

Hydrolysis is a term used rather loosely to indicate the effect of water in decomposing various substances as well as in forming additive compounds; the latter is usually termed *hydration*. Both are really the reverse of the neutralisation of an acid by a base, and may be expressed by the equation:

$$\text{salt} + \text{water} = \underset{\text{acid}}{(\text{acid radicle} - \text{H})} \underset{+}{+} \underset{\text{base}}{(\text{basic radicle} - \text{OH})}$$

Thus, in hydrolysis, a salt splits up, forming an acid and a base or, occasionally, a basic salt. The duration of the exposure to water and the enormous masses involved enable changes to take place in nature which cannot be accomplished in the chemical laboratory, with the result that most silicates and oxides can be hydrolysed and compounds, such as clays, formed, which cannot be produced by any artificial means. In the same way, the hydrolysis of iron compounds in clays and other rocks sometimes results in the formation of a particularly strong ferruginous cement. The yellow films or stains widely distributed among rocks are due to the production of limonite (see p. 275) by the oxidation and hydrolysis of other iron compounds.

Base Exchange. One of the most important consequences of weathering processes is the changes in the exchangeable bases. The influence of the type and amount of absorbed cation is discussed more fully in a later section of this Chapter.

Other Physical Processes due to Water.

1. *Slaking and Swelling.* When a 'lump' of clay is immersed in a large volume of water it gradually softens and eventually falls to a soft 'mushy' mass. This change is known as 'slaking'; it has an important practical significance, because those clays

which slake rapidly can be easily tempered or mixed with water to produce a homogeneous paste.

The time required to slake a clay varies according to the porosity of the material and the cohesion of the particles. A soft, porous, lean clay may be completely slaked in a few minutes, but a hard clay-shale or a highly plastic material may not be fully disintegrated after several weeks. The speed of slaking is increased if the water used to effect it is slightly alkaline.

The slaking is usually accompanied by an increase in the volume of the mass and a small rise in temperature.

2. *Swelling*, is the term applied to the increase in volume when a clay is immersed in water. It is caused by the absorption of water within the crystal lattice of certain minerals, particularly those of the montmorillonite and related layer-lattice crystals (see pp. 147 and 454).

Although the absorption of water is theoretically reversible, and the separation between layers of the crystals returns to the normal value when the water is removed, a permanent increase in the overall volume always remains on drying the swollen mass: this is probably the result of disruption within the mass caused by unsymmetrical drying.

The amount of swelling which occurs when a clay is treated with water depends on many factors which include (*a*) the type and amount of clay minerals which are present; (*b*) the exchangeable ions associated with the clay; (*c*) the particle shape and size of the clay; (*d*) the texture of the mass; (*e*) the soluble salts in the water, and (*f*) any superimposed load on the swelling mass.

Of the clay minerals, the montmorillonites swell to a far greater extent than other types. Sodium montmorillonite expands much more than the same clay mineral saturated with other ions and the rate of swelling is maintained for long periods of time (up to several weeks). Such effects can have serious consequences when buildings or roads are constructed on soils rich in these minerals and there are many examples of disruption being caused when water percolates to the foundation materials in such cases. Mielenz and Okeson[1] studied the foundation displacements which occurred along the Malheur River siphon and found a progressive uplift which sometimes exceeded 1 foot in three years. The constructions were over a sodium bentonite and uplift pressures of the order of 7,200 pounds per square foot were recorded, which lifted not only the siphon but 38 feet of alluvial sand and silt as well.

The water swelling capacity of montmorillonite finds application in oil-drilling processes in which the swollen clay is used to seal porous rock-formations.

Swelling also occurs when dry clays are immersed in various organic liquids. The expansion of the basal spacing of montmorillonite by ethylene glycol,[2] from 14 Å to 17 Å is distinctive and is a valuable means of identifying this mineral, but aliphatic amines are even more effective,[3] provided that they contain a chain of 10 carbon atoms or more. The lattice is expanded by about 8 Å, suggesting that

[1] Mielenz, R. C. and Okeson, C. J., *Econ. Geol.*, 41, 266, 1946.

[2] MacEwan, D. M. C., *X-ray Identification and Crystal Structure of Clay Minerals* (Mineralogical Society, London, 1951), Chapter 5.

[3] Jordan, J. W., Hook, B. J. and Finlayson, C. M., *J. Phys. and Coll. Chem.*, 54, 1196, 1950.

two molecular layers can be absorbed. An expanded montmorillonite is highly absorbent and has other characteristic and valuable properties.

The mechanism of the water-swelling of montmorillonites and some other layer lattice minerals has not been established with certainty, although several interesting theories have been advanced. Forslind[1] suggests that water molecules are arranged in the form of the ice structure which can fit between the layers of montmorillonite crystals provided that the Edelman-Favejee structure (see p. 141) is accepted. Hendricks and Jefferson[2] consider that the extra water molecules in hydrated halloysite $Al_2(Si_2O_5)(OH)_4.2H_2O$ have a definite structural arrangement with linkages to successive sheets of the kaolin-type. Méring[3] is of the opinion, however, that in montmorillonites, the absorbed water is associated with the exchangeable base. Water molecules are considered to co-ordinate around the cations which are present between the layers. Mono- or di-molecular sheets can be recognised in calcium-bearing montmorillonite, but the sodium salt shows progressive hydration with increasing humidity. There is some support for Méring's conclusions from thermal analysis experiments on the hydrated forms of montmorillonite saturated with various ions. The peak corresponding to the loss of absorbed water occurs at different temperatures, which can be related to the exchangeable cation present. If this theory is accepted, then absorbed water is obviously closely related to water of hydration or crystallisation of salts.

3. *Heat of Wetting*. When dry clay is mixed with water, alcohol or various organic liquids, a rise in temperature occurs due to evolution of heat. This 'heat of wetting' is commonly called Pouillet's[4] effect, but Leslie[5] observed it twenty years earlier. According to Adam[6] it may be represented by the equation:

$$u = W - T\frac{dW}{dt}$$

where u is the heat of wetting per unit surface, W the work required to separate the two surfaces (i.e. work of adhesion), and T the temperature in absolute degrees. The term

$$-T\frac{dW}{dt}$$

is sometimes known as the *latent heat of the surface*; it is the amount of heat which has to be applied to the surface to maintain it at a constant temperature.

Although many other factors are involved, the heat of wetting can be related directly to the surface area of the powder by the expression

$$H = h.S_A$$

where h is a constant depending on the nature of the liquid and surface (i.e. the heat evolved when 1 sq. cm. of surface is wetted with a specified agent) and S_A is the total

[1] Forslind, E., *Meddelande*, No. 11, *Svenska*.
[2] Hendricks, S. B. and Jefferson, M. E., *Amer. Min.*, 23, 863, 1938.
[3] Méring, J., *Faraday Soc. Trans.*, 42B, 205, 1946.
[4] Pouillet, *Ann. Chim. Phys.*, 20, 141, 1822.
[5] Leslie, W. B. *Tilloch's Phil. Mag.*, 14, 201, 1802.
[6] Adam, N. K. *Physics and Chemistry of Surfaces* (Oxford, 1930).

surface area which is wetted. Attempts have been made to measure the surface area of powders by the use of such methods, but owing to the uncertainty in the value of *h* and the amount of heat generated, the method has only a limited application[1].

Harkins and Boyd[2] have attempted to deduce values of *h* for several solids and liquids, and find values lying between 100–500 ergs cm.$^{-1}$. On this basis, the heat of wetting of powders with an average particle size of 1 micron would lie between 0·07 and 0·35 cal. per gm.

The method of determination presents considerable experimental difficulty. Normal calorimetric methods using a Beckmann thermometer have been employed, although better results can be secured by measuring the temperature rise with a set of thermocouples,[3] arranged as in a thermo-pile, but with a very small heat capacity.

The heat of wetting of a relatively pure china clay is about 2 calories per gram, and 600 calories per gram of plastic clay, whereas Parks[4] found that the heat of wetting of silica powders by water is only 0·00105 calories per gram.

Detergent action. A property of clay which is seldom realised is its power of reducing the surface tension of water. By this means it enables water to wet surfaces more readily than when the water is used alone. One technical advantage derived from this property is that sometimes mixtures of clay and soap have the same detergent action as soap alone, so that clay added to soap to cheapen it, is not merely a diluent, but has a definite cleansing power.

In former times when only crude and very alkaline soaps were available, the use of clay and water for cleaning (fulling) wool was very common and yielded a softer and plumper product than when soap was used. Hence the term *fullers' earth* for certain montmorillonite clays which have long been used for this purpose.

Emulsifying action. Clay has a powerful emulsifying action, i.e. a mixture of oil and water which, alone, would rapidly separate into its two constituents, remains as an emulsion in the presence of about 1 per cent of clay. Plastic clays are the most effective for this purpose.

The principal effect of adding water to clay is, however, the plasticity or stickiness which is induced. It is in this sense that clay behaves as a true colloid and this physical phenomenon can be explained only by a consideration of such systems.

THE COLLOIDAL STATE

IN the last few decades the study of colloidal suspensions has been advanced to such an extent that it is now a science in its own right. Within the limits of this volume it is possible to give only the briefest description but it must be remembered that clays are only one of many colloidal systems, each of which has certain characteristics.

The existence of colloids was first demonstrated by Graham[5] in 1861. He showed that, whereas a true solution of a soluble salt would pass freely through a

[1] Gregg, S. J., 'Symposium on Particle Size', *Inst. of Chem. Eng. and Soc. of Chem. Ind.*, 1947, p. 40.
[2] Harkins, W. D. and Boyd, G. E., *J. Amer. Chem. Soc.*, 64, 1195, 1942.
[3] Clark, A. and Thomas, B. D., *J. Phys. Chem.*, 43, 579, 1939.
[4] Parks, G. S. *Phil. Mag.*, 4, 240, 1902.
[5] Graham, T., *Phil. Trans. Roy. Soc.*, London, 151, 183, 1861.

parchment membrane, other substances such as glue, starch, and silicic acid, although they appeared to go into solution, would not penetrate the membrane. Graham called these bodies which would not dialyse, *colloids*, after the Greek, *kolla* (glue) and *eidos* (the form).

A colloidal solution may be defined in general terms as a suspension of particles in a liquid, where the grain size of the solid is so small that they will not settle, yet at the same time they cannot be regarded as being in true solution.

Various types of colloidal solutions are known within the three physical states of matter, gas, liquid and solid. Thus, mist, fog and steam are examples of a suspension of liquid in a gas (air); smoke is a suspension of solids in gas; foams and emulsions are gaseous components in liquids; the colloidal solution of a solid in liquid is exemplified by a clay-water system. Although the ultimate particles in a true colloid are too small to be seen with the naked eye or even under the magnifications of an optical microscope, they are clearly revealed by an ultra-microscope in which a narrow beam of light, shining through a colloidal solution, is reflected from the particles and thus characteristic turbidity is produced. A comparable effect may often be observed when a shaft of sunlight picks out fine dust particles in a still atmosphere. This is known as the *Tyndall effect*.

The grain size of particles in a true colloid is within the range of Brownian movement, or in other words, they are small enough to be influenced by collisions with the molecules of the liquid (or suspending) medium, and are in a constant state of agitation. There are no fundamental limits to the size of such particles, but W. Ostwald has suggested that the maximum size is 0·001 mm. and the minimum size 0·000001 mm. Particles much less than this minimum may have colloidal properties, but they chiefly occur in true solution as ions, atoms or molecules, though the difference between some colloidal solutions of very low concentration and some 'true solutions' is not very distinctive.

The present practice is to regard 0·2μ (0·0002 mm.) as the maximum size of particles of colloidal dimension to a minimum of 0·005μ (0·000005 mm.) though some investigators prefer to use other figures.

It is important to realise that some clays, whilst consisting of particles larger than those included as colloids, have so open and spongy a structure that they behave like colloidal particles. This is particularly the case with some clays of the montmorillonite type, including some fullers' earths.

The properties and differentiating characteristics of a colloidal suspension are:

1. The particles are not visible except in an ultramicroscope.
2. The particles will not dialyse (i.e. pass through a membrane).
3. The grain size can apparently be altered by electrolytes. The addition of certain soluble salts will cause *coagulation*, or settling out of the suspended particles.
4. The viscosity of a colloidal suspension is invariably greater than that of the suspending medium, which is not always the case in a true solution.

Many substances show unusual properties in the colloidal state. For example, the exceptional turbidity of zinc oxide when in the form of particles of colloidal dimensions is of value in the production of some opaque glazes and in the paint industry. Similarly an aggregate of carbon particles has been shown to have a maximum hardness value when the grains are of colloidal size.

To be a true colloid the ultimate particles must be smaller than 0·2 microns (1 micron = 10^{-4} cms.) in size, and it is only when grains of these dimensions are present that suspensions show anomalous properties. Some answer to the reasons why exceptionally small particles behave differently from larger-sized ones is shown by a study of their crystal form. Crystallographers have concluded that all crystals are built of fundamental units which are termed *unit-cells* (see Chapter II). The more complex the atomic arrangement, the larger will be the unit cell, and therefore, the less will be the number of units required to form crystals of appreciable size.

Taking calcium fluoride, CaF_2, as an example, the unit cell of which is shown in Fig. VIII.2, there are fourteen atoms of calcium and eight of fluorine. If a cube containing two unit cells in each dimension is considered, each face contact shares five calcium atoms, so the ratio of calcium atoms to fluorine atoms in the complete cube is 63 : 64. As more units are added, the ratio of the two kinds of atoms approaches the stoichiometric 1 : 2 as predicted from the formula CaF_2. However, many units have to be linked before this condition is reached, as shown in Table VIII.1, where *n* represents the number of unit cells in each dimension of a cubical array.

TABLE VIII—I. RATIO OF NUMBER OF CALCIUM ATOMS TO FLUORINE ATOMS IN CRYSTALS OF CALCIUM FLUORITE, CaF_2 WITH INCREASING CRYSTALLITE DIMENSIONS

Number of unit cells per side ♂	*Number of calcium atoms* N_{Ca}	*Number of fluorine atoms* N_F	$\frac{N_F}{N_{Ca}}$	*Length of side of crystallite in millimicrons*
1	14	8	0·572	0·62
5	663	1,000	1·510	3·10
10	4,631	8,000	1·729	6·2
1,000	4,060,301	8,000,000	1·971	62

Thus, in a comparatively simple crystal, the stoichiometric ratio is not exactly obeyed until the crystallite size is quite large (0·1 microns). When more complex crystals are considered, where the unit cells are large, there is great deviation from the true atomic ratios, because particles of colloidal dimensions are, in some cases, composed of only a few unit cells. Consequently, normal chemical laws cannot be applied to colloidal systems, and the concept of electrically-balanced crystals does not hold. Balarew has carefully analysed highly-dispersed precipitates, and has proved to some extent the failure of stoichiometric laws when applied to colloidal suspensions.

The most important feature of colloidal particles is the extremely high surface area relative to their volume. It has been shown in Fig. VIII.1 that the outermost layers of atoms or ions of a crystal are not completely balanced electrically. It might be expected that the properties of such surfaces would differ from the inner balanced layers of crystals. This is in fact so and Weyl has demonstrated the abnormal properties of surfaces due to the inadequate screening of atoms and ions.

In particles where the distorted surface layers are an appreciable proportion of the whole mass (i.e. colloidal particles) it is to be expected that they will be abnormal in behaviour.

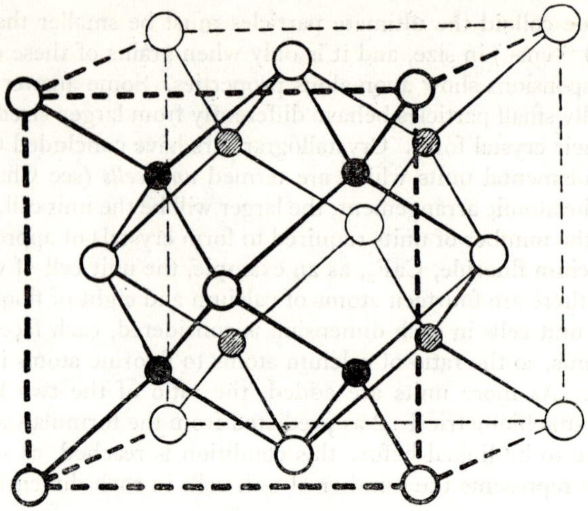

FIG. VIII.2. THE UNIT CELL OF CALCIUM FLUORIDE, CaF_2
Large circles represent calcium ions and small ones fluorine

Colloidal solutions are generally recognised as being of two types: (*a*) **lyophilic** (meaning 'liquid loving'), and (*b*) **lyophobic** (implying 'liquid hating'). When water is the suspending medium then these two forms are frequently called *hydrophilic* and *hydrophobic* sols respectively. The properties of the two types are very different, although many colloidal solutions will, under varied conditions show properties associated with both groups.

The main features of lyophilic and lyophobic sols are compared in Table VIII.II.

TABLE VIII—II. PRINCIPAL PROPERTIES OF LYOPHILIC AND LYOPHOBIC SOLS

Lyophilic	*Lyophobic*
1. The surface tension is often much lower than that of the dispersion medium.	1. The surface tension of the suspension is similar to that of the dispersion medium.
2. The viscosity is much higher than that of the dispersion medium.	2. The viscosity is not very different from that of the dispersion medium.
3. Small quantities of electrolyte have little effect, although larger amounts may cause 'salting-out'.	3. Precipitation is achieved even with small amounts of electrolyte addition.
4. Particles are not easily detected.	4. Particles are readily 'observed' in an ultra-microscope.
5. Particles may migrate in either direction in an electrical field (or not at all) and this may be pH dependent.	5. Particles are all of one charge and migrate readily in one direction in an electrical field.
6. The suspension is easily reversible after drying.	6. After drying, the suspension is not readily reconverted.

No hard and fast rule can be laid down as to whether a particular colloid will be hydrophobic or hydrophilic. In general, metal sols are of the first group, whilst complex organic materials, e.g. starch, glue, are the best examples of the latter.

For the most part, clay minerals form hydrophilic sols, but their properties are not always consistent with those expected from suspensions of this type. Colloidal particles of clay minerals are mostly negatively charged and in addition they are sensitive to small additions of electrolytes. Clay minerals have the capacity to adsorb and absorb other substances from suspensions and these materials often have a pronounced influence on the colloidal properties of the clay. Aliphatic amines, particularly those with a long carbon chain, e.g. dodecylamine $C_{12}H_{25}NH_2$, are absorbed by montmorillonite clays and form true hydrophobic sols; in this form, clay minerals are water resistant, but have a greater capacity for absorbing organic substances.

Forces Acting on Colloidal Particles

Various forces act on colloidal particles suspended in a medium, and these determine the behaviour of slips and slurries which are essentially fine-particled suspensions.

(*a*) **Van der Waal forces** set up an attraction between particles; in a system where the grain size is small, they will not predominate unless the particles approach closely to each other.

(*b*) **Gravitational forces** tend to cause sedimentation; the larger the particles, the greater will be the effect of these gravitational forces.

(*c*) **Thermal movement** of particles opposes the forces of attraction, although it is appreciable only when the particle size is extremely small and it rapidly decreases with increase in size, but it predominates in highly dispersed sols.

Little is known about the relative effect of these three forces, but von Buzagh suggests that if these were the only forces acting, then with particles of colloidal dimensions there would be a strong tendency for coagulation to take place; at the size of colloidal particles, the net attractive force will be at a maximum, implying that particles will more readily come together and aggregate. The fact that stable colloidal suspensions do exist, indicates that other forces must be operative.

(*d*) Colloidal particles, in the majority of cases, possess an **electrical charge**. In an electrical field, therefore, such particles behave in a similar fashion to simple ions—they migrate in a particular direction. This phenomenon is known as *electrophoresis*, and when all particles are negatively charged, the phenomenon is termed *kataphoresis*. Such particles mutually repel and prevent coagulation.

When suspended in a liquid medium, many particles, by virtue of their inherent charge, will adsorb a film of the liquid. Although its thickness is not very great, it acts as a cushion and prevents the close approach of the individual grains of solid suspended matter. The nature of the forces which are present in the *lyosphere*, as the liquid film is termed, will be outlined in the next section, but its existence can be proved by a simple experiment. Finely-grained quartz (less than 200 B.S. sieve size) is a suitable raw material for demonstration purposes and it should be thoroughly dried out prior to testing.

A fixed weight (about 20 gm.) should be placed in each of several identical cylinders. To each of these should be added a different liquid—a useful series is water, ether, toluene and carbon tetrachloride. Excess of liquid should be used to ensure that after the solid has been thoroughly wetted and saturated, some free liquid remains above the sedimented powder. After prolonged shaking, the powder in each cylinder is allowed to settle and its sedimentation volume then measured; a different value will be observed with each liquid. The sedimentation volume per gram of the particular powdered quartz crystal in each liquid is as follows:

TABLE VIII—III. VOLUME OF SEDIMENT IN VARIOUS LIQUIDS

Liquid	*Water*	*Ether*	*Toluene*	*Carbon tetrachloride*
Sedimentation vol. per gm.	0·83	1·15	1·22	1·28

The packing density of the powder in air, when presumably all particles were in contact was 1·39, which gives a sedimentation volume per gram of 0·72.

The increase in the sedimentation volume of the solid particles on settling in the various immersion liquids is due to a film which has been formed around the particles. The thickness of the lyosphere depends on the nature of the liquid used. Of the liquids shown in Table VIII.III the lyosphere is thickest with carbon tetrachloride.

The formation of a lyosphere is a characteristic of the immersing liquid and also of the surface of the solid particles and consequently, the smaller the particle size the greater will be the volume of the lyosphere per unit weight of sample. With grains of colloidal dimensions, the thickness of the adsorbed liquid film is so great that in many respects the lyosphere behaves as though it were a rigid layer preventing the close approach of individual particles. When water is the suspending medium, the liquid film surrounding particles is termed a *hydrosphere* and this hydrated layer is responsible for many of the intriguing properties of colloids.

The Colloidal Micelle

Colloidal particles when suspended in water are charged and are surrounded by a water shell. The molecules of water within this shell are rigidly held by the particle and have a different character and properties from those in the remainder of the dispersion medium. The water shell may also contain cations and anions which play an important part in determining the behaviour of the colloidal suspension. Because of the intimacy of the association of the solid particle with its hydrosphere, it is convenient to regard these *colloidal micelles*, as the basic units in colloidal solutions.

The origin of the charge on particles can, to some extent, be explained from a study of the nature of the raw materials. The crystal structure of clays and other silicate minerals has been illustrated in Chapter III, but it is sufficient here to recall that the essential components are large anions (negatively-charged ions) of oxygen and hydroxyl units held together by much smaller cations (positively-charged ions).

A cleavage section of a kaolinite crystal (see p. 129) will be along the basal plane, exposing layers of oxygen and hydroxyl units, although only weak bonds are broken in the process. Fracture across a plate of these layer minerals, however, breaks valency bonds between the cationic and anionic components. As the negatively-charged ions are very much larger than the positively-charged ones, the former predominate in the external layer. The net result of this, and the fact that hydroxyl units themselves are also adsorbed, is that the surface of clay particles and related minerals is negatively charged when suspended in a water medium. Consequently, it will attract cations and as a direct result the surrounding regions of the water medium in the immediate vicinity of the particles will be rich in positively charged ions which counter-balance the surface charge and maintain electrical neutrality. Helmholtz[1] was the first to recognise this important principle, and he postulated that a *rigid double-layer* was formed in close proximity to the clay surface as illustrated in Fig. VIII.3. Although Helmholtz's theory was a logical deduction, there were many shortcomings in the concept of a rigid monomolecular layer associated with the surface.

Gouy[2] and Freundlich[3] suggested that the layer of cations was not so sharply defined in thickness as Helmholtz supposed and introduced the *diffuse double-layer theory*. They maintain that the double layer is not limited to the layers of the

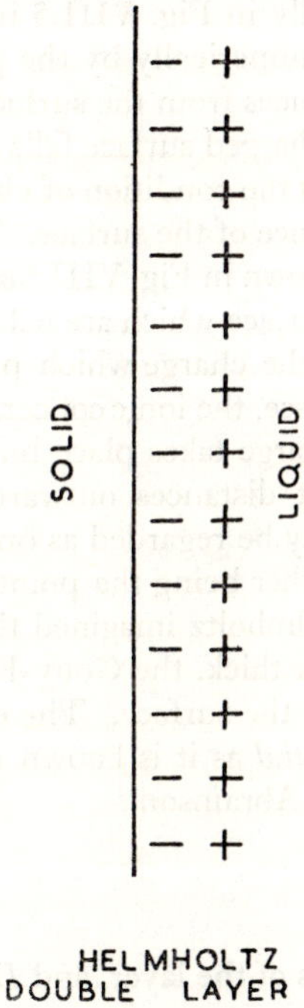

FIG. VIII.3.
THE HELMHOLTZ RIGID
DOUBLE LAYER THEORY

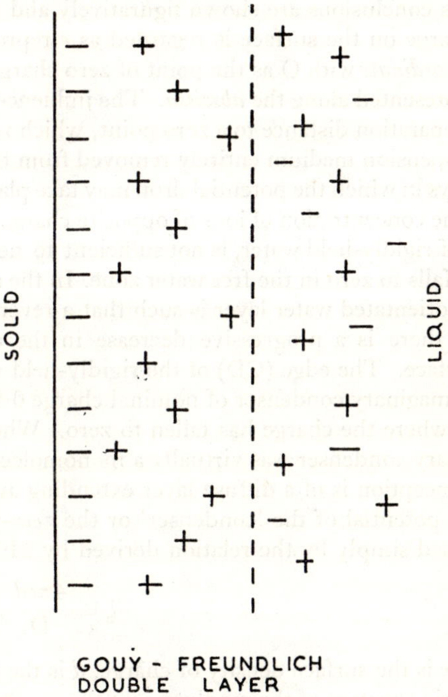

FIG. VIII.4.
THE GOUY-FREUNDLICH DIFFUSE
DOUBLE LAYER THEORY

[1] Helmholtz, H. V., *Wied. Ann.*, 7, 337, 1879.
[2] Gouy, G., *J. Phys. Radium.*, 9, 457, 1910.
[3] Freundlich, H., *Kapillarchemie*, Vol. 1, 4th edition (Leipzig, 1930.)

hydrated shell in intimate contact with the surface, but extends some distance into the liquid medium. In other words, the counter-ions in the liquid are not statically held in close contact with the surface but are diffused throughout a finite volume of the liquid medium (Fig. VIII.4). Gouy considers that the cationic distribution falls off exponentially as the distance from the surface is increased. Thus a type of atmosphere is set up round the surface.

In the particular case of a clay suspended in water this ionic concentration gives rise to a secondary effect which limits the close approach of the cation to the surface of particles and gives rise to the diffuse double-layer. As a result of the intense electrical forces in the vicinity of the surface the molecules of the suspension medium are themselves oriented, giving rise to a layer of rigidly-held liquid. As the distance from the surface is increased, the degree of orientation of the molecules of the suspension medium decreases at the same rate as the concentration of counter-cations. The variation in these properties is a function of the surface and the type and concentration of the ions in solution.

Freundlich has investigated the electrokinetical laws which govern the variation in charge density at points extending outwards from the surface of charged particles and his conclusions are shown figuratively and graphically in Fig. VIII.5 in which the charge on the surface is regarded as *e* represented numerically by the point A on the *ordinate* with O as the point of zero charge. Distances from the surface layer are represented along the *abscissa*. The influence of the charged surface falls rapidly with separation distance to a zero point, which represents the condition of charge in the suspension medium entirely removed from the influence of the surface. Two of the ways in which the potential drop may take place are shown in Fig. VIII.5B. In the first, the concentration of ions of opposite charge to the surface which are held in the layer of rigidly-held water, is not sufficient to neutralise the charge which progressively falls to zero in the free water zone. In the second case, the ionic concentration in the orientated water layer is such that a reversal of charge takes place but, once again, there is a progressive decrease in the charge at distances outward from the surface. The edge (CD) of the rigidly-held water may be regarded as one plate of an imaginary condenser of nominal charge 0·5*e*, the other being the point in the liquid where the charge has fallen to zero. Whereas Helmholtz imagined that this imaginary condenser was virtually a monomolecular layer thick, the Gouy-Freundlich conception is of a diffuse layer extending away from the surface. The electrokinetic potential of the 'condenser' or the *zeta-*(ζ)*-potential* as it is known may be expressed simply by the relation derived by Müller and Abramson:[1]

$$\zeta = \frac{4\pi ed}{D}$$

where *e* is the surface density of charge, *d* is the thickness of the layer, and *D* is the dielectric constant of the medium.

The zeta-potential is a measure of the magnitude and extent of the forces arising as a result of the electrical charges on a particle. The surface density of charge (*e*) is a function of the type of surface and the ions which are present in solution and is known as the Nernst potential.

[1] See Alexander, A. E. and Johnson, P., *Colloid Science* (Oxford University Press, 1950).

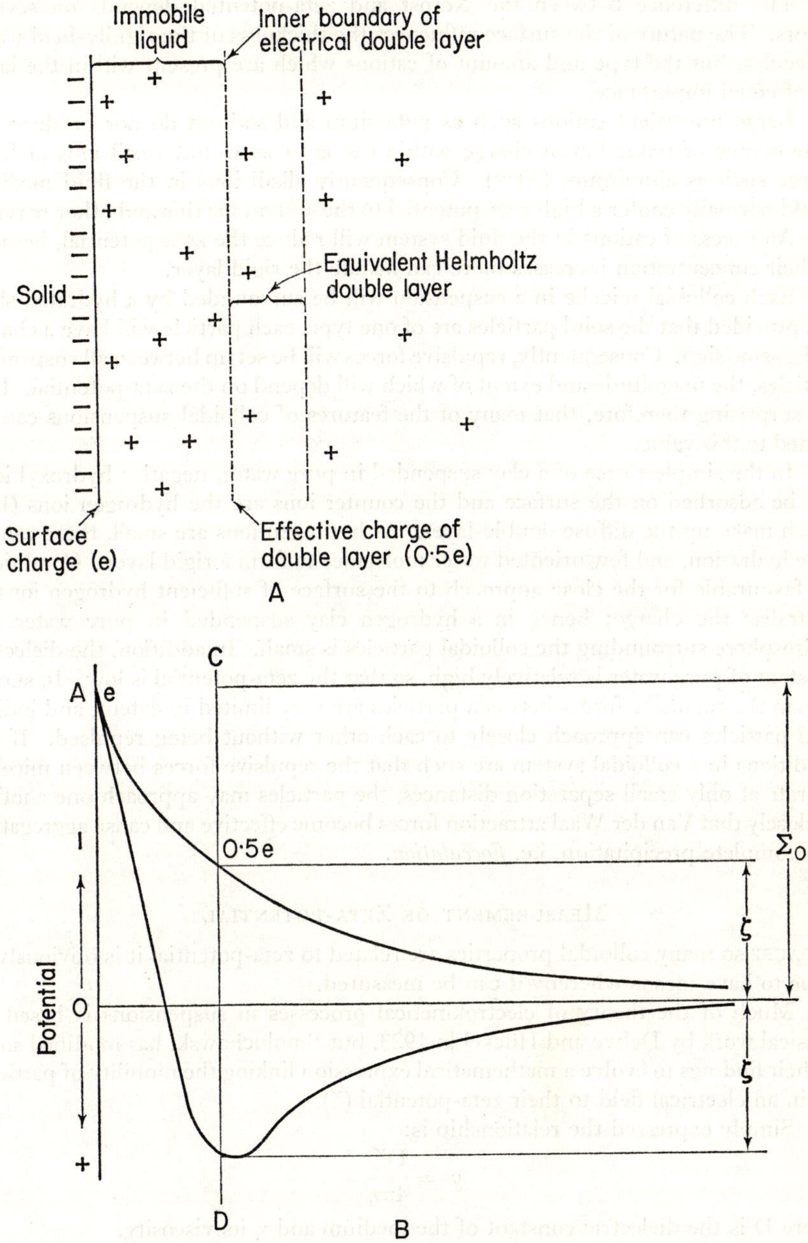

FIG. VIII.5. THE POTENTIAL ENERGY ASSOCIATED WITH A CHARGED SURFACE

The difference between the Nernst and zeta-potential depends on several factors. The nature of the surface influences the thickness of the rigidly-held water molecules, but the type and amount of cations which are present within the layer are of equal importance.

Large uni-valent cations such as potassium and sodium do not produce the same degree of reduction of charge within the layer as would small ions of large charge such as aluminium (Al^{3+}). Consequently alkali ions in the fluid medium would normally confer a high zeta-potential to the system for this and other reasons.

An excess of cations in the fluid system will reduce the zeta-potential, because as their concentration increases more can invade the rigid layer.

Each colloidal micelle in a suspension will be surrounded by a hydrated shell and provided that the solid particles are of one type, each particle will have a charge of the same sign. Consequently, repulsive forces will be set up between all suspended particles, the magnitude and extent of which will depend on the zeta-potential. It is not surprising therefore, that many of the features of colloidal suspensions can be related to this value.

In the simplest case of a clay suspended in pure water, negative hydroxyl ions will be adsorbed on the surface and the counter ions are the hydrogen ions (H^+) which make up the diffuse double-layer. As hydrogen ions are small, there can be little hydration, and few oriented water molecules to form a rigid layer. Conditions are favourable for the close approach to the surface of sufficient hydrogen ions to neutralise the charge; hence in a hydrogen clay suspended in pure water the hydrosphere surrounding the colloidal particles is small. In addition, the dielectric constant of pure water is relatively high, so that the zeta-potential is low. In such a system the repulsive forces between particles are very limited in extent, and individual particles can approach closely to each other without being repulsed. If the conditions in a colloidal system are such that the repulsive forces between micelles operate at only small separation distances, the particles may approach one another so closely that Van der Waal attraction forces become effective and cause aggregation and complete precipitation, i.e. *flocculation.*

Measurement of Zeta-potential

Because so many colloidal properties are related to zeta-potential it is obviously of value to have means whereby it can be measured.

Much of the theory of electrokinetical processes in suspensions is based on classical work by Debye and Hückel in 1923, but Smoluchowski has modified some of their findings to evolve a mathematical expression linking the mobility of particles (μ) in an electrical field to their zeta-potential (ζ).

Simply expressed the relationship is:

$$\mu = \frac{D\zeta}{4\pi\eta}$$

where D is the dielectric constant of the medium and η its viscosity.

This applies strictly only to particles of colloidal size, but it does afford means whereby the zeta-potential of particles within this size range can be measured.

Electrophoresis. The most obvious technique of application of Smoluchowski's equation is where particles suspended in a medium are induced to move in an electrical field. The rate of flow, which can be observed under a microscope or ultra-microscope, can be related to the zeta-potential. In a normal suspension, such as a clay, all particles carry the same charge and hence move in the same direction.

Electro-osmosis differs from electrophoresis only in that in the former the movement of liquid through a bed or plug of particles is measured whilst in the latter the particles move in a static bed of liquid.

Streaming potentials is a simple but elegant method of measuring the zeta-potential of a fluid suspension. In electro-osmosis, a potential field induces movement of the suspension fluid through a mass of particles, but in streaming potentials techniques, a potential difference is developed across a compact of particles by the movement of liquid across it. Gortner[1] first described this method which has been widely applied to clays and other ceramic materials. The diagrammatical representation of the apparatus is shown in Fig. VIII.6. The sample is packed between two platinum gauze electrodes linked to a high resistance circuit capable of measuring very small potential differences. Fluids can be forced through the particulate mass at different rates simply by adjusting the hydrostatic head; if necessary, various electrolytes can be used in the determination.

The zeta-potential can be calculated from the expression

$$\zeta = \frac{4\pi\eta K_s E_s}{PD}$$

where E_s is the streaming potential, P is the pressure forcing liquid through the capillary, and K_s is the specific conductivity of the liquid itself.

Flocculation and Coagulation

The term *flocculation* implies the gathering of particles into 'flocks' which settle rapidly, leaving a clear liquid. This distinguishes it from *coagulation* in which particles unite to form larger ones which do not necessarily settle rapidly. Both terms are used rather loosely.

Deflocculation is the reverse phenomenon, i.e. the dispersion of the aggregation into small particles which remain in permanent suspension.

The zeta-potential is the main factor which governs the stability of suspensions of clay particles. Fig. VIII.7 shows the characteristic variation in fluidity of a kaolinite suspension and its zeta-potential as sodium carbonate is added. In the absence of electrolyte or at low levels of concentration, the hydrogen-form of the clay is flocculated in that it settles fairly rapidly to a low sedimentation volume. Additional electrolyte brings about dispersion and a fully-deflocculated condition of high zeta-potential and low viscosity in which a permanent suspension results. At high concentration of electrolyte, coagulation occurs in which the clay collects into large flocs which settle only slowly to a large-volumed sediment. This is accompanied by a reduction in zeta-potential and an increase in viscosity.

Sols and Gels. A stable colloidal suspension is termed a *sol*, and the particles in them, although solvated, have independent existence and show Brownian movement.

[1] Gortner, J., *Outlines of Biochemistry* (Wiley, 1938).

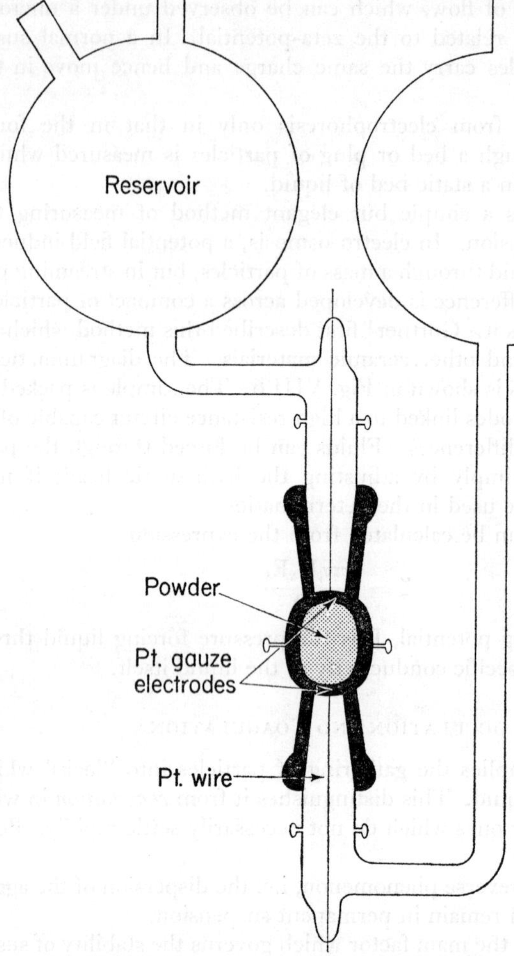

FIG. VIII.6. APPARATUS FOR MEASURING ZETA POTENTIAL BY THE METHOD OF STREAMING POTENTIALS

If such a sol is precipitated or coagulated the particles do not settle to form a dense layer, but the whole colloidal suspension sets to a semi-solid state called a *gel*. The amount of water contained in such a body is very large, but the nature of gels and the intimacy with which the particles and the water are associated is largely unknown although there are many theories, most of which have been outlined by Williamson.[1] Gels are formed only by small-particled materials and some authorities[2] consider that they are the ultimate cause of plasticity (see p. 494) in clays. When a gel forms Brownian Movement ceases, which suggests that a type of balanced structure has

[1] Williamson, W. O., *Trans. Brit. Ceram. Soc.*, 50, 10, 1951.
[2] Webb, H. W., *Trans. Brit. Ceram. Soc.*, 33, 129, 1934.

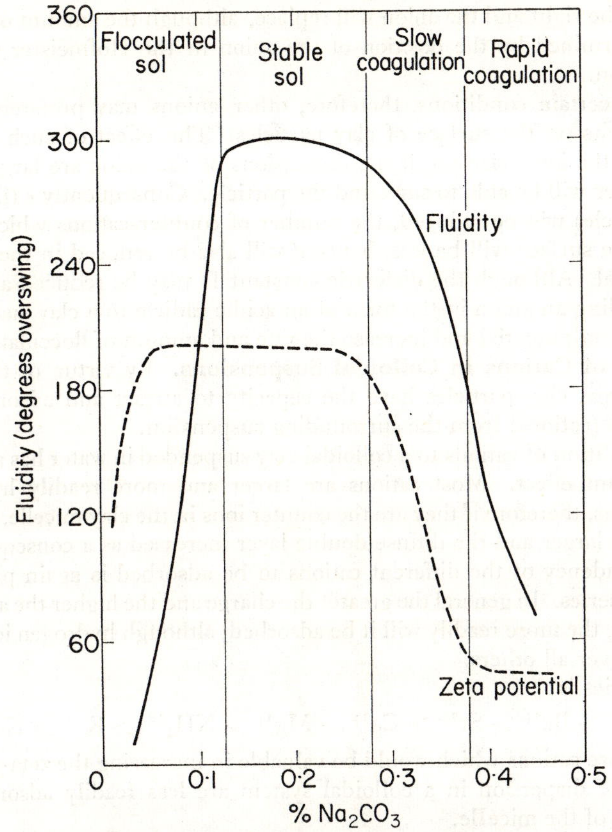

FIG. VIII.7. THE RELATIONSHIP BETWEEN THE FLUIDITY AND ZETA-POTENTIAL OF A KAOLINITE SUSPENSION WITH INCREASING ADDITIONS OF SODIUM CARBONATE

formed; Hauser and Reed[1] postulate that the solid particles are held in an equilibrium position by attractive and repulsive forces and that water molecules occupy the holes in such a structure.

The Effect of Anions in Colloidal Suspensions. When other anions are introduced into a colloidal system, there will be a tendency for them to be adsorbed by the clay particles provided they are in excess. Particles which are capable of negative adsorption, preferentially select different anions according to the Hofmeister series.

$$OH^{1-} > CNS^{1-} > I^{1-} > Br^{1-} > Cl^{1-} > NO_3^{1-} > F^{1-} > SO_4^{2-}$$

However, the ionic exchange is an equilibrium process which may be represented quite generally as:

$$(\text{Clay} - \text{A})^- + \text{B}^- \rightleftharpoons (\text{Clay} - \text{B})^- + \text{A}^-$$

where A and B are any two different anions. Thus, if B is in excess the reaction will

[1] Hauser, E. A. and Reed, C. E., *J. Phys. Chem. N.Y.*, 41, 911, 1937.

proceed to the right and the anion will replace, although the amount of replacement will be determined by the position of the anion in the Hofmeister series and its concentration.

Under certain conditions, therefore, other anions may preferentially replace hydroxyl units on the surface of clay particles. The effect of such replacements depends on the ionic size for, if the dimensions of the anion are large, a relatively small number will be able to surround the particle. Consequently *e* (i.e. the charge on the particle) will be reduced, the number of counter-cations which are able to approach the surface will be less, hence *d* will also be reduced in the equation for zeta-potential. Although the dielectric constant D may be reduced also, the usual effect of adding an anion in the form of an acidic radicle to a clay suspension is to reduce the zeta-potential and increase the rate and amount of flocculation.

The Effect of Cations in Colloidal Suspensions. By virtue of their negative surface charge, clay particles have the capacity to attract and adsorb positively-charged ions (cations) from the surrounding suspension.

The addition of cations to a colloidal clay suspended in water has a pronounced and important effect. Most cations are larger and more readily hydrated than hydrogen ions, therefore if they are the counter ions in the clay micelle, the hydrated layer will be larger and the diffuse double layer increased as a consequence.

The tendency of the different cations to be adsorbed is again predicted by a Hofmeister series. In general the greater the charge and the higher the atomic weight of the cation, the more readily will it be adsorbed, although hydrogen ion (H^+) takes preference over all others.

The series is:

$$H^+ > Al^{3+} > Ba^{2+} > Sr^{2+} > Ca^{2+} > Mg^{2+} > NH_4^{1+} > K^{1+} > Na^{1+} > Li^{1+}$$

Therefore cations which would be valuable in increasing the zeta-potential and the degree of dispersion in a colloidal system are less readily adsorbed into the hydrosphere of the micelle.

This difficulty can be surmounted by adding an excess of the required cation because the replacement reaction is an equilibrium process thus:

$$(\text{clay}^- - X^+) + Y^+ \rightleftharpoons (\text{clay}^- - Y^+) + X^+$$

where X^+ and Y^+ are any two different cations. Thus a sodium clay can be produced from a calcium clay by adding a sufficient excess of sodium ion and the amount of replacement is enhanced if at the same time the calcium ions can be removed from the system.

Monovalent cations such as potassium and sodium are large and highly hydrolysed. For a given charge on the adsorbing surface many more cations of this type can be attracted than (say) aluminium (Al^{3+}). For this reason and also because the number of oriented water molecules is much greater, the hydrosphere of particles with sodium or potassium as the counter-cations is extremely large. The zeta-potential of clays in this state is comparatively very large (Fig. VIII.8) and, consequently, the force of repulsion between particles is such as to produce maximum separation, or, in other words, a *deflocculated* system.

Clays which contain sodium or potassium as the counter ions in the colloidal micelle are, therefore, superior to all others in the production of a disperse or de-

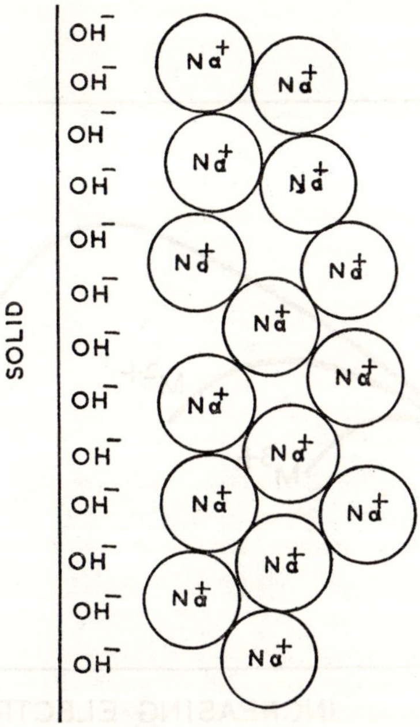

FIG. VIII.8. THE DIFFUSE DOUBLE LAYER WHEN SODIUM IONS ARE THE COUNTER-BALANCING CATIONS

flocculated system. This is shown by Fig. VIII.9 where the zeta-potential is plotted as a function of increasing electrolyte. As the zeta-potential may be regarded as a measure of the degree of deflocculation, the relative effectiveness of uni-, di- and trivalent cations may be directly compared.[1]

At high concentrations of electrolyte or cation concentration the zeta-potential decreases because if the number of cations is increased excessively, they 'crowd' into the diffuse double layer, with the result that the field of influence of the negatively-charged surface is diminished, as shown in Fig. VIII.10. The thickness of the double layer (d) is reduced and, in the limiting case, cations may invade the surface layer itself and so reduce the surface charge (e). The reduction in zeta-potential which results, produces a lower degree of dispersion or deflocculation and will eventually cause coagulation.

The production of a deflocculated system or sol with univalent cations of sodium or potassium from a clay originally saturated with other cations is not an easy matter for, as has been shown, the former type of cations are less readily adsorbed. Provided sufficient univalent cations are present, however, the equilibrium system can be made to proceed to the right thus:

$$(\text{clay OH})_2^- - Ca^{2+} + \underset{(\text{excess})}{2Na^+} \rightarrow 2((\text{clay OH}) - Na^+) + Ca^{2+}$$

[1] Phelps, G. W., *Ceram. Age*, 49, 162, 1947.

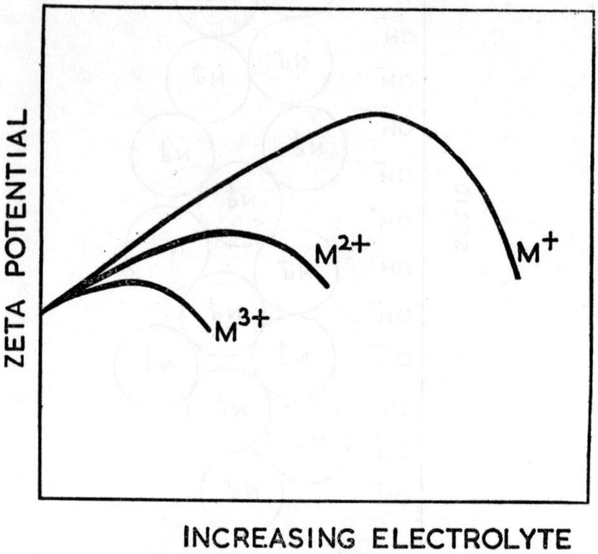

FIG. VIII.9. THE INFLUENCE OF CATIONS OF DIFFERENT VALENCIES ON THE ZETA-POTENTIAL

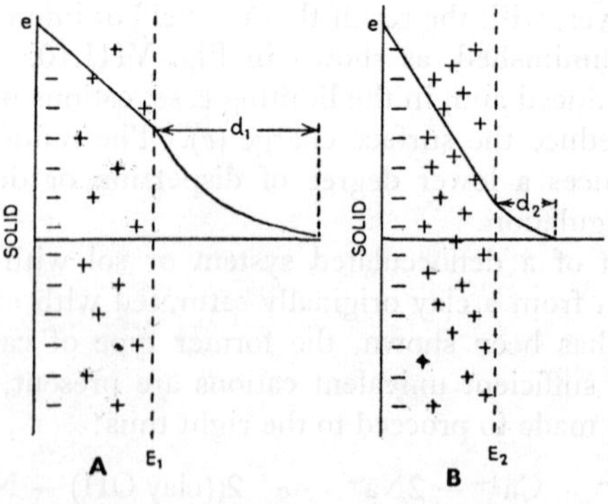

FIG. VIII.10. SCHEMATIC REPRESENTATION OF THE CROWDING OF THE DOUBLE LAYER BY AN EXCESS OF CATIONS

Although the production of a sodium clay is possible by such treatment, the excessive concentration of cations required may result in an overall reduction of the zeta-potential and caustic soda (NaOH) is not always effective in deflocculating a clay.

An alternative method for shifting the above equilibrium state to the right is to remove the calcium ions from the system by adding an anionic component which causes precipitation of an insoluble calcium salt. Carbonate, silicate and phosphate radicles are effective for each forms a calcium salt of low solubility. The reaction then proceeds:

$$\text{(clay OH)}_2^- - Ca^{2+} + \underset{\text{(sodium carbonate)}}{Na_2CO_3} \rightarrow 2(\text{(clay OH)}^- - Na^+) + \underset{\text{(precipitated)}}{CaCO_3}$$

Only small amounts of the sodium salt are required—sufficient to cause the precipitation of the total calcium content of the clay.

In the more general case of a clay saturated with any cation, the most effective precipitant is the phosphate radicle which forms insoluble salts with most cations. Sodium hexametaphosphate is, for this reason, widely used as a deflocculating agent for clays, but its mode of action involves more than simple ion exchange.

The effects of increasing the deflocculation of clay are to decrease the apparent viscosity (which increases the ease of flow), and to stabilise the suspension.

The relative effect of sodium hydroxide and sodium silicate on the deflocculation of a clay is shown in Fig. VIII.11. The influence of each on the viscosity of a 16 per cent kaolinite suspension previously saturated with calcium ions is clearly demonstrated.

As Holdridge[1] has pointed out, the presence of additional salts in a clay may produce abnormal behaviour when electrolytes are added. He cites the case where sodium silicates actually caused flocculation of a clay which was contaminated with calcium sulphate. A double decomposition between the electrolyte and the salt presumably took place with the formation of insoluble calcium silicate and highly soluble sodium sulphate thus:

$$\underset{\substack{\text{calc. sulphate}\\\text{(slightly soluble)}}}{CaSO_4} + \underset{\substack{\text{sod.}\\\text{silicate}}}{Na_2SiO_3} \rightarrow \underset{\substack{\text{calc. silicate}\\\text{(insoluble)}}}{CaSiO_3} + \underset{\substack{\text{sod.}\\\text{sulphate}}}{Na_2SO_4}$$

As a result of such reactions free sulphate ions $(SO_4)^{2-}$ are liberated into the dispersion medium and can to some extent replace hydroxyl ions in the double layer. The zeta-potential will be lowered with consequent flocculation. This is avoided by the prior addition of a little barium hydroxide ($Ba(OH)_2$) to precipitate the sulphate radicle as insoluble barium sulphate.

Anomalous flocculation also occurs when a mixture of caustic soda and sodium chloride is added to a hydrogen clay. Increasing additions of this mixture will cause first a flocculation, then dispersion and finally a second flocculation or coagulation. The first flocculation is due to the invasion of the hydrospheres of the clay micelles

[1] Holdridge, D. A., *Trans. Brit. Ceram. Soc.*, 49, 286, 1950.

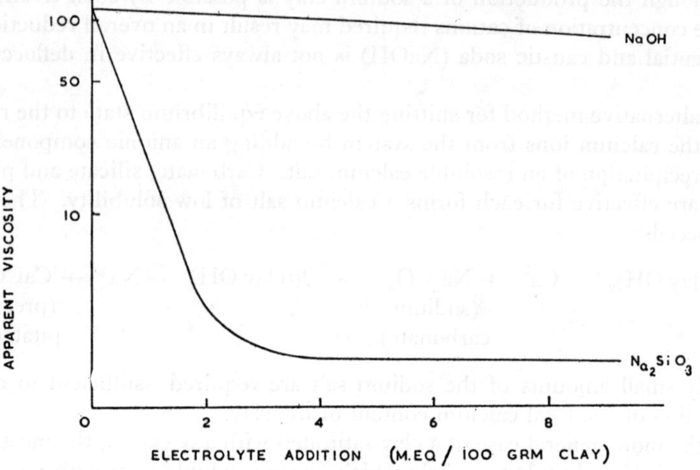

FIG. VIII.11. THE INFLUENCE OF SODIUM HYDROXIDE AND SODIUM SILICATE ON THE DEFLOCCULATION OF A CALCIUM CLAY

by chloride ions which replace the hydroxyl units, thereby reducing the zeta-potential and causing flocculation. As the concentration of sodium ions from the sodium hydroxide is increased, the zeta-potential is raised and deflocculation occurs. Finally at high concentrations, excess sodium ions reduce the thickness of the double layer and the surface charge of the clay particle and so the final coagulated state is formed. This is illustrated graphically by Fig. VIII.12, where the degree of dispersion is judged from the amount of clay remaining in suspension after one hour.

Flocculation or the reduction in zeta-potential may also be caused by adding a colloid of opposite electrical charge. Thus ferric hydroxide which is a positively-

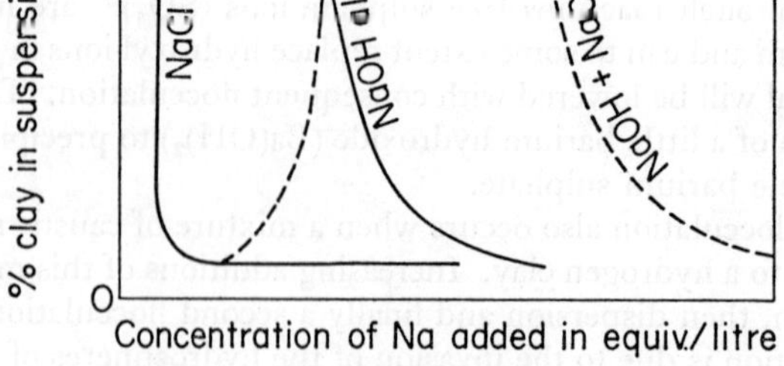

FIG. VIII.12. ABNORMAL FLOCCULATION OF CLAY (Oakley)

charged colloid will cause precipitation of the negative clay colloids. Stremme has suggested that some natural clays may have been produced by the mutual precipitation of colloidal alumina and silica, these substances bearing opposite charges and producing mixed gels. The derivation of the fullers' earth clays of Surrey and Bath may have been of this type.

The **conditions favouring deflocculation** are:

(*a*) The presence of monovalent cations which are capable of being heavily hydrated.

(*b*) The elimination of other cations preferably as insoluble salts.

(*c*) An excess supply of free hydroxyl units $(OH)^-$ in the dispersion medium.

(*d*) The removal of other anions which tend to replace hydroxyl units in the double layer.

(*e*) Not too great an excess of electrolyte in the dispersion medium, which is usually controlled by maintaining the pH between 10–12 depending on the type of electrolyte used.

Protective colloids have the effect of stabilising a clay suspension and preventing or restricting flocculation. Probably the best natural examples of protective colloids are the humus compounds in soils and colloidal lignite in ball clays. Hauser has suggested that humus, lignin and similar materials contain large-sized cations which take part in the double layer and prevent the entry of deleterious ions. The same effect can be produced artificially by adding suitable substances to the clay. Thus crushed lignite is often incorporated into casting slips and the salts of large-sized organic cations are being used extensively to improve the qualities of soils. The value of sodium hexametaphosphate as a deflocculant is thought to be due in part to its capacity to form a protective colloid.

The concept of zeta-potential is valuable for explaining many of the unusual features of slip formation and casting properties and many of the physical phenomena associated with these systems may be readily explained by considering the principles outlined in the previous pages.

Some unusual phenomena of clay suspensions are:

(*a*) **Thixotropy.** It is well known that when a suspension of certain clays is charged with electrolytes, the system on standing appears to *set* and to have many of the characteristics of a *gel*. This state can readily be destroyed by shaking or stirring the material, when immediately a free-flowing liquid *sol* is restored. If the disturbing influence is removed the gelation reappears. To describe this phenomenon the word *thixotropy* is used; it is derived from the Greek *thixis*—the touch, and *trope*—a change.

Freundlich has studied this interesting phenomenon in detail and enumerated many of the contributing factors. Along with other co-workers, he discovered that alkali ions were an essential requirement in the production of a thixotropic suspension of montmorillonite; a dialysed clay (i.e. hydrogen), showed no tendency to gel. Hauser[1] has shown that although there is little or no volume change when thixotropic gels form, all Brownian Movement stops even when the particle size is extremely small indicating that a structure of some type must be formed in which the

[1] Hauser, E. A., *J. Amer. Ceram. Soc.*, 25, 233, 1942.

clay particles, the excess cations and the dispersion medium itself must all play some part. He concludes that the colloidal micelles are locked in an equilibrium position in a type of balanced electrical lattice, where cations in the medium maintain the negative clay particles at a certain separation distance. There is almost certainly a high degree of induced particle orientation which if disturbed will destroy the thixotropic nature—a point beautifully demonstrated by Hofmann, Fahn and Weiss.[1]

Thixotropy may be regarded as a half-way state in the process of coagulation. The addition of more electrolyte to a thixotropic system causes more cations to 'crowd' into the colloidal micelle, with a consequent reduction in the zeta-potential. A coagulated gel is, for this reason, of an entirely different origin from a flocculated one, and although these are often confused they are entirely different in character. A flocculated system will sediment rapidly usually to a small volume; but coagulated particles settle slowly and invariably form a bulky deposit which is troublesome when casting ware.

Freundlich's explanation of thixotropy is probably over-simplified and there may be more than one mechanism involved. Thixotropy is a phenomenon associated with an excess of cations in a suspension of small particles but it can also be observed in the sedimented product of under-dispersed slips. Scott-Blair[2] has proposed two distinct mechanisms for thixotropy, but considerable uncertainty still exists over the complete physical interpretation of the phenomenon (see also p. 495).

(*b*) **Rheopexy.** Rheopexy is closely allied to thixotropy and although the term is not strictly accurate it has been called 'induced thixotropy'. The thixotropic gelation of some suspensions may be slow to develop especially when the particles present are rather large (0·03 microns). If, however, the suspension is vibrated gently or subjected to some other rhythmical movement, gel formation may be produced much more rapidly. Hauser and Reed[3] have demonstrated that in certain montmorillonite sols where normal thixotropy was not obvious in 100 minutes, an effective gel could be induced in less than a minute by vibrating the suspension. Rheopectic behaviour is not general in all suspensions which exhibit thixotropy, but where the particle shape is such that orientation can be developed as a result of vibration, rheopexy will be at a maximum. This is particularly so in the case of clays where the plate-like particles are readily aligned parallel to their basal section on the application of slight forces.

The importance of rheopexy in ceramic suspensions cannot be over-estimated for many of the problems in slip casting may be related to this phenomenon. Some clay slips show little thixotropic behaviour under normal conditions, but, when they are cast in plaster-moulds, the draining effect may cause orientation and rheopexy. As a result an undesirable low-density body will be formed. Rheopexy may also be induced during the extrusion of plastic clays.

(*c*) **Dilatancy.** Almost directly opposed to thixotropy in effect is *dilatancy*, which is exhibited so readily by sand and other substances where the grain size of particles present is comparatively large.

[1] Hofmann, U., Fahn, R. and Weiss, A., *Clay Min. Bull.*, 10, 70, 1953.
[2] Scott-Blair, G. W., *A Survey of General and Applied Rheology* (Pitman London, 1944).
[3] Hauser, E. A. and Reed. C. E., *J. Phys. Chem.*, 41, 911, 1937.

The phenomenon will be known to all who have walked on a sandy beach near to the water's edge as the tide is ebbing. Sand, which is apparently wet and moist, will become dry immediately pressure is applied to it; in other words, an apparently liquid suspension becomes solid when stirred or agitated.

Dilatancy is particularly troublesome when finely-grained powders of non-plastic materials are being made up into slips or suspensions. These tend to adsorb small amounts of water and any attempts to stir or agitate the mixture are opposed by the dilatant effect. Whereas thixotropic systems are readily broken down by an applied force, dilatant suspensions form a more rigid and retentive structure. When the pressure is removed the system returns to its former state.

The earliest explanation of the phenomenon of dilatancy was attempted by Reynolds[1] in 1886 and his views are still generally accepted. When particles of wet sand are closely packed there is a certain amount of void space. If pressure is applied to this mass, the closeness of packing will be destroyed and a larger volume of voids and interstices will form into which any surplus liquid can penetrate. Because the hydrosphere around each individual grain is reduced the mass becomes hard and dry and will not flow with the same ease as before.

(*d*) **The Acidic Nature of Clay Suspensions.** Because hydrogen ions are preferentially absorbed by clay particles, most natural clays contain hydrogen, in part at least, as their exchangeable cation (see p. 266). Consequently, when such clays are suspended in water, dissociation occurs and the hydrogen ion concentration in the suspension is increased.

The acidity of a solution is defined in terms of the dissociated hydrogen ions which are present. In pure water the amount of dissociation of the molecule, H_2O, into the ions, hydrogen (H^+) and hydroxyl (OH^-) is small and only 10^{-7} gram-molecules of hydrogen ion per litre are present. The equilibrium in water can be written

$$H_2O \rightleftharpoons H^+ + OH^-$$

If an acid is added to water the content of free hydrogen ions is increased in proportion to the amount of acid added and to the degree to which it is dissociated. A strong acid is defined as one in which the degree of dissociation is very great; hydrochloric and nitric acids are almost entirely ionised thus:

$$HCl \rightleftharpoons H^+ + Cl^-$$

Some acids, e.g. acetic, are feebly ionised, and in equivalent concentrations yield far less hydrogen ions than strong acids.

Basic substances are those which form solutions of lower hydrogen ion concentration than pure water. Their mode of action may be simply conceived, by visualising the base dissociating into a cation and hydroxyl ions thus:

$$NaOH \rightleftharpoons Na^+ + OH^-$$

When the base is dissolved in water, the concentration of hydroxyl ions in solution is increased above that normally found in pure water. Consequently by

[1] Reynolds, O., *Nature*, 33, 429, 1886.

the Law of Mass Action, the concentration of hydrogen ions will be reduced below that found in pure water.

The hydrogen ion concentration is defined as the number of gram-ions of hydrogen in a litre of solution. As this amount is extremely small it is usually expressed as the logarithm of its reciprocal which is known as the pH value. The number of gram-ions of hydrogen in pure water is 1×10^{-7}; hence its pH is 7.

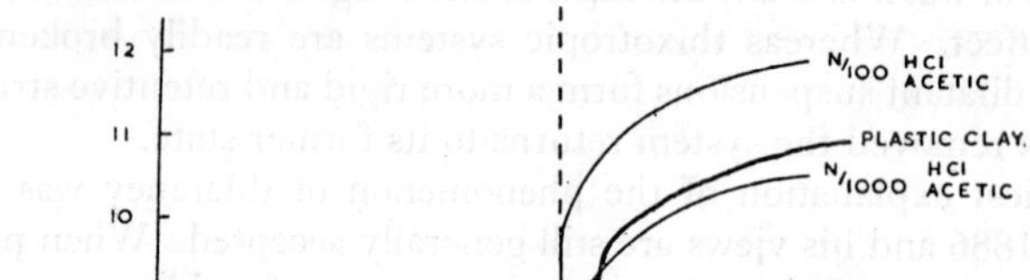

FIG. VIII.13. THE ACIDIC NATURE OF CLAY SUSPENSIONS

As all the solutions with a pH less than 7 are acids (i.e. have an excess of hydrogen ion) and all with a pH greater than 7 are alkaline (i.e. are deficient in hydrogen ion) the pH is a means of assessing the acidity or alkalinity of a liquid.

The pH value of most clay suspensions lies between 4·2 and 7·0 indicating that they behave as weak acids.

This is borne out by titration curves of clay suspensions with bases, the pH changes in a manner comparable to that of a weak acid (Fig. VIII.13).

Many clays, however, owe their acidity not only to the exchangeable hydrogen ions which they have absorbed but to associated organic matter. Humic and other soil acids are frequently associated with clays as 'protective colloids' and these, even in small amounts, may produce marked acidity in clay suspensions.

The most important characteristics of suspensions are the flow properties. These may conveniently be considered under two headings:

(*a*) Viscosity when the suspension is sufficiently fluid to flow under its own weight.

(*b*) Plasticity, in which condition the body can sustain its own weight but can be induced to flow under an applied pressure.

VISCOSITY AND RHEOLOGY

THE viscosity of a liquid is an indication of its ease of flow. In ideal cases, the rate of flow depends only on the pressure or force applied to cause it to flow.

The term *rheological properties* (devised by Bingham in 1928) is used to indicate all properties relating to the flow of all pastes and non-ideal liquids. It is specially important in connection with the flow of slips and the behaviour of clay-pastes.

As the concentration of a clay suspension increases, its properties change from true viscous flow to plastic flow. The character of the clay/water mixtures changes from suspensions and slips, through pastes and muds, to plastic masses which are deformed only by high pressures.

Suspensions and Slips

As the limiting case, the viscosity relationships in a liquid free from any suspended material must be considered. The **viscosity of a liquid** is defined as the ratio between the force or stress applied to a liquid and the shearing strain or rate which is produced, and represents the resistance of a fluid to change of shape or motion. This is more clearly appreciated by considering a simple type of viscometer (Couette design) illustrated in Fig. VIII.14. A fluid is placed in the space between two co-axial cylinders C_1 and C_2. The internal cylinder can be rotated at different velocities and, in order to keep the outer one C_2 at rest, a tangential force F has to be applied.

The viscosity is related to the ratio of F (the shearing force applied) to v the velocity of rotation (which is related to the shearing rate).

The complete expression for an ideal liquid is

$$F = kA\frac{v}{d}$$

where A is the interfacial area of the rotating cylinder, d is the separation between the cylinders and k is a constant. If v/d and A are reduced to unity then $k = F = \eta$, i.e. the coefficient of viscosity.

The unit of viscosity is defined (in c.g.s. units) as the number of dynes per sq. cm. required to maintain a velocity of 1 cm. per sec. between two surfaces, each of 1 sq. cm. area and 1 cm. apart and is termed a *poise* which is represented by η. As the poise is often inconveniently large the *centipoise* ($\eta \times 10^{-2}$) and the *millipoise* ($\eta \times 10^{-3}$) are frequently used. Thus, water has a viscosity of about 1 centipoise and glycerine about 10 poises.

Viscosity is also stated in terms of *stokes* or *centistokes*; this is the unit of kinematic viscosity and the poise the unit of dynamic viscosity. The viscosity in stokes can be found by dividing the viscosity in poises by the specific gravity of the liquid.

The physical concept of viscosity is the frictional resistance of molecules of the liquid to the movement of similar molecules in their immediate vicinity. It is occasionally expressed as a reciprocal or $1/\eta$ when it is known as the *fluidity*.

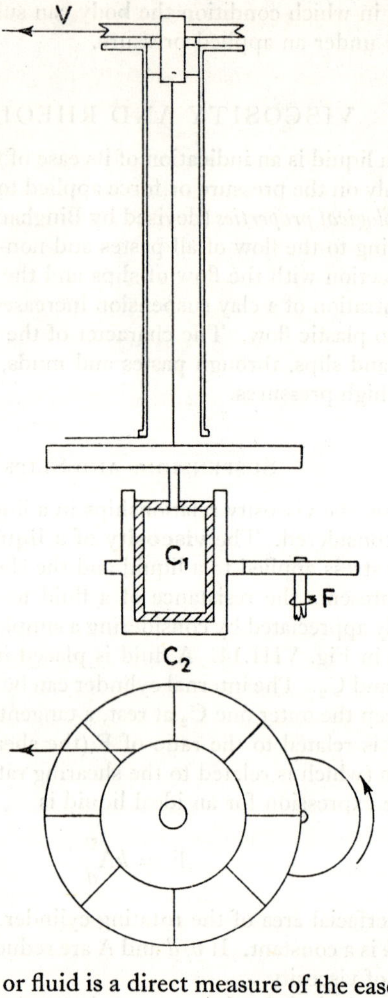

FIG. VIII.14.
PRINCIPLE OF THE
COUETTE VISCOMETER

The viscosity of a liquid or fluid is a direct measure of the ease of flow; hence it is an important and often critical value of clay suspensions and casting slips. It is measured by means of a *viscometer*.

These are of many designs[1] although those of importance in ceramic practice are of four groups.

A. Rotating Cylinder Viscometer. The principles of the method have already been introduced in the previous section where the well-known Couette viscometer (Fig. VIII.14) was illustrated. Several other designs embody these same principles with but minor embellishments. Most modern instruments of this type are designed so that the outer cylinder rotates and the shearing stress on the internal one is measured by observing the angular twist on a fibre or spring suspension. The shearing

[1] Philippof (see p. 484) has described almost 200 different types of viscometers and subdivides them into eight groups.

force can be varied by changing the speed of rotation of the outer cylinder or the space between the two.

The complete expression relating the viscosity of a liquid with the angular twist and the velocity of rotation of the cylinder, depends also on the characteristics of the torsion wire. Normally in a commercially-manufactured instrument of the type illustrated in Fig. VIII.15 the makers supply a list of correction factors to relate the angular twist to the apparent viscosity under a particular set of conditions. Clark and Hodsman[1] have derived a mathematical expression of general application to all rotating cylinder viscometers.

The viscometer illustrated in Fig. VIII.15 is a portable instrument which is very serviceable for the routine control of slips. The reading on the dial can be directly converted to the apparent viscosity and it is also possible by its use to measure the thixotropic behaviour of a slip.

B. Flow of Liquid in a Tube. A useful method of assessing the viscosity of slips is to measure the rate of flow of liquid in a tube when subjected to a certain pressure head. If the bore of the tube is known, the time taken for a given volume to pass a certain point may be calculated and hence the rate of flow determined.

Viscometers of this type are based on an original method by Ostwald,[2] whose apparatus is shown in Fig. VIII.16A. The slip or suspension is poured into the left-hand limb of the U-tube, which has a narrow capillary section along part of its length. Under standardised conditions, the time taken for a constant volume of liquid to flow through the capillary section is related to its viscosity. The usual practice is to measure the time taken for the level of liquid in the tube to fall between two arbitrary marks R_1 and R_2 and to calibrate the apparatus with liquids of known viscosity.

C. Falling Sphere Viscometer. A simple, but nevertheless effective type of viscometer involves measuring the rate of fall of a sphere of known dimensions when it is allowed to fall freely in a column of the liquid.

Although under carefully controlled conditions of experiment the viscosity of the liquid can be theoretically predicted from Stokes's Law (see p. 379) provided that the velocity of fall and the diameter and density of the sphere are measured, correction factors have to be applied in practice. These are necessary because the sphere has to be sufficiently large to enable it to be seen and consequently its fall is not streamlined. Because of these drawbacks, the falling sphere viscometer is more commonly used after being calibrated with liquids of known viscosity.

A small steel ball (S) is allowed to fall in a cylinder (Fig. VIII.16B) containing a liquid of known viscosity (η_1) and known specific gravity (ρ_1); the time (t_1) taken for it to fall between two marks (R_1 and R_2) is noted. The same test is repeated with the liquid or suspension, the viscosity of which is to be measured, and the time of fall (t_2) between the same two marks again noted. If the specific gravity of the test liquid is ρ_2, the following relationship can be applied to determine its viscosity (η_2):

$$\frac{\eta_1}{\eta_2} = \frac{t_1}{t_2} \frac{(\rho - \rho_1)}{(\rho - \rho_2)}$$

where ρ is the specific gravity of the steel ball.

[1] Clark, A. A. and Hodsman, H. J., *J. Soc. Chem. Ind.*, 56, 67, 1937.
[2] Ostwald, W., *Kolloid Zeit.*, 12, 213, 1913.

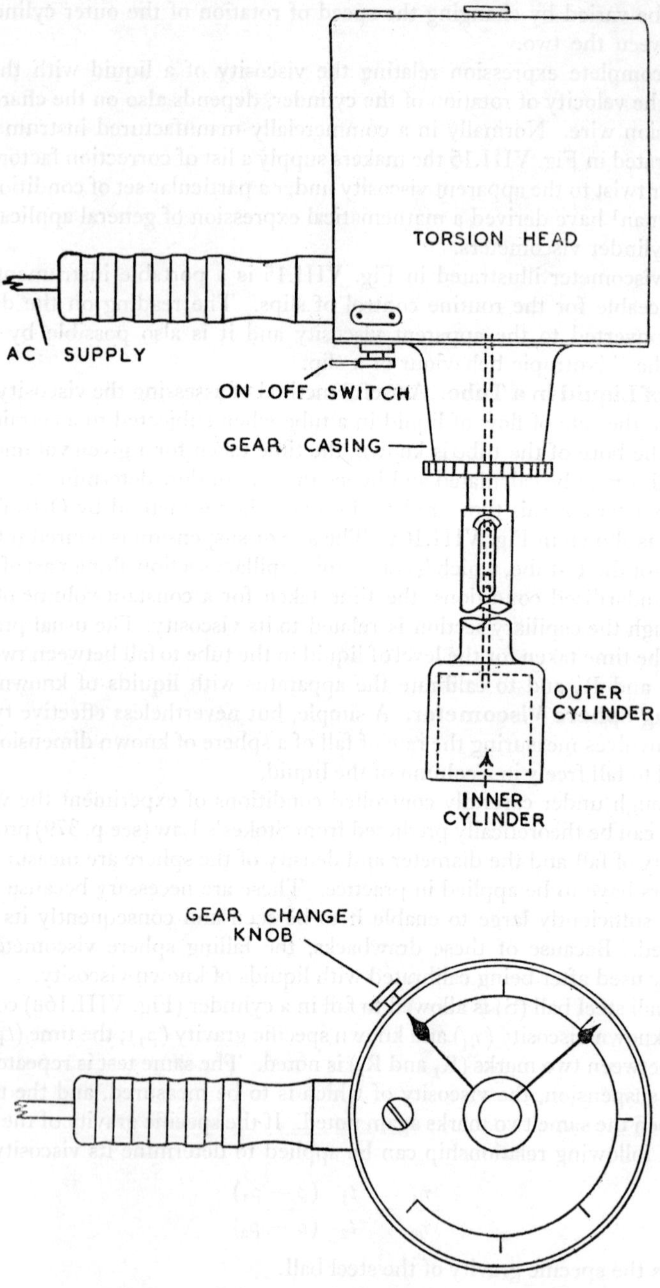

FIG. VIII.15. A PORTABLE ROTATING CYLINDER VISCOMETER
(Supplied by Ferranti Ltd.)

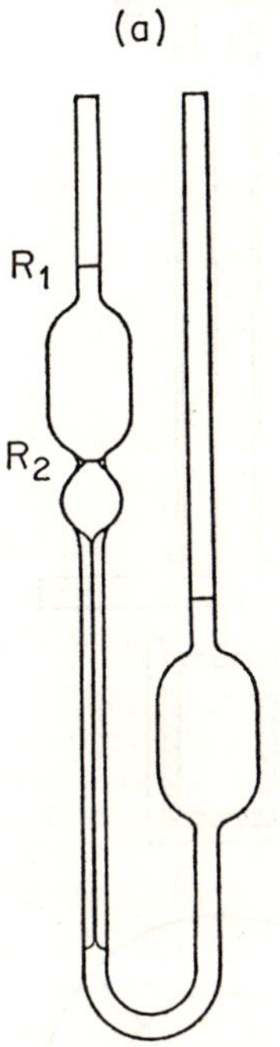

FIG. VIII.16A.
THE OSTWALD VISCOMETER

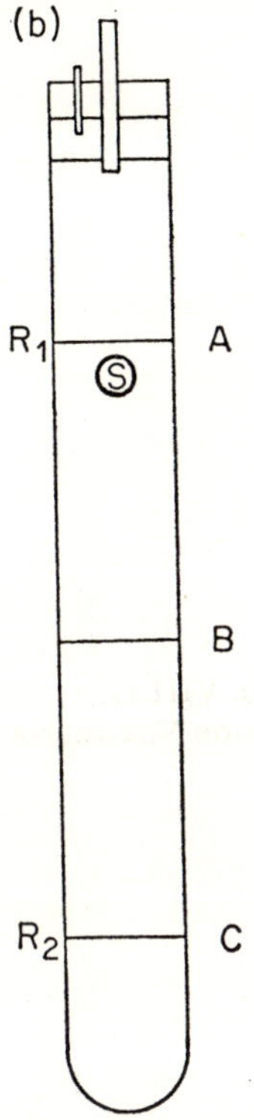

FIG. VIII.16B.
THE FALLING SPHERE VISCOMETER

This method is liable to error if the particles in suspension settle appreciably during the time taken for the test. The falling sphere viscometer is most effective when the viscosity of the fluid being tested is high, and when it is non-opaque.

D. Torsion Viscometers (Fig. VIII.17) have been used for determining the viscosity of clay suspensions (slips) since 1908 in which year A. V. Bleininger[1] immersed a heavy horizontal disc or cylinder suspended by a long wire in a tank containing a clay slip, turned it through an angle of 180° or 360°, released it and measured the amplitude of the swings several times. At a fixed temperature, the ratio between two consecutive amplitudes of swing is constant for a given slip and the relative viscosity (η) can be calculated from the equation:

$$\eta = \frac{t_2 \log r_1}{t_1 \log r_2}$$

[1] Bleininger, A. V., *Trans. Amer. Ceram. Soc.*, 10, 389, 1908.

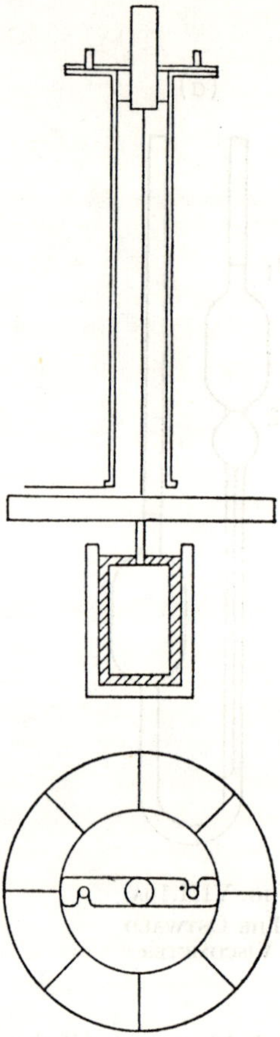

FIG. VIII.17.
THE TORSION VISCOMETER

where t_1 and t_2 are the periods of oscillation in water and the slip respectively, whilst r_1 and r_2 are the ratios of the amplitudes of any two consecutive swings in the same direction in the same liquid at the same temperature. Numerous modifications have been made in the design of torsion viscometers, but the principle of all is the same.

An apparatus developed by Norton[1] is relatively simple to construct and is widely employed in the ceramic industries.

The Couette and torsion viscometers are the most commonly used in the ceramic industries because of their convenience and because they do enable a reasonable amount of control over the properties of slips to be exercised. It should never be forgotten, however, that the property of viscosity as applied to suspensions

[1] Norton, F. H., *J. Amer. Ceram. Soc.*, 21, 33, 1938.

is so complex and its measurement involves so many factors that no single figure can accurately represent it.

VISCOUS FLOW PROPERTIES OF LIQUIDS AND SUSPENSIONS

WHEN an ideal liquid is being tested, the ratio of the shearing rate to the shearing stress is constant and independent of the force applied. In other words the viscosity of such a liquid is constant at all rates of shear, so that its coefficient can be assessed from the slope of a graph of shearing rate against shearing stress. This type of behaviour is known as *Newtonian flow* and it is illustrated in Fig. VIII.18A.

This is rarely, if ever, attained in practice, because in an ideal liquid the molecules are capable of streamlined flow. For all practical purposes, however, water and some simple organic liquids may be so regarded for their coefficient of viscosity is sensibly constant except at extremely high shearing rates.

If a small concentration of spherical particles is added to a liquid the coefficient of viscosity is increased, although the suspension may still behave as though it were an ideal liquid. The ratio of the shearing rate to the stress imposed still remains sensibly constant at all values thereby implying that the viscosity is independent of the applied force.

As the concentration of particles is increased however the flow properties of the suspension depart from those in ideal *Newtonian flow*. As the shearing rate is increased, the stress imposed does not increase linearly. The coefficient of viscosity increases with the applied force. This is expressed graphically in Figs. VIII.18A and B. The former shows the variation of shear rate with stress for an ideal liquid and the latter for a suspension; Fig. VIII.19 indicates the variation in apparent viscosity with the shearing rate and demonstrates the deviation from ideal conditions as the concentration of the suspension is increased.

When non-spherical particles, e.g. clay minerals, are added to the liquid phase, the deviation from Newtonian flow is more pronounced than when all the particles are true spheres. This is due to the fact that when a driving force is applied to a suspension the rate of flow induced is influenced by the following:

(*a*) The resistance of the molecules of the medium itself.

(*b*) The presence of particles in the medium which have to be moved along with the liquid. Not only is there a translatory movement involved but when the particles are irregularly shaped there may be rotational movement in the velocity-gradient of the sheared liquid.

(*c*) At high concentrations, the particles may be sufficiently close to permit mutual interference of their electrical fields. Consequently there will be serious resistance to flow.

For these reasons, the rate of flow of a suspension becomes progressively more restricted as the concentration of solid matter is increased. Furthermore, the amount of restriction depends on the velocity of flow, i.e. the shearing rate, and a linear relation between this factor and the stress imposed does not hold (Fig. VIII.18B). Whenever the viscosities of suspensions have to be compared, it is vitally important

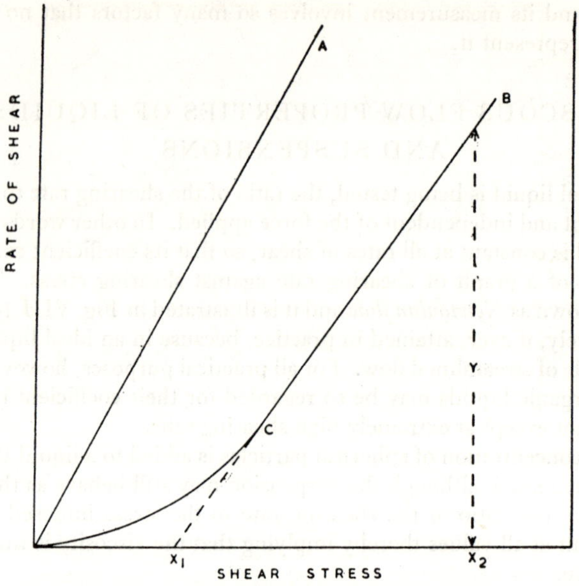

FIG. VIII.18.

(A) NEWTONIAN, AND (B) NON-NEWTONIAN FLOW

In the former, the slope of the stress/strain line, i.e. the viscosity, is constant at all stress values, but in the latter, the relationship is non-linear. The apparent viscosity in non-Newtonian flow is given by X_2/Y and will decrease with increasing stress. The value X_1 which is the extrapolation of BC to zero strain is sometimes called the Yield Value

to carry out all tests under the same conditions of shearing rate. The value so determined is termed the *apparent viscosity* and the conditions under which it is measured should always be stated.

According to Philippof[1] many factors contribute to the apparent viscosity of a suspension and it is not an easy matter to investigate the influence of one whilst maintaining the others constant. These include:

(*a*) The concentration of suspended material.
(*b*) The size of particles, their distribution and the degree of dispersion.
(*c*) The shape of particles.
(*d*) The electrolyte type and concentration, or, in other words the zeta-potential.
(*e*) The flexibility or deformability of the colloidal micelles in the suspension medium.
(*f*) Anomalous behaviour, e.g. thixotropy, rheopexy, dilatancy, etc.
(*g*) The temperature of the suspension.

[1] Philippof, W., *Viskosität der Kolloide* (Steinkopff, Dresden and Leipzig, 1942).

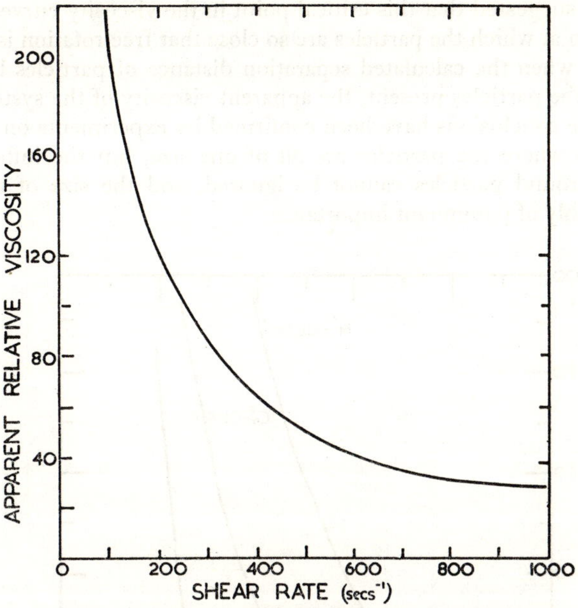

FIG. VIII.19. THE VARIATION OF THE APPARENT VISCOSITY OF A CLAY SUSPENSION WITH INCREASING SHEAR RATE

The **dependence of viscosity on the concentration of particles** has been intensively studied and many writers have evolved mathematical relationships to describe particular systems. It is however only with very dilute suspensions that a general relationship can be applied. The *Einstein equation*[1] holds when the concentration is very low, so that the suspended particles present are widely separated and their electrical fields do not interfere. Viscosity is then a linear function of the *volume* concentration of particles and may be expressed as

$$\eta_a = \eta_m(1 + k\varphi)$$

where η_a and η_m are respectively the observed viscosity and the viscosity of the suspension medium, φ is the volume concentration (expressed as a fraction) and k is a constant which for spherical particles has the value 2·5.

Smoluchowski[2] arrived at a similar conclusion but others have suggested that the value of the constant may be higher, particularly when particles of less symmetrical shapes are considered.

This equation applies only at extremely small concentrations, in fact, clay suspensions with more than 2 per cent of solid matter may show considerable deviation. The mutual interference of particles becomes the predominating influence at higher concentrations and as Fig. VIII.20 shows, above a certain critical concentration the rate of increase in viscosity with concentration is extremely rapid.

[1] Einstein, A., *Ann. Physik.*, 19, 301, 1906 and 34, 592, 1911.
[2] Smoluchowski, M., *Kolloid Zeit.*, 18, 190, 1916.

Norton has suggested that this critical point in the viscosity curve is related to the concentration at which the particles are so close that free rotation is impossible. In other words, when the calculated separation distance of particles becomes less than the size of the particles present, the apparent viscosity of the system is greatly increased. These conclusions have been confirmed by experiments on fractionated clay suspensions where the particles are all of one size, but the influence of the electrical field around particles cannot be ignored, and the size of the colloidal micelles is probably of paramount importance.

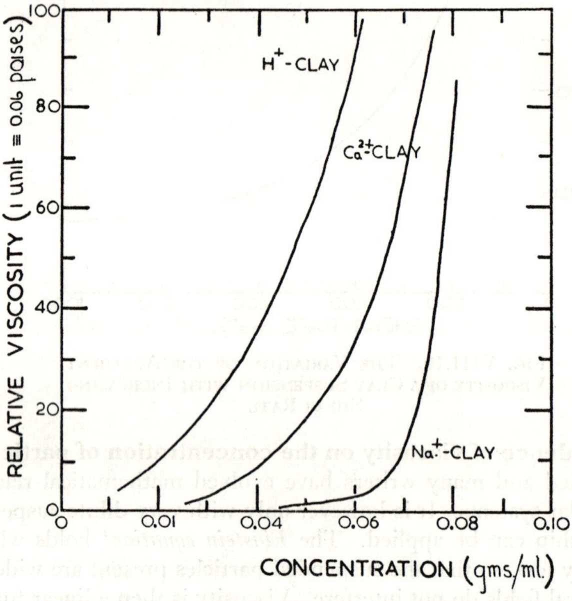

FIG. VIII.20. THE EFFECT OF THE CONCENTRATION OF A CLAY SUSPENSION ON THE APPARENT VISCOSITY
The effect of different saturating cations is also shown

The mathematical treatment of the viscosity/concentration relationship at high concentrations is far from established and many different equations have been advanced. Some writers have simply introduced an additional term into the Einstein equation, which would be the predominating influence when the amount of suspended material was large, thus:

$$\eta_a = \eta_m(1 + k_1\varphi + k_2\varphi^n)$$

Norton, *et al.*,[1] for example, applied such an equation in studies on mono-dispersed kaolinite suspensions and found that in a flocculated system n was near to 3 whilst with clay in the deflocculated state n approximated to 12.

[1] Norton, F. H., Johnson, A. L. and Lawrence, W. G., *J. Amer. Ceram. Soc.*, 27, 149, 1944.

A more general form of the equation has been suggested by Baker[1] as follows:

$$\eta_a = \eta_m(1 + k\varphi)^n$$

On this basis the equation of Norton would be correct to a first approximation. Similarly, if φ were very small, the equation would simplify to the Einstein form.

Mooney[2] has predicted from theoretical reasoning that the viscosity of a suspension of spherical particles should obey the following equation

$$\eta_{rel} = \text{exponential}\ \frac{(a\varphi)}{(1 - k\varphi)}$$

where η_{rel} is the apparent viscosity relative to that of the medium, φ the volume concentration of particles and a and k are constants, the former depending on the shape of the particle and the latter on the 'crowding' or packing. To a first approximation and providing the concentration is not excessive this will simplify to

$$\text{lon}\ \eta_{rel} = \frac{a\varphi}{1 - k\varphi}$$

T. E. Currie[3] has shown that this equation can be applied to deflocculated clays over a comparatively large range of concentration (up to 30 per cent by weight).

The **influence of grain size on the viscosity of clay suspensions** is most pronounced. The finer the particle size the greater will be the viscosity for a given slip concentration as Fig. VIII.21 shows. Undoubtedly the *surface area of the particles* in a suspension is a determining factor on the viscosity and, in this sense, the *particle shape* is an important contributor.[4] Plate-like and fibrous particles have a larger surface factor than spheres or cubes and consequently the viscosity of such particles is larger for an equivalent concentration.

The **electrolyte type and concentration** have a great influence on the apparent viscosity of clay slips and suspensions and their behaviour is most probably related to the changes which they produce in the zeta-potential. The effect of adding an electrolyte to a suspension of a flocculated clay has been demonstrated by Henry and Taylor[5] and their results are shown in Fig. VIII. 22. If a weak solution of sodium hydroxide is added to a suspension of a hydrogen clay, sodium ions progressively replace the hydrogen and increase the zeta-potential. Although the acidity of the suspension (as shown by changes in pH) is gradually lowered, the change in the viscosity is small until the suspension becomes just basic when an amazing increase in fluidity takes place; the viscosity changes to about 1/200 of its original value. Further additions of alkali produce a more alkaline suspension, but the viscosity—which is not much different from that of the suspending medium—does not change again until the amount of alkali added becomes excessive, when the cations invade the double layer and cause coagulation. According to Henry and Taylor the amount of alkali which is required to bring out the reduction in viscosity of the hydrogen clay is equivalent to its base exchange capacity (see p. 263). All clay

[1] Baker, F., *J. Chem. Soc.*, London, 103, 1653, 1913.
[2] Mooney, M., *J. Coll. Sci.*, 6, 164, 1951.
[3] Currie, T. E., Leeds University Ph.D. thesis, 1954.
[4] Ormsby, W. C. and Marcus, J. H., *J. Amer. Ceram. Soc.*, 50, 190, 1967.
[5] Henry, E. C. and Taylor, N. W., *J. Amer. Ceram. Soc.*, 21, 165, 1938.

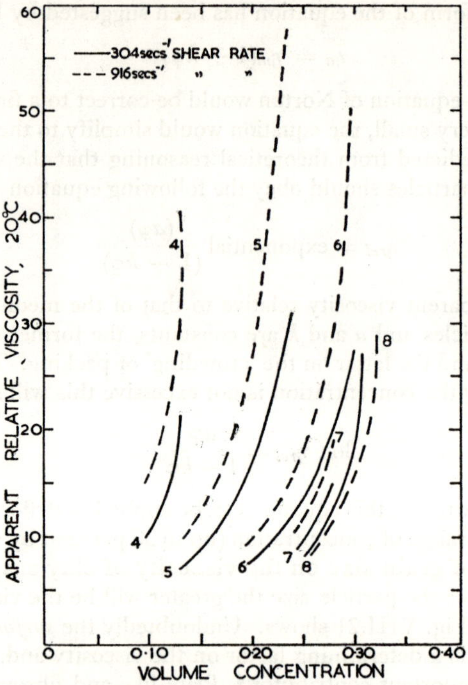

FIG. VIII.21. THE EFFECT OF PARTICLE-SIZE ON THE APPARENT VISCOSITY OF KAOLINITE SUSPENSIONS

Particle-ranges: (4) 0·15–0·20μ; (5) 0·20–0·30μ; (6) 0·30–0·50μ; (7) 0·50–1·0μ; (8) 1·0–2·0μ.
The relationship at two rates of shear is shown

suspensions are susceptible to changes in viscosity under conditions of varying electrolyte type and amount. The presence of monovalent cations, such as sodium and potassium favours lower viscosities at a particular concentration than are given by di- and trivalent cations and also hydrogen ion. The viscous behaviour of kaolinite, montmorillonite and attapulgite clays are compared in Fig. VIII.23 which shows the viscosity/concentration relationship and the influence of sodium hexametaphosphate and hydrochloric acid additions on the natural clays.

The physical concepts of the behaviour of clay suspensions in the presence of electrolytes have not been completely explained. It has been suggested that electrolytes reduce the 'drag' or surface tension of the particles in water. This theory is supported to some extent by the fact that some camphorated products and other detergents reduce the viscosity of clay suspensions without influencing the cation exchange. They may act, however, by forming *protective colloids* (see p. 473), so more is involved than a simple surface tension change in these cases. Norton[1] has

[1] Norton, F. H., *Elements of Ceramics* (Addison-Wesley Press Inc., Cambridge, Mass., 1952).

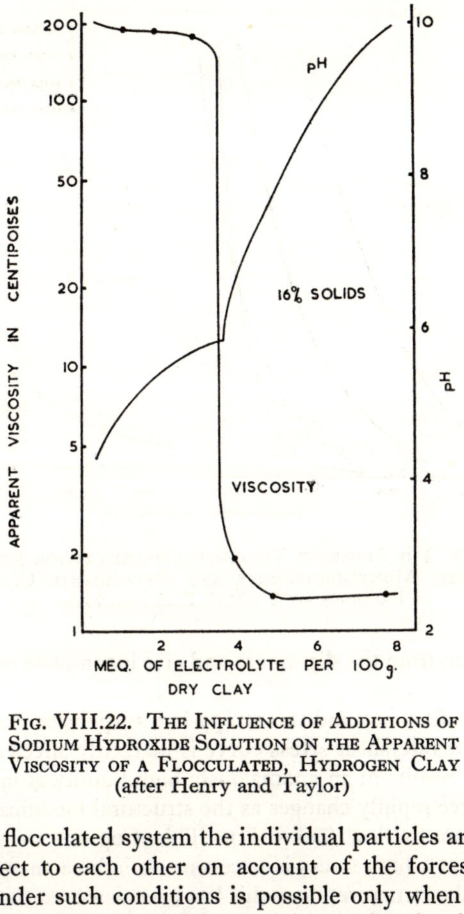

Fig. VIII.22. The Influence of Additions of Sodium Hydroxide Solution on the Apparent Viscosity of a Flocculated, Hydrogen Clay (after Henry and Taylor)

suggested that in a flocculated system the individual particles are in an equilibrium position with respect to each other on account of the forces of attraction and repulsion. Flow under such conditions is possible only when force is applied to disturb the equilibrium disposition of the particles, so that the viscosity will be high. In deflocculated masses, Norton considers that there are no attractive forces between particles, so that they are easily disturbed and the suspension has a much lower viscosity.

Neither of these two theories to explain the behaviour of electrolytes on clay suspensions is generally accepted, but no suitable alternative has so far been advanced which accounts for all the known facts.

The **nature of the colloidal micelles** has an influence on the viscosity of suspensions, but there is little conclusive evidence on this point. The water molecules which are held by the clay particles are influenced both in their amount and the rigidity of their bonding by associated cations. Weakly-held hydrospheres are easily distorted and 'flow' more easily than those of a more rigid nature.

Thixotropic, rheopectic, and dilatant behaviour in suspensions influence the viscosity and so affect the flow properties. Thixotropy is particularly troublesome in ceramic slip-casting processes, because if it is appreciable it may hamper the smooth

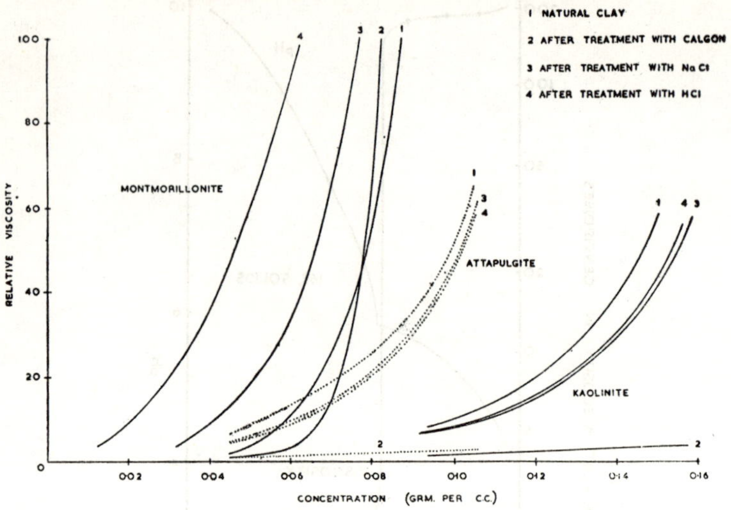

FIG. VIII.23. THE APPARENT VISCOSITY/CONCENTRATION RELATIONSHIP OF KAOLINITE, MONTMORILLONITE AND ATTAPULGITE CLAYS IN THE PRESENCE OF VARIOUS ELECTROLYTES

flow of the suspension from the slip-house and give incomplete or poor shapes in the mould.

If the effects of dilatancy and thixotropy in a suspension are appreciably large, the form of the shear rate/shear stress curve is markedly affected, as shown in Fig. VIII.24. Dilatancy results in an anomalously high fluidity at low stress values but the shape of the curve rapidly changes as the structural modifications in the system impose an opposition to flow (curve D). Thixotropic systems show an almost reversed effect in that a high stress is necessary to induce an appreciable flow rate and to break down the structure; once this has been achieved the flow curve returns to normal linearity (curve T).

Fig. VIII.24 also shows the reversal curves or the changes in stress/strain features as the stress is progressively reduced from high values. Both dilatant and thixotropy systems show marked hysteresis features but of opposite type.

The **temperature of a suspension** has a small effect in so far as the viscosity of the suspending medium is changed. Other factors are of so much greater importance that the temperature changes normally encountered in practice are of little consequence in ceramic slip practice.

VISCOUS AND PLASTIC FLOW

As the water content of a suspension is reduced relative to the amount of solid matter, the lubricating effect of the liquid medium is lessened. The separation between individual particles becomes so small that free flow is prevented; the resistance to shearing forces increases and when it is sufficiently great and the mass is then

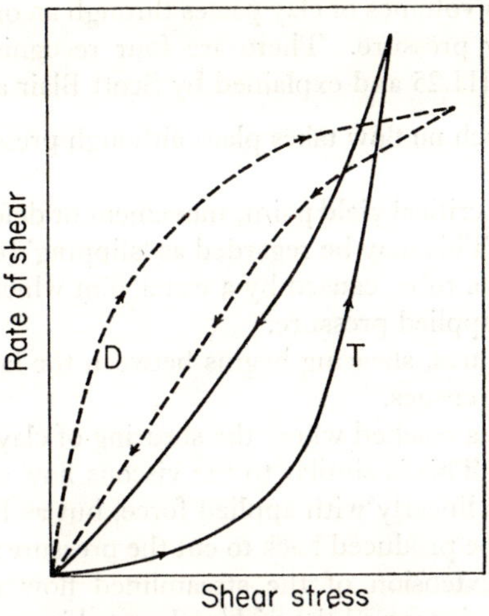

FIG. VIII.24. THE EFFECTS OF DILATANCY (D) AND THIXOTROPY (T) ON THE SHEAR RATE/SHEAR STRESS CURVES OF CLAY SUSPENSIONS

deformed, the change in shape is permanent. Unlike true *viscous flow* which is a feature of dilute slips and suspensions, *plastic flow* implies that a certain yield value must be exceeded before any deformation can take place. Viscous flow is reversible, i.e. the distortion produced by an applied force is relieved immediately the force is removed and the body reverts to its former condition. In plastic flow this is not the case—flow does not occur until the distorting force reaches a certain amount and when the force is removed, a plastic body has a permanent set and does not return to its original shape. There is no sharp dividing line between the two types of flow but a gradual transition as the concentration of solid material in suspension increases.

PLASTIC BODIES

PLASTIC pastes, such as are commonly employed in the clay-working industries, may be regarded as intermediate between solids and viscous liquids. When pressure is applied to a liquid the latter at once commences to flow and continues to do so at a velocity proportional to the pressure. A solid undergoes no appreciable change until its structure is broken down by the pressure. A plastic paste has an intermediate behaviour; on the application of a steadily increasing pressure it appears to be immobile until a critical pressure is applied, after which it behaves like a liquid and —so long as the pressure is maintained—it flows with a velocity proportional to the pressure.

Scott Blair[1] and Bingham[2] have studied the characteristics of plastic flow by

[1] Scott Blair, G. W., *Trans. Eng. Ceram. Soc.*, 30, 138, 1931.
[2] Bingham, E. C., *U.S. Bur. Stand. Sci. Paper*, No. 278, Vol. 13, 309, 1916–17.

comparing the extrusion volumes of clay pastes through an orifice on the application of a steadily increasing pressure. There are four recognisable stages which are demonstrated in Fig. VIII.25 and explained by Scott Blair as follows.

(*a*) A region in which no flow takes place although pressure is being applied to the mass.

(*b*) Above a certain critical yield point, movement or deformation of the mass of clay as a whole occurs. This may be regarded as 'slipping' between the plug of clay paste and the walls of the tube, caused by a water film which may have been forced out of the 'plug' by the applied pressure.

(*c*) At higher pressures, shearing begins between the layers of clay particles—a region of 'mixed flow' ensues.

(*d*) Finally, a state is reached where the shearing of clay layers is complete and streamline flow occurs. This is similar to the viscous flow of a true liquid because the shearing ratio varies linearly with applied force, but as Fig. VIII.25 shows, the straight portion cannot be produced back to cut the pressure axis at zero. The point where the downward extension of the streamlined flow region (shown dotted) touches the pressure line is termed the *yield value*, and is regarded by Bingham as a direct measure of the plasticity of the material—a hypothesis which is not accepted by several well-known authorities, although it is a frequently quoted figure to describe the properties of plastic clay masses.

If the rate of flow (v) of a truly viscous liquid is expressed as proportional to the pressure applied (F) the flow of plastic bodies might be similarly represented as a function of $F - f$ where f is the *yield value*.

Thus the Bingham equation for plastic flow is expressed as

$$v = k\mu(F - f)$$

where k is a constant and μ is termed the *coefficient of mobility*. In the limiting case when $f = 0$ then the expression simplifies to the viscosity relationship for the flow of ideal liquids.

$$v = \mu F \quad (\text{cf. } F = \eta v)$$

Therefore in this case, the coefficient of mobility becomes synonymous with the coefficient of fluidity (φ) or the reciprocal of viscosity.

Scott Blair has suggested that the 'Bingham yield value' should be termed the *static rigidity value* and he considers that the term *yield point* should be reserved for the pressure at which flow just commences. He regards the slope of the pressure/flow line in the region of streamline flow as the *pseudo-viscosity* of the system.

The 'Bingham yield value' or the 'static rigidity' is regarded by some writers as a measure of the coefficient of thixotropy, but, although many clays which exhibit thixotropic behaviour have a high coefficient, this is not the only factor contributing to the absolute value.

The 'Bingham yield value' increases with the concentration of solid particles and is pronouncedly influenced by the presence of electrolytes. Thus, some clays which exhibit a marked degree of plastic behaviour in the natural state, may be converted to a liquid with virtually true viscous behaviour on the addition of a small percentage of electrolyte.

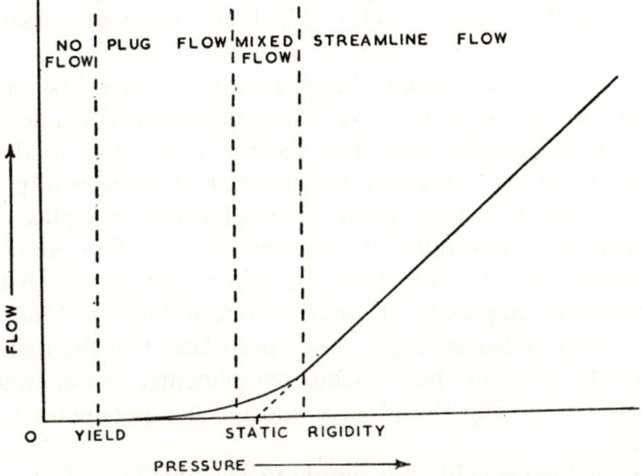

FIG. VIII.25. THE STAGES IN PLASTIC FLOW
(after Scott Blair)

Norton, Johnson and Lawrence[1] have investigated the influence of particle-size on the initial yield point of a hydrogen-ion saturated kaolinite. Their results suggest that an equation of the type

$$\text{Yield point} = kC^n$$

adequately describes the behaviour, where C is the concentration of solid particles and n is 3 for all the size-fractions considered. The value of k increases rapidly as the particle size is reduced.

The property of plastic flow is due to the cohesion between the solid and liquid constituents, which together form an apparently uniform mass in which the liquid acts as a lubricant thereby facilitating the movement of the solid particles and simultaneously restraining them from being separated far from each other. The extent of this cohesion depends on the affinity between the particles of solid and liquid; it is great in plastic materials like clay, which have a high capacity for being deformed by pressure without the individual particles being separated from each other, whilst materials in which the cohesion is low do not form good pastes and are termed non-plastic materials.

Care must be taken, however, not to confuse 'plastic flow' with the 'plasticity' of a clay paste. They are related in some senses, but they are not identical.

PLASTICITY

THE phenomenon of plasticity in all its aspects is the most intriguing study in clay science and technology.

[1] Norton, F. H., Johnson, A. L. and Lawrence, W. G., *Trans. Amer. Ceram. Soc.*, 27, 149, 1944.

It may be defined as the behaviour of clays when they are mixed with water but in insufficient amounts to permit flow without external applied stresses.

There are so many aspects of plasticity that it is impossible to describe it adequately or to ascribe to it parameters which are sufficiently distinctive.

The clay worker defines the plasticity of a clay in terms of 'shortness' or 'fatness', properties dependent on the feel of the material. In other senses, the words 'strong' and 'lean' are applied to clays possessing plasticity which define the capacity for moulding. Green and dry strengths, the rate and extent of deformation, the ability to be extruded or to be pressed into shape are all aspects of plasticity.

Plastic bodies are far from simple in their behaviour to applied stresses. They show elastic characteristics as well as the features of plastic flow, and it is this double nature—which has been fully recognised only in recent years—that confers upon plasticity its intriguing complexity. From the work of Astbury, Moore and Lockett[1] a new model of plasticity has emerged which postulates that the main feature is the breakdown of elastic elements into viscous components. An energy term U_c can be regarded as characterising the process and given a parameter which has been related to plasticity.

Even this model is probably too simple to account for all features of plasticity and it must still be accepted that a craftsman's thumb can often be a better assessor of plasticity than the most complex scientific measurement.

The Nature of Plasticity

It is generally accepted that plasticity is associated mainly with particles which are of colloidal size, therefore it is to be expected that it will be related to the fundamental concepts of the colloidal micelle.

Clay particles when suspended in water are surrounded by a hydrosphere within which are contained ions of different charges. Around the ultimate particle there is a layer of negatively-charged ions and these are balanced by a 'swarm' of cations which are diffused throughout the hydrosphere, which varies in size depending on the number, size, charge, and degree of hydration of the cations (see p. 468). When the hydrosphere is large, the particles in suspensions are at the maximum separation and free, viscous flow is possible as a result of the excess water content. On the other hand, if the hydrosphere is reduced, the separating water film becomes less and more cohesion results.

The counter-cations provide links between the colloidal particles and rigidity in linkage or 'plasticity' develops. The application of a shearing force causes movement of the charged particles, but an equally stable system is readily formed, which retains its new shape after the removal of the force. Consequently small, unhydrated cations adsorbed around a clay particle give greater plasticity than larger, hydrated ones, all else being equal.

Plasticity is associated with the formation of water films of a certain order of thickness around each grain. If more than the required amount is present the 'free water' permits flow with a reduction in the plastic properties. At much lower contents of water, the individual particles touch and plasticity is destroyed. Plasticity is

[1] Astbury, N. F., Moore, F. and Lockett, J. A., *Trans. Brit. Ceram. Soc.*, 65, 435, 1966.

thus a function of the water content, the peak value depending on the surface area capable of being wetted. These adsorbed water films appear to have an induced structure by virtue of the surface with which they are in contact, and a continuum is thus formed which surrounds all the particles and introduces a degree of rigidity and at the same time an ease of deformation.

The thickness of the water film which is present in a clay at maximum plasticity is difficult to calculate with certainty but Norton and Hodgdon[1] have deduced that under a pressure of 8 kg. cm.$^{-2}$ the film thickness varies slightly with different clays as follows:

Clay	Film thickness (1 Å = 10^{-8} cm.) Å
China clay	2100
Ball clay	2400
Brick clay	3100
Bentonite	3400

Mattson[2] has estimated that in soil clay the thickness of the water film is about 700 Å although this may not have been at maximum plasticity.

It is interesting to record that if water separates two parallel plates the attractive force between them when the separation is 2000 Å is 8 kg. cm.$^{-2}$. For this reason alone the plasticity of particles would be expected to be at a maximum when the water film is of that order of thickness.

Norton,[3] however, suggests that the only forces between particles in a plastic mass are (*a*) the repulsive forces between particles and (*b*) the surface tension of the water between them which tends to draw them together. At maximum plasticity, the water film fills only the narrow capillaries between the individual particles and so behaves as a *stretched membrane*. At higher water contents the tension is relieved, hence the force holding the particles together is lessened. With less water, the envelope surrounding all the particles will be broken and the plasticity destroyed.

The dependence of plasticity on both the water content and the nature of the associated cations must lead to the conclusion that the property is associated with the complex electrical relationships surrounding the colloidal micelle. When the zeta-potential of a system is of low order, the counter-balancing cations influence the system only at short distances from the mineral surfaces. Nevertheless, when the counter-balancing cations are of small size and high charge they impose a more rigid structure on water molecules in their vicinity than do those of large size and small charge.

At low water contents, the hydrospheres around each particle merge and the mass attains a stability or structure formation in which continuity is developed between rigidly bound water molecules in adjoining particles.

Such a system resembles thixotropy as described on p. 473, but with the difference that in normal thixotropy as manifested by montmorillonite the suspensions are dilute with an excess of large mono-valent cations.

[1] Norton, F. H. and Hodgdon, F. B., *J. Amer. Ceram. Soc.*, 15, 191, 1932.
[2] Mattson, S., *Soil Sci.*, 33, 301, 1931.
[3] Norton, F. H., *J. Amer. Ceram. Soc.*, 31, 236, 1948.

The thixotropy of concentrated, flocculated clay/water systems is one of the prime causes of plasticity. Elasticity is a natural consequence of distorting the electrical balance which will return to its equilibrium position on the removal of the disturbing force. Plastic flow is possible only when the applied stress is sufficient to break down the force structure; ultimately, yield will occur through shearing between layers of rigidly-bound water molecules when pressures exceed the ultimate strength of the bonds which bind them.

There is thus a parallel between the malleable properties of metals and the plastic properties of clay/water systems, but the analogy must not be taken too far.

Factors Affecting Plasticity

Clays differ greatly in the amount and quality of their plasticity. The 'fat' or 'strong' clays undergo a large change in shape on the application of a small force; 'lean' or 'short' clays undergo a much smaller change of shape and require a greater force to make the change appreciable; some clays are both 'strong' and 'short', i.e. they require a large force to deform them and change only slightly in shape before being ruptured. Hence, in comparing the plasticities of clays, it is necessary to include the amount of change of shape as well as of the force necessary to produce it.

Omitting, for the present, those materials in which the plasticity is due to the use of oil or of added materials other than water, the plasticity of clays is affected chiefly by:

1. The effect of water or other fluid medium on the solid particles and colloidal phenomena connected therewith.
2. The size of the solid particles.
3. The composition of the solid particles.
4. The size distribution of the particles.
5. The shape of the solid particles and their internal structure.
6. The aggregation of the solid particles.
7. The surface area of the solid particles and their intermolecular attraction.
8. The presence of other materials which may have an influence on the properties of the particles.
9. The orientation of particles in the mass.
10. The previous history of the material.

Influence of Water Content. Plasticity is exhibited by clays within a comparatively narrow range of water content. Above a certain value the clay mass becomes a mud or a slip whilst a dry clay is virtually devoid of plasticity. The maximum plasticity of a clay is developed at a specific water content which corresponds with a film around each particle about 2000 Å thick. For most clays, this would be in the water content range 15–25 per cent by weight.

There is a range, however, over which the plasticity is reasonably high, which might be termed the range of workability. In some clays, this may extend over a wide range of moisture content, but, in others, the range is exceedingly small.

The plastic flow diagram Fig. VIII.25 indicates that the flow characteristics of a clay/water system are dependent on the pressure applied. Consequently, a prepared body with a small amount of water may appear dry and non-plastic to the hand, but will flow readily and exhibit aspects of plasticity when deformed under high pressure.

The interdependence of water content and applied pressure in relationship to mouldability cannot be stressed too highly, but in its normal sense, plasticity has always implied the properties of clay pastes as assessed by the hand of the clayworker.

For some purposes it is better to express the proportion of water volumetrically instead of by weight, and plasticity values so obtained are important in civil engineering practice.

The relative volumes of water and solid matter in a moist mass of clay may be represented by the symbol ε.

The ratio $\varepsilon = \dfrac{\text{volume of water}}{\text{volume of solid}}$ in a plastic clay is invaluable to constructional and waterworks engineers who have to deal with the effect of pressure on wet clay. It is most conveniently determined by (i) weighing the clay in air (W_1); (ii) weighing it when suspended in paraffin (W_2); and (iii) weighing after drying at 110° C. (W_3).

If the specific gravity of the paraffin is S:

$$\varepsilon = \frac{S(W_1 - W_3)}{(W_1 - W_2) - (W_1 - W_3)S}$$

When ε is less than 0·35, flow cannot take place without rupture of the particles. Many surface clays with $\varepsilon = 0{\cdot}75$ have a safe load carrying capacity of 1 ton per sq. ft. When $\varepsilon = 2{\cdot}70$ the weights of water and solid matters are equal and the mass is 'sticky' and useless for ceramic purposes; its bearing power is nil and it is practically a viscous fluid.

Effect of Size of the Particles. Various investigators have attributed plasticity to the extreme smallness of the grains of a material, because various materials, usually considered to be 'non-plastic', may, when very finely ground with water, develop a small amount of plasticity. Quartz and limestone, when ground to pass a 200-mesh sieve and then mixed with a small proportion of water, have a slight plasticity, but do not hold together on drying. Orton found a similar effect with ground glass, and Cohn and Atterberg found that precipitated barium sulphate and calcium fluoride are plastic when wetted. Wet-ground felspar which has been allowed to stand for some time has a slight plasticity.

The plasticity developed by grinding non-plastic materials is so small that, whilst it must be admitted that the size of the particles certainly is in some way connected with plasticity, it is only one of the factors involved.

The removal of fine particles from a clay will destroy its plasticity, but calcination to 600°–800° C. will have a similar effect although the ultimate particle size is not unduly influenced.

The effect of the size distribution of particles. Plasticity is related to the degree of packing as discussed on p. 407. Moore and Lockett[1] have shown that a mass of particles of a good packing density gives a better plastometer value than a similar system of higher bulk volume. Some clays composed of wholly fine particles are improved in plasticity or 'body' by the addition of larger grains, but, in general, coarse additives reduce plasticity in a well-balanced clay.

[1] Moore, F. and Lockett, J. A., *Trans. Brit. Ceram. Soc.*, 65, 423, 1966.

Effect of Composition of the Particles. It is sometimes stated that as, plasticity is largely a mechanical property, it must be independent of the composition of the clay. This is incorrect, for though there is no simple relationship between the major constituents of a clay and its plasticity, there is a fairly definite connection between the amount of colloidal matter and the plasticity.

The influence of mineral type on plasticity has not been established for several reasons. In order to be able to compare the properties of two mineral entities not only have particles of similar size to be prepared, but the shape of such particles has to be identical. In plasticity measurements where so many other characteristics exert an influence the elimination of all variables other than mineral type is well nigh impossible.

Currie[1] has attempted to compare the plasticity of various kaolin minerals in fractions where the surface area is kept constant. In these cases although the different minerals showed variations in behaviour, all other factors being constant, the order of the differences was slight compared with the influence of grain-size, shape or electrolyte content. It is not unlikely that although some substances when added to others are known to increase the plasticity, it may be more the result of the increase in the fineness of grain rather than the nature of the mineral itself.

Bentonites are commonly added to non-plastic bodies to increase their plasticity. Although, undoubtedly, such clays increase both the green and dry strengths of bodies to which they are added, materials which include them have a short working range before fracture. The so-called livesite, illite or sericite clays are better in this respect because they deform readily and can withstand considerable distortion before rupturing; they do not impart such great strengths as do the bentonite clays.

The type of cations associated with a clay has a pronounced influence on the plasticity. This aspect is considered more fully in the section on the *influence of electrolytes*, but, in the broadest sense, the exchangeable cations are part of the composition of a clay and, as such, affect the plasticity.

Various organic colloids, including tannin, lignin, and humus may be present in a clay and affect the plasticity. Their mode of action is to form *protective colloids* around the clay particles and to enhance the colloidal properties. Many ball clays owe their high plastic properties to associated lignin, but other clays, which contain no organic matter are equally plastic.

The Effect of the Shape and Structure of the Particles. The plasticity of clays is frequently attributed to their lamellar or platey nature. This is supported by the fact that some minerals with a characteristic fibrous or platey nature, e.g. glauconite, mica, pyrophyllite and talc, develop plasticity when they are finely ground, whereas other minerals which have a more regular crystal habit, e.g. cubic and rhombohedral, are not nearly so plastic, even when crushed to about the same particle-size.

Whilst a lamellar nature in particles is useful in producing plasticity it is not the only suitable shape. Fibrous crystallites, such as those found in attapulgite, and halloysite, are if anything, capable of producing more plastic masses.

Undoubtedly, the surface area per unit weight of particles is a contributing factor to the plasticity and for this reason fibrous and platey particles would be superior to those of spherical or cubic shape, but the orientation or packing of the crystallites is probably of equal or greater importance.

[1] Currie, T. E., Leeds University B.Sc. thesis, 1951.

Effect of Aggregation of the Particles. Evidence is accumulating which seems to show that the structure of the ultimate particles and their aggregation into larger ones has an important influence on plasticity, and it has even been suggested that plastic clays are composed of 'crumbs' each consisting of a felted mass of lath-shaped or needle-shaped crystals. Mellor[1] has suggested that a distribution of grains of various sizes which permits the closest possible packing gives a maximum plasticity, whilst Schurecht[2] concluded that:

(*a*) Loosely cemented aggregates of clay grains are more plastic than closely compacted ones.
(*b*) Fine-grained aggregates are more plastic than coarse-grained ones.
(*c*) Flocculated aggregates are more plastic than deflocculated ones.
(*d*) Flocculated aggregates are more plastic than cemented ones.

The fact must not be ignored that the larger particles of clay are often agglomerations of the smaller particles and possess the same properties, though some aggregates break down more readily than others.

Effect of Surface Area and Intermolecular Attraction. On account of their fineness, grains of plastic clay have a large surface area, and, consequently, exhibit surface phenomena (including intermolecular attraction and colloidal properties) to a very marked degree. It is not unreasonable therefore to conclude that the surface area of the particles present in a plastic mass contributes in no small measure to the ultimate plasticity.

In a mass of particles two factors contribute to the total surface area, viz. the fineness of grain-size and the particle shape.

1. The influence of grain-size may be demonstrated by considering fixed weights of spheres of two different sizes, having radii of 10R cm. and R cm. Their volumes are:

$$\tfrac{4}{3}\pi(10R)^3 \text{ c.c. and } \tfrac{4}{3}\pi R^3 \text{ c.c., respectively.}$$

If the specific gravity of the material composing the spheres is ρ, the weight of each sphere is simply its volume multiplied by ρ.

The *number of spheres* in 1 gm. will therefore be

$$\frac{1}{\text{weight of each sphere}} \quad \text{i.e.} \quad \frac{1}{\frac{4}{3}\pi(10R)^3\rho} \quad \text{and} \quad \frac{1}{\frac{4}{3}\pi R^3\rho}$$

in the two cases.

As the surface area of a sphere is equal to $4\pi(\text{radius})^2$, the total surface area in 1 gm. of spheres of 10R and R cm. radius will be

$$\frac{1}{\frac{4}{3}\pi(10R)^3\rho} \cdot 4\pi(10R)^2 \text{ sq. cm.} \quad \text{and} \quad \frac{1}{\frac{4}{3}\pi R^3\rho} \cdot 4\pi R^2 \text{ sq. cm.}$$

respectively, i.e.

$$\frac{3}{10R} \quad \text{and} \quad \frac{3}{R}$$

[1] Mellor, J. W., *Trans. Eng. Ceram. Soc.*, 21, 91, 1921.
[2] Schurecht, H. G., *Bull. Amer. Ceram. Soc.*, 153, 1923.

Thus the result of decreasing the size of particles by 10 times increases the *specific surface area* by the same factor.

2. The influence of particle shape on the surface area may be assessed on a sphere, cube, plate and needle or fibre of comparable sizes. In Fig. VIII.26, the long dimension in all cases (diameter in the case of the sphere) is 2R cm., and the short dimension is XR where X is considerably less than unity. The density of the material in all cases is ρ.

In a case where the plates and fibres have a thickness to size ratio of 1:10, $R = 1$ and $X = 0{\cdot}1$, so that the specific surface area, i.e. surface area per gm. of the four shapes of comparable sizes is in the following ratio:

Sphere : cube : plate : fibre :: 3 : 3 : 22 : 41

For this reason, platey and fibrous particles exhibit greater plasticity than do spherical or cubical ones.

The values derived from the data in Fig. VIII.26 are for ideal crystals, but the ultimate crystallites in clays and other ceramic materials are often far from being so perfect. Transported particles, either through the medium of air, water or glaciers, are invariably roughened and fractured and the specific surface area may be very much greater than would be predicted from particle-size measurement. R. H. S. Robertson suggests that the term *rugosity* be used to describe the deviation of the surface area of a particle from the ideal.

The Effect of Electrolyte and Flocculants. The addition of soluble salts to clays may influence the plasticity considerably. Not only may the repulsive force between particles in the mass be altered as a result of cationic or anionic absorption but the surface tension of the liquid medium may be changed. Both effects have a pronounced influence on the plasticity of the clay mass. As an example of the former the action of alkalies and acids may be cited. When only 6 lb. of a mixture of sodium carbonate and sodium silicate is added to a ton of some kinds of clay, it can be made into a liquid with less than 20 per cent of its weight of water. If just sufficient acid is added to neutralise the 6 lb. of alkali, the whole mass becomes solid immediately.

The work of Speil[1] and Sullivan and Graham[2] suggests that the greater the cationic concentration near the surface of the colloidal particle the more strongly will water be held. Small cations of high valency (e.g. Ca^{2+}, H^+, etc.) when absorbed will tend to give a higher proportion of bonded water than those of large size and small charge. In a plastic mass where the particles are very close, the absorbed water films around each colloidal grain merge together, so the force required to rupture such masses will depend on the degree of bonding of the water. The plasticity in terms of yield point and breaking strength in torsion are close to the theoretical predictions of the Hofmeister series as shown in Fig. VIII.27. Because the water films surrounding clay particles are more flexible when alkali cations are absorbed, such clays are capable of greater deformation although they rupture more readily than do hydrogen or calcium clays.

More water is required to produce a plastic mass of a particular strength with a hydrogen and calcium clay than with one containing sodium as the exchangeable

[1] Speil, S., *J. Amer. Ceram. Soc.*, 23, 33, 1940.
[2] Sullivan, J. D. and Graham, R. P., *J. Amer. Ceram. Soc.*, 23, 39, 1940.

	Sphere	*Cube*	*Plate*	*Fibre*
		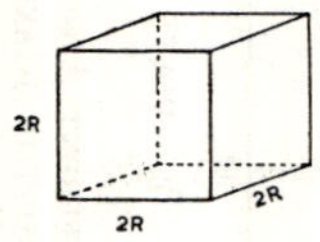	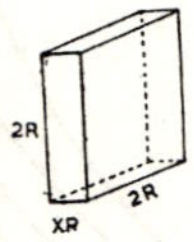	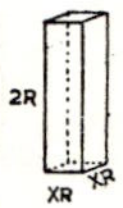
Volume of one particle	$\frac{4}{3}\pi R^3$	$8R^3$	$2R \times 2R \times XR = 4XR^3$	$2R \times XR \times XR = 2X^2R^3$
No. of particles per gm.	$\frac{1}{\frac{4}{3}\pi R^3 \rho}$	$\frac{1}{8R^3 \rho}$	$\frac{1}{4XR^3 \rho}$	$\frac{1}{2X^2R^3 \rho}$
Surface area of one particle	$4\pi R^2$	$6 \times (2R \times 2R) = 24R^2$	$2(2R)^2 + 4(2R \times XR) = 8R^2 + 8XR^2$	$4(2R \times XR) + 2(XR)^2 = 8XR^2 + 2X^2R^2$
Surface area per gm.	$\frac{4\pi R^2}{\frac{4}{3}\pi R^3 \rho} = \frac{3}{R\rho}$	$\frac{24R^2}{8R^3 \rho} = \frac{3}{R\rho}$	$\frac{8R^2(1 + X)}{4XR^3 \rho} = \frac{2(1 + X)}{XR\rho}$	$\frac{2XR^2(4 + X)}{2X^2R^3 \rho} = \frac{4 + X}{XR\rho}$

FIG. VIII.26. VARIATION OF SURFACE AREA WITH DIFFERENT PARTICLE SHAPES

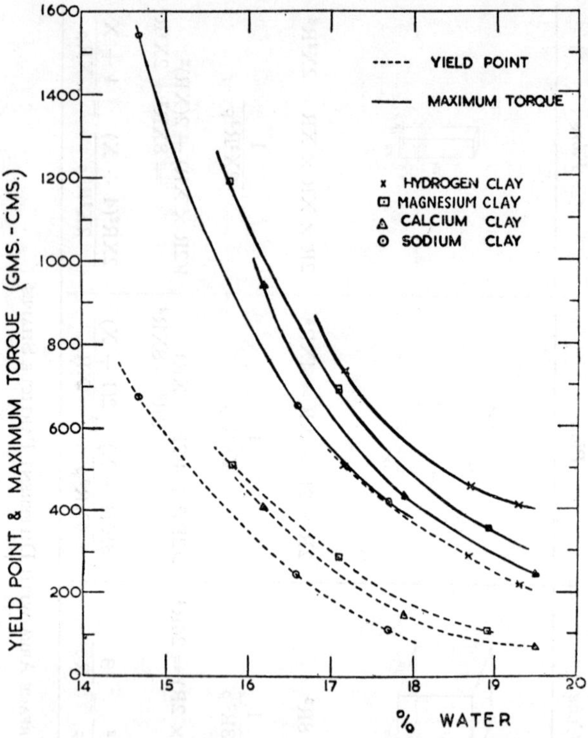

FIG. VIII.27. THE EFFECT OF DIFFERENT ADSORBED CATIONS ON THE YIELD POINT AND THE MAXIMUM TORQUE OF CLAY MASSES AT VARIOUS WATER CONTENTS

cation; so for a particular water content, hydrogen and calcium clays require a greater moulding pressure, but the green body is stronger and less liable to loss of shape than sodium clays.

In general, the range of water content over which the clay is workable is greater with a sodium than with a hydrogen or calcium clay although this is affected by other factors also.

The addition of a material which reduces the surface tension of water also liquefies a plastic mass. Such agents, which permit more concentrated suspensions to be used as casting slips, yet without introducing salts deleterious to refractory properties, are finding increasing use in industrial applications.

The Effect of the Presence or Addition of Other Materials. The plasticity of clays can often be increased or decreased by the addition of other substances than those mentioned above, but the mode of action of many of these is far from clearly understood.

It is not difficult to appreciate that the addition of colloids increases plasticity, although artificial mixtures are seldom as plastic as a natural clay of the same composition. The addition of a *non-plastic*—such as quartz or fired clay (grog)—reduces the plastic behaviour.

Air incorporated in a clay-paste has an adverse effect on its plasticity, so that such a clay appears to be short, but recovers its plasticity when the air has been removed by 'wedging', 'rolling', or by de-airing under low pressure.

The amount of air worked into clay during the preparation of a paste or body is much larger than is generally realised, and its adverse influence is often very serious.

The proportion of air in a plastic clay may be ascertained by placing a weighed quantity of the clay in a vessel of water, the clay being placed beneath a large inverted glass funnel and burette, previously filled with water by suction. On heating the water to boiling, the clay disintegrates and the air is released in the form of small bubbles which rise into the burette and can there be measured. A more accurate eudiometer may be used if required, but for routine tests is not necessary.

The Effect of the Orientation of Particles in the Mass. The plasticity of clay masses is undoubtedly influenced by the alignment of the particles of the predominant minerals particularly those which are plate-like. This may be demonstrated very readily with an oriented mass prepared by settling an air-free clay suspension and removing excess water by decantation or suction, followed by careful drying under vacuum. The mineral plates in such a mass will be arranged so that their basal section is at right angles to the settling direction.

If a force is applied to such a plastic mass at right angles to the oriented plates (i.e. downwards) the mass has a high yield value. Should the force be applied along the direction of 'bedding' (i.e. horizontally) flow commences at much lower applied pressures, because the plates slide on each other. A randomly-oriented mass will normally be intermediate in behaviour. Macey[1] has shown that if a plastic clay mass is continuously subjected to alternating tensional and compressive stresses, a looseness of structure develops and eventually small stresses produce large deformations. This phenomenon is attributed to the gradual orientation of mineral flakes and has been termed *back-lash.* When the clay contains few small particles, the mass exhibits *shear hardening.* The application of alternating forces causes orientation which permits slipping in a direction parallel to the mineral plates; after some time, however, dilatant conditions are set up and a hardening of the structure takes place. This has an important bearing on the shaping of clay by extrusion processes, e.g. wire-cut bricks and salt-glazed pipes.

Effect of Previous History. The treatment (either natural or artificial) which a clay has undergone has a large effect on its plasticity. The primary clays are only feebly plastic because they have not been subjected to conditions which would develop plasticity. Many secondary clays, on the contrary, are highly plastic in a natural state but the degree to which they have been compressed has an important bearing. When a clay is air-dried it loses its plasticity, but this is readily restored when the clay is mixed with a suitable proportion of water. If a clay is dried by artificial heat, or, sometimes if it is exposed to tropical conditions, its plasticity after mixing with water is low because of a partial decomposition of the clay or a partial conversion of the colloidal matter into an inert material. For the same reason, clays which have been heated for a long time at any temperature above 500° C. lose some

[1] Macey, H. H., *Trans. Brit. Ceram. Soc.*, 47, 259, 1948.

of their plasticity permanently. Any treatment which alters the colloidal matter affects its plasticity. All these changes are commonly summarised in the statement that the previous history of a clay has an important influence on its plasticity.

Potential and Actual Plasticity

In considering the plasticity of clays it is necessary to distinguish the *actual* (or *active*) from the *potential* (or *dormant*) plasticity. A dry clay is non-plastic, but on being mixed with a suitable quantity of water it will become plastic, the amount of plasticity depending on the previous history of the clay.

The **potential plasticity** of a clay is an inherent property and is the maximum plasticity which can be developed by suitable treatment of the clay. The **actual plasticity**, on the contrary, is the plasticity at the moment of inspection; it is reduced by drying and may be increased (if this is possible) by mixing the clay with water, or by grinding the clay or by subjecting it to any treatment which will increase the proportion of suitable colloidal matter.

To increase the plasticity of a clay until it equals the maximum potential plasticity is extremely difficult and apparently requires a far longer time than is practicable when artificial means are employed.

It is not generally appreciated that clays differ in the rate at which they absorb water to become plastic. This property may be termed the *imbibitional character* of the clay and it affects many aspects of mixing and forming. Some clays readily take up water whilst others do so only slowly. This property is not fully understood, but the following factors contribute:

(*a*) The particle size of the material and the capillary nature of the compacted mass.

(*b*) The hygroscopic nature of the minerals present.

(*c*) The presence of soluble salts which may change the surface tension of the water.

(*d*) The presence or otherwise of organic constituents in the clay which retard wetting by water and hence its easy penetration.

The rate of imbibition is increased by heating either the water or the clay body.

In the clay-working industries it is important to develop the plasticity to a suitable extent, this being part of the 'preparation' of the clay for use.

Increase and Reduction of Plasticity. Many natural clays are potentially, rather than actually, plastic; their plasticity may, therefore, be increased, up to a limit, by mixing them with a suitable proportion of water or in other ways. Some clays, on the contrary, are too plastic or too sticky to be used in their natural state, and must, therefore, be treated in some way to reduce their plasticity.

Increasing Plasticity. The investigation of methods of increasing the plasticity of clays and ceramic materials is beset with many difficulties; among others is the difficulty of distinguishing between the actual plasticity of a mass and the potential plasticity which may be developed by various means.

The plasticity of a ceramic material may be increased:

(*a*) By the addition or removal of a suitable proportion of water.

(*b*) By the more thorough incorporation of the water and plastic materials present.

(*c*) By the removal of some of the non-plastic material present (usually by converting the material into a slip, allowing some of the non-plastic matter to 'settle out' and then removing the excess of water from the decanted slip).

(*d*) By the addition of a flocculating agent to a slip. This may be achieved in several ways such as:

(i) Using a substance which on fermentation or other decomposition, produces a free acid, e.g. peat, cellulose extract, etc. The decomposition of carbonaceous constituents contributes to the increase in plasticity of some clays when they are weathered.
(ii) Adding a weak acid such as humic, tannic or acetic acid.
(iii) Incorporating other flocculating or coagulating agents (see p. 472.)

(*e*) By the addition of a substance of a colloidal nature. This would include other clays or gels or a substance which could be converted into a *gel* by subsequent treatment with an acid or on prolonged storage (e.g. colloidal silica, alumina, soluble iron salts and a variety of organic complexes).

(*f*) By an electrical treatment which will reverse the electrical charge on the material.

(*g*) By grinding and pugging with or without water, which thereby reduces the particle size and increases the amount of hydrolysis.

(*h*) By developing thixotropic conditions in the clay body.

(*i*) By drying and grinding the clay and using oil instead of water. This is sometimes done when making crucibles and small articles which are shaped by mechanical pressure.

(*j*) By removing air from the clay paste by vacuum treatment. Such a process, in addition to developing increased plasticity also results in a better textured body.

(*k*) By the process known as *ageing* or *souring*. This is a change which takes place in clay pastes and is dependent to some extent on their colloidal nature. When a clay paste is kept moist and cool for a sufficiently long time, its actual plasticity is increased, the water in it is more uniformly distributed, and the homogeneity and working properties of the clay are greatly improved. Hence, where financial considerations permit, clay pastes are greatly improved if they are allowed to 'sour'. In order that the souring may be effective, an ample quantity of water must be present, and to ensure this, and to prevent surface drying, the clay is usually covered with wet cloths during the souring period, the cloths being renewed or rewetted at frequent intervals. Sometimes the clay is placed in pits or in sumps and covered with water for the required period. The surplus water is afterwards run off and the paste is dug out. Wedgwood and some other famous pottery manufacturers have aged the moist clay by keeping it stored in air-tight boxes for several years.

In modern clay practice, it is usual to sour clays in silos or bunkers for periods of time up to 48 hours. Clays which are received in a fairly dry condition from the pit or quarry benefit from this treatment, which is usually carried out immediately prior to the forming process.

After grinding and blending have been completed, a proportion of water is added to the clay which is then transferred to the holding silo. It is frequently observed that the plasticity of the clay, as measured by deformation tests, increases although the water content remains unchanged. Tests of this nature, taken at suitable time intervals, will establish the most profitable souring time required. Steam or hot water accelerates the process of souring to a considerable extent.

The effect of ageing or souring is probably chiefly due to the decomposition of the organic matter present, with the formation of dilute acids which coagulate the fine grains of clay and thus increase the plasticity. A similar effect may sometimes be obtained by adding acid direct to a fresh clay; but the results are not quite the same, as when decomposition occurs *in situ* the acid is more uniformly distributed, whereas when added artificially it must be allowed a considerable time before it is thoroughly distributed through the mass.

The presence of an excess of free alkali in clays limits or may even prevent their improvement by souring. Any such excess should be neutralised by the addition of a suitable acid before any increase in the plasticity of the clay can occur. Seger suggested the addition of a suitable proportion of acetic acid to clays before souring, whilst the addition of old vinegar and of wine was, at one time, a common practice amongst some pottery manufacturers.

It has been suggested that the effect of bacteria or moulds present in a clay has an important effect in the souring,[1] but the extent of their influence is not definitely known. In order to increase the proportion of fermented matter some manufacturers have added sugar or honey to their pottery mixtures before souring.

The increase in plasticity of many clays on ageing is undoubtedly a purely physical phenomenon in which many factors may play a part; a more uniform distribution of water takes place and orientation of particles develops. If a clay paste is allowed to stand the charge distribution on the particles becomes more evenly distributed with a consequent increase in plasticity.

Plasticity may be reduced by the converse of the methods used to increase it. One of the most efficient methods is to heat the clays to about 200° C. Suitable heat treatment can be employed to reduce plasticity to any desired value and it can virtually be completely destroyed by heat treatment at about 500° C.

Pseudo-Plasticity. A modified form of plasticity may be developed in some non-plastic materials by the addition of a suitable viscous fluid, such as oil, or, preferably, an aqueous solution of a gelatinous or gummy material, such as hot starch, dextrin, rubber, sumach, inulin, caramel, gelatin, gum, glycogen, various ferments and enzymes and other colloidal materials.

Oiliness is sometimes regarded as a form of pseudo-plasticity, though it is really distinct from true plasticity and from stickiness. It is often due to the presence of oils or oleaginous substances, and, where necessary, may usually be removed by extracting these materials with benzene or other suitable solvent. The oleaginous nature of some clays is destroyed when they are boiled with a solution of alkali which forms a soap; the latter increases the pseudo-plasticity of the clay. The 'oiliness' of some clays is more readily observed when an attempt is made to mix them with water.

[1] Komuro-Hideo, *J. Kyôto Women's Univ.*, 1, 1, 1952.

Measurement of Plasticity

As there is no generally accepted definition of plasticity, no one method of measurement can be fully descriptive. The measurement of the change in shape of a mass of a certain size when subjected to a definite force is insufficient, as also is the rate of such change.

Many methods have been proposed for the measurement of so-called plasticity. They can be divided into two groups; those in the first determine a property which is indirectly related to plasticity, and those in the second include attempts at a more direct approach.

Indirect methods include measurements of:

(*a*) The *proportion of water required to develop a paste* of recognisable consistency, as suggested by Atterberg and others.

Atterberg[1] has summarised some of the various factors which have been proposed as a basis for the measurement of plasticity, viz.:

(i) The upper limit of fluidity, at which the clay flows almost like water.

(ii) The lower limit of fluidity, at which two portions of the material will not flow together and unite when jerked and shaken under standard conditions.

(iii) The adhesion limit, at which clay ceases to stick to other objects, such as the hands when the clay is squeezed.

(iv) The 'rolling-out limit', at which the clay ceases to be capable of being rolled out into thin cylinders or threads.

(v) The 'cohesion limit' at which the grains cease to stick to one another.

Atterberg has stated that the range of suitable consistency for working lies between (ii) and (iv). He found that the moisture content of the flow-limit and rolling-out limit coincides in non-plastic loams but, as the plasticity of a clay increases, these two limits become more and more separated and so constitute the working range of plasticity.

Atterberg Number. From a consideration of the foregoing, Atterberg suggested that the plasticity of a clay mixture may be expressed numerically by deducting the percentage of water corresponding to the rollling-out limit from the percentage of water corresponding to the flow-limit. Thus, if 40 per cent of water is required to cause the material to flow and 20 per cent is necessary to enable it to be rolled out into cylinders, its plasticity number (*Atterberg Number*) is $40 - 20 = 20$.

He proposed to group clays into four classes:

	Atterberg Number
Class I	17–27
Class II	5–15
Class III	4–7
Class IV	0–1

This method attributes to some substances, such as barium sulphate, a high plasticity, though they are not generally considered to be highly plastic. Atterberg's

[1] Atterberg, A., *Inst. Mitt. Bodens*, 1, 10, 1911 and 2, 149, 1912.

numbers also correspond very closely to the binding power of clays (p. 514), rather than to their true plasticity, and, as R. Rieke[1] has pointed out, the Atterberg's ranges are not always reliable, the plasticity being also due to properties other than those he has considered. The Atterberg number does not fully describe plasticity but defines only the range of water content over which the property exists.

In addition to being indirect, Atterberg's method leaves a large margin for the personal element and is, therefore, undesirable, though it is convenient and has been largely used for comparative tests. It is based on the idea that the plasticity of a material is proportional to the amount of water present, but as it is not the water alone, but the nature of the viscous fluid (of which the water is only one constituent), and the nature of the solid particles as well, which determine the plasticity of a mass, Atterberg's method is fundamentally unsound, but it is a simple and convenient determination for some purposes, e.g. assessing the properties of soils.

(*b*) The *proportion of colloidal material present.* These methods assume that plasticity is dependent only on the amount of colloidal material present and ignore all other variables.

(*c*) The *viscosity* of the mass.

(*d*) The *penetrability* of a 'needle' or steel sphere; these indicate only the softness of the clay mass.

In the extrusion of clay columns, the measurement of penetrability can be a useful control indicator. It can be carried out automatically and used as a measure of the correct moisture content to be included, or a hand instrument can be employed which measures the depth of penetration into the column by a spring-loaded plunger.

(*e*) The *measurement of associated properties* is commonly used to indicate plasticity but the results must be interpreted with caution. The most widely employed property is the dried strength of bars of the clay under test. This is a complex property depending on packing characteristics and the nature of the colloidal fraction, but if the green strength of identical bars, but formed at different moisture contents, is compared, the maximum strength value often indicates the moisture content at optimum plasticity. If several samples are to be compared, their relative plasticities may be in the same order as the maximum strength values.

Direct methods of measuring the plasticity of a material are necessarily of a mechanical nature; they involve three variables—force, shape, and time.

(*a*) *Subjective Methods.* One of the earliest of these methods, which, in spite of its crudity, is still largely used and is very convenient, consists in pressing a small piece of the material in the hand. A paste of a satisfactory consistency for moulding should not adhere to the fingers or palm, yet it should retain the marks produced on it by the lines of the hand. An excess of water will cause stickiness, whilst too little water will produce a mass which will not properly take a clear impress of the fine lines of the hand. This method of determining the plasticity of materials by their 'feel' is satisfactorily employed by those who are continually handling clay, but it is unsatisfactory as a scientific test, as it introduces too much of the personal element, and the essential factors of pressure, changes of shape, and duration of pressure are

[1] Rieke, R., *Sprechsaal*, **44**, 597, 1911.

not accurately defined. The same is true of the 'rolling-out test', in which balls of clay of a pre-arranged size or weight are rolled into strips of convenient size and then wound into a coil. The less plastic clays soon crack and may thus be distinguished from the more plastic ones. Strictly, this test depends more on the binding power than on the true plasticity of the clay. It may also be regarded as a measure of the consistency of a clay-paste.

(*b*) *Extrusion methods*, in which the clay-paste is forced by pressure through an aperture and the quantity extruded is measured, are not really direct methods, for what they measure is the viscosity or flow and not the extent to which the shape of the mass has been changed, and this is an essential in the measurement of plasticity. However, a series of graphs, such as the one shown in Fig. VIII.25 may be very useful in comparing the yield points and secondary flow of several samples. A similar series compiled at different water contents, may also be informative.

(*c*) *Extension Methods*. Several investigators have suggested that either the amount or the rate of extensibility or the tensile strength of a paste are proportional to its plasticity. None of these is alone sufficient, but a useful and consistent figure may be obtained by multiplying the percentage of extensibility by the tensile strength of the material. Macey has pointed out the limitations which arise with this method as the result of shear hardening and induced dilatancy.

Nevertheless the extrusion and extension properties of plastic bodies have such an important bearing on their behaviour in use that although tests of this kind may not give a true measure of plasticity, their importance should not be underestimated.

(*d*) *Compressibility methods* have been suggested which are based on the amount of pressure required to effect a definite change of shape or to deform a sample of a given size to such an extent that it cracks.

A method frequently employed is to apply a fixed load to a standard test piece and to measure the amount of deformation. The load may then be increased and the test continued until total failure occurs.

A simple means of comparing plasticities devised by McDowell[1] consists of a vertical rod mounted in guides so as to slide very easily up and down. The lower end of the rod has a flat plate attached and the upper end a similar, but larger, plate to support weights. A block of plastic clay precisely 2-inch cube is placed on the table, whilst the weighted rod is raised and then lowered gently on to the block of clay. After 30 seconds the height of the block is measured. The reduction in height is regarded as a measure of the plasticity. A weight of 2000 gm. is convenient. McDowell claims that this device gives much better results than the 'feel' of the paste and that more uniform manufacturing conditions can be obtained by it, particularly for clay-paste used for pressing.

Pfefferkorn[2] prefers to drop a known weight through a known height on to a cylinder of moist clay of known dimensions and to measure the deformation. Bowmaker[3] modified the foregoing by allowing a known weight (2 lb.) to fall 6 inches on to a clay cylinder, 3 cm. diam. and 4 cm. high, and measured the additional distance travelled by the weight in compressing the clays.

[1] McDowell, S. J., *J. Amer. Ceram. Soc.*, **11**, 99, 1928.
[2] Pfefferkorn, K., *Sprechsaal*, **57**, 297, 1924 and **58**, 183, 1925.
[3] Bowmaker, *J. Soc. Glass Tech.*, **14**, 330, 1930.

The value of tests based on measuring the plastic deformation under compressive forces has been increasingly recognised in recent years. Such determinations indicate the probable behaviour of plastic masses which are being shaped under pressure and they afford a rapid and reliable means of controlling the consistency of clay pastes. The term *workability* is widely recognised in the U.S.A. as implying the relative ease whereby plastic masses are deformed by pressure and it gives an excellent indication of plastic properties under practical conditions of moulding. The American Society for Testing Materials has adopted a standard procedure for measuring this property and such a method is of value for controlling the shaping or moulding of clay articles.

In the standard test (ASTM. Designation C.181–47) a test piece of 300 gms. is prepared by ramming the clay-water mass into a cylindrical mould of 2 in. diameter. The ramming operation is carried out by using a modified form of the sand rammer recommended by the American Foundrymen's Association.[1] The specimen is subjected to 10 blows on each end from a 14 lb. cylindrical weight and then extruded from the mould by means of a suitable plunger. The length of the test specimen is carefully measured, placed underneath the rammer and given three blows. The percentage deformation in length is the *workability index*.

The workability index is a good guide to the correct consistency of plastic masses for various methods of forming, and is much more rapid and precise than methods based on measuring the grading and the water content.

The following ranges of workability index have been found suitable in the manufacture of fireclay mixes:

Method of Moulding	*Workability Index*
Hand Moulding	35–50
Extrusion	28–34
Machine-made Bricks	15–25
Dry Pressing	5–8

The measurement of the workability index over a small range of moisture contents gives an indication of whether a clay is 'short' or 'fat'. If a change in the workability index from 25–35 is brought about by less than 1·0 per cent of extra water, the clay will be too 'short' for most practical purposes.

Moore[2] has described a plastometer test based on compressive methods which defines true plasticity more accurately. He takes account of the fact that a body of good plasticity should keep its shape at low stress values, but should deform readily once the yield point has been exceeded. A graph of the stress required to produce deformation in small cylindrical specimens is obtained and the plasticity index assessed from the ratio of the stress at 10 per cent strain to that required to produce 50 per cent strain. Values so obtained on different types of clay[3] have given indices in the accepted order of their plasticities, and the method has further advantages in

[1] It is manufactured in Great Britain by J. Ridsdale & Co., Ltd.
[2] Moore, F., *J. Sci. Instru.*, 40, 228, 1963.
[3] Moore, F., *Rheology of Ceramic Systems*, Inst. of Ceramics Textbook Series (McLaren & Sons, London, 1965).

being reasonably independent of moisture content and of the rate of application of stress.

(*e*) *Shear and Torsional Methods.* Many attempts have been made to measure the plasticity of clay bodies using shearing or torsional methods.

The specimen is mounted in a torsion head, which is geared to a motor allowing various speeds of rotation. The stress developed is measured by the movement of an arm located on the specimen and recorded directly by a stylus on a smoked-glass plate. The torque in the specimen is plotted as a function of the angle of twist; a typical result is shown in Fig. VIII.28.[1]

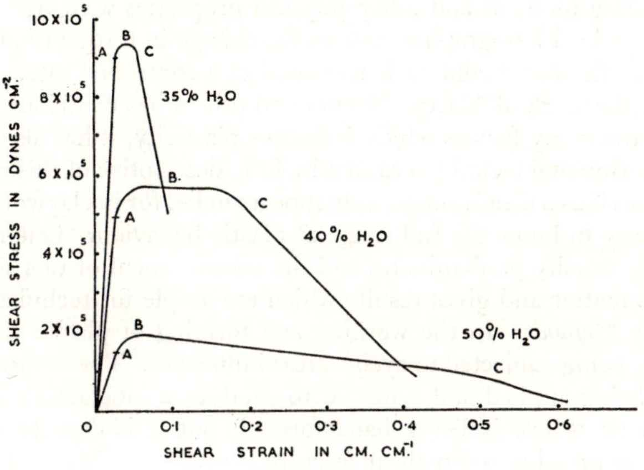

FIG. VIII.28.
THE STRESS/STRAIN RELATIONSHIP IN PLASTIC MASSES
(after Norton)

Three important characteristics of the curve are clearly shown:

A. The yield point of the clay, or the limit of elastic behaviour.
B. The maximum strength of the clay.
C. The point of failure where rupture of the body occurs.

The region between A and C is where the clay is subject to plastic flow and represents the degree of deformation which the body can undergo without breaking. Norton considers that both the *yield point* (which determines the force which the body can withstand without change of shape) and the *maximum extension* or *deformation* should be at a maximum in a plastic body. Thus, if the yield point is low, distortion or slumping may occur after the shape has been formed; a low extensibility indicates that the body does not readily mould, or has a short working range. An increase in yield point invariably results in a reduction in extensibility, but the product of the two usually has a maximum at a particular water content. Norton defines this optimum value as the *maximum workability* point of the clay and Whittaker[2] uses such values as a basis of comparison for the *plasticity* of various clay fractions.

[1] Norton, F. H., *J. Amer. Ceram. Soc.*, 21, 33, 1938.
[2] Whittaker, H., *J. Amer. Ceram. Soc.*, 22, 16, 1939.

Although such a method of measurement gives a useful indication of the working range of clay bodies, there are many factors which influence the results. The time through which the shearing force is applied may influence considerably both the yield point and the extensibility. Thixotropic and dilatant materials often show anomalous behaviour which may give rise to misleading conclusions.

Various other methods have been suggested which are claimed to measure plasticity. Linseis,[1] for example, considers that a useful index is given by recording the tensile breaking strength as a function of the resistance to extrusion ('shear resistance') at different water contents. Other workers confuse binding strength, adhesion, bending moment and other physical properties with true plasticity.

The Brabender Plastograph measures the change in torque required to deform a test sample as the water content is increased at a controlled rate.

Although the work of Macey, Norton and others has contributed much to the knowledge of the many factors which influence plasticity, it has also indicated that one single experimental technique cannot be fully descriptive of this property. This may not be as serious a disadvantage as it appears to be, for in clayworks practice it is seldom necessary to know the full range of plastic behaviour. One aspect or associated property usually predominates and the measurement of this single feature is often a simple matter and gives results which are ample for technical control.

(*f*) *Cyclic Methods*. In the working and forming of plastic clay masses, the body is usually being subjected to cyclic stress influences. The deforming forces are continuously being applied and removed to produce a satisfactory shape. This is obvious when an article is being hand-formed, but it is equally true in screw-extrusion and in pressing when shear fracture is involved. Many of the characteristics of plastic clays are due as much to their behaviour on the removal of stresses as they are to features developing on the application of stresses.

Astbury developed this theme in the 1963 Mellor Memorial Lecture[2] and described his experimental work on cyclic testing at a later date.[3] In this investigation torsional methods were adopted.

A. Hai and the author, in work as yet unpublished, have investigated similar relationships under extrusion conditions where much higher stress values can be employed. The elastic recovery in the enclosed space of the extrusion barrel is also more precisely recorded. A clay body can be investigated at different moisture contents and at various stress values cycled between maximum and minimum values. The use of an Instron Testing Machine[4] permits cycling over a wide range of rates to be studied throughout the elastic and plastic ranges.

Typical results on a clay known to extrude satisfactorily in practice are shown in Fig. VIII.29. Parameters corresponding to elastic and plastic flow can be derived and related to water content. By assessing a wide range of clays, the extrusion characteristics of which are known, it has been possible to define acceptance limits for clays and in some cases to predict modifications to both bodies and machinery to improve extrusion characteristics.

[1] Linseis, M., *Sprechsaal*, 84, 259, 1951.
[2] Astbury, N. F., *Trans. Brit. Ceram. Soc.*, 62, 1, 1963.
[3] Astbury, N. F., Moore, F. and Lockett, J. A., *Trans. Brit. Ceram. Soc.*, 65, 435, 1966.
[4] Manufactured by Instron Ltd., High Wycombe, Bucks.

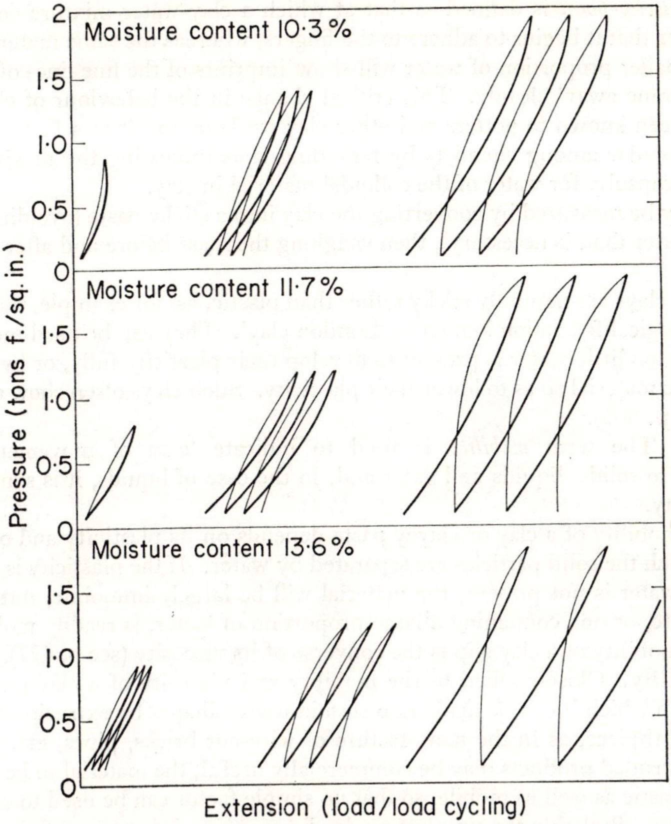

FIG. VIII.29. LOAD/LOAD CYCLING IN THE STUDY OF ELASTIC AND PLASTIC FLOW

PROPERTIES ASSOCIATED WITH PLASTICITY

THE many misconceptions of the nature of plasticity have led to numerous unsound conclusions, and many investigations of the effects of various substances on plasticity have been rendered nugatory because the results have not been correlated to plasticity, but to some other property mistaken for it. Some of these mistakes have been described in the section on the Measurement of Plasticity, whilst others are referred to in the following pages.

Stickiness. When even a small proportion of water in excess of that required to develop the maximum plasticity is added to a clay-paste, the mass adheres to the fingers and becomes 'sticky'. The proportion of water required to create this condition in a clay is a measurement of definite physical significance and is one of the important 'constants' of clays. Its determination is easy to those accustomed to working clay-pastes, but not to others, and for this reason many chemists and physicists have failed to recognise its value.

The *sticky-point* is defined as that at which a clay-water mixture contains so much water that it begins to adhere to the fingers, whereas the same material with a slightly smaller proportion of water will show imprints of the fine lines of the skin, but will 'come away' cleanly. This critical change in the behaviour of clay-pastes has long been known to potters and other clay workers, but it was F. Hardy[1] who made it popular among chemists by regarding it as indicating the maximum imbibitional capacity for water of the colloidal material in clay.

It may be measured by converting the clay into a sticky paste (avoiding the use of more water than is necessary), then weighing the mass before and after drying at 110° C.

Some clays are naturally sticky rather than plastic, as, for example, those found in the geological formation known as 'London clay'. They can be used industrially only when too little water is present to develop their plasticity fully, or by adding a non-plastic material so as to lower their plasticity. Such clays often show a pseudo-plasticity (p. 506).

Mobility. The term *mobility* is used to indicate 'ease of movement'; it is applicable to solids, liquids and gases and, in the case of liquids, it is synonymous with fluidity.

The mobility of a clay or clayey paste depends on its plasticity and on the extent to which the solid particles are separated by water. If the plasticity is low, or if sufficient water is not present, the material will be largely immobile, but a highly plastic paste, or one containing a large proportion of water, is readily mobile.

The mobility of a clay slip is the converse of its viscosity (see p. 477).

Extrudability. Closely allied to the mobility and viscosity of a clay paste is the facility with which it can be made into certain useful shapes by extrusion through a die or mouthpiece, as in the manufacture of wire-cut bricks, pipes, etc. In order that the extruded products may be commercially useful, the material to be extruded must be plastic as well as mobile, so that no simple factor can be used to express its extrudability. Probably the simplest mode of expression is in terms of viscosity, i.e. in the ratio between the quantity extruded in unit time and the pressure required to extrude; but the size and shape of the aperture have so important an influence that it is almost impossible to reduce the idea of extrudability to simple terms. In practice, the relative extrudability of various clay pastes must be expressed in terms of viscosity, the shape and size of the aperture being strictly defined. Thus, a paste which is quite satisfactory for the production of bricks by extrusion may be almost useless for the production of pipes or blocks of special shape by the same process.

Binding Power. The binding power of a clay is measured by the amount of non-plastic material which can be incorporated without the plasticity of the mixture being reduced below that which is necessary for good modelling or moulding properties.

The binding power of a clay must not be confused with its plasticity, however, as these are quite distinct properties, though, as both are dependent on the nature and proportion of the colloidal gel matter present, they are, to some extent, related to one another. Thus some highly plastic clays have a high binding power, but other clays have a high plasticity and a comparatively low binding power.

[1] Hardy, F., *J. Agric. Sci.*, 18, 740, 1928.

The binding power of a clay is analogous to the property possessed by various colloidal gels of retaining particles in their mass in a state of suspension. It appears to be due almost wholly to the viscous liquid constituent of the plastic paste, but the amount of non-plastic material which can be incorporated with any given clay is limited by the amount of such material already present. For this reason, any treatment which increases the proportion of viscous material in the mass also increases its binding power.

The chief effect of water on the binding power of a clay, is to increase the volume of the colloidal gel matter present and, therefore, to enable this matter to cover a large surface of inert material.

The binding strength of clays is dependent on many factors although the surface area, or amount of colloidal fraction, is probably the most important. Thus montmorillonite clays are superior to those of kaolinite type as shown in Table VIII.IV.

TABLE VIII—IV. THE BONDING STRENGTH OF CLAYS

Clay type	*Cross breaking strength (gms. per sq. cm.)*
Kaolinite	200–250
Fireclays	150–450
Ball clays	400–580
Bond clays	600–800
Montmorillonite	1500–2500

Clays, such as ball clays, with a high binding power are very valuable in the manufacture of articles containing, by necessity, a large proportion of non-plastic grains, as grog bricks, porcelain, earthenware and various refractory articles of special kinds. Bonding clays (such as bentonites) which, in general, give a higher wet and dry strength than ball clays, are used in the manufacture of sand moulds for steel castings.

The same property of binding power is involved when clays are used as agents of agglomeration in pelletising practice. In this method finely-divided material is induced to bond together to form nodules or pellets, which are strong enough to hold together. Many materials are insufficiently strong within themselves to pelletise of their own accord, but the addition of a small percentage of plastic clay is of great advantage.

The binding power of clays may be manifested in both the wet (or *green*) state and in the dried material. It is usually measured by determining the breaking (either compressive or tensile) strength of standard-sized blocks made up of various small proportions of the clay with a standard sand, after mixing with various amounts of water.

The *green strength* of clay/sand mixtures resembles plasticity in that a maximum value is recorded at a certain moisture content and, as a result, some writers confuse the two properties and define plasticity in terms of maximum wet strength. For the reasons given previously this is incorrect because binding strength is but one aspect of plasticity.

Grim and Cuthbert[1] have investigated the bonding action of clays in wet sand mixtures. They conclude that the water molecules are associated with the clay mineral particles which, in turn, coat the surface of the sand grains. Furthermore, they claim that the maximum green strength is developed when the adsorbed water molecules are in a rigid condition induced by the surface structure of the clay minerals themselves.

Ideally, a thin film of clay is thought to cover each sand grain and when maximum green compression strength is developed a definite layer of water is associated with the surface of the clay mineral flakes. With montmorillonite-type clays, the water film thickness at maximum strength is about three molecules or 8 Å thick and is fairly uniform throughout the mass, but clays of the kaolin or illite-type are believed to form irregular coatings made up of large clay aggregates formed around the sand grains. In both cases, the clay mineral films run into each other and form *wedge-blocks* which unite adjacent sand particles.

In the 'wedge-block' theory the water molecules are considered to have a rigid structure which is continuous throughout the mass. Grim and Cuthbert favour the concept of a type of ice structure being developed. Although this theory adequately accounts for the concept of green strength in clay/sand mixtures, several objections have been advanced. Macey, for example, has pointed out that Grim and Cuthbert assume that all the water present is associated with the clay and the possibility of some filling the capillaries has been ignored. It may further be argued that under the conditions of mixing, water molecules would be unlikely to penetrate between all montmorillonite flakes and give a uniform layer. The concept of a virtually continuous clay film with wedge-block linkages is, however, most valuable and explains many of the features of this type of bonding.

Organic Bonds for Clays, etc. Organic bonds or *binders* have been added to clays since remote times—gum arabic is added to porcelain and bone-china bodies to increase the plasticity and facilitate the production of delicate porcelain baskets, flowers, etc.

In the manufacture of refractory bricks made largely of non-plastic materials, about 2 per cent of sulphite lye (a by-product of the paper industry) is often used as a binder.

The most important binders or plasticisers are:

(*a*) Flour and starches.

(*b*) Gums—gum arabic, gum tragacanth, gum acacia, dextrin.

(*c*) Alginates—made from seaweed—effective but too expensive except for special wares.

(*d*) Polyvinyl and other alcohols and water-soluble resins.

(*e*) Silicon esters (silicones) which are organic compounds of silica.

(*f*) Methyl ethyl cellulose, sodium carbomethyl cellulose and other cellulose compounds.

(*g*) Casein-glue, glue and gelatin—chiefly used for binding glazes.

(*h*) Water soluble paraffin waxes—chiefly used for small steatite goods.

(*i*) Waste oil-residues, tar, pitch, etc.—used chiefly for firebricks of special shapes.

[1] Grim, R. E. and Cuthbert, F. L., *Illinois State Geol. Survey*, No. 102, 1945.

All these binders create a *pseudo-plasticity* and they are quite distinct in their nature and action from bond clays and plastic clays.

INDUSTRIAL APPLICATIONS OF CLAY/WATER SYSTEMS

1. Slips, Slurries and Suspensions

These are prepared by mixing the solid materials with a suitable proportion of water so that the body has fluid properties and will flow under its own weight. They are used for various purposes, the most important of which are:

(*a*) To ensure a more thorough mixing of ceramic materials than is possible when they are mixed in a dry or plastic state.

(*b*) To enable the materials to be applied more easily than would otherwise be the case, as in coating ware with an engobe or glaze or in decorating ware by painting.

(*c*) To enable articles to be made by the process known as casting (p. 520).

(*d*) To effect the purification of clays and other ceramic materials.

(*e*) To convey fine-particled material in a convenient manner.

Homogeneous mixtures of ceramic materials are best prepared by mixing the dry substances in suitable proportions, soaking the mixture in water until it is thoroughly slaked, and then stirring mechanically until it is of uniform consistency throughout. An alternative method, which is often preferred, consists in making each solid constituent into a separate slip, which is then passed through a sieve in order to separate undesirably coarse material. The consistency of each slip is then adjusted so that it has a definite 'weight per pint' (p. 431). The slips are afterwards mixed in suitable proportions when required. This method ensures the better soaking of the solid materials and the removal of coarse matter before the proportions of the various materials are measured and so leads to a superior product. If the mixed materials are to be converted into the form of a paste or powder, the excess of water may be removed by means of a filter-press or by sedimentation, or by evaporating the water by means of heat, or by a combination of these methods. When a slip contains only a small proportion of solid matter, a large part of the water may sometimes be removed by allowing the slip to 'settle' for some time and then running off the clear supernatant water; the objection to this method is that the heavier and denser particles of solid matter may settle first and so cause an undesirable amount of 'un-mixing'.

This method of preparing pastes is largely used in the manufacture of fine earthenware, porcelain, etc., as it ensures a product of much greater homogeneity than any other method. Its chief drawbacks are its cost and the necessity of increasing the compactness of the paste by a further treatment in a pug-mill or other machine.

Slips for engobes and glazes are prepared by mixing the various solid materials and water as described above, and the slip is then applied direct to the ware by dipping, painting or spraying. The slips are generally made up to a definite weight per pint, as this is a convenient means for measuring the amount of solid matter they contain.

Table VIII.V shows the proportion of solid matter and water in slips in common use in the Potteries.

Adequate *control of the viscosity* of glazes and enamels is very important as it affects the amount retained on the surface of the ware and also the manner in which the surplus slip drains away—too viscous a slip will leave tear-drops or lines and too fluid a slip may drain off to such an extent as to leave too thin a coating on the ware. When glazes and engobes are applied by brushes the correct viscosity is particularly important as a means of preventing bubbles being formed during the brushing: these may later form pinholes in the ware.

TABLE VIII—V. SLIPS USED BY POTTERS

	Standard slop weight oz. per pint	*Weight of solid per standard pint*	*Weight of water per standard pint*	*Solid matter per cent*	*Water per cent*
		oz.	oz.		
Earthenware body	35	24·20	10·80	69·15	30·85
(Casting slip)	36	25·81	10·19	71·68	28·32
	37	27·43	9·57	74·14	25·86
Ball clay (mixed)	24	6·70	17·30	27·91	72·09
China clay (mixed)	26	9·75	16·25	37·50	62·50
Flint slop	31	18·16	12·84	58·58	41·42
Cornish stone slop	31	17·57	13·43	56·69	43·31
Earthenware glaze	33	19·96	13·04	60·48	39·52

One means of controlling the viscosity is to add a suitable proportion of a colloidal material. Clay is very suitable, but it may increase the temperature at which the engobe, glaze or enamel matures and the necessary adjustments may be troublesome. Bentonite is increasingly used for this purpose, the viscosity being controlled by the subsequent addition of a solution of calcium chloride. Instead of clay, 1–5 per cent of starch, dextrin or gum (previously dissolved in hot water) may be added to the slip.

Thixotropy (see p. 473) is an important consideration in glaze and engobe slips. Suspensions which have this property thicken or gel on standing, but when they are stirred free-flowing conditions are restored. Some degree of thixotropy is an advantage in slips especially when heavy or large particles are present which would otherwise tend to settle out on standing. The structure which develops within such suspensions is strong enough to prevent sedimentation yet, when stirred, the slip may be readily transported through pipes or poured into moulds.

Beech and Francis[1] have demonstrated that a thixotropic glaze does not 'run' from sharp edges of ware and yet is sufficiently free-flowing to penetrate all recesses.

Thixotropic bodies are liable to develop faults, however, which arise principally from the large amount of contained water, which leaves air spaces on drying out; 'pin-holing' and similar faults are, in some cases, attributable to this reason. The sedimentation density of slips is drastically reduced when thixotropy is present.

To produce the most satisfactory slip from suitable raw materials, a compromise

[1] Beech, D. G. and Francis, M., *Trans. Brit. Ceram. Soc.*, 44, 25, 1945.

in properties is usually necessary. Ideally, a slip should contain the maximum amount of solid matter consistent with a good flowability. In attaining this, the thixotropy invariably reaches unworkable proportions—the slip 'solidifies' if not kept constantly stirred, draining of moulds becomes difficult and the cast body shrinks excessively on drying.

The best compromise between a high slip density and a low thixotropy is usually secured in practice by using a mixture of sodium carbonate and sodium silicate as deflocculant; the former is an excellent dispersant whilst the latter inhibits thixotropy.

The most suitable amounts and ratio of the two electrolytes can be determined only by experimentation.

The consistency of a slip is dependent on numerous factors, although the amount of colloidal material present is the chief determinant. Consequently the viscosity will be related to the degree of flocculation. Should the required slip be too viscous for pouring, the addition of deflocculants or electrolytes may be of value; on the other hand an increase in viscosity may be effected by adding small quantities of flocculants such as hydrochloric acid, or a salt in which the acid radicle predominates, e.g. zinc chloride, stannic chloride, aluminium chloride or calcium chloride.

Many factors influence the viscosity or 'flowability' of clay slips, and every clay or mixture has its own characteristics and the correct conditions for a suitable slip can be found only by experience. Many of the problems in making-up a clay slip are related to the ease of dispersion of the clays being used—a factor which is often overlooked. The dispersion of prepared china clay and ball clays is a relatively simple matter and provided that slips containing these materials are adequately blunged, the correct consistency is easy to secure.

The principal suppliers of china and ball clays exercise a remarkable control over their products for the industries dependent on casting slips. Particle-size distributions and added components for deflocculation purposes are maintained precisely constant and guaranteed to give absolute reproducibility of slip performance.

In the slip-casting of coarse earthenware and stoneware bodies, however, shaley clays are frequently employed and the breakdown of these materials is often difficult and seldom complete after long periods of blunging. The degree of weathering of such clays is of great importance and casting slips made from materials which have had different treatments, may vary considerably in consistency and other properties.

To secure adequate control over a casting slip, the clays must be at maximum dispersion which may be secured by:

(i) Using only well-weathered clays.

(ii) Grinding the clays and other raw materials as finely as possible prior to use.

(iii) Blunging vigorously and for a long period. Hot water may be used with advantage in the dispersion of some clays.

(iv) Ageing the suspension before use. Some firms allow their slips to stand for a week or more.

It is well known that some slips are considerably improved by additions of weak acids and for this reason humic or tannic acid is often added in small amounts.

Shaley clays frequently break down more readily in an acid solution and their value in casting slips can be enhanced by blunging them, prior to deflocculation, in water to which has been added a small amount (about 0·1 per cent is usually sufficient) of sulphuric or hydrochloric acid.

The dispersion of a clay and the consistency of a slip may be improved by passing it through a fine sieve or lawn after blunging.

The *slip-casting process* (Fig. VIII.30) consists in pouring a suspension of suitable composition and concentration into a preshaped plaster mould, the highly porous nature of which removes water from the slip mainly by capillary action. In the neighbourhood of the plaster wall the suspension becomes locally concentrated and a type of skin formation develops. It is not unlikely that flocculation of the suspension occurs at the surface of the mould due to the exchange of soluble

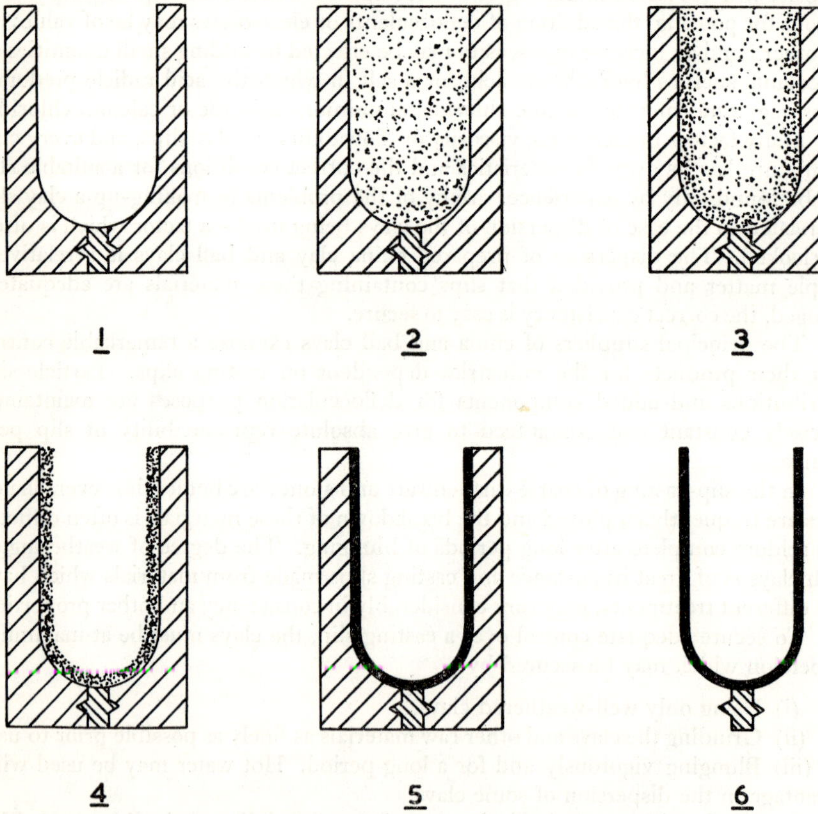

FIG. VIII.30. STAGES IN THE SLIP CASTING PROCESS

1. The prepared plaster mould
2. The mould filled with slip
3. The formation of a 'skin'
4. The 'skin' remaining after emptying the mould
5. The dried 'skin'
6. The article removed from the mould

calcium ions which would considerably assist localised flocculation in a dispersed clay.

The longer the slip is left in the mould the more water is removed and the 'skin' or compacted layer becomes thicker. The process can be stopped at any time by pouring off the excess liquid, or alternatively, more slip may be added to compensate for the water removed, until such time as the complete space within the plaster mould is filled with a solid concentrate. The former process is termed *drain casting* whilst the latter is known as *solid casting*; both methods are used although the former is more frequently employed. Provided that a suitable suspension is prepared, slip casting requires very little skill and yet by its use, articles of complex shape and with extremely thin walls may be produced. It is, therefore, a very valuable method for the manufacture of articles of fine earthenware, porcelain, etc. The method is finding increased application in many ceramic processes where a homogeneous material is required from fine-grained components. Similarly, large thin-walled objects such as crucibles, retorts and sanitary ware are often made by these methods.

Refractory tubular ware is made by spinning a slip inside a cylinder of the required diameter. The suspended particles sediment round the walls by virtue of the centrifugal force and the excess water may be removed by passing a current of warm air through the inside of the mould.

In the drain-casting process, the time required for a slip to remain in the mould before the surplus is poured off varies according to its nature. A slip made without any added electrolyte may require several days, but if a suitable amount of alkali is added the time is considerably reduced. If ware with very thick walls is made by this process, a long time is essential even in the presence of an electrolyte. Thus, in the manufacture of gas retorts, a period of fifteen to twenty hours may be necessary. When a vacuum mould is used much less time is required, as the water is absorbed more rapidly by the mould.

The *rate of 'skin-formation'* is related to the properties of the suspension and many casting difficulties arise through lack of careful control. The sedimentation volume and the solid concentration of the 'skin' are subject to considerable variation, due, in the main, to suspension conditions. Very little work has been done on this perplexing problem but one cause of poor casting quality is structure formation or advanced thixotropy which develops as the suspension begins to concentrate. This can be minimised by ensuring that the electrolyte added to increase fluidity is not present in excess. In other cases, the initial 'skin' is so dense that water cannot readily penetrate it and the increase in layer thickness is exceptionally slow.

The Mechanism of Slip Casting. Although there has been much speculation as to the factors which govern the formation of a 'skin' when a slip is cast in a plaster mould, there is very little reliable information whereby a safe prediction can be made.

It is now generally accepted that the principal mechanism in 'skin' formation is the simple physical reduction in the water content of the slip near to its contact surface with the plaster. The rate of development of the rigid layer proceeds at a uniform rate which suggests that the rate of removal of water is the all-important factor.

Hind[1] demonstrated that the solid layer during casting increased in thickness at a rate which was dependent on the following factors (*a*) the permeability to water of the cast layer or, in other words, its filtration rate, (*b*) the specific gravity of the slip and (*c*) the suction pressure of the mould. If these were maintained constant, Hind showed that the thickness of the solid layer could be related to the square root of the casting time.

Carman[2] applied the Kozeny equation to the slip casting process and assuming that the mechanism is simply one of dewatering he deduced the equation

$$\frac{L^2}{T} = \frac{2PE^3{}_g}{5S_p{}^2\eta\,(y-1)\,(1-E)^2}$$

where L is the thickness developed in time T, P is the suction pressure, E is the percentage of voids in the cast; η is the viscosity of the fluid; S_p is the specific surface area of the solids and y is the volume concentration of the slip.

The filtration rate of the solid layer is governed principally by the proportion of fine-particled material in the slip but Kocatopcu[3] has shown that the packing of the particles and the ratio of plastic to non-plastic components are also important.

The nature of the plaster mould is known to influence the casting rate; the finer the capillaries the greater will be the suction pressure and hence the more rapidly will water be withdrawn.

The condition of the plaster is another influencing factor, a dry mould is more effective than one where the initial water content is high.

In preparing a good casting slip and a good product from it, the following are essential:

(i) Carefully selected and prepared clays and non-plastic materials.
(ii) The correct proportions of each component.
(iii) A minimum proportion of water.
(iv) The appropriate type and amount of electrolyte as a deflocculant.
(v) The best consistency in the slip.
(vi) A suitable mode of preparation and an efficient method of control.
(vii) A well-prepared mould (usually made of plaster of Paris).
(viii) A suitable time in the mould.
(ix) A controlled temperature during casting.

To secure the best results in casting, it is necessary to use a *suitable combination of materials* so as to produce a moderately plastic material of sufficient binding power. Slightly plastic materials are preferable to highly plastic ones, as they require less alkali to maintain a suitable fluidity and the articles can be more readily removed from the moulds. Clays are frequently added to fine-grained components which in themselves are not sufficiently plastic to produce a strong cast. *China clay* and *kaolins* produce a very porous product when cast, but they have not sufficient binding power to produce a strong mass. *Ball clays*, on the other hand, have the requisite binding power and increase the fluidity of slips, but give too dense a product, which makes it very difficult to secure the formation of a sufficiently thick layer in the

[1] Hind, S. R., *Trans. Eng. Ceram. Soc.*, 22, 90, 1923.
[2] Carman, P. C., *Trans. Inst. Chem. Eng.* (*London*), 16, 168, 1938.
[3] Kocatopcu, S. S., *J. Amer. Ceram. Soc.*, 29, 99, 1946

mould. Ball clays are deflocculated at a low pH (on the acid side of the neutral point), but have a higher absorption factor, whilst china clays deflocculate at a higher pH (on the alkaline side), but have lower absorption power, i.e. there are opposite effects and the amount of alkali required does not depend on a single factor. For these reasons, the proportion of ball clay should not exceed 20 per cent of the body, and a large excess of china clay should be avoided. Fireclays vary in their properties in accordance with their nature, but are usually intermediate between kaolins and ball clays.

In mixtures of several materials, serious segregation of the components can take place during casting. As water is drawn through the compacted layer to the mould surface, finely-grained components such as clay may be transported and deposited near to the wall of the mould, so that the external layers may have a much greater concentration of such material than those in the interior. The physical properties of the two regions may be so different that differential shrinkage, during drying and firing, may lead to failures.

The *proportion of water* in a slip used for casting should be as small as possible so as to avoid undue shrinkage of the ware. In estimating the amount of water required, it is necessary to use only dry materials or to take into account the moisture contained in the raw materials. The use of *hot* water is sometimes advantageous.

The fluidity of a slip is largely dependent on the degree of deflocculation, hence the addition of a *small amount of electrolyte* may considerably reduce the amount of water required to produce a fluid suspension. The addition of about 0·25 per cent of sodium silicate or of soda ash, or a mixture of these two substances, will increase the mobility of a slip weighing about 36 oz. per pint to that of an untreated slip weighing only about 26 oz. per pint. If no deflocculant is used, most dry clays will require almost an equal weight of water to convert them into a slip, but with a deflocculant good slips may usually be produced, which contain only about 15–25 per cent of water. Non-plastic materials may usually be made into dense slips by the aid of a deflocculant; thus, slips for casting zirconia crucibles may weigh about 30 oz. per pint.

Normally, casting slips are used in a deflocculated condition because this gives a dense cast body of high strength. Such slips tend to have a slow casting rate and the cast shapes are more difficult to cut and fettle. For these reasons, when complex shapes have to be cast, slight under-deflocculation is often useful although under these conditions it is difficult to avoid a poor run-off condition.

Various deflocculants for clay have been described on page 473, but all are not equally suitable. Sodium carbonate or sodium silicate, or a mixture of these two, are usually the most satisfactory, but when sodium carbonate alone is used a high surface tension is produced and the solid particles tend to 'ball up' and cause the entanglement of air bubbles. Sodium silicate alone, on the other hand, tends to produce a 'stringy' slip and it does not hold the clay in suspension so well as sodium carbonate. If only a single electrolyte can be used, sodium carbonate appears to be the best. Sodium hydrate is much more active than sodium carbonate and silicate, but it is not used to any great extent partly because of its corrosiveness. Sodium carbonate (soda) alone produces a slip with a tendency to be thixotropic.

For the reasons given earlier in this chapter, the presence of soluble salts may interfere with slip formation. Such an example is calcium sulphate which is so undesirable that it may be necessary to pretreat the suspension water with a solution of a barium salt to ensure its complete removal.[1]

The proportion of deflocculant which is the most suitable for any given purpose must, of course, be found by trial. The effect of electrolytes on a ceramic slip may be determined by blunging together definite quantities of the clay or other solid material and water with different proportions of various electrolytes for a period of about two hours. The resultant mixtures are allowed to remain at rest for a suitable time and are then examined. The resulting slip which most closely corresponds with what is required is then noted.

The *correct consistency* in a slip is a major factor in promoting good quality castings. If the slip is too thick, it does not flow readily and imperfections may occur especially if the mould is of complicated shape. A thick slip does not drain successfully and forms a loose-textured 'skin' which is uneven and has an excessive drying and firing shrinkage. A slip which is too thin in consistency is slow to form a 'skin' of the required thickness and the coarser material tends to settle, thereby producing an uneven body.

Although not always realised, the *mode of preparation* of a slip influences its casting properties. The clays and other body materials should always be added to the water and not vice versa. It is equally necessary to use all the water required at once and not in successive quantities. If the latter procedure is adopted, 10–25 per cent more water will be needed, and the slip will be of inferior quality. *Air-bubbles* in slips may cause several serious kinds of defects in the goods, including pin-holes and blisters. They may be avoided by care in mixing (blunging) the various materials with water and may be removed by treating the slip under reduced pressure or even by prolonged storage.

In making slips of various materials it is most important that they should be added to the required amount of water in the following order: (*a*) deflocculant (if any); (*b*) clays; (*c*) non-plastic materials. This procedure is necessary because when all the materials are added at once an excessive quantity of water is required.

Control of the casting-slip is usually effected by determining (1) the 'weight per pint' (p. 431) or the specific gravity at which satisfactory results are secured, (ii) the viscosity of the slip, (iii) measuring the rate of 'skin-formation' in the mould, or (iv) thixotropy.

In accurate work is it necessary to determine both the viscosity (p. 477) of the slip and its 'weight per pint' or specific gravity, as these two properties are not necessarily related. If the viscosity is too low, partial settlement of the suspended matter will occur and may spoil the goods.

The *composition and manufacture of the moulds* into which the slip is cast are of great importance. The material most commonly used is plaster of Paris which has the property of combining with water to form needle-shaped crystals of gypsum which interlock and intergrow to form a strong yet porous mass. The reaction may be simply represented as:

[1] When scraps of broken articles are to be included in the casting slip, it is particularly important to remove (or render harmless by adding barium chloride or carbonate) any calcium sulphate derived from the moulds.

$$\underset{\text{plaster of Paris}}{2CaSO_4.H_2O} + 3H_2O = \underset{\text{gypsum}}{2(CaSO_4.2H_2O)}$$

A mould must have the following properties:

1. Sufficient mechanical strength to withstand the weight of added slip.

2. Perfection in construction so as to give faithful reproducibility in the cast ware.

3. A smooth interior surface finish to avoid imperfections and irregularities in the final shape, particularly at the joints.

4. Sufficient porosity to allow for the rapid removal of water yet with a pore size small enough to prevent penetration of the casting-slip itself.

A plaster of Paris mould is prepared by mixing the finely-powdered raw material to a thick slip with water, pouring it around a former, model or pattern of the desired shape and then allowing the paste to set. The slip must be of a consistency suitable for pouring and, at the same time, the concentration must be high to give a dense hard set. Normally plaster of Paris will stiffen rapidly when mixed with water, although the rate may be influenced by the addition of suitable salts. Borax is of value as a retarder, whilst potassium sulphate or Rochelle salt increase the rate of setting, but reduce the overall shrinkage. Borax is a common addition to pottery plaster, for by its use a more concentrated slip can be made up without undue hastening of the setting time.

In practice, concentrations of up to 66 per cent plaster to water are used for making moulds, although greater dilution may be necessary if a complex shape has to be reproduced.

Plaster moulds are used in dentistry both for casting porcelains and for fabricating metal insertions. In the latter case particularly, the large contraction of the metal on cooling is troublesome and it is not possible to reproduce faithfully the required shape by using normal plaster moulds. Some thermally-expanding material is normally added to the plaster before the mould is made which causes an increase of the internal dimensions when molten metal is added. If the mould mixture is adjusted correctly true compensation for the contraction of the metal shape can be made. Both cristobalite and quartz are used for this purpose and the required amount to be added to plaster must be determined experimentally for a particular metal or alloy.

The number of re-entrant angles in the shape to be cast complicate the slip process. Where these occur, the mould must be made in several pieces to facilitate the removal of the finished cast. Thus, the mould for a complex shape may be composed of many parts, each of which has to be placed into position before the clay slip is added.

The *best length of time in the mould* can be found only by trial; too short a time will produce thin-walled and weak ware; an excessively long time is uneconomical and may produce over-thick ware, but otherwise does little harm.

The *temperature during casting* is important as it affects the stability and viscosity of the slip. A temperature range of about 3° C. is permissible but a large range may yield irregular results.

Improving and Purifying Clays. As the particles of clay minerals are much smaller than those of most others in commercial clays, a considerable amount of improvement and purification can be made by converting such clays into a slip and then passing it through a fine sieve. A still further purification can be obtained by allowing the slip to rest and then pouring off the liquid (which will contain most of the 'true clay') and leave most of the impurities behind as a sediment. The slips must be dilute and for adequate purification the second slip must not usually contain more than 5 per cent of suspended matter; stronger slips should be diluted with water to ensure this.

Clays are frequently pretreated before being used as casting-slips. It is common practice to purify them although this may involve nothing more complex than the removal of coarse or large-sized particles.

Considerable commercial and technical advantages may be secured by making use of the colloidal properties possessed by clays, by means of which they can be suspended in a much smaller proportion of water and yet effectively separated from other (non-colloidal) materials which cannot be so readily suspended. To effect the purification of clays by this means, use is made of the fact that clays, like other colloidal materials, can be dispersed (see p. 473) by means of a suitable electrolyte, and when so dispersed or deflocculated they remain permanently in suspension until their state is changed (by the addition of a coagulant or flocculating agent); the principal impurities found in clay, such as quartz, mica, felspar, pyrites, calcium carbonate, and various iron compounds, are not so affected. Consequently when stirred well with alkaline water, these impurities remain suspended for only a short time and soon settle out, so that the purified clay suspended in water may readily be removed by decantation.

There are many deflocculating agents which are effective and the most suitable for any particular clay can be found only by trial and error.

An alkaline salt is the most suitable and sodium phosphate, silicate and carbonate are in general use. The amount added is usually small and invariably less than 7 lb. per ton of dried clay.

The separation process is best carried out in large tanks which permit the free sedimentation of the coarser grains. The separation of the clay from the greater part of the impurities is fairly sharp, provided that a suitable proportion of deflocculant has been added, but extremely small particles of silica and colloidal components cannot be removed. The chief difficulties connected with the process are found when it is required to separate the purified clay from the water in which is it suspended. To effect this, four general methods are available:

1. *Removal of the water by evaporation*, usually by the aid of heat. This is very effective, but usually too costly. It also causes any soluble salts in the water to remain in the clay. Where waste heat is available, this method is still worth consideration. Care must be taken not to over-heat the clay or its active properties may be destroyed.

2. *Filter-pressing* (see also p. 541).

3. *Flocculation or precipitation* of the suspended clay particles by the addition of acids or other electro-negatively charged substance.

Any of the flocculants mentioned on page 505 may be used though the most convenient are hydrochloric acid, acetic acid, aluminium sulphate or aluminium chloride.

A wide range of organic flocculants is available which are much more powerful in action and introduce less impurity than the inorganic forms.

The size of floc which forms can be controlled within narrow limits by selecting the correct organic components which are usually long-chained molecules with complexing groupings.

The only disadvantage of the use of a flocculant is that any colloidal impurities present or any impurities which may be adsorbed will contaminate the clay.

4. *Electro-osmosis* (p. 542), i.e. by the use of a current of electricity to separate the water and suspended matter. This method is interesting as it can simultaneously effect a further slight purification, especially from iron compounds, but it is very costly.

Various statements have been made from time to time to the effect that crude clays may be purified by electro-osmosis. This is scarcely correct, because the electro-osmosis does not, in itself, effect the chief separation of the impurities from clay, but only separates much of the water. The finest particles of the most important impurities, such as quartz, mica, felspar, etc., bear the same electric charge as the clay particles and so cannot be completely separated, though some oppositely-charged impurities may be separated by electro-osmosis.

5. *Centrifugal action* may be employed to separate particles of almost any size, as it has a force which can be made a thousand times as great as that of gravity and can be varied within narrow limits.

The commonest dewatering appliance used in industrial practice is the hydrocyclone (see p. 536).

Centrifuges are not so convenient in dewatering as hydrocyclones and are more costly to operate, but because of the considerably higher gravitational force which they employ, they are effective with particles of much smaller size.

2. Pastes and Plastic Masses

For the manufacture of bricks and most solid ceramic bodies, the raw material usually contains insufficient water to allow free flow and the mass is moulded to the shape required by the application of pressure. The amount of mixing water varies and is dependent on the nature and complexity of the shape to be made and the raw material employed.

The shaping of plastic masses involves applying a pressure to cause the body to flow until it takes the required shape. By this means the particles of the body are compressed and packed together by the external pressure which is applied, either by hand or by power, to the mass.

As considerable movement of material has to occur it is to be expected that less pressure will be required to form a material which is capable of easy flow than one in which movement is restricted.

Plastic masses, or those in which the water content is high, are easily moulded by pressure, because the moisture behaves as a cushion or lubricant and permits free flow. A drier mass does not flow so readily, hence higher forming pressures are necessary.

The shaping of bodies composed of plate-like particles is further complicated because the individual grains become orientated and offer increased resistance to the applied force (see Fig. VIII.31). Particle orientation is most likely to occur when the water is present in large proportions and when a considerable amount of displacement takes place on the application of the forming pressure. A body so prepared is frequently laminated and may develop cracks and fractures during drying or firing.

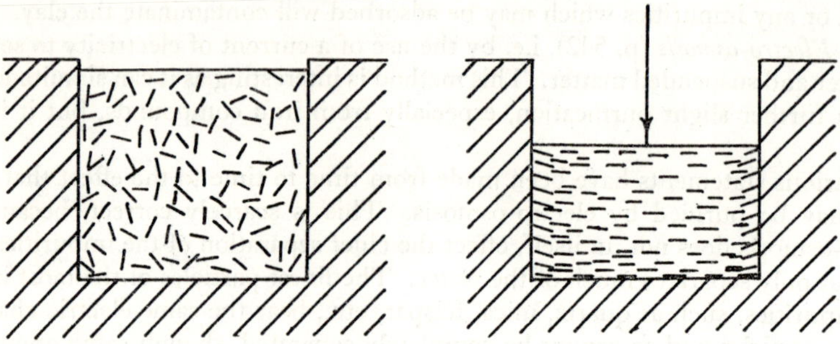

FIG. VIII.31.
THE ORIENTATION OF PLATEY PARTICLES IN PLASTIC MASSES AFTER PRESSING

A manufacturer has, therefore, to effect a compromise; a mass with a high water-content can be shaped under a low pressure but gives a less dense body with a tendency towards laminations; a drier material will pack tighter and with less particle orientation, but only on the application of a much greater pressure.

The pressure to cause flow in a clay/water body is related to the shape of the plastic flow curves over the range of moisture contents. In most forming processes the shearing rate is sufficiently high to result in streamline flow or a close approach to it. The stress required to produce an adequate flow rate is related to the water content in the body and a typical graph of the relationship is shown in Fig. VIII.32.

The choice of the correct moulding pressure for a clay product depends on several factors to which due consideration must be given. These include:

1. The capital cost of equipment. High forming pressures require expensive machines far greater in price than those used when the material is soft.

2. Precision of shape, which is possible only in shapes formed at low moisture content and at high pressure.

3. Permissible drying contraction. The shrinkage occurring in a shape when water is lost depends on the initial forming water content.

4. Load-bearing characteristics of moulded articles. Dry pressed articles can be handled more readily than those formed at low pressures with higher moisture contents.

5. The rate of production which is greater when soft bodies are used in manufacture.

6. The complexity of shape required. Softer clay bodies flow more readily and require smaller forming pressures.

By and large, low-priced products, such as building bricks and most sewer pipes, are formed at low pressures as also are complex ceramic articles. Pressures range

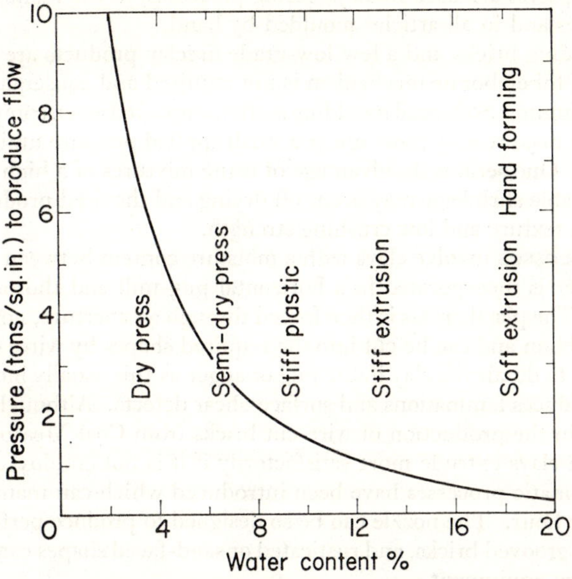

FIG. VIII.32. THE DEPENDENCE OF PRESSURE TO CAUSE FLOW ON THE WATER CONTENT OF A BUILDING BRICK CLAY

from those applied by hand—as in clay-throwing and the potter's wheel—to about 1 ton per sq. in. in extrusion practice.

Refractory bricks to fit accurately into furnace settings are usually made from drier mixes and formed at pressures up to 10 tons per sq. in.

The various methods of moulding used in manufacturing processes may be classified as follows:

The Soft Mud or Soft Plastic Process, where shapes are formed from clay or other raw material containing 15 per cent or more of water. Such mixtures have a soft consistency which allows them to be shapèd by hand for only a little pressure is required to cause free flow within the material.

From the earliest times this process has been used for shaping bricks, tiles and many kinds of pottery and it is still used almost exclusively in the manufacture of earthenware, stoneware and porcelain. The shaping is effected in moulds of iron, brass, wood or plaster of Paris. For simple shapes a one-piece (or sometimes a two-piece) mould is used. It is filled with the soft paste which is either pumped or pressed by the workers' thumbs and fingers so that all the corners are compactly filled. Any surplus clay is then removed by a taut wire. For more complex shapes each portion of a multiple piece mould is filled separately with the paste, then pieces are joined and fastened together and compacted as before. The moulds with their contents are either emptied at once or are set aside for their contents to stiffen and can then be turned out of the mould.

Soft-mud pastes are used in all jiggering processes, in the manufacture of flat-ware of all types and in all articles moulded by hand.

Some building bricks and a few low-grade fireclay products are manufactured by this process, for elaborate mechanism is not required and consequently costs can be kept to a minimum. Soft-mud moulding is often employed when complex shapes are to be made; the mass readily flows under a small applied pressure until it completely fills the mould. One serious disadvantage of using mixtures of a high water content is that considerable shrinkage may occur on drying and the fired product has a comparatively open texture and low crushing strength.

Extrusion Processes involve clays with a moisture content between 10 and 18 per cent. The water is incorporated in a horizontal pug-mill and thoroughly blended with the clay. The plastic mass is then forced through an aperture, nozzle or die and emerges as a ribbon and can be cut into the required shapes by wires or thin blades. It is customary to de-air the clay in the pug or auger as this usually increases the wet strength and reduces laminations and surface-shear defects. Although de-airing is a vital necessity in the production of wire-cut bricks from Coal Measure clays, some of the Wealden clays extrude more satisfactorily if it is not employed.

Fully automatic processes have been introduced which can manufacture up to 7000 bricks per hour. The nozzle can be so designed to produce perforated, hollow or tongued and grooved bricks, and rusticated or sand-faced shapes can be made with simple accessory equipment.

There are two broad classifications of extrusion processes, particularly in the manufacture of building bricks and shapes. These are *soft extrusion* involving bodies of higher moisture content (14–18 per cent) and pressures of about 1000 lb. per sq. in.; and *stiff extrusion* for clays with about 10–14 per cent moisture content requiring pressures of about 1 ton per sq. in.

The former process requires less powerful and hence cheaper equipment and more complicated shapes can be made than is the case with the latter. However, soft extruded bodies shrink appreciably more and are usually too soft to be set directly in a kiln in contrast to those manufactured by stiff extrusion.

Salt-glazed pipes, quarry and tile-bats are also largely extruded products using a plastic body.

The Stiff Plastic or Stiff Mud Process involves the use of materials with a water content of between 8–15 per cent, i.e. there is just sufficient moisture present or added to permit the material to flow under high pressure.

Many building and most fireclay refractory bricks are made in this way. The stiff plastic machine consists of a rotating table containing a series of moulds which are filled in turn by paste forced into them from a vertical pug mill and, after the table has made a partial rotation, the paste in the mould is subjected to pressure. After a further partial rotation of this table, the bricks or other articles are released from the mould and removed from the machine. They are then usually pressed a second time to improve their shape and finish. Most articles made by this process can be sent direct to the kiln without needing any time in a dryer, but for some shapes a partial drying is necessary.

The Semi-dry Process is used where the associated water is about 5–9 per cent. Many clays and shales occur in the natural state with a moisture content in this range and if they are soft in consistency they can be formed directly into bricks or other shapes without further treatment. This is particularly so with some building brick clays of which Oxford Clay is perhaps the best known.

Considerable pressure (up to 10,000 lb. per sq. in.) is required to produce the required shape from masses in the semi-dry state and in order to reduce lamination effects and produce a uniform density throughout, the pressure should be applied both from above and below.

Dry Press Methods are more commonly employed in the manufacture of non-plastic ceramic bodies. The moisture content is about 3–5 per cent and is just sufficient to permit the powdered body to hold together when tightly squeezed.

Although very great forming pressures are required (up to 100,000 lb. per sq. in. for special refractory shapes), the body so formed has an exceptionally high density and the drying and firing shrinkages are at a minimum because the void space or porosity is less than that in a body formed by other methods. Unless precautions are taken to remove entrapped air prior to or during the forming, pressure laminations may develop; they are caused by small pockets of air being compressed and then released during pressing—this is most serious when the individual grains are angular or platey.

There is always a tendency, when clay and other masses are pressed, for some degree of particle orientation to occur and it is a frequent source of weakness and failure particularly in clay bodies, for in them laminations and shear fractures are easily developed. Such faults occur more frequently where the water content is sufficient to allow free flow and where the forming pressure is high; consequently bodies formed by the soft mud and stiff plastic methods are particularly liable to such faults.

Dry pressed bodies, particularly those made from clay, have their own characteristic tendency to faults. Water may concentrate along certain planes at right angles to the pressing direction thus producing a type of lamination. Perhaps the most serious problem is the differential stresses normally developing in high pressure forming. The clay and associated particles are reluctant to flow at the low moisture contents involved so that localised shear failure can occur conferring a type of anisotropy on the whole body.

It is difficult to persuade particles to pack tightly in dry pressing processes, because the overall flow is very low. This can be improved by applying pressure from two faces or by vibrating the body prior to developing full pressure. Even then it is difficult to ensure a uniform and consistent pressure at all points in the pressed body so that a characteristic 'fabric' tends to develop.

A detailed study of the 'fabric' of a body which has been subjected to pressure in various ways has been undertaken by Williamson[1] and also by Haefeli and Amberg.[2] Both groups of workers used a similar technique in which the test pieces were made of alternate layers of differently-coloured wet clay. The resulting masses were then subjected to pressure applied in several ways and the changes in the coloured pattern were observed. The stresses and laminations which are produced in an extruded bar of clay have been examined in detail by Williamson and his results are of the greatest value in explaining many of the defects which occur in clay

[1] Williamson, W. O., *Amer. J. Sci.*, 245, 645, 1947.
[2] Haefeli, R. and Amberg, G., *Schweiz. Ton.-Industr.*, Nos. 1–6, 1949.

masses. The flow properties of the plastic body cause typical particle orientation in certain well-defined directions; in addition the authors have been able to show that stress fractures occur in two planes mutually at right angles.

Such investigations into the 'fabric' of a clay body when subjected to forming pressures will undoubtedly lead to a more detailed appreciation of the many contributory factors involved in the shaping of plastic masses.

Isostatic Pressing is not new, but its application to clays and other ceramic materials is comparatively recent. In this process a dry powder (2–4 per cent water) is sealed into moulds of rubber or plastic in an appropriate container. The whole is submerged in a suitable fluid in a pressure chamber. When pressure is developed it is distributed by the fluid so that it is applied equally as the shape is compacted.

Pressures of 7–10 tons per sq. in. are normally employed and there are several methods of tooling to make the process capable of large-scale production.[1]

Isostatic pressing was originally used in the ceramic industries in the manufacture of spark plugs and electrical insulators, but it is now employed for shaping in the specialised refractories industry. Complicated shapes of high density and low drying and firing contraction are manufactured by these techniques.[2]

An interesting development is the manufacture of precision sewer pipe by isostatic methods. Although rates of production are relatively slow the vastly improved qualities, particularly the shape, have much to commend the technique.

Hot Pressing is not strictly a plastic forming process and does not involve water as a plasticising agent, nevertheless its value and application to ceramic practice is becoming increasingly important.[3]

When subjected to high pressures (*circa* 10 tons per sq. in.) and to moderately high temperatures (700–1200° C.), ceramic bodies can be induced to flow and so to compact and harden.

The processes of crystal shear along lines of cleavage and weakness are involved and so too is conventional plastic flow arising from liquid phases which develop at high temperatures.

So far, hot pressing is restricted to sophisticated materials and shapes, such as carbides and oxides, but the method has great potential in the manufacture of all types of ceramic bodies, including the common brick. The construction and design of moulds present serious problems, but the advantages of 'instant forming' of a ceramic shape are obvious.

Future Developments in Forming Processes

Modern technology in ceramic practice demands a high degree of automation. No longer is the potter able to devote time and labour to controlling his slips and plastic bodies. It is important that the same characteristics must be present in clay/water systems at all times, otherwise high wastage will occur.

The features of viscosity and plasticity which have led to the intensive utilisation of clays since the beginning of civilisation are now proving a virtual embarrassment because of the difficulties attendant on their control.

[1] Catchpole, C., *J. Brit. Ceram. Soc.*, 63, 385, 1964.
[2] Wehrenberg, T. M., Begley, E. R. and Patrick, R. F., *Bull. Amer. Ceram. Soc.*, 47, 642, 1968.
[3] Carruthers, T. G. and Wheat, T. A., *Proc. Brit. Ceram. Soc.*, 3, 259, 1965.

The manufacturers of china clays and ball clays have achieved limits of control undreamed of a decade ago, but common clays used in the majority of ceramic wares cannot be similarly pre-treated on economic grounds alone. Small variations in mineralogical composition, differences in texture—the result of variable weathering or comminution—cause major changes in plastic characteristics with attendant faults in manufacture. In the days of small production units and where craftsmen were available to make frequent necessary adjustment, the problem was not particularly serious; but this is far from true in present-day production lines.

The difficulties inherent in drying and firing of clay bodies are aggravated by the modern need for standardisation and rapidity of turn-round, so that many of the former advantages of clay as a mineral of commerce are no longer pertinent.

New forming methods such as dry and isostatic pressing help in many cases without being the complete answer. Hot forming is still of limited application and may never be of use for large and complex shapes.

An increasing amount of research is being devoted to body preparation in which the complex nature of clays is destroyed by prior heating. The resulting 'grog' is then ground to precise limits and a controlled amount of viscous or plastic character re-introduced either by a prepared clay or by organic additives. Such bodies can be cast, hand or machine moulded, extruded or dry pressed under fully automated conditions. The shaped articles can be dried and fired rapidly and without undue problems and with much greater flexibility than normal clay bodies.

CHAPTER IX

CHANGES IN THE PHYSICAL STATE OF CERAMIC MATERIALS BROUGHT ABOUT BY THE REMOVAL OF FREE WATER

THE clay minerals are mostly hydrated alumino-silicates and contain OH-groups which, when the clay is heated to a sufficiently high temperature, escape in the form of water vapour—known as combined water, constitutional water or crystal water. They also contain water in other forms—some being adsorbed and in a state of slight combination, whilst other water lies between the layers forming the mineral (inter-layer water or interstitial water); in most clays there is water occupying some or all of the open pores.

Water is usually added to all ceramic bodies to assist in the moulding of shapes. The amount used depends upon the material and the method of shaping and varies from around 5 per cent in dry pressing to 30 per cent and above (by weight) in slip casting. Such water is termed free water and its removal must be complete before the ware is fired, otherwise serious faults may arise.

In many clay industries and others where clay is involved as a contaminant, the raw material is thoroughly dispersed in water to form a fine slurry or suspension. In this way coarse minerals can be removed by sedimentation from the fine clay which remains in suspension. There then remains the problem of how to remove the clay from the watery medium which may be 60 per cent or more by weight of the total.

Water removal is thus a prime ceramic operation and its successful and economical completion is an important aspect of scientific and technological study.

The term *dewatering* is commonly used for describing the removal of large quantities of water from slips, etc., the term *drying* being usually restricted to the removal of water from pastes and solids; both terms are used rather loosely.

The term *dehydration* is being increasingly used for the removal of water particularly when the precise means by which the water is removed is not stated; this term includes the removal of water by means other than drying by heat though that is the most widely used method of dehydration.

The removal of water may be effected in various ways, of which the most important are:

(1) *Sedimentation* or settlement.

(2) *Centrifugal separation* by hydrocyclones or centrifuges.

(3) *Filtration* including pressure filtration and absorption into plaster moulds.

(4) *Electro-osmosis* in which the charged particles are caused to migrate in an electric field.

(5) *Evaporation* usually by means of heat or other forms of radiation.

(6) *Desiccation* in which water is removed by absorbing it into a more receptive medium.

The first four techniques are more applicable to dewatering in that none of them dry the solid material thoroughly. Evaporation is used mainly to dry bodies which are already in a plastic condition. Except in very hot, dry climates where solar heat can be harnessed, evaporation is an expensive process and not to be recommended where large quantities of water are involved.

Desiccation is an expensive operation and used only to dry clays on a small laboratory scale.

DEWATERING[1]

THE removal of large quantities of water from dilute slips and suspensions is common to the processing of high purity clays, such as china clays, and to the disposal of clay tailings from several mineral purification operations. These latter include waste disposal from coal washery plants and slimes treatment from various physical and chemical processes. In all cases an essential requirement is that the clarified water should be sufficiently clean either to return to the circuit without reducing efficiency or to be discarded into a public water supply. A requirement of clear water with a maximum solids content in the range 5–20 parts per million is a common specification.

Sedimentaton Methods

If solid particles are suspended in a fluid medium of lower density, they will settle at a rate governed mainly by the size of the particles and the difference in density between the solid and the medium. This is strictly true only when the particles in suspension are small (less than 50 microns) but in dewatering exercises where clay is involved this condition invariably is satisfied.

Sedimentation obeys Stokes' Law which has been developed on p. 379 and which may be written as

$$V = \frac{2}{9}\frac{(\rho_1-\rho_2)gr^2}{\eta}$$

where ρ_1 and ρ_2 are respectively the densities of solid and liquid components and r is the particle size, V is the rate of sedimentation, g is the gravity constant and η is the viscosity of the medium.

When the particle size is small, the rate of sedimentation is slow. Particles of 1 micron size, for example, will sink less than 0·4 cm. in 1 hour. It is not uncommon in some clays to find particles as small as 0·1 micron which would require 100 hours to sediment 0·4 cm., even if other factors were not operative to keep them in permanent suspension.

For a deflocculated clay, therefore, simple sedimentation techniques are not the answer to complete dewatering, although large-particled impurities or components are effectively removed by these means.

In practice, clays are flocculated by some suitable inorganic or organic chemical prior to entering the sedimentation vessel. The clay agglomerates into large flocs

[1] An excellent account of the principles and practice of industrial dewatering units has been given by H. K. Suttle, *Process Engineering Techniques Evaluation, Filtration* (Morgan, Grampian Ltd., London, 1969).

which settle much more rapidly. Flocculated particles are, by their very nature, of low density because of entrapped water. The sedimented volume of such a system is therefore large and although clear water can be removed, the percentage of water in the sediment itself is much larger than in the settled product from a deflocculated suspension.

Much research has been carried out on both the type of flocculant to give optimum results and on the design of thickeners. Dell and Sinha[1] have shown that deep vessels produce more consolidated sediments; Hassett[2] has developed mathematical models to predict sizes and designs of sedimentation tanks. Two types of sedimentation units are in common use in the clay industries.

(*a*) *Lagoon type* in which the suspension is allowed to pass under gravity through a series of lagoons each of which is larger than the one preceding it. The flow rate is thus progressively reduced and finer particles settle in each tank, (Fig. IX.I). This technique was commonly employed in china clay production but has now been largely superseded by other methods.

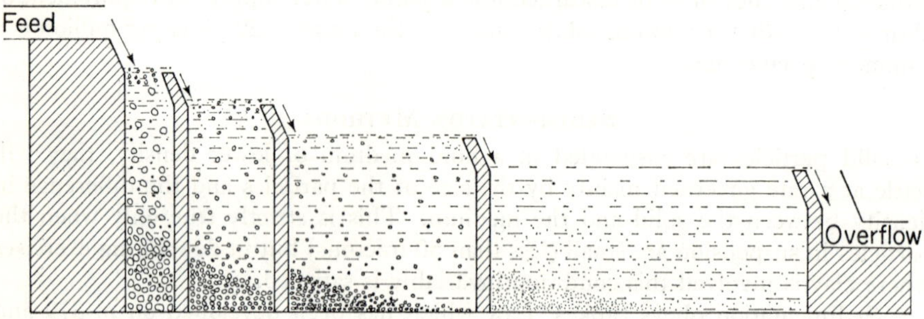

FIG. IX.1. THE PRINCIPLES OF SEPARATION OF PARTICLES IN LAGOONS

(*b*) *Sedimentation cones*, in which a continuous flow of suspension feed enters the centre of a circular pond of large diameter. Clear water overflows at the periphery of the vessel whilst concentrated product is withdrawn continuously from the base of the unit, which is frequently cone-shaped (Fig. IX.2).

A tailings pond fulfils the same purpose. These are reservoirs of large capacity which are operated until solids fill the volume. They are then withdrawn from line and the suspension diverted to another reservoir whilst the solid sediment is being removed from the first pond.

Centrifugal Separation to dewater slips has replaced sedimentation methods in many industries. Hydrocyclones are extremely efficient and simple units which are capable of removing particles down to 5 microns; to extract particles of smaller sizes, the more expensive and cumbersome centrifuges are necessary.

As the title implies, in both methods the normal gravitational forces acting on sedimenting particles are increased, so that the velocity of fall is accelerated.

Hydrocyclones for dewatering purposes involve projecting tangentially at high velocities a jet of suspension around the walls of a cylinder. Two exits are provided;

[1] Dell, C. C. and Sinha, J., *Trans. Brit. Ceram. Soc.*, 63, 603, 1964.
[2] Hassett, N. J., *Trans. Inst. Min. Met.*, 74, 627, 1964.

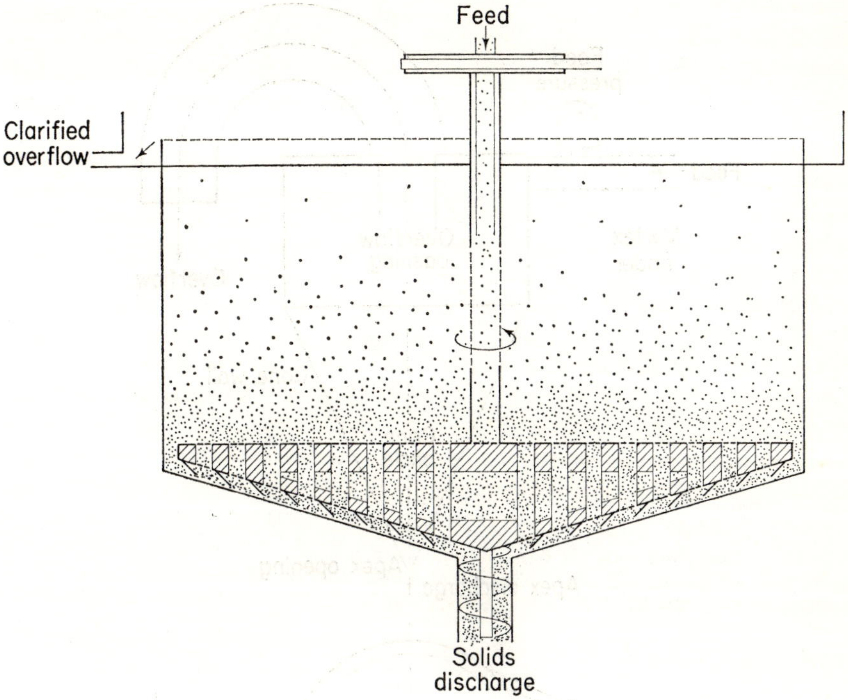

FIG. IX.2. A SEDIMENTATION CONE

one for the clarified overflow (or purified clay) and the second, the apex opening, for the concentrated large-particled solids. The apparatus which was developed by the Dutch State Mines is shown diagrammatically in Fig. IX.3.

The size of particles which can be separated depends on the velocity of feed and its pressure and the diameter and length of the hydrocyclone. Curves of the type shown in Fig. IX.4 illustrate the type of separation which can be achieved.

In the preparation of high quality china clays, batteries of hydrocyclones are commonly employed. The first unit is comparatively large—about 100 mm. diameter which rejects material greater than 20 microns and allows the fine-particled clay to pass in suspension to a small unit. A 15 mm. hydrocyclone will cut at about 5 microns and the overflow product (−5 micron china clay) forms a high quality material, which is then usually treated in a centrifuge often after it has been flocculated.

Further details of hydrocyclone operation are given by F. J. Fontein and C. Dijksman.[1]

Centrifuges have been described in Chapter VII. They are capable of developing high values of gravitational constant, but they are not extensively used in

[1] Fontein, F. J. and Dijksman, C., *Inst. Mining & Met. Symp.*, p. 229, Sept. 1950.

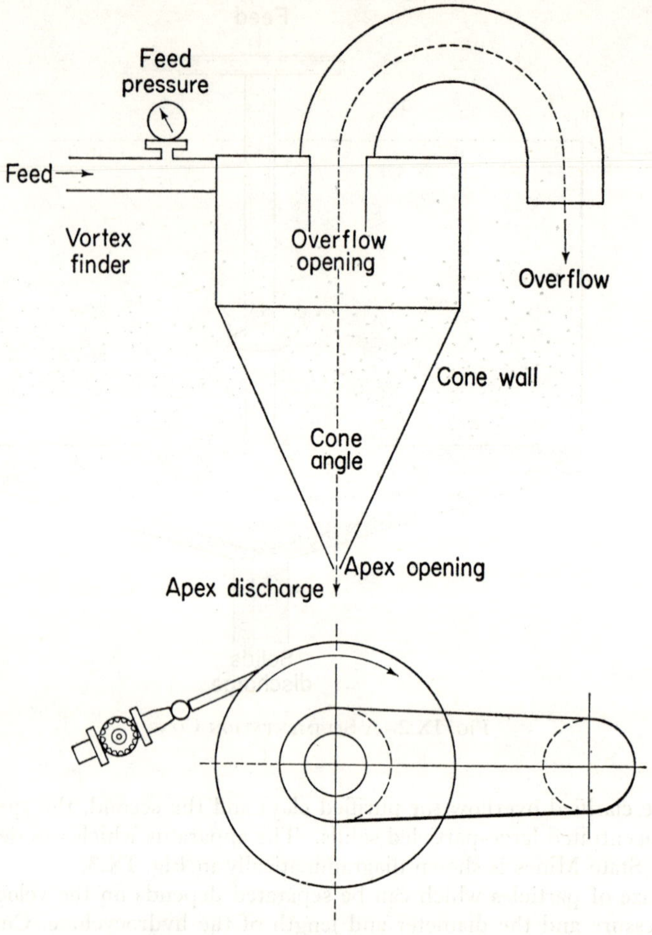

Fig. IX.3. Diagrammatical Representation of a Hydrocyclone

clay industrial practice because of their cost and the fact that they are not easy to operate on a continuous basis.

Filtration Techniques

The removal of water by filtration involves exposing the suspension to a porous material which selectively permits the fluid medium to penetrate into and through itself, whilst retaining the solid particles. It follows that the pore size of the material must be smaller than the particles to be removed but for greatest efficiency the total pore area exposed to the suspension must be as large as possible.

As filtration proceeds, a layer of solids builds up on the interface and provides

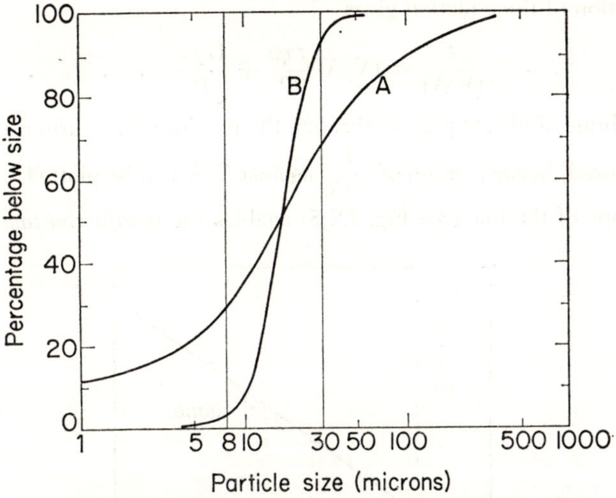

FIG. IX.4. HYDROCYCLONE SEPARATION OF A FINE SAND

(A) Feed material from a washing plant −60-mesh B.S.S., heavily contaminated with clay;
(B) Product recovered from a double hydrocyclone circuit
(i) Rejecting oversize at 30 microns
(ii) Rejecting clay slimes at less than 8 microns
Recovery 44 per cent. Efficiency 92 per cent

a barrier restricting further water removal. The rate of percolation of water through the filtered layer is thus another important factor in this technique of water remcval.

If pressure can be applied to the system, the filtration rate often increases markedly and the whole process can be described by complex mathematical expressions.

The Darcy equation predicts the rate of filtration as follows:

$$\frac{dV}{dt} = \frac{A.P}{r\eta(L_1 + L_2)} \quad . \quad . \quad . \quad . \quad . \quad . \quad . \quad 1$$

where V is the volume of filtrate, A the area of filter cake which forms and L_1 its thickness at any time t. L_2 is a term representing the equivalent cake resistance of the filtration medium. P is the pressure differential and r an important parameter termed the *specific cake resistance*, η is the viscosity of the liquid medium.

From a particular volume of filtrate V, a volume of cake Vv can form which will equal $A.L_1$ when $V = \dfrac{\text{volume of cake}}{\text{volume of filtrate}}$

So equation 1 can be re-written

$$\frac{dV}{dt} = \frac{A.P}{r\eta\left(\dfrac{Vv}{A} + L_2\right)} \quad . \quad . \quad . \quad . \quad . \quad . \quad . \quad 2$$

Integration of the equation gives

$$\frac{t}{(V/A)} = (V/A)\,\frac{r\eta v}{2P} + \frac{L_2 r\eta}{P} \quad . \quad . \quad . \quad . \quad . \quad . \quad . \qquad 3$$

The volume of filtrate passing through the medium after various time intervals can be measured, hence a graph of $\frac{t}{V/A}$ against V/A can be plotted.

The slope of the line (see Fig. IX.5) enables the *specific resistance of the cake*

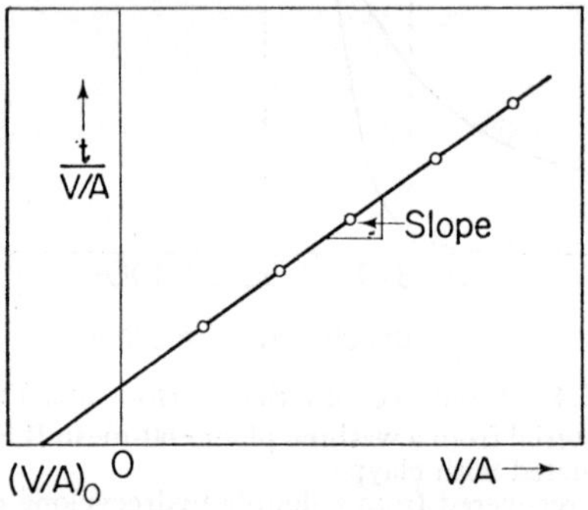

FIG. IX.5. GRAPH FOR CALCULATING THE IMPORTANT PARAMETERS IN FILTRATION PROCESSES

r, to be calculated provided that the thickness of cake at any known time can be measured. From the intercept of the graph with the horizontal axis, the term

$$\frac{L_2 r\eta}{P}$$

can be measured and hence L_2 the *resistance of the filtration medium* can also be calculated.

The specific resistance of the cake is usually relatively high for clays but it depends markedly on the degree of dispersion. Fully deflocculated clays have a very high resistance and consequently they are slow to filter and as the cake builds up the rate of further filtration falls dramatically. The resistance can be lowered by flocculating the clay to a partial extent. Organic chemicals have been developed enabling filtration to be completed in only a fraction of the time of a deflocculated body, but without producing the typical flabby cake of a body flocculated with inorganic reagents or conventional flocculants.

The resistance of the filtration medium can be controlled by carefully selecting the pore size and content. It is of advantage that it should be low but if the pore size is too large the effectiveness of filtration is reduced.

In addition to enabling two important filtration parameters to be deduced, which govern the rate of water removal, the above equations predict that filtration rate is directly related to the filter area and to the pressure differential across it and that it is inversely related to the thickness of cake which forms.

Filtration techniques of clay suspensions both under laboratory conditions and in industrial practice can be classified under three headings:[1]

(*a*) *Absorption into a solid porous medium.* When relatively small amounts of suspensions are to be dewatered, the process is conveniently carried out by absorbing excess water in a Plaster of Paris mould or on a porous tile. This has been described under slip casting processes (p. 520). In dewatering operations where body-finish is not so important, filtration can be accelerated by subjecting the slip to pressure or by applying suction to the mould.

(*b*) *Filter Pressing.* The suspension is forced under pressure through fine-textured cloths; a cake of dewatered solid builds up on each cloth whilst the clarified water passes to waste. This is a batch process and is continued until a cake of predetermined thickness and consistency has developed. Each cloth must then be cleaned before a new cycle can commence. Fig. IX.6A is a schematic representation

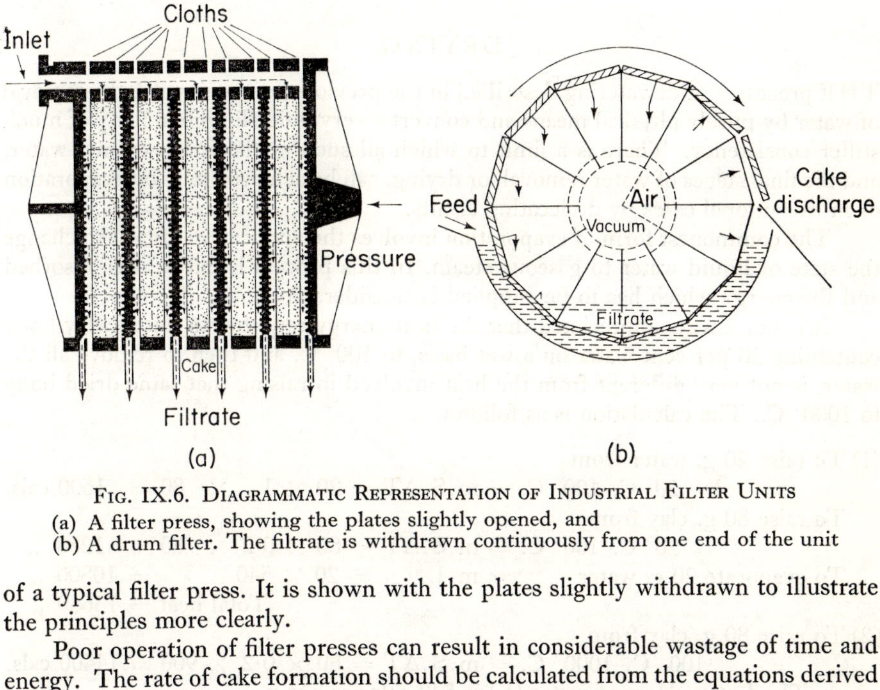

Fig. IX.6. Diagrammatic Representation of Industrial Filter Units
(a) A filter press, showing the plates slightly opened, and
(b) A drum filter. The filtrate is withdrawn continuously from one end of the unit

of a typical filter press. It is shown with the plates slightly withdrawn to illustrate the principles more clearly.

Poor operation of filter presses can result in considerable wastage of time and energy. The rate of cake formation should be calculated from the equations derived in the previous section and the cloths should be cleaned at time intervals consistent with the most efficient operation, taking into account the time and man-power factors.

(*c*) *Drum Filters.* A continuous method of filter pressing, in which the cake is removed progressively, is provided by drum filters. The usual method of operation is to mount filtration cloths in sections on a drum which is revolved through a bath of the suspension to be dewatered. As each section of the drum passes through

[1] An excellent description of the theory and design of industrial filters is given by D. B. Purchas, *Industrial Filtration of Solids* (Leonard Hill, London, 1967).

the suspension, suction is applied to each one in turn to draw water through the cloth and to build up a cake. Once clear of the suspension the cake dries slightly as it rotates in air and finally the cake is blown off the section onto a collection belt, so that a clean cloth enters the suspension each time. The principle is illustrated in Fig.IX.6B.

For the most efficient operation, conditions of immersion time, suction pressure, etc., should be adjusted to conform to Darcy's equation.

Electro-osmosis. Although potentially important in dewatering, electro-osmosis is rarely used for this purpose. It involves causing solid particles in a suspension to move in one direction by virtue of their charge when subjected to a direct current potential field. The process is slow and costly and appears to have no advantages over filter pressing. Under certain conditions, electro-osmosis can simultaneously bleach and dewater and it could be of value in some clay treatments.

DRYING

THE processes of dewatering described in the previous section involve the removal of water by purely physical means and convert a very fluid slip into a body of much stiffer consistency. There is a limit to which all such methods can extract water, and the final stages of water removal, or drying, can be achieved only by evaporation or in exceptional cases by desiccating agents.

The commonest form of evaporation involves the application of heat to change the state of liquid water to gaseous steam. In this process latent heat is absorbed and the energy which has to be supplied is considerable.

It is not always appreciated that the heat energy required to raise a clay body containing 20 per cent water on a wet basis, to 100° C. and then to remove all the water, is not very different from the heat involved in raising that same dried body to 1000° C. The calculation is as follows:

(1) To raise 20 g. water from

$$20° \text{ C.}–100° \text{ C.} = m.S.\Delta T = 20 \times 1 \times 80 = 1600 \text{ cals.}$$

To raise 80 g. clay from

$$20° \text{ C.}–100° \text{ C.} = m.S.\Delta T = 80 \times 0{\cdot}2 \times 80 = 1280 \text{ ,,}$$

To evaporate 20 g. water $= m.L = 20 \times 540 = 10800$,,

Total heat $= 13680$,,

(2) To raise 80 g. clay from

$$100° \text{ C.}–1000° \text{ C.} = m.S.\Delta T = 80 \times 0{\cdot}2 \times 900 = 14400 \text{ cals.}$$

The Latent Heat of water (L) is 540 calories per g.

The Specific Heat of clay (S) is about 0·2.

Measurement of Free Water Content

THE assessment of the amount of water which is present in a clay is subject to many difficulties which have been discussed in Chapter VIII. In drying clays or similar materials the temperature chosen for the determination may remove water which is combined in some way or alternatively some so-called free water may remain in the body. Clays which possess a content of colloidal, amorphous matter progressively lose water and there is no plateau of constant weight as in other materials. Although is has limitations of this nature, it is now generally accepted that 'free water' in clay bodies ought to be defined as that lost on heating to constant weight at 110° C.

Not all authorities agree on how it should be expressed and there are two conventions in common use:

(1) *Wet basis measurements* refer the weight loss to the initial total weight of the wet body.

(2) *Dry basis measurements* refer the weight loss to the weight of the body after drying.

So if W_1 is the weight of the original clay + water mixture and W_2 is the weight of the same material after drying

$$\text{Weight loss (on wet basis)} = \frac{W_1 - W_2}{W_1} \times 100\%$$

$$\text{Weight loss (on dry basis)} = \frac{W_1 - W_2}{W_2} \times 100\%$$

In any determination it is important to define which convention is being used.

In a clay paste free water may be present in one or more of three forms:

(*a*) A superficial film surrounding each solid particle;

(*b*) As interstitial water, i.e. as water occupying the pores or interstices between the solid particles so that each solid grain is separated from every other by a superficial film of water and by the interstitial water;

(*c*) The water may be absorbed into the colloidal material coating the grains, causing the colloidal matter to swell like soaked glue and each solid particle to be separated from every other by the colloidal gel.

As long as sufficient water is present, the clay remains in a swollen state, but as it dries the particles are drawn closer together and the mass shrinks or contracts.

When the water is gradually withdrawn from a paste in this manner, each form in which it occurs is the cause of certain physical changes in the material, and the drying therefore appears to occur in several distinct stages, though these overlap to some extent.

The Mechanism of Drying. During the early stages of drying, evaporation from the surface of a clay body occurs at almost exactly the same rate as from a water surface (e.g. a pond) and continues at that rate until there is an insufficient supply of water from the interior of the mass.

The physical state of fine-grained material when a sufficiently large amount of water is present may be imagined to be as shown in Fig. IX.7A. Each particle is separated from its immediate neighbours by a water film of a relatively appreciable thickness. In this state, the clay or ceramic body can be easily moulded because the 'free water' surrounding all particles acts as a cushion and permits the mass to flow, either unaided or under slight pressure.

The water film is a *continuum* throughout the mass and hence the water surface

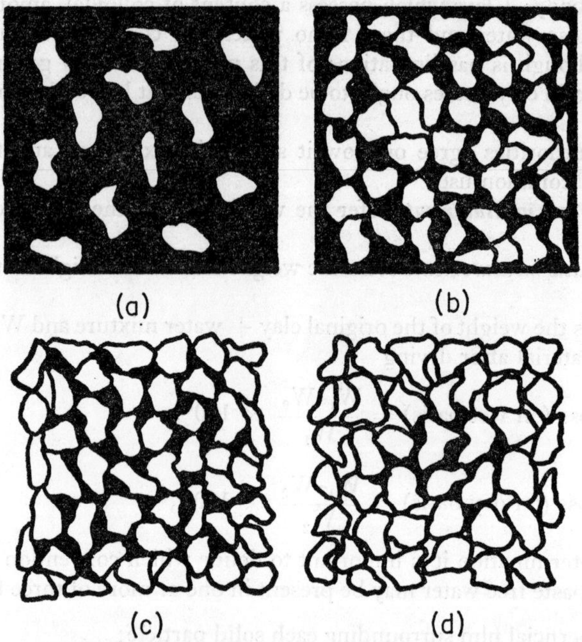

Fig. IX.7. Stages in the Drying of a Plastic Clay Body
(a) The initial state with excess, free water; (b) particles just touching each other—the critical point; (c) the destruction of the continuous surface-water film—water can evaporate only from the ends of capillaries; (d) the final stage of water contained only in isolated pockets in the dried mass

exposed to the atmosphere is unbroken and evaporation can proceed at a rate similar to that from a volume of water with the same surface area.[1]

As water is lost by surface evaporation, the continuum in the mass is preserved by a drawing together of the particles and an over-all *shrinkage* occurs. The clay mass will lose water as though there were a free, surrounding water surface until the solid particles touch each other and the water within the mass can no longer maintain a continuous film (Fig. IX.7b).

In this initial stage of water-loss, the rate of drying, if conditions are kept uniform, is constant throughout. The rate of shrinkage of the mass also very closely parallels the rate of loss of water and is constant in the region. For these reasons, this first period of drying is known as the *constant rate period of drying*.

The rate of drying in this period depends on that at which the water leaves the surface of the article and is absorbed and carried away by the air in the dryer. If this air is completely saturated (i.e. its humidity is 100 per cent) there will be no

[1] Macey, H. H., *Trans. Brit. Ceram. Soc.*, 37, 131, 1937.
Sherwood, T. K. and Comings, E. W., *Ind. Eng. Chem.*, 25, 311, 1933.

drying at all. If the air is stationary, drying will continue until the air is saturated and will then cease. Only if the air is removed, either mechanically or naturally, will drying occur and the rate of drying will depend chiefly on that movement. The hotter the air, the more water will it hold and the more rapidly will it dry the goods, so long as it is not fully saturated, but air which is too hot and too dry may cause cracking.

The water removed from the articles in this state is sometimes termed the *shrinkage water* and the amount of water remaining in the body at the end of this stage is termed the *critical moisture content.*

At the end of the first stage, the articles will then be in a *leather-hard* or *black-hard* state and will contain about 7–10 per cent of water (based on the fully-dried weight).

In the *second stage of drying*, which commences when the solid particles in the mass are in contact with one another, a capillary system is formed. The surface of the mass can no longer behave as a continuously evaporating film, because the water will be drawn into the pores between the solid particles of material.

Precisely what occurs during the removal of the remaining water has not yet been fully established. Morgan and Hursh[1] consider that the process is mainly one of vapour-diffusion to the surface from the ends of the narrow capillaries of water. Ceaglske and Hougen[2] suggest that water is drawn through the capillary system by surface tension forces and evaporated from the surface. Macey,[3] on the basis of his studies on the transmission rate of liquids through porous solids, has advanced the theory that water is removed as a result of the rise in external pressure caused by the repulsion of particle surfaces which are in close proximity. The rate of evaporation depends only on the number of exposed ends of narrow columns of liquid in the mass (see Fig. IX.7C) and as these will become progressively less in size and smaller in number the rate of water loss will fall. Eventually only small isolated pockets of moisture will remain which are so completely enclosed that the rate of evaporation will fall almost to zero (Fig. IX.7D).

This final stage in the drying is known as the *falling rate period* and in many bodies the change-over in drying mechanism is so well defined that the *critical point* can be accurately established (see Fig. IX.8).

The drying behaviour of all ceramic bodies which are agglomerations of small particles follows the same general pattern although the ultimate rates in the two periods and the position of the critical point depend upon many factors.

The Shrinkage on Drying. Fig. IX.9 shows the linear shrinkage as a function of water content for three typical materials, a highly plastic clay (A), a feebly plastic kaolin (B) and a non-plastic sand-aggregate (C).

In all ceramic materials where the moisture content is high, the initial rate of shrinkage on drying is virtually constant. The position of the critical point (i.e. the water content at which free surface evaporation is no longer possible), varies from about 20 per cent in a plastic clay to 4 per cent in a non-plastic sand. The change-over point is more clearly defined in the latter material; in clays there is sometimes

[1] Morgan, W. R. and Hursh, R. K., *J. Amer. Ceram. Soc.*, 22, 271, 1939.
[2] Ceaglske, N. H. and Hougen, O. H., *Ind. Eng. Chem.*, 29, 805, 1937.
[3] Macey, H. H., *Trans. Brit. Ceram. Soc.*, 41, 73, 1942.

no well-marked transition and appreciable shrinkage occurs at a water content below the critical value.

Ford[1] has pointed out that some clays exhibit two well-defined falling-rate periods; in the first, which follows the constant rate period, the rate declines as a linear function of moisture, but in the second stage the relationship is curvilinear.

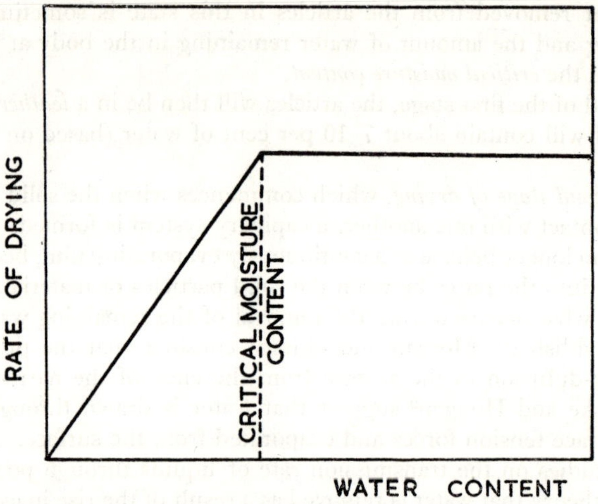

FIG. IX.8. THE RATE OF DRYING IN CERAMIC BODIES

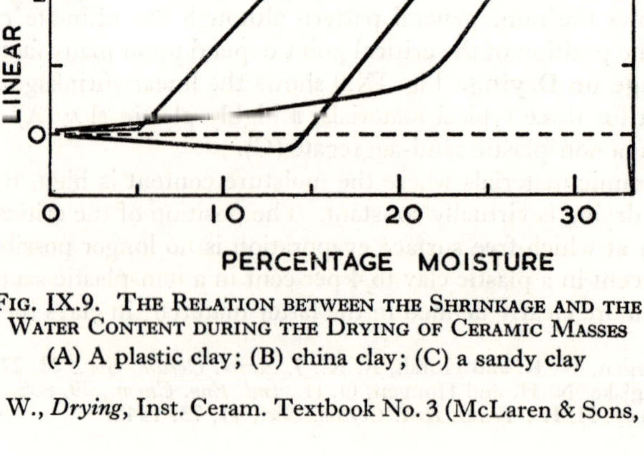

FIG. IX.9. THE RELATION BETWEEN THE SHRINKAGE AND THE WATER CONTENT DURING THE DRYING OF CERAMIC MASSES

(A) A plastic clay; (B) china clay; (C) a sandy clay

[1] Ford, R. W., *Drying*, Inst. Ceram. Textbook No. 3 (McLaren & Sons, London, 1964).

Clays are composed of inherently fine-grained particles which would be expected to have a fine capillary pore-system, hence the amount of retained water will be greater than in sand where the particles are larger and more rounded in shape. The shrinkage in clay bodies at water contents below the critical point has been attributed to a rearrangement of colloidal particles and to the loss of water contained within the structure of certain minerals. The amount of shrinkage below the critical point is, however, relatively small compared with the amount in the constant period of drying.

It is useful to represent drying rate in terms of the volume of a specimen in relationship to the moisture content present. The idealised form of the graph is as shown in Fig. IX.10 where the vertical axis represents the volume occupied by

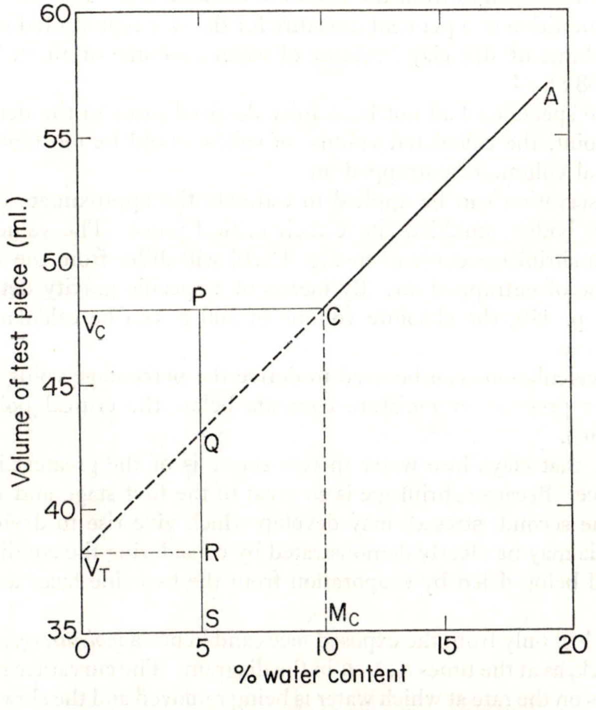

FIG. IX.10. THE IDEALISED VARIATION IN THE VOLUME OF A SPECIMEN WITH ITS WATER CONTENT ON DRYING

100 g. of clay calculated on a dry basis. The bulk density of the test piece at any moisture content can be derived by the relationship (100 + moisture content %) ÷ volume.

The curve establishes several important aspects of drying behaviour:

(*a*) The initial part of the curve, at higher moisture contents, shows a linear relationship between the volume change and the moisture content; so it may be concluded that in the early stages of drying the volume of water lost is compensated by an equivalent shrinkage in the body.

(*b*) At a characteristic critical point C, contraction ceases and the specimen retains the same volume (V_c) throughout the remainder of its drying cycle. At this moisture content M_c the body is in a leather-hard condition.

(*c*) If the linear curve of the constant drying rate period AC is projected to cut the vertical axis, the intercept V_T is the apparent true solid volume of 100 g. of dry clay and the density calculated from this value is approximately the true specific gravity of the clay itself. This is true only if the specimen in the wet state is fully de-aired.

(*d*) At any point on the drying curve the relationship between the volume of dry clay : volume of water : volume of air in pores can be calculated quite simply by drawing a vertical line from the moisture content axis. The line PQRS would indicate the condition at 5 per cent moisture for the clay represented in the diagram.

Thus volume of dry clay : volume of water : volume of air ::: V_T : RQ : QP. In this case 38 : 5 : 4.

(*e*) If the specimen had not been fully de-aired prior to the determination of the critical point, the calculated volume of solids would be too high and in error by the original volume of entrapped air.

This observation can be applied to estimate the approximate content of entrapped air in bodies moulded above their critical point. The value of V_T determined from a shrinkage curve as in Fig. IX.10 will differ from the absolute value by the volume of entrapped air. By means of a specific gravity determination as described on p. 429, the absolute volume of 100 g. can be calculated and hence the void volume.

Similar calculations can be used to derive the percentage voids in clay bodies formed under pressure at moisture contents below the critical point (see Ford, R. W., loc. cit.).

The fact that clays lose water in two stages is of the greatest importance in drying practice. Because shrinkage is so great in the first stage and virtually non-existent in the second, stresses may develop which give rise to drying cracks and warping. This may be clearly demonstrated by considering the conditions in a slab of moist solid being dried by evaporation from the two side faces as illustrated in Fig. IX.11.

Water is lost only from the exposed faces and hence a *moisture-gradient* is set up across the brick, as at the times t_1, t_2, t_3 in the diagram. The curvature in the gradient curve depends on the rate at which water is being removed and the slower the drying-rate the less will be the gradient.

As more water is lost, the moisture in all parts of the slab is reduced but at all times the percentage in the external areas of the two side-faces will be less than in the centre of the slab. Consequently a condition will exist, at some time, t_3, when the outside portions of the slab will have a moisture content below the critical point, yet in the interior the amount of water may be much above this point. The outside layers will, eventually, become almost dry and it is only then that the moisture gradient will be reduced as water from the inside of the slab continues to evaporate.

The moisture gradient at any time in a cylinder being dried from one face can be predicted mathematically; the equation is similar to the one representing the temperature fall across a slab which has been heated to a certain temperature and allowed to cool with one face only exposed. The mathematical expression is

$$\frac{dm}{d\theta} = k\frac{d^2m}{dx^2}$$

where m is the volume concentration of water; θ is the time interval, x is the distance from the cool face and k is a constant.

This equation has been applied by Gilliland and Sherwood[1] to interpret the condition in a slab of drying clay.

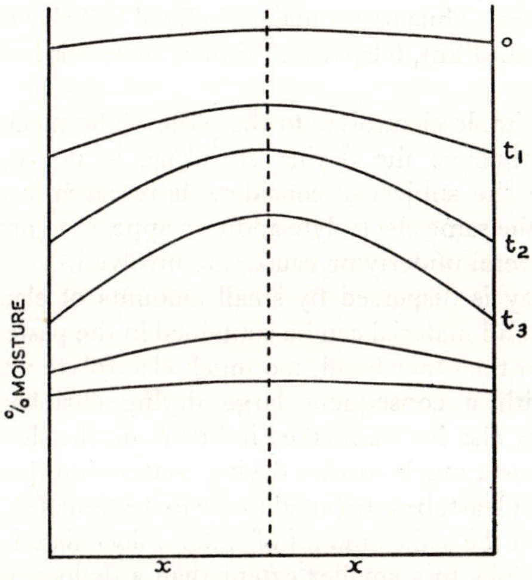

Fig. IX.11. The Moisture Gradients in a Slab of Moist Clay being Dried from the Two Side Faces

The shrinkage in a composite body which is being dried is great above the critical point, but is substantially reduced at low moisture contents. As long as all parts of the slab have a moisture content above the critical point, and, provided the rate of loss is uniform throughout, the shrinkage will be the same at all parts of the body. Once the external areas have dried, however, and contain less water than the critical point of the mass, no further appreciable shrinkage can take place, although the centre region continues to shrink.

The critical point corresponds to the condition when the body becomes *leather hard* or *black hard* and, consequently, is less able to accommodate stresses. If the external parts of the drying slab are in this condition whilst internal areas are still shrinking considerably, there will be a tendency for faults and cracks to develop. Hence, whenever the moisture gradient across the block or article being dried is great, faults and cracks are liable to form and safe drying can only be accomplished

[1] Gilliland, E. R. and Sherwood, T. K., *Ind. Eng. Chem.*, 25, 1134, 1933.

either by (*a*) reducing the drying shrinkage, or (*b*) reducing the moisture gradient.

The Shrinkage may be reduced by:

(i) Making the articles from a stiffer paste. The amount of water to be lost before the critical point is reached is thereby reduced and hence also the shrinkage.

(ii) Using a higher proportion of non-plastic large-grained material in the raw material batch. Such additions have a dual beneficial action; they reduce the overall content of colloidal material and they also build up a rigid skeleton throughout the mass which reduces shrinkage and preserves an open structure which facilitates the drying. Sand and grog are frequently employed for this purpose.

(iii) Using clays and materials which are less colloidal in character, because although other factors are involved, bodies with a high proportion of colloidal matter shrink more than others. Thus pastes made wholly of ball clay may shrink more than 20 per cent when dried. Flint, felspar and similar materials have a very low shrinkage.

(iv) Adding a suitable electrolyte to the paste. The mechanism whereby ions of different kinds influence the drying shrinkage is not yet fully understood, although it has been the subject of considerable research in recent years. Under different conditions, the same electrolyte addition appears to produce varying effects and it may be that several underlying causes are involved.

A flocculated clay is dispersed by small amounts of electrolyte and hence a larger proportion of solid material can be contained in the paste; this will reduce the drying shrinkage. On the other hand, too much electrolyte may give rise to thixotropic conditions with a consequent large drying shrinkage. The individual absorbed ions in a clay also have a marked influence on the shrinkage; calcium bentonite, for example, has a much smaller drying contraction than the sodium variety and this may, in part at least, be attributed to the hydration of the cations themselves.

When mixed with the same amount of water a flocculated clay can be dried far more quickly and shrinks to a smaller extent than a deflocculated specimen. The agglomerated mass is more open in texture and permits the freer removal of water to the surface. Fewer drying defects are encountered with flocculated clays but such bodies have a lower strength than those prepared from a deflocculated mass.

Chlorides, sulphates, and some acids are frequently added to clays to flocculate them and facilitate drying. The amounts added must be small and they should be rigidly controlled otherwise deleterious effects will be introduced.

The moisture gradient across a drying slab may be reduced by:

(i) Slowing down the drying rate by carefully controlled hydrometric conditions. Water contained in the centre portions of a slab will be transferred to the surface at a rate depending principally on capillary and surface tension forces. If the permeability of the mass is low the diffusion of moisture in this way is bound to be slow. If the rate of surface evaporation is reduced, the diffusion process will more nearly equate the moisture gradients.

(ii) Increasing the permeability of the mass. If many of the pores in the material are large, water can readily escape, but where the pores are minute, liquid or vapour transference is not easy. Thus, a lean clay, such as china clay may safely be dried more rapidly than a highly plastic clay, although both contain the same initial amount of water, because the former is more permeable. Sand and similar coarse-textured material have an even greater permeability and are far less susceptible to severe moisture gradients.

(iii) Evaporating the water at a more uniform rate throughout the mass. The rate of evaporation depends upon the relative humidity of the surrounding atmosphere and also on the temperature to which the mass is heated. In a body which is at a constant temperature throughout, the external portions exposed to a drier atmosphere must lose water at a faster rate than the interior where the atmosphere will be more saturated, but, if the inside of a drying block can be raised to a higher temperature than the outside, the rate of internal evaporation can be increased and the moisture gradient reduced.

It is this principle which is utilised in tunnel and humidity dryers.

METHODS OF DRYING

THE drying of clay bodies involves applying heat to them from some external source and thus raising the temperature in order to evaporate the water contained in them.

For this reason, methods of drying are related to the form of heating and the fundamental mechanism of its transfer to the body; this may involve conduction, convection and radiation and all three principles of heat transfer are used in dryer practice.

Conduction Methods

If an article to be dried is placed in direct contact with a hot surface, heat is transferred by conduction and penetrates into the body, thus causing a migration of water to the cooler parts of the shape. It will ultimately escape to the atmosphere from the opposite face.

The theoretical condition from such drying techniques would produce an undesirable moisture gradient and such effects occur if large shapes are to be dried. Conduction methods are successful only if the operation proceeds very slowly or where very thin ware is to be dried.

Hot Floor Dryers. Before humidity dryers were introduced, it was customary to dry all the ware after forming, on *hot floors* or *shelves* heated by steam or by waste gases from the kilns. In this way, water was evaporated entirely from the surface layers exposed to the atmosphere in the dryers, thereby creating large moisture gradients throughout the ware. Excessive losses through differential shrinkage could be avoided only by very slow drying entailing an operation of many days.

Hot floor drying is still in common use in this country in the manufacture of building bricks and articles of special shape such as architectural terra cotta and large refractory blocks. It is being superseded in most sections of the industry by controlled humidity and other methods of drying.

In some porcelain and earthenware factories, the ware is dried in the moulds in a few minutes, the moulds resting on tubular metal shelves which convey hot air to small jets above each mould.[1] The control of temperature and humidity of the air is very important and only thin ware can be dried in this way.

CONVECTION METHODS

THE circulation of warm air around the articles to be dried, serves a dual purpose in heating the ware and in removing the evaporated water by convection circulation.

Only if the rate of circulation of air is rapid can drying proceed quickly, because two effects tend to inhibit evaporation:

(1) When water is converted into steam, latent heat is absorbed which reduces the temperature of ware in the immediate vicinity of the surface.

(2) As waters evaporates from the article, its partial pressure in the air layers in immediate contact rises to saturation point. Unless it is removed progressively, the drying rate will be substantially lowered and may even stop.

The advantage of convection heating in drying practice is that water is evaporated from all exposed surfaces and as the outer layers of moisture are removed, migration from the centre through capillaries can occur, although this may be a slow process in thick ware because heat is slow to penetrate the article.

If the heat applied to surface layers is too great, vaporisation of water can drive moisture back into interior parts of the shape and prevent its escaping.

Many of the corridor, chamber and tunnel dryers of industrial practice operate on the convection principle and by their use, drying can be completed in a few hours or days depending on the size and complexity of shape.

Humidity Dryers involve the convection principle, but they also exploit the concept that a wet article will not lose water, irrespective of its temperature, if the surrounding air is fully saturated with water vapour. In this way the whole volume of large shapes can be raised in temperature quickly and efficiently without water loss occurring and by conduction through the water which is present. Hence there is less tendency to drive moisture into the interior which is an undesirable feature of most ordinary convection dryers.

FIG. IX.12. DIAGRAMMATIC REPRESENTATION OF A TUNNEL DRYER

As it is obviously not feasible to raise the inside of a block to a higher temperature than the outside by a single operation, the whole article must first be heated to a uniform temperature and then the outside cooled to achieve the desired results for humidity drying. It is of the utmost importance to ensure that no water is lost during the initial heating otherwise differential shrinkage is bound to occur. Consequently the surrounding atmosphere during the heating process has to be kept fully saturated

[1] Hancock, W., *Ceramics*, 5, 408, 1953.

with water vapour, and reduced in humidity only after the desired temperature has been attained.

Tunnel dryers usually operate on the so-called counter-flow principle as illustrated in Fig. IX.12.

The ware to be dried is loaded on cars which move slowly through a tunnel usually about 100 ft. long. Warm air, which may be drawn from the cooling kilns, enters at the opposite end and during its passage down the tunnel becomes saturated in moisture derived from the ware which has proceeded some distance into the tunnel. The most efficient humidity dryers incorporate a heating unit near to the ware entrance, hence the wet articles encounter an immediate hot, saturated atmosphere. No water is lost initially, but the temperature is raised throughout the complete body.

As the ware proceeds along the tunnel, the surrounding atmosphere becomes progressively cooler and drier, thereby reducing the outer parts of the ware in temperature and permitting evaporation. The interior of the ware remains at a higher temperature and if the rate of passage of the cars through the dryer is carefully controlled, the moisture gradient across the body may be kept down to a minimum.

Finally, the loaded cars pass out at one end of the dryer, with their contents practically dry, whilst the warm air saturated with moisture, passes out through a chimney at the other end of the dryer.

Many of the tunnel dryers used in the clay industries operate under somewhat different conditions. They make use of waste heat from kilns and it is customary to feed this through the tunnel dryer in the opposite direction to the ware. Consequently the ware, in its transit through the dryer, encounters increasingly *warmer* and drier air. Despite this fact, if conditions are carefully controlled, entering ware is surrounded by saturated air and can be dried uniformly. It is important, however, to secure a correct balance between the temperature and velocity of warm air and the rate of passage of ware through the tunnel dryer.

A similar method, but taking the air through a series of loaded *chambers*, the goods to be dried remaining stationary, also gives very satisfactory results. Such dryers are known as *Keller* or *chamber* dryers.

These units may operate singly or in banks of three or more. The ware must be inserted in a cool chamber and evaporation avoided in the early stages by admitting live steam or by covering the racks with wet sacks. The temperature can be raised quite rapidly provided that the air is maintained in a fully saturated condition. When the whole chamber is at a uniformly high temperature, drier and slightly cooler air is allowed to sweep through the setting, thus drying the ware in a way which avoids moisture gradients.

Humidity dryers not only reduce the losses in the drying process but also considerably improve the space and time factors.

Radiation Methods

The third principle of heat transfer is being applied with increasing importance in ceramic drying practice.

Heat is a form of energy which has a wave-form similar to visible light but of longer wavelength. The radiation involved belongs in the infra-red region of the spectrum in the wavelength range 8×10^{-5} to 4×10^{-2} cm.

Heat transfer by radiation from a hot source to a colder body is much more efficient than it is by conduction or convection means. In addition, water molecules absorb strongly in certain regions of the infra-red radiation range and preferentially retain energy which might otherwise be lost by reflection. For these reasons, heat transfer coefficients from infra-red radiation sources can be 3–4 times as high as those from conduction and convection, and bodies can be heated more rapidly and uniformly. The absorption of characteristic wavelengths by water molecules is particularly important in that it can accelerate heat transfer within the body itself and so avoids large temperature and moisture gradients.

Infra-red drying reduces costs and cuts down drying time. It is ideal for surface drying and has been used extensively for glazed articles and wall-tiles. When drying large articles care must be taken not to volatilise a greater volume of steam than can escape through surface capillaries. Experimental units for drying bricks suggest that shapes made from some clays containing 10–15 per cent water can be dried by infra-red means in about 30 minutes.

Di-electric or high frequency drying processes operate on the same principles as infra-red, but radio-waves of high intensity are the source of radiation.[1] In theory, this wavelength is absorbed by water molecules, which are then excited to volatilisation point. The clay body does not absorb such radiation, but, unfortunately, it acts as an efficient screen to penetration in depth. This is also a problem in infra-red methods.

Much research is being carried out on drying of articles by these methods and the future may see startling developments along these lines.

Electrical Current Drying. Jefremow[2] in 1936 claimed that small ceramic articles could be dried by passing a D.C. electric current through the mass. The author had remarkable success in drying large back-slabs of size 48″ × 24″ × 2″ by these means. The articles were slip-cast and under the most carefully-controlled conventional drying techniques they tended to crack and warp.

A D.C. current of low voltage and high current strength applied at the narrow end edges proved a most efficient drying unit. Conductivity took place through the water continuum which thus became heated to evaporation point. Moisture gradients could not occur because if one part were preferentially dried, less current passed through this zone. It was difficult maintaining a good electrical contact throughout a drying cycle, but cracking and warping were entirely eliminated and the ware could be dried to below its critical moisture point quite rapidly.

All methods of drying used on the industrial scale aim at reducing the water content below the critical point. This ensures that further excessive shrinkage will not take place during the initial stages of kiln firing, hence there is no necessity to waste time and fuel in more thorough drying. In heavily-grogged articles and in others in which the proportion of non-plastic material is high, it may be necessary to reduce the water content below 5 per cent, but, in other cases, appreciably more may be present.

[1] Vassiliou, B. and White, J., *Trans. Brit. Ceram. Soc.*, 52, 377, 1953.
[2] Jefremow, G. L., *Ogneupory*, 1 S, 33, 1936.

That this statement should not be applied generally is well known to all who have had experience of the firing of clayware by rapid cycle methods. In the early stages of firing the fast rate of temperature rise induces a high temperature gradient across articles particularly those of large dimensions. Small quantities of entrapped water in such bodies are driven into the cooler interior and there may accumulate. When the water finally volatilises it may do so with disruptive violence and either shatter completely the block in which it was contained or cause flaking or blistering of the outside layers.

Extruded articles and those made from finely divided clays are particularly prone to this type of fault which may occur even though the body had been dried to a moisture content well below the critical point.

The 'blowing' of articles in this way can be eliminated either by ensuring that a maximum of 1 per cent water is present when articles enter the kiln, or by so adjusting the pre-heating of the kiln that adequate time at a comparatively low temperature (below 200° C.) is available for the complete removal of water.

PROPERTIES OF DRIED ARTICLES

Dry Strength and Hardness. When ceramic bodies are dried the strength of the article usually increases considerably. Clay ware shows a pronounced increase, common brick clays and many fireclays are hard enough to withstand considerable mechanical handling after the drying process.

The underlying mechanism of dry strength is still a perplexing, unsolved problem although several interesting theories have been advanced. It is well established that the specific surface area of the particles in a dried mass is directly related to the strength developed. Furthermore, it is known that the last traces of water adsorbed by a clay are intimately held and are driven off only with difficulty even at high temperatures. One explanation for the dry strength of clay bodies is based on this latter fact. The retained water is regarded as forming a thin film covering the surface of all particles and having a structure induced within it. When the particles are touching, such films may blend together thereby forming a *continuum* which extends throughout the mass; the more points of contact of particles, the greater will be the strength in the mass.

Another theory has been suggested by Weyl, which is based on the distortion of ions in the surface layers of particles. A simplified picture of the ions in a crystal is shown in Fig. VIII.1 (p. 445). A cation which is situated at some distance from the external layers is in equilibrium and balanced by evenly-spaced charges of opposite sign, but, a cation in the surface of the same crystal does not have a full quota of opposite, balancing charges surrounding it, hence there must be some distortion. Weyl suggests that the ions in the surface layers are not at the normal separation distance and that this contraction may extend over several ionic layers in the immediate vicinity of the surface.

This concept of surface distortion is most valuable and some aspects of the unusual behaviour of small particles can be more fully understood on the basis of the irregular properties of the surface (see p. 457). By assuming that such distortion exists, a mechanism which accounts for the dry strength of agglomerated masses can be advanced.

Because ions in the surface layers are unbalanced electrically, they will have a natural potential for attracting other ions, which will be rigidly held by electrical bonding. If another surface, also with distorted external groupings makes contact with the former the two adjacent surfaces can share their unbalanced electrical fields as illustrated in Fig. IX.13. The forces between such particles will be partially ionic in character and considerable pressure will be required to force them apart. The greater the surface areas in contact the higher will be the strength of the aggregated mass. Therefore, clays and similar materials in which the particles are, for the most part, small and plate-shaped are expected to have large dry strengths.

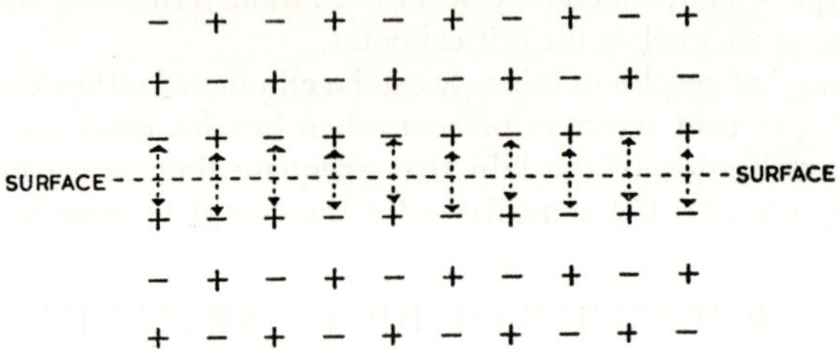

FIG. IX.13. A SUGGESTED MECHANISM FOR THE BONDING BETWEEN TWO ADJACENT CONTACT SURFACES

Although clay bodies are usually stronger after drying, their elasticity decreases and they become more brittle. As long as a small amount of water is present in a clay mass, flow is possible so that the body is able to 'give' and thereby accommodate stresses; but, in a completely dry body, the bonding is rigid and there is virtually no deformation before fracture takes place. For this reason, articles are better cut and drilled or otherwise shaped when they are in the 'green' state than when they are dry.

Hygroscopicity is a term applied to dry materials which absorb water when exposed to moist air. Clays which have been thoroughly dried possess this property but the proportion of water so absorbed is usually very small and never sufficient to make the mass fully plastic. Montmorillonite-type clays can reabsorb up to 15 per cent of water under ordinary conditions of humidity and in doing so undergo considerable expansion.

It is for this reason that drying of clays by desiccation has only a limited application. A fully dried clay is of itself a powerful desiccant capable of extracting water from most other materials and absorbing it onto itself. This can be a serious problem in analysis particularly in determining the free water content by drying at 110° C. Most balances weigh satisfactorily only at room temperature, therefore the sample being tested must be withdrawn from the oven and cooled to room temperature without being allowed to reabsorb water. This is virtually impossible and for truly accurate results the sample should be weighed at 110° C.

DEFECTS DUE TO INCORRECT DRYING PRACTICE

UNDOUBTEDLY many defects in ceramic articles can be attributed to incorrect methods of drying. The faults, in some cases, are not apparent until the ware has been fired, when they have been accentuated by changes brought about by the heat treatment.

In general, the defects which arise as the result of bad drying methods are of four types: (*a*) those due to differential shrinkage and the consequent induced stresses in the ware; (*b*) those due to the migration of the finer particles of the body by moisture transference to the drying surface; (*c*) those which are due to the accentuation of differential stresses arising initially in the forming process; and (*d*) those due to too rapid volatilisation of water within the body of the article which is unable to escape through the fine pore structure at a sufficiently rapid rate.

The most obvious effect of *differential shrinkage* on drying is the *cracking* of ware. This may take many forms but it is most usual for the faults to appear parallel to a face, which is drying more rapidly than other parts of the body. Whilst the material has a water content above the critical point, all parts of the body are plastic and should one area shrink more rapidly than others the strain can be accommodated. Once the leather-hard condition has been reached in part of the ware, deformation is less easy and areas with a high moisture content tend to shrink away from those which are drier. Cracks develop along inherent lines of weakness which in pressed ware are at right angles to the direction at which pressure is applied. Bricks tend to show lamination cracks running parallel to their large face; extruded blocks occasionally show S-shaped cracks along lines of weakness induced by helical movement of the body during forming.

In bodies of irregular shape, the thinnest sections dry more readily and cracks appear at the junction with thicker parts of the ware.

Moulded or cast bodies, especially hollow-ware articles may crack at mould junctions during drying, where there is a line of weakness caused by irregular particle arrangement.

Another defect in ware which is associated with uneven shrinkage is *warpage* or *deformation*. This is most common in thin-ware, but it is also found in large bodies where movement during drying has been restricted by contact with a floor or shelving. Warpage is caused by stresses which arise from uneven drying shrinkage in the body, but where the elastic or plastic nature of the material has permitted movement without cracks developing.

Many faults, normally attributed to incorrect firing, may have their origin in incorrect drying practice. *Dunting* in glazed ware is often due to hair-cracks which have developed at the edges of the body during drying. This is a not too uncommon phenomenon in glazed terra cotta which has been dried on hot floors. The dunting in such cases is limited to one edge or face which has been dried too rapidly.

Shelling may also be caused by improper drying of ware after the application of the glaze slip. Such faults are easily remedied by the correct control of the consistency of the glaze slip, but the defect is apparent after drying and should be clearly differentiated from true shelling, which is due to the incorrect balance between the thermal expansion of the glaze and body and is apparent only after firing.

When drying involves the migration of water from the interior of the article to the surface, fine-particled material may also be transported and deposited in the surface layers. This leads to *skin-formation* and an alteration in physical and sometimes chemical properties of the outside layers relative to the internal areas.

Because of the difference in body composition in various parts of the ware as a result of migration, *cracks* and *warpage* are liable to develop on firing, and frequently surface layers peel off during kiln firing.

The skin formation may be so serious that an impervious layer forms which retards further drying. In such cases, the article is liable to *disrupt* during kiln firing due to the vaporisation of the uneliminated water. An impervious skin formed during drying increases the tendency to '*black-heart*' *formation* by restricting the free access of oxygen during the cleaning period.

A common type of black-heart is caused by bad drying practice alone and has little to do with the carbon content of the clay. When water is entrapped within a body and not fully removed before a moderate heat is attained in kiln firing, volatilisation will occur before or at the criticial temperature of water (374° C.) and high pressures will be involved. Any iron compounds present in the clay tend to become reduced under these conditions and they do not reoxidise very readily as the temperature is increased. The effect is manifested by the fired ware having a characteristic steely grey core of high vitrification and usually with a sharp line of demarcation between itself and the unaffected body. If the body is overfired these steel-grey areas will vitrify excessively, because reduced iron is a powerful flux. Should this occur, the body will blister or bloat due to gas liberation within the core at the vitrification point.

If such faults are present in fired ware they can be cured only by the more thorough and careful elimination of water at low temperatures.

Unless considerable care is taken, it is not unusual for drying faults in clayware to increase during spells of cold weather. It is at such times that dryer cycles are apt to be difficult to maintain, but, no matter how careful the control, if ice is present in the raw clay, or in the mixing water or if it develops in the moulded article, cracking will inevitably occur.

Soluble salts which may be present in the body are readily transported to the surface layers during drying and if they are large in amount they can produce serious *discoloration*. The surface of red-burning ware is frequently unpleasantly discoloured by calcium salts which have been transported during drying and produce yellow or buff areas on firing. Much of the *scumming* on bricks and other ware has a similar origin, but does not show up until the articles have been fired.

In Chapter VIII, the problems of differential stress development in ceramic bodies during their shaping were considered.

When these stresses are of high magnitude they cause cracking or similar defects which are often magnified during the drying process even though shrinkage may not be involved. A fully dried clay article is brittle and has little strength in tension, whilst even when partially wet, the tensile strength may be considerable.

Articles which are formed at moisture contents near to or below the critical point commonly exhibit such stresses and they are most difficult to dry without cracks and other faults occurring.

The only safe way to dry articles which have been formed under high pressures and low moisture contents, is to store them at room temperature in a draught-free atmosphere for a period of time up to 48 hours. Drying should be avoided during this period. Under these conditions inherent strains sort themselves out, and the body can then be dried out rapidly with safety.

The higher the forming pressures involved, the longer the articles should be left in an ambient condition.

Pressure-formed articles are also liable to develop a more impervious outer face than those manufactured with the clay at a soft consistency. In the latter type of shaping, as the water dries from surface layers, a pore structure develops which facilitates further water removal.

Contrary to expected behaviour, the author has noted that some stiff-extruded clay bricks are more difficult to dry and require a longer time than the same clay formed by soft extrusion methods at 3–4 per cent more water content.

The complexities of drying practices are thus considerable and many factors must be taken into account so that defects do not spoil the ware.

Provided that the defects which occur are clearly recognised and their cause ascertained, they can usually be eliminated by correcting the drying procedure. In theory, an efficient, well-regulated humidity dryer can dry all types of ware without inducing differential shrinkage or the transportation of fine particles. Many drying faults, however, are brought about by the incorrect use of such methods or by the failure on the part of the operator to recognise the principles involved.

CHAPTER X

THE EQUILIBRIUM STATE AND PHYSICO-CHEMICAL REACTIONS

In any system of minerals or chemical components, an *equilibrium* state will exist which will be specific to any particular temperature and pressure. Equilibrium may be reached quickly in some mixtures of components but in others it may take a geological age.

When conditions influencing an equilibrium state are changed either by adding another chemical component or by changing the proportions of one or by altering the temperature or pressure, a change or *reaction* may occur which will proceed until another equilibrium state is reached.

Such reactions are of two general types:

(*a*) **Physical reactions,** in which there is no chemical change, but the substance may be changed from a solid to a liquid or gas (or *vice versa*) or into crystals of a different habit, as quartz into cristobalite or tridymite. Physical changes are produced by such processes as melting or fusing, solidifying, volatilisation or vaporisation, sublimation, solution, flocculation or precipitation and inversions and conversions.

(*b*) **Chemical reactions** in which one or more new products of different composition from the original substance are formed, as when copper is heated in air and is converted from a piece of bright metal into a mass of black copper oxide, or when carbon is heated in air and is converted from a black solid into a gas (carbon dioxide), or when a current of electricity is passed through water and converts it into two gases—hydrogen and oxygen.

There is frequently no sharp distinction between chemical and physical reactions in practice—though theoretically they are quite distinct—because a physical change such as a great reduction in the size of the particles, may bring about a chemical reaction at a much lower temperature. A molten mass may have the same composition as the crystals which form on cooling it (physical change), but many molten ceramic materials are not pure, and the crystals may have a composition different from the original liquid and still more different from any of the liquid which forms a slag or glass when cold, the latter case may be regarded as a chemical change on crystallisation or as a physical separation of crystals of one composition from a liquid of another composition.

In consequence of this ambiguity, it is usual to refer to many of these changes as the result of *physico-chemical reactions*, without making any closer distinction between them and to regard such reactions as causing the initial substances to form products of a different type.

Physical Changes. Physical changes in the state of a substance or mixture are usually due to the effects of pressure or heat and, more rarely, to the effect of light, magnetism, electricity and other physical forces.

They are chiefly concerned with:

1. Changes in form, e.g. from solid to liquid or to gas or *vice versa* without any decomposition.

2. Changes in volume, e.g. expansion, contraction (shrinkage), which may lead to changes in texture, hardness, porosity, density and crystalline form.

3. Changes in other physical properties such as viscosity, plasticity, 'strength', and optical and electrical properties.

Some of these apparently physical changes, especially when caused by heat, may be affected by chemical changes which accompany them, such as the reactions which occur when, as the result of heating, some portions of a solid mixture begin to fuse and the resulting liquid then reacts with some of the remaining solid and forms new chemical products instead of acting merely as a solvent. For this reason, readers should refer to Chapters XI and XII which deal with the effect of heat on ceramic materials as well as to page 692 in the present Chapter.

Chemical Reactions. The chief difference between chemical and physical reactions is that the product of a chemical reaction is of a different composition from that of the original material, so that chemical changes appear to be due to forces of a different nature though some physical forces may also be involved.

The most characteristic force associated with chemical reactions is that known as *chemical affinity*—a term which is not very appropriate as *dis*similar atoms attract one another rather than similar ones, but it is still so largely used that it must be retained.

Modern chemists regard the force which produces chemical reactions as the tendency of each atom to surround itself with a balanced shell of electrons (see Chapter II) and in so doing, it effects the combination or rearrangement of atoms, ions and molecules of which the material is composed. In this sense, it is the unbalanced nature of any atomic grouping which largely determines its reactivity and the possibility and extent of any chemical reaction.

The extent to which various substances will react with one another also depends largely on their physical state (this, in turn, being related to their temperature); thus, liquids react more readily than solids, and gases more readily than liquids. In each case, this is partly a result of the freedom of the molecules and of their constituent atoms, the ultimate particles in a gas having the maximum mobility, whilst those in some solids have scarcely any freedom of movement, or any contact with other reacting molecules.

The ease and rapidity with which compounds react with one another also depend on their stability. A very stable substance in which the valencies or affinities of the various atoms for each other are well balanced will not be easily decomposed, but an unstable, unsaturated compound, or one in which the elements are only loosely united, will react readily.

All reactions, when once started, proceed until a state of equilibrium is reached. This state depends on a number of physical factors, such as temperature, pressure, surface tension, electric charges, etc. and if any one of these factors is changed the state of equilibrium will be destroyed and the reaction will proceed further either in its original direction or in the opposite direction. For example, when calcium carbonate (limestone) is heated to 1000° C. in a closed vessel, it is decomposed, forming lime and carbon dioxide gas. This decomposition continues until the pressure of the gas reaches a definite critical value. When the pressure reaches this value, the decomposition will cease, but will start again if a current of air is drawn through the vessel. By continuing to remove the carbon dioxide gas as fast as it is formed, the whole of the calcium carbonate may be decomposed. If, on the contrary, the lime so produced is subjected to the action of carbon dioxide, the original reaction will be reversed and calcium carbonate will be reformed.

The reaction and its reversal may be shown by two equations:

$$(a) \quad CaCO_3 = CaO + CO_2$$
$$(b) \quad CaO + CO_2 = CaCO_3$$

For convenience these may be combined, half arrows being substituted for the sign of equality:

$$CaCO_3 \rightleftharpoons CaO + CO_2$$

The most stable equilibrium which can be produced exists when two substances, each having a great affinity for each other, are combined together. Thus, silicon and oxygen when combined to form silica, produce an exceptionally stable compound, which is highly inert at temperatures below a red heat, though both its constituents (silicon and oxygen) are active elements.

Hence, when acids and bases react with each other, the final equilibrium product will be a substance (salt) in which both the reacting substances are present in maximum proportions. Intermediate states of equilibrium may occur when there is not sufficient of one material to combine with the whole of the other, and under such circumstances a basic salt may be formed if there is a deficiency of acid, and an acid salt if there is a deficiency of base.

Chemical reactions always tend to form the most stable compounds possible under the prevailing conditions. Where several substances are present, those which have the greatest affinity for each other will react in proportion to their affinities until a state of equilibrium is reached. In the case of a single acid reacting with a single base, equilibrium (in this case neutrality) is reached when the maximum amount of neutral substance is formed.

Associated with such changes are energy effects which are usually manifested as heat. Thus some reactions take place with the liberation of heat and are termed *exothermic*, whilst others need to absorb heat from their surroundings and are therefore referred to as *endothermic*. The energy change may not be developed solely as heat; for example, atomic transformations are manifested by light, heat and explosive forces. It is more correct therefore to regard a reaction as a system where the initial energy state is different from the final one. Physical chemists refer to the energy factor as a change in *entropy*.

The total energy change when a reaction takes place may be written as

$$\Delta U = U_{final} - U_{initial}$$

Enthalpy Changes. Sometimes part of the energy is expended in doing work, such as building up a pressure to drive machinery, developing electricity, etc., but usually the energy change of a reaction involves a heat change (ΔH), called the *enthalpy* change of the reacting system. In a mixture of elements and compounds the internal motion of atoms and the interaction between atoms and molecules contribute to the internal energy of the system. When reactions take place, a different set of conditions apply—new chemical or physical entities will form which will have a different amount of internal energy. It is this change in energy which is important in reactions and the change in enthalpy is defined as the heat absorbed or liberated when a reaction takes place at constant temperature and pressure.

By definition an endothermic process is one in which the enthalpy change (ΔH) is positive, whilst for an exothermic change, in which heat is given out, ΔH is negative, i.e. the system has a lower heat energy.

The enthalpy change is of the greatest importance when studying chemical reactions. Not only will it predict the heat liberated or absorbed, but it also indicates the probable direction of a change. A large negative ΔH of a reaction at a particular temperature will usually imply that the reaction will proceed spontaneously and, conversely, a large positive value would indicate that a very stable configuration was already present and that a large amount of external energy would be required to activate it.

It is usual in chemical reactions to represent the enthalpy change in kilocalories.

Thus
$$\underset{\text{(solid)}}{C} + \underset{\text{(gas)}}{O_2} = \underset{\text{(gas)}}{CO_2} \qquad \Delta H = -94{\cdot}1 \text{ kcal.}$$

and
$$\underset{\text{(gas)}}{2H_2} + \underset{\text{(gas)}}{O_2} = \underset{\text{(liquid)}}{2H_2O} \qquad \Delta H = -136{\cdot}64 \text{ kcal.}$$

$$\underset{\text{(gas)}}{H_2} + \underset{\text{(solid)}}{I_2} = \underset{\text{(gas)}}{2HI} \qquad \Delta H = +12{\cdot}4 \text{ kcal.}$$

It is important to remember that the heat changes are usually referred to the condition of the reactants and the products at room temperature and pressure and the values may be different under changed conditions. This is particularly true when a phase change may be involved as in the example of the formation of water quoted in the second equation. In this, two gaseous reactants yield a liquid so that the heat value quoted included the latent heat term, thus,

$$\underset{\text{(liquid)}}{H_2O} = \underset{\text{(gas)}}{H_2O} \qquad \Delta H = +10{\cdot}52 \text{ kcal.}$$

So that for the reaction involving the formation of steam (gas) from component gases, the equation with its energy change would be written

$$\underset{\text{(gas)}}{2H_2} + \underset{\text{(gas)}}{O_2} = \underset{\text{(gas)}}{2H_2O} \qquad \Delta H = -115{\cdot}60 \text{ kcal.}$$

In the reactions quoted, the change can be made to proceed in the reverse direction under certain conditions. When this occurs, the enthalpy change is the same in value but reversed in sign. The changes above involve a *heat of reaction*, but they can also be termed *heats of formation* reactions because there is only one molecular product from single atomic species. In such cases, it is customary to refer the enthalpy change to one mole of the product and to its normal stable state at 25° C. and 1 atmosphere pressure.

Thus, the enthalpy change of formation of the compounds could be referred to one mole of product and hence the heat change would be the heat of formation.

Thus	*Compound*	$\Delta H/mole$
	CO_2 (gas)	$-94{\cdot}1$ kcal.
	H_2O (liquid)	$-68{\cdot}32$ kcal.
	HI (gas)	$+\ 6{\cdot}2$ kcal.

Other examples of heats of formation for more complex substances are given on p. 696.

Free Energy Changes. Although the change in enthalpy—the ΔH term—indicates the heat involved in a reaction and the probable direction of change under specific conditions, it does not indicate the tendency for the reaction to occur or the rate at which it will take place. In these factors another energy term is involved which might be expressed as the limiting amount of energy required to perform the work of initiating and continuing the reaction. It is called the *Free Energy* of the reaction and is given the symbol ΔG.

In ceramic studies the value of ΔG is important because by its use, it is possible to predict the rate at which reactions will proceed and also how far they will continue before coming to equilibrium. ΔG is dependent on pressure and time, whereas ΔH is not, and if in a particular system of components there is a possibility of several reactions taking place, the ΔG value of each combination will establish which change is likely under the experimental conditions.

Unfortunately, most ceramic reactions are complex and take place only at high temperatures where it is difficult to collect precise energy data. Consequently values of both ΔH and ΔG are known with accuracy only for simple reactions, and a few more complex systems.

The potential usefulness of ΔG diagrams in ceramic systems is illustrated in Fig. X.1 where values of the Free Energy of iron oxide reactions are shown along with those in the reaction of hydrogen with oxygen and carbon with oxygen.

If the value of ΔG is negative the reaction will have an equilibrium value towards the right hand component. The greater the negative value of ΔG the more strongly the reaction tends to the right.

When reactions can take place simultaneously, the one preferred is that with the highest negative value of ΔG. So that if Fe_2O_3 is heated in the presence of carbon and the supply of oxygen is limited—as it often is in the firing of clays—at temperatures above 260° C., there will be a tendency for reduction to Fe_3O_4. Above 570° C., further reduction to FeO is likely and at about 700° C. complete reduction to metallic iron can occur. In these cases the ΔG values of the various iron oxide reactions are being considered in relation to the

$$2C + O_2 \rightarrow 2CO \text{ change.}$$

Carbon monoxide itself is a powerful reducer of iron oxides and if it is present in kiln gases it will favour the formation of FeO at temperatures above about 800° C.

If the supply of free oxygen is limited, water vapour will reduce iron oxides successfully, because the ΔG values of the formation of water vapour are considerably more negative at all temperatures than the change.

$$4\,Fe_3O_4 + O_2 \rightarrow 6\,Fe_2O_3$$

and more negative than the change

$$6\ FeO + O_2 \rightarrow 3\ Fe_3O_4$$

at temperatures above 500° C. Reduction of iron oxide is thus predicted by water itself under specific experimental conditions.

Simple thermodynamic principles would thus predict the various types of black-coring of clay bricks which have been discussed on p. 558.

For the most part, ΔG curves are linear and their slope is related to the *entropy* change involved in the reaction. Entropy may be defined as the total internal energy of the components of the system.

If the ΔG curve is projected to absolute zero (−273° C.), its intercept is the *total energy* change of the reaction (ΔH) (see Fig. X.1). Thus for the formula $2C + O_2 \rightarrow 2CO$ the ΔH value is −52·8 kcals. for the reaction, but as ΔH is normally expressed on the basis of g. formula of the product, the value has to be divided by 2 in this case. Thus the energy of formation of carbon monoxide from carbon and oxygen is −26·4 kcal/g. formula.

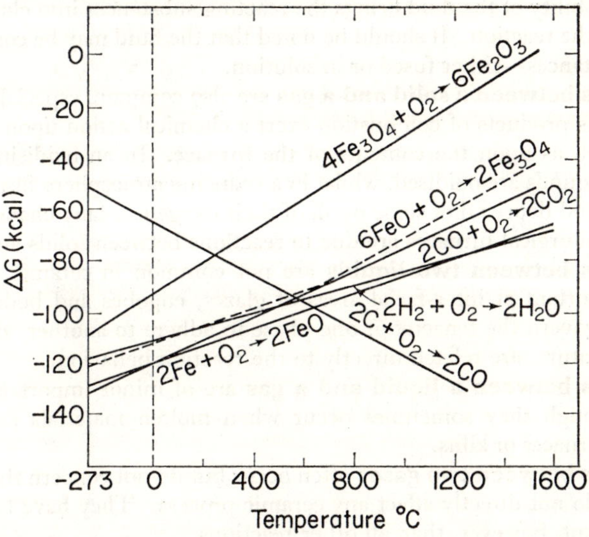

FIG. X.1. ΔG CURVES OF SIMPLE REACTIONS ILLUSTRATING THE PROBABLE BEHAVIOUR OF OXIDES OF IRON IN THE PRESENCE OF CARBON MONOXIDE AND WATER VAPOUR AT DIFFERENT TEMPERATURES

It must not be assumed that if the ΔG values of two alternative equations are compared, reaction in the direction of the one of greater negative value will immediately proceed. ΔG values indicate only the stable conditions at equilibrium and physical conditions can often slow down a reaction or even prevent its taking place. A lump of iron exposed to a damp atmosphere will oxidise only slowly at room temperature although free-energy considerations would strongly favour the oxide.

Finely-divided iron reacts much more quickly and if energy in the form of heat is supplied to initiate the reaction, the oxide will form instantaneously and quantitatively with the liberation of a considerable amount of heat energy.

Chemical reactions may occur between matter in the form of either solids, liquids, or gases. Thus, the following reactions are possible:

1. **Reaction between two solids** is not common at ordinary temperatures, and in most cases, if it occurs at all, it is so slow as to be negligible unless great pressure is applied, as when a mixed explosive is fired by a blow from a hammer. At high temperatures, many solids react with one another; thus lime and silica, when finely powdered and heated to about 600° C. combine to form calcium silicate, the action commencing before either of the substances is in a fluid state. As solid substances are so immobile, at least one of them must usually be in a very fine state of division and in close contact with the other before any combination can occur. The application of pressure, by bringing the particles closer together, facilitates the reaction.

2. **Reactions between a solid and a liquid** are very common and may be exemplified by the corrosive action of acids and alkalies upon various solid substances and by that of fluid slags, etc., upon firebricks and other ceramic materials. The great mobility of the fluid brings the reacting substances into close contact and so facilitates the reaction. It should be noted that the fluid may be composed of one or more substances—either fused or in solution.

3. **Reactions between a solid and a gas** are also common, especially in furnaces, as the gaseous products of combustion exert a chemical action upon the refractory linings as well as upon the contents of the furnace. In an oxidising atmosphere ferrous compounds are oxidised, whilst in a reducing atmosphere ferric compounds are reduced and deprived of some or all of their oxygen. Many changes which take place in metallurgical furnaces are due to reactions between solids and gases.

4. **Reactions between two liquids** are not common in ceramic processes, but they are important at inter-facial areas of glazes, engobes and bodies. Diffusion rates which govern the tenacity of one phase to adhere to another, after cooling to room temperature, are related directly to the reaction behaviour.

5. **Reactions between a liquid and a gas** are of minor importance in ceramic processes though they sometimes occur when molten materials are oxidised or reduced in furnaces or kilns.

6. **Reactions between two gases** often occur but do not concern the present subject as they do not directly affect any ceramic process. They have been studied to a greater extent, however, than all other reactions.

TYPES OF CHEMICAL ACTION

THERE are five chief kinds of chemical action:

1. *Direct combination*, such as when magnesium and oxygen combine:

$$Mg + O = MgO$$

or when silica combines with lime:

$$2CaO + SiO_2 = 2CaOSiO_2\ (Ca_2SiO_4)$$

This type of action is very common in ceramic processes, especially during the burning of goods. Other more complex reactions of the same type also take place, e.g. the combination of fluxes such as lime, soda, potash, magnesia, etc., with silica and alumina. These, together with reactions occurring during oxidation and reduction in ovens, kilns and furnaces, are considered in greater detail in Chapter XI.

2. *Displacement by an element or group of elements*; as when magnesium is attacked by hydrochloric acid:

$$Mg + 2HCl = MgCl_2 + H_2$$

Reactions involving displacement are not very common in ceramic processes, but are common in metallurgical operations.

3. *Mutual exchange or double decomposition*, as when calcium sulphate and sodium carbonate interact:

$$CaSO_4 + Na_2CO_3 = CaCO_3 + Na_2SO_4$$

In all such reactions, the acid radicle or ion of one base leaves it and becomes combined with another base, the acid radicle or ion previously combined with the latter then combining with the first base. The general equation for all mutual exchanges of this character is:

$$AB + CD = AD + CB$$

4. *A rearrangement* is said to occur when the atoms in a compound are recombined in a different manner. The classical example of this is the conversion of ammonium cyanate into urea:

$$(CN)O(NH_4) = N_2H_4(CO)$$

A modified form of 'rearrangement' probably occurs when alumina and other substances become polymerised on heating or when one form of silica such as quartz is converted into another, such as cristobalite, or *vice versa*.

5. *Dissociation* is said to occur when a substance is decomposed into its constituents, as when calcium carbonate is decomposed into lime and carbon dioxide:

$$CaCO_3 = CaO + CO_2$$

or kaolinite is decomposed into free silica, alumina and water:

$$Al_2(Si_2O_5)(OH)_4 = 2SiO_2 + Al_2O_3 + 2H_2O$$

Many instances of dissociation occur in ceramic processes. Thus clay, bauxite, limestone, magnesite, dolomite, etc., are all dissociated when heated. Many of the impurities in ceramic materials are also dissociated by heat; thus, sulphides and sulphates are decomposed with the evolution of gas, limonite (ferric hydroxide) loses water and forms ferric oxide, and many hydroxides behave in a similar manner.

Although all chemical reactions can be included in one of the foregoing types, it frequently happens that several reactions succeed one another in such rapid succession that two or more types of reaction may be involved. The types just mentioned should, therefore, be regarded as guides and not as indicating a very marked dividing line between each type.

SOME FACTORS INFLUENCING CHEMICAL REACTIONS

THE factors which influence chemical reactions are very numerous and need to be carefully considered if the reactions are to be properly understood. The number of these factors, the difficulty in distinguishing completely the influence of each, and especially the fact that most ceramic reactions occur at high temperatures, are largely responsible for the small amount of definite knowledge concerning the reactions of clays and other ceramic materials. It is, therefore, very desirable to study each known factor fully, in order to determine its effect on the other factors and on the course of reaction, but the complexity of many ceramic reactions is so great that it is almost impossible, at present, to investigate them completely.

The principal factors which affect the course of chemical reactions between materials used in the ceramic industries are as follows:

(*a*) Composition (chemical and mineralogical).
(*b*) Sizes of particles and their distribution.
(*c*) Temperature.
(*d*) Time (duration).
(*e*) Mass of material.
(*f*) Pressure.
(*g*) Vapour pressure.
(*h*) Atmosphere.
(*i*) Surface phenomena.
(*j*) Crystallographic structure and properties.
(*k*) Viscosity.
(*l*) Solubility.
(*m*) Selective action.
(*n*) Catalytic action.
(*o*) Thermal conductivity.
(*p*) Electricity.
(*q*) Light.
(*r*) Change of state (solution, fusion, volatilisation).
(*s*) Intimacy of association.
(*t*) Relative quantity of each substance present.

Naturally, the important factors which govern a reaction are the energy changes involved in terms of the heat of the change (ΔH) and the Free Energy (ΔG) or work required to cause it to take place. If these conditions are not conducive to the reaction, then it will not take place in any circumstances. If the energy factors are favourable, then the above conditions can be important.

The *composition* of many ceramic materials has been described in Chapter V. The affinity of materials for each other and their tendency to react is described in greater detail in subsequent chapters.

The *size of particles* and their distribution may be a principal determinant of their reactivity. Reaction between two materials commences at their points of contact and many other reactions commence at surfaces of crystals. These factors are described more fully in Chapters XI and XII.

The *temperature* to which materials are heated influences the reactions which occur within them. Many substances which are stable under ordinary conditions react vigorously at elevated temperatures. Temperature is a principal consideration of the rate at which reactions proceed and of the nature of the equilibrium product.

In some cases, if a particular temperature is exceeded, the products of the reaction may take on a different character. The burning of bricks affords a good example, for in such cases too high a temperature will lead to the formation of an excessive amount of a liquid phase, which will reduce or even eliminate the combination and rearrangement reactions which proceed at a lower temperature. This point has to be closely studied particularly when firing clay bodies, and to secure a suitable product the 'working range' must be carefully established.

Time is almost as important as temperature in enabling a chemical reaction to take place. Although the combination of two single atoms may be instantaneous, yet when large numbers of atoms are involved, time must be allowed in order that they may come into contact with each other. Consequently, if the temperature attained is above the minimum required for the reactions to proceed, a sufficiently prolonged heating at that temperature will enable the reaction to progress to completion equally as well as it would do at a much higher temperature, only in the latter case a much shorter time would suffice.

It is a curious fact that no factor in the reactions of ceramic materials is less considered and less widely used than *time*. In the elementary books on chemistry and physics the experiments suggested are all selected for the speed at which they can be performed, so that most people with only an elementary knowledge of chemistry and physics overlook the importance of the effect of time on all reactions particularly on those of a colloidal nature or on those which involve solid components.

The dispersion or flocculation of a reversible colloid often requires a considerable time and is rarely instantaneous. Hence tests or experiments of too short a duration are liable to be seriously misleading. This is also clearly shown in the case of a fusible clay, or a mixture of clay with some fusible material such as calcium silicate. If the material is heated for only a short time, it will be necessary to raise the temperature very greatly before any signs of fusion can be recognised, whereas if the material be maintained sufficiently long at a much lower temperature, the same amount of fusion will occur.

When relatively *large masses* of material are concerned, the effect of time is even more marked.

An appreciable time interval may be required to raise the mass to uniform temperature, by virtue of the thermal conductivity. If a reaction takes place involving a heat change, the energy liberated or absorbed will take longer to become dissipated in a larger mass than a smaller one.

Similarly, in reactions such as the burning-out of carbonaceous matter or the removal of structural water, a much greater time is involved with large bodies than with smaller samples under laboratory conditions. In kiln-firing practice, the complete mass of a setting in a chamber will be involved in the reactions which occur and it is important in practice to secure a constant and consistent condition throughout.

This prolonged heating is known in the ceramic industries as 'soaking'. It is often much more costly than working the kiln at a higher temperature, but the latter would effect changes in the ware so rapidly that distortion and loss of shape would occur. By allowing the reactions to proceed more slowly (i.e. at a lower temperature), these risks are reduced and, in many cases, are wholly avoided. Where the shape of the product is not important, as in melting metal or glass or burning lime, it is usually desirable to work at as high a temperature as can be obtained without damaging the materials, as the reactions then proceed more rapidly and losses due to radiation, etc., are less serious.

A large part of the burner's skill consists in deciding what is the most suitable duration of heating as well as the maximum temperature to be attained. He has to adjust the temperature so as to complete the desired reactions as rapidly as possible, yet without undue risk of causing distortion of the ware by allowing the reactions to proceed too far.

The mistake, however, of trying to achieve by a short period of high temperature firing what is obtained only by long soaking at lower temperatures, must be carefully avoided; this is a serious fault of the test method known as Permanent Linear Contraction applied to assess the probable behaviour of refractory materials under long service conditions. In it, a sample of refractory material is heated for two hours at 1410° C. and conclusions are derived from the change in volume which occurs, as to the probable behaviour of that material when maintained for longer periods (perhaps several years) at considerably lower temperatures. It must be remembered that at higher temperatures reactions will occur which will never take place no matter how long the material is maintained at lower temperatures, so that the results of such tests are frequently invalidated and may often be misleading.

When the temperature of a kiln or furnace is rising slowly it is fairly easy to stop the heating at any desired point. If the temperature is rising rapidly considerable damage may be done after the heating has ceased because of the accumulated heat in the goods and in the structure.

Pressure usually accelerates chemical action, as it brings the reacting particles in closer contact with each other. In some cases (as where a firebrick is heated under a constant pressure until it collapses), the pressure does not greatly affect the reaction, but merely indicates the extent of the changes which have occurred.

An increase in pressure can, however, greatly accelerate chemical reactions in materials, particularly where water vapour is concerned. Clay minerals are known to rehydrate even though they have been heated to a temperature of about 800° C. The process may take several years to complete under ordinary conditions of water soaking but if steam is employed at high pressures, the reaction may be well advanced after a few hours' treatment.

The effects of high pressure on ceramic bodies has been appreciated only in recent years, but already quite startling discoveries have been made, some of which are discussed in later sections. It is now known that high density forms of silica, e.g. coesite, can develop when the amorphous oxide is subjected to high pressures (*circa* 20 kilobars). Kyanite and sillimanite have stability ranges in systems which have been subjected to high pressures.

Modern fabrication techniques utilise the principle of hot pressing. Reactions can be made to proceed more rapidly at comparatively low temperatures when the body is simultaneously subjected to high pressure. For reasons which are not yet fully explained, such reactions proceed most rapidly at or about a transition point in one of the phases.

Pressure also has an important effect on the area of contact or intimacy of association of the particles.

Vapour pressure, which is the pressure at which the substance and its vapour are in equilibrium at a definite temperature, often exercises a great influence on the progress of a chemical reaction. The vapour pressure of liquids is much greater than that of solids; it increases rapidly as the temperature rises. When a substance which is decomposed on heating evolves a vapour or gas, which is unable to escape, the decomposition or dissociation continues until the vapour pressure of the gaseous product reaches a critical value. The decomposition then ceases no matter how long the heating is continued under the same conditions. If the temperature is raised, decomposition will commence and will continue until the pressure-equilibrium at that temperature is established. Thus, if limestone (calcium carbonate) is heated at 800° C. in an iron tube which has been exhausted of air, the stone will decompose until the vapour pressure of the carbon dioxide is equal to 85 mm. of mercury, and will then cease. On heating to 1040° C. further decomposition occurs until a vapour pressure equivalent of 520 mm. of mercury is reached.

On cooling, carbon dioxide will be reabsorbed and calcium carbonate reformed as the temperature drops until a vacuum is again established.

The changes in the vapour- or sublimation-pressure is expressed by the following equation:

$$\frac{dp}{dT} = \frac{pL_s}{RT^2}$$

where L_s is the molecular latent heat of decomposition, and the vapour formed has the volume RT/p at the temperature T.

The influence of the vapour pressure on the course of a reaction is very important in the lime-burning industry, for unless provision is made in a lime kiln to remove the carbon dioxide as it is formed, the limestone can never, for the reason just mentioned be completely converted into lime.

The vapour pressure of clays with different proportions of water varies with the clay and is thermodynamically related to the 'heat of wetting' (p. 454; see also Dehydration on p. 534).

The surrounding *atmosphere* often has an important influence and in some cases the form and rate of a reaction can be modified by changing the atmosphere.

In those reactions which involve oxidation, such as the burning of carbonaceous matter or the conversion of ferrous salts to ferric components, an excess of oxygen in the atmosphere is necessary, otherwise the reaction will proceed only partially or not at all if the atmosphere is reducing. Quite often the products of combustion are sufficient to retard or prevent further reaction because they surround the reacting material and prevent the free access of oxygen; for this reason, an adequate circulation of air during the cleaning period (300–750° C.) of the kiln firing of clay ware is vitally important.

It is in the colour of clay and glazed products that atmosphere has its greatest effect. The colouring agents, both those that occur naturally in clays and those that are added to it to produce desirable effects, are frequently influenced by gases which surround the body at various stages of firing.

The range of compounds added to glazes to produce the beautiful coloured effects is very great, but they mostly depend on the presence of transition elements which are capable of variable valency (see p. 65). The ultimate colour or shade of many glazes often depends on the gaseous atmosphere which surrounds the article at the maturing temperature. From different parts of the same kiln chamber or from adjacent parts of the setting on the same kiln car, the colour of a particular glaze may be noticeably different due to localised variations in the surrounding atmosphere.

The modern studio potter with small kilns can utilise the atmosphere effects with certain glazes to achieve remarkable but not always predictable results. The ancient craftsman was adept in this technique—the Chinese ox-blood finish has never been satisfactorily reproduced in modern European practice. It has been attributed to the atmosphere in the kiln produced by wood, characteristic of the region, which was used in the firing.

Even with the most sophisticated modern kiln, the potter has to limit the range of his glaze colours to ensure that variable atmosphere does not spoil his ware.

The most pronounced differences in colour in glazes and clay bodies are due to the degree of oxidation or reduction in the kiln gases, but undoubtedly the nature of the gaseous products themselves can produce subtle differences in tint or shade, sometimes with advantage but often with disastrous results.

It is only in recent years that the influence on certain reactions of small amounts of gases either in the surrounding atmosphere or adsorbed on the surfaces of particles, has been appreciated. Although it has been known for many years that if certain glasses are allowed to cool in an atmosphere of burning sulphur, the surface properties are enhanced, it was not until 1940 that Hüttig[1] showed that sulphur trioxide could be adsorbed on the surface of glasses and so retard its reactivity and tendency to crystallise.

Hedvall[2] has shown that the reactions between the crystalline modifications of silica and lime are influenced by pretreatment with gases and he also proved that sulphur trioxide pronouncedly reduced the rate of reaction.

Small traces of adsorbed moisture or the presence of water vapour are known to catalyse certain reactions which would otherwise not take place. The role of water as a surface catalyst has been described in detail by Marshall, Enright and Weyl.[3]

Surface Phenomena. Within the last decade it has been established that the surface layers of a crystal are deformed and not to be compared in atomic arrangement with layers well within the crystal. Verwey, Stranski, Fajans, Weyl and others have recognised that many unusual properties exhibited by small-sized particles are due to the large surface area of distorted layers.

It is not within the scope of this book to detail the causes and implications of

[1] Hüttig, G., *Z. anorg. Chem.*, 53, 35, 1940.
[2] Hedvall, J. A., *Trans. Brit. Ceram. Soc.*, 55, 1, 1956.
[3] Marshall, P. A., Enright, D. P. and Weyl, W. A., *Proc. Int. Sym.* on the Reactivity of Solids (Gothenburg, 1952), Part 1, 273.

surface chemistry and how small crystals with large surface areas differ in reactivity from the classically represented structures. Briefly, it may be stated that the outermost ions or atoms of a crystal are not surrounded on all sides by 'screening' groups. The surface layers are, therefore, distorted and the electrical balance is not maintained. Hence, there is an increased possibility of reactivity or ease of combination with other chemical units in the immediate vicinity, and when reaction occurs it alleviates the insufficient surface 'screening'. (See p. 555.)

The crystal growth of many ceramic materials is attributed to the surface reactivity and so are many of the colloidal aspects of clay suspensions. The surface layers which behave as defective lattice structures often provide nuclei from which reactions can spread through the mass of a crystal.

The speed of reaction, is governed, in many cases, by the surface areas of reactants because it is only at the external, exposed faces that contact between reacting components can occur. Thus, a porous mass presents a much larger reacting surface than the external faces of a solid body of similar dimensions and will usually be attacked more rapidly by liquid slags or corrosive gases.

Intimately linked with the surface layers of particles in a porous mass is the *angle of contact* or *surface tension* of fluids which may become associated with it. This, in turn, affects the chemical reactivity, for it governs the rate of penetration. If the corrosive liquid 'wets' the solid readily and penetrates the pores, the chemical attack will be much more rapid than if the fluid and solid were in less intimate contact.

Surface tension also has an important effect on physical changes in ceramic materials, particularly those relating to plasticity and slips (p. 502). Surface tension decreases with a rise in temperature but is increased by the addition of soluble salts. It is surface tension which causes water and other liquids to rise in capillary tubes and to penetrate ceramic and other materials, against the force of gravity. Water and mercury have a far larger surface tension than most liquids.

Crystallographic Structure and Properties. Reaction between solid materials is slow and proceeds by the diffusion of ions of one component through the other crystalline phases. The mechanism of diffusion is discussed in more detail later in this chapter, but it may be noted here that reactions of this type proceed more readily if the structure is open and contains a maximum of defects in the lattice.

Viscosity has an important effect on chemical action, because changes can only occur when the reacting substances are in intimate contact and a viscous material flows so slowly that it requires a long time to bring the reacting substances into contact. A mobile liquid will react much more vigorously than a viscous liquid with a solid (assuming other factors to be constant).

There are many reactions in ceramic processes which exemplify the effect of viscosity. Thus, slags at their fusing-point are often very viscous and exert little corrosive action upon the bricks with which they come in contact, but if the temperature is raised, many slags become quite mobile and then exert a powerful corrosive action as they flow more readily into the pores of the brick and so expose a greater surface to their action. Most alkaline silicates are very mobile when fused, and consequently, they readily combine with silica or fireclay. Magnesium and ferrous silicates are much more viscous, and as they flow more slowly and penetrate pores less readily, they are much less actively corrosive. Alumino-silicates are usually more viscous than simple silicates and are not generally so corrosive.

As vapours are still more mobile than liquids they may cause a serious amount of corrosion by penetrating into the smallest pores of bricks, etc., where liquids can reach only slowly.

The viscosity of slips and clay-pastes has a noteworthy effect on their behaviour.

The *solubility* of one of the reacting substances in the other or in a third substance usually increases the rate of reaction by bringing them into more intimate contact than would otherwise be the case. As most ceramic materials are insoluble in ordinary liquids, the chief case where solubility is concerned is that of the various substances in the fused material produced when the goods are in the kiln or are heated whilst in use. Such molten material—especially if rich in alkaline or calcareous silicates—has a powerful solvent action and will, in time, effect the destruction of almost all ceramic articles. It is to this action that the corrosion of firebricks, crucibles, etc., by slags is due, and the same action is the cause of most of the distortion which occurs when ceramic materials are 'overheated'. As the fused material dissolves silica or other substances, it becomes increasingly viscous and eventually becomes saturated. After this has occurred, it loses its corrosive action so long as the temperature is not increased. Upon a rise of temperature, however, the equilibrium is destroyed and solution will continue until a new state of equilibrium corresponding to the higher temperature is established.

A very small solubility is sometimes sufficient to cause extensive chemical reactions, provided large masses or long periods of time are involved (see Law of Mass Action).

In certain solvents, of which water is a characteristic example, many substances dissociate into ions of unlike charge (see Chapter II). Chemical reactions which take place between solutions are generally regarded as proceeding through the interaction of ions, and in many cases the reaction cannot take place unless ions form. Thus, hydrogen chloride (HCl) dissolves in water to form hydrochloric acid which readily attacks limestone ($CaCO_3$); if, however, the gas is dissolved in dry benzene there is no interaction. The reaction is, therefore, said to proceed through the agency of hydrogen and chlorine ions into which hydrogen chloride dissociates when dissolved in water.

In the presence of water, many interesting reactions occur, particularly in connection with the flocculation of clays. Whilst these are mainly physical in nature, they also depend on the presence of soluble substances which are chemically active.

The solubility of the product of the reaction also affects the reaction (see Selective Action) as an insoluble product is largely removed out of the sphere of action and so prevents a reversal of reaction taking place.

Selective Action. Where two or more chemical reactions are possible, various conditions will determine which one will take place. If some of the substances present have different affinities for the remainder, the reaction will occur between those which have the greatest affinity for each other. If, by the interaction of two substances, present with several others, one possible product is volatile, that reaction which yields such a product will generally occur in preference to others which do not involve a change of state. Similarly, if one of the possible products is insoluble in the liquid under the prevailing conditions, the reaction which yields that product will usually take place in preference to any other reaction in which all the products are soluble. The presence of a catalyst (below) may also exercise a selective action.

In some cases, chemical action will not occur in the presence of weak acids or bases, but will do so when strong acids or bases are present. This action is termed *disposing affinity*. Thus, the oxides of such a substance may be difficult to form, and when formed may be readily decomposable, whilst the salts of strong acids or the salts produced by the action of strong bases may be quite stable. The precise cause of this mode of action is not known.

Selective action of components is important in determining the products formed on firing commercial magnesites or other basic materials, where the five common constituents are CaO, MgO, Al_2O_3, Fe_2O_3, SiO_2. The preferential combination will be between the components exhibiting respectively the greatest alkalinity and the greatest acidity.

Catalytic Action. The presence of a small proportion of certain substances (catalysts) sometimes causes the interaction of two other materials, whereas without it no reaction would occur under the same conditions.

Indeed, many pure materials are known to be very inert although in the presence of a minute amount of impurity they react readily.

A *catalyst* or *catalyser* is (ideally) a substance which accelerates the rate of reaction but is unchanged itself, although in practice this rarely occurs. The most usual mechanism is for the catalytic agent to form an intermediate, transitional phase which then reacts to form the final product, thereby excluding the catalyst, which is restored to its initial state. The amount of catalyst which is required is often very small (a fraction of 1 per cent), so that it is difficult to investigate its precise action.

Water or water vapour is a frequent catalytic agent in many reactions. Thus, carbon monoxide and oxygen can be safely mixed without reaction taking place provided they are perfectly dry. The least amount of water, however, results in instantaneous and violent combination. 'Spots' of iron oxide catalyse the decomposition reaction of carbon monoxide into carbon and carbon dioxide. This reaction does not proceed to any great extent unless the iron oxide is maintained at about red heat as well as being in close contact with the gas. The reaction is then so active that if the iron spots are in a dense firebrick the carbon deposited may cause the brick to crack seriously.

Catalysts are also important in the conversion of quartz to other allotropic forms of silica. Oxides such as Fe_2O_3, CaO, MgO, etc., greatly accelerate the transformation although, as will be described later in this chapter, the mechanism is such that it is doubtful whether the term 'catalyst' is strictly applicable.

Other catalytic reactions of importance in ceramics include the increase in the rate of crystallisation of periclase from raw magnesia in the presence of alumina or iron oxide.

Clays and diatomaceous earths are well-known catalysers of many reactions involving gaseous components. The large surface area provides nuclei at which reactions occur and when such 'spots' are exhausted, the catalytic activity is usually reduced and the surface is said to be 'poisoned'.

It should be clearly understood that a catalyst will not cause two substances to combine which could not otherwise do so; it merely facilitates the combination or the decomposition, as the case may be, by enabling it to take place under more convenient conditions than otherwise. So far as is known, all the reactions which are effected by the aid of a catalyst can take place without its aid, but only under conditions of temperature and pressure which are much more difficult to secure.

The *thermal conductivity* or ease with which a substance can transmit heat through its particles often determines the rate at which a chemical reaction or change in state can occur. Even an apparently small difference in thermal conductivity (such as that of fireclay bricks and silica bricks) makes a noteworthy difference in the retorts used for the decomposition of coal, zinc, blende, etc.

A low thermal conductivity may lead to undesirable results in a ceramic material containing a large proportion of an inverting material. This is particularly so where silica modifications of quartz, cristobalite and tridymite are present, for these invert with large expansion effects. If the conductivity is low, parts of the material will undergo changes in length whilst others remain unchanged and disruption may result.

The *electrical conductivity* of the substances often affects the rate at which they react upon one another. If they are all non-conductors, any reaction which may occur will take place very slowly. In the presence of an electrolyte the rate of reaction will increase, and if all of the substances are electrolytes, the reaction will usually take place very rapidly. This is in conformity with the ionisation theory, which assumes that all substances are capable, under certain conditions, of splitting up into ions, each of which bears an electric charge (see p. 195).

Electrolytic changes include all changes in which an electric current is involved, including those in which an electric current is passed through a mass and those in which the particles are electrically charged even though no current has been applied.

The reactions in which an electric current is used are: (*a*) *electrolysis* or decomposition by current; (*b*) *kataphoresis*, or those due to the current separating one kind of particles from the others; and (*c*) those in which the electric current is merely a heating agent which eventually fuses a portion of the material and so enables reactions to occur which would not do so at a lower temperature; a typical example of this is the formation of carbides.

Changes in ceramic materials due to the passage of an electric current are of minor interest, as such materials are usually poor conductors of electricity.

The reactions which occur between electrically-charged particles in the absence of an obvious electric current are known as ionic reactions and they are very important in connection with the so-called colloidal properties as well as the acidic and basic properties of clays.

Light is a very important factor in many chemical reactions, such as those which take place in photography and in the oxidation processes occurring in plants and in other ways. The action of light on the chemical action of ceramic materials is not very marked, although sunlight is considered to facilitate the oxidation of impurities during the weathering of clays.

A change of state of one or more substances will often affect its power of reaction, especially if such change is accompanied by the formation of a liquid or gas. Many substances which do not react, or only slightly when in the solid state, will, if one or more is converted into a liquid (either by fusion or solution) or into a gas, react readily. Physico-chemical actions such as the production of solid solutions also occur more readily in the presence of a liquid.

One of the chief reasons for heating solid substances which it is desired should react with one another is that the heating, if sufficiently intense, may cause a change of state of one or more of the substances. Sometimes a small rise in temperature will greatly increase the intensity of chemical action.

The increase in chemical action which follows a change of state is almost wholly due to the greater ease with which the particles may come into contact with each other.

The *intimacy of association* or extent to which the particles are in actual contact with each other is a highly important factor. It is generally agreed that unless the atoms (or ions) of reacting substances come within a very minute distance of each other, so as practically to be in contact, no reaction can occur. In most cases, actual contact appears to be essential, and the greater the area of the surfaces of the materials in contact the greater will be the speed of the reaction. Consequently, the intimacy of association is dependent upon (*a*) the shape of the particles, (*b*) the size of the particles, (*c*) the homogeneity or heterogeneity of the mixture, i.e. the extent to which they are mixed together, (*d*) the porosity in the case of massive materials, and (*e*) the area of surface in actual contact.

If the particles are small and are in the form of thin flat plates, needle-shaped crystals or other shapes with a large surface area and a very small cross-sectional area, reactions will take place more rapidly than with more compact pieces such as spheres, as the former present a greater surface for the same mass of substance and, consequently, they can enter into more intimate contact. For this reason, thin flakes of mica are more harmful as an impurity in clays than the more compact grains of felspar.

If the area of contact between two materials is increased by compressing them (see Pressure, p. 570), the possibilities of chemical reaction will be correspondingly increased.

The *relative quantities of the reacting substances* exert an important influence on the extent to which a chemical reaction can take place, and if large quantities are involved, the result may be quite different from that obtained with small quantities. This difference is due to a variety of causes, such as the much greater heat required to permeate the centre of a large mass of low conductivity and the consequent slower rate of reaction of large masses.

If a material is in large compact pieces of approximately cubic or spherical shape, the rate at which they can react and the extent to which the reaction can proceed must be very small, unless the conditions are such that the products of the reaction are removed as soon as they are formed. Thus, iron filings can be completely oxidised by steam only when a very large surface of iron is exposed, and conversely, iron oxide can be reduced completely to metallic iron only when a very large volume of hydrogen—much in excess of that actually needed to combine with the oxygen—is present. The mass of the available materials appears to affect the reaction quite apart from the penetrability of the solid material concerned. Thus, a typical example of mass action is shown by the fact that, whilst many silicates are unaffected by powerful acids in the laboratory, they are readily decomposed by feeble acids (such as a solution of carbon dioxide in water) when exposed in large quantities, as in the 'weathering' of natural rocks.

What is sometimes termed the *law of mass action* states that broadly a reaction will proceed further or more easily if the mass of one of the reacting substances is in excess, but the results in any given case are subject to various other factors, such as those mentioned in the preceding pages. Not only the total mass, but the relative proportion of the product exerts an important influence on the progress of the reaction.

STATES OF EQUILIBRIUM AND THEIR DISTURBANCES

WHEN a material is 'at rest' both chemically and physically it is said to be in a state of equilibrium. This state can be destroyed by various means of which heating is one. When such a disturbance occurs, there is always a tendency for a fresh equilibrium to be established. This is particularly the case when the disturbance starts a chemical reaction between two or more substances or on the withdrawal of one or more of the interacting substances which maintain the equilibrium.

The Law of Mass Action. Berthollet (1799) pointed out that a chemical reaction does not only proceed in one direction ('to the right') but may do so in the opposite direction ('to the left'). Thus the equation: $A + BC = AB + C$ may proceed only to a certain extent because the opposed reaction $AB + C = A + BC$ can occur under the same conditions and at the same time. A state of equilibrium is then reached in which the opposing reactions balance each other and proceed at equal speeds. Thus reversibility is denoted by substituting the sign $\rightleftharpoons$ for the customary sign of equality (=) as in

$$AB + CD \rightleftharpoons AD + CB$$

Berthollet also showed that the extent to which a reaction proceeds depends on the concentration of each of the reactants present in a given volume and, that if an excess of one reactant is added to a mixture which is in equilibrium, the concentration of that reactant will be reduced.

The Law of Mass Reaction (Guldberg and Waage (1864–7)) states that the extent to which a substance reacts is proportional to its concentration and therefore to the products of the concentrations of each of the reactants.

When representing this mathematically, it is customary to use the chemical symbol for each reactant, enclosed in square brackets, to represent one gram molecule (mole) so that [A] represents one mole of A whilst the symbols *a*, *b*, *c*, *d* . . . represent the number of moles in the reaction. The state of equilibrium is then represented by

$$aA + bB \rightleftharpoons cC + dD$$

and the extent to which a reaction can proceed is proportional to the products of the reactants, i.e.

$$\frac{[A]^a}{[B]^b} = \frac{k[C]^c}{[D]^d}$$

where k is always constant for a particular reaction.

If only one gram molecule of each reactant is involved, $a = b = c = d = 1$ and the ratio of the concentrations of the reactants to that of the products is always constant.

The velocity of the reaction towards the right represented by V_1 is related to the partial concentrations of the reactants by the relationship:

$$V_1 = k_1[A]^a[B]^b$$

Similarly if the reaction is reversible there will be a positive velocity V_2 to the left, viz.:

$$V_2 = k_2[C]^c[D]$$

At equilibrium

$$V_1 = V_2$$

hence

$$k_1[A]^a[B]^b = k_2[C]^c[D]^d$$

or

$$\frac{k_1}{k_2} = \frac{[C]^c[D]^d}{[A]^a[B]^b}$$

As k_1 and k_2 are both constant terms the Law of Mass Action can be stated in general terms as[1]

$$\frac{[A]^a[B]^b}{[C]^c[D]^d} = K$$

The law of mass action may be used to control the results of chemical reactions. If, for example, it is desired to increase the amount of C at equilibrium it can be done in three ways: (i) by increasing the amount of A; (ii) by increasing the amount of B; and (iii) by decreasing the amount of D.

The law of mass action applies to all chemical reactions, but it is sometimes difficult to use because the 'active mass' of one or more of the reacting substances cannot be determined with sufficient accuracy.

Although the implications of the Law of Mass Action are quite general they apply more particularly to reactions which take place in solution. When gas is formed from a solid phase a different condition arises. Thus in the system

$$CaCO_3 \rightleftharpoons CaO + CO_2$$

if the vapour pressures of the components are a, b and c respectively, then, since the concentrations are proportional to the pressure,

$$K = \frac{bc}{a}$$

but a and b are constant because they relate to solid $CaCO_3$ and CaO, hence the equation becomes

$$K = c$$

Hence the pressure of CO_2 has a fixed value at every temperature. This is known as the dissociation value of $CaCO_3$.

According to Guldberg and Waage, 'in a system of reacting bodies, the effect of each substance is proportional to its concentration, and the velocity of the reaction

[1] The constants k_1 and k_2 are termed *velocity constants* and K is the *equilibrium constant*.

is proportional to the product of the molecular concentration of the reacting substances'.

$$\text{i.e.}\quad V = k[A]^a[B]^b$$

where V = velocity of reaction, A and B are the molecular weights of the reactants (in g.) *a* and *b* are the amounts of the products of the reaction and *k* is a constant.

Mass action plays a very important part in some processes connected with the ceramic industries; in the formation of some of the raw materials, it enables weak acids to do what cannot, on a small scale, be accomplished by much stronger acids (see p. 578). It largely determines the progress of reactions involving the decomposition of rock masses by water, such hydrolysis occurring in accordance with the law of mass action and its extent depending on the equilibrium expression and the concentration of the ions of the reacting substances.

In the case of the substances produced by methods of manufacture or during the use of ceramic materials, the final products are largely dependent on the effect of mass action, as the temperatures to which ceramic materials are heated are seldom sufficient to produce complete fusion, in which state the reactions between the various substances take place with the greatest rapidity. Consequently, the reactions are seldom complete and their progress depends upon the temperature and duration of the heating, as well as on the quantities of the various materials employed. In a thin tile or vase exposed on all sides to the heat of the kiln, the amount of fused material and of the various reaction products will be much greater in proportion to the size of the article than they would be in a solid cubic mass of the same material having the same total volume, because the heat will penetrate the former article much more readily than the latter. In this case a change in the shape of the mass has the effect of delaying the progress of the reaction.

In solid systems, the law of mass action still applies although it is usually not possible to apply it in the strictest sense. The overall concentration of one component in the mass may be very small but if it is present in discrete particles, the localised concentration over a small area may be very great. Hence the products in such regions may be totally unpredicted yet they may play a large part in determining the properties of the mass.

Reversible and Irreversible Reactions

The terms reversible and irreversible reactions are used to denote chemical reactions which will or will not proceed in opposite directions when the relative masses of the reacting constituents are changed. In an *irreversible* reaction, the products cannot by any simple change in conditions, such as by varying the temperature or adding an excess of one of the products, be reconverted into the original materials which existed prior to the reaction taking place. Thus, oxidation reactions involving complete destruction of the original substance, as with sugars and other organic materials, may be termed irreversible.

Many reactions which were formerly thought to be irreversible are now known to be otherwise, e.g. the decomposition of kaolinite at moderately high temperatures.

$$Al_2Si_2O_5(OH)_4 \rightleftharpoons Al_2O_3 + 2SiO_2 + 2H_2O$$

which is strictly reversible, for kaolinite can be reformed provided that the pressure of water vapour is sufficiently high. Under normal conditions, however, the reaction may be regarded as irreversible.

In a *reversible* reaction, on the contrary, the original substances may be reformed by a simple change in the conditions or in the proportions of the products. Thus, when a chemical action occurs and the products of the reaction are not removed, a certain point of equilibrium may be reached at which the action may proceed indifferently in either direction, though no equilibrium-point may be apparent at ordinary temperatures. For instance, at 100° C., the reaction between barium carbonate and sodium sulphate is reversible:

$$BaCO_3 + Na_2SO_4 \rightleftharpoons BaSO_4 + Na_2CO_3$$

The system is in equilibrium when there are five parts of sodium carbonate to one part of barium sulphate. With greater concentrations of barium sulphate the latter is decomposed, and with smaller concentrations barium carbonate is decomposed. Thus, in a solution of these salts, either heavy spar or witherite may be deposited, according to the relative proportions of each and the temperature of the solution.

The action of heat upon calcium carbonate also involves a reversible reaction:

$$CaCO_3 \rightleftharpoons CaO + CO_2$$

The dissociation of the calcium carbonate occurs only when its dissociation tendency exceeds the vapour pressure of the carbon dioxide present in the system. As soon as the latter becomes greater than the former, lime and carbon dioxide recombine to form calcium carbonate.

Numerous reversible reactions occur in nature, many of which are of special interest in connection with ceramic materials. The well-known silica inversions and conversions are typical examples.

The Le Chatelier-Braun Principle. The maintenance of a state of equilibrium does not depend only upon the law of mass action. As stated in the Le Chatelier-Braun principle (1884) if *any* factor in a reaction is changed, the whole system will tend to form a fresh state of equilibrium and so annuls the disturbing effect, either partially or completely. In other words the state of equilibrium in a system of reactants and products depends on the relative concentrations, on the temperature, pressure and possibly other variable conditions.

Whenever a reversible reaction can occur and the products of the reaction cannot escape, the condition of equilibrium persists and such a reaction may, therefore, be termed a *balanced reaction.*

The distribution of the reacting masses, when in equilibrium, is determined by the relative concentration of the changing substances, as previously explained.

Until comparatively recent times the existence of equilibrium reactions was unknown; at present the tendency is to regard all reactions as equilibrium reactions, and that even where they appear to be complete, to consider them as only in apparent completion, i.e. in reality, the equilibrium merely lies so far to one side as not easily to be disturbed. If this view is adopted, it will be seen that balanced or equilibrium reactions are especially common in the case of ceramic and other materials which may exist in two or more allotropic forms such as silica, magnesia, etc., the particular form present being then largely dependent on the temperature and pressure. Balanced reactions are equally important in the case of many ceramic products, especially those in which a considerable amount of vitrification has occurred, as in porcelain, stoneware, etc., and in glazes, because the composition of such material depends very largely on the nature of the mixture and its state of equilibrium. This subject is of the greatest importance and is known as *Phase Equilibria.*

Phase Equilibria

When two or more substances are brought into contact, reactions may occur which result in the formation of entirely different chemical entities. The final products will depend on many factors such as the concentration and intimacy of mix of the reactants and the temperature at which the reaction proceeds.

Many physical and chemical changes progress at a finite rate and eventually reach a state of *equilibrium*. A *phase equilibrium* is the final steady state of reacting materials under specific external conditions.

In order to describe a set of reactants and products, the following terminology is usually accepted:

A **System** is the whole of the chemical field under investigation and comprises all the reactants and products under given equilibrium conditions.

A **Phase** is defined as any portion of a system which is physically homogeneous within itself and is bounded by a surface so that it is clearly separated from other parts. Thus, matter which is known to exist in three forms, solid, liquid and gaseous can, under certain conditions, exist in all these three phases. The compound, expressed simply as H_2O, can be solid (ice), liquid (water) or gaseous (steam, water vapour) depending on the conditions under which it is examined. A mixture of substances may form a single phase; sodium chloride dissolves in water, for example, to form a homogeneous, single phase solution. Many liquids are mutually soluble to give a single phase system; alcohol and water mix readily, whilst oil and water is an example of two liquids which do not mix and hence form two phases.

Components are the smallest number of independent variables by which the composition of a mixture may be conveniently expressed. These may be elements, oxides or more complex compounds depending upon the simplicity or otherwise of the system. In ceramic studies, the individual units are usually oxides. The reason for this is more obvious if a system containing calcium, silicon and oxygen is considered. If elements were to be regarded as the basic components, then three composition variables are permissible, viz. calcium, silicon and oxygen. As it is known that the oxygen content under normal conditions is always governed entirely by the relative amounts of calcium and silicon (the total amount being expressed as $CaO + SiO_2$), the oxides are more conveniently regarded as the basic components and thus only two composition variables (lime and silica) are involved. Where an element of variable valency is present, such as Fe, it may be of advantage to consider oxygen as a separate component.

The degrees of freedom or **variance** in a system are the number of independent variables which must be fixed before the system can be clearly defined. In general, only three factors are involved and these are temperature, pressure and concentration. **Equilibrium** exists when the phases do not undergo any further change either in type or amount, under the fixed conditions.

The conditions at equilibrium in any system of components may be accurately forecast by the Phase Rule—deduced by Willard Gibbs—in accordance with which the number of phases that can exist in any system is limited and given by the equation:

$$P + F = C + 2$$

where P is the number of phases, F is the number of degrees of freedom, and C is the number of components.

As a simple illustration, *the single component system* H_2O contains only one component and the equation may be reduced to:

$$P + F = 1 + 2 = 3$$

If one phase is present, the number of degrees of freedom is 2: i.e. the system is bivariant. In other words the temperature and pressure are completely independent variables.

Now if two phases are present then $F = 1$ and the system is univariant. If the phases present are to be kept the same, the alteration of one variable must be accompanied by an equivalent alteration in each other variable. This would be the condition where liquid water is in equilibrium with vapour, as it is over the range 0–100° C.

If three phases, solid, liquid and vapour are all present then $F = 0$; i.e. no degrees of freedom are permissible. The system is *invariant* at such a *triple point.* Should the slightest change occur in the conditions of the system, i.e. in the temperature or pressure, one of the phases must disappear.

In the ice-water-steam system there is only one component, but Gibbs' rule is equally applicable where there are several components. Thus, the conditions of equilibrium of calcium carbonate involve three phases, namely, solid (calcium carbonate), solid (lime), and gaseous (carbon dioxide), so that calcium carbonate forms a three-phase system. Calcium carbonate may be resolved into two components—calcium oxide and carbon dioxide—so that $C = 2$, and on substituting this in Gibbs' rule, the equation becomes

$$P + F = 2 + 2$$

i.e.

$$P + F = 4$$

If only one condition is fixed, i.e. $F = 1$, then $P = 3$, and the three phases can exist only at one definite temperature or pressure. For example, at 600° C. and at all pressures, all three phases can exist simultaneously; but if both the temperature and the pressure are fixed, $F = 2$ and $P = 2$, only two phases could exist under those conditions.

Granite, composed of quartz, felspar, and mica, has three components (alumina, silica, potash) and three solid phases (mica, quartz and felspar). It is, therefore, univariant. It is also in equilibrium because, not being at a transition-point, it can survive small variations in temperature without changing the state of the system.

The phase rule is very valuable because:

1. It permits the classification of systems of similar behaviour.
2. It shows whether the phases of a heterogeneous system are those necessary for equilibrium.
3. It assists in identifying chemical individuals among a series of basic salts or solid solutions.

Gibbs' rule is capable of very wide application and is not limited by the complexity of the factors represented by the symbols P, C, and F. It is, however, qualitative rather than quantitative in its nature, because it does not show which phases will occur, where only a limited number are possible, and it only states the maximum number of phases which could possibly occur. Its value is, therefore, limited to showing the maximum value for each of the symbols P, C and F in any system. When the particular nature of the factors represented by the symbols is required to be known, another method must be adopted, namely, an equilibrium diagram or phase diagram.

The chief value of the phase rule consists in its power of indicating the maximum value of one or more of the symbols in a given system. In some cases, where this value is unity, no further investigation is necessary, but if, for instance, the rule is employed to determine how many phases can occur in a given system and that number is greater than unity, the use of a phase diagram may be essential if it is desired to ascertain the nature of the possible phases.

The chief difficulty experienced in using the phase rule lies in ascertaining what are the precise components, as, in complex cases, their nature is very difficult to determine.

EQUILIBRIUM OR PHASE DIAGRAMS

AN equilibrium or phase diagram is one which shows what phases can exist under given conditions. In any one given system, all equilibrium conditions can be referred to three possible variables, namely temperature, pressure and composition. In simple cases, e.g. a single component system, two variables only need be considered, hence a plane diagram with pressure and temperature as the two axes will completely describe the system. These can conveniently be regarded respectively as the *ordinate* (vertical axis) and *abscissa* (horizontal axis) of a plane diagram.

One-component Systems

A GENERAL form of the diagram is shown in Fig. X.2, where the solid form of the component can exist in two modifications A and B. Solid modification A can exist in equilibrium with vapour along the line WX, which is called a *phase boundary line*. At any point along its length two phases can be present, hence by reference to the phase rule F must equal 1, i.e. the system is univariant.

This is also the condition along the other phase boundary lines shown in the diagram. There are also two *triple points* at W and Z. At such positions three phases can exist simultaneously, hence the number of degrees of freedom if the phase rule is to be obeyed must equal 0. The points W and Z are known as *invariant* points because neither the temperature nor pressure can be altered without causing one phase to disappear.

One-component systems are not of great importance in ceramic studies, although a notable example is that of *silica*, SiO_2, the phase diagram of which is shown in Fig. X.3 (A & B). Both the inversions and conversions are shown and the dotted portions represent the unstable phases under equilibrium conditions.

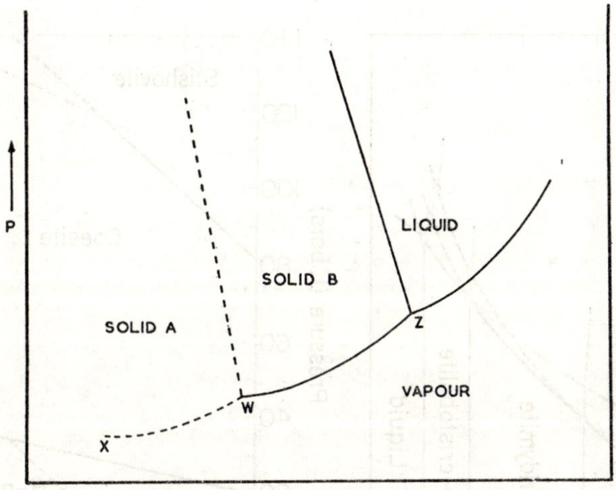

FIG. X.2. PHASE DIAGRAM OF A ONE-COMPONENT SYSTEM SHOWING TWO SOLID MODIFICATIONS, A LIQUID AND A GASEOUS PHASE

Fig. X.3 (Diagram A) is a figurative representation of the well-known changes in the silica minerals. It implies that the vapour pressure of silica increases with temperature—which is true—but the values on the pressure ordinate cannot be defined precisely.

The position of the liquid line and its variation with temperature is speculative. Quartz and tridymite can both be melted under considerable pressure without prior conversion to cristobalite; the fusion temperatures are thought to be at lower temperatures than that of cristobalite.

Fig. X.3 (Diagram B) illustrates the phase relationships of the dense forms of silica, coesite and stishovite, in relation to quartz. Very high pressures are involved and at these values cristobalite and tridymite have virtually no equilibrium stability.

In systems where there is more than one component, the construction of phase equilibrium diagrams becomes more complex. A two-component mixture requires an additional axis before it can be described completely, and as the number of components increases, so does the difficulty of representation.

Reactions between ceramic materials, however, usually involve only solid and liquid phases. The gaseous state is of little consequence and for most purposes can conveniently be disregarded. Such 'condensed' systems no longer have vapour pressure as a variable or degree of freedom hence a modified form of the Phase Rule can be applied to predict the possible phases. This is written:

$$P + F = C + 1$$

where once again, P is the number of phases possible, C is the number of components, and F is the number of variables or degrees of freedom.

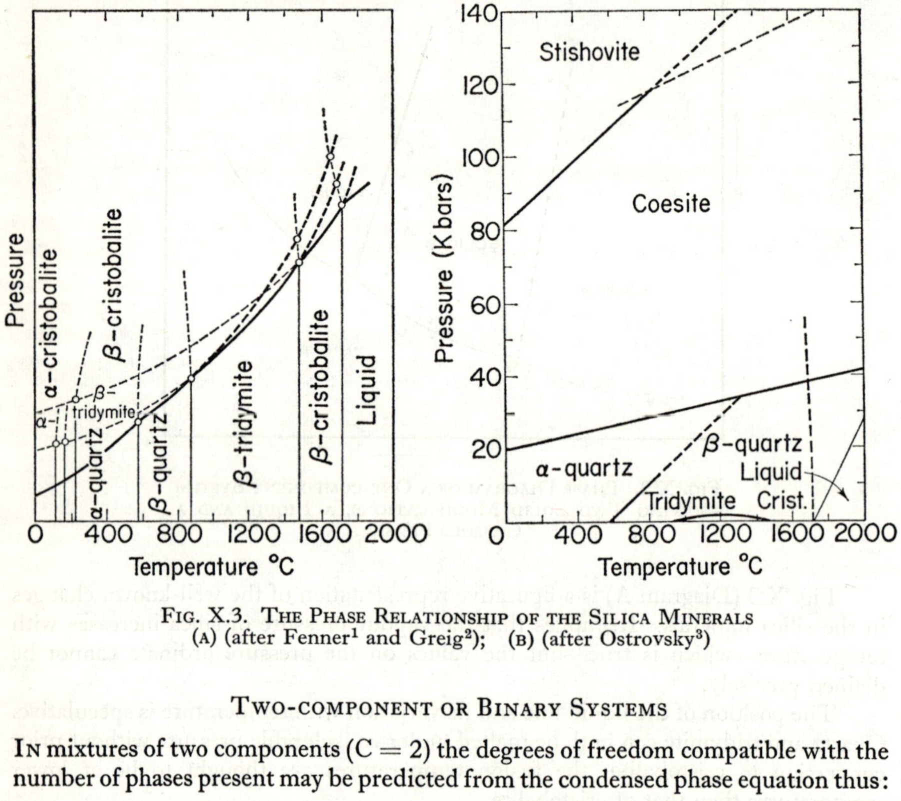

FIG. X.3. THE PHASE RELATIONSHIP OF THE SILICA MINERALS
(A) (after Fenner[1] and Greig[2]); (B) (after Ostrovsky[3])

TWO-COMPONENT OR BINARY SYSTEMS

IN mixtures of two components (C = 2) the degrees of freedom compatible with the number of phases present may be predicted from the condensed phase equation thus:

when $P = 1$ $F = 2$
when $P = 2$ $F = 1$
when $P = 3$ $F = 0$

There are only two variables to consider, namely temperature and composition and as the latter can be represented by one axis, a simple plane diagram can be constructed with temperature as the ordinate and composition as the abscissa. Any mixture of two components can be represented by a point along the horizontal axis.

[1] Fenner, C. N., *Am. J. Sci.*, 36, 383, 1913.
[2] Greig, J. W., *Am. J. Sci.*, 13, 1, 1927.
[3] Ostrovsky, I. A., *J. Geol.*, 5, 325, 1967.

The composition of a mixture can be represented on a weight or molar basis. The diagrams will have different characteristics in so far as the composition axis is involved. Both systems are used and care must be taken in interpreting diagrams to ensure which representation is being used.

Interaction between two components may be of more than one type and consequently several equilibrium diagrams are obtained which may be divided into four general classes as follows:

1. Where the two components do not mix (immiscible) in either the solid or liquid states.

2. Where the components are completely miscible in the liquid region, but are immiscible in the solid phase.

3. Where compound or complex formation between the two components takes place.

4. Where the components are mutually soluble or miscible in both the solid and liquid states.

All four groups are represented in ceramic mixtures and frequently two component systems may, at different compositions or temperatures, exhibit more than one behaviour. Ideally, however, the four groups may be treated separately and the typical diagram for each deduced.

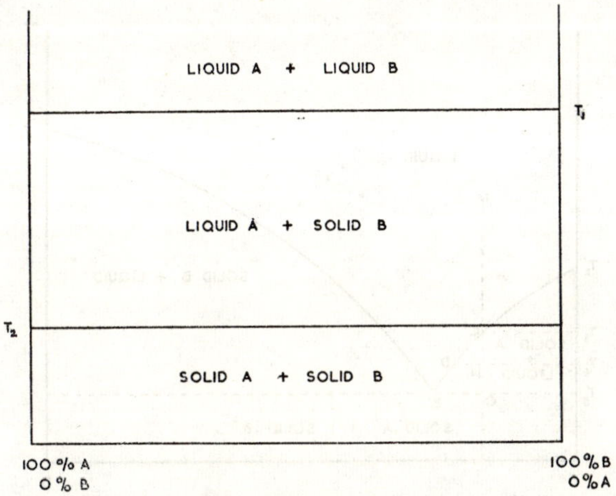

FIG. X.4. THE PHASE DIAGRAM OF TWO COMPONENTS WHICH ARE IMMISCIBLE IN BOTH THE SOLID AND LIQUID FORMS

1. **Components immiscible in the solid and liquid states.** In such a case, when crystallisation takes place, the two components form separate distinct phases and similarly in the liquid region the individual components do not mix but remain separate as do oil and water.

The phase diagram is simple and easily predicted (Fig. X.4).

The two phases A and B melt at temperatures T_2 and T_1 where T_1 is higher than T_2. At temperatures above T_1 the system will consist of the two liquid phases A and B. From the condensed phase rule equation ($P + F = C + 1 = 3$) the system is univariant ($F = 1$); therefore, cooling can take place without any alteration of the phases present.

At a temperature of T_1, however, solid B begins to crystallise. There are now three phases possible, liquids A and B and solid B, hence the system is invariant and before the temperature can be changed, one phase must disappear. This condition applies at all compositions, and so a horizontal *phase boundary line* may be drawn at the temperature of T_1. At temperatures above this point, only two liquid phases can be present; below this temperature, liquid B cannot be present and this component will have completely solidified.

At T_2 a similar three phase co-existence of liquid A, solid A and solid B is possible. This system would be invariant and one phase would have to disappear before the temperature could be changed. Another *phase boundary line* must, therefore, be drawn horizontally across the diagram at T_2. Below this last temperature only solid A and solid B can be present and no other phase change is possible.

Systems where there is complete immiscibility in both liquid and solid states are not common, but there are numerous examples of this phenomenon over a limited range of composition. An example is the silica-rich region of the calcium oxide/silica diagram shown in Fig. X.18.

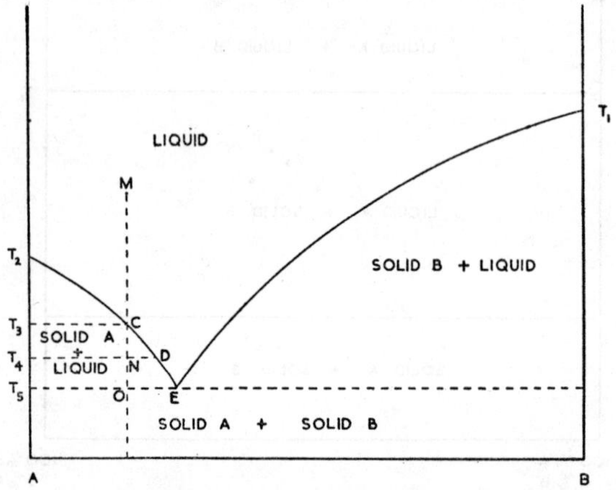

FIG. X.5. THE PHASE DIAGRAM OF TWO COMPONENTS WHICH ARE MISCIBLE WHEN LIQUID, BUT IMMISCIBLE IN THE SOLID STATE

2. Two components miscible in the liquid, but immiscible in the solid states. In the liquid state only one homogeneous phase exists, as in mixtures of water and alcohol. On cooling, however, each crystalline phase separates independently, and, although they may be intimately mixed, the two components are chemically and physically separate.

As shown in Fig. X.5 the two components A and B with melting points of T_2 and T_1 mix in all proportions when both are liquid at temperatures greater than T_1.

The effect of adding one component to the other, however, is to lower the melting point of the latter.

This is a basic thermodynamical principle established by van't Hoff[1] who predicted that the lowering of the freezing temperature (ΔT) resulting from adding 1 mole of any substance to 100 gm. of solvent was:

$$\Delta T = 0{\cdot}02\frac{T^2}{Q}$$

where T is the absolute temperature of melting of the pure component and Q is the latent heat in calories per gram.

The depression of the freezing point is thus independent of the nature of added substance and is a solvent characteristic, provided that the solvent does not form crystalline or solid solutions with the added phase, but separates in a pure form.

As increasing amounts of A are added to B the temperature at which initial crystallisation begins falls progressively along the line T_1E (see Fig. X.5). Similarly, additions of component B to A, cause a lowering of the melting point of A along the line T_2E.

At all points on the line T_1E solid component B is the only phase to crystallise. Along T_2E, only solid A can be formed. At the point E, however, which is the lowest temperature at which liquid can be present, both solid A and B crystallise together. Three phases can now be theoretically present, liquid, solid A and solid B. Hence, this must be an invariant point because, as predicted by the phase equation, the number of degrees of freedom is zero, and, before the temperature can be altered, one phase must disappear. The point E is termed the *eutectic point* of the system.

The system is more clearly appreciated by considering the changes which occur on cooling a specific mixture, such as M, which is a mixture rich in component A. Above the liquidus line, only one phase is present but on cooling to the temperature (T_3) the composition line of the mixture meets the phase boundary line T_2E. Solid component A begins to crystallise. At this point (C), two phases are present, hence the degree of freedom predicted by the 'condensed' phase equation can only be 1, i.e. the system is univariant.

As crystalline A separates, the liquid phase remaining becomes richer in component B and its composition follows the line T_2E.

This progressive crystallisation of solid A and the alteration of the composition of the liquid phase continues until the temperature T_5 is reached, at which point the liquid remaining has a composition corresponding to E. Solid B must also separate at this temperature and the temperature must remain constant until all the liquid phase vanishes. The system then contains only two solid phases which do not undergo further alteration on cooling.

At any temperature above the eutectic point the relative amounts of liquid and solid phase may be assessed by applying the 'lever' rule. At a temperature T_4, for example, which is between the temperature of initial (T_3) and final (T_5) crystallisation, a certain amount of solid A will have crystallised and the liquid remaining will have the composition indicated by the point D. If a horizontal line is drawn across the diagram at the temperature T_4 cutting the mixture composition vertical at N,

[1] J. H. van't Hoff, *Vorlesungen uber theoretische und physikalische Chem.*, Vol. II (Braunschweig, 1903).

then the ratio of solid to liquid is given by the relative lengths of ND and T_4N. That is

$$\frac{\text{amount of solid A}}{\text{amount of liquid (composition D)}} = \frac{\text{length of ND}}{\text{length of } T_4\text{N}}$$

The cooling of a liquid of composition M results in progressive crystallisation. Conversely if a solid mixture of the composition M were heated some liquid would be formed at the eutectic temperature T_5. The amount would be predicted by the lever principle as

$$\frac{\text{amount of solid A}}{\text{amount of liquid (composition E)}} = \frac{\text{length OE}}{\text{length of } T_5\text{O}}$$

Similar conditions of solidification would be observed in a liquid mixture rich in component B. The phase to crystallise would, in this case, be solid B, and the liquid phase would become progressively enriched with A. No solid A could separate until the eutectic temperature (T_5) was reached, when instantaneous solidification of all remaining liquid would take place.

In the interesting case of a mixture with a composition corresponding to that of the eutectic E, the single liquid phase would persist at all temperatures above T_5. At this temperature, however, three phases can co-exist and the system is invariant. requiring that the temperature must remain constant until one phase vanishes, Hence, complete crystallisation can take place at one temperature or, on the reverse cycle, a definite melting point would be observed. This is one of the few cases known in ceramic materials where a mixture may be said to have an established, single melting point.

3. **Compound or complex formation** of three types is known, each of which has a distinct phase equilibria diagram.

A. When a compound is formed which decomposes below the eutectic temperature of the two components, the equilibrium diagram is as shown in Fig. X.6.

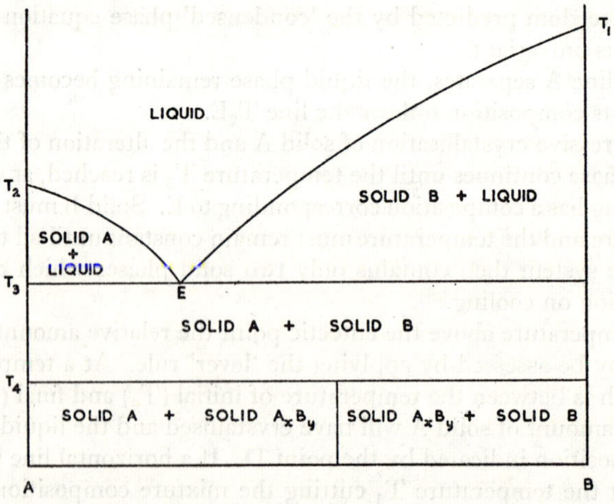

FIG. X.6. A BINARY MIXTURE WHERE A COMPOUND FORMS WHICH IS DECOMPOSED BELOW THE EUTECTIC TEMPERATURE

The changes between compound A_xB_y and the components A and B all occur in the solid state and as the temperature is raised above T_4, the complex decomposes into its equivalent amounts of A and B.

B. A compound may be formed which decomposes before melting, but at a temperature above the eutectic point. This is known as incongruent melting and a peritectic point (P) is produced as in Fig. X.7.

In the two-component system of A + B, the compound A_xB_y is stable in all solid mixtures. On heating to T_p, however, the compound dissociates into a liquid phase (P) and crystals of component B.

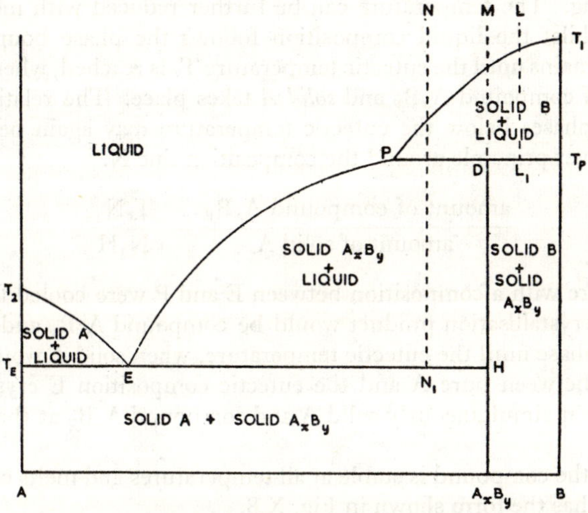

FIG. X.7. A BINARY MIXTURE CONTAINING A COMPOUND WHICH DISSOCIATES BEFORE MELTING, BUT ABOVE THE EUTECTIC TEMPERATURE

The cooling phenomena of three mixtures mroe clearly illustrate the salient features of the diagram. These are mixture L (richer in B than the compound A_xB_y), mixture M (of the composition A_xB_y) and mixture N (with an excess of A over A_xB_y). All are liquid at points above the phase boundary line T_1P but begin to crystallise at the temperature where the compositionline intersects the liquidus line. Solid B is first formed in all cases and continues to crystallise whilst the liquid composition follows the path of the phase boundary line T_1P. As the temperature reaches T_p, the solid component B which has already formed, begins to redissolve in the liquid of composition P and at the same time the compound A_xB_y is crystallised. The point P is thus invariant and the temperature must remain constant until one phase disappears.

In mixture L, solid B is not completely redissolved before all component A in the liquid phase combines to form the compound A_xB_y; so the liquid phase vanishes and the solid system contains crystals of A_xB_y and also B. The relative proportions of the two solid phases are given by applying the lever principle around L_1 with the composition line A_xB_y as one extreme and the B ordinate as the other, i.e.

$$\frac{\text{amount of solid } A_xB_y}{\text{amount of solid B}} = \frac{L_1T_p}{DL_1}$$

In mixture M there is just sufficient liquid phase, containing component A, to redissolve the solid B and precipitate a new solid phase consisting entirely of the compound A_xB_y.

Mixture N is richer in A and therefore after all the originally-crystallised B has been redissolved with formation of solid compound A_xB_y there is still some liquid phase remaining. The temperature can be further reduced with more solid A_xB_y crystallising whilst the liquid composition follows the phase boundary line PE. Some liquid remains until the eutectic temperature T_e is reached, when simultaneous precipitation of compound A_xB_y and *solid A* takes place. The relative amounts of the two solid phases below the eutectic temperature may again be calculated by applying the lever principle around the composition line N.

$$\frac{\text{amount of compound } A_xB_y}{\text{amount of solid A}} = \frac{T_eN_1}{N_1H}$$

If a mixture with a composition between E and P were cooled from the liquid state, the first crystallisation product would be compound A_xB_y and this would be the only solid phase until the eutectic temperature, when solid A would also form.

Mixtures between pure A and the eutectic composition E crystallise solid A initially and then simultaneously solid A and compound A_xB_y at the eutectic temperature.

C. When the compound is stable at all temperatures and melts congruently the phase diagram has the form shown in Fig. X.8.

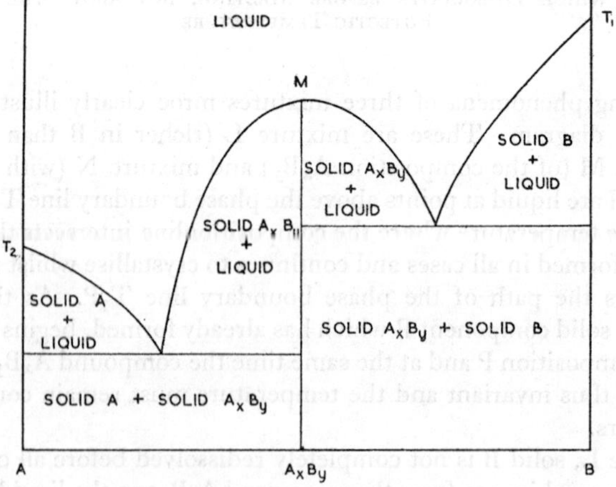

Fig. X.8. The Phase Diagram of a Binary Mixture which forms a Congruent Compound

The diagram is characterised by a maximum in the phase boundary curve corresponding to the composition of the compound A_xB_y. For most purposes, such diagrams may conveniently be separated into two parts divided by the composition line A_xB_y. It will be seen that the parts have each the form of a simple binary mixture with a eutectic point; the components of the systems are A and compound A_xB_y and B and compound A_xB_y respectively. All the laws of crystallisation and melting which apply in the binary system where the two components are immiscible in the solid state, are also operative in this case.

If the pure compound A_xB_y is heated it will melt at M, which is an invariant point, because the system contains only one component (the compound itself) and hence two phases give rise to no degrees of freedom, as predicted by the condensed Phase Rule.

4. **Two components miscible in both liquid and solid states.** In such systems both the solid and liquid phases are homogeneous and the two components form solid or crystalline solutions.

A solid solution is where two crystalline substances merge their chemical entities. Three examples are known:

(*a*) Where two crystalline materials of similar habit and lattice dimensions form a single crystal. Such an example is the mineral dolomite which is a double carbonate of calcium and magnesium. The crystal habit and lattice of dolomite are not unlike those of the single carbonates, the only difference being that calcium and magnesium cations are both present in the lattice at the same time and are mutually interchangeable. The compound must not be confused with a mixture of the two components. Each part of the crystal is identical with all other parts. The formula of dolomite should be written $(Ca.Mg)CO_3$ and not expressed as a mixed carbonate. Solid solution of this kind is very common in ceramic minerals; the spinels, olivine and many common slag minerals are typical examples and even substitution in montmorillonite and micaceous minerals may be so regarded. Ideally, this type of crystalline solution will most readily occur when the interchangeable components or ions are almost identical in size and charge. This is usually the case, but there are many anomalies; for example, one of the best known naturally-occurring solid solutions is the plagioclase felspars, where the interchange is $(Na^+ + Si^{4+})$ for $(Ca^{2+} + Al^{3+})$. A whole series of minerals is known between the two extremes:

$$\underset{\text{albite}}{Na_2[Al_2Si_6]O_{16}} \quad \text{and} \quad \underset{\text{anorthite}}{Ca_2[Al_4Si_4]O_{16}}$$

(*b*) Where foreign ions are contained in holes or defects within the crystal lattice of a mineral. Occasionally, two entirely different crystalline components form solid solutions, but usually in only a limited range of composition. Ionic substitution or replacement is impossible yet the foreign ion is held firmly within the lattice.

Little is known of this kind of solid solution, but its formation has been established with certainty. The formation of such complexes is most common when the crystalline structure is open. This is the case in silicates which are based on the framework structure (see Chapter III) and it is here that solid solutions of this type are most common.

(*c*) Diffusion-type complexes. Many solid substances, when placed in contact with others, will slowly diffuse into the crystal lattice of the other component. This process is usually greatly accelerated at high temperatures and the extent of penetration may, in some cases, cause a breakdown in the crystal lattice. The complex formed may be random in structure with dissimilar ions occupying equivalent positions in the same lattice. Eventually, such diffusion-type complexes may lead to compound formation, but, in the initial stages, they have all the characteristics of true solid solutions.

Although they may be extremely variable in composition, solid solutions are not to be confused with glass. Unlike the latter, they have all the properties of true crystalline solids. On the other hand the term solid solution does not imply compound formation where the combination of components is in a fixed invariable stoichiometric ratio. The term infers that one solid phase is incorporated into another in such a way that a continuum or single phase is formed in which the properties of the parent substances are modified, but not completely changed. The main distinction between a solid solution and a compound is that the melting point in the former extends over a range of temperature whilst in the latter it is usually sharp. There are, however, exceptions to this rule, for some well-defined compounds are known which decompose in the solid state and so have an extended melting range.

The simplest equilibrium diagram for components with a complete range of solid and liquid solution is illustrated in Fig. X.9A. There are two phase boundary lines; the upper of which is known as the liquidus and represents the maximum temperature at which solid phase can exist at all compositions; the lower curve or solidus indicates the minimum temperature at which liquid can form in any particular mixture.

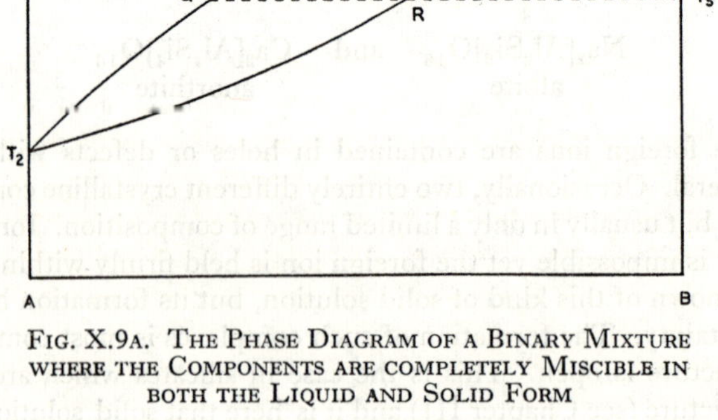

Fig. X.9A. The Phase Diagram of a Binary Mixture where the Components are completely Miscible in both the Liquid and Solid Form

The addition of component A (of lower melting point) to component B depresses its melting point, although amounts of B increase the melting temperature of A.

The region between the two phase boundary lines is known as the field of heterogeneous melting where liquid and solid phase can exist together. The extreme ends of the horizontal lines, known as *tie lines* (e.g. OP and QR), drawn across this region represent the composition of the liquid and solid phases which exist in equilibrium with each other at particular temperatures.

The phases which form on cooling are clearly indicated by a mixture of composition N which, when cooled from a temperature above the liquidus line, has a single uniform phase until the temperature T_3 is reached. Crystallisation commences, but the solid phase first formed has a composition represented by the point M, i.e. richer in the component of higher melting point than is the initial mixture. At this point only two phases are present hence the system is univariant and the temperature can be progressively lowered.

As the cooling continues, the composition of the liquid phase changes (due to crystallisation of an excess of component B), and follows the path of the liquidus curve. At temperature T_4 for example the liquid has the composition of the point O. The solid which is in equilibrium with this liquid has the composition given by the point P, the point of intersection of a horizontal line through O and the solidus phase boundary curve. Hence, not only does the liquid phase change in composition as the temperature is lowered, but so does the solid being crystallised. Ideally, if the rate of cooling were sufficiently slow, the solid phase formed would constantly be redissolved and recrystallised with an ever-changing composition. This cooling process would continue with more solid being formed of a progressively increasing concentration of component A until the temperature T_5 was reached. Here liquid of composition Q is in equilibrium with solid R and as R is along the *isopleth* (line of constant composition) of N, all liquid must disappear and only a solid solution of uniform composition will be present.

The relative amounts of liquid and solid phase at any stage of the crystallisation are given by applying the 'lever principle' about the point of initial composition. At T_4 for example, liquid O is in equilibrium with solid P, and the relative amounts of each are expressed as

$$\frac{\text{amount of liquid O}}{\text{amount of solid P}} = \frac{\text{PS}}{\text{SO}}$$

Thus at the temperature T_5 the amount of liquid of composition Q will be infinitesimal.

Systems of two completely miscible components in both the liquid and solid states are common in both rock-forming minerals and in slag compositions. In general, such mixtures do not cool sufficiently slowly to allow equilibrium to be attained at all temperatures. In some igneous rocks there is evidence that, although cooling may have extended over a geological age, there has been insufficient time for the resolution and solid diffusion reactions to proceed to completion and form homogeneous crystals. Many minerals which occur in eruptive rocks exhibit characteristic *zoning* or *fractional crystallisation* and this phenomenon is even more common in the typical silicate minerals found in slags.

The formation of zoned structures is clearly demonstrated by considering the cooling of a mixture of composition N (Fig. X.9A), but under more rapid non-equilibrium conditions. Once again crystallisation begins at a temperature T_3 and the composition of the first solid phase to appear is represented by the point M. The liquid phase becomes deficient in component B and its composition follows the path of the liquidus curve. If conditions were ideally those of equilibrium cooling, the solid phase would be constantly changing in composition due to resolution and re-crystallisation or to solid diffusion, but, under practical conditions, these processes do not proceed rapidly enough and crystallisation of material of ever-changing composition continues with the original deposits acting as seeds for later crystallisation. This process continues until all the liquid disappears with the result that the final solid consists of zones of material, the centre core of which is comparatively rich in component B but with the composition changing and being more enriched in A towards the outer edges. The overall composition is equivalent to the original composition of the liquid mixture N.

Zoning of this type is a typical feature of the plagioclase felspars where albite ($Na_2O.Al_2O_3.6SiO_2$) and anorthite ($CaO.Al_2O_3.2SiO_2$) are the extreme end members of a solid solution series. It is also found in the olivine, pyroxene and melilite series. In every case the 'core' of zoned crystals is richer in the component of higher melting point.

The system illustrated in Fig. X.9A, is the simple case of two components which are completely miscible in the solid and liquid states. Other cases are known where a maximum or minimum point occurs in the diagram as shown in Figs. X.9B and C. These maxima or minima are of interest because they are examples of invariant points where apparently the Phase Rule does not so predict. At the points M in both diagrams, only two phases can co-exist, namely a homogeneous liquid phase and a solid phase of identical composition with complete miscibility of the components. Hence, with $P = 2$ and $F = 1$, the points are theoretically *univariant*.

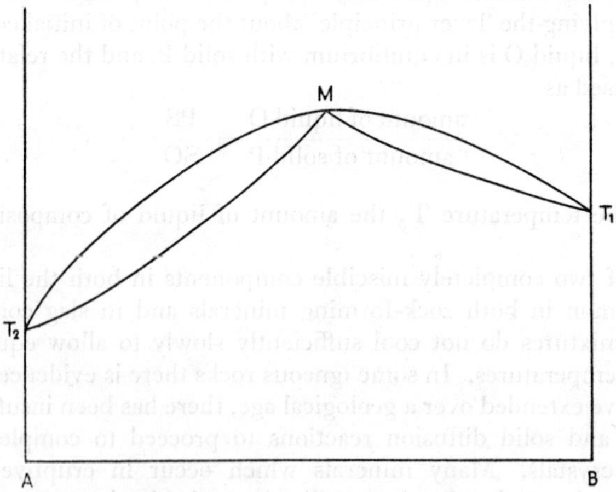

FIG. X.9B. THE PHASE DIAGRAM OF MISCIBLE COMPONENTS WITH A MAXIMUM POINT

A concise explanation of this apparent anomaly is outside the scope of this volume—it arises as a result of an over-simplification of the Phase equation. Terms which appear in the original form cancel out in the vast majority of systems, but in this one case they have to be considered and contribute to the equation to make the maximum and minimum points invariant.

Examples of a maximum-point diagram (Fig. X.9B) are not common in ceramic systems, but they are the usual form of constant melting or boiling point systems, such as alcohol and water. There are, however, some important mineral systems with a minimum point, and the iron-calcium monticellite $2(CaO,FeO)SiO_2$ is an example.

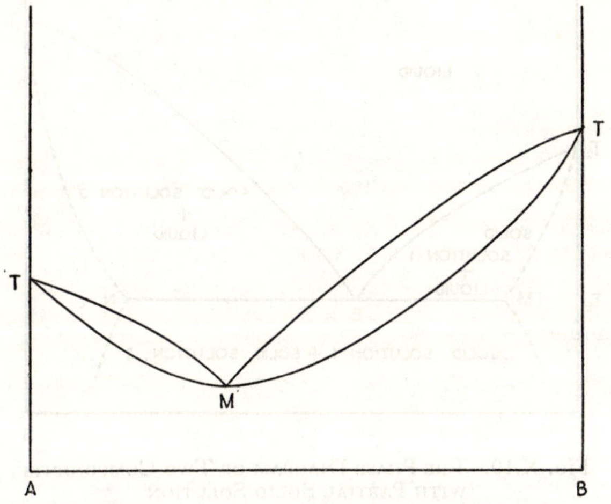

FIG. X.9C. THE PHASE DIAGRAM OF MISCIBLE COMPONENTS WITH A MINIMUM POINT

Partial miscibility in the solid state is of common occurrence in silicate systems. Many components will mix together to form a completely miscible liquid but a homogeneous, solid, crystalline solution is produced over only a limited range of compositions.

The simplest type is illustrated in Fig. X.10 where the two components form a eutectic mixture. Limited solid solution formation is encountered in mixtures rich in either component, but intermediate mixtures do not form a homogeneous solid phase.

The addition of component A to component B causes a reduction in the melting point of the latter and, similarly, additions of B to A also produce a lowering. When a melt of any particular composition is cooled, crystallisation will begin at the temperature of the liquidus line in that region. The composition of the first solid phase to form is given by drawing a horizontal tie-line (p. 595) to cut the solidus line. The composition of the liquid phase follows the phase boundary line and, ideally, the solid solution changes in composition progressively. This process will continue until all liquid disappears when one solid homogeneous phase will be the result or until the liquid reaches the composition of the eutectic mixture E. At this point liquid can be in equilibrium with two solid solutions of composition M and N. True invariant conditions exist and the two solid phases are distinctly separate.

Any mixture with a composition between the points M and N will solidify to give two separate crystalline solutions. Mixtures richer in component A than the point M or richer in B than is mixture N will finally crystallise as one homogeneous solid solution phase, provided equilibrium cooling has taken place. The composition of this single phase will be that of the original mixture.

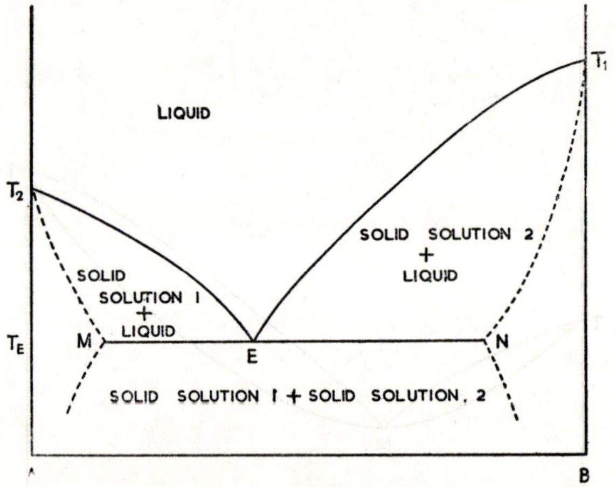

FIG. X.10. THE PHASE DIAGRAM OF TWO COMPONENTS WITH PARTIAL SOLID SOLUTION

Solid solutions may also be associated with both stable and incongruent compound formation. The establishment of the amount of solid solution in silicate systems is difficult and it is only in recent years that exhaustive studies using specialised techniques have established their existence.

Polymorphic Changes in Binary Systems. The changes described in the various two component systems may be further complicated by the presence of inversion or conversion phenomena, where a crystalline entity changes into another modification. Many minerals are known which are polymorphic or are capable of existing in more than one form. Silica is perhaps the best known example; three well-defined minerals are known, each of which has separate characteristic inversional phenomena (p. 160).

When changes of this type take place below the lowest temperature at which liquid may exist and, where there is no interaction or miscibility between the two components of a binary system, the polymorphic transformation will proceed at a definite temperature, which is independent of the composition of the mixture. Beyond the introduction of a horizontal line at the appropriate temperature (T_3) in the solidus region, the equilibrium diagram in such cases is unchanged (Fig. X.11A).

When a transformation occurs above the eutectic temperature an inflection point usually appears in the diagram as in Fig. X.11B. This is to be expected as the latent heat of fusion of the two forms will most probably be different. In Fig. X.11B, the polymorphic change takes place in component A at a temperature of T_t. Such

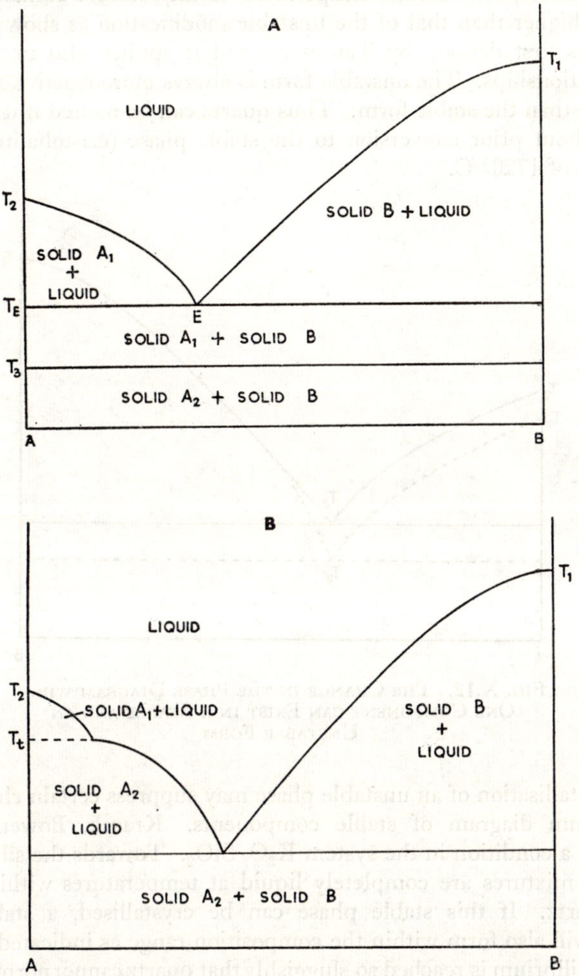

FIG. X.11. THE PHASE DIAGRAM OF A BINARY MIXTURE IN WHICH ONE COMPONENT UNDERGOES A POLYMORPHIC CHANGE
A below and B above the eutectic temperature

a diagram represents the equilibrium conditions of mixtures of silica and alkaline oxides where cristobalite, tridymite and quartz may all have stability ranges in certain regions of composition and temperature.

Examples are common in silicate systems, in which an unstable crystalline phase may separate from a liquid melt on cooling. In such cases, the equilibrium diagram will be different from that where the stable phase is produced. If two components form a eutectic mixture, but one component can be present as either a stable or unstable variety, the eutectic temperature of the mixture containing the stable form will be higher than that of the unstable modification as shown in Fig. X.12. This rule was first defined by Tammann and it applies also to the fusion and solubility relationships. The unstable form is always more reactive and has a lower melting point than the stable form. Thus quartz can be melted if it is very pure at 1670° C. without prior conversion to the stable phase (cristobalite) which has a melting point of 1720° C.

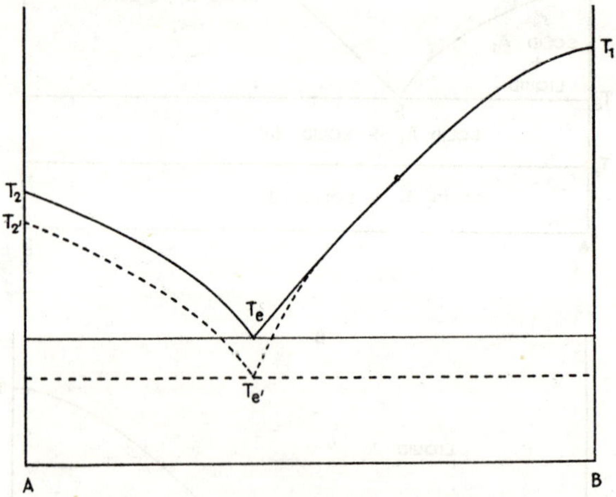

Fig. X.12. The Change in the Phase Diagram when One Component can Exist in a Stable and an Unstable Form

The crystallisation of an unstable phase may suppress certain characteristics of the equilibrium diagram of stable components. Kracek, Bowen and Morey[1] describe such a condition in the system K_2O/SiO_2. Towards the silica-rich end of the diagram, mixtures are completely liquid at temperatures within the stability range of quartz. If this stable phase can be crystallised, a stable compound $K_2O.4SiO_2$ will also form within the composition range as indicated by Fig. X.13. However, equilibrium is reached so sluggishly that quartz cannot normally be crystallised even though the temperature is below 870° C.; instead tridymite or cristobalite crystallise from the melt. Under these conditions, the crystal phase $K_2O.4SiO_2$ does not appear and a low temperature eutectic with the silica phase and $K_2O.2SiO_2$ may be observed.

Polymorphous changes in solid solution. The transformation of a pure mineral into a polymorphic variety invariably takes place at a definite temperature although such changes are sometimes very sluggish. When solid solution occurs, however, an

[1] Kracek, F. C., Bowen, N. L. and Morey, G. W., *J. Phys. Chem.*, 41, 1188, 1937.

inversion-interval may replace the invariant point and the change or changes take place at temperatures removed from those in the pure material.

Miscibility in the solid state is most likely when the components are similar in crystalline form. When polymorphic changes in crystal structure take place in one component of such a mixture, the ratio of the two components which can take part in the high temperature solid solution will probably be different from that in the original mixture. Furthermore, the formation of solid solutions usually results in an alteration of the temperature of inversion in a comparable way to the change in the melting or boiling point produced by mixing components. Solid phase relationships in the region of polymorphic changes are complex but can be represented by characteristic equilibrium diagrams.

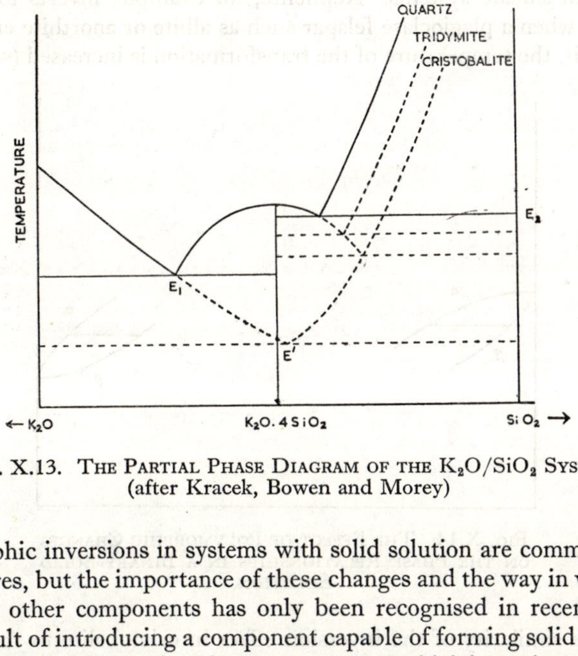

FIG. X.13. THE PARTIAL PHASE DIAGRAM OF THE K_2O/SiO_2 SYSTEM (after Kracek, Bowen and Morey)

Polymorphic inversions in systems with solid solution are commonly found in silicate mixtures, but the importance of these changes and the way in which they are influenced by other components has only been recognised in recent years. The invariable result of introducing a component capable of forming solid solutions with a polymorphic material is to alter the temperature at which inversions take place. As in the case of the melting of a binary solid solution, a heterogeneous field is present, bounded by two phase-boundary lines. The upper line represents the transformation of a high to a low form in the presence of varying amounts of the second miscible component, whilst the lower phase-boundary curve indicates the variation in the low to high transformation with the degree of solid solution. At any one particular temperature, therefore, a solid solution of the high temperature modification of an inverting mineral can be in equilibrium with a solid solution of the low temperature form. The inversion effect thus takes place over a temperature interval and not at one specific point.

There are two well-established cases covered by Konowalow Rules which state:

If the low temperature or α-phase of the inverting material contains *less* of the additional miscible component than the high or β-form, the inversion temperature of the resulting solid solution will be *lower* than that of the pure mineral.

Conversely, if the high, or β-variety is *less rich* in the other component than the low form, the inversion temperature in the solid solution will be *raised* with respect to that of the pure mineral.

These two cases are shown in Fig. X.14, the former case being on the left-hand side of the diagram, whilst the latter is on the right. Changes of this type are frequently interposed on already familiar diagrams such as the two illustrated in Fig. X.15A and B. The diagrams are of two components with partial miscibility in the solid state, where one component (A in both cases) has an inversion effect. Fig. X.15A is commonly encountered in metallurgical studies, whilst Fig. X.15B is often associated with silicate systems. Nephelite, for example, inverts to carnegieite at 1254° C., but when a plagioclase felspar such as albite or anorthite enters into solid solution with it, the temperature of the transformation is increased (see Fig. X.30).

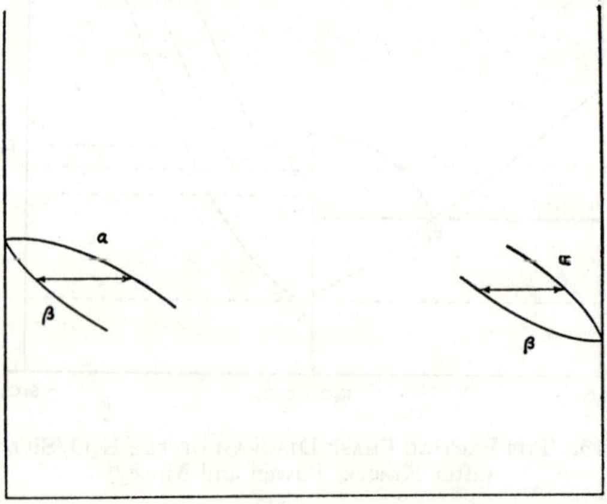

FIG. X.14. THE EFFECT OF POLYMORPHIC CHANGES ON THE PHASE RELATIONSHIPS IN A BINARY SOLID SOLUTION

The close similarity of the inversion effect in cristobalite to the example just cited, suggests that the well-known variation in temperature of the transformation of the α- to the β-variety of this mineral, may be due to solid solution effects. The crystal structure of cristobalite is sufficiently open to contain cations such as Ca^{2+}, Mg^{2+}, Na^{+}. Furthermore, Barth, Hummel and other workers have shown that silicon atoms in the structure can be replaced by aluminium with the inclusion of electrical-balancing cations without changing the atomic arrangement. Evidence which favours this theory is that the inversion of cristobalite is at a higher temperature when the sample has been subjected to prolonged or high temperature heat treatment. This would also have the effect of increasing the amount of solid solution which in turn should increase the inversion temperature, provided the equilibrium diagram is of the form indicated in Fig. X.15B.

The solid solution theory of the cristobalite inversion effect is difficult to establish with certainty. Solid solution within the cristobalite lattice cannot be confirmed by X-rays when the amounts of substituted atoms are small and randomly situated. It has, however, been clearly demonstrated (by thermal and microscopic evidence) that cristobalite with an apparently high inversion-temperature transforms over a range or interval and not at one specific temperature.

The transformations within the silica minerals are further complicated by pressure. Kracek[1] has suggested that at pressures in excess of 1500 atmospheres, quartz is stable to its melting point and, furthermore, it would be the crystallisation product from liquid silica at these pressures.

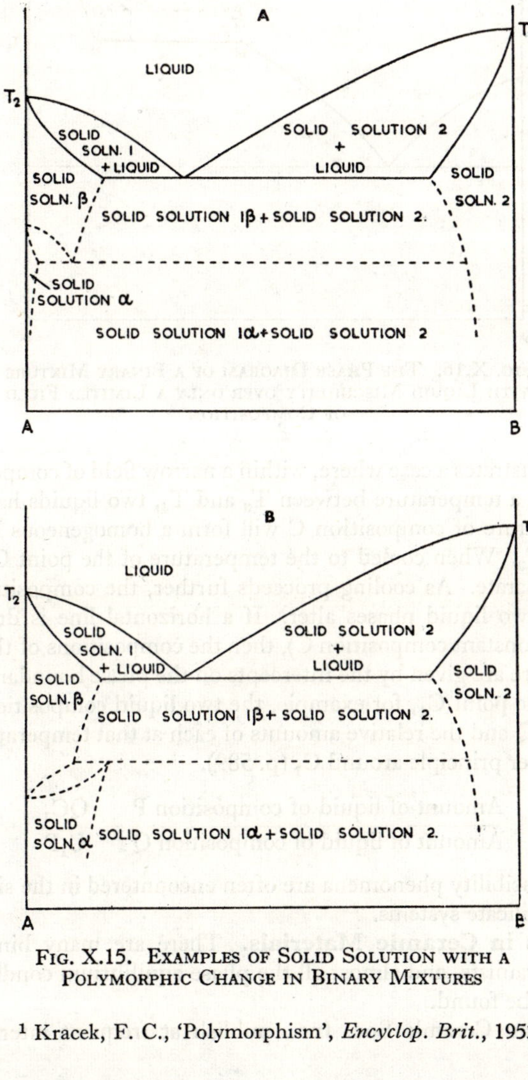

FIG. X.15. EXAMPLES OF SOLID SOLUTION WITH A POLYMORPHIC CHANGE IN BINARY MIXTURES

[1] Kracek, F. C., 'Polymorphism', *Encyclop. Brit.*, 1953.

Partial Miscibility in the Liquid Phase. In almost all the theoretical cases considered so far, the assumption has been made that the liquid phase is homogeneous and the two components are miscible in all proportions. Examples are not common of components which are completely immiscible in the liquid state, but many binary systems have partial fields of composition where the two components exist as independent liquid phases.

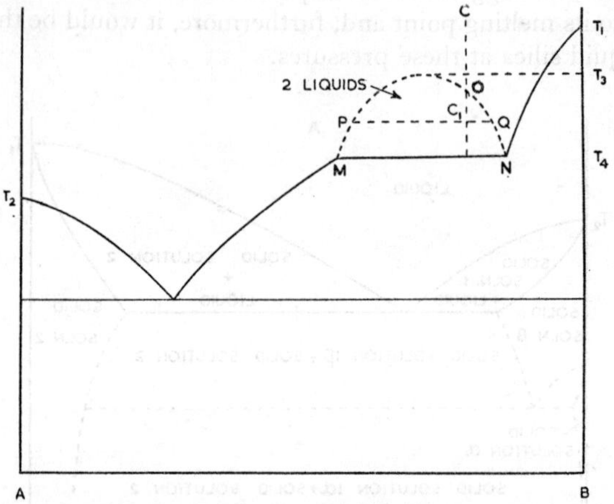

FIG. X.16. THE PHASE DIAGRAM OF A BINARY MIXTURE WITH LIQUID MISCIBILITY OVER ONLY A LIMITED FIELD OF COMPOSITION

Fig. X.16 illustrates a case where, within a narrow field of composition bounded by M and N and a temperature between T_3 and T_4, two liquids have independent existence. A mixture of composition C will form a homogeneous liquid at a temperature above T_3. When cooled to the temperature of the point O, however, two liquid phases separate. As cooling proceeds further, the composition and relative amounts of the two liquid phases alter. If a horizontal line is drawn to cut the isopleth (line of constant composition C), then the compositions of the liquid phases at that temperature are given by the intercepts on the phase boundary curve. At the temperature of the point C_1, for example, the two liquid compositions are given by the points P and Q and the relative amounts of each at that temperature may be calculated by the lever principle around C_1 (p. 589).

$$\frac{\text{Amount of liquid of composition P}}{\text{Amount of liquid of composition Q}} = \frac{QC_1}{C_1P}$$

Partial immiscibility phenomena are often encountered in the silica-rich region of alkaline earth silicate systems.

Binary Systems in Ceramic Materials. There are many binary systems of importance to ceramists, and almost all the phase equilibrium conditions described previously are to be found.

The American Ceramic Soc. Inc., publish at frequent intervals up-to-date information on phase equilibrium diagrams of importance in ceramic studies. Entitled 'Phase Diagrams for Ceramics' by E. M. Levin, H. F. McMurdie, and F. P. Hall, editions were published in 1956, 1962, 1964 and 1969.

The *alumina/silica* diagram is perhaps the most commonly encountered binary system. Seger, in 1893, was the first to attempt its construction by noting the temperature at which a liquid phase appeared in mixtures of different composition. Bowen and Greig[1] studied the system in more detail and deduced a phase diagram in which the salient features included a eutectic point at a temperature of 1545° C. at a composition of 5·5 per cent alumina, 94·5 per cent silica. A peritectic point at 1810° C. suggested that a compound was formed which was not stable to its melting point. This is the well-known mineral, mullite ($3Al_2O_3.2SiO_2$) which is produced by heating almost all alumino-silicate mixtures and which at 1810° C was thought to dissociate into corundum (α-Al_2O_3) and a liquid phase.

Since the earlier work was published, several authors have studied sections of the diagram in great detail. It has been suggested that limited solid solution exists and that compounds other than mullite are stable in the solidus region. The modifications most generally accepted are those due to Toropov and Galakhov[2] and Aramaki and Roy[3] who proved that mullite is in fact a congruently melting compound —but only just. The new phase diagram is shown in Fig. X.17.

The inter-relationship of the alumino-silicate minerals, kyanite, andalusite, sillimanite and mullite has been established as the result of the work of Newton[4] and Haskell and DeVries.[5] They have studied the pressure/temperature relationship up to 50kbars pressure and shown that whereas sillimanite is the favoured phase at high temperatures, kyanite is stable at higher pressures; andalusite has a limited stability range but mullite will form at equilibrium at normal pressures, from all other types.

The *lime-silica* system has been examined in detail by Rankin and Wright[6] and modified by Muan and Osborn.[7] There are three eutectic points as follows:

1. At a temperature of 1436° C. a eutectic mixture consisting of tridymite and pseudo-wollastonite $\alpha CaO.SiO_2$ is formed.

2. At 1460° C. a eutectic containing 53 per cent of lime and 47 per cent of silica consisting of pseudo-wollastonite and $3CaO.2SiO_2$ is formed.

3. At 2065° C. a eutectic containing 67·5 per cent of lime and 32·5 per cent of silica and consisting of calcium orthosilicate and lime is formed.

Four definite binary compounds are shown in the equilibrium diagram (Fig. X.18), namely, calcium orthosilicate ($2CaOSiO_2$) which melts at 2130° C., calcium metasilicate ($CaOSiO_2$) which melts at 1544° C.; a compound with the formula $3CaO.2SiO_2$, which melts incongruently at 1464° C.; and tricalcium silicate, $3CaO.SiO_2$, which dissociates into CaO and $2CaO.SiO_2$ at 1900 °C. which is below the eutectic temperature (2065° C.).

[1] Bowen, N. L. and Greig, J. W., *J. Amer. Ceram. Soc.*, 7, 242, 1924.
[2] Toropov, N. A. and Galakhov, F. Ya., *Doklady Akad. Nauk. S.S.S.R.*, 78, 301, 1951.
[3] Aramaki, S. and Roy, R., *J. Amer. Ceram. Soc.*, 42, 644, 1959, and 45, 239, 1962.
[4] Newton, R. C., *Science*, 151, 1223, 1966.
[5] Haskell, R. W. and DeVries, R. C., *J. Amer. Ceram. Soc.*, 47, 203 and 236, 1964.
[6] Rankin, G. A. and Wright, F. E., *Amer. J. Sci.*, 39, 5, 1915.
[7] Muan, A. and Osborn, E. F., *J. Amer. Iron and Steel*, 1951.

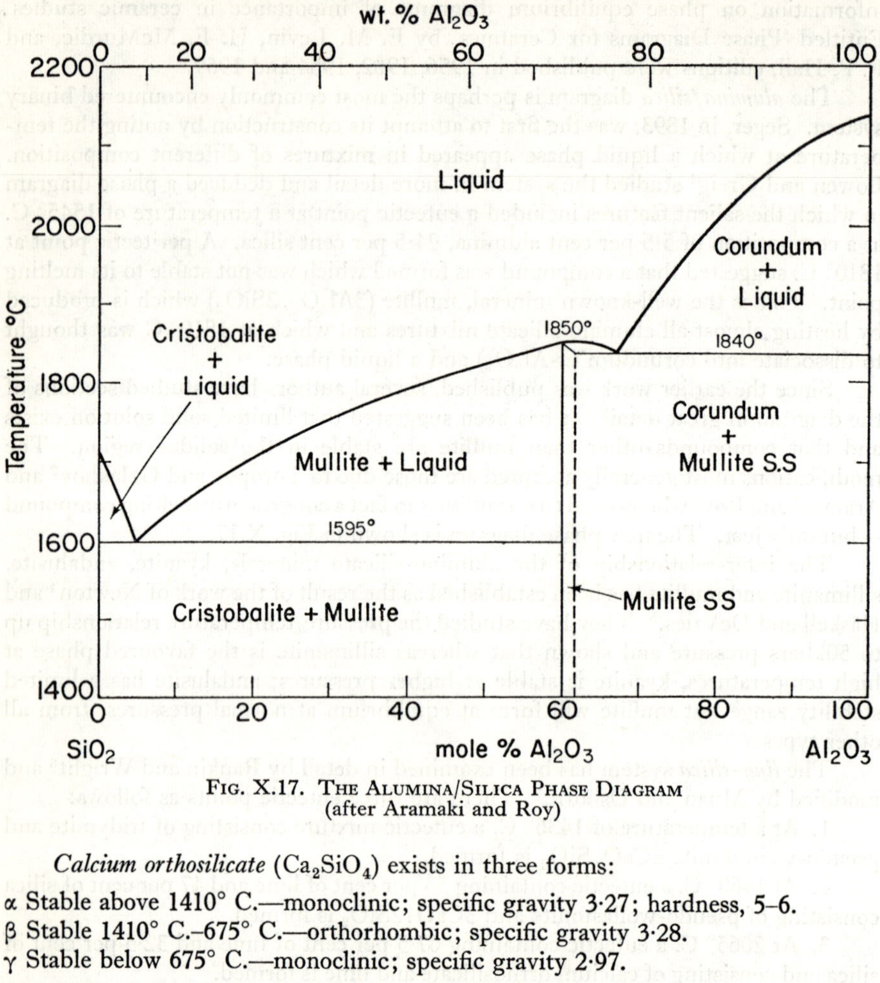

FIG. X.17. THE ALUMINA/SILICA PHASE DIAGRAM
(after Aramaki and Roy)

Calcium orthosilicate (Ca_2SiO_4) exists in three forms:

α Stable above 1410° C.—monoclinic; specific gravity 3·27; hardness, 5–6.
β Stable 1410° C.–675° C.—orthorhombic; specific gravity 3·28.
γ Stable below 675° C.—monoclinic; specific gravity 2·97.

Calcium metasilicate (wollastonite) ($CaSiO_3$) exists in two forms:

α Stable between 1125° C. and the melting-point (1540° C.).
β Stable up to 1125° C.

In the *magnesia-silica system* examined by Bowen and Anderson,[1] Fig. X.19, there are two compounds (*a*) clinoenstatite $MgSiO_3$ (incongruent, melting point 1557° C.) and (*b*) forsterite Mg_2SiO_4 (melting point 1890° C.). This is a particularly difficult system to investigate as the products are so viscous, even at high temperatures, that there is always uncertainty as to whether equilibrium has been reached.

The phase diagrams of divalent oxides with silica show interesting and similar features in the important silica-rich region. A close correlation with the size of the

[1] Bowen, N. L. and Anderson, O., *Amer. J. Sci.*, 37, 488, 1914.

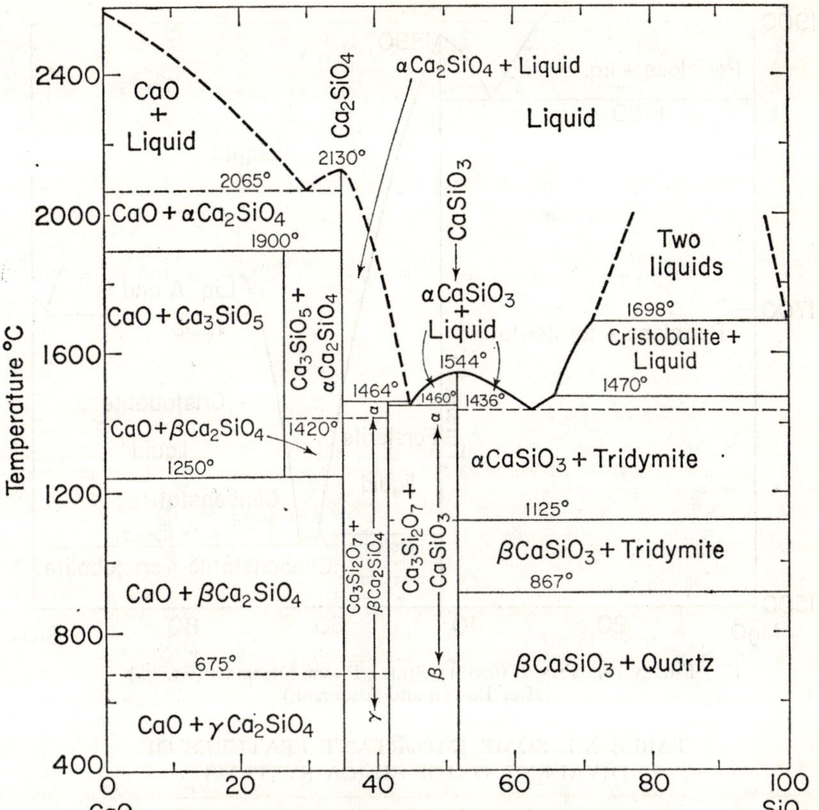

Fig. X.18. Phase Equilibrium Diagram of CaO-SiO$_2$ (wt. %)
(after Rankin and Wright and Muan and Osborn)

cation is observed if the range of immiscible liquids and the composition of the first eutectic are compared. This is illustrated by the data in Table X.I.

It is of interest to note that with the exception of barium oxide, all alkaline earth oxide-silica systems have a range in the silica-rich part of the diagram where the two components form immiscible liquids. This is of importance in basic open hearth practice, where silica bricks are used as roof bricks. The dusts rich in free lime are not exceptionally corrosive if the silica is of high purity because there is little tendency to mix. Should a third component, such as alumina, be present, even in small amounts, this desirable feature is destroyed and complete homogeneity in the liquid phase develops. The slagging action of the lime on the refractory is then greatly enhanced.

All the divalent oxide/silica systems contain a great complexity of compounds and eutectics, but it is also of interest to observe that oxides of elements of large cationic size (e.g. Ba) have the greatest complexity whilst the smaller ones (e.g. Mg)

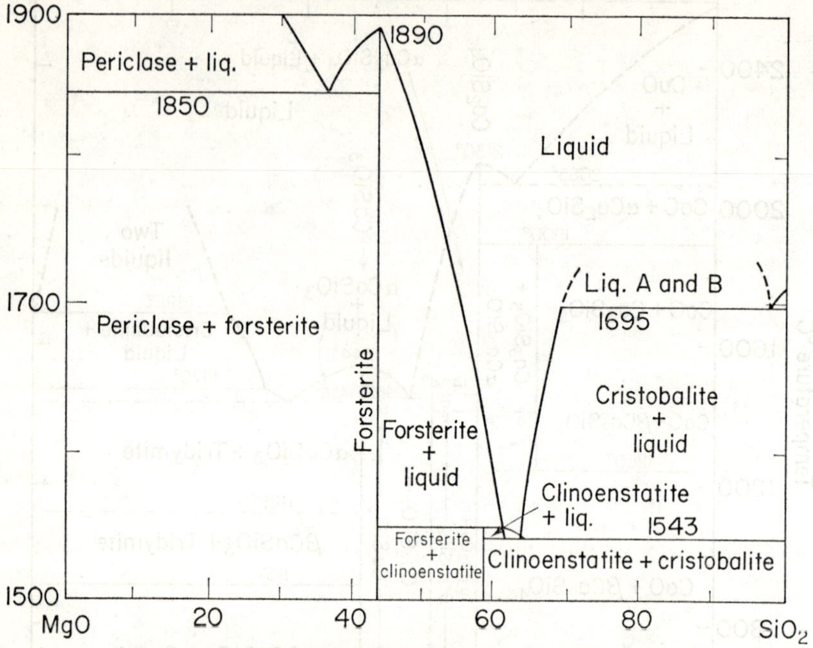

FIG. X.19. THE MAGNESIA/SILICA PHASE DIAGRAM (wt. %)
(after Bowen and Anderson)

TABLE X.I. SOME IMPORTANT FEATURES OF DIVALENT OXIDE/SILICA SYSTEMS

Oxide	*Cation Size* Å	*Immiscible Liquid Range*	*SiO_2-rich eutectic*	
		SiO_2 molar range	*Molar % SiO_2*	*Temp. °C.*
BaO	1·35	—	73	1380
SrO	1·13	80–97	67	1360
CaO	0·90	72–97	63	1440
MgO	0·65	59–97	45	1550
PbO	1·20	40–97	29	728
MnO	0·80	52–97	46	1200
FeO	0·76	62–97	38	1180
ZnO	0·74	66–97	52	1432
CoO	0·74	48–97	39	1381
NiO	0·72	57–97	41	1650

are relatively simple. The beryllium oxide/silica system[1] where the cation (Be^{2+}) is extremely small (0·31Å) has only a single eutectic in the liquidus range at 1670° C. and with 94 per cent molar concentration of SiO_2. A compound $2BeO.SiO_2$ forms but only in the solid phase; it dissociates into the component oxides at 1560° C.—well below the eutectic point.

[1] Morgan, R. A. and Hummel, F. A., *J. Amer. Ceram. Soc.*, 32, 255, 1949.

The *iron oxide-silica systems* are complicated by the existence of three iron oxides and by the comparative ease with which the red ferric and the black magnetic oxides can be reduced to form the dark ferrous oxide. At temperatures in excess of about 1000° C., however, all forms convert more or less readily into *wustite*, FeO, and Bowen and Schairer[1] have deduced the relationships between this component and silica as shown in Fig. X.20.

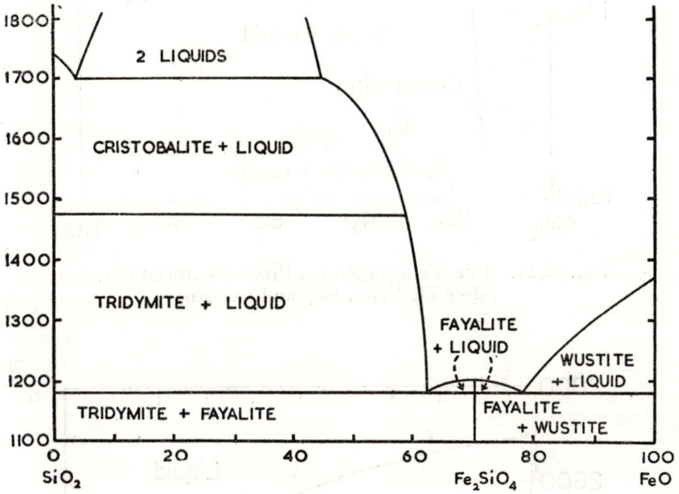

FIG. X.20. THE IRON OXIDE/SILICA PHASE DIAGRAM (wt. %)
(after Bowen and Schairer)

One compound, *fayalite*, Fe_2SiO_4, has been found and there are eutectic points at 1183° C. and 1181° C. containing 62 per cent and 77 per cent FeO respectively. Two immiscible liquid phases form in the silica-rich region in a similar way to alkaline earth-silica mixtures.

The *titania-silica system* was studied by DeVries, Roy and Osborn;[2] they found only one eutectic point at 1550° C., but showed that titania and silica were immiscible in the liquid form over most of the composition range. The phase diagram is shown in Fig. X.21.

The *zirconia-silica system* has been investigated by Geller and Yavorsky,[3] and as Fig. X.22 shows, contains the compound $ZrSiO_4$, zircon, which melts incongruently and gives rise to a peritectic point at 1775° C. The eutectic at a little below 1700° C. contains only about 1 per cent ZrO_2. Solid solution is also established.

The *alkali oxide-silica* mixtures give complex diagrams, characterised by many compounds and extremely low melting point eutectics. The principal features of the Li_2O, Na_2O, K_2O—silica systems are summarised in Table X.II, and the phase diagram of soda/silica is in Fig. X.23.

[1] Bowen, N. L. and Schairer, J. F., *Amer. J. Sci.*, 24, 200, 1932.
[2] DeVries, R. C., Roy, R. and Osborn, E. F., *Trans. Brit. Ceram. Soc.*, 53, 525, 1954.
[3] Geller, R. F. and Yavorsky, P. J., *J. Res. Nat. Bur. Stand.*, 35, 109, 1945.

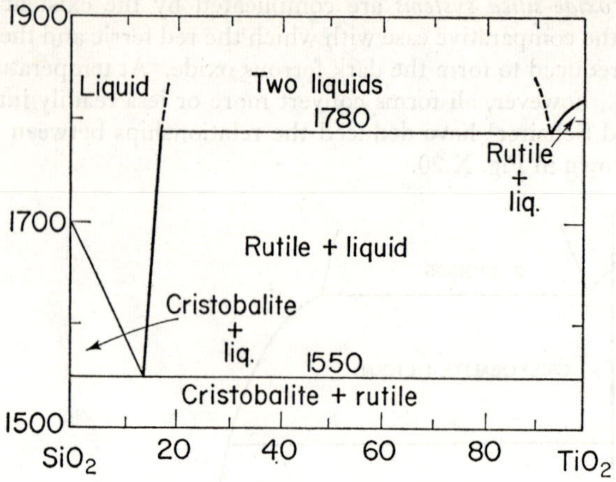

FIG. X.21. THE TITANIA/SILICA PHASE DIAGRAM (wt. %)
(after De Vries, Roy and Osborn)

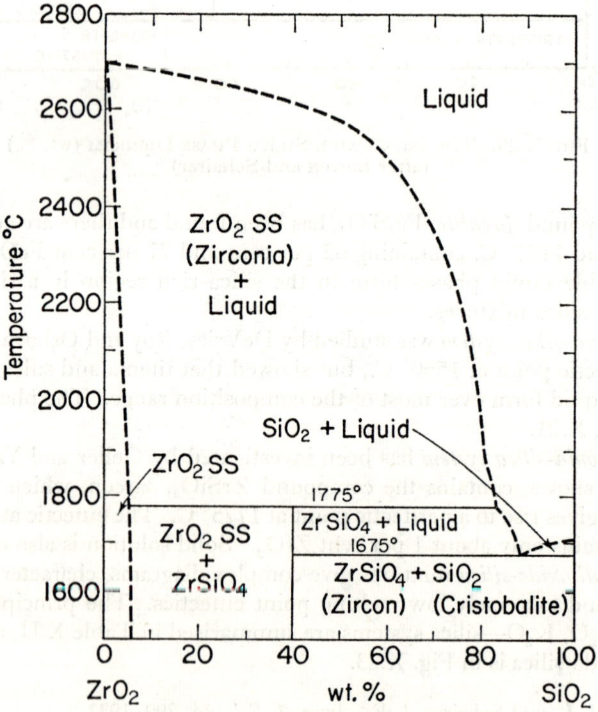

FIG. X.22. THE ZIRCONIA/SILICA PHASE DIAGRAM (wt. %)
(after Geller and Yavorsky)

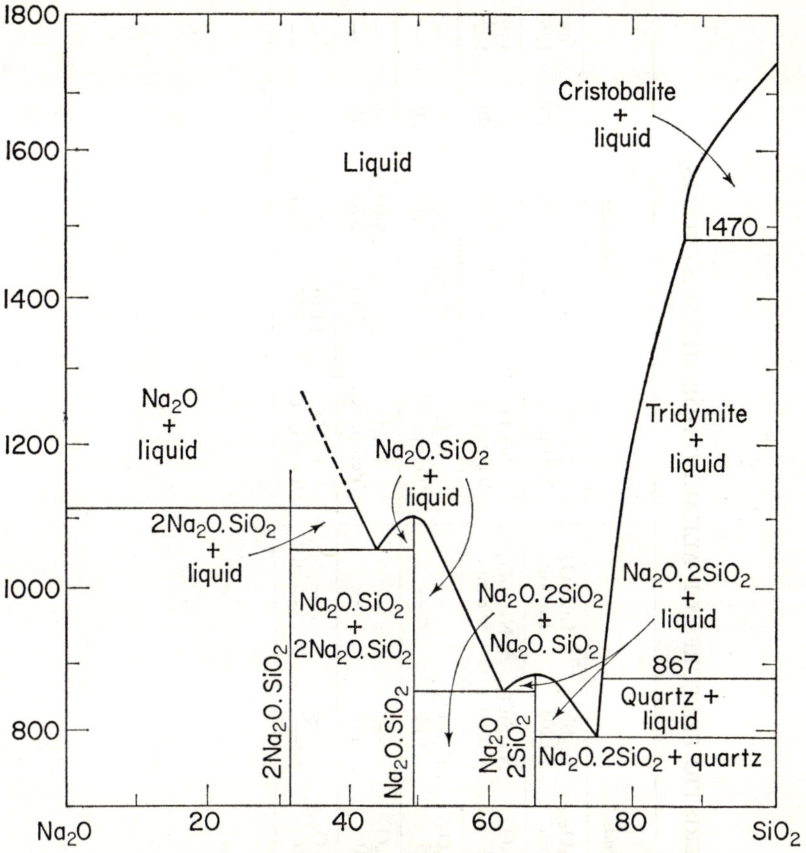

FIG. X.23. THE SODIUM OXIDE/SILICA PHASE DIAGRAM
(after Kracek)

The full diagram is probably even more complicated. Williamson and Glasser[1] have proved that an incongruently melting compound $Na_6Si_8O_{19}$ exists in the region of the 799° C. eutectic.

Much interesting information can be derived from a comparison of the various binary systems with silica as one component. The amount of liquid formed as the temperature is raised when a small amount of impurity is present in silica may be predicted.

If 1 per cent molar concentration of the various oxides is mixed with pure silica, a liquid will first form at the appropriate eutectic point and its amount is given by applying the lever principle around the composition point (see p. 589). The amount of liquid at all temperatures above the eutectic may be similarly assessed until the liquidus line is reached, when no solid matter remains.

[1] Williamson, J. and Glasser, F. P., *Science*, 148, 1589, 1965.

TABLE X—II. CHARACTERISTIC FEATURES OF ALKALI OXIDE-SILICA SYSTEMS

Alkali Oxide		*Compounds*					*Eutectics*			
Li_2O	Formulae	—	$Li_2Si_2O_5$	Li_2SiO_3	Li_4SiO_4	Li_2O	—	17%	44%	49%
	M. pt. °C.		Incong.	1201	Incong.	Temp. °C.	—	1028	1024	1255
Na_2O	Formulae	—	$Na_2Si_2O_5$	Na_2SiO_3	Na_4SiO_4	Na_2O	—	28%	37%	55%
	M. pt. °C.		874	1080	Incong.	Temp. °C.	—	799	840	1020
K_2O	Formulae	$K_2Si_4O_9$	$K_2Si_2O_5$	K_2SiO_3	—	K_2O	26·4%	32·4%	45·0%	—
	M. pt. °C.	770	1045	976	—	Temp. °C.	767	752	780	—
Rb_2O	Formulae	$Rb_2Si_4O_9$	$Rb_2Si_2O_5$	—	—	Rb_2O	16·0%	25·0%	43·0%	—
	M. pt. °C.	870	1090	—	—	Temp. °C.	770	876	745	—
Cs_2O	Formulae	$Cs_2Si_4O_9$	$Cs_2Si_2O_5$	—	—	Cs_2O	14·0%	23·0%	—	—
	M. pt. °C.	950	1070	—	—	Temp. °C.	875	910	—	—

Fig. X.24 shows the amount of liquid forming as the temperature is raised with several 1 per cent oxide-silica mixtures. It is significant that although with alumina, liquid does not form until 1545° C., the amount present at 1650° C. is in excess of that in the sodium oxide-silica mixture. For this reason, alumina is an undesirable impurity in the raw materials for the manufacture of silica bricks.

The *lime-alumina system* includes various calcium aluminates which are much more fusible than either lime or alumina.

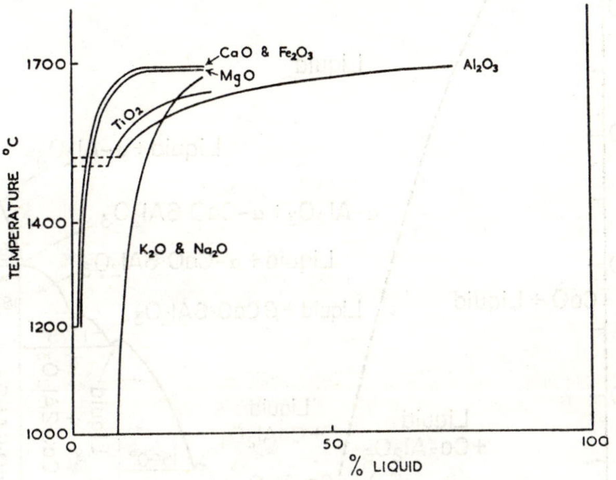

FIG. X.24. THE AMOUNT OF LIQUID FORMED WITH MIXTURES OF SILICA WITH 1 PER CENT OF VARIOUS OXIDES AT DIFFERENT TEMPERATURES

Free lime and alumina begin to react at about 800° C. and equilibrium, completely in the solid state, is reached rapidly at 1300° C. No liquid forms in the system below 1400° C.

There are many compounds known between the two components—some are congruent so that there are several eutectics. There are others which melt incongruently. The phase diagram shown in Fig. X.25 is a simplified form attributed to Rankin and Wright[1] and Rolin and Thanh.[2]

The *magnesia-alumina system*[3] yields one compound—spinel ($MgOAl_2O_3$), with a melting point of 2135° C. and the three eutectics shown in Table X.III.

TABLE X—III

Constituents	*Magnesia per cent*	*Alumina per cent*	*Melting point*
Spinel/magnesia	45	55	2030° C.
Spinel/α-alumina	98	2	1925 ± 40° C.
Spinel/β-alumina	92	8	1925 ± 40° C.

[1] Rankin, G. A. and Wright, F. E., *Amer. J. Sci.*, 39, 11, 1915.
[2] Rolin, M. and Thanh, P. H., Rev. Hautes. Temp. Refr. 2, 181, 1965.
[3] Rankin, G. A. and Merwin, H. E., *Amer. J. Sci.*, 45, 304, 1918.

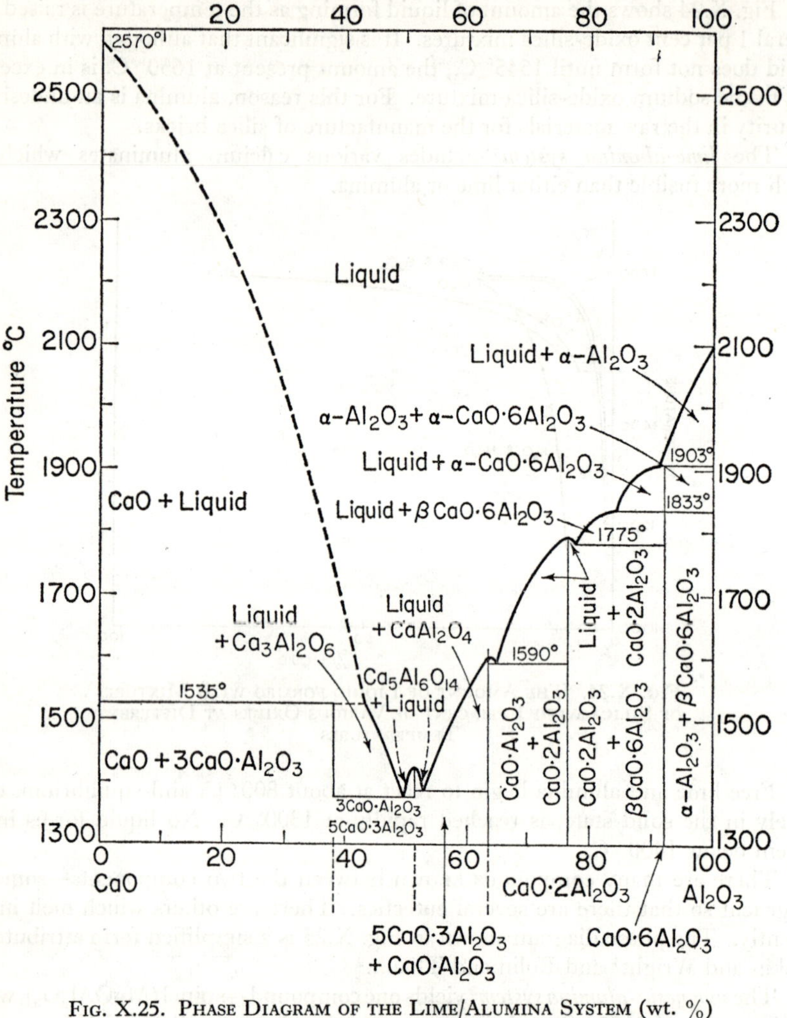

FIG. X.25. PHASE DIAGRAM OF THE LIME/ALUMINA SYSTEM (wt. %) (after Rankin and Wright and Rolin and Thanh)

Spinel forms a series of solid solutions with α-alumina.[1] The phase diagram is shown in Fig. X.26.

The *lime-magnesia system*[2] contains one eutectic composed of 67 per cent of lime and 33 per cent of magnesia; it melts at 2300° C. (see Fig. X.27). Tröjer and Konopicky[3] have suggested that partial solid solution of magnesia in lime takes

[1] Roy, D. M., Roy, R. and Osborn, E. F., *J. Amer. Ceram. Soc.*, 36, 149, 1953.
[2] Rankin, G. A. and Merwin, H. E., *J. Amer. Chem. Soc.*, 38, 571, 1916.
[3] Tröjer, F. and Konopicky, K., Radex-Rundschau, 4, 161, 1949.

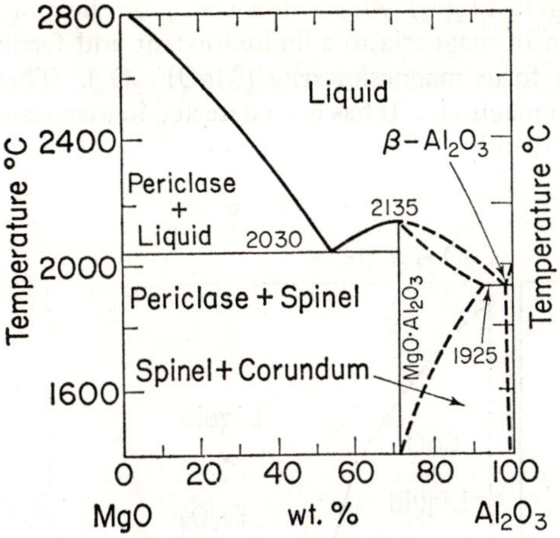

FIG. X.26. PHASE DIAGRAM OF THE MAGNESIA-ALUMINA SYSTEM (wt. %) (after Roy and Osborn)

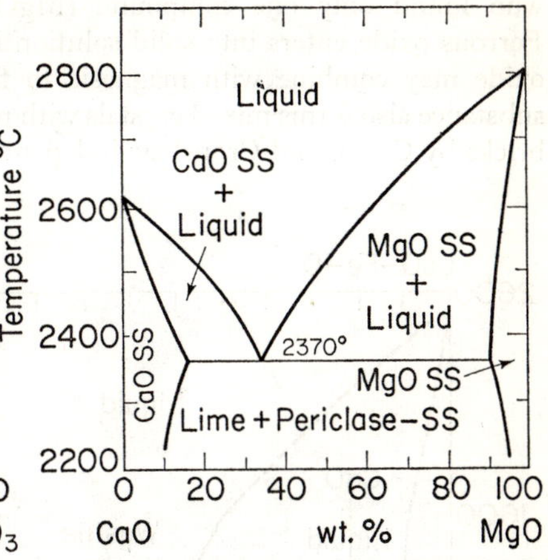

FIG. X.27. TEMPERATURE/CONCENTRATION DIAGRAM OF THE LIME-MAGNESIA SYSTEM (after Doman *et al.*)

place, and this view has been confirmed by Doman *et al.*,[1] whose diagram is shown in Fig. X.27.

The *iron oxide-alumina system* like the corresponding silica system is complicated by the existence of three iron oxides (p. 609). So far as the ferrous oxide-alumina system is concerned the chief compound is *hercynite* ($FeOAl_2O_3$). The iron oxide-alumina system has not, however, been fully investigated.

The *iron oxide-lime system*[2] is complicated like the corresponding ones containing alumina and silica respectively. When lime and iron oxide are heated together ferrites may be formed. The two chief compounds are calcium metaferrite ($CaOFe_2O_3$), which melts at 1215° C., and dissociates at the same temperature, forming long, black, needle-shaped crystals, whilst the second compound calcium orthoferrite ($2CaOFe_2O_3$) melts at 1436° C. and almost immediately dissociates, forming black crystals having a yellowish-brown tinge by reflected light. The eutectic contains 8 per cent of lime and 92 per cent of ferric oxide; it melts at 1203° C. The phase diagram, due to Sosman,[3] is shown in Fig. X.28B.

Very little calcium meta- or ortho-ferrite is formed when clays are burned as the proportion of iron oxide is not usually sufficient.

When the iron oxide is in the partially reduced state, which is so often the case when burning clays and other ceramic materials, wustite (FeO) is one component of the system and the phase diagram can be entirely different in form[4] (Fig. X.28A).

The *iron oxide-magnesia system* has been investigated by Phillips and Muan[5]

[1] Doman, R. C., Barr, J. B., McNally, R. N. and Alper A. M., *J. Amer. Ceram. Soc.*, 46, 313, 1963.
[2] Sosman, R. B. and Merwin, H. E., *J. Wash. Acad. Sci.*, 6, 532, 1916.
[3] *Amer. J. Sci.*, 39, 11, 1915.
[4] Johnson, R. E. and Muan, A., *J. Amer. Ceram. Soc.*, 48, 360, 1965.
[5] Phillips, B. and Muan, A., *J. Amer. Ceram. Soc.*, 45, 590, 1962.

who found only one compound ($MgO. Fe_2O_3$) which dissociates on heating. Ferrous oxide enters into solid solution in magnesia to a limited extent and ferric oxide may combine with magnesia to form magnesioferrite ($MgOFe_2O_3$). This substance also forms mixed crystals with magnesia. It has been detected in magnesia bricks by Cornu and Cronshaw independently.

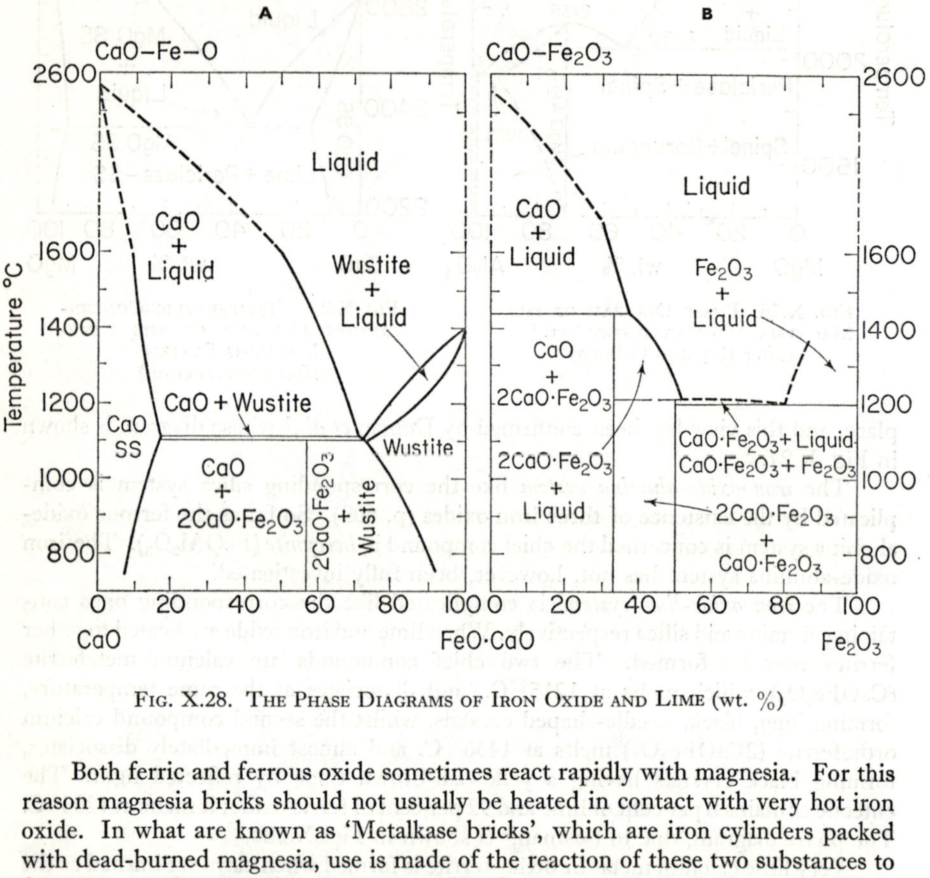

FIG. X.28. THE PHASE DIAGRAMS OF IRON OXIDE AND LIME (wt. %)

Both ferric and ferrous oxide sometimes react rapidly with magnesia. For this reason magnesia bricks should not usually be heated in contact with very hot iron oxide. In what are known as 'Metalkase bricks', which are iron cylinders packed with dead-burned magnesia, use is made of the reaction of these two substances to form a magnesic mass with a very strong bond.

Binary Systems of Complex Components. Although it is most usual to consider equilibrium diagrams in terms of the simplest components (namely oxides), it is frequently convenient to select complex compounds for this purpose. A small region of a ternary or multi-component system can be more clearly appreciated by constructing the binary field between two compounds which are formed in the mixture; in fact, a ternary equilibrium diagram is frequently built up from information given by firing mixtures of two compounds in definite proportions.

A simple example of the use of complex components is shown in Fig. X.29 where the phase relationships between sodium disilicate (Na_2O2SiO_2) and corundum

(Al_2O_3) in the ternary system Na_2O—Al_2O_3—SiO_2 are illustrated. The comparatively simple binary diagram shows congruent compound formation of the mineral carnegieite $Na_2O.Al_2O_3.2SiO_2$ which transforms to nephelite at 1254° C. Binary eutectics are formed at 768° C. and 1475° C.

Another example from the same ternary mixture but where two complex compounds are taken as components is shown in Fig. X.30. If carnegieite (or nephelite) $Na_2O.Al_2O_3.2SiO_2$ and the sodium felspar, albite, $Na_2O.Al_2O_3.6SiO_2$ are regarded

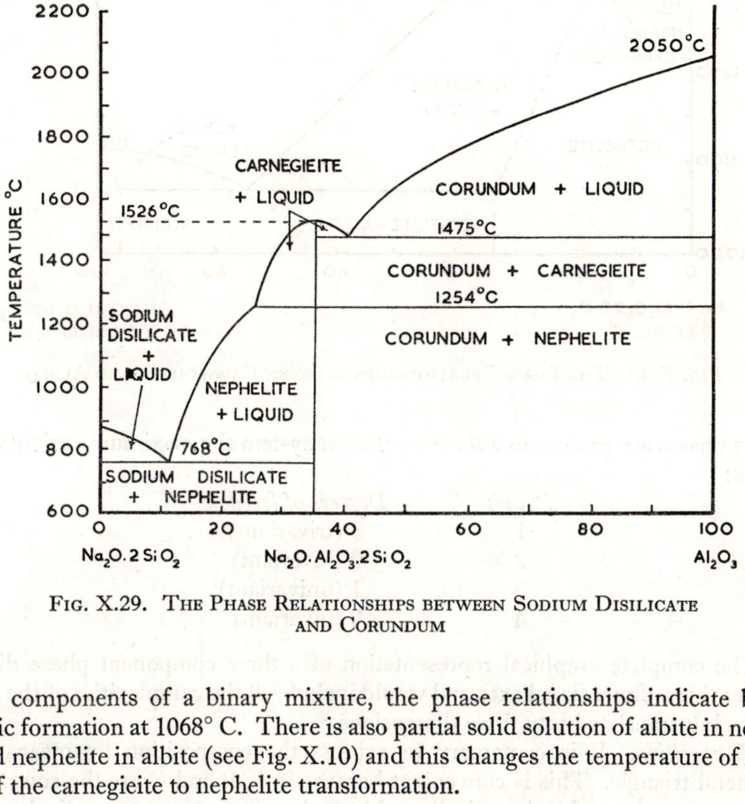

FIG. X.29. THE PHASE RELATIONSHIPS BETWEEN SODIUM DISILICATE AND CORUNDUM

as the components of a binary mixture, the phase relationships indicate binary eutectic formation at 1068° C. There is also partial solid solution of albite in nephelite and nephelite in albite (see Fig. X.10) and this changes the temperature of inversion of the carnegieite to nephelite transformation.

THREE-COMPONENT OR TERNARY DIAGRAMS

WHEN the number of components in a system is increased to three, the possible total of independent variables is *four*. These are pressure, temperature and two concentration degrees of freedom. In a ternary system, provided the concentrations of two components are known, the third can be implied; thus the composition can be accurately defined with only two concentration variables.

In ceramic systems, vapour pressure, in most cases, can be disregarded and a 'condensed' form of the Phase Rule used to predict the degrees of freedom. When

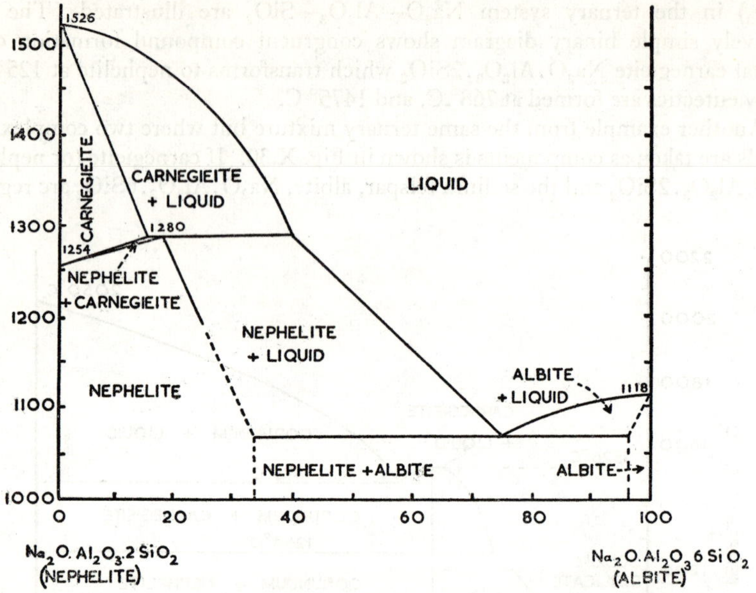

FIG. X.30. THE PHASE RELATIONSHIPS BETWEEN CARNEGIEITE AND ALBITE

various phases are present in a *three-component* system the maximum variants are as follows:

Phases	*Degrees of freedom*
1	3 (trivariant)
2	2 (bivariant)
3	1 (univariant)
4	0 (invariant)

The complete graphical representation of a three component phase diagram requires three-dimensional axes and would include all the complexities of the binary systems but translated into three dimensions.

Composition. It is customary to express the composition by means of an equilateral triangle. This is convenient because if A, B and C are the components and the proportions of each are indicated by a, b and c per cent respectively, then,

$$a + b + c = 100 \text{ per cent}$$

in every case. If a and b are regarded as independent variables, c can always be deduced provided the first two are known. The composition of any mixture may conveniently be expressed by three planar axes which intersect at angles of 60° and are parallel to the sides of the triangle.

The apices of the equilateral triangle represent 100 per cent of each pure component, either on a weight or molar basis. The opposite side of the triangle to any particular apex will be the composition line of mixtures containing zero percentage of the component. Mixtures with intermediate concentrations of any component fall at an appropriate distance between the apex and opposite side. A triangular composition diagram is shown in Fig. X.31. The proportion of the components A, B and C in any particular mixture is easily deduced by drawing lines at convenient intervals across the diagram parallel to the three sides. Thus, point P contains 20 per cent of component A, 40 per cent of component B and 40 per cent of component C.

A triangular diagram of this kind offers many advantages, one of which is that if amounts of one compound are progressively added to or subtracted from a ternary mixture whilst the other two components are kept in a fixed ratio, the ensuing change in composition may be represented by a straight line drawn from the appropriate apex through the point of original composition and produced if

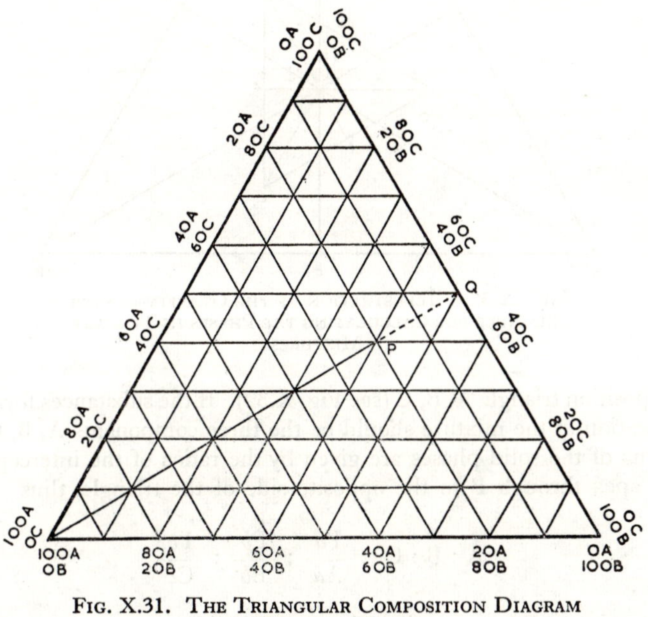

FIG. X.31. THE TRIANGULAR COMPOSITION DIAGRAM OF A TERNARY MIXTURE

necessary. This is shown in Fig. X.31, where the line PA indicates the change in composition brought about by adding amounts of component A to a mixture of composition P (consisting of 20 per cent A; 40 per cent B; 40 per cent C); PQ represents the change in composition of the same mixture when component A is progressively removed. Similar additions or subtractions of the other components of the mixture would result in the composition altering along a line joining the original point P to the corresponding apex of the triangle or to the extension of such a line beyond point P. This is an important fundamental concept which greatly assists in interpreting ternary diagrams.

Another important aspect of the triangular composition diagram is the *gravity-centre principle*[1] for determining the quantitative composition when several phases form from a given mixture. This may be applied to ternary systems in a similar manner to the lever rule in binary mixtures. The point P, for example, is a mixture

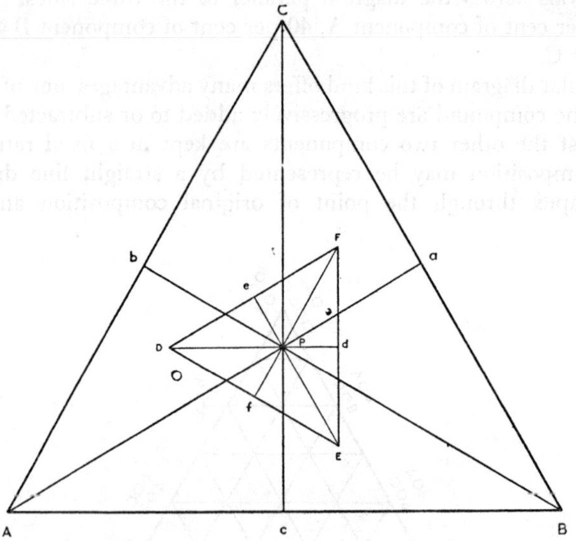

FIG. X.32. ILLUSTRATION OF THE GRAVITY-CENTRE PRINCIPLE FOR CALCULATING THE PHASES IN TERNARY MIXTURES

in the composition triangle A, B, C (see Fig. X.32). If the substances formed during the solidification of the mixture should be the three components A, B, C, the relative amounts of the solid phases are given by the ratios of the intercepts on lines from each apex through P to the opposite side of the triangle, thus

$$A : B : C :: \frac{Pa}{Aa} : \frac{Pb}{Bb} : \frac{Pc}{Cc}$$

If, however, mixture P separates into three other solid components, say D, E and F, the relative proportions are given by constructing a composition triangle and taking intercepts on the lines from the apices to P produced to the opposite sides of the triangle DEF. Thus

$$D : E : F :: \frac{Pd}{Dd} : \frac{Pe}{Ee} : \frac{Pf}{Ff}$$

The gravity-centre principle is of value in assessing the relative amounts of phases when compound-formation occurs in ternary mixtures.

Temperature. There remains the problem of representing the third variable in

[1] That is by using the same geometrical construction as that employed to find the centre of gravity of a plane surface.

ternary mixtures, namely temperature. This may be solved by introducing an additional axis projecting at right angles from the plane of the triangular diagram. The complete ternary diagram using such a model would be a solid figure bounded by three faces each one based on one side of the composition equilateral triangle and projecting vertically upwards.

Planar diagrams of such solid models can be constructed by representing the temperature as contour lines in a similar way to which height is shown on survey maps.

Phase boundary lines are normally shown by heavy lines, whilst zones of constant temperature or isotherms are in lighter or dotted lines. With a little experience it is not a difficult matter to become familiar with ternary phase diagrams in projection on the equilateral base.

1. **Ternary Eutectic Formation**. The simplest example of a ternary system is where three components are completely miscible when liquid, but do not mix in any proportion in the solid state. In such a case a *ternary eutectic* is formed which is exactly analogous to that of the binary system on page 588 except that three components are involved.

A perspective drawing of the resulting solid diagram is shown in Fig. X.33A, and the corresponding projection in Fig. X.33B. The addition of any one component to another causes a reduction in the melting point of the latter. In addition, the third component will give rise to a lowering of the melting point of any mixture of the other two components. Hence, the solid diagram is shaped like a well with three troughs falling into the middle (point E_4). Each trough arises at the eutectic point of a binary mixture (points E_1, E_2, E_3), and they correspond to phase boundary lines which effectively divide the diagram into three distinct regions. The field bounded by the points A–E_1–E_4–E_3 represents the stability range of component A in equilibrium with liquid. When any mixture, the composition of which falls in this field, is cooled the first product of crystallisation will be component A. Similarly component B will be formed first from a mixture in the field B–E_1–E_4–E_2 and crystals of C will be the first solid phase to form on cooling a composition in the C–E_2–E_4–E_3 field.

In each stability field the isotherms fall towards the eutectic point (E_4) and along the phase boundary lines E_1–E_4, E_2–E_4, and E_3–E_4 the temperature also decreases in the direction of the ternary eutectic, E_4.

A further simplification in representing ternary diagrams consists in dispensing with the isotherms and indicating changing temperatures by arrows on the phase boundary lines, as in Fig. X.33C, which shows the arrows pointing in the direction of decreasing temperature.

The phase boundary lines in the ternary diagram shown in Fig. X.33C, are the positions at which two solid phases can be in equilibrium with liquid. Along the line 1–4, solid A and B can co-exist, B and C may be present together along 2–4 and A and C form simultaneously along the line 3–4. In each of these conditions the system is univariant and has one degree of freedom. At point 4, however, three solid phases A, B and C may be present along with a liquid, hence this is an invariant point and before the temperature can change, one phase must vanish. The ternary eutectic is thus the point of lowest temperature in the system at which a liquid phase can exist.

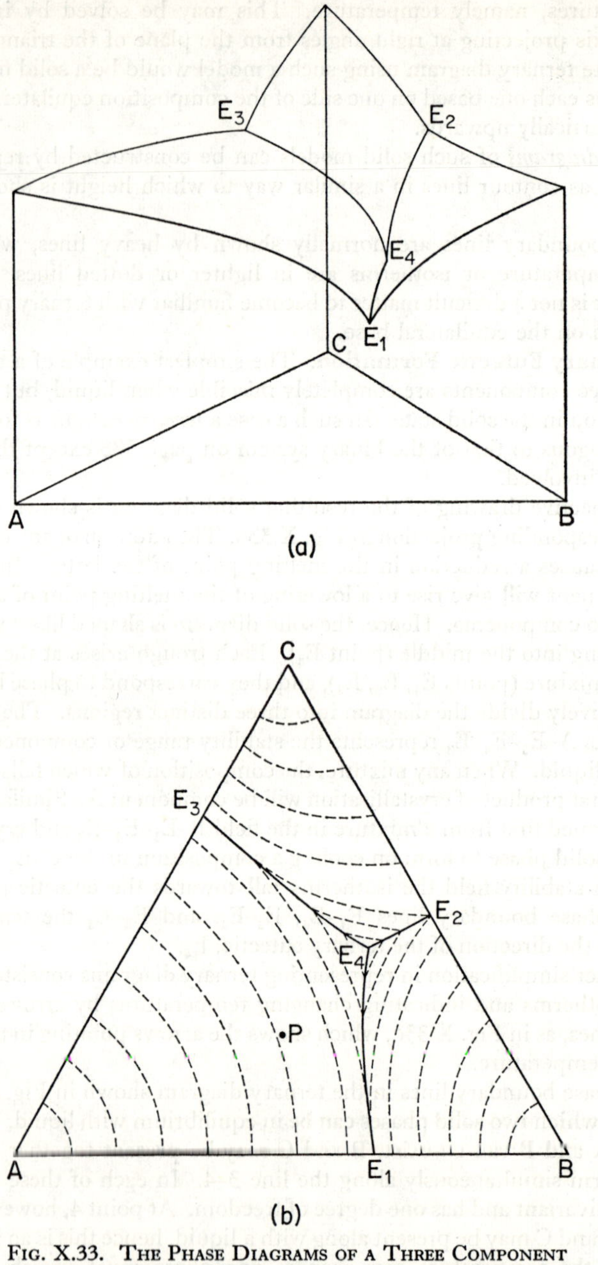

FIG. X.33. THE PHASE DIAGRAMS OF A THREE COMPONENT SYSTEM SHOWING THE FORMATION OF A TERNARY EUTECTIC
(a) A solid model
(b) A planar representation showing isotherms

The crystallisation paths when any mixture is cooled are similar in all parts of a ternary-eutectic system and may be illustrated by a composition N in the stability field of component A (see Fig. X.33c).

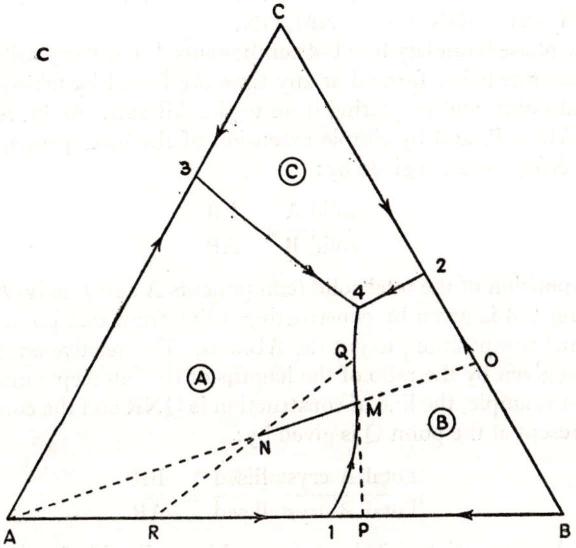

Fig. X.33c. The Phase Diagrams of a Three Component System showing the Formation of a Ternary Eutectic
A planar representation with arrows indicating the direction of falling temperature

The mixture is liquid at all temperatures above the solidus curve. Crystals of A are the first to separate from the cooling melt and in accordance with the principle laid down on page 620 the composition of the liquid phase changes along the line NM which is AN produced. When the liquid composition becomes that of the point M on the phase boundary line 1–4, both solid A and solid B can form together and they do so with the liquid composition changing along the line M–4. At the eutectic point 4, crystals of C also separate and eventually all liquid will vanish as the cooling proceeds. The final state is, therefore, a mass of crystals with each solid component a separate entity.

At any temperature, the relative proportions of each phase present may be calculated by applying the gravity-centre principle. The determination of the amount of component A which crystallises between the initial solidification point N and the phase boundary line at M demonstrates its use.

If ANM is produced it will meet BC at O. The amount of component A in a mixture represented by the point N is given by the ratio ON/OA; similarly the amount of A in mixture M is OM/OA. The difference in amount of A in the two mixtures would be equivalent to the amount crystallised during the cooling of the initial mixture from the temperature at N to that at M. This may be expressed as:

$$\frac{ON}{OA} - \frac{OM}{OA} = \frac{NM}{OA}$$

The ratio of the amount crystallising to the amount of component A originally present is NM/OA ÷ ON/OA = NM/ON.

Along the phase boundary line 1–4 components A and B crystallise together and the relative amounts being formed at any time are found by taking the tangent to the line at that point, and projecting it on to the AB axis. At M, for example, the tangent cuts AB at P, and by simple extension of the lever principle, the ratio of solid A and B *being formed* is given by:

$$\frac{\text{solid A}}{\text{solid B}} = \frac{PB}{AP}$$

The composition of the total solid (components A and B only) *having formed* at any point along 1–4 is given by constructing a line from that point through N the point of original composition, to cut the AB axis. The relative amounts of solid A and B are then given by the ratio of the lengths of the intercepts made on this axis. At point Q, for example, the line of construction is QNR and the composition of the solid phase present at the point Q is given by:

$$\frac{\text{Total A crystallised}}{\text{Total B crystallised}} = \frac{BR}{AR}$$

This is so because mixture N has separated into a liquid of composition Q and a combined solid phase. The composition of the solid must lie along QN produced and as it contains only components A and B, it must also fall on the line AB, hence point R is the composition point of the solid. The percentage amount of component A in such a mixture is given by BR/100 and that of B by AR/100. Hence, the ratio of A to B must equal BR/AR.

The *total amount* of components which have crystallised at any point, or conversely the composition of the liquid phase remaining, may be readily calculated from first principles. The percentages of solid A and solid B which have crystallised at point Q from an original mixture N may be deduced as follows:

	A	B	C
Composition of liquid N	*a* parts	*b* parts	*c* parts
At Q the liquid remaining has the composition	*d* ,,	*e* ,,	*f* ,,

As component C does not crystallise up to the point Q, there must be the same amount of this component in the liquid phase at Q as in the original mixture. Hence, *f* must be reduced to the value of *c*, and *d* and *e* must be altered in the same proportion thus:

	A	B	C
Composition of liquid phase at Q	$d\frac{c}{f}$ parts	$e\frac{c}{f}$ parts	*c* parts

The amounts of A and B which crystallise on cooling to the point Q are:

$$\text{A:}\quad a - d\frac{c}{f} \text{ parts} \qquad\qquad \text{B:}\quad b - e\frac{c}{f} \text{ parts}$$

or expressed as a percentage of the original amount of component.

$$\text{A:}\quad \left(1 - \frac{dc}{af}\right)100 \qquad\qquad \text{B:}\quad \left(1 - \frac{ec}{bf}\right)100$$

Alternatively, the relative amounts of phases at the point Q can be determined by the gravity-centre principle. The three phases present have the composition A, B and that of the point Q, hence, a triangle can be constructed with each composition as an apex (Fig. X.32). The relative amounts of the three phases is given by the intercepts on the lines from each apex through P produced on to the opposite side of the triangle. Thus in Fig. X.34:

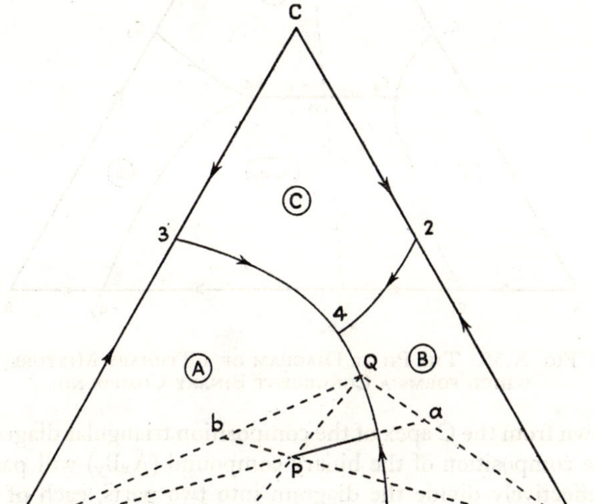

FIG. X.34. THE TERNARY EUTECTIC PHASE DIAGRAM SHOWING THE APPLICATION OF THE GRAVITY-CENTRE PRINCIPLE TO DETERMINE THE COMPOSITION OF PHASES

Ratio of solid A : solid B : liquid of composition Q :: Pa/Aa : Pb/Bb : Pc/Qc

2. **Binary Compound Formation without Solid Solution.** In many ternary mixtures, two of the components combine together to form a binary compound. There are several variants depending on the nature of the compound:

(*a*) Where the *binary compound melts congruently* (i.e. without decomposition) over the *whole* field of its existence. The equilibrium diagram is as shown in Fig. X.35. A solid diagram would show a *solidus* (p. 594) declining in temperature from the C apex to the phase boundary line E_4–E_5–D–E_6–E_3. This line has as extremities the binary eutectics of mixtures A–C and B–C; there are two points of minimum temperature E_5 and E_6 separated by a 'col' of which D is the maximum point. The centre dome-shaped portion of the E_5–E_6 phase-boundary line is maintained towards the composition line AB but with temperatures gradually increasing. The phase equilibria conditions along the line AB are of a binary mixture with congruent compound formation as illustrated in Fig. X.8 (p. 592).

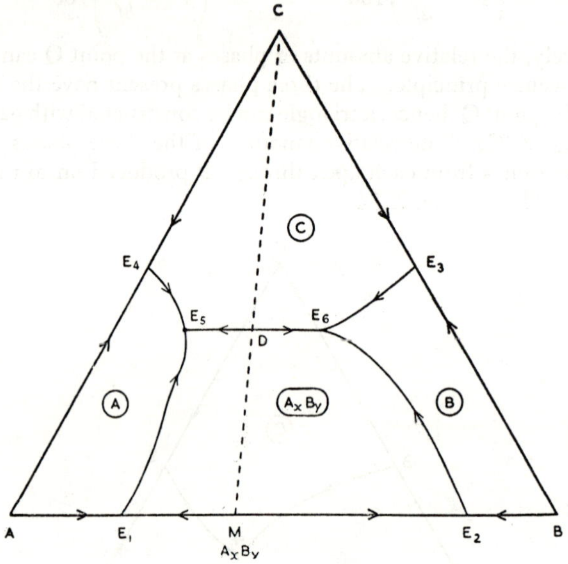

FIG. X.35. THE PHASE DIAGRAM OF A TERNARY MIXTURE WHICH FORMS A CONGRUENT BINARY COMPOUND

A line drawn from the C apex of the composition triangular diagram to the point M which is the composition of the binary compound (A_xB_y) will pass through the point D and effectively divide the diagram into two parts, each of which has the characteristics typical of ternary eutectic formation. The components of the first triangle are A, C, and the compound A_xB_y, the second triangle has B, C and A_xB_y as components. CDM is known as a *conjugation line* and the temperature varies along its length as in binary eutectic mixtures. The point D is at a minimum temperature with respect to C and M.

The *crystallisation-paths* of any mixture may be interpreted by applying the principles relating to ternary eutectic mixtures. The final products of crystallisation of any mixture are, however, determined by the *conjugation triangle* in which it occurs. Mixtures to the left of the conjugation line CDM can have only A, C and compound A_xB_y as solid products; whilst a mixture in the composition triangle bounded by CMB will crystallise only to solid B, C, and A_xB_y.

(*b*) Where the *binary compound is stable to its melting point in only a limited field of composition.* As the equilibrium diagram in Fig. X.36 shows, the binary field AB is that of normal, stable-compound formation. The dome-shaped portion does not continue into the ternary field but falls to a *ternary peritectic inflection* P. The phase boundary lines from this point and from the binary eutectic points E_2 and E_3 meet at a ternary eutectic or minimum temperature point E_5.

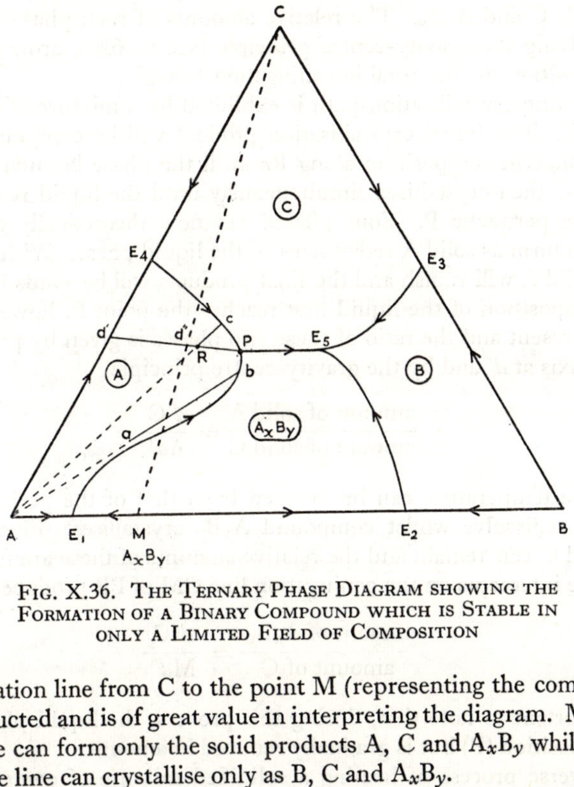

FIG. X.36. THE TERNARY PHASE DIAGRAM SHOWING THE FORMATION OF A BINARY COMPOUND WHICH IS STABLE IN ONLY A LIMITED FIELD OF COMPOSITION

A conjugation line from C to the point M (representing the compound A_xB_y), may be constructed and is of great value in interpreting the diagram. Mixtures to the left of this line can form only the solid products A, C and A_xB_y whilst mixtures to the right of the line can crystallise only as B, C and A_xB_y.

The crystallisation paths of various mixtures can be deduced from this axiom as several examples show. Thus, a mixture of composition *a* will begin to crystallise at the corresponding temperature of the solidus. This is in the stability field of A, hence this component will be the first to crystallise and the composition of the liquid will follow the line A*a* produced (see p. 620). This process will continue until the liquid composition reaches the phase boundary line E_1P at the point *b*. Now the solid phases A and A_xB_y will crystallise together with the liquid phase changing in composition from *b* towards P. At P, component C can also crystallise; hence four phases will be in equilibrium, namely solids A, C and AxBy and a liquid; the point *p* is, therefore, invariant and one phase must vanish before further reduction of temperature can take place. Solid component A begins to react with the liquid phase with simultaneous crystallisation of the compound A_xB_y. If the original mixture lies in the conjugation triangle AMC, all the liquid will crystallise before solid A is completely redissolved. On the other hand, from a mixture whose composition lies in the triangle BMC but to the left of the phase boundary line E_1P, A would still be the first phase to crystallise, but at the peritectic point P it would be completely redissolved before all the liquid crystallised. At this stage, the liquid would change in composition along P–E_5 with simultaneous crystallisation of A_xB_y and C until at E_5, a ternary eutectic, solid B would also form. The final solid would be composed only of solid B, C and A_xB_y. The relative amounts of each phase in all cases are found by applying the gravity-centre principle (see p. 620), around the point of original composition in the suitable conjugation triangle.

An interesting crystallisation-path is exhibited by a mixture of composition R (see Fig. X.36). The initial crystallisation product will be component A and the liquid will change in composition along R*c* until the phase boundary line E_4–P is joined. Solid C then crystallises simultaneously until the liquid reaches the composition of the peritectic P. Four phases can now theoretically exist, for solid A_xB_y begins to form as solid A redissolves in the liquid phase. With this particular mixture, all solid A will vanish and the final products will be solids B, C and A_xB_y. When the composition of the liquid first reaches the point P, however, only solid A and C are present and the ratio of these two phases is given by producing PR to meet the AC axis at d' and by the gravity-centre principle

$$\frac{\text{amount of solid A}}{\text{amount of solid C}} = \frac{d'\text{C}}{\text{A}d'}$$

Before the temperature can be lowered from that of the peritectic point, all solid A must redissolve whilst compound A_xB_y crystallises. Finally, only solid A_xB_y and solid C can remain and the relative amounts of these are found by taking the ratio of the intercepts on the conjugation line CM of PR produced to d'', thus

$$\frac{\text{amount of } A_xB_y}{\text{amount of C}} = \frac{d''\text{C}}{\text{M}d''}$$

It is important to note that, during this process, the liquid phase remains at constant composition (P) but is gradually reduced in amount.

The converse process of heating a solidified mixture of composition R well illustrates the meaning of the ternary peritectic point. Solids B, C and A_xB_y which are the solid phases of mixture R begin to melt at the temperature of E_5. At the temperature of point P, the compound A_xB_y dissociates into solid A and liquid and thus decomposes at temperatures below its true melting point M.

(*c*) Where the *binary compound is unstable to its melting point over all ranges of composition*. In other words there is both *binary and ternary peritectic formation*; the equilibrium diagram is shown in Fig. X.37.

The conjugation line CM lies outside the field of stability of the compound A_xB_y at all points and the binary system AB which contains a peritectic point P_1 is identical with the system illustrated in Fig. X.7 (p. 591). The point P_2 is a ternary peritectic point similar in all respects to the point P in the previous section (Fig. X.36).

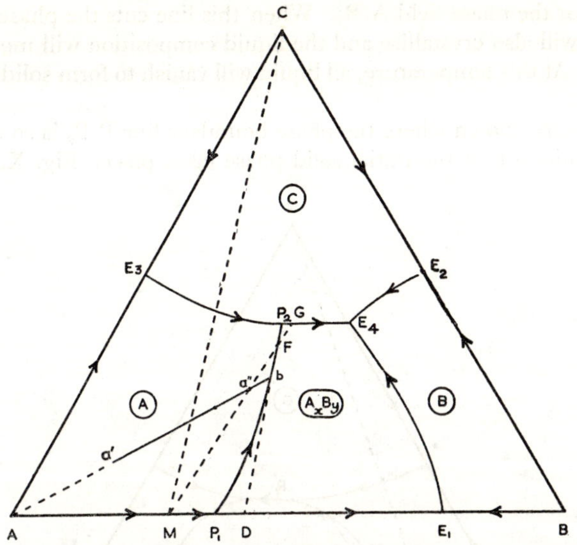

FIG. X.37. THE TERNARY DIAGRAM SHOWING THE FORMATION OF A BINARY COMPOUND WHICH IS INCONGRUENT AT ALL COMPOSITIONS

Once again the position of a mixture relative to the conjugation line CM determines the final products of crystallisation although the processes involved during equilibrium cooling of certain mixtures are rather more complex.

The crystallisation path of a mixture of composition *a*′ to the left of the conjugation line CM, illustrates many of the principal features.

As '*a*'' lies in the stability field of component A, this will be the first product of crystallisation and the liquid composition will proceed along the line A*a*′ produced until the phase boundary line P_1P_2 is reached at *b*. The compound A_xB_y will then form also, but the solid A already crystallised will begin to redissolve. This is so because the tangent to the phase boundary line P_1P_2 at *b* cuts the AB axis at D which is outside the A/A_xB_y stability line. The re-solution of A continues until the liquid composition reaches P_2 which is an invariant point. As the mixture '*a*'' lies within the A–C–A_xB_y conjugation triangle, some solid A must remain after all the liquid phase has crystallised into solid C and compound A_xB_y.

If the original mixture had the composition of the point *a*″, however, although solid A would still be the original phase to form, it would completely redissolve whilst the liquid composition was progressing along bP_2. The position at which all solid A disappears may be predicted by producing a line from M (the composition of the compound A_xB_y), through a″ to F on the phase boundary line P_1P_2. Fa″ produced thus intersects the A–A_xB_y composition line at a point corresponding to 100 per cent of the A_xB_y phase which must be the composition of the solid which has crystallised when the liquid corresponds to the point F. Further cooling of a mixture of original composition *a*″ will result in crystallisation of more A_xB_y and no other solid form. Hence, the liquid phase will follow the composition of the line MF produced *across* the phase field A_xB_y. When this line cuts the phase boundary line P_2E_4, solid C will also crystallise and the liquid composition will move to E_4, a ternary eutectic. At this temperature, all liquid will vanish to form solid B, C and compound A_xB_y.

Examples are known where the phase boundary line P_1P_2 is so shaped that recurrent crystallisation of the initial solid phase takes place. Fig. X.38 illustrates a

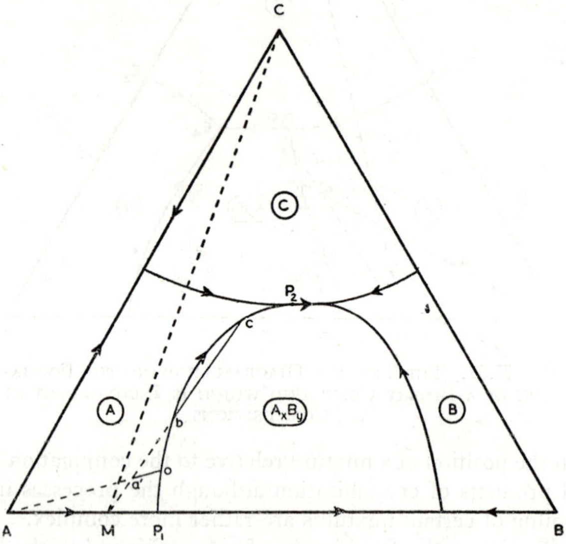

FIG. X.38. A TERNARY PHASE DIAGRAM SHOWING RECURRENT CRYSTALLISATION OF ONE PHASE

system which would exhibit this phenomenon. The crystallisation path of a mixture *a'* would be along *a'o* with solid A separating. The liquid composition would then follow the phase boundary line P_1P_2 to the point *b* with A_xB_y only being crystallised and solid A redissolving. At *b* no crystals of A remain and the crystallisation path follows along M*a'b* produced, with only solid A_xB_y being formed. This line, however, again cuts the phase boundary line P_1P_2 at *c*. Solid A can now be reformed because the tangent to P_1P_2 at *c* cuts the AB axis between A and A_xB_y indicating that both solid forms can co-exist. It is only when the liquid composition reaches the peritectic point P_2 that solid A becomes unstable again and redissolves in the liquid.

(*d*) Where a *binary compound is stable only when it is formed above a certain minimum temperature*, as shown in Fig. X.39.

The compound A_xB_y melts congruently in the binary field AB at M, but in the ternary field *below* the temperature of the point D it is not a stable phase.

(*e*) The converse case where a *binary compound dissociates above a certain temperature* is also known and the consequent equilibrium diagram is shown in Fig. X.40.

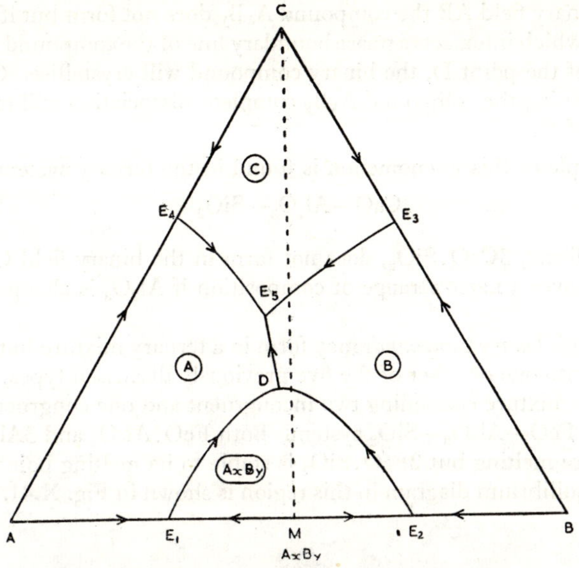

Fig. X.39. A Ternary Phase Diagram showing the Formation of a Binary Compound which is Stable only above a Certain Minimum Temperature

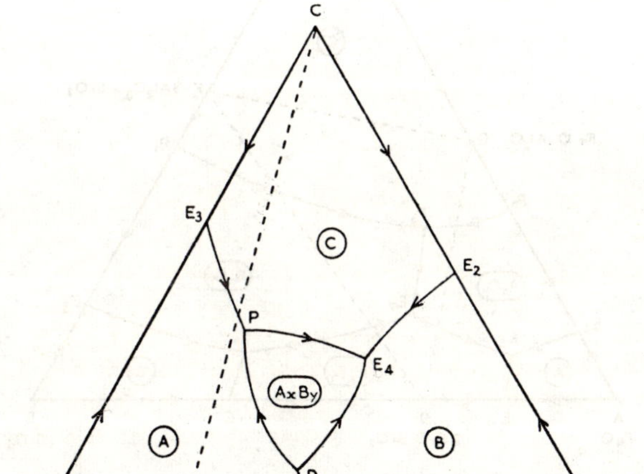

Fig. X.40. A Ternary Phase Diagram showing the Formation of a Binary Compound which Dissociates above a Certain Temperature

In the binary field AB the compound A_xB_y does not form but if a ternary mixture is cooled which intersects a phase boundary line of the compound A_xB_y below the temperature of the point D, the binary compound will crystallise. On heating any mixture containing the compound A_xB_y complete dissociation will take place at the temperature of D.

An example of this phenomenon is found in the ternary system

$$CaO—Al_2O_3—SiO_2$$

Tricalcium silicate, $3CaO.SiO_2$, does not form in the binary field $CaO—SiO_2$ but does separate over a narrow range of composition if Al_2O_3 is also present (see Fig. X.50).

(*f*) *Multiple binary compounds* may form in a ternary mixture but each form can be separated into one or other of the five previously discussed types.

A ternary mixture containing two incongruent and one congruent binary compounds is the $FeO—Al_2O_3—SiO_2$ system. Both $FeO.Al_2O_3$ and $3Al_2O_3.2SiO_2$ decompose before melting but $2FeO.SiO_2$ is stable to its melting point. A simplified form of the equilibrium diagram in this region is shown in Fig. X.41.

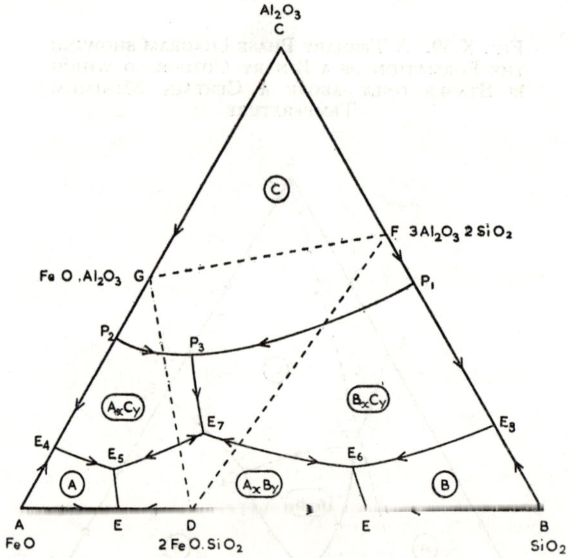

FIG. X.41. A SIMPLIFIED FORM OF THE PHASE DIAGRAM OF THE COMPONENTS $FeO/Al_2O_3/SiO_2$ SHOWING COMPOUND FORMATION

The conjugation triangles enable the final, solid phases after equilibrium cooling to be predicted. For example, a mixture, the composition of which lies in the triangle bounded by DGF, will produce only the three compounds in ratios which may be determined by the gravity-centre principle.

3. Ternary Compound Formation without Solid Solution. When three components are present in a mixture, ternary compounds of the general composition $A_xB_yC_z$ may be formed. There are two general cases to be considered depending on whether the compound is stable to its melting point (i.e. congruent) or whether it dissociates prior to melting (incongruent).

(*a*) *Congruently melting compound formation* is characterised by an equilibrium diagram which in the solid form would have a dome-shaped centre. From the maximum temperature point the isotherms would fall away to three ternary eutectic points. The solid form and projection models with and without isotherms are shown in Fig. X.42A, B, C.

The insertion of the conjugation lines AM, BM and CM divides the diagram into three simple areas of ternary eutectic formation. The three components of each conjugation triangle are, in each case, two of the primary components plus the ternary compound. There are three binary eutectics E_1, E_2 and E_3 and three ternary eutectics at E_4, E_5 and E_6.

The final, solid phases crystallising from any ternary mixture are determined from the appropriate conjugation triangle and the relative amounts present by applying the gravity-centre principle within that triangle.

The crystallisation path of any mixture is similar to that in a ternary eutectic system. Mixture R, for example, in the stability field of component A (Fig. X.42c) will, at the solidus temperature, form solid A. The liquid composition will follow the line AR produced to S on the phase boundary line E_4E_6; the compound $A_xB_yC_z$ will then also form and the liquid will proceed in composition along E_6E_4 to the ternary eutectic invariant point where the liquid phase will vanish to leave solid A, B and compound $A_xB_yC_z$.

(*b*) When the *ternary compound melts incongruently* the composition point of the compound falls outside its field of stability as shown in Fig. X.43.

A line from A through M (the composition of the compound) will meet the phase boundary line FE_4 at D. This point will be at a maximum temperature and the isotherms will fall away to the ternary eutectic E_4 and 'transformation point F'. A similar maximum occurs on the phase boundary E_4–E_5 at the intersection of CM, but as BM does not cut FE_5 there is no maximum temperature point along this phase boundary.

The crystallisation-paths in a ternary system with an incongruent, triple compound resemble those in the ternary system with a binary compound which dissociates below its melting point over all the composition range (see Fig. X.39, p. 631). A mixture of composition a' within the conjugation triangle AMB forms solid A as the initial phase and the liquid composition proceeds to b which lies between D and F on the phase boundary curve E_4F. The tangent to this curve at b indicates that solid A is an unstable phase, hence crystallisation proceeds with the formation of compound $A_xB_yC_z$ and re-solution of solid A. At F, solid B can also form which would be a fourth phase; hence F is an invariant point. Because a', the composition of the original mixture, lies in the conjugation triangle AMB, all liquid must vanish at the temperature of F because only solid A, B and $A_xB_yC_z$ can be present in the final crystallised mixture.

Should a composition a'' within the conjugation triangle CMB be crystallised, all

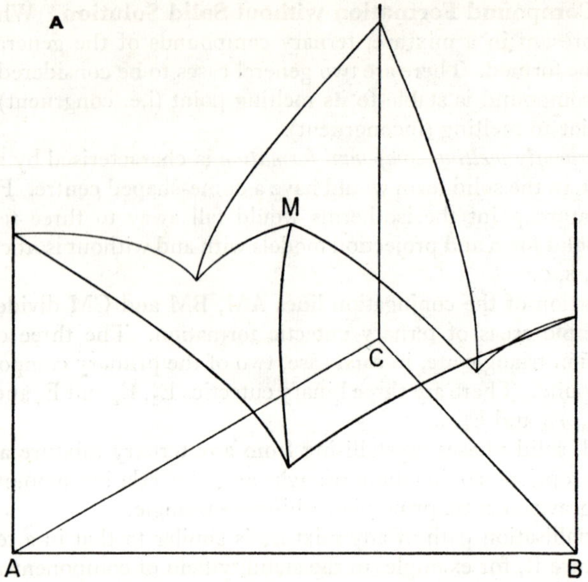

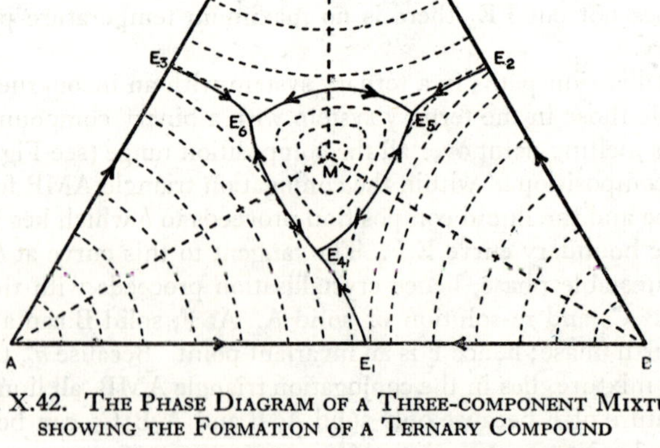

FIG. X.42. THE PHASE DIAGRAM OF A THREE-COMPONENT MIXTURE SHOWING THE FORMATION OF A TERNARY COMPOUND

A A solid model

B A planar representation with isotherms

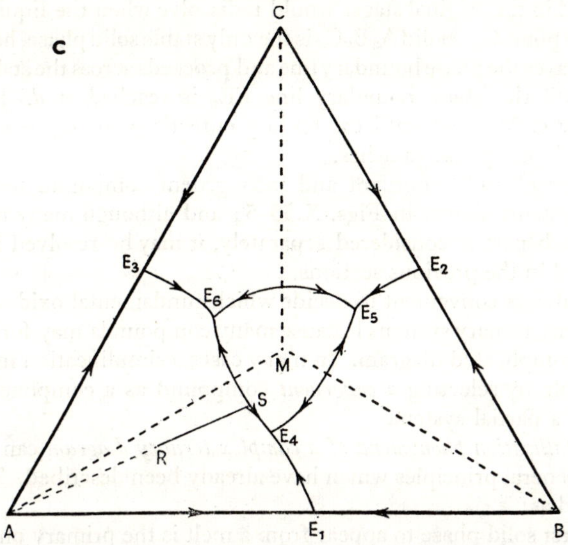

Fig. X.42. The Phase Diagram of a Three-component Mixture showing the Formation of a Ternary Compound

c A planar representation with arrows indicating the direction of falling temperature

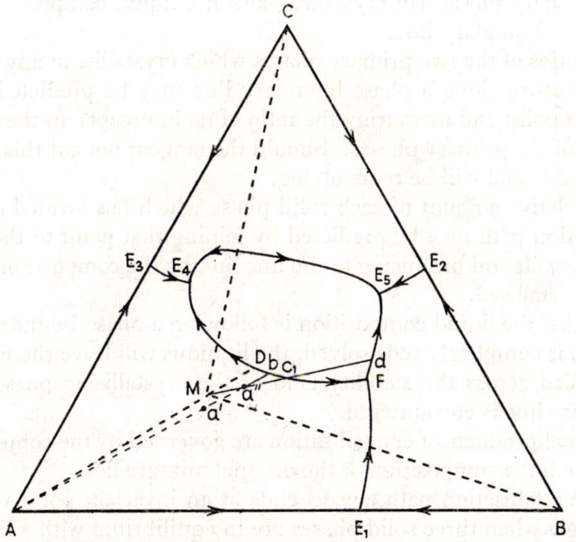

Fig. X.43. A Ternary Phase Diagram with a Ternary Compound which Melts Incongruently

solid A formed in the original stages would redissolve when the liquid had the composition of the point C_1. Solid $A_xB_yC_z$ is the only stable solid phase, hence the crystallisation path leaves the phase boundary line and proceeds across the stability field of the compound until the phase boundary line FE_5 is reached at *d.* Solid B is then simultaneously crystallised until the ternary eutectic point E_5, where solid C also forms and the liquid phase vanishes.

Many examples of congruent and incongruent compound formation will be given in the systems shown in Figs. X.50–53 and although many mixtures appear complex, if each part is considered separately, it may be resolved into one of the types described in the previous sections.

It is not always convenient to decide which fundamental oxides form the components of many ternary systems because many compounds may form which would give rise to a complicated diagram. In many cases a simplification in representation may be possible by selecting a *congruent* compound as a component and thereby studying only a partial system.

The *crystallisation phenomena of a complex ternary diagram* can be deduced by applying the general principles which have already been described. These may now be summarised as:

1. The first solid phase to appear from a melt is the primary phase for the part of the system in which the melt composition lies.

2. The liquid phase changes along the extension of a line joining the composition point of the primary crystallising phase to that of the composition of the original melt.

3. Only one phase will crystallise until the liquid composition line meets a phase boundary line.

4. At a phase boundary line which divides the stability field of two constituents, the second primary phase will crystallise, and the liquid composition will proceed along the phase boundary line.

5. The ratios of the two primary phases which crystallise at any one instant at a certain temperature along a phase boundary line may be predicted by taking the tangent at that point and measuring the ratio of its intercepts on the line joining the compositions of the primary phases. Should the tangent not cut this line, one of the phases is unstable and will be redissolving.

6. The relative amount of each solid phase which has formed at any point on the crystallisation path may be predicted by joining that point to the original composition of the melt and producing to the line linking the compositions of the phases which have crystallised.

7. If, whilst the liquid composition is following a phase boundary, an unstable primary phase is completely redissolved, the liquidus will leave the phase boundary line and proceed across the stability field of the crystallising phase until another phase boundary line is encountered.

8. The final products of crystallisation are governed by the conjugation triangle (p. 626) in which the composition of the original mixture lies.

9. The crystallisation path always ends at an invariant point within the conjugation triangle when three solid phases are in equilibrium with a liquid.

10. The relative proportions of the final phases on crystallisation are found by applying the gravity-centre principle.

Ternary Systems with Solid Solutions. Ternary systems in which solid or crystalline solutions are possible between all three components present many complex problems. Many examples of this type of phenomena are known in silicate systems; compound formation frequently occurs simultaneously and adds further complications.

The quantitative assessment of solid solution in ternary silicate systems is so difficult that relatively few mixtures have been fully investigated. Most of the present day information has been implied from research into metallurgical systems and in his classical treatise F. A. H. Schreinemakers has discussed the most important types of crystalline-solution ternary diagrams. Although it is more than likely that complex solid solutions form in many ternary silicate and other ceramic systems, most investigators have not been able to consider this possibility and in most cases all deductions have been made on the basis of a completely immiscible solid state.

It is not within the scope of this volume to describe all the phenomena associated with ternary systems containing solid solutions, but a few of the main features may be mentioned briefly:

1. *Three components completely miscible in both liquid and solid states* and no complex formation.

This simple case is where the binary system illustrated in Fig. X.9A (p. 594), is projected into three dimensions. The perspective diagram is illustrated in Fig. X.44A and, as would be expected, there is a solidus- and liquidus-boundary throughout the whole range of ternary composition. Thus, there is a particular solid composition in equilibrium with each liquid mixture at each solidification temperature.

Above the liquidus field, only one phase is present, namely a completely homogeneous melt; whilst below the solidus field the only phase is a homogeneous solid solution. Only between these two fields can two phases co-exist at one temperature. For any particular temperature between the fusion points of the highest and lowest melting components, an isotherm can be drawn which encircles the liquidus and solidus areas. Such a section indicates the composition relationships between two co-existing solid and liquid phases. A line drawn through an isothermal section from a certain liquid composition to the solid in equilibrium with it is called a *conode* or *tie-line*.

The principal features of the system may be more clearly demonstrated by considering the cooling phenomena associated with a specific melt.

A mixture of composition P will be a homogeneous liquid at temperatures above T_1 on the liquidus field. At this temperature, a solid phase will begin to separate with a composition S_1 indicated by the tie line l_1–S_1. The liquid phase will then change in composition and follow a path on the liquidus area to l_2. As the temperature is lowered the solid phase in equilibrium with the liquid will also change in composition and ideally, if equilibrium conditions are maintained, the solid phase will be continually redissolving and precipitated again with a slightly different composition. The relative amounts of solid and liquid phases at any temperature are found by applying the *lever principle* (see p. 589) around the vertical line through P to the equivalent solidus and liquidus points. When the solid being formed has assumed the composition of the initial melt, the liquid phase will completely disappear, leaving only a solid phase.

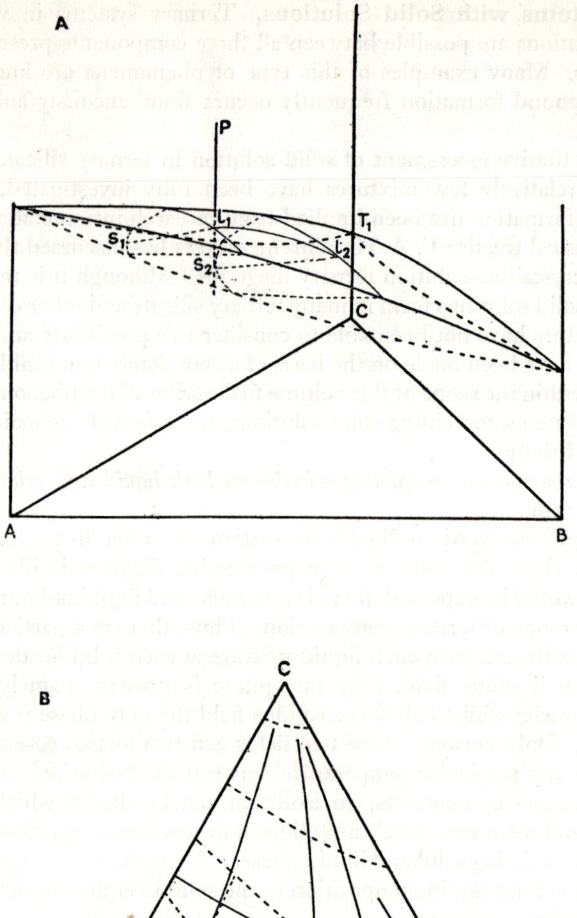

FIG. X.44. A TERNARY PHASE DIAGRAM WHERE THE THREE COMPONENTS ARE COMPLETELY MISCIBLE IN BOTH THE SOLID AND LIQUID FORMS

A A solid representation
B A planar representation

If, as is most usual in silicate systems, cooling does not proceed under equilibrium conditions, the solid phase will not completely readjust its composition as the temperature decreases. As a result, successive layers of solid, each with a different composition, are formed and give rise to *zoned crystals*.

It is difficult to represent completely a ternary system with solid solution in a planar diagram when the temperatures are shown as 'contour lines'. Not only is it essential to have some means of indicating liquidus temperatures for all compositions, but tie-lines giving the liquidus and solidus equilibrium conditions must be included at numerous points. It is also advantageous to know the crystallisation paths, or the change in composition of the liquid phase, at many positions on the diagram. At the same time, the planar diagram must indicate any phase boundary lines which arise as a result of complex formation.

A projected model of the system of three components with complete miscibility in the solid state is shown in Fig. X.44B. The isotherms are shown as dotted lines and indicate that the pure component C has the highest melting point with B intermediate and A with the lowest fusion point. Crystallisation paths are shown in heavy type and denote the change in composition of specific melts when they crystallise. Two sets of tie-lines are also shown, liquid phase l_1 for example will be in equilibrium with solid S_1; liquid l_2 with solid S_2. For a complete interpretation of cooling phenomena, information of this type would be required for the entire diagram.

In silicate systems, ternary mixtures with complete solid miscibility of simple type are rare, although examples are known where a maximum or minimum point occurs. The precise physical meaning of such phenomena has not yet been established with certainty, but inflection points might represent a change-over in solid solution type (see p. 596). At a maximum or minimum, the solid and liquid phases in equilibrium have the same composition and should a melt of that particular composition be cooled, solidification will be complete at one temperature.

The best-known example of this kind of behaviour occurs in *melilites* which are frequently found in basic slags. These are part of the quaternary system CaO, MgO, Al_2O_3, SiO_2 but the three congruent compounds $2CaO.Al_2O_3.SiO_2$ (*gehlenite*), $2CaO.MgO.2SiO_2$ (*akermanite*) and $3CaO.Al_2O_3.3SiO_2$ (*sarcolite*), may be regarded as the components of a ternary system. In the separate binary fields, maxima occur between gehlenite and sarcolite, and akermanite and sarcolite, but a minimum point is present in the gehlenite and akermanite series. A combination of these effects translated into the ternary system results in 'saddle-formation' as in Fig. X.45A; the projected drawing showing isotherms and crystallisation paths is in Fig. X.45B.

2. *Binary Solid Solution in Ternary Systems.* It is not an unusual occurrence for only two components of a ternary system to admix in the solid state whilst the third component separates on cooling, as an independent phase. The simplest case is where components A and B form a homogeneous solid phase over the whole composition range, but both exhibit immiscible binary eutectic formation with the third component C.

A spacial equilibrium diagram for such a system is shown in Fig. X.46A, with the projection giving crystallisation paths and the phase boundary line in Fig. X.46B.

Within the composition field bounded by CE_1E_2 the stable solid phase is

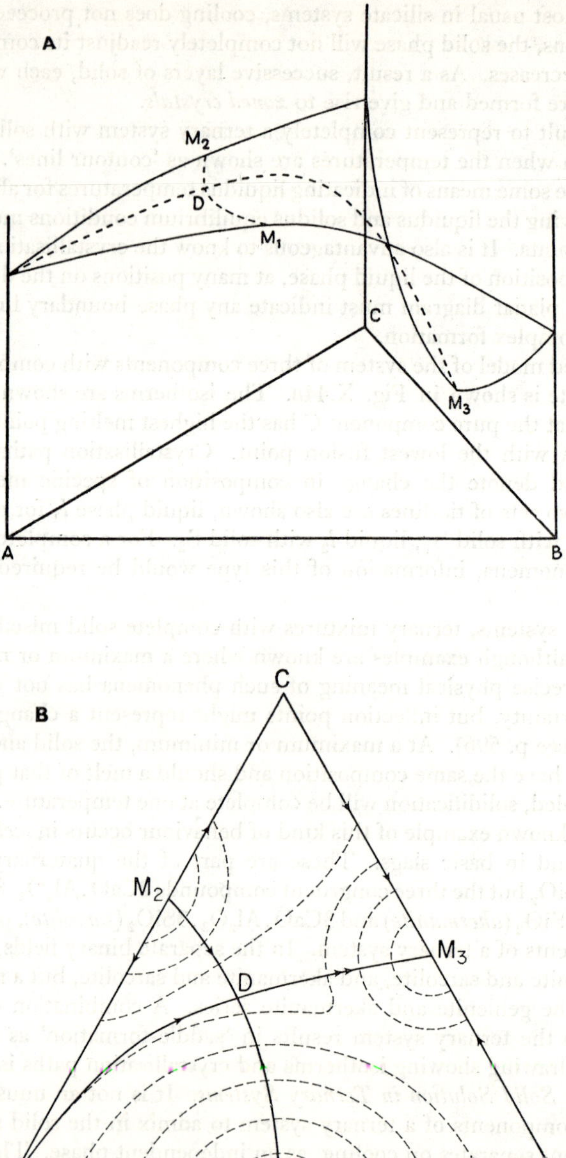

FIG. X.45. A TERNARY PHASE DIAGRAM WHERE THE THREE COMPONENTS ARE COMPLETELY MISCIBLE IN BOTH THE LIQUID AND SOLID STATES, BUT WITH A MINIMUM POINT

A The solid model showing only the liquidus lines

B A planar representation

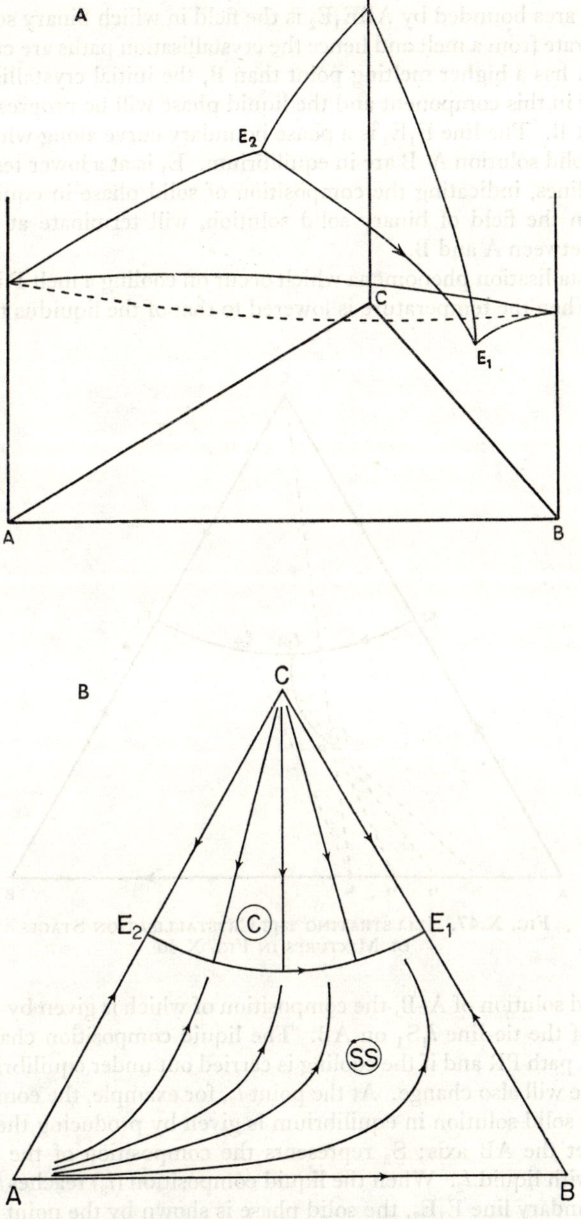

FIG. X.46. THE PHASE DIAGRAM OF A TERNARY MIXTURE WHERE TWO COMPONENTS FORM SOLID SOLUTIONS

A A solid model

B A planar representation

component C and when this crystallises, the liquid composition changes linearly (see p. 620). The area bounded by ABE_1E_2 is the field in which binary solid solutions of A and B separate from a melt and hence the crystallisation paths are curved. Because component A has a higher melting point than B, the initial crystallisation products will be richer in this component and the liquid phase will be progressively enriched in component B. The line E_1E_2 is a phase boundary curve along which two phases, solid C and solid solution A–B are in equilibrium. E_1 is at a lower temperature than E_2. The tie-lines, indicating the composition of solid phase in equilibrium with a liquid melt in the field of binary solid solution, will terminate at a point on the solidus line between A and B.

The crystallisation phenomena which occur on cooling a melt P is illustrated by Fig. X.47. When the temperature is lowered to that of the liquidus the crystallising

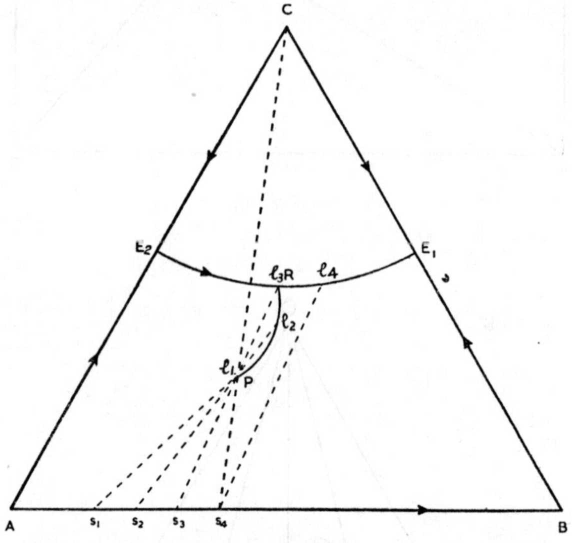

FIG. X.47. ILLUSTRATING THE CRYSTALLISATION STAGES OF MIXTURES IN FIG. X.46

phase is a solid solution of A–B, the composition of which is given by S_1 which is the intersection of the tie line l_1S_1 on AB. The liquid composition changes along the crystallisation path PR and if the cooling is carried out under equilibrium conditions the solid phase will also change. At the point l_2, for example, the composition of the homogeneous solid solution in equilibrium is given by producing the line joining l_2 and P to meet the AB axis; S_2 represents the composition of the solid phase in equilibrium with liquid l_2. When the liquid composition (l_3) reaches the point R on the phase boundary line E_1E_2, the solid phase is shown by the point S_3. Now solid component C also separates and the final solidified state will be composed of this component and a solid solution of A–B. The lever-principle (p. 620) applied around the point of original composition P shows that the solid solution S_4 and component C will be the final states and if the tie-lines of the system are known, the composition of the final liquid, prior to complete crystallisation may be predicted as l_4.

The ternary system *diopside, albite* and *anorthite* provides a typical example of binary solid solution in a three-component mixture. The two plagioclase felspars are considered by Bowen to be completely miscible in the solid state over the whole range of composition, but form binary eutectics with diopside.

Binary Compound Formation with Solid Solution. An important extension of ternary solid solution diagrams is where a binary compound A_xB_y is formed which is miscible in the solid state with component C. The type of diagram depends upon the nature of the compound, which may be congruent (melts stably) or incongruent (decomposes prior to melting).

When a congruent compound forms a solid solution with the other component, the phase diagram is relatively simple and is as illustrated in Fig. X.48. The solid solution phase is composed of varying amounts of component C and compound A_xB_y, so the composition will have the general form: $aC.bA_xB_y$, where $a + b = 1$. The ratio of components A and B in solid solution is thus always the same and equal to $x:y$.

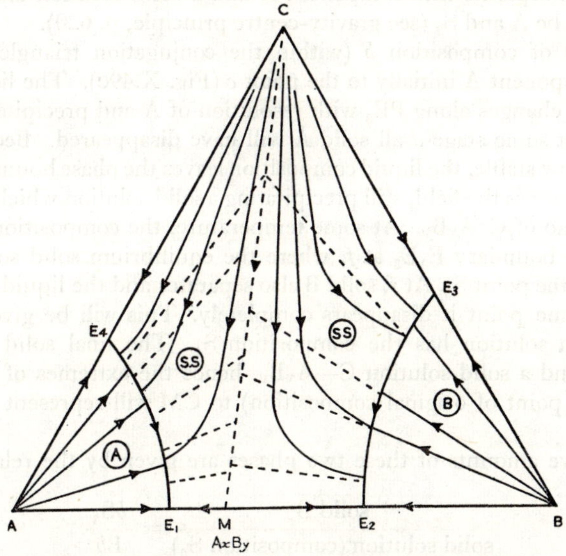

FIG. X.48. THE PHASE DIAGRAM OF A TERNARY MIXTURE WHERE A BINARY COMPOUND IS PRODUCED WHICH FORMS SOLID SOLUTIONS WITH THE THIRD COMPONENT

The conjugation-line CM effectively divides the diagram into two systems each of which is that of a binary solid solution in a ternary mixture (Fig. X.46).

Mixtures to the left of CM in the conjugation-triangle ACM will crystallise to solid A and a solid solution of C—A_xB_y, whilst mixtures in the triangle BCM form solid B and a solid solution of C—A_xB_y.

When the binary compound (A_xB_y) is *incongruent* the equilibrium diagram of the ternary system is as shown in Fig. X.49A, in which the binary field AB has a peritectic point P and there are two phase-boundary curves where solid solution C—A_xB_y crystallises simultaneously with either the A or B component. The conjugation line CM, however, is not entirely within the stability field of the solid solution and over the range MD, the C—A_xB_y complex would dissociate before melting by virtue of the decomposition of the compound A_xB_y.

The crystallisation of certain mixtures making use of the crystallisation paths and tie-lines of Fig. X.49A, is shown in Fig. X.49B. A mixture of composition '*a*' (within the conjugation triangle AMC) will initially form solid component A. The liquid composition will follow A*a* produced until the phase boundary curve PE_3 is joined at *c* where a solid solution of composition S_1 also separates. At this point, solid A is an unstable phase (p. 636) hence it is continually being redissolved as the liquid changes in composition along PE_3. The solid solution being formed also changes composition until at some point *d* on the phase boundary curve, solid solution S_2 is the equilibrium composition. At this point, all remaining liquid must vanish because if, as predicted by the conjugation triangle ACM, a melt of composition '*a*' is to separate into component A and a solid solution along CM, these two phases will be A and S_2 (see gravity-centre principle, p. 620).

A mixture of composition *b* (within the conjugation triangle CMB) again crystallises component A initially to the point *c* (Fig. X.49C). The liquid composition, as before, changes along PE_3 with resolution of A and precipitation of a solid solution until at some stage *d* all solid A will have disappeared. Because only one solid phase is now stable, the liquid composition leaves the phase boundary curve PE_3 and progresses across the field, still precipitating a solid solution which is continually changing its ratio of C:A_xB_y. At some temperature, the composition of the liquid joins the phase boundary E_1E_2 at *f*, where the equilibrium solid solution has the composition of the point S_3. At *f*, solid B also separates and the liquid changes along fE_2 until at some point it disappears completely. This will be given by *g* where the equilibrium solution has the composition S_4. The final solid contains only component B and a solid solution C—A_xB_y, hence the extremes of a line from B through *b* (the point of original composition) to CM will represent the final solid phases.

The relative amounts of these two phases are given by the relationship:

$$\frac{\text{solid B}}{\text{solid solution (composition } S_4)} = \frac{bS_4}{Bb}$$

Ternary systems with solid solution phenomena are known where complex formation is even more involved than the examples already quoted. It is beyond the scope of this volume to include such systems although it may be emphasised that many silicate systems fall into the category. The reader who wishes to pursue this subject further is referred to the works of Schreinemakers, Bakhuis and Roozeboom, Vogel, Eitel and others.

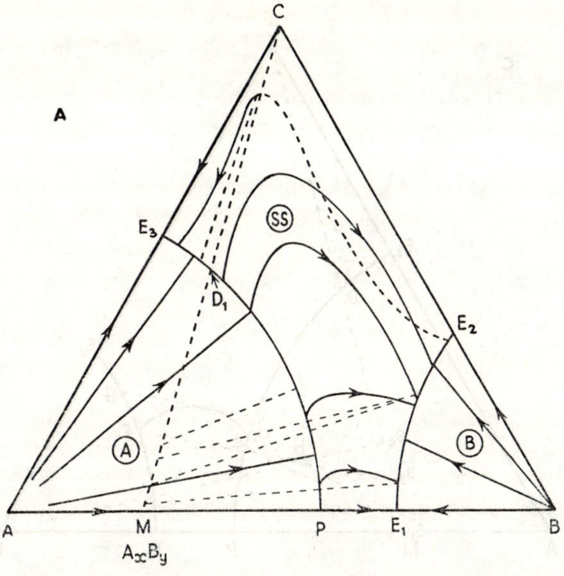

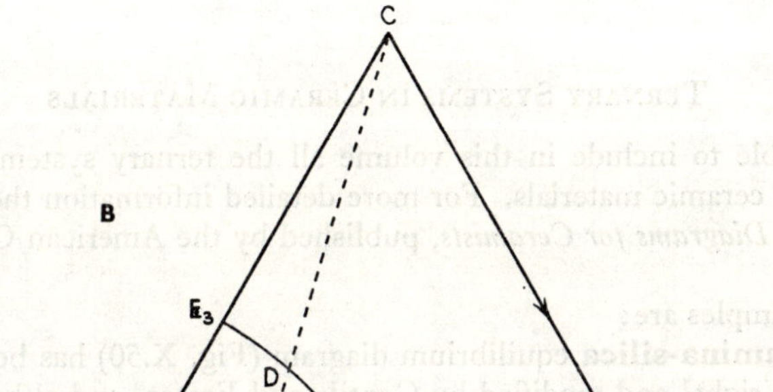

FIG. X.49. THE PHASE DIAGRAM OF A TERNARY MIXTURE WHERE A BINARY COMPOUND WHICH FORMS SOLID SOLUTIONS WITH THE THIRD COMPONENT, MELTS INCONGRUENTLY

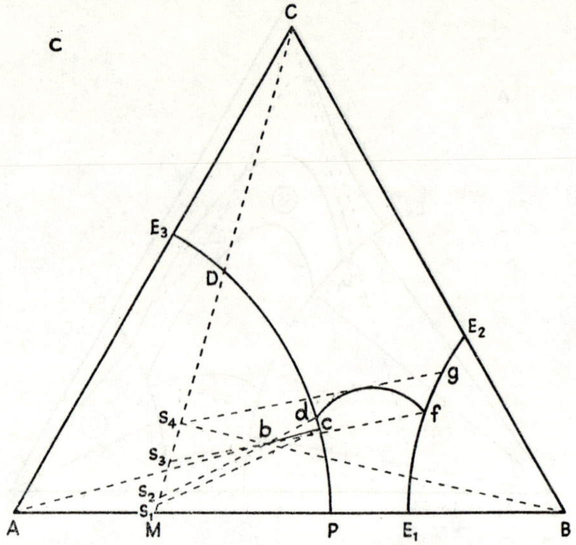

FIG. X.49. THE PHASE DIAGRAM OF A TERNARY MIXTURE WHERE A BINARY COMPOUND WHICH FORMS SOLID SOLUTIONS WITH THE THIRD COMPONENT, MELTS INCONGRUENTLY

TERNARY SYSTEMS IN CERAMIC MATERIALS

IT is impossible to include in this volume all the ternary systems which are of importance in ceramic materials. For more detailed information the reader should consult *Phase Diagrams for Ceramists*, published by the American Ceramic Society (see p. 605).

Some examples are:

The **lime-alumina-silica** equilibrium diagram (Fig. X.50) has been deduced by Rankin and Wright[1] and modified by Gentile and Foster[2] and although it appears highly complex, the phases which would crystallise from any mixture of components can be deduced by drawing the conjugation triangle embracing three adjacent phases and applying the principles laid down in the preceding pages.

There are ten binary and two ternary compounds formed in the system as shown in Table X.IV.

The lowest temperature at which liquid can be present is 1170° C. corresponding to a mixture of 23 parts CaO, 15 parts Al_2O_3, 62 parts SiO_2.

The **soda-alumina-silica** system has been studied by Schairer and Bowen[3] and

[1] Rankin, G. A. and Wright, F. E., *Amer. J. Sci.*, 39, 52, 1915.
[2] Gentile, A. L. and Foster, W. R., *J. Amer. Ceram. Soc.*, 46, 76, 1963.
[3] Schairer, J. F. and Bowen, N. L., *Amer. J. Sci.*, 245, 196, 1947.

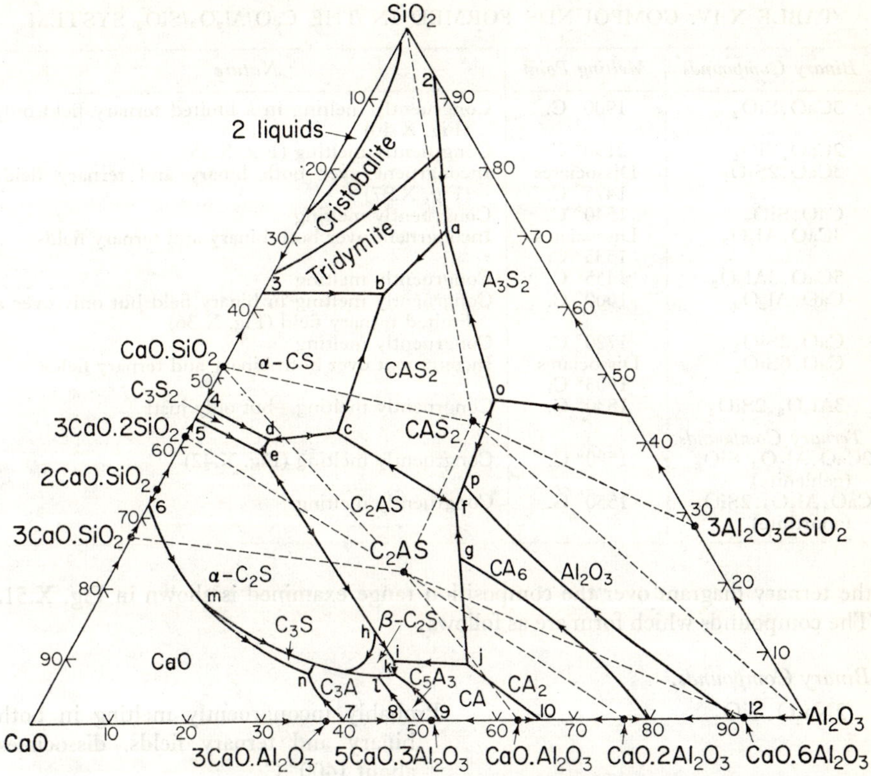

Binary points						Ternary points					
1	1800° C	P	8	1395° C	E	a	1345° C	P	i	1360° C	P
2	1545	E	9	1400	E	b	1170	E	j	1505	P
3	1436	E	10	1590	E	c	1265	E	k	1335	E
4	1455	E	11	1775	E	d	1310	E	l	1335	P
5	1475	P	12	1903	P	e	1335	P	m	1900	P
6	2065	E				f	1380	E	n	1470	P
7	1535	P				g	1475	P	o	1512	E
						h	1450	P	p	1500	P

E — Eutectic P — Peritectic

FIG. X.50. THE PHASE DIAGRAM OF LIME-ALUMINA-SILICA (wt. %)

TABLE X.IV. COMPOUNDS FORMED IN THE $CaO/Al_2O_3/SiO_2$ SYSTEM

Binary Compounds	*Melting Point*	*Nature*
$3CaO.SiO_2$	1900° C.	Congruently melting in a limited ternary field only (Fig. X.40)
$2CaO.SiO_2$	2130° C.	Congruently melting (Fig. X.35)
$3CaO.2SiO_2$	Dissociates 1475° C.	Incongruent over both binary and ternary fields (Fig. X.37)
$CaO.SiO_2$	1540° C.	Congruently melting
$3CaO.Al_2O_3$	Dissociates 1535° C.	Incongruent over both binary and ternary fields
$5CaO.3Al_2O_3$	1455° C.	Congruently melting
$CaO.Al_2O_3$	1600° C.	Congruently melting in binary field but only over a limited ternary field (Fig. X.36)
$CaO.2SiO_2$	1720° C.	Congruently melting
$CaO.6SiO_2$	Dissociates 1903° C.	Incongruent over both binary and ternary fields
$3Al_2O_3.2SiO_2$	1830° C.	Congruently melting—but only just!
Ternary Compounds		
$2CaO.Al_2O_3.SiO_2$ (gehlenite)	1590° C.	Congruently melting (Fig. X.42)
$CaO.Al_2O_3.2SiO_2$ (anorthite)	1550° C.	Congruently melting

the ternary diagram over the composition range examined is shown in Fig. X.51. The compounds which form are as follows:

Binary Compounds.

$2Na_2O.SiO_2$	—	Probably incongruently melting in both binary and ternary fields, dissociates about 1400° C.
$Na_2O.SiO_2$	M.P. 1089° C.	Congruently melting.
$Na_2O.2SiO_2$	M.P. 874° C.	Congruently melting.
$3Al_2O_3.2SiO_2$	Dissociates at 1810° C.	Incongruently melting in both binary and ternary fields.
$Na_2O.Al_2O_3$	—	Not examined.

Ternary Compounds.

$Na_2O.Al_2O_3.6SiO_2$ (albite)	M.P. 1118° C.	Congruently melting.
$Na_2O.Al_2O_3.2SiO_2$ (carnegieite, nephelite)	M.P. 1526° C.	Probably congruently melting.

The lowest temperature at which liquid can be present is 732° C. corresponding to a mixture of 26 parts Na_2O, 13 parts Al_2O_3, 61 parts SiO_2.

The **potash-alumina-silica** system has also been studied by Schairer and Bowen[1] and the diagram (Fig. X.52) is similar to that of the soda-alumina-silica system. The

[1] Schairer, J. F. and Bowen, N. L., *Amer. J. Sci.*, 245, 196, 1947.

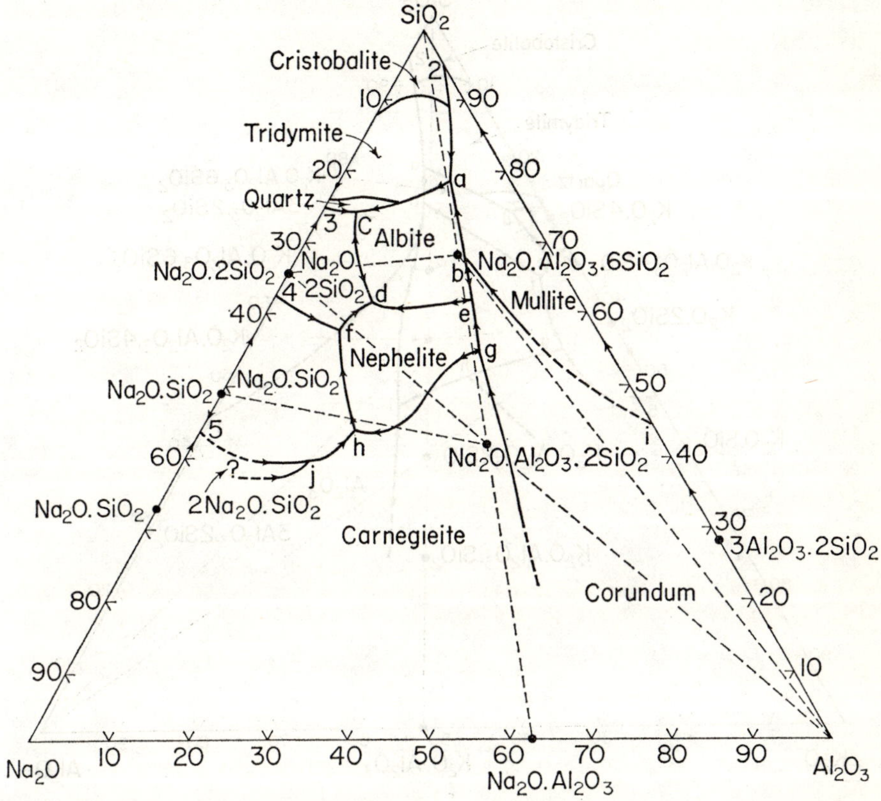

Binary points			Ternary points					
1	1810 °C	P	a	1050 °C	E	f	760 °C	E
2	1545	E	b	1104	P	g	1270	P
3	789	E	c	740	E	h	915	P
4	837	E	d	732	E	i	955	P
5	1022	E	e	1063	E			
E – Eutectic			P – Peritectic					

Fig. X.51. The Phase Diagram of Soda-Alumina-Silica (wt. %)

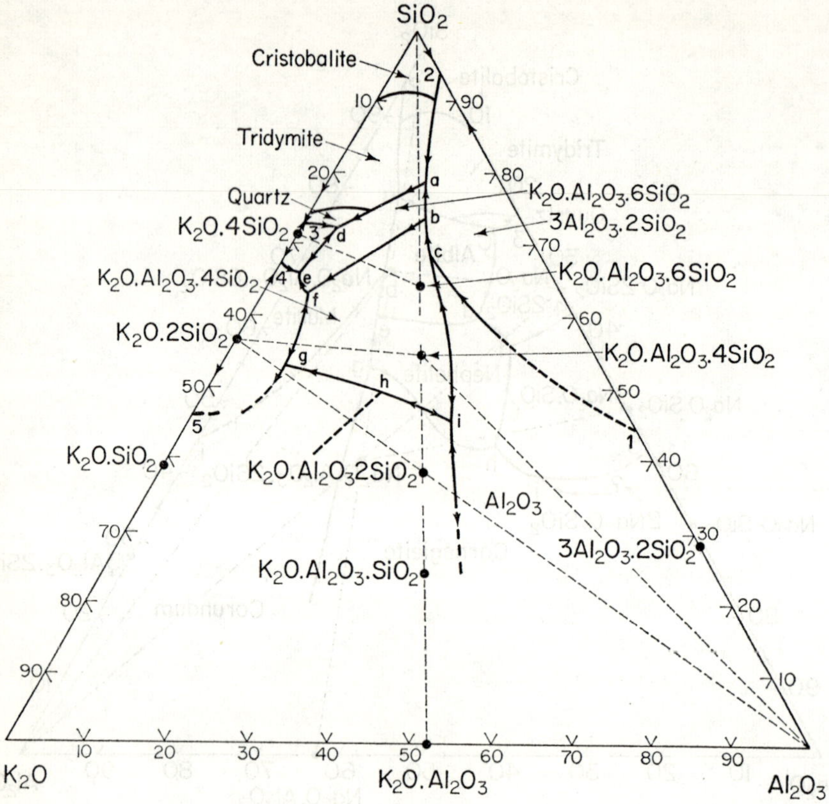

Compounds		Points					
Binary	Ternary	Binary			Ternary		
$3Al_2O_3.2SiO_2$	$K_2O.Al_2O_3.6SiO_2$	1	1810°C	P	a	985°C	E
mullite diss. 1810	felspar diss. 1100	2	1545	E	b	1140	P
$K_2O.4SiO_2$	$K_2O.Al_2O_3.4SiO_2$	3	769	E	c	1315	P
m.pt. 770	leucite m.pt. 1686	4	742	E	d	710	E
$K_2O.2SiO_2$	$K_2O.Al_2O_3.2SiO_2$	5	780	E?	e	695	E
m.pt. 1045	kaliophilite 1730				f	810	P
$K_2O.SiO_2$	$K_2O.Al_2O_3.SiO_2$				g	905	E
m.pt 976					h	1540	P
$K_2O.Al_2O_3$					i	1553	E
m.pt. ?		E – Eutectic			P – Peritectic		

FIG. X.52. THE PHASE DIAGRAM OF POTASH-ALUMINA-SILICA (wt. %)

ternary compound $K_2O.Al_2O_3.6SiO_2$ (orthoclase felspar), however, melts incongruently except in a limited field of composition (Fig. X.43). A further ternary compound $K_2O.Al_2O_3.4SiO_2$ (leucite) is also formed and this melts congruently at 1686° C. The lowest temperature at which liquid can be present is 695° C. corresponding to a mixture of 30 parts K_2O, 4 parts Al_2O_3, 66 parts SiO_2.

Ternary mixtures within the components **magnesia-lime-alumina-iron oxide-chromic oxide-silica** have been extensively studied because of their importance in the manufacture of basic refractories and the changes which occur in them when used in contact with basic slags.

An excellent summary of the complex phase relationships in these systems has been given by E. F. Osborn.[1] Fig. X.53, illustrates a part of the CaO—MgO—SiO_2 system within the conjugation triangle of the compounds $CaO.MgO.2SiO_2$ (diopside), $2MgO.SiO_2$ (forsterite) and SiO_2. The phase conditions represent the case of a binary compound, $MgO.SiO_2$ which is not stable to its melting point and which forms a series of solid solutions with the third component, diopside. These solid solutions which have the general formula xMgO.SiO_2. yCaO.MgO.2SiO_2 are known as the *pyroxenes*. The crystallisation paths of such mixtures have been outlined on page 646 (Fig. X.49).

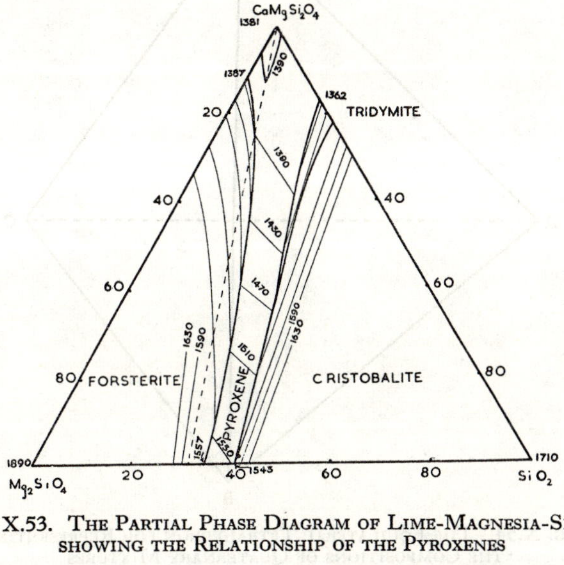

FIG. X.53. THE PARTIAL PHASE DIAGRAM OF LIME-MAGNESIA-SILICA SHOWING THE RELATIONSHIP OF THE PYROXENES

QUATERNARY AND MULTI-COMPONENT SYSTEMS

As the number of components in a system is increased the possible equilibrium conditions extend rapidly. It has been shown that ternary representation requires three binary systems to be translated into three dimensions and that increasing the components to four involves no less than four ternary and six binary systems combined together.

[1] Osborn, E. F., American Iron and Steel Institute, Technical Meeting, October 1954.

Because of the difficulty of representation when so many components are present, it is best wherever possible to consider multi-component systems as partial systems with congruent compounds represented as components. Thus, for example, the important MgO—CaO—FeO—SiO_2 system which has been investigated in detail by Osborn and Schairer[1] may, over much of the composition range, be studied as partial ternary or binary systems.

Quaternary systems and even those with a higher number of components can, however, be graphically constructed. For example, the component-composition in a quaternary mixture may be represented along the four edges of an *equilateral tetrahedron*. Each corner corresponds to 100 per cent of one component and, thus, the composition of any four-component mixture can be fully described by such a spacial model (Fig. X.54). There is no axis available for temperature representation, but usually it is indicated by three-dimensional isotherms or more simply by arrows on phase boundary curves, pointing in the direction of falling temperature.

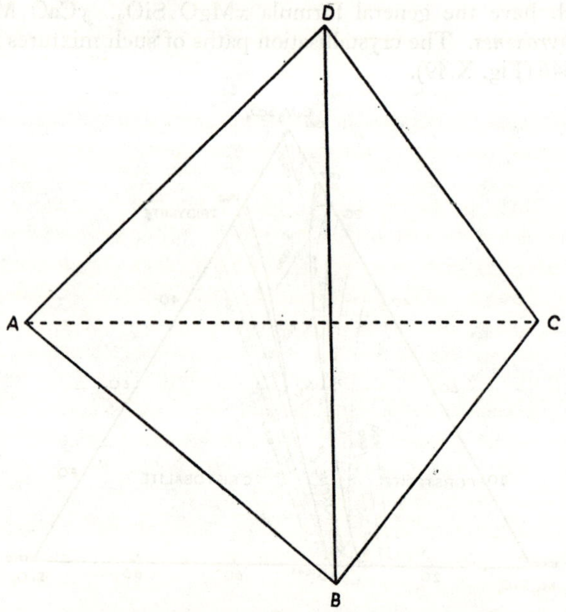

FIG. X.54. THE EQUILATERAL TETRAHEDRON FOR REPRESENTING THE COMPOSITIONS OF QUATERNARY MIXTURES

EXPERIMENTAL METHODS FOR THE DETERMINATION OF EQUILIBRIUM DIAGRAMS

Equilibrium implies that all reactions between the components of any mixture have proceeded to completion under the specific experimental conditions and that no

[1] Schairer, J. F. and Osborn, E. F., *J. Amer. Ceram. Soc.*, 33, 160, 1950.

further change would take place in the products if the same conditions were maintained for an infinite period of time. Reactions between solid components are frequently very slow, particularly where the rates of diffusion are low, as in many silicate mixtures. Consequently, it is doubtful whether equilibrium conditions would be reached, in many instances, in the periods of time feasible for laboratory experimentation; the simple heating of mixed components to specific temperatures, even for long periods, is insufficient to reach equilibrium in many ceramic systems.

Heating Curves. In metallurgical practice the appearance of a phase in a certain mixture is so rapid (sometimes instantaneous) that a *thermal arrest* can be detected at a well-defined temperature on either a heating or a cooling curve. This is produced by the changes in latent heat which accompany the melting or crystallisation of phases. Phase boundary lines can be established with certainty and there is no necessity to identify the equilibrium products in a wide range of mixtures in order to derive the complete equilibrium diagram.

Ceramic mixtures do not, however, readily lend themselves to such treatment. Crystal phases form too slowly to produce a pronounced 'arrest' and *super-cooling* is a common phenomenon. Silicate melts can often be cooled to a temperature well below that at which a solid phase should appear without any crystallisation taking place.

Quenching Methods. When silicate systems are being studied another very troublesome feature is that a glassy phase is easily formed and it may persist and crystallise only with difficulty even under prolonged conditions of test. Hence the determinations of phase equilibrium diagrams for ceramic systems require special techniques; the most widely used of which is that described by Morey.[1] The finely-divided components are intimately mixed in known proportions and then fired in suitable containers to the desired temperature. After prolonged, carefully controlled soaking, the samples are instantaneously quenched to room temperature in a suitable liquid. This usually has the effect of *freezing* the phases which were present at the temperature of the test and thus they may be identified and often estimated by suitable methods (see Chapter V). It is advisable to ensure that true equilibrium has been reached by regrinding the quenched specimen and again heating and soaking it at the same temperature. If after requenching, the phases have not been materially altered, equilibrium may be assumed; otherwise, further treatment is necessary.

If tests are being carried out at temperatures where little or no liquid can form, considerable time may be saved by first heating the mixed components to a temperature in excess of that at which the phase determination is to be made. The mass will react more rapidly and become more uniform and thereby reach equilibrium at the desired temperature more rapidly than would otherwise be the case.

A quenched body will show crystals developed at the temperature of soaking and also a glassy, homogeneous phase which has been formed from any liquid present in the heated mass. A microscopic examination in conjunction with X-ray analysis are the most valuable methods of phase identification, although others may be used with advantage in some instances.

Should any crystal phase have a rapid transition at temperatures below that of the test, obviously, quenching may give misleading results; it is usually possible,

[1] Morey, G. W., *J. Wash. Acad. Sci.*, 13, 325, 1923.

however, to make allowance for such changes and, if the mechanism of transition is known, to deduce how much of the original phase was present.

Newkirk[1] has refined the technique for ternary and more complex mixes by centrifuging the equilibrium mixes to separate crystalline phases. The liquid is removed from the system by filtering through a platinum sponge. After quenching, the resultant glass can be chemically analysed.

The establishment of phase boundary lines by this method is a comparatively simple matter. If the same mixture of components is soaked at a series of ever-increasing temperatures and the phases present at each stage are carefully assessed, the range of stability of each crystal form can be established. Similar experiments may be carried out, using mixtures of different compositions and phase boundary lines can be constructed for the whole diagram.

When the components form solid solutions, the determination of the phases is greatly complicated. In most cases neither microscopic nor X-ray determinations are adequate to indicate the composition of the solid solution and it is usually necessary to separate the phases and to analyse each individually.

Studies of Model Systems. Silicate mixtures and other compositions of interest to ceramists melt only at high temperatures where the difficulties of observation and measurement are very great. Goldschmidt[2] has shown that similar phase relationships exist between some fluorides and oxides, but the changes in the fluorides take place at much lower temperatures, and can thus be studied more conveniently.

The fluoride ion is of similar size to oxygen but is of unit charge. Monovalent cations form fluorides which have a similar structure and properties to oxides of divalent cations of the same size, but, because the bonding between monovalent ions is much weaker than between divalent ions, the melting point is correspondingly lower (see p. 687). Table X.V shows the equivalent fluoride and oxide compounds and their melting points.

TABLE X—V. THE MELTING POINTS OF SOME OXIDES AND THEIR CORRESPONDING FLUORIDE MODELS

Fluoride	*Cation Size* Å	*Melting Point* °*C.*	*Melting Point* *K°**	*Oxide*	*Cation Size* Å	*Melting Point* °*C.*	*Melting Point* *K°**	$\frac{T_o}{T_f}$
LiF	0·60	870	1143	MgO and	0·65	2820	3093	2·71
				ZnO	0·74	1975	2248	1·97
NaF	0·95	988	1261	CaO	0·99	2572	2845	2·25
KF	1·33	846	1119	BaO	1·35	1923	2196	1·96
BeF_2	0·31	543	816	SiO_2	0·41	1728	2001	2·45
MgF_2	0·65	1240	1513	TiO_2	0·68	1825	2098	1·39
CaF_2	0·99	1330	1603	ZrO_2 and	0·80	2720	2993	1·87
				ThO_2	1·02	2800	3073	1·91

* K° indicates absolute temperature.

[1] Newkirk, A., see Bogue, R. H., *3rd Int. Symp. Chem. Cement*, 62, 1952.
[2] Goldschmidt, V. M., *Skr. Nor. Videnskaps-Akad.*, Oslo 1, *Mat-Naturv. Klasse*, 8, 7, 1927.

The relationship between the absolute temperature (K°) of melting of the oxides and the corresponding fluoride is about 2 to 1, as shown by the last column in the Table.

Provided that the limitations of model systems are appreciated, their study often reveals much valuable information of the equilibria relationship of high melting point oxides.

The **lithium fluoride-sodium fluoride** mixture is very similar to the **calcium oxide-magnesium oxide** system; a eutectic is formed in both which corresponds in composition to about 40 per cent molecular ratio of the lower melting component. The similarity between the **lithium fluoride-beryllium fluoride**[1] and the **zinc oxide-silica** mixtures is shown in Fig. X.55 A and B. The liquidus curves are remarkably similar but the compound Li_2BeF_4 which resembles Zn_2SiO_4 (willemite) dissociates at temperatures below that of the eutectic. The study of the model system **calcium fluoride-beryllium fluoride** enabled the phase relationships in **thoria-silica** mixtures to be predicted.[2]

An excellent survey of the value and use of fluoride models in equilibrium studies of refractory oxides has been compiled by E. F. Osborn.[3]

The Importance of Equilibrium Diagrams in Ceramic Studies

Although it is undoubtedly a fact that most ceramic bodies do not reach equilibrium under normal manufacturing conditions, equilibrium diagrams are of very great value and increasing use is being made of the information they contain. They must be used with care, however, because—in addition to rarely attaining equilibrium in practice—raw materials are often too impure to be regarded as simple mixtures of components. Despite serious objections of this nature, if due precautions are taken, important deductions can be made from equilibrium studies even of highly complex mixtures.

Equilibrium diagrams are valuable for:

1. Predicting the changes which are liable to occur in a body on prolonged firing.
2. Indicating the lowest temperature at which liquid will be found in a mixture of components and also the amount of such liquid.
3. Determining the nature of the bond in a fired ceramic material or article, i.e. whether it is glassy or crystalline.
4. Suggesting the approximate refractoriness or melting point of the body.
5. Developing new body compositions which possess unusual and desirable properties in practice. Examples which may be quoted, from a considerable number where phase equilibrium studies have been largely instrumental in evolving new ceramic products include:

Bodies composed of mixtures of talc and clay are known to possess excellent electro-ceramic properties. That this was due mainly to the presence of cordierite was established, but the composition for optimum properties was evolved from phase equilibrium studies of the $MgO—Al_2O_3—SiO_2$ diagram.

[1] Roy, D. M., Roy, R. and Osborn, E. F., *J. Amer. Ceram. Soc.*, 37, 300, 1954.
[2] Roy, D. M., Roy, R. and Osborn, E. F., *J. Amer. Ceram. Soc.*, 36, 185, 1953.
[3] Osborn, E. F., *Ceram. Age*, 60, 41, 1952.

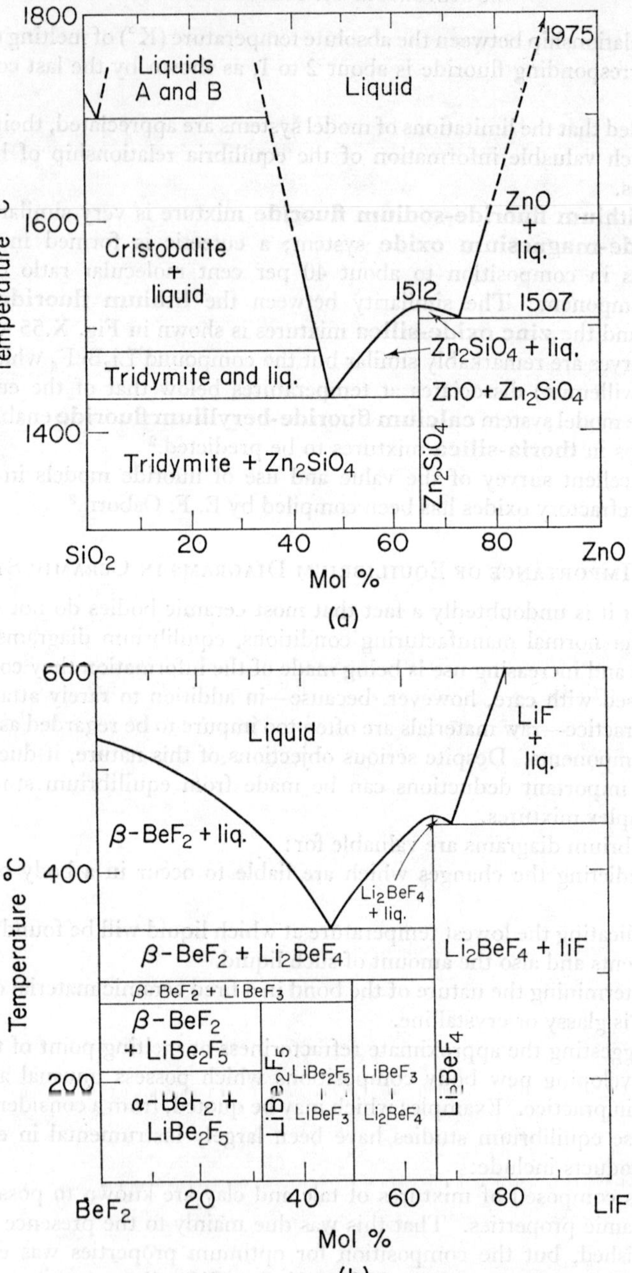

FIG. X.55. A COMPARISON OF (a) THE ZINC OXIDE-SILICA PHASE DIAGRAM AND (b) ITS FLUORIDE MODEL, THE LITHIUM FLUORIDE-BERYLLIUM FLUORIDE DIAGRAM

In the same way wollastonite bodies (containing $CaSiO_3$) have been brought to a high state of perfection by examining the appropriate and related phase equilibrium systems.

A wide range of low expansile bodies within the Li_2O—Al_2O_3—SiO_2 system was predicted and then developed commercially.

The developments of the newer ceramics for magnetic and electrical purposes are mainly the result of phase equilibria studies. Ferrites, titanates and related bodies have been evolved and their properties brought to the required levels by these means.

6. Avoiding compositions or components which might lead to defects in ceramic ware.

The most obvious examples in this category are the inversional and conversional changes within the silica minerals. The selection of fired compositions and the industrial application of such bodies depends to a large extent on phase changes which occur.

Of equal importance in other systems, have been the solution of dusting problems in stabilised dolomite bricks, and the techniques of overcoming transitional changes in zirconia and zircon bodies.

7. Enabling slag compositions to be predicted, which whilst conferring the maximum benefit to the smelting process, are not excessively corrosive to refractories.

8. By studying solid solution behaviour and controlling it, properties such as thermal expansion, conductivity, magnetic and electrical characteristics, can frequently be modified to a remarkable degree. This is one aspect of phase equilibria studies which is exciting considerable attention in the semi-conduction field and as a means of reducing brittleness and the tendency to thermal shock.

Until recent years, it has been customary to disregard the effects of pressure when studying equilibria in ceramic systems. That this is not an entirely valid assumption has been demonstrated in many systems and several examples have been mentioned in previous pages. Gaseous components which surround articles during firing or when they are in use are now known to have a marked effect on equilibrium conditions and on the subsequent properties and behaviour.

These aspects are being studied with great interest and as the technology of ceramic production improves, such principles will undoubtedly play a major role in improving existing ceramic compositions and in developing new ones.

Much information has been derived from a study of equilibrium diagrams in systems which are mainly liquid at the temperature of firing, such as glass or slags, where the phases present often approach true equilibrium. In silicate systems, for example, good permanent glasses are liable to form in the composition ranges where eutectics with low melting-points or incongruent compounds are formed. Crystal formation and other changes are more liable to occur when there is easy movement of the reacting material and the respective atoms. Silicate melts have a fairly low viscosity at high temperatures, but this rapidly increases as the temperature is lowered. Hence mixtures which cannot contain solid phases until the temperature is low, will virtually freeze and prevent crystal formation. The formation of incongruent compounds always implies a change which takes place largely in the solid state. Diffusion reactions of this type are invariably slow and are easily suppressed if the viscosity of the system is high and the cooling relatively rapid.

THE VELOCITY OF REACTIONS AND CHEMICAL KINETICS

ALTHOUGH equilibrium studies on ceramic materials have been actively pursued only in comparatively recent years, even less attention has been devoted to the speed or rate at which such reactions occur. This subject is termed *chemical kinetics* and although the processes or reactions in many gaseous or solution systems have been deduced, the fundamental mechanisms of solid transformations are not well established.

Some reactions proceed so rapidly that they are explosive in character, whilst others change at a rate undetectable by normal methods of measurement. Many oxidation reactions such as those of sulphur and carbon in the presence of potassium nitrate are complete in a fraction of a second, but the change from cristobalite to quartz at low temperatures is so slow that it requires a geological age for completion.

For a reaction to proceed from an equilibrium state, it is necessary that energy be applied to the system. This should be sufficient in intensity to permit atoms or ions in the crystal lattices or in molecular forms to vibrate from their normal position and be available to reform into other combinations. The minimal energy which is required for this purpose is termed the *activation energy*; it is most usually in the form of heat, but electricity, light and sometimes sound or vibration can supply it.

Even when the activation energy has been exceeded, the progress of a reaction in terms of its direction and velocity is governed by the Free Energies of the possible combinations which can occur and the energy changes (represented by ΔG) which would take place if they did occur (see p. 564).

The activation energy must not be confused with the energy change which occurs when a reaction takes place. When two substances are mixed which theoretically should react together, they frequently will not do so, because the atoms in the various original forms are strongly held and are unable to escape or release themselves from their environment. Energy needs to be applied to such a system to lift it over a 'threshold' to enable the change to take place freely. This is the *activation energy*.

Once the reaction has been initiated it will proceed, provided that thermodynamical conditions are satisfied, either exothermically if heat is given out or endothermically if heat has to be supplied continuously from an external source to cause it to proceed. The heat or energy change during the course of the reaction is the *enthalpy* or *heat* (or *energy*) *of reaction*.

In ceramic systems, the prediction of reactions which are likely to occur and of their rate of progression is far from easy from theoretical thermodynamic principles. For simpler systems, relationships have been evolved which predict the behaviour of reacting substances. Although different reaction principles must be involved in ceramic systems, it is surprising that the simple laws of chemical kinetics can often be applied.

Reactions between Gases and Liquids

Arrhenius[1] was the first to recognise that reactions which took place entirely in the gaseous or liquid phase depended on the collision-frequency of reacting molecules, atoms or ions. If this were the only factor involved, the reaction rate would depend only on the concentration, for this would determine the number of collisions in a given time. Arrhenius therefore concluded that some molecules in a gas must be more 'activated' and are able to react more readily when they come into contact with others. These 'active' molecules may be considered to have a higher kinetic energy than the average. On this basis, the reaction rate for different types of *gaseous reactions* may be predicted as depending on the number of molecules which take part in the change.

First Order Reactions. When only one molecular type is involved in the change, the rate of reaction must be proportional to its concentration at any time. Supposing for example, the reaction is of the type $A \rightarrow B + C$ (as in decomposition reactions), the rate of reaction $-(dc/dt)$ will be directly proportional to the concentration (c) of the reactant A at that particular time; thus

$$\frac{-dc}{dt} = kc \quad . \quad . \quad . \quad . \quad . \quad (1)$$

where k is a constant depending on the collision rate of 'active' molecules. It is known as the *rate constant*. Simple integration of this expression gives:

$$ln\ c = -kt \quad . \quad . \quad . \quad (2)$$

$$\text{or } c = C_o e^{-k(t-t_o)} \quad . \quad . \quad . \quad . \quad (3)$$

where C_o is the initial concentration at time t_o.

Equation (2) predicts that, if the reaction is first order or unimolecular, the concentration of the reactant will fall logarithmically with respect to time.

Second Order Reactions. If two molecules are involved in a reaction then it is of *second order* and is of the type:

$$A + B \rightarrow C + D$$

The rate of such a reaction will depend on the relative concentrations of both the reactants and these will each decrease at the same rate and the rate of reaction may be expressed as:

$$\frac{-dC_A}{dt} = \frac{-dC_B}{dt} = kC_A . C_B$$

where C_A and C_B are the respective concentrations of components A and B at time t.

If $$C_A = C_B = c$$

then $$\frac{-dc}{dt} = kc^2 \quad . \quad . \quad . \quad . \quad . \quad (4)$$

By integration $$\frac{1}{c} = kt \quad . \quad . \quad . \quad . \quad . \quad (5)$$

[1] Arrhenius, *Z. physik. Chem.*, 4, 226, 1899.

Therefore in *bimolecular* or *second order reactions* the concentration of reactants varies inversely with time.

Third Order Reactions. Reactions between three components are known although they are rare. The rate may be predicted by similar reasoning as:

$$\frac{-dc}{dt} = kc^3 \quad \text{if } C_A = C_B = C_C = c \quad . \quad . \quad . \quad (6)$$

Such simple behaviour may be complicated by reverse, side or consecutive reactions which may be simple in themselves, but, when taken together, may give rise to a complex rate expression, unless one stage predominates or is *rate determining*.

The *rate constant* (*k*) of any reaction is thus the important rate-determining factor and this is related to the *activation energy* (E) and also the temperature (T) of the system by the relationship:

$$k = Ae^{-E/RT} \quad . \quad . \quad . \quad . \quad (7)$$

This is an exponential equation which can be represented in a simpler fashion by taking logarithms of both sides so that:

$$ln\ k = \frac{-E}{RT} + ln\ A \quad . \quad . \quad . \quad (8)$$

In this equation T is in degrees absolute, E is in calories/mole (or gram-formula) R and A are constants, with the former (R) being the gas constant per g. molecule which has a value of 1·99 cal./°C/mole, and the latter (A) being related to free energy changes in the reacting system.

The simple kinetical equations derived in the previous sections apply only to gaseous or liquid reactions where the individual molecules are capable of free movement and reactions proceed mainly through collisions. The molecular energy term is predominantly that of *translation*.

Reactions Involving Solid Substances

When reactions involving solid phases are considered, the individual reacting units of atoms or ions are rigidly held within a crystal lattice and, consequently, before a rearrangement can take place the equilibrium lattice positions have to be disturbed. Consequently *vibrational* movement of the bonds becomes the dominant factor and reaction can occur only when this has exceeded a certain amount. Even when sufficient energy is available to break bonds, the released units (unless they are gaseous), have no freedom of movement, but can only transfer to an adjacent lattice-site. Such reactions can proceed only through the mechanism of *diffusion*.

Reactions in the solid state are, consequently, slow, but they nevertheless proceed with finite velocities. For example, pure kaolinite ($Al_2O_3.2SiO_2.2H_2O$), will not form a liquid phase below 1545° C., yet at considerably lower temperatures, mullite and cristobalite are formed from it.

The kinetical mechanism of most reactions involving solids is difficult to derive because the *rate-determining factor* is not always obvious.

In a simple liquid reaction, such as $CaCl_2 + Na_2CO_3 = CaCO_3 + 2NaCl$ the reaction rate is governed by the rate of formation of insoluble $CaCO_3$.

In the decomposition of calcium carbonate

$$CaCO_3 = CaO + CO_2$$

the partial pressure of CO_2 over the reactants is the rate-determining factor.

In most solid state changes, no such obvious governing mechanism is present. Solid reaction proceeds because the *free energy* of the system is lowered as a consequence; thus free energy may be expressed in terms of the lattice or crystal energies of the solid phases. If a decrease in total energy can be brought about by a reaction between the components, that change will tend to take place.

Thus, in the simple alkali halides, sodium chloride reacts with potassium fluoride to give sodium fluoride and potassium chloride because such a reaction involves a free energy decrease of about 9,000 cals. per mole.

Thus $NaCl + KF = NaF + KCl + 9000$ cals. per mole.

If the finely-powdered reactants are intimately mixed the reaction proceeds readily. Unfortunately, the lattice energy of crystals is not easy to calculate and the value for only a relatively few simple crystals is known with certainty, but this concept of a thermodynamical basis for solid reactions is valuable.

Solid reactions may be generally classified into two groups (*a*) internal re-arrangements, and (*b*) external interactions. The first type simply involves a 'change-round' in an existing atomic arrangement at a specific temperature without the introduction of foreign agencies; whilst the latter type implies transference of the atoms or ions of one crystal form across a surface into a lattice of another type. Although these two types are fundamentally different in character the underlying mechanisms and driving forces are similar. In each case, atoms or ions must leave their original lattice-sites and transfer to vacant positions in the immediate vicinity. As would be expected, the type of lattice determines the relative ease of such a transfer; massive, tightly-packed structures are not conducive to ionic movements, but spongy, cellular and framework structures are particularly adaptable to the process of *diffusion*.

(*a*) **Internal Rearrangement Reactions.** When no atomic or ionic material crosses the crystal boundaries, yet a reaction takes place at a certain temperature, the chemical composition of the initial and final states must be the same. Such reactions are termed *polymorphic transformations* and two clearly defined types are recognised:

1. The first involves a slight rearrangement and an alteration in the degree of order or symmetry of the crystal. No bonds within the crystal are broken; only a spacial reorientation of adjacent atoms occurs. Little or no heat is involved in the transformation, but the specific heats of the initial and final states may be substantially different. The reaction takes place at a clearly defined temperature and is instantaneously reversible. This type of change is known as an *order-disorder change* and is typified by the various silica *in*versions. In these cases, the lattice of the β- or high temperature modification is of the highest symmetry but, on cooling, the contraction is so great that the angles between linkages cannot be maintained. The change from β- to α-quartz is considered to involve only a change in the —Si—O—Si— bond angle from 180°, although in cristobalite and tridymite a

rotational movement may also occur (see Chapter III). The heat change in every case is small, being of the order of 1–4 cal/g.[1]

2. A second type of internal rearrangement reaction requires a migration of atoms or groupings and a disruption and re-forming of bond linkages. The final state has a completely different structure and symmetry system. The reaction involves a considerable energy change and although a definite temperature can be defined below which the transformation cannot proceed, the reaction, even well above this critical point is often sluggish. In contrast to the order-disorder type of change, the reverse reaction below the critical point is also slow and may require a geological age for completion. Changes in silica minerals also illustrate this type of reaction but, in this case, the *con*versions provide the example. Thus, quartz with a pseudo-hexagonal lattice converts to cristobalite with a cubic lattice, or to tridymite which is tetragonal, at temperatures above 870° C. The energy change, in each case, is comparatively high being of the order of 6,000 cals. per mole.[2,3] Fig. X.56 is an idealised picture of the structural difference between the two types of polymorphic transformations. The order-disorder inversion (*a*) involves only a bond angle change, the conversion reaction (*b*) changes the atomic arrangement completely.

Changes of the latter type, although producing a complete alteration of structure, involve little migrational or translatory movement of atoms or ions. The structure may simply glide or twist about an axis, or ions may migrate along a channel in the framework structure in such a way that crystal discontinuities do not arise. The change from meta-kaolinite to mullite could be considered as of this type. At temperatures above about 1,000° C., the silicon ions or silica groupings $(SiO_4)^{4-}$ partly migrate and take up other positions around the octahedral alumina layers which remain virtually unchanged spacially throughout the reaction.

(*b*) **External Interactions.** When two unlike materials are in contact and reaction occurs, the transfer of ions across the adjacent surfaces proceeds by a mechanism similar to that involved in the rearrangement of ions in polymorphic transformations. Ions transfer to vacant lattice-sites across the contact-surfaces in such a way that only a minimum translatory movement is involved. If the opposing crystal lattices are close-packed, little migration is possible and ionic transference will usually be towards the medium containing the maximum vacant lattice-positions or the more open structure.

Reactions between two or more solid substances are of two general types as follows:

(*a*) Additive—where $A + B = AB$

and the compound AB forms a separate phase, or

(*b*) Exchange—where $A + BC = AC + B$

or $AB + CD = AD + CB$

The mechanism is similar in both cases, i.e. *diffusion.*

[1] Sabatier, G., *Bull. Soc. fr. Miner. Crist.*, 80, 444, 1957.
[2] Bernal, J., *Trans. Farad. Soc.*, 34, 6, 1938.
[3] The heat change on the conversion of TiO_2 (anatase) → TiO_2 (rutile) has been calculated as +24,000 cal/mole.

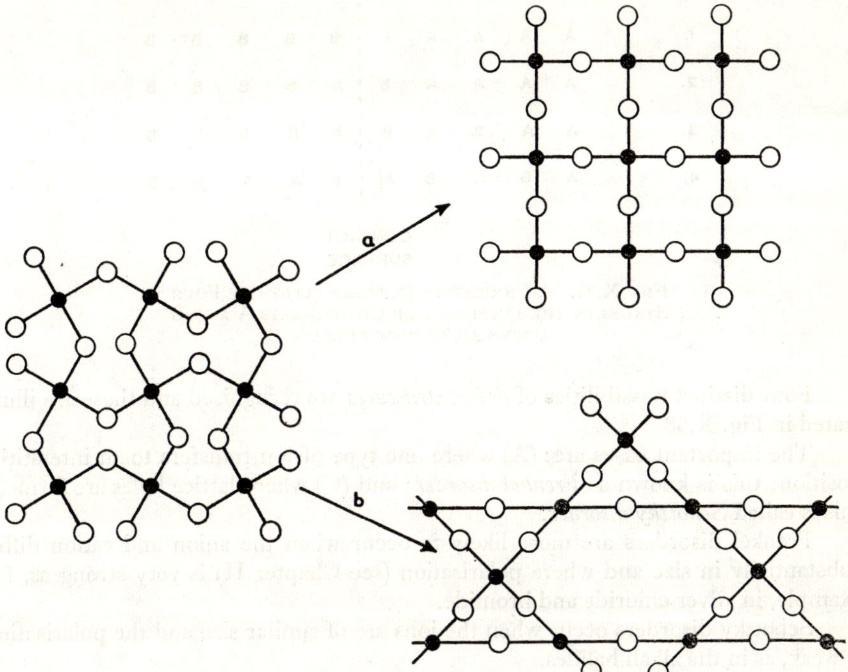

FIG. X.56. AN IDEALISED REPRESENTATION OF THE DIFFERENCE BETWEEN (a) THE INVERSION; AND (b) CONVERSION REACTIONS

MECHANISM OF DIFFUSION REACTIONS

THE physical mechanism of diffusion is directly related to the properties of the surfaces in contact. Weyl's suggestion that the outermost layers of atoms in a crystal have insufficient 'screening' and hence must be distorted, has already been demonstrated (p. 555). In these dynamically unstable surface regions there is always the possibility of ionic transference, especially if the vibrational movement is accelerated by heat, or if a suitably receptive surface is brought into close association.

The mutual diffusion stages of components A and B might be represented as in Fig. X.57, although the individual rates into the opposite medium may not be identical. There is, however, another factor which has an important bearing on the rate of diffusion reactions. Defects are common in crystal lattices and these facilitate the rate of transfer of ions. Schottky and Wagner have calculated that in any crystal lattice at temperatures above absolute zero, a definite number of ions will have left their normal sites and be occupying positions of disorder. Defects of this type do not occur on a large scale—perhaps not more than one ion in 100,000 will be out of place—yet this is sufficient to permit diffusion.

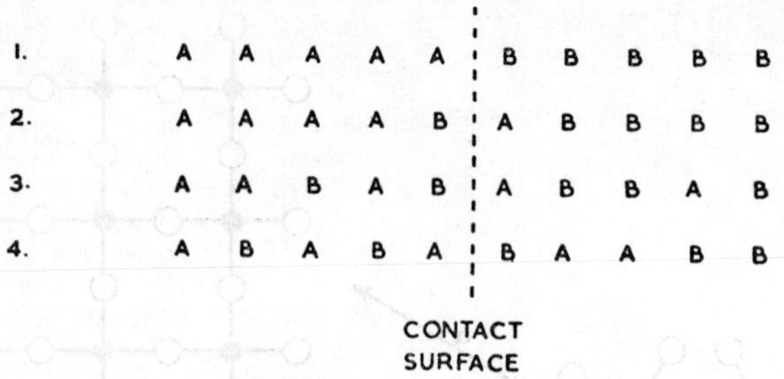

FIG. X.57. A FIGURATIVE REPRESENTATION OF FOUR STAGES IN THE DIFFUSION OF COMPONENTS A AND B ACROSS A CONTACT SURFACE

Four distinct possibilities of *defect structures* are recognised and these are illustrated in Fig. X.58.

The important cases are: (A) where one type of ion transfers to an interstitial position; this is known as *Frenkel disorder*; and (C) where lattice holes are formed; this is called *Schottky disorder*.

Frenkel disorders are more likely to occur when the anion and cation differ substantially in size and where polarisation (see Chapter II) is very strong as, for example, in silver chloride and bromide.

Schottky disorders occur when the ions are of similar size and the polarisation is weak, as in the alkali halides.

The number of lattice imperfections in a crystal depends on the temperature, hence the diffusion rate increases likewise.

Although diffusion proceeds through the transference of ions through defects in a crystal or through the spaces in an open-structured lattice, the detailed mechanism is a matter for conjecture.

Jander[1] suggests that the reaction

$$A + B = AB$$

proceeds in stages as illustrated in Fig. X.59.

A weak bond between the reactive components A and B is first formed at contact surfaces (stage 1). Mutual ionic transfer coupled with bonding in the surface layer then proceeds (stage 2) until a coherent layer of compound is formed (stage 3). The final phase involves molecular development (stage 4) inside the crystal which eventually resolves into an ordered crystal lattice (stage 5). Diffusion proceeds through the agency of a layer of product which migrates away from the surface. The nature of the product or diffusion layer may not correspond to the final product, but may be of an intermediate or transitory character.[2]

The Nature of Diffusion Layers. Although it is well established that diffusion reactions propagate through an advancing interface, the true nature of this 'diffusion layer' can only be surmised. It may be that the condition in such regions is a random distribution of diffusing ions which distort the original structure, until eventually a

[1] Jander, W., *Z. Angew, Chem.*, 49, 88, 1936.
[2] Hedvall, J. A. and Sjoman, P., *Z. Electrochem.*, 37, 130, 1931.

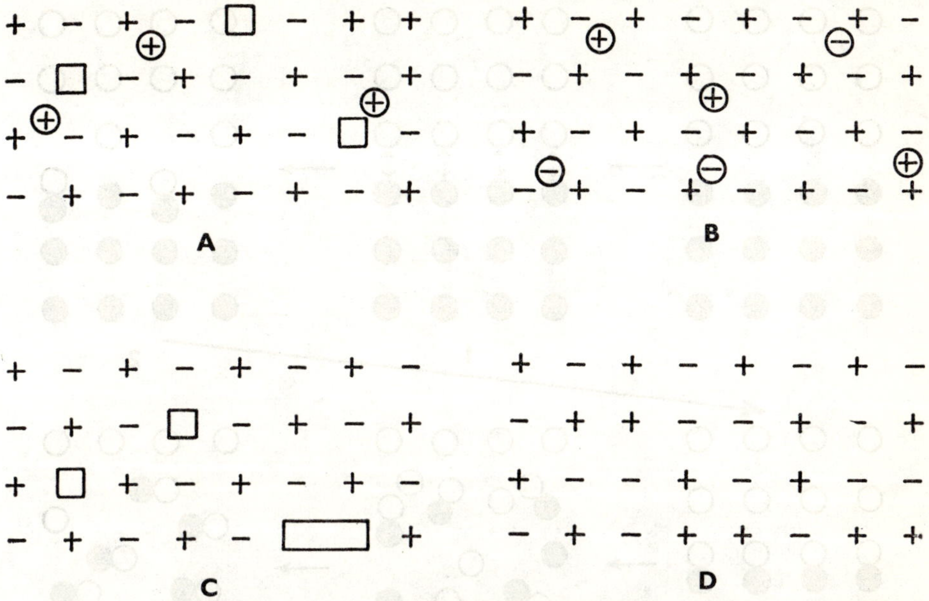

FIG. X.58. FOUR TYPES OF DEFECTS IN CRYSTAL LATTICES
A The transfer of ions to interstitial lattice positions; B the inclusion of both anions and cations in interstitial lattice positions; C vacant lattice sites; and D inverted ionic positions

new crystal phase will form which Jander has termed a *hybrid compound* and other authorities have referred to a *transitional form*. Diffusion proceeds more rapidly between two components under conditions where unstable compounds are more liable to form, although this generalisation cannot be carried too far.

Solid reactions commence at points of contact of opposite components. Hence, the 'active' concentration of reactants in such places will depend on their relative diffusion rates and may not be related to the overall ratio in which they are mixed. Jander has shown, for example, that in a mixture of silica and calcium oxide the first product of solid reaction is always the orthosilicate and it is only after prolonged heating that the metasilicate may form. Tricalcium silicate also appears at some stage of the reaction, as shown in Fig. X.60.

The formation of a diffusion layer is dependent on several factors and more than one underlying mechanism is involved. Where one component diffuses into another of different type, the mobility of the transferring ions will be the principal determinant. Therefore the formation and spread of a diffusion layer will depend on the availability of foreign ions and on the ease of transfer through the receptive lattice.

Diffusion proceeds mainly through crystal defects or along channels in a crystal with a framework structure. An increase in temperature will produce more defects and, at the same time, increase the mobility of the diffusing ions. Hence, the rate of diffusion is largely dependent on the temperature of the surroundings. Hedvall has

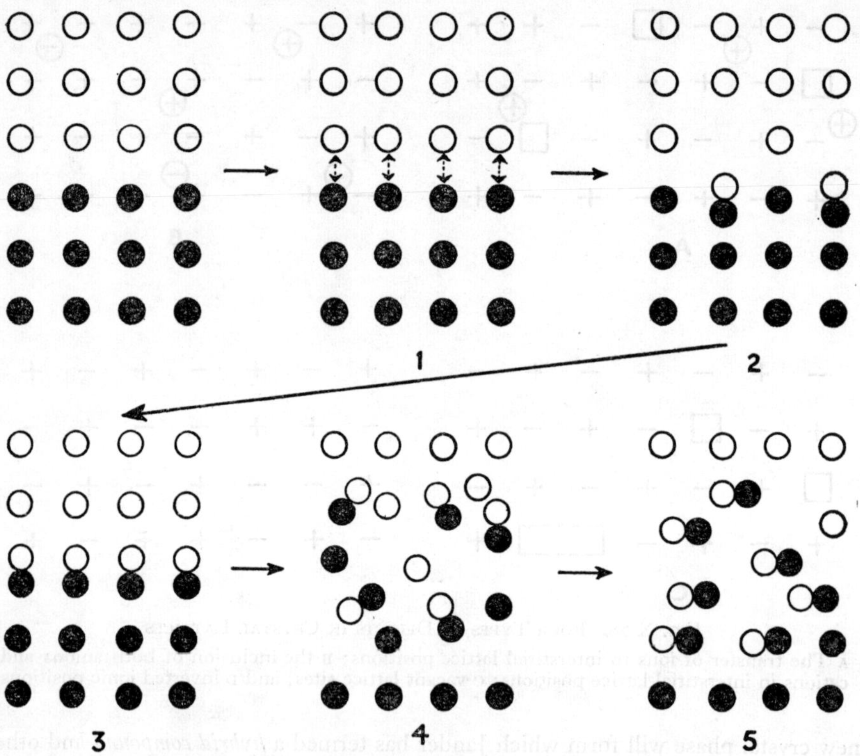

FIG. X.59. STAGES IN THE DEVELOPMENT OF A DIFFUSION REACTION

shown that the diffusion rate is also considerably accelerated when the receptive medium is undergoing a transitional change. Iron oxide, for example, can be induced to enter the quartz lattice with an appreciable velocity at about 573° C. (the inversion temperature); the rate does not increase to any extent above this temperature until the lattice becomes unstable when the conversion to cristobalite commences.

An altogether different mechanism of diffusion-layer formation and progression must be dominant in reactions which do not involve ionic transfer. Polymorphic transformation (p. 661) or decomposition reactions require that the linkages between adjacent ions should be altered in direction or completely broken. This can only arise as the result of an increase in the vibrational energy of the lattice which, if it becomes sufficiently large, will permit an ion to move out of the spheres of influence of its immediate counterbalancing neighbours. Here again, an increasing temperature will be an important factor as also will be a defective structure or an open lattice. Weyl has pointed out that the presence of foreign ions in a disturbed lattice may considerably increase the ease of rupture of linkages.

Even in changes which do not involve ionic transfer, the reaction proceeds by the propagation of a diffusion-type layer. At some point or points in the lattice,

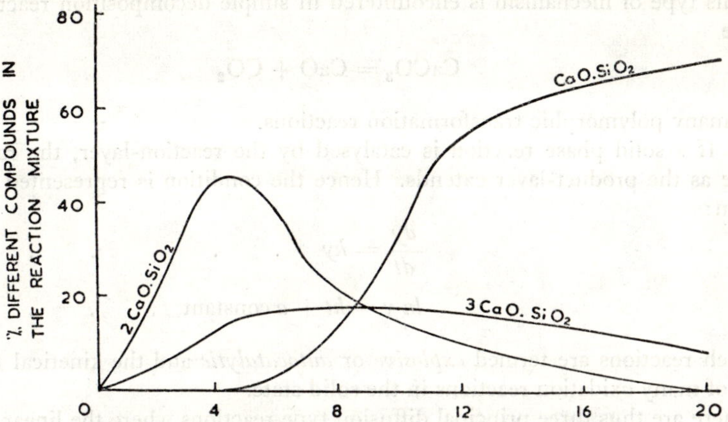

FIG. X.60. THE PRODUCTS OF FIRING LIME-SILICA MIXTURES FOR VARIOUS PERIODS OF TIME AT 1300° C. (after Jander)

conditions are more favourable to the commencement of a reaction which then spreads from nuclei of this type throughout the entire mass. This diffusion layer exerts a pronounced influence on the progress of the solid reaction; it may behave as (*a*) a barrier which slows down the rate, or (*b*) as an intermediary which propagates at a steady linear rate, or (*c*) as a catalyst which accelerates the process.

(*a*) Jander has made a comprehensive analysis of the kinetics of a reaction involving a diffusion barrier and he deduced that, in a cylindrical specimen reacting from one end-face under isothermal conditions, the rate of reaction (dy/dt) was inversely proportional to the thickness (y) of the layer formed, so that:

$$\frac{dy}{dt} = \frac{k}{y} \qquad (9)$$

so by integration

$$y^2 = 2kt + \text{a constant} \qquad (9a)$$

Therefore the rate of reaction will progressively decrease as the thickness of the reaction layer increases. This is a feature of many ceramic reactions in the solid state.

(*b*) When the reaction-layer boundary moves at a constant velocity and the product has neither a retarding nor a catalytic effect, the rate of reaction must be constant throughout. Hence the rate equation will be:

$$\frac{dy}{dt} = k \qquad (10)$$

or, by integration:

$$y = kt + \text{a constant (A)} \qquad (10a)$$

This type of mechanism is encountered in simple decomposition reactions of the type

$$CaCO_3 = CaO + CO_2$$

and in many polymorphic transformation reactions.

(*c*) If a solid phase reaction is catalysed by the reaction-layer, the rate will increase as the product-layer extends. Hence the condition is represented by the equation:

$$\frac{dy}{dt} = ky \quad . \quad . \quad . \quad . \quad . \quad (11)$$

or

$$ln\ y = kt + a \text{ constant} \quad . \quad . \quad . \quad (11a)$$

Such reactions are termed *explosive* or *autocatalytic* and the kinetical form is typical of many oxidation reactions in the solid state.

There are thus three principal diffusion type reactions where the linear rate of *propagation* may be expressed by standard equations (see 9, 10 and 11 above). In ceramic masses, however, it is exceptional for a reaction to be capable of being so simply represented; usually a shape factor has to be taken into consideration and, in addition, the initial formation of a surface layer (*nucleation*) may modify the kinetical mechanism in the initial stages.

Influence of Shape Factor on Propagation Rate. The equations (9, 10 and 11) in the previous section apply to the condition where a cylindrical specimen reacts on one end-face only and when the diffusion layer so formed advances along the length of the specimen. Thus, the area of new surface presented to the reacting layer is the same at all instances. This condition does not frequently occur and, more usually, the area available for reaction changes as the reaction proceeds. A typical example is a solid sphere reacting from all points on its surface; the diameter and, hence, the reacting surface, progressively decrease as the reaction proceeds.

In the sphere shown in Fig. X.61, the initial radius is *r* and after a time *t*, the reacting surface has advanced a distance *y* from the initial surface and *x* per cent of the sphere has reacted.

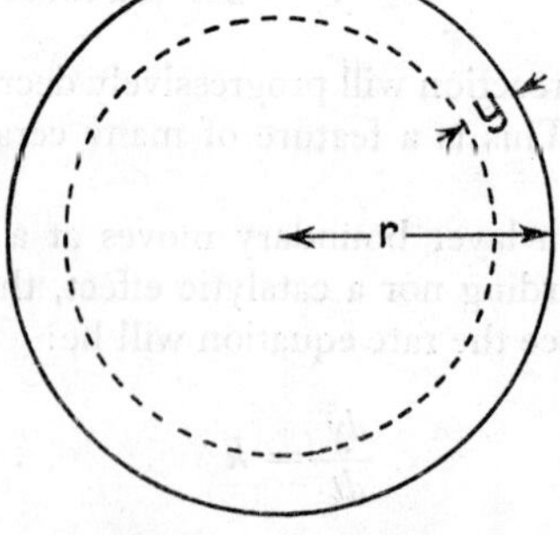

FIG. X.61. THE RATE OF THE PROPAGATION REACTION IN A SPHERE

The relationship between x and the thickness of the reacted layer y is given by:

$$\frac{x}{100} = \frac{\text{volume of original sphere} - \text{volume unchanged after time } t}{\text{volume of original sphere}}$$

$$= \frac{\dfrac{4\pi r^3}{3} - \dfrac{4\pi(r-y)^3}{3}}{\dfrac{4\pi r^3}{3}}$$

$$= 1 - \frac{(r-y)^3}{r^3}$$

or

$$\frac{(r-y)^3}{r^3} = \frac{100-x}{100}$$

$$\frac{r-y}{r} = \sqrt[3]{\frac{100-x}{100}}$$

$$\frac{y}{r} = 1 - \sqrt[3]{\frac{100-x}{100}}$$

$$y = r\left(1 - \sqrt[3]{\frac{100-x}{100}}\right) \quad . \quad . \quad . \quad . \quad . \quad (12)$$

As y is the distance advanced by the interface in time t, if the reaction is retarded by the diffusion-barrier, equation 9a can be substituted, i.e.

$y^2 = 2kt$ (the constant A is equal to zero because $y = 0$ when $t = 0$)

Therefore the expression (12) becomes

$$2kt = r^2\left(1 - \sqrt[3]{\frac{100-x}{100}}\right)^2$$

or

$$\left(1 - \sqrt[3]{\frac{100-x}{100}}\right)^2 = \frac{2kt}{r^2} = \mathrm{K}t \quad . \quad . \quad . \quad (13)$$

The constant K contains the diffusion-coefficient D and is also inversely related to the square of the initial radius of the sphere, or to the surface area factor (surface area of a sphere $= 4\pi r^2$).

Equation 13 derived from Jander's mathematical reasoning has been questioned in its detail by several physicists who have deduced their own equations of which those of Dünwald and Wagner[1] and Ginstling and Brounshteĭn[2] deserve special mention. The latter authors have developed an equation which is thought to be a more accurate representation of typical ceramic systems than that of Jander. Their equation is

$$1 - \frac{2x}{3} - \frac{\sqrt[3]{(100-x)^2}}{100} = \mathrm{K}_{\mathrm{GB}}t \quad . \quad . \quad . \quad (13b)$$

[1] Dünwald, H. and Wagner, C., *Z. Physik. Chem.*, B24, 53, 1934.
[2] Ginstling, H. M. and Brounshteĭn, B. I., *J. App. Chem. U.S.S.R.*, 23, 1327, 1950.

The various equations have been examined in detail by Carter[1] and Giess[2] and they conclude that activation energies as calculated by the Ginstling-Brounshteĭn equation are more in line with predicted values.

Equation 13 is a rate expression for it relates the percentage of sphere decomposed in unit time; hence K is, in every sense, a rate-factor, and its value governs the speed of reaction. As in simple gas reactions (p. 659), K is related to absolute temperature (T) by the standard Arrhenius equation

$$K = Ae^{-E/RT}$$

where R is the gas constant; A is a constant depending on the ease of diffusion, the number of mobile particles and the physical state of the material; and E is a term which expresses the energy required to activate a gram-molecule of particles so that they can move between points in the lattice; in other words, E is the *energy of activation.*

The general equation (13) expressing the rate of a reaction in a sphere is of general application. It describes the conditions under which a solid sphere will dissolve in a surrounding liquid or slag, provided that the diffusing layer behaves as a barrier or retarder to the progress of the reaction.

When the reaction layer transmits the reaction at constant velocity and neither retards or catalyses the rate of transfer (as in case (*b*) page 667), equation (10a) must be substituted in the expression

$$y = r\left(1 - \sqrt[3]{\frac{100 - x}{100}}\right)$$

Then $$1 - \sqrt[3]{\frac{100 - x}{100}} = \frac{kt}{r} \quad . \quad . \quad . \quad (14)$$

k is usually termed the *linear rate of propagation* and is given the conventional symbol μ. This equation may be expressed in another form as

$$1 - a = \left(1 - \frac{ut}{r}\right)^3$$

where *a* is the proportion of material reacted in time *t*, which, when completely expanded, is

$$1 - a = 1 - \frac{3ut}{r} + \frac{3(ut)^2}{r} - \frac{(ut)^3}{r}$$

To a first approximation, this expression may be regarded as the expansion of the equation:

$$1 - a = e^{\frac{-3ut}{r}}$$

or $$ln\,(1 - a) = \frac{-3u.t}{r} \quad . \quad . \quad . \quad . \quad (14a)$$

[1] Carter, R. E., *J. Chem. Phys.*, 34, 2010, 1961, and 35, 1137, 1961.
[2] Giess, E. A., *J. Amer. Ceram. Soc.*, 46, 374, 1963.

Now a is the proportion of material which has reacted in time t so if α is the proportion of original material remaining then

$$ln\ \alpha = \frac{-3u.t}{r} \quad . \quad . \quad . \quad . \quad (14b)$$

This last equation has the form of a first order equation (p. 659) where $3u/r$ is equivalent to the rate constant k.

Similar equations can be derived for particles of other shapes and where the diffusion mechanism is of one of the three general types as outlined on p. 667.

The Nucleation Rate. In the diffusion reactions, so far considered, only the rate of propagation of an interface has been derived, but the rate at which such a layer is initially formed is of equal importance.

If two solid substances react, they can do so only at those surface areas which are in contact. The reaction must be initiated by the transfer of atoms or ions of one type over the surface layers of both phases and into the lattice of the other component. It is to be expected that the energy required and the mechanism of such transfer will be dissimilar from the propagation of foreign ions which have already been incorporated in another lattice.

Diffusion proceeds by the transfer of ions from a lattice-site into a defect or vacant position elsewhere in either the same crystal or in an entirely different crystal. Defects in lattices are limited and occur mainly in surface layers (p. 663), hence, if two crystals are in contact and a defect or 'activation centre' in one component is in close proximity to a receptive condition in the other, transference of one phase into another can proceed. A reaction will commence, therefore, at nuclei on the surface and will then radiate by normal diffusion until a complete interface is formed as shown in Fig. X.62. The first diagram (A) shows two contact surfaces with defect points; the second (B) shows several nucleated areas which spread with a spherical interface; and the third (C) shows how several such points may merge to form a linear interface.

Some reactions, involving a polymorphic change without requiring any transference of ions across a contact surface, can also commence on a surface defect which nucleates readily and then merges with others to form a continuous interface, as in the conversion of pure quartz which is discussed more fully in a later section.

The rate of formation of nuclei is governed by the normal probability law as follows:

$$n_t = n_o e^{-kt} \quad . \quad . \quad . \quad . \quad (15)$$

where n_o and n_t are respectively the total number of possible activation centres and the number which remain after time t, and k is a constant, equivalent in all respects to a normal rate constant.

If it is assumed that the process of nucleation involves an appreciable molecular change, as often appears to be the case, and α is the percentage of the original phase remaining after time t

$$\alpha = e^{-kt} \quad . \quad . \quad . \quad . \quad . \quad (15a)$$

or

$$ln\, \alpha = -kt \quad . \quad . \quad . \quad . \quad . \quad (15b)$$

Thus the rate of a purely nucleation reaction has a form equivalent to that of a first order reaction (p. 659).

Obviously, the number of nucleation spots will influence the reaction rate, hence if surface defects are the main cause, the rate will be proportional to the surface area of the reacting particles.

The rate of nucleation (k) is also dependent on temperature and related by the Arrhenius equation (p. 660):

$$k = Ae^{-E_N/RT}$$

where E_N is the *energy of nucleation.*

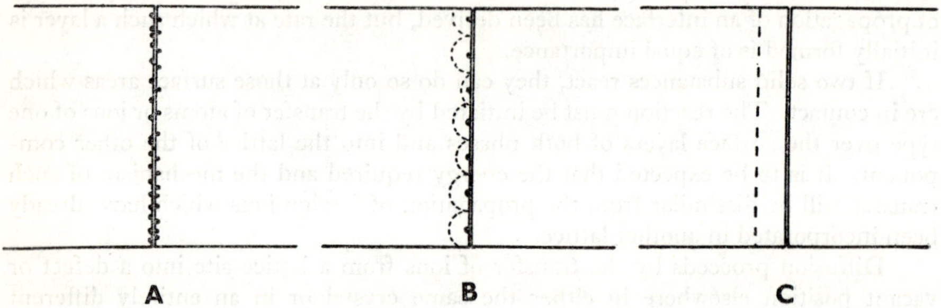

FIG. X.62. THE PROGRESS OF A SOLID REACTION WHICH PROCEEDS ACROSS AN INTERFACE
A Defect points in the surface layers; B the formation of nucleated areas; and C the formation of a linear reaction layer

Relationship between the Nucleation and Propagation Rates in Solid Reactions. Solid reactions may be divided into two stages (*a*) *nucleation*—when mainly surface defects are being nucleated, and (*b*) *propagation*—when such nucleated regions spread by normal lattice diffusion through the mass of the material. The form of reaction can be mathematically deduced and the rate at which each stage proceeds is governed by constant terms which are known respectively as (*a*) *the nucleation constant* (see equation 15b), and (*b*) the *propagation constant* (see equations 13 and 14).

The complete form of the solid reaction depends entirely on the relative magnitudes of these two terms and, if one greatly exceeds the other, it will be the predominant mechanism. There are three important cases:

1. Where the nucleation rate is rapid, but the rate of propagation is slow. Nucleation centres form so rapidly that the entire exposed layers of surface become completely covered.

The initial stage will be completed rapidly and the reaction will proceed through the transfer of a continuous layer into the interior of the crystal. The rate of propagation, which will depend on the nature of the diffusion layer and shape factor, will, after the rapid initial stage, be the *rate-determining factor.*

2. Where the rate of propagation is rapid and the nucleation rate is slow. In the limiting condition when one nucleation spot forms on the contact surface the reaction spreads from this centre and decomposes the entire particle. Thus, the

rate at which nucleation occurs will be the *rate-determining factor* in such cases. Such a mechanism is to be expected in many solid-state reactions involving two components. Nucleation can take place only at the points of contact between the two reactants, hence diffusion will proceed into the interior of each crystal from a relatively small number of surface points.

3. Where both nucleation and propagation rates are similar. A complex reaction curve may result which will be a function of the two individual rates. For a certain initial period of time, the rate may be mainly dependent on the number of nucleation spots and the rate at which each spreads. If the particles are of sufficient size, the second mechanism may predominate in the latter stages. The general form of such reaction curves is typically sigmoid-shaped.

Examples of Solid State Reactions in Ceramic Materials

1. **Decomposition of Clay.** Murray and White[1] have studied the kinetical mechanism of the decomposition of clays and have concluded that, as the evolution of water is logarithmically related to time, the form is *first order*. Unfortunately, these authors were not able to assess accurately the loss in weight in the initial stages of the reaction because of experimental difficulties, hence the existence of an induction or nucleation period was not established. The authors suggest that propagation must be rapid relative to the nucleation rate as 'only then would a first order law be obeyed'. This is not necessarily correct for if the particles are assumed to be spherical, with the diffusion layer behaving purely as a linear agent of transfer (case *b*, p. 667), decomposition by propagation will be logarithmically related to time, as in equation (14b) irrespective of nucleation conditions.

Murray and White have deduced the energy of activation of various clays as follows:

Clay	*Energy of Activation (cals/mole)*
Supreme kaolin	37,970
Newton Abbot ball clay	34,310
Wyoming bentonite	57,640
Eureka halloysite	36,740

2. **Dehydroxylation of Muscovite.** Kodama and Bryden[2] investigated the breakdown of muscovite as a function of temperature and concluded that the reaction was of first order form. The activation energy was calculated as 54,000 cal/mole.

3. **The Inversion of Silica Minerals.** Chaklader and Roberts used an elegant technique of high-speed cine-photomicrography to study the inversional change of cristobalite. The β-form of cristobalite is isotropic under polarised light whereas the strained lattice of the α-form (p. 661) gives the crystal a degree of birefringence. A magnified image of a section was photographed as it was raised slowly in temperature on a hot-stage microscope. Individual crystallites of cristobalite could be observed in the process of inverting. The change commenced at a point on the

[1] Murray, P. and White, J., *Trans. Brit. Ceram. Soc.*, 48, 187, 1949.
[2] Kodama, H. and Bryden, J. E., *Trans. Farad. Soc.*, 64, 3112, 1968.

surface of a grain or along a crack and proceeded rapidly from this point throughout the grain. Only rarely was more than one nucleating centre observed on a single grain, so it may be concluded that the rate-determining factor in the inversion of cristobalite is the formation of nuclei.

4. The Quartz Transformation. Grimshaw, Hargreaves and Roberts[1] have studied the kinetical mechanism of the destruction of quartz at high temperatures and have shown that two mechanisms can be ascribed depending on the presence, or otherwise, of catalytic agents.

(*a*) *Pure quartz.* When pure quartz, in the form of small particles, is heated to a sufficiently high temperature, a reaction takes place which results eventually in the total decomposition of the mineral. The reaction starts at centres on the surface of particles and these eventually merge together so that the reaction continues through the advance of an interface into the interior of each crystal. In large particles the reaction rate is complex whilst in fine-grained material (less than 60 microns) the decomposition proceeds at a logarithmical rate throughout.

Careful experimental tests have revealed that both the nucleation rate and propagation rates are important but, whereas the former depends inversely on surface area ($1/r^2$) the latter is related to $1/r$ (p. 669). In coarse-grained material the propagation rate is more rapid and hence a slow initial induction or nucleation period is distinguished, whilst, in fine material, the two rates are comparable in velocity and the two regions are indistinguishable.

The diffusion layer formed in this transformation is of type *b* (p. 667) and it neither retards nor catalyses the reaction rate, which is to be expected in a polymorphic change.

The reaction rates in the nucleation and propagation periods have been deduced from the two equations:

Nucleation $$ln\ \alpha = -kt \text{ (equation 15b)}$$

Propagation $$ln\ \alpha = \frac{-3u}{r} t \text{ (equation 14b)}$$

and, by calculating the appropriate rate constants at several temperatures, the energy change in the two periods may be derived from the Arrhenius equation. This gave

Energy of nucleation: 124,000 cals. per mole.
Energy of activation: 96,000 cals. per mole.

(*b*) *Quartz and catalysts.* The transformation of quartz is greatly accelerated by the presence of certain oxides which have been termed catalysts. Many of these (including the alkali oxides and iron oxide), react by forming a liquid phase with silica and, in such cases, the reaction proceeds rapidly to completion. Other catalysts, of which the alkaline earth oxides are examples, react only in the solid phase. The kinetical mechanism of these latter catalysts was studied in detail. The relative effectiveness of the various agents was assessed by comparing mixtures containing an *equal number of similar-sized particles* of each oxide mixed with quartz

[1] Grimshaw, R. W., Hargreaves, J. and Roberts, A. L., *Trans. Brit. Ceram. Soc.*, 55, 36, 1956.

grains; the number of contact points at which reaction must start was thereby kept constant in every case. The form of the reaction was similar in most cases. Initially, the rate of quartz transformation was very rapid but, after a brief period the velocity decreased rapidly and became comparable with the rate in the final stages of transformation in pure quartz.

The authors concluded that the particles of catalyst nucleate the quartz grains with which they are in contact and the transformation reaction is propagated very rapidly through those crystals. The catalytic effect is probably transmitted into a second ring of quartz grains which touch those in 'primary contact', but its influence is then dissipated and no further grains are attacked. Further reaction can then only proceed by the pure quartz grains reacting independently of catalytic activity of any kind.

This type of behaviour was found when calcium, magnesium and titanium oxide were added to quartz. Alumina was anomalous and appeared to retard the transformation at all stages; not only was there no initial period of pronounced catalytic activity, but the latter stages were also reduced in rate. Weyl has suggested that aluminium ions, being of high charge and small in size, more effectively screen highly-polarised oxygen ions, and therefore inhibit the nucleation process. Furthermore aluminium more readily diffuses through a quartz lattice than other cations and can therefore affect more quartz particles.

Chaklader and Roberts[1] have studied the formation of cristobalite following the quartz transformation and shown that a different kinetical mechanism applies.

There is no direct conversion of quartz into cristobalite and reaction proceeds through an intermediate phase which is dependent on catalytic ions. Although the predominant kinetical mechanism is of first order form, nucleation and propagation rates are of similar order which gives rise to anomalies when considering reaction in materials of different particle size.

5. **Reactions between Mullite, Silica and Alumina and certain Carbonates.** Pole and Taylor[2] measured the reaction rates between these substances at several temperatures by calculating the amount of carbon dioxide evolved. The mechanism was shown to correspond to the case of a diffusion layer retarding the rate, as in equation (9) (p. 667). Sodium carbonate was the most effective reactant used and, in general, mullite was more readily attacked than quartz with corundum least of all.

6. **Reaction between Lime and Silica.** Jander and Hoffman[3] studied the reaction between lime and silica as a function of relative concentrations, time and thoroughness of mixing and found that the orthosilicate ($2CaO.SiO_2$) was, apparently, the first product in all cases. When the proportion of lime was large, tricalcium silicate developed after long periods of firing, whilst the metasilicate $CaO.SiO_2$ appeared in those mixtures with a lower concentration of lime. Water vapour was found to be an activator of the reaction.

7. **Transformation of γ- to α-alumina.** Cannon and White[4] examined the

[1] Chaklader, A. C. and Roberts, A. L., *J. Amer. Ceram. Soc.*, 44, 35, 1961, and 46, 192, 1963.
[2] Pole, G. R. and Taylor, N. W., *Penn State Coll. M.I. Exp. Stat. Tech.*, Paper No. 20, 1935.
[3] Jander, W. and Hoffman, E., *Z. Anorg. Allgem. Chem.*, 218, 211, 1934.
[4] Cannon, J. H. and White, J., *Trans. Brit. Ceram. Soc.*, 55, 82, 1956.

crystallographic transformation of oxides of alumina, by observing the density change which accompanies the γ- to α-change when the material was heated isothermally for various time intervals.

At all temperatures, the reaction obeyed first order laws in which the concentration of the reacting phase varied inversely with time.

It would appear that the reaction is similar in kinetical form to the quartz transformation, but because the crystallite size was small, nucleation and propagation regions were not clearly defined. Nucleation was probably the rate-determining factor.

The calculated activation energy of the reaction was 79,000 cal/mole.

Heterogeneous Reactions

CLAYS, zeolites and many other ceramic materials are being used on an ever-increasing scale as catalytic agents to promote gaseous and liquid reactions. To quote an example, ethylene and bromine can be mixed together without reaction, although if they are contained in a glass vessel, reaction will proceed on the surface; however, if a small amount of clay is added, the velocity of the reaction is greatly accelerated.

Dupont and Dulou[1] have described the polymerisation reactions of pinene at clay surfaces. Many condensation and alkylation reactions of the Friedel-Crafts type progress rapidly if a clay is used as a catalyst and Sachanen and Caesar[2] have summarised many of the important applications in this field.

Reactions of this type are termed *heterogeneous* and always involve a surface which is capable of adsorbing the gaseous or liquid phase. The forms of such reactions are of interest and can be deduced as follows:

In a system composed of a surface and gas molecules, the latter strike the former, condense, remain on the surface for a time and then evaporate; eventually an equilibrium will be set up between the molecules impinging and those evaporating.

The number of molecules n which strike unit area of surface in unit time must be proportional to the pressure (p) of the gas and so, if a is the proportion adhering, an is the total number of molecules colliding with and being temporarily retained by a unit area in unit time.

Furthermore if S is the total proportion of surface covered at any instant then $1 - S$ is the amount of surface available for the condensation of more particles, hence, assuming monomolecular adsorption, the rate of condensation is $(1 - S)\,an$ moles. per sq. cm./sec.

The rate of evaporation is proportional to the surface covered (i.e. S) therefore it can be expressed as VS where V is a constant.

At equilibrium, the rates of condensation and evaporation are equal, hence

$$(1 - S)an = VS \qquad (16a)$$

or

$$S = \frac{an}{an + V} \qquad (16b)$$

[1] Dupont, G. and Dulou, R., *Verre Silicates Ind.*, 12, 8, COBEA, 1947.
[2] Sachanen, A. N. and Caesar, P. D., *Ind. Eng. Chem.*, 38, 43, 1946.

The adsorbed gas reacts on the receptive surface and, if the reaction velocity is small compared with the ratio of condensation and evaporation, the decomposition of some molecules will not appreciably alter the equilibrium condition.

The rate of reaction will be proportional to the concentration of molecules on the surface, raised to a certain power depending on the order of the reaction (see p. 659).

Hence
$$\frac{dx}{dt} = kS^m$$

where k is a constant depending on the rate of evaporation of product.

In unimolecular reactions $m = 1$ so

$$\frac{dx}{dt} = kS$$

Substituting in equation (16b):

$$\frac{dx}{dt} = \frac{k.an}{an + V} = \frac{k.\frac{an}{V}}{\frac{an}{V} + 1} \quad . \quad . \quad . \quad (17a)$$

Now n is proportional to the pressure of gas p and ka/V and a/V are constant under particular conditions, hence equation (17a) can be written as

$$\frac{dx}{dt} = \frac{K_1 p}{K_2 p + 1} \quad . \quad . \quad . \quad . \quad (17b)$$

So a reaction which involves a unimolecular change does not obey first order kinetics when it takes place where a surface layer serves as a catalyst.

There are two limiting cases:

1. When the surface is sparsely covered and molecules do not condense easily but evaporate readily. Then a is small, but V is high and S (the proportion of surface covered) is small. Therefore, $1 - S \doteq 1$.

So equation (16a) becomes $an = VS$ and substituting the unimolecular form

$$\frac{dx}{dt} = kS \quad \text{as above,}$$

$$\frac{dx}{dt} = \frac{kan}{V} = Kp \quad . \quad . \quad . \quad . \quad (18a)$$

i.e. rate is dependent on pressure so the reaction is truly first order in such cases, but only when the surface adsorption approaches zero.

2. When the surface becomes almost entirely covered with adsorbed molecules $S \doteq 1$ so the rate equation is:

$$\frac{dx}{dt} = kS \doteq k \quad . \quad . \quad . \quad . \quad (18b)$$

The rate of reaction is independent of pressure and is of *zero order*. In other words the gas particles are retained immediately they collide with the surface and are not easily released. This is approximately the condition when organic molecules react at a clay surface.

There are thus two extreme cases of heterogeneous reactions given by the equations.

$$(18a)\ldots \frac{dx}{dt} = kp \quad \text{and} \quad (18b)\ldots \frac{dx}{dt} = k$$

Therefore the general case is

$$\frac{dx}{dt} = kp^x \qquad . \quad . \qquad (18c)$$

where x lies between 0 and 1. Hence such reactions have *fractional orders.*

The kinetical form of gas/surface reactions is complicated by the behaviour of the compound formed. In many cases, the product is strongly held by the surface and so retards the reaction, especially in the advanced stages, thus poisoning the catalyst.

When two reacting gases are involved, the rate of reaction depends on the relative adsorption capacity of the molecules of each type.

The mechanism of heterogeneous reactions or surface catalysis is not fully understood. Formerly, it was thought that, as the gaseous molecules are adsorbed on the surface, there is a localised concentration of reacting particles. It is now concluded that the surface layers hold the adsorbed molecules rigidly by virtue of the unsatisfied valency conditions in the outermost atomic groupings. The electron field of the gaseous molecule is, thereby, distorted and reactions can proceed which would, otherwise, be impossible.

This view is confirmed, to some extent, by reactions which lead to different end-products when different solids are used as catalytic surfaces. An example of this is the dissociation of ethyl alcohol vapour when passed over metallic surfaces such as copper and nickel, and over non-metallic materials like alumina. The change proceeds as follows:

$$\underset{\text{ethyl alcohol}}{CH_3.CH_2OH} \xrightarrow[\text{over Cu,Ni, etc., at 400° C.}]{} \underset{\text{acetaldehyde + hydrogen}}{CH_3CHO + H_2}$$

$$\underset{\text{ethyl alcohol}}{CH_3.CH_2OH} \xrightarrow[\text{over } Al_2O_3 \text{ at 400° C.}]{} \underset{\text{ethylene + water}}{CH_2 = CH_2 + H_2O}$$

The theory has been advanced that the catalytic surfaces attract different configurations or groupings of the alcohol molecule. The metallic surfaces containing atoms of a single type tend to attract hydrogen atoms only, whilst the alumina surface attracts alternate hydrogen and oxygen atoms as shown in Fig. X.63

Some surfaces contain active adsorption centres of more than one type. Thus platinum wire catalyses the decomposition of ammonia and also the dissociation of hydrogen iodide. If it is used for the former reaction it rapidly becomes 'poisoned' and inactive; but in this 'poisoned' state it will readily catalyse the latter change.

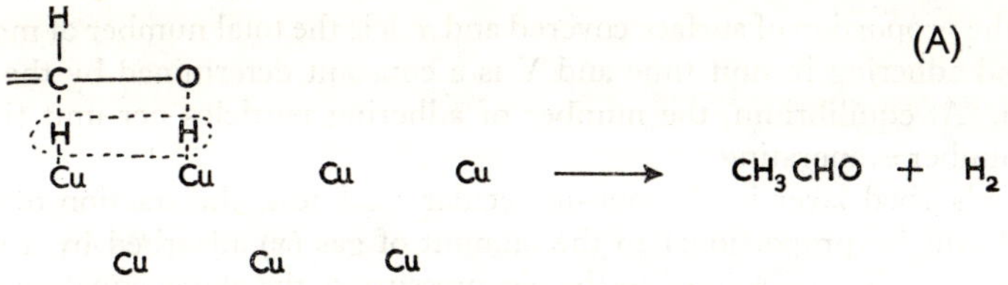

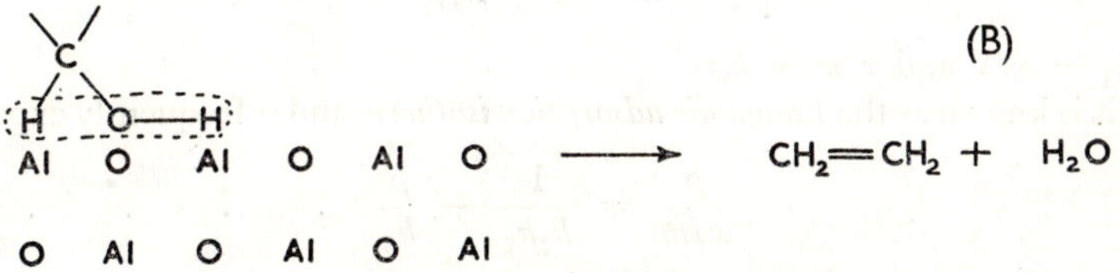

FIG. X.63. THE MECHANISM OF THE DISSOCIATION OF ETHYL ALCOHOL
At (A) a metallic surface; and (B) the surface of alumina powder

KINETICS OF ADSORPTION

HETEROGENEOUS reactions proceed by adsorption and as this phenomenon is a characteristic feature of clays, it is of special interest to ceramics.

The term *adsorption* implies a concentration of a particular component at the surface of another phase and, in theory, should be distinguished from *absorption* which implies internal penetration (see p. 448).

The amount of adsorption depends upon the surface area exposed, the concentration of so-called 'active centres' in the surface, the temperature, the concentration of adsorbent or the pressure, if a gaseous phase is involved.

The adsorption of gases or other molecules by a surface undoubtedly involves attractive forces, the nature of which can only be conjectured. Van der Waal forces (see Chapter II) have been suggested but, in recent years, a more tenable theory has been advanced by Fajans and developed by Weyl and others. Briefly, atoms and ions in the outermost surface layers are electrically unbalanced and the electron fields are distorted or 'unscreened'. Such units will readily link with molecules to reduce the unbalanced 'screening'. The rigid bonding between surface layers and adsorbed particles can only involve one molecular thickness because such forces act over a small distance. When deformation of the adsorbed molecule is high, however, extra layers may build up but they can do so only if the pressure or the concentration is high and even then they are much more loosely held than the first layer.

From a kinetical stand-point, adsorption can be treated in a similar fashion to heterogeneous catalysis. Equations 16a and b apply namely:

$$(1 - S)an = VS \quad \text{or} \quad S = \frac{an}{V + an}$$

where S is the proportion of surface covered and $a.n$ is the total number of molecules colliding and adhering in unit time and V is a constant determined by the rate of evaporation. At equilibrium, the number of adhering particles per unit time will equal the number evaporating.

If the adsorbed layer is of mono-molecular thickness, the fraction of surface covered (S) will be proportional to the amount of gas (x) adsorbed by a mass of solid (m), and as n is proportional to the gas pressure p, the above equation may be written

$$\frac{x}{m} = \frac{k_1 k_2 p}{1 + k_1 p} \quad . \quad . \quad . \quad . \quad (19a)$$

where $k_1 = a/V$ and $x/m = k_2 S$.

This is known as the *Langmuir adsorption isotherm* and is frequently expressed as

$$\frac{p}{x/m} = \frac{1}{k_1 k_2} + \frac{p}{k_2} \quad . \quad . \quad . \quad . \quad (19b)$$

Hence a graph of $\frac{p}{x/m}$ against p will give a straight line and under standard conditions the surface area of powders can readily be compared (see Chapter VII).

There are two limiting cases:

1. When either the pressure or the degree of adsorption is low, only a small portion of the surface is covered hence $S \doteqdot 0$ and $1 - S \doteqdot 1$ and equation 16a becomes $an = VS$ which substituting as in equation (19a)

$$\frac{x}{m} = k_1 k_2 p \quad . \quad . \quad . \quad . \quad (19c)$$

i.e. the gas adsorbed is directly proportional to pressure.

2. When the adsorption is large $S \doteqdot 1$ so equation (16a) becomes

$$(1 - S)an = V$$

$$S = 1 - \frac{V}{an}$$

or

$$\frac{x}{m} = k_2 - \frac{k_2}{k_1 p} \quad . \quad . \quad . \quad . \quad (19d)$$

As the pressure increases $k_2/k_1 p$ becomes smaller and x/m approaches the value k_2. So at low pressures the adsorption is given by $x/m = kp$ and at high pressures by $x/m = k$.

At intermediate pressures or adsorptions the general equation $x/m = kp^n$ applies where n lies between 0 and 1.

Adsorptions from solutions by solid surfaces obey similar laws; the general equation is

$$\frac{x}{m} = kc^n$$

where c is the concentration.

The adsorption of dye-stuffs from solutions by clays is a typical example of this type of adsorption.

CRYSTAL GROWTH

IN the ceramic processes which have been considered so far, there has been a change in the physical or chemical nature of the original components. At high temperatures, however, many crystal forms increase markedly in size without any change in crystallographic type. In a similar way, compacted powders shrink when heated and form bonds which are purely the result of a change in crystal dimensions or of a development of a continuum between adjacent particles. The lattice arrangement in the parts of the crystal which form on heating is indistinguishable from that of the initial crystallites and, in addition, particles are linked together by the intermerging of their atomic or ionic configurations. Such crystal growth is termed *sintering* and is of especial interest in ceramic studies.

An important consideration in sintering processes is that the crystal growth takes place at temperatures well below the point of liquid formation.[1] For instance, periclase, MgO, does not fuse below a temperature of about 2,800° C. yet when fired in a rotary kiln at only 1650° C., minute crystallites will grow readily to an appreciable size. In a similar way mullite can be detected in kaolinite fired at around 1000° C. only by X-ray analysis. If the temperature is maintained at 1250° C., the crystals become large enough to be visible under a low-power microscope, yet the amount of the mineral present has not been substantially increased. The melting point of mullite ($3Al_2O_3 . 2SiO_2$) is 1810° C., so that a liquid phase cannot have been involved.

In recent years, considerable progress has been made in the manufacture of high-purity oxide refractories. Sintering plays an important role in their formation because a hard, dense body can be prepared by firing well below the liquidus temperature.

SINTERING

THE fundamental mechanism of sintering is not yet fully understood, although many authorities consider that diffusion phenomena are involved. Crystal growth is undoubtedly intimately linked with the distorted surface layers of crystals and with the degree of compactness or number of contact surfaces within the mass. Three possible mechanisms have been advanced to account for the transfer of material (*a*) vaporisation and condensation, (*b*) plastic flow of material, and (*c*) solid diffusion.

Huttig[2] has demonstrated that significant changes can be detected in the cohesion and electrical conductivity of compacted metal powders at temperatures far below their melting point. This conclusion is supported by Clark, Cannon and

[1] The term sintering is frequently used in the ceramic industries to describe a certain condition in the firing of ware (see also p. 686). It represents a state of partial fusion when the mass has developed a degree of strength but is still porous. The individual particles have either grown, fused or been fused together but the bonding is confined to surfaces or points which are in contact. In the specialised usage of the term *sintered ware*, fusion is not implied because the crystal growth takes place at temperatures well below the lowest possible point of liquid formation.

[2] Huttig, G. F., *Z. Anorg. Chem.*, 247, 221, 1941.

White[1] who have demonstrated that an alumina compact develops strength at about 600° C. This, they attribute to the removal of a hydrate layer and the formation of weak surface linkages which increase the 'screening' of the exposed oxygen atoms.

True sintering, however, which is manifested by the growth of crystallites and an increase in density of the compact, is usually acknowledged to start at the Tammann temperature which is about 0·55 to 0·60 that of the melting point of the component.

When sintering occurs, the mass of small crystallites changes slowly into a compact of a few large intergrown crystals. Microscopic observation suggests that each initial particle merges with its neighbours and the spaces between the particles progressively decrease and finally disappear, until eventually large, perfectly-developed crystals are formed.

There are two phenomena which are not easy to explain when crystals grow in sintered material.

1. Despite an initial random distribution, the small crystallites grow into well-developed crystals which are optically oriented throughout.

2. The void space between the sintering crystallites vanishes entirely without apparent diffusion of a gaseous phase.

Theories of the Mechanism of Sintering.

1. *Vapour phase transference.* Until a few years ago the most commonly accepted explanation of sintering was that of the transference of mass through the vapour phase. The vapour pressure of solid ceramic materials is known to be low at normal temperatures, but it is appreciable, and, as the temperature is raised, the pressure is substantially increased.

Any small particles with a large surface area will form a vapour phase more readily than large ones. Hence, material will be lost from smaller particles and will condense on larger grains. A compact of small particles will thus tend to change into a mass of larger grains and eventually into a single crystal. On this basis the formation of large uniform crystals is readily explained because the condensing phase would take up the lattice configuration of the 'accepting' crystal.

This theory receives little support at the present time because microscopic examination does not indicate that small particles decrease in size at the expense of larger ones, but that adjacent grains grow together and so form enlarged crystals.

2. *Diffusion mechanism.* A mechanism similar to that of diffusion has been suggested to explain sintering or crystal growth. Diffusion of ions from one lattice through a crystal of dissimilar type has been clearly established and it is to be expected that if vacant lattice sites or interstitial ions are suitably present, transfer between two similar crystal lattices is feasible. Undoubtedly transfer of ions in this way does occur in sintering processes, but this may not be the principal mechanism involved. The calculated energy of activation required for self-diffusion reactions is much higher than the energy factor involved in sintering processes.

3. *Viscous and Plastic Flow.* Many authorities on sintering conclude that the transfer of material must be brought about by a shearing of layers by capillary forces between contact surfaces in the original compact.

[1] Clark, P. W., Cannon, J. H. and White, J., *Trans. Brit. Ceram. Soc.*, 52, 1, 1953.

The flow-process has been systematically investigated in the field of metallurgy[1] and it is concluded that the mechanism is mainly 'slip' within the lattice structure of the minerals. Crystals are known to be far from perfectly developed, and along certain planes of atomic configuration, dislocations may occur which, over a localised region, may destroy the symmetrical atomic arrangement. In such positions, the force required to move the atoms is far less than would be required over a normal part of the lattice and the weak forces which develop in the region of surfaces in contact may be sufficient to produce a movement, especially at high temperatures.

Clark and White[2] and Mackenzie and Shuttleworth[3] have independently concluded that 'flow' must be the principal mechanism of sintering; the former as the result of experimental evidence from pure oxides; the latter from their work in metallurgical systems.

The equations which both groups of workers have derived are essentially similar and indicate that *plastic flow* with a definite yield point must be the main mechanism in the systems studied. This is in contrast to glass in which viscous flow (see Chapter XIII, p 902.) occurs. Clark and White suggest that in the initial stages of sintering, capillary forces in the region of contact surfaces produce flow in the material and thus build up a 'lens' between two adjacent particles.

It may be that whilst plastic flow provides the best explanation of sintering in its initial stages and up to a densification of 0·80, in the later or final stages, as the body approaches zero porosity, diffusion becomes more important. Ford[4] considers that true solid state sintering takes place in three overlapping, but well-defined, stages.

In the first stage, the contact areas between adjacent particles build-up through the formation of 'necks'. As a result, the volume of the compact decreases at a relatively rapid rate. This continues until the neck diameters have increased to between one-fifth and one-third of the diameters of the particles themselves.

During this process it can be predicted theoretically and verified experimentally that under constant temperature conditions:

$$\frac{dL}{L} = kt^{2/5}$$

where $\frac{dL}{L}$ is the fractional linear shrinkage in time t, k is a constant which depends on temperature and from it, an energy value for the process can be determined.

In the initial stage of sintering there is virtually no increase in the size of crystallites, and densification proceeds by virtue of the drawing together of particles as contact necks increase in size.

The second stage of sintering, to a relative density of about 0·95, proceeds by a mechanism in which the pores are decreasing in size although they remain inter-connected. Some grain growth takes place in this period so that the rate of densification is usually complex.

[1] Mott, N. F. and Nabarro, F. R. N., Bristol Conf. on the Strength of Solids (*Phy. Soc., London*), 1948.
[2] Clark, P. W. and White, J., *Trans. Brit. Ceram. Soc.*, 49, 305, 1950.
[3] Mackenzie, J. K. and Shuttleworth, R., *Proc. Physical Soc.*, B62, 833, 1949.
[4] Ford, W. F., 'The Effect of Heat on Ceramics', *Inst. Ceram. Text Book No. 4*, 1967.

In the third and final stage of sintering, densification proceeds by the elimination of unconnected or closed pores by grain growth. In most cases, complete densification cannot be achieved because the pores become trapped and are unable to migrate to a boundary. In this stage the porosity decreases in linear relation with log time. The slope of the relationship depends on the rate of grain growth.

The rate of sintering decreases dramatically with time and is also a function of temperature. These effects are well-brought out in Fig. X.64, which shows the effect of temperature and time on the densification of pure titania compacts.[1]

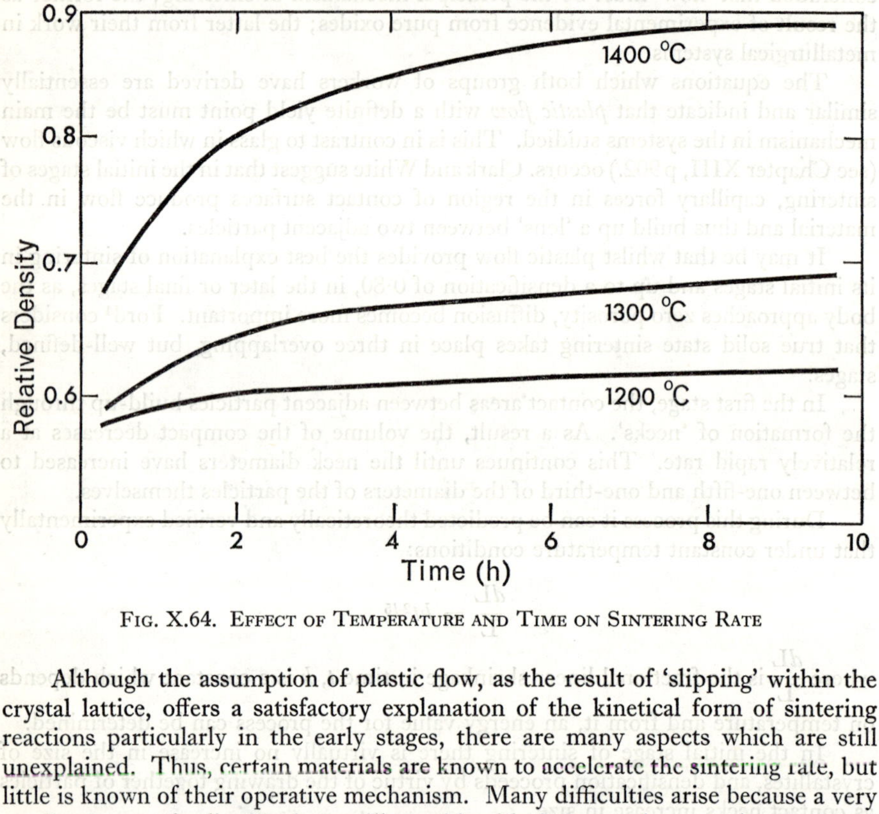

FIG. X.64. EFFECT OF TEMPERATURE AND TIME ON SINTERING RATE

Although the assumption of plastic flow, as the result of 'slipping' within the crystal lattice, offers a satisfactory explanation of the kinetical form of sintering reactions particularly in the early stages, there are many aspects which are still unexplained. Thus, certain materials are known to accelerate the sintering rate, but little is known of their operative mechanism. Many difficulties arise because a very small amount of a liquid phase will considerably increase the crystal growth and although most workers have used materials of high purity in their investigations, trace impurities, especially under the pressures developed by capillary forces could lead to liquid formation at temperatures well below the melting point of the mass. Even though amounts were small, traces of liquid would facilitate the transfer of material between crystallites.

The phenomenon of crystal growth in the presence of small amounts of

[1] Cannon, J. H. and White, J., *Trans. Brit. Ceram. Soc.*, 55, 82, 1956.

impurity is a fascinating new field of ceramics. One aspect of it is *epitaxy* or the formation of whiskers of a particular phase by growth at high temperatures.[1]

Hot-pressing techniques depend on sintering and crystal growth for densification. Plastic flow is increased by the application of high stress values to a compacted powder maintained at high temperature (see also p. 532).[2]

Vitrification

Densification in ceramic articles, when they are raised to high temperatures, is frequently the result of a liquid phase developing at crystal interfaces by chemical reaction within the body. When the amount of liquid is small, the predominant mechanism within the grains of the body is a rapid increase in crystallite size. This is presumably by virtue of phase transfer within the liquid medium and may be compared with the growth of crystals in an aqueous medium. Small particles, because of their different surface energies dissolve and the material is reprecipitated on larger grains.

Grain growth in aqueous media is generally inversely proportional to grain size so that

$$\frac{dD}{dt} = \frac{k}{D} \text{ where D is grain diameter.}$$

By integration $D^2 - D_0^2 = k(t - t_0)$ where D_0 is the initial size of particles (at time t_0).

Buist has shown that in vitrification studies involving the grain growth of lime, magnesia and alumina in the presence of small amounts of liquid, the relationship is better represented by:

$$D^3 - D_0^3 = k(t - t_0)$$

The grain diameter is thus proportional to the cube root of time at constant temperature.

By carrying out experiments in which grain growth was measured as a function of time, the rate constant was measured. Tests at different temperatures enable an activation energy to be calculated. Buist has derived a value of about 40,000 cal/mole for the examples of grain growth quoted above.

Grain growth in a liquid phase is a complex phenomenon, particularly when the fluid does not surround the grains completely. Contact between the solid particles and liquid medium then depends on surface tension forces and the degree of wetting. A contact angle (called the *dihedral angle*) between fluid and solid is established, which may be compared with the angle of the meniscus when a liquid rises in a glass tube.

In ceramic bodies, the lower the value of the dihedral angle the more readily will the fluid make contact with or 'wet' the solid which, in turn, leads to a faster rate of grain growth.

This is a fascinating field of research and one of great importance in the control and development of ceramic bodies. It has been applied to the control of particle

[1] Knippenberg, W. F. and Verspui, G., *Philips Tech. Rev.*, 29, 252, 1968.
[2] Ubbelohde, A. R., *Nature*, 220, 434, 1968.

size in alumina-cutting tools. Small amounts of Cr_2O_3 additions produce a liquid of low dihedral angle, so that large crystallites develop quite rapidly at temperatures about the eutectic point.

FUSION AND CRYSTALLISATION

WHEN most ceramic materials are raised to a sufficiently high temperature they are gradually converted from a solid to a liquid material, often undergoing some decomposition in the process.[1] According to the temperature attained the duration of the heating and some of the physical and chemical properties of such materials, the amount of fluid produced may be very slight and the term *incipient fusion* is then applied; a larger proportion of liquid may be produced and such a material on cooling is then said to be *sintered*, particularly if the liquid has been almost wholly superficial, i.e. when it forms a thin coating on many of the particles. When a large part (or the whole) of the material has been converted into a liquid the change is known as *fusion* or *melting*. *Vitrification* is an intermediate stage between sintering and complete fusion, but, it is becoming increasingly common in the ceramic industries to include all stages of partial melting as vitrification. The term sintering is seldom used except to describe the products from solid state reactions (see previous section) and occasionally in the production of frits and of some cements.

The complete change from the solid to the liquid state may be described as occurring in four stages:

Solid → sintered → vitrified → fused or melted (liquid)

These stages tend to overlap because some fused material is produced in almost every stage after the heating has commenced and fusion occurs throughout the whole series of changes. These stages are, however, of great industrial and technical importance and should be clearly understood and appreciated.

Fusion is defined as the change in state of a substance from the solid to a liquid phase. It is usually brought about through the agency of heat, although pressure may also be an influencing factor. In ceramic and allied materials the term 'fusion' is preferred to 'melting' because, unlike many pure materials, the change of state takes place over a temperature range and a melting point cannot be established.

The change from a solid to liquid state can be explained on the basis of modern theories of the structure of materials. A solid crystalline material is composed of atoms or ions which are arranged spacially in characteristic groupings to give a symmetrical structure (see Chapter II). The distribution of electrical charge is such that each unit is maintained in an equilibrium position and although small vibrational movements may be possible they are insufficient to disturb the structure. As the temperature is increased, however, the atoms or ions are 'excited' by the heat energy supplied and the intensity of movement becomes greater. At some point the oscillation is sufficient to permit atoms to move out of their equilibrium position and hence the structure may break down completely. When this occurs the individual atoms

[1] Some substances, which are so highly resistant to high temperatures that they undergo scarcely any noticeable change, are said to be *refractory* and to have the property of *refractoriness* (see p. 777). Hence, broadly, refractoriness is the converse of *fusibility*.

or ions are virtually independent entities,[1] no bonds exist and the mass is capable of easy deformation. A liquid state has thus been formed.

The Melting Point of Simple Compounds

Although the process of melting involves the separation of the cations and anions and the destruction of its structure, it is difficult to predict the temperature at which substances will melt except for the most simple compounds.

There have been many attempts to correlate the melting point of a crystal with a specific property such as the lattice energy, interatomic forces, type and character of the bonding, but none has been completely satisfactory and of general application. Weyl[2] has developed an 'atomistic approach to the melting of simple compounds' in which he suggests that as the temperature is increased, the defects in the structure increase and some cations occupy interstitial positions. Consequently some cations approach more closely to others and cation to cation repulsion forces are developed. Before a completely disordered structure can form, these repulsion forces have to be overcome and Weyl suggests that the melting point of a crystal is characterised by the thermal energy at which the cations can pass one another.

The theory advanced by Weyl satisfactorily accounts for the melting point of simple inorganic compounds and for a detailed explanation the reader is advised to consult the original publication.

Expressed simply, the following factors influence the cation to cation repulsion forces and hence the melting point of crystals.

1. The size and charge of the cation (see Chapter II). The smaller the size and the larger its valency the greater will be the repulsion between cations. This factor may be expressed as Z/r^2 where Z is the charge and r the radius of the cation.

2. The relative size of the anion. Large anions 'screen' small cations and so restrict their close approach. The repulsion forces which can develop are reduced and hence compounds with large anions tend to have a lower melting point than those of the same cation with smaller anions (Table X.VI).

TABLE X—VI. MELTING POINT OF ALKALI HALIDES

Element	*Fluoride* (1·36 Å)	*Chloride* (1·81 Å)	*Bromide* (1·95 Å)	*Iodide* (2·16 Å)
Lithium (0·60 Å)	842° C.	606° C.	549° C.	446° C.
Sodium (0·95 Å)	988	800	740	660
Potassium (1·33 Å)	846	768	728	693
Rubidium (1·48 Å)	775	713	681	642
Caesium (1·69 Å)	684	626	627	621

With the exception of the lithium compounds and sodium iodide, the melting points of the alkali halides decrease with increasing cation size and also with increasing anionic size.

[1] It is not strictly accurate to regard a liquid as being composed of individual units, because there is undoubtedly grouping of some type in most cases. However, if linkages do occur they are irregular and non-continuous and the structure is completely random.

[2] Weyl, W. A., *O.N.R. Tech. Report*, No. 52, 1952.

3. The numerical ratio of anions to cations. When the number of cations in a crystal is small, the possibility of cations approaching each other is also small, and hence the melting point will be low. This is borne out by the following series:

Boron	Fluoride (BF_3)	oxide (B_2O_3)	nitride (BN)	carbide (B_4C_3)
Melting point	−128° C.	450° C.	decomp. at 2000° C.	2350° C.
Anion to cation ratio	3:1	1·5:1	1:1	0·75:1

This same conclusion also applies to those compounds where the anion is of one type, but where the anion:cation ratio is increased by raising the charge on the cation. For example, in the following series of oxides, the cations are of comparable size but the melting point decreases as the anion:cation numerical ratio increases.

Oxide	MgO	Al_2O_3	SiO_2	P_2O_5	SO_3
Melting point	2800° C.	2030° C.	1713° C.	570° C.	45° C.
Anion:cation ratio	1:1	1·5:1	2:1	2·5:1	3:1

There are, however, two opposing effects which give rise to apparent anomalies in the melting point of compounds. As the cationic size becomes smaller and its valency increases the repulsion force (Z/r^2) becomes greater, thereby tending to increase the melting point, but at the same time the anion:cation size and numerical ratio is also increased—a factor leading to a lowering of the melting point.

The influence of the two factors is well demonstrated in Table X.VII where the melting points of some oxides are compared on the basis of the order of elements in the Periodic Table (p. 58).

TABLE X—VII. THE MELTING POINT OF SOME OXIDES

	Group 1	*Group* 2	*Group* 3	*Group* 4
Period I	Li_2O > 1700° C.	BeO 2570° C.	B_2O_3 577° C.	CO_2 gaseous
Period II	Na_2O 1275 (subl.)	MgO 2820	Al_2O_3 2050	SiO_2 1728
Period III	K_2O 430	CaO 2572	Sc_2O_3 2430	TiO_2 1850
Period IV	Rb_2O 400	SrO 2430	Y_2O_3 2410	ZrO_2 2720
Period V	Cs_2O 400	BaO 1923	La_2O_3 2315	HfO_2 2812

The ions of the alkali elements in Group 1 are comparatively large so the 'screening' effect of the anions is not very great. The principal effect is the repulsion term Z/r^2 and as this decreases with increasing size of the cation, the melting point decreases accordingly. The alkaline earth oxides of Group 2 with the exception of beryllia follow the same general pattern of decreasing melting point with increasing cationic size. Because the beryllium ion is small (0·31 Å) the repulsion term (Z/r^2) is large, but the possibility of the close approach of the cations is not great, due to the large anion: cation size ratio. The melting point of the oxide is, therefore, lower than magnesium oxide. For most of Group 3 and all of Group 4 the melting point increases with increasing size of the cation indicating a greater screening by the anions of the smaller cations, and this factor outweighs the repulsive force of the cations.

In the halide series of Table X.VI the anions are larger than oxygen and their greater screening effect (i.e. the influence of the anion to cation size) accounts for the apparently low melting points of all the lithium halides and sodium iodide.

With the exception of the alkalies the oxides of Periods 1, II, and III decrease in melting point with increasing cationic charge, showing the effect of increasing the numerical ratio of anions to cations. This trend is, however, reversed in Periods IV and V where with the increased relative cationic size, the repulsion term (Z/r^2) becomes dominant and so the melting point is increased.

This atomistic concept of melting is of value in explaining many anomalies in the fusion of ceramic materials. If the atomic structure of a crystal is simple and all linkages and bonds between the component ions are identical, the energy required to cause breakdown will be similar at all parts and the material will have a definite melting point. In more complicated structures, involving several different kinds of atoms or ions and more than one type of linkage, some bonds will be broken at lower temperatures than others, and the fusion will occur gradually over a range of temperature even though the crystal is free from impurities.

Nature of the Liquid Phase in Silicates

The transition from the solid to the liquid state in clays and related silicates is a complex phenomenon and has presented formidable thermodynamic problems of interpretation. The latent heats of fusion, the viscosity of the liquid phase, refractive index changes have not followed predicted laws. Scientists have recognised that the liquids which result from melting silicate materials must be much more complex than those from other inorganic forms.

Toop and Samis[1] applied thermodynamic calculations similar to those used in polymer chemistry to analyse specific characteristics of silicate liquids. They deduced that macro-molecules of large size and complexity must exist in such melts. Furthermore the nature of the silicate molecule was dependent on the associated cations.

Masson[2] has extended their ideas and calculated parameters for certain cations, which are a form of rate constant. In a silicate melt containing divalent cations, it can be assumed that ionic forms will be present of the type

$$M^{2+},\ O^{2-},\ Si_nO_{3n+1}{}^{2(n+1)-}$$

[1] Toop, G. W. and Samis, C. S., *Can. Met. Quart.*, 1, 129, 1962.
[2] Masson, C. R., *J. Amer. Ceram. Soc.*, 51, 134, 1968.

The proportion of the complex ion and its nature can be adduced by measuring certain physical characteristics and by applying equilibrium calculations to measure the rate constant *k*. The smaller the value of the parameter *k*, the lower the proportion and the less complicated in structure are the complex ions in the melt.

In a pure silica liquid, for example, *k* approaches zero, so that the predominating silicate ion type is SiO_4^{4-}. Values of *k* determined for several divalent cation/silicate glasses were as follows

Ca^{2+}	0·003	Fe^{2+}	1·4
Pb^{2+}	0·2	Co^{2+}	3·3
Mn^{2+}	0·75	Ni^{2+}	(46)

One important practical advantage of this conclusion relates to the viscosity of the melt. Small, simple ionic forms flow readily whilst those of a complex nature give sluggish melts of high viscosity. Calcium-containing liquid silicates are much more fluid than those containing the other cations which were investigated by Masson.

In some of the divalent cation silicate melts with high *k* values, ionic complexes containing up to 5 atoms of silicon were thought to be present.

Latent Heat

The process of melting involves an energy change which is virtually equivalent to the force required to break the bonds within the crystal. Consequently when liquid is formed, energy is absorbed from the surroundings and this is usually manifested as a heat effect. If a substance is steadily raised in temperature an 'arrest' invariably occurs in its heating curve and if the melting point is sharply defined, no further increase in temperature will occur until the substance has been entirely liquefied. The energy absorbed during melting or, conversely, that evolved during crystallisation, is termed the *latent heat of fusion.*

When the liquid form changes to a vapour the same phenomenon occurs. Heat has to be supplied to change the physical state of the material whilst the temperature remains constant and this is known as the *latent heat of evaporation.*

The latent heat of a substance is defined as the amount of heat in calories to change the state of a substance but without changing its temperature. The values for certain simple, well-known materials are given in Table X.VIII.

Most ceramic materials do not behave as simple compounds; on heating, the bond types are usually so varied that a long range of fusion occurs. In addition, the liquid formed is often so extremely viscous at temperatures near to the fusion point that complete disruption of the structure is not instantaneous. Furthermore, when the crystal lattice is mainly ionic, the energy required to separate the individual ions is invariably in excess of the bond strength itself and superheating effects are common. Similarly, on cooling, the crystallisation may be delayed or inhibited by a variety of causes which produce supercooling or 'retarded crystallisation'. For these reasons the temperature of fusion and the latent heat of ceramic materials are difficult to determine.

When another component is added to a substance the melting point of the

TABLE X—VIII. THE LATENT HEATS OF SOME SUBSTANCES

Substance	*Formula*	*Latent Heat of Fusion*			*Latent Heat of Evaporation*		
		Temp °C.	*ΔH cal.* *g.*	*mole*	*Temp ° C.*	*ΔH cal.* *g.*	*mole*
Carbon monoxide	CO	− 205·1	7·1	200	− 191·5	51·4	1440
Oxygen	O_2	− 218·4	3·3	106	− 183·0	50·9	1630
Methane	CH_4	− 182·5	14·1	225	− 161·5	122·6	1960
Sulphur dioxide	SO_2	− 72·7	32·2	2060	− 10·0	93·1	5960
Benzene	C_6H_6	+ 5·5	30·4	2360	+ 80·1	94·7	7400
Water	H_2O	0	80·0	1436	+ 100·0	540·0	9820
Aluminium	Al	+ 659	95·0	2555	+2467		
Sodium chloride	NaCl	+ 801	115·9	6800	+1413		
Iron	Fe	+1539	66·2	3700	+3000		
Silica	SiO_2	+1730	113·3	6800	+2200	—	—

latter is lowered unless a solid solution is formed (p. 588). This phenomenon is termed Raoult's law of the *depression of the freezing point* and the extent of the lowering (*dt*) is given by the formula

$$dt = \frac{RT^2M}{L}$$

where T is the absolute temperature; M is the molar concentration of the impurity; L is the latent heat of fusion; and R is a constant. This explains why eutectics, or points of lowest melting are formed between components.

E. W. Washburn[1] calculated the latent heat of fusion of cristobalite from the formula:

$$L = \frac{RT^2x}{\Delta t}$$

where L is the molar heat of fusion, Δt the lowering of the freezing point in some solution, such as the eutectic between cristobalite and alumina (= 160° C.) and x the mole-fraction of alumina in the solution which is 0·0882. From these figures, Washburn estimated that the latent heat of cristobalite is 6,800 calories per mole, or 110 calories per gram.

He gave the following figures for the latent heat of fusion of lime and magnesia:

TABLE X—IX. LATENT HEAT OF FUSION

Substance	*Fusion-point (Kanolt), °C.*	*Latent Heat of Fusion (cals. per gram)*	*Latent Heat of Fusion (cals. per mol.)*
Lime	2570 ± 3	490 ± 20 per cent	30,000
Magnesia	2800 ± 3	700 ± 15 per cent	30,000

[1] Washburn, E. W., *J. Amer. Ceram. Soc.*, 2, 1006, 1919.

The latent heat of fusion of some complex materials is very high, possibly on account of their decomposition at the fusion-point; in that case, the heats of decomposition would be included in the latent heat of fusion.

Table X.X shows some values of complex materials which have been derived by various means.

TABLE X.X. HEATS OF FUSION OF COMPLEX MATERIALS

Substance	*Melting-point* °*C.*	*Latent Heat* (*cal/g.*)	*Heat of Fusion* *cal/mole*	*Observer*
Anorthite	1532	100	+28,000	Vogt
Diopside	1225	100	22,000	Vogt and White
Enstatite	1375	125	25,000	Vogt
Olivine	1400	130	18,000	Vogt
Albite	1118	45	11,890	Waldbaum, D.R.
Sanadine	1200	43	11,100	Waldbaum, D.R.
Akermanite	—	90	—	
Calcium borate	—	49	—	
Leucite	—	26	—	

There is a relationship between the temperature of melting and the latent heat of a substance, but the parallelism is not entirely complete because the formation of complex groupings in the liquid may affect the determination of the absolute latent heat of fusion.

There is a linear relationship between the latent heat of evaporation expressed in calories/g.formula and the boiling point of substances, but this is rarely significant in ceramic studies.

The Mechanism of Fusion in Ceramic Materials

THERE are three principal reasons why ceramic materials fuse over a range of temperature:

1. Several different minerals may be present in any one sample, each of which may have a different melting temperature.
2. Many bond types may occur in one particular crystal.
3. Eutectics may be formed.

Sharp melting points are known in ceramic mixtures but they are rare and can only occur in the following cases:

1. A pure component or congruently melting compound with only one type of linkage throughout its structure.
2. A mixture of the composition of a eutectic point.
3. Specific mixtures corresponding to a maximum or minimum in a solid solution diagram (see Fig. X.9, p. 596).

Even in these cases, melting will take place at one particular temperature only when the rate of heating is extremely slow, and when the quantity of material involved is very small in amount. When large quantities are heated, especially if the substance has a low thermal conductivity, or contains impurities, signs of fusion may appear at a temperature much below that at which liquefaction is complete.

The nature of the physico-chemical reactions which take place during the fusion of one or more ceramic materials can best be learned from a careful study of the appropriate equilibrium diagrams (p. 584ff.), but it is important to realise that there is very little evidence as to the nature of the substances which are produced in the *earlier* stages of fusion, as most of the information which has been obtained relates to substances which have been either partially or completely fused and afterwards cooled prior to investigation.

The fusion point of ceramic substances is greatly complicated by the heterogeneous nature of these substances. Localised concentrations of individual components may give rise to partial fusion at temperatures well below those predicted by equilibrium studies. Conversely, a mass that is non-homogeneous may not liquefy until a temperature well above the theoretical fusion point because equilibrium is attained very slowly.

Rate of Fusion. It is impossible to devise simple formulae which will indicate the rate of fusion in terms of time and temperature because such fusion also depends very largely on the size of the grains of material in the surface exposed to heat or chemical action, and on the viscosity of the fused products; this viscosity depends on the nature of these products, some substances producing much more viscous fluids than others. Thus, magnesia produces a very viscous slag or glass and, consequently, the range of temperature through which fusion occurs (sometimes termed the *vitrification range*) is much longer where magnesia is the chief flux than when lime is present in considerable proportion. Thus, articles containing magnesia as a flux are less likely to soften and lose their shape than those in which lime is used.

The rate of fusion also depends on the solubility of the less fusible materials in the liquid portion. This, in turn, depends partly on their mutual reactivities—a fusible mobile silicate, rich in bases having a much greater solvent action than one—such as felspar—of a more neutral character. The action of felspar is very slow, so that it does not cause the ware to lose shape very quickly as is the case with some other fluxing minerals.

The vaporisation point of a substance is that temperature at which it is converted into a vapour at normal atmospheric pressure. Many substances vaporise at temperatures well below their boiling point. Thus, iron oxide does so at about 1250° C. and more rapidly as the temperature is increased. Chromium oxide, copper oxide, barium oxide, and zinc oxide also vaporise freely at the temperatures reached in firing stoneware and porcelain. At the temperature of the electric arc most ceramic materials vaporise.

The rate at which vaporisation occurs depends partly on the temperature and partly on the vapour pressure (p. 571), each substance acting independently.

The boiling point of a substance is the temperature at which it is converted into vapour at such a rate that its vapour pressure is that of a column of mercury 760 mm. in length i.e. 1 atmosphere. Most ceramic materials have boiling points which are so high as to be outside the range of attainable temperatures, so that their boiling points are not of great importance. The chief exception to this is water, which, when pure, has a boiling point of 100° C. or 212° F. The boiling point of any substance is reduced if the pressure of the atmosphere in which it is heated is also reduced, so that water can be boiled rapidly *in vacuo* at about 60° C. and slowly at lower temperatures.

If a liquid is heated above its boiling point in a closed vessel so that the vapour of the liquid cannot escape, the vapour pressure will increase in proportion to the heat absorbed. A typical example of this is the pressure of steam in an ordinary boiler. Water heated under pressure has more solvent action than that at the ordinary boiling point, and this is probably one cause of the decomposition of felspar and other minerals from which clay is derived. Any substances which dissolve in a liquid will increase the boiling point of the latter in proportion to the concentration of the solution.

The Solidification of Molten Masses

When a molten mass cools, it eventually becomes solid, but the physical and chemical nature of the product may vary greatly, according to the nature of the materials present and the rate of cooling.

The liquid state is a collection of atoms, ions or isolated groups of these with so great a vibrational energy that continuous linkages cannot form; hence *flow* can readily take place. As the temperature is decreased the movement within the liquid is reduced, more linkages can be formed and the mass becomes more viscous.

The atoms or ions in the liquid endeavour to combine with other units to preserve an electron balance, and at the same time, to form the most stable co-ordination (see Chapter II). The change from the virtually random arrangement of the liquid state to the symmetrical lattice of the crystalline medium, may involve considerable displacement. In most cases, the liquid phase is sufficiently fluid to permit the complete crystallisation at a definite temperature, but, particularly in silicate systems, crystallisation is sluggish, because of increasing viscosity, and the liquid may 'freeze' with an unsymmetrical structure and form a *supercooled liquid* which is known as *glass*; the structure of such material has been described in Chapter III.

Crystallisation is a nucleation reaction (p. 671). At particular points in the cooling liquid, centres of incipient crystallisation begin to develop at the melting temperature and from them solidification progresses throughout the entire mass.

The number of nuclei which form determines the crystallinity of the solid. If many centres are nucleated at about the same time, the resulting solid is likely to be a mass of small crystallites; but if the number of nuclei can be kept to a minimum, only large crystals will be produced.

Nucleation depends, to some extent, on the rate of cooling; if this is rapid, many more centres are likely to develop before crystallisation is complete, than if the temperature had been slowly reduced.

In many cases, the formation of nucleation-centres does not readily occur and the liquid may supercool. If this is undesirable, crystallisation may be induced by 'seeding' the liquid with solid particles which provide centres for solidification.

Therefore, depending upon the composition of the melt and the conditions under which it is cooled, the solid formed may be (*a*) a homogeneous non-crystalline glass, (*b*) a wholly crystalline mass of either large or small particles, or (*c*) a heterogeneous mixture of both crystals and glass. If the cooling is sufficiently slow, all liquids will crystallise but, in some cases, the process is so slow that it cannot be completed in a finite time.

Quite often a glassy phase can be *devitrified* (i.e. caused to crystallise) by maintaining the solid at a high temperature, yet below the melting point. Thus, vitreous or fused silica is a true glass, but if it is held at a temperature in excess of 1200° C. it slowly crystallises to either tridymite or cristobalite.

When, on cooling, the components of a liquid can form solid solutions, the composition of the crystals depends on the rate of cooling (see p. 595). Thus, slow cooling will more nearly simulate equilibrium conditions and the solid phase will be homogeneous throughout; rapid cooling and crystallisation results in 'zoned crystals' where the internal composition is different from the external layers.

Crystallisation is accompanied by the liberation of heat which is equivalent to the *latent heat of fusion* absorbed in the melting process. The energy change is the difference in 'entropy' between the random state of the liquid and the ordered crystalline lattice of the solid.

If a liquid is slowly cooled and crystallisation is instantaneous at one temperature an 'arrest point' will occur on the cooling curve. The temperature will not fall further until crystallisation is complete. Should a glassy non-crystalline phase form, however, no 'kink' is noticeable and the cooling curve is perfectly regular.

THERMAL EFFECTS ACCOMPANYING REACTIONS

AS stated in the earlier part of this Chapter, reactions are accompanied by an energy change which is usually manifested as a heat effect. Thus a heat change may occur when substances form, dissolve, combine, dissociate, melt or crystallise. The amount of energy involved in such changes depends on the initial and final states of the material. If the final state has a lower energy content than the original, heat will be evolved and the reaction is termed *exothermic*; conversely, if the final state has a higher energy content, energy will be absorbed and the change is *endothermic*.

Heats of Reaction and Formation

If sufficient energy in the form of heat is available to cause the vibrational amplitude of a bond in a crystal to be exceeded, disruption of the lattice will occur, one or more new compounds may be formed, and the energy change is known as the *heat of reaction*; on the other hand, the material may melt with an energy change known as the *latent heat of fusion*; should the material be completely dispersed into individual atoms or ions then the accompanying energy effect is termed the *heat* (*or energy*) *of dissociation*, which is equal to but of opposite value to the *heat of formation*.

The heat of reaction is usually further subdivided and is known by the type of reaction which occurs. Thus, when minerals hydrate, the energy liberated is known as the *heat of hydration*; when they crystallise it is known as the *heat of crystallisation*; the *heat of solution* is the energy change when substances dissolve in a liquid medium; the *heat of combination* refers to the specific case when two or more materials react to form another.

The energy change on reaction is equivalent to the difference in lattice energies between the initial and final states. It is not easy to calculate on this basis alone although it can often be derived if the heats of formation of the components are known.

The heat of formation is defined as the energy change accompanying the formation of a compound from its individual constituent atoms (see p. 563). Thus, when water (H_2O) forms from two atoms of hydrogen and one of oxygen, a heat change of 57,826 calories per gram molecule is involved. The heats of formation of various other compounds from their elements and of various silicates are shown in Table X.XI.

TABLE X—XI. HEATS OF FORMATION OF SOME SIMPLE SUBSTANCES

Compounds	*Calories per g. mole*	*Compounds*	*Calories per g. mole*
H_2O	−57,826	Na_2O	−99,160
CO	−26,400	K_2O	−86,260
CO_2	−94,100	FeO	−64,040
SO_2	−70,900	Fe_2O_3	−196,500
SO_3	−94,500	Fe_3O_4	−266,910
SiO_2	−217,500	$CaCO_3$	−288,500
MgO	−143,700	$MgCO_3$	−267,390
BaO	−133,100	$CaSO_4$	−342,400
CaO	−151,900	$MgSO_4$	−305,500
Al_2O_3	−399,100	UO_2	−267,000
B_2O_3	−279,810		

According to Hess's law, the quantity of heat evolved or absorbed by a reaction which takes place in two or more stages, is the same as would have been evolved or absorbed if the reaction had occurred in one stage. Thus, the formation of carbon monoxide from carbon and oxygen is accompanied by the evolution of 26,400 calories, the formation of carbon dioxide from carbon monoxide by the evolution of 67,700 calories; in the formation of carbon dioxide from carbon, the same amount of heat (viz. 94,100 calories) is evolved as if the formation had taken place in the two stages just mentioned. Hence:

1. The heat of formation of a compound is independent of its mode of formation.

2. The thermal value of a reaction is independent of the time occupied by the process.

3. The thermal value of a reaction is the sum of the heats of formation of the final products of the reaction less the heats of formation of the initial products of the reaction.

The amount of heat rendered latent when a compound is dissociated is often termed its *heat-content*. When heat is required to effect such a dissociation it is often represented by a figure (preceded by a minus sign) *after* the equation describing the reaction. Similarly, when heat is evolved in a reaction, it is represented by a figure preceded by a plus sign.

The heat of reaction may sometimes be calculated from the heats of formation of the components and the products.

As an example the decomposition of calcium carbonate may be cited:

				ΔH values
$CaCO_3$	→	Ca + C + 3O	− 288,500 calories	+ 288,500 calories
Ca + O	→	CaO	+ 151,900 ,,	− 151,900 ,,
C + 2O	→	CO_2	+ 94,100 ,,	− 94,100 ,,

Therefore the total reaction which may be conceived as involving (*a*) a complete breakdown of $CaCO_3$, and (*b*) the re-formation of CaO and CO_2, has an overall energy change of:

$$\Delta H + 288{,}500 - (151{,}900 + 94{,}100) \text{ calories}$$

i.e. +42,500 calories per g. molecule.

The difference between the initial and final states is thus 42,500 calories per gram molecule and the reaction may be fully represented as:

$$CaCO_3 \rightarrow CaO + CO_2 \quad -42{,}500 \text{ cals.}$$

The heat change is the enthalpy of the system (see p. 658).

The equation indicates that, in the decomposition of calcium carbonate, heat is absorbed and the reaction is thus endothermic. When calcium is oxidised to form lime or when carbon is burnt and forms carbon dioxide, heat is evolved and the reaction is exothermic.

It is not easy to calculate heats of formation of clays and related minerals from direct measurements. Barany deduced the heat of formation of muscovite mica by dissolving it in hydrofluoric acid, measuring the heat change and deducting the heats of formation of the fluorides which form. He postulates a value of −1,400,000 cal/mole formula.

The latent heats of fusion of materials have been discussed on p. 690.

CHAPTER XI

CHEMICAL CHANGES IN CERAMIC MATERIALS

CERAMIC materials in an *unfired, unburned* or *green* condition are available in three states:

1. As raw materials—chiefly natural minerals, such as clays, sand, felspar, water, etc.

2. As prepared materials, such as those used as *bodies, engobes, frits* and for *glazes, colours*, etc.

3. As ceramic articles such as bricks, tiles, pottery, etc., prior to their being heated in an oven or kiln.

The chemical and physical changes which occur in such materials and articles are so numerous and varied that they are conveniently divided into two groups:

(*a*) Those which occur at 'ordinary' temperatures—a term used rather loosely to include temperatures below 110° C.

(*b*) Those which occur at higher temperatures, and particularly such as are reached in pottery ovens, kilns and furnaces.

Some physical and chemical reactions in and between ceramic materials start at or below ordinary atmospheric temperatures, but most of them then proceed so slowly that it is convenient to take a rather higher temperature (105°–110° C.) because this is the one used in chemical and physical laboratories for drying many materials and test-pieces. The reactions proceeding between 15° C. and 110° C. are almost wholly those which proceed at ordinary atmospheric temperatures though more rapidly at about 110° C.

Whilst there is no sharp dividing line between ordinary and high temperatures, the figure of 110° C. is particularly convenient in considering ceramic materials.

CHEMICAL CHANGES AT ORDINARY TEMPERATURES

THE *chemical reactions* to which ceramic materials are subjected at ordinary temperatures are chiefly due to: (*a*) water; (*b*) acids; and (*c*) alkalies.

EFFECTS ON RAW MATERIALS

Water has no rapid chemical effect upon most ceramic materials at ordinary temperatures, with the exception of certain metallic oxides such as lime, magnesia and calcined dolomite, which are hydrolysed more or less rapidly by it.

Magnesia, when calcined at high temperatures, is converted into the stable oxide, periclase, and in this form, it is relatively unattacked by water. Lime, however, cannot be stabilised in this manner, but hydrolysis can be reduced by coating the particles with tar; alternatively, the lime may be combined with silica to form tricalcium silicate, which does not react with water.

Caustic magnesia is very soluble in water and still more so in slightly acidified water, but magnesite bricks made of intensely calcined magnesia or periclase are practically insoluble in water or dilute acids, except on prolonged exposure.

Most silicates, when subjected to the action of water for very long periods of time, tend to hydrolyse and although some change fairly rapidly, the reaction of most can be observed only in a geological age.

Water has a pronounced influence on many other minerals some of which are common constituents of naturally-occurring clays. Probably the most important effect of water on clay is its ability to form ions from soluble components. The base or cationic exchange capacity (see p. 263) is of great importance in a clay, because it can greatly influence the texture. The type and concentration of absorbed cations is largely determined by the conditions during the deposition of the clay but they can be markedly altered if at some stage the clay has been subjected to leaching by percolating water, particularly if it is charged with cations of a different type. Calcium-bearing solutions have been known to assist in the consolidation of a shaley deposit, solely by a cationic exchange altering the colloidal properties of the clay.

The leaching power of water can also break down mineral components, e.g. felspars, which 'weather' to form clay minerals under the influence of climatic conditions. Most alkali-bearing minerals are weakly ionised, and prolonged exposure to moisture is sufficient to cause a complete alteration of the mineral composition. It is likely that by a similar process some hydrous micas are produced by the leaching of primary micas which are richer in alkali.

Water is an intermediary through which other constituents may become active. Bacteria thrive in moisture and assist in the formation of carbonaceous components in soils and clays. Their action may, in turn, liberate acids which then attack other minerals which may be present.

The action of *carbonated water* (i.e. a solution of *carbon dioxide* in water) is, in some cases, relatively rapid, as is the action of rain water (containing carbon dioxide) upon limestone, basalts and olivines.

Water has a considerable influence on many other minerals, some of which are common impurities in natural clays. Thus, iron oxides may be converted into hydrates. Pyrite, FeS_2, readily oxidises in the presence of moisture and forms soluble sulphate. The physical effects of water in clays, after freezing, are described under 'Weathering' in Chapter VIII.

Acids may come into contact with ceramic materials in a variety of ways. Peaty or organic acids are frequently produced by bacterial action on carbonaceous constituents and can directly attack clays with which they are in contact. Rain water falling in an atmosphere rich in sulphurous and similar fumes may absorb sufficient gases to become distinctly acid. Many ceramic products are made specifically to withstand the attack of acids and are used as containers and transporters of such products in industrial practice.

Many acids will dissolve certain substances in clays and will alter their chemical composition as well as their physical form. Alkali-, alkaline earth-, and iron compounds are more readily attacked, although when in sufficient concentrations, strong acids decompose some of the clay minerals.

Hydrochloric acid attacks many calcareous and ferruginous substances, including dolomite, magnesite, calcite, iron oxides, olivine, serpentine, chlorite, nepheline, epidote, leucite and apatite; some other phosphates, monazite, etc., are attacked to some extent. Some sulphides are also dissolved.

Sulphuric acid when hot and concentrated attacks clays and other hydrated alumino-silicates and also most basic materials. When clay is heated for a long time with concentrated sulphuric acid the clay is decomposed, with the separation of silica and the formation of aluminium sulphate.

Sulphuric acid also has the power of improving the working qualities of some shales. Only a very small proportion of acid is required—about 1 per cent of the water used to soften the clay increases the modulus of rupture and the speed at which the clay-paste can be extruded through a die. No scum is formed as a result of adding the acid.

Nitric acid behaves in a similar manner to hydrochloric acid. In addition, it decomposes sulphides forming nitrates.

Phosphoric acid attacks silica glass and many silicates at temperatures above 400° C., forming the corresponding phosphates and liberating silica. Calcined and crystalline silica is not attacked by phosphoric acid below 300° C., but precipitated silica is attacked at lower temperatures.

Hydrofluoric acid decomposes many siliceous minerals which are unaffected by other acids, silicon fluoride being volatilised.

Schwarz[1] examined the relative rates of attack by hydrofluoric acid on the various silica minerals. Samples of quartz, cristobalite, tridymite and vitreous silica were prepared to a standard equivalent particle diameter of 0·04 mm. They were treated for $\frac{1}{2}$ or 1 hour at 100° C., in solutions of various strengths and the loss in weight recorded. The results are shown in Table XI.I along with values for a similar treatment with sodium carbonate solution.

TABLE XI.—I. SOLUBILITY OF SILICA MATERIALS (%) (After Schwarz)

	5% Na_2CO_3 ($\frac{1}{2}$ hour)	1% HF (1 *hour*)	5% HF ($\frac{1}{2}$ *hour*)
Quartz	2·11	5·2	30·1
Tridymite	2·77	20·3	76·3
Cristobalite	—	25·8	74·3
Vitreous silica	—	52·9	96·6

Hydrofluoric acid or a similar derivative is occasionally employed to dissociate a complex silicate mixture and so facilitates its analysis. A few drops of concentrated sulphuric acid are added at the same time so that other radicals are converted to sulphates and the silica evaporates as silicon tetra-fluoride which is volatile and escapes to the atmosphere. This treatment should be used with care, as many minerals are not decomposed under these conditions and others only slightly so. Mullite and felspars, both of common occurrence in clays and fired ceramic materials, are scarcely attacked by such reagents.

Sodium (and *potassium*) *bisulphate* when fused at temperatures of a dull red heat readily dissolve many silicates, but have no effect on quartz or the other allotropic

[1] Schwarz, R. Z., *Z. Anorg. Chem.*, 76, 422, 1912.

forms of silica. This is the basis of the Trostel and Wynne[1] method for the chemical estimation of quartz in clays.

Silica is very resistant to all acids, except hydrofluoric acid and phosphoric acid at high temperatures.

Silicon carbide is slightly attacked by hydrofluoric acid, but not to any great extent.

Alkalies usually combine with acid substances, but their action in the cold upon silicates and alumino-silicates is slight, and it is very feeble in the case of crystalline silicates. Freshly-precipitated hydrated silica is readily soluble in alkaline solutions. Flint is readily soluble in alkaline solution at 200° C., this fact being used in one method of making sodium silicate.

Some glasses and slags are readily attacked by hot solutions of alkalies, though some 'chemical glassware' and some glassware used for cooking utensils are highly resistant to such solutions.

Alkalies do not greatly affect silicates in the absence of water at temperatures below red heat or until the alkali salts themselves begin to melt. Reaction may then be quite rapid and most minerals are decomposed. Sodium carbonate is particularly reactive and is the material used for the fusion of silicates prior to a chemical analysis.

The effect of reagents on most ceramic materials is not usually very distinct on account of the heterogeneous nature of the matter and the limited extent to which they may be attacked. Only in isolated instances can one mineral be separated from the rest, as before one is completely dissolved others will have begun to decompose.

Effect of Pressure on Aqueous Reactions

Whilst water, dilute acids and alkalies attack clays and similar materials only to a small extent under normal conditions, the behaviour under high pressures can be much more pronounced.

Brindley, Roy, Tuttle and their co-workers at the Pennsylvania State University, Holmquist and many other workers have shown that many reactions of decomposition and synthesis within the clay mineral groups can take place under pressurised hydrothermal conditions. Much of this work has been summarised in Chapter III.

That water under high pressure can have a marked influence on many other materials has been demonstrated by Tuttle and England[2] who proved that cristobalite could be melted hydrothermally at 1475° C., under a pressure of 300 atmospheres, whilst quartz melted at 1025° C., under 1400 atmospheres in saturated water vapour.

Holmquist[3] has studied the effect of water vapour on the transitions between the silica minerals. At pressures between 100–800 atmospheres, water induced disordered states in cristobalite and tridymite and brought about transformations after relatively short periods at 400° C.

[1] Trostel, L. J. and Wynne, D. J., *J. Amer. Ceram. Soc.*, 23, 18, 1940.
[2] Tuttle, O. F. and England, J. L., *Bull. Geol. Soc. Amer.*, 66, 149, 1955.
[3] Holmquist, S. B., *J. Amer. Ceram. Soc.*, 49, 397, 1966.

In studying the conversion of tridymite to quartz under the above experimental conditions, Holmquist noted that maximum reaction occurred at about 200 atmospheres pressure and was almost complete in 2 hours.

There is no doubt that water at high temperatures and pressures can markedly affect the surface structural characteristics of many silica minerals and thus has a pronounced effect on their properties.

The enhanced reaction of bases with acid materials in the presence of water at elevated temperatures and pressures is utilised in the manufacture of sand-lime bricks.

The raw materials used are graded silica in the form of sandstone or, even better, flints and calcined lime which for convenience is frequently slaked before being used.

The composite mixture is pressed into the required shapes which are then processed in an autoclave operating at temperatures slightly in excess of 100° C., and at fully saturated steam pressures up to several atmospheres. Under these conditions, calcium silicate develops at contact points between the sand and lime particles so that a hard hydraulic bond develops throughout the brick shape. The following reactions may occur:

$$Ca(OH)_2 + SiO_2 = CaOSiO_2H_2O$$

$$CaOSiO_2H_2O + Ca(OH)_2 = 2CaO.SiO_2 + 2H_2O$$

$$Ca(OH)_2 + 2SiO_2 = CaO2SiO_2H_2O$$

The briquetting of ores with lime, soda, and water-glass are similar instances of the action of alkalies upon silica in the presence of water.

Chemical Reactions of Calcined Materials

The production of new chemical and physical states when clays are heated often results in a material with a very diffierent chemical reactivity from that of the original form.

When clays are heated to different temperatures, the reactivity of the product when it is cooled to room temperature shows many unusual characteristics.

Acid Leaching. Jaffé[1] and other workers at Leeds University have investigated the extent to which kaolinite and other clays can be leached by acids following calcination. After treatment at various temperatures, decomposition reactions took place which liberated alumina from its normal fixed lattice positions and enabled it to be taken up freely into acid solutions. Fig. XI.1 illustrates the changing activity of kaolinite towards acid as a function of pre-calcination treatment.

The raw clay was attacked only slightly at room temperature, but the reaction was greatly enhanced when the clay was heated in the region of its decomposition temperature (*circa* 500° C.). The temperature and duration of heating were significant and, provided that maximum dissociation had been achieved, the lower the temperature of calcination the greater the content of alumina which could be

[1] Jaffé, D., 'The Effect of Heat Treatment on the Reactivity of Minerals', Leeds, University, Ph.D. Thesis, 1969.

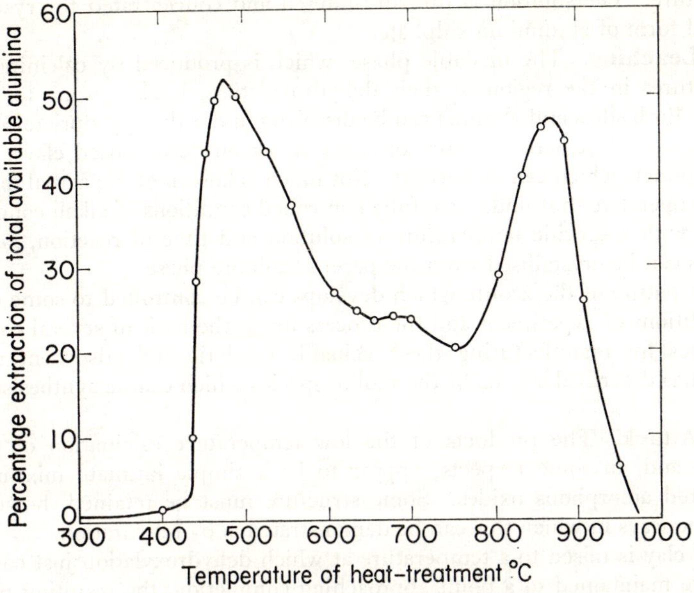

FIG. XI.1. THE EXTRACTION OF ALUMINA FROM HEAT TREATED KAOLINITE

Conditions of experiment (a) Pre-calcination for 5 hours
(b) Leaching with 5N HCl for 2 hours at 50° C.
When 5N H_2SO_4 was used, 95–100% Al_2O_3 was extracted from the samples precalcined at 460–540° C.

extracted. Under certain conditions of experiment virtually all the alumina could be removed from the calcined clay.

Above the decomposition point, further heat treatment caused a marked lowering in activity of the calcined clay. Jaffé proved that this was due to the partial recrystallisation of alumina to the γ-form. When kaolinite was heated to about 940° C.—approaching the temperature of the exotherm noted in differential thermal analysis—a further region of pronounced chemical reactivity was noted.

The appearance of mullite as a crystalline phase which occurred after treatment at higher temperatures signalled the disappearance of most of the chemical activity.

Jaffé also observed similar effects when testing fireclays and in these cases, the solution of iron oxide by acid media closely paralleled that of alumina—even in the region of the peak at 940° C. He theorised that in these cases, iron was probably an integral part of the kaolin mineral lattice where it replaced aluminium. There was further evidence to suggest that the iron was in a ferrous form.

The ease of removal of alumina from clays after calcination in the temperature region 500–700° C., has been utilised by the Laporte Industries Limited in the Spence process for manufacturing aluminium sulphates. Clays of high alumina content are extracted by sulphuric acid after pre-heating to the most convenient temperature. The solution is pressure-filtered and concentrated to crystallise the hydrated form of aluminium sulphate.

Alkali Leaching. The unstable phase, which is produced by calcining clays to temperatures in the region of their dehydroxylation, is also prone to attack by alkalies. Both silica and alumina can be dissolved under the appropriate conditions.

The most fascinating feature of alkali attack on decomposed clays, however, is the products which can be derived. Not much is known of the actual mechanism which is operative, but under carefully controlled conditions of alkali concentration coupled with a specific temperature of solution and time of reaction, *zeolite-type* products can be crystallised from the parent kaolinite phase.

The nature of the zeolite which develops can be controlled to some extent by the conditions of experiment and the process forms the basis of several commercial techniques for manufacturing these valuable catalytic and adsorbent materials. Barrer[1] has described several of the zeolite species which can be synthesised in this way.

Water Attack. The products of the low temperature calcination of clays are unstable and, in some respects, appear to be a simple intimate mixture of the dehydrated amorphous oxides. Some structure must be retained, however, and one of the ways in which this can be demonstrated is by rehydration.

If a clay is raised to a temperature at which dehydroxylation just commences and there maintained to a point approaching completion, the resulting product is extremely hygroscopic. It is impossible to cool down such a material to room temperature without water being taken up from the surroundings. The author has not yet found a desiccant more powerful than clay which has been pre-calcined to the critical dehydroxylation temperature.

If a sample is transferred to a balance pan immediately on removing from the furnace, an almost instantaneous weight gain can be recorded as soon as the material cools to below 100° C.

The gain in weight on exposure to atmosphere is rapid in the initial stages but continues at an appreciable rate for several hours and can still be detected after several days. Table XI.II shows the type of results which are obtained and the criticality of the pre-calcination temperatures. In every case in these experiments the sample was pre-calcined for 5 hours.

TABLE XI—II. REHYDRATION STUDIES OF PRE-CALCINED KAOLINITE

Temperature of Preheat ° C.		110	460	480	500	1000
Total weight loss (from 110° C.) %		0	6·4	13·3	13·4	13·4
Rehydration weight gain %	10 min.	0·5	1·8	2·9	1·1	0·4
	20 min.	0·7	2·5	3·8	2·0	0·5
	1 hour	0·9	2·8	4·8	2·3	0·6
	8 hours	1·0	3·4	5·2	2·5	0·6
	24 hours	1·1	3·7	5·4	2·7	0·7
	168 hours	1·2	4·1	5·6	2·8	0·7
Autoclaved		1·5	6·1	10·7	3·7	0·9

[1] Barrer, R. M., *Trans. Brit. Ceram. Soc.*, 56, 155, 1957.

Differential thermal analysis tests on the rehydrated products of the sample precalcined at 480° C., showed a peak at the normal kaolinite dehydroxylation temperature, but this accounted for only one-third of the water gain. Other peaks in the 175–280° C., region—associated with amorphous alumina and silica—accounted for the remainder.

The same specimens were autoclaved at 150° C./54 lbf/in^2 steam pressure and after eight hours' treatment a larger weight gain was recorded, as is shown in the table. These values were recorded after the samples had been dried at 110° C.

The material which had been pre-calcined at 460° C., had regained almost all its original loss in weight; that heated at 480° C., had recovered a high proportion. Differential thermal analysis tests on these materials showed a peak in the normal dehydroxylation region of kaolinite although it was not of the same shape nor did it occur at the identical temperature. It corresponded in size to about 85 per cent of the original kaolinite value in the 460° C., pre-calcined sample, and 64 per cent in the case of the 480° C., pre-calcined sample.

Similar results have been reported for other alumino-silicates by several authors.

Solubilities of Glasses and Glazes

Because of their random crystallographic nature (p. 170) glasses and glazes have an inherent tendency to be more soluble than other silicates of mineral form. A feature of glass structures is the large holes in the lattice which can admit foreign ions and expose a large surface area of silica groups to attack.

By suitable combinations of alumina and silica and the appropriate content and balance of fluxing elements, most glasses and glazes can be manufactured with low solubilities and a high resistance to both acid and alkali attack (p. 344). In some cases, however, as in high-gloss glazes, some compromise is essential and an increased solubility is frequently unavoidable in order to achieve aesthetic acceptability.

One of the most serious problems with high-gloss glazes was lead solubility but this was overcome by ensuring that the lead oxide/silica ratio was kept in the correct proportions. It has recently been reported in the U.S.A., that some lead glazes were being attacked by acetic acid and similar components of table condiments and sauces. It is extremely unlikely that a sufficient quantity of lead would be released in these circumstances to cause harm, but such glazes frequently contain small quantities of cadmium and other trace elements which might be more injurious to health. These would be slightly soluble under acid conditions and could be leached from certain glaze compositions.

Moore[1] has examined some of the factors which affect the resistance to chemical attack of boron glazes. He lists the ratio of $B_2O_3 : SiO_2$ as being of the greatest importance in assessing solubility; the proportions should not exceed 1 : 8 for good chemical resistance and, in addition, the ratio of alkalies + alkaline earth oxides to boric oxide should not be less than 1 : 5. The presence of Al_2O_3, ZnO and BeO increases chemical resistance and permits the addition of more alkalies and boric acid to a glaze composition.

[1] Moore, H., *Trans. Brit. Ceram. Soc.*, 55, 589, 1956.

Leachable Glasses. Douglas and El-Shamy[1] have studied the leaching characteristics of certain typical glasses and glazes and proved that hydroxyl ions from the aqueous medium can replace alkali and alkaline earth cations in the lattices. The reaction is diffusion-controlled and the amount of cations extracted is proportional to the square root of time.

Everest and his colleagues[2] have designed glass compositions which are specifically prone to leaching by mild acidic conditions. The purpose of the exercise was to convert a mineral species, which was most difficult to process by normal means, into a form from which the valuable elements could be extracted chemically.

The most successful operation of this nature was the treatment of certain beryllium ores[3] which were made into glasses by being fused with calcium carbonate together with amounts of alumina and silica. After cooling the glass, it was crushed and then leached with dilute acid. Beryllium ions rapidly diffused out of the glass phase and were concentrated in a soluble form in the aqueous medium.

A Polish process has been described for extracting pure alumina from clays by preparing a lime or soda-lime sinter from which the alumina can be extracted by mild acid leach. The product is used in the preparation of metallic aluminium.

Leachable glasses are the base material for a new series of silica compositions of low thermal expansion, high melting point and of light-weight. The composition of the glass is carefully chosen so that rapid and uniform leaching can proceed easily.

Undoubtedly, the techniques of leachable glasses will be used on an increasing scale both in mineral processing operations and to prepare new ceramic bodies.

Moisture Expansion. Because of the tendency to be attacked by water, the presence of glass in fired ceramic bodies can present serious problems. When reactions of this type occur, the body expands slightly, often with disastrous results.

The direct reaction between water and fired ceramic ware is slow but nevertheless appreciable. Terra-cotta roofing tiles slowly rehydrate when exposed to atmospheric weathering and after several years' service may have reabsorbed considerable moisture.

The action of absorbed water in fired clay ware is frequently the cause of defects often attributable to other sources. Many of the faults which develop in glazed tiles and terra cotta, such as cracking or shelling of the glaze and some cases of delayed crazing, are due to the absorption of water by the body. If the absorption is of a chemical nature, the body itself expands, but more often defects arise either because of soluble salt action or due to the physical changes in water on freezing.

Delayed dunting (or cracking of a ceramic body either during storage or in use) is often due to the action of water which is absorbed by the body from humid air, even though the ware is stored under cover.

Most clay products, after being fired, will expand permanently on being exposed to the atmosphere. In normal well-fired products the amount of expansion

[1] Douglas, R. W. and El-Shamy, T. M. M., *J. Amer. Ceram. Soc.*, **50**, 1, 1967.
[2] Everest, D. A., Napier, E. and Wells, R. A., 93rd Annual Meeting A.I.M.E., New York, 1964.
[3] Napier, E. and Lee, D. A., Scientific Report N.C.L./AE 219, 1965.

from this cause is slight and even after several years' exposure the increase in size is very small. Nevertheless as Hosking and Heuber[1] have shown, the expansion of brickwork by moisture absorption can be a cause of serious structural defects.

The amount of expansion varies between 0·05 and 0·20 per cent and is thought to be due to the rehydration of amorphous crystalline phases in fired clays or to the absorption of water in alkali-bearing glasses formed at low temperatures.[2] The amount of moisture expansion can be estimated by autoclave tests.

This samc type of effect is known to cause serious faults in glazed ware; *crazing*, which results in a multitude of fine hair-cracks in glazes, is due to the expansion changes arising from water attack on the body of the ware. The importance of moisture absorption in glazed ware has been discussed by Vaughan and Dinsdale[3] (see p. 795).

Wilkinson and Dinsdale[4] are of the opinion that a serious fault of some enamel glazed ware, known as 'spit-out', is due, at least in part, to the readsorption of water into a fired body. When such bodies are refired the adsorbed moisture is not released before the glaze becomes liquefied. Greatest retention occurs in defects and cracks at the body surface and these residual traces of water tend to escape in the final stages of firing when they bubble through the glaze and burst from the surface giving the characteristic crater-like hole.

Chemically Resistant Ware

Despite the adverse comments of the previous pages, fired ceramic ware in general is remarkably resistant to chemical attack of all types and is superior to most other materials. Without ceramic materials in construction, liquid transport, and container applications, much industrial and domestic development would be seriously hampered.

Brickwork is remarkably weather-resistant and well-fired ware is capable of withstanding all elemental conditions for indefinite periods. Clay-pipes are unsurpassed in their excellence for resisting chemical attack in water, acid and sewage transportation. Sanitary ware in ceramic materials is accepted as being superior to competitive materials in terms of its resistance to chemical attack.

Where conditions are likely to be severe, special clay-ware ceramics have been designed to withstand the action of most chemicals under atmospheric conditions.

Silicates and alumino-silicates are generally resistant to most acids. The attack of alkalies is more severe in that silica tends to dissolve slowly.

The ideal chemically-resistant article is one in which the porosity and permeability are at a minimum in order to reduce the surface area exposed to attack. The glass content of such ware should also be kept to a minimum or should be chemically adjusted to be resistant to corrosive agents. These two effects are not always

[1] Hosking, J. S. and Heuber, H. V., *Trans. 7th Int. Ceram. Congr.*, 311, 1960.
[2] Young, J. E. and Brownell, W. E., *J. Amer. Ceram. Soc.*, 42, 571, 1959.
[3] Vaughan, F. and Dinsdale, A., *Trans. Brit. Ceram. Soc.*, 61, 1, 1962.
[4] Wilkinson, W. T. and Dinsdale, A., *A. T. Green Book*, p. 255, B.C.R.A. publication, 1959.

compatible but nevertheless excellent chemically-resistant ware is manufactured.

The most acid-resisting of ceramic materials are hard porcelains, some stoneware which closely resembles hard porcelain, and Staffordshire blue bricks.

Where only a moderate resistance to acids suffices, glazed ware is usually satisfactory but to resist hot concentrated acids, the body of the ware should be fully vitrified and should contain a low proportion of alkali silicates. It should also have a minimum porosity.

Class A engineering bricks with a water absorption of less than 3 per cent are used extensively in the chemical industry for lining acid-tanks operating at high concentrations and temperatures. The ware produced at Accrington, Lancashire, is world-famous in this respect. It is made in a variety of shapes, some of which are of interlocking form to reduce jointing problems. Careful body preparation is essential for such ware, but it is possible to produce articles of excellent shape and dimensional tolerances with water absorptions not exceeding 1 per cent.

Acid-resistant tiles are made from clays which can be well-vitrified at comparatively low firing temperatures. The degree of acid resistance is largely dependent on the porosity or water absorption of the body. Those with a water absorption of less than 1 per cent will not be attacked by most concentrated acids (except hydrofluoric or its derivatives) even at comparatively high temperatures. They are also little attacked by concentrated alkali solutions although some clays may be bleached by such reagents.

The main problem in acid resistant ware is the jointing to be used between individual ceramic blocks. In some cases, this can be overcome only by using expensive resin compositions. It is one of the ironies of the ceramic trade that in some chemical vessels lined with fired clay-ware, the jointing may involve ten or more times the cost of the ceramic ware. Even then the life of the vessel may be limited by the preferential wear on the jointing!

Chemical Behaviour Induced by Grinding

In an earlier chapter, the influence of grinding on the surface properties of minerals has been discussed. The distorted layers which develop have a chemical reactivity far in excess of that which can be attributed to the increase in surface area.

Harris[1] has investigated in detail the release of octahedral ions from the lattice of chrysotile after various grinding operations and his results can be compared with observations on kaolinite and quartz by other workers. The results shown in Table XI.III are not directly comparable, because the conditions of grinding and leaching were not the same throughout, but at least they represent the order of magnitude of the effect. In each case the amount of cation released is calculated as a percentage of the original content of that cation.

In both chrysotile and kaolinite, the amount of silica capable of being dissolved under all conditions of experiment was low and never amounted to more than 2 per cent of the total unless the magnesium or aluminium were prior-leached by acid. After this treatment large amounts of silica could be dissolved by caustic soda.

[1] Harris, A. M., private communication.

TABLE XI—III. THE LEACHING OF BALL-MILLED MINERALS

		Unground	*Short duration ball-milling* (1–12 *hours*)	*Long duration ball-milling* (24–168 *hours*)
Chrysotile	pH 9	1·6	9·2	9·9
% magnesium dissolved at	pH 7	2·4	58·7	79·4
	pH 5	2·6	60·2	79·6
Kaolinite	pH 9	0·2	3·1	4·1
% aluminium dissolved at	pH 7	0·3	6·1	7·1
	pH 5	1·2	32·3	40·6
Quartz	pH 9	0·1	22·7	36·8
% silica dissolved at	pH 7	0·0	6·9	11·5
	pH 5	0·0	0·3	0·4

The results are thought to have a significant bearing on the health hazards, silicosis and asbestosis. Lung fluids never reach high acidic or basic values, but they are known to be capable of dissolving colloidal or highly reactive silica. The distorted layers induced on quartz grains by grinding are thus in an active state and capable of reacting in the lung.

The magnesium in ground chrysotile is readily leached by neutral solutions and thus leaves a highly active silica phase on the residual mineral body.

Kaolinite, although it is heavily distorted by grinding, is not significantly attacked except by solutions of comparatively high acidity—a condition not likely to occur in lung fluids. The alumina may dissolve but under neutral pH conditions it will immediately precipitate as a colloidal hydroxide gel. It is speculated that this coats active silica grains, so preventing interaction between the silica and lung fluids.

It is a comforting fact that silicosis and related lung diseases are comparatively rare in the clay industries, even when clays rich in silica are being processed. The alumina does appear to have some kind of suppressing action and the explanation advanced above may be the reason.

REACTIONS AT HIGH TEMPERATURES

THE chemical changes which occur when unfired ceramic materials are heated depend on many factors; these include:

(*a*) The mineralogical composition of the material.
(*b*) The chemical composition.
(*c*) The size-distribution of the minerals.
(*d*) The areas of contact of the reacting particles.
(*e*) The rate of heating at each stage or period.
(*f*) The maximum temperature attained during the heating.
(*g*) The duration of the heating at various stages or periods.
(*h*) The atmosphere surrounding the material during the heating.

The chief chemical reactions which occur when ceramic materials are heated have been described in general terms in Chapter X. They may be grouped as follows:

1. Those which depend upon the effect of heat alone on a simple substance, and include: (*a*) dissociation; (*b*) rearrangement of atoms; and (*c*) decomposition.

2. Those which depend on the reaction of two or more constituents of a mixture, including substances present as impurities or purposely added, as in a fine earthenware body, or as a catalyst. These include: (*a*) displacement; (*b*) double decomposition; and (*c*) direct combination.

3. Those which depend on the presence of a gas (such as oxygen) in the atmosphere causing *oxidation*; or on carbon monoxide in the kiln gases which may cause *reduction*. The gas may be formed by the action of the air on one or more of the constituents of the ceramic material (as carbonaceous matter in clay) or on the fuel or it may be introduced directly in the gaseous form.

It is often difficult to decide which reactions occur and, in some cases, there is some doubt as to whether a change should be regarded as *physical* (e.g. the loss of moisture) or *chemical* (e.g. the loss of hydrate-water). R. E. Grim[1] includes the loss of all kinds of 'water' (whether 'free' or 'combined') on heating under the term *dehydration* but most ceramic chemists prefer to restrict dehydration to water lost at temperatures below 110° C.[2] and all 'water' lost at higher temperatures as due to dissociation or decomposition, though there is some doubt as to how to describe the 'water' lost by some clays between 110° and about 640° C. There is probably no sharp distinction, though the assumption that the 'water' is lost at different distinct stages is convenient (see p. 446).

Changes in Composition on Heating Clays

The reactions which take place when clays are fired may be classified as follows:

1. Loss of hygroscopic and hydrate water.
2. Decomposition reactions:
   (*a*) loss of combined water;
   (*b*) decomposition of the oxy-salts;
   (*c*) oxidation processes;
   (*d*) reduction processes.
3. Re-forming reactions:
   (*a*) recrystallisation.
   (*b*) recombination.
4. Liquid formation.
5. Changes on cooling.

[1] Grim, R. E., *Clay Mineralogy* (McGraw-Hill, London, 1953), p. 190.

[2] The loss of water at its boiling point under normal pressure (i.e. at 100°) is theoretically useful, but is seldom complete in reasonable periods of heating so that a rather higher temperature (105°–110° C.) is generally used as the dividing line.

The extent to which these reactions occur in any one clay depends on the nature of the minerals present and the temperature to which the sample is heated.

1. **Loss of hygroscopic and hydrate water.** The *loss of hygroscopic water* is usually complete at a temperature of about 100° C. The amount to be liberated depends largely on the fineness of grain-size of the minerals in the clay, because the moisture is associated with the surface area. Adsorbed water is not strongly held (see p. 448) but some clay minerals, particularly those of the montmorillonite type may absorb water between the plate-like layers of which the ultimate crystal is composed. Such absorbed moisture is more tightly retained and is not released until a higher temperature is reached. Most montmorillonites will lose their inter-layer water at about 150° C., but the temperature depends on the type of absorbed cations which are present.

The amount of hygroscopic water retained by a clay is a function of the humidity of the atmosphere. Like all deliquescent substances, clays have a positive vapour pressure, hence the amount of moisture they can adsorb will be related to the partial pressure of water vapour in the surrounding atmosphere, or in other words, the relative humidity.

The amount of adsorbed moisture lost on heating to 200° C. by different types of clays which had previously been exposed to saturated air at 20° C. is given in Table XI.IV.

TABLE XI—IV. HYGROSCOPIC WATER CONTENT OF SOME CLAYS

Type	*Percentage weight lost*
Cornish china clay	1·2
Stourbridge fire clay	3·8
Brick clay	2·0
Ball clay (Dorset)	5·7
Bonding clay (Scottish)	7·4
Eureka halloysite	6·3
Wyoming bentonite (montmorillonite)	14·2

Water which is reversibly adsorbed may or may not be partially removed during the drying process. As outlined in Chapter IX, the principal aim in drying is to reduce the water used to facilitate moulding, to an amount below which no further shrinkage occurs. The amount of moisture still retained beyond this point may be considerable and it is not lost until the initial stages of firing in an oven or kiln. Articles made of clay are very friable when fully dry and it is, therefore, a technical advantage to place them in the kiln or oven whilst a small proportion of moisture is still present and the articles are in a black-hard rather than in a white-hard condition.

Hydrate water is defined as the water of crystallisation associated with salts and cations contained in the clay. The temperature at which it is removed is variable and there can be no sharp distinction between this form and the true combined water of typical clay minerals.

Two of the commonest examples of hydrates which occur in clays are the sulphates of calcium and magnesium. These minerals readily absorb water from their surroundings and form, respectively, gypsum $CaSO_4.2H_2O$ and epsomite $MgSO_4.7H_2O$, but revert to the dehydrated form at temperatures in the region of 200° C. The absorption of hydrate water is usually accompanied by a volume expansion and the converse process, i.e. the removal of water, by a shrinkage. The amounts of hydrating minerals which are present in clays are insufficient to produce a noticeable effect but in the fired products, calcium and magnesium sulphates are frequently troublesome in producing efflorescence and also internal disruption due to the expansion caused by hydration.

In ceramic usage, the initial period of kiln firing up to about 150° C. is termed the *smoking period* (see p. 731).

2. Decomposition Reactions

(*a*) Many of the minerals associated with clays contain water as an inherent part of their structure. As the temperature is progressively raised such minerals decompose at specific temperatures with the evolution of water. Reactions of this type are not usually rapid and the loss of water frequently takes place over a range of temperature depending on the rate at which the temperature rises. Kaolinite, for example, begins to decompose at about 415° C., but at this temperature the reaction is so slow that it is partially complete only after several weeks of maintained heat; at 550° C., however, complete loss of water is effected in about 30 minutes.

The amount of water lost by each mineral depends on its crystal structure and the number of hydroxyl units contained in the lattice. The theoretical amounts held in some typical hydrous clay minerals are given in Table XI.V which also includes the temperature at which this water is given off at a rapid rate corresponding to the 'peak' in the differential thermal analysis curve for that mineral. This is not the initial temperature of decomposition, which may be 150° C. lower, but this point is difficult to establish with precision.

TABLE XI—V. THE DECOMPOSITION OF SOME CLAY MINERALS

Mineral	*Chemical Formula*	*Temperature of rapid decomposition*	*Percentage water lost*
Kaolinite	$Al_2O_3.2SiO_2.2H_2O$	585° C.	14·1
Halloysite	$Al_2O_3.2SiO_2.2H_2O$	558° C.	14·1
Montmorillonite	$Al_2O_3.4SiO_2.H_2O$ (ideal)	678° C.	5·0
Muscovite mica	$K_2O.3Al_2O_3.6SiO_2.2H_2O$ (ideal)	856° C.	4·5
Hydrous mica	As above but less alkali and more water	550° C. (upwards)	6·0–10·0
Gibbsite	$Al(OH)_3$	325° C.	34·6
Goethite	$FeO(OH)$	350° C.	10·1

Each of these minerals gives a pronounced endothermic heat effect on decomposition, so much so that they can be identified and frequently estimated by differential thermal methods (see Chapter V).

When such minerals decompose with the evolution of water, the solid products which remain are virtually an intimate mixture of the component oxides. The internal arrangement still has many of the characteristics of the original mineral and can be rehydrated to some extent, but, chemically, the oxide components have separate entities.

(*b*) Other minerals which are present in clays may also decompose when heated. These are limited almost entirely to the oxy-salts such as carbonates ($''CO_3$) and sulphates ($''SO_4$), which tend to lose carbon dioxide (CO_2) or sulphur trioxide (SO_3) at some stage of firing and are then converted to the simple oxides.

Calcium carbonate, calcite ($CaCO_3$) is the most common mineral of this type in clays and it is a frequent constituent of deposits of marine origin, where it is present as shells or the remains of sea-water organisms. It decomposes at a temperature of 880° C. with a weight loss of 44 per cent.

Magnesium carbonate, $MgCO_3$, is associated with some clays although rarely in appreciable amounts, except in the form of the mineral dolomite $(Ca,Mg)CO_3$ Carbon dioxide is evolved from magnesium carbonate at about 790° C. with a weight loss of 52·4 per cent. Dolomite, on heating, behaves as though it were a mixture of calcium and magnesium carbonates; under certain conditions, the loss of carbon dioxide takes place in two stages.

Sulphates of calcium and magnesium are not of common occurrence in clays although they may frequently form on firing by the direct combination of sulphurous gases in the kiln with other alkaline-earth bearing minerals. Both these sulphates dissociate to some extent at high temperatures (*circa* 1100–1300° C.) and may contribute to the *bloating* frequently associated with clays which vitrify at low temperatures.

The mineral *alunite* (see p. 321), has been reported in clays in Egypt and North America. It dissociates above 550° C. and liberates sulphur trioxide and water; the former is not entirely removed even after firing at a high temperature. Clay containing small amounts of alunite is used in the United States as a raw material for the manufacture of ladle bricks. Liberation of sulphur dioxide at the temperature of service, gives rise to a large 'after expansion' which seals the joints between adjacent bricks and prevents the penetration of molten metal into them.

(*c*) There are several mineral components which are of frequent occurrence in clays that readily oxidise at high temperatures.

Oxidation may be defined as causing oxygen (or its equivalent) to combine with a substance, forming compounds which, in the simplest cases, are known as oxides. More strictly, oxidation is a change whereby the proportion of the electro-negative constituents in a compound is increased as, for example, in the conversion of Cu_2O to CuO or of $FeCl_2$ to $FeCl_3$. The conversions of -ous to -ic compounds are all examples of oxidation reactions.

Oxidation plays an important part in the various reactions involved in the production of ceramic articles. The chief technical objects of oxidation are:

(*i*) The conversion of impurities during the burning process (i.e. carbon, sulphur, etc.) into a form which can be easily removed, as by their conversion into gases or soluble substances.

(*ii*) The oxidation of iron compounds either to render them more refractory or for the production of a desirable colour, as in the manufacture of red bricks, terra-cotta, etc.

The removal of *carbonaceous matter* is most readily effected by oxidation, i.e. by heating it in a suitable atmosphere in which it combines with oxygen, forming carbon dioxide and water, both of which escape as gases. The temperature of reaction varies with the type of carbonaceous material (see Chapter VI). Algae and the carbonaceous matter in oil shales burn readily at temperatures around 200° C. When these are present, as they are to a considerable extent in clays of the Oxford and Kimmeridge beds, their oxidation reaction contributes to a more economical and uniform firing of the kilns. Ligneous or peaty components are oxidised at a higher temperature, but not as high as that required for bituminous or coaly material which does not burn readily until the kiln temperature has been raised to about 700° C. Most fireclays contain a proportion of the last-mentioned type of carbonaceous material which, if it is present in appreciable quantities, is difficult to burn away. When anthracitic material is present in a clay, the oxidation reaction is extremely slow and such carbonaceous material is removed only slowly at temperatures of about 1000° C., in an atmosphere rich in oxygen.

In many cases, it is of great technical importance that all the carbonaceous material should be oxidised in the early stages of burning, otherwise superficial fusion may occur, sealing the pores, thus preventing the access of air to the interior and so making it impossible to burn off all the carbon. If this should occur—as is liable in clays containing bituminous and anthracitic materials—a 'black-heart' may be formed which could cause disruption of the fired shape when in service. This may be avoided only by raising the temperature slowly—particularly between 350° C. and 800° C. and ensuring a free circulation of air.

Sulphides are decomposed when heated, if the atmosphere is oxidising, sulphates or sulphur dioxide and an oxide being formed. Pyrite (FeS_2) loses 60–75 per cent of its sulphur at a temperature below 800° C. and the remainder above this temperature. The latter is removed very slowly, the evolution decreasing in rapidity as the temperature rises; usually some combination occurs and a corresponding proportion of sulphur is never expelled.

Ferrous compounds are oxidised to the corresponding ferric compounds or to red ferric oxide when they are heated in a current of air. A temperature of 900° C. is usually required for rapid oxidation, but if the temperature is very high the red ferric oxide (Fe_2O_3) may be decomposed and magnetic iron oxide (Fe_3O_4) produced.

Certain hydrates of iron which are of the ferrous form oxidise readily immediately the water is evolved. Siderite, $FeCO_3$, is also oxidised on decomposition.

(*d*) *Reduction*, as a chemical term, may be defined as the removal of oxygen (or its equivalent[1]) from a compound; it is, therefore, the converse of oxidation. More strictly, the reduction of a substance consists in an increase of the electro-positive charge or a decrease in its electro-negative charge.

Reduction plays both a useful and a harmful part in the chemical reactions

[1] The conversion of $FeCl_3$ to $FeCl_2$ is regarded as a reduction, though no oxygen is removed.

connected with ceramic processes. The chief reducing agents used in connection with ceramic materials are carbon monoxide and hydrogen (generally produced by burning coal with an insufficient amount of air), but finely divided carbon and several metals have a powerful reducing action when at a sufficiently high temperature. The presence of sulphur dioxide in the kiln gases, as occurs when a sulphurous coal is used, has a reducing effect upon any iron oxide present and may produce ferrous compounds. Sulphur trioxide, if present in the kiln gases, may attack the bases present in the clay and form sulphates. If an oxidising atmosphere is maintained these sulphates will persist and may produce a scum on the goods, but if reducing conditions arise they will be decomposed again.

Silica is reduced by carbon at temperatures above 1800° C., forming the element silicon and carbon monoxide. The reaction is more rapid in the presence of iron and some other metallic substances. In some cases, silicon carbide or silicon monoxide may be produced. Zirconia, when heated to very high temperatures, tends to be reduced by its combination with carbon, forming zirconium carbide.

The successive reduction of iron compounds may be represented by the following equations:

$$6Fe_2O_3 - O_2 = 4Fe_3O_4 = 4Fe_2O_3 + 4FeO$$
$$2Fe_2O_3 - O_2 = 4FeO$$
$$2FeO - O_2 = 2Fe$$

An instance of a *desirable* reducing action[1] is in the manufacture of biue bricks, where a vitrified mass of requisite strength is produced and also the blue colour which is characteristic of ferrous silicates. Some colours used in glazes require to be fired in a reducing atmosphere in order to produce the desired effect.

Reduction processes are sometimes used in the manufacture of refractory materials, such as silicon carbide by the reduction of silica in an electric arc:

$$SiO_2 + 3C = SiC + 2CO$$

The reaction begins, according to Lampen, and also to Tucker, Gillet and Saunders, at a temperature of about 1820°–1920° C.; other carbides and carboxides are prepared in a similar manner.

A reducing action may be *harmful* when it is produced accidentally, as it may result in the decomposition of a useful material and the production of undesirable fusible compounds. Thus, if fully oxidised iron compounds, such as ferric oxide, are reduced to the ferrous state, in contact with clay or silica, they immediately lower the refractoriness of the bodies in which they occur, because ferrous silicate is readily fusible and may, therefore, cause much trouble and loss. Carbon has a stronger affinity for oxygen than iron, and, consequently, little or no oxidation of the iron compounds can occur in the presence of carbonaceous matter. It is, therefore, necessary to use a highly oxidising atmosphere in the earlier stages of firing clays, etc., containing carbonaceous matter, in order to oxidise and remove it. At a later stage, when all the carbonaceous matter has been burned away, some or all of the iron compounds may be reduced to the ferrous state without damaging the ware,

[1] Other instances of desirable reducing actions are the manufacture of *multicoloured bricks*, *black bricks*, and the *sang-de-bœuf* colour of some glazed ware.

but great care is needed in the control of the furnace or kiln, as in a reducing atmosphere at high temperatures the ferrous compounds will attack the silica present and form slag-like masses of fusible iron silicate.

At temperatures above 1000° C., ferric oxide has an increasing tendency to dissociate and it will do so unless a considerable excess of oxygen is maintained. In many kilns this is not feasible nor is it always desirable. A brick clay will often develop a more attractive fired colour and will vitrify at a lower temperature if the amount of excess oxygen in the kiln gases is kept low. Great care is needed, however, to ensure that some excess oxygen is always present, otherwise the iron compounds will be reduced with consequent damage to the ware.

When ironstone particles are present in discrete grains in a brick and reduction conditions persist during firing, as they may when carbonaceous matter is present, the iron compounds may be reduced to metallic iron or to iron carbide. Many of the black specks observed when fireclays and coarse earthenware are fired are of this form.

This can be a most objectionable fault, not least when the body has to be glazed and then refired. This is a common procedure in the manufacture of glazed bricks, faience and glazed earthenware. The water from the glaze slip reacts with the reduced iron or the iron carbide and releases gases which cause 'blow-offs'. A large section of glaze may fly off the surface of the body, sometimes before the ware enters the kiln for the second fire. More frequently, the fault is detected only after the glaze coat has been fired, when an unpleasant crater will result.

Many of the 'black-cores' in fired clayware are due, not to undecomposed carbonaceous matter, but to iron in a reduced state. The latter can be recognised by a steely appearance and usually the dividing-line between the iron core and the clean body is sharply defined.

Carbon has a greater affinity for oxygen than sulphur; consequently, it may retard the oxidation of the latter when they are present as sulphides.

Water at high temperatures has a very strong reducing action and also lowers the refractoriness of any fireclay or silica in contact with it, the refractoriness being lowered roughly in proportion to the amount of iron oxide present.

The reaction processes which take place in the initial stages of firing serve to decompose most minerals which are present and in effect reduce the clay to an intima mixture of component oxides. In the latter stages of the burning cycle, crystalline compounds of an entirely new nature begin to form and spread throughout the mass.

3. Crystalline Formation at High Temperatures

(*a*) *Recrystallisation Reactions* are those in which the individual oxides undergo atomic transformations to form crystalline compounds.

Alumina, Al_2O_3, is an excellent example; the amorphous form crystallises to the γ- (gamma) crystalline modification at a temperature between 500°–1000° C. and at higher temperatures it is converted to the α- (alpha) form, known as corundum (see p. 184).

Silica recrystallises from an amorphous form to either *cristobalite* or *tridymite* at a temperature of about 1250° C. This change is sufficiently rapid to be detectable by differential thermal methods, provided a sensitive recording equipment is employed. In most, if not every case, cristobalite is the first crystallisation product of amorphous silica at high temperatures. Tridymite may re-form rapidly from the cristobalite provided conditions are favourable, but there is no evidence to suggest that this phase may be the first crystallisation product.

Many natural forms of *amorphous silica* exhibit rapid crystallisation at about 1250°. This is particularly so of chalcedonic or flinty varieties and, for this reason, such materials are not suitable for the manufacture of silica bricks.

Crystalline *quartz* is one of the few minerals of common occurrence in clays, which does not break down to an amorphous form at some stage of the heating process. This mineral tends to convert to one of the other crystalline modifications of silica on prolonged firing, especially if suitable catalysts are present. The high alumina content of most clays, however, exerts a retarding action, so that fireclays, even after firing to 1250° and above, may still be rich in unchanged quartz.

Some other oxides may also crystallise at high temperatures although such changes are usually insignificant in clays. It is more than likely that such oxides would combine with other components or form a liquid phase at temperatures well below their crystallisation point. Thus, *magnesia* slowly transforms at high temperatures from an amorphous non-crystalline variety to the cubic form, periclase, but, when in contact with silica and the other component oxides of clays, it combines readily to form either stable silicates or, at higher temperatures, a liquid phase.

Anatase and, less commonly, *brookite*, are crystalline modifications of titania which are occasionally found in clays. These, like quartz, are not decomposed by heat although they are readily converted into the more stable variety, *rutile*, at temperatures as low as 400° C. In the presence of associated minerals, the transformation temperature may be appreciably altered.

(*b*) *Recombination reactions* are those which take place between two or more of the amorphous or crystallised components of heated clays. They are, in the main, processes which take place almost entirely in the solid state and so, like recrystallisation phenomena, are slow in reaching completion.

The best known example in clay systems is the combination of alumina with silica to form *mullite* in accordance with the equation

$$\underset{\text{alumina}}{3Al_2O_3} + \underset{\text{silica}}{2SiO_2} = \underset{\text{mullite}}{3Al_2O_3.2SiO_2}$$

The amount of mullite which forms when a raw clay is fired depends largely on the alumina content. It should theoretically be at a maximum when pure kaolin clays are fired, thus

$$3(Al_2O_3.2SiO_2.2H_2O) = \underset{\text{(mullite)}}{3Al_2O_3.2SiO_2} + \underset{\text{(cristobalite)}}{4SiO_2} + 6H_2O$$

The idealised composition of a well-fired pure kaolinite body would be

mullite 64·0 per cent by weight
cristobalite 36·0 per cent by weight

Mullite itself consists of 71·8 per cent Al_2O_3 and 28·2 per cent SiO_2 by weight.

The temperature at which this reaction takes place depends upon many factors, such as the intimacy of the mixture and the presence of other components some of which accelerate and others which retard or even prevent the reaction.[1] Mullite can be detected in some clays which have been heated to only 800° C., but the rate of reaction is not rapid below about 1000° C. Comeforo, Fischer and Bradley[2] have shown by electron microphotographs that mullite develops in pure kaolinite crystals at the approximate temperature of the exothermic thermal 'peak' at 980° C. When it is first formed, mullite is of extremely small crystallite size and cannot be resolved even under the highest magnifications of an optical microscope. On prolonged heating, especially at a high temperature, long needle-shaped crystals are formed which interlock and intergrow throughout the mass.

Mullite is a very stable phase in alumino-silicate mixtures of all types and is usually the first product to form in any mixtures where the two components are present.

Navias[3] has reported that the theoretical amounts of mullite were formed in all cases when a range of ball clays were fired to Cone 10. At higher temperatures side-reactions and liquid formation reduce the amount of mullite but it is always the principal crystalline component of fired alumino-silicates (see also p. 743).

Many other crystalline products may be formed when clays are heated under favourable conditions, although in many cases they are a transient form which disappears into the glassy matrix on prolonged firing.

The reactions of *iron oxide* in a clay body are of especial interest because they are the main factors which contribute to the colour of the fired body. The most usual forms of iron compounds in clays dissociate at temperatures below 500° C. and if the atmosphere has been oxidising, ferric oxide only will remain. The colour of the mass at this stage will be red, but the tone and shade will depend on the amount of ferric oxide, its grain-size and the evenness of its distribution through the mass. At higher temperatures, reaction with silica occurs; if conditions are favourable, alumina may also be present in the crystalline phase and the mass then undergoes a pronounced change in colour. Definite crystalline modifications are difficult to identify because of the ease with which solid solutions are formed; furthermore, their existence is limited to a narrow range of temperature above which they melt to form liquid.

Clay minerals rich in iron oxide, such as the *nontronites* or *chlorite* types, form *fayalite* or iron-rich *olivines* (Fe_2SiO_4). *Pyroxenes* approximating to the composition $FeO.SiO_2$ are also produced at temperatures around 900° C. possibly from the decomposition of less stable fayalite. Rigby[4] has reported that *iron cordierite*, ($2FeO.2Al_2O_3.5SiO_2$) occurs in the form of very fine crystals in the black cores of imperfectly-fired fireclay bricks.

Clays which contain nodular ironstone and carbonaceous material may form iron carbonyl if they are burned too rapidly or under non-oxidising conditions. The presence of this compound is a source of blowholes or 'blow-outs' in

[1] Matthews, K. J. D., *Trans. Brit. Ceram. Soc.*, 68, 97, 1969.

[2] Comeforo, J. E., Fischer, R. B. and Bradley, W. F., *J. Amer. Ceram. Soc.*, 31, 254, 1948.

[3] Navias, L., *J. Amer. Ceram. Soc.*, 8, 296, 1925.

[4] Rigby, G. R., 'The Thin-Section Mineralogy of Ceramic Materials' (*B.C.R.A.*, 1953).

glazed ware which is bisque-fired prior to the glaze being applied; iron carbonyl gives off gas when in contact with water or when heated in an atmosphere of oxygen and there is frequently sufficient near to the surface to be glazed to cause serious defects.

Crystalline modifications of *calcium-bearing minerals* are also formed, especially in clays which are rich in limestone or chalk. *Calcium sulphate* frequently occurs on firing when calcium oxide is subjected to vapours or kiln gases rich in sulphur trioxide derived from sulphurous coal or other fuel, or from the decomposition of pyrites in the clay itself.

Silicates of calcium may, occasionally, be produced, although usually any free calcium oxide will combine with alumina and silica to form a liquid phase at temperatures below those at which crystalline modifications form rapidly. *α-dicalcium silicate*, $2CaO.SiO_2$ is a constituent of Portland cement and is the result of firing clay with lime; it is also present in some fired brick clays of marine origin where the lime content is high. *Gehlenite*, $2CaO.Al_2O_3.SiO_2$, has been identified in engineering bricks although it may in such cases be a crystallisation product from a liquid phase rather than the result of a direct combination of the individual components. *Felspar*, corresponding closely to *anorthite* ($CaO.Al_2O_3.2SiO_2$) is present in some fired clays, but more particularly when such products have been subjected to calcium-bearing slags at high temperatures.

Magnesium-rich clays occasionally form *forsterite* $2MgO.SiO_2$ or *enstatite* $MgO.SiO_2$ when heated around 1000° C. These minerals are not very stable in the presence of alumina and readily decompose or form a liquid. Grim and Bradley[1] have shown that magnesium-containing montmorillonites frequently exhibit a spinel phase (mainly $MgO.Al_2O_3$) as a transient product when heated to about 850° C.

Alkali-bearing minerals are not common in fired clays, principally because both sodium and potassium oxides form liquids at temperatures well below those at which recombination reactions occur. After slow cooling or on prolonged exposure to a moderate temperature, alkali-containing minerals may crystallise from the glassy phase. *Leucite*, $K_2O.Al_2O_3.4SiO_2$, for example, may be present in fireclays used as blast furnace linings and *nepheline*, $Na_2O.Al_2O_3.2SiO_2$, can be formed by the interaction of clays with soda vapour at high temperatures.

Crystallisation usually begins at small nuclei and hence the initial crystallites are extremely small. After long periods of firing at high temperatures these tend to grow or sinter unless otherwise destroyed by side-reactions (see also p. 748).

4. **Liquid Formation.** Liquid formation in clays commences at the temperature of the lowest eutectic point of the components in the mixture; then, with progressively increasing temperature, the amount of liquid phase increases rapidly. Alkali-bearing minerals or clays with a high content of adsorbed, exchangeable, alkali cations begin to fuse at low temperatures; a mixture of sodium, alumina and silica, for example, partially melts at 732° C.; the equivalent mixture with potassium replacing sodium shows incipient fusion at 695° C.

In most ceramic materials, little glass is produced below a temperature of 1000° C. unless a considerable proportion of alkali is present. The glassy matter then usually begins to develop at about 750° C., and increases fairly rapidly up to

[1] Grim, R. E. and Bradley, W. F., *J. Amer. Ceram. Soc.*, 23, 242, 1940.

about 1150° C., any felspar fusing readily about this temperature. The partial fusion continues rapidly up to about 1300° C., the fused material dissolving the smaller particles and thus increasing the volume of the fused product. After this, the fusion proceeds more slowly, with a steady rise in temperature, because the alumino-silicates are more viscous and flow less readily than the simple silicates and so require a longer time to effect an equal amount of corrosion.

Other cations also induce fusion in alumino-silicate mixtures at comparatively low temperatures. The eutectic temperatures for many of the common impurities in clay, when associated with alumina and silica are shown in Table XI.VI.

TABLE XI—VI. FUSION TEMPERATURES OF FLUXES IN ASSOCIATION WITH ALUMINA AND SILICA

Oxide	*Lowest Eutectic Temperature °C.*
Na_2O	732
K_2O	695
CaO	1170
MgO	1345
Fe_2O_3	1073

Although some liquid usually forms in most mixtures at about the eutectic temperature, the increase in amount depends on many factors which include:

(*a*) The temperature to which the mass is raised.
(*b*) The amount of fluxing elements.
(*c*) The grain sizes of the various minerals present and the evenness of distribution throughout the mass.
(*d*) The rate of increase in temperature.
(*e*) The time that the maximum temperature of heating is maintained.
(*f*) The viscosity of the liquid formed.

(*a*) From phase equilibria studies the increase in the amount of liquid with the *temperature* of heating may be predicted. In some cases, small variations in temperature do not cause a pronounced change, but in other cases the quantity of liquid may substantially increase over a small temperature range (see also p. 613).

(*b*) The *amount of fluxing elements* present in a clay will influence the extent of liquid formation as will be evident from a study of the appropriate equilibrium diagrams (see Chapter X). There is, however, a further complication in that some liquids readily dissolve other components at temperatures well below those predicted by equilibrium studies. Free silica, for example, may be incorporated into a liquid phase in large amounts, especially if it is of fine grain-size and where alkali oxides are the fluxing oxides. Under favourable conditions, one molar percentage of alkali oxide may liquefy up to 20 molar parts of silica and even more may be incorporated if small amounts of alumina are also present.

(*c*) The *grain size of the minerals* containing the fluxing elements is important because it determines the distribution of the molten liquid throughout the clay mass. Alkali-bearing micaceous minerals are of frequent occurrence in clays and the particle size of this component largely determines the vitrification rate at a particular temperature. If the mica is of small grain-size the alkali ions contained in the lattice can react more readily with the clay and form the maximum of liquid in the shortest possible time. On the other hand, coarsely grained mica tends to form isolated regions of liquid around the edges of the grains but the spread of liquid throughout the body is not very rapid.

(*d*) The *rate at which the temperature rises* affects the amount of liquid formed because combination reactions are slow and require a considerable time to reach completion. A slow increase in temperature affords more time for the liquid phase to diffuse through the body and therefore to cause maximum reaction.

(*e*) *The time that the maximum temperature of heating is maintained* determines to a large extent, the uniformity of the product. If insufficient time is allowed at the maximum temperature, the amount of liquid which forms in the exterior of the article may be greater than that formed in the interior. It is advisable that ware should be held at top temperature for as long as is practicably possible, especially if large shapes are being fired.

(*f*) The *viscosity of the liquid phase* produced influences the total amount formed because a highly mobile liquid penetrates and interdisperses throughout a body more quickly than one which is viscous and flows slowly and irregularly. The more mobile liquid will come into contact with a greater proportion of soluble components than the more viscous one and will form a greater quantity of liquid in the same time interval.

The viscosity of a silicate melt is dependent on its composition and temperature. In general, the viscosity at a particular temperature, will decrease as the amount of fluxing elements increases. Alumina and, to a lesser extent, titania exerts an unusual influence on the liquid produced on heating silicates. Both aluminium and titanium cations can replace silicon as a network former in glass structures (see p. 173). In small amounts these may reduce the viscosity of the liquid phase of such glasses, but if larger amounts are incorporated the mobility is usually reduced because of 'induced' structural forms in the liquid phase. Magnesium oxide and silica form extremely viscous liquids, but when alumina is added to such mixtures, it increases the mobility.

The liquid state is always a temporary one, so far as ceramic materials are concerned. As soon as the liquid begins to cool its physical properties change, its viscosity increases and any chemical reactions are slowed down and eventually cease.

Many changes occur in the cooling stage—chiefly resulting in the formation of crystals and of glass or a mixture of these. Most of the crystals are much too small to be visible to the unaided eye and many ceramic materials therefore appear to be amorphous.

5. **Cooling Reactions.** The chemical change which can take place on cooling in fired clays and other ceramic bodies has never received the attention which it deserves. It is still common practice in the industry to cool ware as fast as the kiln will permit provided that the ware does not dunt or shell or become discoloured.

The changes that can and do take place on cooling can have a significant influence on the properties of ware from many aspects.

At the maximum temperature of firing, a ceramic body may contain a large content of liquid arising from the chemical combination of fluxing elements with alumina and silica. The amount which forms is not easy to calculate, but Table XI.VII shows the percentage which might be anticipated in simple Na_2O–Al_2O_3–SiO_2 systems at different temperatures with varying amounts of alkali, provided that equilibrium has been reached.

TABLE XI—VII. PREDICTED AMOUNTS OF LIQUID AT VARIOUS TEMPERATURES IN Na_2O–Al_2O_3–SiO_2 SYSTEMS

% Na_2O	*Firing Temperature*			
	800° C.	1000° C.	1200° C.	1400° C.
1	5·2%	6·7%	14·3%	33·0%
2	10·5	13·4	28·6	66·0
3	15·2	20·0	42·9	100·0
4	20·8	26·7	55·6	100·0
5	25·2	33·0	71·4	100·0

These liquids are rich in silica; consequently on cooling they would tend to form glasses of the same composition as the liquid.

This is not necessarily the situation and if the cooling is sufficiently slow, crystallisation may take place forming cristobalite or tridymite and some mullite.

McGee[1] has studied certain aspects of mineralogical changes in clays which were fired to high temperatures and then cooled in various ways. The samples were all refractory clays ranging from a 35 per cent alumina body with 2·4 per cent total alkali oxide and 1·9 per cent alkaline earths, to a kaolinite with 44 per cent alumina, with 0·4 per cent alkalies and 1·4 per cent alkaline earths. After firing briquettes of these to 1350° C., they were cooled in a variety of ways and the mineralogical composition then calculated by X-ray diffraction methods. Some of the results are summarised in Table XI.VIII.

TABLE XI—VIII. MINERALOGICAL CONTENT OF FIRED FIRECLAYS

	Mullite	*Cristobalite*
Plastic Fireclay (34% Al_2O_3)		
100° C./hour cool	27%	16·5%
800° C. quench	28	14·0
Air quench	28	1·4
Flint Fireclay (44% Al_2O_3)		
100° C./hour cool	33	10·5
800° C. quench	33	9·0
Air quench	33	2·2
Kaolinite (44% Al_2O_3)		
100° C./hour cool	29	15
800° C. quench	28	14
Air quench	28	15

[1] McGee, T. D., *J. Amer. Ceram. Soc.*, 49, 83, 1966.

The bodies with the high liquid contents showed marked differences in cristobalite content after various cooling techniques, although the mullite phase was not unduly affected. Rapid cooling resulted in a high glass content in contrast to the more crystalline system developing under slow conditions.

This effect is apparent in kiln-firing practice. Modern rapid-firing processes involve heating clay products to a higher temperature than that formerly used when a long soaking time was acceptable. More liquid will form as the body vitrifies and this, coupled with rapid cooling practice, will result in a more glassy body.

As far as the author is aware there has not been any systematic study of the relative performance of products which were fired and cooled slowly and those produced from a rapid kiln programme.

In brick manufacture, fast-fired products are obviously more vitreous and brittle than their counterparts fired under slower conditions.

Some glazes tend to be more soluble when fired and cooled rapidly.

In some refractory products a high glass phase may be an advantage in that the problems of silica mineral inversions are largely eliminated. On the other hand, recrystallisation under service conditions might result in serious volume changes.

The Mechanisms of Thermal Changes in Clays and Related Minerals

Although the changes which take place on firing clays are well-recognised, the fundamental processes involved in the clay minerals, in particular, are far from fully understood.

1. Kaolinite.

Dehydroxylation. The water which is lost when clay minerals decompose is derived from the hydroxyl groups which are an integral part of their crystalline lattice. These have been fully discussed in Chapter III.

It is generally accepted that hydroxyl groups are removed from a crystal lattice by virtue of the vibrational energy which is imparted to them by heating. When the group attains a sufficient degree of agitation, a proton can split from one group and migrate to another thus producing a molecular unit of water. This is then lost from the lattice which retains one oxygen anion in place of the two original hydroxyl groups.

The temperature at which decomposition occurs depends on the affinity of the cationic forms which are attached to the hydroxyl groups and on the bond strength.

In most clay minerals, all the hydroxyl groups are associated in an identical fashion in the crystal lattice, but a notable exception is kaolinite. The structure of this mineral has been described fully in Chapter III, but for convenience a structural representation is reproduced in Fig. XI.2.

Hydroxyl groups in kaolinite have two specific and different lattice locations. In a unit layer, 6 OH groups are on the surface, associated only with aluminium atoms whilst a further two groups are at the centre of the unit, linked only to aluminium atoms but coordinated in a very different manner.

There has been considerable speculation in scientific literature as to whether

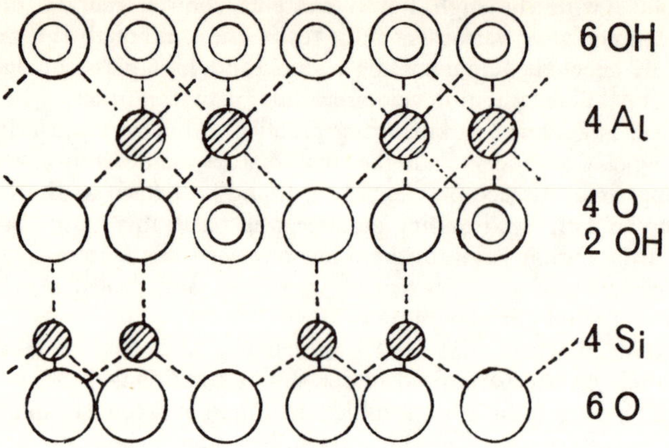

a

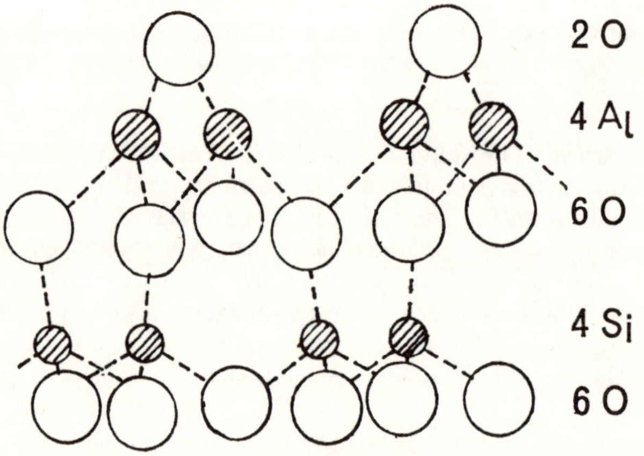

b

FIG. XI.2. THE STRUCTURE OF KAOLINITE
(a) Before Dehydroxylation
(b) Theoretical Condition after Dehydroxylation

the two types of hydroxyl groups behave differently on dehydroxylation. Most of the information suggests that they cannot be differentiated but there are some well-authenticated opposite views.

Murray and White[1] studied the kinetics of the kaolinite decomposition and

[1] Murray, P. and White, J., *Trans. Brit. Ceram. Soc.*, 54, 137, 1955.

concluded that the reaction had a first-order form. However these conclusions could not be applied to later stages of the reaction where considerable deviations from the linear plot were observed. It was suggested that when decomposition had approached 90 per cent totality, the lattice collapsed, which restricted the removal of the last traces of water.

Jaffé[1] obtained well-pronounced and reproducible humps when he investigated the weight loss of kaolinite when heated to near to the dehydroxylation temperature. In samples which were maintained at a constant temperature for 20–100 hours before being raised to a slightly higher value, there was a definite suggestion that a two-stage process was involved (see Fig. XI.3).

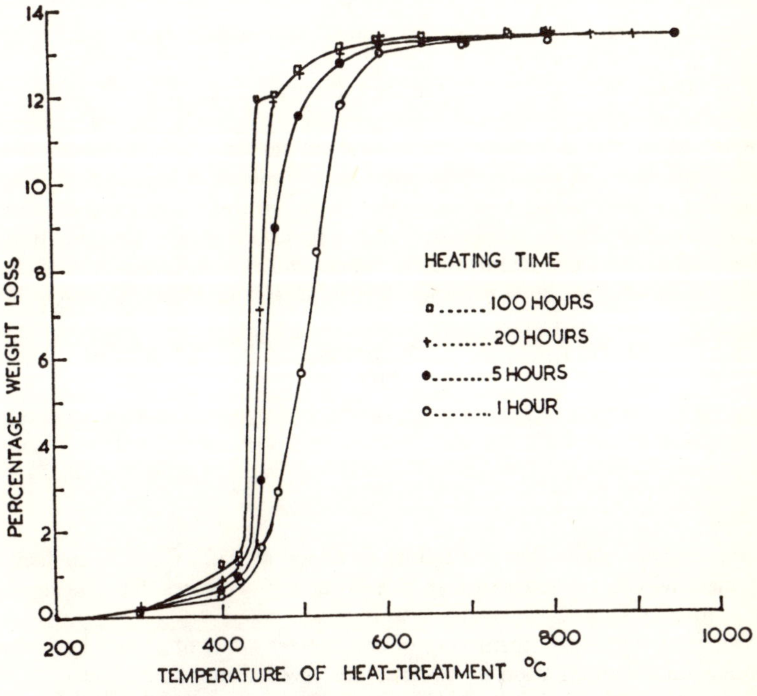

FIG. XI.3. THE STRUCTURAL DEHYDRATION CURVES OF KAOLINITE

The Decomposition Product. When water has been removed from the kaolinite lattice, the breakdown product does not change drastically in the spacial positions of cationic and oxygen groups. A transitional-type structure called *metakaolin* is thought to develop. In it, the structural bonds must be weakened to permit the enhanced chemical attack, but at the same time, the essential structural order of the

[1] Jaffé, D., 'The Effect of Heat Treatment on the Reactivity of Minerals', Leeds University, Ph.D. Thesis, 1969.

original kaolinite must be preserved because of the ease with which rehydration can take place, with its consequent reversion to a kaolinite-type mineral—albeit, not kaolinite itself.

The detailed structure of metakaolin has been examined by several authors, but they are not agreed on the nature of this transient phase. Brindley and Nakahira[1] favour a structure in which aluminium and silicon retain their lattice positions but co-ordinate in a different manner with oxygen. Taylor[2] prefers the idea of a specific oxygen ion arrangement with a random distribution of aluminium and silicon cations. He further considers that proton migration is an essential feature of the structure.

Freund[3] has proved that the metakaolin structure must contain a high concentration of defects coupled with tetrahedral co-ordination of aluminium, so that it may not be correct to ascribe any particular structure to the phase.

It is generally agreed that the new phase preserves much of the orientation of the original kaolinite and in mineralogical terminology it is called a *topotactic* transformation.

Recrystallisation Reactions. Most modern theories for explaining recrystallisation phenomena of metakaolin preserve the concept of topotactic reaction and suggest that a defect spinel-type phase develops. In this structure both aluminium and silicon occupy lattice positions, but the structure is only marginally different in X-ray characteristics from the well-established γ-alumina lattice (p. 185). Brindley and Nakahira have proposed the following schematic change:

Kaolinite		*Metakaolin*		*Spinel*
$6(OH)^{1-}$		$2O^{2-}$		
$4Al^{3+}$		$4Al^{3+}$		$4Al^{3+}$
$2(OH)^{1-} + 4O^{2-}$	$\rightarrow$	$6O^{2-}$	$\rightarrow$	$6O^{2-}$
$4Si^{4+}$		$4Si^{4+}$		$3Si^{4+} + 2\square$
$6O^{2-}$		$6O^{2-}$		$6O^{2-}$

The authors claim that the spinel develops at 980° C., but, in fact, defect spinel material can be detected at much lower temperatures if the heating rate is slow.

The final stage—the formation of mullite—is thought to be due to cationic migration but the orientation of layers is preserved throughout.

From chemical reactivity experiments, Jaffé has formulated somewhat different principles based on combined chemical and X-ray evidence. The idea of metakaolin is accepted where Si–O–Al bonds still exist although they are much weaker than in kaolinite itself. Hence aluminium can be leached from the phase quite readily.

As the temperature of metakaolin is increased in the range 500–900° C., the vibrational energy tends to disrupt the weakened Si–O–Al bonds and when this occurs, true γ-alumina can be formed. Long periods of heating at a maintained

[1] Brindley, G. W. and Nakahira, M., *J. Amer. Ceram. Soc.*, 42, 311, 1959.
[2] Taylor, H. F. W., *Clay Minerals Bull.*, 5, 45, 1962.
[3] Freund, F., *Ber. deut. keram. Ges.*, 44, 5, 1967.

temperature favour the maximum formation of this phase and by heating specimens for up to 100 hours at 800° C., a large content of γ-alumina can be identified.

Jaffé showed that in the 750–800° C. region an unexpected maximum region of chemical activity could be detected, which was more pronounced in samples subjected to rapid heating cycles. This he concluded was due to a rearrangement in bond angles.

At temperatures above 920° C., mullite tended to form, provided that the heating rate had been reasonably rapid. The Si–O–Al bonding of the metakaolin would be highly unstable at these temperatures, but mullite is thermodynamically preferred to the spinel structure of γ-alumina. It was noted that if much γ-alumina had developed at intermediate firing temperatures, mullite did not appear below 980° C., and the D.T.A. curve showed no exothermic peak at 950° C. Mullite only developed in large quantities in such mixtures at higher temperatures (*circa* 1100–1200° C.), and then at the expense of γ-alumina.

The strong exothermic peak at 950° C., typical of kaolinite was produced only if the sample were raised in temperature with sufficient rapidity to avoid forming appreciable amounts of γ-alumina; in other words, by subduing widespread Si–O–Al disruption until the higher temperatures. The exothermic peak could then be attributed to the catastrophic, self-accelerated dissociation of the weak bonding of the metakaolin form. From the dissociation of these groups, mullite would be the natural and immediate phase to form.

2. **Other Layer Lattice Minerals.** None of the other clay mineral groups has been studied in such detail as that of kaolinite, probably because most of them are not used industrially in high temperature processing. When they do occur in commercial bricks or fireclays, it is only as minor amounts, so that their influence on overall firing performance is marginal.

Pyrophyllite. $Al_2Si_4O_{10}(OH)_2$ usually occurs in a very compact form, free from fluxing impurities and with a low loss on ignition. It dehydroxylates at about 800° C., to give a metastable phase which, on heating in excess of 1100° C., forms a felted mass of cristobalite and mullite needles and suffers negligible shrinkage. For these reasons, pyrophyllite is a valuable material for tile bodies and electrical porcelains and also as an additive conferring volume stability on many ceramic compositions.

Montmorillonites contain so many substituted cations that their break-down behaviour is complicated by the appearance of liquid at low temperatures.

Muscovite and other micas dehydroxylate at about 700° C., and can readily reabsorb water to revert to their original form.[1] Structural break-down occurs above about 1000° C., with the formation of γ-alumina (spinel?), mullite and several other minor phases, such as leucite which contains alkali. Liquid becomes a predominant phase and some of the products thought to occur on firing micas may indeed be crystallising from the liquid on cooling.

Magnesium and iron layer-lattice silicates generally undergo dehydroxylation at somewhat higher temperatures than the corresponding aluminium derivatives. Almost immediately following the water loss, the stable ortho-silicate tends to form rapidly, unless impurities are present.

[1] Vedder, W. and Wilkins, R. W. T., *Amer. Mineral.*, 54, 482, 1969.

Chrysotile $Mg_3Si_2O_5(OH)_4$ loses its hydroxyl groups at 640° C., and the D.T.A. curve shows an immediate large exothermic effect which presages the formation of forsterite. At higher temperatures (about 1100° C.), enstatite also develops.

Talc $3MgO.4SiO_2.H_2O$ loses lattice water at 900° C., and crystallises immediately to enstatite Mg SiO_3. In pure talc some cristobalite also forms, but this is small in amount and quickly dissolves in a liquid phase.

3. Hydroxides

The hydroxides of aluminium form a highly complex reaction system. Both mono- and trihydroxides are known and each of these exists in more than one crystal variety. Dehydroxylation of the trihydrate usually takes place in two stages. The first in the range 250–300° C., results in a monohydrate form which subsequently dehydrates completely at about 550° C. From this stage onwards, several modifications of alumina can crystallise depending upon the raw material, the conditions of heating, the presence of impurities, the particle size, etc. Lippens and De Boer[1] have described 5 individual species given the Greek symbols γ, η, δ, θ and χ as prefixes.

At temperatures in excess of 1200° C., all revert to the α-form, corundum.

Duncan and Creyke[2] have pointed out that β-alumina can also form when hydroxides of alumina are heated. Small amounts of impurities are necessary for its formation, particularly sodium and magnesium oxides.

In pure alumina bodies, recrystallisation commences immediately following the final dehydroxylation, indicating easy reorientation of the cationic and anionic groupings. This is in marked contrast to the behaviour of alumina in metakaolin where the silicon bonds exert a restraining force to prevent recrystallisation.

The greater activity of alumina and the converse influence of Si–O–Al bonding is well illustrated by the reaction of various alumina-containing minerals with lime. Grim, Machin and Bradley[3] have shown the following interesting sequence

$$\left.\begin{array}{l}\text{gibbsite}\\ (Al_2O_3.3H_2O) + SiO_2\\ \text{kaolinite}\\ (Al_2O_3.2SiO_2.2H_2O)\\ \text{montmorillonite}\\ (Al_2O_3.4SiO_2.H_2O)\end{array}\right\} + \begin{array}{c}\text{calc. carbonate}\\ (CaCO_3)\end{array} \left.\begin{array}{c}\xrightarrow{890^\circ\text{ C.}}\\ \xrightarrow{930^\circ\text{ C.}}\\ \xrightarrow{1200^\circ\text{ C.}}\end{array}\right\} \begin{array}{l}\text{gehlenite}\\ (2CaO.Al_2O_3.SiO_2)\end{array}$$

With calcined gibbsite the reaction commenced at the dissociation temperature of calcium carbonate; kaolinite did not react until its own critical dissociation temperature (930° C.), was reached; montmorillonite was even less reactive and it might be presumed that the Si–O–Al bonding is stronger in this mineral after dehydroxylation than it is in kaolinite.

The hydroxides of iron show a bewildering reaction sequence on heating which has not been fully elucidated. There may be as many varieties of transient oxide

[1] Lippens, B. C. and De Boer, J. H., *Acta. Cryst.*, 17, 1312, 1964.
[2] Duncan, J. H. and Creyke, W. E. C., *Trans. Brit., Ceram. Soc.*, 68, 137, 1969.
[3] Grim, R. E., Machin, J. S. and Bradley, W. F., State Geol. Survey Ill. Report No. 69, 1945.

phases as there are in alumina, but superimposed on these changes are the dissociations into different oxidation or valency states of iron itself. Above 400° C., the normal ferric oxide, haematite, tends to convert to magnetite, Fe_3O_4, and at higher temperatures wustite, FeO, is the stable form. Prior to 400° C., several forms of monohydrate have been recognised and also two varieties of Fe_2O_3.[1,2]

The hydroxide of magnesium does not present such a complex pattern of changes as the hydroxides of aluminium and iron. Brucite, $Mg(OH)_2$, is the only known form and this dehydroxylates at about 435° C. The resulting amorphous oxide shows no well-defined features until the temperature is raised to 1200° C., or above, when cubic periclase develops by growing from centres of nucleation.

The main reactions of clays and associated minerals have been summarised by Brett, MacKenzie and Sharp.[3] Table XI.IX shows a simplified reaction cycle for some of the more important mineral varieties.

CHEMICAL REACTIONS IN KILNS AND OVENS

THE most important chemical reactions between ceramic materials occur during the process of burning (or firing) the articles in an oven or kiln.[4]

Industrially, three different terms are applied to the heating of ceramic articles:

(*i*) *Baking* refers to heating to a moderate temperature, e.g. not exceeding a bright red heat. In many cases it may be said to correspond to the end of the decomposition stage (p. 712).

(ii) *Burning* refers to the heating, independently of the temperature or the duration. It differs from 'baking' in implying that a higher temperature has been reached and that further changes (in addition to decomposition) have occurred. The term 'burning' is also used to imply destruction of the material heated, as when fuel is burnt; to avoid misunderstanding many ceramists prefer to use the word 'firing' instead of 'burning' when heating the ware is implied.

(iii) *Firing* is a term which should be restricted to the placing of the fuel on the fire and all the actions necessary to keep it properly burning, but it is often applied to the whole process of heating the goods, as in the phrase 'the kiln is fired in order to bake or burn the goods'.

All these terms are used loosely, but the term 'firing' is gradually replacing the term 'burning'. A distinction should still be made between 'burned', meaning heated and 'burnt' meaning damaged or destroyed by heat; thus china-ware is burned, but fuel is burnt.

[1] Rooksby, H. P., *X-ray Identification and Crystal Structures of Clay Minerals*, 2nd edition (Mineralogical Society, 1961), p. 354.

[2] Mackenzie, R. C. and Berggren, G., *Differential Thermal Analysis* (Academic Press, 1970), p. 272.

[3] Brett, N. H., MacKenzie, K. J. D. and Sharp, J. H., *Quart. Rev.*, 24, 185, 1970.

[4] In the ceramic industries the distinction between an oven and a kiln is that in a *kiln* the goods are heated in contact with the flames or hot gases from the fuel (though in some cases they may be protected by being placed in saggars) whereas in an *oven* or *muffle-kiln* the goods do not come in (obvious) contact with the hot gases. The distinction is often disregarded and tunnel kilns with cars are sometimes known as tunnel ovens and pottery kilns in which the goods are placed in saggars are often known as ovens.

TABLE XI—IX. THERMAL DECOMPOSITION REACTIONS OF CLAYS AND ASSOCIATED MINERALS

Mineral	*Dehydroxylation Temperature ° C. (D.T.A. values)*	*Intermediate Crystal Phases*	*Final Crystal Phase and Temperature*
Gibbsite $Al(OH)_3$	300	Boehmite AlO.OH $\xrightarrow{450} \gamma \xrightarrow{750} \delta \xrightarrow{1000} \theta + \alpha Al_2O_3$	1200° C. corundum $\alpha \cdot Al_2O_3$
Iron hydroxides	200–300	Goethite FeO.OH → complex forms →	400° C. haematite $\alpha . Fe_2O_3$
Brucite $Mg(OH)_2$	435	Unstable oxide →	1200° C. periclase αMgO
Kaolinite $Al_2Si_2O_5(OH)_4$	550	Metakaolin $\xrightarrow{700\text{–}900}$ γ alumina or spinel →	980° C. mullite $3Al_2O_3 . 2SiO_2$ + cristobalite SiO_2
Pyrophyllite $Al_2Si_4O_{10}(OH)_2$	750	Dehydroxylate →	1200° C. mullite + cristobalite
Montmorillonite (see p. 140)	700	Dehydroxylate $\xrightarrow{1000}$ spinel →	1200° C. mullite + cristobalite
Muscovite $K_2Al_4(Al_2Si_6)O_{20}(OH)_4$	700	Dehydroxylate $\xrightarrow{1050}$ spinel + sanadine → $KAlSi_3O_8$	1300° C. mullite + (leucite)
Chrysotile $Mg_3Si_2O_5(OH)_4$	635	Transitional immediately reverting to forsterite → Mg_2SiO_4	1100° C. forsterite + enstatite $MgSiO_3$
Talc $Mg_3Si_4O_{10}(OH)_2$	950	Transitional immediately reverting to enstatite → $MgSiO_3$	900° C. enstatite
Vermiculite (see p. 154)	800–900	Transitional immediately reverting to enstatite →	900° C. enstatite

The chemical reactions which occur during the heating in ovens and kilns are usually described in five groups or stages:

Drying or *Smoking*, in which the articles are heated to a temperature not exceeding about 150° C.

Preheating, which includes some decomposition, the oxidation and the *water-smoking* of the goods, i.e. the period in which the clay is partially decomposed and its combined water liberated and passed out through the chimney.[1] The preheating stage of firing normally covers the temperature range 300–800° C.

The Full Fire Stage includes the cycle within the temperature range above 800° C. to the maximum temperature attained. In this stage, the kiln or oven is usually heated at its maximum rate.

The Finishing Stage or the *Soaking Period* represents the period during which the maximum temperature of heating is maintained. The principal recrystallisation, recombination and liquid formation reactions take place during this period.

The Cooling Stage is the period of time during which the burned articles are reduced in temperature from the maximum temperature of heating to a point at which it is safe and convenient to remove them from the kiln or oven.

The effects of heat in burning or firing ceramic articles are most conveniently studied by considering separately each stage in the process, commencing with the effect of a very gentle heat—technically termed 'smoking'—and proceeding until the final or 'finishing stage' of the firing is reached. It must, of course, be understood that there is no sharp line of demarcation between these various stages, the effects of heat being continuous, so that the division of the firing into several 'stages' is wholly empirical.

Smoking. When a piece of clay or an article made of clay and allied materials which has been dried to the extent customary in manufacture is placed in a kiln or oven, the first effect of the heat is to remove any moisture still remaining in the material. It is by no means unusual for 5 per cent of water, equivalent to 1 cwt. or 3,000 cubic feet of steam per ton of goods, to be present in the 'dried' articles taken to the kiln and to be evolved as steam during the first stage of heating. For the steady evolution of this enormous volume of vapour without any damage to the contents of the kiln, it is very important to provide ample ventilation and to take precautions that the temperature shall not rise too rapidly, as, otherwise, the goods may be cracked by being unable to withstand the pressure of the steam produced in their interior and unable to escape with sufficient rapidity through the small pores in the articles.

The first or 'smoking' stage of the burning of ceramic materials is finished at a temperature of about 300° C.[2] When this temperature is attained throughout the kiln and its contents, the whole of the water should have been vaporised and the second stage of heating may then be commenced.

[1] The steam or 'white smoke' which issues from the chimney at this stage of the firing is produced by the completion of the drying of the goods and by the water liberated on the decomposition of the clay.

[2] It is difficult to remove all free water from ceramic ware in kilns at temperatures nominally below 300° C. The rate of evaporation from within bricks or blocks is extremely slow and although it can theoretically be completed at 100–120° C., the time factor would be too great. In practice, therefore, the kiln is raised slowly to 300° C. to allow all water to escape without harming the ware.

At so low a temperature as 300° C. very few chemical changes occur, other than those produced when the heating is effected by waste gases bearing products of combustion from another part of the kiln. Such gases often contain sulphur trioxide which can combine with any lime present in the clay and, in the presence of moisture, may form calcium sulphate in the form of a white deposit or 'scum' on the surface of the goods.

The **decomposition stage** is the second stage in the burning or firing of ceramic materials; its range of temperature extends from about 300° C. to about 800° C., and in this interval many important chemical and physical actions occur (p. 712). It is usually essential at this stage that the temperature shall rise very steadily, though the rate of heating may vary within wide limits with different materials.

At a temperature between 300° and 800° C., many chemical changes commence which are not completed until a higher temperature is attained, but having once started they proceed gradually and are completed (as far as this is permitted) at a higher temperature.

It is important in this stage of burning to complete the decomposition of minerals which lose water, carbon dioxide or sulphur dioxide before the commencement of vitrification, which would seal the surface layers. As in the smoking stage, a large volume of gases has to be removed during this operation. Weight losses of 1–2 cwts. per ton have commonly to be accepted and most of this is evolved as steam.

Other changes which occur at this stage are oxidation processes which depend for their satisfactory completion on the presence of an ample supply of air. As these changes occur in the order of their affinity for oxygen, the heating must be properly controlled or undesirable reactions may occur whilst desirable ones remain incomplete. Thus, carbon has the greatest affinity for oxygen, so that in a highly porous and heat-resistant body the oxidation of compounds of iron, sulphur, etc., will not be completed until all the carbonaceous matter has been fully oxidised, and where the latter is present in large quantities it is important that the temperature should rise sufficiently slowly to remove the carbonaceous matter entirely before too high a temperature is reached. Otherwise, as the heat is nearly always applied externally to the materials,[1] the exterior may be raised to such a temperature that partial fusion occurs and the exterior pores are thereby sealed before all the carbonaceous gases can escape. The result of such sealing will be shown at a later stage by the articles being swollen or bloated by the pressure of gases in their interior. The appearance of the articles may also be spoiled by the charred material remaining behind and being incapable of oxidation because the air in the kiln cannot gain access to it on account of the sealed pores.

The bloating effects resulting from incomplete combustion in the decomposition range can be put to good effect in preparing lightweight aggregates from certain clays and shales. If some clays are heated to about 550–600° C., to remove most of the hydrate water and then raised quickly to about 1250° C., bloating occurs due to the liberation of sulphur dioxide and carbon monoxide when the

[1] The only exceptions being clamp bricks and some articles in which sawdust or other combustible matter forms an essential part of the material of which they are made.

body is in a partially vitrified state. Stamboliev[1] has reported that aggregates possessing a bulk density of 0·45–0·60 ml./g. can be prepared from certain American clays and shales.

Herath[2] considers that the addition of about 2 per cent finely divided calcium carbonate to the body is of advantage in forming lightweight aggregates from Ceylon clays.

The **full-fire stage** in burning ceramic ware commences at that temperature at which the kiln or oven may be heated as rapidly as is possible without serious risk of damage to its contents. In the case of some clays, it probably commences at about 800° C., but it is usually safer to regard it as commencing at 900° C. with some reservation with respect to the more fusible clays, glazes, etc.

The 'full-fire' stage is in every respect an intermediate one and in it the various chemical and physical changes which commenced in the prior, or decomposition stage continue at a more rapid rate, some of them reaching completion. Recombination reactions, partial fusion and some recrystallisation occur and, if the temperature is sufficiently high or sufficiently prolonged, the molten liquid readily penetrates the pores in the more refractory material, dissolving some of the latter and so producing a still further quantity of the fused glass, slag or 'vitrified matter'.

The chemical changes which occur in this stage are not essentially different from those in the latter part of the previous stage. They consist chiefly in the combination of the bases or basic silicates with a further amount of silica. The higher the temperature attained in this stage, the more complete will be the chemical reactions between the basic and acid materials present and the greater the amount of silica entering into solution in the fused portion of the material. These reactions are more conveniently considered separately (see *Vitrification*, p. 736).

The **finishing stage** of the firing of ceramic ware is that in which the desired reactions and other changes are completed or have progressed to such an extent as to produce articles or materials having the requisite properties. In this stage the temperature of the kiln or oven does not rise appreciably above that of the full-fire stage. On the contrary, it should usually be maintained as constant as possible. For this reason, this period is often known as the *soaking stage*, the materials or articles being regarded as 'being soaked in the heat' until it has penetrated completely through them in much the same manner as water will penetrate a porous material immersed in it for some time.

The purpose of this prolonged heating or soaking at an almost constant temperature is to enable the various changes which have commenced in previous stages to be completed or at any rate to progress to an extent which will produce the desired properties in the product. If the temperature is allowed to rise rapidly until the firing is finished there is always a danger—except with the most refractory materials—that the articles will be distorted or their colour or other desirable properties will be spoiled by overheating. This serious defect is largely avoided by maintaining the temperature at an almost constant level during the last stage of firing, as this procedure enables the process to be more satisfactorily and readily controlled.

[1] Stamboliev, H., *J. Amer. Ceram. Soc.*, 44, 577, 1961.
[2] Herath, S. W. and Grimshaw, R. W., *J. Brit. Ceram. Soc.*, 5, 436, 1970.

An adequate soaking period at 1100–1200° C. (i.e. prior to the final finishing at a higher temperature) is very important when firing china ware, porcelain and vitrified ware; it allows the necessary time for reactions to proceed as fully as may be required.

This prolonged heating or 'soaking' at a suitable high temperature may have the following effects:

(*a*) It may increase the amount of fused matter and, consequently, the amount of chemical action between the fluxes and the more refractory constituents, thus producing a larger amount of vitrified or glassy material which will gradually fill the interstices between the other particles, rendering the whole impermeable, as in the case of stoneware, or even translucent, as in porcelain.

By allowing ample time for the various chemical reactions to occur, prolonged heating of a ceramic material at the close of the firing imparts great stability to the material as a whole and enables a state of stable equilibrium to be attained. This is very important, when the product is to be reheated in use, as an unstable product—due to insufficient 'soaking'—will continue to undergo various physical and chemical changes which may have very serious consequences. The heating must not, of course, be prolonged to such an extent that undue distortion or loss of shape (see *Squatting*, p. 828) occurs. It is in the avoidance of this defect that the skill of the fireman is revealed, for the chemical reactions which are facilitated by the burning process must be stopped before they have proceeded too far. In other words, in the production of ceramic ware, most of the chemical reactions possible between the various constituents can never be completed, as that would render the material quite useless for the purpose for which it is intended. In this sense, as J. W. Mellor has pointed out, the chemistry of the firing of ceramic substances is a chemistry of arrested reactions. Unless the arrest takes place within a very narrow range of time or temperature, the product will be spoiled, hence, the enormous importance of as complete a control as possible over the kilns or ovens in which the burning is effected.[1]

The extent to which the burning of ware in a kiln has been completed is still often judged by observing the squatting of cones or bars[2] of known refractoriness value, set in a convenient position in the kiln. Although such methods are undoubtedly of value, they can be misleading if their full significance is not appreciated.

A cone squats because it has become sufficiently vitrified to flow under its own weight; it will be influenced not only by the temperature of its surroundings but also by the rate at which the temperature has been increased. The point at which a particular cone bends depends upon the amount of *heat work* it has received which is an integration of temperature and time. Seger Cone 7, for example, bends when the surrounding temperature is 1230° C. provided that the rate of rise of temperature

[1] The extraordinary success with which the firemen in pottery and other works control the heating within the requisite limits has long evoked the wonder of all who have some understanding of the subject. Considering the thousands of years during which clay has been heated industrially it seems equally remarkable that automatic methods for controlling the temperatures of ceramic kilns have only been in use for a few years, and in chamber and arch kilns their value is still questionable.

[2] In U.K. practice, it is usual to employ the Seger type cone (p. 777) or the Holdcroft bar depending on the convenience of location in the kiln. In the U.S.A. the Orton Cone is used.

has been 5° C. per minute; when the temperature is raised at 1° C. per minute cone 7 will bend at 1140° C., and at still lower temperatures under slower rates of temperature increase.

Each standard cone has a different chemical composition (see p. 776) and when they are used as temperature indicators in kiln firing, each may have different vitrifying characteristics. It is not unusual to see during the latter stages of slow firing in a kiln that a cone of higher value will bend before one of nominally lower value; this is more likely to be due to anomalous softening behaviour—the cone of higher value having a greater softening range—than to deficiencies in the composition of cones—which are checked under a standard heating rate of 5° C. per minute.

The vitrification behaviour of ware being fired alongside standard cones may be such that two separate kiln firings, each to the same nominal cone value, may produce a very different degree of burning in the ware. This is because the rate of heating had been different in both cases.

On the other hand, the alternatives to cones or bars are thermocouples, but these measure temperature at one point only and that usually not in the ware itself. Even in the best regulated kilns, large temperature differences throughout a setting often occur which cannot be detected by a single thermocouple.

If the obstacle of temperature variability can be overcome, there still remains the problem that thermocouples measure only the temperature at a particular time whereas the firing of ceramic ware is an integration of temperature/time effects. A cone or bar effectively records this characteristic—within limits—but when using thermocouples, a pre-determined heating rate and soaking time must be adopted and strictly followed.

In the firing of bricks in continuous chamber kilns, the correct use of contraction rods is still the most effective means of controlling the firing.

Whenever it is suspected that temperature variation is occurring in a kiln chamber or on a tunnel kiln setting, Buller's Rings afford an excellent assessment. These small circular discs can be placed in many strategic positions as the ware is set and after firing they can be measured for contraction so that a complete survey of the firing conditions can be carried out. A Buller's Ring Survey of a kiln is a worthwhile exercise at regular intervals in that it can reveal many unsuspected faults and deficiences of firing. A set of Buller's Ring values is given in Chapter XII.

In most of the chemical processes employed in other industries the principal object is to complete the reaction as rapidly as possible consistent with obtaining the maximum yield of the desired product. This is not the case in firing ceramic wares, as the possible reactions cannot be allowed to proceed to completion, but must be stopped at a point which enables the product to possess the complex series of properties which are essential to its use to the best advantage. Anyone with even a small knowledge of chemical reactions knows how difficult many of them are to arrest even when they take place at the temperature of boiling water, and can realise how much more difficult it is to arrest such reactions when occurring at 1000° C. or above without damage to the product; yet such a stoppage must be effected promptly and at precisely the right time, especially in the manufacture of some of the most delicate wares.

(*b*) It may permit the crystallisation of some of the constituents. Thus mullite may be formed in clay goods, cristobalite or tridymite in siliceous materials, periclase in magnesite bricks, etc. In some cases, the production of a mass of felted crystals gives the mass an added strength and a greater resistance to sudden changes in temperature. In others, the crystals are important because they are the most stable form of material under the conditions in which it will be used. In glazes and sometimes in binding materials, the production of crystals is undesirable and is commonly regarded as a defect, though some crystalline glazes are very beautiful.

(*c*) It may cause the volatilisation of alkalies and thus increase the refractoriness of the residual material, though this can only occur to a very small extent in the soaking stage of burning; it is more important when ceramic products are subjected to prolonged heating at a high temperature during use.

Vitrification commences when the fusion point of the least refractory constituents (or of the most fusible product of any reactions which may have occurred) is reached, but owing to the complex nature of ceramic materials no single temperature can be stated as that of the commencement of vitrification. With some very fusible clays and glazes it is as low as 450° C., whilst some highly refractory materials, such as magnesia and alumina, show no signs of fusion below 2000° C. With most of the crude clays used in the ceramic industries the commencement of vitrification occurs at about 750°–800° C.

A readily fusible substance present in the material may be the first to fuse, but it is more usual for a sodium or potassium salt present as an impurity to decompose and combine with the silica, forming a fusible silicate, which will, as the temperature rises, act as a solvent for some of the other silicates and bases present; the molten material thus forms a liquid in which the various substances can react far more rapidly than when all are in the solid state. By this means, fused complex compounds, eutectics and solid solutions are formed. These changes, which have been described in Chapter X, increase in extent and velocity as the amount of vitrified material increases and the temperature rises, until—unless its progress is previously arrested—a point is reached when there is so much fused material present that the mass is unable to retain its shape and loss of shape occurs, and if the heating is still further continued at a suitable temperature, the whole of the material is reduced to a molten or liquid state and forms, when cooled, a glass or slag.

The rate of vitrification is very slow at low temperatures, but increases rapidly at temperatures above 1200° C. (sometimes at a much lower temperature), except with the most refractory materials, some of which can only be melted in the intense heat of the electric arc.

The chemical changes which occur during the production of vitrified material cause various physical changes to take place simultaneously. Thus, the volume of the material changes as the interstices are filled and the solid material is dissolved, and also as products of different specific gravity are formed. The porosity and permeability decrease as the pores are filled with the molten material. The strength of the hot material decreases as the amount of fused material increases, the mass becoming more mobile, but when cold the strength is increased as a result of the larger amount of fused glassy matter produced acting as a bond uniting the other particles firmly together. These effects are described in Chapter XII.

The control of vitrification involves not only the gradual production of a fused material, but the retention of the shape of the article as a result of the presence of sufficient refractory particles, so distributed that they act as a skeleton which is not affected by the movement of the molten portion of the body. Unduly prolonged heating will change the properties of the ware by altering the proportion of glass and skeletal matter and the structure of the latter.

The term *vitrification range* is applied to the range of temperature between the commencement of fusion and the loss of shape due to the production of vitrified material. It varies according to the nature of the substances present; in some clays the range may be as low as 30° C., whilst in some refractory clays the range may be 300° C. or more.

In clays and in some other materials, the vitrification range depends chiefly on the nature of fluxes present. Magnesia is usually regarded as producing the longest range of vitrification obtainable with siliceous materials, especially if clay is present, as the product is more viscous than that of the corresponding lime- or alkali-compounds, and as it does not penetrate the pores so readily its rate of attack is much slower. Some Keuper Marls probably owe their long vitrification range to the presence of magnesia which may occur in amounts up to 6 per cent, although at the normal temperatures of firing it is not appreciably fusible in theory at least.

Clays or ceramic bodies which contain several fluxing elements in small proportions have long vitrification ranges; the alkali elements react with the alumina and silica at comparatively low temperatures (*circa* 700° C.) and the amount of liquid gradually increases until at 1100° C. or thereabouts, lime and magnesia also react.

A long vitrification range is also given by clays in which the minerals which contain the fluxing elements are present in a wide size range. Small particles of mica, for example, react quickly, but larger particles are slow to form a liquid.

Finishing Temperature. The temperature at which the firing of ceramic materials or articles is finished depends on the nature of the contents, but in any case it will be below that at which the vitrification of the product is so far advanced that serious loss of shape occurs.

Different materials and articles require different finishing temperatures for reasons already stated, and even articles of one class, such as building and firebricks, cannot all be finished at the same temperature on account of variations in the composition of the materials of which they are composed. The following notes are intended to give some idea of the temperatures used for different articles:

Building bricks, *roofing tiles*, and *terra-cotta* are burned at various temperatures according to their composition and the type of the product required.

Clays used in the U.K. for the manufacture of building bricks cover a wide geological and mineralogical distribution. A broad classification of finishing temperatures is as follows:

Carboniferous clays	1000–1100° C.	Jurassic clays	950–1000°C.
Etruria marls	1100–1150	Wealden clays	1020–1060
Keuper marls	1000–1020	Glacial clays	950–1100

Bricks of engineering quality are fired to higher temperatures to develop a greater degree of vitrification.

The temperature of finishing depends also on the length of the firing cycle.

The modern trends for rapid turn-round often involve increasing the top temperature by up to 100° C., but cutting down the length of soaking.

Colour is affected to a considerable extent by the finishing heat and this consideration often determines the maximum temperature sometimes to the detriment of strength.

Salt-glazed articles must be fired at a higher temperature than building bricks, because the salt and clay will not combine satisfactorily at temperatures below 1180° C. (Cone 5*a*).

Earthenware is a term applied to articles made of so many kinds of clay and mixtures of clay with other materials that no single figure can be given for its finishing temperature. Some of the most crude earthenwares and some majolica wares are finished at 790° C. (Cone 015*a*), whilst white ware is finished at 1140°–1300° C. when in the biscuit state and at 1080°–1300° C. when glazed. Between these wide limits it is impracticable to give closer figures unless all the required properties are known with great exactitude.

Earthenware is required to have a porous body to which the glaze will adhere readily. The ware should also be strong enough to withstand ordinary usage, but it need not be vitrified. It is, therefore, burned at a lower temperature than porcelain ware, though the temperature reached in some kilns in which the best qualities of earthenware are burned, approaches very closely to the finishing temperature of porcelain.

In the manufacture of earthenware and porcelain, the ware is first fired in the unglazed state, producing *biscuit*, which is afterwards glazed and refired.

Stoneware is required to contain more vitreous material than earthenware and must, therefore, be fired under conditions which ensure the requisite amount of vitrification. The best qualities of stoneware are quite devoid of porosity even apart from the glaze. The nature of the glaze sometimes affects the temperature at which stoneware is finished as it is customary to apply the glaze to the dried ware, whilst earthenware is made into biscuit and then glazed. The term 'stoneware' includes many varieties of ware; that to which it is chiefly applied in this country is usually finished at a temperature of about 1200°–1350° C. (Cones 6*a*–12).

China and other kinds of *porcelain* are heated so as to secure the maximum amount of vitrification without loss of shape. The finishing temperature is largely influenced by the nature of the materials used and this accounts for the great variations shown in Table XI.X. Like the better qualities of earthenware, it is usually necessary to fire porcelain articles at least twice—once in the unglazed state and afterwards when glazed. The latter temperature may be lower than the former if the purposes for which the ware is to be used permit this.

Glazed ware may be fired in either one or two periods; in the so-called 'single-firing' process, the glaze is applied to the unfired ware and the burning must then be controlled in a manner similar to that described for earthenware, but continued to such a higher temperature as may be necessary to fuse the glaze completely and cause it to spread uniformly over the glazed surface. In the so-called 'glost-firing', the glaze is applied to the fired or biscuit ware and the glost-firing can, therefore, be effected much more rapidly as the only changes of importance which are required to take place are those which occur in the glaze.

Glazes are required to be completely fused and to become sufficiently mobile that they distribute themselves uniformly except in those cases where an irregular distribution is required as a means of decoration. As the various chemical reactions should proceed to completion in the fused glaze, the temperature may usually be raised fairly rapidly, but care must be taken not to spoil the glaze by overheating it especially if it has been applied to unfired ware.

The retention of a suitable oxidising or reducing atmosphere during the firing is very important in glazed ware and especially in the case of coloured glazes, for many substances used as glaze-stains are profoundly modified in colour by the nature of the atmosphere in which they are heated. Thus, chromium compounds are green in a reducing atmosphere, but red or buff in an oxidising one; lead glazes may be blackened in a reducing atmosphere, and manganese and cobalt compounds form bubbles of oxygen if the glaze is heated too long in an oxidising one. Hence, it is impossible to fire all kinds of colours at one time in a kiln. Steady firing is essential for glazes, or the ware may be blistered or 'feathered'. Very slow firing is detrimental to glaze, so that the firing should be as rapid as possible without damaging the ware.

Table XI.X shows the finishing temperatures for various classes of goods.

TABLE XI—X. FINISHING TEMPERATURES OF EARTHENWARE, ETC.

Type of Ware	*Finishing Temperature* *Seger Cone*[1]	°*C.*
Whiteware, biscuit	3*a*–10	1140–1300
Whiteware, glost	01*a*–10	1080–1300
Easy earthenware, biscuit	4*a*–5*a*	1160–1180
Hard earthenware, biscuit	5*a*–6*a*	1180–1200
Easy earthenware, glost	03*a*–1*a*	1040–1100
Hard earthenware, glost	2*a*–3*a*	1120–1140
White stoneware (soft glaze)	09*a*–03*a*	920–1040
White stoneware (hard glaze)	1*a*–10	1100–1300
Stoneware with salt glaze	5*a*–10	1180–1300
Stoneware (chemical)	6*a*–12	1200–1350
Majolica ware	015*a*–05*a*	790–1000
China biscuit ware	9–10	1280–1300
Soft porcelain ware	8–10	1250–1300
Hard porcelain ware	7–20	1230–1530
Hard porcelain glaze	13–16	1380–1460
Glass colours	022–021	600–650
Easy enamel kiln	020–018	670–710
Medium enamel kiln	017–016	730–750
Hard enamel kiln	016–015*a*	750–790
Porcelain colours and lustres	022–010*a*	600–900

Refractory articles such as firebricks, crucibles, retorts, etc., should usually be fired at a temperature higher than that at which they are to be used. If this is done, any further heating to a lower temperature cannot have very much adverse effect.

A mistake which is commonly made is to reach a high temperature, but not to

[1] For a description of Seger Cones see p. 776.

maintain it for a sufficiently long time, the firing being stopped before the heat has had time to penetrate the articles sufficiently.

In *refractory bricks*, there is usually a very small proportion of fused material. The firing of these materials should therefore be such that crystal development reactions are able to proceed; otherwise the ware would have insufficient strength. In practice, a small amount of liquid is desirable at the temperatures at which refractory ware is fired.

Low grade fireclay bricks are frequently finished at about Cone 5*a* (1180° C.). The better qualities should be fired to Cones 7–12 (1230°–1350° C.), and high alumina products are fired to Cone 14 (1410° C.) or still higher temperatures.

Cooling. When the supply of heat to an oven or kiln containing ceramic articles (or to a furnace or other structure of which they form a part) ceases, the articles gradually cool down. The heat will be lost slowly if the kiln or oven is tightly sealed, but it may be withdrawn rapidly by passing a sufficient quantity of cold air through the chamber, provided that the goods do not crack or disintegrate, as most ceramic materials, when fired, are sensitive to sudden changes in temperature. To avoid damage, it is necessary to ascertain what are the best conditions for cooling. These naturally differ with the nature of the product, but, as a general rule, the more porous the material, the more rapidly can it be cooled with safety. Glass, porcelain and other articles containing a large proportion of vitreous matter, on the contrary, must be cooled skilfully, and it is usually necessary with such materials to vary the rate of cooling at different temperatures; this variation or control of the cooling is known as *annealing*. Unless the proportion of vitreous matter is large it is usually possible and desirable to allow the ware to cool rapidly from the finishing temperature to about 900° C., but below the latter temperature the cooling must usually be much slower, and even above it it an excessive rate of cooling may be dangerous with large, thick pieces of highly vitrified ware.

Many fused and vitreous materials tend to crystallise if maintained too long at a temperature above 900° C., so that rapid cooling through the 'crystallisation zone' is essential where a vitreous structure is desired as in glazes and in most ceramic ware. Such crystallisation may be beneficial or objectionable according to the crystals formed. If the crystals form a felted mass (e.g. mullite) they act as a reinforcement to the glassy matter and produce very strong ware (e.g. porcelain and stoneware). Isolated crystals and those arranged in small bundles like faggots are a source of weakness rather than strength. In glazes, the formation of crystals is usually regarded as a defect (*devitrification*), but occasionally they are considered to be decorative, as in *crystalline* and *aventurine glazes*.

Cooling the kilns improperly is a fruitful source of many defects, especially 'crazing', 'cracks', 'dunts' and 'feathering' or crystallisation, and great skill and care are needed to avoid them.

The few investigations which have been made on the effect of cooling ware at different rates during various stages of cooling show that there is scope for a very careful study of this subject. At present there is not sufficient information for any precise statement to be made on the various rates which should be adopted; they appear to differ with each class of ware.

Firing Cycles

In order to avoid certain defects, it is customary to raise and lower the temperature of kilns containing clay-ware very slowly. In many works, bricks, tiles, and terra-cotta do not attain the maximum temperature in less than ten days (i.e. 240 hours) and a further six days or more is allowed for cooling. Pottery and other hollow-ware can be heated and cooled much more rapidly, but for much of it more than 168 hours is allowed. These very low rates are partly due to the limitations imposed by the kilns employed and investigations on a laboratory scale have indicated that much more rapid rates of burning are practicable. With the advent of tunnel kiln methods of firing, the burning cycle of some types of ware has been reduced to 40–80 hours. Unless great care is exercised during firing under such conditions, the quality of the ware will be inferior to that produced from slower fired kilns. It is important that the ware entering rapid-cycle tunnel kilns should be perfectly dry and that the temperature gradients should be so adjusted that an excessive rate of change of temperature is avoided during critical stages in the firing. Clays with a high content of free quartz and those containing carbonaceous matter cannot be burned too rapidly without developing serious faults.

In such kilns, bricks, tiles and pottery can be fired, cooled and drawn in 48–72 hours, whereas in an intermittent or Hoffmann-type of continuous kiln two or three weeks would be required.

More research has been put into kiln construction and design in recent years than in most other aspects of ceramic production.

It has been recognised that firing methods suffer from two general faults:

(*a*) Non-uniform heating throughout the setting due to poor heat transfer or aerodynamic flow.

(*b*) Inefficient removal of products of combustion coupled with an inadequate supply of oxygen at critical periods.

Slow firing rates which were essential in former practice to achieve the above two objectives are wasteful in fuel, labour and efficiency of turn-round.

Firing cycles of an almost unbelievable rapidity are now being proposed. Thin glazed ware, such as tiles, is being processed in minutes, but perhaps the most startling proposal is that of the General Shale Corp. of America[1] who claim to burn building bricks on an 8-hour cycle. A special kiln has been designed in which the bricks stand one-high. The whole production plant is fully automated and manufactures extruded bricks with a large number of holes.

Many British bricks can be fired on rapid cycles. The author has had most success with Keuper marls, but other clays can be processed satisfactorily in about 10–12 hours provided that a longer soak in the region of 700–850° C. is allowed. The advantage of small settings during the cooling cycle is obvious. Heat can be lost more readily to the surroundings, so large temperature gradients are avoided.

Fast-fired ware is generally of good quality, but it is frequently more brittle than the same article produced more slowly. Because a slow soak at top heat is not possible in rapid cycles, a higher firing temperature is necessary. More liquid is thereby produced. Because cooling is also rapid, less crystallisation takes place and

[1] Sells, G. C. and Postel, J., *Brick and Clay Record*, (1967), p. 48.

the product contains a greater proportion of vitreous glass than its slow-fired counterpart.

Some decorated, glazed wares must be fired several times because some of the colours used are spoiled by overheating and because some glazes cannot be applied to unfired ware. This use of *two or more firings* was first introduced into England by Enoch Booth of Tunstall, Staffs. (about 1750); he fired a white stoneware body and produced *biscuit ware* which was then glazed and re-fired at a lower temperature, producing *glost ware*. In *coloured or decorated ware*, each colour may require a different temperature, so that they must be applied in an order of diminishing temperatures, with a separate firing after each application. The position in which decorated articles are placed in a glost-oven is very important and much skill and experience are needed to ensure the correct setting of such wares.

THE COMPOSITION OF FIRED CLAYS

THERE are two entirely distinct phenomena occurring in clays heated at high temperatures. These are (*a*) crystal formation due to the recombination and recrystallisation reactions, and (*b*) liquid formation. This latter material may crystallise on cooling, but siliceous melts are prone to supercool and often 'freeze' to a homogeneous glass or a partially crystallised system.

The relative rates of crystal formation and of liquid development in a particular clay cannot be predicted because of the complexity and number of factors involved. The composition, grain-size and distribution of minerals, firing rate and maximum temperature all contribute to both types of reaction and, in addition, the two phases may mutually interact after they have been produced. Thus the liquid which develops during firing may dissolve some of the crystals formed by solid reaction processes and, conversely, some liquid may crystallise on contact with the solid by reason of the change in composition produced, or by the 'seeding' of a supercooled phase.

The factors which influence glass formation on cooling have been outlined in Chapter X and these may be summarised as:

(*a*) The rate of cooling.

(*b*) The viscosity of the liquid in the temperature region of incipient crystallisation.

(*c*) The composition of the liquid.

Crystallisation in silicate melts is usually slow and consequently a rapid rate of cooling may inhibit the appearance of a crystalline phase which under slower rates of decreasing temperature might crystallise completely.

Some silicate melts are viscous even at high temperatures—particularly when magnesium oxide is the principal fluxing element. Such mixtures do not crystallise readily because the free movement of the ions or atoms of the liquid is retarded and they are unable to rearrange to form an ordered crystalline lattice. Other elements, for example the alkali metals, form liquids which do not crystallise until the temperature is below 700° C., and the viscosity is invariably high. These mixtures crystallise only with extreme difficulty and readily form glasses.

Fired clay mixtures which are low in alumina and high in silica will contain more glass on cooling than those with a higher proportion of alumina, even though the fluxing elements, firing temperature and cooling rate are kept constant in both cases.

As the glassy phase must be derived from a parent liquid, the amount which is present in a vitreous body will depend on the extent of liquid development under the particular firing conditions. It is, therefore, to be expected that, in addition to the temperature and rate of heating, the amount and availability of fluxing elements contribute to the proportion of glass in the cooled ceramic material.

The changes in composition which determine the constituents of a clay when it is fired and then cooled may be summarised as:

(*a*) Decomposition reactions occurring at comparatively low temperatures (p. 712).

(*b*) Primary crystallisation through reactions proceeding in the solid state (p. 716).

(*c*) Liquid formation which on cooling forms:

1. glass or
2. secondary crystallisation products.

Low Refractory Clays

INFERIOR or low-grade clays are usually rich in fluxing elements principally of the alkali type and have a high silica content, due to the proportion of sandy constituents. With such compositions, liquid forms in large amounts even at relatively low temperatures and it does not crystallise readily on cooling. *Building bricks, roofing tiles, terra-cotta ware*, and similar materials are principally composed of a crude glass which surrounds isolated crystals mainly of unchanged quartz. Primary crystallisation products (except mullite) are rarely detectable because, if they form at all, they are quickly dissolved in the liquid phase. Secondary crystallisation is not extensive especially when the fluxing elements are principally alkaline. Bricks rich in lime occasionally contain felspar produced by crystallisation from the liquid during the cooling.

The *silica minerals* are commonly present in fired clay bodies. Quartz is frequently a principal mineral in clays and unless the temperature of firing is excessively high it does not convert to cristobalite or tridymite to a great extent. This change is slow at temperatures below 1250° C., especially when alumina is present to act as a retarder (see p. 674). Flinty or chalcedonic particles of silica associated with sedimentary clays are readily converted at above 1000° C. and are usually changed on firing to one of the more stable crystalline forms of silica. Free silica is rapidly attacked by alkalies or liquids rich in alkalies and is dissolved in the melt.

In brick clays and other clay products which have not been fired above 1100° C., quartz will persist throughout the burning and will be a constituent of the final product. As such, it can present problems in the cooling of ware, because at its inversional point of 573° C., the consequent large contraction is a frequent source of dunting cracks.

The complete fired composition of building bricks, sewer pipes, roofing and quarry tiles, coarse earthenware, faience and stoneware, all of which are derived from raw clays, is difficult to calculate because it depends on the initial mineral forms, the particle size, the temperature and duration of firing and the cooling procedure.

The phase present in greatest amounts in all of these bodies is glass, rich in silica and alkalies, but containing alumina, iron oxide, lime and magnesia. This surrounds the crystal phases of mullite and quartz and occasionally cristobalite which is probably derived from the liquid phase on cooling.

The author has examined the fired composition of a brick clay of the Etruria Marl series, samples of which were processed in two different kilns. The first was a transverse arch kiln with a total firing cycle of about 3 weeks, a maximum temperature of 1100° C. with 8–12 hours of soaking followed by slow cooling. The second was a modern fast-fired tunnel kiln with a complete cycle of 52 hours. The maximum temperature was 1150° C. and cooling was extremely rapid. The results are given in Table XI.XI.

TABLE XI—XI. THE COMPARISON OF THE FIRED COMPOSITION OF BRICKS OF ETRURIA MARL BURNED IN DIFFERENT KILNS

					Fired Composition	
Chemical Composition of Clay					*Slow firing max.* 1100° C.	*Fast Firing max.* 1150° C.
SiO_2	58·5%	CaO	0·4%			
Al_2O_3	21·8	MgO	1·1	Mullite	18%	22%
TiO_2	1·2	K_2O	1·3	Quartz	25	23
Fe_2O_3	7·6	Na_2O	0·2	Cristobalite	15	4
Loss on ignition 7·6%				Non-crystalline phase	42	51

The non-crystalline phase was estimated by difference. It is not correct to term it glass because it contains titanium, iron compounds, etc., which are not easy to identify.

Some stoneware and highly vitreous tiles contain only small amounts of crystalline phases.

Although they are present in only small amounts, some of the impurities commonly associated with brick clays undergo changes on firing which persist in the fired brick and can have consequences in service which seriously affect quality.

Calcium-bearing minerals are amongst the most important in this respect and two species warrant special mention.

Calcium carbonate is of fairly common occurrence in brick clays. If it is present in disseminated particles which are soft in texture its only influence on firing is to change the colour of the body from red to buff, yellow or grey. In this form it can be present in amounts up to 25 per cent without causing undue harm. In harder, larger sizes, however, calcium carbonate is a serious contaminant.

Calcining calcium carbonate at about 900° C. removes carbon dioxide and leaves a highly active calcium oxide grain. If the particle size is small, the lime reacts with silica or alumino-silicates or it may enter the liquid phase. As such the calcium oxide becomes stabilised in the fired clay body. When the particle size of the original carbonate is large, the oxide is not able to react fully and may remain as free lime in the brick body. In this form, lime is highly reactive and on being cooled and exposed to the atmosphere, it will rehydrate or re-carbonate with an expansion sufficient to cause blow-outs in surface layers or, if limestone is present in large amounts, to shatter the bricks completely.

Laird and Worcester[1] suggest that increasing the vitrification of clay bricks is beneficial, but in the author's experience this is not always true. A soft under-fired brick has low strength and hence it has low resistance to stresses; on the other hand, a highly vitrified brick may be brittle and prone to conchoidal fracture from internal stresses.

Fast-firing kiln cycles seem to aggravate lime-blowing problems probably for two reasons:

(*a*) there is less time for the lime to combine with the other components of clays,

(*b*) the fired body is more glassy and brittle.

The only positive cures for lime-blowing are either to remove the limestone from the clay by selection in the quarry or to grind the clay to a sufficient fineness to ensure reaction on burning.

If the second course is adopted, the material must be reduced in size so that it will all pass a 30 B.S. sieve (500 microns). A good guide to the potential lime-blowing characteristics of a clay is to disperse a known weight of it in water and to pass it as a slurry through a 30-mesh screen. The weighed residue on the screen should be treated with an excess of dilute hydrochloric acid and filtered. The loss in weight indicates the content of limestone *liable to give lime-blowing*. Even when large-sized limestone is present in amounts of only 1 per cent (expressed as the weight of limestone retained on a 30-mesh screen against the total weight of clay) lime-blowing will occur in noticeable amounts. It is likely to be serious if the content exceeds 2 per cent. The magnitude of the effect will also be related to the individual sizes and hardness of the calcite grains and the texture of the fired brick.

It has been suggested that lime-blowing can be reduced (*a*) by immersing the bricks in cold water immediately they leave the kiln. This is claimed to cause rapid rehydration which is less harmful, or (*b*) by adding 0·2–0·5 per cent sodium chloride (common salt) to the brick batch. It is difficult to theorise why this should help the problem unless it is because a greater degree of vitrification will be induced in the brick body. Nevertheless, cases where salt does help are indisputably established but it certainly does not apply in every instance.

Lime-blowing is a particularly worrying problem in that it often does not show up until the bricks have been in service for several weeks or even months. Even a small amount of limestone can produce an unsightly surface.

Sulphates of Calcium and Magnesium. Gypsum and epsomite bands are not common in clays but they are liable to occur in those of the Etruria and Keuper

[1] Laird, R. T. and Worcester, M., *Trans. Brit. Ceram. Soc.*, 55, 545, 1956.

Marl series of the U.K. Sulphates often form from sulphurous fumes in the dryer or kiln during drying or firing.

Both calcium and magnesium sulphates are stable on firing up to about 1100° C. As such these minerals persist in the clay body where they are sources of *scumming* and *efflorescence.*

In the former fault the sulphates migrate to the surface of ware during drying. There they are concentrated, become dehydrated and fired into the clay. They combine with surface iron compounds to give a buff-coloured bleached coloration which is often localised or blotched because of irregular drying.

Magnesium and calcium sulphates are both appreciably soluble in water, so that if bricks containing them become saturated in use and then dry out, the salts from within the bricks gradually migrate to the surface. There they cause an unpleasant white furry deposit which persists despite all efforts to wash it off, simply because it is soluble and is absorbed into the brick more quickly than it can be removed. The crystallisation of magnesium and calcium sulphates is accompanied by a large volume expansion. In bricks which have been exposed to the weather for many years the successive cycles of solution and crystallisation can cause shattering and disruption of surface layers.

If the sulphates are present in the original clay both scumming and efflorescence can be avoided by adding a proportion of barium carbonate to the brick batch. This reacts to give insoluble barium sulphate and the harmless carbonates of calcium and magnesium thus:

$$\left.\begin{matrix}\text{Ca}\\\text{Mg}\end{matrix}\right\}SO_4 + BaCO_3 = BaSO_4 + \left.\begin{matrix}\text{Ca}\\\text{Mg}\end{matrix}\right\}CO_3$$

(soluble) (soluble) (insoluble) insoluble)

When sulphates form during drying or firing nothing can be done to cure efflorescence, but hard firing helps. The unpleasant effects on walls can be reduced by spraying the surface with a silicone preparation, but this requires frequent renewal.

Refractory Clays

As the proportion of fluxing elements in clays is decreased, the amount of liquid which forms at any particular high temperature is reduced in amount. Refractory bodies are normally fired to much higher temperatures than building bricks, so the proportion of liquid at the maximum temperature may not be dissimilar in many clay products which have been vitrified to the same extent, but at different temperatures.

Only in low-grade refractory bricks, with alumina contents of 30 per cent or less, can any quartz be detected. Firebricks should be fired to temperatures at which all quartz converts into cristobalite (or tridymite). The liquid phase which develops in fired fireclays is richer in alumina than that in common bricks, so there is a greater tendency for crystallisation if the cooling rate is reasonably slow.

Konopicky[1] has surveyed a wide range of German and other Continental commercial firebricks and represented their constitution on a ternary diagram.

[1] Konopicky, K., *Ber. Deut. Keram. Ges.*, 35, 367, 1959.

The predominant phases which he found were mullite, cristobalite and glass. All mineral phases were identified and estimated by X-ray methods and the glass was assessed by difference. Konopicky's results have been summarised in Table XI.XII.

TABLE XI—XII. FIRED COMPOSITION OF ALUMINO-SILICATE REFRACTORIES

	Mineral Phases %		
Alumina Content Range %	*Mullite*	*Cristobalite*	*Glass*
39–45	43–55	5–30	12–45
32–39	30–44	8–40	20–60
20–32	20–33	10–50	25–70

It is interesting to note that the mullite content was consistently related to the total alumina content of the brick, but the relative ratio of cristobalite and glass was associated with the content of fluxes.

Johnson and Andrews[1] have examined a range of British alumino-silicate fired products containing 39–85 per cent Al_2O_3. Some of their materials were made from high grade fireclays whilst others were low quality fireclays beneficiated with alumina; others were manufactured from calcined kyanite or sillimanite and those of the highest alumina content contained essentially calcined bauxite as a raw material. The compositions of some of these products are listed in Table XI.XIII.

TABLE XI—XIII. THE FIRED COMPOSITION OF SOME BRITISH ALUMINO-SILICATE REFRACTORIES

Chemical Composition				*Fired Composition*				
Al_2O_3	SiO_2	*Total alkalies*	*Total alkaline earth*	*Mullite*	*Alumina**	*Cristobalite*	*Quartz*	*Glass*
39	54	0·4	1·1	52	—	31	3	14
41	56	2·1	1·1	51	—	—	—	49
44	49	0·7	1·1	63	—	24	—	13
54	39	1·2	1·0	37	23	11	—	30
64	26	0·8	0·4	70	33	—	—	7
81	13	0·3	0·4	51	36	—	—	13
83	12	1·0	0·9	34	44	—	—	22

* This is presumably corundum.

The importance of the content of fluxing elements and of silica on liquid formation is clearly shown. The sample with 64 per cent alumina was obviously manufactured from calcined kyanite which conferred a high mullite value.

McGee[2] has presented a quantitative analysis of some American alumino-

[1] Johnson, W. and Andrews, K. W., *Trans. Brit. Ceram. Soc.*, 61, 724, 1962
[2] McGee, T. D., *J. Amer. Ceram. Soc.*, 49, 83, 1966.

silicate materials which were fired and cooled under laboratory conditions (see also p. 722). His materials were a plastic- and flint-fireclay and also kaolin. They were fired to 1350° C., and maintained there for 5 hours, after which they were cooled in different ways. Mullite was the predominant crystalline phase detected by X-rays and in the three samples, which were all high in alumina, the content was in the range 27–33 per cent—lower than in the commercially-fired products of equivalent composition reported by Konopicky and Johnson and Andrew.

When the samples were cooled slowly 10–17 per cent cristobalite was produced, but this value was markedly altered by quenching. The cristobalite content in the plastic fireclay block reduced from 16·5 per cent to 1·1 per cent if the sample were quenched. Presumably this was due to the different rate of crystallisation from a liquid phase, because the kaolin samples showed no appreciable change in cristobalite content when subjected to the same treatment.

China Ware and Porcelains

China ware and *Porcelains* vary considerably in composition (p. 334). Soft porcelains closely resemble opaque yet translucent glasses. A glass of high melting point is incorporated as a frit and the other ingredients fuse at the temperature of firing and form an impervious body. The particles of frit provide a skeleton which prevents excessive liquid formation and consequent distortion. The amount of crystalline material in soft porcelain is very small.

Hard porcelains contain a much greater amount of crystalline phase than do soft porcelains, although in most commercial porcelains, the proportion rarely exceeds 35 per cent.

The changes on firing in a hard porcelain body are such that the body consists of a felted mass of crystals that are mainly mullite and these form a skeleton throughout the body, the voids of which are filled with glass which may be partially crystallised to mullite or felspar.

The changes on heating and the fired composition of *bone china* have been studied by St. Pierre.[1] Little change other than the evolution of water and carbon dioxide occur up to about 1000° C., but above this temperature the bone ash begins to decompose giving β-tricalcium phosphate ($Ca_3(PO_4)_2$). The lime which is liberated during this reaction combines with the clay body to form anorthite. A highly fluid liquid is formed which reacts with other raw ingredients and gradually dissolves them—quartz is least readily attacked but in well-fired bone china no free quartz can be detected. The final phases on cooling are β-tricalcium phosphate, anorthite and glass, the amounts of which are determined by the proportions of bone ash and lime in the raw materials of which the body is composed.

The composition of other ceramic materials including steatite, cordierite, rutile and zircon bodies have been described in Chapter VI.

In *glazed ware manufacture* the correct temperature conditions during the glost-firing must be maintained to permit the fusible layer of glaze to react partially and adhere to the clay body, but excessive fluidity must be avoided.

From the foregoing it will be seen that in burning ceramic articles made from

[1] St. Pierre, P. D. S., *J. Amer. Ceram. Soc.*, 38, 217, 1955.

clay, crystalline components and a liquid phase form simultaneously at speeds governed principally by the composition of the body and rate of temperature rise. The liquid phase, especially at high temperatures, becomes dominant, dissolving the crystalline products, permeating the mass and eventually, if the heating were sufficiently prolonged and intense, the whole of the material would change into a molten liquid.

As this would involve complete destruction of the article, the reactions which occur must be arrested at a point at which sufficient vitreous material has been formed to impart to the article (when cold) the desired properties, of which the chief are usually the strength and porosity, though other properties dependent thereon, such as resistance to acids as well as other independent properties, may require to be considered in determining the finishing temperature and the duration (if any) of the soaking period. Thus, if colour is of great importance, as in some pottery and terra-cotta, the finishing temperature must be low and the proportion of vitreous material small; if the strength of the product is to be the predominant property, the finishing temperature must be higher and the soaking period prolonged so as to produce a larger proportion of vitrified material; whilst if complete impermeability, resistance to acids, hardness, or translucency, or any of these are to be predominant, the finishing temperature must usually be very high and the duration of the soaking very prolonged in order that the reactions which produce the fused vitreous material may proceed as far as possible without causing loss of shape in the product.

To anyone acquainted, even to a minor degree, with the nature of the reactions involved and the necessity of arresting them at precisely the right time, the firing of porcelain and some other forms of ceramic ware, under the conditions and with the restrictions imposed in commercial work, must be a continual source of wonder. Were we less accustomed to it we should find it difficult to believe that such complex reactions could be controlled so effectively, with such crude and imperfect means as are used regularly in the manufacture of earthenware, china-ware, and other pottery of the highest quality. In probably no other industries does the art of controlling chemical reactions, whose very nature is largely unknown, rise so high as in the firing of ceramic wares.

Changes in Composition of Fired Clays in Use

In the foregoing sections, it will have become apparent that the various mineral phases in fired clay bodies are not in their equilibrium state. Changes in composition are therefore to be expected when the materials are in use, particularly if they are maintained at high temperature.

Changes at Ordinary Temperature. When fired clays are cooled to room temperature, the solid phases have all the appearance of being absolutely stable, and within the normal expected life-span of such materials this is the case.

That changes of a remarkable degree must occur over extended periods can be inferred from studying pottery of earlier eras. Ware produced in Roman times, such as tiles and bricks, contain a much more complex crystalline composition than their modern counterparts. The presence of felspars, such as anorthite, and felspathoids, such as leucite, has been reported and these are probably the result of recrystallisation from the glass phase taking place slowly over many centuries. Cristobalite is also present in such bodies in amounts greater than would be expected unless changes had occurred.

Other changes in fired clay ware can be observed. Quartz and cristobalite show characteristic low temperature inversions which are manifested on cooling. Frequently, textural faults arise in ceramic ware due to these transformations, but they develop only after long service life.

Moisture absorption (see p. 706) is another effect which is liable to occur in fired ceramic articles which are exposed to atmospheric conditions. This is due to chemical rehydroxylation of glass or crystalline components.

Changes at Maintained High Temperatures. Many refractory clay products are designed for continuous high temperature use in coke-ovens, steel-plants and similar installations. Normally, such materials are fired to a higher temperature than their service use so that extended vitrification should not occur, but, nevertheless, the system tends to an equilibrium condition. The situation is further complicated, because certain chemical components have an appreciable volatility when maintained at a high temperature. A ceramic composition will frequently lose alkali elements and hence become more refractory and lower in liquid phase.

After long periods at high temperatures, refractories reach virtual equilibrium as predicted from phase equilibria. Mullite, usually with iron oxide in solid solution, can be present in massive crystal form and in amounts corresponding to theoretical predictions. The alkalies frequently crystallise as the felspars, albite, anorthite, or orthoclase, and leucite is a mineral frequently capable of identification. Iron minerals, such as the silicate, fayalite, or in the form of pyroxenes, are often present when refractories have been subjected to dusts containing iron oxide. Lime dusts form anorthite in considerable quantity, usually at surface interfaces.

The amount of crystalline phases in 30 per cent Al_2O_3 refractories which have seen continuous service in coke ovens at 1200° C. for several years is remarkably high.

Glasses are particularly prone to crystallise when they are maintained at high temperature, even though this may be well below the softening point.

Cristobalite forms quite rapidly from vitreous silica when it is heated at 1200° C. and the same phase forms at 1000° C. if the temperature is maintained for long periods of time. The crystalline phase destroys many of the valuable properties of the vitreous form; it increases the thermal expansion and consequent tendency to thermal shock and also reduces the transparent qualities.

Commercial glasses designed for oven-ware, slowly crystallise after frequent use and eventually lose their characteristics. Surface scratches appear to enhance the tendency to crystallise. Cristobalite is the main phase to form in most cases.

Changes in Fired Clays in Contact with Slags. Where fired clay bricks are used in contact with molten slags, hot dusts or molten metals, chemical reactions are of common occurrence. An alumino-silicate composition readily forms low melting-point eutectics with many of the chemical ingredients of slags. High silica refractories are remarkably resistant to both iron oxide and lime because over a wide range of composition in the silica-rich region, two immiscible liquids exist. The presence of alumina destroys this stability and liquid formation at eutectic compositions is readily developed.

High alumina bodies have a better resistance to most slags; the liquids which are produced in them are more viscous and do not penetrate the brick very readily. As liquid forms at the contact surface between brick and slag, an interface develops in which crystals often form and grow. This may be due to the preferential diffusion of certain elements from the slag. The calcium felspar, anorthite, is frequently observed when lime has attacked alumino-silicate refractories. This type of reaction often increases the rate of corrosion or wear on the refractory, because the crystalline development within the brick layers gives rise to a plane of weakness. Break-down occurs by the shelling-off of layers.

On the other hand, some alumino-silicate refractories develop a high resistance to alkali slags, such as sodium carbonate, because of the development of albite which grows and swells within the pores of the brick, thus producing a non-permeable skin. At the normal temperature of service, the alkali-bearing minerals have sufficient fluidity to withstand stresses.

CHANGES IN COMPOSITION ON FIRING SILICEOUS MATERIALS

SILICA may exist in a variety of crystalline forms, each of which is stable over a fixed range of temperature as described in Chapters III and X.

Two quite distinct types of thermal change are associated with the changes in the silica minerals; these are (*a*) conversions, and (*b*) inversions.

Conversions involve a pronounced change in crystal type (see p. 158); they proceed at a sluggish rate and, although reversible, the change takes place so slowly at the critical temperature that it is not observable under ordinary conditions.

The stability ranges of the minerals have been quoted by Fenner[1] as

$$\text{Quartz} \underset{870^\circ\text{ C.}}{\rightleftharpoons} \text{tridymite} \underset{1470^\circ\text{ C.}}{\rightleftharpoons} \text{cristobalite}$$

The transformation of quartz is not as simple as the above scheme suggests, nor is that of tridymite to cristobalite.

The changes are so slow that pure quartz can be maintained at temperatures well above 870° C. without pronounced change and, once the high temperature varieties are formed, they can be cooled and maintained at room temperature for an indefinite period of time without reconverting into quartz.

Pure quartz crystals are highly stable even at temperatures well above 870°. Grimshaw, Hargreaves and Roberts[2] have measured the amount of Madagascar crystal which is decomposed at various temperatures after long periods of firing. Their results are summarised in Table XI.XIV.

The critical temperature for the change was deduced by assessing the rate constant at various temperatures and extrapolating back to zero conversion. The initial product of the decomposition of quartz is probably a transitional, unstable entity which then slowly reverts to one of the other crystalline modifications.

[1] Fenner, C. N., *Amer. J. Sci.*, 36, 331, 1913.
[2] Grimshaw, R. W., Hargreaves, J. and Roberts, A. L., *Trans. Brit. Ceram. Soc.*, 55, 36, 1956.

Almost invariably cristobalite forms in preference to tridymite even in the temperature range of stability of the latter mineral. It is only after prolonged firing that tridymite appears as a product of heating pure quartz and unless some catalytic agent is present the change may take a geological age if it proceeds at all.

Silica brick manufacturers must ensure that the firing is sufficiently intense to destroy as much as possible of the quartzitic material in the raw silica rock and convert it into tridymite or cristobalite. Only by this means can bricks be produced which are stable and free from further large volume changes (see Chapter XII) when in use.

TABLE XI—XIV. THE PERCENTAGE DECOMPOSITION OF PURE MADAGASCAR QUARTZ CRYSTAL AFTER FIRING AT DIFFERENT TEMPERATURES

Firing Temperature	1270° C.		1320° C.		1370° C.		1450° C.	
Grain-size range* B.S.S.	−100 +120	−240 +300	−100 +120	−240 +300	−100 +120	−240 +300	−100 +120	−240 +300
Firing time (hours)								
½	0	0	0	0	0	0·3	1·0	9·8
2	0	0	0	4·1	0	4·1	1·9	16·8
4	0	0·7	0	4·5	0·4	6·7	9·7	44·1
12	0	2·3	0	7·2	1·5	16·8	59·5	80·0
24	0·2	3·6	0·2	14·9	7·5	29·2	83·6	95·5
48	0·4	6·2	0·6	25·9	27·4	49·3	100	100
96	1·0	13·5	3·5	45 1	55·5	87·2	100	100
240	8·7	28·0	61·5	80·0	92·0	100	100	100

* The centre grain-sizes of the two fractions were 0·138 mm. and 0·053 mm. respectively.

Fortunately, the slow change of crystalline type can be greatly accelerated by catalytic agents (see p. 162). Many materials are suitable and have been employed; some promote the formation of tridymite whilst others accelerate the production of cristobalite, and the reaction can proceed either entirely in the solid state or through the medium of a liquid phase.

The mode of action of the solid state catalysts, or *promoters*, appears to be the establishment of nucleated surface regions where contact occurs with a quartz grain. These rapidly spread until the whole particle is decomposed when reaction either stops, or under favourable conditions continues into other grains in contact with the first (see p. 674).

When a liquid phase is capable of being formed between the catalyst and silica at the temperature of firing, the activity or rate of decomposition is greatly increased as the mobility of the promoter is greater when in the liquid form.

Many attempts have been made to explain the fundamental mechanism of the catalytic behaviour. In some, the size and electrical nature of the cation of the promoter have been regarded as all-important. This is by no means a complete explanation, because the anionic component may also influence the change. Potassium oxide for example, or the corresponding tungstate or phosphate are extremely rapid in action, whilst potassium chloride added in the same molecular amounts is much less effective. In a similar way, the rate of change in the presence of some catalysts is influenced by the atmospheric conditions inside the kiln or furnace. Reducing conditions or an excess of carbon monoxide assist the change, although nitrogen has the opposite effect.

Salts of the alkali metals (lithium, sodium and potassium) are the catalytic agents which, in general, produce the most rapid changes. This is to be expected, as these components combine readily with silica, even at low temperatures to form a mobile liquid. Iron oxide is another commonly-occurring material which forms a eutectic with silica at comparatively low temperatures and it too is an energetic catalyst. Lime and magnesia react entirely in the solid form, but are quite active especially in the initial stages of firing (see p. 158).

Alumina is anomalous in behaviour because it not only fails to catalyse the decomposition of quartz, but retards the reaction and furthermore inhibits the activity of other normal catalysts when it is mixed with them.

Titania is of doubtful behaviour; under oxidising conditions it is a weak catalyst, but, in a reducing atmosphere, it may exert a retarding influence, but to a lesser extent than alumina.

Relatively small amounts of a catalyst are required to cause conversion, especially if a liquid phase is formed during the heating. When the reaction proceeds entirely in the solid state, however, the rate of reaction depends mainly on the contact points between the catalyst and the quartz, hence the speed of the change may be increased by adding larger quantities of the promoter.

The relative effect of various oxide catalysts on the decomposition of quartz is shown in Table XI.XV. The amounts added in each case were calculated to give the same number of contact points for reaction between the quartz and oxide (see p. 674). The particle-number ratios of quartz to catalyst were about 200:1 in all cases which corresponds to an addition of about 1 per cent of oxide in each case. Three additions of calcium oxide are included, with progressively increasing amounts, to show the effect of concentration.

The alkali oxides cannot be included in the Table because at 1370° C., quartz is completely decomposed in their presence in a few minutes. Even at 1100° C., 1 per cent of either sodium or potassium oxide will completely convert quartz of the same grain size in four hours.

The temperature of heating also influences the rate of reaction in the presence of catalysts in much the same way as with pure quartz.

It is of great importance that when the added promoter does not form a liquid but catalyses entirely in the solid state, the reaction is rapid only in the initial stages of kiln firing and eventually, if there is an insufficient supply of the accelerating agent, the rate falls to that of pure quartz (see p. 674). This is related to industrial practice—where the normal addition is milk of lime; the bulk of the quartz is rapidly converted (as shown by density measurements), but the final traces can only be transformed after prolonged firing at a very high temperature. Should alumina be present, even in small amounts, the rate of conversion is much less and as this component has other undesirable influences on the properties of the finished bricks, quartzites with the minimum quantity of alumina are preferred.

In addition to influencing markedly the conversion rate, catalytic agents also determine, to a large extent, the nature of the product formed.

The alkali metals, especially when combined with highly polarisable cations such as phosphate or tungstate, favour the production of tridymite. These are the only agents which are accepted generally as tridymite promoters, although it has been claimed that iron oxide added as boiler scale behaves similarly. Whilst it is true that silica bricks rich in tridymite can undoubtedly be produced by the addition of 5 per cent of this material, it is a matter of conjecture whether the catalytic influence should be attributed to the iron oxide or the alkali salts which are also present in boiler scale.

Most other oxides or compounds either catalyse the formation of cristobalite or do not influence the conversion of quartz to any appreciable extent.

TABLE XI—XV. THE PERCENTAGE DECOMPOSITION OF MADAGASCAR QUARTZ CRYSTAL IN THE PRESENCE OF VARIOUS CATALYSTS AT 1370° C.

Mixture	*Particle No. Ratio quartz:catalyst*	*Time of firing (hours):* ½	1	2	4	8	16
Pure quartz		1·0	3·0	4·0	6·5	11·0	20·0
+ Fe_2O_3	200:1	11·9	15·7	24·9	38·0	62·7	95·4
+ CaO	200:1	13·9	20·8	25·0	27·4	30·2	39·7
	100:1	20·0	32·6	41·3	47·4	51·2	59·3
	50:1	30·0	41·9	50·2	59·5	63·9	80·0
+ CaO + Al_2O_3	200:1:1	5·5	13·1	18·6	22·7	26·5	28·5
+ Al_2O_3	200:1	1·0	3·0	4·5	6·5	6·5	7·4
+ MgO	200:1	5·0	10·0	17·5	21·5	25·5	33·5
+ TiO_2	200:1	4·2	10·0	18·2	20·5	25·0	30·0

The grain-size of quartz was in all cases − 100 + 120 B.S.S.

The alkali earth oxides are particularly efficient as catalysts and as they do not produce a liquid phase with silica at the temperatures of firing and service, they are the catalysts most commonly employed in the manufacture of high-grade products. Calcium oxide (as milk of lime) is frequently used in amounts up to 2 per cent. The commercial use of barium salts as catalysts in India has been reported.

The raw materials used in the manufacture of silica and siliceous refractories differ from pure quartz crystals in composition and texture and their firing characteristics are correspondingly different. Silica occurs in nature in a variety of forms (see Chapter VI), which include the truly amorphous types, such as opal, and crypto-crystalline varieties (of which flint and chalcedony are examples), in addition to crystalline quartz. Natural quartzites are mixtures of these various forms and the relative proportions determine the nature of the material and of the fired product.

The true amorphous silica minerals are of extremely small particle-size—in fact, the ultimate grains may be of molecular size. Crypto-crystalline material is thought to be composed of particles of quartz of submicroscopic size; X-ray and thermal tests indicate that the proportion of true quartz in them may be as low as

20 per cent, but whether this is the result of the extreme fineness of the particles or to associated colloidal material is not clearly established.

Crystalline quartzites are those with larger grains easily detectable under a microscope, but many samples contain associated flint or chalcedonic grains and most show evidence of secondary silicification (see Chapter I) indicating a proportion of amorphous or non-quartzitic material.

All the naturally-occurring siliceous materials contain small proportions of a wide variety of minerals present as impurities. Sericite or hydrous mica is common in British silica rocks and this is usually the source of alumina in the material. Silcrete, a crypto-crystalline silica rock, mined in Germany, Italy and South Africa frequently contains up to 5 per cent of titanium-bearing minerals.

On firing, such materials are converted to the high temperature modifications but, usually at a far faster rate than are crystals of pure quartz. This is mainly because of the smaller particle-size and the presence of colloidal or crypto-crystalline material which changes rapidly and then acts as 'seeds' for the reaction of the more crystalline varieties. The presence of an amount of such rapidly-converting material is probably essential if a quartzite is to be suitable for the manufacture of siliceous refractories. Pure quartz reacts too slowly under commercial conditions; but, on the other hand, an excess of amorphous silica is also unsuitable. So rapidly does this substance convert that a shape prepared from it may be shattered by the excessive rate of change in volume on heating. Under fairly rapid heating conditions, the transformation in amorphous silica may be detected by observing the thermal change at about 1250° C.

Quartzites of good quality may contain up to 98 per cent silica, but the small percentage of impurity minerals is sufficient to produce an appreciable amount of liquid on firing. This is augmented by the action of the catalyst or promoter added to the raw material to accelerate the rate of conversion. *Lime* or *iron oxide* (the two principal additives) do not readily form a liquid with silica, but the presence of a small proportion of alumina is sufficient to cause fluidity at a kiln-firing temperature of 1450° C.[1]

Ford[2] has calculated the proportion of liquid which can form in silica mixtures when fired to various temperatures. The effect of varying amounts of alumina on the liquid formed by 2 per cent additions of lime and magnesia is shown in Table XI.XVI.

Most of this liquid which is produced on firing becomes a glass on cooling which surrounds the crystals of cristobalite and tridymite which have formed. The properties of this glass are, however, unusual and it may be that, as it is highly siliceous, a structure can be induced in it by the embedded silica crystals.

Alumina when present in siliceous materials, not only gives rise to undesirable liquid formation at high temperatures but also inhibits the degree of conversion. This is especially noticeable in siliceous clays where even at high temperatures and on prolonged heating, quartz grains are converted only slowly to one of the high

[1] It has been estimated that 1 per cent of alumina in a quartzite is sufficient to produce 10 per cent liquid in a brick at a temperature about 1620° C. in the roof of a basic open-hearth furnace.

[2] Ford, W. F., *Refract. J.*, 33, 76, 1957.

temperature varieties of silica. Quartz can often be detected in fireclay refractories which have been in constant service at temperatures over 1250° C. for periods of many years.

Crypto-crystalline quartzites usually contain a higher proportion of impurities than the more coarsely grained varieties, yet, for reasons which have not yet been definitely established, the amount of liquid formed on firing is not as large as would be expected from their composition. This may be due to the rapidly-inverting material trapping the impurities which are more evenly distributed than in the coarse-grained varieties. The liquid is thus limited to small isolated pockets and is not able to disseminate throughout the body as when a coarse material is used.

Silica bricks must be fired to ensure that all the quartz has been converted and that no further change will occur in service. Those of the best quality ought to be fired at Cones 14–18 (1410–1500° C.), including a soaking period (p. 733) at or near the maximum temperature of at least 8–10 hours.

TABLE XI—XVI. THE PROPORTIONS OF LIQUID FORMED IN MIXTURES OF SILICA WITH ALUMINA AND LIME OR MAGNESIA

	2 *per cent lime*			2 *per cent magnesia*	
Alumina (*per cent*)	1·00	0·67	0·30	1·00	0·67
1400° C.	9·4	8·0	6·4	7·0	4·7
1500° C.	11·6	9·3	7·4	9·1	7·8
1600° C.	16·7	12·0	9·0	10·0	8·7
(1650° C.)	(40)	(23)	(16)	(20)	(17)
1700° C.	85	60	30	60	40

The figures in brackets are estimated values.

At temperatures slightly in excess of 1700° C., the crystalline modifications of silica melt into a highly viscous fluid. Pure quartz, if heated in the absence of air, can be melted without excessive conversion at 1690° C., tridymite similarly at 1710° C., whilst cristobalite (the most stable phase) requires to be heated to 1725° C.; in each case the product is virtually identical. Once liquefied, silica does not readily recrystallise and can be supercooled to form a glass which is known commercially as *fused silica*. This can be 'devitrified' or recrystallised by prolonged heating above 1200° C. when cristobalite forms.

Although, after suitable firing treatment at about 1450° C., little or no quartz remains in silica refractories, the combined proportions of tridymite and cristobalite may not be greater than 50 per cent and rarely exceed 75 per cent. Quartz is converted into an intermediate, unstable form of silica from which cristobalite develops. If promoters which catalyse the formation of tridymite are present, the primary cristobalite, in turn, is converted into a transitional phase from which tridymite crystallises.[1] This succession of changes is slow and the catalysts and promoters tend

[1] Gaskell, J. A., Grimshaw, R. W. and Roberts, A. L., *Gas Research Board Publication*, 1956.

to become exhausted as the reactions proceed. Consequently in fired silica bricks, it is usual to find a considerable amount of non-crystalline silica over and above that which could be present as a glass. Firing at high temperatures (*circa* 1650° C.) causes the amorphous silica to crystallise as cristobalite[1] and it may well be that the remarkable ability of silica refractories to withstand temperatures to within a few degrees of their melting point is due to this effect which increases the proportion of crystalline components by as much as 30 per cent.

Inversions. The silica minerals each undergo additional changes when heated which do not involve a pronounced change in crystallographic form (see p. 160). These small rearrangements in structure are termed *inversions* and take place at well-defined temperatures. In contrast to the conversions these reactions are rapid and instantaneously reversible.

They may be summarised as:

Quartz	α	$\rightleftharpoons \beta$	573° C.
Cristobalite	α	$\rightleftharpoons \beta$	210–280° C.
Tridymite	α_1	$\rightleftharpoons \beta_1$	116° C.
	α_2	$\rightleftharpoons \beta_2$	163° C.

It is now known that quartz inverts over a short range of temperature although in most cases it occurs within a degree of 573° C. The variations in the inversion temperature are attributed by Keith and Tuttle[2] to possible defects in the structure or the incorporation of foreign ions. The variability of the inversion temperature of cristobalite is most probably due to a similar cause. The higher the temperature and the longer the duration of firing, the higher is the inversion temperature of cristobalite, but, the presence of foreign ions also exerts a marked influence; thus, sodium catalyses the formation of cristobalite from quartz and the material so produced after heat treatment at 1350° C. inverts at 268° C. Under the same conditions, calcium, magnesium and other cations form cristobalite with an inversion at 232° C.; but cristobalite formed on heating pure kaolin at 1250° C. has an inversion point at about 216° C.

The inversion effect may be measured by the small thermal change which accompanies the reaction; the total heat change is about +2·8 calories per g. and the reaction is thus an endothermic effect on heating but, on cooling, is exothermic, i.e. will liberate heat.

Cristobalite prepared in different ways is frequently different in thermal behaviour; the presence of magnesium gives a cristobalite in which the inversion is sharp and the heat is liberated over only a small range of temperature; lime and, to an even greater extent sodium compounds catalyse the formation of a form which inverts over a range of up to 10° C.

The range of temperature in the inversional effect of cristobalite samples is due to many individual crystallites which each invert instantaneously but at different temperatures. Chaklader and Roberts[3] have demonstrated this fact by cine-microphotography on a thin-section of a silica brick which was slowly heated or

[1] Chaklader, A. C. D. and Roberts, A. L., *J. Amer. Ceram. Soc.*, 44, 35, 1961.
[2] Keith, M. L. and Tuttle, O. F., *Amer. J. Sci.*, Bowen Vol., Part 1, 203, 1952.
[3] Chaklader, A. S. and Roberts, A. L., *British Ceram. Soc.*, Autumn Meeting, 1956.

cooled through the inversional range of cristobalite on a hot-stage microscope; the change from $\alpha \rightleftharpoons \beta$ cristobalite in a crystallite was accompanied by an instantaneous increase in transparency.

Tridymite is unusual in having inversions at two temperatures but it is probable that two forms of the mineral exist each of which occurs in both the low/high or α/β varieties; the difference in type may be due to solid solution. Tridymite formed from pure quartz by firing with potassium oxide for short periods at 1350° C. exhibits only the lower inversion effect. Prolonged firing increases the magnitude of the change at 163° C. until eventually it approaches in size that at 116° C. which has progressively diminished.[1]

The amorphous varieties of silica have no inversion effects and the cryptocrystalline types show the quartz change on a very reduced scale. Finely-ground quartz crystals have an inversion effect which is very much less than that in the untreated material. This is due to the production of a distorted surface layer in which the configuration of quartz has been destroyed.

High temperature or β-cristobalite has been reported as existing in a metastable condition at room temperatures. In partially devitrified siliceous glasses, a crystal form has been detected which gives an X-ray pattern corresponding to the theoretical spacings of β-cristobalite. It has been postulated that the surrounding glass phase exerts a restraining influence on the crystal and prevents the inversion. β-quartz may also exist in the metastable form when silica glass is devitrified at temperatures below 870° C.

The stabilisation of the high temperature forms is not of common occurrence, however, and they are readily destroyed by prolonged heat treatment and by fine grinding.

The Composition of Silica Refractories. A silica brick of good quality must contain less than 1 per cent of alumina and a total alkali content of less than 0·2 per cent. Calcium oxide in amounts up to 2 per cent or iron oxide in amounts up to 5 per cent, are added as mineralisers.

At high temperatures, liquid is produced, the amount of which depends on the maximum temperature and the composition of the bricks. Invariably, all the alumina present enters the liquid phase and remains as a glass on cooling; associated with it are alkali and alkaline earth elements and a considerable proportion of silica, but the amount of liquid at a particular temperature in a silica brick is largely dependent on the proportion of alumina.

The crystalline phases in a fired silica refractory are usually restricted to cristobalite and tridymite. The former is the main product if lime is used as a promoter, whilst the latter is favoured by alkali elements or boiler scale which is principally iron oxide. Because the amount of calcium oxide which is required to convert raw quartzite in a reasonably short period of time is small and because, in small amounts, it does not injure the valuable properties of silica refractories, lime is the usual promoter used in the manufacture of high-grade silica bricks, which are used in furnaces to withstand temperatures of about 1650° C. Cristobalite is the only crystalline mineral formed under such conditions and because of its inversion at 220–280° C. bricks which contain it are liable to thermal shock (p. 948) if subjected

[1] Plumat, E., *Verre Sil. Ind.*, 13, 81, 1948.

to temperature fluctuations within this range of temperature. Bricks composed largely of tridymite are less liable to this defect and in all cases where silica bricks have to withstand being heated or cooled at low temperatures, tridymite bricks are preferred, e.g. in gas retorts. Because the impurity content of such bricks is relatively large they cannot be used at high temperatures. As stated on p. 756, commercial silica refractories frequently contain a high proportion of non-crystalline silica due to the arrested reactions in the cycle:

Quartz → transitional phase → cristobalite → transitional phase → tridymite

The proportion of non-crystalline or 'transitional phase' silica is usually higher in bricks of the greatest purity.

The Calcining of Flint. Flint is an important source of highly reactive silica, of particular value in pottery bodies and in glazes. The origin of flint has been discussed on p. 312; it occurs in nodular form, probably arising from the precipitation of silica from solutions around an organic nucleus. Mineralogically it consists of fine crystallites of quartz, but of such a size as to give a diffuse X-ray pattern and a low differential thermal analysis effect. Both combined water and organic matter are associated with the pebbles and it is these which cause problems in use.

If raw flint were to be used in ceramic bodies, it would shatter on heating due to the explosive release of combined water between 300–400° C. Articles in which the raw flint was incorporated would also tend to suffer damage during the inversional change of quartz at 573° C.

Before it can be used in pottery bodies, therefore, flint has to be calcined to eliminate the shattering tendency, but care is essential to avoid overfiring. Water and carbonaceous matter can linger in the calcined flint even up to 1100° C. At this temperature, although the density has fallen from 2·60 to 2·48, cristobalite is not usually detectable, although this crystalline silica phase does develop rapidly when flint is heated in the range 1150–1250° C.

Cristobalite is a harmful phase to incorporate into a pottery body, because it is both slow to react and undergoes volume changes at low temperature. Therefore, the successful calcination of flint involves critical control of the kiln to attain complete loss of water at about 1100° C., but without overfiring to produce cristobalite.

The calcining of flint is carried out without prior grinding. The natural form is extremely hard, but heating shatters the grain making size reduction very much easier.

FIRING CHANGES IN OTHER ALUMINO-SILICATES

THE principal alumino-silicates of commercial importance apart from clays, are the kyanite, andalusite, sillimanite group which have the general formula $Al_2O_3.SiO_2$. These minerals occur in a relatively pure state in Nature and are used in the manufacture of high alumina refractories. They are stable when heated until temperatures in excess of 1000° C. are reached; above this, dissociation into mullite and cristobalite occurs as shown by the following equation:

$$3(Al_2O_3.SiO_2) = \underset{\text{(mullite)}}{3Al_2O_3.2SiO_2} + \underset{\text{(cristobalite)}}{SiO_2}$$

The true temperature of reaction has not been established with certainty, because the change occurs entirely in the solid state and is therefore slow. In practice the raw material is heated as high as possible to develop large crystals of mullite which confer exceptional strength on the body.

A true sillimanite (so-called) brick prepared from one of the highly-calcined ortho-silicates is rich in mullite (see p. 111). Attempts have been made to compound a similar product by adding calcined bauxite to a fireclay. By and large, such bricks are vastly inferior in properties to a true mullite because corundum tends to separate as the primary crystalline phase with mullite forming only in relatively small amounts as the result of secondary development from a liquid phase. McGee and Dodd[1] have shown that such bricks contain only 23 per cent mullite even after firing to 1600° C. A better product resulted when small amounts of sodium fluoride were added to the brick batch.

Bricks produced from calcined kyanite, sillimanite or andalusite are generally called *sillimanite bricks* even though this mineral is not present. This is an old terminology which became current before the mineralogical changes which occur on firing had been adequately explained.

Synthetic mullite[2] and articles made from it, must not be confused with the alumina-enriched fireclays. A commercial product is manufactured from pure calcined bauxite with a china clay addition to give the ideal mullite formulation. Suitable catalytic additives and firing to about 1750° C. enable a body to be produced which is virtually pure mullite in a well-developed crystalline form of low porosity. As an ingredient of refractory bodies, this material is superior even to the best calcined kyanites in that it possesses excellent load-bearing characteristics at high temperatures and is remarkably resistant to abrasion and corrosion.

Fused mullite articles are made by melting calcined diaspore or bauxite, grinding it to a powder, mixing it with sufficient fireclay to produce a so-called 'synthetic mullite', fusing the mixture in an electric arc furnace and casting into suitable shapes. The production is too costly for most purposes, but such bricks for glass-melting furnaces are very satisfactory. These blocks are extremely dense and resistant to erosion; they are composed of a felted mosaic of mullite needles and dendrites with glass filling voids and interstices.

Alumina bricks containing more than 92 per cent alumina contain a high proportion of corundum and some mullite. Because a clay is usually added as a bond, there is a proportion of glass, but most of the liquid formed on firing tends to crystallise on cooling as corundum or mullite.

FIRING CHANGES IN OTHER CERAMIC MATERIALS

Magnesia and Magnesite Refractories

The most important basic raw material for use as a refractory is magnesium oxide. It does not occur in Nature but is easily prepared by calcining the carbonate or hydroxide, both of which are available commercially (see Chapter VI). These

[1] McGee, T. D. and Dodd, C. M., *J. Amer. Ceram. Soc.*, 44, 277, 1961.
[2] Hawkes, W. H., *Trans. Brit. Ceram. Soc.*, 61, 689, 1962.

minerals dissociate at temperatures below 800° C. to form amorphous magnesium oxide, which, in this condition, will readily reabsorb carbon dioxide or water vapour on exposure to the atmosphere.

On being calcined at a sufficiently high temperature, magnesia slowly crystallises to *periclase* which is stable and can be cooled and maintained at room temperatures without changing form. This material which is produced by firing the magnesium mineral in a rotary kiln at about 1650° C. is known as *dead-burned magnesite.*

The formation of periclase from the amorphous oxide is a slow process even at high temperatures, but the reaction may be accelerated by the addition of certain oxides which increase the sintering rate; iron oxide is commonly employed for this purpose as also is alumina. Although such additions are necessary, they increase the proportion of impurity in the raw magnesite which in its natural state is usually contaminated with varying amounts of calcium, silicon, iron and aluminium minerals.

The total impurity content in a dead-burned magnesite, is comparatively small, yet it exerts a pronounced influence on the properties of the fired product. At high temperatures, the various components combine to form specific minerals, the nature of which is governed purely by the chemical composition. The predominant component is magnesia, and its high temperature form, periclase, may constitute 90 per cent or more of the fired product The periclase crystals are cubic which neither intergrow nor interlock and do not, as a consequence, impart strength to the material. The bonding between the periclase grains is due to the additional minerals which form through the combination of the impurities in the raw material.

The complete mineralogical composition of a fired magnesite can only be computed accurately by referring to the complex equilibrium diagram in which each component is represented. Rigby *et al.*[1] have, however, evolved a simple scheme which enables the phases present in any mixture to be forecast.

The components of most magnesite materials are principally calcium oxide, magnesium oxide (in excess), iron oxide, alumina and silica in order of decreasing basicity. At the high temperatures of firing, reactions take place at a rate which depends principally on the relative affinity of the various components. Hence, the more basic components will combine most readily with those of strongest acidity, though the relative concentrations also largely determine the ultimate products. As a result of deductions made from the X-ray study of many magnesites—the chemical composition of which had been accurately assessed—Rigby *et al.* formulated general rules for predicting the fired composition. There are two general cases.

1. When the $CaO:SiO_2$ molecular ratio is less than 2:1

(*a*) When the ratio of the *molecular concentrations* of $CaO:SiO_2$ is *less than 2, but greater than 1·5.* Of all the components, lime has the greatest affinity for silica, hence these two components will combine to form the stable *dicalcium silicate* $2CaO.SiO_2$. The silica present in excess will combine with part of the lime and magnesia to give *merwinite* ($3CaO.MgO.2SiO_2$). The relative proportions of these two minerals will depend on the $CaO:SiO_2$ ratio, the former predominating if this is near 2:1, but the latter becomes increasingly prevalent in mixtures where the ratio is nearer 1·5:1.

[1] Rigby, G. R., Richardson, H. M. and Ball, F., *Trans. Brit. Ceram. Soc.*, 46, 311, 1947.

(*b*) Where the $CaO:SiO_2$ ratio lies *between 1·5:1 and 1·0*, the silica combines with the available calcium and some magnesium. *Merwinite* is the principal phase in compositions near the higher ratio of lime: silica but *monticellite* ($CaO.MgO.SiO_2$) also forms at lower ratios.

(*c*) When the ratio of $CaO:SiO_2$ is *less than 1:1* all the calcium is combined as monticellite and the excess of silica reacts with magnesia to form *forsterite* $2MgO.SiO_2$.

In all cases considered above any alumina or iron oxide which may be present combines with magnesia to form *spinel* $MgO.Al_2O_3$ and *magnesioferrite* $MgO.Fe_2O_3$.

2. **When the $CaO:SiO_2$ ratio is equal to or greater than 2:1**

(*a*) *If the ratio is exactly 2:1*, all the lime and silica present will, theoretically, combine together to form *dicalcium silicate* and the alumina and iron oxide react with magnesia as before.

(*b*) *If the ratio is greater than 2:1*, all the silica combines with lime to give *dicalcium silicate.* The other phases are determined by the relative excess of lime. *Brownmillerite* ($4CaO. Fe_2O_3.Al_2O_3$) forms in most mixtures, although under favourable conditions *dicalcium ferrite* ($2CaO.Fe_2O_3$), *tricalcium silicate* ($3CaO.SiO_2$), *tricalcium aluminate* ($3CaO.Al_2O_3$), or *penta-calcium trialuminate* ($5CaO.3Al_2O_3$) and the other crystalline phases previously mentioned, may be present.

Many of these compounds have a comparatively low melting point (Table XI.XVII), or can form low melting point eutectics which cause the failure under stresses of magnesite refractories at comparatively low temperatures.

TABLE XI—XVII. MELTING OR SOFTENING POINTS OF SOME COMPOUNDS FOUND IN MAGNESITE REFRACTORIES

Mineral	*Composition*	*Melting Point °C.*
Periclase	MgO	2800
Forsterite	$2MgO.SiO_2$	1910
Monticellite	$CaO.MgO.SiO_2$	1500*
Merwinite	$3CaO.MgO.2SiO_2$	1575*
Dicalcium silicate	$2CaO.SiO_2$	2130
Tricalcium silicate	$3CaO.SiO_2$	1900
Brownmillerite	$4CaO.Fe_2O_3.Al_2O_3$	1415
Calcium ferrite	$CaO.Fe_2O_3$	1205*
Dicalcium ferrite	$2CaO.Fe_2O_3$	1436*
Tricalcium aluminate	$3CaO.Al_2O_3$	1535
Pentacalcium trialuminate	$5CaO.3Al_2O_3$	1455
Magnesioferrite	$MgO.Fe_2O_3$	1751
Spinel	$MgO.Al_2O_3$	2135

* Dissociates.

The **fired compositions** of many magnesite products have been calculated by Rait[1] from phase equilibria studies and confirmed by X-ray and thin-section microscopy determinations. He showed that the nature and amounts of the minerals which were present were related to the composition of the raw material. The

[1] Rait, J. R., *Basic Refractories* (Iliffe and Sons Ltd., London, 1950).

principal minerals which occur in magnesite refractories and the proportions in which they are liable to be present are summarised in Table XI.XVIII.

Most British magnesite refractories are made from sea-water magnesite (p. 325) and in general they contain more lime than those made from naturally-occurring carbonates. In bricks made from a material low in lime, e.g. Austrian magnesite, *monticellite* in amounts up to about 5 per cent is present. Siliceous magnesites may contain a high proportion (up to 20 per cent) of forsterite, but there is a consequent reduction in the content of periclase. The amount of glass which is found in commercial magnesites is always small in amount.

TABLE XI—XVIII. THE PRINCIPAL MINERALS IN BRITISH MAGNESITE REFRACTORIES

(after Rait)

Periclase	73–88 per cent	(average 86 per cent)
Dicalcium ferrite	0–11 per cent	(average 1 per cent)
Magnesioferrite	1–8 per cent	(average 4 per cent)
Spinel	0–7 per cent	(average 2 per cent)
Brownmillerite	0–8 per cent	(average 4 per cent)
Merwinite	0–9 per cent	(average 1 per cent)

Talc-containing bodies are of considerable interest in ceramic studies because many of them have excellent thermal and electrical properties coupled with low thermal expansions.

Talc itself has the composition $3MgO.4SiO_2.H_2O$ and is a layer-lattice silicate (p. 140). The complete reaction on heating may be represented by the equation:

$$Mg_3Si_4O_{10}(OH)_2 = 3(MgSiO_3) + SiO_2 + SH_2O$$

enstatite cristobalite

Enstatite does not form readily, except at high temperatures and the presence of silica introduces undesirable properties.

Steatite bodies are formed by combining talc with appropriate amounts of kaolinite to produce a fired body containing mainly enstatite and cordierite ($2MgO.2Al_2O_3.5SiO_2$), with only a minimum quantity of free silica.

Steatite bodies have a very short firing range but have excellent dielectric properties. The addition of small amounts of barium carbonate to a talc body containing 10–15 per cent china clay produces a body which has better firing characteristics.

Cordierite bodies are manufactured from talc to which appropriate amounts of clay and alumina or sillimanite have been added to correspond to the cordierite composition ($2MgO.2Al_2O_3.5SiO_2$). Firing of bodies of this composition is difficult in that liquid develops at a rapid rate over a small temperature range. Cordierite bodies have extremely low thermal expansion characteristics and in addition to possessing excellent thermal expansion characteristics they also have stable electrical properties at high temperatures.

Lime and Calcium Containing Ceramic Bodies

Calcium, either as carbonate, sulphate, or alumino-silicate is a common constituent of clays where it behaves as a powerful flux or liquid-former particularly at high temperatures.

Calcium oxide either in the pure form or associated with other basic materials, such as magnesia, is a potential source of valuable refractory material. The melting point of lime is in the region of 2600° C. and the eutectic composition with magnesia does not fuse below 2200° C.

Lime does not occur in nature but is readily produced by calcining chalk or limestone—both commonly-occurring carbonates (see p. 328)—at about 900° C. The mixed carbonate of calcium and magnesium, found as *dolomite* or *magnesian limestone* is also a common rock formation; on calcination at a high temperature, it is converted into an intimate mixture of the two oxides, lime and magnesia.

Calcium oxide, however, unlike magnesium oxide, cannot be stabilised so as to avoid atmospheric rehydration or recarbonisation merely by firing at a high temperature. Sintering may take place and the decrease of surface area thereby resulting reduces the rate of attack of water vapour or carbon dioxide, but even then deterioration and eventual disintegration proceed at a sufficient speed to prevent calcium oxide from being used as a refractory for all except small laboratory articles. Stabilisation of a temporary nature can be secured by coating the individual grains with tar to exclude air and, in this form, lime (and more particularly dolomite) is used to line basic furnaces. On calcination, however, the tar is burnt away and the lime again becomes active and would disintegrate on exposure to cold humid air.

Complete stabilisation is possible by converting the lime into a chemical compound which is not susceptible to rehydration or recarbonisation. The most obvious choice is to form a calcium silicate, but this leads to difficulties because the ortho-variety of dicalcium silicate ($2CaO.SiO_2$), (which is the one most liable to be produced by the direct combination of lime with silica), exists in three forms (see p. 112). The α- and β-varieties, which are high temperature modifications with closely similar properties, develop on calcination but, on cooling, the γ-variety forms with a large volume expansion of the order of 10 per cent. This would be sufficient to disrupt any brick or other shape which was composed of the mineral and, even when present in small amounts, dicalcium silicate causes *dusting* which is a frequent defect in dolomite bricks.

Efficient stabilisation of lime is possible by converting it into tricalcium silicate which also has inversions but these do not bring about a drastic change in property. The mineral is, however, stable only within a small range of composition (see Fig. X.50, p. 647) and any chemical additions have to be carefully controlled and blended to ensure the desired reaction.

In practice, magnesium silicate, as talc or serpentine, is normally added to the dolomite to provide silica and to introduce a higher proportion of magnesia. The complete reaction is as follows:

$$\underset{\text{dolomite}}{6(MgCO_3.CaCO_3)} + \underset{\text{talc or serpentine mineral}}{3MgO.2SiO_2.2H_2O} \rightarrow \underset{\text{tricalcium silicate}}{2(3CaO.SiO_2)} + \underset{\text{periclase}}{9MgO} + 12CO_2 + 6H_2O$$

It is considered desirable to have an excess of lime rather than silica in the fired product.

The manufacture of dolomite refractories has been considerably eased and the properties improved by stabilising α- or β-dicalcium silicate by small additions of boric oxide, chromic oxide or phosphorus pentoxide.[1] The change to γ-dicalcium silicate below 675° C. then does not occur or does so only slowly, so that the large volume expansion and consequent disruption of a shape which contains the mineral can be largely avoided.

Although bricks containing 30 per cent of stabilised α-dicalcium silicate are manufactured and do not seriously deteriorate during storage, a dolomite refractory with such a relatively high proportion would have other undesirable properties which would make it unsuitable in service. More liquid would be formed at temperatures exceeding 1600° C. and furthermore it would be less resistant to the action of slags rich in silica, lime and iron oxide, than would a brick containing chiefly tricalcium silicate. In British practice, it is usual to keep the content of dicalcium silicate below 5 per cent.[2]

Stabilised dolomite refractories contain a considerable proportion of impurities, similar to those in magnesite. These arise from the contaminants in the raw ores and from the additions made to promote sintering. Dolomite bricks are comparable in property to those of magnesite, but they require careful chemical control during manufacture and some have a tendency towards atmospheric deterioration.

The firing changes which take place during the burning of dolomite refractories are complex and have been studied exhaustively by Rait.[3] The presence of alumina and iron oxide leads to the initial formation of a liquid at a low temperature, so that the proportion of these two impurities must be kept as low as possible. However, should alumina and iron oxide be entirely absent, liquid would not form at the usual temperatures of firing (*circa* 1450° C.) and the bricks would be of low strength. Rait has shown that the alumina and iron oxide present in a stabilised dolomite refractory usually form brownmillerite ($4CaO.Fe_2O_3.Al_2O_3$) and he has investigated the effects of this component on the production of liquid at high temperatures. From the analyses of typical dolomite refractories, Rait concluded that liquid would be formed initially at about 1280° C. and, at 1400° C., 12–16 per cent could be expected. On cooling, most of the liquid crystallises and the nature and amounts of the minerals present are shown in Table XI.XIX.

Semi-stable dolomite is calcined in a rotary kiln and the lime 'stabilised' by

TABLE XI—XIX. CONSTITUTION OF STABILISED DOLOMITE REFRACTORIES

Periclase	40–45 per cent
Tricalcium silicate	45–38 per cent
Dicalcium silicate	4–12 per cent
Brownmillerite	6–9 per cent
Spinel	1–3 per cent

[1] Bates, P. H. and Klein, A. A., *Bur. Stand. Tech. Paper No.* 78, 1917.
[2] Swinden, T. and Chesters, J. H., *J. Iron & Steel Inst.*, 144, 105, 1941.
[3] Rait, J. R., *Basic Refractories* (Iliffe and Sons, Ltd., London, 1950).

coating the crushed, sintered mass with tar to exclude the atmosphere. A proportion of impurities is necessary to accelerate the sintering rate and to provide a flux which gives strength to the mass and to some extent 'bottles' the lime and so retards its rehydration. Free lime, periclase and a small proportion of tricalcium silicate are the principal minerals present, but amounts of brownmillerite and dicalcium ferrite which total between 5–8 per cent are invariably present.

Modern furnaces for melting steel operate under vicious conditions of temperature (in excess of 1700° C.) and basic slag attack. Live oxygen is blown into the melt to accelerate the formation of steel. A charge of over 200 tons of steel can be prepared in about 20 minutes.

Such furnaces are generally lined with lime or dolomite of high purity. Tar-bonded bricks made by sealing the calcined raw material with tar by forming them under high pressure are commonly used.

Wollastonite is the metasilicate of calcium $CaSiO_3$ which frequently crystallises as an undesirable phase in glass compositions but is valued as a ceramic body in its own right. It is a major component of some electrical ceramics which combine a low dielectric loss with a high mechanical strength.

Wollastonite forms readily from mixtures of lime and silica; it melts congruently at 1544° C. (see Fig. X.18. p. 607), but exists in two solid forms, (i) pseudo—or α–wollastonite which is stable above 1125° C. and (ii) the normal — or β–wollastonite, which is the stable form below 1125° C. The transition is very slow under ordinary conditions, and this can present problems in use.

Chrome Products

The essential mineral component of chrome ores is *spinel* which has the general composition $R^{2+}O.R_2{}^{3+}O_3$ where R^{2+} is a divalent element, such as Mg and Fe^{2+} and R^{3+} is a trivalent element, for example Al, Fe^{3+}, or Cr (for more details see p. 191). Associated with this mineral are various impurities such as hydrated silicates of magnesium and iron.

On firing, the spinel grains do not change appreciably in composition, though there may be some oxidation of ferrous iron and some solid solution effects. The impurities which are more prevalent than in either raw magnesite or calcium-bearing materials, recombine to give the typical mineral components which form the inter-crystalline linking in fired magnesites. Spinel crystals are cubic, like periclase, hence the composition of a fired chrome ore brick is similar to that of a magnesite with spinel replacing periclase and a greater predominance of crystalline impurities. Because silica is always present in appreciable amounts in chrome ores, the gangue is rich in low-melting silicates which reduce the resistance of the material to high-temperature stresses.

This drawback in chrome refractories can be largely offset by introducing finely-divided magnesia to react with the excess silica and form a proportion of forsterite which has a high melting point. Such composite materials containing 70 per cent chrome ore with 30 per cent dead-burnt magnesite are termed *chrome-magnesite* and because of their enhanced properties at high temperatures they find considerable application in steel-melting furnaces.

Chrome-magnesite refractories are composed essentially of spinel grains surrounded by a matrix which is mostly crystalline. The composition of the spinel grains depends chiefly on the constitution of the raw ore (p. 328) although some changes in the type of solid solution may occur on firing. The matrix, which comprises about 25 per cent by weight of the brick, is rich in magnesia and contains between 50–70 per cent of finely-divided periclase, forsterite is present in amounts between 12–17 per cent, whilst 10–20 per cent of monticellite ($CaO.MgO.SiO_2$) is also to be found. The only other constituent which may occur in any quantity in the matrix is finely-divided spinel.

The behaviour of bricks under stress at high temperatures is largely dependent on the proportion of monticellite as this is a low-melting constituent which combines readily with slags rich in silica, alumina and iron. The amount of monticellite in a chrome-magnesite refractory depends on the content of lime in the raw material, hence this component should be kept as low as possible. There is, however, a limiting value because, otherwise, too little liquid would form during kiln-firing and bricks of sufficient strength could not be produced.

CHEMICAL PROPERTIES OF FIRED CERAMIC PRODUCTS

THE chief chemical properties of fired ceramic articles and other products are those which enable them to resist the chemical action of acids, alkalies and slags—the latter being a serious cause of *corrosion*.

Resistance to acids and alkalies is chiefly obtained in highly vitrified products such as some engineering bricks, stoneware and porcelains. When resistance to alkalies, lime, magnesia and slags is required at very high temperatures it is less feasible to use highly vitrified materials as they tend to fuse too readily; in such cases the density must be obtained by skilful grading of the raw materials so as to ensure close packing and so controlling the firing that no more glass is produced than is necessary.

Most ceramic materials are used because of their *inertness* to their contents (if hollow) or to their immediate surroundings. If they are unduly affected, they are either used unsuitably or are defective and should be dealt with accordingly.

CORROSION

CORROSION may be defined as a chemical, physical or electrolytic reaction between two or more components which results in the solution or disintegration of at least one of those components. The nature of the reactions may be similar to those involved in the firing of the raw material, but corrosion effects are, in general, undesirable and imply a reduction in the valuable properties of the ceramic material involved. Thus, the action of iron oxide on magnesia or of lime on silica may be useful, if in a small degree, in producing a suitable bond, but when these substances in excessive amounts attack refractory materials whilst the latter are in use, the effects are undesirable and may be included under the term ‘corrosion’.

Many factors influence corrosion, some of which have been described on pp. 698–709 and on p. 750. The principal ones are:

(*a*) The available area of contact between the corroding agent and refractory.

(*b*) The chemical affinity between the two reactants.

(*c*) The mobility of the corroding agent.

(*d*) The permeability or ease of penetration of the corroding agent into the ceramic material.

In some cases, the continued action of a corroding agent may, in time, prevent or hinder further action taking place. Thus, when calcium compounds attack fireclay a calcium alumino-silicate is first formed which is high in calcium and has a low fusing point. As more clay is dissolved, the material formed becomes richer in alumina and silica and at the same time more refractory, so that in time a protective coating is formed and further corrosion is prevented or at least proceeds very slowly. There arc six chief groups of substances which have a corrosive action on ceramic materials:

(*a*) Free bases which combine directly with clay, silica, etc.

(*b*) Silica, clay, and other materials which react like acids with basic refractory materials such as magnesite, etc.

(*c*) Salts which, on heating, are decomposed, liberating a base and an acid, the base uniting with the fireclay, silica, or analogous refractory material and producing a fusiblc compound. To this class belong substances such as calcium sulphate, sodium chloride, etc. Thus, the soluble salts present in some coals used for making gas or coke cause much trouble by corroding the material of which the retort or coke oven is made. Sodium chloride is usually the most important of these salts, though others may also be present.

(*d*) Direct-acting salts which act like free bases, though to a less degree. Felspar, mica, Cornish stone, and the carbonates are of this class.

(*e*) Fly ash and other corrosive substances carried in the hot gases in kilns, furnaces and ovens. These may be derived from the fuel or they may be picked up by the high velocity gas stream.

(*f*) Corrosive gases of especial chemical nature which preferentially attack specific components in the ceramic material.

Fluorine is particularly troublesome in that it volatilises readily at high temperature and is a powerful flux. It will combine with silica to form gaseous silicon tetrafluoride.

Carbon monoxide is not specifically a corrosive agent but it is troublesome in that it will react at the surface of iron nodules contained in ceramic bodies and there liberate carbon which can weaken the refractory to further attack.

The chief corrosive agents to which clay, silica, and analogous materials are likely to be subjected are molten metals, slags, such as those produced in open-hearth, reverberatory, converting, and cupelling furnaces, consisting chiefly of basic materials or silicates (such as the slags produced in assaying), and metallic oxides. The action of volatilised substances may also be serious in some cases.

Lime should not be brought into contact with clay or silica at high temperatures or partial fusion may take place.

Iron compounds form ferrous or ferric alumino-silicates, the former being more fusible and, therefore, more corrosive than the ferric compounds. Iron oxide corrodes fireclay bricks more severely than it does silica bricks. Sulphides are also very corrosive and rapidly penetrate clay products and silica. Iron sulphides are especially harmful in their action on these materials, because they combine with silica to form fayalite ($2FeOSiO_2$) and sulphur dioxide which escapes as a gas. Any sulphur which may be present in the fuel used in furnaces may exert a corrosive action upon the brickwork at high temperatures, the sulphur being converted first into sulphur dioxide and then into sulphuric acid which attacks the bricks.

The presence of steam often facilitates the corrosive action of other substances; thus, its presence is necessary for the decomposition of some of the soluble salts in coal, and, for this reason, the preliminary drying of a coal may reduce the amount of corrosion of any firebricks with which it may be heated.

Fireclay bricks which are high in alumina are usually less easily corroded by fluxes, etc., than those rich in silica, but the amount or rate of corrosion is not always proportional to the total amount of alumina present. The solution of alumina by a slag containing a metasilicate may increase the fluidity of the slag and, therefore, increase its solvent action on bricks. With silicates richer in flux than metasilicates, however, the solution of alumina decreases the viscosity and increases the refractoriness of the bricks.

Table XI.XX shows the refractoriness of various mixtures of slag and refractory material, obtained in the slag test devised by R. M. Howe, S. M. Phelps and R. F. Ferguson, in which the lowering of the refractoriness of a powdered mixture of slag and refractory material is regarded as the index of the intensity of the reaction.

TABLE XI—XX. REFRACTORINESS OF MIXTURES OF SLAGS AND FIRECLAY BRICKS*

(The results are expressed in Seger Cones)

Type of Slag	*Percentage of Slag in Mixture*								
	0	4	8	12	16	20	30	40	50
Acid open-hearth	32	32	32	30	28	26	13	11	8
Blast furnace	32	30	29	20	17	13	13	12	9
Basic open-hearth	32	30	26	16	15	13	9	8	5
Heating furnace	32	31	30	26	19	18	10	9	8
Coal ash (clinker)	32	32	32	31	31	31	30	29	20

* The fireclay had an alumina content of 33 per cent.

Metallic vapours sometimes cause the destruction of firebricks. Thus, when vapours of metallic zinc penetrate the bricks of the blast furnaces in which the ores are smelted, they are oxidised by the CO_2 present and are solidified in the pores. This may cause disintegration if the bricks are not very strong. A similar corrosion occurs when zinc is smelted in retorts.

The penetration of vapours of sodium chloride into fireclay retorts may cause the formation of ferrous chloride which will, in turn, volatilise and act as a corrosive agent.

Flue dust has a complex action upon firebricks producing a fusible mixture of various silicates and alumino-silicates, according to the composition of the dust.

Mellor and Emery found that the oxides of copper, zinc, and iron in a finely divided condition have a very high penetrative power under reducing conditions at very low temperatures. Copper oxide will penetrate at 600° C. and iron at 1100° C,

The texture of **fireclays** largely influences the degree of resistance to erosion. A low-grade fireclay, for example, which has been fired to a temperature at which extensive vitrification occurs may be more resistant to metal or slag penetration than a clay of higher grade fired at the same temperature. The low refractory clay readily forms a protective glassy coat on its surface when in contact with the slag and thereby prevents the penetration of the corrosive agent along the pores of the material. Similarly, a brick with an initially low permeability is less liable to corrosive action than one where the penetration may be high.

Clays which bloat at high temperatures are often extremely resistant to corrosion probably because the increase in size is sufficient to seal surface pores and to reduce the sizes of the joints between adjacent bricks. Such bricks are commonly used, especially in the U.S.A. as linings for steel ladles. In general, siliceous clays are more readily corroded than others due to the greater ease with which they form a liquid.

Aluminous clays are less liable to be attacked and this may be due in no small part to the ease with which they form *mullite*, which is extremely stable and is not readily decomposed.

Silica bricks are in theory more reactive to basic slags at high temperatures than are fireclays, but the amount of attack depends on the texture. The penetration of slags into a siliceous refractory is through the glass which surrounds the crystallites of cristobalite and tridymite, or along pores and fissures leading from the surface; a close-textured brick and one containing little glass is much less readily attacked. The action of lime and iron oxide on silica at high temperatures is not as pronounced as would be expected from a combination of bases and acids. As shown in Figs. X.18, 19, 20 (p. 607ff.) oxides of iron, calcium and magnesium form an immiscible liquid phase with silica and for this reason a large proportion of these oxides can be absorbed in a silica brick without an undue deterioration in property occurring. If small amounts of alumina are present as an impurity, the reaction is very rapid and the brick readily disintegrates.

The permeability of a silica brick has a pronounced influence on its corrosion resistance to dusts and slags at high temperature. Bricks made from silcrete (p. 316) contain a larger proportion of impurities than do refractories made from silica rock. The porosity of the former is only about half that of the bricks made from quartzites, consequently the slagging action is confined largely to the surface and does not readily penetrate along the pores. For this reason, silcrete bricks are far more resistant to corrosion than those made from high quality quartzite.

Fused silica has a high resistance to attack by most chemicals and because it also has a high resistance to thermal shock it is a valuable material in chemical plants and in laboratory apparatus. It is rapidly attacked by hydrofluoric acid and should not be kept in contact with alkali.

Magnesite, dolomite, chrome and chrome-magnesite bricks are basic in character and are consequently more resistant to basic slags than are silica and fireclay bricks. Components of such slags, however, combine readily with the matrix minerals of such bricks and form low melting compounds (p. 762) which cause disintegration.

Chrome-magnesite refractories, because the gangue material is essentially forsterite are less liable to the penetration of slags. The spinel crystals which form the greater part of the refractory are attacked by iron oxide. The action is still not completely explained although apparently iron oxide Fe_3O_4 is isomorphous with certain spinels and can form solid solutions giving rise to a large volume expansion. This 'iron-oxide bursting' may be very severe and limits the use of some chrome-magnesite bricks in many industrial and metallurgical applications.

Aluminous bricks and those made from kyanite are moderately resistant to most slags—the rate of attack is very dependent on the texture.

Zircon and **zirconia** bricks and crucibles are highly resistant to slags and consequently they are very valuable refractories. They withstand the corrosive action of acid slags, liquid glasses of various kinds, most molten metals and some basic slags. Zircon refractories have been used in gas retorts where the attack by iron oxide at high temperatures is very severe and have proved successful, but their high cost limits their use.

Carbon bricks are extremely resistant to molten slags and metals of all types. They are serviceable at high temperatures, however, only in the absence of air.

Silicon carbide is also highly resistant to most slags and molten metals.

Measurement of Corrosion. The usual method for measuring the resistance of ceramic materials to slags is to drill a hole 1 in. diameter and $\frac{1}{2}$ in. deep in a block 3 in. × 2 in. × 2 in. cut from a brick. This recess is filled with a powdered sample of the slag and the block fired to the normal contact temperature between the slag and the refractory material. After the maximum temperature has been maintained for two hours, the furnace is cooled and the block cut vertically to show the depth of penetration.

Chesters[1] has described a slag spray test in which powdered slag is fed into the flame from a gas burner and then sprayed on to a panel of the bricks under test. The extent of penetration and the wear on the bricks are then measured.

Howe, Phelps and Ferguson[2] suggest that the resistance of refractory materials to the action of slag may be tested by powdering the bricks and slag so that they will pass completely through an 80-mesh sieve, mixing them in various proportions, and determining the refractoriness of the mixtures. They claim that this method more nearly agrees with the conditions observed in works' practice, the chief of these being (*a*) increased intensity of action with increased slag concentration; (*b*) the different action of different slags; and (*c*) the effect of any slag on different refractory materials.

Although this method is of value in indicating the probable nature of the chemical attack during corrosion it takes no account of the texture of the ceramic material which is often more important than its chemical nature.

Partridge and others measure the corrosive effect of glass on clay by immersing a specimen of the clay (15 mm. sq. section) in a porcelain crucible containing the

[1] Chesters, J. H., *Steelplant Refractories* (The United Steel Companies Ltd., Sheffield, 1957).

[2] Howe, R. M., Phelps, S. M. and Ferguson, R. F., *J. Amer. Ceram. Soc.*, 6, 589, 1923.

molten glass for a suitable time, afterwards cutting the specimens and polishing. The change in the size of the specimen is measured and the polished surface is examined under a low-power microscope.

Mackenzie[3] has devised a similar simple but highly satisfactory comparison test for measuring the relative erosion rate of ladle materials in the hot metal. A long bar of 1 in. square end section is partly lowered into a crucible of the hot metal. After a certain time interval it may be withdrawn and the amounts of corrosion assessed.

[3] Mackenzie, J., *Trans. Brit. Ceram. Soc.*, 53, 654, 1954.

CHAPTER XII

PHYSICAL CHANGES IN CERAMIC MATERIALS

As with the chemical changes, it is possible to subdivide the physical changes which occur in ceramic materials during processing into (i) those which take place at ordinary temperatures (up to about 120° C.); and (ii) those which occur at elevated temperature.

Changes Occurring at Ordinary Temperatures

In previous chapters, many of the physical changes which occur in ceramic materials have been described in detail: these include (*a*) changes that take place during preparation of the material (Chapter VII); (*b*) those which are the result of the action of water (Chapter VIII); and (*c*) those which accompany the loss of water or during the drying of the body (Chapter IX). This Chapter is concerned mainly with those changes which take place at elevated temperatures, but for completeness those occurring at low temperatures may be summarised.

The physical changes which take place at low temperatures in ceramic materials are chiefly concerned with:

(*a*) A reduction in volume (shrinkage on drying).

(*b*) Increased compactness and density.

(*c*) A loss in weight (chiefly moisture but sometimes due to the evolution of a gas such as carbon dioxide).

(*d*) The removal of one or more substances in solution, as in *weathering* (p. 449).

(*e*) Changes in plasticity and viscosity.

(*f*) Small changes also occur in the crushing strength, transverse strength, toughness, elasticity, porosity and texture, though they are much less than when a higher temperature is reached and the materials are afterwards allowed to cool.

The effects of *pressure* in increasing compactness, of *surface tension* in facilitating the moulding or extrusion of various materials, of *vapour pressure and viscosity* in connection with the extrusion of clay bodies, of *flocculation* and *deflocculation*, all require much attention in various researches as well as in industrial production.

At 'low' temperatures many of these reactions and other changes cannot proceed to completion, though various degrees of stability may be reached and they are of considerable technical importance. Thus the appearance and behaviour of a piece of highly plastic clay are very different from the same piece after it has been allowed to dry at ordinary room temperature and slightly more so after more complete drying at 105°–110° C. Hence the importance of considering such changes as occur at 'low' temperatures apart from those which occur at higher ones.

These physical changes are practically the same as those which occur both at ordinary room temperatures and at any temperature up to 120° C. (though in some cases some decomposition also occurs as the temperature of 100°–120° C. is reached), but the changes usually associated with drying are accelerated by the application of gentle heat or by a current of warmed air.

Physical Changes which occur in Ceramic Materials at Elevated Temperatures

In the previous chapter, the chemical changes which occur on heating ceramic materials have been considered. Firing treatment also causes a pronounced alteration in the purely physical and textural properties either as the direct consequence of a change in the physical properties of the minerals present (e.g. expansion, specific gravity) or through chemical action (e.g. liquid formation, crystalline development).

It is not possible or even desirable to distinguish between the two types of effects in many cases, but their combined action may be observed in the various changes which take place in the textural properties of the body as follows:

(*a*) A change in the temperature of the mass.
(*b*) A change in state of the material, i.e. from a solid to a liquid or to a gas.
(*c*) A change in the volume of the material (expansion or contraction).
(*d*) A change in the porosity of the body.
(*e*) A change in the specific gravity of the material.
(*f*) A change in the strength of the body.
(*g*) A change in weight in the mass.
(*h*) Changes in other physical properties, e.g. optical, electrical, thermal, feel, hardness, etc.
(*i*) A change in colour.

CHANGES IN TEMPERATURE

CHANGES in temperature may be purely physical in character and a simple result of what is termed the absorption of heat (this may simultaneously effect a change in the physical state of the substance, p. 775), or they may be the result of chemical changes which evolve or absorb heat.

The rate and manner in which a mass will increase in temperature depends on several factors, the most important of which are outlined below:

(i) The *difference between the temperature of the environment and that of the body itself*. Heat, being a form of energy, is readily transferred from a hot to a cold source at a rate proportional to the temperature difference.

(ii) The *total heat content of the source*. A hot object behaves as a reservoir of heat; if it is small in size or of a low heat capacity (see p. 929) it will not have much heat to impart to a cold body with which it comes into contact. On the other hand, if a body is placed into a furnace of large heat capacity, it will be quickly raised in temperature without appreciably altering the environmental conditions.

(iii) The *mechanism of heat transfer to the cold body*. A solid body can receive heat either by conduction or radiation. The former process requires that the source of heat be in direct contact and the transfer of energy is from particle to particle of the materials involved. *Conduction* will take place when a cold body touches another at a higher temperature or equally when hot gases are circulated round the mass. Although other factors are involved, the rate of conduction is directly proportional to the difference in temperature.

Radiation is of an entirely different nature. Heat is similar to light in that it can be transmitted in the form of energy waves of a certain frequency; these rays are of longer wavelength than those of the visible part of the spectrum but their mode of transmission is identical. Heat is therefore radiated from a hot body and can be absorbed within a suitable receptive material which is raised in temperature as a consequence, but the intervening medium, usually air or gas which transmits the rays is little changed in temperature. The amount of heat absorption depends on the nature of the material and also its surface, but it is approximately proportional to the difference between the fourth powers of the temperature of the source and receptive materials, i.e. to $T_s^4 - T_o^4$, where T_s is the temperature of the source of radiating heat and T_o is the temperature of the body being heated.

(iv) The *surface of the body* exposed to the source of heat and *the geometrical shape*. The heating of a shape by an external agency proceeds from the surface exposed through the mass of material. It is an obvious fact that a brick will more readily reach a uniform temperature throughout if it is heated on all sides than if only one or two faces were exposed. Similarly, a small specimen will heat up more quickly than one of larger size under the same conditions.

(v) The *specific heat of the material* being heated. Different substances have varying capacities for absorbing heat, some needing far more heat to raise a unit weight to a given temperature than do others (see Chapter XIV).

(vi) The *thermal diffusivity* of the body or the rate at which heat is transmitted through the material. In ceramic bodies which may be regarded as porous aggregates, heat transfer is a highly complex phenomenon and will be considered in detail in Chapter XIV.

(vii) *Exothermal or endothermal changes* within the material being heated. As the temperature of a substance is raised, reactions of a chemical or physical nature may occur which are accompanied by either an absorption or liberation of heat. The former are known as endothermal and the latter as exothermal changes. In addition to purely chemical reactions involving decomposition or recombination, changes of state from solid to liquid or liquid to gas may be accompanied by thermal effects which alter the rate of heating of a substance at a particular temperature. When ceramic clay materials are fired in the kiln, a temperature arrest frequently occurs on the decomposition of the clay minerals; alternatively, if carbonaceous constituents are present, the heat liberated during oxidation may increase the rate of temperature rise to an alarming degree.

CHANGES OF STATE

AN important influence of heat on ceramic materials is the change from a solid to a liquid form. The formation of a liquid is a true physical effect, but in materials of a complex composition, liquefaction is often accompanied by a chemical change.

The factors which influence the melting point of simple compounds have been described in Chapter X, but only if a material is very pure and of a simple structure will it melt at a definite temperature. Even when a substance is almost pure, the change of state proceeds at a rate which depends on the smallness of the individual particles and if the liquid formed is viscous, the melting of a large volume does not proceed readily.

The refractoriness of a ceramic material involves a more complex condition than the melting point of a simple organic or inorganic salt. Most of the articles made from clays contain a complexity of minerals and glasses, some of which may have their own specific melting point, but, in the process of melting, the liquid which forms is so viscous that it is virtually impossible to judge the critical transformation point. Many minerals melt incongruently, i.e. they dissociate into two or more components whilst still in the solid phase and each entity has its own melting characteristics.

Consequently, when they are heated, ceramic materials soften and melt over a wide temperature range and it is usually incorrect to ascribe a particular value as being their melting point. It is customary to refer to a *refractoriness value*, which is the temperature at which the body has attained a viscosity or flowability sufficient to cause a standard shape to deform by a measurable amount.

The point at which this deformation attains prescribed limits can be observed visually or electronically and the temperature measured by pyrometers or thermocouples. Because the melting processes in ceramic materials are so complex, this method can give rise to errors, because time is as important as temperature in reactions between silicate bodies. Consequently it is essential to observe a strict schedule of rate of temperature rise and even then to compare the melting characteristics of a test piece with a standard cone rather than by measuring the temperature by direct means.

For this purpose and also to provide a widely-used method for the control of optimum firing conditions in kilns, series of standard pyrometric cones have been devised. Those used in the U.K., throughout Europe and elsewhere are based on the Seger series whilst in American practice it is usual to use Orton cones.

The materials of which all cones are composed are artificial mixtures similar to porcelains but controlled chemically with great precision. Those of lowest refractoriness contain soda, lead oxide, boric acid and silica, but as cone numbers increase, the fluxing elements are decreased and alumina progressively introduced. The numbers of Seger cones and their Pyrometric Cone Equivalent temperature are shown in Table XII.IA. Orton cone values are included in Table XII.IB and Buller's Ring numbers in Table XII.IC. The last mentioned temperature indicators are not normally used for measuring refractoriness values but they are most useful for assessing variations in optimum firing conditions in kilns.

The equivalent temperature values indicate the point at which the tip of the cone will bend over to touch the base plate if the rate of temperature rise were maintained at 5° C./min. in the softening range.

TABLE XII—Ia. CHARACTERISTICS OF SEGER CONES

Cone No.	Equivalent Temp. °C.	Cone No.	Equivalent Temp. °C.	Cone No.	Equivalent Temp. °C.
022	600	1a	1100	26	1580
021	650	2a	1120	27	1610
020	670	3a	1140	28	1630
019	690	4a	1160	29	1650
018	710	5a	1180	30	1670
017	730	6a	1200	31	1690
016	750	7	1230	32	1710
015	790	8	1250	33	1730
014	815	9	1280	34	1750
013	835	10	1300	35	1770
012	855	11	1320	36	1790
011	880	12	1350	37	1825
010	900	13	1380	38	1850
09	920	14	1410	39	1880
08	940	15	1435	40	1920
07	960	16	1460	41	1960
06	980	17	1480	42	2000
05	1000	18	1500		
04	1020	19	1520		
03	1040	20	1530		
02	1060	21–25	Not manufactured		
01	1080				

TABLE XII—Ib. CHARACTERISTICS OF ORTON CONES*

Cone No.	Bending Temperature ° C.	Bending Temperature ° F.	Cone No.	Bending Temperature ° C.	Bending Temperature ° F.	Cone No.	Bending Temperature ° C.	Bending Temperature ° F.
1	1160	2120	14	1398	2548	31	1683	3061
2	1165	2129	15	1430	2606	31½	1699	3090
3	1170	2138	16	1491	2716	32	1717	3123
4	1190	2174	17	1512	2754	32½	1724	3135
5	1205	2201	18	1522	2772	33	1743	3169
6	1230	2246	19	1541	2806	34	1763	3205
7	1250	2282	20	1564	2847	35	1785	3245
8	1260	2300	23	1605	2921	36	1804	3279
9	1285	2345	26	1621	2950	37	1820	3308
10	1305	2381	27	1640	2984	38	1835	3335
11	1325	2417	28	1646	2995	39	1865	3389
12	1337	2439	29	1659	3018	40	1885	3425
13	1349	2460	30	1665	3029	41	1970	3578
						42	2015	3659

* Heating rate: Cones Nos. 1–20, 150° C. per hour.
Cones Nos. 23–38, 100° C. per hour.

The Measurement of the Refractoriness of Ceramic Materials

The procedure for measuring the refractoriness or pyrometric cone equivalent (P.C.E.) of ceramic materials has been the subject of much controversy, but most

TABLE XII—Ic. CHARACTERISTICS OF BULLER'S RINGS

Ring Value	*Equivalent Temp. °C.*	*Ring Value*	*Equivalent Temp. ° C.*
2–5	980–1000	26–28	1140–1160
6–7	1000–1020	29–30	1160–1180
8–11	1020–1040	31–34	1180–1200
12–13	1040–1060	35–37	1200–1230
14–17	1060–1080	38–41	1230–1250
18–19	1080–1100	42–45	1250–1280
20–23	1100–1120	46–48	1280–1300
24–25	1120–1140		

countries have now standardised their test methods which are all very similar to each other.[1]

Preparation of Test Cones. When testing a clay or other ceramic raw material, careful sampling and size reduction is of paramount importance. A carefully quartered sample (see p. 210) should be crushed to pass 210 microns (70-mesh U.S.; 72-mesh British), then heated to about 1000° C. in an oxidising atmosphere and maintained at heat for 30 minutes to remove water and other undesirable components which might cause distortion and bloating.

The sample, which should weigh 200–300 g., is thoroughly mixed and sufficient water and organic binding agent added, such as dextrin or gum tragacanth (it must be alkali-free), to enable it to be moulded.

Small cones are manufactured from the sample in a metal die producing a truncated trigonal pyramid with its base either making a right angle (U.K.) with the vertical edge or inclined at a small angle to it (U.S.A.). This is to ensure that the cone will always bend in the same predetermined direction when melting commences.

The shape of the test-cone is not standardised and various countries permit different sizes. Results from specimen cones of different sizes do not give markedly varying results. The American cone has a base side measurement of 0·272 in. and a vertical height of 1·081 in. whilst the British sample requires an 0·5 in. base side and a perpendicular vertical length of 1·5 in.

After the cones have been moulded they are removed from the die and dried carefully.

When testing a brick or fired shape under British test procedures, it is permitted to cut a cone to the above dimensions, but the author prefers the American requirement which involves sampling, grinding and fabricating a cone as for clay. If a cut section is used there is a possibility that a large grog grain or individual mineral fragment might invalidate the test.

The sample cones are mounted on a ceramic disc or plaque of a suitable refractory material, interspersing them with standard cones of appropriate refractory value to encompass the anticipated melting temperature of the sample. The vertical edge of the cone should be set so that on melting, the test specimen tilts inward; this is not an absolute essential, but it is usually the best practice.

[1] The U.S. Standard test is given in ASTM Designation C24–56. The British Standard is covered in B.S. 1902, Part 2A, 1966.

Heating Procedure. The furnace can be of any suitable type but a vertical tube construction has many advantages. It must have a reasonably long constant temperature zone, be capable of rising to 1800° C. or above under oxidising conditions and to be able to maintain a rate of temperature rise of at least 5° C./minute up to these temperatures.

In order to ensure absolute constancy, it is an advantage to rotate and to oscillate the specimen plaque within the furnace throughout the heating cycle. A typical mounting technique and furnace assembly is shown in Fig. XII.1.

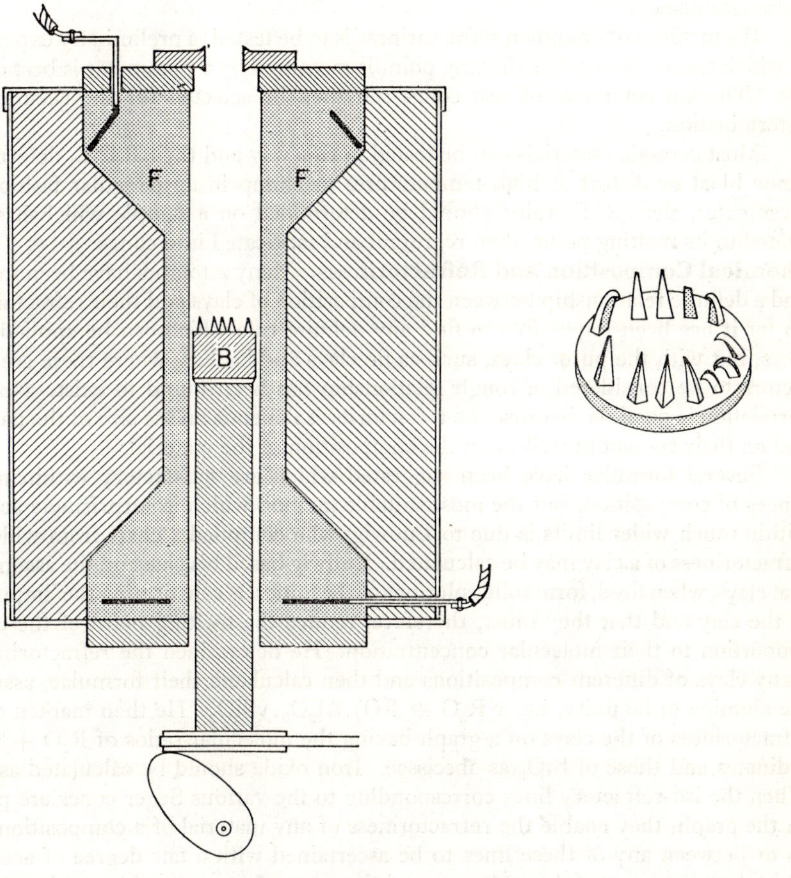

FIG. XII.1. AN APPARATUS FOR MEASURING THE REFRACTORINESS (P.C.E.) OF A CERAMIC MATERIAL

The heating rate must be rigidly controlled. The British Standard recommends 5° C./min. from a point not less than 200° C. from the predicted refractoriness value. The American Standard publishes a heating chart which specifies a heating rate of 2–2½° C./min.

Assessment of Refractoriness Value. As the sample cone commences to melt, its apex slowly tilts inwards until eventually it would touch the base, ideally having described a graceful arc. At a critical point of the test when some tilting of the sample has been observed to be taking place, the plaque should be withdrawn from the hot zone and the amount of bend compared with the deformation on the standard cones. The one nearest in condition to the sample defines the refractoriness value.

The condition of the cones at any time is difficult to judge, but if a silica tube is inserted into the furnace to a point just above the cones and cold air blown in, the tip of the cones will darken just sufficiently and momentarily to allow their condition to be established.

If a material of unknown refractoriness is to be tested, a preliminary experiment in which the approximate softening point is measured by a pyrometer is best carried out. The correct range of test cones can then be selected for the accurate test determination.

Most ceramic materials can be tested in this way and their P.C.E. determined. Some bloat or distort at high temperatures or slump in an irregular fashion. In these cases, the P.C.E. value should be determined on a sample which has been heated to its melting point, then reground and fabricated into cones.

Chemical Composition and Refractoriness. Many attempts have been made to find a definite relationship between the composition of clays and their refractoriness. So far it has been impossible to find any definite rule which can be applied to all clays, but with the purer clays, such as fireclays and kaolins, where there are fewer factors to be considered, a rough relationship has been found to exist. No exact correlation is possible, because the refractoriness depends on the size of the particles and on their contact as well as on the composition of the material.

Several formulae have been suggested which are satisfactory within narrow ranges of composition, but the most suitable method which is surprisingly accurate within much wider limits is due to Ludwig, who compiled a chart from which the refractoriness of a clay may be calculated. Ludwig based his chart on the assumption that clays, when fired, form solid solutions of the minerals constituting the 'impurities' in the clay and that they must, therefore, reduce the melting point of the clay in proportion to their molecular concentration. He determined the refractoriness of many clays of different compositions and then calculated their formulae, assuming the alumina to be unity, i.e. $x(R_2O + RO).Al_2O_3.ySiO_2$. He then marked off the refractoriness of the clays on a graph having the molecular ratios of $R_2O + RO$ as ordinates and those of SiO_2 as abscissae. Iron oxide should be calculated as FeO. When the iso-refractory lines corresponding to the various Seger cones are plotted on the graph, they enable the refractoriness of any material of a composition lying on or between any of these lines to be ascertained with a fair degree of accuracy. This chart is very useful for clays containing up to 6 per cent of bases, but for less pure clays it is useless, on account of the heterogeneous nature of the materials. Ludwig's chart is shown in Fig. XII.2.

Although the presence of impurities greatly increases the fusibility of clays, it is a surprising fact that the refractoriness of alumino-silicates is very largely a property of the ratio of their alumina to silica contents. Mellor and Green[1] have shown that

[1] Mellor, J. W. and Green, A. T., *J. Soc. Chem. Ind.*, 57, 627, 1938.

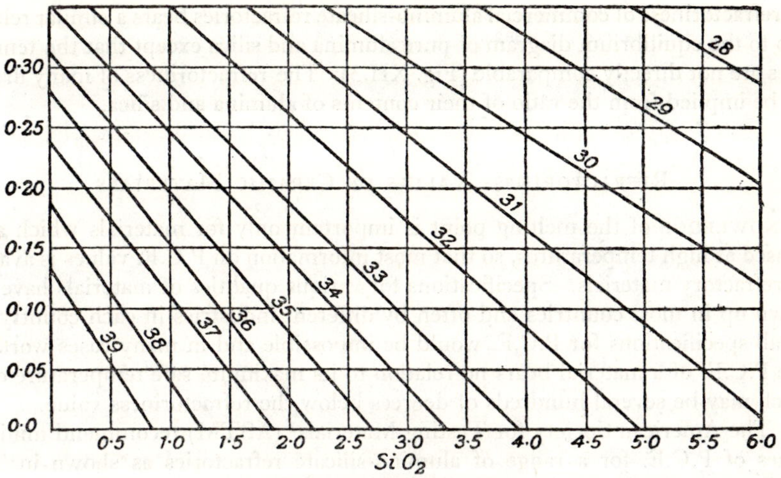

FIG. XII.2. LUDWIG'S CHART

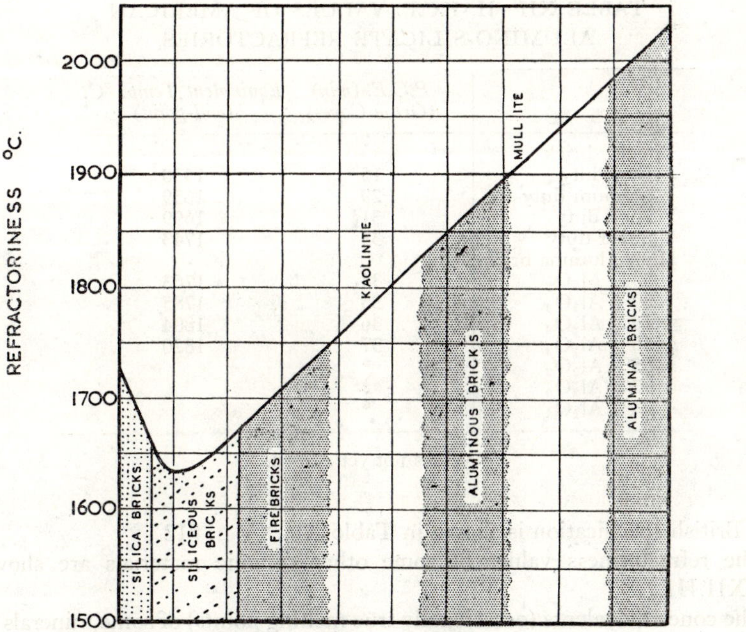

FIG. XII.3. THE REFRACTORINESS OF COMMERCIAL ALUMINO-SILICATE MATERIALS
(after Mellor and Green)

the refractoriness of commercial alumino-silicate refractories bears a similar relationship to the equilibrium diagram of pure alumina and silica except that the temperatures are not directly comparable (Fig. XII.3). The refractoriness of many fireclays can be implied from the ratio of their contents of alumina and silica.

Refractoriness Values of Ceramic Materials

A knowledge of the melting point is important only for materials which are to be used at high temperatures, so that most information on P.C.E. values is available for refractory materials. Specifications for various qualities of materials have been drawn up in most countries and often by different industries in each country. To list all specifications for P.C.E. would be impossible and in many cases worthless. The P.C.E. of a material bears no relation to its maximum safe temperature of use which may be several hundreds of degrees below the refractoriness value.

The American Society for Testing Materials[1] (ASTM) recommend minimum values of P.C.E. for a range of alumino-silicate refractories as shown in Table XII.II.

TABLE XII—II. P.C.E. VALUES OF AMERICAN ALUMINO-SILICATE REFRACTORIES

	P.C.E. (min) (Orton Cones)	*Equivalent Temp. °C. (end-point)*
Fireclay brick		
Low duty	15	1430
Medium duty	29	1659
High duty	$31\frac{1}{2}$	1699
Super duty	33	1743
High Alumina brick		
50% Al_2O_3	34	1763
60% Al_2O_3	35	1785
70% Al_2O_3	36	1804
80% Al_2O_3	37	1820
85% Al_2O_3	*	
90% Al_2O_3	*	
95% Al_2O_3	*	

* values not yet agreed.

A British specification is shown in Table XII.XV p. 812

The refractoriness values of some other ceramic materials are shown in Table XII.III.

The cone equivalents (occasionally true melting points) of some minerals likely to occur in ceramic materials are shown in Table XII.IV.

The **change of ceramic materials into vapours** is seldom regarded as a simple physical change. It is usually one result of chemical decomposition such as the reduction of silica to silicon monoxide or silicon which are readily volatilised at high

[1] ASTM Designation C27–60.

TABLE XII—III. REFRACTORINESS VALUES OF CERAMIC MATERIALS

Mineral	*P.C.E. ° C.*	*Mineral*	*P.C.E. ° C.*
Alumina	2049	Lanthanum oxide	1840
,,	2020	Lime (pure)	2570
,,	2050	Magnesia (pure)	2800
Barium oxide	1925	Magnesite brick	2165
Bauxites	1600–1850	Molybdenum carbide	2700
Bauxite brick	1565–1785	Quartz	1710
Bauxitic clay	1790–1810	Silica brick	1700–1750
Beryllium oxide	2525	Silica sand	1700–1750
Boron carbide	2400	Silicon carbide	2700
Brick clays	1250–1350	Sillimanite	1800
Cerium oxide	2950	Tantalum pentoxide	1875
Chromite	2180	Thoria	2470
Chromite brick	2050	Titanium oxide (rutile)	1600–1700
Chromium oxide	2059	Tungsten carbide	3000
Corundum	2030	Yttria	2400
Dolomite	2100	Zircon	2550
Fireclay brick	1555–1725	Zirconia	2590
Kaolin	1735–1740	Zirconium carbide	3200

temperatures, but are soon re-oxidised and deposited as a white powder. This may appear as though silica itself is volatilised in some furnaces but that is not the case.

The melting of some components in a ceramic material alters many of the physical characteristics of the substance. It may cause a change in the size or shape and, on cooling, articles in which large amounts of liquid have been present will be much changed in texture and surface appearance. They will be vitrified (p. 736), and may have altered drastically in colour.

The Softening Point of Glasses and Glazes. Ceramic materials which are entirely glassy in nature, show softening characteristics at high temperatures which are different from those of bodies of a more crystalline nature. Glasses and glazes are super-cooled liquids which gradually increase in viscosity as the temperature is raised.

The softening point of such bodies is defined as the temperature at which a fibre of the material 0·55–0·75 mm. in diameter elongates under its own weight at a rate of 1 mm. per minute when a 10 cm. length is heated uniformly in a furnace, which is rising in temperature at 5° C./min.[1]

CHANGES IN VOLUME

CHANGES in volume effected by heat result in either an increase or a decrease in the volume of the substance; the former is termed *expansion* and the latter *contraction* or *shrinkage*. These changes are often of great technical importance, especially when substances which undergo such changes are made into articles which are required to be of definite size within very narrow limits. Thus, first-class firebricks are not expected to vary in length by more than 1 per cent, whilst for some electrical fittings made of stoneware or porcelain a much greater degree of accuracy in size is required.

[1] A.S.T.M. Designation C 338–57.

TABLE XII—IV. REFRACTORINESS VALUE OF MINERALS

Mineral	*P.C.E. °C.*	*Mineral*	*P.C.E. °C.*
Albite	1120–1220	Forsterite	1910
Alumina	2050	Galena	1115
Andalusite	1770	Gehlenite	1570
Anorthite	1532	Grossular	1150–1250
Apatite	1270–1300	Hornblende	1180–1220
Asbestos	1285–1310	Kaolin	1745–1790
Augite	1145–1150	Labradorite	1477
Barium metasilicate	1490	Lepidolite	925–945
Barium metaborate	1050	Leucite	1320–1370
Barium orthoborate	1350	Magnesia	2500
Barium pyroborate	1000	Magnesium metasilicate	1565
Bauxite	1820	Magnesium orthosilicate	1900
Beryl	1410–1430	Magnetite	1538
Beryllium metasilicate	2000	Monticellite	1435
Beryllium orthosilicate	2000	Mullite	1810
Biotite	1155–1240	Muscovite	1255–1290
Borax	878	Nepheline	1223
Bronzite	1310–1370	Nephite	1180–1210
Calcium aluminate	1587	Orthoclase	1175–1225
Calcium diborate	1025	Potassium metaborate	947
Calcium metaborate	1095	Rhodonite	1210
Calcium pyroborate	1215	Rutile	1700
Calcium phosphate	1550	Scapolite	1120–1140
Calcium orthosilicate (α)	2130	Silica (quartz)	1685*
Calcium metasilicate (α)	1540	Sillimanite	1816
Tricalcium aluminate	1537	Sodium chloride	819
Tricalcium ferrite	1455	Sodium metaborate	960
Chromite	1850–2180	Sodium metasilicate	1018
Corundum	1750–1800	Spinel	2135
Cristobalite	1720	Spodumene	1380
Cryolite	977	Strontium metasilicate	1287
Diopside	1391	Strontium orthosilicate	1593
Enstatite	1380–1500	Titanite	1200–1300
Epidote	1250	Titanium oxide	1610
Fayalite	1210	Titanium orthosilicate	1650
Ferrous oxide	1419	Tourmaline	1000–1100
Ferrous metasilicate	1100	Wollastonite	1540
Ferric oxide	1548	Zinc metasilicate	1479
Fluorite	1361	Zinc orthosilicate	1484

* α-Quartz is inverted at 573° C. to β-Quartz, which, if finely ground, is converted at 870° C. into tridymite, and at 1470° C. into cristobalite; the last-named melts at 1720° C.

The changes in volume caused by the action of heat may be divided into two groups: (*a*) reversible, and (*b*) irreversible changes.

(*a*) *Reversible volume-changes* are those in which the volume of a substance changes on the application of heat, but the original volume is regained when the heat is withdrawn, i.e. when the substance is cooled to the original temperature. Such changes are usually of a wholly physical nature.

(*b*) *Irreversible changes* are those in which any increase or decrease in volume which may occur is permanent and is not reversed on cooling. Such a change may be the result of the formation of a substance of different specific gravity, such as the conversion of quartz to cristobalite, or some change of a chemical character whereby new compounds are formed. A change of state which results in the formation of a liquid phase will give rise to an irreversible shrinkage of the mass especially if the initial porosity is high.

Reversible Changes on Heating

The *reversible changes on heating* are of two kinds: (i) the normal coefficient of expansion or contraction associated with all materials; and (ii) sudden volume changes accompanying reversible reactions, e.g. the $\alpha \rightleftharpoons \beta$ inversion of quartz or changes of state.

The coefficient of expansion or contraction is usually expressed as a fraction of the length, cross-sectional area, or volume of the substance (the latter being taken as unity) when heated so that its temperature rises 1° C. The coefficient of expansion may be expressed in terms of length, i.e. the coefficient of linear expansion, or in terms of volume, i.e. the coefficient of cubical expansion. If the coefficient of linear expansion is represented by a, the original length of the piece being l_0 at 0° C. and l_t and t° C., then

$$l_t = l_0(1 + at)$$

Hence, if the length prior to expansion be unity, that after expansion will be $(1 + a)$ and the volume after expansion will be $(1 + a)^3$ or $1 + 3a + 3a^2 + a^3$. As a is a small fraction, a^2 and a^3 will often be so small as to be negligible, and it is then sufficient to regard the volume of the expanded material as $(1 + 3a)$. Hence, the coefficient of volume expansion is almost exactly three times the linear coefficient. If the linear expansion or contraction is as high as 10 per cent, omitting the fractions as suggested will produce an error of about 10 per cent of the total volume change.

The expansion of materials on heating may be determined in various ways. Where it is large, as in the case of some metals, a bar is heated with one end fixed and the other in contact with an indicating lever and the amount of deflection noted.

The standard method now in common use for ceramic materials is to heat a square-ended or cylindrical specimen in a silica glass sheath, which, in turn, fits loosely into a tubular wire-wound furnace. The specimen is supported on a silica rod and a distance rod of the same material rests on top of the test bar and projects out of the fused silica sheath which is maintained in a vertical position. The complete assembly is shown in Fig. XII.4, including the dial gauge which is attached to the silica sheath. Its pointer is in contact with the end of the protruding distance piece, hence the relative difference in expansion between the internal assembly and the outside sheath can be accurately measured.

The expansion of the bar under test can be calculated relatively simply from such an apparatus.

If the silica sheath is l cms. in length and the specimen is x cms. long, the expansion recorded on the dial gauge at any particular temperature t is the increase in length of the specimen (of x cms.) plus the expansion of the internal, silica supporting rod and distance piece (of equivalent length $l-x$ cms.) minus the expansion of the outside silica sheath (of length l cms.).[1]

[1] As the dial gauge is supported by the sheath any increase in length in the latter would be indicated as a contraction reading, hence it would have to be subtracted from the overall indication on the gauge.

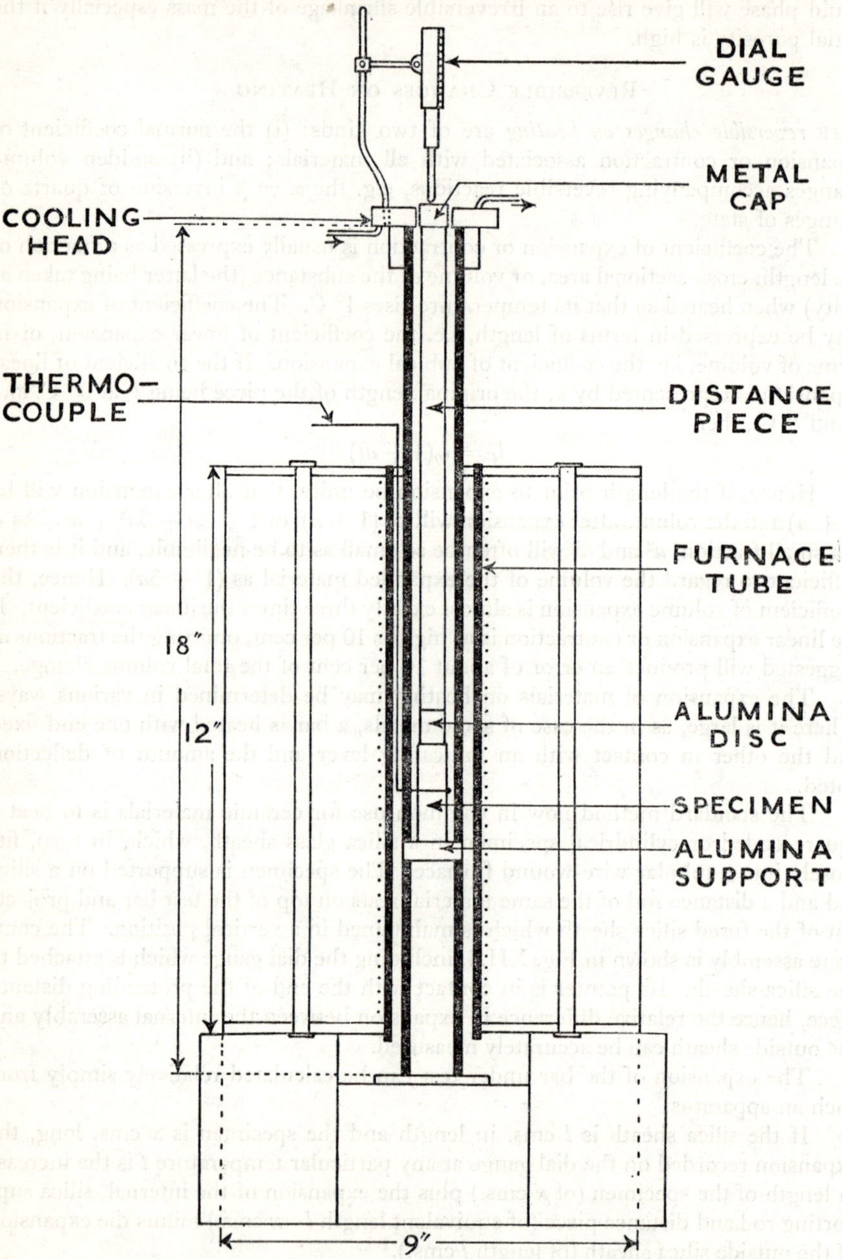

FIG. XII.4. AN APPARATUS FOR MEASURING THE THERMAL EXPANSION OF CERAMIC MATERIALS

If the coefficient of expansion of silica is α_1 and that of the test specimen α_2, the dial gauge reading may be expressed as:

$$\begin{aligned}\text{Reading} &= (x\alpha_2 + (l - x)\alpha_1 - l\alpha_1)t \\ &= (x\alpha_2 - x\alpha_1)t\end{aligned}$$

The expansion of the specimen itself, i.e. $x\alpha_2 t$, is thus the reading on the dial gauge *plus* the expansion of an equivalent length of a fused silica rod. This latter is relatively small and can be calculated by using a material of known expansion characteristics or from data supplied by the makers.

In practice, the expansion test is best carried out using an extremely small rate of temperature rise (about 1° C. per minute should not be exceeded).

Fused silica rods and sheaths are serviceable only for measurements up to 1000° C.; above this temperature mullite should be substituted and, as this material has a much larger expansion than silica, greater care must be taken with the initial calibration.

In recent years, similar equipment has been described for measuring the expansion of small-sized specimens. The dial gauge is then usually replaced by a more sensitive indicating or recording device. An optical lever arrangement has been used by some workers; others prefer to make the top of the distance piece one plate of a condenser with the other fixed and the whole forming part of an electrical circuit. Both these methods enable a continuous record of the expansion curve to be made.

The Ceramics Whiteware and Related Products Committee of the American Society of Testing Materials has adopted a standard method of measuring the linear thermal expansion of fired products by an interferometric method. In it, a small test piece expands against fused silica interferometer plates causing a fringe pattern to move across a field of view.[1]

By such means, the linear change in size of a material may be measured and hence its coefficient of expansion or contraction. In addition, any inflection in the curve may be detected and related to inversional changes and associated phenomena in the minerals present.

Reversible Volume Changes on heating occur to a greater or less extent in all materials. Basically, they are the result of changes within the crystalline lattices of the component minerals due to the heating effect, and on cooling the body returns to its former size. Reversible volume changes depend on the coefficient of expansion (or contraction) of the body, and this varies with different materials according to

(*a*) the mineralogical composition;
(*b*) the texture.

The coefficient of expansion of minerals. Numerous recent attempts to correlate the thermal expansion properties of materials with their crystalline structure have met with a considerable amount of success with simple crystals, but more complex substances—particularly some of those of interest to the ceramist—are still beyond the scope of theoretical treatment.

The present-day concept of matter has been outlined in Chapter II. Crystals

[1] A.S.T.M. Designation C 327–56.

are considered to be composed of atoms or ions, arranged in particular geometrical patterns and forming crystalline lattices. Although it is convenient to regard each unit of such a lattice as having a fixed position in space, this concept is not strictly accurate. The individual atoms or ions have a certain degree of latitude and have freedom of movement about a so-called equilibrium position. The extent of these vibrations is extremely small but is nevertheless appreciable, considering the minuteness of the atomic scale.

The degree of vibrational movement is markedly affected as the temperature is increased. The amplitude of the oscillation about the equilibrium position increases until at some point an ion or atom will escape from the restoring influence of its surroundings and the crystal lattice will collapse.

There are two opposite forces acting on an atom or ion within a crystal lattice: (*a*) attraction, towards a centre of opposite charge in the near vicinity; and (*b*) repulsion or restoration by other groups or charged centres in the same vicinity. The two forces serve to keep each ion in an equilibrium position or at an average separation distance from others which are in close proximity. A simple schematical representation of the action of the interatomic forces is shown in Fig. XII.5. Within a certain separation distance, the repulsion forces predominate, but if the centres are far apart the force of attraction tends to restore them to an equilibrium position indicated by *a*. There is, however, a small distance (*da*) through which ions can vibrate, and so the interatomic distance varies between the extremes $a - da$ and $a + da$ and as these represent points of different energy, for a certain vibrational amplitude (*da*), an energy change dE_1 will be involved.

This energy term is important because it determines the strength of the bond between atoms or ions in the crystal. Where the resultant-force line (see Fig. XII.5)

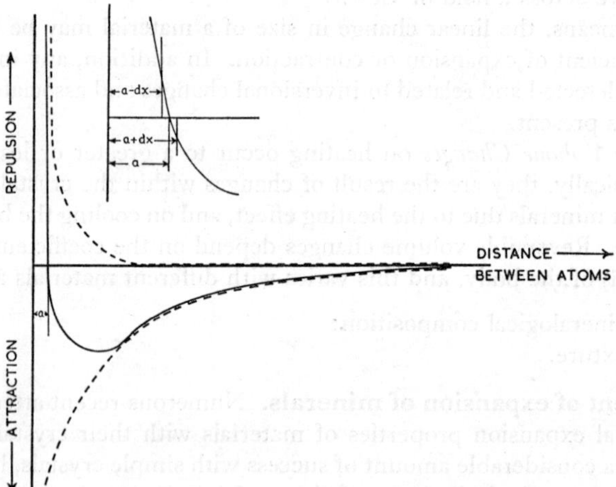

FIG. XII.5. THE FORCES OF ATTRACTION AND REPULSION BETWEEN ADJACENT ATOMS IN A CRYSTAL

The solid line indicates the resultant

is steeply inclined, the bond strength or lattice energy will be larger than in cases when the line has a smaller slope.

In the neighbourhood of the equilibrium position or zero point (*a*) of the resultant-force line of Fig. XII.5, the curve undergoes a change in character. At shorter interatomic distances, the repulsion forces between ions rise steeply whilst, when the equilibrium position is exceeded, the forces of attraction also increase, but at an ever-decreasing rate. Therefore, as the amplitude of vibration increases, the repulsion and attractive forces around the zero position become unsymmetrical and the restoring force is greater from the $a - da$ extreme than from the $a + da$ point. As a result, the mean centre of vibration tends to be displaced towards a greater interatomic distance and this will become increasingly so as the amplitude extends. The crystal, therefore, expands and the volume increase is thus related to the relative slopes of the resultant-force line on each side of the equilibrium position and to the temperature, which governs the magnitude of the vibrations.

Megaw[1] has investigated the thermal expansion of crystals in relation to their structure and has been able to account for the properties of simple compounds on a quantitative basis. She also explains the anomalous behaviour of more complex types on the basis of different bond characteristics within the same crystal, or the possibility of rotational movement of linkages which supplement or disguise true vibrational effects.

The strength of the bond between adjacent atoms or ions in a lattice is related to the charge or valency of the ion divided by the number of ions of opposite charge which surround it. This concept was first advanced by Pauling and the value (*q*) is known as the electrostatic share of the ion. (For a more detailed explanation see Chapter II, p. 84.) Megaw has shown that for simple crystals, the coefficient of expansion (α) is inversely proportional to the square of the electrostatic share, i.e.

$$\alpha = \frac{C}{q^2} \quad \text{or} \quad \alpha q^2 = \text{a constant}$$

This relationship holds only for such minerals as the alkali halides, although, if only one type of bond is present in the crystals, the expansion coefficient of more complex substances decreases with increasing bond strength as shown in Table XII.V and αq^2 is approximately constant.

TABLE XII—V. RELATIONSHIP BETWEEN BOND STRENGTH (ELECTROSTATIC SHARE) AND THE THERMAL EXPANSION COEFFICIENT

Mineral	*q*	$\alpha \times 10^{-6}$	$\alpha q^2 \times 10^{-6}$
NaCl	1/6	40	1·11
CaF_2	1/4	19	1·19
MgO	1/3	10·5	1·17
ZrO_2	1/2	4·5	1·12
BeO	1/2	5·1	1·26
Al_2O_3	1/2	5·1	1·26
SiO_2	1	0·5*	0·50

* Figure for silica glass. The low temperature forms of the silica minerals have a large expansion due to rotational effects and changes in the bond angles.

[1] Megaw, H. D., *Z. Krist.*, 100, 58, 1939.

Hummel[1] has compared the thermal expansion characteristics and the crystal structure of many natural minerals and suggests that those with the same crystallographic symmetry have similar coefficients. Magnesium oxide, for example, has a cubic structure based on rock salt (NaCl) and these two minerals have expansion coefficients of $10{\cdot}5 \times 10^{-6}$ and $40{\cdot}0 \times 10^{-6}$ respectively. CaO, BaO, FeO and many others have a similar structure and all are characterised by a large thermal expansion.

BeO and Al_2O_3, have an hexagonal structure but are still close-packed, hence the expansion coefficient is relatively high (*circa* $5{\cdot}1 \times 10^{-6}$). Silicate minerals are lower, primarily because the SiO_4-grouping, which usually predominates, has a high bond strength. Framework structures based on three-dimensionally linked SiO_4-tetrahedra should have the lowest expansion coefficients, but such lattices are not close-packed and rotational movements and bond angle changes may occur, which cause large volume changes (see Table XII.XIV). At high temperatures, however, when only vibrational effects are involved, such minerals have only slight or even negative expansion coefficients. Hummel[2] concluded that if such high temperature structures could be stabilised, a material of low thermal expansion would result, and he synthesised the lithium-alumino-silicates, β-eucryptite and β-spodumene. Both of these have a structure which is based on that of high or β-quartz with aluminium ions replacing silicon, and lithium ions filling holes in the lattice thus preventing inversions. Both minerals have an extremely low expansion; in fact, the former has a negative coefficient in the temperature range 0–1000° C.

Although the simple concept of heat affecting the vibrational movement of bonds within the crystal lattice is far from a complete explanation, many useful conclusions may be derived. As Austin[3] has pointed out, it offers an explanation as to why the expansion coefficients, heats of formation and melting points of simple materials are related. Furthermore, the variation in thermal properties along different crystallographic axes in a substance may be readily explained on this basis.

The thermal properties of a composite material such as are encountered in ceramic studies are influenced not only by the mineral forms which are present, but also by the texture of the body itself. It is understandable, therefore, that the theoretical considerations which have been deduced for simple crystalline substances, must be used with caution in ceramic materials.

The coefficient of expansion of crystals is constant along all crystallographic directions (p. 90) only in those of the cubic variety. Because most ceramic materials are composed of a mass of small crystallites randomly arranged this has little significance, but where large crystals are present or where there is some degree of orientation, the amount of expansion in different directions may vary.

The coefficient of expansion of bodies. Where more than one type of mineral is present in a material, the coefficient of expansion of the mass is roughly the average of the components, but any porosity in the composite body will have an important effect.

The texture of the body has a pronounced effect on the coefficient of expansion of

[1] Hummel, F. A., *J. Amer. Ceram. Soc.*, 33, 102, 1950.
[2] Ibid. *J. Amer. Ceram. Soc.*, 34, 235, 1951.
[3] Austin, J. B., *J. Amer. Ceram. Soc.*, 35, 10, 1952.

a material, but only insofar as the porosity reduces the linear and volume continuity of the material and hence the overall thermal expansion will be lowered. In some cases, a small amount of pore-space in a body will influence the expansion to a very great extent; this is often true in partially-crystallised glasses particularly when cristobalite has formed. On cooling through its inversion point, cristobalite contracts away from the surrounding glass leaving a small amount of void space. When such a specimen is reheated the large volume change associated with the $\alpha \rightleftharpoons \beta$ cristobalite inversion is not apparent because the crystals expand into the void space formed during cooling and do not influence the overall size of the material to any great extent.

The influence of porosity on the thermal expansion of fired clay bodies is illustrated in Table XII.VII.

The reversible thermal expansion of materials which react during firing can only be measured at temperatures below that at which reaction occurs, otherwise part of the volume change indicated will be permanent.

Clays. Table XII.VI shows the coefficient of expansion of various kinds of clays after calcination at various temperatures.

TABLE XII—VI. COEFFICIENT OF EXPANSION OF CLAYS
(Houldsworth and Cobb), coeff. of Exp. $\times 10^{-7}$

Substance	*Temperature Range, ° C.*	*Temperature of Firing prior to Test*			
		Cone 06	Cone 9	Cone 14	Cone 20
Kaolin	100–250	—	76	57	—
,,	15–1000*	40	53	47	44
Yorkshire fireclay	100–250	—	102	87	—
,, ,,	15–500	48	76	67	—
,, ,,	500–600	133	127	107	—
,, ,,	600–1000*	21	18	25	—
,, ,,	15–1000*	49	58	54	30
Ball clay	15–500	54	—	—	—
,, ,,	500–600	98	—	—	—
,, ,,	600–900	43	—	—	—
,, ,,	15–1000*	55	57	43	—
Ayrshire Bauxitic clay	100–250	—	72	65	—
,, ,,	15–1000*	48	60	56	41
Scottish fireclay	100–250	—	98	80	—
,, ,,	15–500	40	74	54	—
,, ,,	500–600	127	122	88	—
,, ,,	600–1000*	27	29	34	—
,, ,,	15–1000*	45	61	55	39

* 600°–900° C. and 15°–900° C. for the Cone 06 specimens.

The effect of porosity on the coefficient of expansion of fireclay is shown in Table XII.VII due to Houldsworth and Cobb.[1]

The effect of increasing the porosity of a body is to decrease its coefficient of thermal expansion at all temperatures. Fireclay which has been fired to Cone 9 has a

[1] Houldsworth, H. and Cobb, J. W., *J. Soc. Glass Tech.*, 5, 36, 1921.

greater expansion at a particular porosity in the range 15°–500° C. than has the same material fired to Cone 06; this trend is reversed in the expansion coefficient over the range 500–600° C. The amount of residual quartz in fireclay fired to 06 is relatively high and, because of the expansion accompanying the $\alpha \rightleftharpoons \beta$ inversion at 573° C., the coefficient of expansion in the range 500°–600° C. is high, but firing to Cone 9 converts some of the quartz to cristobalite or tridymite, thereby reducing the expansion in the range 500°–600° C. but increasing that at low temperatures. The higher temperature of firing produces a greater proportion of glass which is highly siliceous and hence confers a lower thermal expansion at high temperatures.

TABLE XII—VII. EFFECT OF POROSITY ON THE COEFFICIENT OF EXPANSION OF FIRECLAY

Temperature of Calcination	*Porosity*	*Coefficient of Expansion* $\times 10^{-7}$			
		15°–500° C.	500°–600° C.	600°–1000° C.	15°–1000° C.
Cone 9	28·7	76	127	18	58
	44·6	60	105	28	51
	50·2	55	99	29	49
Cone 06	31·7	48	133	20	45
	47·7	44	122	21	42
	62·7	40	120	19	40

The average expansion of a well-burned firebrick when heated to 1300° C. is about 0·000006 per 1° C. or 60×10^{-7} per 1° C.

The coefficient of other non-vitrified claywares is similar to that of firebricks. Mellor[1] found the coefficient of expansion of various kinds of tiles to be 70×10^{-7} per 1° C. on average.

Vitrified Claywares. The coefficient of expansion of vitrified claywares varies considerably according to their nature and the extent of vitrification, but is generally less than that of non-vitrified claywares.

Much work has been done on the coefficient of expansion of porcelains and the influence of the various components on this property have been studied. Much of the evidence is apparently highly conflicting but many of the anomalies can be resolved by considering the reactions which take place during the firing.

The major constituents of porcelain are: (*a*) felspar; (*b*) flint; and (*c*) clay; at high temperatures these react to form liquid and other mineral varieties, so that, on cooling, the body is composed of crystals which are mainly mullite and cristobalite with some felspar and glass. The proportion of each component is governed by the proportion of the original constituents and the conditions of firing and cooling in the kiln.

In general, the crystalline components of porcelain will have a higher coefficient of expansion than that of the glass, but the glass itself will have varied expansion characteristics which will depend largely upon its composition; glass rich in silica

[1] Mellor, J. W., *Trans. Eng. Cer. Soc.*, 5, 159, 1905–6.

has a low thermal expansion, but as the proportions of alkali elements and alumina in it are increased, the coefficient increases as also does its tendency to crystallise on cooling.

Porcelains low in felspar have a relatively large proportion of crystals in them because the flux content is insufficient to form an extensive amount of liquid at the temperature of firing. Consequently, the expansion coefficient of such bodies is high, but it may be decreased by adding more felspar to reduce the proportion of crystalline phases. Beyond a certain limit of felspar addition, the coefficient increases again due to the formation of a less siliceous glass and an enhanced tendency for the liquid which forms to crystallise on cooling. This is illustrated in Table XII.VIII which shows the change in expansion brought about by increasing the felspar content at the expense of clay, keeping a constant amount of flint.

TABLE XII—VIII. EXPANSION CHARACTERISTICS OF PORCELAIN MIXTURES

Composition by weight			*Coefficient of Expansion*
Flint	*Felspar*	*Clay*	
30	7	63	65×10^{-7}
30	31	39	40×10^{-7}
30	49	21	70×10^{-7}

Additions of clay to a felspar-flint mixture decrease the expansion when small amounts are incorporated, because the alumina in clay increases the tendency to liquid formation and stabilises the glassy phase. Larger amounts of clay increase the coefficient of expansion of porcelain because the total amount of felspar is thereby reduced and more crystals will form. The addition of more clay to a porcelain within the normal composition range tends to increase its coefficient of expansion slightly.

The addition of flint to a porcelain body may either decrease or increase the coefficient of expansion depending on whether the extra flint enters the liquid phase at high temperatures thereby producing more siliceous glass or whether it crystallises to cristobalite.

The hard porcelain with the lowest thermal expansion corresponds to a mixture of 30 per cent flint, 30 per cent felspar and 40 per cent of clay which has a coefficient of 39×10^{-7} (0°–1000° C. temperature range).

The expansion of porcelain is not regular at all temperatures, which is due in part to the presence of cristobalite with an inversion at about 220°–280° C. Where quartz is used instead of flint an increased expansion between 550°–600° C. is often apparent.

Stoneware has a somewhat greater expansion than that of porcelain, but this is due almost entirely to the presence of free unchanged quartz or to a greater proportion of cristobalite.

Glazes must have a coefficient of expansion very similar to that of the body to which they are attached, otherwise serious manufacturing faults will arise. When cooling a glaze from its maturing temperature, there is no sudden arrest or change in characteristics due to crystallisation; the liquid gradually increases in viscosity and only becomes brittle hard in the range 500°–700° C. Consequently any strain which may be imposed upon it by the differential contraction of the body to which it is attached can be readily accommodated by the glaze at temperatures above a dull red heat. Below this temperature, however, should there be a difference in the coefficients of expansion of the body and glaze, stresses will be set up; if the body contracts more than the glaze (i.e. if its coefficient of expansion is greater), the glaze will be put into compression and *peeling* may occur. On the other hand, if the glaze has the greater contraction, it will be put into tension and *crazing*, characterised by fine hair cracks will certainly occur. Glazes are weak in tension but strong in compression, hence it is better to use a glaze which has a slightly smaller expansion than that of the body.

The correct adjustment of a glaze so that its expansion will 'fit' that of a clay body is not an easy matter, because the two materials are very different in both chemical composition and physical characteristics. The use of engobes or several vitreous layers of gradually-changing properties is desirable and necessary in many cases, but, even when these are employed, the thermal expansion characteristics of each vitreous coat must be carefully controlled.

The thermal expansion of a glaze (and of a glass) depends largely upon its chemical constituents; each oxide component contributes a share to the total expansion of the glaze and the total coefficient of expansion (α) can be roughly calculated by adding the contribution of each component thus:

$$a_1x_1 + a_2x_2 + \ldots a_nx_n = \alpha$$

where $a_1, a_2 \ldots a_n$ are the percentage by weight of each oxide component in the glaze and $x_1, x_2 \ldots x_n$ are factors for that particular oxide.

TABLE XII—IX. FACTORS FOR CALCULATING THE THERMAL EXPANSION OF GLAZES

Oxide	*Linear Coefficient of Thermal Expansion* $\times 10^{-7}$	
	Winkelmann and Schott	*English and Turner*
SiO_2	26·7	5·0
Al_2O_3	166·7	166·7
B_2O_3	3·3	− 65·3
CaO	166·7	163·0
MgO	3·3	45·0
BaO	100·0	140·0
PbO	100·0	106·0
ZnO	60·0	70·0
Na_2O	333·3	416·0
K_2O	283·3	390·0
P_2O_5	66·6	—

Winkelmann and Schott[1] in 1894 were the first to propose this method of calculating the thermal expansion and other properties of glazes and they drew up a

[1] Winkelmann, A. A. and Schott, F. O., *Ann. Physik. Chem.*, 51, 730, 1894.

table of factors for the oxides which are most commonly present in glazes. English and Turner[1] derived factors which are of more general application. They are shown in Table XII.IX.

Although it is not possible to calculate the accurate coefficient of thermal expansion of a glaze from such factors, they are useful in indicating the approximate value and also they are a valuable guide towards improving a glaze which 'crazes' or 'peels' when applied to a particular body. Additions of alkali reduce the tendency to peel but enhance crazing, whilst more silica or boric acid should be incorporated in a glaze which is liable to craze. Care must be exercised in making such changes because they will, at the same time, alter other important properties of the glaze, such as the maturing temperature and fired appearance.

The stresses which develop between body and glaze during cooling and their effect upon the finished article have been the subject of many investigations. Three idealised cases are shown in Fig. XII.6. Glazes which harden about 700° C. usually

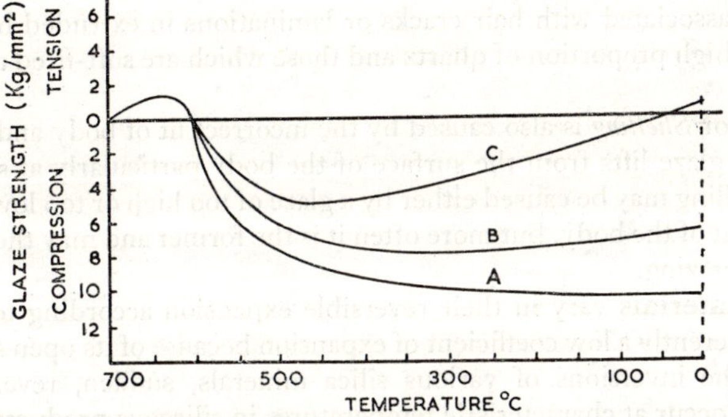

FIG. XII.6.
STRESSES INDUCED IN GLAZES DURING COOLING
A. Where the body has a much greater coefficient of expansion which may cause peeling;
B. where the body has a greater thermal expansion than the glaze, but not an excessive one; and
C. where the body and glaze are of comparable expansion characteristics tending to give crazing

show a small rise into tension on cooling to 600° C., but below this temperature they are put into compression due to the contraction in the body at the quartz inversion. At lower temperatures a high compression may be maintained if the glaze is of low coefficient of expansion (curve A), or if it is of greater expansion, stresses may be reduced, but at room temperature it may still be in slight compression (curve B) or in slight tension (curve C). A body with glaze A would probably 'peel', and glaze C will most certainly 'craze', with glaze B being most suitable. A residual compressive stress of about 5 kg./mm². or slightly above in a glazed body gives the best performance in service.

[1] See Turner, W. E. S., *J. Amer. Ceram. Soc.* 12, 760, 1929.

One of the most serious problems in glazing, especially in ware which is exposed to the atmosphere or to water, is *delayed crazing* which may appear only after many months of use. The cause is considered by Mattyasovsky-Zsolnay[1] to be the expansion of the body due to moisture absorption thereby reducing the compressive stresses in the glaze to a point at which crazing can occur.

The crazing-resistance of a glazed body can be assessed by autoclaving the ware in steam at 50 lbf/in^2, 100 lbf/in^2 or 150 lbf/in^2. Blakely[2] has shown that under such treatments the compressive stresses may be reduced by 5 kg./mm^2., and Mattyasovsky-Zsolnay considers that tiles which withstand steam at 100 lbf/in^2 for two hours will not craze in twenty years.

Dunting is another defect in glazed ware which is the result of a differential expansion between body and glaze or between different parts of the body, but unlike crazing it is more localised and appears as a crack or series of cracks in one or more areas of the article. The most frequent cause of dunting is too rapid surface cooling which induces tensional stresses in the glaze or outside layers of the body; for this reason the cooling of glazed ware in the kiln should not be unduly hurried. Dunting may also be associated with hair cracks or laminations in extruded bodies. Bodies containing a high proportion of quartz and those which are soft-fired are more liable to dunting.

Shilling or *Shelling* is also caused by the incorrect fit of body and glaze. When it occurs the glaze lifts from the surface of the body particularly at sharp edges or corners. Shilling may be caused either by a glaze of too high or too low an expansion relative to that of the body, but more often it is the former and may then be regarded as a type of crazing.

Siliceous materials vary in their reversible expansion according to their origin. Silica has inherently a low coefficient of expansion because of its open structure, but, because of the inversions of various silica minerals, sudden, reversible volume changes may occur at characteristic temperatures in siliceous products (Fig. XII.7).

The coefficient of expansion of natural quartzites, ganisters and silica stones varies between 130–180 $\times$ 10^{-7} over the range 0°–1000° C. Above 600° C., however, these materials are virtually volume stable.

Silica bricks expand between 1·0–1·3 per cent linearly between 0° and 1000° C. but 90 per cent of this expansion occurs below 300° C. and is associated with the inversions of cristobalite and tridymite. Bricks rich in tridymite expand less than those with a large proportion of cristobalite.

Fused silica has a very low thermal coefficient of expansion; it is only 5·4 $\times$ 10^{-7} or one-twelfth that of ordinary soda glass, so that it is very insensitive to sudden changes in temperature and can be quenched from red heat in cold water without fracture.

According to Vergano, Hill and Uhlmann[3], felspars and felspathic glazes and glasses show anomalous expansion characteristics at their transition points in the region of 1100–1200° C.

Potassium felspar changes from a value of 66 to 120 $\times$ 10^{-7} whilst sodium

[1] Mattyasovsky-Zsolnay, L., *J. Amer. Ceram. Soc.*, 29, 200, 1946.
[2] Blakely, A. M., *J. Amer. Ceram. Soc.*, 21, 243, 1938
[3] Vergano, P. S., Hill, D. C., and Uhlmann, D. R., *J. Amer. Ceram. Soc.*, 50, 59, 1967.

felspar shows an increase from 74 to 200 $\times 10^{-7}$ in coefficient of expansion across the transition point.

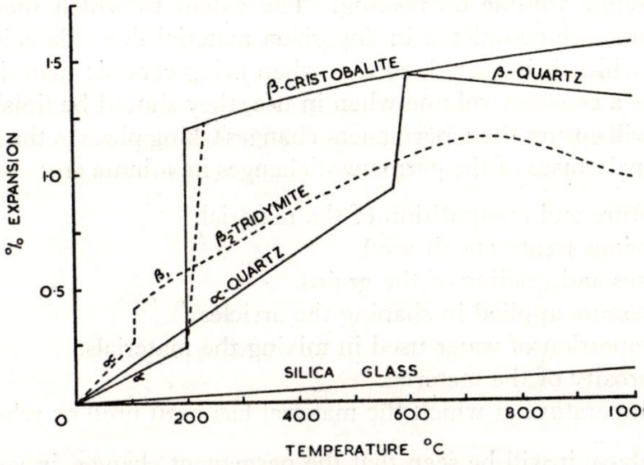

FIG. XII.7. THE EXPANSION CHARACTERISTICS OF THE SILICA MINERALS

Glasses have an expansion which depends on their composition; those rich in alkali elements may have a coefficient of up to 90 $\times 10^{-7}$. Glasses of low expansion are rich in silica, contain magnesium or calcium oxide as fluxes and amounts of boric oxide or phosphorus pentoxide.

Other Ceramic Materials have thermal expansions which depend largely on the minerals which they contain. Table XII.X shows the linear coefficients of expansion of some minerals and ceramic products.

TABLE XII—X. LINEAR COEFFICIENTS OF EXPANSION OF SOME CERAMIC MATERIALS

	Coefficient of expansion $\times 10^{-7}$		*Coefficient of expansion* $\times 10^{-7}$
Alumina	81	Steatite bodies	70
Beryllia	56	Zircon porcelain	48
Mullite	45	Titanate bodies	80–100
Silicon carbide	47	Magnesite bricks	127
Titania (rutile)	70	Dolomite bricks	110
Zirconia	45	Chromite bricks	70
Zircon	42	Chrome-magnesite bricks	77
Electrical porcelain	60	Forsterite bricks	90
Cordierite bodies	30		

PERMANENT VOLUME CHANGES ON HEATING

PERMANENT changes in volume differ with the materials in which they occur. The chief change which occurs when ceramic materials are made into various articles and fired is the contraction or shrinkage which occurs when they are heated. Some materials, such as suitable mixtures of clay and silica, are practically constant in volume, whilst others, such as silica, expand when heated and do not regain their original volume on cooling. The extent to which these permanent changes in volume are completed in any given material depends chiefly upon the temperature to which it is heated, so that when firing ceramic materials which are required to have a constant volume when in use, they should be finished at such a temperature as will ensure these permanent changes taking place to the fullest extent.

The principal causes of the permanent changes in volume are:

(*a*) The nature and composition of the material.
(*b*) Its previous treatment (if any).
(*c*) The sizes and grading of the grains.
(*d*) The pressure applied in shaping the articles.
(*e*) The proportion of water used in mixing the materials.
(*f*) The porosity of the material.
(*g*) The temperature at which the material has been fired or reheated.

From the above it will be seen that the permanent changes in volume may be divided into two groups:

(i) Changes which are dependent upon the inherent qualities of the materials.
(ii) Changes which are dependent on the method of manufacture.

The changes in the first group are chiefly due to: (*a*) decomposition and the removal of some constituent which causes a reduction in volume, as in the case of clay; (*b*) the formation of other allotropic forms of the same material which cause a change in the specific gravity and, consequently, either an expansion or contraction; (*c*) a chemical reaction which produces a new material of greater specific gravity and with consequent volume change; (*d*) the formation of a liquid phase which flows into and fills some of the original pore space; and, (*e*) sintering reactions which again result in a decrease in pore space and an overall volume reduction.

The physical causes of change in volume, such as texture, porosity etc., are considered in the respective sections dealing with these subjects.

Whilst it is very desirable it is not always possible to complete the various changes in volume which occur during the firing of ceramic materials, so that a compromise must usually be made, and—where it is important—the maximum permissible permanent expansion or contraction in use should be specified. This is particularly the case with refractory materials, as it is in these that after-contraction or expansion is most harmful.

Clays, when properly burned, change in size permanently to a varying extent, depending on their composition. This takes place in two stages: (*i*) during drying; and (*ii*) during burning. The shrinkage in drying is due to the removal of water from the surfaces of the grains and has been described in Chapter IX.

The change in shape in clays when fired at high temperatures is mainly the result of the reactions which occur within and between the minerals present. These reactions are of various types as follows:

(*a*) *Decomposition.* Clays and related minerals lose water or carbon dioxide at comparatively low temperatures (p. 712). Although the loss in weight can be considerable, the minerals undergoing decomposition do not change significantly in size during the reaction and then it is usually to give a slight permanent expansion.

(*b*) *Inversional changes* on heating clays are in theory reversible, but when they take place they frequently cause disruption in the surrounding structure of the clay mass, thus producing irreversible changes. Quartz is the mineral which has the greatest influence of this nature, and because it is present in most natural clays it can bring about marked permanent changes in volume as it is heated through its inversional change at 573° C.

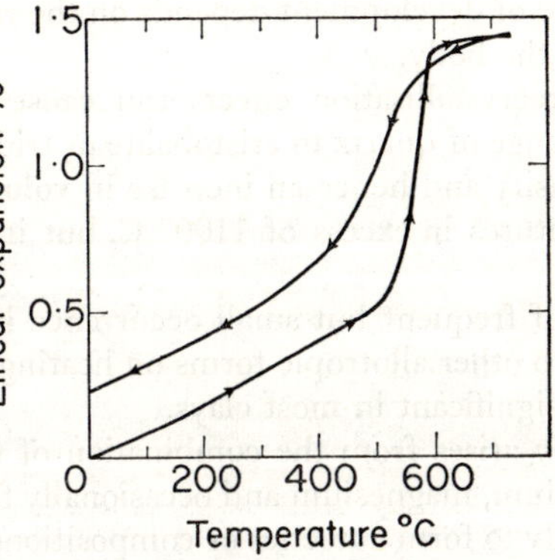

FIG. XII.8. IRREVERSIBLE VOLUME CHANGES ON HEATING CLAY/QUARTZ MIXTURES

This is illustrated in Fig. XII.8 which shows the volume changes in a clay/sand mixture heated to 750° C. and then cooled. Under these conditions, the only changes taking place are the normal reversible thermal expansions of clays and other minerals, the very slight permanent expansion of the clay mineral on decomposition and the large volume change of the quartz undergoing its $\alpha \rightleftharpoons \beta$ transition at 573° C. In theory, this last change should reverse on cooling and it will no doubt do so, but because the clay body is non-elastic, the initial expansion opens up the texture. On cooling, the new packing conditions cannot reverse although the mineral grains of quartz undergo contraction. Thus the final body is of larger size, but with a greater porosity and often of much lower strength.

This is an important aspect of sand ramming compositions. To overcome the deleterious effects of silica inversions in unfired bodies, it is better to use a siliceous material of low quartz crystallinity. Flint is not an adequate substitute because this also undergoes changes of a disruptive nature on heating due to the loss of water and the decomposition of carbonaceous matter. Some chalcedonic forms of silica have particular advantages in these applications.

(*c*) *Crystallisation.* Following the decomposition of the main clay mineral, the unstable dehydroxylation product retains much of the character of the original form (see p. 725). In most cases, however, as the temperature rises, the unstable form (metakaolin in the case of kaolin-type clays), begins to recrystallise and to undergo solid state sintering. Contraction occurs and continues at an increasing rate although frequently no new crystal phase can be detected or identified.

As the temperature continues to rise, the tendency for other mineral phases to develop becomes increasingly great. γ-alumina, or a distorted spinel phase incorporating silicon ions in the lattice is known to form in most clay minerals, commencing at about 750° C. and increasing progressively in amount and crystallite size.

At high temperatures mullite, fosterite, enstatite, corundum may form and in every case dense crystals grow from a porous mass of virtually amorphous components. Contraction is the inevitable result of such changes but the amount which occurs and its rate of development depends on the rate and duration of firing and the composition of the body.

In one instance, recrystallisation effects can cause expansions in clays on heating. This is the change of quartz to cristobalite or tridymite with a consequent reduction in crystal density and hence an increase in volume. This change occurs rapidly only at temperatures in excess of 1100° C. but it has great significance in refractory practice.

Anatase, TiO_2, is of frequent but small occurrence in clays. This mineral is also thought to change to other allotropic forms on heating but the volume changes due to this cause are insignificant in most clays.

(*d*) *Liquid formation* arises from the combination of fluxing elements such as sodium, potassium, calcium, magnesium and occasionally ferrous iron with alumina and silica of the clay body to form eutectics or compositions of low melting point.

Liquid development fuses the body and draws together the solid components. It fills voids in porous masses and is the principal agent of the transfer of mass from one point of a body to another during the firing process.

Consequently as liquid develops on firing clay bodies, contraction always occurs. The rate and extent of contraction depends on

(*a*) the amount and nature of fluxing elements which are present in the body,

(*b*) the overall composition of the body,

(*c*) the particle size of mineral components, particularly those containing the fluxing elements,

(*d*) the maximum temperature attained and the rate of temperature rise.

The alkali elements are the most effective fluxing elements in clays, in that they combine with alumina and silica to form liquid at temperatures only slightly above 700° C. The amount of liquid increases as the temperature is raised, but its quantity and quality depend on other components which can take part in the liquid phase. In highly siliceous bodies, the rate of increase of liquid when alkalies are the fluxing elements is not great even up to 1500° C., but when alumina is present along with silica, large amounts of liquid form at much lower temperatures (p. 720). Consequently, the rate and extent of shrinkage is very much dependent on the composition of the body.

Calcium and magnesium are less powerful fluxes than alkalies at temperatures below 1000° C. although they can take part in ternary, quaternary and higher component mixtures and influence the type and amount of liquid which forms.

Above 1000° C. the alkaline earth elements can form liquids with alumina and silica and thus cause contraction and vitrification. In general, liquids containing calcium are more fluid than those containing magnesium and hence they result in more rapid and uncontrollable shrinkage.

In firing ceramic bodies, rapid volume shrinkage can generally be achieved by the body having a single fluxing element present in appreciable amounts and in a finely-divided state.

Conversely, a slow progressive contraction on firing is best achieved by combining two or more fluxing elements contained in mineral forms which are themselves present throughout a range of particle sizes.

(*e*) *Gas liberation at high temperatures.* When an appreciable quantity of liquid has developed in a ceramic body on firing, its condition might be described as pyro-plastic, i.e. having the ability to flow at high temperature. Normally, in firing clays the degree of pyro-plasticity is controlled so that as contraction takes place there is no excessive distortion due to differential pressures in the setting.

However, if gas is liberated within the body whilst it is in a pyro-plastic condition and if that gas cannot readily escape through the pore system, internal pressures may develop of such a magnitude as to cause overall expansion. When gaseous release at vitrification point is of a considerable order, large volume bloating or surface blistering may occur. Many brick clays show small expansile effects near to their vitrification point which, although they may be insufficient to spoil the bricks, prevent their densification to a point at which they would be suitable for engineering brick quality.

The important characteristics of volume changes on firing are best illustrated by referring to examples. Fig. XII.9 shows the linear contraction on firing which takes place in specimens of ball clays of various mineral compositions. The rate of

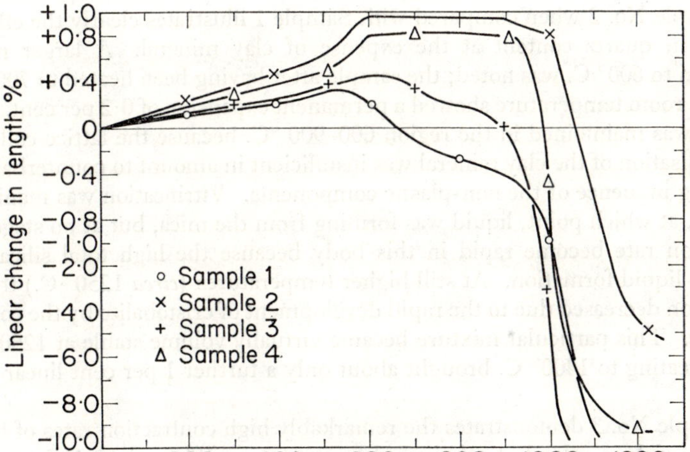

FIG. XII.9. LENGTH CHANGES ON HEATING CLAYS

heating in the tests was maintained at 5° C./min. Four examples have been chosen to emphasise the relative importance of the amount of clay mineral, quartz and mica (or other alkali-bearing) minerals. In the examples, alkalies have been calculated as mica and the particle size range of the various minerals is assumed to be constant. The chemical and calculated mineralogical compositions are in Table XII.XI, but only the major components are listed.

TABLE XII—XI. CHEMICAL AND MINERALOGICAL COMPOSITION OF BALL CLAYS

Sample No.	*Chemical Analysis*				*Mineralogical Analysis*		
	SiO_2	Al_2O_3	K_2O+ Na_2O	*Loss on ignition*	*Kaolin Mineral*	*Quartz*	*Mica*
1	48	35	1·5	12·9	76	6	14
2	71	19	2·0	5·7	31	48	17
3	49	33	2·8	13·0	57	10	25
4	60	28	3·2	7·6	43	24	29

Ball clay No. 1 which is rich in kaolin-clay mineral showed only a small expansion to about 500° C. and this was of a reversible nature. At this point dehydroxylation occurred with a slight although barely detectable irreversible expansion. At 600° C. lattice collapse, accompanied by progressive recrystallisation resulted in a relatively slow contraction. An accelerated rate at about 1000° C. was due to the melting of mica but this was comparatively slight in effect because of the small amount present. If no mica had been present, the slow rate of contraction would have been maintained and 10 per cent linear shrinkage would not have resulted until about 1450° C.

Sample No. 2 when compared with Sample 1 illustrates clearly the effect of an increase in quartz content at the expense of clay mineral. A larger reversible expansion to 600° C. was noted; the sample after having been heated to 700° C. and cooled to room temperature showed a permanent expansion of 0·2 per cent. Volume stability was maintained in the region 600–900 °C. because the lattice collapse and recrystallisation of the clay mineral was insufficient in amount to counterbalance the stabilising influence of the non-plastic components. Vitrification was manifested at 1000° C., at which point, liquid was forming from the mica, but at no stage did the contraction rate become rapid in this body because the high total silica content inhibited liquid formation. At still higher temperatures (*circa* 1250° C.) the rate of contraction decreased due to the rapid development of cristobalite by the conversion of quartz. This particular mixture became virtually volume stable at 1250° C. and further heating to 1400° C. brought about only a further 1 per cent linear contraction.

Sample No. 3 demonstrates the remarkably high contraction rates of ball clays which are rich in mica but relatively low in quartz. The principal feature in the heating curve was the shrinkage between 950° C. and 1025° C. which amounted to about 10 per cent linear. It is also of interest that mica behaves in a somewhat similar fashion to quartz in the temperature range 600–900° C. It stabilises the body and prevents the shrinkage attributable to lattice collapse and recrystallisation of the clay mineral.

Sample No. 4 contained a high content of both quartz and mica. As is to be expected, the former mineral confers high initial expansion to 600° C. and following this point, volume stability was preserved almost to the same extent as in Sample No. 2. Above 950° C., the alkalies from the mica form appreciable amounts of liquid which was manifested by a sharp and rapid contraction. The rate and extent of this shrinkage was not as great as in Sample 3 because of the stabilising effect of the high quartz content. At about 1200° C., the conversion of quartz to cristobalite caused a marked slowing down of the contraction rate.

The samples which were tested are typical British ball clays supplied commercially for a variety of purposes. Sample No. 1 is a high quality product used as a constituent of domestic pottery to confer strength and good casting characteristics. Sample No. 2 is of high silica content used as a major component of stoneware articles. Sample No. 3 is typical of material favoured by tile and electrical porcelain manufacturers where high strength and a rapid contraction rate to produce a body of low porosity is required. Sample No. 4 is of particular value where fast firing cycles are preferred, as in sanitary ware.

The effect of the grain size of minerals on the firing contraction is illustrated by the shrinkage curves of a range of medium grade fireclays studied by Carr, Hargreaves, Grimshaw and Roberts,[1] who investigated the shrinkage produced in a clay body by differing amounts and various sizes of alkali-bearing hydrous mica. Provided that the grain-size of this liquid-producing mineral was kept constant, the shrinkage produced at 1250° C. was almost directly proportional to the amount present. Finely-divided mica, however, reacted much more readily than coarser material and the resulting volume decrease in the clay was far more pronounced and developed with greater rapidity (see Fig. XII.10A).

There was little or no fire-shrinkage between 600° C. and 900° C., but above 900° C. the shrinkage increased considerably in amount up to 1100° C. It thus appeared that during the burning of the carbonaceous matter scarcely any shrinkage takes place, so that the temperature may safely be raised fairly rapidly between 600° and 900° C. if little carbonaceous matter is present. Shrinkage continued up to the highest temperature at which the clayware was fired and would still continue, if the heating were prolonged, up to the point at which fusion occurred, unless the conditions were such as to cause bloating, when an expansion would result. Consequently, clays will usually shrink in use if they are then heated to a temperature higher than that attained in the firing. This is very important in connection with refractory clays, which should be fired at a sufficiently high temperature to prevent an excessive contraction in use.

As Fig. XII.10B shows, when the synthetic mixtures of hydrous mica (illite) and clay were refired at 1400° C., the coarse-grained mica reacted causing considerable contraction in the body. The fine micaceous mineral had more closely approached equilibrium by firing at 1250° C., so that reheating to 1400° C. produced considerably less contraction.

[1] Carr, K., Hargreaves, J., Grimshaw, R. W. and Roberts, A. L., *Trans. Brit. Ceram. Soc.*, 51, 345, 1952.

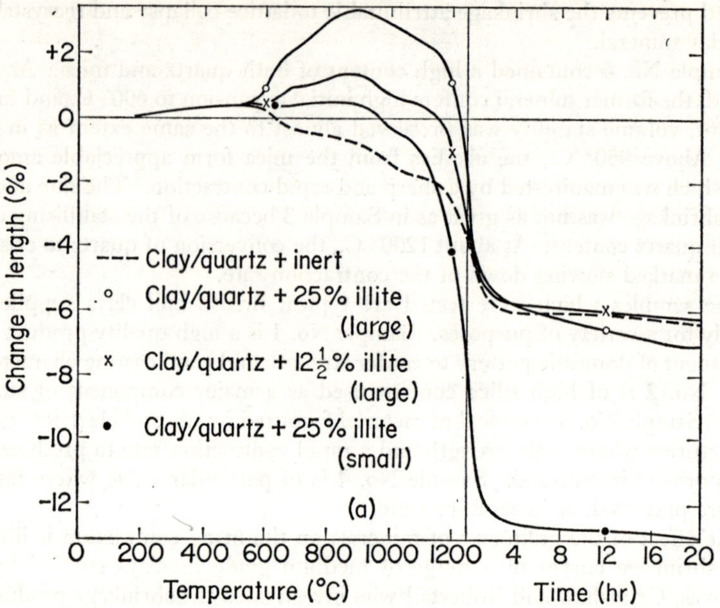

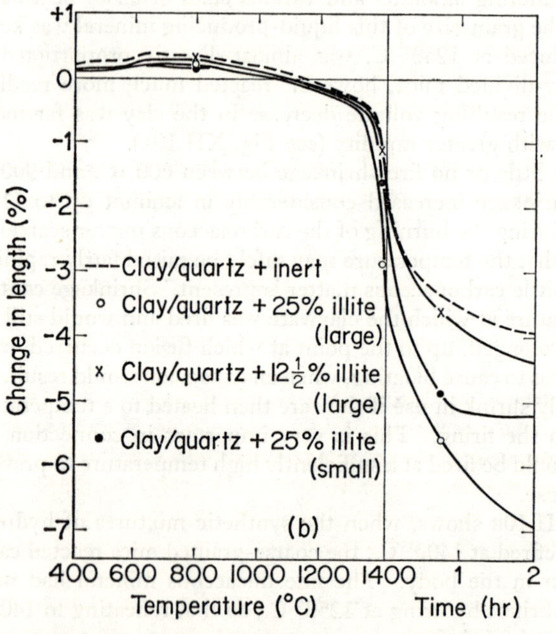

FIG. XII.10. FIRING SHRINKAGE OF FIRECLAYS

(A) Shrinkage on Firing in Synthetic Mixtures of China Clay and Illite
(B) Shrinkage on Reheating Mixtures to 1400° C.

The presence of non-plastic material, such as grog, quartz or alumina greatly reduces the firing shrinkage of clays, apparently by acting as a non-reactive skeleton during heating. By the addition of such materials an excessive shrinkage in a clay body can be avoided.

Semi-silica bricks containing between 10–15 per cent of alumina and about 80–85 per cent silica are made from inferior ganisters or quartzites to which the requisite amount of a clay is added if necessary. These bricks show little or no volume change during firing or in subsequent service at temperatures up to 1350° C. probably because a small proportion of quartz converts to cristobalite and the resulting expansion is sufficient to counterbalance any contraction in the clay.

In Table XII.XII is shown the volume changes at various temperatures in clays; the specimens were maintained for two hours at each temperature.

TABLE XII—XII. VOLUME CHANGES OCCURRING ON HEATING CLAYS

	Temperature °C.						
	200	400	600	800	1000	1200	1400
Plastic clay	+0·1	+0·3	+0·3	+0·0	−4·3	−16·2	—
Plastic clay + 30 per cent sand	+0·3	+0·5	+0·7	+0·8	−0·1	−4·3	—
Fireclay (cone 30)	+0·1	+0·2	+0·3	+0·4	0·0	−9·2	−14·0
Fireclay + 20 per cent grog	+0·1	+0·2	+0·4	+0·4	+0·1	−6·3	−9·5
Fireclay + 40 per cent grog	+0·2	+0·2	+0·5	+0·6	+0·4	−3·0	−6·2

Carr *et al.* found irregular volume changes in fireclays when heated and were able to associate various effects with the presence of particular minerals. They noted an expansion between 550°–600° C. in clays containing appreciable amounts of free quartz and concluded that where expansion occurred between 600°–900° C., it was due to the exfoliation of hydrous mica (see Fig. XII.10A).

Clays used in the manufacture of *common building bricks* are more complex in chemical and mineralogical composition than either ball clays or fireclays, so that a scientific interpretation of their volume changes on firing is not easy.

In the firing of Building Bricks it is only rarely that maximum temperatures in excess of 1100° C. are used and in some cases 950° C. is sufficient. Consequently the only elements which behave as fluxes and can form liquids with alumina and silica are those of the alkali group, sodium and potassium. Other elements such as iron and calcium can be dissolved in the liquid phase, but apart from affecting the colour, they behave as stabilising non-plastic components and tend to reduce shrinkage. Magnesium also dissolves in the liquid phase and when the content of this element is large, as it is in some Keuper marls, it can markedly reduce the fluidity of the liquids which develop at high temperatures. In these cases, the shrinkage at about the maturing point often proceeds at a slower and more controlled rate.

Apart from the above exceptions, therefore, the volume changes in common brick clays at high temperatures depend largely on the amount of alkali-bearing minerals which are present and the particle size of these minerals. Mostly the minerals involved are of the mica-type, but some clays of recent origin have more complex species.

It is implicit that in clays of this nature there will always be sufficient available silica and alumina to develop liquids, but this is rarely otherwise.

In Chapter V (p. 306), typical compositions of building brick clays of the U.K. have been listed. If these values are compared with the maximum temperature of firing of bricks of different geological eras in the U.K. (Chapter XI, p. 737), the relationship between maturing temperature and total content of alkalies is clearly established.

Keuper marls containing 4–5 per cent total alkalies often mature at 1000–1020° C., whilst Etruria marls with only 2–2·5 per cent alkalies require to be burned up to 1100–1150° C.

This is far from the complete story, however; the rate at which vitrification and its consequent firing shrinkage develop is at least as important. If a clay has a short firing range it is difficult to control in the kiln and wastage may be high because of under- and over-burning. At the best, the size range of the bricks from kiln will be outside acceptable limits if shrinkage is occurring rapidly near to the maturing temperature.

A comprehensive survey of the firing characteristics of typical brick clays has been undertaken by Freeman.[1] He suggests that it is possible to classify most of the clays used for brickmaking in the U.K. into 4 sub-divisions. Within these groups the volume changes which occur can be related to the content of calcium carbonate and to sand, which behave largely as non-plastic components.

Clays used in brickmaking mostly show similar characteristics up to 750° C. as those of the ball clays illustrated in Fig. XII.9. An overall expansion occurs which is most marked in the 550–600° C. region and its magnitude depends on the content of quartz. Above 750° C. contraction commences at a rate and amount determined by the alkali mineral content and size.

Superimposed on the general pattern, however, are effects due to other minerals.

If considerable lime is present as carbonate, the rate of contraction is slowed down in the temperature range 850–1000° C. and then shrinkage occurs suddenly and often dramatically. Between 800–850° C. in bricks containing large amounts of calcium carbonate a sharp shrinkage occurs which Freeman suggests is due to the structural collapse associated with the loss of carbon dioxide.

Free silica, as quartz, also reduces the vitrification shrinkage in building brick clays. It tends to shorten the temperature range over which the contraction occurs in these clays which mature at high temperatures, but to lengthen it in clays which would otherwise have low vitrification temperatures. Grog, as pre-calcined brick or cinders behaves in a similar fashion.

Many brick clays show a pronounced tendency to bloat at temperatures near to the full maturing point. This causes a reduction in the shrinkage rate but prevents complete densification of the bricks. Gaseous liberation from sulphides, sulphates and other minerals is thought to be the cause of bloating, but an insufficient cleaning period can be a contributory factor.

Most clays with a low maturing temperature tend to bloat as the top temperature is reached and conversely those of higher refractoriness are often, but not always, less troublesome.

[1] Freeman, I. L., *Trans. Brit. Ceram. Soc.*, 57, 346, 1958.

The shrinkage of clays is dependent on the physical texture of the material as well as on the mineralogical composition. The individual grains composing a compacted material may vitrify and shrink within themselves at high temperatures but an overall shrinkage of the article can occur only when reactions between adjacent grains take place. Consequently the shrinkage of clays depends upon the intimacy of mixing and forming and on the surface area of contacts between grains. In large-particled material, there are few contact points, so that shrinkage normally takes place at slower rates and at higher temperatures than when the material is finely divided and subjected to high pressures.

This is well-illustrated in Fig. XII.11 which shows the volume shrinkage

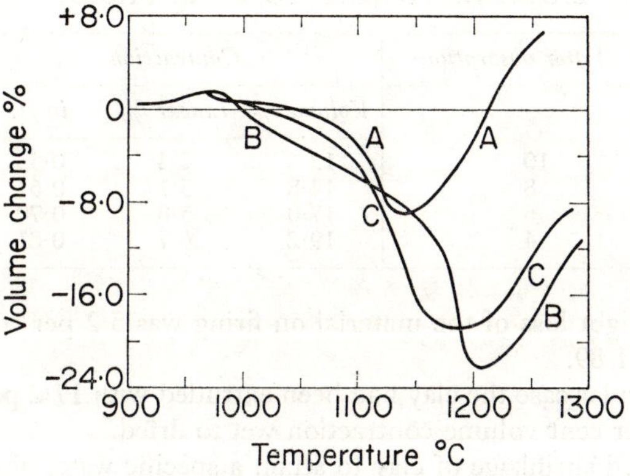

FIG. XII.11. THE FIRING CHARACTERISTICS OF A CLAY PREPARED IN DIFFERENT WAYS

(A) Ground – 200 B.S. and dry pressed
(B) Ground – 6 B.S. and semi-dry pressed
(C) Composite body, semi-dry pressed

characteristics of the same clay but prepared in different ways. Curve A is of material ground through 200 mesh and pressed into tile shapes at 10 tonsf/sq.in. The product, on firing to 1120° C., was suitable for chemical engineering tiles with a water absorption of less than 0·1 per cent. The total volume shrinkage was small because of the high initial pressed density, but it mostly occurred over a fairly narrow temperature range.

Curve B shows the shrinkage of the identical clay, but crushed to pass 6 mesh and then pressed into ladle brick shapes at about 3 tonsf/sq.in. A greater contraction was noted and this also occurred over a small temperature range but its maximum was at 1200° C. where the minimum water absorption was 3 per cent. At higher temperatures, the bricks bloated to a great extent.

Curve C shows the volume change of an experimental mixture prepared by adding 20 per cent of the finely ground 200 mesh clay to the 6 mesh quality. The form of the shrinkage curve was markedly different in that a longer firing range resulted with a lowered maturing point, but at the same time an earlier incidence of bloating was unavoidable.

This is a general characteristic of many clays. The rate of firing shrinkage can be better controlled by grinding a proportion to very fine size and then reblending it with the coarser-ground main body. The technique is successful only with clays which mature at about 1000° C. and above and is less effective with extruded bodies than with those that are pressed.

In the firing of clays to a dense product satisfactory for use, the total contraction depends on the method of manufacture and the density of the finished product.

A typical carboniferous brick clay showed the following total shrinkage *dry* to *fired* when burned to specific water absorptions (Table XII.XIII).

TABLE XII—XIII. TOTAL FIRING CONTRACTION OF A TYPICAL BRICK CLAY (DRY TO FULL FIRED)

% *Water absorption*	*Contraction*		
	Volume %	*Linear* %	*in/ft.*
10	12·6	4·4	0·53
8	14·8	5·1	0·61
6	17·0	5·8	0·70
4	19·2	6·7	0·81

The total weight loss of the material on firing was 5·2 per cent and the dried bulk density was 1·89.

In this particular case the clay had been extruded with 17·2 per cent water and had given 13·1 per cent volume contraction wet to dried.

The total fired shrinkage of clay to attain a specific water absorption depends on the total loss in weight provided that the same method of manufacture into shapes has been used.

This is to be expected because of the physical and chemical principles which are involved and which have been discussed earlier. When water or carbon dioxide is lost from a clay on heating, no shrinkage occurs, so that the porosity will increase in direct relation to the volume of gas lost. When vitrification takes place shrinkage must firstly counterbalance the increase in porosity which has occurred on calcination and then further shrinkage is involved in drawing the grains together.

The advantage of grog additions to reduce shrinkage is thus apparent, because this will not undergo any increase in porosity during the early stages of firing.

It must also be remembered that not all clays can be fired to bodies of low water absorption without serious distortion and bloating occurring.

Porcelain bodies shrink considerably on firing because the finished article has to be of such a low porosity as to be virtually impervious. The shrinkage is the result of liquid formation developing within the body. Felspar is normally added to provide alkalies which react with the silica of the calcined flint and the alumino-silicates of the china and ball clays. The control of the rate of shrinkage in porcelain bodies is difficult and one of the many skills of the potter lies in the setting of ware in the kiln.

Differential shrinkages in various parts of an article must be reduced to a minimum to keep distortion and warpage within acceptable limits.

Pipe bodies are frequently manufactured from clays which are similar to those used for common bricks or low-grade firebricks, but the material is generally ground to a fine-particled size. The rate of contraction at the maturing temperature, coupled with the overall maturing range, is of even greater significance than it is in bricks, because accuracy of shape is of paramount importance.

Fully vitrified pipes with a water absorption between 3–6 per cent, should have a maturing range of at least 50° C. within which the body conforms to the above porosity range and during which the linear shrinkage should not exceed 1·3 per cent.

To achieve these characteristics, the clay should be low in its content of free quartz and if this mineral is present, the body must be ground finely. The minerals which contain the fluxing elements (mainly the micas and felspars) should be distributed over a wide range of particle size. The clays which are most satisfactory contain 2·5–4·5 per cent total alkalies as oxides, and the mica or felspar is present in all size ranges from 53 microns to 1·8 microns (wet analysis). The percentage of material exceeding 53 microns (300 B.S. mesh size)—calculated on a dry basis—should not exceed 12 per cent.

Silica bricks expand considerably during the burning, in course of manufacture, on account of the conversion of quartz to tridymite or cristobalite (Table XII.XIV), which involves a volume expansion of almost 20 per cent or at least $\frac{7}{8}$ in. per linear foot. The complete conversion is never entirely effected in commercial practice during the firing of the bricks, but a slow and constant expansion continues whilst the bricks are in use at high temperatures, so that if the bricks are to be as constant in volume as possible when in use, they must be heated sufficiently during the first firing to effect as much as possible of the total expansion. Bricks which have not been heated sufficiently during the firing may fail—as the result of continued expansion—when in use.

In the manufacture of high-grade silica refractories for use in steel plant practice, the quartz content after firing must be reduced to a minimum. Burning should continue until the specific gravity does not exceed 2·34.

TABLE XII—XIV. SPECIFIC GRAVITY AND VOLUME CHANGES ACCOMPANYING THE INVERSIONS AND CONVERSIONS OF SILICA

Temperature °C.	*Variety*	*Specific gravity*	*Changes into*	*Specific gravity*	*Volume increase per cent*
Inversions:					
573	α-quartz	2·65	β-quartz	2·60	2
210–280	α-cristobalite	2·32	β-cristobalite	2·21	5
117	α_1-tridymite	2·28	β_1-tridymite	2·27	0·5
163	α_2-tridymite	2·28	β_2-tridymite	2·27	0·5
Conversions:					
870	β-quartz	2·60	β-tridymite	2·27	13
870	β-quartz	2·60	β-cristobalite	2·21	18
1680	β-quartz	2·60	Fused silica	2·21	18
1470	β-tridymite	2·27	β-cristobalite	2·21	5
1605	β-tridymite	2·27	Fused silica	2·21	5
1719	β-cristobalite	2·21	Fused silica	2·21	0

Bauxitic and aluminous clays shrink greatly when fired during their course of manufacture even though the raw material has been pre-calcined at a very high temperature.

Sintered alumina ware and other oxide compacts may show up to 15 per cent volume reduction during firing due mainly to the decrease in porosity resulting from crystal growth.

Sillimanite bricks usually have a small shrinkage unless the amount of clay added as a binder is excessive. The raw material from which such refractories are made, however, has, in most cases, to be pre-calcined prior to shaping. Three minerals are in common use for the manufacture of sillimanite ware. These are kyanite, andalusite and sillimanite all of which, on high temperature calcination, change into a mixture of mullite and cristobalite. Sillimanite undergoes little change in density on conversion but andalusite and kyanite show marked increases in volume, the former of 3 per cent and the latter of 20 per cent.

Other refractory wares, such as bricks, crucibles, etc., made of carbon, magnesia, carbides, chromite, zirconia, etc., contract very little during the firing process and have a constant volume when in use. The kiln contraction during manufacture is almost wholly due to the changes in volume of the bond.

The measurement of the permanent volume changes in ceramic materials can be carried out in a similar apparatus to that for the reversible changes (see p. 786). It may also be readily determined from the length or volume changes of a test piece after treatment.

The provisional method for testing the contraction of clay consists in drying the clay at a temperature not exceeding 70° C., crushing it so that it will pass through a 28-mesh sieve, mixing it with water and shaping it into a suitable form, by means of a mould. The test piece is marked with vertical lines 9 cm. apart and allowed to dry, first at the ordinary temperature, then for four to five hours at 70°–80° C., and finally at 110° C. The drying shrinkage is then measured. The test piece is then heated in a furnace at the rate of not more than 100° C. per hour to 900° C., and afterwards at the rate of not more than 4 cones (about 80° C.) per hour to the end of the firing. The test piece is then allowed to cool and is measured. The difference between the vertical lines (*a*) before and after drying gives the drying shrinkage, (*b*) after firing and after drying gives the kiln shrinkage, and (*c*) before drying and after firing gives the total shrinkage. These results are usually expressed as a percentage of the distance between the fiduciary marks before drying the test piece.

Permanent Volume Changes in Fired Bodies may occur if the articles are subjected to heat treatment in service which is sufficient to cause further reactions, e.g. liquid formation. They may occur if the ware is used at a higher temperature than that at which they have been fired in the kiln, or if the soaking period during manufacture has not been sufficiently prolonged.

The *after-contraction* or *-expansion*, as such changes are termed, are important, because unless they are extremely small, excessive length changes may occur on heating in structures made from refractory bricks, thus resulting in cracking or disruption.

Although the importance of volume changes in fired refractory articles on reheating and under service conditions is recognised, there is considerable divergence of opinion as to how the amount of such changes can be predicted prior to the bricks being used. Many firebricks are manufactured to withstand temperatures of 1150°–1250° C. when in use in furnaces, muffles or oven settings; such bricks are normally fired to about 1250°–1280° C. and soaked for 12–24 hours within that temperature range. The user of the refractories must have some means of predicting the amount of volume change which the materials will undergo when maintained perhaps for many months at the operating temperature.

Both the British Standards Committee[1] and the American Society for Testing Materials[2] have specified a Reheat Test for measuring the After-Contraction or After-Expansion of refractories which involves heating the material to a considerably higher temperature than would be encountered in practice. It is then assumed that a short period of firing at this elevated temperature would be equivalent to a much longer period of maintained heat at the normal operating temperatures. For some clays, useful conclusions may be derived from such a test, but for a great many others, it is of doubtful value. At 1400° C., reactions proceed which do not occur at 1300° C. because the equilibrium conditions are quite different. The liquid phase which forms at elevated temperatures is less viscous, so that, especially in porous materials, a shrinkage will occur which would not develop at lower temperatures. Furthermore, at 1400° C., low-grade fireclays tend to bloat, thereby masking any tendency to contract.

Despite these objections, the Reheat Test is widely used as an indication of the serviceability of a refractory and most users specify a maximum permissible volume or length change.

In British practice the Reheat Test is normally carried out on test pieces $2\frac{1}{2}$ in. × 2 in. × 2 in. cut from a brick or shape. The square ends must be ground as near parallel as possible; a line is drawn down the centre of one long side and the length of the specimen along this line measured with a vernier caliper or some other means which must read to the nearest 1/1000 inch. The test pieces are then placed in a suitable furnace with the long face vertical and fired either to 1350° C. or to 1410° C. and maintained there for two hours. Silica and basic bricks are usually fired to a higher temperature, either 1500° C. or 1650° C. being specified. After cooling, the test pieces are re-measured and the linear change in length calculated.

The A.S.T.M. test requires that a full brick be used and recommends a five-hour soak at the maximum temperature. Fireclay bricks should be tested at 1200° C., 1290° C., 1350° C., 1400° C., 1500° C. or 1600° C. depending upon their proposed application. Normally a shrinkage of more than 1·5 per cent is not permitted.

When the shrinkage by volume is to be determined accurately, a volumeter such as is used for porosity determinations (p. 434) must be used, the volume of the sample being determined before and after the test and the percentage of volume change calculated from these figures. The expansion of bloating bricks used in steel ladles and as casting-pit refractories can be determined in a similar manner.

Standard Specifications for alumino-silicate refractories as used in the British Practice require that a maximum after-contraction should not be exceeded in the

[1] British Standard 1902, Part 1A, 1966.
[2] A.S.T.M. Designation C 113–61.

various grades of materials.[1] Details of these specifications are given in Table XII.XV. In the U.S.A. every industry specifies its own conditions for refractories.[2]

TABLE XII—XV. BRITISH SPECIFICATIONS FOR ALUMINO-SILICATE REFRACTORIES (1966)

	Grade A	*Grade D*	*Grade L*†	*Grade M*†
Composition	Silica not less than 94% SiO_2	Semi-silica 78–85% SiO_2	Firebrick less than 78% SiO_2	Firebrick less than 78% SiO_2
Reheat test	1450° C. 2 h.	1350° C. 2 h.	1350° C. 2 h.	1410° C. 2 h.
Maximum change in size permitted (per cent)	Exp. less than 0·5	−0·50* +1·00	S. 1·0 L. 1·2	S. 1·0 L. 1·2
Minimum refractoriness (Cone No.)	31	28	28	29

	Grade N	*Grade P*‡ (1 *and* 2)	*Grade Q*1‡	*Grade Q*‡ (2 *and* 3)
Composition	Firebrick 28–35% Al_2O_3	Firebrick 35–40% Al_2O_3	Firebrick 40–45% Al_2O_3	Firebrick 40–45% Al_2O_3
Reheat test	1410° C. 2 h.	1410° C. 2 h.	1410° C. 2 h.	1500° C. 2 h.
Maximum change in size permitted (per cent)	S. 1·0 L. 1·2	S. 0·8(2) 1·0(1) L. 1·2(1)	S. 0·5 L. 0·8	0·8(2) 0·5(3)
Minimum refractoriness (Cone No.)	29	31	33	33

* Semi-silica for coke oven usage must not change linearly by more than ±0·5.
† In Grade L and M material the specifications for large (i.e. where each dimension is at least 6 in.) and small sized blocks are different.
‡ Grades P & Q allow more than one value depending on the maximum apparent porosity.

Although the Reheat Test affords a rapid and simple method of measuring the probable behaviour of refractories when in use, a more accurate and reliable indication is given by observing the changes in length of a specimen when maintained at a suitable temperature (corresponding to that encountered in practice) for varying periods of time. The rate of the change in length can thus be calculated and extra-

[1] British Standard 1758:1966.
[2] A.S.T.M. C106–51 (Incinerators).
C 64–61 (Heavy Duty Boilers).
C 63–61 (Malleable Iron Furnaces).
C435–63 (Steel Pouring Pits).

polated to infinite time. An apparatus such as that described on p. 786 is most suitable for these tests.

Mellor has shown that if it is assumed that the change of contraction during the firing is proportional to the square of the contraction which has yet to take place, the effect of repeated heating at 1120°–1160° C. (Cones 2*a*–4*a*) approximately follows the law for bimolecular reactions and the contraction may be calculated as follows:

$$x = \frac{a^2kt}{1 + akt}$$

where x is the contraction after any number of firings t, a is the maximum contraction after an indefinite number of firings, and k is a constant depending on the nature of the material. He found that fireclay bricks contracted more in a reducing atmosphere than in an oxidising one. The increased contraction is probably due to the greater fluxing effect of theiron in a reducing atmosphere.

B.S. 1758: 1966 also classifies ladle bricks as Grade K in which the requirements are a maximum silica content of 78 per cent and a minimum P.C.E. of 18. The maximum permitted Permanent Linear Change is not specified because some of these products bloat to a considerable extent although the recommended test temperature is 1350° C. for 2 hours.

Bloating in ladle bricks is caused by gaseous release—probably sulphur dioxide from sulphates—when the body is in a pyroplastic condition. It is essential that the bloating shall take place in a uniform fashion and not form cavities or blisters. The bloated product must have a honeycomb structure which is possible only with certain natural clays.

Bloating bricks have been must successful in steel ladles because their expansile characteristics tend to seal joint irregularities and develop a monolithic structure. They virtually eliminate cobble-stone attack in ladles which is caused by the corrosion or erosion of joints.

Silica bricks occasionally show an after-expansion due to the conversion of some quartz which was unchanged during kiln firing, and sometimes to a change from cristobalite to tridymite. The after-expansion measured at 1500° C. may be as high as 6 per cent in some bricks and it is important that this should be determined so that an expansion allowance can be made when constructing a furnace from them.

Basic bricks, provided that they have been well-fired and that they are made from high-quality materials, rarely show a pronounced volume change when heated to 1500° C. for two hours. A small contraction (*circa*. 0·3 per cent) is not harmful and can be attributed to the sintering of one or more of the crystalline components such as periclase or spinel to form a mass of lower porosity.

CHANGES IN POROSITY ON HEATING

THE porosity of a material has been defined in Chapter VII, as the total proportion of the pores, voids or interstices which occur in a body or agglomerated mass.

In the unfired material, the porosity is related solely to the size and shape of the grains and the manner in which they are blended together, but on heating, the nature of the materials comprising the mass has a very important influence because of the chemical reactions and partial fusing which occurs. The fineness of grain is an added factor in porosity changes at high temperatures but only in so far as it increases the rate at which reactions take place. In this respect, it is interesting to compare the influence of particle-size on the porosity in the raw and fired states. Mixtures of fine materials have a large surface factor and are often more porous in the raw state than those containing coarse materials. At high temperatures, however, small-particled material fuses more readily than coarser grained bodies. The liquid formed fills the interstices and voids in the mass thus reducing the porosity.

The porosity of a body may increase or decrease as the temperature is raised, depending largely on the composition and minerals which are present.

Materials Increasing the Porosity. In general, the porosity of a mass will increase if it contains components which volatilise, evaporate or decompose with gas liberation at a particular temperature. Thus, in the drying of moulded ceramic bodies, water is driven off from the pores between the solid particles. Although the removal of water may cause minor rearrangements in the dispositions of individual particles, provided the moisture content is below the critical point (see p. 545), the water lost on drying and firing is entirely replaced by air.

If carbonaceous or other organic constituents are present the porosity will increase when oxidation or burning occurs. Light, porous bodies can be artificially produced by incorporating some combustible, such as sawdust or finely powdered coke or coal, prior to shaping. Naturally-occurring porous materials, such as expanded vermiculite, kieselguhr or diatomaceous earth may also be admixed with clay to produce a similar effect.

Decomposition reactions involving gas liberation can also cause an increase in the porosity of the mass. The destruction of the clay minerals and other hydrous substances, combined with the dissociation of carbonates tends to give a maximum void space in clay bodies at about 800° C.

Materials Reducing the Porosity. The materials which cause a reduction in the porosity of fired articles, partly or wholly composed of clay, are those which are fusible at the temperature of firing, or they are *fluxes* which combine with some of the other constituents to form a liquid, which fills the pores of the materials. The amount by which the porosity is decreased depends upon the type of fluxing elements, the form in which they are present, the grain-size of the minerals which contain them and the temperature at which the body is fired.

Sintering and crystal growth also reduce the overall porosity.

Because alkalies form a liquid with alumina and silica at much lower temperatures than other fluxes, it is to be expected that they would be the most effective in reducing porosity, but because they are usually present in clay bodies as mica or felspar they do not readily react unless the minerals are finely divided.

The temperature of firing largely determines the amount of liquid which forms in a clay body containing fluxes, hence the porosity of an article tends to diminish when it is fired to higher temperatures.

Table XII.XVI shows the effect on the porosity of china clay to which were added varying amounts of felspar of three grades and then fired to different temperatures.

The reduction in porosity was greatest with large amounts of felspar of small grain-size and after firing to the highest temperature.

TABLE XII—XVI. EFFECT OF FLUXES ON POROSITY
(after J. Keele)*

Size of Particles of Felspar	*Felspar*	*Porosity in dry State*	*Temperature of Firing in Seger Cones*						
			05	1	3	5	7	9	12
	Per cent								
40-mesh	40	31·00	34·50	30·95	27·55	23·65	19·60	10·95	4·96
	30	33·63	37·30	31·70	28·55	25·40	21·70	14·40	9·33
	20	36·60	40·00	37·15	32·00	30·00	25·40	18·45	14·21
	10	39·50	43·00	40·00	35·70	32·35	30·50	23·00	18·65
100-mesh	40	32·85	36·50	30·80	23·80	17·10	13·20	6·60	1·40
	30	35·20	39·00	38·30	29·40	25·05	18·30	10·80	4·92
	20	37·00	41·00	38·40	30·60	27·50	27·10	14·40	8·87
	10	39·20	42·70	38·15	34·55	32·25	25·30	23·95	16·37
200-mesh	40	35·10	38·60	22·65	18·50	14·80	6·00	—	—
	30	36·45	40·00	28·80	20·50	15·80	13·65	4·40	—
	20	38·05	41·55	33·60	28·45	23·70	21·90	10·50	7·11
	10	41·10	43·80	36·70	33·15	28·25	27·08	17·20	12·75

* *Trans. Amer. Ceram. Soc.*, 13, 731, 1911.

The heat treatment to which materials are subjected also influences their porosity. Burned articles usually have a greater porosity than unfired materials up to the temperatures at which vitrification commences. Articles which have been fired to higher temperatures may be less porous and in some cases are impermeable.

Once a high temperature has been reached, the reduction in porosity of a clay closely follows the shrinkage because both depend on the rate at which liquid develops or at which recrystallisation occurs.

Porosity, however, can be measured in two ways (see p. 419) (*a*) on a weight basis where the total amount of water which can be absorbed is expressed as a percentage of the dried weight of the specimen, and, (*b*) on a volume basis, where the volume of air contained in voids in the specimen is expressed as a percentage of the total volume of the specimen. If the specimen is fully permeable and free from closed pores, the water absorption equals the volume porosity divided by 2·6–2·8. When fusion is well-developed, sealed pores increase in number so that the water absorption may decrease at a much greater rate than the overall porosity. Some bloated clays have virtually zero water absorption but they may have up to 50 per cent volume porosity.

Many ball clays can be fired to a low porosity after treatment at moderate temperatures. The rate at which this is achieved depends on the mineralogical composition in much the same way as the changes in volume.

Brick clays cannot usually be fired to very low porosities because they tend to bloat and distort at temperatures just above the maturing point. Some are better than others in this respect; good ones give water absorption values of less than 7 per cent (Class B Engineering) and the best can have values lower than 5 per cent (Class A Engineering) under optimum firing conditions.

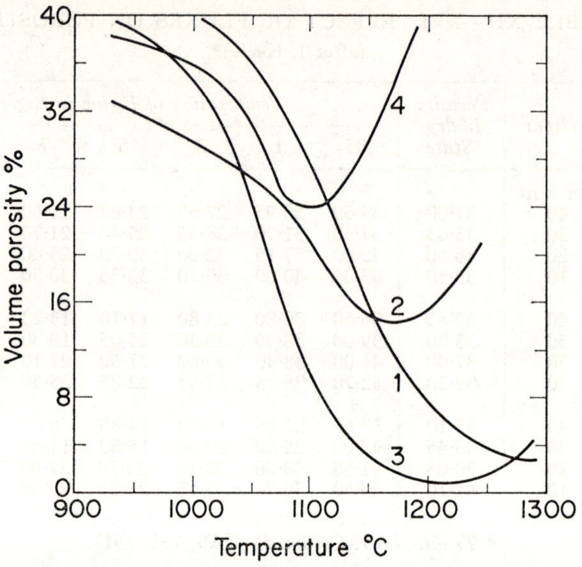

FIG. XII.12. CHANGES IN THE POROSITY ON FIRING TYPICAL CLAYS

Fig. XII.12 illustrates the changes in porosity of four clays of different characteristics. Samples 1 and 3 are the ball clays listed in Table XII.XI under the same designations. The high mica content in Sample 3 produces a low porosity body at 1160° C., whilst the high clay content and small amount of fluxes in Sample 1 result in a much slower reduction of porosity.

Sample 2 is a brick clay which can be fired to Engineering Class B quality at 1150° C. whilst Sample 4 is a brick clay which bloats excessively above its maturing temperature of 1080° C.

The ability of a clay to fire to a mass of low porosity is of particular value when preparing grog or pre-calcined material for use in castable and mouldable compositions and for additions to refractory clay batches. If grog of virtually zero porosity is incorporated, it cannot shrink further in service and furthermore it is remarkably resistant to slag and metal penetration and to erosion. Flint clay grogs are particularly valuable in this respect and they combine the property of low porosity with high refractoriness values.

Brick clays which fire to a low porosity have the added advantage of being very strong and resistant to frost attack.

The porosity of articles made of clay increases as they are heated until a temperature of about 750°–850° C. is reached. During this heating the clay expands slightly and the loss of water and carbonaceous matter, which occurs at the same time, causes the porosity to increase. At the moment when clay is dehydrated (about 450°–500° C.), the dissociation of the material and the liberation of the combined water may increase the porosity by about 10 per cent. At 700°–800° C., the porosity depends partly on the initial porosity of the mass and partly on the amount of carbonaceous matter present. When the temperature is increased above 900° C. partial fusion occurs and the molten matter so produced begins to flow into the pores. If the temperature is sufficiently high and the heating is sufficiently prolonged the fused matter will fill all except the closed pores and a non-porous product will be formed. With most clays the article will lose its shape or bloat before this stage is reached, but not in the case of vitrified engineering bricks, porcelain, and some stoneware.

Certain types of clays are reduced in porosity after moderate firing temperatures but, as the temperature is raised further, bloating of the mass occurs with an increase in porosity (see p. 801). Although the amount of void space in a bloated brick is large, most of the pores are sealed, so that the amount of water absorption may be very small.

Nodular ironstone is a serious contaminant of clays because at high temperatures, especially in a reducing atmosphere, it forms localised centres of a highly-bloated, fusible mass, which gives an unpleasing appearance to the brick, in addition to forming areas of high porosity.

Some types of clay increase in porosity after firing to high temperatures because they *exfoliate*. This behaviour is typical of hard, laminated fireclays, especially those which contain a high proportion of carbonaceous matter. The evolution of gas during burning forces open the bedded layers and produces a mass of high porosity. For this reason, it is very difficult to produce a dense brick from some Scottish fireclays although considerable beneficiation is afforded by fine-grinding such clays prior to shaping into bricks.

The porosity of fired ceramic ware can be reduced by the careful control of *grading* especially when the material does not readily form a liquid phase at high temperatures. The influence of grading on the porosity has been described in Chapter VII.

By *increasing the pressure* used in shaping bricks a considerable reduction in total porosity throughout the firing cycle can be secured. Higher forming pressures compress the particles of which the body is composed more tightly and at the same time less water is required during shaping. If the pressure during the shaping of bricks is increased from 1,000 lbf/in^2 to 8,000 lbf/in^2 the initial porosity may be reduced by about 20 per cent of its original figure. The use of still greater pressures would further reduce the porosity but for most claywares the additional expense involved would not be justified. Higher forming pressures are used in the manufacture of high-grade refractory products.

The onset of vitrification or the development of a liquid phase is not the only factor causing a reduction in porosity on firing. Crystal formation may produce a similar effect although higher temperatures are usually required before a sizeable decrease in void space is observed. At elevated temperatures, crystalline minerals begin to form in most ceramic bodies. The first products are often small crystallites hardly visible even under high magnifications; on prolonged heating, however, unless side reactions occur to destroy them, these crystallites start to grow and the mass *sinters* (see p. 681).

Sintering or crystal growth resulting in large decreases in the porosity of the mass becomes important, however, at elevated temperatures of the order of 1500° C.

and above, as in the manufacture of pure oxide refractories for special purposes. Compacts of materials such as finely-divided magnesia, alumina, zirconia, beryllia, etc., may be fired at temperatures well below their melting point and form bodies of low porosity and great strength. The void space within such bodies is reduced by a physical process involving only solid state reactions (see p. 682), which increase in activity as the temperature is further raised or maintained.

Burned clay-wares may be divided into two classes, according to their porosity, as follows:

1. *Porous ware*, including bricks, tiles, terra-cotta, refractory ware, coarse pottery, and earthenware.
2. *Impermeable ware*, including stoneware, china ware, and porcelain.

These divisions are only general, however, as some bricks are quite impermeable to water, and some china ware is appreciably porous.

Building bricks should be moderately non-porous as otherwise when moisture condenses from the air, the resultant water collects in drops on the inside of the walls, producing damp spots. If the pores are sufficiently minute, the bricks may be more porous to allow the walls to 'breathe'. Bricks with large, coarse pores should be avoided, as they admit rainwater too easily, but do not always part with it readily, so that walls in which they are used often remain permanently damp. The porosity of ordinary building bricks should not usually exceed 12 per cent by weight or 30 per cent by volume.[1] Common red bricks and facing bricks usually have a porosity of 5–10 per cent by weight or 12–20 per cent by volume, whilst wire-cut bricks, rubbers, and gault bricks usually have a porosity of 12–20 per cent by weight and 30–55 per cent by volume.

In the south of England, the average amount of water absorbed by bricks during complete immersion is about 12 per cent of the weight of the brick. North of the Trent and in Wales, denser bricks are usual and the average water absorption is seldom above 8 per cent of the weight of the brick. Bricks made by the semi-dry process—that is, by compressing the clay in the form of a damp dust—usually absorb only about 5 per cent of water, though they vary greatly in this respect.

Bonnell and Butterworth[2] have measured the water absorption of many typical building bricks of the U.K. on the basis of their geological origin and method of manufacture. Clews[3] has also produced excellent summary tables of the properties of building bricks and tiles, including their porosity values.

Bricks and other articles which are required to resist chemical action at ordinary or moderately low temperatures, or which must possess great strength, must be denser than ordinary architectural materials. Blue bricks of good quality usually have a porosity of only 0·1–2 per cent; it should not generally exceed 3 per cent by weight.

The large hollow blocks used in fireproof floors, etc., are usually made of a highly porous material. This reduces the cost of carriage and the weight of material in the structure, and provides insulation.

Tiles, terra-cotta and other unglazed argillaceous building materials should have

[1] For definitions and methods of measuring porosity and water absorption see p. 419.
[2] Bonnell, D. G. R. and Butterworth, B., 'Clay Building Bricks of the United Kingdom', *Nat. Brick Advis. Council Paper 5*, London, H.M.S.O., 1950.
[3] Clews, F. H., *Heavy Clay Technology* (Academic Press, London, 1969).

a similar porosity to building bricks, but the exposed surface should be non-porous. The best roofing tiles made from Etruria marls have a porosity between 3 and 8 per cent, light red tiles may have a porosity of 10–18 per cent and the darkest and densest red tiles a porosity of 3–5 per cent by weight (see B.S. 402, 1945).

With roofing tiles, the porosity is of minor importance compared with the permeability. So long as the tiles will not allow water to pass through them in the heaviest storm, the more porous they are the better, as porous tiles dry more rapidly than dense ones and are more resistant to sudden changes in temperature.

Other *unglazed wares* should have a porosity corresponding to the purpose for which they are to be used. Such articles as flower-pots, filters, etc., must be very porous.

Refractory Articles. The porosity of *fireclay bricks* is often much more important than in bricks which are not heated to a high temperature. It varies very considerably according to the purposes for which the bricks are to be used, but is generally between 8 and 24 per cent by volume or 3–9 per cent by weight. Where there is little corrosion or abrasion, a highly porous brick may be quite satisfactory and, even where these actions do occur, a porous brick may be necessary to give the requisite resistance to sudden changes in temperature.

Porous fireclay bricks which contain iron spots, when heated to redness, decompose carbon monoxide liberating free carbon. This is undesirable especially in gas producers, etc., used in the production and utilisation of carbon monoxide gas.

In general, when the articles are to be heated to a high temperature they should be as porous as is consistent with the other properties required.

Saggars should generally have a porosity of 10–40 per cent by volume or even greater, provided that the pores are very small and do not allow the fire-gases to enter into the interior of the saggar and so discolour the goods, and that the saggars possess sufficient strength.

Muffles should have a reasonably porous structure so that they will permit sudden heating or cooling. Muffles which are too dense are not economical in fuel, as they require a much longer time to heat, as well as to fire the goods.

Glass-house pots and *Tank-blocks* should be close textured and dense so as to resist the corrosive action of the molten glass and the flames.

Retorts should have as high a porosity as is consistent with the other requirements, so that they may be resistant to changes in temperature. Too great a porosity must be avoided, however, where resistance is required to fluxes, ash, etc.

The porosity of *silica bricks* may vary very considerably, though the usual limits are 15–32 per cent by volume.

Basic bricks should have a low porosity and a dense structure; this low porosity is necessary because such bricks are used largely under conditions where there is a considerable risk of corrosion by basic slags. Where they are not subject to corrosion, bricks with a porosity up to 40 per cent by volume may be employed. Ordinary magnesite bricks have a porosity of 12–30 per cent by volume. After a prolonged use at a high temperature, the porosity may be reduced to as low as 10 per cent.

Asbestos bricks are highly porous and, therefore, withstand sudden changes of temperature, but they are not resistant to the action of fire-gases, slags, and other corrosive agencies.

Stoneware should be devoid of porosity, but this is seldom attained, and the commercial articles absorb 0·5–2 per cent of their weight of water. They should not exceed the latter amount.

Porous bisque or *biscuit ware* is an intermediate product in the manufacture of glazed ware. It is fired prior to glazing to give it greater strength in the dipping and decorating processes and to enable the glaze to adhere well to the surface. A suitable porosity is 10–15 per cent by weight. Excessive porosity is undesirable, as too much of the glaze is then absorbed by the ware; this not only wastes glaze, but tends to cause warping and produces a dull surface.

The porosity of some fired ceramic products along with their specific gravity and bulk density is shown in Table XII.XVII.

TABLE XII—XVII. PHYSICAL PROPERTIES OF FIRED CERAMIC PRODUCTS

	True S.G.	*Bulk Density*	*Porosity (by volume)*
Building bricks	2·4–2·8	1·6–1·95	10–35
Engineering Bricks Class B	2·4–2·8	1·7–2·0	15–19
Class A	2·4–2·8	1·8–2·2	5–13
Tiles	2·4–2·7	1·4–2·0	3–18
Fireclay bricks	2·5–2·7	1·9–2·1	20–30
Porous fireclay bricks	2·5–2·7	1·0+	up to 60
Semi-porcelain bisque	2·45–2·65	1·9–2·2	13–23
Hard porcelain bisque	2·45–2·65	2·25–2·45	3–9
China ware bisque	2·45–2·65	2·25–2·45	2–11
Earthenware bodies	2·4–2·6	1·8–2·0	20–25
Stoneware	2·3–2·5	2·2–2·4	0·5–2·0
Silica brick (cristobalite)	2·34	1·92	21
Silica brick (tridymite)	2·31	1·85	22
Magnesite brick	3·4–3·6	2·6–2·8	12–21
Chrome magnesite brick	3·85	2·8–3·0	12–26
Chrome brick	3·95	3·15–3·27	14–21
Dolomite brick	3·0	2·52	23
Diatomaceous earth	2·62	0·6	73
Fired kyanite brick	3·2	2·0–2·3	22–30

CHANGES IN PERMEABILITY ON FIRING

PERMEABILITY is closely associated with porosity but the two properties are quite distinct and should not be confused. A permeable body is one which contains a large number of open pores so that gases and sometimes liquids can pass through it, whilst in a porous body the pores may be sealed or open. A body may be porous which is not necessarily permeable.

A permeable body is usually produced if the amount of liquid formation at the temperature of firing has not been very great, or where the grading of the body is poor and large void spaces are present. Permeability may also be induced by in-porating wood fibres or thin splinters in the body which, on burning, leave open channels in the structure.

If a body is to be impermeable but to have a high porosity, the majority of pores must be sealed; whilst some porous articles are often highly permeable, there is no definite relation between the porosity and permeability of a material and a tile may have a true porosity of 25 per cent and yet be impermeable even to gases.

Ordinary building bricks should be permeable to air, so that they may 'breathe', i.e. allow air to pass through them, but they should not be permeable to water, or rain will enter and make the wall 'damp'. If a house is built with solid walls of impermeable bricks it will appear to be damp, because the impervious walls cannot absorb the moisture condensed on them during cold weather. Other argillaceous building materials, such as terra-cotta, etc., should usually have similar properties. Paving bricks should be quite impermeable.

Roofing tiles should be permeable to air, but not to water, or they will allow rain to pass through them and accumulate on or even drop from the underside of the tiles, thereby causing damage.

Filters and land tiles should be very permeable so that they allow water or other liquid to pass through them at a suitable rapid rate. The pores must, however, be sufficiently small to prevent any 'dirt' or other undesirable solid matter passing through them. A filter may be tested for permeability by noting the amount of water passing through it in a given time (say twenty-four hours).

Saggars and muffles should be porous but not very permeable, or they will not protect the goods from flame, smoke, etc., produced by the fire-gases. This object is attained by having the pores as small as possible consistent with the articles being resistant to sudden changes of temperature.

Retorts should be porous, but not permeable, or much of the vapour or gas produced in them may be lost. This loss may be reduced by regulating the withdrawal of the vapour or gas, so as to avoid the creation of an appreciable pressure in the interior of the retort. In new gas-retorts, permeability is seldom serious, as the formation of a deposit of carbon on the interior soon prevents any loss of gas by permeation.

Blast furnace linings, ladle bricks and casting pit refractories should have a very low permeability to molten metals and slags. For the most part, they should also be impermeable to gases so as to preserve maximum thermal efficiency; sand moulds, however, should be impermeable to liquid metals, but permeable to vapour, in order that occluded gases and entrapped air may readily excape. The permeability of some refractory materials is shown in Table XII.XVIII.

TABLE XII.XVIII. PERMEABILITY OF CERAMIC MATERIALS

Type of Brick	*Permeability: ml. per sec. per sq. cm. through slab 1 cm. thick*	*Type of Brick*	*Permeability: ml. per sec. per sq. cm. through slab 1 cm. thick*
Fireclay	0·025—0·063	Magnesite	0·065—0·092
Silica	0·040—0·120	Chrome-magnesite	0·030—0·064
Sillimanite	0·010—0·030	Diatomite	0·008—0·080
Alumina	0·028—0·042	Porous insulating fireclay:	0·10 —1·2
Ladle Bricks	0·003—0·062	Glasstank blocks	0·002—0·013

The permeability may vary considerably in different directions in a pressed brick; it is usually greatest in the direction at right angles to the applied pressure.

CHANGES IN THE SPECIFIC GRAVITY

WHEN a material is heated changes also occur in the specific gravity or the density of the body. The true specific gravity of the material of which the body is composed may alter and in addition there may be variations in the apparent specific gravity or bulk density of the shape itself (see p. 432). These latter are the result of changes in the exterior volume and mass of the body.

True Specific Gravity

THE true specific gravity of a material is the average value of the true density of the minerals present. If the proportion of each mineral type can be deduced, the average specific gravity may be calculated by taking the arithmetical mean of the value of the components so that changes which occur on firing are related to those changes in the actual mineral components themselves. Thus a particular mineral may decompose and then recrystallise or recombine with other components to form entirely different minerals; or liquid may form at high temperatures which solidifies to a glass of a different specific gravity from that of the initial components.

Kaolinite, for example, has a specific gravity of 2·615. On heating it dissociates and finally forms mullite (s.g. 3·03) and cristobalite (s.g. 2·32) in accordance with the following equation:

$$\underset{\text{kaolinite}}{3(Al_2O_3.2SiO_2.2H_2O)} = \underset{\text{mullite}}{3Al_2O_3.2SiO_2} + \underset{\text{cristobalite}}{4SiO_2} + 6H_2O$$

The weights of mullite and cristobalite in the fired product are in the ratio of 426 to 240. Therefore the final specific gravity is

$$\frac{426 \times 3{\cdot}03 + 240 \times 2{\cdot}32}{666} = 2{\cdot}77$$

Silica is one of the few materials which decreases in density on heating. This is the result of the allotropic changes and the production of low density forms. The recrystallisation of other oxides also results in a change in specific gravity as shown in Table XII.XIX.

TABLE XII—XIX. CHANGES IN SPECIFIC GRAVITY ON HEATING

	450° C.	1000° C.
Ferric oxide	5·102	5·195
Alumina	3·401	3·729
Chromic oxide	4·933	5·050
Magnesia	2·950	3·674

The true specific gravity of minerals, especially those rich in silica is altered when they are melted and then cooled to a glassy form. Even when only partial fusion occurs the changes in the specific gravity are often much more rapid than at a slightly lower temperature, and the product may be quite different in character, even though its composition (as shown by analysis) is unchanged. Thus, if crystalline quartz is heated to any temperature below its fusing-point its specific gravity cannot be reduced below about 2·3, and that of other siliceous materials of a crystalline nature cannot be reduced below 2·32 or 2·33 even by prolonged heating, provided partial fusion does not occur, yet as soon as sufficient fused material is formed the specific gravity drops rapidly to 2·2. This change in behaviour is due to the transformation, during fusion, from crystalline to the amorphous state; the explanation is confirmed by the fact that flint and other amorphous forms of silica may have a specific gravity as low as 2·2 before fusing, and no sudden change occurs in their specific gravity when the mass melts. The change in specific gravity which occurs on melting various siliceous materials is shown in Table XII.XX, due to Purdy and Moore.[1]

TABLE XII—XX. EFFECT OF MELTING ON SPECIFIC GRAVITY

Material	*Specific Gravity in Crystalline Form*	*Specific Gravity in Glassy Form*	*Reduction in Specific Gravity, per cent*
Quartz	2·6630	2·2800	16·3
,,	2·6500	2·1900	17·3
Mica	3·0719	2·2405	27·0
Orthoclase	2·5740	2·3280	9·6
,,	2·5883	2·3073	10·9
Microcline	2·5393	2·3069	9·1
Albite	2·6040	2·0410	21·6
Oligoclase	2·6600	2·2580	15·1
,,	2·6061	2·3621	9·1
,,	2·6141	2·1765	16·7
Labradorite	2·7333	2·5673	6·1

The rate of cooling often has an important effect on the specific gravity of a material; slow cooling increases the amount of crystallisation from a liquid and thus usually results in an increase in the true specific gravity of the material.

Because the reactions which occur in ceramic materials are slow, even at high temperatures, the maximum temperature attained and the duration of heating also affects the specific gravity.

When a complex or impure ceramic material is heated, the impurities present may play an important part in changing its true specific gravity by combining with a portion of it to form a fusible material of a different nature and specific gravity from that of the original substance. For example, clays from various sources differ in specific gravity on account of variations in their composition and structure. On heating, these variations cause the clays to act in different ways. Table XII.XXI shows that whilst ordinary fireclays and most red-burning clays decrease in true specific gravity on firing, some highly-aluminous clays increase in specific gravity.

It will be seen that in the last four clays in the Table, the specific gravity decreases most in the ones containing the largest proportion of flux. In the first seven

[1] Purdy, R. C. and Moore, A. H., *Trans. Amer. Ceram. Soc.*, 9, 226, 1907.

TABLE XII—XXI. SPECIFIC GRAVITY OF BURNED CLAYS

	Flux per cent	600° *C.*	750° *C.*	900° *C.*	1000° *C.*	1100° *C.*	1200° *C.*	1300° *C.*	1400° *C.*	*Change*
Ruabon aluminous fireclay	4·24	2·55	2·57	2·56	2·59	2·58	2·52	2·49	2·37	−0·18
Kilwinning	3·05	2·46	2·52	2·55	2·66	2·55	2·56	2·55	2·50	+0·04
Mansfield	2·58	2·45	2·53	2·61	2·65	2·65	2·63	2·53	2·50	+0·05
Stourbridge	2·02	2·50	2·53	2·60	2·64	2·57	2·57	2·63	2·62	+0·12
Kilwinning aluminous shale	1·57	2·53	2·63	2·74	2·72	2·70	2·71	2·72	2·72	+0·19
Halifax	1·14	2·50	2·54	2·50	2·53	2·54	2·57	2·61	2·54	+0·04
Ayrshire bauxitic shale	0·69	2·57	2·64	2·67	2·82	2·75	2·79	2·88	2·88	+0·31
Stourbridge	2·62	2·48	2·51	2·60	2·62	2·52	2·50	2·46	2·33	−0·15
,,	1·80	2·50	2·52	2·55	2·60	2·56	2·51	2·49	2·38	−0·12
Halifax	2·42	2·47	2·48	2·37	2·41	2·43	2·48	2·45	2·34	−0·13
Grossalmerode	1·74	2·51	2·64	2·63	2·61	2·60	2·56	2·53	2·41	−0·10

clays, however, the presence of a large amount of alumina causes variations and in all cases but one, an increase in the specific gravity. The continued decrease in the specific gravity of overburned clays shows that molecular changes are still going on and that equilibrium is not reached at any convenient stage of vitrification.

The specimens were tested after they had been maintained at the various temperatures of firing for two hours.

The true specific gravity of *porcelain* varies from 2·24 to 2·35; the more complete the vitrification the lower will be the specific gravity. The specific gravity of porcelain is also markedly influenced by its composition.

The specific gravity of *earthenware* and *stoneware* bodies also depends on their composition and duration of heating. Earthenware bodies vary between 2·4 and 2·6 whilst stonewares, because they are more vitrified, have a lower value (see Table XII.XVII).

Siliceous Materials. It has previously been shown that the temperature attained, and more especially the duration of heating, is of great importance in the conversion of quartz to the low specific gravity forms of silica. Table XII.XXII shows the effect of heating various siliceous materials to a temperature of 1430° C., for various time intervals.

The rate of reduction in the specific gravity of pure siliceous materials on firing depends on the size of the quartz grains, and the relative amount of crypto-crystalline or amorphous silica. Pure quartz converts very slowly, whilst amorphous silica is rapidly changed to cristobalite at 1250° C. The addition of small amounts of impurities, e.g. lime, increases the rate of conversion of quartz (p. 162).

Sillimanite refractories may change appreciably in specific gravity on firing depending on whether kyanite, andalusite or sillimanite is the principal mineral variety in the unfired body. At high temperatures, all these minerals form mullite and cristobalite and the fired material has a true specific gravity of 2·99, so the greatest change will be in kyanite (sp. gr. = 3·59) and less in andalusite (sp. gr. = 3·18) and sillimanite (sp. gr. = 3·08).

Although kyanite has to be pre-calcined before bricks can be made from it, the raw material is a valuable ingredient in some high temperature cements, mortars and linings where the increase in volume caused by its change in specific gravity is of value.

Basic materials such as magnesite and dolomite undergo considerable change in true specific gravity on firing. The amorphous oxides of magnesium and calcium which usually form after low temperature calcination, change to crystalline varieties at high temperatures. Magnesite (magnesium carbonate) in the unfired state has a specific gravity of 2·9–3·1; at about 700° C. it loses carbon dioxide to form amorphous magnesia (sp. gr. 3·23); intense burning in a rotary kiln produces periclase with a true specific gravity of 3·57–3·62 depending on the presence or otherwise of impurities. The change in raw dolomite is less pronounced from 2·85 to about 3·2.

Chrome ores do not change appreciably in specific gravity on firing unless the type of spinel (p. 193) alters due to oxidation or solid solution.

The true specific gravities of some ceramic products are shown in Table XII.XVII.

TABLE XII—XXII. EFFECT OF REPEATED HEATING ON SPECIFIC GRAVITY OF SILICEOUS MATERIALS

Material	Specific Gravity											
	Period of Maintained Heat in Kiln at 1430° C. (hours)											
	Raw	2	4	6	8	10	12	14	16	18	20	22
Quartz from Norwegian pegmatite (in pieces)	2·65	2·38	2·33	2·325	2·33	2·32	2·31	2·32	2·33	—	—	—
Quartz from Norwegian pegmatite (in powder)	2·65	2·37	2·34	2·34	2·335	2·33	—	—	—	—	—	—
Quartz sand	2·651	2·591	2·502	2·456	2·450	2·442	2·428	2·386	2·369	2·343	2·338	2·328
Pure geyserite from Taurus	2·651	2·555	2·492	2·478	2·391	2·394	2·366	2·344	2·321	2·316	2·313	—
Average geyserite (ground)	2·651	2·456	2·361	2·349	2·333	2·317	2·306	2·307	—	—	—	—
Quartz crystal (Brazilian)	2·65	—	—	—	—	2·64	—	—	—	—	2·63	—
Quartz crystal (Madagascar) (in powder)	2·65	2·568	2·553	2·547	2·519	2·475	—	—	—	—	—	—
Flint (in pieces)	2·632	2·231	2·241	2·248	2·239	2·233	2·255	2·251	2·231	2·230	—	—
Chalcedony (in pieces)	2·607	2·16	2·17	2·17	2·19	—	—	—	—	—	—	—
Hydrated silica (Kahlbaum)	—	2·322	2·319	2·312	2·316	2·317	—	—	—	—	—	—
Quartz glass (Heræus)	2·21	2·327	2·328	2·33	—	—	—	—	—	—	—	—

The Apparent or Bulk Density

CHANGES in the bulk density of ceramic materials on heating are dependent on changes in the true specific gravity and the porosity of the mass as a whole; changes in either of these factors will change the bulk density.

The bulk densities of some fired ceramic products are shown in Table XII.XVII.

CHANGES IN WEIGHT

MOST ceramic materials change in weight during heating. This is the direct result of reactions which take place with the liberation or absorption of gases.

Weight is lost from ceramic materials in the following ways:

(*a*) During drying when adsorbed water is removed.

(*b*) In the early stages of kiln firing, when water which is more strongly retained is evaporated.

(*c*) By the oxidation or burning of carbonaceous material.

(*d*) Through the decomposition of clay and other hydrous minerals. The temperature range over which such reactions occur depends on the mineral type, but is usually between 400–700° C.

(*e*) On the decomposition of other minerals, such as carbonates or sulphates.

The amount of weight loss in clays is variable and depends on the relative concentrations of decomposable minerals. Purified china clay loses about 14 per cent by weight at the temperature of reaction of kaolinite. Fireclays and ball clays being less rich in hydrous minerals lose less, in general, although if the concentration of carbonaceous material is high or should hydrates of alumina be present, the average figure of 6–8 per cent may be considerably exceeded. Brick clays, in which the proportion of quartz is usually high, average about a 5 per cent loss, although this may be greater if carbonaceous matter or calcium carbonate is present.

Siliceous materials, once thoroughly dried, undergo little change in weight on heating. The common starting materials for basic refractories are either the carbonate or hydroxide of magnesium or calcium where weight losses of up to 50 per cent occur, hence pre-calcination is necessary before they can be used.

A gain in weight infrequently occurs during the calcination of ceramic materials. This is the result of oxidation of a particular component as, for example, ferrous iron minerals to the ferric form. Sulphides under favourable conditions may be oxidised to sulphates which also leads to an overall increase in weight.

CHANGES IN THE TEXTURE

CHANGES in texture frequently occur on heating ceramic mixtures; in fact, conditions are often designed to effect certain desirable alterations. Texture as defined in Chapter VII is related to the size, shape and grading of the particles which compose the mass. Therefore any changes in these three conditions is tantamount to a change in texture.

A coarsely-grained, open-textured material can be changed through the agency of heat to a homogeneous non-porous structure, e.g. glasses, glazes and porcelain bodies. On the other hand, crystallinity may be enhanced after high temperature heat treatment as in sintered bodies.

Changes in texture are thus related to the chemical and physical processes which occur at high temperatures, i.e. to crystal development and liquid formation.

CHANGES IN OTHER PHYSICAL PROPERTIES

MANY other properties of ceramic and other materials are liable to change as the result of the application of heat. These include simple alteration in the character of the body, such as the feel and hardness (see Chapter XIV) and strength (see Chapter XIII) or they may involve more fundamental features, as for example, thermal, optical and electrical properties (see Chapter XIV).

The changes which occur in the physical properties of a ceramic material are, like the chemical changes which bring them about, slow in reaching completion. This is desirable when, for example, pronounced volume alterations occur, otherwise serious faults might develop in the fired ware. For this reason the rate of heating and cooling must be rigidly controlled, as also must be the maximum temperature of firing and the period of time that the ware is soaked. Failure to observe the conditions will lead to faults in the ware.

EFFECTS OF EXCESSIVE HEATING

WHEN any ceramic material or article is heated under such conditions that it loses its shape, swells undesirably, or undergoes any other changes which reduce its usefulness for the purposes for which it is intended, it is said to be *over-burned*. Over-burning is also said to occur when a material which should be highly porous becomes less so, even though it does not lose its shape. If an undesirable colour is produced as a result of excessive heating, the article or material is also said to be over-burned.

Over-heating may, therefore, be described as any result of heating which enables the reactions which lead to fusion to progress further than is desired in the particular case under consideration. Hence, what may be a sign of over-burning when a material is required for some purposes may indicate the opposite (under-burning) of the material for other purposes.

Over-heating may be due either to the material or article being raised to too high a temperature, to the heating being unduly prolonged or to the heating being repeated too frequently.

Distortion or **squatting** occurs when the amount of fused material present in a mass is sufficient to cause the latter to change its shape. The cause of this distortion is the mobility acquired by the solid particles in the presence of the molten material, whereby the mass as a whole cannot resist the pressure due to its own weight and, therefore, becomes reduced in height and increased in width until a shape is reached at which the whole mass remains in a state of equilibrium. If the temperature rises still further, or if the heating is prolonged at the critical temperature, more molten material will be formed and still further changes of shape will occur, until eventually the whole mass becomes fluid. The increase in mobility may be due either to the material being heated to too high a temperature, to the heating being too prolonged or the pressure too great.

As particles immersed in or floating on a fluid move more readily when pressure is applied to them, the amount of squatting is greater when a ceramic material is heated under pressure, or, alternatively, the change in shape occurs at a lower temperature because of the increased mobility of the solid particles when under pressure. Hence, the resistance to heat or refractoriness of a material is lower when it is under pressure or supports a load (p. 886) than when it is quite free.

The shape of a mass also affects the temperature at which squatting first appears, a tall mass changing its shape at a lower temperature than a short one, as the pressure on a unit area in the lower part of the latter is much less. The shape of the mass before heating also determines its behaviour after squatting has commenced; thus, Seger cones bend over before becoming wholly liquid.

Boiling, or the conversion of a liquid into vapour or gas, will occur if the liquid is raised to a sufficiently high temperature without it undergoing decomposition. This change seldom occurs in ceramic materials as their boiling points are usually unattainable in commercial furnaces. Glazes sometimes present the appearance of having been suddenly solidified whilst in a state of ebullition, but this is due to a wholly different cause and is quite distinct from true boiling, being caused by the partial escape of moisture or other gases from the ware after the glaze has been fused. A spurious boiling also occurs when clays rich in carbonaceous matter are heated too rapidly, the surface of the clay being sealed with fused material before the gases formed by the burning carbonaceous matter have escaped from the mass.

The **volatilisation** of some of the constituents of ceramic materials may occur at very high temperatures. Thus, the alkalies and some of the silica in clay may be partially volatilised so that when clays are heated repeatedly at high temperatures they become slightly more refractory (see also p. 736).

Mellor[1] found that nearly 20 per cent of the alkalies in an earthenware body were volatilised by overfiring it at 1400° C. Under reducing conditions and in the presence of water vapour, the loss is still greater. Mellor[1] also found that a saggar lost 22 per cent of its alkalies after firing it seven times at 1200° C.

The volatilisation of alkali from some glazes and the resultant dulling of the glaze is mentioned on p. 830.

Silica appears to be appreciably volatile at high temperatures, especially in the presence of carbon, as the latter reduces it to silicon monoxide which volatilises and is again oxidised and re-forms silica. This change of composition does not occur to any very great extent below 1650° C., but at higher temperatures quartz crystals, large enough to be readily seen with the naked eye may be completely volatilised in a reducing atmosphere. In the absence of carbon, silica is thought to boil at 2200° C.[2]

Effect of Prolonged Heating

The effect of a prolonged 'soaking' during the firing of ceramic materials has been described on p. 733. A moderately long 'soaking' is often beneficial, but if it is unduly prolonged, the effect may be deleterious instead of useful, as the production

[1] Mellor, J. W., *Trans. Eng. Ceram. Soc.*, 5, 75, 1906.
[2] Sosman, R. B., *The Phases of Silica* (Rutgers Univ. Press, N.J., 1965).

of an excessive amount of fused matter may endanger the stability of the mass or crystallisation may cause a decrease in strength and may render the material brittle or, as in the case of devitrification of silica glass, it may spoil the material for the purpose for which it is intended.

Prolonged heating may also cause excessive shrinkage in clay wares or excessive expansion in siliceous materials, if the finishing temperature during the burning of the ware was not sufficiently high, and these changes in volume may endanger the stability of the mass.

Swelling, bloating, and similar defects may occur as a result of prolonged heating at high temperatures, and if any fluxes are present which can combine with silica or bases, an increased amount of fused material may be produced, so that the effect of excessively prolonged heating may be similar to that of heating the material to too high a temperature and so over-burning it.

As volatilisation of part of the soda and potash may occur if the heating of a glaze is prolonged, this may lead to the glaze becoming dull instead of glossy; undue prolonging of the heating at the finishing temperature of glazes should be avoided.

Excessive heating of a molten glaze in contact with clay will usually enable reactions to occur between them with the result that an unfused product is formed. This destroys the glossy appearance of the glaze and so makes the ware defective. Such reactions are precisely the same as those which occur when a crude clay or pottery body is heated excessively, but the effect being confined more closely to the surface of the ware is more easily observed.

When clay is subject to continuous high temperature for many months or even years, remarkable changes take place in its physical nature.

An examination of brickwork from gas retorts, coke-ovens and regenerators which have been in service for a normal campaign life will show a pronounced change in the physical nature of the refractory. The most obvious characteristic is that the glassy bond of the original refractory has almost entirely disappeared. The product is virtually entirely crystalline with a vastly different texture from the original. Mullite usually develops in quantities approaching the theoretical maximum and in large crystallite sizes. Most of the alkalies evaporate, at least from surface layers, leaving a much more refractory product but usually with a considerable reduction in strength.

When slagging agents, such as iron oxide and lime, have been present in the firing process, these penetrate to considerable distances into the brickwork, forming crystalline components, such as iron cordierite, anorthite, gehlenite and several others.

Effect of Repeated Heating

When ceramic materials are repeatedly heated and cooled the effect is similar in many respects to that of prolonged heating (p. 829), the same kind of changes taking place in both cases, but a much higher temperature or a much longer time is required compared with that needed if there has been no cooling.

The effects of repeated heating and cooling also differ according as the treatment is (*a*) at a rapid rate; and (*b*) at a slow rate.

The effect of rapid heating and cooling when repeated several times is one cause of the defects known as spalling and shattering (p. 948) due to thermal shock.

The chief materials which are subjected to repeated heating and cooling are the walls of some kilns and furnaces, saggars, glass-melting pots, retorts, crucibles, etc.

The temperature range through which the material is cycled during heating and cooling is most important. The main physical effect which is occurring during the cycling is the change in volume due to the reversible thermal expansion. Not only does this occur in the brick structure as a whole, but also in individual grains within the body itself. In a composite material composed of many mineral species, this constant subjection to differential stresses causes a weakening of the bonds and the development of a multitude of fine cracks.

These effects are most serious when large expansion coefficients are involved or when minerals are present which undergo phase changes in the temperature range at which cycling occurs.

Basic materials, which are rich in magnesia, have large thermal expansion coefficients at high temperatures and tend to deteriorate to a serious degree when subjected to cycling conditions of temperature, as, for example, was the case in open-hearth practice.

Silica materials have virtually no volume changes above their inversional temperatures and they are remarkably resistant to cycling conditions provided that temperatures are not allowed to fall to the transition point of the major silica mineral which is present.

Silica bricks, rich in cristobalite, used to be preferred in steel practice where reversal cycling between about 700–1700° C. was to be encountered, whilst in gas plant and coke oven practice, tridymite bricks were used because cycling between 200–1200° C. was a likely condition.

Clay bricks normally resist temperature cycling extremely well, unless quartz is present in quantity, when catastrophic results might ensue if repeated heating and cooling through 573° C. occurs.

In the case of all bodies which are repeatedly heated and then cooled to room temperature, thermal shock or serious reduction in strength can be avoided only by slow, controlled heating particularly at low temperatures.

Clay bodies such as crucibles and saggars are subjected to this type of cycling continuously throughout their life and they become increasingly prone to cracking and fracture with frequent use, possibly because of secondary crystallisation from the glassy phase.

A considerable improvement in the life of saggars in recent years has been effected by introducing amounts of talc or steatite in the clay body used in their preparation. Such additions lower the thermal expansion of the fired body and so reduce the tendency to thermal failure, but other factors are also probably involved.

Crucibles, when repeatedly heated, become hard and vitreous as a result of the action of the contents upon the crucible. The resistance of the crucibles to sudden changes of temperature is correspondingly decreased so that after a time crucibles 'perish' and can no longer be used.

When *fused silica* is repeatedly heated at about 1200° C. it shrinks and devitrifies forming tridymite or cristobalite, especially in the presence of basic or alkaline dust or ash.

CHAPTER XIII

STRENGTH AND ALLIED PROPERTIES

As with so many other properties of ceramic materials, *strength* is an apparently simple concept which on closer study reveals a field of great complexity.

Associated with the ability of a material to withstand pressure being applied to it without rupturing, are (i) its capacity to be elongated or compressed without fracture, (ii) its elastic recovery on the removal of pressure, (iii) its inclination to deform progressively with time and temperature under applied pressure, (iv) its resistance to abrasion and (v) several other important but differing aspects.

Even when ultimate rupture is implied there are several very different principles which may be involved and it is important to distinguish between them.

PRINCIPLES OF STRESS[1]

WHEN pressure is applied to an object, rupture can occur only when differential stresses of sufficient magnitude are developed. If one part of the body is restrained or fixed relatively to the applied stress, fracture will occur when the resultant stress exceeds a particular value and imparts an excessive strain. The direction of the applied stress in relation to the fixed point or plane determines the mode of fracture.

In general six different main types of fracture are encountered in ceramic systems. These are compressional, tensile, shear, torsional, transverse and impact and Fig. XIII.1 illustrates the different principles involved.

1. **Compression Stresses** are those commonly encountered in load-bearing structures such as walls constructed from bricks. The fixed plane is the base and load is applied vertically downwards and evenly distributed over the top face. True compression occurs only when the bearing surface is not penetrated and when the side walls are in restraint to prevent sideways thrust. In brittle materials, such as ceramics, compressive failure is usually manifested as a shear fracture.

2. **Tensile Stresses** are those which tend to pull a body apart. They are those which in plastic materials produce elongation along the direction in which the stress is applied.

3. **Shear Stresses** cause one part of a body to slip or slide relative to another. In shear it is implied that the displacing force is acting at an angle to the fixed plane other than perpendicularly towards or away from it. The direction may be parallel to the fixed plane but more often an obtuse angle is implied. Shear forces frequently develop as the resultant of two or more stresses applied to a body in different directions.

new list of standard conventions (1970) adopted in British practice requires that ressure should be represented as pounds *force* per sq. in. (lbf./in.^{2}).

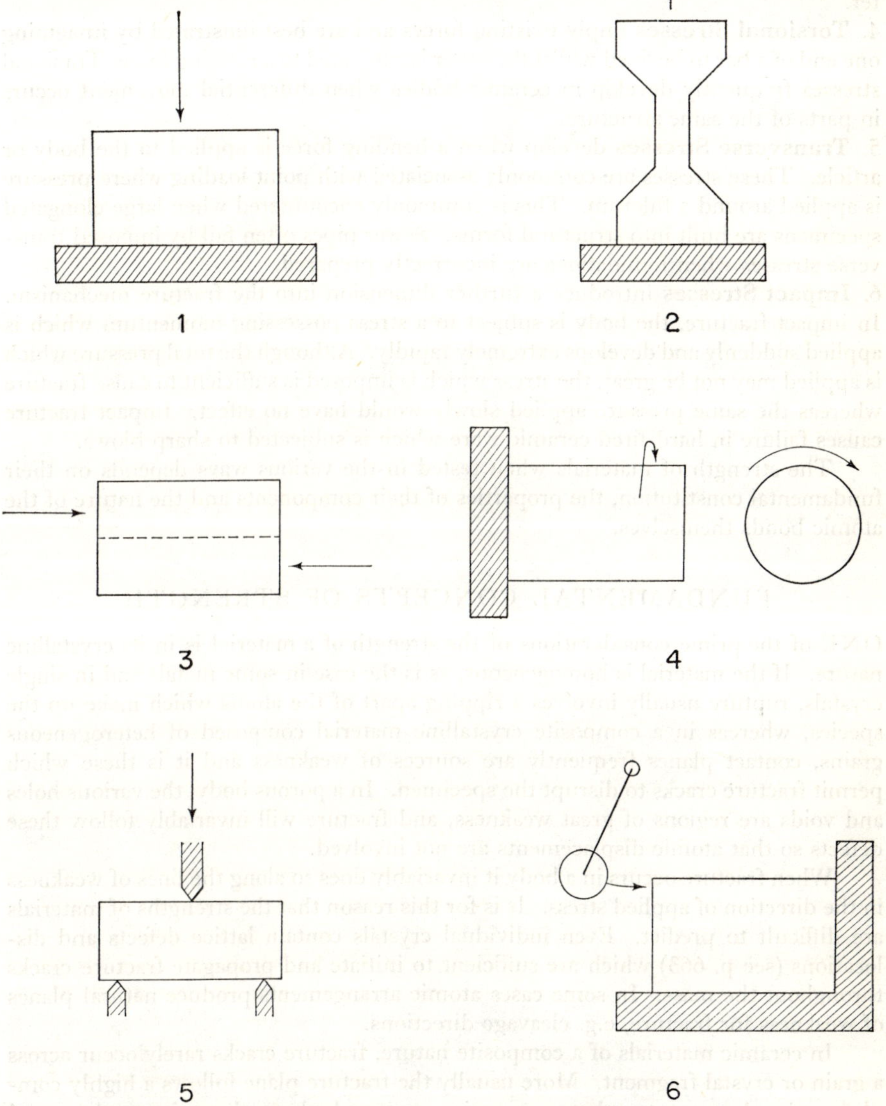

FIG. XIII.1. THE PRINCIPAL METHODS OF APPLYING STRESS

1. Compression
2. Tensile
3. Shear
4. Torsion
5. Transverse
6. Impact

The directions of principal stresses are indicated by arrows. The shaded portions represent the fixed planes.

Clays in a plastic condition usually fail in shear; bricks subjected to crushing on their principal plane also fail in shear for reasons given later in this Chapter.

4. **Torsional Stresses** imply twisting forces and are best illustrated by imagining one end of a bar to be fixed whilst the other is subjected to a rotating force. Torsional stresses frequently develop in ceramic bodies when differential movement occurs in parts of the same structure.

5. **Transverse Stresses** develop when a bending force is applied to the body or article. These stresses are commonly associated with point loading where pressure is applied around a fulcrum. This is commonly encountered when large elongated specimens are built into structural forms. Sewer pipes often fail by imposed transverse stresses when foundations are incorrectly prepared.

6. **Impact Stresses** introduce a further dimension into the fracture mechanism. In impact fracture, the body is subject to a stress possessing momentum which is applied suddenly and develops extremely rapidly. Although the total pressure which is applied may not be great, the stress which is imposed is sufficient to cause fracture whereas the same pressure applied slowly would have no effect. Impact fracture causes failure in hard-fired ceramic ware which is subjected to sharp blows.

The strength of materials when tested in the various ways depends on their fundamental constitution, the properties of their components and the nature of the atomic bonds themselves.

FUNDAMENTAL CONCEPTS OF STRENGTH

ONE of the prime considerations of the strength of a material is in its crystalline nature. If the material is homogeneous, as is the case in some metals and in single crystals, rupture usually involves a ripping apart of the atoms which make up the species, whereas in a composite crystalline material composed of heterogeneous grains, contact planes frequently are sources of weakness and it is these which permit fracture cracks to disrupt the specimen. In a porous body, the various holes and voids are regions of great weakness, and fracture will invariably follow these defects so that atomic displacements are not involved.

When fracture occurs in a body it invariably does so along the lines of weakness in the direction of applied stress. It is for this reason that the strengths of materials are difficult to predict. Even individual crystals contain lattice defects and dislocations (see p. 663) which are sufficient to initiate and propagate fracture cracks throughout the mass. In some cases atomic arrangements produce natural planes of weakness for fracture, e.g. cleavage directions.

In ceramic materials of a composite nature, fracture cracks rarely occur across a grain or crystal fragment. More usually the fracture plane follows a highly complex path where contact between various mineral phases is at its weakest and particularly where it is possible to link up voids in the structure itself.

A theoretical consideration of the strength of ceramic materials is of little value, therefore, except in the case of individual crystals of perfect form, but, nevertheless, the basic principles of crystal structure and bonding afford a better understanding of the different strength characteristics between metals and non-metallic materials. These same principles are useful guides for researches into means for improving certain aspects of the strength of ceramic materials.

The bonds which are developed between atoms in a crystal lattice have been described fully in Chapter II. The importance of these in relationship to the melting point of crystals has been discussed in Chapter X. When a crystal melts, sufficient thermal energy has to be supplied to it to enable the electrostatic field between adjacent atoms to be broken. In the fracturing of a crystal, mechanical energy has to be developed which is sufficiently strong to split individual atoms from their neighbours along the plane of rupture. It is not surprising, therefore, that in many crystals, melting point and strength are closely inter-related. There is not an absolute correlation because fracture involves only a single plane where there may be inherent weakness, whereas melting involves the whole mass of material; in addition, some crystals do not dissociate entirely on melting and molecular groupings persist in the melt. Nevertheless as Table XIII.I shows, the atomistic approach to both strength and melting point is soundly based for simple species.

Unfortunately, accurate values of compression or tensile strengths are not readily available for a sufficiently wide range of minerals to be compared. The hardness values (Mohs' Scale) provide a reasonably good guide to strength values and these are established for a range of simple elements, oxides and compounds which melt congruently into a simple molecular liquid state. These are the figures which are compared in Table XIII.I.

TABLE XIII—I. A COMPARISON OF THE MELTING POINT AND THE HARDNESS OF SOME SIMPLE MATERIALS

Material	*Formula*	*Hardness Mohs' Scale*	*Melting Point °C*	*Material*	*Formula*	*Hardness Mohs' Scale*	*Melting Point °C*
Potassium	K	0·5	62	Iron	Fe	5	1539
Sulphur	S	1·0	119	Haematite	Fe_2O_3	6	1550
Lead	Pb	1·5	327	Quartz	SiO_2	7	1760
Rock Salt	NaCl	2·0	801	Corundum	Al_2O_3	9	2030
Aluminium	Al	2·5	659	Boron	B	9	2300
Fluorite	CaF_2	4	1360	Diamond	C	10	+3000

The most significant feature revealed by the atomistic approach to the strength of materials is in the nature of the fracture which occurs in perfect crystals.

The two main types of bonds which are present in crystal lattices are the covalent/ionic form, in which electrons are transferred or shared between adjacent atoms, and the metallic form in which cations are symmetrically distributed in a semi-mobile 'electron sea'. The electrons in metal structures are not rigidly bonded to particular atoms and have freedom of movement within the strictures of normal interlinked orbitals.

In general, both covalent/ionic and metallic bonds are relatively strong, whilst other types such as hydrogen, hydroxyl and Van der Waal are inherently weak. When these latter bonds are present in crystals, lines or planes of weakness are the inevitable result and the crystal will fracture readily along such layers. This is the case in kaolinite where hydrogen bonding links adjacent layers and in graphite where Van der Waal linkages are present. Both these crystal forms are inherently flaky, because of the relative ease with which fracture occurs in one particular plane.

The covalent/ionic forms of crystals show marked differences in strength characteristics from those which possess mainly metallic bonds. In the former the electrons are rigidly bound in discrete orbits and although slight displacements in equilibrium ionic positions do occur, restoring forces prevent large disturbances unless these disrupt the lattice entirely. In the case of metallic bonds, the electrons are not so rigidly bound, so that ionic movements of considerable magnitude may occur. The cations can slide or slip in planes relative to each other without rupture occurring until the displacement is very great.

Consequently, whereas crystals with covalent/ionic bonds are rigid, develop stresses with great rapidity, show little or no elastic or plastic flow and rupture after only small displacements, metallic crystals often flow readily on the application of low pressures, they absorb stresses easily, show marked elastic and plastic flow and will usually distort markedly before rupturing.

Most natural minerals and metals exhibit their typical characteristics to a greater or lesser extent. Metals such as copper and gold are extremely malleable and represent the ideal metallic state; on the other hand, molybdenum and tungsten, contain so-called 'metallic electron bonds', which confer greater rigidity and ultimate strength and a higher melting point. So also in non-metallic crystals; some, such as diamond and quartz, are extremely rigid, but with very high strengths and melting points, whereas others such as silicon carbide, certain nitrides and beryllia, have free electrons in their structure and have a small degree of malleability.

This fascinating field of study has excited great interest in ceramic studies. It has great bearing on thermal shock characteristics (see p. 948) and it has a strong influence on the valuable properties of semi-conduction.

Although most ceramic materials, particularly those made from clay, are mostly rigid at room temperature, those with a high content of glass show marked plasticity and even malleability at high temperatures. The fact that glass is present, even at low temperatures, can often result in appreciable yield and distortion without rupturing, especially when the applied stress is maintained for long periods of time. Even some crystals can flow under stress along defects or dislocations within the lattice. This is the phenomenon referred to as *creep*.

The atomistic concept of strength explains one other interesting feature of ceramic materials and other related inorganic species. When a crystal of the covalent/ionic bond type is subjected to compression, the ionic components are being forced into closer juxtaposition. The forces of repulsion which then develop are large and one atom can only be separated from its neighbours by forces which are sufficient to squeeze it from its normal position. If a similar material is subjected to tensional or elongational stresses or to shear forces the principles involved are quite different. As atoms or ions are pulled from their equilibrium position, restoring forces develop but these are more readily overcome and do not increase with distance from the equilibrium position to the same extent as those in compression. A covalent/ionic bonded material is strong in compression, but weak in tension and shear and this is a feature of most ceramic materials.

Metallic-bonded bodies have a less rigid electron structure and this is manifested particularly in their resistance to tensional and stretching forces. Stresses are more readily accommodated and bodies with this type of bonding are generally higher in tensile strength than those with electron bonds of a more rigid nature.

THE NATURE OF STRESS/STRAIN RELATIONSHIPS

THE general forms of stress/strain curve for two types of bodies are shown in Fig. XIII.2 where the specimens are being subjected to stretching forces which

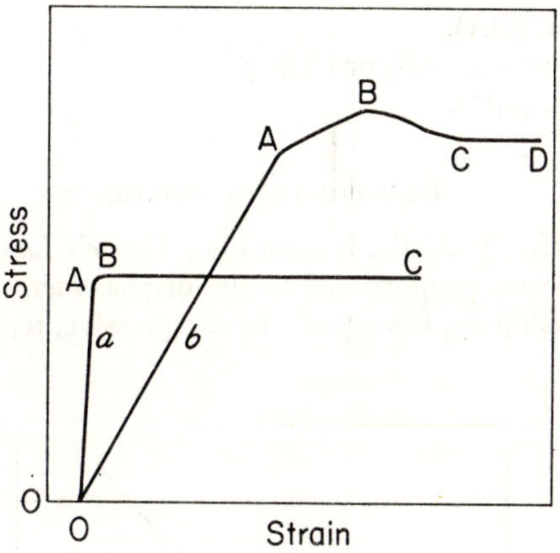

FIG. XIII.2. TYPICAL STRESS/STRAIN CURVES
(*a*) brittle body; (*b*) metallic specimen

are tending to elongate the test piece. Curve *a* is for a typical crystalline inorganic mineral or fired ceramic material and shows that the application of stress produces little or no yield (OA) until a fairly high value of applied pressure is reached. Some elongation then occurs (AB), but this is usually small in amount before final fracture (BC). In the region AB, a type of plastic flow is occurring which is due to gliding and slipping along dislocations or natural cleavage planes or to localised internal failures at discrete points. Most natural or fired products are never as strong as would be predicted from theoretical principles because of such internal weaknesses. Curve *b* is the stress/strain relationship of a characteristic metal body. In this case the portion OA is clearly defined as of linear form; when stress is applied to a value not exceeding A, the body is perfectly elastic and will return to its former length on removal of pressure. The AB region of the curve represents plastic flow and indicates that the body is undergoing permanent irreversible changes in shape. Under the conditions of test, namely elongation, most metals would show pronounced 'necking' prior to fracture so the cross-sectional area over parts of the test-piece would be reduced. This accounts for the apparent reversal in the BC region of the curve prior to the fracture load indicated by CD.

In essential details the curves for rigidly electron-bonded ceramic bodies (*a*) and for the 'free electron-cloud' metals (*b*) are the same, but they differ markedly in detail. The same features are exhibited also by wet clays and by ceramic bodies which are tested at temperatures above that at which liquid forms. In these cases, the form of curve shows a larger plastic region (AB) before fracture and there is also a more clearly defined elastic region but only at low stress values. In these cases, a comparison with metal structures can be made in which water or liquid glass fulfils similar functions to the electron cloud. Such composite bodies do not exhibit the same elastic characteristics, however, and there are important differences in flow principles so that the comparison is not absolute.

The three basic features exhibited in all stress/strain characteristics are as follows:

(i) elastic region (OA),
(ii) plastic flow region (AB) or (ABC)
(iii) final fracture (CD).

ELASTICITY OF SOLIDS

FOR a body to be perfectly elastic, it must obey *Hooke's Law* which states that the change of load is linearly proportional to the displacement which occurs in such bodies. This relationship applies equally to compressive, tensile, torsional or shear forces.

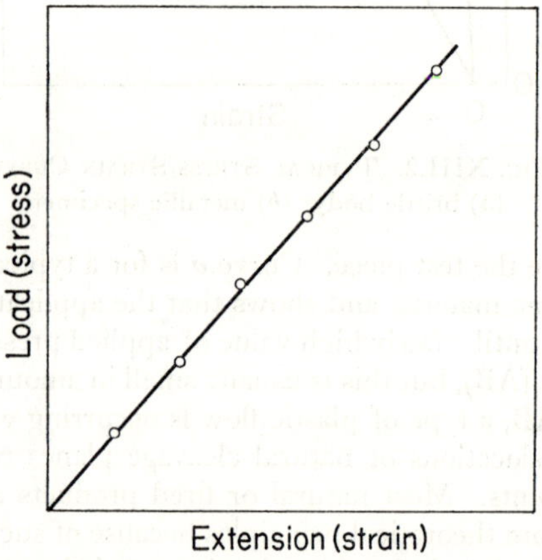

FIG. XIII.3. THE ELASTICITY OF A SOLID

Expressed graphically the relationship is shown in Fig. XIII.3. As the line passes through the origin, it follows that

$$\frac{\text{load}}{\text{extension}} = \text{constant}$$

and because stress is defined as load per unit area and strain as extension (or change of length) per unit length, the above equation can be written more generally as

$$\frac{\text{stress}}{\text{strain}} = \text{constant}$$

When tensional or compressive testing is involved the constant is known as *Young's Modulus* or the *modulus of elasticity* and commonly given the symbol E.

In shear testing the linear relationship also applies but the constant is known as the *modulus of rigidity* and given the symbol E_R (C, G and N are used by other authors for this value). E is usually greater than E_R by a factor of 2–4.

Some typical values for these constants are included in Table XIII.XV (p. 895).

Metals usually have a greater modulus of elasticity than ceramic materials; steels are in the range $23–32 \times 10^6$ lbf./in.^{2}, copper and aluminium have respectively values of 15·0 and $10{\cdot}3 \times 10^6$ but fired clays are of the order $0{\cdot}5–3{\cdot}0 \times 10^6$ lbf/in^2.

Elasticity and Rigidity Moduli in ceramic ware depend very much on porosity and on the amount of glassy phase which is present. True glasses have elasticity moduli in the range $10–25 \times 10^6$ lbf./in.^{2}.

There is a further important constant which can be applied to a body undergoing elastic strain. When a body is undergoing tensile (elongation) or compressive stress, the longitudinal strain which results produces in turn a strain of opposite sign in a perpendicular direction, i.e. laterally. The ratio of the lateral to the longitudinal strains is called *Poisson's Ratio.*

The significance of elasticity and rigidity moduli and of Poisson's Ratio will be discussed later in this Chapter where methods for measuring their values are described.

Plasticity of Solids

When subjected to stress values above the elastic limit (Point A Fig. XIII.2), materials distort permanently and will ultimately fail by fracture. The region of irreversible flow in bodies, i.e. when they are undergoing permanent distortion or deformation but before fracture takes place, is termed the plastic condition of the body.

The aspects of plasticity which relate to the flow of clay/water systems have been discussed fully in Chapter VIII and similar principles apply to single crystals, to mineral bodies and to fired ceramic articles although the fundamental mechanisms are different.

One of the important features of plastic deformation, or 'creep', as it is often termed, is that it is time dependent. A body subjected to stress above its elastic limit and there maintained, will continue to distort slowly but progressively.[1] Most ceramic bodies are so rigid at room temperatures that the region of stress in which plastic flow occurs is small. No marked creep can be observed without actual rupture taking place.

Creep is evident in clays made plastic by water and also in those which are maintained at high temperature under stress for long periods of time. The slumping and collapse of kiln structures is one manifestation of creep where bricks exhibit plastic flow so as to distort under their own weight. This occurs only after long

[1] Creep can actually occur in bodies which are subjected to stress below the so-called elastic limit provided that the load is applied for a sufficient period of time. This is probably of little significance in ceramic bodies, but it is an important feature of rock strata.

periods of maintained heat and stress. The same bricks could probably be heated to 100–200° C. higher in temperature without any noticeable yield if they were raised in temperature at 5° C./minute.

The refractoriness-under-load test is designed to reveal creep characteristics in ceramic ware to be used at high temperature (see p. 883).

Fig. XIII.4 shows some interesting results of a particular fireclay brick (35

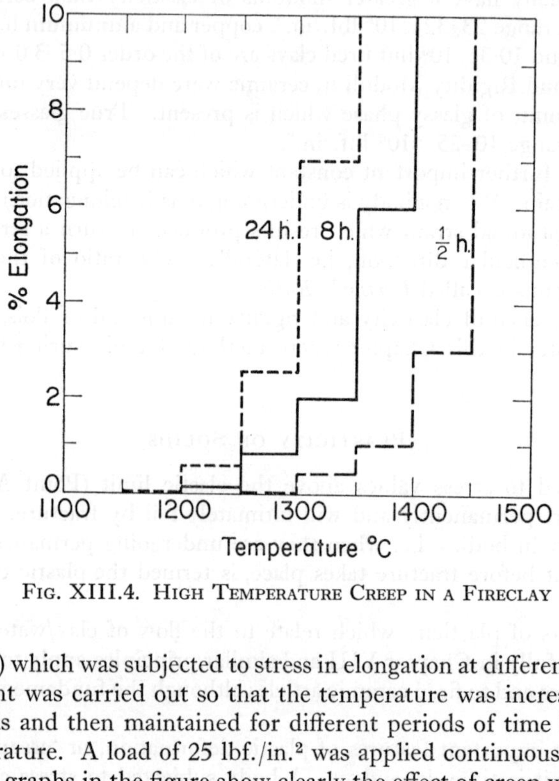

FIG. XIII.4. HIGH TEMPERATURE CREEP IN A FIRECLAY

per cent Al_2O_3) which was subjected to stress in elongation at different temperatures. The experiment was carried out so that the temperature was increased in steps of 50° C. intervals and then maintained for different periods of time before the next raise in temperature. A load of 25 lbf./in.^{2} was applied continuously.

The three graphs in the figure show clearly the effect of creep when comparing the results of maintaining temperatures from $\frac{1}{2}$, 8 to 24 hours at particular temperatures. Considerably more creep was obvious in those samples which were maintained at temperatures for long periods of time.

Although ultimate fracture occurred at lower temperatures in specimens which were subjected to the greatest creep, this should not be taken as an indication that a slow rate of heating causes absolute failure at lower temperatures than would be produced by a rapid rate. The samples showing the greatest creep developed a distinct 'neck' of smaller cross-sectional area, hence the actual stress (lbf./in.^{2}) at the fracture point was considerably greater than the initial stress on the test-piece of uniform cross-section.

Equally, it cannot be implied that if a load were applied slowly to a ceramic body it would fracture in elongation at lower pressure than when the stress was developed more rapidly. In fact, the reverse is the case, as will be discussed later.

When articles are tested in a similar fashion but in compression, the creep behaviour follows the same pattern, but at higher temperatures the rate of flow appears to slow down and ultimate fracture may occur at a higher temperature in samples maintained for long periods under load. This is because the sample is 'bellying' and the overall cross-sectional area increases, thereby producing a lower stress.

The causes of creep in ceramic materials, particularly that developing at low temperatures, are associated with complexities within the crystal structures and textures of the bodies themselves. When glass is present in quantity and testing is carried out at elevated temperatures, undoubtedly the partial conversion to a pseudo-liquid is the prime cause of irreversible flow, but such a mechanism does not account for the changes at low temperature or for those in a perfect crystalline body.

Creep in ceramic materials, at temperatures below those at which glass becomes appreciably soft, can be attributed to two mechanisms:

(*a*) Flow in crystals themselves along lattice defects, along natural lines of cleavage, or along contact planes between one crystal and another and

(*b*) Permanent deformation due to internal fractures which do not propagate throughout the mass but are nevertheless sufficient in magnitude to cause localised failure and yielding. This is probably a predominant mechanism in porous bodies and multi-component mineral masses. It is significant that bodies which have been subjected to high stresses for considerable periods of time, invariably have an appreciably lower ultimate fracture strength than the original unstressed article.

Ultimate Strength of Solids

The theoretical considerations which govern the stress/strain relationship of ceramic and other materials have already been outlined. Fracture occurs in different circumstances depending upon the direction of applied stress. A body may break in compression, in elongation (tensile stresses), in transverse strain, in torsion, in shear or by impact.

In rigid or brittle bodies which are typified by most ceramic materials, the actual mode of fracture is similar independently of the direction or type of stress. Tensile fracture, where the body is literally stretched or thrust apart, affords the simplest picture. To cause fracture, a sufficiently high stress must be applied to pull apart adjacent layers along the plane of greatest weakness at right angles to the direction of stress (Fig. XIII.5/2).

In compressive fracture, layers of the body are squeezed together in the direction of applied pressure, but this sets up a strain of opposite sign in the direction at right angles. This resultant strain, once it reaches sufficient magnitude, causes fracture, in effect, by elongation in a direction at an angle to the applied compressive stress. A ceramic block subjected to a crushing pressure shows a characteristic cone of fracture ideally with two planes of elongational fracture as shown in Fig. XIII.5/1.

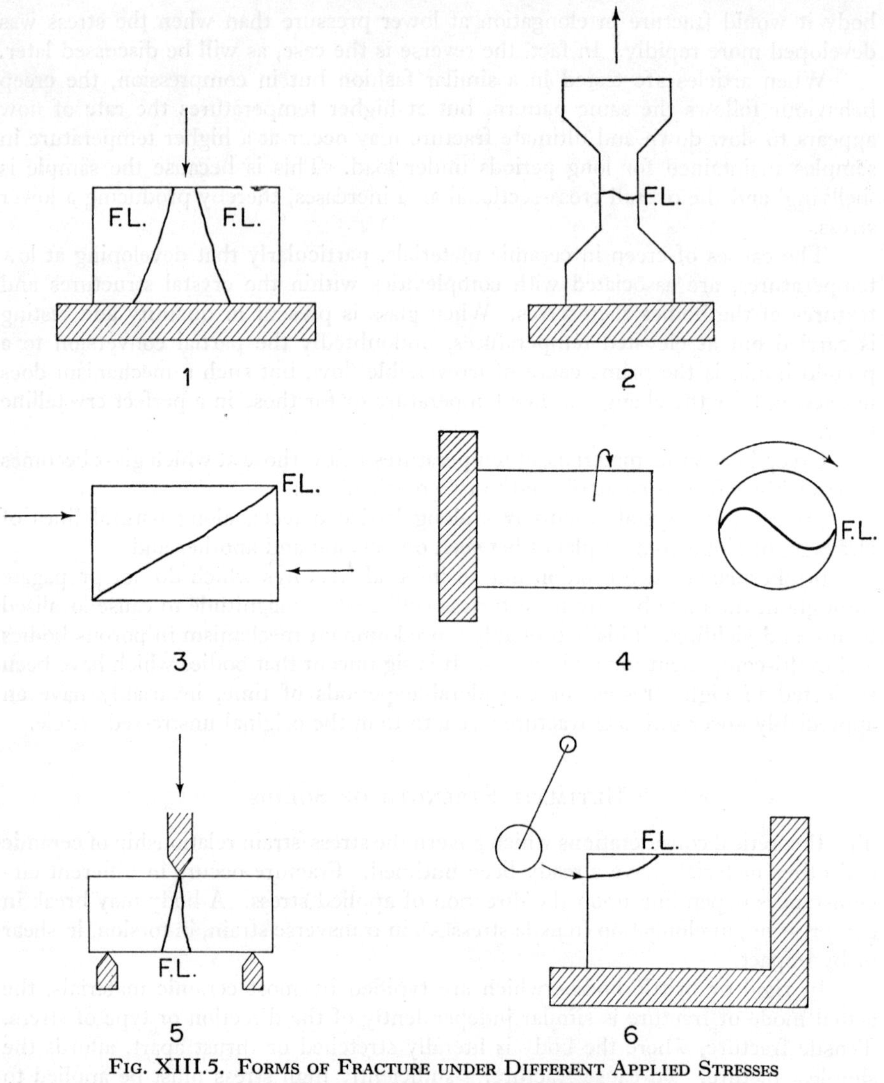

Fig. XIII.5. Forms of Fracture under Different Applied Stresses
1. Compression 4. Torsion
2. Tensile 5. Transverse
3. Shear 6. Impact
The directions of principal stresses are indicated by arrows

Shear fracture is not unlike torsional rupture; tensile forces develop at an angle of approximately 45° to the direction of the shearing force. The body then splits along the plane, parallel to the applied force (Fig. XIII.5/3).

Fracture by torsional forces involves similar principles. Layers of the body are forced apart but the greatest pressure is on the outside of the test specimen.

Tensile stresses develop at an angle to the direction of shear so that failure can often produce a characteristic S-shaped shear fracture line (Fig. XIII.5/4).

In transverse fracture, the test piece is subjected to a bending stress, so that the lower layers are in tension. Fracture commences on the lower surface and is propagated upwards throughout the block along planes of greatest weakness (Fig. XIII.5/5). As the crack develops the point of maximum compaction in the upper planes acts as a type of fulcrum.

The fracture by impact stresses is a more complex phenomenon. Impact stress implies subjecting a body to a pressure applied almost instantaneously and with considerable momentum. Brittle bodies are unable to transmit such pressures with sufficient rapidity to dissipate them throughout the structure. High stress values develop locally which initiate fracture and these follow the direction of the momentum of the impacting force. Consequently impact fracture frequently has a characteristic conchoidal shape (Fig. XIII.5/6). Essentially, however, the initial impact fracture is by tensile stress.

In considering the fracture of brittle bodies, therefore, by pressure applied in any direction, the important factor is the resultant shear or tensile stress which must be developed in a particular plane. When this exceeds the maximum stress that the body can withstand, rupture will occur.

The Development of Strains in Ceramic Shapes

When stress is applied to a body, a strain is created provided that there is a fixed plane or axis against which the stress can develop. From simple geometrical concepts, it is possible to calculate the strain which results when the pressure or stress is applied in different ways.

1. **Tensile stress.** Consider the case of a simple cube which is being subjected to a tensile stress p_X in the direction parallel to OX (or in the X direction). The resulting strain e_X is given from the equation

$$\frac{\text{stress}}{\text{strain}} = \text{E (modulus of elasticity)}$$

$$\text{So } e_X = \frac{p_X}{E}$$

At the same time, there will be resulting compressive strains being developed to a maximum in directions at right angles to that of the tensile pressure. These can be represented as e_Y and e_Z along the Y and Z directions, but in a symmetrical object, such as a cube, they will be equal. Their value is given by the relationship.

$$e_Y = e_Z = -\frac{p_X \nu}{E}$$

where ν is Poisson's Ratio (p. 839).

Strain has no dimensions and can be represented as $\frac{\Delta l}{l}$ to indicate the fractional contraction (or expansion) which would take place when stress is applied in one particular direction. Rigid materials, such as concrete, fired bricks and most other ceramic materials cannot be elongated by tensile forces more than 0·0003–0·0004 in. per linear inch.

In common bricks, the elastic modulus is in the range $0{\cdot}7\text{–}3{\cdot}0 \times 10^6$ lbf./in.², so the order of tensile stress required to cause fracture in elongation can be calculated as

$$\begin{aligned} p_X &= e_X \times E \\ &= (3-4) \times 10^{-4} \times (0{\cdot}7\text{–}3{\cdot}0) \times 10^6 = (2{\cdot}1 - 12{\cdot}0) \times 10^2 \\ &= 210\text{–}1200 \text{ lbf./in.}^2 \end{aligned}$$

Shear fracture values are larger by amounts up to 50 per cent, but compressive strengths can be 10 or more times higher for reasons discussed later.

2. **Compressive Stress.** The opposite case of an object being subjected to a crushing pressure p_X along the X dimension, would produce a negative strain (contraction) given by

$$e_X = \frac{-p_X}{E}$$

but a resulting tensile strain at right angles of e_Y or e_Z where

$$e_Y = e_Z = \frac{p_X \nu}{E}$$

The total resultant strain has the characteristics of a shear strain so that fracture occurs at approximately 45° to the direction of the compressive stress. The calculation of absolute strain in these cases is complex and the problem is best analysed from the aspects of stress as in the next section. Fracture in brittle bodies, however, is the direct result of elongation in some direction and once a critical value has been exceeded, failure will occur.

Fracture in transverse, torsional and impact stresses requires even more complex treatment. The value has to be calculated in terms of the geometry of the shape and then the resultant fracture stress of either a shear or tensile character worked out. This is illustrated later in the chapter.

Basically, however, a brittle body can fracture only by tensile (elongation) strain or by shear strain irrespective of the nature of the stresses which are applied. Only when the critical tensile or shear strength in a particular direction is exceeded will the body fracture.

In the case of a brick (Engineering Class B) subjected to crushing stresses on its large face, fracture will occur in the form of a shear crack at an angle of less than 45° to the applied stress when the shear stress exceeds about 2000 lbf./in.². To achieve this a compressive stress of 7000 lbf./in.² has to be applied.

THEORETICAL ASPECTS OF FRACTURE IN BRITTLE MATERIALS

THE assessment of strains which are set up in a body when it is subject to external stresses is highly complex and involves complicated mathematical vector analysis.

The solution involves three-dimensional theory and an absolute proof is not always possible.

A useful compromise is to consider a two-dimensional model which is much easier in concept, but nevertheless affords useful pointers to the stress relationship in simple bodies when they are subjected to one or more external stresses applied in different ways. A more complicated model, but of similar type can be constructed for the assessment of the resulting strains, but this has to take account of the added complication of Poisson's Ratio.

Consider a small cube of infinitesimal size with axes X, Y, Z and liable to stresses on each of its major sides. To simplify the treatment the action of forces acting along the Z axis perpendicular to the plane of the paper must be disregarded. The model simplifies therefore to a two-dimensional system with stresses being applied from the X and Y directions only, as shown in Fig. XIII.6.

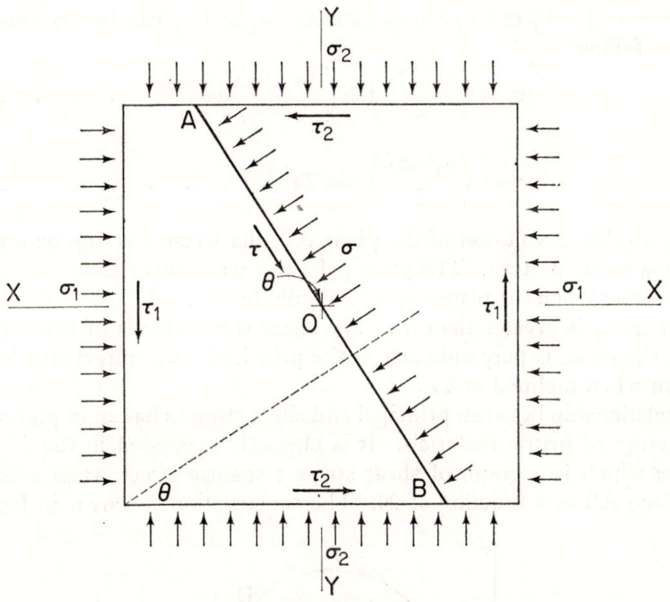

FIG. XIII.6. A TWO-DIMENSIONAL SYSTEM OF STRESSES AND THEIR RESULTANTS

The analysis of the total stress relationships in such a model can be effected by considering the resultant stress acting on a small point source in a line AB as it makes a complete revolution around the centre point O within the square of applied forces.

Suppose that stresses of different magnitude σ_1 and σ_2 are being applied to the X and Y faces respectively. The calculation is not basically affected whether these are regarded as compressive or tensile stresses but in detailed theory it is necessary to adopt a convention of giving opposite signs to compressive and tensile stresses. The tensile stresses must also be given a sign to indicate direction in relation to the compressive stresses, but there is no standard convention yet adopted and workers differ in their terminology.

The stresses σ_1 and σ_2 are the principal stresses of the system. When the plane *AB* is perpendicular to either of the two principal stress directions, any small unit area along its length will be subject only to that pure principal stress acting vertically upon it. At intermediate positions of inclination of the plane *AB*, there will be a resultant stress σ acting in the direction indicated by the arrows in the diagram and intermediate in value between σ_1 and σ_2.

The application of a direct stress induces a corresponding shear stress acting in a perpendicular direction. Shear stress values are normally indicated by the symbol τ. Consequently, the principal stresses acting on a small area of the line AB (Fig. XIII.6) will induce also a resultant shear stress, the magnitude of which will determine the tendency of the block to glide or shear along AB.

The components of direct and shear stress in a unit area of a plane inclined at an angle θ and subject to principal stresses σ_1 and σ_2 can be expressed by two formulae as follows

$$\sigma = \left(\frac{\sigma_1+\sigma_2}{2}\right)+\left(\frac{\sigma_1-\sigma_2}{2}\right)\cos 2\theta \quad . \quad . \quad . \quad . \quad . \quad 1$$

$$\tau = \left(\frac{\sigma_1-\sigma_2}{2}\right)\sin 2\theta \quad . \quad . \quad . \quad . \quad . \quad . \quad . \quad . \quad . \quad . \quad 2$$

In a complete revolution of the plane AB, the stresses acting on a small area of it follow a cyclic pattern. The principal stress component shows maximum and minimum values when the plane is perpendicularly oriented to each stress direction, assuming that σ_1 is greater than σ_2. The shear stress also is of cyclic form; it is zero when the plane is perpendicular to the principal stress directions, but reaches a maximum when inclined at 45°.

The relationship between principal and shear stresses has an important bearing on the fracture of brittle materials. It is elegantly expressed in the *Mohr's circle construction* which is a graph of shear stress τ against direct stress σ acting on a point in plane AB as a function of 2θ. The construction is shown in Fig. XIII.7.

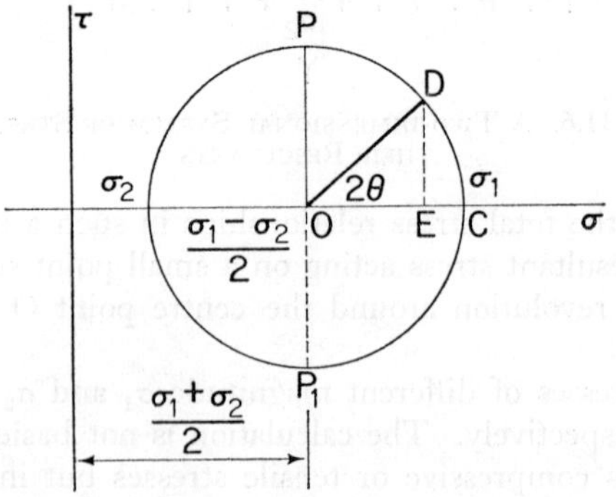

FIG. XIII.7. A MOHR'S CIRCLE CONSTRUCTION OF STRESS RELATIONSHIPS

The normal stress axis σ is represented horizontally and on it lengths are marked off to represent σ_1 and σ_2. If a circle is constructed on this axis it will have its centre as the point $\frac{\sigma_1+\sigma_2}{2}$ and a radius of $\frac{\sigma_1-\sigma_2}{2}$; this is the value of τ at P and P_1 when the angle $2\theta = 90°$ or $270°$, i.e. at an inclined angle of plane AB of 45° and 135°. The circle thus represents the values of σ and τ at any inclination θ of the plane AB.

It must be remembered that the stress path from σ_1 to σ_2, which is a semi-circle sweeping an angle of 180°, represents the actual stress relationships as the relative inclination of the plane AB in Fig. XIII.6 changes by only 90°. Consequently in Mohr's circle representation, an angle of 2θ to the normal indicates an angle of inclination of the plane AB of θ to the direction of principal stress.

From the diagram, the maximum values of shear stress occur when the angle $2\theta = 90°$ (and 270°), indicating that when the plane AB is inclined at 45° (and 135°) the greatest shear stress is to be expected.

The point D in the diagram would illustrate the stress conditions on a plane inclined at an angle θ to the principal stress direction. The value of σ in this plane would be given by the position of point E on the σ_1 σ_2 axis and the value of τ, the shear stress, by the length of the line DE. The values of shear stress can have a positive or negative value depending on the angle of inclination θ of the plane AB to the direction of the principal stresses.

The diagram illustrated in Fig. XIII.7 is for a condition where σ_1 and σ_2 are both stresses of compression, but where σ_1 is greater than σ_2. If σ_2 were to equal 0, the circle would be constructed to touch the vertical axis. When σ_2 is a stress of opposite sign to σ_1 (i.e. tensile as opposed to compressive) the diameter embraces the origin, but it has exactly the same meaning otherwise.

When $\sigma_1 = \sigma_2$ no circle can be constructed, which indicates that no shear stress (τ) can be present at any inclination of a plane to the principal stress directions.

In practice, when a body fails in stress, the angle of the fracture plane θ is always steeper than 45°. This is explained by assuming that the imparted shear stress (τ) to cause fracture has not only to exceed the theoretical or direct shear strength (S_0) of the material but also to overcome a frictional force which opposes the shearing action. This latter term is related to the mean direct stress σ of the system. Failure will occur only when

$$\tau \geqslant S_0+\mu\sigma \text{ or } S_0 \leqslant \tau-\mu\sigma \quad . \quad . \quad . \quad . \quad . \quad . \quad 3$$

where μ is termed the *coefficient of internal friction.*

In the above equation, the values of τ and σ can be replaced by their appropriate values in σ_1, σ_2 and θ in equations (1) and (2) to give

$$S_0 = \tau-\mu\sigma = \frac{\sigma_1-\sigma_2}{2}\sin 2\theta-\mu\left[\frac{(\sigma_1+\sigma_2)}{2}+\frac{(\sigma_1-\sigma_2)}{2}\cos 2\theta\right]$$

$$= -\tfrac{1}{2}\mu(\sigma_1+\sigma_2)+\tfrac{1}{2}(\sigma_1-\sigma_2)(\sin 2\theta-\mu\cos 2\theta) \quad . \quad . \quad . \quad . \quad . \quad 4$$

If this is differentiated with respect to θ a maximum value occurs when

$$\tan 2\theta = -\frac{1}{\mu} \quad . \quad . \quad . \quad . \quad . \quad . \quad 5$$

and the value of the maximum is

$$\tfrac{1}{2}\mu(\sigma_1+\sigma_2)-\tfrac{1}{2}(\sigma_1-\sigma_2)\,(1+\mu^2)^{\frac{1}{2}} \quad . \quad . \quad . \quad . \quad . \quad 6$$

An interesting calculation can be derived from this observation by constructing a Mohr's circle corresponding to a fracture condition. This is illustrated in Fig. XIII.8 and the fracture plane indicated by FO where the angle 2θ is greater than

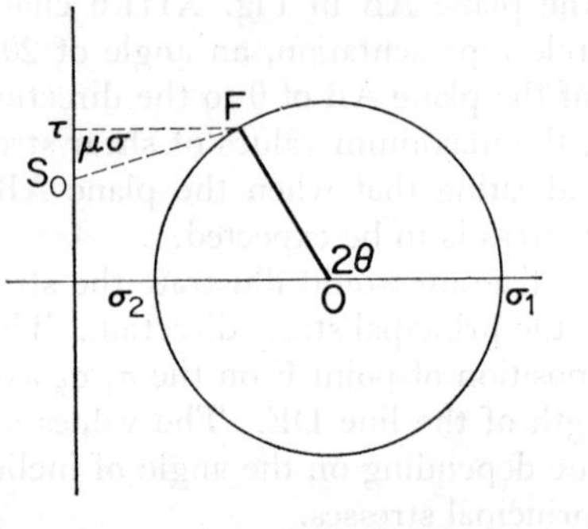

FIG. XIII.8. A FRACTURE CONDITION (F) INDICATED ON A MOHR'S CIRCLE CONSTRUCTION

90°. The value of τ as indicated by the point F exceeds the stress at failure S_o under direct shear conditions by $\mu\sigma$ (equation 3). The significance of S_o is illustrated more clearly in Fig. XIII.9 and on p. 851.

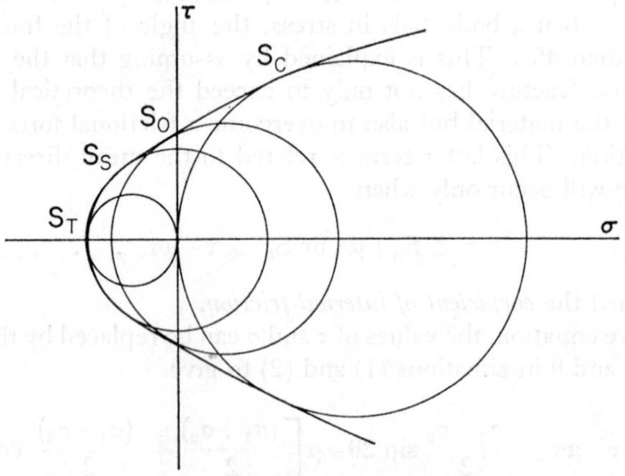

FIG. XIII.9. THE FRACTURE ENVELOPE CONSTRUCTED FROM A SERIES OF MOHR'S CIRCLES

So two equations can be evolved for the shear stress at F (by utilising equations 1 and 2).

(i) Substituting equation 1 in 3

$$\tau = S_o + \mu\left[\frac{(\sigma_1+\sigma_2)}{2} + \frac{(\sigma_1-\sigma_2)}{2}\cos 2\theta\right] \quad . \quad . \quad . \quad . \quad 7$$

(ii) Restating equation 2

$$\tau = \frac{(\sigma_1-\sigma_2)}{2}\sin 2\theta \quad . \quad . \quad . \quad . \quad . \quad . \quad 8$$

At point F, $\tan 2\theta = \dfrac{-1}{\mu}$

So $$\sin 2\theta = \frac{1}{(1+\mu^2)^{\frac{1}{2}}} \text{ and } \cos 2\theta = -\frac{\mu}{(1+\mu^2)^{\frac{1}{2}}} \quad . \quad . \quad . \quad . \quad 9$$

Equating (7) and (8) above and substituting the values for sin 2θ and cos 2θ gives

$$S_o + \mu\left[\frac{(\sigma_1+\sigma_2)}{2} - \frac{(\sigma_1-\sigma_2)}{2} \times \frac{\mu}{(1+\mu^2)^{\frac{1}{2}}}\right] = \frac{(\sigma_1-\sigma_2)}{2} \times \frac{1}{(1+\mu^2)^{\frac{1}{2}}} \quad . \quad 10$$

This simplifies to

$$2S_o = \sigma_1[(1+\mu^2)^{\frac{1}{2}}-\mu] - \sigma_2[(1+\mu^2)^{\frac{1}{2}}+\mu] \quad . \quad . \quad . \quad . \quad 11$$

This equation represents a criterion for failure under the effect of combined stresses, and although it should strictly apply only to a compressive condition producing shear fracture, it holds over a range of tensile stresses also.

A. In the case of failure under compressive stress applied only to one face of a symmetrical block

$$\sigma_1 = S_C \text{ and } \sigma_2 = 0$$

Therefore $$2S_o = S_C[(1+\mu^2)^{\frac{1}{2}}-\mu] \quad . \quad . \quad . \quad . \quad . \quad . \quad . \quad . \quad 12$$

B. When tensile stress only is involved in the opposite direction

$$\sigma_1 = 0 \text{ and } \sigma_2 = -S_T$$

Therefore $$2S_o = S_T\,[(1+\mu^2)^{\frac{1}{2}}+\mu] \quad . \quad . \quad . \quad . \quad . \quad . \quad . \quad 13$$

where S_C is the compressive strength and S_T is the tensile strength of the material.

From equations (12) and (13)

$$\frac{S_C}{S_T} = \frac{[(1+\mu^2)^{\frac{1}{2}}+\mu]}{[(1+\mu^2)^{\frac{1}{2}}-\mu]} \quad . \quad . \quad . \quad . \quad . \quad . \quad . \quad 14$$

For many minerals normally found in ceramic materials—in fact for most of those which contain oxygen atoms as their main anions, μ approaches the value of 1. In the equation above, therefore,

$$\frac{S_C}{S_T} = \frac{[\sqrt{2}+1]}{[\sqrt{2}-1]} = \frac{2{\cdot}414}{0{\cdot}414} = 5{\cdot}85$$

$$S_C = 5{\cdot}85\ S_T \quad . \quad . \quad . \quad . \quad . \quad . \quad . \quad . \quad . \quad . \quad 15$$

In fact, the value of μ varies with different materials and it is often taken as a measure of the elastic character of brittle minerals. Non-elastic incompetent rocks (often of sedimentary character), such as shales, have a μ value as low as 0·7 giving $\frac{S_C}{S_T} \doteqdot 4$; harder sedimentary rocks, such as limestones and single crystal minerals are termed semi-elastic and have $\mu \doteqdot 1{\cdot}0$ and $\frac{S_C}{S_T} \doteqdot 6$; some igneous rocks and materials of a glassy nature are quasi-elastic with $\mu \doteqdot 1{\cdot}5$ and $\frac{S_C}{S_T} \doteqdot 10$. As Table XIII.II shows, this relationship is of the predicted order.

TABLE XIII—II. RELATIONSHIP BETWEEN TENSILE AND COMPRESSIVE STRENGTHS OF SOME BRITTLE MATERIALS

Material	*Tensile Strength* (S_T) (*lbf./in.*^{2})	*Compressive Strength* (S_C) (*lbf./in.*^{2})	*Ratio* $\frac{S_C}{S_T}$	μ
Granite	3,000	22,000	7·3	1·17
Marble	3,300	20,000	6·0	1·03
Limestone	2,200	14,000	6·4	1·07
Sandstone	610	2,800	4·6	0·87
Shale	120	500	4·2	0·82
Silica Brick	600	3,000	5·0	0·92
Common Brick	680	5,000	7·3	1·17
Engineering Brick	1,030	10,000	9·7	1·45
Quartz crystal	6,000	35,000	5·8	1·00

The value S_S, defined as the *pure or simple shear strength*, represents the condition of failure, when the body is acted upon by compressive and tensile stresses in perpendicular directions and of equal magnitudes. By a similar calculation to that linking compressive and tensile strengths, so also can the pure shear strength be related to the other values. Once again the ratio depends on the value of μ, the coefficient of internal friction, but the following values are useful guides

$S_S \doteqdot S_T \doteqdot 0{\cdot}25\, S_C$ for non-elastic bodies ($\mu \doteqdot 0{\cdot}7$) . . . 16

$S_S \doteqdot 1{\cdot}2S_T \doteqdot 0{\cdot}20\, S_C$ for semi-elastic bodies ($\mu \doteqdot 1{\cdot}0$) . . . 17

$S_S \doteqdot 1{\cdot}5S_T \doteqdot 0{\cdot}15\, S_C$ for quasi-elastic bodies ($\mu \doteqdot 1{\cdot}5$) . . . 18

The relationship $S_0 = \tau - \mu\sigma$ for the fracture relationship of brittle materials forms the basis of the **Coulomb-Navier criterion of failure** and by using it with appropriate values inserted, a graph relating the stress in fracture to the mean applied principal stress can be derived.

In **Mohr's criterion of failure** a fracture envelope is constructed from a series of Mohr's circles as shown in Fig. XIII.9. In this diagram only four circles are shown, but in each, stresses are represented which cause fracture to occur. The four circles are significant in that they represent respectively, from right to left, (1) shear fracture in pure compression when σ has a positive value; (2) direct shear failure where the fracture point coincides with τ axis at S_0; (3) fracture in pure shear and (4) fracture in mean tension with negative values of σ.

On the various stress circles which are constructed, failure does not occur at maximum values of τ because, due to internal friction forces, the fracture plane is not inclined at 45°.

A fracture envelope can thus be constructed from a series of Mohr's circles which is a true representation of the relationship between the principal stresses and the point at which failure will occur.

When overall tension stresses operate, the form of the envelope undergoes a marked change and some authorities consider it unwise to place too much significance on the shape of the failure curve in this region. Strictly the values of τ at failure represent the stress condition only when shear fracture occurs. When $\tau = 0$, $S_0 = -\mu\sigma$ which is the condition for failure in pure tension.

Mohr's representation, although empirical in its derivation, is more accurate than the Coulomb–Navier mathematical model. In the latter, the fracture envelope would be linear, whereas it is more correctly indicated by a parabola. The relationship $S_0 = \tau - \mu\sigma$ does not strictly hold true and is inaccurate at high values of σ.

The **Griffith criterion of brittle fracture** is a mathematical approach to the problem in which it is postulated that tensile fracture is initiated at microcracks and flaws in the material. The treatment has been of the greatest value in predicting the strength relationships of glasses and it has also been used in problems of rock mechanics.

Basically, the theory involves triaxial stress considerations (hence σ_3 is employed) and their influence on cracks or flaws which intersect the main stress fields. The mathematical expressions are of a complex nature but under certain conditions, a simplification to an equation which represents a Mohr's circle can be derived thus·

$$\left(\sigma - \frac{\sigma_1+\sigma_3}{2}\right)^2 + \tau^2 = \left(\frac{\sigma_1-\sigma_3}{2}\right)^2 \quad . \quad . \quad . \quad . \quad . \quad 19$$

From this expression, a generalised equation for the Mohr's envelope can be adduced which may be written as

$$\tau^2 - 4\,S_T\sigma = 4\,S_T{}^2 \quad . \quad . \quad . \quad . \quad . \quad . \quad . \quad . \quad 20$$

where S_T is the tensile strength of the material under test.

These types of construction and calculation are of important application to failure problems in structures of all kinds in the ceramic industry. Structural clay products are frequently used where units are in compression from overlying materials but where stresses at right angles often develop. Failure conditions can be adduced from a knowledge of their conventional crushing strength (p. 892) and by applying Mohr's circle theory to the stresses which can be anticipated.

Refractory furnace construction often requires that blocks and bricks are subject to a variety of compressive and tensile stresses both by nature of the design and by virtue of structural movement during heating. A study of the compressive, tensile and shear stresses which are likely to develop can help in overcoming potential failure positions.

The foregoing treatment has established the relationship between compressive, tensile and shear stress, but in ceramic practice, transverse bending stresses and torsional stresses are also involved.

Transverse Bending Stress. When a body is held rigid at one or more points along its length and subjected to stresses at positions on its opposite surface, the tendency is for the body to bend until, if the applied stresses are sufficiently great, fracture occurs by tensile cracking. In the simple example of Fig. XIII.10, a load

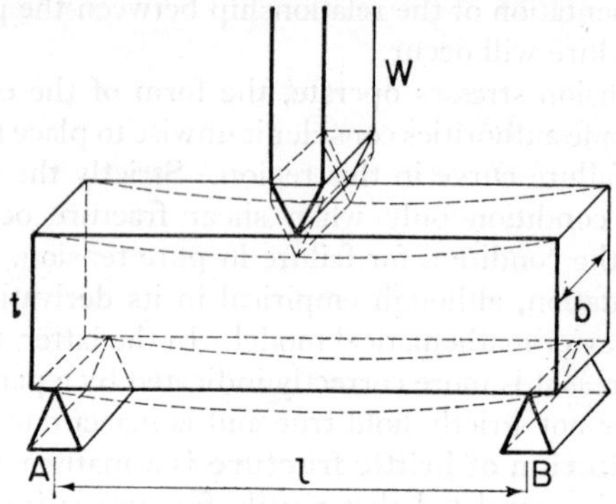

FIG. XIII.10. CALCULATION OF THE TRANSVERSE STRENGTH OF A BAR

W is applied at the centre of a beam which is supported on its lower surface at its extreme ends A, B.

The applied force W induces bending in the beam as shown by the dotted lines in the Figure, and, in the process, the lower planes of the beam are put into tension whilst those in contact with the loading point are in compression. Fracture commences in the lower surface of the beam by tensile failure when the stresses which are induced exceed the tensile strength of the material.

The stresses which will cause fracture in such a body are those which will develop excessive tensile stresses at the point of maximum bending.

It can be calculated quite simply that

$$T = \frac{Mt}{2I} \qquad \ldots\ldots\ldots \qquad 21$$

where T is the transverse strength also called the modulus of rupture, M the maximum bending moment, t the thickness of the bar and I the moment of inertia of the section around the axis of the bar.

The maximum bending moment for the example given in Fig. XIII.10 is

$$M = \frac{W}{2} \times \frac{l}{2}$$

where l is the distance between the lower load points.

The moment of inertia depends on the cross-sectional geometry of the bar or shape subjected to transverse forces. For a bar of thickness t, breadth b the relationship is

$$I = \frac{bt^3}{12} \qquad \text{so } T = \frac{3}{2} \cdot \frac{W.l}{bt^2}$$

For a circular bar diameter d

$$I = \frac{\pi d^4}{64} \qquad \text{so } T = 8 \,.\, \frac{Wl}{\pi d^3}$$

A pipe-section with an external diameter of D and an internal diameter d would have

$$I = \frac{\pi}{64}(D^4 - d^4) \quad \text{so } T = 8\frac{Wl}{\pi(D-d)^3}$$

Transverse stresses are frequently encountered in ceramic bodies in use. Sewer pipes are usually subjected to this type of stress when they are suspended by two adjacent sockets in a trench and back-fill piled on top of them, some of which may be large stones which virtually impart point loading. Beam sections in building structures and in kilns are also liable to transverse stresses.

Torsional Stresses. When a bar or rod of material is subjected to torsional stress, it has to resist a twisting action and will ultimately fail when the tensile stress is exceeded in a direction at right angles to the tangential direction of the twist.

The calculation of breaking stress is possible by use of the formula

$$S_o = \frac{Tt}{2J} \quad \ldots\ldots\ldots\ldots \quad 22$$

where S_o is the breaking stress, T the torque applied in lb.ft. or other appropriate units, t is the thickness or diameter of the bar and J the *polar moment of inertia.*

The value of J can be calculated by similar principles to those used for assessing I in the previous section, but the moment involves a different axis so the values are not the same. In fact J is twice the value of I for bars or rods of the same shape and dimensions.

Torsional fracture is not uncommon in pipes and beams which are subjected to twist in service due to differential movements.

Impact Stresses. When a moving weight impinges on a body, the resulting tensile or shear stress which may develop is considerably in excess of that which results when the same weight is applied slowly or without its momentum.

Suppose, as in Fig. XIII.11, a bar AB is maintained *rigidly* at point A. A weight W is allowed to fall from height h on to the bar near to its unsuspended end B so that a stress develops along the length of the bar. The bar has a thickness t and an area of cross-section a.

When the weight strikes the bar it will induce a stress σ and a strain (in compression) of Δt. The work done by the falling weight will equal the energy absorbed by the bar, so that,

$$W(h + \Delta t) = \tfrac{1}{2}\sigma . a . \Delta t \quad \ldots\ldots\ldots \quad 23$$

Now $$\frac{\text{stress}}{\text{strain}} = E \text{ (elasticity modulus)}$$

So that $$\frac{\sigma . t}{\Delta t} = E$$

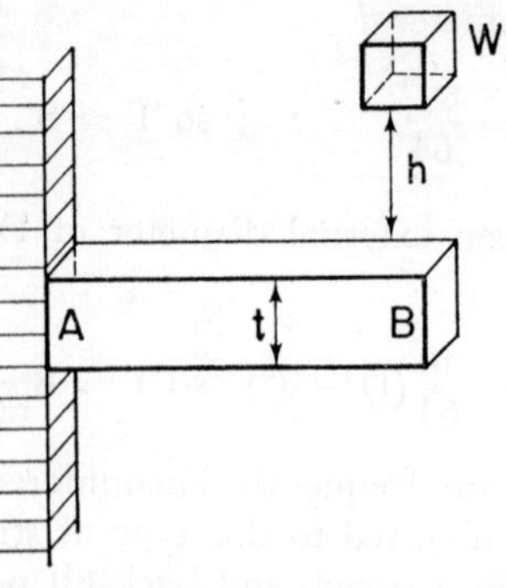

FIG. XIII.11. CONDITIONS OF IMPACT FRACTURE IN A BAR

Substituting the value of Δt in equation 23 gives

$$\frac{W(h+\sigma.t)}{E}=\frac{\sigma^2 at}{2E} \quad . \quad . \quad . \quad . \quad . \quad . \quad . \quad . \quad 24$$

This equation can be solved for σ and the positive root is

$$\sigma=\frac{W}{a}+\sqrt{\left(\frac{W}{a}\right)^2+\frac{2WhE}{at}} \quad . \quad . \quad . \quad . \quad . \quad . \quad 25$$

This is an important equation because it enables induced stresses to be compared (*a*) when the weight is dropped on to the bar; (*b*) when the weight is applied suddenly; and (*c*) when the weight is applied gradually.

In case (*a*) the term $\sqrt{\frac{2WhE}{at}}$ dominates and all others can be ignored.

In case (*b*) $h = 0$ so

$$\sigma=\frac{W}{a}+\frac{W}{a}=\frac{2W}{a} \quad . \quad . \quad . \quad . \quad . \quad . \quad . \quad . \quad 26$$

but in case (*c*) the stress simply equals $\frac{W}{a}$ 27

A sudden application of pressure produces a stress twice as great as when the pressure is applied slowly, but the stress induced by impact is many times greater.

This is best illustrated by an example:

Suppose a weight of 5 lb. is dropped from a height of 6 in. on to a bar of material 1 in. thick and having a cross-sectional area of 1 sq. in. E, the modulus of elasticity, equals 1×10^6 lbf./in.^{2}.

From equation (25)

$$\sigma=\frac{5}{1}+\sqrt{\left(\frac{5}{1}\right)^2+\frac{2\times5\times6\times10^6}{1\times1}}$$

$$=5+\sqrt{25+60\times10^6}$$

$$=7{,}750 \text{ lbf./in.}^2 \text{ (approx.)}$$

If conditions were kept the same, but the pressure were to be applied suddenly, but without impact $h = 0$, so

$$\sigma = 5 + \sqrt{5^2 + 0}$$
$$= 5 + 5 = 10 \text{ lbf./in.}^2$$

Finally when the weight is applied gradually

$$\sigma = \frac{W}{a} = 5 \text{ lbf./in.}^2$$

The stress values induced by impact under these conditions of experiment are many times greater than those when the same load is applied in direct contact.

The calculation of the actual failure conditions in impact can present problems. In the ideal conditions illustrated in Fig. XIII.11, the bar is maintained in an absolutely rigid condition and the applied force acts perpendicularly to the impacted surface. The principal and only stress may then be regarded as one of compression acting downwards. The mathematical treatment is that of a Mohr's circle with σ_1 = maximum impact stress and $\sigma_2 = 0$. Failure is ideally by shear, and absolute values for fracture can be calculated.

Only rarely do ideal conditions apply. The impacted body may acquire kinetic energy to absorb part of the impact stress, or the angle of impact may not be 90°. This latter factor can have considerable significance; if the impact stress impinges at an obtuse angle to the surface, failure conditions tend towards tensile fracture where breakage is more likely to occur.

The concept of impact fracture is of vital importance in many aspects of ceramic practice. It is invoked in crushing of the original raw material in hammer mills, and in the fracture of articles which are dropped or struck by a weight.

The calculations in the previous section under equations (25, 26, 27) illustrate another important aspect of the strength of ceramic and other brittle materials. The resultant stresses which develop within a body subjected to external pressures depend on the rate of application of such pressures.

Equations 26 and 27 indicate that the resultant stress in a bar can be doubled when compressive stresses are applied instantaneously as against the condition of slow development.

This fact can have far-reaching significance in both the testing and use of materials of a brittle nature.

The physical concept of the above conclusion is that stresses are not transmitted instantaneously throughout a body. Sudden applications of external stresses produce localised areas of intense strain, thereby causing incipient fracture which is transmitted throughout the body.

Circumferential Stresses in Cylindrical Shapes Subjected to Internal Pressure. A special case of stress development in ceramic articles is to be found when pipes, tubes, containers or tanks are subjected to internal pressure. A typical case is illustrated in Fig. XIII.12 where the cross-section of a pipe of diameter *d* and

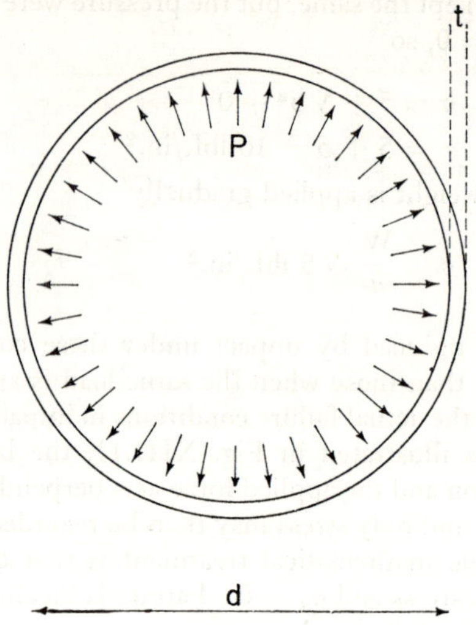

FIG. XIII.12. THE STRESS CONDITION IN A PIPE SECTION

wall thickness t, is being subjected to an internal pressure of P. This may arise due to a head of water.

The resultant stress on the circumference is exhibited in tensile and it can be expressed quite simply as

$$T_o = \frac{Pd}{4t} \quad . \quad . \quad . \quad . \quad . \quad . \quad . \quad . \quad 28$$

This relationship is strictly true only if the ratio of $t : d$ is less than 1 : 20, but it gives a good guide to the safe limit of use of glassware and of sewer pipes. A glass tube with a tensile strength of 7000 lbf./in.^{2} of 10 in. diameter and a wall thickness of 0·025 in. could withstand an internal pressure given by

$$P = \frac{7000 \times 4 \times 0{\cdot}025 \text{ lbf./in.}^2}{10}$$

$$= 70 \text{ lbf./in.}^2$$

Hardness. A property closely related to the ultimate strength of a material is its surface hardness. This may be defined as the ability of a substance to resist scratching or abrasion or the penetration of other bodies into its surface.

Because the measurement of hardness is relatively easy, this property is often used as an indication of strength, but care must be exercised in the interpretation of results more so than with any other method of strength measurement.

Basically, when measuring the hardness of a material, the property under test is the capacity of a point, ball-bearing or similar contact area to penetrate into the surface. When the penetrating device is large relative to the units comprising the material, the main stresses being exerted are compressive, but because there is no localised restraint there may be some complementary flow in shear or tensile strain. When measuring the hardness of a mineral surface of perfect crystallinity, the test may be regarded as compressive, because the individual displaced units are atoms of the lattice—extremely small in relation to the penetrating device.

When testing materials of a composite nature, particularly those of a porous character, hardness measurements can be confusing. The penetrating device might encounter a crack or void and thus cause deformation by tensile or shear fracture.

Because hardness is measured as a function of the depth of penetration (or the amount of surface abraded) in most cases over a very small area, results can be variable and often unreliable.

It is better, when measuring hardness values, to subject as large an area of surface as possible to the penetrating load. In so doing, it follows that unless extremely high stresses are employed the amount of penetration will often be small and difficult to measure with accuracy.

Although there have been numerous attempts to correlate hardness values, determined in a variety of ways, with a particular strength characteristic, the approaches are essentially empirical. Details of several hardness testing devices are given later in the Chapter.

THE MEASUREMENT OF STRENGTH AND ALLIED PROPERTIES

THE principal methods of determining the strength of clays and other ceramic materials are naturally based on the various forms in which the 'strength' of these materials is exhibited. According to the information required, a method is used which determines one or more of the following: (1) tensile strength; (2) compression or crushing strength; (3) transverse strength or modulus of rupture; (4) resistance to impact; (5) torsional strength; (6) modulus of elasticity; (7) hardness and deformability; (8) miscellaneous methods.

Some of these methods may be applied either to the cold material or to it at any suitable temperature. The strength of a material or article at the temperature at which it is used is, in many cases, of much greater importance than its strength at atmospheric temperatures. This is especially the case with refractory materials.

Tensile Strength. The failure of a body in tensile is the least complicated of all strength tests. Pressure is applied to one face to elongate or stretch the test piece until fracture occurs. As the Mohr's circle construction of Fig. XIII.9 shows, failure in pure tension involves no resultant or complicating shear stresses; the body fractures when the induced strain in the direction of the principal stress becomes too great to be accommodated. The size and shape of the test piece has little or no significance and the tensile strength is simply defined as the pressure applied divided by the area of cross-section at the fracture plane.

Unfortunately simple tensile testing is not an easy procedure. Test pieces of special shape are required so that they can be held securely in the clips of the testing machine; it is also necessary to ensure that the stress develops exactly at right angles to the smallest area of cross-section.

When measuring the tensile strength of a wet clay, it is an easy matter to press a shape in a mould of the correct dimensions, but dried and fired articles need careful sizing and shaping to ensure that test conditions are correct. It may also be argued that fired ceramic ware is rarely subjected to pure tensile forces, hence this method of test is used mainly for measuring green properties.

Tensile testing is used widely in the electrical ceramic ware industry and in specialised applications.

The test-piece usually employed for making tensile tests consists of a figure-8-shaped block of material. The dumb-bell shaped ends must fit securely into the chucks or clips of the test machine, an example of which is shown in Fig. XIII.13.

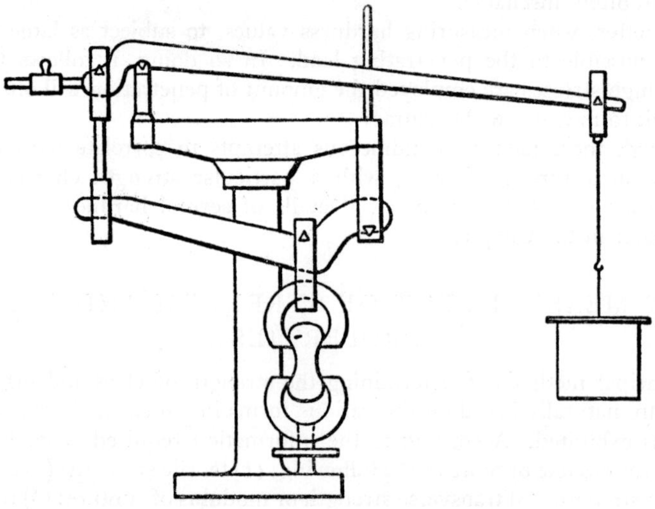

FIG. XIII.13. A TENSILE TESTING MACHINE

This is a simple model for green clays in which the pressure is applied through a fulcrum from the container on the right which is slowly filled with water, sand or mercury. More complex units apply the load by mechanical or hydraulic means.

It is important that the rate of increase of stress should be carefully controlled and standardised. Normally this is about 10 lbf./in.^{2} per second for specimens which fracture in the range 200–1200 lbf./in.^{2}, but it should be slower for specimens of lower strength.

Various other methods have been devised for determining the tensile strength, one of which consists in measuring the length of a clay column which can be extruded through an aperture of a certain definite size before the column breaks. This test is conveniently made by placing the clay in a vertical cylinder with a $\frac{3}{8}$-inch aperture in the base, and a piston with screw-down motion at the top. On turning the screw slowly and regularly the paste is extruded through the aperture and hangs vertically until its weight is greater than its tensile strength, when the column breaks. The tensile strength is found by dividing the weight of the broken piece of paste by the area of the aperture through which it was extruded.

The strength of dried pottery is sometimes determined by means of a tensile test, but the transverse test, which is much simpler, is increasingly used for this purpose.

The *Extensibility* may be measured by placing the two portions of the broken test piece together and measuring the increase in its length. For this purpose, it is desirable before making the tensile test to make two small dots or thin lines on the test piece at a convenient distance apart and to measure the distance between them before and after the test. The increase in length multiplied by 100 and divided by the original length or distance between the marks will be the percentage of extensibility. Extensibility tests are preferably made on rods or cylindrical test pieces.

The compression or crushing strength test is much more widely used and it is the standard method for assessing the load-bearing capacity of building and engineering bricks, refractory shapes, concrete blocks, etc.

Fracture under compressive stresses is a complex phenomenon and the strength so determined depends on many factors other than the simple intergranular cohesion. When a compressive stress is applied to one face of a block, a complex resultant stress and strain pattern is set up, so that fracture occurs in shear (p. 844).

The ultimate compressive stress which is required depends on the geometry of the shape and, to an important degree, on the rate of application of the stress. Shear fracture lines tend to follow flaws and cracks in the test-piece and large grains of grog or hard minerals can act as barriers. The specimen to be tested must be as large as possible and appreciably greater than the maximum size of the individual grains of which it is composed. The smallest dimension of the test-piece should be at least 10 times that of the maximum dimension of the largest grains, if consistent and reliable results are to be obtained. It is obviously better to test as large a sample as possible but this may involve a powerful machine capable of developing several hundred tons pressure.

When testing concrete specimens, it is the usual practice to employ cubical test-pieces. Provided that the blocks are sufficiently large, the crushing strength is consistent irrespective of size; small-sized blocks, however, even though they are cubical, give a wider scatter of results due to grain effects.

As the dimensions are changed, marked differences occur in the compressive strength (expressed as pressure applied per unit area of bearing surface). For cylindrical specimens the following relationship holds over a wide size and shape range

$$S_c = S_{co}\left(0{\cdot}8 + \frac{0{\cdot}2}{L/D}\right) \quad . \quad . \quad . \quad . \quad . \quad . \quad . \quad . \quad . \quad 29$$

where S_{co} is the compressive strength when $L/D = 1$ and L is the length between the bearer plates and D the diameter. S_c is the actual recorded strength.

Thus a specimen which has a large L/D ratio, i.e. an elongated cylinder, will have a reduced compressive strength, whereas a disc with a L/D ratio of less than 1 will have a greater strength.

Empirical relationships have been suggested for rectangular shapes.

The expression $S_c = S_{co}\left(0{\cdot}8 + \dfrac{0{\cdot}2}{L/\sqrt{A}}\right)$ 30

gives fair agreement for blocks which have an approximately square end-section of area A. In this case the value S_{c_0} would be for a perfect cube.

Where one of the end dimensions is smaller than the other, the shear failure tends to predominate along the smaller axis, so that an approximate formula can be used, thus

$$S_c = S_{c_0}\left(0{\cdot}8 + \frac{0{\cdot}2}{L/W}\right) \quad . \quad . \quad . \quad . \quad . \quad . \quad . \quad . \quad 31$$

where W is the shortest dimension of the cross-section.

On this basis, a whole brick compressed on its 9 in. × 4½ in. face would give a similar strength to a half-brick of the same material because in both cases the thickness L = 3 in. and the shortest dimension equals 4½ in., but usually the smaller test-piece gives a slightly higher average value.

The ratio of compressive strengths (lbf./in.²) of a brick tested on its three principal faces should be in the following ratios

Bearing-face	L/W	*Relative compressive strength*
9 in. × 4½ in.	3/4½	1·0
9 in. × 3 in.	4½/3	0·85
4½ in. × 3 in.	9/3	0·75

The greatest problem in compressive testing is to produce perfectly flat and rigid bearing-surfaces. If the former condition is not achieved, localised areas of high pressure will develop, resulting in premature failure. Solid bricks should be rubbed free of imperfections with sand-paper; bricks with frogs or perforations should be filled with mortar. A material which will not flow and penetrate the brick surface during testing should be used as a seating between upper and lower press platens and the specimen. In British practice three-ply wood sheet, 3 mm. thick is used.

In former practice, a type of felt material was used; this was squeezed by pressure into pores in bricks, and fracture occurred at much lower values than expected due to a tensile stress component being developed.

Building bricks are normally tested when fully saturated with water. This results in a markedly lower compressive strength than when dry bricks are tested, but it simulates strength values under the worst weather conditions.

The testing machine normally should have one floating ball-seated platen to take up shape irregularities. In British practice for building bricks a loading rate of 2,000 lbf./in.²/min. is specified from a value half that of the anticipated failure value.

The test procedure for common building bricks varies in the specifications for different countries.

The U.K. requires tests on whole bricks, pressure to be applied on the face on which they are normally set in service, i.e. the 9 in. × 4½ in.—10 samples are to be tested.[1]

[1] B.S. 3921, 1965.

The American standard[1] requires 5 half bricks; the bearing surfaces are thus $4\frac{1}{2}$ in. square.

German requirements[2] are that bricks should be sawn in half and the two halves bedded together with thin layers of mortar to form an approximate cube.

Even with the best practice, results of duplicate determinations differ by as much as ± 20 per cent from a mean value. Hence, it is necessary to have multiple testing.

The crushing tests on refractory bricks are normally carried out in similar fashion, but specimens may be tested on the flat or on end at a loading rate of 500 lbf./in.^{2}/min. in British practice.[3] The American requirement[4] is for the tests to be carried out with the end-faces ($4\frac{1}{2}$ in. $\times$ 3 in. or $2\frac{1}{2}$ in.) as the bearing surfaces. Loading rates of up to 20,000 lbf./min. are recommended (this is over $13\frac{1}{2}$ or $11\frac{1}{2}$ sq. in.).

The strength of refractory materials at high temperatures may be determined in two ways:

(1) The material may be heated to the desired temperature in a tubular furnace and a load applied through refractory thrust rods until the specimen fractures, or

(2) A definite load or pressure may be applied to the test piece, which is then heated under this pressure (*a*) to a fixed temperature at which the rate of deformation is measured, or (*b*) at a pre-determined rate of temperature rise until failure occurs. These tests are a measure of the *refractoriness-under-load* property which is described on p. 883.

Transverse Tests have, for some purposes, superseded compression or tensile tests, as they can readily be made without the use of large and expensive apparatus.

In making a transverse or cross-breaking test, the test piece is supported upon two triangular 'knife-edges'[5] and a pressure is applied by means of a third knife-edge on top of the test piece, resting on a sheet of glass or steel to prevent cutting. The pressure is increased at a pre-arranged rate, until the test piece breaks. The pressure at the moment of fracture is a measure of the strength of the test piece. The modulus of rupture (in lbf./in.^{2}) may then be found from the following formula:

$$T = \frac{3Wl}{2bt^2}$$

Where W is the breaking load in pounds, l is the span in inches, and b and t are the width and depth of the test piece, respectively, in inches.

A simple but effective transverse tester for small moulded or cut specimens is shown in Fig. XIII.14. A test piece which may be of wet, dry or fired material is placed on the two lower knife-edges and pressure applied to the platform above the upper knife-edge. For specimens of low strength the breaking point can be determined by slowly running mercury or water into a beaker placed on the platform. Stronger specimens are best tested by fitting the tester between the plates of a hydraulic press and slowly applying a load until fracture occurs.

The American Society for Testing Materials requires that transverse tests should be carried out on whole bricks (C133–55), that the bearing edges should be rounded to a $\frac{5}{8}$ in. radius and that the loading should be applied at the rate of 4000 lbf.

[1] A.S.T.M., C62-35T. [2] D.I.N. Standards. [3] B.S. 1902, Part 1A, 1966.
[4] A.S.T.M., C133-55. [5] Roller-bearings are also in common use.

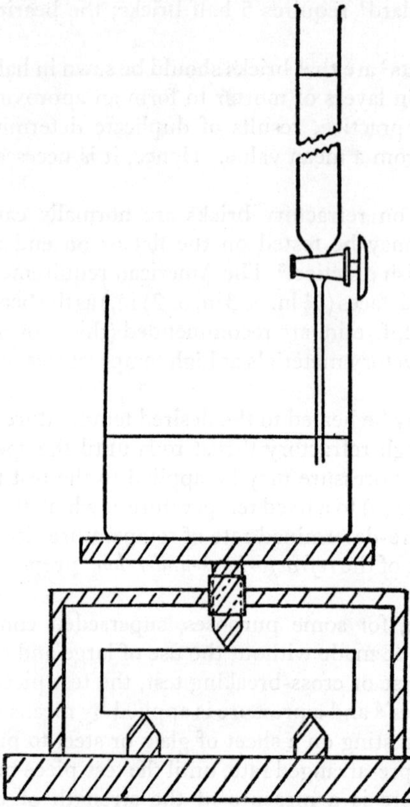

FIG. XIII.14. AN APPARATUS FOR MEASURING THE TRANSVERSE STRENGTH OF SMALL SPECIMENS

per min. The lower bearing edges are 7 in. apart and across the $4\frac{1}{2}$ in. face of the brick, and the load is applied at mid-span. At least 10 test pieces must be used.

It is most important that the test-pieces for the transverse test should be perfectly straight and free from flaws, otherwise unreliable results may be obtained. Even with the greatest care, transverse tests give very irregular results with dried clays.

The British Standard method[1] for measuring the Modulus of Rupture or Transverse Strength is identical with the American recommended procedure except that a loading rate of 1500 lbf./min. is required.

The determination of the *strength of pipes* recommended in British standards[2] involves laying a whole pipe horizontally and supporting it on two bearings spaced 1 in. apart for each foot of pipe diameter along the length of axis. Pressure is applied at the rate of 2000 lbf./ft. run through another bearing midway along the top of the pipe. The bearing surfaces are rigidly specified.

[1] B.S. 1902 Part 1A, 1966. [2] B.S. 65, 1963.

The strength is normally reported in terms of lbf. per linear foot of pipe at its fracture point.

Most manufacturers and users of pipes insist also on a hydraulic proof test for products in which a sealed end pipe is subjected to a maximum pressure of 20 lbf./in.^{2} applied at a rate not exceeding 2 lbf./in.^{2}/sec. If the pipe will withstand a maintained pressure for 5 seconds it is considered acceptable.

Torsion Tests. The value of torsion testing of ceramic materials has been appreciated only in recent years. The method of determining the *workability* of a plastic clay by torsional methods developed by Norton[1] is described in Chapter VIII. Roberts *et al.*[2] have applied the method to high temperature testing of refractory materials with great success and by its means to measure the modulus of rigidity.

Their apparatus is illustrated in Fig. XIII.15.

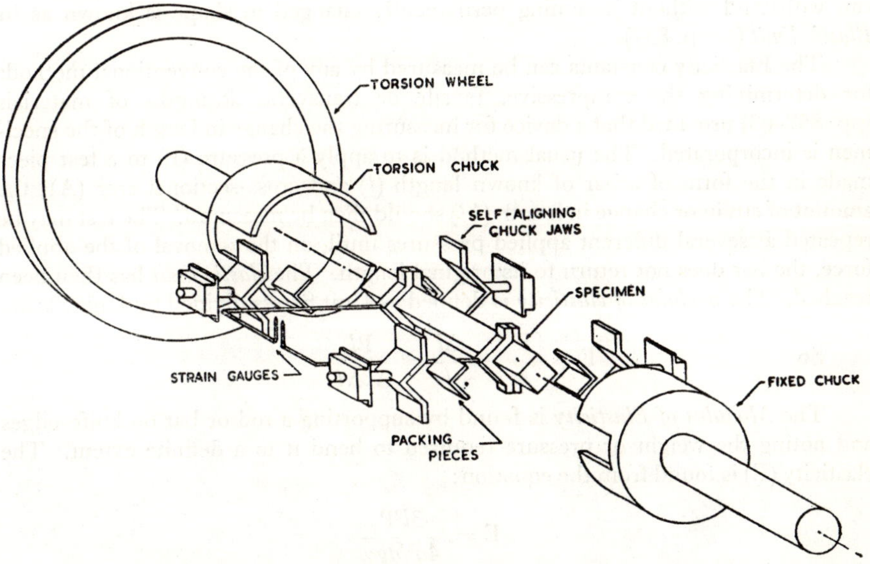

FIG. XIII.15. AN APPARATUS FOR DETERMINING THE TORSIONAL PROPERTIES OF CERAMIC MATERIALS (after Roberts)

A specially-shaped test rod, cylindrical in the centre but with square end-sections is clamped at both ends in a torsion head. A load is applied at one end to exert a twisting force on the specimen, which may be surrounded and heated, if desired, by a tubular furnace (see Fig. XIII.15). Torsion tests not only permit the measurement of the progressive deformation under load, but also the elasticity of the specimen at any desired temperatures (p. 838).

The **Modulus of Elasticity** is a measure of the capacity of a body to withstand

[1] Norton, F. H., *J. Amer. Ceram. Soc.*, 31, 236, 1948.

[2] Roberts, A. L. and Cobb, J. W., *Trans. Brit. Ceram. Soc.*, 32, 22, 1933 and 35, 186, 1938.

strains without becoming permanently deformed or fractured. It is an important property of many ceramic materials, particularly of glazes which may be put into considerable stress during the cooling of ware. The Modulus of Elasticity may also have an important influence on the ability of a body to withstand thermal shock (p. 948).

When a bar or block of a material is subjected to a pressure, it will be reduced or increased in length depending on whether the force is compressive or tensile. Provided that the applied force has not been too great the bar will return to its original length once the pressure has been released. The ratio of

$$\frac{\text{Unit Stress}}{\text{Unit Strain}}$$

is defined as the Modulus of Elasticity (E) and the maximum stress which a material can withstand without becoming permanently changed in shape is known as its *Elastic limit* (see p. 839).

The Elasticity constants can be measured by any of the conventional methods for determining the compressive, tensile or transverse strengths of materials (pp. 857–63) provided that a device for measuring the change in length of the specimen is incorporated. The usual method is to apply a pressure (P) to a test piece made in the form of a bar of known length (l) and cross-sectional area (A); the amount of strain or change in length (Δl) should then be measured. The test may be repeated at several different applied pressures until, on the removal of the applied force, the bar does not return to its original length. The *elastic limit* has then been reached. The *modulus of elasticity* is defined as Unit Stress divided by Unit Strain.

So $$E = \frac{P}{A} \div \frac{\Delta l}{l} = \frac{Pl}{A.\Delta l}$$

The *Modulus of Elasticity* is found by supporting a rod or bar on knife-edges and noting the weight or pressure required to bend it to a definite extent. The elasticity (E) is found from the equation:

$$E = \frac{3l^2P}{4a^3btg\alpha}$$

where a is the thickness of the test-piece, b its width, l the distance between supports, P the load or pressure and $tg\alpha$ a factor based on the angle of rise of one end of the bent test piece and on the length of the rising portion.

When a successively increasing load is applied and removed, the load which is just too great to permit the test piece to recover its normal position after the removal of the load is taken as the elastic limit. The elastic limit diminishes as the temperature of the test-pieces increases above a certain minimum. The load must be applied for a sufficient time (8 minutes or more) or erratic results will be obtained.

The modulus of elasticity of mineral specimens and ceramic materials is readily measured by using modern hydraulic testing machines fitted with accurate load cells and strain gauges. Batchelor[1] has described a method for determining the elastic moduli of electrical porcelains and glazes.

[1] Batchelor, R. W., *Trans. Brit. Ceram. Soc.*, 54, 40, 1955.

The sonic method has been used by Lakin and West[1] and Soga and Andersen[2] in a study of the characteristics of refractories at high temperatures. The method depends upon the fact that solid bodies vibrate with a characteristic resonant frequency which depends on their modulus of elasticity, their density and dimensions.

A vibrator imparting a wave of known frequency is applied to one end of a brick and the position of fundamental modes determined by a gramophone crystal pick-up connected to a double beam oscillograph which also receives the signal from the vibrator. Longitudinal, torsional and flexural vibrations are set up within the brick and these can all be measured, so that, by applying suitable formulae, the modulus of elasticity can be measured.

The sonic method gives a more accurate measure of the modulus of elasticity than do torsional measurements but, because the brick is not subjected to any strain, the development of true plastic flow cannot be observed. The development of liquid is, however, indicated by a change in the characteristic frequencies of vibration. Sonic methods of measuring the elastic constants of a ceramic material are especially valuable in their application to studies of thermal shock.

Impact Tests are used for determining the resistance of articles which are subjected to rough handling or are liable to be struck against other articles or to be dropped. They are also applicable to domestic table ware, electrical porcelain, and other articles in which brittleness (p. 843) is objectionable. The tensile strength of materials is sometimes determined as an indication of the resistance of articles to impact, but in this respect it is not always reliable.

An impact method for testing porcelain plates consists of a pendulum 24 inches long, at the end of which is a $1\frac{1}{4}$ lb. flattened steel ball with a spherical striking surface, having a radius of $1\frac{1}{8}$ inches. The sample to be tested is placed loosely against a $\frac{5}{8}$-inch pine board and the pendulum is allowed to swing through an angle of about 30 degrees and strike the test piece, the blows being repeated (with the pendulum moving through a greater angle at each blow) until the test-piece is broken. The impact strength is then calculated from the following formula:

$$E = HW \quad \text{or} \quad E = W(1 - L\cos\theta) \qquad 32$$

where E is the energy of impact in foot-pounds, H is the height of drop in feet $= 1 - L\cos\theta$, where $\theta =$ the angle through which the pendulum swings, and L $=$ the length of the arm in feet. W is the weight of the hammer in pounds.

Hardness Testing. The hardness of a material can be more significant than its ultimate strength in some instances. For example, in crushing and grinding operations, ore bodies tend to be abraded, i.e. their surface layers pared off; so too in their resistance to erosion by dusts, high velocity gas streams, molten slags and metals, it is the surface hardness of ceramic materials which is the important property.

Most methods of measuring hardness rely upon the observation of an indentation which is made in the surface of the material to be tested when it is

[1] Lakin, J. R. and West, C. S., *Trans. Brit. Ceram. Soc.*, 56, 8, 1957.
[2] Soga, N. and Andersen, O. L., *J. Amer. Ceram. Soc.*, 49, 318, 1966.

subjected to an applied force. Others, which may have more significance, measure abrasion resistance or the amount removed by techniques resembling milling, cutting or sand-blasting, although these involve more complex equipment and are time-consuming.

There may be little or no relationship between the results of indentation tests and abrasion tests, because the former is most usually associated with the hardness of the mineral components of a composite material whilst the latter is an indication of the strength of the bonds which hold the grains together.

Indentation Methods

A. *Mohs' Hardness Scale*. The simplest hardness test and the one most commonly used in rock and mineral testing, is where the substance is scratched in turn by a series of 'standard' minerals which increase progressively in hardness.

The hardness value of the material under test is that of the mineral which will just produce a positive indentation or scratch when viewed under a lens.

A Mohs' hardness testing kit is readily available from many companies who supply mineral specimens and accessory equipment for measuring their properties.

The test must be used with discretion—as must all hardness methods. Crystal imperfections can give false results and some minerals have different hardness values in different crystallographic directions.

A typical series of minerals used for the purpose of Mohs' hardness testing is shown in Table XIII.III along with some convenient substitutes.

TABLE XIII—III. MOHS' SCALE
(substitutes shown in Italics)

Hardness No.	*Material*
1	Foliated talc
2	Rock salt or gypsum, or *finger nail*
3	Transparent calcite or *copper wire*
4	Fluorspar, *scratches copper wire*
4–5	*Ductile iron; window glass*
5	Transparent apatite
5–6	*Blade of good pocket-knife*
6	Orthoclase felspar
6·5	*File*
7	Transparent quartz
7–8	*Will scratch a knife*
8	Transparent topaz
9	Sapphire or corundum
10	Diamond

B. *Brinell Hardness Testing*. In this method, an indentation is made in a polished specimen under test by applying a static load to a hardened steel ball of diameter D placed on the surface. The load is applied gradually and maintained at full pressure for 15 seconds; the indentation is measured in terms of its depth (y) below the original surface and of its mean diameter (d); both measurements involve using a micrometer microscope. A schematic representation is shown in Fig. XIII.16.

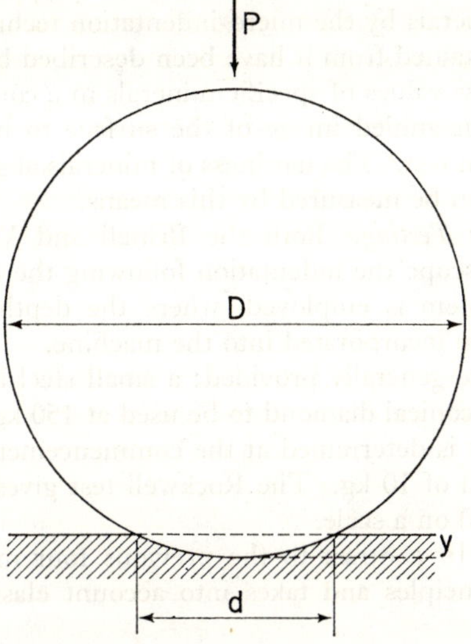

FIG. XIII.16. THE CALCULATION OF THE BRINELL HARDNESS NUMBER

The Brinell Hardness Number (H) is defined as the applied load in kilogrammes divided by the spherical area of the indentation in square millimetres.

The area of indentation is given by

$$A = \pi D y = \tfrac{1}{2}\pi D(D - \sqrt{D^2 - d^2})$$

so $$H(BHN) = \frac{P}{A} = \frac{P}{\tfrac{1}{2}\pi D(D - \sqrt{D^2 - d^2})} \quad . \quad . \quad . \quad . \quad 33$$

Full details of the method, the apparatus and recommended ball sizes and loads for different materials have been given in B.S. 240 Part 1, 1937.

C. *Vickers Hardness Testing*. The Brinell method is useful and widely used, but very hard materials may themselves indent the steel ball and thus invalidate the results. In the Vickers method the steel ball is replaced by a square-based *pyramidal diamond*, the faces of which make an angle of 136° with each other.

The test is similar in most respects to that of Brinell but the area is calculated from the measured diagonal (d) of the square which is the usual shape of the indentation when viewed under a microscope

so $$H(VHN) = \frac{P}{A} = \frac{2P \sin \tfrac{1}{2}\theta}{d^2}$$

If $\theta = 136°$ this simplifies to

$$H(VHN) = 1{\cdot}854 \frac{P}{d^2} \quad . \quad . \quad . \quad . \quad 34$$

The diamond pyramid method has been widely applied to the measurement of the hardness of minerals by the micro-indentation technique. The method and some of the results obtained from it have been described by Young and Millman.[1] By its use, the hardness values of specific minerals in a composite body can be determined, because a magnified image of the surface to be tested can be viewed prior to the indentation test. The hardness of minerals aligned in different crystallographic directions can be measured by this means.

D. *Rockwell Hardness Testing*. Both the Brinell and Vickers methods involve measuring by a microscope the indentation following the test but in the Rockwell method a simpler system is employed, where the depth of indentation is read directly on a dial gauge incorporated into the machine.

Two indenters are generally provided; a small steel ball operating at 100 kg. for softer materials; a conical diamond to be used at 150 kg. for harder substances. A reference base point is determined at the commencement and conclusion of the test by applying a load of 10 kg. The Rockwell test gives results in terms of the deflection (in divisions) on a scale.

E. *Shore Scleroscope*. In contrast to the previous four methods, the scleroscope relies on dynamic principles and takes into account elasticity factors as well as absolute hardness.

A cylindrical test piece with an accurately polished upper surface is placed at the lower end of a vertical glass tube of about 1 in. diameter. A small hammer or plunger with a diamond striking point and weighing about $\frac{1}{12}$ ounce slides freely inside the tube and is allowed to fall on to the specimen from a height of 10 inches. The height of the rebound observed on a calibrated scale is a measure of the Scleroscope Hardness Number.

The apparatus is simple to operate and depends on the principle that the energy required for the rebound is reduced by the work done in producing the indentation. Harder materials produce the greatest rebound which has a maximum of 140 scale divisions.

The various methods of measuring hardness find favour in different industries. The Brinell method is preferred in metallurgical practice largely because a simple relation can be derived between the number (BHN) and the tensile strength (T_o) of the metal

$$T_o = k \times (\text{BHN})$$

where k is reasonably constant and has a value of 515 when the Brinell value is less than 175 and 496 when it is above. Tensile strength is measured in lbf./in.^{2}.

The Vickers diamond pyramid method is widely used for metals, minerals and hard ceramic materials, particularly in conjunction with the micro-indentation refinement described above.

The Rockwell methods are mostly confined to metals but they have been applied to mineral systems—particularly single crystals.

The scleroscope is one of the best techniques for measuring the hardness of minerals and ceramic materials. The coefficient is easy to measure in materials with a Mohs' Hardness greater than 7 and the results given can be directly related to the crushing and grinding characteristics of mineral bodies.

[1] Young, B. B. and Millman, A. P., *Trans. Inst. Min. Metal.*, 73, 437, 1963–4.

It is difficult and hazardous to compare the hardness values given by the different methods, because so many principles are involved. Nevertheless, a useful guide can be compiled as in Table XIII.IV.

These values were obtained in different laboratories at the University of Leeds, but there was an inter-relationship in that common specimens were used in comparing one technique with another. The Brinell, Rockwell and Vickers diamond pyramid methods were in use on metal specimens, so that a direct comparison was feasible up to about 1200 VHN values. The scleroscope and micro-indentation techniques were compared on mineral and rock specimens of all types varying in hardness on the Mohs' scale from 4–9.

It is on this basis, that the comparison values in Table XIII.IV must be regarded, but in some cases, figures have been averaged and others have been extrapolated and may not be exact.

TABLE XIII—IV. A COMPARISON OF HARDNESS VALUES

Vickers	*Brinell*	*Rockwell C*	*Rockwell B*	*Scleroscope*	*Mohs' Scale*
+4000	—	—	—	140	10
2600	—	—	—	133	9
1600	940	82	—	126	8
1400	880	78	—	119	7
1200	810	73	—	110	
1000	740	68	—	100	
800	630	62	—	87	6
600	530	53	—	73	
400	390	41	114	55	5
200	200	—	93	30	4
100	100	—	58	—	3
50	50	—	27	—	2
25	25	—	—	—	
15	15	—	—	—	1

The values of the hardness of various minerals given on the Mohs' Scale are included in the Appendix, but the above Table enables a Vickers and a Scleroscope value to be compared with reasonable accuracy.

F. *Penetrometers* are similar to the indentation units described in previous sections but they are designed principally to test soft materials, such as clays prepared for forming processes. A cylindrical plunger is forced into the compacted clay against a spring-loaded upper container; the harder the clay, the greater the stress required for penetration, hence the contact cylinder does more work against the spring. This is indicated by a calibrated gauge which provides an excellent relative indication of different consistencies.

Abrasion Tests

A. *The Rattler Tests* are chiefly used for testing bricks for use in floors and roads in order to determine (*a*) whether the bricks are tough enough to resist the forces of impact and abrasion to which they are subjected when in use, and (*b*) whether the material is uniform in quality. The toughness is determined by the average loss of weight which the pieces to be tested undergo as the result of chipping or breaking, and the uniformity by the variation in the weight of the individual bricks or other articles after the completion of the test.

The American Society for Testing Materials has specified conditions under which the Rattler Test should be carried out (C7–52).

Briefly, a representative number of bricks or other specimens are rolled in a large cylindrical vessel along with a fixed number of cast iron balls. At the conclusion of a specified time, the charge is removed, debris sieved and weighed to give the total abrasion and each test-piece weighed separately to indicate the uniformity of quality.

For paving bricks, the total loss in weight under standard conditions should not exceed 26 per cent and no individual brick should lose more than 30 per cent of its weight.

B. *Abradability Tests.* Numerous methods have been described for assessing the degree of abrasion of ceramic products by measuring the total weight loss when a known area of sample is subjected to a standard grinding procedure. A graded gritstone, emery powder or diamond dust is used as the abrasive medium which is carried on a revolving plate whilst the specimen is held at a known pressure in contact with it. After a predetermined test period, the loss in weight or change in surface character is calculated.[1]

No matter how carefully such experiments are carried out, there are too many uncontrollable variables involved for accurate measurement.

A rather better test is to measure the rate of penetration of a standard grinding wheel or diamond cylindrical cutter into a block of the material or to measure the amount eroded by a standard surface grinder wheel.[2] It is never easy to ensure constant pressure or to take account of variable angles of cut, but nevertheless as a rough guide to abradability such tests are quite satisfactory.

C. *Abradability Index Tests.* More accurate methods for determining abradability values as a comparative index involve measuring the amount eroded after subjecting a known area of specimen to a blast of abrasive particles which are controlled in size and velocity of impact.

Aldred, Elliott and Cowling[3] have developed an apparatus to measure the Abradability Index which is very reproducible and is capable of general application. A regulated air blast projects a stream of graded fused alumina particles at the sample arranged so that it is at a fixed angle to the stream. A pre-determined weight of fused alumina particles (14 lb.) is used in each test and the Index is calculated by measuring the amount of specimen abraded and dividing this by its bulk density. The method has been adopted as a standard test by several laboratories and is known as the Morgan-Marshall Abrasion Test. The Abradability Index of some materials is given in Table XIII.XVII (p. 902).

Storey and Mackenzie[4] have suggested that there is a relationship between the Abradability Index and the Modulus of Elasticity of refractory materials.

[1] A.S.T.M. (for porcelain enamels) C448–64.
[2] A.S.T.M. (for tiles) C501–62T.
[3] Aldred, F. H., Elliott, A. and Cowling, K., *Trans. Brit. Ceram. Soc.*, 54, 239, 1955.
[4] Storey, C. and Mackenzie, J., *Trans., Brit. Ceram. Soc.*, 56, 14, 1957.

FACTORS AFFECTING THE STRENGTH OF CERAMIC MATERIALS

MANY factors affect the strength of ceramic materials, the most important of which may be grouped under the following heads:

(*a*) The chemical or mineralogical composition of the material.
(*b*) The physical properties of the material.
(*c*) The mode of preparation of the material.
(*d*) The mode of manufacture of the article.
(*e*) The conditions of drying.
(*f*) The conditions of burning.
(*g*) The temperature at which the article or material is used, or at which its strength is determined.
(*h*) Other conditions to which the article or material is or has been subjected, including weathering, sudden changes of temperature, prolonged heating, etc.

The strength of raw clay will depend chiefly on factors (*a*), (*b*), and (*h*), and chiefly on the second; the strength of a dried clay will depend on factors (*a*) to (*e*); the strength of freshly-burned products on (*a*) to (*g*); whilst the strength of articles which have been in use for some time may be further complicated.

In order to ascertain the effect of these various factors upon the strength of ceramic materials they should first be studied separately, and afterwards in combination.

The **Chemical** or **Mineralogical Composition** frequently has a very important effect on the strength of ceramic articles, but care must be taken not to exaggerate its influence.

The mineralogical nature of the raw clay may have a pronounced influence on its physical features and in this sense this property is all-important. Clay minerals, such as montmorillonites, have remarkable cohesion and bonding power and are thus capable of imparting great strength, whilst non-plastic components, such as quartz, confer hardly any strength whatsoever.

The colloidal properties of clays are largely influenced by the electric charges which develop on minerals of small particle size and in this sense also, the mineralogical and chemical properties of raw clays make a major contribution to strength in a green and dry state. These properties and their influence have been described in Chapter VIII.

When clays and other ceramic materials are fired, chemical reactions take place between mineral components. The liquid which forms surrounds more refractory grains and on cooling, the resultant glassy phase provides a continuous bond. Crystal growth may also provide a mass of interlocking particles of high strength. These characteristics also depend on the chemical and mineralogical nature of the material.

The **physical properties** of ceramic materials have a very important influence on their strength. The chief physical properties to be considered are (*a*) the size and shape of the article; (*b*) the texture of the material; (*c*) the porosity; (*d*) the coefficients of expansion or contraction; and (*e*) the adhesive power of the bond.

Size and Shape. The crushing strength is to some extent dependent on the size of an article or mass, and usually the larger an article the greater will be the total power required to crush it, though it need not necessarily have a high crushing strength, as the latter is expressed in terms of unit area (lbf./in.^{2}). The shape has an important effect on its strength, for if it is designed badly it may be unduly weak in certain parts and so bend or crack when subjected to a great stress. For instance, sharp corners and sudden changes in the thickness of an article are sources of weakness, and, where possible, these should be avoided. The apparently abnormally great strength of hollow columns as compared with solid ones is well known and it is worthy of note that perforated bricks are stronger in *proportion to their area* than solid ones, as shown in Table XIII.V, which compares bricks made from different clays.

TABLE XIII—V. PERFORATED *v.* SOLID BRICKS

Crushing Strength, lbf./in.^{2}		*No. of Holes*	*Diameter of Holes in.*
Solid Bricks	*Perforated Bricks*		
3,200	6,400	14	0·6
2,500	3,100	14	0·5
7,200	9,700	3	1·0
9,100	12,200	10	0·7
4,200	7,800	3	0·8
5,200	6,300	3	0·5

Texture is concerned with (*a*) the shapes and sizes of the individual particles, and (*b*) the arrangement and size of the pores or interstices between the grains. In ceramic materials, the maximum strength is obtained by the use of irregular, angular grains of numerous sizes, which interlock freely; rounded grains produce a weak mass as they cannot interlock properly. Thus, bricks made of powdered calcined flint or of sea-sand are usually weak, because the grains are rounded and, consequently, do not hold well together. The strength of a ceramic mass, both in the dried and fired state, usually increases with the fineness of the grains, but an excessive proportion of fine grains is undesirable.

One of the most important factors which influences the green and dry strength of raw clays is the specific surface area or surface area per unit weight of material. This aspect has been studied by T. E. Currie[1] who found that the dry strength of a wide variety of kaolin-type clays was related to the available surface area. Some results are shown in Table XIII.VI.

The relationship between dry strength and surface area is approximately logarithmic and of the form:

$$\log S = aA + b$$

where S is the cross-breaking strength of the dried material, A is the surface area per gram of clay and *a* and b are constants.

The presence of small particles facilitates reactions on firing and is the probable reason for the increase in strength when such material is present. With too many

[1] Currie, T. E., B.Sc. thesis, Leeds Univ., 1951.

TABLE XIII—VI. THE RELATION BETWEEN THE TRANSVERSE STRENGTH AND THE SURFACE AREA OF SOME KAOLIN-TYPE CLAYS

Clay	*Transverse Strength* (*gm./sq. cm.*)	*Specific Surface Area* (*sq. mt./g.*)
Bond clay	1140	11·64
Devon ballclay	805	9·20
Dorset ballclay	660	7·15
Scottish fireclay	480	1·81
Stourbridge fireclay	440	1·78
Yorkshire fireclay	650	2·57
Durham fireclay	385	1·17
Shropshire fireclay	435	1·67

fine particles, excessive shrinkage may occur and then warping, distortion and loss of strength will certainly result.

The addition of sand, grog, or other material to reduce the shrinkage and facilitate drying, lessens the hardness of the fired articles unless the heating is sufficient to cause partial vitrification.

In addition to angular grains, some ceramic products (particularly porcelains and some refractory wares) contain a felted mass of crystals of mullite. This is probably the reason why porcelains rich in mullite are superior to others in mechanical strength and in resistance to thermal shock. The size of the grains of grog in fireclay mixtures has an important influence on the resistance of the fired material to transverse loads at high temperatures. Large ceramic slabs or other articles which are required to carry heavy loads should be made of comparatively coarse material. Various investigators are agreed that about 50 per cent of grog sizes corresponding to 4–20-mesh should be used and all fine materials should previously have been removed from it.

The strongest unfired mass consists of a suitably graded mixture of angular grains of various sizes selected so as to produce as compact a mass as possible and provided with a sufficient amount of binding material to cover each particle of aggregate and to unite them together. The production of graded aggregates of this kind is described on p. 418. The proper interlocking of the particles of aggregate is of very great importance, and the low transverse strength of many fireclay and other slabs is due to their being made of badly-graded mixtures. Similarly, the strongest burned material consists of a mass of interlocking crystals formed *in situ* and united by a glassy cement. Such a mass is stronger and more compact than can be obtained by any artificial means of assembling the particles.

It should be observed that the texture of a mass is not always uniform in every direction on account of the method used in shaping it. Thus, if a tile is made by applying pressure to its upper and lower surfaces, its structure when laid flat will not be quite the same as that when it is on edge. Hence, in all machine-made products, and to some extent with hand-made ones, the structure varies in different directions, and this should be taken into consideration when investigating the strength of an article.

Porosity. The size and arrangement of the pores in a ceramic mass affect its strength, inasmuch as the larger and more numerous are the pores, the thinner must be the enclosing 'walls' of solid material. Hence, a highly porous material—if the pores are large—must be weak. When the pores are extremely small, their effect on the strength of the material is less noticeable.

In many fired ceramic materials, the tensile and crushing strengths are roughly inversely proportional to the porosity. Thus, if certain bricks having a porosity of 6–12 per cent have a crushing strength of 7000–15,000 lbf./in.^{2}, other bricks of a similar character, but with a porosity of 14–25 per cent, will probably have little more than half the strength of the former ones.

In ceramic materials, it appears that great strength is incompatible with high porosity, though there is only a very indefinite relation between the porosity and strength of such materials.

In this sense the crushing strength of a ceramic material is definitely related to its physical rather than to its chemical properties. In other words it conforms to the general Mariotte-Boyle law that

$$pv = c$$

where p and v are suitably selected variables, and c a constant. If p (the pressure corresponding to the crushing strength) is taken as one variable, v may be a variable dependent on any other suitable property, such as density, porosity, or volume, and then

$$p = \frac{c}{v}$$

Griffith[1] has plotted the crushing strengths against the porosity (water absorption) of a very large number of ceramic articles and has found that the characteristic curve is a smooth one accurately conforming to the above expression, though he prefers to split the denominator into two parts making:

$$p = \frac{cs}{A + A_o}(1 \pm e)$$

where p is the crushing strength in lbf./in.^{2}, c the efficiency factor, s the parameter or factor of dimensional stress, A the porosity (expressed as weight of water absorbed), A_o another parameter which includes sealed pore-space, and e the probable error for a single datum. The working equivalent based on the tests is

$$p = \frac{80{,}000}{A + 4}(1 \pm 0{\cdot}25)$$

This is equivalent to saying that the strength of a ceramic mass is determined solely by its density, and if a brick were wholly non-porous, so that $A + A_o = 4$, the crushing strength would be 20,000 lbf./in.^{2}, or 1,285 tons f per sq. ft.

Hasselman[2] has evolved empirical relationships relating the porosity (P) of

[1] Griffith, J. H., *J. Amer. Ceram. Soc.*, 14, 325, 1931.
[2] Hasselman, D. P. H., *J. Amer. Ceram. Soc.*, 46, 564, 1963.

various ceramic bodies to their tensile strength (S_T) and Young's Modulus (E).

His general relationships are

$$S_T = S_0 (1 - A_S P)$$
$$E = E_0 (1 - A_E P)$$

where A_S and A_E are constants and S_0 and E_0 are the strength and elasticity values at zero porosity.

The values quoted for sintered alumina are

$$A_S = 6{\cdot}0 - 6{\cdot}8;\ A_E = 4{\cdot}3 - 4{\cdot}4;\ E_0 = 59{\cdot}5 \times 10^6 \text{ p.s.i.}$$

Spriggs[1] investigated a range of polycrystalline refractories and proposed

$$E = E_0 e^{-bP}$$

to describe the relationship of the elastic moduli to the porosity P expressed as a volume fraction; *b* is a constant. The expansion of this expression and neglecting all multiple power terms would conform with the equation of Hasselman.

Spriggs' equation is more general and is a better relationship for materials of high porosity.

Knudsen[2] has also discussed similar relationships as also have Kingery and Coble.[3]

Power of the Bond. In ceramic materials five types of bond occur: (*a*) bonds of a plastic nature, such as clays; (*b*) bonds of an adhesive nature, such as glue, dextrin, molasses, etc.; (*c*) bonds of a hydraulic nature, such as mortar and Portland cement; (*d*) glassy bonds such as occur in vitrified ware; and (*e*) crystalline bonds where the body is held together by a felted mass of intergrowing and interlocking crystals.

Each of these bonds increases the strength of the mass as a whole, though to very different extents and in very different ways (see also Chapter VII).

A *plastic agent*, such as clay, possesses a peculiar binding power (p. 890), whereby it is able to extend its plastic properties to limited proportions of non-plastic materials which may be mixed with it, and when the mixture is dried and afterwards burned, the mass becomes stronger than before, although it has lost its plasticity.

Shales and indurated clays produce masses which are weaker in the raw and dried state, but develop great strength on firing, so that the strength finally attained may not be greatly different from that reached by more plastic clays and non-plastic material. The chief difference in this respect is that whilst plastic clays will develop great strength when heated to a relatively low temperature (900°–1000° C.), lean clays must be heated to a higher temperature (1200°–1600° C.), so that they undergo considerable vitrification in order to attain the same strength.

An *adhesive*, such as glue, tar, viscous mucilage, dextrin, flour or starch paste, heavy mineral oils, cellulose or molasses, forms a coating around each of the particles and so unites them together. The resultant mass will be moderately strong when dry, but as such adhesives are destroyed by heat they can only be used to a limited

[1] Spriggs, R. M., *J. Amer. Ceram. Soc.*, 44, 628, 1961.
[2] Knudsen, F. F., *J. Amer. Ceram. Soc.*, 45, 94, 1962.
[3] Kingery, W. D. and Coble, R. L., 'Review of Effect of Microstructure on Mechanical Behaviour of Polycrystalline Ceramics', *Nat. Bur. Std.*, Monograph 59, 113, 1963.

extent in connection with ceramic materials. When non-plastic material is to be made into bricks an adhesive is often added to confer strength prior to firing. After such treatment, the dried mass will be strong enough to handle and stack and although there will be a rapid reduction in tenacity when the gum begins to burn away, provided that the setting has been carefully carried out there should be no collapse.

Cellulose sulphite lye and molasses are common adhesives in refractory practice, the former being of particular value in silica and basic production. Although adhesive bonds are burnt away at moderate temperatures, they leave no residue and, for this reason, are preferred to clay bonds which alter the chemical composition of the product.

A *hydraulic bond*, such as Portland cement, lime, etc., behaves like an adhesive, so far as the dried materials are concerned. On heating the mixture, the hydraulic bond is usually converted into a glassy bond.

A *glassy or vitrified bond* consists of a glass- or slag-like mass of molten material which, when the ceramic material containing it is at a sufficiently high temperature, melts and flows into the interstices between the solid particles. At that stage the mass as a whole is weak, but when it has cooled and the bond is solidified, an extremely strong bond is obtained. A glassy bond can seldom be used when the material is in the moist or dried state—unless water-glass (a sodium silicate) is included in this type of bond—as it becomes adherent only when fused and afterwards allowed to solidify.

In most ceramic materials, the glassy bond is produced when the articles are in the kiln and, consequently, such materials attain their maximum strength only after they have been 'burned' or 'fired'.

A *crystalline bond* is, as the name implies, one in which the body is essentially composed of crystals which grow together; the best analogy is a box of pins, although a better representation would be given if each join or point of contact between the pins were welded.

The best conditions for crystalline bonding are when needle-shaped crystals are produced in large quantities at the temperature of firing. Thus the development of mullite in alumino-silicate mixtures may give rise to a felted mass of acicular crystals.

In general a crystalline-bonded material is not as strong in the cold as one which is essentially glassy, but in the former the strength is maintained virtually unchanged until the crystals themselves melt. Thus, mullite brick, free from any glass, will suffer little change in shape under a load up to about 1750° C.

The **mode of preparation** of ceramic materials often has a very important influence on the strength of articles made from them.

The *grinding* of the materials determines the sizes and shapes of the various particles and so directly affects the strength of the mass (p. 872).

The *amount of water* present or added to the material largely influences the strength of the mass, both directly in the damp material and indirectly in the dried and fired product.

When less water is present in the ground raw material than will form a thin film around each particle, the addition of a further quantity of water will increase the strength of the material. When each particle is completely surrounded by a film of water of the required thickness, the addition of further water will effect a reduction in the strength of the dried and fired articles due mainly to the increase in porosity which results (Table XIII.VII shows these effects with refractory clay blocks).

TABLE XIII—VII. EFFECT OF AMOUNT OF WATER ON STRENGTH

Water, per cent	*Compressive Strength after drying kg. per sq. cm.*	*Compressive Strength after burning at 1300° C., kg. per sq. cm.*
15	12	180
16	13	195
17	15	190
18	15	220
19	17	230
20	17	220
21	15	205

A deficiency of water is usually preferable to an excess in the preparation of ceramic materials.

The *proportion of added materials*, such as the bonding agent, etc., may seriously affect the strength of the prepared material.

Sometimes a very small proportion of an added material will have a very marked effect on the strength of the product. This is particularly noticeable in the case of *electrolytes*, which, when added to clay in the form of pastes and slips, considerably increase the strength of some dry and fired goods made from them.

The mechanism by which the strength of materials is increased by the addition of electrolytes is probably intimately linked to the deflocculating effects described in Chapter VIII. Colloidal particles are more readily dispersed in water when electrolytes are present and a better distribution of plastic material is thereby achieved.

The addition of 1 per cent of sodium hydroxide increased the dry strength of a stoneware clay by 79·22 per cent on the unground material, and nearly trebled it after wet grinding. A flint clay which showed a modulus of rupture of 305 lbf./in.^{2} after wet grinding with plain water had its modulus of rupture increased to 439 lbf./in.^{2} when the material was ground with a 1 per cent solution of sodium hydroxide.

The *method of mixing* ceramic materials has a very great influence on the strength of the products. If, as is often the case, the mixing is incomplete, the strength of the articles will vary in different parts of their structure. The process of 'pugging' is a far less effective method of mixing than 'tempering' the material in an edge-runner mill[1] with revolving pan (the so-called wet-pan mill), though pugging is much cheaper than tempering. Not only is a sufficient period of mixing very

[1] Care should be exercised during tempering to ensure that no reduction in particle-size occurs, otherwise the grading and other properties of the material may be drastically changed. This can be a serious disadvantage when using an edge-runner mill and for the best results, these machines should be operated with the rollers raised and an extra paddle or churn introduced.

necessary to secure a uniform mixture of the materials, but it will also be found that imperfectly mixed ceramic materials are a constant source of annoyance and loss.

Effect of Shaping on Strength. The method of shaping articles has an important influence on their strength. Hand-moulded articles are seldom as strong as machine-made ones, provided that the machine is of a suitable character for the material. It is not always easy to ensure the suitability of the machinery, as so much depends on apparently trivial properties in the material. Thus, in making articles by extrusion or expression, a defective die or mouthpiece puts an excessive strain on the column of clay as it issues from the opening and so the material may crack. The cracks may not be serious and sometimes are not visible when the articles are in the freshly-made state, but during drying and firing, cracks or other defects may develop and seriously weaken the goods. A laminated structure is sometimes imparted to a column of clay made by extrusion; it is usually due to lack of sufficient space between the bridge which carries the knives of the machine and the end of the mouthpiece, but it may also be due to the lamellar structure of the original clay (p. 54).

The various methods of shaping ceramic articles afford many other opportunities for reducing the strength of the material. Thus, if they are shaped by compression in a press which does not apply a uniform pressure over the whole surface, the resulting article will be subject to internal strains and will have a low strength, no matter how great a pressure may be applied to some parts of it.

When the pressure is applied uniformly, its amount affects the strength of the product. Under normal conditions, the greater the pressure applied the greater will be the strength of the article produced. Table XIII.VIII shows the effect on their tensile strength of the pressure applied in producing tiles by compression of an almost dry clay-dust.

An increase of moulding pressure has a pronounced effect on the strength of refractory bricks. Shapes made by dry press methods may be more than twice as strong as those which are wet moulded.

Although the appearance of bricks and tiles may be improved by double pressing, they may suffer a reduction in strength unless care is taken to ensure that a disturbance of the mass does not occur in the repress; if the box is too large a heterogeneous body may be formed.

Effect of Drying on Strength. The manner in which articles made of clay and allied materials are dried has a great influence on their strength when in the dry and also in the fired state. To avoid rupture, all materials having plastic or kindred properties must be dried under conditions which will permit the water to be removed at a uniform rate throughout the mass and without the formation of an impervious skin or crust through which water from the interior of the mass cannot penetrate. The necessity of drying under suitable conditions (see Chapter IX) is particularly great with articles containing a large proportion of plastic clay, as these are specially liable to form a hard surface-skin which, at a later stage in the drying, is very liable to crack under the pressure of the water vapour which cannot escape through it. The partly-dried material is very weak and cannot withstand the strains produced in it by the pressure of water vapour in the interior and by the uneven contraction. Joints in the articles are usually very weak at this stage, and so need special care in drying, and also when drying articles which vary greatly in thickness in different parts. When articles are dried before being placed in the kiln it is often desirable not to remove the whole of the water, as this causes them to become very friable and difficult to handle without rubbing the edges and fine mouldings, and so causing irreparable damage. A very small proportion of water is sufficient to overcome this difficulty.

TABLE XIII—VIII. EFFECT OF PRESSURE ON STRENGTH

Pressure lbf./in.^{2}	*Tensile Strength lbf./in.*^{2}	*Pressure lbf./in.*^{2}	*Tensile Strength lbf./in.*^{2}
4000	1400	2000	850
3750 }		1750	850
3500 }	980	1500	960
3250 }		1250	960
3000	1150	1000	940
2750	—	750	810
2500	1060	500	790
2250	1060		

Materials consisting largely of plastic clay increase in strength and rigidity when dried slowly and, under favourable conditions, the more thorough the drying the greater is the crushing strength.

Effect of Burning on Strength. The conditions to which ceramic articles are subjected in the kiln form one of the most important factors in determining their final strength. If they have been heated at a sufficiently slow rate, so as not to crack them, they will be much stronger after firing than in the plastic or dry state, the final strength depending chiefly on (i) the nature and amount of the bonding material produced during the heating; (ii) the particles of aggregate united by the bond; and (iii) the better consolidation of the grains of material which results from the shrinkage of the material during the firing.

The *bond* affects the strength of the fired material after the latter has been allowed to cool, because the bond is, as its name implies, the agent which unites the other particles together. Consequently, the stronger the bond and the larger the proportion of it present, the stronger will be the articles. For this reason, vitrified masses—in which all interstices between the particles of aggregate are filled with the bond—are stronger than porous ones; the vitrified material not only contains a larger proportion of bond, which unites the other particles together more securely, but the bond itself also possesses great intrinsic strength. Clays containing at least 6 per cent of ferric oxide, or its equivalent, when fired in a reducing atmosphere, form a strong product, because the reduced iron oxide acts as a flux, combines with the clay and so produces a mobile fluid of fused matter which penetrates and fills the pores (see Chapter XI). When the same clays are fired in an oxidising atmosphere, the oxidised iron oxide does not fuse, and the small proportion of fused material formed does not fill the pores, so that a weaker and less vitrified article is produced. Porcelain, china, stoneware, and other articles of a vitrified nature are all much stronger than porous materials.

Refractory materials, in general, contain a much smaller proportion of glass than a vitrified ceramic product; consequently they are not as strong when tested in the cold, for their porosity is comparatively high. At elevated temperatures, however, refractory articles will retain their strength and suffer little deformation, because there is insufficient glass to liquefy. Glassy ceramic materials, on the other hand, melt and lose strength at much lower temperatures.

The *temperature attained in the firing* affects the strength of the finished and cold articles, because it determines, to a large extent, the amount of fusible matter or bond produced, and this, in turn, controls the strength of the mass.

Grimshaw and Sutton have measured the effect of the firing temperature on a series of Pennsylvanian shale clays used for brick-making. The transverse strength increased in remarkable fashion with the development of a glassy phase as shown in Table XIII.IX.

TABLE XIII—IX. EFFECT OF FIRING TEMPERATURE ON THE COLD TRANSVERSE STRENGTH OF PENNSYLVANIAN SHALE CLAYS

Sample No.	*Modulus of Rupture (lbf./in.²)*									
	Room Temp.	110° C.	300° C.	500° C.	600° C.	700° C.	800° C.	900° C.	1000° C.	1100°C.
1	87·2	127·4	90·4	108·3	91·1	108·0	147·7	935·2	3,279	6,540
2	73·3	115·0	64·4	72·5	69·6	74·5	89·9	764·3	2,759	8,450
3	45·0	87·0	57·7	49·3	47·2	n.d.	58·8	332·5	1,469	4,708
4	91·6	170·9	106·1	109·2	119·7	128·9	170·2	1,400	6,379	bloat
5	102·9	146·2	100·1	98·6	102·8	110·9	123·4	1,212	2,533	6,164

The strength of clays which have been heated to points within the range 110° C. to 325° C. is practically constant, but then begins to change as the temperature of firing is increased. At temperatures of between 400°–600° C. most clay minerals undergo decomposition and in this region a pronounced reduction in tenacity may occur, because of the loss of strength in the bonding material.

Above 1000° C. the amount of liquid formed increases considerably. On cooling, this liquid solidifies mainly to a glass which acts as a cement serving to bind the mass together, thereby conferring great strength on the body when tested *at room temperatures.*

The effect of liquid formation on the strength of clay products *at high temperatures* is shown by heating test pieces under the influence of a tensile stress to elevated temperatures. A gradual increase in length is observed until at a certain point the extension increases rapidly and eventually the test-piece breaks. A typical temperature-extension curve of fireclays is shown in Fig. XIII.17 in which the point A corresponds to a temperature of about 1500° C. If the tensile stress is not too large (under 550 gm. per sq. cm.) the resultant curve shows a slight contraction immediately before the rupture of the test-piece.

Das and Roberts[1] have studied the variations which occur in the rigidity of

[1] Das, S. S. and Roberts, A. L., *Trans. Brit. Ceram. Soc.*, 48, 215, 1949.

silica refractories when heated up to 1620° C. and subjected to torsional strain. A pronounced decrease in modulus was recorded in the region of the inversional changes of tridymite and cristobalite but the value increased rapidly above this temperature up to about 1200° C. In certain cases, the rigidity attained a maximum which was five times greater than that recorded at the inversion temperatures. At still higher temperatures, the rigidity decreased slowly, but even at 1400° C., it tended to be greater than that at room temperature. No plastic flow was detected below 1400° C.

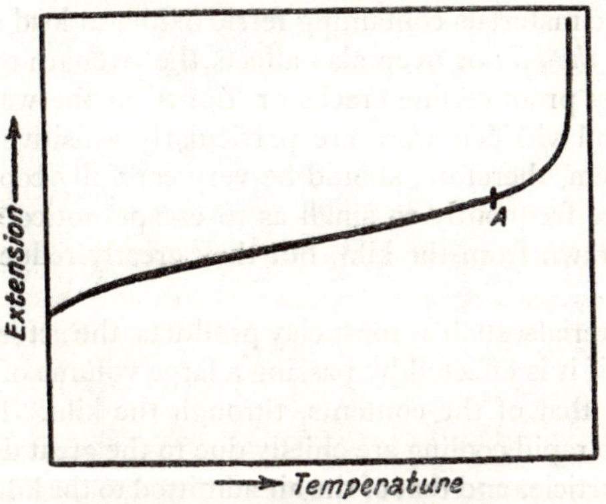

FIG. XIII.17. TYPICAL TEMPERATURE-EXTENSION CURVE FOR FIRECLAYS

Wiechula and Roberts[1] carried out similar tests on a range of alumino-silicate refractories. They showed that where the material tested was rich in crystalline silica a marked increase in rigidity was recorded up to temperatures of about 700° C. When mullite or corundum was the dominant crystalline phase the change in rigidity was less or non-existent. In all the samples tested, a transition occurred between 700°–800° C. from complete rigidity to incipient plasticity irrespective of the composition of the material. The rate of 'flow' was less with material of high alumina content and the authors considered that this was related to the increase in the amount and probably the grain size of mullite and corundum crystals.

In general, the higher the temperature of firing the greater is the cold crushing strength of the products because the amount of glass and the size and degree of interlocking of the crystals is increased. The *rate of firing* in the kiln may also affect the strength; if too rapid it may cause cracks or 'bloating' in the articles.

The *duration of the firing* also affects the strength of the articles when cold, because prolonged heating at a sufficient temperature produces an increase in the proportion of fusible bonding material. Hence it is not only necessary for the final temperature attained in the burning to be sufficiently high, but in many cases it is equally necessary to maintain the kiln at that temperature for a sufficient time to enable the maximum strength of the contents to be developed. This is especially the case with porcelain (also with silica and magnesia bricks) in which special reactions

[1] Wiechula, B. A. and Roberts, A. L., *Trans. Brit. Ceram. Soc.* 51, 173, 1952.

must occur before the maximum strength is obtained, but it is also important with many other ceramic materials.

The *atmosphere* in which the goods are fired may influence the strength; some, such as red bricks, firebricks, and most refractory materials are fired, as far as possible, in an oxidising atmosphere, as this ensures a material which will have the greatest strength at high temperatures, but if some of the same materials (which contain ferric oxide or its equivalent) were fired in a reducing atmosphere, their strength, when cold, would be greatly increased, because of the larger proportion of bond formed. A reducing atmosphere lowers the resistance of refractory materials, and of most ceramic materials containing ferric oxide, to load at high temperatures.

The *cooling of the kiln* or oven also affects the strength of the contents, as too rapid a cooling may produce fine cracks or 'dunts' in the ware. Articles made of silica, magnesia, and vitrified ware are particularly sensitive in this respect, and articles made of them, therefore, should be very carefully cooled after firing. The cracks produced are frequently so small as to escape notice when the articles are immediately withdrawn from the kiln, but they greatly reduce the strength of the ware.

With some materials, such as most clay products, the actual rate of cooling need not be detrimental if it is effected by passing a large volume of air, at a temperature only slightly below that of the contents, through the kiln. The cracks and other defects attributed to rapid cooling are chiefly due to the great difference between the temperature of the articles and that of the air admitted to the kiln. In other materials, however, the rate of cooling is important, especially where a critical range of cooling has to be passed through, as in the case of silica; the rate of cooling must then be regulated so as to effect all the necessary physical or chemical changes desired and yet prevent any undesirable ones. Such regulation can only be learned by constant inquiry. An investigation of the critical range of cooling of various porcelains and other ceramic materials would probably result in a decrease of the total time required for cooling the kilns and would also reduce the proportion of cracked and dunted ware.

Repeated changes of temperature and repeated heating, especially in the case of refractory materials, gradually reduces their strength to an extent depending on (*a*) the coefficient of expansion or contraction of the material, and (*b*) the chemical and physical changes which may take place during the repeated heating and cooling. Materials such as silica and magnesia, which are very sensitive to sudden changes in temperature, soon lose so much of their strength as to become useless unless carefully treated (see also Chapter XII).

The **temperature during use** has an important influence on the strength of ceramic materials, because their strength is usually much lower at high temperatures. They retain their original (cold) strength until a temperature is reached at which the bonding material begins to soften and yield under pressure, or at which, as previously mentioned (p. 734), the fluxes present in the material begin to soften or fuse and so decrease the strength of the mass at the high temperature to which it is exposed. In this way, fluxes play a double part, as they reduce the strength of a ceramic material at high temperatures and increase it at lower ones.

The strength of refractory materials when hot is often more important than the strength when cold, as the latter is usually ample, whilst the former is not merely lower, but is often an unknown factor. Moreover, as the strength when cold is usually much greater than the strength when heated, any material which is sufficiently strong at the highest temperature attained during its use, is almost certain to be perfectly satisfactory at ordinary temperatures.

As most ceramic materials are not quite pure, they do not usually possess a sharply-defined softening or yielding point when heated, and consequently their crushing strength diminishes gradually over a long range of temperature; this is still more noticeable if the material be subjected to a considerable pressure or load which is not intense enough to crush the cold material, but causes it to lose its shape rapidly when any softening of the bond or other fusible constituent occurs. If the material under examination consists of a single pure substance, such as pure silica, however, the strength remains fairly constant until the fusion-point is reached and loss of shape then occurs suddenly, especially if the material is under pressure.

The long range of temperature through which some materials lose strength progressively is due solely to the gradual production of fused or partially fused material. This is clearly shown by varying the pressure or tensional stress applied to the material, when it will be found that loss of shape occurs at a lower temperature when the pressure or tension is greater and *vice versa.* When only a very small amount of mobile materials has been formed a much greater stress is required to deform the mass and to cause the unfused particles to move through or in the fused portion. As the temperature increases, or the time of heating is prolonged, more and more fused and mobile material is formed and less stress is needed to move the remaining particles.

The fact that most ceramic materials have a much lower strength at high temperatures than when in the cold state, is very important in connection with their use in the construction of furnaces, retorts, crucibles, and other refractory articles.

When an essentially crystalline material is being used, a high strength is generally maintained up to a comparatively high temperature; on the other hand, glassy-bonded refractories begin to soften at low temperatures and progressively lose strength on further heating. An important property of refractories is their *refractoriness-under-load*, i.e. their ability to withstand high temperatures without undergoing deformation under an applied force.

The refractoriness-under-load of a refractory material is usually carried out on a cylindrical specimen of 2 in. diameter and 2½ in. long cut from a brick or other shape.[1] This is placed on alumina or silicon carbide blocks B, in a carbon granule resistance furnace F, and a load L applied to it by weights through a system of levers to an alumina thrust rod *R* (Fig. XIII.18). The specimen is heated at a fixed rate (10° C. per min. is recommended for fireclays) and the change in length measured on a suitable recording instrument. This rate of *rising temperature* may be maintained until the test piece fails, i.e. shears or suffers a contraction of 10 per cent or more, or a certain *temperature may be maintained* and the time required for the specimen to fail is recorded. In the former test it is usual to record the initial softening point, the point of rapid failure (corresponding to a shrinkage of 1 per cent per min.) and the temperature of complete failure. (See B.S. 1902, Part 1A, 1966.)

[1] Blocks 2½ in. × 1¾ in. × 1¾ in. are preferred by some workers.

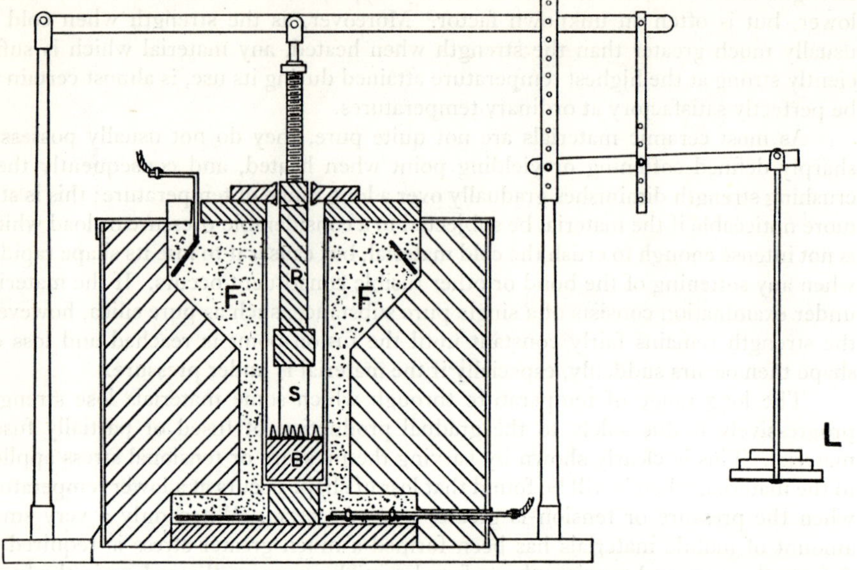

FIG. XIII.18. AN APPARATUS FOR MEASURING THE REFRACTORINESS-UNDER-LOAD

A most important consideration is to ensure that the specimen is heated uniformly and evenly throughout and that oxidising conditions prevail at all temperatures. Carbon granule resistance furnaces are much more serviceable than gas-fired ones, especially at the high temperatures which are encountered in testing high-grade fireclays and siliceous products; the measuring thermocouple should be in contact with the specimen.[1]

Lahr and Hardy[2] have described an ingenious modification to the apparatus in which the specimen or an inner lining to the furnace is not only rotated (a refinement introduced by Cross[3]) but oscillated in a vertical direction. They claim that the refractoriness-under-load of silica specimens thus measured is very reproducible and can be correlated with the amount of alumina present as impurity, provided that a slow rate of temperature rise (1° C. per min.) is employed.

The test is normally carried out under a load of 28 lbf./in.^{2} (2 kg. per sq. cm.) or 50 lbf./in.^{2}.

The American Society for Testing Materials specifies that the refractoriness-under-load should be measured on a full brick under an applied force of 25 lbf./in.^{2}; a slow heating rate is also recommended.

The refractoriness-under-load properties of some refractory materials under a load of 28 lbf./in.^{2} are shown in Fig. XIII.19 and Table XIII.X.

[1] Some workers prefer to use Seger cones to measure the temperature or duration of heating. These should be positioned near to the specimen (see Fig. XIII.18).
[2] Lahr, H. R. and Hardy, C. W., *Trans. Brit. Ceram. Soc.*, 56, 369, 1957
[3] Cross, A. B., *Trans. Brit. Ceram. Soc.*, 54, 461, 1955.

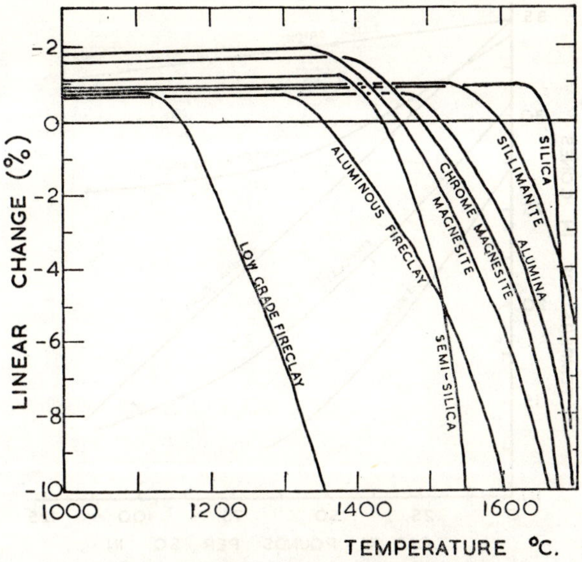

FIG. XIII.19. THE REFRACTORINESS-UNDER-LOAD OF VARIOUS REFRACTORY MATERIALS

TABLE XIII—X. REFRACTORINESS-UNDER-LOAD CHARACTERISTICS OF SOME REFRACTORY BRICKS

	Fireclay (low grade)	*Fireclay (42 per cent Al_2O_3)*	*Silica*	*Dolomite (semi-stable)*	*Magnesite*	*Chrome-magnesite*	*Synthetic Mullite*
Initial softening	1135° C.	1320° C.	1670° C.	1350° C.	1400° C.	1420° C.	1690° C.
Rapid softening	1260	1450	1690	1520	1540	1555	1760
Final failure	1350	1590	1710	1600	1620	1680	1780

The effect of different applied pressures on the failure of some materials is shown in Fig. XIII.20. Glassy-bonded materials, such as fireclays, show pronounced changes in the refractoriness-under-load properties with different applied loads, but silica bricks show little change.

The refractoriness-under-load properties of alumino-silicate refractories depend on their chemical and mineralogical composition. Such refractories yield when a sufficient quantity of liquid has formed which flows when the body is subjected to pressure. Not only is the amount of liquid important, but also its physical nature (which depends largely on its chemical composition) and also the rheological behaviour of the crystal and liquid system.

In general, refractory materials with a high alumina content have a high yield point under load, but the type and amount of fluxes (compounds of iron, calcium, magnesium, sodium and potassium) which may be present are important factors. Well-developed crystals or hard-fired grog can increase considerably the yield point

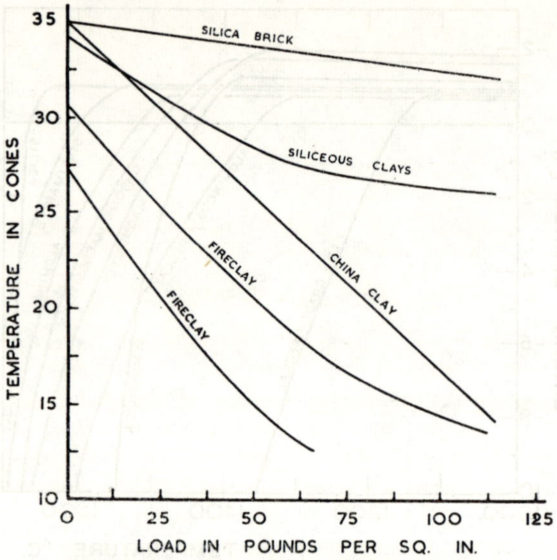

FIG. XIII.20. THE REFRACTORINESS OF SOME MATERIALS UNDER VARIOUS APPLIED PRESSURES

of an alumino-silicate refractory; hence an article prepared from kyanite or sillimanite containing large crystals of mullite is often superior to bodies containing a higher proportion of alumina but prepared from bauxitic clays.

Quartz behaves in a similar way so that the load-bearing properties of some siliceous fireclays at high temperatures are remarkable. Coarsely-grained quartz is unreactive in alumino-silicate bodies up to about 1350° C. and it has valuable stabilising properties to these temperatures. At high temperatures, however, quartz is rapidly attacked and although it may increase the initial yield point, the rate of subsidence above this point is extremely rapid. This effect is illustrated by the figures in Table XIII.XI which were obtained on alumino-silicate refractories when tested under a load of 28 lbf/in.^{2}.

Alumino-silicate refractories with a high proportion of crystalline material are less prone to differences in behaviour when tested under different applied pressures than are those with a high proportion of glassy bond. The initial yield point of silica, sillimanite and alumina refractories are little different when tested under a load of 28 or 50 lbf./in.^{2}, but with fireclays, particularly the non-siliceous variety, differences in yield point of up to 100° C. may be recorded between the different pressures.

It is common practice to test all alumino-silicate refractories under a load of 28 lbf./in.^{2}, but 50 lbf./in.^{2} is more favoured for silica.

The loss of strength at high temperatures may be due to:

(*a*) Chemical rearrangement of the constituents of the mass with the production of liquid.

(*b*) Physical breakdown due to melting of crystal components.

(*c*) Structural defects which develop during the heating.

TABLE XIII—XI. REFRACTORINESS-UNDER-LOAD PROPERTIES OF SOME ALUMINO-SILICATE ARTICLES

Type of Refractory:	*Silica brick*	*Semi-silica*	*Siliceous fireclay*	*Low alkali fireclay*	*High alkali fireclay*	*Sillimanite*	*Aluminous*	*Alumina*
Percentage of Alumina:	0·3	15	27	42	40	63	72	90
Initial softening	1690	1445	1400	1400	1350	1530	1490	1550
Rapid softening (5% subsidence)	1700	1490	1480	1500	1440	1610	1580	1620
Final failure (10% subsidence)	1720	1520	1510	1610	1525	1745	1695	1730

Effect of Weathering on Strength. Exposure to weather greatly decreases the strength of most ceramic articles, those composed of clay which has merely been dried being the most affected. Most building bricks and roofing tiles are very resistant to weathering, but this is not the case with firebricks which may lose from 1–21 per cent of their crushing strength after six months, and 11–28 per cent after twelve months' exposure. One brand of specially strong machine-made bricks was found to withstand six months' exposure to the weather practically without loss in strength. Some very porous hand-made fireclay bricks are greatly reduced in strength on exposure and may be made practically worthless after six months' exposure.

Silica bricks may lose as much as 39 per cent of their crushing strength when cold if they are exposed to the weather for twelve months, whilst some magnesite bricks decrease in crushing strength by 15 per cent after six months' exposure and 33 per cent after twelve months' exposure. The strength of magnesite bricks when heated under load is also affected; some bricks, after being weathered for twelve months, failing at a temperature of 40° C. lower than new bricks under a pressure of 25 lbf./in.^{2} Some magnesite bricks are more seriously affected by the weather than almost any other kind of refractory brick and much greater care is therefore required in their storage.

The reduction in strength of ceramic articles on exposure to weathering conditions is intimately related to their permeability or the ease with which water can penetrate. The slightly solvating action of rain water is sufficient to cause a loosening of the bonds within the mass thereby resulting in an overall loss of strength.

Effect of Frost on Crushing Strength. The crushing strength of articles is reduced by subjecting them to the action of frost, sometimes by as much as 30 per cent in a single cycle.

The effect of freezing does not appear to be related to the original strength of the brick, but to depend upon other factors such as texture, porosity, temperature and duration of burning, etc. Some idea of the probable effect of exposure to frost may be gained by exposing the goods or materials to a 'freezing test', such as is frequently applied to building materials. In one form of this test, the article or material to be tested is dried at 100° C., weighed, and boiled in water for thirty minutes. It is then allowed to cool and is placed in ice-cold water for one hour, after which it is placed in a refrigerator and kept at about −7° C. or 20° F. for twenty-four hours. It is then thawed and boiled again. This treatment is repeated twenty times, the loss in weight being noted each time and expressed as a percentage of the total dry weight of the article. It was at one time supposed that there is a relationship between the porosity of burned-clay goods and their resistance to freezing, but this is not the case.

A more rapid 'freezing test' consists in immersing the article for forty-eight hours in a 15 per cent solution of sodium sulphate in water at 21° C., withdrawing it, and then placing it in a dryer at 110° C. for about seven hours. This treatment is repeated as often as may be desired. The pressure of the crystals of sodium sulphate is supposed to be similar in action to that of ice.

The freezing test of the American Society for Testing Materials consists in saturating five bricks by placing them in cold water and then heating the water to a temperature of 200° F. in thirty minutes. The soaked samples are then placed in ice water for at least one hour, after which they are weighed and then frozen in a refrigerator with all surfaces exposed at a temperature of less than 15° F. for at least five hours. The samples are then removed and placed in water at a temperature of 150°–200° F. for one hour, the freezing and thawing being repeated twenty times. After this the bricks are re-weighed and the loss in weight, as well as any visible alterations, are noted.

In another form of freezing test, the bricks or other articles, after being frozen and thawed for a prescribed number of times, are dried, and their crushing strength is then determined and compared with that of the original bricks. The effects of the freezing are then expressed in terms of the percentage loss in crushing strength.

Notwithstanding the large amount of work done both in Great Britain and in the United States, no satisfactory freezing test has yet been devised. As saturated bricks undergo a permanent expansion during freezing a measurement of this expansion may prove a better guide to the resistance of the bricks to frost than the crude methods of inspection or measuring the crushing strength of different samples before and after freezing.

Freezing tests can often be of value in revealing machine and lamination faults in extruded clay columns. A suitable length cut from a column should be placed immediately in a refrigerator unit operating at 10–15° F.; after 12 hours of such treatment the piece will show cracks or other defects if the flow of clay has been irregular in any particular direction.

W. N. Thomas[1] has investigated the mechanism of frost action on building bricks and has deduced empirical relationships which link the probable influence of the total water absorption, the saturation coefficient, and the compressive strength on the resistance to frost. The saturation coefficient is defined as the ratio between the amount of water absorbed (C) after 24 hours' immersion in cold water to that absorbed (B) after boiling the brick for 5 hours in water and allowing it to cool whilst still immersed.

Bonnell and Butterworth[2] have suggested that if the total water absorption (B)

[1] Thomas, W. N., *Building Res. Stn. Tech. Paper*, No. 17 (H.M.S.O., London, 1938).

[2] Bonnell, D. G. R. and Butterworth, B., 'Clay Building Bricks of the United Kingdom', *Nat. Brick, Advis. Council*, Paper 5 (H.M.S.O., London, 1950).

is not greater than 7·0 per cent by weight, or if the saturation coefficient (C/B) is less than 0·60 or if the compressive strength of the fired brick is greater than 7,000 lbf./in.[2] then even severe exposure conditions will not appreciably reduce the durability of the brick. Even when none of these conditions is fulfilled the brick will still be resistant provided that B + 100C/B is not greater than 72. Should this relation be between 72 and 93, the durability of the bricks is doubtful. Bricks with a figure of greater than 93 would be most liable to break up under severe conditions although under normal conditions they might be quite satisfactory if they had been well fired and had only a low content of soluble salts.

Butterworth and Baldwin have retracted the above statement and shown that it has not the general application which was previously thought to be true. It errs mostly in that it is too severe and condemns many bricks which have proved to be acceptable.

An alternative method is to measure the change in dimensions resulting when fully saturated bricks are completely frozen at −3° C. and then thawed out. Bricks which are liable to frost attack undergo permanent expansion in this test and successive treatments cause further volume changes.[1]

The frost resistant characteristics of bricks and the strength changes which can occur in structures over several winters are worrying factors and a correct assessment is often tempered by commercial considerations.

The climate of the U.K. is more searching than that of most other areas in being wet and variable in temperature. The freezing/thawing cycle at full saturation of bricks occurs quite frequently in a normal year, especially in the northern counties. The author has seen so many examples of serious frost attack in bricks in exposed conditions that anything less than the Bonnell/Butterworth specifications should be viewed with great caution.

Effect of External Chemical Agents on the Strength

(*a*) *Deposited carbon.* Some fireclay bricks are exceptionally prone to lose strength when exposed to furnace gases, rich in carbon monoxide. The cause of failure is attributed to the deposition of carbon in the region of iron spots within the structure of the fired material.[2]

The carbon fills pores and voids in the vicinity of the iron spots and forces the bricks apart, and characteristic flaking may then occur. The reduction in strength takes place only in alumino-silicates rich in iron spots. In well-fired materials little trouble is encountered because the iron is mostly converted to a silicate which does not catalyse the formation of carbon.

(*b*) *Slags and Flue-dust.* When in contact with molten slags at high temperatures, most ceramic materials are subjected to considerable erosion. In addition, the slag may penetrate the pores of the material and cause a considerable reduction in strength, especially at high temperatures. The slag may force apart the grains of the brick or it may chemically combine with the material of the brick and form low-melting compounds or liquid. Dusts which are carried in furnace gases may behave in a similar fashion.

[1] Watson, A., *Trans. Brit. Ceram. Soc.*, 63, 681, 1964.
[2] Davis, W. R. and Rigby, G. R., *Trans. Brit. Ceram. Soc.*, 53, 511, 1954.

THE STRENGTH CHARACTERISTICS OF CERAMIC MATERIALS

IN the previous sections of this chapter, various methods have been described for measuring different aspects of the strengths of ceramic materials. Numerous parameters can be derived which may define the behaviour of a material under certain specified conditions. The actual values of strength and related properties may conveniently be considered under three headings, (*a*) failure tests, (*b*) elastic moduli, and (*c*) hardness values.

FAILURE TESTS

THE stress required to cause fracture in a ceramic material depends on the manner in which the test is carried out. Normally a test is devised which relates most closely to the conditions encountered in actual practice.

Strength of Raw Clay and Clay Pastes. The strength of raw clays and clay pastes, i.e. of clays in the plastic state, varies very greatly. Some lean clays will break under a tension of a few ounces, whilst other highly plastic clays such as bentonites may withstand a tension of several pounds per square inch. The duration of the application of the force also affects the result, as a small force acting throughout a long period will have an effect similar to that of a much larger force acting for a much shorter time.

The *green strength* (i.e. the wet strength) of a clay or other ceramic material is an important consideration, especially when the moulded shape has to be handled or transported prior to drying. Plastic clays are usually much stronger in the raw state than shales and indurated clays.

Strength of Dry Clays. The strength of ceramic materials in the dried condition is often important, for, obviously, the dried articles must be sufficiently strong to withstand the amount of handling to which they are likely to be subjected and also the weight of any articles placed above them when they are set in the kiln. Such articles as tiles must be particularly strong when in the dry state. Some pieces of pottery also require to be made of material of considerable strength when dry.

Like clays in the plastic state, dried clays vary greatly in tensile strength; in some, it is as low as a few pounds per square inch, whilst in others it may be as high as 400 lbf./in.^{2}. Brick clays in the dry state have usually a tensile strength of about 100 lbf./in.^{2}. Clays which have the greatest plasticity when in the form of a paste usually have the greatest strength when dry.

Some clays are of value principally for the strength they impart to other materials with which they are mixed before the body is moulded and dried. Examples of this type are the bond clays, which are used to impart strength to sand mouldings for metal casting purposes, and ball clays, which are used to strengthen whiteware bodies and at the same time produce a white, fired product. The strength of brick clays and fireclays is variable, but, in general, it is less than that of the bond or ball clays, but greater than that of the residual kaolins.

The main factor influencing the dry strength of clays is undoubtedly the fineness of particle-size or the available surface area per unit weight of material. Thus, montmorillonite-type clays, such as the bentonites, where the particle-size is extremely small, are far more adhesive than any other type of clay. Similarly, clays containing the mineral, *livesite*, which is inherently fine-grained, are capable of imparting greater strength than true kaolinites. Table XIII.XII shows the tensile strength of test pieces made of silver sand, bonded, in each case, with 5 per cent of clay of different type. Compressive strengths are 4–6 times higher.

The dry strength required will depend on the nature of the articles to be made; exceptionally thin products, such as tiles or pottery require to be strong when dry in order to avoid damaging in handling. As the cohesive power of most materials in the raw and dried states is low compared with that in the burned condition, they must be carefully handled so as to avoid damage.

Dried articles are usually very brittle and readily damaged if incorrectly handled. On the other hand, a green or wet clay body is able to accommodate sudden slight knocks or stresses more readily because of its more plastic nature. For this reason, it is frequently safer to handle articles in their wet state and load them into kilns or on to cars for drying and firing, rather than to set dry ware.

TABLE XIII—XII. TENSILE STRENGTH OF SAND BONDED WITH 5 PER CENT OF CLAY OF DIFFERENT TYPES

Type of Clay	*Tensile Strength*	*Type of Clay*	*Tensile Strength*
	lbf./in.^{2}		*lbf./in.*^{2}
Montmorillonite clay	823	Fireclay (Shropshire)	170
Bonding clay (Irish)	220	,, (Yorkshire)	50–92
(Scottish)	225	Brick clays	40–120
Ball clay (Devon)	175	Brick shales	35–91
(Dorset)	187	Sedimentary kaolins	50–105
		Residual kaolin (Cornwall)	30

Strength of Burned Clay Wares. The strength of articles, etc., made of burned clay varies according to the materials used and the mode of manufacture, very great variations sometimes occurring in articles made under apparently identical conditions. The need for careful control is, therefore, very evident.

The strength of *building bricks* varies through wide limits and depends entirely on the amount of glass which has developed under the conditions of firing, and on the final porosity.

From a purely structural load-bearing point-of-view, a brick does not have to be excessively strong in compression to withstand the normal loading of overlying brick-work in a typical house construction.

If in a normal construction of 9 in. × 4½ in. × 3 in. units each weighing 9 lb., bricks were laid flat, one above another, to a vertical height of 100 ft. (400 bricks), the maximum compressive pressure would not exceed 100 lbf./in.^{2}. (Actually it would be $(400\times 9) \div 40\frac{1}{2} = 89$ lbf./in.^{2}).

Greater stresses would develop where major floor and roof tie beams were let into the load-bearing brickwork but even in these cases the total compressive strength would not exceed a few hundred pounds force per sq. in.

However, much higher strengths are generally preferred to avoid excessive damage during transport and handling and also to ensure against failure in severe weather conditions. Low strength and high porosity or water absorption generally go hand-in-hand, so where the latter is undesirable, bricks of high strength are generally specified. In certain positions in normal building constructions, bricks may be subject to abnormally high compressive and tensile stresses, so that shapes of less than 500 lbf./in.² are rarely acceptable and strengths of +7000 lbf./in.² are often specified.

British Standard Specification 3921, 1965, lays down the following strength and water absorptions for different classes of bricks in accordance with CP 111.

TABLE XIII—XIII. STRENGTH AND ABSORPTION OF BRICKS

Designation	*Class*	*Minimum Average Compressive Strength (lbf./in.²)*	*Minimum Average Water Absorption %*
Engineering Brick	A	10,000	4·5
	B	7,000	7·0
Load Bearing Brick	15	15,000	No specific requirements
	10	10,000	
	7	7,000	
	5	5,000	
	4	4,000	
	3	3,000	
	2	2,000	
	1	1,000	
Bricks for damp-proof courses	D.P.C.	As required	4·5

In the same standard, facing and common bricks can be accepted above 750 lbf./in.² and blocks of ordinary quality may be deemed satisfactory if they will withstand a compressive strength exceeding 400 lbf./in.².

Butterworth[1] has compiled a comprehensive list of the strengths of bricks manufactured from different clay strata of the U.K. London Stocks, Flettons and Gault bricks are generally less strong than those made from Keuper Marl and Coal Measure Shales.

The American Society for Testing Materials classifies facing and building bricks into three groups depending on the degree of frost and weathering they will have to withstand (Code C216–50). Each must conform to certain minimum strength requirements when tested by standard methods. The grades and specifications are as follows:

Grade	*Exposure conditions*	*Minimum compressive strength (flatwise) (lbf./in.²)* — *Average of 5*	*Individual*
SW	Severe frost and weather	3,000	2,500
MW	Medium weathering	2,500	2,200
NW	Inside wall, no exposure	1,500	1,200

[1] Butterworth, B., 'Clay Building Bricks', *Nat. Build. Stud. Bull.*, London (H.M.S.O., 1948).

Roofing Tiles are usually tested by transverse methods because it is by such stresses that failure is most likely to occur in practice. B.S. 402: 1945 recommends mean values of 1900 and 1000 lbf./in.^{2} depending on the thickness.

The strength of porous clay wares, such as **porous pipes, tiles, terra-cotta,** etc., is generally about the same as that of building bricks. Articles containing a larger proportion of vitrified material are stronger, and highly vitrified ware, such as stoneware, has a strength approximating to that of blue or vitrified bricks.

Earthenware and **china** vary considerably in strength, the former being usually much weaker than the latter and the latter being usually rather weaker than other forms of porcelain.

Porcelain invariably has a high strength because of the glassy nature of the fired material, although different samples vary because of the raw materials used and the mode of manufacture. Compressive strengths of about 40,000 lbf./in.^{2} are quite common in ordinary felspathic porcelains, whilst the best high-tension electrical porcelains may withstand a load of 65,000 lbf./in.^{2}. The tensile strength which for some purposes is more important varies from 1500 to 13,000 lbf./in.^{2}.

Electrical porcelains made of steatite have compression strengths of 60,000–150,000 lbf./in.^{2}; those of zirconia 75,000–95,000 lbf./in.^{2}; rutile and titanate bodies 50,000–130,000 lbf./in.^{2}. Cordierite and porcelain bodies have much lower tensile strengths (of the order $2\text{–}6 \times 10^3$ lbf./in.^{2}) than zircon porcelain (13×10^3 lbf./in.^{2}), steatite ($8\text{–}12 \times 10^3$ lbf./in.^{2}), and titanates ($6\text{–}10 \times 10^3$ lbf./in.^{2}). Glazed porcelains are usually stronger than unglazed ones.

The crushing strength of **fireclay bricks** varies according to the temperature at which they have been burned and depends on the extent to which the particles are bonded together with fused matter. As a general rule, the higher the temperature at which such bricks are burned, the greater will be their strength when cold. Firebricks should be at least strong enough to withstand mechanical handling and to resist the weight of any superimposed structure.

The Standard Specification for Refractory Materials for the Gas and Coking Industries required cold crushing strengths of at least 1800 lbf./in.^{2} for E. and F grade bricks and 2000 lbf./in.^{2} for G and H grades, but in the new specification B.S. 1758: 1966, no values have been quoted.

The American Society for Testing Materials (A.S.T.M. C27–60) lays down minimum modulus of rupture values for fireclay which range from 500–1200 lbf./in.^{2} with the maximum requirement being reserved for those materials which have to resist slags.

Fireclay bricks and other alumino-silicate refractories should be selected so that there is little or no deformation or loss of strength at the temperatures of use, as when they are built into a setting and have to withstand the weight of the overlying structure.

The transverse strength of firebricks usually lies between 500 and 700 lbf./in.^{2}.

Salt-Glazed Pipes are tested by special techniques described on p. 862. The minimum values which pipes of different internal diameters should withstand are designated in B.S. 65: 1963 as in Table XIII.XIV.

Glass, being a uniform, non-porous body with a continuous structure, is stronger than a non-vitrified mass of packed particles; though it is a relatively weak material

TABLE XIII—XIV. CRUSHING STRENGTHS OF SALT-GLAZED PIPES

Nominal Internal Diameter (in.)	*lbs./foot of effective length*	
	Standard	*Extra Strength*
4	1350	1500
6	1350	1500
9	1350	1850
12	1450	2250

compared to metals, it is usually *brittle* and is extremely variable in strength. Turner and Holland,[1] for example, found that the modulus of rupture of several identical specimens cut from the same sample of sheet glass varied from 750 Kg/cm^2 to 1600 Kg/cm^2 and differences of a similar order have been reported by other investigators.

The mechanical strength of glass depends on many factors, some of which are:

(*a*) The composition and crystalline linkages of the material.

(*b*) The degree of perfection in the glass. Microcracks are a frequent source of low strength.

(*c*) The 'flow-orientation' in the sample. Glasses which have been extrusion cast or drawn may have different strengths when tested along several directions due to the differences in crystal development.

(*d*) The method of applying the breaking force. Glass is a super-cooled liquid and deforms progressively under an applied load.

(*e*) The temperature at which the test is carried out. As the temperature is raised the viscosity of glass falls and a gradual reduction in strength takes place.

Glass is to some extent elastic, that is, it can be deformed by an applied force but it readily regains its original shape when the force is removed. It is this property which is of importance in *glazes*, which frequently are subjected to strains arising from the expansion or contraction of the body. Like all brittle solids, glasses have a much greater compression than tensile strength. The usual range is: Compression strength 110×10^3 to 130×10^3 lbf./in.^{2} and tensile strength 7×10^3 to 13×10^3 lbf./in.^{2}.

The tensile strength of glasses depends greatly on the diameter of the test-piece, glass-fibres have tensile strengths from 15×10^3 to 900×10^3 lbf./in.^{2}. By *tempering* (i.e. heating well above the annealing temperature and then blowing cold air uniformly over both surfaces) the surface layers are in strong compression and the interior is in tension so that the total strength is greatly increased. Thus, a glass which has been well annealed and has a transverse breaking strength of 7,000 lbf./in.^{2} will probably have a transverse breaking strength of 25,000–30,000 lb.f/in.^{2} after being tempered.

The brittleness of glass is greatly reduced by *lamination*, i.e. causing two sheets to adhere by the insertion of polyvinyl butyral or other suitable, light-stable resin which, on the breaking of the glass by sudden impact, leads to small radiating cracks and avoids the 'flying' of the glass.

[1] Holland, A. J. and Turner, W. E. S., *J. Soc. Glass. Tech.*, 20, 72, 1936.

Quartz-glass or vitreous silica has a crushing strength of 200,000–300,000 lbf./in.^{2} and a tensile strength of 11,000–13,500 lbf./in.^{2}; these it retains up to about 1400° C. but softens rapidly at higher temperatures.

The strength of **glazes** has not been investigated to any great extent, yet it and the elastic modulus are important in connection with the avoidance of cracks and 'peeling'.

The elastic modulus of stoneware glazes is generally between 5700 and 6800 kg. per sq. cm.; that of lead glazes is lower than for leadless ones though the variation in the figures obtained by different observers makes it uncertain whether any real relationship exists between them.

The tensile strength of glazes is of more importance than the crushing strength. The tensile strength of stoneware glazes is about 5 to 8 kg. per sq. mm., though some glazes have much lower strengths.

Elastic Moduli

The values of the Elasticity and of the Rigidity moduli depend on two main factors: (*a*) the mineralogical composition; and (*b*) the texture, of which the former is the more important. The atomic structure of the minerals and the ease whereby movement can occur within the lattice on the application of stresses determine the modulus of elasticity. For the most part ceramic materials have a low elasticity compared with metals; glass and glassy-bonded bodies have a higher modulus of elasticity than those which are mainly crystalline provided that the materials are of similar composition. The modulus of elasticity (also known as Young's modulus) of some ceramic bodies is given in Table XIII.XV.

TABLE XIII—XV. THE MODULUS OF ELASTICITY OF SOME CERAMIC BODIES (lbf./in.^{2})

	$E \times 10^6$		$E \times 10^6$
Fireclays	0·5–1·0	Leadless glaze	8
Silica bricks	0·4–0·6	Lead glaze	12
China ware	10	Glasses	10–25
Building bricks	0·7–3·0	Porcelains	8–12
Engineering bricks	3–5	Cordierite	7–15
Mullite bricks	1·5	Steatite	12–16
Magnesite refractories	50	Zircon porcelain	21–24
Earthenware body	5	Titania	13–16
Tile body	2	Asbestos	1·5–1·8

The modulus of rigidity of ceramic materials usually has a value of about two-fifths that of the modulus of elasticity.

The change in the modulus of rigidity at different temperatures in silica,[1] fireclay[2] and basic refractories[3] has been studied by Roberts and his co-workers. A

[1] Das, S. S. and Roberts, A. L., *Trans. Brit. Ceram. Soc.*, 48, 215, 1949.
[2] Wiechula, B. A. and Roberts, A. L., *Trans. Brit. Ceram. Soc.*, 51, 173, 1952.
[3] Chung, H. S. and Roberts, A. L., *Brit. Ceram. Soc. Meeting* (London, 1953) (as yet unpublished).

rise in modulus up to about 800° C. commonly occurs, although the cristobalite inversion point in silica refractories is marked by a sharp decrease in elasticity. The production of liquid in fireclay and basic refractories marks the beginning of a progressive decrease in modulus and this occurs at temperatures which depend on the composition of the brick. Fig. XIII.21 shows the variation with temperature of the rigidity modulus of alumino-silicate refractories. Brick A is a siliceous fireclay; the quality of fireclay as measured by the alumina content increases through bricks B, C and E, this last containing 79·50 per cent alumina; Brick D is a sillimanite refractory. The dotted line shows the changes in a silica brick under the same test conditions.

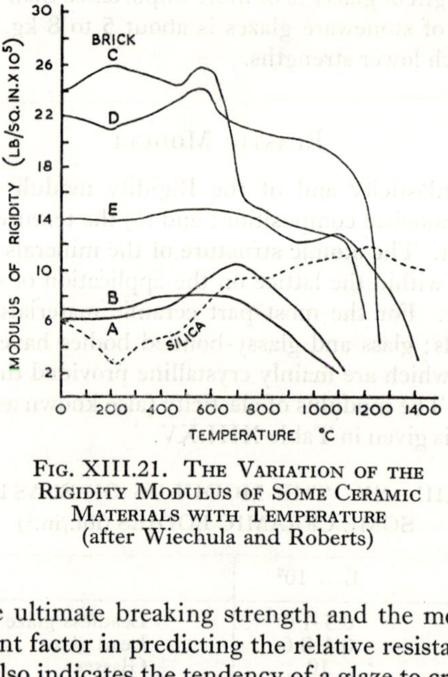

FIG. XIII.21. THE VARIATION OF THE RIGIDITY MODULUS OF SOME CERAMIC MATERIALS WITH TEMPERATURE (after Wiechula and Roberts)

The ratio of the ultimate breaking strength and the modulus of elasticity or rigidity is an important factor in predicting the relative resistance of ceramic bodies to thermal shock; it also indicates the tendency of a glaze to craze or peel.

When tested at room temperature, most ceramic materials behave elastically until their ultimate breaking point, but at elevated temperatures the test piece 'creeps' under high pressures and permanent deformation occurs. Roberts and Wiechula recognise three different types of stress/strain curves as shown in Fig. XIII.22; (i) is of a perfectly elastic material in which the line AB represents the stress/strain relationship on both loading and unloading. A*b* is the strain at the maximum applied stress. Curve (ii) shows an 'imperfectly elastic' deformation where the loading (AB) and recovering curves (BC) do not coincide—a 'permanent set' of AC occurs. Within the plastic range of the material, curve (iii) is obtained, which shows 'creep' or deformation (BB′) whilst a fixed stress is maintained; AB again represents the stress/strain relationship during loading and B′C that during unloading.

In each case the tests were carried out by applying or removing the load in

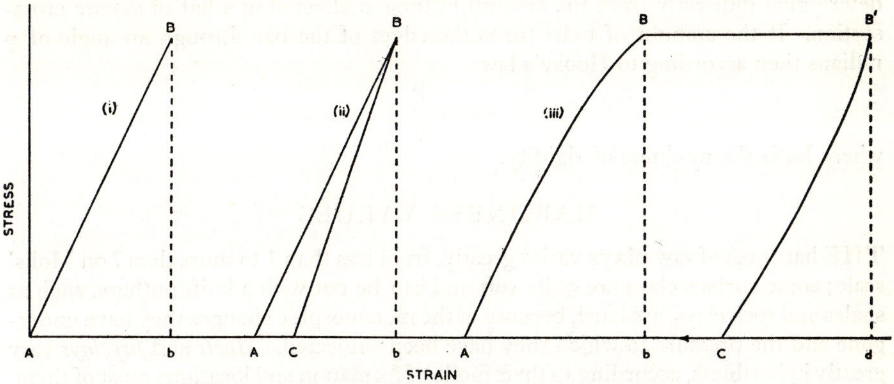

FIG. XIII.22. DIFFERENT FORMS OF STRESS/STRAIN CURVES
(i) Elastic material; (ii) imperfectly elastic material; (iii) showing plastic flow (after Wiechula and Roberts)

fixed increments and maintaining at each value for 2 minutes. The maximum stress was maintained for 30 minutes.

The variation of elastic moduli at high temperatures has also been studied by Davies and Brough,[1] by Spriggs,[2] and by Davis,[3] using dynamic resonance methods. Similar changes to those reported by Roberts and his co-workers were noted. The Young's Modulus of vitreous silica, for example, changes by 11 per cent when heated from 0–1300° C.

Poisson's Ratio is an important elastic constant of ceramic materials; it is defined as the ratio between the decrease in width and the increase in length of a bar of material under tensile stress. It is related to the moduli of elasticity and rigidity by the equation.

$$\nu = \frac{E}{2E_R} - 1$$

where ν = Poisson's Ratio; E = modulus of elasticity; and E_R = modulus of rigidity.

The value of ν is relatively constant for ceramic materials, being about 0·30 for most articles made largely of clay. Glasses have a somewhat lower value and for fused silica the value is 0·15.

The Poisson's ratio decreases at higher temperatures, but not considerably unless plastic flow occurs; it is an important constant which relates to the resistance of a body to thermal stress (p. 949).

Rigidity is defined as resistance to change in shape (deformation); if the force applied is P and the amount of deformation (or strain) is *a*; $P/a = E_R$ where E_R is the modulus of rigidity. It may be measured directly but in clay-pastes it is usually

[1] Davies, L. J. and Brough, R., *Trans. Brit. Ceram. Soc.*, 62, 451, 1963.
[2] Spriggs, R. M., *J. Amer. Ceram. Soc.*, 44, 628, 1961.
[3] Davis, W. R., *Trans. Brit. Ceram. Soc.*, 67, 515, 1968.

determined indirectly from the amount of torsion effected in a bar of square cross-section. If the amount of twist turns the edges of the bar through an angle of φ radians then according to Hooke's law

$$E_R = \frac{P}{\varphi}$$

where E_R is the modulus of rigidity.

HARDNESS VALUES

THE hardness of raw **clays** varies greatly, from less than 1 to more than 7 on Mohs' scale; some surface clays are quite soft and can be cut with a knife; others, such as shales and rock clays, are hard, because of the metamorphic changes they have undergone and the pressure to which they have been subjected. *Marls* and *fireclays* vary greatly in hardness, according to their mode of formation and location; most of them, when in the dry state, can be easily cut with a knife, but some are hard enough to scratch glass. This scratching is largely confined to siliceous (quartzose) impurities in the clay and not to the clay itself. Ayrshire bauxitic clay and some flint clays are extremely hard due to their mode of origin and subsequent alteration.

As clays are usually converted into a plastic paste, or into a slip or cream before use, their hardness is chiefly of importance in so far as it affects the methods to be employed in grinding them to powder or otherwise preparing them for use. Clays which can readily be crushed to the required fineness are naturally easier and cheaper to prepare and are, therefore, more desirable than very hard materials which require much power to reduce them.

The Mohs' hardness values of some minerals associated with ceramic materials are shown in Table XIII.XVI.

TABLE XIII—XVI. HARDNESS OF CERAMIC MATERIALS ON MOHS' SCALE[1]

Material	*Hardness*	*Material*	*Hardness*
Agate	7	Kaolinite	1–2·5
Apatite	5	Kyanite	7
Baddeleyite	6·5	Magnesite	3–5
Bauxite	1–3	Magnetite	5·5–6·5
Boron carbide	9–10	Mica	3
Brookite	5·5–6	Monazite	5–5·5
Calcite	3	Periclase	6
Chromite	5·5	Quartz	7
Coal	2·3	Rutile	6–6·5
Common clay (dry)	0–4	Sand	7
Corundum	9	Silicon carbide	9–9·5
Dolomite	3·5–4	Sillimanite	6–7
Felspar (orthoclase)	6	Spinels	7–8·5
Fluorspar	4	Talc	1
Graphite	1–2	Tantalite	6
Gypsum	2	Titanite	5–5·5
Haematite	5·5–6·5	Tridymite	7
Hydromagnesite	3–5	Zircon	7·5
Ilmenite	5–6		

[1] For other values see Appendix.

Silica has a hardness of 7 on Mohs' scale, but many siliceous materials in which individual grains are very hard may be easily crushed, as the bond which unites these particles is quite weak. For the same reason, the individual grains of silica in articles made of that material are much harder than the average hardness of the articles, on account of the softness of the bond.

Aluminous materials vary greatly in hardness; bauxite usually has a hardness of 1–3 on Mohs' scale, whilst corundum has a hardness of 9–10 and is one of the hardest substances known.

Magnesite varies in hardness according to the form in which it occurs. Coarse crystalline magnesite usually has a hardness of about 4 on Mohs' scale, whilst crypto-crystalline magnesite has a hardness of 3–5. Brucite also has a hardness of 3–5.

Carbides, nitrides and carboxides are extremely hard, corresponding to 9–10 on Mohs' scale, being harder than crystalline alumina (9), but not so hard as diamond (10).

Sintered Oxides and similar materials can be very hard, so much so that they have been used to replace the more expensive carbide tools in steel-turning and for other applications.

Hardness of Burned Ceramic Materials

Burned clay is much harder than raw clay and articles composed of it, either alone or mixed with other materials, may conveniently be arranged in two groups, according to their hardness:

Group I.—Soft wares, which can be scratched by iron, including fired sandy-clayey bodies, such as bricks, cooking utensils, crucibles, jars, unglazed earthenware, faience, roofing tiles, and most refractory materials.

Group II.—Hardwares, which cannot easily be scratched by steel, including fine earthenware, ceramic stoneware, pipeclay ware, flint ware, and hard china.

The resistance of ceramic materials to abrasion is usually more important, especially as many of them are subjected to a considerable amount of abrasion when in use. Thus, domestic pottery requires to resist the scratching and rubbing action of knives, forks, etc.; paving bricks, floor tiles, etc., require to be resistant to traffic; and many refractory materials are required to resist the abrasive and corrosive dust contained in the hot gases in furnaces, kilns, etc., which rapidly wear away any soft portions of refractory material.

The bricks which form the lining of vertical shaft furnaces, such as blast furnaces, cupolas, lime, magnesia, and other calcining kilns, etc., are subject to great abrasion by the descending charges, and it is essential that such bricks should have the necessary resistance to this action and to the differential movements which often occur in shaft kilns and also exercise a considerable abrasive action.

The manner in which gas retorts and some other appliances made of ceramic materials are charged and discharged also calls for the use of a material which is highly resistant to abrasion both in the hot and cold states. 'Rough usage' also has a great abrasive effect, as well as necessitating the use of a material which is resistant to blows and shocks.

The resistance of a ceramic material to abrasion depends upon one or more of the following:

(*a*) The nature of the material, and especially its texture and hardness.
(*b*) The mode of its preparation.
(*c*) The nature of the bond (if any).
(*d*) The amount of bond (if any).
(*e*) The extent of vitrification.
(*f*) The temperature of the material when it is examined.

Building Bricks. The hardness of building bricks is seldom of much importance, as, in the ordinary way, they are not required to be highly resistant to abrasion. Most well-burned bricks are rather harder than sandstone. Very soft bricks should not be used, except, possibly, as panels in interior work, as they are generally underburned and deficient in strength and resistance to the weather.

Bricks are not of strictly uniform hardness throughout their mass and the surface is usually harder than the interior on account of the greater pressure applied to it, and to the migration of fine particles during drying.

There appears to be no definite correlation between the hardness and the compressive strength of bricks and similar articles.

Vitrified bricks, when used for roads, etc., require to be specially resistant to abrasion, and, such bricks should not lose more than 14 per cent of their weight when subjected to the Standard Rattler Test, and no one brick should lose more than 18 per cent of its weight.

Roofing tiles should be sufficiently hard to be handled without chipping and to prevent them from being damaged by frost. They should not be vitreous or they will 'sweat' when in use. A good roofing tile should usually be difficult to scratch with a piece of steel; roofing tiles which are harder than steel are generally too hard and vitreous. The desired hardness is largely determined by the manner in which the tiles are fired in the kiln.

Floor tiles should be sufficiently hard to resist any abrasion to which they may be subjected.

Fireclay bricks vary greatly in hardness, some being sufficiently soft to crumble when cut with a steel blade, whilst others are harder than steel. The hardness depends to a large extent upon the temperature at which the bricks are fired; the higher the finishing temperature the harder will be the bricks, and soft bricks are usually underburned.

Silica bricks are composed of hard grains, but the bond is soft, so that the bricks are readily rubbed down, and they resist abrasion only to a small extent. The hardness of the individual grains is equal to that of quartz (p. 898). The greater the proportion of lime used in making the bricks the harder will they be, as their resistance to abrasion is entirely dependent on the bond formed by the combination of lime and silica, and unless a sufficient amount of this bond is produced the particles of quartz, etc., are only feebly held together.

Silica bricks which have not been burned at a sufficiently high temperature are very soft and easily abraded. Well-burned bricks are harder and emit a good 'ring' when struck.

Fused silica has a hardness of about 5 (Mohs' Scale).
Magnesite bricks are fairly resistant to abrasion at atmospheric temperatures, but they are very soft at high temperatures.
Fused magnesia has a hardness of 5–6 according to Mohs' scale.
Silicon carbide bricks are extremely hard, and an angular fragment from them will readily cut glass. At high temperatures some silicon carbide bricks appear to be extremely resistant to abrasion, more so than any other form of refractory brick.
Chrome bricks are fairly resistant to abrasion in the cold, but at high temperatures they are rather soft.
Glazed ware should be sufficiently hard to resist the abrasion to which it is subjected in ordinary use. Domestic earthenware is scarcely hard enough for severe use, as it is too easily scratched by knives and forks. Bone china is much better in this respect, and the 'hard porcelain' of the Continent is the hardest and most resistant of all such ware.

For many ceramic articles, **resistance to abrasion** is of minor importance, but in grinding-wheels and some cutting-tools a high abrading power is of great value. These articles are made of particles (of selected sizes) of diamonds, emery, silicon carbide, quartz (sand), glass and various synthetic materials such as fused alumina. The particles (known technically as *grit*) are mixed with a suitable bond such as rubber shellac, magnesium oxychloride, sodium silicate, a vitreous bond similar in composition to hard porcelain, or various resinoid materials such as bakelite. The articles are moulded from a stiff plastic paste, dried and baked or fired at a temperature corresponding to the nature of the bond and the hardness or grinding power desired.

An abrasive material must not only be hard—7–10 on Mohs' scale; it must also be tough and strong so that when the particles break they do so sharply and leave angular fragments behind to provide fresh cutting edges. (The effectiveness of most abrasives depends also on the pressure with which they are applied, the speed at which the abraded material is removed and, in the case of some metals, efficient cooling at the point of abrasion.)

Resistance to Abrasion is an important property of some ceramic materials; well-known instances are paving bricks, balls used in ball grinding mills, linings for grinding cylinders and the resistance of furnace linings to the downward movement of the charge. In all these instances, *toughness* rather than hardness is the most important factor. A finer texture is also desirable so that relatively large pieces are not torn out of their surroundings.

Resistance to abrasion can be ascertained only by full-scale trials as it is not proportional to the hardness scale nor has it any simple relation to any other property, though vaguely related to several, such as the strength of the bond uniting the grains together, and the modulus of rupture. There appears to be no co-relation between 'Hardness' ('scratch tests'), porosity tests and resistance to abrasion.

The resistance of plates and saucers to the abrasion of knives and forks is very important, as such ware when scratched is unattractive.

The abradability of some ceramic materials as determined by the Morgan-Marshall Test (p. 870) is shown in Table XIII.XVII.

TABLE XIII—XVII
ABRADABILITY INDEX OF SOME CERAMIC MATERIALS.

Material	*Abradability Index*	*Material*	*Abradability Index*
Fused basalt tiles	34	Fireclay bricks (22% porosity)	80—120
Vitreous fireclay tiles	36—37	Insulating Refractory Concrete	90—140
Red flooring tiles	43—56	Silica bricks	70—100
Dense fireclay bricks	50—65	Carbon bricks	130—180
Molochite blast furnace bricks	45—55	Insulating fireclay bricks	200—300

VISCOUS AND PLASTIC FLOW AT HIGH TEMPERATURES

IN Chapter VIII, the flow properties of suspensions and the various stages in the deformation of plastic masses have been described. At high temperatures, solid bodies may behave in a similar manner and their properties in this condition are exactly paralleled by the behaviour of plastic masses, slips and suspensions.

When ceramic materials, particularly those containing clays, are heated, the fluxes present combine with other ingredients to form a liquid. Over a certain temperature range, the body is composed of solid particles suspended in a liquid, but if a sufficiently high temperature is reached, the mass will usually become converted into a perfectly homogeneous fluid.

On cooling, some solid particles will develop, or, in glasses and glazes, associated groups form; in both cases, the mass becomes progressively more rigid and finally solidifies completely as a crystalline mass, or as a glass, or as a mixture of the two forms.

A solid mass, particularly if it is crystalline, behaves elastically, i.e. if a stress is applied, a change in shape of the body occurs, but it returns to its original shape when the stress is removed. The elasticity is maintained until the stress becomes sufficiently large to fracture the specimen.

When a proportion of liquid forms or when any glass present begins to soften, the body exhibits plastic flow (p. 491) and, when the amount of liquid is high, the flow becomes viscous.

The flow properties of clay bodies at high temperatures is well brought out by measuring the torsional properties which are described on pp. 895–898.

Glasses. The viscosity or flow properties at high temperatures of glasses and glazes is particularly important. A glassy material can be induced to flow at low temperatures, but a large pressure is required and it has to be maintained for a long period of time before any noticeable deformation occurs. At higher temperatures, less force is required to cause movement, and flow becomes progressively easier and eventually a very fluid liquid may be formed. The ratio between the applied force and the resultant deformation is related to the viscosity (p. 477), so the effect of increasing the temperature of a glass or glaze is to decrease its viscosity.

The change of viscosity with temperature of a glass is shown in Fig. XIII.23. The important regions of the curve are:

1. *Rigid glass*, above 10^{14} poises, where little noticeable flow occurs.
2. *The Annealing range* $10^{12 \cdot 5}$–$10^{13 \cdot 4}$ poises, over which internal stresses can be relieved and some degree of atomic movement is possible.
3. *The Working range*, 10^4–$10^{7 \cdot 6}$ poises, represents a condition where the glass is handleable and can be blown or rolled easily without it being too fluid. The upper limit of the working range is known as the *Softening Point*.
4. *The Melting range*, $10^{1 \cdot 5}$–$10^{2 \cdot 5}$ poises, is the range of viscosity over which the glass flows readily and corresponds to its condition in the melting tank.

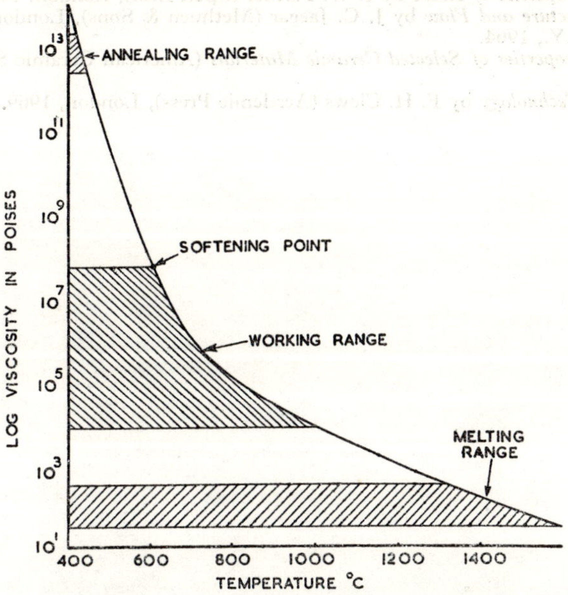

FIG. XIII.23. THE RELATIONSHIP BETWEEN THE VISCOSITY AND THE TEMPERATURE OF A GLASS

The temperatures over which these ranges occur differ with the composition of the glass. Alkali glasses are more viscous at a particular temperature than those containing lime and magnesia and, in general, alkali glasses have a shorter working range.

Glazes exhibit a similar behaviour to glasses but usually at a higher temperature. The most critical viscosity range in glazes is, however, between $10^{2 \cdot 5}$–$10^{4 \cdot 3}$ poises,[1] because this represents the condition in which the glaze would flow with sufficient freedom to cover the surface of the body, but would be insufficiently fluid to run off the ware or to develop bubbles or pin-holes.

The viscosity of glazes and its variation with temperature is influenced by the chemical composition. Glazes rich in silica and alumina mature at higher temperatures than those with appreciable amounts of fluxes and the burning temperature must be adjusted to suit the glaze being used. Of very great importance is the

[1] Bremond, P., *Bull. Soc. Franc. Ceram.*, 11, 4, 1951.

maturing range of a glaze, i.e. the temperature range over which the viscosity changes from $10^{2 \cdot 5}$ to $10^{4 \cdot 3}$; this should be as large as possible to allow for possible temperature variations in the kilns or muffles used for burning the ware. A long maturing range can be achieved by correct adjustment of the composition and using more than one fluxing element.

BIBLIOGRAPHY

1. *Engineering Properties of Rocks* by I. W. Farmer (Spon Ltd.), London, 1968.
2. *Elasticity, Fracture and Flow* by J. C. Jaegar (Methuen & Sons), London, 1964; (Wiley & Sons), N.Y., 1964.
3. *Engineering Properties of Selected Ceramic Materials* (American Ceramic Society Publication), 1966.
4. *Heavy Clay Technology* by F. H. Clews (Academic Press), London, 1969.

CHAPTER XIV

MISCELLANEOUS PROPERTIES OF CERAMIC MATERIALS

CERAMIC materials are of many kinds, differing in composition and property. In many cases, a product is so manufactured that one particular property is emphasised which is of value in service. The present chapter is concerned with such properties, their physical nature and their causes and effects.

COLOUR

OF all the properties exhibited by ceramic materials, none is more fascinating than colour. It is probably also true that no other property is so unpredictable or so difficult to reproduce with precision in fired ware.

Colour is a manifestation of light which is one of the forms of energy. It involves radiation of a characteristic wavelength range to which the retina of the eye is particularly sensitive. Why the eye should discriminate between radiations of different wavelengths is not known with certainty and it is possible that some other living species may 'see' a different range of wavelength from that of human beings. However, the total range of radiation visible to man occurs over only a relatively small interval in the complete sequence of radiation shown in Fig. XIV.1. Within the narrow band of visible radiation the eye can distinguish specific colours of characteristic wavelengths as in Table XIV.I.

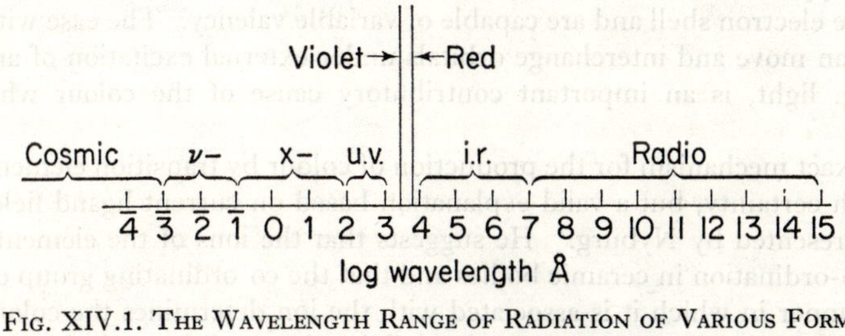

FIG. XIV.1. THE WAVELENGTH RANGE OF RADIATION OF VARIOUS FORMS

TABLE XIV—I. CHARACTERISTIC WAVELENGTHS[1] OF VISIBLE LIGHT RADIATION

Red	6562×10^{-8} cm.	Blue	4854×10^{-8} cm.
Orange	6072	Indigo	4618
Yellow	5853	Violet	4330
Green	5230		

[1] Angstrom Unit (Å) = 10^{-8} cm. and the wavelength of light is frequently expressed in these units.

White light is a composite radiation of wavelengths throughout the whole of the visible spectrum, but it can be split into its components quite readily.

Because light is a form of radiation, it can be transmitted, reflected, refracted or diffracted by surfaces on to which it is allowed to fall. The property of refraction, i.e. the capacity of a substance to change the velocity of radiation which passes through it, is the major method by which white light can be subdivided into its component coloured radiations. Because of the range of wavelengths involved in white light, rays of longer wavelength are refracted more than shorter ones so that after passing through a suitable medium, a spectrum band ranging from red to violet is produced.

Basically, the elementary theory of transmission, reflection and refraction accounts for the colour of everyday objects. White light from the sun or other source falls on the object and is partly transmitted and partly reflected at the surface. If it is completely transmitted the object appears black, but total reflection would be manifested by a white object. Coloured objects are those which transmit light of certain wavelengths, refract others and reflect only a certain narrow band or bands to the observer. The nature of the reflecting substance determines the colour and it is related to its atomic or molecular structure.

The main mechanisms for the absorption of infra-red radiations of specific wavelengths have been discussed in Chapter V (p. 246). In a similar fashion, some electron and atomic configurations of mineral species absorb visible radiations of characteristic wavelength, so that the reflected or transmitted light is deficient in certain radiations.

The theoretical reasoning of the colours produced by materials and chemical compounds is highly complex and often largely speculative. In the inorganic field of colour production, which is that most usually associated with ceramic materials, the presence of elements of the transition group (see p. 351) has an important significance. Many of these elements when present in compound form are coloured, e.g. iron, copper, cadmium, mercury, chromium. These elements have an unfilled penultimate electron shell and are capable of variable valency. The ease with which electrons can move and interchange orbitals under external excitation of an energy source, e.g. light, is an important contributory cause of the colour which they produce.

The exact mechanism for the production of colour by transition elements is not known with certainty, but a valid explanation based on current ligand field theory has been presented by Nyburg.[1] He suggests that the ions of the elements are in complex co-ordination in ceramic bodies and that the co-ordinating group or ligand and the manner in which it is associated with the ion determines the colour. The presence of ligands affects the energy state of the elemental ion and the absorption bands which it can exhibit.

Weyl[2] has considered various factors to account for the colour which develops in glasses. On the basis of polarisation effects in ions and their relative attraction for oxygen anions when they are in different valency states, he offers a plausible explanation for the colours observed in practice. Yates has applied this reasoning

[1] Nyburg, S. C., *Trans. Brit. Ceram. Soc.*, 58, 565, 1959.
[2] Weyl, W. A., 'Coloured Glasses', Society of Glass Technology Monograph 1958.

to account for the colour produced in glasses of different compositions[1] on the basis of the role of cations in the glass lattice.

Colour is exhibited in various ways, so that an object may show different colours under different conditions, viz.:

1. The colour by reflected light.
2. The colour by transmitted light if the object is transparent.
3. The colour of the powder, as distinct from the mass.
4. The appearance of special types of colour, such as (*a*) *play of colours*, caused by the unequal bending of rays of light of different wavelengths giving a vari-coloured effect; (*b*) *opalescence*, which gives a pearly or milky appearance, as shown by opals; (*c*) *iridescence*, or display of colours, due to inclusions of air or liquid in transparent crystals, or to peculiar irregularities in the surface of the material, as in some specimens of quartz, calcite and mica; (*d*) *fluorescence*, which is the ability of a material to absorb energy and then re-emit it with its wavelength altered to one in the visible part of the spectrum; (*e*) *phosphorescence*, the ability to absorb light energy and then to emit radiation when the source of excitation has been removed. Objects which exhibit phosphorescence continue to glow in the dark after being exposed to light. Normally the intensity of phosphorescence will decline with time, but it can be reactivated on exposure to light. Some materials, for example, white phosphorus, and radio-active substances, glow permanently due to chemical reactions within themselves which provide the excitation energy, and (*f*) *pleochroism*, a quality possessed by some transparent minerals, by which they exhibit different colours when viewed by polarised light passing through them in different directions (see also Chapter V).

Lustre is a property which may be classed with colour; it is due to the manner in which light is reflected from an object, rays of different wavelengths being reflected at different angles and so producing diffraction effects. The lustre may be (*a*) *metallic lustre*, such as is exhibited by many metals; (*b*) *vitreous* or *glassy*, as in glass, quartz, etc.; (*c*) *resinous*, as in resin; (*d*) *pearly*, as in pearls; (*e*) *silky*, as in satin spar; and (*f*) *adamantine*, as in diamond and some forms of silicon carbide.

Colour is a much more constant feature in opaque substances than in transparent ones; the latter often vary greatly in colour, because of the presence of minute traces of other substances as impurities. Many substances have a characteristic colour, which is of value as indicating their nature, e.g. compounds of iron, chromium, copper, uranium, etc.

The colour of most raw ceramic materials is usually a minor property and chiefly of value as an indication of the impurities present. In finished articles, it is often a very desirable property and one which contributes greatly to their beauty. In some ornamental ware and even in some utilitarian articles the colour is more important than other properties. Thus, an architect may select red bricks on account of their colour and may have to disregard the fact that the ones he chooses may not be as strong as some others. It is often easy by firing bricks at a higher temperature to make them much stronger, but, if the colour is less pleasing, they will usually be more difficult to sell.

Where a definite colour is desirable in the finished articles, it may be produced by various means such as:

[1] Yates, D. J., *Trans. Brit. Ceram. Soc.*, 58, 573, 1959.

(*a*) By heating the material or articles at a suitable temperature, and in a controlled atmosphere, so as to produce the desired colour. If the composition of the raw materials is suitable, no other treatment may be required. If certain colour-producing impurities are present in the raw material they may prevent the formation of the desired colour, in which case some other method must be employed. A typical example of this method of producing colour is a ferruginous clay, which, if burned at a certain temperature (depending on the clay) in an oxidising atmosphere, may produce a pleasant red colour, but if it is burned in a reducing atmosphere a dark, dull 'blue' colour is obtained.

(*b*) Introducing a coloured substance into the raw material, so as to mask any undesirable colour and give the finished goods the desired tint. Thus, the addition of manganese dioxide will produce a dark brown or nearly black article, whilst cobalt compounds will give blue shades and chromium compounds green ones. The lighter the natural colour of the burned material, the greater will be the variety of colours which can be produced in this manner.

(*c*) If the colour of a fired material is undesirable it may be masked by coating the article with an engobe of another material which burns to a desirable colour. The engobes used for this purpose are usually white-burning clays or a mixture of such clays with flint, Cornish stone, or felspar, together with a suitable colouring agent if required, the engobe being adjusted so as to have the same shrinkage as the article to which it is applied, as well as the desired colour when burned.

(*d*) Covering the engobe with a transparent glaze, to protect it and render the article impervious to water. Sometimes, instead of a coloured engobe, a coloured glaze is used and in some faience and majolica ware a coloured glaze is used over a white engobe.

When the colour is required to be in the form of a pattern, it may be applied directly to the article or to the engobe by means of a brush, stencil, transfer, or any other convenient method.

Sources of Colour in Ceramic Materials

Many of the pure minerals and chemical components of ceramic materials are perfectly white. Alumina, silica, magnesia, lime, zirconia, china clay, talc and many others, reflect light of all wavelengths and show no characteristic coloration. Carbon, in various forms which occur in natural clays, is black so that when these components are present the resulting colour is grey of various shades.

It is rare, however, for minerals to be pure in a natural state. Iron is a common impurity which is rarely completely absent in any mineral deposit.

Iron compounds in their various forms give rise to yellows, reds, greys, buffs, browns of different degrees and intensities in most natural and fired clays and other ceramic materials.

It is important to recognise that colour can be due to both (*a*) the physical, and (*b*) the chemical condition of a material.

Physical Causes of Colour. Because colour is the result of complex transmission, reflection and refraction of light within a substance, it is not surprising that the physical condition of the material can have a marked effect.

The particle size is of considerable significance. Large crystals or grains may be very different in colour from the same material ground to fine size. A substance which is transparent or translucent may appear dark or even black when of large grain, but pure white when reduced in size. Flint provides a good example in that in nodular form or along a fracture surface it may appear black, but in a crushed form it is valued as a white filler.

Internal reflections and refractions within a crystal often produce coloured effects of great beauty. Opals owe their desirable characteristics to the presence of intense stresses within the crystal which refract and reflect the light in a myriad of colours. The same specimen is usually dull and uninteresting when viewed in transmitted light.

The gorgeous brilliance of diamonds, zircons and similar gemstones is due to their high refractive indices in particular directions. The resulting birefringence (p. 225) gives the crystal its clarity and brilliance.

The refractive index of minerals is a major contributory factor to the intensity of their whiteness and brightness when in small particle sizes. Titanium oxide, zinc oxide and tin oxide are valuable in this respect.

Chemical Causes of Colour. Certain substances have the property of possessing colours other than perfect whiteness. These materials may be of an organic or an inorganic nature.

Organic colouring materials are of great importance in the paint, plastic and similar industries but are not so significant in ceramic products, because they usually burn out completely on firing. Coloured organic substances are frequently employed in engobe and glaze slips to indicate to the potter the number of coats which have been applied. Each application is coloured differently, but because the source is organic it disappears on firing without affecting the ceramic body itself.

Natural organic materials, ranging from humic acids and algae to high-rank coals and graphite are frequent components of natural clays. They produce yellow, grey, brown, and black colorations which burn-off at moderately high temperatures. They can be undesirable in clays which are used in their natural state as a source of whiteness; for example, china clays in paper-making and for medical purposes.

Inorganic colouring materials are retained in the body even after it has been fired to high temperatures and hence these are the sources of colour in fired ceramic bodies.

Iron, in one or other of its compound forms, is the most frequent inorganic source of colour and because of its variable valency states, it can produce a wide range of colours from yellow, red, blue, brown, grey even to black. The fineness of size of iron compounds is also of importance in the colour which is produced.

Natural clays rarely contain sufficient amounts of inorganic colouring agents, other than iron, to have a noticeable effect. Some elements, such as calcium, may modify the iron colour, but there are few which confer colour in their own right.

Chrome ores and products made from them are usually characteristically red or brown (due partly to iron impurity); occasionally chrome/serpentines are green.

The range of artificial inorganic colouring agents added to clays, but more especially to glazes and glasses is very large and the most important ones have been listed in Chapter VI (p. 351).

Most inorganic colouring agents are compounds of elements of the transition groups. The more prominent colours tend to be given by the elements which are towards the middle of each particular transition sequence of the Periodic Table (p. 58). Table XIV.II illustrates the characteristic for the more common elements of Period IV.

TABLE XIV—II. NORMAL COLOUR PRODUCED BY SOME TRANSITION ELEMENTS

Period IV	Sc	Ti	V	Cr	Mn	Fe	Co	Ni	Cu	Zn
	white	white	green	green	brown	red	blue	green	green	white

Elements in transition groups of higher Periods also produce coloured compounds but they are frequently too expensive to use. Exceptions are cadmium (yellow) gold (red, green) mercury (red), tungsten and uranium (yellow). Selenium is exceptional in not belonging to a group of transition elements and yet producing an attractive red colour. Other elements are employed in glaze colouring to produce specific effects.

The colours listed above are for the single element usually combined in the oxide form. Combinations of these elements and their various compound forms can often result in very different effects. When various elemental forms are used for colouring glazes, the composition of the glaze itself often has a marked influence on the final tint or shade. In some instances, colour develops as a result of the element forming part of the glass or glaze lattice itself, whereas in others a type of colloidal suspension results.

Application of Colour in Ceramic Bodies

In many branches of the ceramic industry, the manufacturer has to accept the natural colour which develops in his raw material and so adjust methods of preparation and firing to bring out the best effects. Occasionally, additives can be made to the body to produce different colours, but these are of limited application principally because of economic factors.

The production of ware of aesthetic quality for a comparatively high-priced market permits the manufacturer to employ colouring agents of exotic qualities but applied in thin coatings.

The colouring of ceramic ware may thus conveniently be sub-divided into three important classes, (*a*) body colouring, (*b*) veneer facings, (*c*) glaze and slip decoration.

Body Colouring. Most ceramic materials in their natural state contain a sufficient quantity of colouring elements (usually iron), so that at all stages of their processing and including the finished fired shapes, the whole body will be of a non-white appearance. Unless there has been some migration in drying or some localised atmospheric condition in firing, the colour will be uniformly distributed throughout the whole body.

Clays containing iron minerals will fire to pink, red, yellow or brown depending on the amount and size distribution of the iron, on other minerals which may be present, and on the temperature and atmospheric conditions during burning. Only occasionally, as in china clays and some ball clays, is the iron content sufficiently low to yield a white body.

For some purposes, bodies have to be stained to a uniform colour throughout their mass. An article, such as a floor tile, selected for its aesthetic appearance but which is liable to become worn with use must have a uniform colour throughout. Body staining is often necessary for similar reasons in facing bricks.

For the most part, staining compounds are expensive and require to be added in considerable proportions to produce a desired or even noticeable colour. Consequently, body staining, with all but the cheapest ingredients, is confined to ceramic articles of reasonably high market value.

Bricks and other structural clay products are rarely artificially body coloured, because of cost considerations. It is usual to select natural clays which will give the requisite colour on firing without further additions. This is not always possible and there are limited modifications in terms of additives which have a certain measure of success.

Red colours in fired clays are generally preferred. They are the most common colours to which natural clays will fire, but the tint and intensity which can be developed are not always ideal. Red is a critical colour in firing and changes in maximum temperature of only a few degrees can often markedly affect the final shade. When a deep red colour cannot be obtained in a particular clay, it is most difficult, if not impossible, to develop such colours by synthetic additions. Iron compounds in natural clays are present in colloidal sizes, evenly distributed by nature throughout the mass. No amount of mixing of oxides or other compounds of iron produce the same effects; the synthetic colours are invariably browns or buffs.

The author had some success in producing a reasonably uniform red coloration in a ball clay body, by blunging a slip of the clay with ferric chloride solution. After filter-pressing, the body was carefully dried, reground and pressed into tile shapes. After firing to 1020° C. the article was a pleasant light red of uniform distribution, but attempts to intensify it usually gave a brown colour.

The exercise can have only academic interest, as the cost of such treatment would be prohibitive for most commercial articles. No other means of adding iron compounds to produce true red colours appears to have been evolved.

Yellow body colours in fired clays are relatively easy to produce. Some clays burn to this colour naturally; normal red-burning clays will burn yellow if a proportion of finely-divided lime or limestone is added to the brick batch. It is relatively easy to convert a red-burning clay into a yellow or buff body, but practically impossible to achieve the reverse effect.

Some Keuper marls burn to a red colour at 950° C. but at 1000° C. they are yellow or grey. Brown colours in clays can be induced either by overburning normal reds or by adding small amounts (0·5 per cent) of *manganese dioxide* to the brick batch. The latter process is preferred because the product is easier to control. A greater quantity (1·5 per cent) of manganese dioxide in combination with iron can produce a black body. Blue bricks can frequently be produced by completing the burning of the ware in a reducing atmosphere, by excluding air and creating an excess of carbon monoxide or hydrogen in the kiln. This has the effect of converting iron compounds into magnetite, Fe_3O_4, or wustite FeO and sometimes into the carbide or the elemental form.

Bricks, made by the old clamp process, frequently exhibit a wide and attractive range of body colourings due to differences in atmospheric conditions during their burning. In any one batch, colours may range from red to blue whilst others have distinct violet and purple hues.

It is only in rare circumstances that bricks with other body colours are produced. Industrial effluents or waste may be of interest and a green brick with a certain degree of aesthetic attraction was produced by blending a low grade buff-burning fireclay with sludge from a chromium-plating plant. The same waste added to other clays produced an unsightly mess; just as in the colour blending of paints, it is so easy with ceramic colours to end up with a predominantly unpleasant greyish base.

Floor tiles and architectural ceramics frequently are body coloured to give pleasing effects. It is not easy to produce an acceptable colour with a natural clay as a starting material, because its own colour tends to mask the effect of the additive or to produce a composite shade of doubtful value.

White-burning clays generally form the basis for body-coloured ware of good quality. Additions of 1–2 per cent of oxides of cobalt (blue), chromium (green), cadmium (yellow), copper (blue-green) or specially compounded blends of several elements are commonly employed. Some colours are atmosphere and temperature sensitive to a marked degree so that care is necessary in firing.

Veneer Facings. Rather than incorporate an expensive colouring additive to a complete brick batch, it is frequently preferable to apply a coating to the faces of the article. The veneer of colour need only be thin and is often applied in conjunction with clay, sand or grog to the article whilst it is still in a wet state. Many extruded brick shapes are readily treated in this way and a remarkably wide range of colour effects can now be produced economically. This veneer colouring process is often combined with texturing or 'rustication' where a pattern is scratched or rolled into the surface by some simple device.

In floor tile production, the addition of a veneer coating has been developed to a fine art by automatic sequential operations. The main body material of the tile is fed from one hopper and pressed lightly into shape. The bottom press platen is then dropped slightly to permit a further layer of coloured body to be added to the upper surface. This is supplied from a second hopper/feeder and mechanisms are available to produce flamed, marbled and variegated patterns. The veneer face is normally about 1–3 mm. thick. The whole unit is then pressed to form the complete tile shape.

Veneer facings of both bricks and tiles have added a new dimension of beauty to structural ceramics. They are limited to applications where abrasion and erosion are not severe, but they are remarkably durable and weather resistant.

The most usual colours which have been developed as veneers or slip coatings to be used with clays are as follows:

Black articles or the nearest approach to a true black which can be obtained may be produced by:

(*a*) The addition of a mixture of manganese dioxide and iron oxide to the clay. Where there is enough iron oxide in the clay, the addition of a suitable proportion of manganese dioxide is usually sufficient. The precise colour produced depends on the relative proportion and fineness of the two colouring agents. A red-burning clay with the addition of about 1½ per cent of manganese dioxide will usually produce a beautiful jet black. It is important to use fine precipitated manganese dioxide; that produced from spent material used in the manufacture of chlorine, etc., is much less satisfactory. If the colour is too brown or violet, the proportion of manganese dioxide should be increased or the iron oxide reduced.

(*b*) The addition of a mixture of iron and cobalt oxides.

(*c*) The addition of iron, manganese, cobalt, and chromium oxides to the clay.

Grey may be produced by the use of smaller proportions of the same materials as are used for blacks (*supra*), but the iron must not be present in very large proportions when light shades of grey are required. Greys may also be produced by means of iridium oxide or platinum chloride or both, but these are so costly that their use is confined to the most expensive pieces of art ware.

Blue colours are produced in white wares by the addition of cobalt oxide to the clay. The shade of colour is largely dependent upon the alumina present, a highly-aluminous body giving a sky-blue tint, whilst a siliceous mass, or one containing zinc oxide, usually assumes more of an indigo shade. Blues may be modified by varying the proportion of cobalt and also by the addition of other substances, such as alum or frits.

Violet colours are extremely difficult to obtain; the best are usually produced by a mixture of chromium and cobalt oxides. Under suitable conditions, precipitated manganese dioxide produces an excellent violet.

Green colours are generally produced by the addition of chromium or nickel oxides to bodies containing only a very small proportion of iron oxide, as otherwise the colour will not be pure green. Chromium greens are the most easily produced, and may be modified towards blue by the addition of cobalt oxide. Nickel greens are somewhat uncertain, especially at temperatures above 1100° C., when no other ingredient is used, but with cobalt oxide a fairly reliable olive-green is obtained.

Copper oxide also produces a series of green colours, but they are more easily spoiled at high temperatures than chromium greens.

At temperatures above 1200° C., iron silicates impart a greenish-yellow colour to the ware, whilst when such silicates fuse they become quite green and produce the colour of green bottles.

Yellow colours are, where possible, produced almost entirely by iron oxide, the tints varying from yellow through orange to yellowish brown. For high-class pottery, titanium and antimony oxides or lead chromate may be used. For orange-yellow tints, uranium oxides may be employed. For these tints an oxidising atmosphere is essential, as a reducing atmosphere gives a greenish-grey shade.

Red colours, produced at comparatively low temperatures, are usually due to iron oxide. In vitrified wares, iron oxide seldom produces a pleasing red shade. A rose or pink colour may be produced by the addition of a finely powdered frit of bichromate of potash and alumina, whilst a lilac tint may be obtained by the addition of a little cobalt oxide.

Occasionally bricks are made to appear red by dipping them in, or painting them with a slip of Venetian red with water just prior to sale.

Glaze and Slip Decoration. The best-known examples of colour being applied to ceramic ware are in domestic pottery and china ware, which usually have a glaze finish. Coloured compounds may be applied by dipping, spraying or transfer either to dried ware or more frequently to fired ware.

Compounds which burn to a white colour, such as the oxides of titanium, zinc and tin are frequently used to cover a brown or buff-coloured body. These are generally applied as an engobe layer (p. 341).

Colour may be applied to ceramic ware in two ways,

(*a*) directly on to the body or engobe and then a glaze applied to cover the whole. This is called *underglaze decoration* which ensures a more durable effect although the pattern is less distinct and a narrow range of colours has to be accepted,

(*b*) on to the glaze itself after the coating has been prior-fired. The coloured pattern can then include a wide range of colouring agents and after the article is refired the pattern is more vivid and lustrous, but not as resistant or durable. This is termed *on-glazing* or *over-glazing*. The principal colouring compounds and combinations used in glazing have been described on p. 351.

The colours used in glazing are mostly inorganic salts in combination with gums and oils for ease of application. Decorative ceramic ware may often be coated either in its entirety or in pattern-form with a slip composed of clay and added colouring matter. In this sense, slip coatings are not unlike veneers but they tend to be more fusible and to be applied in much thinner layers.

In ceramic practice, the colour develops after a firing process. Reactions occur at high temperatures to produce a stable colour which is often markedly different from its appearance in the unfired form.

Factors which affect Colour in Ceramic Ware

ALTHOUGH oxides and compounds of certain metals may produce a specific colour when added to clays and glazes, the shade and tint depend upon many factors. These include:

1. The mineral form of the colouring agent.
2. The state of sub-division.
3. The amount of the compound which is added.
4. The firing conditions, including maximum temperature, duration of heat and the atmosphere in the kiln at all stages of firing.
5. The presence of other components in the body which can modify the colour.
6. The nature of fluxes and the extent of vitrification.

The mineral form of the colouring agent can influence colour in both the raw and fired states. Iron compounds provide the best examples.

In natural clays, iron compounds of many types may be present (p. 275). Limonite is the name given to a group of hydroxides of iron which occur in disseminated form in clays and confer a yellow colour in the raw state. Natural red clays generally contain oxides such as haematite.

On firing, iron compounds lose water and oxidise to haematite Fe_2O_3 at fairly low temperatures if sufficient oxygen is available. At higher temperatures magnetite, Fe_3O_4 or wustite, FeO may form which are grey or blue. At even higher temperatures iron silicates or alumino-silicates form which are frequently yellow or buff in colour, but, when they fuse, a brown body generally results.

The state of sub-division of the agent determines whether the effect will be of even distribution or of discrete spots of intense colour on a much lighter background. Iron as a natural colouring agent exhibits the extremes; when it is present in clay in a colloidal form, the fired colour is uniform and evenly distributed, but individual nodules of iron-stone produce blotched areas which can be most unpleasant.

When artificial colours are used as additions to clay bodies, discrete grains can often be observed when inspecting the ware under a microscope, but the overall effect may be quite pleasant.

The amount of coloured compound which is present obviously has an effect on the tint or shade of the fired article. The addition of artificial colouring agents can be so adjusted in amount to produce a range from delicate pastel shades to intense strident colourings.

The amount of iron calculated as oxide in clays influences the colour markedly. In amounts below 3 per cent, bricks are unlikely to be more intense than flesh pink, 6–9 per cent Fe_2O_3 is generally necessary to give a pleasing deep red colour, but even this depends on other factors such as the impurity content and the temperature of firing.

The firing conditions can effect colour of ware to a considerable extent. The elements which give rise to colours in ceramic ware are mostly capable of existing in more than one valency state, each of which may have a characteristic colour. The stable valency state of an element depends on the temperature and the oxidation condition of the surrounding atmosphere. Colour matching from kilns is often difficult, particularly with certain colour combinations because even on the same car and in the same chamber, firing conditions are not sufficiently uniform.

High class coloured ware must be fired in muffle kilns where gaseous atmospheres other than air are excluded. Such ware is invariably pre-fired before the colouring ingredients are applied to avoid the potential adverse influence of water vapour or decomposing organic matter.

Impurities in the clay body frequently react with added colouring compounds to produce an entirely different effect from that anticipated. Both lime and magnesia are powerful bleaching agents; the former when present in amounts up to 5 per cent can eliminate completely the normal red colour of a clay containing 6–9 per cent Fe_2O_3. Even smaller amounts of soluble compounds of calcium and magnesium can have a marked effect on surface colorations because they migrate during drying and produce scum.

The nature of fluxes affect colour in so far as they influence the amount and quality of liquid which forms in a clay at high temperature. Some liquid glass compositions readily dissolve certain oxides or compounds of metals whereas others will maintain them in colloidal form.

The Measurement of Colour

It is an undoubted fact that the human eye, when in perfect condition, is the best measuring instrument for colour. The retina is remarkably sensitive and can often detect slight differences in shade and intensity in certain colour ranges which are beyond the resolution of the best instruments. However, colour response varies with individuals; errors due to environmental factors are common and the eye quickly becomes fatigued.

The assessment of colour involves complex factors. Basically, a colour effect can be identified by the wavelength range of light which is transmitted through, or reflected from a surface, coupled with the intensity or amount of the radiation over that spectrum range.

It has been explained earlier in the chapter that when light impinges on a surface, part is transmitted, part is reflected and part is refracted. Some bodies preferentially reflect specific wavelengths; a small-particled material will have a greater number of potentially reflecting surfaces than one with large crystals; a porous surface texture will cause more light scattering than one which is smooth and solid. Even the incident light in which the specimen is examined can have a marked effect on the colour of a specimen. If the incident light is deficient in certain wavelengths in comparison with sunlight, and the specimen would normally reflect some of these wavelengths, its apparent colour will be different from that which it shows in sunlight.

The position is further complicated because the human eye, by which the aesthetic or appeal quality of colour is judged, scans a wide area of surface in assessing a colour effect. It often disregards microscopic detail in tiny areas, but will immediately detect slight overall differences in shade or tint at various parts of the specimen. Instruments for colour measurement, by their very nature, tend to measure opposite factors; they scan small areas with precision but cannot encompass a complete tile or brick surface in a single determination.

For these reasons, the most usual technique for assessing colour is by comparators, in which the eye examines a particular specimen and compares its intensity, shade, gloss, etc., with a series of pre-formed standards. Comparisons must be made under standardised lighting conditions, which in wavelength spread are similar to sunlight but which can be maintained at a fixed intensity. Human error and deficiencies in individual retinal reactions are not uncommon in such methods, but a skilful and practised operator can classify accurately and quickly for short periods of time. Small differences in shade, tint and gloss are unavoidable in fired ceramic ware, but a good and experienced colour assessor can judge the acceptable range of such variations.

Instrumental methods for measuring colour in terms of its shade and intensity depend upon measuring the spectral range of light either reflected or transmitted when the specimen is illuminated by a standard light source. In *spectrophotometric methods* the intensity of radiation over the whole wavelength range of visible light is usually assessed, but where specific colours are involved it may be sufficient to examine only a limited wavelength range. Such measurements are useful but not easy to apply to the definition of a specific colour.

The Nature of Colour. Light is a complex phenomenon; it is defined in terms of radiations within a wavelength range which produce a sensitive reaction in the retina of the eye. The light emitted by the sun and received on earth contains radiation throughout the whole wavelength range which is sensitive to the eye, but the relative intensities in different parts of the spectrum range are not equal and these vary within themselves in different parts of the world and at different times of the day.

Furthermore, the eye itself is not equally sensitive to all parts of the visible spectrum band. There has been much research and speculation on the response characteristics of the human eye to various spectral colours and it is now generally accepted that a curve such as that shown in Fig. XIV.2 represents the sensitivity of the 'average eye' to radiations of various wavelengths within the visible range.

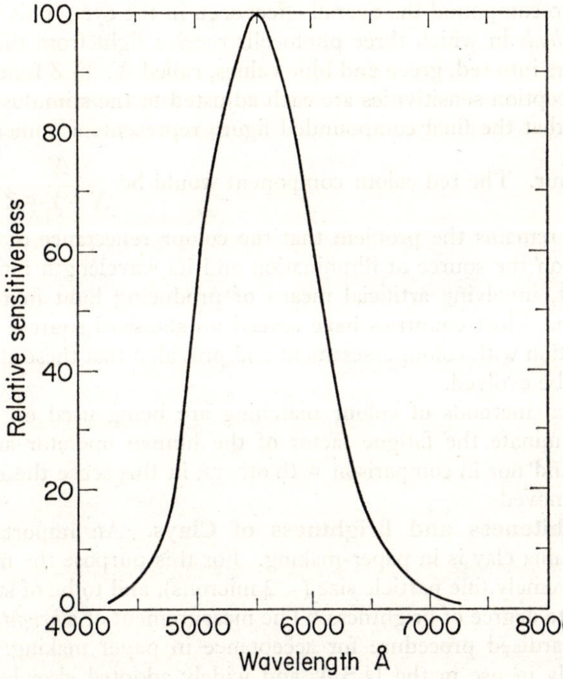

FIG. XIV.2. THE SENSITIVITY OF THE AVERAGE EYE TO LIGHT OF DIFFERENT WAVELENGTHS

The sensation of colour communicated to the brain from the eye is the result of its reaction to various parts of the spectrum and as the figure shows, the greatest sensitivity occurs in the yellow-green region.

This type of curve has led to the concept that the sensation of colour can be defined in terms of the sensitivity of the eye to various specific colours and then to assume that the overall sensation is the summation of the amounts of the individual components.

It is considered sufficient to divide the eye response into three wavelength forms and to ascribe values to these, called *tristimulus values*, which are in the red, green and blue region of the spectrum respectively. The colour that the eye registers is a component from each of the three coefficients. White light for example would correspond to

18·7 units red; 65·5 units green; 9·8 units blue.

These values have been accepted by definition (Commission Internationale de l'Eclairage 1931) and are commonly accepted to ascribe numerical values to light of a particular colour. Standard Chromaticity Diagrams,[1,2] have been drawn up which are types of colour charts.

The instrumental method most commonly used for measuring colour takes account of these values and integrates the individual coefficients for red, green and blue emissions to compound the overall effect seen in the eye. This is achieved by *trichromatic methods* in which three photocells receive light from the source, subdivide it by filters into red, green and blue values, called X, Y, Z functions respectively. Their reception sensitivities are each adjusted to the stimulus response of a normal eye, so that the final compounded figure represents a numerical value for a particular colour. The red colour component would be $\frac{X}{X+Y+Z}$ and so on.

There still remains the problem that the colour reflectance of light from an object depends on the source of illumination and its wavelength range. This is a complex subject, involving artificial means of producing light in relation to the visible spectrum. Most countries have agreed on standard sources of light to be used in conjunction with colour assessment and provided that these are used, useful parameters can be evolved.

Instrumental methods of colour matching are being used on an increasing scale. They eliminate the fatigue factor of the human operator and measure a colour directly and not in comparison with others; in this sense the environmental factor is also removed.

Measuring Whiteness and Brightness of Clays. An important industrial application of china clay is in paper-making. For this purpose the material has to be white, of extremely fine particle size (<2 microns), and to be of smooth texture to give a requisite degree of brightness. The measurement of a *brightness factor* has become a standardised procedure for acceptance in paper making. Details of a method currently in use in the U.S.A. and widely adopted elsewhere have been given by Millman.[3]

The clay to be tested is thoroughly dried (but not over-dried) at 105° C. and then brushed through a 200-mesh screen. A cylinder of the clay $1\frac{1}{2}$–$1\frac{5}{8}$ in. diameter and weighing about 15 grams is pressed under a load of about 40 lbf/in^2 against a glass plate.

The compacted clay cylinder is carefully removed from the glass plate, which

[1] German DIN standard 5033.
[2] American Standards Association Z 58.7.2. 1951.
[3] Millman, N., *Paper Trade Journal*, p. 43, Sept. 1948.

has given it a smooth surface, and tested in a reflection spectrophotometer. It is important that the surface of the specimen which has been formed against the glass plate should not be touched or rubbed in any way; it is equally important that the specimen should be tested as soon as possible before it absorbs too much water. When reabsorption to the extent of more than 2 per cent by weight of water has occurred, the whiteness value is impaired.

The reflection spectrophotometer should be capable of accepting reflected light over a reasonably wide angle to allow for the scatter which normally occurs. If the specimen is inclined at 45°, the angle of collection should not be less than 38–52°.

The total reflected light is concentrated in a prism or lens system and focused on a photocell where its intensity is measured. It is standard practice to measure the intensity at a specific wavelength by inserting a filter of known characteristic between the light and the photocell. The four filters in common use and the wavelength which they transmit are shown in Table XIV.III.

TABLE XIV—III. FILTERS USED IN BRIGHTNESS DETERMINATION

Colour	*Wavelength*
Blue	$4{,}800 \times 10^{-8}$ cm.
Green 1.	5,200
2.	5,500
Yellow	5,800

The intensity of the reflected beam received on the photocell is compared with that given by a standard sample of either barium sulphate, magnesium oxide or carbonate which is assumed to be 100 per cent. The value of brightness can be measured at several different wavelength values and is expressed as a percentage of the standard sample.

To assess *whiteness*, the whole curve of reflectivity must be examined as shown in Fig. XIV.3, which shows results of several materials tested by the author.

The brightness of the material is indicated by the vertical reading at different wavelengths. The whiteness is assessed by observing the horizontal inclination of the curve.

The pure magnesia powder used as standard is assumed to give a horizontal line at 100 per cent throughout the entire spectrum range. The precipitated titanium dioxide has a high brightness value over the whole range, but at wavelengths less than 4000 Angstrom Units, its intensity falls appreciably, indicating that the material has a slight yellow tinge. To be acceptable in paper making, china clays and talcs should have reflectance values in excess of 88 per cent over the range 4000–6000 Angstrom Units indicating that they are both bright and white.

The Removal of Colour from Ceramic Materials

In the foregoing sections of this chapter it has been shown that natural clays and other ceramic materials may be coloured by the presence of iron minerals or by carbonaceous matter. Occasionally titanium compounds may also be present and contribute to colour. In some industries, white raw materials and fired products are essential so that colour, if it exists, must be removed prior to use.

The presence of iron and the colour which it imparts to clay and fired products is frequently objectionable. Various methods are available to counteract the undesirable effects, some of which involve the removal of the iron mineral itself, others counter the unpleasant red by a complementary colour addition, whilst yet others rely upon the bleaching action of additives, usually by converting ferric forms to the ferrous state.

The Removal of Iron Minerals is applicable mainly to materials of large size. Four processes are claimed to be successful:

(*a*) *Washing*, in which the clay is dispersed in water and the slurry passed through a fine lawn (60-mesh). Large ironstone particles are retained along with coal and stones and may be discarded. The purified clay is freed from objectionable iron-spotting which would normally have developed on firing.

(*b*) *Magnetic separation* can often be effective in clays which have been prior-fired to between 500–750° C. Under these conditions, some iron minerals are converted into magnetite Fe_3O_4 and can be removed by magnetic methods.

(*c*) *Chemical solution* usually by acids, although most minerals are not readily attacked and the process is expensive.

(*d*) *Flotation methods* in which iron and other minerals which have a darkening effect are removed from the clay by froth flotation.[1]

Complementary Colours to the brown, reds and yellows of iron compounds are those which are predominantly blue. Clays and glasses which have a slight iron tinge can be made to appear much whiter by adding a small proportion of a blue cobalt compound in finely-divided form. The same effect can be achieved by adding a blue organic colour, but this is effective only in products which have not to be fired. It is an interesting psychological point that a body with a slight yellow or brown tint appears dirty, whilst a blue shade is considered clean—almost 'whiter than white'.

The addition of cobalt oxide to a sedimentary china clay produced the reflectance effects given by curve (*e*) in Fig. XIV.3. The blue intensification caused a lowering of reflectance in the 6000 Angstrom Unit region, but the overall curve became more horizontal, indicating a whiter (or greyer) product. The untreated product shown in curve (*d*) had a greater predominance of red and yellow, but it was brighter in total effect.

Small amounts of cobalt oxide are frequently added to glazes and glasses, to hide a natural brown tint. The amount required must be found by experimentation, but it is rarely effective if the iron oxide content exceeds 0·02 per cent.

Bleaching Agents of many types are employed to disguise iron coloration. In fired products, lime is by far the most effective and when added to clays containing up to 0·5 per cent iron oxide, a fired product of an acceptable whiteness for most purposes can be produced. The lime must be added as a very fine powder and intimately mixed.

Other bleaching agents have been used with considerable success to improve the whiteness of natural china clays, ball clays, talcs, etc. Hydrosulphites of sodium,

[1] U.S. Patent No. 2, 569, 680, 1951.

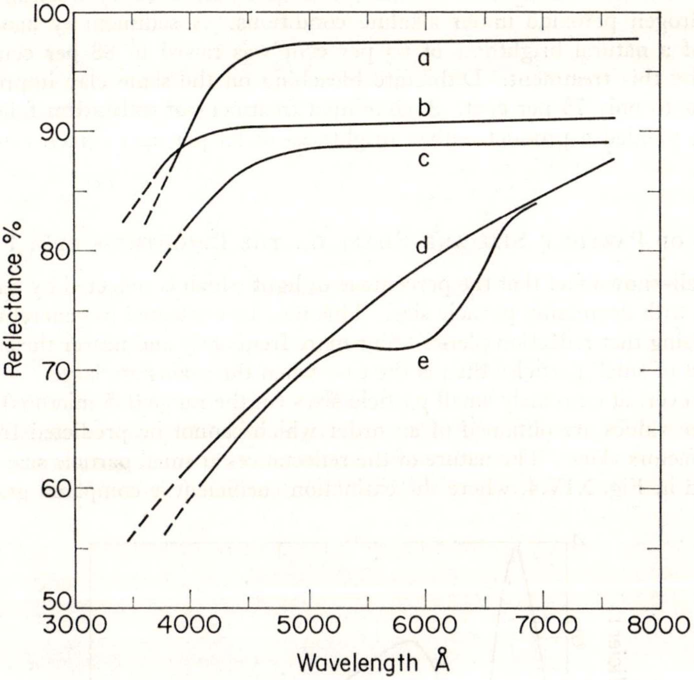

FIG. XIV.3. THE REFLECTIVITY OF VARIOUS MATERIALS TO VISIBLE LIGHT
(a) Rutile filler. (b) Ceylon china clay (– 2 microns), bleached and oxidised. (c) Ceylon china clay (– 2 microns) oxidised only. (d) Ceylon china clay (– 2 microns) untreated. (e) Ceylon china clay (– 2 microns) plus cobalt oxide.

zinc and calcium in amounts up to 0·5 per cent of the raw clay are often effective.[1,2] It is claimed that the greatest benefit results from a double treatment with a stage of filtration and redispersion following the first treatment. Ferric salts are converted to ferrous form by this treatment, with a consequent reduction in the brown coloration. Hydrosulphites are usually called dithionites in the U.K.; they are salts of the organic radicle $(S_2O_4)^{2-}$.

Organic Matter is a common cause of discoloration in raw clays. Low rank materials, such as algaeic or ligniceous forms, are usually of a colloidal nature and give rise to an evenly distributed grey or brown colour. Higher rank materials of a coaly character are more frequently present as discrete grains of a typically black appearance, but fine disseminated forms are also common.

Sedimentary clays commonly contain organic matter and this is a frequent source of discoloration. It is sometimes confused with iron compounds, but, unlike them, it cannot be removed by bleaching.

[1] Huber Corporation, British Patent No. 1,043,252, 1966.
[2] Georgia Kaolin Corp., British Patent 924,536, 1963.

Carbonaceous matter of low rank can often be removed by treating the clay with hydrogen peroxide under alkaline conditions. A sedimentary kaolin from Ceylon of a natural brightness of 65 per cent was raised to 88 per cent (paper quality) by this treatment. Dithionite bleaching on the same clay improved the brightness to only 73 per cent. A combined treatment of **oxidation** followed by bleaching yielded a product with a brightness of 93 per cent (green filter) (Fig. XIV.3).

Effect of Particle Size and Shape on the Brightness of a Powder

It is a well-known fact that the percentage of light which is reflected by a material, increases with decreasing particle size. This may be explained in general terms by presupposing that reflection planes occur more frequently and nearer the surface in a compact of small particles than is the case when the grains are large.

However, at extremely small particle sizes (in the range 0–5 microns) unusual reflectance values are obtained of an order which cannot be predicted from geometrical factors alone. The nature of the reflectances at small particle size values is illustrated in Fig. XIV.4, where the extinction coefficient is compared graphically

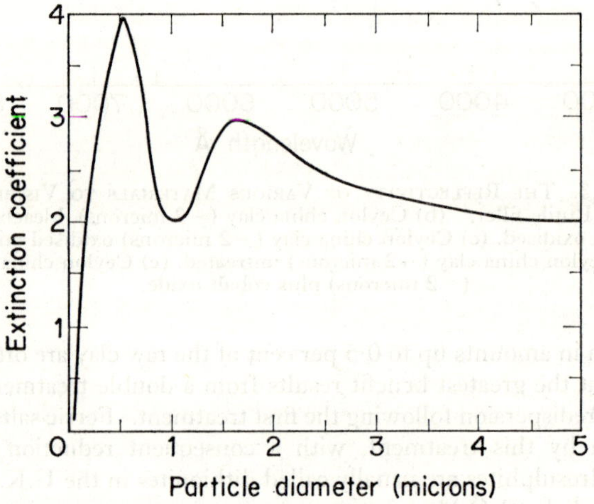

Fig. XIV.4. The Variation of Extinction Coefficient with the Particle size of a Powder

with the diameter of uni-sized spheres of a powder. All powders, irrespective of their composition, show similar characteristics but the absolute values of the extinction coefficients differ. These depend on the refractive index, shape, colour and other properties. The maxima on the curves are also not fixed at particular particle diameters, but depend on the wavelength of light which is being reflected. In the graph illustrated, the abscissa should represent the term

$$\varkappa = \frac{\pi d}{\lambda}$$

because the positions of maximum peak extinction (or reflectance) change slightly with respect to particle diameter (*d*) when tests are carried out with light of different wavelengths (λ).

The principal maximum reflectance value always occurs with particles which are between 0·3–1·0 microns diameter in size, and for obvious reasons the peak is less sharp when white light is used.

The sharp decline in extinction coefficient at very small particle sizes is also of major significance in selecting the most suitable size range to secure maximum brightness in a powder.

Particles of a size smaller than 2 microns are most beneficial and a high percentage of ultimate grains in the size range 0·3–0·6 microns produces optimum brightness. Any crystallites smaller than these diameters markedly reduce brightness and should be avoided at all costs.

These principles apply to the selection of fillers for the paint, plastic and paper industries in particular, but the same problems occur in selecting white additives to produce high gloss in glazes and glasses.

In the selection of china clays for the manufacture of gloss paper, careful particle size selection is essential and two-stage centrifuging is usually involved. Particles greater than 2 microns are rejected in the first unit, whilst the second selects the +0·3 micron fraction from the undesirable debris.

Lyons[1] has shown that the opacity or brightness value is closely related to the percentage of material which is less than 2 microns in size. His results were produced from a Georgia clay and the percentage of material less than 0·3 microns did not exceed 10 per cent. Fernando[2] has examined the brightness factors of Ceylon china clays and shown that when the percentage of particles less than 0·3 microns exceeds 15 per cent by weight, the brightness decreases considerably (Fig. XIV.5).

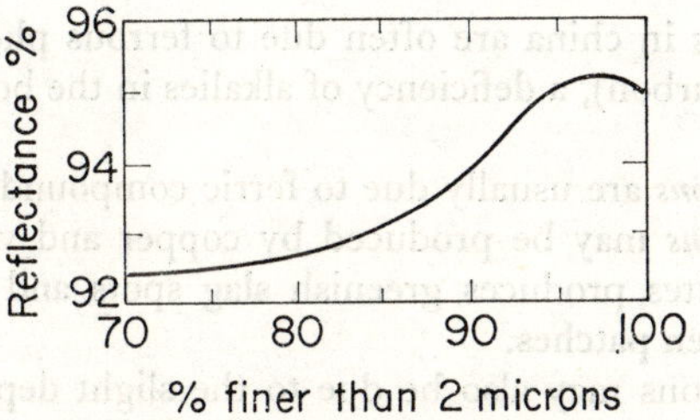

FIG. XIV.5. CHANGES IN REFLECTANCE OF CHINA CLAY WITH PARTICLE SIZE

Kaolinite is an ideal material for use as a filler in paper-making. The plate-like crystallites orient themselves parallel to the plane of the paper during forming and so produce a perfectly flat surface for optimum reflection. Lyons has also shown that particles of kaolinite less than 2 microns in size are mostly ultimate grains and

[1] Lyons, S. C., Paper Coating Pigments, Tappi Monograph Series No. 20, 1958.
[2] Fernando, W. L. W., Ceylon Ceramics Corporation, private communication.

ideal flakes. Larger sized material tends to be present as irregularly stacked layers which do not orient so perfectly.

Well-ordered kaolinites are preferred because they are reported to impart a higher gloss in paper-making than those of poor-order.[1] This may be entirely due to the fact that disordered kaolins tend to contain a high percentage of very fine particles, because when these are removed from Ceylon china clay (which is distinctly disordered) excellent reflectance values are obtained.

The particle shape has an appreciable effect on the brightness (sometimes termed whiteness) values of paint fillers. Titanium dioxide, zinc oxide and tin oxide are valued for this purpose and they are most effective when present in small particle sizes (about 0·5 microns) and when the particles are in the form of perfect spheres. Chemically precipitated qualities are generally preferred.

Discoloration of Ceramic Materials

The term 'discoloration' is applied to materials or articles possessing colours of an undesirable nature upon their surface or in their interior. The principal causes of discoloration are:

(*a*) Insoluble substances.

(*b*) Soluble substances or 'scum'.

(*c*) Substances derived from the fuel and present in the kiln gases.

(*d*) Substances volatilised from articles adjacent to the ones which are discoloured.

Discoloration by insoluble substances includes several colouring agents mentioned in preceding pages which produce an undesirable appearance.

Black spots are usually due to ferrous or manganese compounds, but sometimes white ware fired in silicon carbide saggars is discoloured by grey, red, or black stains.

Blue discolorations in china are often due to ferrous phosphate (derived from bone ash containing carbon), a deficiency of alkalies in the body, or to the action of reducing gases.

Brown discolorations are usually due to ferric compounds.

Green discolorations may be produced by copper and vanadium compounds; thus, cupriferous pyrites produces greenish slag spots and vanadium molybdate produces irregular green patches.

Green discolorations may also be due to the slight deposition of soot on the goods.

Green spot on glazed ware can be particularly troublesome. If it is of regular and consistent occurrence it is probably caused by minerals which contain copper being present in the clay. On the other hand, a sudden incidence of the fault is more likely to be due to adventitious copper contaminating the ingredients. Copper wire used in blasting operations in the mine or quarry, severed ends of cable introduced by careless electricians, residual copper scrap in lorries used for transporting clay, are likely sources of trouble.

On one occasion, the author counted over one hundred serious green spots on

[1] Murray, H. H., and Lyons, S. C., 4th Nat. Clay Conf., Oct. 1955.

sanitary fireclay articles which were later attributed to a 12 in. length of welding rod which had been dropped into the clay grinding pan.

Pink discolorations on biscuit ware or buff terra-cotta often indicates that the ware has been heated too rapidly below 700° C., so that the combined water has not been driven off properly. This fault is known as *steaming*.

The red discoloration in some hard porcelain, known as *la malade jaune* or *jaune de cuisson*, is due, according to B. Moore and J. W. Mellor,[1] to the presence of ferric oxide and of oxidising conditions in the first stages of burning.

Yellow discolorations are often due to ferric compounds. For means of correcting a yellowish tinge in china and white earthenware, see p. 920.

Scum is a defect known by a variety of names, such as *whitewash*, *nitre*, *salt*, *mould* (all of which are incorrect), and *florescence*, *efflorescence*, *wall-white*, *kiln-white*, etc., which more accurately describe it.

In recent years it has become customary both in Great Britain and the United States to reserve the term *scum* for substances actually formed or deposited during processing as in the dryers or kilns, and *efflorescence* for effects caused by materials originally present in the clay or formed during firing which show up only after periods of exposure to the weather.

Scum. The chalky-white or vari-coloured film or deposit which is occasionally observed on ware at some stage of the processing is referred to as scum. It has been attributed to one of the following:

(*a*) Soluble salts present in the clay being transported to the surface of the ware during the removal of water. These are chiefly calcium, magnesium, potassium, sodium, ferrous or aluminium sulphates and occasionally chlorides and nitrates.

(*b*) Substances deposited on or formed on the surface during the drying or burning, either by condensation products from the hot gases or by some chemical reaction taking place within the kiln. Sulphurous fumes are the most objectionable because these readily form sulphates in the presence of moisture.

(*c*) The migration to the surface due to incorrect drying practice of fine-particled material contained in the clay. The surface layers of bricks dried on hot floors are frequently enriched in clay in this way but this may not cause a serious deterioration in fired colours; finely-divided calcium carbonate, usually present as chalk may also be concentrated and in such clays the fired colour is always affected. Fine silica sand in a clay can also cause troublesome effects.

White scums are the commonest and include 'kiln white' and 'dryer white'. The former consists chiefly of calcium sulphate with small amounts of magnesium and alkaline sulphates, and occasionally a trace of alum. It is more usually attributable to the action of sulphurous fumes in the kiln gases which act on the lime, etc., in the clay.

A *brown scum* on some fired goods may be due to the presence of soluble iron salts formed by the oxidation of pyrites in the clay or rock to ferric sulphate which, being soluble, rises to the surface of the moist material by capillary action during the drying and is converted to ferric oxide during the firing.

A *grey scum* is sometimes formed if calcareous water comes into contact with a

[1] *Trans. Eng. Ceram. Soc.*, 16, 58, 1917.

clay. Calcium and magnesium sulphates, when present in a clay, also tend to impart a drab appearance.

A *yellowish scum* may be formed by the interaction of sulphuric acid in the kiln gases with alumina, lime, and silica in the clay at temperatures approaching that of vitrification. Seger found a yellowish-green scum to be produced by the presence of soluble vanadium and molybdenum compounds.

When calcium or magnesium sulphates are present in the clay itself, scum can usually be prevented by adding barium carbonate in accordance with the equation

$$\left.\begin{matrix}Ca\\Mg\end{matrix}\right\}SO_4 + BaCO_3 = \left.\begin{matrix}Ca\\Mg\end{matrix}\right\}CO_3 + BaSO_4$$

soluble soluble insoluble insoluble

The calcium and magnesium sulphate content of clays can be estimated by agitating a known amount (about 100 g.) in a fixed volume of water. After two hours' stirring, a small quantity can be filtered by pressure methods. An aliquot portion of the filtrate can be tested for calcium and magnesium by E.D.T.A. methods, and the sulphate estimated gravimetrically by precipitation with barium chloride.

A content of soluble sulphate less than 0·1 per cent in amount does not usually cause an objectionable amount of scum. 1·0 per cent or more will frequently bleach the whole surface layers of brick and change the colour entirely. The most serious concentration is in the range 0·2–0·5 per cent of soluble sulphate which will form a scum or discoloration on corners, edges and exposed surfaces in an irregular fashion.

If the content of soluble sulphate is about 0·2 per cent, the objectionable scumming effects can be eliminated by incorporating barium carbonate with the clay mix at the rate of 5 lb. per ton.

The prevention of scum is often effected by drying the articles correctly and thoroughly in an atmosphere free from sulphurous fumes. During the initial firing the sulphur content of the gases should also be kept to a minimum in order to reduce 'kiln white'. 'Dryer white' can be reduced only by careful humidity drying or by reducing the content of soluble salts in the raw clay (see also under efflorescence).

Scum is undesirable in that it leads to an unsightly appearance of the ware. The localised concentrations of such salts as calcium and magnesium in the surface layers of the body may seriously affect the colour on firing and so destroy the required red shade of many facing bricks.

Efflorescence. Some bricks are apparently perfect in colour after firing, but when exposed to the weather for a period of time, an unpleasant film or scurf forms which is usually white or grey with a characteristic crystalline appearance. This is efflorescence and it is due to a variety of causes:

(*a*) Soluble salts occurring naturally in the clay itself.

(*b*) Soluble salts produced from other ingredients in the material as a result of weathering.

(*c*) Soluble salts produced by interaction between the component minerals in the clay during firing, or by the action of furnace gases on some component within the clay.

(*d*) Soluble salts contained in the mortar used in erecting a building of the bricks.

(*e*) Soluble salts developed by interaction of the mortar and bricks or tiles.

(*f*) Soluble salts in the water used in the manufacture of the ware, or which later gains access to the structure.

(*g*) Soluble salts in the ground on which the goods are stacked, or in ashes or other materials which come into contact with the ware.

Similar colorations may appear as in scumming and the same salts may cause both these defects, though the formation of each is rather different. In a brick which exhibits efflorescence, the soluble salts are initially uniformly distributed throughout the body. Successive impregnations and dryings when exposed to the weather, result in the soluble components being brought to the surface by capillary action. Many salts, for example, calcium and magnesium sulphate, absorb water from the atmosphere and form a hydrate, and in the process undergo a large volume expansion. When this occurs within the surface of the brick, especially if the strength is low, crumbling and disruption may result. Calcium sulphate, in amounts as low as 0·1 per cent may cause serious efflorescence, but magnesium sulphate is even more deleterious and 0·01 per cent of this component may produce undesirable results.

Soluble sulphates are not of frequent occurrence in raw clays. They are liable to form, however, when pyrites and carbonates are present together and when the firing is not wholly under oxidising conditions. Sulphurous gases in the kiln may also react on the carbonates and other minerals to form the corresponding sulphates.

Efflorescence may be prevented or reduced in the following ways:

1. By adding barium carbonate or chloride to the clay; this converts any soluble sulphates into insoluble barium sulphate and renders them harmless. This treatment is not effective for other salts which cause efflorescence.
2. By maintaining oxidising conditions during the initial stages of burning and reducing the sulphur content of the kiln gases to a minimum.
3. By avoiding the use of water or mortar containing soluble salts.
4. By avoiding contamination of the material at all stages of manufacture.
5. By painting the surface of brickwork with a solution of a silicone or other organic silicate compound. Although this is an expensive method it is most effective.

The *efflorescence tendency* of a brick can be measured by the Wick test or a modification of it. A whole brick is set on end in a bath of distilled water to a depth of 1 in. and left for seven days. The level of water must be maintained constant throughout this time. Water is taken up in the pores of the brick, dissolves any soluble salts and transports them through the brick to the upper parts where surface evaporation occurs. After seven days, the brick is thoroughly dried in an oven and afterwards examined for discoloration effects. The American Society for Testing Materials specifies that at least five bricks should be tested, but on no account must bricks of more than one type be placed in the same pan of water otherwise contamination of one by others may occur.

Mottling in silica bricks is a form of scumming where acid fumes in the kiln gases react with iron-bearing minerals in the moulded brick. This occurs only in certain regions so that the surface has a characteristic mottled appearance with large patches of red coloration on a yellow or buff background. These stains are restricted to the surface layers and do not influence the properties of the brick in any way.

Mottled surfaces on glazed wares are due to a variety of causes, including defective protection when placing in the oven or kiln, imperfect mixing of the ingredients in the preparation of the body, engobe or glaze, and the shape of an article which causes a coloured glaze to flow irregularly. After firing in the kiln, bricks which normally burn to a light colour, sometimes have a surface discoloration which may be red or brown. This effect may be confined to one or two bricks or to parts of bricks, but in some cases a complete setting may be influenced. This is known as *steaming* and the discoloration is usually characteristically streaked, so that it is often mistaken for 'flame-flash'.

Steaming is caused by (*a*) the ware being set in a wet condition and heated too rapidly in the initial stages of kiln firing, or (*b*) warm, humid gases circulating round the ware during firing and water condensing on the cold articles. Although it has not been definitely established, it is most likely that sulphurous fumes in the kiln gases combine with iron compounds in the clay when water is present, render them soluble and then they are drawn to the surface during drying out.

The discoloration effects of steaming spoil the appearance of the ware, but it is indicative of a serious condition in the kiln. When the amount of water present in the kiln in the early stages of firing, becomes excessive, the bricks are liable to become fused together or to crack. In serious cases, the complete setting in a kiln may slump or collapse.

MINOR PROPERTIES

Ring. The clear ringing note or 'ring' which some burned ceramic materials emit when struck is often a good indication of the extent of the burning and the absence of cracks. A dull 'ring' is usually due to the presence of small cracks, some of which may cause serious trouble if used under exacting conditions, whilst others are so minute as to be of little or no importance.

Whilst of considerable use as a rough test, too much reliance must not be placed upon the sound emitted when a sample is struck, because the note depends on so many factors, some of which may affect the 'ring' without being necessarily harmful.

Feel. Most ceramic materials and many others have a characteristic 'feel', which may be (*a*) smooth, (*b*) rough, (*c*) meagre or harsh, (*d*) greasy, soapy, silky, or unctuous. Most refractory materials belong to the first three groups, but many clays are included in the last one. Some materials may be classed in two groups simultaneously; thus, china clay and dry ball clay are smooth and unctuous; many fireclays feel 'rough', but a freshly-cut surface, when rubbed with the fingers, has a slightly greasy feel. As the tongue is often more sensitive than the fingers, it has long been the custom to compare some fine clays by placing a small portion in the mouth and 'working' it with the tongue. By this means the presence of a very small proportion of grit in an otherwise impalpable material is readily recognised.

Experts can readily distinguish different varieties of porcelain and other ware by the 'feel', and this property is often very useful as a supplementary indication of the nature of a material.

Odour. Many clays, when moist, have a characteristic earthy odour, which probably is due to carbonaceous matter present. When the clay is heated to redness, the characteristic odour is lost. Impurities may sometimes be detected by the odour they emit, especially when heated. Thus, some clays and alum shales have a sulphurous odour on account of the pyrites present in them. This is particularly noticeable when the clay is freshly cut or heated to redness in a closed vessel.

Sectility and Fissibility are two closely related terms, indicating that a material to which they are applied can readily be cut, in at least one direction. Shales are of this character, though the 'cutting' is possibly more in the nature of 'splitting', or separating the existing laminae of which the material is composed. In order to be sectile a material must usually be moderately soft, but a hard material composed of thin sheets united by a soft cement will be fissile in a direction parallel to the sheets.

The most sectile clays are ball and china clays when almost dry, and also some of the surface clays. Harder or leaner clays are friable rather than sectile, so that when a knife is applied to them they are crushed rather than cut.

Most ceramic materials lose their sectility when heated to redness, but this property is retained to a large extent by certain rather soft bricks (known as *cutters*), and by many tiles.

Friability may be regarded as the converse of resistance to abrasion; it is considered more fully in Chapter XIII.

Acoustics and Sound Absorption. Although the walls of most dwelling houses are built of stone or bricks and the interior ones are covered with plaster they are not efficient absorbents of sound—seldom absorbing more than about 3 per cent of the sound reaching their surface. Thus brickwork walls 9 in. thick will allow conversation to be heard through them but it is seldom intelligible to the listener. Walls $13\frac{1}{2}$–$31\frac{1}{2}$ in. permit sound to be transmitted to the extent that a noise can be heard but conversation is unintelligible unless the voices are raised. It requires brick walls at least 36 in. thick to render conversation on one side wholly inaudible on the other. Hollow blocks tend to transmit sounds much more readily than solid walls as the insulation of sound depends chiefly on the mass of the insulator.

In sound-insulation the structure, rigidity and hardness of the surface are the most important factors and non-ceramic materials can provide these more effectively than ceramic ones (see also p. 441).

THERMAL PROPERTIES

MANY of the thermal properties of ceramic materials have been described in Chapter XII, in connection with the physical changes which are brought about by the effect of heat. There are three other thermal properties of materials which are important in ceramic practice; these are (*a*) the thermal capacity or the specific heat, (*b*) the thermal conductivity and diffusivity, and (*c*) the thermal shock resistance.

The Thermal Capacity or Specific Heat

VARIOUS materials appear to have different capacities for heat, i.e. equal masses of different substances which are all at the same temperature absorb different quantities of heat before they are raised to a given temperature. The ability of a substance to absorb and retain heat is termed the thermal capacity of that substance, and in order to compare the thermal capacities of different substances a standard of measurement is used, termed the *specific heat*, which may be defined as the amount of heat required to effect an increase of one degree in the temperature of unit mass of the substance under a definite pressure (usually one atmosphere). For convenience the specific heat of water is assumed to be unity, and other substances are expressed in relation to this.

The physical explanation of the specific heat of a pure crystal is afforded by a study of the forces acting between adjacent atoms or ions in the lattice. As Fig. XIV.6 shows, there is a certain separation distance *a* between ions in a crystal lattice at which the forces of attraction and repulsion exactly balance. As they approach one another more closely, repulsion forces tend to restore the equilibrium position, as do attraction forces when the separation distance becomes greater than *a*. There is, however, a small distance through which the ions vibrate and this becomes greater as the temperature increases, but in this process heat is absorbed equivalent to the amount of energy expended. Those crystals in which the resultant force line, Fig. XIV.6, is steeply inclined in the neighbourhood of the equilibrium position will absorb more energy in vibrating over a certain distance than those where the resultant force line is less steep. So when heated through a certain temperature range some crystals absorb more heat than others, i.e. their specific heat is greater.

Crystals with the greatest specific heat are those in which the ions are close-packed and those in which the cations and anions have widely-different electronegativities (Chapter II).

The specific heat depends on whether the substance is in the crystalline or glassy state, or, if the former, what is its crystalline form and internal molecular arrangement. Thus, crystalline quartz has a specific heat of 0·185 between 12° and 100° C., whilst the specific heat of amorphous silica (opal) between 12° and 100° C. is 0·2375.

The specific heat is not constant at all temperatures, so that it is necessary to specify the particular rise in temperature employed in the definition; this may be from 0°–1° C., 32°–33° F., 60°–61° F., or any other convenient range. The temperature should be stated in any report of the results.

The specific heat is usually fairly constant between 10° C. and 200°–300° C., but boron, carbon, and silicon are exceptions, and vary very greatly between these temperatures. At higher temperatures the specific heat of ceramic materials increases somewhat more rapidly than the corresponding rise in temperature.

Vogt has stated that the specific heat at different temperatures may be calculated from the formula:

$$C_t = C_o(1 + 0{\cdot}000078t)$$

where C_t is the specific heat at temperature t° C., and C_o is the specific heat at 0°C.

The specific heat of substances is generally greater when they are in the liquid than in the solid and gaseous states.

The specific heat of kaolin over the range 0°–1000° C. averages 0·5, and that of ball clay 0·45; for smaller ranges above 1000° C. it tends to be about 0·2.

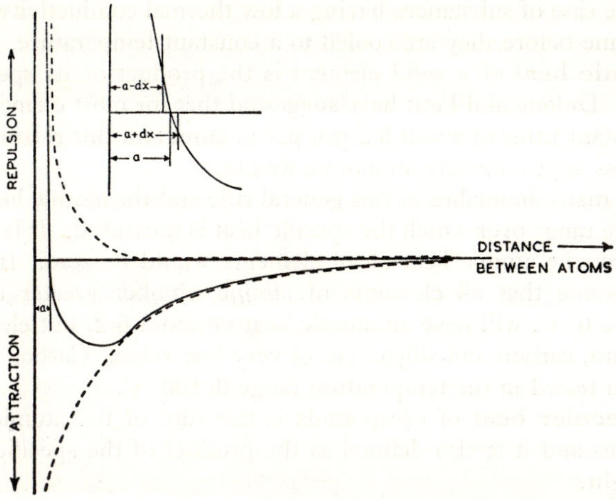

FIG. XIV.6. THE FORCES OF REPULSION AND ATTRACTION BETWEEN ADJACENT ATOMS IN A CRYSTAL
The resultant is shown as a solid line

A *determination of the specific heat* of a substance is usually made by means of the method of mixtures,[1] by heating a weighed quantity of the substance to the required temperature for some time until it is uniformly heated. It is then quickly transferred to a suitable calorimeter containing a weighed quantity of cold water at a known temperature, and the latter is stirred so as to distribute the heat. The temperature of the water is noted at intervals of half a minute or so, until a constant temperature is reached. The specific heat is then calculated from the following formula:

$$\text{Specific heat} = \frac{W_2 \times (T_3 - T_2)}{W_1 \times (T_1 - T_3)}$$

where W_1 is the weight of the substance in grams.
W_2 is the weight of the water in grams.
T_1 is the maximum temperature of the substance.
T_2 is the original temperature of the water.
T_3 is the final temperature of the water.

The above formula is based on the assumption that the quantity of heat in the substance at the given temperature is W_1T_1s, where W_1 and T_1 have the same significance as in the formula and s is the specific heat at the temperature T_1.

The heat in the water prior to the determination is W_2T_2, and after the determination it is W_2T_3, and as the whole of the heat lost by the substance is assumed to be transferred to the water $W_1(T_1 - T_3)s = W_2(T_3 - T_2)$.

Where accurate results are required, a special calorimeter must be used and numerous precautions taken to avoid the loss of heat by radiation. This is especially

[1] Cf. Cohn, W. M., *J. Amer. Ceram. Soc.*, 10, 347, 1927.

necessary in the case of substances having a low thermal conductivity, which take a considerable time before they are cooled to a constant temperature.

The **atomic heat** of a solid element is the product of its specific heat and atomic weight. Dulong and Petit have suggested that for most elements the atomic heat has a constant value of about 6·2 cal. per g. atom and this relationship is often utilised to assess approximately an atomic weight.

There are many anomalies to this general rule and the atomic heat depends on the temperature range over which the specific heat is measured. It is predicted that at absolute zero, the atomic heat of all elements would be zero. It is reasonably accurate to assume that all elements of atomic number greater than 35 when measured above 0° C. will have an atomic heat of about 6·2, but elements such as boron, beryllium, carbon and silicon are of very low value. Carbon has a value of about 1·5 when tested in the temperature range 0–100° C.

The **molecular heat** of compounds is the sum of the atomic heats of the individual atoms and it is also defined as the product of the specific heat and the molecular weight.

To a first approximation the molecular heat is equal to 6·2 N where N is the total number of atoms in the molecular formula. This is a useful concept in determining the atomic weight of complex minerals and compounds.

The heat capacity of ceramic materials is often very important, especially when they form part of the walls of furnaces, etc., as all the heat absorbed by such material is, in a sense, wasted and ought to be used in heating the contents of the furnace.

The specific heat of most materials increases with the temperature, so that the fuel consumption when heating ceramic materials at high temperatures is much greater than at low temperatures, the same weight of material being used in each case. Thus, in fireclay bricks, the relation between the fuel consumption at 1200°–1400° C. and that at 100°–300° C. is nearly 3:2.

Clays. The specific heat of raw clay is about 0·237, that of clay heated to 650° C. 0·204, and that burned at 1050° C. is 0·200 C.G.S. units, the essential difference is the deyhdration of clay which causes a decrease in the specific heat. The specific heat at 1050° C. is practically the same as that at 650° C. The specific heat of kaolin at different temperatures is shown in Table XIV.IV.

TABLE XIV—IV. CHANGES IN SPECIFIC HEAT OF KAOLIN

Temperature, ° C.	*Specific Heat*
—	0·235
22–98	0·2242
440–1000	0·235
650	0·204
1050	0·200

The specific heats between 500–1050° C. of a number of materials of a refractory nature have been measured by Clements.[1] He suggests that the relationship

[1] Clements, J. F., *Trans. Brit. Ceram. Soc.*, 61, 452, 1962.

TABLE XIV—V. MEAN SPECIFIC HEAT OF SOME REFRACTORY MATERIALS (c.g.s. units)

Temperature ° C.	*Firebrick*	*Silica Brick*	42% Al_2O_3 *Brick*	*Mullite Brick*	*Alumina Brick*	*Magnesite Brick*	*Chrome Magnesite*	*Dolomite Brick*	*Zircon Brick*	*Diatomite Brick*
20	0·167	0·211	0·164	0·195	0·192	0·233	0·189	0·203	0·140	0·185
500	0·228	0·240	0·230	0·241	0·245	0·263	0·224	0·237	0·165	0·232
700	0·245	0·249	0·247	0·254	0·259	0·272	0·235	0·247	0·172	0·245
900	0·256	0·257	0·258	0·264	0·268	0·281	0·245	0·256	0·178	0·254
1050	0·261	0·262	0·263	0·270	0·272	0·286	0·250	0·261	0·181	0·259

between mean specific heat and temperature can be represented by an equation of the form

$$S = a + bT - cT^2$$

The value of S is defined as the *mean value of the specific heat* between 20° C. and the temperature T in °C. of the test, and a, b and c are constants.

Some values of specific heats determined by Clements are shown in Table XIV.V.

The values of a are those given for the specific heat at 20° C.; b also varies with the material, and ranges from $0{\cdot}5\text{–}1{\cdot}7 \times 10^{-4}$ but Clements' figures throw no light on its significance. In alumino-silicates, it appears to increase to a maximum with 42 per cent alumina bodies. The constant c has less significance, except at very high temperatures, because its value is in the range $0{\cdot}8\text{–}7{\cdot}0 \times 10^{-8}$, but it varies in a similar fashion to b.

The values quoted are mean specific heats, but a *true specific heat* at a particular temperature is defined as the calories per g. to effect a change of 1° C. at that temperature.

Clements has also derived constants from which true specific heats can be calculated over a range of temperatures up to 1050° C. The equation can be represented as

$$S_{True} = a + b(2T - 20) - c(40T + 3T^2)$$

where a, b and c have the same values as before.

Table XIV.VI shows the variation in true specific heat of some forms of pure silica. The inversional change in quartz brings about a pronounced reduction in value.

TABLE XIV—VI. SPECIFIC HEAT OF VARIOUS FORMS OF SILICA

Temperature ° C.	*Quartz Glass*	*α-Quartz*	*β-Quartz*	*Cristobalite*
100	0·202	0·204	—	—
250	0·236	0·244	—	—
500	0·266	0·294	—	—
550	—	0·313	—	—
750	0·280	—	0·277	0·278
1000	0·290	—	0·288	0·285
1100	—	—	—	0·287

The Transmission of Heat

When the surface of a body is heated, the rise in temperature is not restricted to the external layer, but the heat is transmitted through the mass. Similarly, if a hot body is placed in contact with a colder one, heat is transferred to the latter.

The transmission of heat from one body to another may be effected in various ways:

(*a*) By **conduction**, in which the heat is transferred by bond vibration of the hotter substance to the cooler substance and so setting up a corresponding motion amongst the atomic structure in the latter; the acceleration of motion progresses at a definite rate through the second substance. The ease at which the heat passes through a material—as measured by the rise in temperature of the latter—is termed *thermal conductivity*.

As a result of their different atomic structures and electron configurations, materials have different powers of conducting heat. Most metals have a higher thermal conductivity than non-metals or compounds, i.e. they conduct heat rapidly through their mass. Ceramic materials are, on the whole, very poor conductors of heat, though some, such as silicon carbide, have a higher thermal conductivity than others.

The fundamental equation governing the conduction of heat may be expressed as:

$$Q = -kA\frac{dT}{dx}$$

where Q is the amount of heat flowing through a body of cross-sectional area A, and dT/dx is the temperature gradient in the direction perpendicular to A.

(*b*) **convection**, in which the heated particles move away from the hotter to the cooler parts of the mass and carry 'heat' with them. Convection can only occur in fluids, as the particles in a solid are not sufficiently mobile.

(*c*) **radiation**, in which the heat is carried neither by conduction nor convection, but in a manner comparable to the transmission of light. Heat may be radiated instantaneously through space as well as through air and other gases, and when so radiated it scarcely affects the temperature of the medium through which the 'rays of heat' are passed. There is a very close relationship between radiated heat and light. Both can be reflected by mirrors and deflected or refracted by prisms and lenses; in fact, the chief difference between them appears to lie in the difference in wavelength.

Hence, when a body is sufficiently heated, the heat rays emit a form of light and the body is said to be 'incandescent'. When heat rays are absorbed by any substance the latter is heated and its temperature increased in proportion to the amount of heat absorbed, but the air through which the heat rays pass may remain at a much lower temperature. This is due to the fact that radiated heat is absorbed more by some surfaces than others, the nature of the surface being of greater importance, in this respect, than the composition of the heat absorbing material. Similarly, the amount of heat radiated from a body varies according to the colour and nature of the radiating surface, being low for polished metal and high for rough, black surfaces. In the latter, it is proportional to the fourth power of the absolute temperature:

$$E = k(T_1^4 - T_o^4)$$

where E denotes the radiation from a body at T_1, to one at T_o and k is a constant. The radiation loss, apart from convection, is usually (for a temperature difference of 100° C.) 0·015 gram-cal. per second for each cm. of heat-radiating surface.

Factors Influencing Thermal Conductivity. The chief factors to be considered in connection with the rate of passage of heat through ceramic materials are:

(*a*) The chemical composition of the material used.

(*b*) Its previous heat treatment.

(*c*) The texture or physical condition of the material.
(*d*) The porosity of the material and the sizes and distribution of the pores.
(*e*) The temperature at which the material is used or tested.

Thermal conductivity is a property which is, as yet, not fully explained on the basis of theoretical reasoning. Debye[1] considers that the atomic or ionic groupings within a crystal act as oscillators or vibrators and if these frequencies correspond to those of the thermal waves, conduction will readily proceed. On this basis, highly symmetrical crystals where the vibrations are harmonic will be better conductors of heat than those which are disordered or contain many types of bond. Whilst this simple concept applies in general terms, there are many discrepancies in experimental findings which suggest that other factors are involved.

Thermal and electrical conductivity are closely related properties, so it is a safe presumption that the nature of the electron linkages and orbital transfers play a large part in the transfer of heat as they undoubtedly do in electrical conduction.

Materials differ with regard to their thermal conductivity, which also changes with the heat treatment to which they have been subjected, just as the temperature and duration of firing during manufacture modifies other properties of such materials. It is also important to know the condition of the material during use, as the thermal conductivity of a brick or block may be quite different from that of the same substance when in the powdered state. Thus, magnesia when in the form of a refractory brick is a moderately good conductor of heat, whilst in the form of a powder it is extremely resistant to heat and has a high insulatory value.

The influence of texture and porosity may be considered together, as the principal effect on the thermal conductivity is due to the relation between the amount of solid and of air which the heat has to traverse in passing through the material. As air is a much better insulator than any solid material, the larger the proportion of air the greater will be the thermal insulating power of the material. Hence, a fine-grained, close-textured material has a much greater thermal conductivity than one with a coarser open texture. The relation between insulating power and texture or porosity cannot, however, be expressed in very simple terms, as it is modified by (*a*) temperature, (*b*) the size, and (*c*) the shape of the pores or interstices, (*d*) the position of the interstices relative to each other and to the solid matter.

If the rate of radiation increases until it equals the rate of conduction through a solid the pore spaces will cease to act as insulators. With pores 0·01 cm. diameter this equality of heat transfer should occur at 3600° C. At lower temperatures, or with wider pores, the rate of radiation is lower than that at which the heat passes through the solid material, so that the presence of pores in materials used at any temperature ordinarily attainable decreases the thermal conductivity.[2] A. T. Green[3] has, however, stated that at much lower temperatures (e.g. 1400°–1500° C., and even at 1150° C.) some of the pore spaces lose their insulating properties and transmit heat at nearly the same rate as the solid matter. If this statement is correct, pore spaces in ceramic materials do not have so great an insulating power at high temperatures as is usually assumed.

[1] Debye, P., *Amer. Physik.*, 39, 789, 1912.
[2] There are some exceptions to this statement.
[3] *Trans. Eng. Ceram. Soc.*, 21, 394, 1921–22.

It might well be that convection plays a greater part in the transmission of heat through a porous material than has hitherto been suspected. Pores are filled with air and if there is a considerable degree of interlinking, air movement will be a natural sequence when temperature gradients occur. By this means, thermal insulation will be decreased.

An important distinction must be made between open and closed pores (see p. 419), because the former permit the passage of hot gases and thus may increase the effective thermal conductivity. A permeable brick is invariably not as good in thermal insulation characteristics as a non-permeable one of the same porosity.

In comparing the thermal conductivity of different substances, the unit is the amount of heat which passes in one second through a mass of the material of unit thickness and area when the difference in temperature of opposite faces is one degree. If the C.G.S. unit is adopted the thermal conductivity will be expressed as gram-calories per second per sq. cm. of material 1 cm. thick for a temperature-difference of 1° C. The British Unit of thermal conductivity is the number of B.T.U. per second per sq. in. for material 1 in. thick for a temperature-difference of 1° F. Another convenient British Unit is the B.T.U. transmitted per sq. ft. of material 1 in. thick in one hour for a temperature-difference of 1° F.

The smaller British unit can be obtained by dividing the conductivity in C.G.S. units by 178 and the larger British unit by multiplying the C.G.S. figure by 2903. It is standard American practice to use the larger unit (B.T.U. basis).

The reciprocal of thermal conductivity or resistance offered by unit mass of a substance to the passage of heat is termed its *resistivity* or *insulating power* (p. 947).

The thermal conductivity of a solid substance, such as a ceramic material, is usually determined by exposing one face to a constant source of heat and measuring the temperature at opposite faces, or at the hot face and at a point in the material a convenient distance from it. The greatest source of error occurs in applying the heat uniformly to the hot face and ensuring its uniform distribution through the material to be tested, so the various methods which have been devised for determining the thermal conductivity differ in the means used to avoid this source of error.

One of the earliest references to a method of measuring thermal conductivity was by Wologdine[1] in 1909, and the same basic principles are involved in the more common types of apparatus in use at the present time.

The methods commonly used in English practice are based on the apparatus developed by Blakeley and Cobb.[2] The design is shown in Fig. XIV.7.

For the purposes of the test, a whole brick (9 in. × 4½ in. × 3 in. or 2½ in.) is used. It is placed face downwards on an electric heater comprising a flat steel or brass plate fitted over a tile wound with resistance wire or tape. The top surface of the brick is covered with a blackened metal plate into the centre of which is fitted a thermocouple T_3. Further thermocouples (T_2) are placed on the upper surface and (T_1) beneath the lower surface of the brick in contact with the heater plate and it is advisable to recess this into the brick, otherwise there may not be contact between the brick and heating element at all points.

[1] Wologdine, C. T., *Bull. Soc. Encour.*, 3. 879, 1909.
[2] Blakeley, T. H. and Cobb, J. W., *J. Soc. Chem. Ind.*, 51, 237, 1932.

The brick and heater assembly is enclosed in a box lined with diatomite blocks, and diatomite or alumina powder filled into the gap between the brick and the box.

A steady current is applied to the heater element and measurements of the temperature on the top and bottom faces of the brick (as indicated by the thermo-couples T_2 and T_1) and measurement of T_3 (to indicate the heat loss) should be taken when they have reached a constant value. As the thickness of the brick and the temperature gradient across it are known, the thermal conductivity can be calculated. Determinations of the conductivity at various hot face temperatures can be carried out by regulating the current supplied to the heater.

The American Society for Testing Materials (1947)[1] has adopted the method described by Norton in 1942.[2] The heating element must be capable of giving a hot face temperature over the range 400–2800° F. and be equipped with controlling devices to ensure a constancy of ±2·5° F. A water calorimeter is employed to measure heat flow and end losses are reduced by using similar bricks along the sides of the test brick and soap bricks against the end faces.

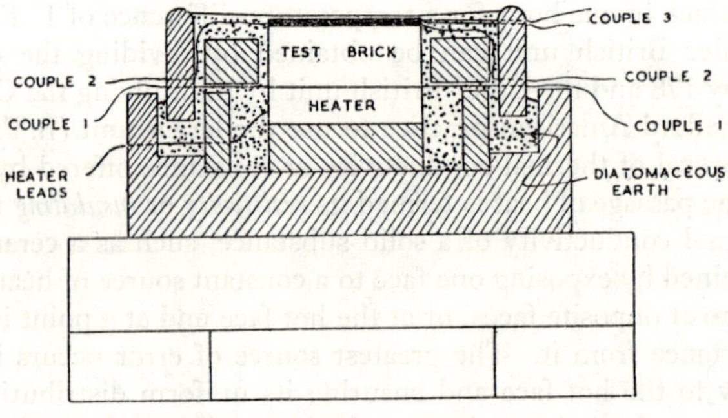

FIG. XIV.7. AN APPARATUS FOR MEASURING THE THERMAL CONDUCTIVITY OF CERAMIC MATERIALS

Clements and Vyse[3] have described in detail an apparatus for measuring thermal conductivity values of bricks using similar techniques to those outlined above, but suggesting several precautions which enable more accurate results to be obtained.

The thermal conductivity of common bricks is often measured by the hot wire method in which a heating element is inserted longitudinally down a hole drilled in the centre of the brick. Thermo-couples are used to measure the temperature gradient throughout the brick at different hot wire temperatures.

Thermal conductivity is a measure, under steady state conditions, of the heat flow per degree of temperature through a unit area of cross-section of specimen of unit length in unit time.

[1] *A.S.T.M.*, Designation C.201–47.
[2] Norton, F. H., *Refractories*, 2nd edn. (McGraw-Hill, New York, 1942).
[3] Clements, J. F. and Vyse, J., *Trans. Brit. Ceram. Soc.*, 53, 134, 1954.

It may be represented by the formula

$$k = \frac{QL}{A(T_1 - T_2)}$$

where Q is heat flow per unit time, L and A are respectively the thickness and cross-sectional area of the specimen, and T_1 and T_2 are respectively the higher and lower temperatures of the test specimen.

The cross-sectional area is in fact the area of plate or water calorimeter used to measure the heat flow.

When a water calorimeter is used, Q is simply a measure of the temperature rise of water from the inlet to exit points multiplied by the weight of water which is flowing through in unit time (usually 1 hour).

The thermal conductivity varies with the actual temperature conditions within the specimen, so that the mean temperature must always be calculated and reported when assessing the results. Some workers quote hot face temperatures, particularly when tests on whole bricks are involved. It is useful to represent *k* values at several mean temperatures, but in standard specification practice one determination at a mean temperature of 1400° F. (760° C.) is considered satisfactory.

It is extremely difficult when testing whole bricks or any other shape of specimen to preserve a linear rate of heat flow through any fixed cross-sectional area. The problem is basically one of size of apparatus, because it is at the end sections of specimens where heat losses occur and these must be reduced to a minimum in order to obtain accurate results.

If too large a test piece is selected, the problem of maintaining a uniform temperature becomes serious. A compromise solution is to select a large test piece (a full brick), insulate it thoroughly, heat the full top face as uniformly as modern techniques will permit and measure the heat flow through a small centrally-positioned area of cross-section.

For more accurate determinations of the thermal conductivity and for entirely eliminating end effects, it is better to use a specimen which completely surrounds the heat source. Adams and Loeb[1] have described a highly satisfactory apparatus using a specimen in the form of a prolate spheroid moulded round a heater core. Kingery[2] has used spherical and cylindrical specimens.

The thermal conductivity is a very important property in ceramic materials. Thus, the walls of a furnace which is internally heated are required to have a low thermal conductivity so that as little heat as possible will be lost through them. The walls of a muffle, retort, saggar, or similar appliance, on the contrary, must have a high thermal conductivity in order that the heat may pass through them in order to heat their contents.

Factors which Affect the Thermal Conductivity of Ceramic Materials. The flow of heat through a ceramic material depends upon the relative ease of transfer through the solid phase and through the intervening air spaces. Basically, therefore, *k* values are influenced by:

(*a*) The porosity of the material.

[1] Adams, M. and Loeb, A. L. *J. Amer. Ceram. Soc.*, 37, 73, 1954.
[2] Kingery, W. D., *et al.*, *J. Amer. Ceram. Soc.*, 37, 67, 1954.

(*b*) The material of which the body is composed.

(*c*) The temperature at which the test is carried out.

(*a*) The *porosity* of the material is a major factor in influencing its thermal conductivity, but no simple relationship can be postulated because of the complexity of the various factors involved. Air is a vastly superior insulating material to most solid bodies, so that a highly porous body will, *per se*, be less conductive than a solid body of the same chemical composition.

The actual conductivity value will depend not only on the total void space, but also on the size and the nature of the voids—whether they are closed or interlinked. These factors are important because heat can be transferred across an air space by conduction, radiation and convection.[1] Although some authors question whether convection can be an acceptable mechanism of heat transfer in porous bodies, it cannot be denied that heat differentials will result in air flow through a body if continuous pores are present.

Although, at first sight, a body with large pores might be expected to be more insulating than one with small pores but with the same total porosity, this is not the case. There are many explanations advanced for this, not all of which are agreed by various authorities. In a system of small holes, however, the path length through the solid from hot face to cold face must be far greater than is the case when only a few large holes are present. For this reason alone, the conductivity of small-pored material would be less than a large-holed specimen of equal total porosity. Clements[2] has demonstrated this point clearly and deduced a relative series of values for the pore size/thermal conductivity values as illustrated in Fig. XIV.8.

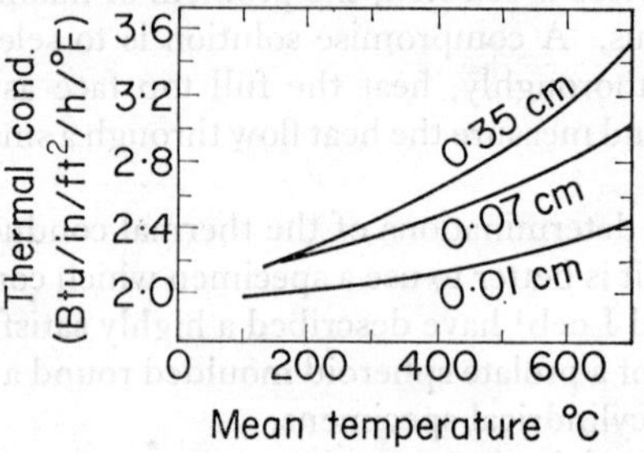

FIG. XIV.8. THERMAL CONDUCTIVITY/PORE SIZE RELATIONSHIP

The values have been equated to a bulk density value of 40 lb/ft^3 in every case. The average pore size calculated for each specimen is shown in the diagram.

As Clements points out, however, it is possible to obtain truly high porosities in ceramic materials only when pore sizes are large, and the best insulating materials —with k values as low as 1 B.T.U's in/ft^2/h./°F. and bulk densities of about 25 lb/ft^3—must of necessity contain large pores.

Because of the complicating factor of pore size, the relationship between thermal conductivity and total porosity is best represented by a band spread of the type shown in Fig. XIV.9, for a range of alumino-silicate products. The actual

[1] Barrett, L. R., *Trans. Brit. Ceram. Soc.*, 48, 235, 1949.
[2] Clements, J. F., *Trans. Brit., Ceram. Soc.*, 65, 479, 1966.

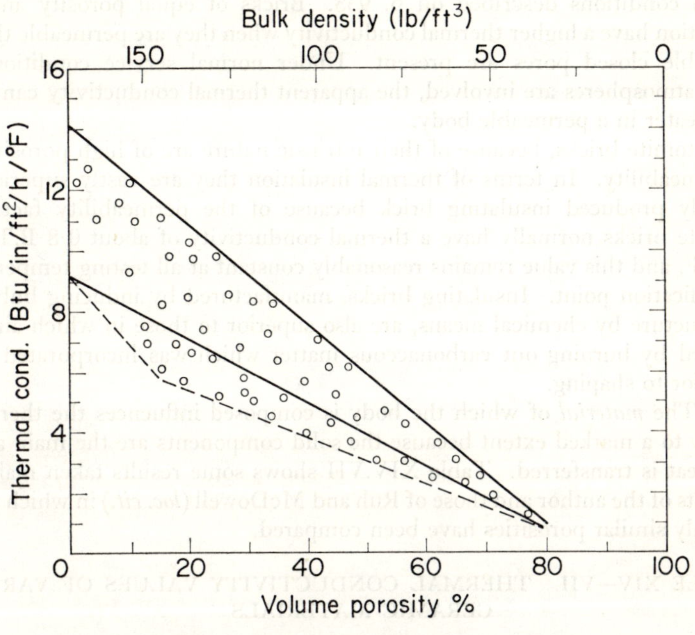

FIG. XIV.9. THE RELATIONSHIP BETWEEN THERMAL CONDUCTIVITY AND VOLUME POROSITY FOR SOME ALUMINO-SILICATE BODIES

chemical compositions varied between 22–42 per cent Al_2O_3; firing temperatures to which the bodies had been subjected were between 1000–1450° C. Hence, variables other than total porosity are involved, but these would not result in a major order of difference.

The results shown were obtained in the author's own laboratories, but figures published by Clements (*loc. cit.*) and by Ruh and McDowell[2] correspond closely with these values.

Bricks of lower alumina content and those containing the highest percentage of glassy phase usually fall nearer to the line of lowest values. Some of the bricks tested and included in Fig. XIV.9 were actually building bricks. Those which were hard-fired and composed of Coal Measure shales, fell within the band spread of the full lines—but towards the lower limit. Common bricks of high porosity, particularly those rich in lime, fell outside the normal band spread for refractory materials of similar porosity. The dotted line shown in Fig. XIV.9 indicates the lower limit of thermal conductivities found by the author for common bricks manufactured from various strata levels in the U.K. Not all types are included because some tend to vitrify under the conditions of this test. Common bricks vary in thermal conductivity values between 3·0–10·0 B.T.U's in/ft²/h.°F.

The permeability of bricks to air has a marked influence on thermal conductivity values. To some extent, this can be established by testing under the

[1] Ruh, E. and McDowell, J. S., *Trans. Amer. Ceram. Soc.*, 45, 189, 1962.

standard conditions described on p. 938. Bricks of equal porosity and similar composition have a higher thermal conductivity when they are permeable than when appreciable closed pores are present. Under normal service conditions, when gaseous atmospheres are involved, the apparent thermal conductivity can be many times greater in a permeable body.

Diatomite bricks, because of their intrinsic nature are of high porosity, but of low permeability. In terms of thermal insulation they are vastly superior to any artificially produced insulating brick because of the permeability factor alone. Diatomite bricks normally have a thermal conductivity of about 0·8 B.T.U's/in./ft.^{2}/h./°F. and this value remains reasonably constant at all testing temperatures to the vitrification point. Insulating bricks, manufactured by inducing bubbles into their structure by chemical means, are also superior to those in which air space is developed by burning out carbonaceous matter which was incorporated with the body prior to shaping.

(*b*) The *material* of which the body is composed influences the thermal conductivity to a marked extent because the solid components are the main agents by which heat is transferred. Table XIV.VII shows some results taken mainly from the results of the author and those of Ruh and McDowell (*loc. cit.*) in which materials of roughly similar porosities have been compared.

TABLE XIV—VII. THERMAL CONDUCTIVITY VALUES OF VARIOUS CERAMIC MATERIALS

Material (Brick forms)	% *Porosity*	*Thermal Conductivity* (*B.T.U's/in./ft.*^{2}*/h./*°F) 700° F. (371° C.)	1000° F. (537° C.)	1400° F (760° C.)	1832°F. (1000° C.)
Firebrick (28% Al_2O_3)	22	5·8	7·9	9·8	11·9
Firebrick (42% Al_2O_3)	19	8·5	9·0	9·4	9·8
Sillimanite	23	6·5	8·1	10·9	12·2
Alumina (72%)	22	10·6	10·4	10·2	10·0
Alumina (90%)	21	14·6	13·9	12·9	13·0
Alumina (99%)	24	26·2	23·3	18·2	17·1
Silica brick	23	9·3	10·0	11·2	12·1
Magnesia (94% MgO)	20	47·6	37·3	29·8	19·2
Chrome-magnesite	22	12·0	12·2	12·3	12·4
Zircon brick	17	19·2	17·7	16·6	16·6

The temperatures quoted are all mean values.

The high conductivity of magnesia is worthy of especial comment as also is the remarkable reduction in its value when tested at high temperatures.

(*c*) The *temperature* of testing of materials affects the thermal conductivity as is also illustrated in Table XIV.VII.

Most materials of the alumino-silicate range increase in thermal conductivity values as the temperature of testing is raised, but this tendency is less noticeable as the alumina content becomes greater and actually reverses at very high alumina values.

Magnesia shows remarkable characteristics and its thermal conductivity at a mean temperature of 1000° C. is less than one-half that at 371° C. No other common ceramic material shows such a marked change, which is due to the crystal structure of the main phase, periclase, and to its anomalous co-ordination condition at high temperatures (p. 183).

Several authors have attempted to derive empirical relationships to relate thermal conductivity to the temperature of testing and also to the porosity of the test piece, but most of these are of doubtful merit.

Some ceramic materials show remarkable changes in thermal conductivity values at very low temperatures.[1]

Powdered material generally has a lower thermal conductivity than when in a compacted form. This is to be expected from the data shown in Fig. XIV.8.

Ruh and Renkey[2] have made an exhaustive survey of the thermal conductivities of refractory castable materials. When tested at low temperatures, the presence of unfired cement resulted in comparatively high thermal conductivity values, but above 500° C. (mean temperature) the values were similar to those obtained from conventional refractories.

This is an interesting observation because the presence of water either in a combined or free form invariably increases thermal conductivity values. This phenomenon is noted when common bricks are being measured. A fully saturated brick has 2–3 times the conductivity value of the same material when dried.[3]

The thermal conductivities of various *electrical ceramic materials* are shown in Table XIV—VIII.

TABLE XIV—VIII. THERMAL CONDUCTIVITIES OF ELECTRICAL CERAMIC BODIES

	Thermal Conductivities	
	g.cals. /cm./cm.²/sec./°C.	*B.T.U./in./ft.²/h./°F.*
Porcelain	0·002–0·004	5·8–12·0
Steatite	0·005–0·006	15·0–18·0
Cordierite	0·003–0·005	9·0–15·0
Zircon	0·011–0·012	30·0–36·0
Rutile	0·007–0·010	20·0–30·0
Titanate	0·008–0·0095	24·0–30·0

Cermets, or composite metal-ceramic bodies usually have a high thermal conductivity and some have similar values to those of metals, which are in the range 0·5–1·2 g.cals./cm./cm.²/sec./°C.

Non-siliceous materials usually have a high thermal conductivity and Table XIV.IX, due mainly to Kingery[4] shows values for several pure ceramic materials which were either single crystals or were sintered to a mass of low porosity.

Thermal conductivity measurements are particularly liable to variations and are subject to numerous errors. There are many discrepancies between the results

[1] Burk, M., *J. Amer. Ceram. Soc.*, 46, 150, 1963.
[2] Ruh, E. and Renkey, A. L., *J. Amer. Ceram. Soc.*, 46, 89, 1963.
[3] Owen, W. J. and Clews, F. H., *Trans. Brit. Ceram. Soc.*, 65, 559, 1966.
[4] Kingery, W. D., *J. Amer. Ceram. Soc.*, 38, 3, 1955.

TABLE XIV—IX. THERMAL CONDUCTIVITIES OF SOME CERAMIC MATERIALS

Material	*Thermal Conductivity (g. cals./cm./cm.²/sec./°C.)*		
	100° C.	400° C.	1000° C.
Al_2O_3	0·072	0·031	0·015
BeO	0·525	0·222	0·049
Graphite	0·426	0·268	0·149
MgO	0·086	0·039	0·017
Spinel	0·036	0·024	0·014
ThO_2	0·025	0·014	0·008
Zircon	(0·016)	0·012	0·010
TiC	0·060	0·032	0·014
TiC cermet	0·083	(0·04)	(0·02)
Mullite	0·015	0·011	0·010
Zirconia	0·0047	0·0049	0·0055
SiC	0·480	0·372	0·192

Bracketed figures are estimated values.

reported by different authors and it is not easy to establish absolute values. Ideally the specimen to be tested should be in the form of a sphere or cylinder, but this is not always either convenient or satisfactory. The results obtained by any one worker on a series of materials are comparative within themselves but should not be used in conjunction with other figures obtained under different experimental conditions.

Thermal conductivity measurements are made after equilibrium in the heat flow conditions have been attained and take no account of the rate at which heat flows through the material. This latter property is the *Thermal Diffusivity* which may be defined as the rise in temperature produced in 1 ml. of the substance by 1 calorie acting during 1 second through 1 sq. cm. of a layer 1 cm. thick, having a temperature difference of 1° C. between its faces.

The coefficient of diffusion is represented by:

$$\frac{K}{dh}$$

where K is the amount of heat in g.-cals. which is transmitted in 1 second through a plate 1 cm. thick per sq. cm. of its surface when the difference in temperature between the two sides is 1° C., d is the specific gravity, and h the specific heat of the material.

A. T. Green[1] has found the average temperature-diffusivity of six silica and six fireclay products to be those shown in Fig. XIV.10; the thermal conductivity of the same materials is shown in Fig. XIV.11. Individual materials differ so greatly from the average that no comparison in general terms is very reliable.

The coefficient of thermal conductivity (K) at t° C is

$$K = kcs$$

where k is the coefficient of thermal diffusivity at t° C., c the apparent density, and s the true specific gravity at t° C.

[1] Green, A. T., *Trans. Brit. Ceram. Soc.*, 26, 171, 1926-7.

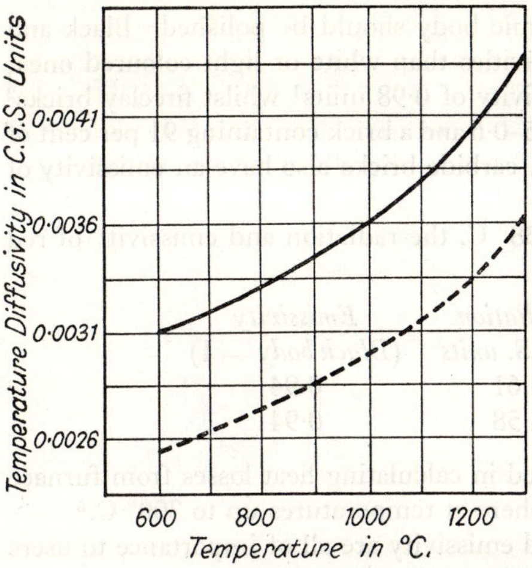

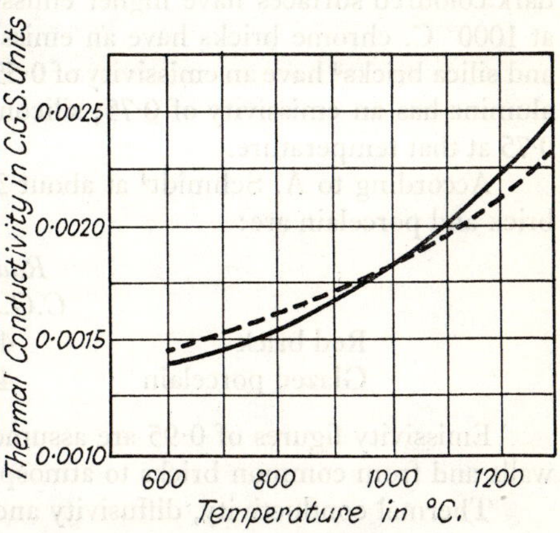

Average for six Silica Products ————
Average for six Fireclay Products - - - - - - -

FIG. XIV.10. TEMPERATURE/THERMAL DIFFUSIVITY CURVES OF SILICA AND FIRECLAY PRODUCTS

FIG. XIV.11. TEMPERATURE/THERMAL CONDUCTIVITY CURVES FOR SILICA AND FIRECLAY PRODUCTS

Fourier's law for the linear flow of heat is:

$$\frac{d\theta^2}{dx^2} = k\frac{d\theta}{dt}$$

where θ denotes the temperature, *x* the distance from a hot surface, *t* the time and *k* the diffusivity. Calculations based on this formula, made by Heyn, Bauer and Wetzel, give much lower results for conductivity than those obtained by means of calorimeters.

Thermal diffusivity apparatus is difficult to design but can be based on the standard conductivity equipment.[1] Two special techniques have been described; the first due to Fitzsimmons[2] involves immersing the specimen in a lead bath and measuring the rate of rise of temperature at the centre of the body; Van Zee and Babcock[3] have measured the thermal diffusivity of molten glass by superimposing a small temperature variation (58° C.) on the mean temperature of the sample which could be maintained at several points between 700°–1400° C. The amount of time taken for the impulse to travel to the centre of a cylindrical specimen was judged to be a measure of the diffusivity.

Emissivity is defined as the amount of heat or other energy radiated or absorbed by a substance; that emitted by an ideal black body being taken as unity.[4] To be

[1] Plummer, W. A., Campbell D. E. and Comstock, A. A., *J. Amer. Ceram. Soc.*, 45, 310, 1962.

[2] Fitzsimmons, E. S., *J. Amer. Ceram. Soc.*, 33, 327, 1950.

[3] Van Zee, A. F. and Babcock, C. L., *J. Amer. Ceram. Soc.*, 34, 244, 1951.

[4] The radiation of an ideal black body is assumed to be $4{\cdot}9 \times 10^{-8}$ kilo-calories per sq. m. per hour per °K^4. In furnaces and kilns working under normal conditions and with the doors closed, the radiation from the walls and contents conforms broadly to that of a hot black body.

truly comparable the surface of the ceramic body should be polished. Black and dark coloured surfaces have higher emissivities than white or light coloured ones; at 1000° C. chrome bricks have an emissivity of 0·98 units[1] whilst fireclay bricks[2] and silica bricks[3] have an emissivity of 0·55–0·6 and a brick containing 91 per cent of alumina has an emissivity of 0·75; silicon carbide bricks also have an emissivity of 0·75 at that temperature.

According to A. Schmidt[4] at about 38° C. the radiation and emissivity of red brick and porcelain are:

	Radiation C.G.S. units	*Emissivity (Black body = 1)*
Red brick	4·61	0·94
Glazed porcelain	4·58	0·94

Emissivity figures of 0·95 are assumed in calculating heat losses from furnace walls and from common bricks to atmosphere at temperatures up to 300° C.[5]

Thermal conductivity, diffusivity and emissivity are all of importance to users of ceramic materials.

For example, the loss of heat through a furnace wall is given by the expression:

$$H = K(T_1 - T_0)\frac{S}{l}$$

where H represents the flow of heat, T_0 the temperature of the cold or outer surface, T_1 that of the hot or inner surface of the wall, S the area, l the thickness, and K the mean thermal conductivity between the temperature T_0 and T_1.

If s is the inner area of the wall and S the outer area, the geometric mean is $\sqrt{sS}$ and

$$H = K(T_1 - T_0)\frac{\sqrt{sS}}{l}$$

In electrical furnaces the term *thermal mho* is generally used to express in watts the heat radiated in g.-cals. per second per cm. cube for a difference of 1° C., a watt being equivalent to 0·2388 calories per second. The reciprocal value of this, the *thermal ohm*, represents the difference of temperature divided by the flow of heat in watts per cm. cube.

The thermal conductivity expressed in C.G.S. units may be converted into thermal ohms by multiplying the reciprocal of the conductivity by 0·2388. To reduce g.-cals to watts, the reciprocal of the conductivity is multiplied by 4·186. The watts may be resolved into power as follows:

Watts × 0·0013411 = horsepower
Watts × 0·0568776 = B.T.U. per minute
Watts × 0·0143329 = calories per minute

[1] Heilman, R. H., *Trans. Amer. Inst. Chem. Eng.*, 31, 165, 1934.
[2] Barrett, L. R., *Trans. Brit. Ceram. Soc.*, 48, 235, 1949.
[3] Baritel, A., *Chal et. Indust.*, 19, 237, 299, 1938.
[4] Schmidt, A., *Stahl u Eisen*, 1927, p. 1336.
[5] A.S.T.M. Standards, Proposed Procedures, p. 551, 1965.

The Insulation of Heat or the Thermal Resistivity

The converse of the transmission of heat is termed *insulation* and substances which prevent the transmission of heat or have a very low thermal conductivity are known as *insulators*. They are particularly used for retaining heat in kilns, furnaces, and other heating appliances.

The loss of heat by radiation and convection from a boiler plant frequently reaches 10 per cent of the fuel used; in gas-retorts it is usually nearer 20 per cent, and, with other types of plant, such as metal-heating furnaces and pottery kilns, as much as 60–70 per cent of the heat supplied may be lost. In storage tanks, drying chambers, and refrigerating plants, the loss is less noticeable, but is equally important.

The materials available for heat insulation are all characterised by a relatively high proportion of air space, but the smallness of the pores and their lack of communication with each other are even more important than their total volume. If the pores are large and communicate freely with one another, heat may be transmitted by convection across them, even at comparatively low temperatures.

Heat is also transmitted across the pores by radiation and, as the temperature rises, the quantity radiated increases with great rapidity, until, at 1500° C. it almost equals that conducted through the solid substance, and the material ceases to act as an insulator.

At relatively low temperatures, wood, cork, rubber, rope, mill-board, foamed plastic and various textile materials, are good insulators. For rather higher temperatures aluminium-foil, asbestos, vermiculite, diatomite, porous concrete, slag wool, and glass-silk are suitable, but for still higher temperatures a refractory ceramic material must be used. No insulator is equally suitable for all temperatures.

The insulating power of a material is ascertained by determining the thermal conductivity and then calculating its reciprocal, or by comparing the loss of heat when various insulators are used.

In selecting an insulating material for use at high temperatures it is important to choose only those which are sufficiently refractory, have a low shrinkage when heated to the temperature at which they will be used, and have sufficient strength to support any load or pressure likely to be applied to them.

Air, when motionless, is often regarded as a heat-insulating material, but Mellor[1] has shown that for a given size of pore-space, or volume of air, there is a temperature at which the quantity of heat passing through it by radiation equals that which would be carried by a solid substance occupying the same space. For an air-distance of 0·5 cm. the temperature is 730° C.; for 0·1 cm., 1400° C., and for 0·01 cm. the temperature is 3000° C. Hence, the smaller the pore-space the greater is its insulating power.

The most important materials used for heat-insulation are:

(*a*) *Powdered and Fibrous Materials*, such as slag-wool, glass-silk (fibre glass), expanded plastic compositions, asbestos, magnesia, diatomite, or kieselguhr. Most of these are suitable only for use at temperatures below 800° C.

[1] *J. Soc. Chem. Ind.*, 38, 140, R., 1919.

(*b*) *Diatomite*, kieselguhr, and the like made into bricks or other articles. They have a low crushing strength (seldom exceeding 300 lb. per sq. in.), a low refractoriness, and cannot be used above 800–850° C.

(*c*) *Vermiculite* which is a hydrated magnesium silicate capable of exfoliation on heating (see p. 154). The material has a high degree of insulating power and can withstand a higher temperature than diatomite.

(*d*) *Refractory materials* composed of fireclay mixed with paper-pulp, saw-dust or other combustible material which burn away, leaving a highly porous mass the insulating power of which is lower than that of diatomite, but the bricks can resist a much higher temperature. Porosity can also be induced by developing bubbles of air or gas in the body during forming.

Thermal Shock Resistance

Many ceramic materials are unable to withstand sudden changes in temperature without flaking, dunting, spalling, cracking or other form of disintegration. The ability of a substance to resist such treatment is known as its *thermal shock resistance.* Failure of the brick as a result of its inability to withstand sudden temperature changes is manifested in a characteristic corner fracture, cracking or spalling.

Thermal shock is a direct consequence of internal stresses set up in the brick or shape on the application of heat or on too rapid cooling. Most ceramic bodies are poor conductors of heat and, in consequence, unless careful control of the rate of temperature rise is maintained, thermal gradients may be set up between adjacent areas. If the body expands appreciably on heating, some parts of it will increase in volume to a greater extent than others due to the differences in temperature; stresses will arise which, if they exceed the elastic limit of the body, will cause spalling.

The principal factors which contribute to a low thermal shock resistance are:

(*a*) A high coefficient of thermal expansion.
(*b*) A low thermal conductivity or diffusivity.
(*c*) An inability to accommodate stresses, or in other words, a low modulus of elasticity.
(*d*) A low tensile strength.

Although thermal spalling and some of the factors which contribute to it have been studied for many years, the inter-relation of the various properties of an article with its thermal shock resistance have been elucidated only recently, and even now, except for some pure oxides, the conclusions are only empirical.

Kingery[1] has shown that an important factor which influences the thermal shock resistance of a material is its surface heat-transfer coefficient, i.e. its ability to absorb or give up heat to the surrounding medium.

If the surface heat-transfer coefficient is large, i.e. if heat is exchanged readily, the thermal shock resistance (R) can be expressed as:

$$R = S_t(1 - \nu)/E\alpha \qquad (a)$$

[1] Kingery, W. D., *J. Amer. Ceram. Soc.*, 38, 3, 1955.

but, if the heat transfer coefficient is low, the thermal shock resistance (R′) is more correctly represented by the equation:

$$R' = kS_t(1 - \nu)/E\alpha \quad . \quad . \quad . \quad . \quad (b)$$

where S_t = the tensile strength of the material at the temperature of the test (t).
ν = Poisson's ratio (see p. 843).
E = the modulus of elasticity.
α = the thermal expansion.
k = the thermal conductivity.

For most ceramic materials, Poisson's ratio is reasonably constant (p. 897) and does not change much with temperature and although the strength of a body can vary between wide limits, the elasticity modulus E changes similarly, so that the ratio S_t/E is not very different for many materials. At high temperatures, however, the strength may be reduced and this tends to increase the thermal shock tendency, but when liquid forms, there is a considerable increase in elasticity, so that spalling resistance increases.

The thermal conductivity and expansion have a considerable effect on the thermal shock resistance and the ratio of these two values is often an indication of the spalling tendency of a material.

Kingery has derived thermal stress resistance factors for some ceramic materials and these are shown in Table XIV.X.

TABLE XIV—X. THERMAL STRESS RESISTANCE FACTORS FOR SOME CERAMIC MATERIALS
(after Kingery)

Material	100° *C.*		400° *C.*		1000° *C.*	
	R	*R′*	*R*	*R′*	*R*	*R′*
Al_2O_3	37	2·7	36	1·1	40	0·6
BeO	33	17·3	31	6·9	14	0·7
MgO	22	1·9	24	0·9	26	0·5
Mullite	75	1·1	68	0·8	84	0·8
Spinel	32	1·2	32	0·8	32	0·5
ThO_2	51	1·3	47	0·7	37	0·3
Zircon	137	(2·2)	115	1·4	73	0·7
ZrO_2	66	0·3	62	0·3	53	0·3
Fused silica	2500	9·5	2400	10·1	—	—
Glass (soda-lime-silica)	94	0·4	94	0·4	—	—
TiC	42	2·5	40	1·3	(40)	0·6
Porcelain	116	0·5	103	0·4	116	0·5
Fireclay	38	0·1	41	0·1	178	0·5
TiC cermet	208	(17·3)	200	(8)	198	(4)

Bracketed figures are estimated values.

Before the thermal shock behaviour of a material can be assessed, the surface heat-transfer coefficient must be determined. This is not an easy matter because so many factors can have an important influence; these include (i) the size and shape of the body; (ii) its surface texture; (iii) the nature of the surrounding medium. Bodies quenched in air or subjected to the impingement of hot gases on their surface, generally have a much lower surface heat transfer than when the quenching or heating medium is a liquid.

So important is the property of surface heat transfer, that research into methods of improving this factor and the so-called *Biot's modulus* or heat transfer ratio would obviously be of paramount importance in improving the thermal shock resistance of ceramic materials. Colour, texture, surface coatings, all can influence this characteristic to a remarkable degree.

The modulus of elasticity of brittle materials is generally accepted as being low and, therefore, another unsurmountable factor which confers on ceramic materials their poor resistance to thermal shock. It is true that conventional materials, at low temperatures particularly, have low values of elasticity modulus, but this may be due more to methods of manufacture and to imperfections on a macro- and micro-scale than to basic properties. Although Griffith published his classical work on the failure of brittle materials in 1920,[1] it has been given its due regard only in recent years. The recognition that flaws or cracks within a body initiate failure at stress values much below those which could be predicted from theoretical reasoning, has led to improvements in the thermal shock resistance of highly sophisticated ceramic materials to a remarkable extent. In recent discussions with the author, Professor Pask[2] of the University of California, mentioned that single crystals of refractory oxides had been prepared in the U.S.A. which were free from flaws and had appreciable ductility, as well as extremely high tensile strengths.

There are obviously several promising lines of development which are leading to vast improvements in the thermal shock resistance of ceramic materials.

White[3] has summarised many of the important factors which contribute to thermal shock characteristics of brittle materials.

The reversible thermal expansion frequently determines the tendency of a ceramic material to spall. When the expansion coefficient is high, as in silica and siliceous materials, the rate of heating should be slow so as to reduce the temperature gradients to a minimum. Silica minerals, however, have a high expansion coefficient only at comparatively low temperatures as a result of the inversions (see p. 796). Once this critical temperature has been exceeded only a negligible increase in volume occurs and the tendency to thermal shock is considerably reduced.

Materials with a high thermal diffusivity have little tendency to spalling even though the coefficient of expansion may be appreciable. Such materials are not common in ceramic studies although magnesia provides an interesting case. As will be seen in Table XII.X (p. 797), bricks made of this material have a comparatively large thermal expansion, but because the conductivity is also high at low temperatures (see Table XIV.VII, p. 942), they have a moderately good resistance to thermal shock. At higher temperatures, however, the expansion is maintained but the conductivity is markedly decreased; hence, the tendency to spall becomes a serious factor.

[1] Griffith, A. A., *Phil. Trans. Roy. Soc.* 221, 163, 1920.

[2] See also Gorum, A. E., Parker, E. R. and Pask, J. A., *J. Amer. Ceram. Soc.*, 41, 161, 1958.

[3] White, J., *Trans. Brit. Ceram. Soc.*, 57, 591, 1958.

The ability to accommodate stresses which may be set up from thermal gradients in a body is not easy to measure. A vitrified mass is more liable to spall than one where crystalline development predominates, but probably the greatest factor is the porosity of the body.

There are many anomalies in the literature on the subject of the effect of porosity on the spalling tendency of the body. When a high porosity is induced in a ceramic body by underfiring, the composition may also be quite different so that a direct comparison is valueless. Furthermore, a spalling crack in a porous body may be less obvious than one in a dense article, so unless careful measurements are made, a misleading conclusion may be reached.

A porous refractory block usually shows less tendency to disintegrate under rapid heating and cooling conditions than does a dense brick, but measurements of the elasticity coefficient by sonic methods have shown that spalling occurs readily in porous shapes, but the cracks which result are not continuous and frequently do not involve a loss of material, but they do involve a loss of strength.

Coble and Kingery[1] have studied the influence of porosity on the thermal shock resistance of sintered alumina and conclude that the spalling tendency is *increased* by a factor between three and six times for an increase in porosity of 50 per cent.

Eusner and Debenham[2] concluded that small increases in porosity in a dense fireclay brick cause a *reduction* in the thermal shock resistance, but that there was no direct correlation between the two properties if the porosity was above 10 per cent. They found that the method of forming the bricks contributed to the spalling tendency; induced laminations were frequent sources of spalling cracks.

The spalling resistance can be measured in various ways. Two types of standard test are used, (*a*) where small blocks cut from whole bricks or blocks are subjected to frequent and rapid temperature changes, and (*b*) panel spalling tests, where whole bricks are laid together to form a complete furnace door which is subjected to heating and cooling.

The small block test is described in British Standard Specification B.S. 1902, Part 1A, 1966. Test pieces of prism shape 3 in. × 2 in. × 2 in. are cut from the block to be tested and placed in a cold furnace which is then raised slowly in temperature over a three hour period. If silica bricks are being tested, a maximum temperature of 450° C. is sufficient but for all others 1000° C. is preferred. Maximum temperature is maintained for 30 minutes and the blocks should then be withdrawn and placed on end on a brick floor. After 10 minutes cooling they are returned to the furnace which has been maintained at the maximum temperature and left there for a further 10 minutes. The cycle of cooling and heating for 10 minute intervals should be repeated until the specimen can be pulled apart or until it has withstood ten cycles. The thermal shock is reported as the number of cycles which the specimen will endure without fracturing.

This test has obvious limitations and has been severely criticised, but, nevertheless, it is easy to carry out, requires no expensive or large scale equipment and provides at least a rough guide to the tendency of a material to thermal shock under fairly extreme conditions.

[1] Coble, R. L. and Kingery, W. D., *J. Amer. Ceram. Soc.*, **38**, 33, 1955.
[2] Eusner, G. R. and Debenham, W. S., *Bull. Amer. Ceram. Soc.*, **31**, 489, 1952.

The American Society for Testing Materials[1] recommends the panel spalling method in which twelve or more bricks are built into a furnace door so that a 9 in. × 2½ in. or 3 in. face will be exposed to the heat treatment. Before setting in place, each brick should be marked and weighed. The bricks are suitably insulated at the rear and side to maintain constant temperature gradients across the specimens. In the full test the panel should be set in position as the door of a large furnace which is heated to a specified temperature in 5–8 hours and there maintained for 24 hours. After switching off the furnace, the panel must be left in place for 8 hours before being removed for examination for shrinkage, bloating, etc.

The spalling test is then carried out on the same panel which is placed in position and raised to the prescribed temperature in a fixed period of time as indicated in Table XIV.XI.

TABLE XIV—XI. INITIAL HEATING RATES FOR THERMAL SHOCK TEST (A.S.T.M.)

Temp.	*Time*	*Temp.*	*Time*
1830° F.	20–25 min.	2370° F.	75–80 min.
2190° F.	50–55 min.	2550° F.	120–125 min.

After 10 minutes at the maximum temperature, the test panel is withdrawn and replaced by another sample or by a dummy. The hot brick face should be cooled by a blast of air and water mist for ten minutes and returned to the heating furnace. The cycling operation should be repeated for a specified number of tests, usually 12, after which the panel should be cooled, each brick cleaned and re-weighed and the spalling result reported in terms of the percentage loss in weight.

A positive method of measuring the thermal shock tendency of materials is to establish the temperature difference between the surroundings and the body which is just sufficient to cause spalling. The maximum stresses which the body can withstand are thus established and the results on different materials are directly comparable. The simplest method using this principle is to quench a cylinder or block of material from a controlled furnace temperature into air, water or other suitable medium. The temperature of the specimen prior to quenching can be raised by small increments until the critical temperature difference to cause spalling is reached.

Buessem and Bush[2] have devised a method for measuring thermal shock resistance in which a pile of rings of the material under test are heated from the inside by an electric element and cooled on the outside. The temperature gradient across the rings can be raised gradually until spalling occurs.

Thermal shock measurements by conventional means have limitations in that they rely upon an actual observation of a fracture condition, whereas an article may fail to a sufficient degree to weaken it, but the crack does not fully propagate to cause disruption.

The 'soundness' of an article from kiln can often be judged by its ring and many

[1] A.S.T.M. C38—1958.
[2] Buessem, W. R. and Bush, E. A., *J. Amer. Ceram. Soc.*, 38, 27, 1955.

manufacturers employ the 'penny-test' to remarkably good effect. An experienced operator can detect a dunt in a salt-glazed pipe or a slab of faience which is not otherwise visible or detectable.

Clements and Davis[1] applied this characteristic on a scientific basis in their sonic spalling test. The test-piece, usually a section cut from a brick or block, is placed in a heater assembly and connected to a vibrator unit. A crystal pick-up cartridge, attached to a wire in contact with the block, feeds the vibrations through the block to an oscillograph screen. By means of a tuning circuit, a resonance frequency for the block can be established which would be indicated by a trace of large amplitude.

The heater current is then switched on and a thermal gradient established across the block. This can be increased until at some point a sudden change in the amplitude or wavelength of the resonance frequency occurs. This is an indication of a thermal shock condition which can be related to a specific thermal gradient.

By these means, safe heating rates for a variety of refractory articles have been predicted, although the authors qualify their conclusions and urge caution because of scale-up factors.

It is generally accepted that silica bricks should not be heated at a greater rate than $2\frac{1}{2}$° C./minute up to 300° C. High alumina fireclays can withstand rates of 10–15° C./minute, but if quartz is present care must be taken through the inversional temperature.

The resistance of *fireclay* bricks and other clay products to thermal shock depends largely on the proportion of quartz or free silica in them. Grog bricks are less likely to crack than fireclay bricks made wholly of clays, and aluminous bricks —in which most of the clay has been converted into crystalline mullite—are still more resistant to sudden changes in temperature.

Large blocks are more susceptible to sudden changes in temperature than are smaller ones, but if properly made they should resist all ordinary conditions of heating and cooling. To secure the necessary resistance they should have a more open texture than is needed for bricks or small blocks, so that the pores may take up any strains which occur.

With many fireclay articles, the author has found that slow cooling in the kiln is extremely beneficial to thermal shock resistance. Annealing at temperatures in the range 700–900° C. also has a remarkably good effect. The introduction of fast-firing cycles into the production of ceramic ware of all types undoubtedly results in the development of high stresses within the body on cooling. Thermal shock problems have increased dramatically as a result, but manufacturers will find that an annealing treatment of a few hours at a point at which slight pyro-plastic conditions still pertain will usually be of marked advantage.

Glass-melting pots and **retorts** require to be very resistant to sudden changes in temperature, but it is difficult to secure this property at the same time as a high resistance to corrosion. Moderately porous materials are the most suitable for the purpose where the corrosion is not excessive.

Crucibles must be very resistant to sudden changes of temperature, as when in use they are withdrawn rapidly—from a furnace (which may have a temperature up to

[1] Clements, J. F. and Davis, W. R., *Trans. Brit. Ceram. Soc.*, 57, 624, 1958.

1800° C.) into the open air. This resistance is attained if the crucible has an open, porous texture which may be obtained in the case of fireclay crucibles by introducing a sufficiently large proportion of grog into the mixture of which the crucibles are made. The number of times which a crucible can be reheated varies according to its contents and the manner in which it is used.

Silica bricks are very susceptible to sudden changes in temperature and soon develop fine cracks, which later increase in size and cause the bricks to become weak or to spall and flake. Most silica bricks are so sensitive to changes in temperature, that if a current of air is drawn through them during the cooling of the kiln they are liable to crack. For this reason, after the firing of the bricks is completed, the kilns are closed and allowed to cool in such a manner that the whole of the heat is lost by radiation through the walls and floor of the kiln.

The great liability of silica bricks spalling at temperatures below 500° C. is due to the α-β transition range described on p. 796.

Silica bricks are only liable to spall at low temperatures (see p. 950) and at high temperatures are extremely resistant to thermal shock. A good quality product may be used quite safely in the roof of a basic open hearth furnace where, on the reversal of the firing cycle, the temperature may fluctuate between 1650° C. and 1200° C. in a matter of seconds. The crystal type of silica present governs the safe lower limit of temperature. If quartz is present, spalling is always liable below 573° C.; when cristobalite is the principal modification present, the brick is 'safe' above about 300° C.; tridymite bricks can be used with comparative safety at temperatures above 170° C. Under normal testing conditions as described on p. 951, however, silica bricks rarely withstand more than one cycle.

Magnesite and **chrome bricks** are particularly subject to spalling but only at high temperatures. The reasons for this unusual behaviour have been outlined on p. 950.

Zirconia and **zircon** refractories are very resistant to temperature changes and for this reason, their use may increase in the near future.

Carbon and silicon carbide bricks are quite insensitive to sudden changes of temperature as their coefficient of expansion is very low, and they have high thermal conductivities.

Crazing, peeling and allied phenomena are specialised types of thermal shock. If the glaze and body compositions are not correctly adjusted so that the coefficients of expansion are closely similar, one will increase in length at a greater rate than the other when heated. If the elastic limit of the glaze is exceeded as a consequence, then it will fracture. In many cases it is impossible to adjust the expansion of the glaze to that of the body without losing other desirable characteristics. Engobes are therefore used, which act as a bridge between the expansion of the body and glaze; these are materials which have intermediate expansion coefficients and so lessen the possibilities of cracking (see p. 341).

Dunting is a fault also associated with thermal shock which occurs when large articles are cooled too quickly or unevenly in a kiln. It is commonly ascribed to the inversional change of quartz which is present in the body and in fireclay articles this may well be the case. Dunting does occur in non-siliceous material, so any factor leading to high thermal shock tendency can cause it.

It is a particularly serious fault in sewer and drain pipe production where it is exhibited as a longitudinal crack which may be revealed only in service.

Delayed dunting is particularly troublesome, because this fault may occur after days or even weeks of storage. Presumably, strains due to thermal stresses do not instantaneously trigger off fracture cracks, but the changing atmospheric conditions provide the extra stresses that are needed.

With the advent of jet and rocket propulsion and atomic energy, an urgent need has arisen for materials capable of resisting high temperatures and very rapid heating rates. All refractory bodies so far developed disintegrate rapidly under these conditions because of their comparatively low resistance to thermal shock, and the metals which are available melt under the conditions of operation. For these reasons, attempts have been made to 'wed' metals and ceramics in the hopes that such a 'marriage' would produce a substance with the high elasticity coefficient of the former and the high softening point of the latter. Such developments are outside the scope of this volume, but *cermets* (as the metal-ceramic products have been named) have opened up a new era of high temperature operations.

ELECTRICAL PROPERTIES

THE electrical properties of ceramic materials are important in connection with both their manufacture and use. These properties include:

(1) The capacity of clays when suspended in water to conduct an electric current. The nature of the colloidal minerals in clays and their ionisation tendency (p. 460) determines their conductibility to a very large extent. When subjected to a direct current, the solid particles in a clay suspension migrate towards the positive electrode. This property, known as *electro-osmosis* and kataphoresis is occasionally used in the purification of clays and as a means of dewatering them.

(2) The high electrical resistance possessed by many ceramic materials. For this reason, porcelains are commonly employed as resistors and insulators in electrical installations.

(3) Their dielectric capacity. Ceramic materials are widely employed as the dielectric medium in condensers and as cores in transformers.

Clay Slips possess electrical properties which are often important in connection with the purification of these materials and in the production of ware by the casting process. Thus, whilst the electrical conductivity of pure, distilled water is 10^{-6} reciprocal ohms (or *mhos*) at 18° C., the presence of even minute quantities of soluble salts causes large differences in the conductivity. When larger proportions are present, the conductivity does not conform to simple rules, but is dependent on several complex considerations.

The conductivity of solutions at different temperatures varies considerably; it may be calculated from the formula:

$$C_t = C_{18}(1 + k(t + 18))$$

where C_t is the conductivity at any given temperature t in ° C., C_{18} is the conductivity at 18° C., and k is the temperature coefficient. For salts, the temperature coefficient varies from 0·02–0·023; for acids, and for some acid salts it is 0·009–0·016; and for caustic alkalies about 0·02 ohms per ml.

The electrical conductivity of clay slips depends chiefly on the proportion of soluble salts present in the water but also on the type of clay.

Electrical Conductivity and Resistivity of Ceramic Materials

The conduction of electricity in materials is the result of ionic or electronic movement throughout the mass of the body. In solutions and some melts, the freedom of movement of the individual ions is the determining factor, but in solid materials this is, for the most part non-existent, and conductivity depends mainly on the ease with which electrons can move about the body. Some materials, such as the metals, have so-called *free electrons* which can wander, almost without restriction, between the electron orbitals of the ions; these electrons facilitate the passage of electricity and impart a high conductivity to the mass.

Modern concepts of electrical conduction are based on an atomistic approach and the quantum states of electrons as predicted by the Pauli exclusion principle. A given quantum state (p. 63) or electron orbital position cannot be occupied by more than two electrons and even these must have opposite spins. The electrons associated with particular atoms or ions in a crystal lattice are confined to 'bands' and associated with specific electron shells. If a sufficient amount of energy can be supplied to the crystal lattice, electrons can be raised to bands of higher energy and some of these interlink to form a continuous 'conduction band' throughout the structure. In metal structures, conduction bands exist at energy levels only slightly greater than the normal valency state. So electrons are easily excited into the condition where they can wander at random throughout the structure. When potential energy is applied across such a material, the electrons acquire directional energy; they can be excited to higher levels within the conduction band and hence move in the direction of the applied field. The energy levels between the so-called valency bands and the conduction bands virtually merge in some metals (Fig. XIV.12a).

In a material with covalent/ionic bonds, the position is very different. A large energy gap (Fig. XIV.12c exists between the valency electrons—those linking

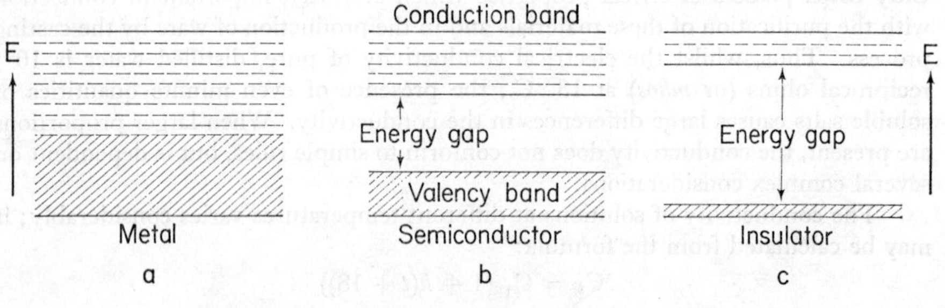

Fig. XIV.12. Electron Energy Levels in Solids

(a) Metal conducting bodies. (b) Semi-conductors. (c) Insulating or non-conducting bodies.

The cross-hatched levels represent bound or valency states; the horizontal hatching indicates the conduction bands.

atoms or ions in the structure—and the conduction bands. In order to raise electrons to the conduction band, a large amount of energy has to be supplied, which is not always feasible.

There exists a class of materials of intermediate character known as semi-conductors where the energy gap between the valency and conduction bands is not very great (Fig. XIV.12b). The application of external energy such as heat or light is often sufficient to excite some electrons into the conduction band, so that conductivity increases to a considerable degree.

The position is further complicated because impurity ions or lattice defects can provide electrons to enter the conduction bands, or 'positive holes' to serve as carriers.

This simple concept of electrons as carriers of electric current is most useful in explaining the difference between conductors, semi-conductors and non-conductors. It can be taken further to account for actual conductivity values and the changes which occur when materials are heated.

In a metal, the conduction band is readily filled with electrons which develop directional motivation in the presence of an applied electrical field. The ions or nuclei (including valency electrons) hinder the passage of 'free electrons' which progress with a wave motion. At elevated temperatures, the ionic vibrations within the lattice increase, so that the 'free electrons' have to take a greater path length. Although more valency electrons may be excited to the conduction band, it is invariably true that metals decrease in electrical conductivity at high temperatures. Conversely, most metals increase in conductivity at low temperatures and some become super-conducting as their temperature approaches absolute zero. Some metals are less conducting than others largely because of the different tendencies for electrons to enter the conduction band and because of the differences in the freedom of movement of electrons within the band itself, but these detailed considerations are outside the scope of this book.

In semi- and non-conductors a different condition applies. External energy sources increase the tendency for electrons to enter the conduction band and heat also increases the proportion of lattice defects, so that such materials, when raised in temperature, tend to increase in conductivity and often to a remarkable degree. The semi-conducting metals such as selenium and germanium are the best-known examples of materials which change in conductivity when exposed to external energy sources. The former is particularly sensitive to light and is the basis of most photo-electric devices; the latter is affected by heat and its electrical conductivity changes quite dramatically over a small temperature range.

So-called non-conductive materials, which comprise most ceramic materials, exhibit different electrical conductivities largely because of their electron configuration and defects within their lattices.[1] They also show different behaviour for the same reasons when subjected to heating.

Some minerals show anomalous behaviour in their electrical conductivity when passing through a phase change.

At very high applied electrical voltages, the valency electron bands of non-conductors can become sufficiently excited for some electrons to be raised to the

[1] Ford, W. F. and White, J., *Trans. Brit. Ceram. Soc.*, 51, 1, 1952.

conduction band level with a consequent complete break-down in the insulating character of the material.

When cold, most ceramic materials are practically non-conductors, but, owing to the resistance they offer to the current, they become heated in time and the conductivity rises rapidly with the temperature. When in a molten state, most silicates behave as electrolytes and as dissociated substances in solution. The softening point of the principal constituent (eutectic) can often be determined by the change in the conductivity at the point of fusion.

The electrical conductivity and resistivity of ceramic materials are expressed in different ways according to convenience:

1. The *electrical conductivity* may be expressed directly in C.G.S. units as *mho* (or reciprocal ohms) per centimetre.

2. The *electrical resistance* may also be expressed in C.G.S. units as *ohms* per centimetre (the temperature being also stated both in this and the electrical conductivity).

3. The *breakdown* or *puncture voltage*, is the voltage at which a piece of the material of definite thickness is unable to act as an insulator, and, therefore, allows the electric current to pass through it.

4. The *arc-over voltage* is that at which an electric current 'jumps the gap' between an insulator and an electrode and depends on the distance between them, as well as on the resistance of the insulator. The minimum permissible distance (in inches) is usually taken to be numerically equal to the voltage divided by 3000.

British Standard Specification No. 116 specifies the arc-over voltage as at least 10 per cent more than the *one-minute* test, which consists in exposing the insulator for one minute to a voltage equal to $(2\frac{1}{2}\,V + 2000)$ where V is the line-voltage.

5. The *di-electric* or *electric strength* is another term used to indicate puncture voltage. It is expressed by the number of volts per mm. of thickness of the insulator which must be passed at 250° C. before a current will flow. For pieces 0·25 in. thick, the electric strength of a porcelain insulator should be at least 200 volts per mm. The di-electric strength diminishes rapidly as the temperature increases.

6. The *'Te' value* or the temperature at which the electrical resistance is reduced to 1 megohm per cm.

7. The *specific inductive capacity* or the relative dielectric strength of the article or material, that of an equal thickness of air being taken as unity.

The general factors influencing the electrical conductivity and resistivity of ceramic materials are:

1. The composition of the material.
2. The size of the article or test piece at the time.
3. The density and texture of the material.
4. The manner of heating and the extent to which the material has been heated during the 'burning' of the ware.
5. The piezo-electric effect.
6. The temperature of the ware.
7. The length of exposure to the electric current.

The electric current can take two paths through a ceramic material (i) directly through a mass as a whole—*volume conductivity*. It is measured by placing two equal-sized conductor plates at opposite ends of the material so as to form a circuit and passing the current through it. (ii) Chiefly (or wholly) through the surface layers—*surface conductivity*. It varies with the humidity of the atmosphere and is probably due, in many cases, to the condensation or absorption of oil or moisture; the latter may, in time, dissolve small quantities of electrolytes (alkalies, soluble salts, etc.) from the material and so produce a conducting film.

Some materials are used as *conductors* but others, including some ceramic articles, are used to resist or wholly prevent the passage of an electric current, i.e. as *insulators*. Thus resistance may be regarded as the opposite of conductivity and expressed as the *reciprocal* of conductivity. Thus if the conductivity is represented by the letter ε, the resistance or resistivity (τ) would be represented by the expression $1/\varepsilon$ so that

$$\tau = \frac{1}{\varepsilon}$$

When considering conductors it is simple to refer to their conductivity; when insulators are under consideration it is often better to refer to their resistance or resistivity.

The *composition* of the ceramic material largely determines its electrical properties, some materials being much better conductors than others. In composite materials, such as porcelain, both the chemical composition and texture are of equal importance.

Fig. XIV.13 shows the variation with temperature of the resistivity of some

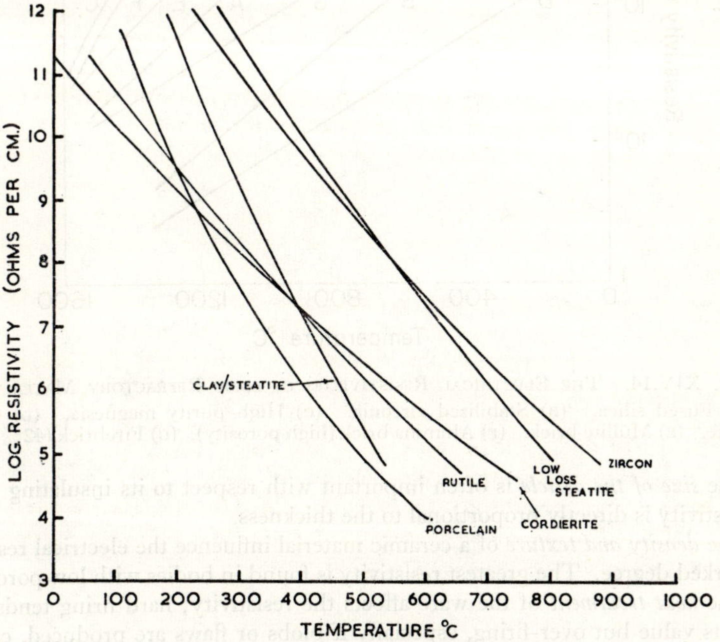

Fig. XIV.13. The Variation of the Resistivity of some Electrical Ceramic Materials with Temperature

electrical ceramic materials; steatite and zircon bodies have exceptionally high values and these are maintained up to high temperatures. Electrical porcelains made of clay, felspar and flint, have a high resistivity at room temperature, but it falls rapidly with increasing temperature, and the Te value is between 300°–500° C.

The Te value can be increased by replacing felspar with other fluxes which do not contain alkali. The presence of sodium and potassium in a porcelain body reduces the resistivity especially at high temperatures.

Wallace and Ruh[1] have examined the electrical resistivity of a number of refractory materials and the manner in which the property changes with temperature. Some of their results are shown in Fig. XIV.14 along with those of some other workers.

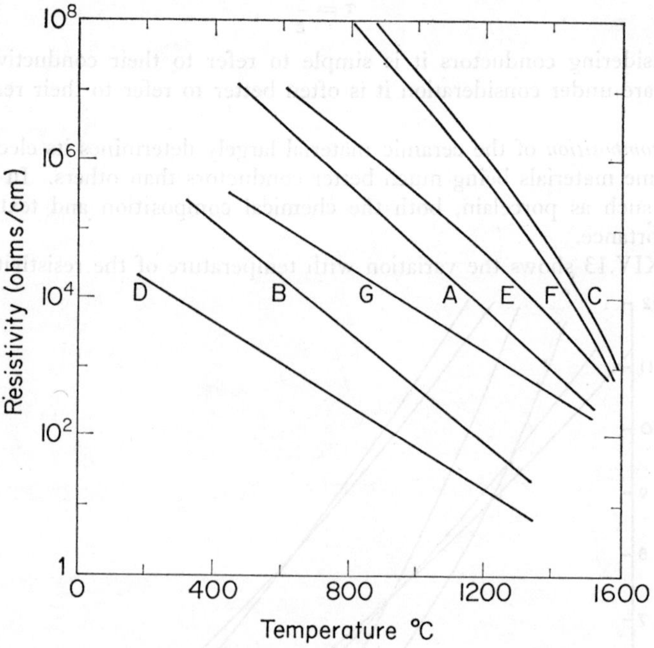

FIG. XIV.14. THE ELECTRICAL RESISTIVITIES OF SOME REFRACTORY MATERIALS

(A) Fused silica. (B) Stabilised zirconia. (C) High purity magnesia. (D) Chrome magnesite. (E) Mullite brick. (F) Alumina brick (high porosity). (G) Firebrick (42% Al_2O_3).

The *size of the article* is often important with respect to its insulating power; the resistivity is directly proportional to the thickness.

The *density and texture* of a ceramic material influence the electrical resistivity to a marked degree. The greatest resistivity is found in bodies with low porosity.

The *heat treatment* of the ware affects the resistivity; hard firing tends to increase its value but over-firing, especially if blobs or flaws are produced, causes a

[1] Wallace, R. W. and Ruh, E., *J. Amer. Ceram. Soc.*, 50, 358, 1967.

reduction. The Te value of porcelain is affected by the rate of cooling; rapid cooling produces a closer and more vitreous structure than slow cooling.

The *piezo-electric effect*, in which an electric current is produced by pressure, sometimes seriously reduces the electrical resistance of an insulator if the latter is under compression. Some specimens of quartz and porcelain exhibit this phenomenon when the pressure is applied to two diametrically opposite faces parallel to the major axes; a potential difference is then set up in the faces perpendicular to those in which the pressure is applied, this difference varying directly as the pressure. The converse of this may also occur when some quartz crystals are subjected to the prolonged action of an electric current and a change in the dimensions of the crystals may then result, an expansion occurring along one axis and a contraction along another. If these changes are hampered by the surrounding magma, great local stresses may occur; consequently, when a piece of porcelain is placed in an alternating field of electrostatic force, a vibratory movement results, owing to the repeated changes in the dimensions of quartz or other crystals similarly affected, and these vibrations may cause a rupture along the cleavage planes of the crystals and also between the crystals and the matrix. The rupture may result in a leakage of current through the spaces so formed, and the dielectric strength then rapidly deteriorates. The piezo-electric effect appears to be exhibited only in connection with certain crystals; it is, therefore, at a minimum in those porcelains in which the quartz crystals originally present in the raw materials have been most completely dissolved in the felspathic ground mass by prolonged heating at a suitable temperature. So far as can be ascertained, mullite and cristobalite crystals—which are also formed in most well-burned porcelains—do not show this effect.

The *temperature of the ware* when in use has a notable effect on the electrical resistance or insulating power, as most ceramic materials become much better conductors of electricity when at a high temperature than when cold. This is well shown in Fig. XIV.13 and 14.

With most electrical ceramic materials the resistivity/temperature relationship conforms reasonably well to a first order form where

$$R = Ae^{E/kT}$$

where R is the resistivity, T the absolute temperature, A and k are constants and E may be defined as the *activation energy of conductivity*. The value of E varies between 20–40 k.cal./mole.

The curves in Fig. XIV.13 and Fig. XIV.14 show a close approach to linearity if they are plotted in the form of log R against the reciprocal of absolute temperature, because, from the equation quoted above,

$$\log R = \frac{E}{kT} + \log A.$$

The *duration of exposure to an electric current* has a noteworthy effect on the apparent resistivity or dielectric strength of porcelains, as a much lower voltage applied for a long time will puncture the material in the same manner as a current of high voltage applied only for a very short time. In making comparative tests it is, therefore, important to state the time during which the current was applied.

The *electric sensitiveness* of insulators is tested by heating the sample in water to 50° C. for one hour, then placing it in ice-water for an hour, and, after wiping it dry, subjecting it three times in succession to a flash-voltage.

Porcelain. The electrical insulating power of porcelain and stoneware depends largely on the texture and density, and these are controlled by the chemical composition and the manner of burning.

Until comparatively recent years, electrical porcelain bodies were similar in composition to the conventional hard and soft varieties described in Chapter VI. The presence of quartz has been recognised as being undesirable, particularly as the demand for high tensile strength products has increased.

The science of electrical porcelain manufacture has advanced considerably and bodies are now being used which are very different from the standard recipes for conventional porcelains.[1]

The principal advances have been in the introduction of alumina, sillimanite and kyanite into clay compositions at the expense of quartz and felspars. Sparkplugs, in which high strength and resistance to thermal shock are prime requirements, now mostly contain more than 90 per cent alumina.

In all high-strength electrical porcelains, the alumina content is high (in excess of 45 per cent) and this is achieved by adding calcined bauxites ground to a fine-particle size to the mixture. Glaze compositions have to be carefully controlled to avoid faults in manufacture.

A porcelain with a high cristobalite content has been developed in Japan and this has led to the development of bodies of exceptionally high strength, but without the high cost associated with alumina additions. The cristobalite crystallises from highly siliceous bodies in the form of extremely fine particles.

The whole emphasis in modern electrical porcelain manufacture is on fine-textured and crystalline bodies which possess high strength and do not develop excessive localised stresses when in use. Modulus of rupture values in excess of 12,000 lbf/in.^{2} are frequently demanded. Felspars are still the main source of flux, but these are selected with great care to be extremely low in iron and other undesirable impurities. The other major components are ball clays and china clays which are nowadays produced in an extremely fine-particle size, ideal for inclusion in electrical porcelains.

The electrical conductivities of porcelains are low and range from $4 - 7 \times 10^{-16}$ (ohm/cm.) at room temperature. At 600° C. the value falls to $5 - 7 \times 10^{-7}$ but they retain a reasonably high resistivity up to their melting point.

The dielectric strengths of bodies of typical porcelain compositions have been measured by Oliver and Waye.[2] They report values in the range 15–30 KV/mm.

Determination of Electrical Conductivity and Resistivity. The conductivity or resistivity of a material is usually determined on a centimetre cube of the sample which is placed between two electrodes. The current flowing when a high voltage is applied is measured and the resistivity calculated by simple application of Ohm's Law.

It is important to obtain good contacts between the terminal plates—achieved

[1] Bloor, E. C., *J. Brit. Ceram. Soc.*, 7, 77, 1970.
[2] Oliver, J. and Waye, B. E., *Trans. Brit. Ceram. Soc.*, 69, 121, 1970.

by applying a spray of platinum paste before securing the electrodes—and to avoid surface conduction effects.

The *puncture-voltage* (i.e. the voltage required to break down the resistance and allow a current to pass readily) is generally determined by placing a sample of known thickness between two electrodes and applying a gradually increasing voltage until puncture or other breakdown occurs and noting the maximum voltage applied, the test piece being usually immersed in oil, unless the test is made at a high temperature, when oil cannot, of course, be used.

Surface Conductivity. In some cases, a material will have a high resistivity in volume, but will not behave so well in practice because of the ease with which electricity can pass through its surface layers. This arises principally as the result of the adsorption of a thin film of oil or water which acts as conducting layers. Materials with a smooth finish are less liable to break down in this way and for this reason a glaze or surface coating is usually applied. Silicones (organic silicon complexes) are being increasingly used for this purpose, because of their low water adsorption.

Dielectric Constant

An electrical condenser is an arrangement of two parallel plates separated by a medium. The capacity of such a unit is governed largely by the nature of the intervening material and where high frequency alternating current is involved or high capacities required, it is important that the condenser plates be separated by a substance with a large *dielectric constant.*

The Dielectric Constant is the inductive capacity of a material when placed in a potential field, as for example when ceramic bodies are used in electrical condensers.

The dielectric constant of a material is expressed by means of Coulomb's Law, which defines the force F which develops between two point charges q_1 and q_2 which are separated by a distance d as,

$$F = k\frac{q_1 q_2}{d^2}$$

k is a proportionality constant which is expressed in different ways in standard text-books, but it depends on the units which are used for charge, distance and force. The modern convention is to represent these in coulombs, metres and newtons so the value of k equals

$$9 \times 10^9 \; nt\text{-}m^2/\text{coul}^2$$

k is related to the *permittivity* ε_0 of the material between the charged points by the equation

$$k = \frac{1}{4\pi\varepsilon_0}$$

so the net force F is given by

$$F = \frac{1}{4\pi\varepsilon_0}\frac{q_1 q_2}{d^2}$$

and $\varepsilon_0 = \dfrac{1}{4\pi k} = 8{\cdot}85 \times 10^{-12}\ \text{coul}^2/nt\text{-}m^2$

The permittivity of the space between the points or plates of charge varies with the material which is interposed; it is customary to define the permittivity of a material in relation to that of free space. Ideally that of a vacuum is preferred, but, in fact, the value for air is only slightly larger than that of a vacuum and for most purposes by an insignificant amount. So that

$$\varepsilon = K\varepsilon_0$$

where ε is the permittivity of the material between the charged points and K is a dimensionless constant, termed the *dielectric constant* or the *specific inductance capacity* of the material. For vacuum or air the K value is 1.

The value of its dielectric constant, determines the capacity of a material in electrical condenser applications.

Most of the commonly-employed solid electrical insulating materials, including paraffin wax, mica, glass and porcelain have dielectric constants of less than 10 (relative to air which is regarded as unity) and, furthermore, when subjected to alternating currents of high frequency such material becomes heated and the dielectric constant falls appreciably.

The addition of titanium dioxide to clay bodies increases the dielectric constant to about 60 and this discovery has led to the development of titanate bodies; barium titanate has a dielectric constant of about 1200 and when mixed with other ingredients, values of up to 10,000 have been attained. Above a certain temperature, known as the Curie temperature, such bodies lose their dielectric properties but additions of small amounts of zirconates, or stannates can increase their temperature stability.

Steatite bodies have a dielectric constant of about 60 and can withstand temperatures of up to 1000° C. without showing undue breakdown.

When certain substances are placed in an electric field, ionic alignments may be set up, thereby inducing *polarisation.* Usually there is a specific temperature above which such substances lose their *ferroelectric* properties (the Curie temperature), but below this point the electric polarisation may be very high in some cases.

The dielectric constant K of materials may be expressed by the equation:

$$K = 1 + 4\pi\frac{P}{E}$$

where P is the spontaneous electric polarisation and E is the applied electrical field.

The subject of ferro-electrics is outside the scope of this volume but it should be noted that some ceramic materials have remarkable properties in this respect. One such example is *perovskite* or barium titanate $BaTiO_3$; dielectric constants in excess of 10,000 are reported with this material as compared with figures of the order of 5 for rocksalt, quartz and similar materials. Materials of this type have also a large piezo-electric effect (p. 961).

Low Loss Ceramics

MATERIALS which are used as insulators of alternating currents, tend to absorb energy which is dissipated as heat. This represents wasted energy and at the same time the electrical efficiency of the medium is reduced because of the attendant rise in temperature (see p. 964). When exposed to a high frequency alternating current, many of the standard insulating materials, such as porcelains and mica become excessively hot because they have a high *power factor*.

Ceramic bodies have been developed in which much less electrical energy is wasted; these include cordierite (talc + clay), clinoenstatite and steatite (or talc) (p. 763). Bodies containing titania (TiO_2) are also valuable for this purpose especially at higher frequencies; a rutile (TiO_2)—zirconia (ZrO_2) mixture has a power factor which is almost constant over a wide range of radio and electrical frequencies.

The composition of materials of value as insulators and dielectrics has been discussed in Chapter VI and Chapter XI.

MAGNETIC PROPERTIES

FOR the most part, clays and products derived from them are of little or no value in terms of their magnetic properties. Some impurities contained in clays may show magnetic characteristics to a sufficient extent to enable them to be removed.

Some burned fireclays, particularly if the temperature of heating has not been high, may show feeble magnetic characteristics because of the iron impurities which they contain, especially when magnetite, Fe_3O_4 has been formed. The magnetic properties of some minerals associated with ceramic materials are shown in Table XIV.XII, but these frequently owe their characteristics to the presence of iron minerals either in solid solution or as impurities.

The physical causes of magnetism are not yet fully explained, but it is known that in bodies which exhibit this property, the electron bonding which occurs aligns the spins of electron pairs into a regular order and direction and sets up a 'domain structure'.

In their combination with atoms or ions of opposite valency, certain atoms in a crystal lattice develop magnetic moments on the atomic scale which are not balanced by parallel bands of the same moment but in an opposite direction.

Various types of magnetic effects can occur depending upon the type and direction of the atomic magnetic moments within the crystal lattice.

The structures which can develop these specific conditions are relatively few and most frequently they are of the 'defect form' in which cations are present in non-normal co-ordination situations.

For a detailed description of the known factors which cause magnetic characteristics in bodies, the reader is referred to several interesting specialised publications on the subject.

It is a fascinating subject, more so because synthetic materials of unusual magnetic characteristics, called ferrites, are classed as ceramic materials.

Several important groups of such minerals have been described by Sadler.[1]

Recent developments in the field of electronics necessitate the use of materials which are ferro-magnetic and yet have a high electrical resistivity. The particular applications are as cores for coils and transformers carrying high frequency currents where energy is likely to be dissipated in the form of eddy currents. For a variety of

[1] Sadler, A. G., *J. Canadian Ceram. Soc.*, 33, 1964.

TABLE XIV—XII. MAGNETIC PROPERTIES OF MINERALS

Highly Magnetic	*Moderately Magnetic*	*Feebly Magnetic*	*Almost Non-Magnetic*	
Magnetite	Hypersthene	Chlorite	Zircon	Rutile
Titanoferrite	Augite	Staurolite	Corundum	Barytes
Ilmenite	Garnet	Epidote	Galena	Most iron-free minerals
Pyrrhotite	Siderite	Limonite	Fluorspar	Clay
Haematite	Olivine	Actinolite	Pyrite	Silica
	Hornblende	Kyanite	Cassiterite	
	Chromite	Fayalite		
	Wolfram			
	Monazite			

reasons, ferrites are of great value in these applications and are now being used on an extensive scale. Ferrites differ widely in composition but they have all an identical crystal structure, based on the inverse spinel type discussed in Chapter IV. These materials owe their magnetic properties to the situation of cations in the lattice which behave as carriers of magnetic moments.

For a more detailed description of the properties of these and other ferro-magnetic materials, the reader is referred to other publications including those of Snoek,[1] Neel[2] and Merz.[3]

[1] Snoek, J. L., *New Developments in Ferro-magnetic Materials* (Elsevier, New York, 1947).
[2] Neel, L., *Ann. de Physique*, 3, 137, 1948; *Comptes. Rendus.*, 230, 190, 1950.
[3] Merz., W. J., 'Crystal Chemistry Symposium' (*Ceramic Age*, 1951).

NOTES TO THE APPENDIX

1. The optical values have either been determined by the author or are taken from various collected data.

2. (a) The X-ray data for ceramic minerals are largely from the author's collection, supplemented by data published by the American Society for Testing Materials.

(b) In most cases, the three lines quoted are the strongest of those given by a particular mineral, but where necessary, the characteristic lines by which a mineral may be identified are preferentially quoted. Where minerals have been arranged in families, reflections from equivalent spacings are compared irrespective of their relative intensities.

(c) The values of the intensity of lines are relative to the intensity (10) of the strongest line for a particular mineral.

3. (a) Differential thermal analysis data have been derived by using purified minerals from reputable sources, tested by the standard method described in Chapter V.

(b) The absence of values of thermal effects for certain minerals does not necessarily mean that those minerals are unreactive. Undoubtedly, some minerals undergo polymorphic transformations when heated which would be manifested as a small thermal effect capable of detection by more sensitive equipment. These have not been included in the present survey, nor have thermal effects about which there is a reasonable doubt.

(c) The magnitudes of thermal effects are classified as:

v. large	— producing a temperature difference	greater than 90° C.
large	— ,, ,, ,, ,,	,, ,, 50° C.
medium	— ,, ,, ,, ,,	between 10–50° C.
small	— ,, ,, ,, ,,	less than 10° C.
v. small	— ,, ,, ,, ,,	,, ,, 3° C.

OPTICAL PROPERTIES

	Composition	*Crystal system*	*Specific gravity*	*Hardness*	*Refractive index*	*Birefringence*	*Interference figure*	*Crystal habit*	*Colour*
Clay and Related Minerals									
1. kaolinite	$Al_2(Si_2O_5)(OH)_4$	Triclinic	2·61	2	1·56	0·006	B-neg.	Plates	None
2. halloysite (normal)	$Al_2(Si_2O_5)(OH)_4$	Hexagonal?	2·62	—	1·52	0·030	B-neg.	Small needles	None
3. livesite	$Al_2(Si_2O_5)(OH)_4$	Monoclinic	2·63	—	1·54	0·020	B-neg.	Small plates	Yellow?
4. dickite	$Al_2(Si_2O_5)(OH)_4$	Monoclinic	2·62	2·5	1·56	0·006	B-neg.	Plates	None
5. nacrite	$Al_2(Si_2O_5)(OH)_4$	Orthorhombic	2.54	2·5	1·56	0·006	B-neg.?	Plates	None
6. anauxite	$Al_2O_3.3SiO_2 2H_2O$	—	2·60	—	—	—	—	—	—
7. allophane	$Al_2O_3.1{\cdot}5SiO_2 nH_2O$	—	—	—	1·49	—	Isotropic	—	—
8. montmorillonite	$(Al, Mg)_2Si_4O_{10}(OH)_2$	Monoclinic	2·5	2	1·51	0·021	B-neg.	Small flakes	None
9. beidellite?	$Al_2(Al, Si)_4O_{10}(OH)_2$	Monoclinic	2·5	2	1·51	0·020	B-neg.	Small flakes	Cream
10. nontronite	$(Al, Fe, Mg)_2(Al, Si)_4O_{10}(OH)_2$	Monoclinic	2·6	2	1·56	0·023	B-neg.	Small laths	Brown
11. hectorite	$(Mg, Li, Al)_3Si_4O_{10}(OH)_2$	Monoclinic	2·4	2	1·49	0·022	B-neg.	Small flakes	None
12. saponite	$Mg_3(Al, Si)_4O_{10}(OH)_2$	Monoclinic	2·5	2	1·50	0·015	B-neg.	Small flakes	None
13. talc	$Mg_3Si_4O_{10}(OH)_2$	Monoclinic	2·75	1	1·58	0·040	B-neg.	Plates	None
14. pyrophyllite	$Al_2Si_4O_{10}(OH)_2$	Monoclinic	2·80	1·5	1·57	0·050	B-neg.	Plates	None
15. vermiculite	$(Mg, Fe)_3(Al, Si)_4O_9(OH)_3$	Monoclinic	2·4	1·5	1·56	0·020	B-neg.	Plates	Brown
16. glauconite	$KMg(Fe, Al)_3Si_6O_{18}3H_2O$	Monoclinic	2·4	2	1·60	0·020	B-neg.	Small laminae	(Red) Green
17. sepiolite	$H_4Mg_2Si_3O_{10}.nH_2O$	Orthorhombic	2	2	1·53	0·010	U-neg.	Small fibrous	None
18. palygorskite (attapulgite)	$Mg_5Si_8O_{20}(OH)_2 8H_2O$	Orthorhombic	—	—	1·51	0·020	—	Small fibrous	—

X-RAY ANALYSIS						DIFFERENTIAL THERMAL ANALYSIS (Grimshaw and Roberts Apparatus)				
Three principal lines						Endothermic Peaks (maxima)		Exothermic Peaks (maxima)		
d in Å; I, relative intensity						Temp. °C.	Magnitude	Temp. °C.	Magnitude	
d	I	d	I	d	I					
7·15	10+	3·566	10+	2·331	10	105 585	small large	985	v. large	1
7·2	8	3·566	8	4·422	10+	115 550	medium large	985	large	2
7·15	10	3·566	10	4·455	8	110 200? 553	medium small large	925–945	medium	3
7·14	10	3·576	10+	2·318	8	105 580 695	v. small small large	980	large	4
7·08	10	3·58	9	2·418	10	105 690	v. small large	980	large	5
7·1	10	3·57	10	2·35	10	105 580	v. small large	980	large	6
—		—		—		200 520	medium small	985	medium	7
14·7	10	4·42	5	1·49	4	140 700 875	medium medium small	920	small	8
14·2	10	4·45	4	1·49	4	150 540	medium small	915	small	9
14·0	10	4·48	3	1·51	2	140 490	medium medium	900	small	10
14·3	10	4·44	3	1·51	5	145 840	medium medium	—	—	11
14·0	10	4·51	5	1·52	5	150 870	medium medium	—	—	12
9·30	8	3·10	10	1·53	8	910	medium (broad)	—	—	13
9·14	8	3·06	10	1·50	8	750	medium (broad)	—	—	14
14·1	10	3·54	6	1·53	6	120–200 820	medium small	860	small	15
9·92	10	4·45	9	3·34	9	110 565 895	small medium small	920	medium	16
12·15	10	4·60	10	2·61	8	140 810	medium medium	300–600 880	medium (broad) medium	17
10·5	10	4·49	8	3·23	10	110 260 410 535	small medium medium small	940	medium	18

OPTICAL PROPERTIES

	Composition	Crystal system	Specific gravity	Hardness	Refractive index	Birefringence	Interference figure	Crystal habit	Colour
Micas									
19. muscovite	$KAl_2(Al_1Si_3)O_{10}(OH)_2$	Monoclinic	2·9	2·5	1·59	0·036	B-neg.	Plates	None
20. paragonite	$NaAl_2(Al_1Si_3)O_{10}(OH)_2$	Monoclinic	2·9	2·5	1·60	0·030	B-neg.	Plates	None
21. lepidolite	$K(Li, Al)_3(Al_1Si_3)O_{10}(OH)_2$	Monoclinic	3·0	3	1·56	0·020	B-neg.	Plates	None
22. biotite	$K(Mg, Fe)_3(Al_1Si_3)O_{10}(OH)_2$	Monoclinic	3·1	2·5	1·63	0·050	B-neg.	Plates	Brown
23. margarite	$CaAl_2(Al_2Si_2)O_{10}(OH)_2$	Monoclinic	3·0	4	1·65	0·010	B-neg.	Tabular	None
24. hydrous mica	Variable	Monoclinic	2·75	—	1·57	0·021	B-neg.	Plates	None
Felspars									
25. orthoclase	$KAlSi_3O_8$	Monoclinic	2·55	6	1·52	0·006	B-neg.	Prismatic	None
26. microline	$KAlSi_3O_8$	Triclinic	2.55	6	1·53	0·007	B-neg.	Equant	None
27. celsian	$BaAl_2Si_2O_8$	Monoclinic	3·4	6	1·59	0·010	B-pos.	Prismatic	None
28. albite	$NaAlSi_3O_8$	Triclinic	2·61	6	1·53	0·010	B-pos.	Tabular	None
29. andesine	$(Na, Ca)(Al, Si)_4O_8$	Triclinic	2·67	6	1·55	0·007	B-pos.	Tabular	None
30. anorthite	$CaAl_2Si_2O_8$	Triclinic	2·76	6	1·58	0·013	B-neg.	Tabular	None
Felspathoids									
31. leucite	$KAlSi_2O_6$	Orthorhombic	2·47	6	1·51	0·001?	Isotropic?	Equant	None
32. nephelite	$NaAlSiO_4$	Hexagonal	2·61	6	1·54	0·004	U-neg	Prismatic	None
33. carnegieite	$NaAlSiO_4$	Triclinic	2·51	6	1·51	0·005	B-neg.	Needle	None
34. eucryptite	$LiAlSiO_4$	Hexagonal	2·67	—	1·55	0·002	U-neg.	Prismatic	None
Chlorites									
35. antigorite	$Mg_3Si_2O_5(OH)_4$	Orthorhombic	2·57	3·5	1·57	0·008	B-neg.	Fibrous	Green
36. amesite	$(Mg_2Al)(AlSi)_2O_5(OH)_4$	Monoclinic	2·8	2·5	1·58	0·015	B-pos.	Lamellae	Pale green
37. daphnite	$(Fe_2Al)(Al, Si)_2O_5(OH)_4$	Monoclinic	3·0	2	1·65	0·001	B-neg.	Lamellae	Green

X-RAY ANALYSIS: Three principal lines (d in Å; I, relative intensity)						DIFFERENTIAL THERMAL ANALYSIS (Grimshaw and Roberts Apparatus): Endothermic Peaks (maxima)		Exothermic Peaks (maxima)		
d	I	d	I	d	I	Temp. °C.	Magnitude	Temp. °C.	Magnitude	
9·98	10	3·33	9	2·57	10	600–900	small (broad)	—	—	19
9·97	6	4·41	10	3·10	10	—	—	—	—	20
9·84	5	4·42	10	3·60	10	750	medium	—	—	21
10·1	10	3·36	9	2·65	7	550 700–900	medium small (broad)	—	—	22
—		—		—		—	—	—	—	23
10·0	10	4·45	8	3·32	9	105 550 625–660	v. small small medium	940	medium	24
3·20	10	1·81	9	4·02	8	850	small (broad)	—	—	25
3·21	10	1·80	8	2·16	6	—	—	—	—	26
2·60	10	2·09	8	1·79	7	—	—	—	—	27
3·20	10	1·80	6	4·00	5	820	medium	—	—	28
3·19	10	1·88	6	4·00	5	—	—	—	—	29
3·17	10	2·51	6	4·02	4	780	medium	—	—	30
3·27	10	5·30	9	3·42	8	640	v. small	—	—	31
3·01	10	3·82	9	3·17	8	—	—	—	—	32
4·29	10	2·61	9	1·50	6	690	v. small	—	—	33
3·97	10	2·55	10	6·92	8	440	v. small	—	—	34
7·14	9	3·59	10	2·52	7	105 640	v. small large	840	large	35
7·0	10	3·50	10	2·47	9	105 580	v. small medium	780	small	36
6·9	8	3·51	10	4·69	5	105 460–600	v. small medium (broad)	780	small	37

OPTICAL PROPERTIES

	Composition	*Crystal system*	*Specific gravity*	*Hardness*	*Refractive index*	*Birefringence*	*Interference figure*	*Crystal habit*	*Colour*
38. ferro-antigorite	$Fe_3Si_2O_5(OH)_4$	Monoclinic	3·2	2	1·66	0·007	B-neg.	Lamellae	Dark green
39. penninite	$(Mg, Fe, Al)_3 (Al, Si)_2O_5(OH)_4$	Monoclinic	2·70	2	1·57	0·002	B-pos.?	Lamellae	Green
40. chrysotile (serpentine)	$Mg_3Si_2O_5(OH)_4$	Orthorhombic	2·55	2·5	1·55	0·013	B-pos.	Fibrous	Green
Olivines									
41. forsterite	Mg_2SiO_4	Orthorhombic	3·22	6·5	1·67	0·035	B-pos.	Equant	None
42. fayalite	Fe_2SiO_4	Orthorhombic	4·32	6·5	1·88	0·051	B-neg.	Equant	Green, red
43. chrysolite	$(Mg,Fe)_2SiO_4$	Orthorhombic	3·3	6·5	1·68	0·035	B-pos.	Prismatic	Green
44. monticellite	$CaMgSiO_4$	Orthorhombic	3·2	5	1·64	0·017	B-pos.?	Prismatic	None
Pyroxenes									
45. enstatite	$MgSiO_3$	Orthorhombic	3·18	5·5	1·65	0·004	B-pos.	Prismatic	None
46. clinoenstatite	$MgSiO_3$	Monoclinic	3·19	5·5	1·65	0·004	B-pos.	Prismatic	None
47. diopside	$(CaMg)SiO_3$	Monoclinic	3·28	5	1·66	0·032	B-pos.	Prismatic	None
48. augite	$(Ca,Mg,Fe)SiO_3$	Monoclinic	3·4	5·5	1·71	0·026	B-neg.	Prismatic	Green, brown
49. spodumene	$(LiAl)SiO_3$	Monoclinic	3·15	6·5	1·68	0·027	U-pos.	Prismatic	None
50. wollastonite	$(Fe,Ca)SiO_3$	Triclinic	2·92	4·5	1·61	0·020	B-neg.	Needles	Pale green
Amphiboles									
51. tremolite	$Ca_2Mg_5Si_8O_{22}(OH)_2$	Monoclinic	3·1	5·5	1·62	0·026	B-neg.	Fibrous	None
52. hornblende	$Ca_2Na(Mg,Fe,Al)_5 Al_2Si_6O_{22}(OH)_2$	Monoclinic	3·3	5·5	1·68	0·019	B-neg.	Fibrous	Green, brown
Melilites									
53. akermanite	$Ca_2MgSi_2O_7$	Tetragonal	3·18	5	1·63	0·007	U-pos.	Prismatic	None
54. gehlenite	$Ca_2Al(SiAl)_2O_7$	Tetragonal	3·04	5·5	1·67	0·010	U-neg.	Prismatic	None

X-RAY ANALYSIS						DIFFERENTIAL THERMAL ANALYSIS (Grimshaw and Roberts Apparatus)				
Three principal lines						Endothermic Peaks (maxima)		Exothermic Peaks (maxima)		
d in Å ; I, relative intensity						Temp. °C.	Magnitude	Temp. °C.	Magnitude	
d	I	d	I	d	I					
7·12	10	3·50	10	2·63	4	110 480	small large	760	medium	38
14·3	6	7·17	10	4·78	10	105 650 810	small large small	840	medium	39
7·07	4	3·59	10	2·45	6	105 635 800	v. small large small	850	medium	40
1·74	10	3·89	5	2·49	5	—	—	—	—	41
1·76	4	3·71	4	2·85	10	—	—	—	—	42
1·75	10	2·46	10	2·52	10	—	—	—	—	43
1·81	9	1·59	10	2·65	7	—	—	—	—	44
3·16	10	2·86	7	1·48	6	—	—	—	—	45
3·16	10	2·87	9	1·52	8	—	—	—	—	46
3·0	10	2·51	8	1·62	6	—	—	—	—	47
2·98	10	2·52	9	1·62	9	—	—	—	—	48
2·90	10	2·75	9	4·35	7	—	—	—	—	49
2·98	10	3·30	8	1·75	7	—	—	—	—	50
3·30	10	1·56	10	4·61	8	—	—	—	—	51
8·80	10	1·44	10	2·70	8	—	—	—	—	52
2·85	10	1·76	8	1·38	7	—	—	—	—	53
2·82	10	1·75	7	1·39	6	—	—	—	—	54

OPTICAL PROPERTIES

	Composition	*Crystal system*	*Specific gravity*	*Hardness*	*Refractive index*	*Birefringence*	*Interference figure*	*Crystal habit*	*Colour*
Alumino-silicates									
55. sillimanite	$Al_2O.SiO_4$	Orthorhombic	3·08	7·5	1·66	0·022	B-pos.	Fibrous	None
56. andalusite	$Al_2O.SiO_4$	Orthorhombic	3·18	7·5	1·63	0·010	B-neg.	Prismatic	None
57. kyanite	$Al_2O.SiO_4$	Triclinic	3·59	7·5	1·71	0·011	B-neg.	Tabular	None
58. mullite	$3Al_2O_3.2SiO_2$	Orthorhombic	3·03	7·5	1·65	0·012	B-pos.	Needles	None
Misc. Silicates									
59. zircon	$ZrSiO_4$	Tetragonal	4·7	7·5	1·93	0·053	U-pos.	Prismatic	None
60. phenacite	Be_2SiO_4	Rhombohedral	3·0	7·5	1·65	0·016	U-pos.	Prismatic	None
61. topaz	$(Al,F)_2SiO_4$	Orthorhombic	3·55	8	1·63	0·010	B-pos.	Prismatic	None
62. beryl	$Be_3Al_2Si_6O_{18}$	Hexagonal	2·71	7·5	1·57	0·006	U-neg.	Prismatic	None
63. cordierite	$Al_3Mg_2(Si_5Al)O_{18}$	Orthorhombic	2·6	7	1·54	0·008	B-neg.	Prismatic	None
64. tourmaline	Complex silicate of B+Al+other cations	Hexagonal	3·0	7	1·63	0·029	U-neg.	Prismatic	Blue
65. titanite (sphene)	$CaTiO.SiO_4$	Monoclinic	3·5	5	1·99	0·110	B-pos.	Plates	Brown, green
66. merwinite	$Ca_3Mg(SiO_4)_2$	Monoclinic	3·15	6	1·71	0·010	B-pos.	Prismatic	None
Oxides									
67. quartz (α)	SiO_2	Hexagonal	2·65	7	1·55	0·009	U-pos.	Prismatic	None
68. cristobalite (α)	SiO_2	Tetragonal	2·32	7	1·48	0·003	U-neg.?	Octahedral mosaic	None
69. tridymite (α)	SiO_2	Orthorhombic	2·28	7	1·47	0·004	B-neg.	Laths 'twinned'	None
70. opal	$SiO_2.nH_2O$	Amorphous	2·1	6	1·45	—	Isotropic	Pebbles	Milky
71. chalcedony	SiO_2	Hexagonal?	2·6	6	1·53	0·007	?	Fibrous	None
72. anatase	TiO_2	Tetragonal	3·9	6	2·55	0·073	U-neg.	Prismatic	Brown
73. rutile	TiO_2	Tetragonal	4·2	6	2·89	0·275	U-pos.?	Fibrous	Red
74. brookite	TiO_2	Orthorhombic	4·0	6	2·64	0·108	B-pos.	Prismatic	Brown
75. ilmenite	$FeTiO_3$	Hexagonal	4·7	7	—	—	—	Lamellar	Opaque black

X-RAY ANALYSIS						*DIFFERENTIAL THERMAL ANALYSIS (Grimshaw and Roberts Apparatus)*				
Three principal lines						*Endothermic Peaks (maxima)*		*Exothermic Peaks (maxima)*		
d in Å ; I, relative intensity						*Temp. °C.*	*Magnitude*	*Temp. °C.*	*Magnitude*	
d	*I*	*d*	*I*	*d*	*I*					
3·39	10	2·20	9	1·52	7	—	—	—	—	55
4·58	10	1·48	10	2·17	8	—	—	—	—	56
1·95	10	1·38	10	3·20	10	—	—	—	—	57
3·42	10	2·21	8	1·53	8	—	—	—	—	58
3·29	10	2·52	9	1·71	7	—	—	—	—	59
3·11	10	2·52	8	1·26	8	—	—	—	—	60
2·96	10	1·40	10	1·38	9	—	—	—	—	61
8·10	10	3·33	8	2·92	8	—	—	—	—	62
8·29	7	3·33	8	3·0	10	—	—	—	—	63
2·59	10	6·50	8	3·48	8	—	—	—	—	64
3·20	10	2·59	10	2·98	9	—	—	—	—	65
1 90	10	2·68	6	1·53	6	—	—	—	—	66
3·35	10	1·814	8	4·21	7	573	v. small	—	—	67
4·05	10	2·48	9	2·85	5	220–280	v. small	—	—	68
4·30	10	4·10	8	3·80	7	117 163	v. small v. small	—	—	69
—		—		—		220 1250	small small	—	—	70
—		—		—		1250	small	—	—	71
3·51	10	1·89	9	1·70	7	—	—	—	—	72
3·24	10	1·36	8	1·69	10	—	—	—	—	73
3·47	10	1·88	8	2·90	10	—	—	—	—	74
2·75	10	2.54	8	3·73	7	650	small	—	—	75

OPTICAL PROPERTIES

	Composition	*Crystal system*	*Specific gravity*	*Hardness*	*Refractive index*	*Birefringence*	*Interference figure*	*Crystal habit*	*Colour*
76. corundum	Al_2O_3	Hexagonal	4·03	9	1·77	0·008	U-neg.	Angular	None
77. haematite	Fe_2O_3	Hexagonal	5·2	5	3·0	0·21	U-neg.	Platey	Brown
78. periclase	MgO	Cubic	3·65	6	1·74	—	Iso-tropic	Equant	None
79. baddeleyite	ZrO_2	Monoclinic	5·7	6·5	2·20	0·070	B-neg.	Tabular	None
80. cassiterite	SnO_2	Tetragonal	6·9	6	2·0	0·097	U-pos.	Prismatic	Brown
81. perovskite	$CaTiO_3$	Cubic	4·0	5·5	2·34	—	Iso-tropic	Cubes	Yellow Green
82. chrysoberyl	$BeAl_2O_4$	Orthorhombic	3·64	8·5	1·75	0·004	B-pos.	Tabular	None Green
Spinels									
83. spinel	$MgAl_2O_4$	Cubic	3·6	8	1·72	—	Iso-tropic	Equant	None
84. chromite	$FeCr_2O_4$	Cubic	4·5	5·5	2·12	—	Iso-tropic	Equant	Orange
85. magnetite	$FeFe_2O_4$	Cubic	5·2	6	2·42	—	Iso-tropic	Equant	Opaque black
Hydroxides									
86. diaspore	AlO(OH)	Orthorhombic	3·4	7	1·75	0·048	B-pos.	Prismatic	None
87. gibbsite	$Al(OH)_3$	Monoclinic	2·35	3	1·59	0·020	B-pos.	Tabular	None
88. goethite	FeO(OH)	Orthorhombic	4·28	5	2·40	0·140	B-pos.	Prismatic	Brown
89. brucite	$Mg(OH)_2$	Hexagonal	2·4	2·5	1·58	0·019	U-pos.	Tabular	None
Sulphates									
90. anhydrite	$CaSO_4$	Orthorhombic	2·9	3	1·61	0·044	B-pos.	Prismatic	None
91. celestite	$SrSO_4$	Orthorhombic	4·0	3·5	1·63	0·009	B-pos.	Tabular	None
92. barite	$BaSO_4$	Orthorhombic	4·5	3	1·65	0·012	B-pos.	Tabular	None
93. anglesite	$PbSO_4$	Orthorhombic	6·3	3	1·89	0·017	B-pos.	Tabular	None
94. mirabilite (glauber salt)	$Na_2SO_4.10H_2O$	Monoclinic	1·46	1·5	1·44	0·001	B-neg.	Fibrous	None
95. epsomite	$MgSO_4.7H_2O$	Orthorhombic	1·68	2	1·46	0·028	B-neg	Prismatic	None

X-RAY ANALYSIS						DIFFERENTIAL THERMAL ANALYSIS (Grimshaw and Roberts Apparatus)				
Three principal lines						*Endothermic Peaks (maxima)*		*Exothermic Peaks (maxima)*		
d in Å ; I, relative intensity						*Temp. °C.*	*Magnitude*	*Temp. °C.*	*Magnitude*	
d	*I*	*d*	*I*	*d*	*I*					
2·08	10	1·60	10	2·55	9	—	—	—	—	76
2·69	10	2·51	8	1·69	8	—	—	—	—	77
2·10	10	1·49	7	1·21	2	—	—	—	—	78
3·19	10	2·85	8	1·81	4	—	—	—	—	79
3·32	10	2·62	10	1·75	10	—	—	—	—	80
2·70	10	1·91	8	1·56	7	—	—	—	—	81
2·08	10	1·61	10	3·24	8	—	—	—	—	82
2·41	10	1·42	10	4·62	8	—	—	—	—	83
2·51	10	1·46	8	3·10	4	—	—	—	—	84
2·53	10	1·48	7	2·96	6	325 620	small small	—	—	85
3·98	10	2·31	8	1·63	8	550	large	—	—	86
4·83	10	4·34	6	3·30	3	240 305 520	small v. large medium	—	—	87
4·15	10	2·43	7	2·67	6	335	large	—	—	88
4·75	8	2·36	10	1·79	4	435	large	—	—	89
1·85	10	3·49	10	2·79	7	1020	medium	—	—	90
2·04	10	3·32	10	2·97	10	1025	medium	—	—	91
2·11	10	3·42	8	3·09	8	1045	medium	—	—	92
2·07	10	4·24	9	3·0	7	900	medium	—	—	93
3·10	10	2·77	10	4·62	8	145 880	v. large medium	—	—	94
4·40	10	2·92	6	4·04	3	200 320 920	large medium medium	—	—	95

OPTICAL PROPERTIES

	Composition	*Crystal system*	*Specific gravity*	*Hardness*	*Refractive index*	*Birefringence*	*Interference figure*	*Crystal habit*	*Colour*
96. gypsum	$CaSO_4.2H_2O$	Monoclinic	2·32	2	1·53	0·009	B-pos.	Tabular	None
97. plaster of Paris	$2CaSO_4.H_2O$	Monoclinic	2·5	2	1·55	0·020	B-pos.?	Tabular	None
98. alunite	$KA1_3(OH)_6(SO_4]_2$	Rhombohedral	2·60	4	1·57	0·020	U-pos.	Cubic	None
Carbonates									
99. magnesite	$MgCO_3$	Rhombohedral	2·96	3·5	1·70	0·191	U-neg.	Equant	None
100. siderite	$FeCO_3$	Rhombohedral	3·85	3·5	1·87	0·242	U-neg.	Equant	Grey, brown
101. calcite	$CaCO_3$	Rhombohedral	2·71	3	1·65	0·170	U-neg.	Equant	None
102. dolomite	$(CaMg) CO_3$	Rhombohedral	2·86	3·5	1·68	0·176	U-neg.	Equant	None
103. aragonite	$CaCO_3$	Orthorhombic	2·94	3·5	1·68	0·155	B-neg.	Prismatic	None
104. witherite	$BaCO_3$	Orthorhombic	4·3	3·5	1·68	0·148	B-neg.	Prismatic	None
Sulphides									
105. pyrite	FeS_2	Cubic	5·02	6·5	—	—	—	Cubic	Opaque yellow
106. chalcopyrite	$(Cu,Fe)S_2$	Tetragonal	4·2	3·5	—	—	—	Prismatic	Opaque yellow
107. marcasite	FeS_2	Orthorhombic	4·89	6	—	—	—	Tabular	Opaque yellow
108. galena	PbS	Cubic	7·5	2·5	—	—	—	Cubic	Opaque grey
109. sphalerite	ZnS	Cubic	4·0	3·5	2·36	0·022	—	Cubic	Yellow
Miscell-aneous									
110. diamond	C	Cubic	3·3	10	2·42	—	Iso-tropic	Cubic	Yellow
111. graphite	C	Hexagonal	2·25	1	—	—	—	Lamellar	Opaque black
112. borax	$Na_2B_4O_7.10H_2O$	Monoclinic	1·70	2	1·47	0·025	B-neg.	Prismatic	None
113. apatite	$Ca_5 (F,C1)P_3O_{12}$	Hexagonal	3·2	5	1·64	0·004	U-neg.	Prismatic	Green, brown
114. fluorspar	CaF	Cubic	3·3	4	1.43	—	Iso-tropic	Cubic	None, Variable
115. halite	NaCl	Cubic	2·2	2	1.54	—	Iso-tropic	Cubic	None

X-RAY ANALYSIS						DIFFERENTIAL THERMAL ANALYSIS (Grimshaw and Roberts Apparatus)				
Three principal lines						Endothermic Peaks (maxima)		Exothermic Peaks (maxima)		
d in Å ; I, relative intensity						Temp. °C.	Magnitude	Temp. °C.	Magnitude	
d	I	d	I	d	I					
4·29	10	3·06	7	2·87	7	225 350 960	large medium small	—	—	96
6·05	10	3·01	9	2·80	9	220 1020	large medium	—	—	97
2·98	10	1·89	6	4·93	5	570 795	large large	680	medium	98
2·74	10	1·697	9	2·099	8	760	v. large	—	—	99
2·78	10	1·728	8	2·127	6	545	medium	620	large	100
3·03	10	1·917	9	1·868	8	880	v. large	—	—	101
2·88	10	1·801	7	2·189	6	770 870	large large	—	—	102
3·40	10	1·98	10	2·70	8	820	v. large	—	—	103
3·58	10	1·91	9	2·56	9	840 983	large small	—	—	104
1·63	10	2·42	10	2·70	8	—	—	440	large	105
1·86	9	1·59	2	3·03	10	—	—	440 520	large medium	106
1·76	8	3·44	4	2·72	10	—	—	430	large	107
1·79	9	2·07	8	2·93	10	—	—	385	large	108
1·63	3	3·12	10	1·91	6	—	—	480	medium	109
2·06	10	1·26	8	1·07	6	—	—	—	—	110
3·37	10	2·03	6	1·16	5	—	—	820	v. large	111
2·57	10	2·81	9	4·83	7	150 410	v. large v. large	—	—	112
2·82	10	2·71	8	2·78	5	—	—	—	—	113
3·15	10	1·93	10	1·65	4	—	—	—	—	114
2·82	10	1·99	6	1·63	2	801	v. large			115

INDEX TO AUTHORS

INDEX TO SUBJECTS

Printed in Great Britain by
Western Printing Services Ltd., Bristol

ENCYCLOPAEDIC
TECHNICAL ENCYCLOPAEDIA
Z
ZIRCON TO ZIRCONIA